# Water Quality and Treatment

# Other McGraw-Hill Reference Books of Interest

# Water Quality and Treatment

## A Handbook of Community Water Supplies

## American Water Works Association

### Frederick W. Pontius, *Technical Editor*

### Fourth Edition

**McGraw-Hill, Inc.**

New York  St. Louis  San Francisco  Auckland  Bogotá
Caracas  Hamburg  Lisbon  London  Madrid
Mexico  Milan  Montreal  New Delhi  Paris
San Juan  São Paulo  Singapore
Sydney  Tokyo  Toronto

Library of Congress Cataloging-in-Publication Data

Water quality and treatment: a handbook of community water supplies /
   American Water Works Association.—4th ed.

   p.  cm.
   Includes bibliographical references and index.
   ISBN 0-07-001540-6
   1. Water—Purification.  2. Water quality.  3. Water-supply—
United States.  I. American Water Works Association.
TD430.W365   1990                                          90-6192
628.1'6'0973—dc20                                          CIP

The second edition was published in 1950 by The American Water Works
Association, Inc.

1 2 3 4 5 6 7 8 9 0   DOC/DOC   943210
ISBN  0-07-001540-6

The sponsoring editor for this book was Harold B. Crawford, the
editing supervisor was Rita Margolies, the designer was Naomi
Auerbach, and the production supervisor was Suzanne W. Babeuf. It
was set in Century Schoolbook by McGraw-Hill's Professional and
Reference Division Composition Unit.

Printed and bound by R. R. Donnelley & Sons Company.

For more information about other McGraw-Hill materials,
call 1-800-2-MCGRAW in the United States. In other
countries, call your nearest McGraw-Hill office.

# Contents

## Chapter 8. Filtration    455
### John L. Cleasby, Ph.D., P.E.

## Chapter 9. Ion Exchange and Inorganic Adsorption    561
### Dennis A. Clifford, Ph.D., P.E.

## Chapter 10. Chemical Precipitation    641
### Larry D. Benefield, Ph.D., and Joe S. Morgan, Ph.D.

# Contributors

**Appiah Amirtharajah, Ph.D., P.E.**  *Professor, School of Civil Engineering, Georgia Institute of Technology, Atlanta, Ga.* (CHAP. 6)

**Katherine Fox Arnold**  *Consultant, Tucson, Ariz.* (CHAP. 2)

**Larry D. Benefield, Ph.D.**  *Professor, Department of Civil Engineering, Auburn University, Ala.* (CHAP. 10)

**John L. Cleasby, Ph.D., P.E.**  *Professor, Department of Civil and Construction Engineering, Iowa State University, Ames, Iowa* (CHAP. 8)

**Dennis A. Clifford, Ph.D., P.E.**  *Professor of Environmental Engineering, Department of Civil and Environmental Engineering, University of Houston, Houston, Tex.* (CHAP. 9)

**William J. Conlon, P.E.**  *Vice President, Stone & Webster, Water Technology Services Division, Fort Myers, Fla.* (CHAP. 11)

**David A. Cornwell, Ph.D., P.E.**  *President, Environmental Engineering & Technology, Inc., Newport News, Va.* (CHAP. 5)

**Joseph A. Cotruvo, Ph.D.**  *Director, Criteria and Standards, Office of Drinking Water, U.S. Environmental Protection Agency, Washington, D.C.* (CHAP. 1)

**Peter W. Doe, P.E.**  *Consultant, Havens and Emerson, Inc., Saddle Brook, N.J.* (retired) (CHAP. 16)

**Edwin E. Geldreich**  *Senior Research Microbiologist, U.S. Environmental Protection Agency, Drinking Water Research Division, Cincinnati, Ohio* (CHAP. 18)

**William H. Glaze, Ph.D.**  *University of North Carolina, Department of Environmental Science and Engineering, Chapel Hill, N.C.* (CHAP. 12)

**Ross Gregory**  *WRc Swindon, Blagrove, Swindon, Wiltshire, England* (CHAP. 7)

**Charles N. Haas, Ph.D.**  *Professor, Pritzker Department of Environmental Engineering, Illinois Institute of Technology, Chicago, Ill.* (CHAP. 14)

**Carl L. Hamann, Jr., P.E.**  *Director of Water Treatment and Innovation, CH2M Hill, Inc., Denver, Colo.* (CHAP. 3)

**John A. Hroncich**  *Director of Marketing, Laboratory Resources, Inc., Westwood, N.J.* (CHAP. 4)

**J. Brock McEwen, P.E.**  *Senior Process Consultant, CH2M Hill, Inc., Denver, Colo.* (CHAP. 3)

**Joe S. Morgan, Ph.D.**  *Associate Professor, Department of Civil Engineering, Auburn University, Ala.* (CHAP. 10)

**Anthony G. Myers, P.E.**  *Senior Process Consultant, CH2M Hill, Inc., Milwaukee, Wisc.* (CHAP. 3)

**Charles R. O'Melia, Ph.D., P.E.**  *Professor, Department of Geography & Environmental Engineering, The Johns Hopkins University, Baltimore, Md.* (CHAP. 6)

**Thomas G. Reeves, P.E.**  *National Fluoridation Engineer, U.S. Public Health Service Centers for Disease Control, Atlanta, Ga.* (CHAP. 15)

**Robert H. Reinert, P.E.**  *Vice President, Malcolm Pirnie, Inc., Paramus, N.J.* (CHAP. 4)

**Michael R. Schock**  *Research Chemist, U.S. Environmental Protection Agency, Drinking Water Research Division, Cincinnati, Ohio* (CHAP. 17)

**Vernon L. Snoeyink, Ph.D.**  *Professor of Environmental Engineering, Department of Civil Engineering, University of Illinois at Urbana-Champaign, Urbana, Ill.* (CHAP. 13)

**Carol H. Tate, D. Env., P.E.**  *Vice President, James M. Montgomery Consulting Engineers, Pasadena, Calif.* (CHAP. 2)

**Craig D. Vogt, P.E.**  *Office of Marine and Estuarine Protection, U.S. Environmental Protection Agency, Washington, D.C.* (CHAP. 1)

**Thomas F. Zabel**  *WRc Medmenham, Medmenham, Oxfordshire, England* (CHAP. 7)

# Preface

*Water Quality and Treatment* was written as a companion to the recently published *Water Treatment Plant Design*, a cooperative endeavor of the American Water Works Association and the American Society of Civil Engineers. The books complement each other; *Water Quality and Treatment* emphasizes theory and practice, and *Water Treatment Plant Design* discusses the "nuts and bolts" of design. Those familiar with the third edition of *Water Quality and Treatment* will recognize that this fourth edition is not an update but a complete rewrite. The entire book is new, and each chapter was written by a recognized expert in the subject area.

*Water Quality and Treatment* was written for professionals in the field—engineers, chemists, and advanced operators—and college students who are interested in the theory of water treatment practice and the development of design criteria. The book provides authoritative information on the current knowledge of water quality, water treatment processes, and water quality control, but specific cost data are not included. Information on water treatment chemicals in the third edition is available in other American Water Works Association publications. This edition is more sophisticated than the third edition in that it emphasizes theory, but it was not written for researchers in the field. The style and level of sophistication are appropriate for readers with the goal of increasing knowledge as well as for professionals in the field. Most chapters include example problems with solutions to assist the reader in understanding the theoretical concepts presented.

When this project was in its formative stage, the outline of the book was debated. Should it be organized by contaminant—removal of turbidity, iron, color, and so forth—or by unit process—sedimentation, filtration, oxidation, and so forth? The latter was chosen, but to help readers find which chapters contain information about the control of a specific contaminant, Chapter 3 was prepared as a "roadmap" for the reader. For example, the reader who is interested in the control of taste and odor will be guided to the chapters on Stripping and Aeration, and Adsorption.

*Water Quality and Treatment* was reviewed by an *ad hoc* committee representing the Technical and Professional Council and the Water Quality Division of the American Water Works Association. The Committee comprises two water utility reviewers, two consulting engineering reviewers, and two academic reviewers. The Technical and Professional Council was represented by James M. Symons, University of Houston (academic); Raymond H. Taylor, California Water Service Company (water utility); and Glen H. Abplanalp, Havens and Emerson, Inc. (consulting engineer). The Water Quality Division was represented by John L. Cleasby, Iowa State University (academic); James J. Costello, City of Chicago (water utility); and Carl L. Hamann, CH2M Hill, Inc. (consulting engineer). Because both Hamann and Abplanalp were closely involved with the production of *Water Treatment Plant Design*, overlap was minimized and coordination ensured.

The review committee wishes to thank all the authors for their dedication to completing this important assignment. They also want to thank the word processors, those who worked with the authors and those at AWWA Headquarters, who typed, proofed, and corrected the many drafts of each chapter. In addition, the review committee offers thanks to Mary Kay Kozyra, Senior Technical Editor for the American Water Works Association, for her patience, diligence, and talent throughout the completion of this book. Finally, major thanks go to Frederick W. Pontius, Technical Editor of *Water Quality and Treatment*, without whose tireless efforts and countless hours of work this project would not have been completed.

*James M. Symons, Chairman*
Ad hoc *Review Committee*
*University of Houston*

# Rationale for Water Quality Standards and Goals

**Joseph A. Cotruvo, Ph.D.**

*Director, Criteria and Standards*
*Office of Drinking Water*
*U.S. Environmental Protection Agency*
*Washington, D.C.*


**Craig D. Vogt, P.E.**

*Office of Marine and Estuarine Protection*
*U.S. Environmental Protection Agency*
*Washington, D.C.*

In order to be used as healthful fluid for human consumption, water must be free from organisms that are capable of causing disease and from minerals and organic substances that could produce adverse physiological effects. Drinking water should be aesthetically acceptable; it should be free from apparent turbidity, color, and odor and from any objectionable taste. Drinking water also should have a reasonable temperature. Water meeting these conditions is termed "potable," meaning that it may be consumed in any desired amount without concern for adverse effects on health.

Potable water has been so readily available in the United States that few people consider its abundance a modern miracle. One could travel from coast to coast freely consuming water from any public water supply tap without feeling constrained to ask whether the water was "safe." This general acceptance of the "safety" of public water is a dramatic tribute to the achievements of the waterworks industry, the state and local departments of health, the U.S. Public Health Service (USPHS), and the U.S. Environmental Protection Agency (USEPA).

In recent years, however, increasing population pressures, the evolving industrial-chemical society, and great advances in science have resulted in many questions being raised about the potability or safety of drinking water just as they have been raised about safety in all other aspects of life. The terms "safe" or "unsafe" have given way to the consideration of the concept of risk—what is the risk of illness due to the drinking water? The concept of risk is always present but not always articulated or quantifiable.

## Historical Need and Development of Drinking Water Standards

Historically, civilizations began and centered within regions of abundant water supplies. Water quality was not very well documented, and people knew relatively little about disease as it related to water quality. Early historical treatment was performed only for the improvement of the appearance or taste of the water. No definite standards of quality other than general clarity or palatability were recorded by ancient civilizations.[1]

The first drinking water standards were issued at least 4000 years ago. In *The Quest for Pure Water,* Baker quotes a Sanskrit source[2]:

> ... it is directed to heat foul water by boiling and exposing to sunlight and by dipping seven times into a piece of hot copper, then to filter and cool in an earthen vessel.

Hippocrates, the father of medicine (460 to 354 B.C.), stated that "water contributes much to health."[2] His interest in water centered on the selection of the most health-giving source of supply rather than on purifying the waters that were bad. Apparently, ancient people deduced by observation that certain waters promoted good health, while others produced infection. And, although they knew nothing about the causes of disease, they appeared, at least in some instances, to have been astute enough to recognize the health-giving properties of pure and wholesome water. Unfortunately, such information had to be acquired as a result of illness and death of many people.

By the eighteenth century, filtration of particles from water was established as an effective means of clarifying water. The general practice of making water clean was well recognized by that time, but the degree of clarity was not measurable.[1] The first municipal water filtration plant started operations in 1832 in Paisley, Scotland.[2] Aside from the frequent references of concern for the aesthetic properties of water, historical records indicate that standards for water quality were notably absent up to and including much of the nineteenth century.

With the realization that various epidemics (e.g., cholera and typhoid) had been caused and/or spread by water contamination, people saw that the quality of drinking water could not be accurately judged by sensory perception. Reliance on taste and smell was not an accurate means of judging the acceptability of water; more stringent quality criteria would be a necessary historical development. As a result, in 1852 a law was passed in London stating that all waters should be filtered.[1] This was representative of the new heights that had been reached in the ability to observe and correlate facts. In 1855, epidemiologist Dr. John Snow was able to prove empirically that cholera was a waterborne disease. In the late 1880s, Pasteur demonstrated the particulate germ theory of disease, which was based upon the new science of bacteriology. Only after a century of generalized public observations of deaths due to waterborne disease was this cause-and-effect relationship firmly established.

The growth of community water supply systems in the United States began in Philadelphia, Pa. In 1799, a small section was first served by wooden pipes and water was drawn from the Schuylkill River by steam pumps. By 1860, over 400 major water systems had been developed to serve the nation's major cities and towns.

Although municipal water supplies were growing during this early period of the nation's development, health and sanitary conditions did not begin to improve significantly until the turn of the century. By 1900, an increase in the number of water supply systems to over 3000 contributed to major outbreaks of disease because pumped and piped supplies, when contaminated, provided an efficient vehicle for the delivery of pathogenic bacteria.

In the mid- to late-1800s, acute waterborne disease of biological origin was still prevalent in the United States. Following the lead of European investigators, slow sand filters were introduced in Massachusetts by the mid-1870s. Empirical observations showed that this improved the aesthetics of water quality. In the mid-1890s, the Louisville Water Company, Louisville, Ky., combined coagulation with rapid sand filtration, significantly reducing turbidity and bacteria in the water. The Louisville studies refined the knowledge of the process and showed the essential need for pretreatment, including sedimentation for Ohio River water. The next major milestone in drinking water technology was the use of chlorine as a disinfectant. Chlorination was first used in 1908 and was introduced in a large number of water systems shortly thereafter.

## History of Federal Drinking Water Standards

Federal authority to establish standards for drinking water systems originated with the enactment by Congress in 1893 of the Interstate

Quarantine Act. The act,[3] among other things, authorized the USPHS Director

> ... to make and enforce such regulations as in his judgment are necessary to prevent the introduction, transmission, or spread of communicable disease from foreign countries into the states or possessions, or from one state or possession into any other state or possession.

This provision of the act resulted in promulgation of the interstate quarantine regulations in 1894, but it was not until 1912 that the first water-related regulations were adopted: the use of the common drinking cup on interstate carriers was prohibited.

### 1914 standards

The first formal and comprehensive review of drinking water concerns was launched in 1913. In development of the standards, reviewers quickly realized that "most sanitary drinking water cups" would be of no value if the water placed in them was unsafe. Thus, a maximum level of bacterial contamination, 2 coliforms per 100 milliliters (mL), was recommended. The concept of a maximum permissible, safe limit was introduced. The limit was for interstate systems and did not apply to municipalities that did not serve interstate carriers.

With the promulgation of the 1914 standards by the Department of the Treasury on October 14, 1914,[1] a basis for federal, state, and local cooperation was established. Because local and state officials were responsible for inspecting and supervising community systems, they inspected the carrier systems also. In 1915 a federal commitment was made to review the drinking water regulations on a regular basis.

### 1925 standards

By this time, large cities applying either filtration, chlorination, or both encountered little difficulty in complying with the 2 coliforms per 100 mL limit. Thus, the standards were revised to reflect the experience of systems with excellent records of safety against waterborne diseases.[1] The limit was changed to 1 coliform per 100 mL, and the principle of attainability was established. In addition to bacteriological standards, standards were established for physical and chemical (lead, copper, zinc, excessive soluble mineral substances) constituents.

### 1942 standards

In February 1941, an advisory committee for revision of the 1925 drinking water regulations was appointed by the USPHS. The committee was composed of representatives of federal agencies, scientific associations, and three members at large.

Significant new initiatives in the 1942 standards included:[1]

- Samples for bacteriological examination were to be obtained from points in the distribution system, a minimum number of bacteriological samples for examination each month was established, and the laboratories and procedures used in making these examinations became subject to state or federal inspection at any time.

- Maximum permissible concentrations were established for lead, fluoride, arsenic, and selenium. Salts of barium, hexavalent chromium, heavy metals, or other substances having deleterious physiological effects were not allowed in the water system.

- Maximum concentrations, not to be exceeded, where more suitable, i.e., where alternative water sources were available, were set for copper, iron plus manganese, magnesium, zinc, chloride, sulfate, phenolic compounds, total solids, and alkalinity.

### 1946 standards

The technical provisions of the 1946 standards were essentially the same as those of the 1942 standards except that a maximum permissible concentration was added for hexavalent chromium and wording that excluded the presence of salts of barium, hexavalent chromium, heavy metal glucosides, and other substances was changed to prohibit the use of these compounds in water treatment processes.[1] The 1946 standards were further amended by publication in the *Federal Register* on March 1, 1957, of provisions authorizing the use of the membrane filter procedure in the bacteriological examination of water samples.

### 1962 standards

As of the early 1960s, over 19,000 municipal water systems had been identified, and federal water pollution control efforts had called attention to chemical and industrial wastes polluting many surface waterways. In addition, a concern for radioactive pollutants needed to be acknowledged. A new advisory committee was appointed including members from the USPHS; Food and Drug Administration; U.S. Geological Survey; and 12 national transportation, technical, professional, and trade associations.

The 1962 standards provided[4]

- The addition of recommended maximum limiting concentrations for alkyl benzene sulfonates (synthetic detergents), barium, cadmium, carbon-chloroform extract (a measure of organic residues), cyanide, nitrate, and silver
- The addition of a new section on radioactivity

### Interim drinking water standards

The Safe Drinking Water Act (SDWA) (Public Law 93-523) was passed in 1974 because of congressional concern about organic chemical contaminants in drinking water and uneven (and frequently ineffective) state supervision of public drinking water supplies. Based in part on data collected in the 1969 Community Water Supply Survey, monitoring of drinking water quality, especially in small communities, was determined to be seldom-practiced and compliance with the Public Health Service standards (which were actually nonenforceable guidelines) for drinking water was minimal. Except for the coliform standard under the Interstate Quarantine Act mentioned previously, drinking water standards were not legally binding until the passage of the SDWA.

The SDWA required the USEPA to set enforceable standards for health-related drinking water contaminants to apply to all public water systems (those serving at least 25 persons). Regulations were to be set in two steps: (1) interim regulations, based primarily upon the USPHS 1962 guidelines, were to be set immediately, and (2) these interim regulations were to be revised after a comprehensive National Academy of Sciences' assessment of human exposure via drinking water and the toxicology of contaminants in drinking water.

Interim standards (Table 1.1) were set in 1975[5] and amended in 1976,[6] 1979,[7] and 1980.[8] The standards specify maximum contaminant levels (MCLs) for a variety of substances.

### Secondary drinking water standards

In addition to health-related enforceable standards, the SDWA required the USEPA to set nonenforceable federal guidelines for contaminants that may adversely affect the aesthetic quality of drinking water.[9] Initial secondary drinking water standards (Table 1.2) specifying secondary MCLs (SMCLs) were set in 1979.[10] The SMCL for fluoride was revised in 1986,[11] and additional SMCLs were proposed in 1989.[12] SMCLs will be discussed further later in this chapter.

### Drinking water quality in the 1970s and 1980s

**Synthetic organic chemicals and trihalomethanes.** The focus of the USEPA efforts in the mid- to late-1970s was on (1) synthetic organic chemicals (SOCs) in drinking water resulting from industrial contamination of surface water supplies and on (2) organic contaminants that were produced in the disinfection treatment process, i.e., trihalomethanes (THMs).

**TABLE 1.1    National Interim Primary Drinking Water Regulations**

| Contaminant | MCL (enforceable)[a] |
|---|---|
| Organics | |
| Endrin | 0.0002 mg/L |
| Lindane | 0.0004 mg/L |
| Methoxychlor | 0.1 mg/L |
| Toxaphene | 0.005 mg/L |
| 2,4-D | 0.1 mg/L |
| 2,4,5-TP silvex | 0.01 mg/L |
| Trihalomethanes (chloroform, bromoform, bromodichloromethane, dibromochloromethane) | 0.10 mg/L |
| Inorganics | |
| Arsenic | 0.05 mg/L |
| Barium | 1.0 mg/L |
| Cadmium | 0.010 mg/L |
| Chromium | 0.05 mg/L |
| Fluoride | 1.4–2.4 mg/L[b] (ambient temp) |
| Lead | 0.05 mg/L |
| Mercury | 0.002 mg/L |
| Nitrate (as N) | 10 mg/L |
| Selenium | 0.01 mg/L |
| Silver | 0.05 mg/L |
| Sodium and corrosion | No MCL, monitoring and reporting only |
| Radionuclides | |
| Radium 226 and radium 228 | 5 pCi/L[c] |
| Gross alpha particle activity | 15 pCi/L[d] |
| Beta particle and photon radioactivity | 4 mrem[e] (annual dose equivalent) |
| Microbials | |
| Coliforms | < 1/100 mL |
| Turbidity | 1 TU[f] (up to 5 TU) |

[a]Monitoring and reporting for each contaminant also required.
[b]Revised MCL and MCLG for fluoride are 4 mg/L.
[c]5 pCi/L = ~0.19 Bq/L.
[d]15 pCi/L = ~0.56 Bq/L.
[e]4 mrem = $4\mu$ Sv
[f]TU = turbidity unit.

Over 1000 SOCs have been detected in drinking water at one time or another; most are probably of no consequence, but some may pose a potential health risk to consumers. Some SOCs are considered toxicants as well as suspected human carcinogens. The existence of SOCs in drinking water is not necessarily a new problem, but it was not recognized earlier because of the lack of analytical measurement techniques. Much of the science and technology dealing with analytical methods, potential health risks, and treatment techniques to reduce exposure have been developed since the early 1970s.

TABLE 1.2    National Secondary Drinking Water Regulations

| Contaminant | SMCL (nonenforceable) |
|---|---|
| Chloride | 250    mg/L |
| Color | 15    color units |
| Copper | 1    mg/L |
| Corrosivity | Noncorrosive |
| Fluoride | 2    mg/L |
| Foaming agents | 0.5    mg/L |
| Iron | 0.3    mg/L |
| Manganese | 0.05 mg/L |
| Odor | 3 (threshold odor number) |
| pH | 6.5–8.5 |
| Sulfate | 250    mg/L |
| Total dissolved solids | 500    mg/L |
| Zinc | 5    mg/L |
| | Proposed SCMLs, mg/L |
| Aluminum | 0.05 |
| Dichlorobenzene, o- | 0.01 |
| Dichlorobenzene, p- | 0.005 |
| Ethylbenzene | 0.03 |
| Monochlorobenzene | 0.1 |
| Pentachlorophenol | 0.03 |
| Silver | 0.09 |
| Toluene | 0.04 |
| Xylene | 0.02 |

The extent and significance of organic chemical contamination of drinking water or drinking water sources first came to public attention in 1972, when a report, "Industrial Pollution of the Lower Mississippi River in Louisiana," was published.[13] Although the report did not show that THMs were formed during treatment, it provided the first evidence of the presence of THMs in drinking water supplies as well as showed that SOCs from industrial pollution were present in drinking water at low levels.

The findings in New Orleans prompted follow-up studies in 1974, primarily for the purpose of determining how widespread and serious the SOC contamination of drinking water was. Impetus was added by the passage of the SDWA the same year, that directed the USEPA to conduct a comprehensive study of public water supplies and drinking water sources to determine the nature, extent, sources, and means of control of contamination by substances suspected of being carcinogenic. Work by Rook[14] in the Netherlands and the studies by Bellar, Lichtenberg, and Kroner[15] at the USEPA showed that chloroform and other halogenated methanes are formed during the water chlorination process. At that time, Rook[16] provided insight into the organic precursors that might be responsible for the formation of THMs. The USEPA conducted the National Organics Reconnaissance Survey (NORS) for

Halogenated Organics,[17] or "80-City Study," that was designed to determine the extent of the presence of four THMs: chloroform, bromodichloromethane, dibromochloromethane, and bromoform, along with carbon tetrachloride and 1,2-dichloroethane. In addition, the NORS examined the effect water sources and water treatment practices had on the formation of these compounds.

In 1976 to 1977, the USEPA conducted the National Organics Monitoring Survey,[18] aimed not only at determining the presence of THMs in additional water supplies but also at determining the seasonal variations in concentrations of these substances. The USEPA concluded that drinking water is the major source of human exposure to THMs and that THMs are among the most ubiquitous synthetic organic contaminants found in drinking water in the United States. They were generally found at the highest concentrations of any such chemicals measured at that time. THMs are produced in the course of water treatment as by-products of the chlorination process and thus are controllable. In addition to the feasibility of monitoring, means are available to reduce THM concentrations in drinking water. THMs are also indicative of the presence of a host of other halogenated, oxidized, and potentially harmful by-products of the chlorination process that are concurrently formed in even larger quantities but that cannot be readily chemically characterized.

Initial perceptions were that the contamination of surface waters by SOCs from industrial sources warranted regulation under the SDWA. Further data have shown, however, that although a broad spectrum of SOCs can be found in certain surface water supplies at low levels, their significance from a health perspective and the need for universal regulatory action have not been established. This continues to be an area of investigation. Apparently, SOC concentrations in surface waters have been reduced in more recent years as industrial discharges have had to meet stringent standards under the Clean Water Act.[19]

**Volatile organic chemicals.**   In the late 1970s and early 1980s, concerns about SOCs in drinking water were related to contamination of groundwater supplies by a class of chemicals termed volatile organic chemicals (VOCs) that are commonly used as solvents. Many groundwater supplies across the country, once thought of as pristine, have been found to contain measurable amounts of VOCs. The VOCs pose a possible health risk because a number of them are probable or known human carcinogens. To assess the magnitude of groundwater contamination problems, the USEPA conducted the National Ground Water Supply Survey in 1981 to 1982, which sampled nearly 1000 public water systems using groundwater.[20] The results showed that approximately one-fourth of the systems contained at least one VOC at concentrations above the detection limit (i.e., approximately 0.2 µg/L). The vast majority of VOCs were present at very low levels, but because these are man-made chemicals, their detection indicates that a pollution incident has occurred in a resource that was once thought to be invulnerable.

**Pesticides.** Pesticides in groundwaters have emerged in the early to mid-1980s as an important national issue. Contamination resulting from the legal registered use of pesticides has been detected in water supplies in many states, e.g., California, Wisconsin, Rhode Island, New York, Florida, and Hawaii. Numerous pesticides have been found, including ethylene dibromide (EDB), aldicarb, chlordane, and dibromochloropropane (DBCP). A national survey by the USEPA of over 1000 public and private groundwater supplies will be completed in about 1990, which should provide a comprehensive assessment of the extent of contamination by pesticides. Major surface water supplies were sampled by the USEPA in 1980 for a number of pesticides. Atrazine was detected most frequently [39 out of 169 systems in the range of 0.1 to 4.9 parts per billion (ppb) in source and/or treated water].

**Microbiological.** Despite improvements in disinfection and other types of water treatment, outbreaks of waterborne disease still occur, particularly in smaller communities. From 1972 to 1981, there were 335 reported outbreaks of waterborne disease involving almost 78,000 cases;[21] 50 outbreaks and 20,000 cases occurred in 1980 alone.[22] At least two deaths were reported. Major causes of outbreaks in community water systems were contamination of the distribution system and treatment deficiencies such as inadequate filtration and interruption of disinfection. In noncommunity water systems, contamination of groundwater used without treatment or with treatment deficiencies (usually interruption of or inadequate disinfection) was responsible for most outbreaks and cases. The recent concerns about microbiological quality of drinking water relate to *Giardia lamblia, Cryptosporidium,* and viruses.

*Giardia lamblia* is a protozoan that is a human intestinal parasite and is the cause of giardiasis, a disease that can be mildly or extremely debilitating. *Giardia* infections can be acquired by ingesting viable cysts from food or water or by the fecal-oral route. Both humans and wild and domestic animals have been implicated as hosts. Between 1972 and 1981, 50 reported waterborne outbreaks of giardiasis occurred with about 20,000 reported cases.[21] At the present time, no simple and reliable method exists for assaying *Giardia* cysts in water samples. Microscopic methods for detection and enumeration are tedious and require skill and patience on the part of the examiner.

*Giardia* cysts are relatively resistant to chlorine, especially at higher pH and low temperatures, but ozone appears to be a particularly effective disinfectant. Filtration, whether through diatomaceous earth or granular media if properly done, has been shown to be effective for removing cysts of *Giardia* and another pathogenic protozoan, *Entamoeba histolytica.* The effectiveness of these processes for removing other protozoan organisms, such as *Cryptosporidium,* which are smaller than *Giardia* cysts, has not been fully determined.

Viruses have been implicated in numerous outbreaks of waterborne disease. Between 1972 and 1981, 11 waterborne outbreaks involving about 5000 cases were attributed to viruses.[21] Undoubtedly, the reported number of outbreaks is substantially lower than actual numbers. Moreover, in about half the outbreaks of waterborne disease, the causative agent has not been found. A suspicion is growing that most of these were caused by viruses. These organisms are generally more resistant to disinfection than coliforms, and thus may be present in some drinking waters meeting coliform standards.

**Inorganic chemicals.** Monitoring for inorganic chemicals has shown that possibly 1500 to 3000 systems have concentrations for one or more contaminants above the current standards.[23] Inorganics are mostly problems in groundwaters, and removal of inorganic chemicals can be difficult and relatively expensive on a per capita basis for small public water systems. Problems continue primarily with arsenic, barium, lead (from pipe or solder corrosion), natural fluoride, and to an increasing degree, nitrate. Lead is known to occur widely as a result of lead plumbing materials and the action of corrosive water. First-draw drinking water is the primary concern, especially in newer homes. Many states have banned the use of lead-containing plumbing materials. In 1986, the SDWA was amended such that the use of lead-containing pipes, solders, and fixtures in new installations or repairs is unlawful.

Asbestos occurs frequently in drinking water both from natural mineral sources and from the degradation of asbestos-cement water pipe in contact with aggressive water. Although airborne asbestos is a recognized health hazard, the effect of asbestos ingested from drinking water is not generally considered to be significant. The role of asbestos in the etiology of gastrointestinal (GI) cancer continues to be a matter of scientific controversy. Many aspects of asbestos have been the subject of intensive investigation, including the health effects of ingested asbestos exposure from asbestos-cement pipe and natural sources. In general, human epidemiology and animal studies have not detected adverse health effects from ingested asbestos fibers.

**Radionuclides.** Radionuclide contamination of drinking water is one of the more significant emerging concerns. Contamination is caused primarily by natural sources. Although man-made radioactivity in drinking water has not been determined to be a general problem, the potential for contamination exists throughout the country, as releases from medical facilities or nuclear power plants could result in discharge to surface waters of man-made radionuclides. Natural radionuclides of concern, because of their potential health effects (i.e., carcinogenicity) and widespread occurrence, include radium, uranium, and radon. Radon is known to occur widely in groundwater. Airborne exposure from radon released into the home from water (from sources such as showers, washing clothes, and dishes) is more significant than direct ingestion from drinking water. Transport of radon from water to air in the home results in a water-to-air concentration ratio of about 10,000 to 1 [i.e., 10,000 picocuries per liter (pCi/L) in water contribute about 1 pCi/L to indoor air]. The most significant source of radon in indoor air, however, is, by far, that which migrates directly from the ground into dwellings. A national survey of some 1200 groundwater supplies for radionuclides (and inorganic chemicals) was completed in 1987 [National Inorganics and Radionuclides Survey (NIRS)].[23] The results from the NIRS showed that 72 percent of supplies tested positive for radon with a mean of positives of 881 pCi/L, a median of positives of 289 pCi/L, and a maximum of 26,000 pCi/L. Levels as high as 3 million pCi/L have been detected in private wells. Using data from the survey and other sources and the individual risk rate for radon determined from uranium miner data, radon probably contributes the most significant cancer risk of any substance in drinking water.

## Safe Drinking Water Act

The SDWA (first passed in 1974 with major amendments in 1986)[24] requires the USEPA to publish primary drinking water regulations that

1. Apply to public water systems
2. "Specify(s) contaminants which (sic), in the judgment of the Administrator, may have any adverse effect on the health of persons"
3. Specify for each contaminant either (*a*) maximum contaminant level or (*b*) treatment techniques for the control of the contaminant.

A treatment technique requirement would only be set if "it is *not* economically or technologically feasible" (emphasis added) to ascertain the concentration of a contaminant in drinking water.

The term "public water system" means a system for the provision to the public of piped water for human consumption, if such system has at least 15 service connections or regularly serves at least 25 individuals. The term "maximum contaminant level" means the maximum permissible level (concentration) of a contaminant in water that is delivered to any user of a public water system.

The SDWA required the USEPA to initially set interim regulations that were to protect health to the extent feasible. The interim regulations were to be established within 100 days of the enactment of the SDWA.

Following the interim regulations, the USEPA was to revise those regulations based upon a comprehensive review of the occurrence of contaminants in drinking water and the potential health effects of human exposure. For this purpose, the National Academy of Sciences (NAS) was contracted to assess the toxicology of contaminants in drinking water. The NAS report was to contain

1. A summary of publications on the toxicology of contaminants in drinking water
2. Methodologies and assumptions for use in (a) estimating levels at which adverse health effects may occur and (b) estimating margins of safety
3. Proposals for maximum contaminant level goals (MCLGs)
4. List of contaminants in drinking water that cannot be monitored
5. Recommended future research

In the revised national primary drinking water regulations (NPDWR), MCLGs must also be specified (originally termed recommended MCLs, or RMCLs, the 1986 amendments to the SDWA changed the term to MCLGs). *MCLGs are nonenforceable health goals* for public water systems. MCLGs are to be set at a level at which, in the administrator's judgment, "no known or anticipated adverse effects on the health of persons occur and which (sic) allows an adequate margin of safety." Congressional guidance on MCLGs for carcinogens was contained in House Report 93-1185:

> ... The Administrator must consider the possible impact of synergistic effects, long-term and multi-stage exposures, and the existence of more susceptible groups in the population. Finally, the recommended maximum level must be set to prevent the occurrence of any known or anticipated adverse effect. It must include an adequate margin of safety, unless there is (sic) no safe threshold for a contaminant. In such a case, the recommended maximum contaminant level (i.e., MCLG) should be set at a zero level.

The primary drinking water regulations must also set MCLs; *MCLs are the enforceable standards.* MCLs must be set as close to MCLGs as is feasible. The term "feasible" means feasible with the use of the best technology, treatment techniques, and other means that the administrator finds, after examination for efficacy under field conditions and not solely under laboratory conditions, are available, taking cost into consideration. In addition, the 1986 amendments specified that granular activated carbon (GAC) adsorption is feasible for the control of SOCs, and any technology, treatment technique, or other means found to be the best available for the control of SOCs must be at least as effective in controlling SOCs as granular activated carbon adsorption.

The 1986 amendments required the USEPA to set standards and monitoring requirements for 83 drinking water contaminants by 1989, as shown in Table 1.3. The amendments also required the USEPA to set regulations for nine contaminants by June 1987, 40 by June 1988, and the remainder of the list by June 1989. Seven substitutes in the list were allowed for contaminants that may pose a greater health risk. In addition, by January 1988, the USEPA is to specify a list (i.e., the Drinking Water Priority List) of contaminants that occur in drinking water, or are anticipated to occur in drinking water, that pose a health risk and that may warrant regulation under the SDWA. By January 1991, the USEPA is to set standards and monitoring requirements for at least 25 contaminants from the January 1988 list. The list is to be updated every 3 years and at least 25 contaminants from each revised list are to be regulated.

In addition, the 1986 amendments required the USEPA to specify criteria under which filtration (including coagulation and sedimentation) would be required for water systems using surface water sources. Criteria were to be set by December 1987 and states were to use the criteria to determine which water systems would have to install filtration. The 1986 amendments also specify that the USEPA shall set regulations by June 1989 such that all public water systems are required to practice disinfection.

The SDWA also specifies that primary drinking water regulations must contain criteria and procedures to ensure a supply of water that complies with the MCLs. It authorizes the USEPA to require by regulation any public water supplier to keep records, make reports, conduct monitoring, and provide such other information as may be required to assist in determining compliance with the SDWA. The 1986 amendments also require the USEPA to establish regulations for systems to monitor for unregulated contaminants in drinking water. All public water systems are required to monitor, and monitoring is to be repeated every 5 years.

Variances are allowed if a system cannot comply with MCLs despite installation of the best available technology (BAT). Exemptions are allowed for systems that cannot comply with an MCL for compelling reasons (including economic ones). For systems with 500 connections or less,

**TABLE 1.3    Contaminants Required to Be Regulated Under the SDWA of 1986**

| Volatile Organic Chemicals | |
| --- | --- |
| Trichloroethylene | Benzene |
| Tetrachloroethylene | Chlorobenzene |
| Carbon tetrachloride | Dichlorobenzene |
| 1,1,1-Trichloroethane | Trichlorobenzene |
| 1,2-Dichloroethane | 1,1-dichloroethylene |
| Vinyl chloride | trans-1,2-Dichloroethylene |
| Methylene chloride | cis-1,2-Dichloroethylene |

| Microbiology and Turbidity | |
| --- | --- |
| Total coliforms | Viruses |
| Turbidity | Standard plate count |
| Giardia lamblia | Legionella |

| Inorganics | |
| --- | --- |
| Arsenic | Molybdenum |
| Barium | Asbestos |
| Cadmium | Sulfate |
| Chromium | Copper |
| Lead | Vanadium |
| Mercury | Sodium |
| Nitrate | Nickel |
| Selenium | Zinc |
| Silver | Thallium |
| Fluoride | Beryllium |
| Aluminum | Cyanide |
| Antimony | |

| Organics | |
| --- | --- |
| Endrin | 1,1,2-Trichloroethane |
| Lindane | Vydate |
| Methoxychlor | Simazine |
| Toxaphene | Polynuclear aromatic hydrocarbons (PAH) |
| 2,4-D | Polychlorinated biphenyls (PCB) |
| 2,4,5-TP | Atrazine |
| Aldicarb | Phthalates |
| Chlordane | Acrylamide |
| Dalapon | Dibromochloropropane (DBCP) |
| Diquat | 1,2-Dichloropropane |
| Endothall | Pentachlorophenol |
| Glyphosate | Pichloram |
| Carbofuran | Dinoseb |
| Alachlor | Ethylene dibromide (EDB) |
| Epichlorohydrin | Dibromomethane |
| Toluene | Xylene |
| Adipates | Hexachlorocyclopentadiene |
| 2,3,7,8-Tetrachlorodibenzodioxin (dioxin) | |

TABLE 1.3    Contaminants Required to Be Regulated Under the SDWA of 1986 (*Continued*)

| Radionuclides | |
| --- | --- |
| Radium 226 and 228 | Uranium |
| Gross alpha particle activity | Radon |
| Beta particle and photon radio-activity | |

| Removed from SDWA List of 83 | |
| --- | --- |
| Zinc | Molybdenum |
| Silver | Vanadium |
| Aluminum | Dibromomethane |
| Sodium | |

| Substituted into SDWA List of 83 | |
| --- | --- |
| Aldicarb sulfoxide | Heptachlor epoxide |
| Aldicarb sulfone | Styrene |
| Ethylbenzene | Nitrite |
| Heptachlor | |

exemptions can be renewed. All systems, including those with variances, are to be put on schedules that require eventual compliance.

Public notification requirements provide that any violation of an MCL, failure to comply with an applicable testing provision, or failure to comply with any monitoring required must be reported to the persons served by the water system. Public notice is a fundamental part of achieving compliance with MCLs and other requirements. The concept is that an informed public would bring pressure on water systems to come into compliance with the regulations as well as be more willing to support the costs that may be incurred in doing so.

The 1974 SDWA established the National Drinking Water Advisory Council, which consists of 15 members appointed by the administrator. The council's objectives are to advise, consult with, and make recommendations to the administrator on matters relating to activities, functions, and policies of the agency in implementing the SDWA. Five members are from the general public; five members are appointed from state and local agencies concerned with water hygiene and public water supply; and five members are appointed from private organizations or groups demonstrating an active interest in the field of water hygiene and public water supply. Each member of the council holds office for a term of 3 years.

The SDWA provides that implementation of USEPA regulations should be by states provided that states adopt regulations at least as stringent as USEPA regulations. Federal funds are provided to states to assist in the implementation and enforcement of drinking water regulations.

## Philosophy in Setting Drinking Water Goals and Regulations

The SDWA mandates are typically characterized as preventive or prudent in addressing actions on protection of the consumer against adverse health effects. When insufficient data are available for a clearcut decision, the mandate would direct the USEPA to err on the side of safety.[25]

The USEPA's philosophy in setting drinking water regulations is to initially assess the potential for harm, followed by a determination of the feasibility of attainment.[26,27] Federal drinking water standards are designed to be the benchmark from which the safety (or risk) of drinking water is judged. If water utility monitoring programs find contamination, USEPA standards can be used to determine appropriate actions.

Regulatory actions must be scientifically and legally defensible as well as technically and economically feasible. This can only be accomplished through careful study and analysis and extensive communication with those affected by the regulations (i.e., state agencies, public water suppliers, and the scientific community), as well as other interested parties in the public sector.

### Selection of contaminants for regulation

Regulating every chemical that may appear in drinking water and that theoretically may adversely affect health in some remote circumstances is impractical and irrational. What is needed is some priority determination for regulation so that a reasonable number of contaminants of sufficient concern can be addressed in regulations that will advance the goals of the SDWA and provide definitive guidance to address potential human health effects of exposure to hazardous materials in drinking water. The most relevant considerations in selection of contaminants are (1) the occurrence or potential for occurrence in drinking water and (2) the potential health risk.

A set of selection criteria have been developed that describe the primary factors listed above. Use of a specific formula to apply selection criteria is not appropriate because of the many variables associated with contaminants in drinking water. For each contaminant, the essential factors in the analysis are as follows:

1. Is an analytical method available to detect the contaminant in drinking water? If the USEPA cannot ascertain whether the contaminant can be detected in drinking water, a regulation for that contaminant may not be appropriate. Alternatively, a treatment technique may be specified.
2. Are sufficient health effects data upon which to make a judgment on the potential health effects of human exposure available?
3. Have potential adverse health effects from exposure to the contaminant via ingestion been demonstrated?
4. Does the contaminant occur in drinking water?
   a. Has the contaminant been detected in significant frequencies and in a widespread manner?
   b. If data are limited on the frequency and nature of contamination, does a significant potential for drinking water contamination exist?
   c. Factors considered in the analysis of potential occurrence include the following:
      (1) Occurrence in drinking water other than public water supplies
      (2) Presence in direct or indirect additives
      (3) Presence in ambient surface water or groundwater
      (4) Presence in liquid or solid waste
      (5) Mobility to surface water (runoff) or groundwater (leaching)
      (6) Widespread dispersive use patterns
      (7) Production rates

**Seven substitutes to the list of 83 contaminants[28,29]**

The USEPA chose seven contaminants for removal from the statutory list of 83 contaminants: aluminum, dibromomethane, molybdenum, silver, sodium, vanadium, and zinc. Each contaminant was considered for removal on an individual basis and had one or more of the following characteristics: insufficient health and/or occurrence data to develop a regulation, or sufficient health and/or occurrence data to show that it is not expected to cause any adverse health effect. All seven were added to the Drinking Water Priority List, discussed below.

Seven substances were selected to replace the above chemicals on the statutory list of 83: aldicarb sulfone and aldicarb sulfoxide, ethylbenzene, heptachlor and heptachlor epoxide, nitrite, and styrene. Adequate health effects and/or occurrence data exist to demonstrate that these seven substitute chemicals have a greater potential to pose public health risks than those contaminants removed from the list.

**Drinking Water Priority List (DWPL)[29]**

Published on January 22, 1988,[29] the DWPL (Table 1.4) is based on the following SDWA requirements:

1. Each contaminant removed from the list of 83 contaminants must be included on the DWPL.

2. The DWPL must contain substances that are known or anticipated to occur in public water systems and that may require regulation under the SDWA.

3. The USEPA must consider certain lists of substances.

4. The USEPA must form an advisory working group, including members from the National Toxicology Program (NTP); USEPA Offices of Drinking Water, Pesticides, Toxic Substances, Ground Water, Solid Waste, and Emergency Response; and any other members the administrator deems appropriate.

Candidate substances for the first DWPL for regulation in 1991 represent a significant cross section of the most important or potentially

TABLE 1.4    Drinking Water Priority List

| | |
|---|---|
| 1,1,1,2-Tetrachloroethane | Chloramine |
| 1,1,2,2-Tetrachloroethane | Chlorate |
| 1,1-Dichloroethane | Chlorine |
| 1,2,3-Trichloropropane | Hypochlorite ion |
| 1,3-Dichloropropane | Chlorite |
| 1,3-Dichloropropane | Chloroethane |
| 2,2-Dichloropropane | Chloroform |
| 2,4,5,-T | Chloromethane |
| 2,4-Dinitrotoluene | Chloropicrin |
| Aluminum | Cryptosporidium |
| Ammonia | Cyanazine |
| Boron | Cyanogen chloride |
| Bromobenzene | Dibromoacetonitrile |
| Bromochloroacetonitrile | Dibromochloromethane |
| Bromodichloromethane | Dibromomethane |
| Dichloroacetonitrile | Dicamba |
| Ethylene thiourea | Tribluralin |
| Metolachlor | Vanadium |
| Metribuzin | Zinc |
| Molybdenum | o-chlorotoluene |
| Ozone by-products | p-chlorotoluene |
| Silver | Halogenated acids, alcohols, aldehydes, |
| Sodium |    ketones, and other nitriles |
| Strontium | Isophorone |
| Trichloroacetonitrile | Methyl tertbutyl ether |
| Bromoform | |

important drinking water contaminants taken from the following groups:

*Group 1:*   The seven contaminants proposed for removal from the statutory list of 83

*Group 2:*   Disinfectants and contaminants formed as a result of the disinfection process ("disinfection by-products")

*Group 3:*   The first 50 contaminants on the priority list the USEPA is required to compile under Section 110 of the Superfund Amendments and Reauthorization Act (SARA) of 1986

*Group 4:*   Design analyses of the USEPA National Pesticides Survey and pesticides reported to be present in drinking water in certain federal and state surveys

*Group 5:*   Unregulated contaminants to be monitored under Section 1445 of the SDWA

*Group 6:*   Certain other substances reported frequently in other recent surveys

A new list of contaminants to be considered for regulation will be published triennially. The USEPA must promulgate 25 NDPWRs from the DWPL within 3 years of publishing the triennial list.

### Regulatory basis of MCLGs

MCLGs are to be set at a level at which "no known or anticipated adverse effects on the health of persons occur and which (sic) allows an adequate margin of safety."[26,27]

**Noncarcinogens.**   For toxic agents not considered to have carcinogenic potential, "no effect" levels for chronic and/or lifetime periods of exposure including a margin of safety are referred to commonly as reference doses (RfDs) [previously referred to as acceptable daily intakes (ADIs)].[26] RfDs are considered to be exposure levels estimated to be without significant risk to humans when received daily over a lifetime. For noncarcinogenic toxicity, an organism is assumed to tolerate and detoxify some amount of a toxic agent without ill effect up to a certain dose or threshold. A threshold is defined as that dose of a given substance that is required to elicit a measurable biologic response. As the threshold is exceeded, the extent of the response will be a function of the dose applied and the length of time exposed.

   The intent of a toxicological analysis is to identify the highest no-observed-adverse-effect level (NOAEL) based upon assessment of available human or animal data (usually from animal experiments).

To determine the RfD for regulatory purposes, the NOAEL is divided by (an) appropriate "uncertainty" or "safety" factor(s), or UF. This process accommodates for the extrapolation of animal data to the human, for the existence of weak or insufficient data, and for individual differences in human sensitivity to toxic agents, among other factors.

The NAS[30] recommended an approach for use of uncertainty factors when estimating RfDs for contaminants in drinking water. The NAS guidelines are as follows:

- An uncertainty factor of 10 should be used when good acute or chronic human exposure data are available and supported by acute or chronic data in other species.
- An uncertainty factor of 100 should be used when good acute or chronic data are available for one species but not for humans.
- An uncertainty factor of 1000 should be used when acute or chronic data in all species are limited or incomplete. Other uncertainty factors can be used to account for other variations in the available data.

RfDs traditionally are reported in milligrams per kilograms per day [mg/(kg · day)], but for MCLG purposes, the "no-effect" level needs to be measurable in terms of concentrations in drinking water, i.e., milligrams per liter. An adjustment of the RfD to milligrams per liter is accomplished by factoring in a reference weight of the consumer and a reference amount of drinking water consumed per day. The no-effect level in milligrams per liter has been termed the drinking water equivalent level (DWEL).

DWELs are calculated as follows:

$$\text{DWEL} = \frac{[\text{NOAEL in mg/(kg · day)}] \, (70 \text{ kg})}{(\text{UF}) \, (2 \text{ L/day})}$$

where NOAEL = no-observed-adverse-effect level
70 kg = assumed weight of an adult
2 L/day = assumed amount of water consumed by an adult per day
UF = uncertainty factor (usually 10, 100, or 1000)

To determine the MCLG, the contribution from other sources of exposure, including air and food, is taken into account. When sufficient data are available on the relative contribution of other sources, the MCLG is determined as follows:

MCLG = RfD − contribution from food − contribution from air

This calculation ensures that the total exposure from drinking water, food, and air does not exceed the RfD. Comprehensive data are usually not available on exposures from air and food. In these cases the MCLG is determined as follows:

MCLG = DWEL (percentage drinking water contribution)

The DWEL is the value in milligrams per liter of the RfD assuming that the entire daily exposure occurs from drinking water. The drinking water contribution often used in the absence of specific data is 20 percent. This effectively provides an additional safety factor of 5.

In addition to the MCLGs, the USEPA also issues guidance values, called health advisories, for short- or long-term exposure situations, such as spills or accidents. These evaluations are considered to be exposure levels that would not result in adverse health effects over a roughly specified period [usually 1 day, 10 days, longer-term (several months to several years), and a lifetime]. Health advisories are discussed further below.

**Carcinogens and equivocal evidence of carcinogenicity.[26]**    Determination of MCLG no-effect levels for substances that may possess carcinogenic potential is a two-phase process. In the first phase, the toxicological database for noncarcinogenic toxicity is evaluated in the same manner as described above for noncarcinogens. In the second phase, assessment is made of the evidence that measures directly the carcinogenic potential (e.g., long-term bioassays in rodents) as well as evidence that provides indirect support (e.g., mutagenicity and other short-term test results). This process is difficult because the production of cancer is a multistage event, determined by a multiplicity of mechanisms, the nature of which remain, for the most part, hypothesized rather than identified.

To date, scientists have been unable to demonstrate experimentally a threshold of effect for carcinogens, according to the 1977 report of the NAS Safe Drinking Water Committee.[30] This finding leads to the policy *assumption* that because no zero-effect exposure dose has been experimentally demonstrated for carcinogens, any exposure may theoretically represent some finite level of risk. Depending upon the potency of the specific carcinogen and the level, such a risk could be vanishingly small at very low doses.

Human epidemiology data usually cannot define cause-and-effect relationships and, consequently, are extremely limited in their ability to identify carcinogenic risks. Thus, animal experiments are conducted, from which potential human risk is extrapolated. In the first volume of *Drinking Water and Health*, the NAS Safe Drinking Water

Committee provided principles to serve as guidance to the USEPA when assessing irreversible effects[26,30]:

*Principle 1:*   Effects in animals, properly qualified, are applicable to humans.

*Principle 2:*   Methods do not now exist to establish a threshold for long-term effects of toxic agents.

*Principle 3:*   The exposure of experimental animals to toxic agents in high doses is a necessary and valid method of discovering possible carcinogenic hazards in humans.

*Principle 4:*   Material should be assessed in terms of human risk, rather than "safe" or "unsafe."

The above principles are the subject of continuing discussion, debate, and disagreement within the scientific community. The USEPA and other researchers and scientists believe that the principles, when properly qualified, are reasonable. Consequently, these principles serve as the basis for assessment of irreversible effects, such as carcinogenicity, for regulatory purposes.

Several groups of scientists have attempted to classify chemicals on the basis of available evidence for carcinogenicity. These include the International Agency for Research on Cancer (IARC), the NAS Safe Drinking Water Committee, and the USEPA via its recently promulgated risk assessment guidelines for carcinogenicity.

The IARC is responsible for a program on the evaluation of carcinogenic risk of chemicals to humans.[31] The program involves the preparation and publication of monographs providing a qualitative assessment of the carcinogenic potential of individual chemicals and complex mixtures. The assessments are made by independent, international working groups of experts in cancer research. The program has existed since 1971 and has evaluated about 600 chemicals to date.

Criteria used for evaluating carcinogenic risk to humans were first established in 1971 and were used by the IARC for the preparation of the first 16 volumes of the monographs. These criteria consisted of the terms "sufficient evidence" and "limited evidence" of carcinogenicity, referring to the amount of evidence available and not to the potency of the carcinogenic effect.

The NAS classified chemicals in four categories based upon the strength of the experimental evidence.[30] These categories are human carcinogens, suspected human carcinogens, animal carcinogens, and suspected animal carcinogens.

The USEPA set guidelines for carcinogen risk assessment in August 1986 that contain a classification system for chemicals using the degree of evidence of carcinogenicity. The categorization scheme places chemicals into five groups.[32]

*Group A:*   Human carcinogen (sufficient from epidemiological studies)

*Group B:*   Probable human carcinogen

*Group B1:*   At least limited evidence of carcinogenicity to humans

*Group B2:*   Usually a combination of sufficient evidence in animals and inadequate data in humans

*Group C:*   Possible human carcinogen (limited evidence of carcinogenicity in animals in the absence of human data)

*Group D:*   Not classified (inadequate animal evidence of carcinogenicity)

*Group E:*   No evidence of carcinogenicity for humans (no evidence of carcinogenicity in at least two adequate animal tests in different species or in both epidemiological and animal studies)

The USEPA Office of Drinking Water has evaluated these three (IARC, NAS, USEPA) approaches and developed a three-category approach based upon strength of evidence of carcinogenicity to set MCLGs.[25] Category I includes those chemicals that have sufficient human or animal evidence of carcinogenicity to warrant their regulation as probable human carcinogens. The MCLGs for category I chemicals are set at zero. Zero is not measurable and provides little practical guidance on what levels should be the ultimate target for reduction of a carcinogen in drinking water. In meeting the directive of the SDWA, however, zero is an aspirational goal that expresses the ideal concept that drinking water should not be the cause of adverse health effects and should not contain carcinogens.

Category II includes those substances for which some limited inconclusive evidence of carcinogenicity exists from animal data. These are not regulated as human carcinogens. MCLGs will, however, reflect that some possible evidence of carcinogenicity in animals exists. Thus, they are treated more conservatively than category III substances. Category III includes substances with inadequate or no evidence of carcinogenicity; MCLGs are calculated based upon RfDs. These categories are summarized below:

*Category I:*   Strong evidence of carcinogenicity (USEPA group A or B)

*Category II:*   Equivocal evidence of carcinogenicity (USEPA group C)

*Category III:*   Inadequate or no evidence of carcinogenicity in animals (USEPA group D or E)

The method for determining the MCLGs for category II chemicals is more complex than for the other categories. To be placed in category II, chemicals are not considered to be probable but rather possible carcinogens via ingestion. Thus, these substances should be treated more conservatively than category III "noncarcinogens," yet less conservatively than category I chemicals.

Two options are available for setting the MCLGs for category II chemicals; the first option involves basing the MCLG on the RfD. To account for the possible evidence of carcinogenicity, an additional factor would be applied (e.g., RfD divided by a factor of 10 or some other value). The second option involves basing the MCLG on a lifetime risk calculation in the range of $10^{-5}$ to $10^{-6}$ (i.e., 1 in 100,000 or 1 in 1,000,000 excess cancers in a lifetime of ingestion) using a conservative method that is unlikely to underestimate the actual risk. Both of these approaches are commonly considered to be protective, and in the future, if additional data lead to reconsideration of a chemical's carcinogenicity, the MCLG would still have been set at a level that would represent an extremely low nominal risk. The first option, basing the MCLG upon the RfD, is used if sufficient valid chronic toxicity data are available. If sufficient chronic toxicity data are not available, MCLG is based on a risk calculation if sufficient data are available to perform the calculations.

### Regulatory basis of MCLs

MCLs are to be set "as close to" the MCLGs "as is feasible." The term "feasible" means feasible with the use of the BAT, taking costs into consideration.[25]

The general approach to setting MCLs is to determine the feasibility of controlling contaminants. This requires an evaluation of (1) the availability and cost of analytical methods, (2) the availability and performance of technologies and other factors relative to feasibility and identifying those that are "best," and (3) an assessment of the costs of the application of technologies to achieve various concentrations. Key factors include the following:

1. Technical and economic availability of analytical methods: precision and accuracy of analytical methods that would be acceptable for accurate determination of compliance (i.e., practical quantitation levels, see later discussion), limits of analytical detection, laboratory capabilities, and costs of analytical techniques.
2. Concentrations attainable by application of the BAT.

    *a.* Levels of contamination in drinking water supplies.

    *b.* Feasibility and reliability of reducing contaminants to specific concentrations.

3. Other feasibility factors relating to the best means of treatment such as air pollution and waste disposal and effects on other drinking water quality parameters.

4. Costs of treatment to achieve contaminant removals.

In the VOC rule published on July 8, 1987,[27] the USEPA explained the feasibility basis for its determinations of MCLs for eight VOCs. This included designation of GAC adsorption and packed tower stripping as BATs for all of the VOCs (except vinyl chloride). Packed tower stripping was designated as the BAT for vinyl chloride. This determination was based upon the demonstrated cost and efficiency of the control methods. The USEPA also stated that an MCLG of zero does not imply that actual harm would necessarily occur to humans at levels somewhat above zero, but rather that zero is an aspirational goal that includes a margin of safety within the context of the SDWA.

For substances that were regulated as known or probable carcinogens, the MCLGs were zero, but the MCLs, of course, could not be zero. The USEPA identified a "reference risk range" of $10^{-4}$ to $10^{-6}$ (i.e., theoretical incremental upper bound lifetime risk of one in ten thousand to one in a million). The USEPA considers MCLs that are set within this range to be safe levels and protective of human health. "Safe" generally means unlikely to result in adverse effects on health. All the VOC MCLs based on potential carcinogenic risk fell within that range. The range was consistent with recommendations in the 1984 World Health Organization drinking water guidelines (discussed below).

## Compliance monitoring considerations

The objective of monitoring is to assure compliance with the MCLs and, of course, to indicate the quality of the drinking water. Monitoring requirements vary depending upon the contaminants and vulnerability of water supplies to specific contaminants.

Certain contaminants such as coliforms, turbidity, and some inorganic and organic chemicals are widely detected in drinking water supplies and pose serious, often acute health risks when MCLs are exceeded. Without consistent or frequent oversight, these substances have a high potential for posing health risks. Such contaminants warrant national fixed regular monitoring requirements. States would be required to adopt and apply those regulations as written; states can produce more stringent requirements as needed.

On the other hand, the occurrence of many drinking water contaminants is often predictable based upon geological conditions, source type, and historical record. The occurrence of contaminants such as natural radionuclides, certain pesticides, and some inorganics such as barium may well be predictable; thus, repeated monitoring according to a fixed formula may use resources for nonproductive monitoring, once compliance status has been determined and source conditions are stabilized. Cases such as these warrant conferring maximum discretion to state regulatory agencies so that activities can be tailored to regional conditions.

The fundamental questions that are considered in developing compliance monitoring requirements are the following:

1. What minimum requirements should be set?
2. What distinctions should be made between ground and surface water systems?
   a. What location should be chosen for sampling?
   b. What number of samples per system should be used?
   c. Should monitoring be done once or over a period of time? Should a minimum repeat frequency be established? If so, what frequency should be chosen and upon what basis?
   d. How much time should be allowed for public water systems to complete the monitoring per system?
3. What sampling requirements should be set?
4. What follow-up actions may be needed to assist the public water systems and the states when positives are reported?
   a. Should follow-up confirmation sampling be performed?
   b. What health and treatment advisories should be made?
5. What reporting and public notice requirements should be set?

Thus, states have a very active role in determining appropriate monitoring requirements. The USEPA generally specifies a minimum frequency and provides guidance on factors to take into account when assessing a system's vulnerability to contamination.

Public notice of violations of MCLs, monitoring, and reporting is required so that notification is provided immediately for violations that represent a serious health risk. Depending upon the type of violation, notice must be by publication in newspapers and in water bills and also provided to radio and television stations. Public water systems are required to provide public notice that results of monitoring for the unregulated contaminants are available upon request.

## USEPA Process for Setting Standards

Development of regulations by the USEPA involves interpretation of mandates and directives of the SDWA, technical and scientific assess-

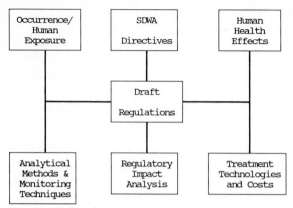

**Figure 1.1** Development of draft USEPA regulations (adapted from the *Federal Register*).

ments to meet the SDWA requirements, preparation of regulations that blend the technical and science aspects with policy considerations, review of draft regulations within the agency, and review by the public of draft and proposed regulations.

### Regulation development within the USEPA

Development of regulations within the USEPA is achieved by first assessing the available technical and scientific information, preparing the draft regulations, and then submitting the regulations to intra-agency review.

**Technical and scientific assessments.** Separate support documents for each contaminant are prepared on (1) occurrence and human exposure, (2) health effects and toxicology, (3) analytical methods and monitoring, and (4) treatment technologies and costs.* In addition, a regulatory impact assessment is prepared for each regulation. These assessments are shown in Fig. 1.1.

The occurrence evaluations of contaminants in drinking water are primarily based upon USEPA-conducted national surveys of public water systems.[28] Surveys completed to date include

Community Water Supply Survey (1969 USPHS)[33]

National Organics Reconnaissance Survey (1975)[17]

National Organics Monitoring Survey (1976–1977)[18]

---

*Many of these documents are available for a fee from the National Technical Information Service, U.S. Department of Commerce, Washington, D.C. (800-426-4791).

National Screening Program for Organics (1977–1981)[34]

Community Water Supply Survey (1978)[33]

National Rural Water Survey (1978)[35]

Ground Water Supply Survey (1980–1981)[20]

National Inorganics and Radionuclides Survey (1986)[23]

National Pesticides Survey (initiated in 1987)[36]

These surveys provide national, statistically based assessments of contaminants in drinking water. These data are used as part of the decision process on which contaminants should be regulated and on the potential impact of regulations. The surveys also provide information from which states can determine monitoring requirements for systems vulnerable to specific contaminants. The occurrence documents also include estimates of human exposure to contaminants by other routes of exposure such as air or food.

The health effects documents provide a comprehensive review of available literature on the potential adverse health effects of human exposure via drinking water to the contaminants. The types of data evaluated include human epidemiology studies, animal studies, and sometimes human clinical studies. The health effects assessment attempts to determine toxicology endpoints. Specific areas addressed include

- Carcinogenicity
- Mutagenicity
- Teratology
- Pharmacokinetics
- Absorption factors
- Metabolism

These evaluations lead to cancer risk estimates (for carcinogens) and RfDs (for noncarcinogens) for each of the contaminants considered.

Analytical methods and monitoring techniques are presented in the methods and monitoring support document. Factors that are considered in specifying which analytical methods should be approved include

- Reliability (precision and accuracy) of the analytical results
- Specificity in the presence of interferences
- Availability and performance of laboratories

- Rapidity of analysis to permit routine use
- Costs of analysis

Guidance for implementation of the monitoring requirements are included in the document.

To account for variabilities in the quality of data, practical quantitation levels (PQLs) are used as one of several factors in assessing what concentration would be enforceable for MCLs.[28] The PQL serves as the performance requirement for laboratories applying for certification to provide compliance data for implementation of the regulations. A method detection limit (MDL) is the minimum concentration of a substance that can be measured and reported with 99 percent confidence that the true value is greater than zero. MDLs are not necessarily reproducible over time even in a given laboratory and even when the same analytical procedures, instrumentation, and sample matrix are used.

The PQL is the lowest concentration that can be achieved within specified limits of precision and accuracy during routine laboratory operating conditions. The PQL thus represents the lowest concentration achievable by good laboratories within specified limits during routine laboratory operating conditions. The PQL is determined through interlaboratory studies and by application of principles of analytical science. Differences between MDLs and PQLs are expected because the MDL represents the lowest achievable concentration under ideal laboratory conditions, whereas the PQL represents the lowest achievable concentration under practical and routine laboratory conditions. PQLs are usually in a range between 5 and 10 times the MDL and can be achieved by well-managed laboratories.

The treatment and costs document summarizes the availability and performance of the treatment technologies that can reduce contaminants in drinking water. Costs of treatment are determined for each technology for many sizes of water systems. The BAT is determined based upon a number of factors, some of which include technologies that

- Have the highest efficiencies of removal
- Are compatible with other types of water treatment processes
- Are available as manufactured items or components
- Are not limited to application in a particular geographic region
- Have integrity for a reasonable service life as a public work
- Are reasonably affordable by large metropolitan or regional systems
- Can be mass-produced and put into operation in time for implementation of regulations

The regulatory impact assessment is required under Presidential Executive Order 12291. The assessment is submitted to the Office of Management and Budget (OMB) for review and contains an evaluation of the potential national cost (and other) impacts and benefits of proposed and final regulations.

### Intra-agency review process

The objective of the extensive review mechanism within the USEPA is to ensure that regulations under development (1) have met their legislative mandates, (2) are scientifically and technically defensible, and (3) address potential cross-program issues in other agency activities. The process is shown in Fig. 1.2 and is discussed briefly below.

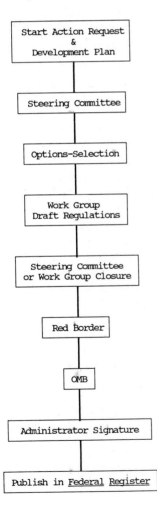

**Figure 1.2** Development and review process for proposed and final USEPA regulations (adapted from the *Federal Register*).

1. *Start action request (SAR) and development plan:* This is the very first step and is a presentation of the need for a regulation and the general concepts of what the regulation would cover. The development plan identifies SDWA mandates, planned actions, issues, and schedules. A work group assists and reviews the SAR and development plan. The work group consists of staff-level representatives of all agency programs interested in the regulations being considered. The work group must meet and consider reviewers' comments and issues between each major step in Fig. 1.2.

2. *Steering committee review of the SAR and development plan:* The steering committee is a standing committee with senior-level representatives of each major program area (e.g., water, air, pesticides and toxics and research).

3. *Review of major issues and options at the options-selection meeting with the deputy administrator:* Assistant administrators from each program office in the USEPA participate in this meeting. The objective is to obtain guidance on major issues prior to the commitment of resources to certain directions in regulation development. This process takes about 3 to 4 months from preparation of briefing memos to final decisions.

4. *Preparation of draft regulations with assistance from the work group.*

5. *Review of draft regulations by the steering committee* (approximately a 4-week process): Review by the work group and closure on all issues during a formal work group meeting can be substituted for steering committee review.

6. *Red border sign-off:* The regulations are circulated to each assistant administrator for final sign-off (4-week process).

7. *OMB review:* The OMB reviews the technical aspects and the potential impacts of proposed and final regulations, and the policy implications as well (minimum of 2 weeks and up to 6 months).

8. *Administrator signature.*

9. *Publication in the* Federal Register.

These steps are repeated twice in the process, initially for preparation of a proposal that is published for public comment in the *Federal Register,* and then again when the final regulations, which are prepared after the public comment process, have been completed. The typical period for the entire process is 2 to 4 years.

### Public participation in regulation development

Fundamental to the successful development of a regulation is the involvement of the public. The USEPA operates under a mandate from

Congress to produce standards for contaminants in drinking water, often under circumstances of incomplete toxicological and technological information. Participation by the scientific community as well as those potentially impacted by the regulations is essential. Public involvement from the inception of the regulation development process through completion of final regulations is actively solicited. Public participation is shown in Fig. 1.3.

Public workshops, meetings, and hearings are conducted from early in the process until completion. These allow informal and formal discussions of issues, and the opportunity to provide data and information to the USEPA. The advance notice of proposed rule making (ANPRM) is an optional step and provides an extra opportunity for public comment. Public comment periods generally last from 30 to 45 days and up to 120 days. The USEPA reviews each submitted comment and uses it in the preparation of the proposed and final regulations. A detailed document is prepared that presents each comment and the USEPA's response to it.

The last block in Fig. 1.3 is litigation, which, if filed, under the SDWA is to be filed in the U.S. Court of Appeals for the District of Columbia Circuit.

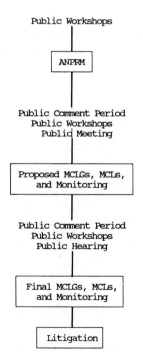

**Figure 1.3**   Public participation in regulation development (adapted from the *Federal Register*).

## Drinking Water Regulations and Goals

This section provides summaries of published standards, guidelines, and goals for drinking water quality. These include

USEPA primary drinking water regulations

USEPA secondary drinking water regulations

Canadian drinking water guidelines

World Health Organization (WHO) *Guidelines for Drinking Water Quality*

European Economic Community (EEC) drinking water directives

USEPA advisories on direct and indirect additives

USEPA health drinking water advisories

In addition, a discussion of the water quality criteria for lakes, rivers, and streams published by USEPA is included in this section.

Terms such as regulations, standards, goals, guidelines, criteria, advisories, limits, levels, and objectives are commonly used to describe numerical or narrative qualities of drinking water that would protect public health. Distinctions between these terms generally fall into either (1) legally enforceable concentrations (regulations or standards) or (2) concentrations that represent desirable water quality but that are not enforceable (goals or criteria). In addition, both regulations and goals usually represent either (1) a health-based no-adverse-effect level or (2) a level representing a balance between health risks and feasibility of achieving those levels.

### USEPA primary drinking water regulations

As stated previously, the USEPA sets both nonenforceable health goals (MCLGs) and enforceable regulations (MCLs). The national interim primary drinking water regulations (NIPDWRs) (Table 1.1) are enforceable MCLs until revised. Currently (1989), final MCLGs and MCLs have been set for a number of contaminants, but many more are in various stages of development. Table 1.5 presents the numerical primary regulations existing or proposed at press time. Table 1.6 presents the primary regulations for surface water treatment (filtration and disinfection)[37] and total coliforms[38] promulgated on June 29, 1989.

Approximately 220,000 public water systems in the United States are regulated under USEPA's standards. Of these, 60,000 systems are community water systems and 160,000 are considered noncommunity water systems. Noncommunity systems serve transient populations

TABLE 1.5    USEPA Primary Drinking Water Regulations

| | Existing | |
|---|---|---|
| Volatile organic chemicals | MCLG, mg/L | MCL, mg/L |
| Trichloroethylene | zero | 0.005 |
| Carbon tetrachloride | zero | 0.005 |
| Vinyl chloride | zero | 0.002 |
| 1,2-Dichloroethane | zero | 0.005 |
| Benzene | zero | 0.005 |
| 1,1-Dichloroethylene | 0.007 | 0.007 |
| 1,1,1-Trichloroethane | 0.2 | 0.2 |
| p-Dichlorobenzene | 0.075 | 0.075 |

| | Proposed | |
|---|---|---|
| Inorganic chemicals | MCLG, mg/L | MCL, mg/L |
| Arsenic | zero | 0.03 |
| Asbestos (fibers/1 > 10 μm) | $7 \times 10^6$ | $7 \times 10^6$ |
| Barium | 5 | 5 |
| Cadmium | 0.005 | 0.005 |
| Chromium | 0.1 | 0.1 |
| Copper | 1.3 | OCC* |
| Lead | zero | OCC |
| Mercury | 0.002 | 0.002 |
| Nitrate (as N) | 10 | 10 |
| Nitrite (as N) | 1 | 1 |
| Selenium | 0.05 | 0.05 |
| Fluoride | 4 | 4 |

| | Proposed | |
|---|---|---|
| Synthetic organic chemicals | MCLG, mg/L | MCL, mg/L |
| Acrylamide | zero | TT† |
| Alachlor | zero | 0.002 |
| Aldicarb | 0.01 | 0.01 |
| Aldicarb sulfoxide | 0.01 | 0.01 |
| Aldicarb sulfone | 0.04 | 0.04 |
| Atrazine | 0.002 | 0.002 |
| Carbofuran | 0.04 | 0.04 |
| Chlordane | zero | 0.02 |
| cis-1,2-Dichloroethylene | 0.07 | 0.07 |
| DBCP | zero | 0.0002 |
| 1,2-Dichloropropane | zero | 0.005 |
| o-Dichlorobenzene | 0.6 | 0.6 |
| 2,4-D | 0.1 | 0.1 |
| EDB | zero | 0.00005 |
| Epichlorohydrin | zero | TT |
| Ethylbenzene | 0.7 | 0.7 |
| Heptachlor | zero | 0.0004 |
| Heptachlor epoxide | zero | 0.0002 |
| Lindane | 0.0002 | 0.0002 |
| Methoxychlor | 0.4 | 0.4 |
| Monochlorobenzene | 0.1 | 0.1 |
| Polychlorinated biphenyls | zero | 0.0005 |
| Pentachlorophenol | 0.2 | 0.2 |

TABLE 1.5    USEPA Primary Drinking Water Regulations (*Continued*)

| Synthetic organic chemicals(*Continued*) | Proposed | |
|---|---|---|
| | MCLG, mg/L | MCL, mg/L |
| Styrene | zero | 0.005 |
| Tetrachloroethylene | zero | 0.005 |
| Toluene | 2 | 2 |
| 2,4,5-TP | 0.05 | 0.05 |
| Toxaphene | zero | 0.005 |
| *Trans*-1,2-Dichloroethylene | 0.1 | 0.1 |
| Xylene | 10 | 10 |

| Radionuclides | Potential | |
|---|---|---|
| | MCLG | MCL |
| Radium 226 | zero | N/A‡ |
| Radium 228 | zero | N/A |
| Uranium | zero | N/A |
| Radon | zero | N/A |
| Gross alpha | zero | N/A |
| Gross beta | zero | N/A |

| Microbiological parameters | Promulgated 6/89 | |
|---|---|---|
| | MCLG, mg/L | MCL, mg/L |
| Total coliform | zero | P/A§ |
| Turbidity | none | FD¶ |
| *Giardia* | zero | FD |
| Viruses | zero | FD |
| HPC | none | FD |
| *Legionella* | zero | FD |

*OCC—optimal corrosion control.

†TT—treatment technology specified polymer addition practices (i.e., limited dosage).

‡N/A—not available.

§P/A—Presence and absence of coliform in samples. Small systems cannot have more than 1 coliform positive sample per month; for large systems, not more than 5 percent of total samples per month may be coliform positive.

¶FD—filtration and disinfection.

Note: Some MCLGs and MCLs have not yet been proposed at press time for other contaminants on Table 1.3.

such as parks, campgrounds, and restaurants. These systems are only required to meet MCLs for acute toxicants such as nitrate, or indicators of contamination such as coliforms and turbidity. A subgroup within the noncommunity group is the 20,000 nontransient, noncommunity systems that consistently serve the same nonresident populations (e.g., schools and factories that produce their own drinking water). This group would have to meet all standards for community water systems.

TABLE 1.6    USEPA Regulations for Surface Water Treatment and Total Coliforms[37,38]

## Surface Water Treatment Requirements

**Coverage.** All public water systems using any surface water or groundwater under direct influence of surface water must disinfect and may be required by the state to filter, unless certain water quality source requirements and site-specific conditions are met.

Treatment technique requirements are established in lieu of MCLs for *Giardia,* viruses, bacteria, *Legionella,* and turbidity. Treatment must achieve at least 99.9 percent removal and/or inactivation of *G. lamblia* cysts and 99.99 percent removal and/or inactivation of viruses.

All systems must be operated by qualified operators as determined by the state.

### Criteria to be met to avoid filtration

**Source water criteria.** Fecal coliform concentration must not exceed 20/100 mL or the total coliform concentration must not exceed 100/100 mL before disinfection in more than 10 percent of the measurements for the previous 6 months, calculated each month.

Minimum sampling frequencies for fecal or total coliform determinations are

| System size, persons | Samples per week |
|---|---|
| < 501 | 1 |
| 501–3300 | 2 |
| 3301–10,000 | 3 |
| 10,001–25,000 | 4 |
| > 25,000 | 5 |

If not already conducted under the above requirements, a coliform test must be made each day that the turbidity exceeds 1 nephalometric turbidity unit (NTU).

Turbidity levels must be measured every 4 h by grab sample or continuous monitoring and may not exceed 5 NTU. If the turbidity exceeds 5 NTU, the system must install filtration unless the state determines that the event is unusual or unpredictable, and the event does not occur more than two periods in any 1 year, or 5 times in any consecutive 10 years. An "event" is one or more consecutive days when at least one turbidity measurement each day exceeds 5 NTU.

TABLE 1.6    USEPA Regulations for Surface Water Treatment and Total
Coliforms[37,38] (*Continued*)

## Site-specific conditions

**Disinfection.**   Disinfection must achieve at least a 99.9 and 99.99 percent inactivation of *Giardia* cysts and viruses, respectively. This must be demonstrated by the system meeting CT values in the rule [CT is the product of residual concentration (mg/L) and contact time (min) measured at peak hourly flow]. C and T must be determined at or prior to the first customer. The total percent inactivation can be calculated based on unlimited disinfectant residual measurements in sequence prior to the first customer. Failure to meet this requirement on more than one day in a month is a violation. Filtration is required if a system has two or more violations in a year, unless the state determines that at least one of these violations was caused by circumstances that were unusual and unpredictable. A third violation in 12 months, regardless of the cause, triggers filtration.

Disinfection systems must (*a*) have redundant components, including alternate power supply, automatic alarm and start-up to ensure continuous disinfection of the water during plant operation; or (*b*) have automatic shutoff of the delivery of water to the distribution system whenever the disinfectant residual is less than 0.2 mg/L, provided that the state determines that a shutoff would not pose a potential health risk to the system.

For systems using chloramines if chlorine is added prior to ammonia, the CT values for achieving 99.9 percent inactivation of *G. lamblia* cysts can also be assumed to achieve 99.99 percent inactivation of viruses. Systems using chloramines and adding ammonia prior to chlorine must demonstrate with on-site studies that they achieve 99.99 percent inactivation of viruses. For systems using disinfectants other than chlorine, the system may demonstrate the other CT values than those specified in the rule, or other disinfection conditions are provided that achieve at least 99.9 and 99.99 percent inactivation of *G. lamblia* and viruses, respectively.

Disinfectant residuals in the distribution system cannot be undetectable in more than 5 percent of the samples, each month, for any two consecutive months. Samples must be taken at the same frequency as total coliforms under the revised coliform rule. A system may measure for HPC in lieu of disinfectant residual. If the HPC measurement is less than 500 colonies per milliliter, the site is considered to have a "detectable" residual for compliance purposes. Systems in violation of this requirement must install filtration unless the state determines that the violation is not caused by a deficiency of treatment

TABLE 1.6    USEPA Regulations for Surface Water Treatment and Total
Coliforms[37,38] (*Continued*)

of the source water. For systems that cannot maintain a residual or practically monitor for HPC, the state can judge whether adequate disinfection is provided in the distribution system.

Systems must maintain a disinfectant residual concentration of at least 0.2 mg/L in the water entering the system, demonstrated by continuous monitoring. If there is a failure in the continuous monitoring, the system may substitute grab sample monitoring every 4 h for up to 5 days. If the disinfectant residual falls below 0.2 mg/L, the system must notify the state as soon as possible but no later than the end of the next business day; notification must include whether or not the residual was restored within 4 h. If the residual is not restored to at least 0.2 mg/L within 4 h, it is a violation and the system must filter, unless the state determines that the violation was caused by unusual and unpredictable circumstances. Systems serving 3300 people or less can take daily grab samples in lieu of continuous monitoring. Minimum grab sampling frequencies are 1/day, <501 people; 2/day, 501–1000 people; 3/day, 1001–2500 people; 4/day, 2501–3300 people. If at any time the residual is below 0.2 mg/L, the system must conduct grab sample monitoring every 4 h until the residual is restored.

**Other conditions.**  Systems must maintain a watershed control program that will minimize the potential for contamination by human enteric viruses and *G. lamblia* cysts. Systems must monitor and control the activities in the watershed that may have an adverse impact on water. Systems must demonstrate through ownership or written agreements with landowners in the watershed that they are able to limit and control all human activities that may have an adverse impact upon water quality. The watershed control program and disinfection treatment must be inspected on-site and approved by the state annually.

Systems must not have had any waterborne disease outbreaks, or if they have, such systems must have been modified to prevent another such occurrence, as determined by the state.

Systems must not be out of compliance with the monthly MCL for total coliforms for any two months in any consecutive 12-month period, unless the state determines that the violations are not due to treatment deficiency of the source water.

Systems serving more than 10,000 people must be in compliance with MCL requirements for total trihalomethanes.

TABLE 1.6    USEPA Regulations for Surface Water Treatment and Total Coliforms[37,38] *(Continued)*

**Criteria for filtered systems**

**Turbidity monitoring.**    Turbidity must be measured every 4 h by grab sample or continuous monitoring. For systems using slow sand filtration, or filtration technologies other than conventional treatment, direct filtration, or diatomaceous earth filtration, the state may reduce the sampling frequency to once per day. The state may reduce monitoring to one grab sample per day for all systems serving less than 500 people.

**Turbidity removal.**    *Conventional filtration or direct filtration* water must achieve a turbidity level in the filtered water at all times less than 5 NTU and not more than 0.5 NTU in more than 5 percent of the measurements taken each month. The state may increase the 0.5 NTU limit up to less than 1 NTU in greater than or equal to 95 percent of the measurements, without any demonstration by the system, if it determines that *overall treatment* with disinfection achieves at least 99.9 and 99.99 percent removal or inactivation of *Giardia* cysts and viruses, respectively.

*Slow sand filtration* must achieve a turbidity level in the filtered water at all times less than 5 NTU and not more than 1 NTU in more than 5 percent of the samples taken each month. The turbidity limit of 1 NTU may be increased by the state (but at no time exceed 5 NTU) if it determines that there is no significant interference with disinfection.

*Diatomaceous earth filtration* must achieve a turbidity level in the filtered water at all times less than 5 NTU and of not more than 1 NTU in more than 5 percent of the samples taken each month.

*Other filtration technologies* may be used if the system demonstrates to the state that they achieve at least 99.9 and 99.99 percent removal or inactivation of *G. lamblia* cysts and viruses, respectively, and are approved by the state. Turbidity limits for these technologies are the same as those for slow sand filtration, including the allowance of increasing the turbidity limit of 1 NTU up to 5 NTU, but at no time exceeding 5 NTU upon approval by the state.

**Disinfection requirements.**    *Disinfection with filtration* must achieve at least 99.9 and 99.99 percent removal or inactivation of *Giardia* cysts and viruses, respectively. The states define the level of disinfection required, depending on technology and source water quality. Guidance on the use of CT values to make these determinations is available in the *Guidance Manual.*[45] Recommended levels of inactivation are based on expected occurrence levels of *Giardia* cysts in the source water and the filtration technology in place. Disinfection requirements for point of entry to the distribution system and within the distribution system are the same as for unfiltered systems.

### Analytical requirements

Except for ozone, testing and sampling must be in accordance with *Standard Methods,* 16th ed.,* or methods approved by the USEPA for total coliforms, fecal coliform, turbidity, disinfectant residuals, temperature, and pH. Residual disinfectant concentrations for ozone must be measured by the indigo method or automated methods, which are calibrated in reference to the results obtained by the indigo method.

### Reporting

All parameters required in the rule must be reported monthly to the state. Unfiltered water systems must also report annually on their watershed control program and on-site inspections.

### Compliance

**Surface water systems.**    Unfiltered systems must meet monitoring requirements beginning 18 months following promulgation, unless the state has determined that filtration is required. Unfiltered systems must meet the criteria to avoid filtration beginning 30 months following promulgation, unless the state has determined that filtration is required, or they are in violation of a treatment technique requirement. Unfiltered systems must install filtration within 18 months following the failure to meet any one of the criteria to avoid filtration, or within 48 months following promulgation, whichever is later, or they are in violation.

Filtered systems must meet monitoring and performance requirements beginning 48 months following promulgation.

The interim turbidity monitoring and MCL requirements will remain in effect for unfiltered systems until 30 months following promulgation, and for filtered systems until 48 months following promul-

---

*Arnold E. Greenberg, R. Rhodes Trussell and Lenore S. Clesceri, *Standard Methods for the Examination of Water and Water Wastewater,* prepared and published by the American Public Health Association, American Water Works Association, and the Water Pollution Control Federation, 1985.

TABLE 1.6    USEPA Regulations for Surface Water Treatment and Total Coliforms.[37,38] (*Continued*)

gation. For systems that prior to 30 months following promulgation the state determines must filter, the interim turbidity requirements will remain in effect until 48 months following promulgation, or until filtration is installed, whichever is later.

**Groundwater systems under direct influence of surface water.**  All systems using groundwater under direct influence of surface water must meet the treatment requirements under the surface water treatment rule (SWTR). States must determine which community and noncommunity groundwater systems are under direct influence of surface water within 5 years and 10 years, respectively, following promulgation. Unfiltered systems under the direct influence of surface water must begin monitoring within 6 months following the determination of direct influence unless the state has determined that filtration is required. Systems under direct influence of surface water must begin meeting the criteria to avoid filtration 18 months after the determination of direct influence, unless the state has determined that filtration is required. Unfiltered systems under direct influence of surface water must install filtration within 18 months following the failure to meet any of the criteria to avoid filtration.

**Variances**

Variances are not applicable.

**Exemptions**

Exemptions are allowed for the requirement to filter. Systems using surface water must disinfect (i.e., no exemptions). Exemptions are allowed for the level of disinfection required.

**Total Coliform Requirements**

**Effective date.**  All public water systems must meet the revised final coliform MCL and monitoring requirements 18 months after promulgation of the regulations. Current rule remains in force until the effective date of the revised rule.

**Maximum contaminant level goal.**  Zero.

**Maximum contaminant level**

- Compliance is based on presence or absence of total coliforms in sample, rather than on an estimate of coliform density.

TABLE 1.6    USEPA Regulations for Surface Water Treatment and Total
Coliforms[37,38] (Continued)

- MCL for systems analyzing at least 40 samples per month: no more than 5.0 percent of the monthly samples may be total coliform positive.
- MCL for systems analyzing less than 40 samples per month: no more than 1 sample per month may be total coliform positive.

## Monitoring requirements for total coliforms

- Each public water system must sample according to a written sample siting plan. Plans are subject to state review and revision. The state must establish a process that ensures the adequacy of the sample siting plan for each system.
- Monthly monitoring requirements are based on population served (see Table 1.6a).
- A system must collect a set of repeat samples for each total coliform-positive routine sample (see Table 1.6b) and have it analyzed for total coliforms. At least one repeat sample must be from the same tap as the original sample; other repeat samples must be collected from within five service connections both upstream and downstream of the original sample. The system must collect all repeat samples within 24 h of being notified of the original result, except where the state waives this requirement on a case-by-case basis.
- If total coliforms are detected in any repeat sample, the system must collect another set of repeat samples, as before, unless the MCL has been violated and the system has notified the state (in which case the state may reduce or eliminate the repeat sampling requirement for the remainder of the month).
- If a system that collects fewer than five routine samples per month detects total coliforms in any routine or repeat sample (and the sample is not invalidated by the state), it must collect a set of five routine samples the next month the system provides water to the public, except that the state may waive this requirements if (1) it performs a site visit to evaluate the contamination problem; or (2) it has determined why the sample was total coliform positive and (a) explains this conclusion in writing, (b) obtains the signature of the supervisor of the state official who draws this conclusion, (c) makes the documentation available to the USEPA and the public, and (d) requires the system to collect at least one sample under certain circumstances described in the rule.
- Unfiltered surface water systems, or systems using unfiltered

TABLE 1.6    USEPA Regulations for Surface Water Treatment and Total
Coliforms.[37,38] (*Continued*)

groundwater under the direct influence of surface water, must analyze one coliform sample each day the turbidity of the source water exceeds 1 NTU.

■ Tables 1.6*a* and 1.6*b* summarize the routine and repeat sample
monitoring requirements for total coliforms.

### Invalidation of total coliform-positive samples

■ All total coliform-positive samples count in compliance calculations,
except for those samples that are invalidated by the state. Invalidated
samples do not count toward the minimum monitoring frequency.

■ A state may invalidate a sample only if (1) the analytical laboratory
acknowledges that improper sample analysis caused the positive result; (2) the system determines that the contamination is a domestic
or other nondistribution system plumbing problem on the basis that
any repeat sample taken at the same tap as the original total
coliform-positive sample is total coliform positive, but all repeat
samples at nearby sampling locations are total coliform negative; or
(3) the state has substantial grounds to believe that a total coliform-
positive result is due to some circumstance or condition not related
to the quality of drinking water in the distribution system, if (*a*) this
judgment is explained in writing, (*b*) the document is signed by the
supervisor of the state official who draws this conclusion, and (*c*) the
documentation is made available to the USEPA and the public.

**Variances and exemptions.**    None allowed.

**Sanitary surveys.**    Periodic sanitary surveys are required for all systems collecting fewer than 5 samples per month, according to the
schedule in Table 1.6*c*.

### Fecal coliforms and *E. coli*; heterotrophic bacteria (HPC)

■ If any routine or repeat sample is total coliform positive, the system
must analyze that total coliform-positive culture to determine if fecal coliforms are present. The system may test for *E. coli* in lieu of
fecal coliforms. If fecal coliforms or *E. coli* are detected, the system
must notify the state before the end of same business day, or, if detected after the close of business for the state, by the end of the next
business day.

**TABLE 1.6     USEPA Regulations for Surface Water Treatment and Total Coliforms.[37,38] (Continued)**

- If any repeat sample is fecal coliform or *E. coli* positive, or if a fecal coliform- or *E. coli*-positive original sample is followed by a total coliform-positive repeat sample, and the original total coliform-positive sample is not invalidated, the system is in violation of the MCL for total coliforms. This is an acute violation of the MCL for total coliforms.

- The state has the discretion to allow a water system, on a case-by-case basis, to forego fecal coliform or *E. coli* testing on total coliform-positive samples, if the system treats every total coliform-positive sample as if it contained fecal coliforms, i.e., the system complies with all sections of the rule which apply when a sample is fecal coliform positive.

- State invalidation of the routine total coliform-positive sample invalidates subsequent fecal coliform or *E. coli*-positive results on the same sample.

- Heterotrophic bacteria can interfere with total coliform analysis. Therefore, if the total coliform sample produces (1) a turbid culture in the absence of gas production using the multiple tube fermentation (MTF) technique; (2) a turbid culture in the absence of an acid reaction using the presence or absence (P/A) coliform test; or (3) confluent growth or a colony number that is "too numerous to count" using the membrane filter (MF) technique, the sample is invalid (unless total coliforms are detected, in which case the sample is valid). The system must collect another sample within 24 h of being notified of the result from the same location as the original sample and have it analyzed for total coliforms. In such cases, the USEPA recommends using media less prone to interference from heterotrophic bacteria for analyzing the replacement sample.

**Analytical methodology**

- Total coliform analyses are to be conducted using the 10-tube MTF technique, the MF technique, the P/A coliform test, or the minimal media ONPG-MUG test (Autoanalysis Colilert System). The system may also use the five-tube MTF technique (20-mL sample portions) or a single-culture bottle containing the MTF medium, as long as the 100-mL water sample is used in the analysis.

- A 100-mL standard sample volume must be used in analyzing for total coliforms, regardless of the analytical method used.

TABLE 1.6*a*    Total Coliform Sampling Requirements, According to Population Served

| Population served | Minimum number of routine samples per month* | Population served | Minimum number of routine samples per month* |
|---|---|---|---|
| 25 to 1,000† | 1‡ | 59,001 to 70,000 | 70 |
| 1,001 to 2,500 | 2 | 70,001 to 83,000 | 80 |
| 2,501 to 3,300 | 3 | 83,001 to 96,000 | 90 |
| 3,301 to 4,100 | 4 | 96,001 to 130,000 | 100 |
| 4,101 to 4,900 | 5 | 130,001 to 220,000 | 120 |
| 4,901 to 5,800 | 6 | 220,001 to 320,000 | 150 |
| 5,801 to 6,700 | 7 | 320,001 to 450,000 | 180 |
| 6,701 to 7,600 | 8 | 450,001 to 600,000 | 210 |
| 7,601 to 8,500 | 9 | 600,001 to 780,000 | 240 |
| 8,501 to 12,900 | 10 | 780,001 to 970,000 | 270 |
| 12,901 to 17,200 | 15 | 970,000 to 1,230,000 | 300 |
| 17,201 to 21,500 | 20 | 1,230,001 to 1,520,000 | 330 |
| 21,501 to 25,000 | 25 | 1,520,001 to 1,850,000 | 360 |
| 25,001 to 33,000 | 30 | 1,850,001 to 2,270,000 | 390 |
| 33,001 to 41,000 | 40 | 2,270,001 to 3,020,000 | 420 |
| 41,001 to 50,000 | 50 | 3,020,001 to 3,960,000 | 450 |
| 50,001 to 59,000 | 60 | 3,960,001 or more | 480 |

*In lieu of the frequency specified in this table, a noncommunity water system using groundwater and serving 1000 persons or fewer may monitor at a lesser frequency specified by the state until a sanitary survey is conducted and the state reviews the results. Thereafter, noncommunity water systems using groundwater and serving 1000 persons or fewer must monitor in each calendar quarter during which the system provides water to the public, unless the state determines that some other frequency is more appropriate and notifies the system (in writing). Five years after promulgation, noncommunity water systems using groundwater and serving 1000 persons or fewer must monitor at least once per year.

A noncommunity water system using surface water, or groundwater under the direct influence of surface water, regardless of the number of persons served, must monitor at the same frequency as a like-sized community public water system. A noncommunity water system using groundwater and serving more than 1000 persons during any month must monitor at the same frequency as a like-sized community water system, except that the state may reduce the monitoring frequency for any month the system serves 1000 persons or fewer.

†Includes public water systems that have at least 15 service connections but serve fewer than 25 persons.

‡For a community water system serving 25 to 1000 persons, the state may reduce this sampling frequency, if a sanitary survey conducted in the last 5 years indicates that the water system is supplied solely by a protected groundwater source and is free of sanitary defects. However, in no case may the state reduce the sampling frequency to less than once per quarter.

**TABLE 1.6b    Monitoring and Repeat Sample Frequency after a Total Coliform-Positive Routine Sample**

| Number of routine samples per month | Number of repeat samples* | Number of routine samples next month |
|---|---|---|
| 1 or fewer | 4 | 5 |
| 2 | 3 | 5 |
| 3 | 3 | 5 |
| 4 | 3 | 5 |
| 5 or greater | 3 | Table 1.6a |

*Number of repeat samples in the same month for each total coliform-positive routine sample.

Except where the state has invalidated the original routine sample, substitutes an on-site evaluation of the problem, or waives the requirement on a case-by-case basis.

**TABLE 1.6c    Sanitary Survey Frequency for Public Water Systems Collecting Fewer Than Five Samples per Month**

| System type | Initial survey completed by | Frequency of subsequent surveys |
|---|---|---|
| Community water system | 5 years after promulgation | Every 5 years |
| Noncommunity water system | 10 years after promulgation | Every 5 years* |

*For a noncommunity water system that uses protected and disinfected groundwater, the sanitary survey may be repeated every 10 years instead of every 5 years.

## USEPA secondary drinking water regulations

Table 1.2 provides USEPA's secondary regulations that set desirable levels for drinking water contaminants that may adversely affect the aesthetic value of drinking water. These are not enforceable by the federal government, but states are encouraged to adopt them. States may establish higher or lower levels that may be appropriate depending upon local conditions such as unavailability of alternative source waters or other compelling factors, provided that public health and welfare are not adversely affected.

## Canadian drinking water guidelines

In Canada, drinking water is a shared federal-provincial responsibility. In general, provincial governments are responsible for an adequate, safe supply, whereas the Federal Department of National Health and Welfare develops quality guidelines and conducts research. Guidelines for Canadian drinking water quality are developed through a joint federal-provincial mechanism and are not legally enforceable unless promulgated as regulations by the appropriate provincial agency. The first comprehensive Canadian drinking water guidelines were published by the Department of National Health and Welfare in 1968.[39] They were completely revised in 1978[40] and again in 1987.[41] The current guidelines for Canadian drinking water quality are listed in Table 1.7.[41]

## World Health Organization Guidelines for Drinking Water Quality[42]

The primary aim of the WHO *Guidelines for Drinking Water Quality* is the protection of public health and thus the elimination, or reduction to a minimum, of constituents in water that are known to be hazardous to the health and well-being of the community.

In presenting this summary of guideline values (Tables 1.8 through 1.12), WHO's intent is that individual values should not necessarily be used directly from the tables. Guideline values should be used and interpreted in conjunction with the information contained in the appropriate section of chapters 2 to 5 of *Guidelines for Drinking Water Quality*[42] and within national policies and authorities.

## European Economic Community drinking water directives[43]

The EEC, having been established by a treaty of the Council of the European Communities, issued a council directive relating to the quality of water intended for human consumption.[43] Specifically, the EEC standards provide for both the setting of standards to apply to toxic chemicals and bacteria that present a health hazard, and the definition of physical, chemical, and biological parameters for different uses of water, specifically for the use of human consumption. The member states are directed to bring into force laws, regulations, and administrative provisions necessary to comply with the directive on water standards (see Tables 1.13 through 1.18).

TABLE 1.7    Canadian Guidelines for Drinking Water Quality (1987)

| Parameter | Type[a] | MAC[b] | IMAC[b] | AO[b] | Notes |
|---|---|---|---|---|---|
| Alachlor | P | — | — | — | 1 |
| Aldicarb | P | 0.009 | — | — | 2(A) |
| Aldrin and dieldrin | P | 0.0007 | — | — | 1 |
| Antimony | I | c | — | — | 1 |
| Arsenic | I | 0.05 | — | — | 1 |
| Asbestos | I | d | — | — | 3 |
| Atrazine | P | — | 0.06 | — | 2(A) |
| Azinphos-methyl | P | 0.02 | — | — | 2(A) |
| Barium | I | 1.0 | — | — | 1 |
| Bendiocarb | P | 0.04 | — | — | 2(A) |
| Benzene | O | 0.005 | — | — | 2(A) |
| Benzo(a)pyrene | O | 0.00001 | — | — | 2(A) |
| Boron | I | 5.0 | — | — | 1 |
| Bromoxynil | P | — | 0.005 | — | 2(A) |
| Cadmium | I | 0.005 | — | — | 3 |
| Carbaryl | P | 0.09 | — | — | 2(R) |
| Carbofuran | P | 0.09 | — | — | 2(A) |
| Carbon tetrachloride | O | 0.005 | — | — | 2(A) |
| Cesium 137 | R | 50 Bq/L | | | |
| Chlordane | P | 0.007 | — | — | 1 |
| Chloride | I | — | — | < 250 | 5 |
| Chlorobenzene; 1,2-di- | O | 0.2 | — | 0.003 | 2(A) |
| Chlorobenzene; 1,4-di- | O | 0.005 | — | 0.001 | 2(A) |
| Chlorophenol; 2,4-di- | O | 0.9 | — | 0.0003 | 2(A) |
| Chlorophenol; penta- | O | 0.06 | — | 0.03 | 2(A) |
| Chlorophenol; 2,3,4,6-tetra- | O | 0.1 | — | 0.001 | 2(A) |
| Chlorophenol; 2,4,6-tri- | O | 0.005 | — | 0.002 | 2(A) |
| Chlorpyrifos | P | 0.09 | — | — | 2(A) |
| Chromium | I | 0.05 | — | — | 3 |
| Coliform organisms | — | e | | | |
| Color | — | — | — | < 15 TCU | 5 |
| Copper | I | — | — | < 1.0 | 5 |
| Cyanazine | P | — | 0.01 | — | 2(A) |
| Cyanide | I | 0.2 | — | — | 4 |
| 2,4-D | P | 0.1 | — | — | 1 |
| DDT and metabolites | P | 0.03 | — | — | 1 |
| Diazinon | P | 0.02 | — | — | 4 |
| Dicamba | P | 0.12 | — | — | 2(A) |
| 1,2-Dichloroethane | O | — | — | — | 1 |
| 1,1-Dichloroethylene | O | — | — | — | 1 |
| Dichloromethane | O | 0.05 | — | — | 2(A) |
| Diclofop-methyl | P | 0.009 | — | — | 2(A) |
| Dieldrin and aldrin | P | 0.0007 | — | — | 1 |
| Dimethoate | P | — | 0.02 | — | 2(A) |
| Dinoseb | P | — | — | — | 1 |
| Dioxins | O | — | — | — | 1 |
| Diquat | P | 0.07 | — | — | 2(A) |
| Diuron | P | 0.15 | — | — | 2(A) |
| Endrin | P | — | — | — | 2(D) |

TABLE 1.7    Canadian Guidelines for Drinking Water Quality (1987) *(Continued)*

| Parameter | Type[a] | MAC[b] | IMAC[b] | AO[b] | Notes |
|---|---|---|---|---|---|
| Ethylbenzene | O | — | — | < 0.0024 | 2(A) |
| Fluoride | I | 1.5[f] | — | — | 4 |
| Furans | O | — | — | — | 1 |
| Gasoline | O | [g] | — | — | 2(A) |
| Glyphosate | P | — | 0.28 | — | 2(A) |
| Hardness | I | — | — | [h] | 4 |
| Heptachlor and heptachlor epoxide | P | 0.003 | — | — | 1 |
| Iodine 131 | R | 10 Bq/L | | | |
| Iron | I | — | — | < 0.3 | 5 |
| Lead | I | 0.05 | — | — | 1 |
| Lindane | P | 0.004 | — | — | 1 |
| Linuron | P | — | — | — | 1 |
| Malathion | P | 0.19 | — | — | 2(A) |
| Manganese | I | — | — | < 0.05 | 5 |
| MCPA[i] | P | — | — | — | 1 |
| Mercury | I | 0.001 | — | — | 3 |
| Methoxychlor | P | 0.9 | — | — | 2(R) |
| Methyl-parathion | P | 0.007 | — | — | 1 |
| Metolachlor | P | — | 0.05 | — | 2(A) |
| Metribuzin | P | 0.08 | — | — | 2(A) |
| Nitrate | I | 10.0[j] | — | — | 1 |
| Nitrilotriacetic acid (NTA) | O | 0.05 | — | — | 4 |
| Nitrite | I | 1.0[j] | — | — | 1 |
| Odor | — | — | — | Inoffensive | 5 |
| Paraquat | P | — | 0.01 | — | 2(A) |
| Parathion | P | 0.05 | — | — | 2(R) |
| PCBs | O | — | — | — | 1 |
| Pesticides (total) | P | 0.1 | — | — | 1 |
| pH | — | — | — | 6.5–8.5[k] | 5 |
| Phenols | O | — | — | — | 2(D) |
| Phorate | P | — | 0.002 | — | 2(A) |
| Picloram | P | — | — | — | 1 |
| Radium 226 | R | 1 Bq/L | | | |
| Selenium | I | 0.01 | — | — | 3 |
| Silver | I | — | — | — | 2(D) |
| Simazine | P | — | 0.01 | — | 2(A) |
| Sodium | I | [l] | — | — | 3 |
| Strontium 90 | R | 10 Bq/L | | | |
| Sulfate | I | 500 | 00 | < 150 | 1 |
| Sulfide (as $H_2S$) | I | — | — | < 0.05 | 5 |
| 2,4,5-T | P | 0.28 | — | < 0.02 | 2(A) |
| 2,4,5-TP | P | — | — | — | 2(D) |
| Taste | — | — | — | Inoffensive | 5 |
| TCA[m] | P | — | — | — | 1 |
| Temephos | P | — | 0.28 | — | 2(A) |
| Temperature | — | — | — | < 15°C | 5 |
| Terbufos | P | — | 0.001 | — | 2(A) |
| Tetrachloroethylene | O | — | — | — | 1 |
| Toluene | O | — | — | < 0.024 | 2(A) |

TABLE 1.7    Canadian Guidelines for Drinking Water Quality (1987) *(Continued)*

| Parameter | Type[a] | MAC[b] | IMAC[b] | AO[b] | Notes |
|-----------|---------|--------|---------|-------|-------|
| Total dissolved solids | I | — | — | < 500 | 5 |
| Toxaphene | P | — | — | — | 2(D) |
| Triallate | P | 0.23 | — | — | 2(A) |
| 1,1,1-Trichloroethane | O | — | — | — | 1 |
| Trichloroethylene | O | — | — | — | 1 |
| Trifluralin | P | — | — | — | 1 |
| Trihalomethanes | O | 0.35 | — | — | 1 |
| Tritium | R | 40,000 Bq/L | | | |
| Turbidity | — | 1 NTU[n] | — | < 5 NTU[o] | 1 |
| Uranium | I | 0.1 | — | — | 2(R) |
| Xylenes | O | — | — | < 0.3 | 2(A) |
| Zinc | I | — | — | < 5.0 | 5 |

[a]I—Inorganic constituent; O—Organic constituent; P—Pesticide; R—Radionuclide.

[b]Unless otherwise specified, units are mg/L; limits apply to the sum of all forms of each substance present. MAC = maximum acceptable concentration; IMAC = interim maximum acceptable concentration; AO = aesthetic objectives.

[c]An objective concentration only was set in 1978, based on health considerations.

[d]Assessment of data indicates no need to set numerical guideline.

[e]No sample should contain more than 10 total coliform organisms per 100 mL; not more than 10 percent of the samples taken in a 30-day period should show the presence of coliform organisms; not more than two consecutive samples from the same site should show the presence of coliform organisms; and none of the coliform organisms detected should be fecal coliform.

[f]*It is recommended, however, that the concentration of fluoride be adjusted to 1.0 mg/L, which is the optimum level for the control of dental caries. Where the annual mean daily temperature is less than 10°C, a concentration of 1.2 mg/L should be maintained.*

[g]Assessment of data indicates no need to set a numerical guideline.

[h]Public acceptance of hardness varies considerably. Generally hardness levels between 80 and 100 mg/L (as $CaCO_3$) are considered acceptable; levels greater than 200 mg/L are considered poor but can be tolerated; those in excess of 500 mg/L are normally considered unacceptable. Where water is softened by sodium-ion exchange, it is recommended that a separate unsoftened supply be retained for culinary and drinking purposes.

[i]2-Methyl-4-chlorophenoxyacetic acid.

[j]As nitrate- or nitrite-nitrogen concentration.

[k]Dimensionless.

[l]It is recommended that sodium be included in routine monitoring programs since levels may be of interest to authorities who wish to prescribe sodium-restricted diets for their patients.

[m]Trichloroacetic acid.

[n]For water entering a distribution system. *A maximum of 5 NTU may be permitted if it can be demonstrated that disinfection is not compromised by the use of this less stringent value.* See also section on microbiological characteristics.

[o]At the point of consumption.

*Notes:*

1. Under review for possible revision, deletion from, or addition to the guidelines.

2. It is proposed that a guideline be added for this parameter for the first time (A); a change be made to the previous guideline (R); or the guideline be deleted (D). If after 1 year, no evidence comes to light that questions the appropriateness of the proposal, it will be adopted as the guideline.

3. Reassessment of data indicates no need to change 1978 recommendation.

4. Adapted from *Guidelines for Canadian Drinking Water Quality: 1978*; reassessment considered unnecessary at this time.

5. Previously listed in *Guidelines for Canadian Drinking Water Quality: 1978* as a maximum acceptable concentration based on aesthetic considerations.

**TABLE 1.8  World Health Organization Guidelines for Microbiological and Biological Quality**

| Organism | Unit | Guideline value | Remarks |
|---|---|---|---|
| **I. Microbiological Quality** | | | |
| **A. Piped Water Supplies** | | | |
| *A.1 Treated water entering the distribution system* | | | |
| Fecal coliforms | number/100 mL | 0 | Turbidity <1 NTU; for disinfection with chlorine, pH preferably |
| Coliform organisms | number/100 mL | 0 | < 8.0; free chlorine residual 0.2–0.5 mg/L following 30 min (minimum) contact |
| *A.2 Untreated water entering the distribution system* | | | |
| Fecal coliforms | number/100 mL | 0 | |
| Coliform organisms | number/100 mL | 0 | In 98% of samples examined throughout the year—in the case of large supplies when sufficient samples are examined |
| Coliform organisms | number/100 mL | 3 | In an occasional sample, but not in consecutive samples |
| *A.3 Water in the distribution system* | | | |
| Fecal coliforms | number/100 mL | 0 | |
| Coliform organisms | number/100 mL | 0 | In 95% of samples examined throughout the year—in the case of large supplies when sufficient samples are examined |
| Coliform organisms | number/100 mL | 3 | In an occasional sample, but not in consecutive samples |
| **B. Unpiped Water Supplies** | | | |
| Fecal coliforms | number/100 mL | 0 | |
| Coliform organisms | number/100 mL | 10 | Should not occur repeatedly; if occurrence is frequent and if sanitary protection cannot be improved, an alternative source must be found if possible |
| **C. Bottled Drinking Water** | | | |
| Fecal coliforms | number/100 mL | 0 | Source should be free from fecal contamination |
| Coliform organisms | number/100 mL | 0 | |
| **D. Emergency Water Supplies** | | | |
| Fecal coliforms | number/100 mL | 0 | Advise public to boil water in case of failure to meet guideline value |
| Coliform organisms | number/100 mL | 0 | |
| Enteroviruses | No guideline value set | | |
| **II. Biological Quality** | | | |
| Protozoa (pathogenic)—no guideline value set | | | |
| Helminths (pathogenic)—no guideline value set | | | |
| Free-living organisms (algae, others)—no guideline value set | | | |

**TABLE 1.9    World Health Organization Guidelines for Inorganic Constituents of Health Significance**

| Constituent | Guideline value* | Remarks |
| --- | --- | --- |
| Arsenic | 0.05 | |
| Asbestos | No guideline value set | |
| Barium | No guideline value set | |
| Beryllium | No guideline value set | |
| Cadmium | 0.005 | |
| Chromium | 0.05 | |
| Cyanide | 0.1 | |
| Fluoride | 1.5 | Includes both natural fluoride and deliberately added fluoride. Local or climatic conditions may necessitate adaptation. |
| Hardness | No health-related guideline value set | |
| Lead | 0.05 | |
| Mercury | 0.001 | |
| Nickel | No guideline value set | |
| Nitrate | 10 | |
| Nitrite | No guideline value set | |
| Selenium | 0.01 | |
| Silver | No guideline value set | |
| Sodium | No guideline value set | |

*All values are in mg/L.

## USEPA advisories on direct and indirect additives

In addition to drinking water contaminants for which national enforceable standards are set, the USEPA has provided nonenforceable guidance to states on the acceptability of compounds "added to" drinking water, either directly or indirectly, in the course of treatment and transport of drinking water.[44] Public water systems use a broad range of chemical products to treat water supplies and for maintaining storage and distribution systems. Systems may directly add chemicals such as chlorine, alum, lime, and coagulant aids in the process of treating water to make it suitable for public consumption. These are known as "direct additives." As a necessary function of maintaining a public water system, storage and distribution systems (including pipes, tanks, and other equipment) may be painted, coated, or treated with chemical products that may leach into or otherwise enter the water. These products are known as "indirect additives."

The USEPA provided technical assistance to states and public water systems on the use of additives through the issuance of advisory opinions on the acceptability of many additive products. In 1985, the National Sanitation Foundation entered into a cooperative agreement

**TABLE 1.10 World Health Organization Guidelines for Organic Constituents of Health Significance**

| Constituent | Guideline value* | Remarks |
|---|---|---|
| Aldrin and dieldrin | 0.03 | |
| Benzene | 10† | |
| Benzo(a)pyrene | 0.01† | |
| Carbon tetrachloride | 3† | Tentative guideline value‡ |
| Chlordane | 0.3 | |
| Chlorobenzenes | No health-related guide-line value set | Odor threshold concentration between 0.1 and 3 μg/L |
| Chloroform | 30† | Disinfection efficiency must not be compromised when controlling chloroform content |
| Chlorophenols | No health-related guide-line value set | Odor threshold concentration 0.1 μg/L |
| 2,4-D | 100§ | |
| DDT | 1 | |
| 1,2-Dichloroethane | 10† | |
| 1,1-Dichloroethene¶ | 0.3† | |
| Heptachlor and heptachlor epoxide | 0.1 | |
| Hexachlorobenzene | 0.01† | |
| Gamma-HCH (lindane) | 3 | |
| Methoxychlor | 30 | |
| Pentachlorophenol | 10 | |
| Tetrachloroethane¶ | 10† | Tentative guideline value‡ |
| Trichloroethane¶ | 30† | Tentative guideline value‡ |
| 2,4,6-Trichlorophenol | 10†§ | Odor threshold concentration, 0.1 μg/L |
| Trihalomethanes | No guideline value set | |

*All values are in μg/L.

†These guideline values were computed from a conservative hypothetical mathematical model which cannot be experimentally verified, and values should therefore be interpreted differently. Uncertainties involved may amount to two orders of magnitude (i.e., from 0.1 to 10 times the number).

‡When the available carcinogenicity data did not support a guideline value, but the compounds were judged to be of importance in drinking water and guidance was considered essential, a tentative guideline value was set on the basis of the available health-related data.

§May be detectable by taste and odor at lower concentrations.

¶These compounds were previously known as 1,1-dichloroethylene, tetrachloroethylene, and trichloroethylene, respectively.

**TABLE 1.11   World Health Organization Guidelines for Aesthetic Quality**

| Constituent | Guideline value* | Remarks |
|---|---|---|
| Aluminum | 0.2 | |
| Chloride | 250 | |
| Chlorobenzenes and chlorophenols | No guideline value set | These compounds may affect taste and odor |
| Color | 15 TCU† | |
| Copper | 1.0 | |
| Detergents | No guideline value set | There should not be any foaming, taste, or odor problem |
| Hardness | 500 (as $CaCO_3$) | |
| Hydrogen sulfide | Not detectable by consumers | |
| Iron | 0.3 | |
| Manganese | 0.1 | |
| Oxygen, dissolved | No guideline value set | |
| pH | 6.5–8.5 | |
| Sodium | 200 | |
| Solids, total dissolved | 1000 | |
| Sulfate | 400 | |
| Taste and odor | Inoffensive to most consumers | |
| Temperature | No guideline value set | |
| Turbidity | 5 NTU‡ | Preferably <1 for disinfection efficiency |
| Zinc | 5.0 | |

*Unless otherwise specified, all units are mg/L.
†TCU—true color unit.
‡NTU—nephelometric turbidity unit.

**TABLE 1.12   World Health Organization Guidelines for Radioactive Constituents**

| Constituent | Guideline value* | Remarks |
|---|---|---|
| Gross alpha activity | 0.1 | If the levels are exceeded, more detailed radionuclide analysis may be necessary. |
| Gross beta activity | 1 | Higher levels do not necessarily imply that the water is unsuitable for human consumption. |

*Units are Bq/L

**TABLE 1.13    European Economic Community Standards for Organoleptic Parameters**

| Parameters | Expression of the results | Guide level | Maximum admissible concentration |
|---|---|---|---|
| Color | mg/L Pt/Co scale | 1 | 20 |
| Turbidity* | mg/L $SiO_2$ | 1 | 10 |
|  | Jackson units | 0.4 | 4 |
| Odor* | Dilution number | 0 | 2 at 12°C |
|  |  |  | 3 at 25°C |
| Taste* | Dilution number | 0 | 2 at 12°C |
|  |  |  | 3 at 25°C |

*Refer to EEC standards for comments.


**TABLE 1.14    European Economic Community Standards for Physicochemical Parameters (in Relation to the Water's Natural Structure)**

| Parameters | Expression of the results | Guide level | Maximum admissible concentration |
|---|---|---|---|
| Temperature | °C | 12 | 25 |
| Hydrogen ion concentration* | pH unit | $6.5 \leq pH \leq 8.5$ |  |
| Conductivity* | μS/cm at 20°C | 400 |  |
| Chlorides* | Cl mg/L | 25 |  |
| Sulfates | $SO_4$ mg/L | 25 | 500 |
| Silica* | $SiO_2$ mg/L |  |  |
| Calcium | Ca mg/L | 100 |  |
| Magnesium | Mg mg/L | 30 | 50 |
| Sodium* | Na mg/L | 20 | 150–175 |
| Potassium | K mg/L | 10 | 12 |
| Aluminum | Al mg/L | 0.05 | 0.2 |
| Total hardness* |  |  |  |
| Dry residues | mg/L after drying at 180°C | 1500 |  |
| Dissolved oxygen* | % $O_2$ saturation |  |  |
| Free carbon dioxide* | $CO_2$ mg/L |  |  |

*Refer to EEC standards for comments.

**TABLE 1.15    European Economic Community Standards for Parameters Concerning Substances Undesirable in Excessive Amounts**

| Parameters | Expression of the results | Guide level | Maximum admissible concentration |
|---|---|---|---|
| Nitrates | $NO_3$ mg/L | 25 | 50 |
| Nitrites | $NO_2$ mg/L | | 0.1 |
| Ammonium | $NH_4$ mg/L | 0.05 | 0.5 |
| Kjeldahl nitrogen* | N mg/L | | 1 |
| $KMnO_4$ oxidizability* | $O_2$ mg/L | 2 | 5 |
| Total organic carbon (TOC)* | C mg/L | | |
| Hydrogen sulfide | S µg/L | | Undetectable organoleptically |
| Substances extractable in chloroform | mg/L dry residue | 0.1 | |
| Dissolved or emulsified hydrocarbons* | µg/L | | 10 |
| Phenols (phenol index)* | $C_6H_5OH$ µg/L | | 0.5 |
| Boron | B µg/L | 1000 | |
| Surfactants | µg/L (lauryl sulfate) | | 200 |
| Other organochlorine compounds* | µg/L | 1 | |
| Iron | Fe µg/L | 50 | 200 |
| Manganese | Mn µg/L | 20 | 50 |
| Copper* | Cu µg/L | 100, 3000 | |
| Zinc* | Zn µg/L | 100, 5000 | |
| Phosphorus | $P_2O_5$ µg/L | 400 | 5000 |
| Fluoride* | F µg/L | | |
| | 8–12°C | | 1500 |
| | 25–30°C | | 700 |
| Cobalt | Co µg/L | | |
| Suspended solids | | None | |
| Residual chlorine* | Cl µg/L | | |
| Barium | Ba µg/L | 100 | |
| Silver* | Ag µg/L | | 10 |

*Note:* Certain of these substances may even be toxic when present in very substantial quantities.

*Refer to EEC standards for comments.

**TABLE 1.16    European Economic Community Standards for Parameters Concerning Toxic Substances**

| Parameters | Expression of the results | Guide level | Maximum admissible concentration |
|---|---|---|---|
| Arsenic | As μg/L | | 50 |
| Beryllium | Be μg/L | | |
| Cadmium | Cd μg/L | | 5 |
| Cyanides | CN μg/L | | 50 |
| Chromium | Cr μg/L | | 50 |
| Mercury | Hg μg/L | | 1 |
| Nickel | Ni μg/L | | 50 |
| Lead* | Pb μg/L | | 50 (in running water) |
| Antimony | Sb μg/L | | 10 |
| Selenium | Se μg/L | | 10 |
| Vanadium | V μg/L | | |
| Pesticides and related products* | μg/L | | |
| Substances considered separately | | | 0.1 |
| Total | | | 0.5 |
| PAH* | μg/L | | 0.2 |

*Refer to EEC standards for comments.

**TABLE 1.17    European Economic Community Standards for Microbiological Parameters**

| Parameters | Results: volume of the sample, mL | Guide level | Maximum admissible concentration | |
|---|---|---|---|---|
| | | | Membrane filter method | Multiple tube method |
| Total coliforms* | 100 | — | 0 | < 1 |
| Fecal coliforms | 100 | — | 0 | < 1 |
| Fecal streptococci | 100 | — | 0 | < 1 |
| Sulfite-reducing Clostridia | 20 | — | — | < 1 |

**TABLE 1.17    European Economic Community Standards for Microbiological Parameters (*Continued*)**

| Parameters | Temperature, °C | Results: size of the sample, mL | Guide level | Maximum admissible concentration |
|---|---|---|---|---|
| Total bacteria counts for water supplied for human consumption† | 37 | 1 | 10 | |
| | 22 | 1 | 100 | |
| Total bacteria counts for water in closed containers† | 37 | 1 | 5 | 20 |
| | 22 | 1 | 20 | 100 |

*Provided a sufficient number of samples is examined (95 percent consistent results).
†Water intended for human consumption should not contain pathogenic organisms.
*Note:* If it is necessary to supplement the microbiological analysis of water intended for human consumption, the samples should be examined not only for the bacteria referred to in this table but also for pathogens including: salmonella, pathogenic staphylococci, fecal bacteriophages, and enteroviruses; nor should such water contain parasites, algaes, or other organisms such as animalcules (worms, larvae).

**TABLE 1.18    European Economic Community Standards for Minimum Required Concentration for Softened Water Intended for Human Consumption**

| Parameters* | Expression of the results | Minimum required concentration (softened water) |
|---|---|---|
| Total hardness | Ca mg/L | 60 |
| Hydrogen ion concentration | pH | |
| Alkalinity | HCO$_3$ mg/L | 30 |
| Dissolved oxygen | | |

*Refer to EEC standards for comments.

with the USEPA to develop a voluntary, third-party, private-sector program for evaluating drinking water additives. The National Sanitation Foundation is leading this effort through a unique consortium approach, which includes the American Water Works Association, American Water Works Association Research Foundation, Association of State Drinking Water Administrators, and the Conference of State Health and Environmental Managers. The USEPA has terminated its additives advisory program. Several third-party certification mechanisms are being developed to implement the consortium standards. These include Underwriter's Laboratories, the Safe Water Additives Institute, and the National Sanitation Foundation.

### USEPA drinking water health advisories [46,47]

The USEPA also provides nonregulatory health guidance to assist in dealing with incidences of drinking water contamination. Health advisories describe concentrations of contaminants in drinking water at which adverse health effects would not be anticipated to occur. A margin of safety is included to protect sensitive members of the population.

Health advisories are developed from data describing carcinogenic and noncarcinogenic toxicity. For chemicals that are known or probable human carcinogens, nonzero 1-day, 10-day, and longer-term advisories may be derived, with attendant caveats. Projected excess lifetime cancer risks are provided to give an estimate of the concentrations of the contaminant that may pose a carcinogenic risk to humans. To date, approximately 200 drinking water health advisories have been prepared.*

### USEPA water quality criteria

Under the directives of the Clean Water Act, the USEPA publishes water quality criteria for lakes, rivers, and streams. The criteria contain two essential types of information:[19] (1) discussions of available scientific data on the effects of pollutants on public health and welfare, aquatic life, and recreation and (2) quantitative concentrations or qualitative assessments of the pollutants in water that will generally ensure water quality adequate to support a specified water use. These criteria are based solely on data and scientific judgments on the relationship between pollutant concentrations and environmental and human health effects. Criteria values do not reflect considerations of economic or technological feasibility.

The water quality criteria are not regulations. Rather, these criteria present scientific data and guidance on the environmental effect of pollutants that can be useful in deriving regulatory requirements based on considerations of water quality impacts.

### Trends for the Future

Legislative mandates, regulatory processes, and policies contribute to the development of national drinking water regulations and guide-

---

*Many of these documents are available for a fee from the National Technical Information Service, U.S. Department of Commerce, Washington, D.C. (800-426-4791).

lines in the United States. Regulations not only apply specifically to the public water supplier, but they also have potentially important roles in other contexts where specifications for protection of surface and groundwater sources, discharge controls, and cleanup requirements are established by federal or state actions.

Overall, the quality of drinking water in the United States is good, but the length and breadth of lists of drinking water standards will continue to grow in the foreseeable future as the technological advancement of analytical science, toxicology, and human hazard assessment continues. Demands upon water suppliers for more frequent monitoring and more sophisticated treatment processes to assure higher-quality drinking water will increase along with the cost of public drinking water. Thus, the responsibility for assuring the safety of drinking water at the tap will be shared by federal, state, and local authorities; the public water supplier; and consumers.

## Acknowledgments

At the time this manuscript was being prepared, Craig D. Vogt was Deputy Director, Criteria and Standards Division, U.S. Environmental Protection Agency, Office of Drinking Water, Washington, D.C.

## References

1. J. A. Borchardt and G. Walton, "Water Quality," in American Water Works Association, *Water Quality and Treatment,* 3d ed., McGraw-Hill, New York, 1971.
2. M. N. Baker, *The Quest for Pure Water,* vol. I, 2d ed., McGraw-Hill and American Water Works Association, New York, 1981.
3. "Interstate Quarantine Act of 1893," *U.S. Statutes at Large,* chap. 114, vol. 27, February 15, 1893, p. 449.
4. *Federal Register,* title 42, chap. 1, part 72, March 6, 1962, p. 2152.
5. *Federal Register,* title 40, part 248, December 24, 1975, p. 59569.
6. *Federal Register,* title 41, part 13, January 20, 1976, p. 2916.
7. *Federal Register,* title 44, part 231, November 29, 1979, p. 68624.
8. *Federal Register,* title 45, part 168, August 27, 1980, p. 57332.
9. *Federal Register,* title 42, part 62, March 31, 1977, p. 17143.
10. *Federal Register,* title 44, part 140, July 19, 1979, p. 42195.
11. *Federal Register,* title 51, part 63, April 2, 1986, p. 11396.
12. *Federal Register,* title 53, part 97, May 22, 1989, p. 22062.
13. USEPA., "Industrial Pollution of the Lower Mississippi River in Louisiana. Region VI," Dallas, Texas, April 1976.
14. J. J. Rook, "Formation of Haloforms During Chlorination of Natural Water," *Water Treatment and Examination,* vol. 23, part 2, 1974, p. 234.
15. T. A. Bellar et al., "The Occurrence of Organohalides in Chlorinated Drinking Water," *Journal AWWA,* vol. 66, no. 12, December 1974, p. 703.
16. J. J. Rook, "Chlorination Reactions of Fulvic Acids in Natural Waters," *Environmental Science and Technology,* vol. 11, 1977, p. 478.
17. J. M. Symons et al., "National Organics Reconnaissance Survey for Halogenated Organics," *Journal AWWA,* vol. 67, no. 11, November 1975, p. 634.
18. H. J. Brass et al., "National Organics Monitoring Survey: Sampling and Analyses

for Partible Organic Compounds," in R. J. Pojasek (ed.), *Drinking Water Quality Enhancement Through Source Protection,* Ann Arbor Science, Ann Arbor, Mich., 1977.

19. "The Clean Water Act," PL-100-4, 1981.

20. J. J. Westrick et al., "The Ground Water Supply Survey," *Journal AWWA,* vol. 76, no. 5, May 1984, p. 52.

21. G. F. Craun and W. Jakubowski, "Status of Waterborne Giardiasis Outbreaks and Monitoring Methods." American Water Resources Association, *Water Related Health Issues Symp.,* Atlanta, Ga., November 1986.

22. E. C. Lippy and S. C. Waltrip, "Waterborne Disease Outbreaks—1946–1980: A Thirty-Five Year Perspective," *Journal AWWA,* vol. 76, no. 2, February 1984, p. 60.

23. J. P. Longtin, *Final Report—National Inorganics and Radionuclides Survey (NIRS),* USEPA, Office of Drinking Water, Technical Support Division, Cincinnati, Ohio (in preparation).

24. "Safe Drinking Water Act," PL-99-339, June 19, 1986.

25. *Federal Register,* title 50, part 219, November 13, 1985, p. 46936.

26. J. A. Cotruvo, "Risk Assessment and Control Decisions for Protecting Drinking Water Quality," in I. H. Suffet and M. Malaiyandi, eds., *Advances in Chemistry,* American Chemical Society, Washington, D.C., 1987.

27. J. A. Cotruvo, "Drinking Water Standards and Risk Assessment," *Regulatory Toxicology and Pharmacology,* vol. 8, no. 3, September 1988, p. 288.

28. *Federal Register,* title 52, part 130, July 8, 1987, p. 25690.

29. *Federal Register,* title 53, part 14, January 22, 1988, p. 1892.

30. *Drinking Water and Health,* vol. I, National Academy of Sciences, Washington, D.C., 1977.

31. International Agency for Research on Cancer (IARC), IARC Monographs on the Evaluation of Carcinogens Risk of Chemicals to Humans, Supplement 4, Lyon, France, 1982.

32. *Federal Register,* title 51, part 185, September 24, 1986, p. 34000.

33. H. J. Brass et al., "Community Water Supply Survey: Sampling and Analysis for Partible Organics and Total Organic Carbon," *Proc. AWWA Annu. Conf.,* St. Louis, Mo., 1981.

34. USEPA, "National Screening Program for Organics, 1977–1981," Final Report, Office of Drinking Water, Washington, D.C., unpublished.

35. USEPA, "National Rural Water Survey," Office of Drinking Water, Washington, D.C., 1978, unpublished.

36. USEPA, "National Pesticide Survey," Ongoing project, Office of Pesticide Programs and Office of Drinking Water, Washington, D.C. (Publishing date in 1991).

37. *Federal Register,* title 54, part 124, June 29, 1989, p. 27486.

38. *Federal Register,* title 54, part 124, June 29, 1989, p. 27544.

39. Health and Welfare Canada. *Canadian Drinking Water Standards and Objectives, 1968,* Minister of Supply and Services, Ottawa, 1969.

40. Health and Welfare Canada, *Guidelines for Canadian Drinking Water Quality,* Minister of Supply and Services, 1978.

41. Health and Welfare Canada, *Guidelines for Canadian Drinking Water Quality,* 1987.

42. World Health Organization, *Guidelines for Drinking Water Quality,* Geneva, 1984.

43. *Official Journal of the European Communities,* August 30, 1980, vol. 23, Official Directive, no. L229/11–L229/23.

44. *Federal Register,* title 53, part 130, July 7, 1988, p. 25586.

45. USEPA, *Guidance Manual for Compliance with the Filtration and Disinfection Requirements for Public Water Supplies Using Surface Waters,* National Technical Information Services, PB 90 148016/AS, October 1989.

46. *Reviews of Environmental Contamination and Toxicolgy,* vols. 104, 106, and 107, Springer-Verlag, New York, 1988, 1989.

47. *Drinking Water Health Advisories: Pesticides,* Lewis Publishers, Inc., 1989.

# Health and Aesthetic Aspects of Water Quality

## Carol H. Tate, D. Env., P.E.

*Vice President*
*James M. Montgomery*
*Consulting Engineers*
*Pasadena, California*

## Katherine Fox Arnold

*Consultant*
*Tucson, Arizona*

Health and aesthetics are the motivation for water treatment. In the late 1800s and early 1900s, acute waterborne diseases, such as cholera and typhoid fever, spurred development and proliferation of filtration and chlorination plants. Subsequent identification in water supplies of additional disease agents, such as *Legionella* and *Giardia,* and contaminants, such as cadmium and lead, resulted in more elaborate pretreatments to enhance filtration and disinfection. Additionally, specialized processes such as granular activated carbon (GAC) adsorption and ion exchange were occasionally applied to water treatment for control of taste and odor and removal of contaminants such as nitrates.

A variety of developments in the water quality field during the last decade and a half and an increasing understanding of health effects are creating an upheaval in the water treatment field. With the identification in water of low levels of potentially harmful organic compounds, a coliform-free and turbidity-free water is no longer sufficient. New information regarding inorganic contaminants, such as lead, is forcing suppliers to tighten control of water quality within distribution systems. Increasing pressures on watersheds have resulted in a heavier incoming load of microorganisms to many treatment plants.

Although a similarly intense reevaluation of the aesthetic aspects of water quality has not occurred, aesthetic quality is important. Problems, such as excessive minerals, fixtures staining, and color, are generally local. Aesthetic quality has not warranted the national attention that health-associated contaminants do, but it does affect consumer acceptance of the water supply. The main exception to the general lack of activity within this area is in the control of taste and odors. Significant advances in the identification of taste- and odor-causing organisms and their metabolites have occurred within the last 2 decades.

This chapter summarizes the current (1988) state of knowledge on health and aesthetic aspects of water quality. Following an introductory discussion of waterborne disease outbreaks and basic concepts of toxicology, separate sections of the chapter are devoted to pathogenic organisms, indicator organisms, inorganic constituents, disinfectants, organic compounds, disinfection by-products, particulates, and radionuclides. Taste and odor, turbidity, color, mineralization, and hardness are discussed in a final section devoted to aesthetic quality.

Caution is necessary when applying the information presented in this chapter. The chapter is intended as an introductory overview of health effects information. The cited references and additional sources must be studied carefully and public health officials consulted prior to making any decisions regarding specific contamination problems.

## Waterborne Disease Outbreaks

In the past, infectious diseases were frequently transmitted through contaminated drinking water. Improvements in wastewater disposal practices and the development, protection, and treatment of water supplies have reduced the prevalence of infectious disease in the United States and other developed countries. To understand the problem of water-related disease, classifying water-related illness into the following four general groups based on epidemiological considerations is helpful: (1) waterborne diseases, (2) water-washed diseases, (3) water-based diseases, and (4) water-vectored diseases.

Waterborne diseases are those transmitted through the ingestion of contaminated water. The water acts as the passive carrier of the infectious or chemical agent. Classic waterborne diseases are cholera and typhoid fever. Chemical poisonings and methemoglobinemia may also be caused by contaminated water supply systems. In addition, chronic ingestion of low levels of some chemical contaminants in drinking water has been associated with adverse health effects, but these associations are poorly understood at present. Note that diseases caused by pathogenic bacteria, viruses, protozoans, and helminths are transmitted through the fecal-oral route from human to human or animal to human, with drinking water being only one of several possible sources of infection. Diarrheal illness is an important cause of infant mortality and morbidity in developing countries.

Water-washed diseases are related to poor hygienic habits and sanitation. Unavailability of water for washing and bathing contributes to diseases that affect the eye and skin. Water-based diseases are those in which the pathogen spends an essential part of its life in water or is dependent upon aquatic organisms for the completion of its life cycle. Schistosomiasis and dracontiasis are examples of water-based diseases. Water-vectored diseases, such as yellow fever and malaria, are transmitted by insects that breed in water or that bite near water. Although important in the protection of public health, water-washed, water-based, and water-vectored diseases will not be considered in this book, but references are available.[1,2]

The introduction of coliform drinking water standards in 1914 underscored the fact that waterborne microorganism disease outbreaks could be dramatically reduced through source control, filtration, and chlorination. A substantial number of outbreaks do still occur, however; and since 1950, there has been a steady increase in the number of reported cases (Fig. 2.1). Much of this increase is presumed to be caused by public awareness and associated increases in reporting of

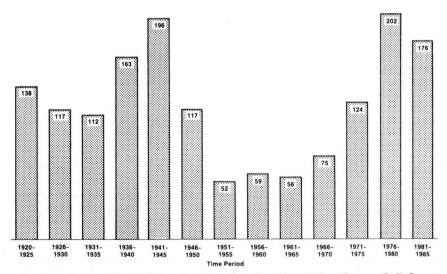

**Figure 2.1**  Number of waterborne disease outbreaks, 1920–1985.  (*Source: G. F. Craun, Surface Water Supplies and Health. Journal AWWA, vol. 80, no. 2, Feb. 1988, p. 40.*)

TABLE 2.1    Etiology of Waterborne Outbreaks in Water Systems, 1971 to 1985[3]

| Illness | Number of outbreaks | Cases of illness |
|---|---|---|
| Gastroenteritis, undefined | 251 | 61,478 |
| Giardiasis | 92 | 24,365 |
| Chemical poisoning | 50 | 3,774 |
| Shigellosis | 33 | 5,783 |
| Hepatitis A | 23 | 737 |
| Gastroenteritis, viral | 20 | 6,524 |
| Campylobacterosis | 11 | 4,983 |
| Salmonellosis | 10 | 2,300 |
| Typhoid | 5 | 282 |
| Yersiniosis | 2 | 103 |
| Gastroenteritis, toxigenic *E. coli* | 1 | 1,000 |
| Cryptosporidiosis | 1 | 117 |
| Cholera | 1 | 17 |
| Dermatitis | 1 | 31 |
| Amebiasis | 1 | 4 |
| Total | 502 | 111,228 |

outbreaks. Along with this have been improvements in sampling and analytical techniques that allow better reporting. Most outbreaks are caused by the use of contaminated, untreated water or by inadequacies in treatment, and the majority tend to occur in small community and noncommunity systems.[3-5]

Organisms associated with recent disease outbreaks include the protozoan *Giardia lamblia,* bacteria such as *Salmonella,* and viral agents such as hepatitis A. Table 2.1 summarizes the etiology of waterborne disease outbreaks reported in the United States from 1971 to 1985. Significantly, approximately half of the reported outbreaks have an undefined etiology. This is traced mostly to difficulties in sampling and culturing of the organisms responsible.[4]

## Understanding Health Effects of Chemicals

Every chemical has an effect on living organisms exposed to it.[6] Toxicology, the study of the adverse effects of chemicals on living organisms, provides a means of evaluating and understanding these effects. Although a complete review of toxicological principles is beyond the scope of this chapter, basic concepts needed to understand the information presented in this chapter are presented below.

Living organisms respond in one of two ways when exposed to a chemical. An organism may show an effect that is of no health consequence to the organism. Alternatively, an organism may show an adverse or deleterious effect that is undesirable and threatens the health of the organism. Some adverse health effects in organisms are immediate [within 24 to 48 hours (h) after exposure], but others are delayed (for example, 10 to 20 years or more for cancer in humans).[6] Adverse effects may be reversible depending upon their nature, the severity of the effect, and the organ affected.[6]

The response of a living organism exposed to a chemical depends upon the chemical dose or exposure level. The higher the dose, the more significant the effect. This is termed the dose-response relationship. Understanding this concept is important because simply knowing that a substance can have a particular toxicological property (i.e., it is carcinogenic) is not sufficient in and of itself to assess human health risk. The dose-response relationship must also be known, as well as information concerning human exposure, before a judgment can be made regarding the public health significance of that substance. The USEPA regulatory process described in Chap. 1 is designed to determine the acceptable levels of substances found in drinking water on a national basis which have known or suspected possible adverse health effects in humans.

A variety of adverse health effects are possible. In this chapter, the following general terms will be used to describe these effects.

*Toxic:*  Causing a deleterious response in a biologic system, seriously injuring function, or producing death. These effects may result from acute conditions (short high-dose exposure), chronic (long-term, low-dose) exposure, or subchronic (intermediate-term and dose) exposure.

*Neurotoxic:*  Exerting a destructive or poisonous effect on nerve tissue.

*Carcinogenic:*  Causing or inducing uncontrolled growth of aberrant cells into malignant tumors.

*Mutagenic:*  Causing heritable alteration of the genetic material within living cells.

*Teratogenic:*  Causing nonhereditary congenital malformations (birth defects) in offspring.

Many tests are used for both short- and long-term assessments of toxicology. Two procedures often employed are the National Cancer Institute (NCI) feeding study for the detection of carcinogens and the Ames test for mutagenicity. In the former test, male and female rats, mice, or hamsters selected for sensitivity to carcinogens, are fed the test chemical for 2 years. Doses are the maximally tolerated dose (MTD) and half the MTD. The incidence of neoplasms in the experimental group is then statistically compared to a control group.[7] The Ames test, also known as the *Salmonella* microsome assay, is one of

TABLE 2.2    Cancer Risk Estimates for VOCs[9]

| | Concentration in drinking water, corresponding to a $10^{-5}$ risk, μg/L | |
| Compound | NAS* | CAG† |
| --- | --- | --- |
| Trichloroethylene | 45 | 26 |
| Tetrachloroethylene | 35 | 6.7 |
| Carbon tetrachloride | 45 | 2.7 |
| 1,2-Dichloroethane | 7.0 | 3.8 |
| Vinyl chloride | 10 | 0.15 |

*National Academy of Sciences.
†Cancer Assessment Group.

many in vitro or in vivo screening systems. Strains of *Salmonella typhimurium,* specially developed to have a histidine requirement, are exposed to the test chemical. Mutagenicity in that test system is measured by comparing reversions to histidine independence in the experimental group to those that would occur spontaneously in a control group.[8] Since many known carcinogens test positive in the Ames test, it has been used as a candidate screening tool for carcinogens.

Health effects information is compiled by the USEPA as a basis for drinking water regulations. The agency uses a number of sources of data and also peer review systems, including the NAS and the IARC, and its own external science advisory board. The assessment framework and associated nomenclature is described in Chap. 1, while the conclusions are reported in this chapter. Because of statutory demands, the task of systematically evaluating the occurrence and health effects of many constituents is sometimes proceeding much more quickly than would be dictated purely by traditional scientific consensus procedures. The SDWA requires that the USEPA make conservative judgments to assure protection of public health. Thus, gaps in data and discrepancies among different groups' conclusions occur. This is most notable in the assignment of cancer risks, where the use of different models and assumptions can lead to significantly different figures. Table 2.2 shows the concentrations of five volatile organic compounds corresponding to a $10^{-5}$ incremental lifetime cancer risk as developed by the NAS and the USEPA cancer assessment group (CAG).[9] Within this chapter, for consistency, CAG values are used. Other discrepancies in conclusions are noted wherever possible; however, precedence is given to USEPA values.

## Pathogenic Organisms

Table 2.3 lists potential waterborne disease-causing organisms. Shown are characteristic elements of the disease each organism

TABLE 2.3    Potential Waterborne Disease-Causing Organisms

| Name of organism or group | Major disease | Major reservoirs and primary sources |
|---|---|---|
| | Bacteria | |
| *Salmonella typhi* | Typhoid fever | Human feces |
| *Salmonella paratyphi* | Paratyphoid fever | Human feces |
| Other salmonella | Salmonellosis | Human and animal feces |
| Shigella | Bacillary dysentery | Human feces |
| *Vibrio cholerae* | Cholera | Human feces |
| Enteropathogenic *E. coli* | Gastroenteritis | Human feces |
| *Yersinia enterocolitica* | Gastroenteritis | Human and animal feces |
| *Campylobacter jejuni* | Gastroenteritis | Human and animal(?) feces |
| *Legionella pneumophila* and related bacteria | Acute respiratory illness (legionellosis) | Thermally enriched waters |
| *Mycobacterium tuberculosis* | Tuberculosis | Human respiratory exudates |
| Other (atypical) mycobacteria | Pulmonary illness | Soil and water |
| Opportunistic bacteria | Variable | Natural waters |
| | Enteric Viruses | |
| Enteroviruses | | |
|   Polioviruses | Poliomyelitis | Human feces |
|   Coxsackieviruses A | Aseptic meningitis | Human feces |
|   Coxsackieviruses B | Aseptic meningitis | Human feces |
|   Echoviruses | Aseptic meningitis | Human feces |
|   Other enteroviruses | Encephalitis | Human feces |
| Reoviruses | Mild upper respiratory and gastrointestinal illness | Human and animal feces |
| Rotaviruses | Gastroenteritis | Human feces |
| Adenoviruses | Upper respiratory and gastrointestinal illness | Human feces |
| Hepatitis A virus | Infectious hepatitis | Human feces |
| Norwalk and related GI viruses | Gastroenteritis | Human feces |
| | Protozoans | |
| *Acanthamoeba castellani* | Amoebic meningoencephalitis | Soil and water |
| *Balantidium coli* | Balantidosis (dysentery) | Human feces |
| *Cryptosporidium* | Cryptosporidiosis | Human and animal feces |
| *Entamoeba histolytica* | Amoebic dysentery | Human feces |
| *Giardia lamblia* | Giardiasis (gastroenteritis) | Human and animal feces |
| *Naegleria fowleri* | Primary amoebic meningoencephalitis | Soil and water |
| Algae (blue-green) | | |
| *Anabaena flos-aquae* | Gastroenteritis | Natural waters |
| *Microcystis aeruginosa* | Gastroenteritis | Natural waters |
| *Alphanizomenon flos-aquae* | Gastroenteritis | Natural waters |
| *Schizothrix calciola* | Gastroenteritis | Natural waters |

causes and major reservoirs or primary sources of each organism. The information given in this table is further detailed in Ref. 10.

According to convention, every biological species bears a latinized name that consists of two words.[11] The first word indicates the taxonomic group of immediately higher order, or genus, to which the species belongs, and the second word identifies it as a particular species of that genus. The first letter of the generic name is capitalized, and the whole phrase is italicized or underlined. The generic name is often abbreviated to its initial letter, e.g., *E. coli.*

Health aspects of organisms most significant to drinking water are discussed below. Many of these organisms are actually a group of intimately related microorganisms distinguished on the basis of antigenic composition, known as serotypes. Space limitations do not allow a complete treatise on all organisms listed in Table 2.3. Additional information on the characteristics of the organisms listed may be found in the cited references.

## Bacteria

Bacteria are single-celled microorganisms. They possess no well-defined nucleus, are devoid of chlorophyll, and reproduce by binary fission.[11] Bacteria exhibit almost all possible variations in morphology, from the simple sphere to very elongated, branching threads. Many serious diseases of humans are caused by bacteria.[11] Bacteria of interest in drinking water are *Salmonella, Shigella, Yersinia enterocolitica, Campylobacter jejuni, Legionella,* enteropathogenic *E. coli, Vibrio cholerae, Mycobacterium,* and opportunistic bacteria.

**Salmonella.** Over 2200 known serotypes of *Salmonella* exist, all of which are pathogenic to humans. Most cause gastroenteritis; however, *S. typhi* and *S. paratyphi* cause typhoid and paratyphoid fevers, respectively. These last two species only infect humans; the others are carried by both humans and animals. At any one time, 0.1 percent of the population will be excreting *Salmonella* (not, however, because of waterborne-derived disease; the bulk of infections are caused by contaminated foods). The 2300 cases of waterborne salmonellosis reported to have occurred between 1971 and 1985 were primarily associated with small systems.[3]

**Shigella.** Four main subgroups exist in this genus, *S. sonnei, S. flexneri, S. boydii,* and *S. dysenteriae.* They infect humans and primates and cause bacillary dysentery. *S. sonnei* causes the bulk of waterborne infections, although all four subgroups have been isolated during different disease outbreaks. Waterborne shigellosis is most often the result of contamination from one identifiable source, such as an improperly disinfected well. The survival of *Shigella* in water and their response to water treatment is similar to that of the coliform bacteria. Systems adequately controlling coliforms should be protected against *Shigella*. Between 1971 and 1985, 33 outbreaks were reported. These outbreaks were attributed to inadequately maintained and/or monitored water systems.[3]

*Yersinia enterocolitica.*   This organism can cause acute gastroenteritis and is carried by humans and a variety of animals. Two outbreaks of waterborne gastroenteritis (one reported in 1974 and one in 1977) were linked to *Yersinia*; however, definitive association was never made. *Yersinia* can grow at temperatures as low as 39°F (4°C) and has been isolated in untreated surface water more frequently during colder months. Field studies and surveys indicate that *Yersinia* is somewhat resistant to chlorine, and correlates poorly with levels of total or fecal coliforms.

*Campylobacter jejuni.*   This organism can infect humans and a variety of animals. In humans it causes acute gastroenteritis for a period of about 1 week. Four outbreaks were reported in the United States between 1971 and 1981, three of which were in municipal systems. Coliform tests were not indicative of contamination in any of the three. Eight more occurred between 1981 and 1985. Laboratory tests indicate that conventional chlorination should control the organism; however, in only one of the three municipal outbreaks was inadequate disinfection implicated.

*Legionella.*   Twenty-six species of *Legionella* have been identified. Seven are etiologic agents of Legionnaires disease; however, *Legionella pneumophila* is the primary causative agent. *Legionella* organisms can cause effects similar to acute pneumonia or a milder, nonpneumonic illness designated Pontiac fever. The former typically is found in immunosuppressed individuals, or those who are susceptible because of age, smoking, or underlying illness.

   *Legionella pneumophila* is a naturally occurring and widely distributed organism. It was isolated from all samples taken in a survey of 67 rivers and lakes in the United States. Higher isolations occurred from warmer water. Some data suggest that *Legionella* are less prevalent in groundwaters.[12] *Legionella* are relatively resistant to chlorine, and small numbers are expected in the finished water of systems employing full treatment.[12] *Legionella* can effectively colonize plumbing systems, especially warm ones, and aerosols from fixtures, such as showerheads, are postulated as the typical routes of infection. Cooling towers have also been suggested as sites of *Legionella* proliferation and sources of transmission.[9] Thus far, little evidence exists to suggest that ingestion of water containing *Legionella* leads to infection.[13]

**Enteropathogenic *E. coli*.** Approximately 11 of the more than 140 existing serotypes of *E. coli* cause gastroenteritis in humans. Enteropathogenic *E. coli* is a prime cause of diarrhea in infants. A number of waterborne enteropathogenic disease outbreaks have been traced to fecal-contaminated water. During the period 1961 to 1970, four outbreaks with 188 associated cases were reported in the United States.[4] Prior to 1961, enteropathogenic *E. coli* had not been implicated in any disease outbreaks, and since 1970 no new outbreaks have been detected. Environmental processes and water treatment are effective in controlling *E. coli;* its presence is an indication of recent fecal contamination from warm-blooded animals.[14]

***Vibrio cholerae*.** This virulent pathogen produces acute intestinal disease with diarrhea, vomiting, suppression of urine, dehydration, reduced blood pressure, subnormal temperature, and complete collapse. Death may occur within a few hours unless medical treatment is given. *Vibrio cholerae* has been associated with epidemics throughout the world. The first recorded cholera epidemic was reported in 1563 in India. Development of protected water supplies, control of sewage discharges, and water treatment dramatically reduced widespread epidemics; however, they still occur. In 1970, epidemics were reported in southern Russia, Africa, Egypt, and Guinea.[14] In the United States one outbreak caused by *V. cholerae* occurred during the period 1971 to 1985. The outbreak was caused by wastewater contamination of an oil rig's potable water system, resulting in 17 cases of severe diarrhea.

**Mycobacteria.** *Mycobacterium tuberculosis* causes tuberculosis in humans. It is typically transmitted via person-to-person contact; however, sewage-contaminated water is a potential pathway.[14] Species of nontubercular mycobacteria, *M. kansasii* and *M. avium-intracellulare,* can cause pulmonary and other diseases in individuals. Those identified as being at highest risk are the very young, the elderly, and persons with preexisting health conditions. Some evidence exists linking drinking water to disease caused by mycobacteria. Potable water can, however, support their growth and transmit them through the distribution system. In addition, waterborne *M. avium-intracellulare* has been epidemiologically linked to infections in hospital patients.

**Opportunistic bacteria.** Opportunistic bacteria comprise a heterogeneous group of gram-negative bacteria that cause disease infrequently, but they can cause significant disease in newborns, the elderly, and individuals with preexisting health problems. Included in this group are *Pseudomonas, Aeromonas hydrophila, Edwardsiella tarda, Flavobacterium, Klebsiella, Enterobacter, Serratia, Proteus, Providencia, Citrobacter,* and *Acinetobacter*. These organisms are ubiquitous in the environment and are usually the predominant group of bacteria in finished water. The heterotrophic plate count (HPC) can, thus, provide a good measure of their occurrence. High-quality supplies, meeting coliform standards, usually contain between 1 and 500 HPC organisms per milliliter.

### Viruses

Viruses are a large group of infectious agents ranging from 10 to 25 nanometers (nm) in diameter. They are not cells, but particles composed of a protein sheath surrounding a nucleic-acid core.[11] Viruses are characterized by their total dependence on living cells for reproduction and by lack of independent metabolism.[11]

Viruses belonging to the group known as enteric viruses infect the GI tract of humans and, in some cases, animals and are excreted in the feces of infected people or animals. Over 100 types of enteric viruses exist. Enteric viruses of particular interest in drinking water are hepatitis A, Norwalk-type viruses, rotaviruses, adenoviruses, enteroviruses, and reoviruses.

**Hepatitis A.**   Although all the enteric viruses are potentially transmitted by drinking water, evidence of this route of infection is strongest for hepatitis A (HAV). HAV causes infectious hepatitis, an illness characterized by inflammation and necrosis of the liver. Symptoms include fever, weakness, nausea, vomiting, and diarrhea. Jaundice may occur. Between 1971 and 1985, 23 outbreaks of disease caused by HAV were documented by the Centers for Disease Control (CDC).[3]

Because of difficulties in culturing HAV (and other enteric viruses), it is only recently that definitive information on its occurrence in drinking water supplies and/or its removal during treatment has been generated. Rao et al. have shown effective removal of HAV through coagulation, flocculation, and filtration. They also confirmed that HAV is somewhat more resistant to removal and inactivation than are some other enteric viruses.[15]

**Norwalk-type viruses.**   The Norwalk-type viruses cause acute epidemic gastroenteritis. At least three serologically distinct viruses exist in this group. Like HAV, these viruses have not been cultured in the laboratory; however, immunochemical methods exist for detecting their antigens and antibodies. Evidence of waterborne disease outbreaks attributed to Norwalk-type viruses is based on serological evidence from individuals who are ill. No information exists on their occurrence in source waters or their removal during treatment.

**Rotaviruses.** Rotaviruses cause acute gastroenteritis, primarily in children. In developing countries, rotavirus infections are a major cause of infant mortality. Immunochemical methods for detection of rotavirus antigens and antibodies are available; culturing rotaviruses is difficult. Rotaviruses have been detected in finished drinking waters. Rao et al. recently demonstrated effective rotavirus removal during pilot-scale coagulation, flocculation, and filtration processes. Removals of greater than 99 percent were consistently achieved.[15]

**Adenoviruses, enteroviruses, and reoviruses.**    These are the three other groups of enteric viruses. Viruses in each group can infect both the enteric and upper respiratory tract. Adenoviruses have been detected in wastewater and contaminated surface water, but not to date in drinking water. Entero- and reoviruses have been detected in wastewater, natural water, and finished drinking water. No outbreaks of illness from drinking water contaminated with these viruses have been documented; therefore, their significance as disease-causing agents in drinking water is uncertain. Drinking water has been implicated as the route of transmission for the poliomyelitis virus (an enterovirus), but the epidemiological evidence is not conclusive.

### Protozoans

Protozoans are unicellular, colorless, generally motile organisms that lack a cell wall.[11] Various groups of protozoans exist, including amoebas, flagellated protozoans, ciliates, and sporozoans. Several protozoans are pathogenic to humans. Pathogenic protozoans of interest in drinking water are *Giardia lamblia, Entamoeba histolytica, Cryptosporidium,* and *Naegleria fowleri.*

*Giardia lamblia.*    This flagellated protozoan can exist as a trophozoite, 9 to 21 μm long, or as an ovoid cyst, approximately 10 μm long and 6 μm wide[16] (Fig. 2.2). When ingested, *Giardia* can cause giardiasis, a GI disease manifested by diarrhea, fatigue, and cramps. Symptoms may last anywhere from a few days to months. Between 1971 and 1985, 92 reported outbreaks of giardiasis occurred, involving 24,365 cases.[3] The infectious dose for giardiasis is low; humans tested at doses between 1 and $10^6$ cysts were infected at all levels above 10 cysts,[17] while mice were infected at doses as low as 1.4 cysts.[9]

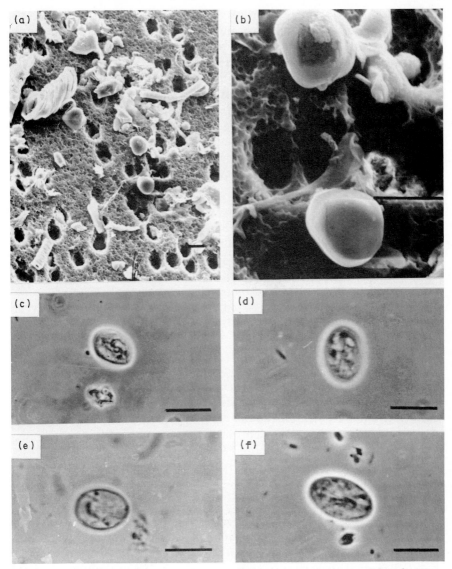

**Figure 2.2**  *Giardia* cysts from humans. (*a*) An SEM micrograph of several *Giardia* cysts from human material on an 8-μm pore membrane at 750 × . (*b*) The same material magnified to 3000 × . (*c*) to (*f*) Photo micrographs of cysts of the same sources at 2000 × . [*Source: W. Jakubowski and J. C. Hoff (eds.), Waterborne Transmission of Giardiasis, USEPA 600/6-79-001, June 1979.*]

Water can be a major vehicle for transmission of giardiasis, although person-to-person contact and other routes are more important. Samples taken from 204 surface waters serving as drinking water sources tested 48 percent positive for *Giardia*.[18] Both humans and animals, particularly beavers and muskrats, are reservoirs for *Giardia*. Significantly, *Giardia* cysts can survive in water for 1 to 3 months.

*Giardia* cysts are considerably more resistant to chlorine than coliforms. Granular media filtration can achieve greater than 99 percent removal of cysts provided that pretreatment is optimized. When coagulant doses are low or suboptimal, or when finished water turbidities are >1.0 NTU, poor cyst removal occurs. Diatomaceous earth filtration can also be effective in removing cysts.[19]

**Entamoeba histolytica.** This protozoan exists either as a fragile trophozoite, 15 to 25 $\mu$m long, or as a more hardy cyst, 10 to 15 $\mu$m in diameter (slightly larger than *Giardia*). When ingested, *Entamoeba* can cause amoebic dysentery, with symptoms ranging from acute diarrhea and fever to mild gastroenteritis. Occasionally the organism can invade the bloodstream, reach other organs (typically the liver), and cause amoebic abscesses.

The last reported waterborne outbreak of disease caused by *E. histolytica* was in 1953; however, 2000 to 3000 cases of amoebiasis typically occur in the United States each year, and the prevalence of *E. histolytica* in human carriers is estimated at 3 to 10 percent of the population. Thus, the potential for additional outbreaks exists. Unlike *Giardia,* however, animals do not serve as reservoirs, so the potential for source water contamination is low.

*Entamoeba histolytica* is more resistant to chlorine than are either enteric bacteria or viruses. Conventional filtration can, however, effectively remove the protozoan.

**Cryptosporidium.** Since 1907, a variety of species of the protozoan *Cryptosporidium* have been identified. Two are associated with infections in mammals, *C. muris* and *C. paryum*. Isolates from animals have been shown to initiate infections in humans. Human cryptosporidiosis was first reported in 1976. Medical interest subsequently increased because of its occurrence as a life-threatening infection in acquired immune deficiency syndrome (AIDS) patients. *Cryptosporidium* infections have now been reported in surveys on six continents.[20] A recent (1987) waterborne outbreak of several thousand cases of cryptosporidiosis occurred in a filtered water system in Carollton, Ga. Prior to that, one other outbreak with 177 cases had been reported.[3] Another recent outbreak has occurred in the vicinity of London, England.

The primary symptom of cryptosporidiosis is acute diarrhea. Other symptoms may include abdominal pain, vomiting, and low-grade fever. Sources of infection include pets and farm animals, person-to-person contacts, and contaminated drinking water. The infectious dose has not been well defined.

An ongoing investigation into the occurrence of *Cryptosporidium* in western United States water has detected the organism a high percentage of the time. Of 74 samples taken from rivers, lakes, streams, and reservoirs, 56 have been positive, with a range of 0.04 to 18 oocysts per liter. The higher concentrations were in streams receiving wastewater discharges. In finished drinking water samples, oocysts were found in two of four unfiltered, chlorinated waters, and in one water receiving filtration without coagulation. No oocysts were detected in water receiving conventional treatment.[21] Specific testing of individual processes for *Cryptosporidium* removal efficiencies has not yet generated definitive data.

**Naegleria fowleri.**   This free-living amoeba is found in soil, water, and decaying vegetation. The trophozoite measures 8 to 15 μm. Exposure is typically from swimming in freshwater lakes having high concentrations of the organism. The route of infection is via the nasopharynx, through the olfactory epithelium, and up the olfactory nerve plexus to the brain. Primary amoebic encephalitis results with symptoms of headache, high fever, and somnolence. Death usually occurs on the fifth or sixth day. All disease incidents have been associated with swimming in natural water; treated drinking water supplies are not a suspected route of transmission. The risk of infection has been estimated at less than 1 in 2.5 million exposures.

## Algae

Algae are unicellular, generally nonmotile plants. They range in size from about 5 to 100 μm. All use photosynthesis as their primary mode of nutrition. Algae are typically not of health concern; however, certain species may produce endo- or exotoxins, which, if ingested at high enough concentrations, may be harmful.

Three species of blue-green algae, *Anabaena flos-aquae, Microcystis aeruginosa,* and *Aphanizomenon flos-aquae*, produce exotoxins. Toxin concentrations generated during algal blooms have resulted in illness or death in mammals, birds, and fish that have ingested a sufficient dose. Some evidence exists that human exposure in recreational and possibly drinking water has caused contact irritation and possibly gastroenteritis.

Endotoxins are a normal component of the outer wall of heterotrophic, gram-negative bacteria and blue-green algae. They are ubiquitous in water. They can cause toxemia and shock in immunosuppressed patients; however, little evidence exists that ambient levels of endotoxin found in most water supplies pose a health risk to the normal population. High concentrations of endotoxin associated with a bloom of *Schizothrix calciola* may have been responsible for an outbreak of gastroenteritis in Sewickley, Pa.

## Indicator Organisms

Isolating and identifying specific pathogenic microorganisms is beyond the capability of most water utility laboratories; thus, protection of public health is achieved by the application of treatment technologies. Because of technical difficulties, and also because the number of pathogens relative to other microorganisms in water can be very small, indicator organisms are typically used to measure the effectiveness of the treatment process, and thus indirectly the apparent safety of the drinking water.

The primary function of an indicator organism is to provide evidence of recent fecal contamination from warm-blooded animals. The general criteria for an indicator organism are

1. The indicator should always be present when the pathogenic organism of concern is present, and absent in clean, uncontaminated water.

2. The indicator should be present in fecal material in large numbers.

3. The indicator should respond to natural environmental conditions and to treatment processes in a manner similar to the pathogens of interest.

4. The indicator should be easy to isolate, identify, and enumerate.

5. The ratio of indicator/pathogen should be high.

6. The indicator and pathogen should come from the same source (GI tract).

A number of microorganisms have been evaluated as indicators, including total coliforms, fecal coliforms, *E. coli*, fecal streptococci, *Pseudomonas aeruginosa*, enterococci, and HPC. Yeasts have also recently been proposed as effective indicators.[22] However, total coliforms have been the indicator of choice for decades, and remain so.

Both fecal coliforms and the HPC have application. Each is briefly described below and in Chap. 18. Information on the specific utility of other indicator organisms is available.[23]

## Total coliform

The coliform group of organisms, referred to as total coliform, is defined in water bacteriology as all the aerobic and facultatively anaerobic, gram-negative, nonspore-forming, rod-shaped bacteria that ferment lactose with gas formation within 48 h at 35°C. This is an operational rather than a taxonomic definition and encompasses a variety of organisms, mostly of intestinal origin. The definition includes *E. coli*, the most numerous facultative bacterium in the feces of warm-blooded animals, plus species belonging to the genera *Enterobacter, Klebsiella,* and *Citrobacter.* These latter organisms are present in wastewater but can be derived from other environmental sources, such as soil and plant materials.

No organism fulfills all the criteria for an indicator organism, but the coliform bacteria fulfill most. Drawbacks to the use of total coliform as an indicator include their regrowth in water, thus becoming part of the natural aquatic flora. Their detection then becomes a false positive. False positives can also occur when the bacteria *Aeromonas,* which can biochemically mimic the coliform group, is present in a sample. On the other hand, false negatives can occur when coliforms are present along with high populations of HPC bacteria. The latter organisms may act to suppress coliform activity.[24] Finally, a number of pathogens have been shown to survive longer in natural water and/or through various treatment processes than coliforms. This is particularly true for protozoans and viral pathogens. Despite these drawbacks, the total coliform measure remains the most useful indicator of drinking water microbial quality, and coliform numbers in finished water are regulated by the USEPA.

## Fecal coliforms

Fecal coliforms provide stronger evidence of the possible presence of fecal pathogens than do total coliforms. Fecal coliforms are a subgroup of the total coliforms, distinguished in the laboratory through elevated temperature tests (43 to 44.5°C, depending on the test). Although the test does determine coliforms of fecal origin, it does not distinguish between human and animal contamination. Fecal coliform numbers are typically much lower than are total coliform numbers, and they are not as widely used as an overall indicator of source water contamination, treatment effectiveness, and posttreatment degradation or contamination.

### Heterotrophic plate count

The HPC measures a broad group of bacteria including nonpathogens, pathogens, and opportunistic pathogens. Because of its lack of specificity, the HPC cannot be correlated with any likelihood of waterborne disease outbreak; drinking water having any positive HPC level might contain many, few, or no pathogens. The significance of the HPC lies in its indication of poor general biological quality of the drinking water. Five hundred colonies per milliliter has been suggested as an upper level above which corrective action should be taken.

## Inorganic Constituents

Inorganic constituents may be present in natural water, in contaminated source water, or, in some cases, in water which has had contact with piping or plumbing materials. Lead, copper, and asbestos are constituents that can derive from distribution and plumbing systems.

Selected inorganics found in drinking water can cause a variety of health concerns. Some are known or suspected carcinogens. Arsenic, lead, and cadmium fall into this group. A number of inorganics are essential to human nutrition at low doses, yet demonstrate adverse health effects at higher doses. These include arsenic, selenium, chromium, copper, molybdenum, nickel, zinc, and sodium. Two inorganics, sodium and barium, have been associated with high blood pressure.[25] Numerous reports have also shown an inverse relationship between water hardness and hypertensive heart disease.[25–27] Each of these relationships is equivocal and under continuing investigation.

Health aspects of inorganic constituents of interest in drinking water are summarized below. Drinking water regulations and health advisories are listed in Table 2.4. Health advisories are periodically reviewed by the USEPA as new data become available. Additional information on health effects of inorganics may be found in the cited references.

### Aluminum

Aluminum occurs naturally in nearly all foods, the average dietary intake being about 20 mg/day. Aluminum is common in treated drinking water, especially those treated with alum. A 1983 survey found a

**TABLE 2.4  USEPA Drinking Water Regulations and Health Advisories for Inorganic Contaminants**

| Chemical | Status reg.* | NIPDWR, mg/L | MCLG, mg/L | MCL, mg/L | Status HA* | 1-day, mg/L (10-kg child) | 10-day, mg/L (10-kg child) | Longer-term, mg/L (10-kg child) | Longer-term, mg/L (70-kg adult) | RfD, mg/(kg·day) | DWEL, mg/L | Life-time, mg/L | $10^{-4}$ Cancer risk, mg/L | USEPA Cancer Group |
|---|---|---|---|---|---|---|---|---|---|---|---|---|---|---|
| Aluminum | L | — | — | — | D | — | — | — | — | — | — | — | — | — |
| Ammonia | L | — | — | — | D | — | — | — | — | — | — | — | — | — |
| Antimony | T | — | 0.003 | 0.01/0.005 | D | 0.015 | 0.015 | 0.015 | 0.015 | $4 \times 10^{-4}$ | 0.015 | 0.003 | — | D |
| Arsenic | T | 0.05 | zero | 0.05 | D | — | — | — | — | 0.001 | — | — | 0.003 | A |
| Asbestos (>10 μm) | P | — | $7 \times 10^6$ fibers/L | $7 \times 10^6$ fibers/L | — | — | — | — | — | — | — | — | — | — |
| Barium | P | 1 | 5 | 5 | F | 5 | 5 | 5 | 5 | — | — | 5 | — | D |
| Beryllium | T | — | zero | 0.001 | D | 30 | 30 | 4 | 200 | 0.005 | 0.2 | — | $7 \times 10^{-4}$ | B2 |
| Boron | L | — | — | — | D | 3 | 2 | 0.9 | 3 | 0.09 | 3 | 0.6 | — | D |
| Cadmium | P | 0.01 | 0.005 | 0.005 | F | 0.04 | 0.04 | 0.005 | 0.02 | $5 \times 10^{-4}$ | 0.02 | 0.005 | — | D |
| Chromium (total) | P | 0.05 | 0.1 | 0.1 | F | 1 | 1 | 0.2 | 0.8 | 0.005 | 0.2 | 0.1 | — | D |
| Copper | P | — | 1.3 | 1.3 | — | — | — | — | — | — | — | — | — | D |
| Cyanide | T | — | 0.2 | 0.2 | F | 0.2 | 0.2 | 0.2 | 0.8 | 0.022 | 0.8 | 0.2 | — | D |
| Fluoride | F | 1.4–2.4 | 4 | 4 | — | — | — | — | — | 0.060 | — | — | — | — |
| Lead (at source) | P | — | zero | 0.005 | — | — | — | — | — | — | — | — | — | B2 |
| Lead (at tap) | P | 0.05 | zero | TT† | — | — | — | — | — | — | — | — | — | B2 |
| Manganese | P | — | — | — | F | — | — | — | — | — | — | — | — | D |
| Mercury | L | 0.002 | 0.002 | 0.002 | D | — | — | — | 0.002 | $3 \times 10^{-4}$ | 0.01 | 0.002 | — | D |
| Molybdenum | T | — | — | — | F | — | — | — | — | — | — | — | — | D |
| Nickel | P | — | 0.1 | 0.1 | F | 1 | 1 | 0.1 | 0.6 | 0.02 | 0.6 | 0.1 | — | D |
| Nitrate (as N) | P | 10 | 10 | 10 | F | 10 | 10 | — | — | — | — | — | — | — |
| Nitrite (as N) | P | — | 1 | 1 | F | 1 | 1 | — | — | — | — | — | — | — |
| Nitrate and Nitrite | P | — | 10 | 10 | — | — | — | — | — | — | — | — | — | — |
| Selenium | P | 0.01 | 0.05 | 0.05 | — | 0.2 | 0.2 | 0.2 | 0.2 | 0.005 | 0.2 | 0.1 | — | D |
| Silver | L | 0.05 | 0.05 | 0.05 | D | — | — | — | — | — | — | — | — | — |
| Sodium | L | — | — | — | D | — | — | — | — | — | 20‡ | — | — | — |

TABLE 2.4 USEPA Drinking Water Regulations and Health Advisories for Inorganic Contaminants (Continued)

| | Regulations | | | | Health advisories | | | | | | | | | |
| | | | | | | 10-kg child | | | 70-kg adult | | | | | |
| Chemical | Status reg.* | NIPDWR, mg/L | MCLG, mg/L | MCL, mg/L | Status HA* | 1-day, mg/L | 10-day, mg/L | Longer-term, mg/L | Longer-term, mg/L | RfD, mg/(kg·day) | DWEL, mg/L | Lifetime, mg/L | $10^{-4}$ Cancer risk, mg/L | USEPA Cancer Group |
|---|---|---|---|---|---|---|---|---|---|---|---|---|---|---|
| Strontium | L | — | — | — | D | 25 | 25 | 25 | 90 | 2.5 | 90 | 17 | — | D |
| Sulfate | T | — | 400 | 400 | D | — | — | — | — | — | — | — | — | — |
| Thallium | T | — | $5 \times 10^{-4}$ | 0.002/0.001 | D | 0.007 | 0.007 | 0.007 | 0.02 | $7 \times 10^{-5}$ | 0.002 | $4 \times 10^{-4}$ | — | — |
| Vanadium | L | — | — | — | D | — | — | — | — | — | — | — | — | D |
| Zinc | L | — | — | — | D | — | — | — | — | — | — | — | — | — |

Information based on USEPA, Office of Drinking Water, Criteria and Standards Division, summary of drinking water regulations and health advisories (April 1990).

Definitions for abbreviations in column heads are as follows:

NIPDWR—National interim primary drinking water regulation. Interim enforceable drinking water regulations first established under the Safe Drinking Water Act that are protective of public health to the extent feasible.

MCLG—Maximum contaminant level goal. A nonenforceable concentration of a drinking water contaminant that is protective of adverse human health effects and allows an adequate margin of safety.

MCL—Maximum contaminant level. Maximum permissible level of a contaminant in water that is delivered to any user of a public water system.

RfD—Reference dose. An estimate of a daily exposure to the human population that is likely to be without appreciable risk of deleterious effects over a lifetime.

DWEL—Drinking water equivalent level. A lifetime exposure concentration protective of adverse, noncancer health effects, that assumes that all the exposure to a contaminant is from a drinking water source.

*The codes for the Status regulations and Status health advisory columns are as follows: F—final; D—draft; L—listed for regulation; P—proposed; T—tentative (to be proposed).

†TT—treatment technique.

‡—Guidance.

NOTE: Large discrepancies between lifetime and longer-term health advisory values may occur because of the USEPA's conservative policies, especially with regard to carcinogenicity, relative source contribution, and less-than-lifetime exposures in chronic toxicity testing. These factors can result in a cumulative uncertainty factor (UF) of 10 to 1000 when calculating a lifetime health advisory.

range of 0.014 mg/L in groundwater with no alum coagulation to 2.57 mg/L in surface water with coagulation.[9]

Aluminum shows low acute toxicity. At lower doses, aluminum administered to laboratory animals is a neurotoxin. Chronic exposure data are limited, but indicate that aluminum likely affects phosphorus absorption that can create weakness, bone pain, and anorexia. Carcinogenicity, mutagenicity, and teratogenicity tests have all been negative. An association between aluminum and two neurological disorders, Alzheimer's disease and dialysis dementia, has been hypothesized.[28–30]

Aluminum was included on the original list of 83 contaminants to be regulated under the 1986 SDWA amendments. The USEPA removed aluminum from the list as one of the seven substitutes allowed (see Chap. 1) because it concluded that no evidence existed at the time (1988) that aluminum ingested in drinking water poses a health threat.[31] Research on the health effects of aluminum continue, and aluminum is included on the USEPA's Drinking Water Priority List. The USEPA has proposed an SMCL of 0.05 mg/L to ensure removal of coagulated material ahead of the distribution system.[32]

### Arsenic

Arsenic concentrations in U.S. drinking water are typically low; McCabe et al.[33] reported arsenic at levels greater than 0.01 mg/L in only 4 percent of drinking water supplies sampled. Data from two surveys found arsenic greater than 0.005 mg/L in 55 of 330 groundwater supplies and in only 2 of 115 surface supplies.[9] Erosion of arsenic-containing surface rocks probably accounts for a significant amount of the arsenic in water supplies. The other major source of environmental arsenic is the smelting of nonferrous metal ores, especially copper.[26]

Arsenic is a likely dietary requirement and is a constituent of many foods such as meat, fish, poultry, grain, and cereals. It is often present in organic arsenical forms, which are less toxic than inorganic arsenic. During 1971 to 1974 the average daily intake of arsenic was 11.4 μg.[27]

In excessive amounts, arsenic causes acute GI damage and cardiac damage. Chronic doses can cause vascular disorders, such as blackfoot disease. Epidemiological studies in Chile and Taiwan have linked arsenic with skin and lung cancer; however, these findings have not been supported by an animal model. Arsenic has been shown to be mutagenic in several bacterial test systems, and sodium arsenate and arsenite have shown teratogenic potential in several mammalian species. The USEPA has classified arsenic as a human carcinogen (group A).[9]

### Asbestos

Asbestos is the name for a group of naturally occurring, hydrated silicate minerals with fibrous morphology. Included in this group are the minerals chrysotile, crocidolite, anthophyllite, and some of the tremolite-actinolite and cummingtonite-grunerite series. All except chrysotile fibers are known as amphibole. Most commercially mined asbestos is chrysotile. Asbestos occurs in water exposed to natural deposits of these minerals, asbestos mining discharges, and asbestos-cement pipe.[26]

Samples taken from 406 cities in 47 states, Puerto Rico, and the District of Columbia showed the following concentrations of asbestos fibers.[9]

| Asbestos concentration, $10^6$ fibers/L | Number of cities |
|---|---|
| Less than detection limit | 117 |
| Less than 1 | 216 |
| 1–10 | 33 |
| Greater than 10 | 40 |

The physical dimensions of asbestos fibers rather than the type are more important in health effects, with the longer, thinner fibers more highly associated with cancers by inhalation. Human occupational and laboratory animal inhalation exposures are associated with lung cancer, pleural and peritoneal mesothelioma, and GI tract cancers.[34] The latter group may be associated with ingestion of some fraction of inhaled fibers; however, only one study has shown an association between asbestos in the diet of laboratory animals and possible carcinogenicity. This 1984 National Toxicology Program (NTP) study found a significant increase in benign epithelial neoplasms in the large intestines of only male rats exposed to intermediate range fibers (65 percent of the fibers greater than 10 μm, 14 percent greater than 100 μm) for their lifetime.[9]

The CAG bases its 1/1,000,000 cancer risk estimate of $7 \times 10^6$ long fibers per liter on the NTP rat study that concluded only long fibers (10 μm or longer) were involved. The NAS estimates a 1/1,000,000 risk of 11,000 fibers per liter based on occupational exposure data, assuming 30 percent of inhaled fibers are ingested. Inhaled asbestos is clearly carcinogenic. The USEPA has classified ingested asbestos as a possible human carcinogen (group C), with a proposed MCLG and a proposed MCL of $7 \times 10^6$ fibers per liter (10 μm or longer).[32]

## Barium

Barium occurs naturally in trace amounts in most surface and groundwaters from their exposure to barium-containing rocks. Industrial release of barium occurs from oil and gas drilling muds, coal-fired power plants, jet fuels, and auto paints.[9] Barium in typical finished water ranges from 1 to 172 μg/L, with a mean of 28.6 μg/L. The drinking water of many communities in Illinois, Kentucky, Pennsylvania, and New Mexico, however, can contain barium at 10 times the current NIPDWR of 1 mg/L.[30]

The acute effects of barium exposure include prolonged stimulation of the cardiac, GI, and neuromuscular systems. Chronic exposure may contribute to hypertension. A 16-month study of rats given 100 mg/L barium in drinking water demonstrated hypertensive effects; however, human epidemiological studies with community drinking water containing 2 to 10 mg/L barium did not provide definitive results.

The USEPA has not classified the carcinogenicity of barium (group D) because of inadequate evidence. An MCLG and a revised MCL for barium of 5 mg/L have been proposed.[32]

## Beryllium

Beryllium is not common in drinking water. Groundwater surveyed between 1962 and 1967 had a 5.4 percent detection frequency with a maximum concentration of 1.22 μg/L. Surface water in the same surveys had a 1.1 percent detection frequency with a maximum concentration of 0.17 μg/L.

Inhalation of beryllium causes pulmonary ailments but has not been shown to cause lung cancers in humans. It can, however, cause osteosarcomas in laboratory animals irrespective of the mode of administration.[26]

Initial analysis showed limited potential for drinking water exposure to cause a significant health risk. No regulation has been proposed; however, data collection is continuing.[9,32] Beryllium is included on the list of contaminants to be regulated under the 1986 SDWA amendments; a proposal is expected in 1990.

## Cadmium

Cadmium enters the environment from a variety of industrial applications, including mining and smelting operations, electroplating, and pigment and plasticizer production. Cadmium occurs as an impurity in zinc and may enter consumers' tap water as a result of galvanized-pipe corrosion.[26] NIPDWR compliance monitoring as of November 1985 shows 25 public water supplies with levels of cadmium greater than 0.010 mg/L (the NIPDWR standard). Federal surveys conducted between 1969 and 1980 showed a mean concentration of 3 µg/L in 707 groundwater supplies, and 3.2 µg/L in 117 surface water supplies.[9]

Cadmium acts as an emetic at ingested doses of 3 to 90 mg, becomes toxic at 10 to 326 mg, and is fatal at 300 to 3500 mg. Chronic exposure results in renal dysfunction. Cadmium has been shown to induce sarcomas at injection sites in laboratory animals and to induce lung tumors in rats exposed to cadmium chloride via aerosol for carcinogenic effects.[9]

The USEPA has classified cadmium as a probable human carcinogen (group B1), based on positive carcinogenicity testing; however, cadmium is being regulated based on its renal toxic effects, because the carcinogenicity occurs via inhalation. An MCLG and an MCL of 0.005 mg/L have been proposed.[32]

### Chromium

Chromium occurs in drinking water in its +3 and +6 valence states with +3 being more common. The valence is affected by the level of disinfection and presence of reducible organics. Primary sources in water are old mining operations, wastes from plating operations, and fossil fuel combustion. NIPDWR compliance monitoring shows 17 groundwater systems and 1 surface water system with chromium levels greater than 50 µg/L. Three of the groundwater systems serve more than 10,000 people, while the surface water system serves more than 100,000 people. Federal surveys conducted between 1969 and 1980 showed a mean of 10 µg/L for surface water systems and a mean of 16 µg/L for groundwater systems.[9]

Chromium III is nutritionally essential, nontoxic, and poorly absorbed. Deficiency results in glucose intolerance, inability to use glucose, and other metabolic disorders. The NAS estimates a safe and adequate intake of 0.05 to 0.20 mg/day.[27]

Chromium VI is toxic, producing liver and kidney damage, internal hemorrhage, and respiratory disorders. Subchronic and chronic effects include dermatitis and skin ulceration.[9] Chromium VI has been shown to cause cancer in humans and animals through inhalation exposure, but it has not been shown to be carcinogenic through ingestion exposure. The USEPA classifies chromium as a human carcinogen (group A), although the proposed MCLG and MCL (for total chromium) of 0.1 mg/L are based upon noncancer toxic effects.[32]

## Copper

Copper is commonly found in drinking water. Low levels (generally below 20 µg/L) can derive from rock weathering. Some industrial contamination also occurs, but the principal sources in water supplies are corrosion of brass and copper pipe and the addition of copper salts during water treatment for algal control.[9] A 1967 survey of 380 finished waters conducted by the U.S. Department of the Interior found copper concentrations from 1 to 1060 µg/L, with a mean of 43 µg/L. A more recent region V USEPA survey of 83 finished waters found copper at less than 5.0 to 200 µg/L, well below the SMCL of 1.0 mg/L. These measurements did not, however, include increases in copper within the distribution system.

Copper is a nutritional requirement. Lack of sufficient copper leads to iron deficiency and reproductive abnormalities. The NAS lists a safe and adequate copper intake of 2 to 3 mg/day.[27]

Copper doses in excess of nutritional requirements are excreted; however, at high doses, copper can cause acute effects such as GI disturbances, damage to the liver and renal systems, and anemia. A dose of 5.3 mg/day was the lowest at which GI tract irritation was seen. Exposure of mice via subcutaneous injection yielded tumors; however, oral exposure in several studies did not. Mutagenicity tests have been negative.

The USEPA has not classified the carcinogenicity of copper (group D) because of inadequate evidence. Proposed regulation of copper is as a primary instead of a secondary contaminant. The proposed MCLG and MCL are 1.3 mg/L.[9]

## Cyanide

Cyanide occurs as an industrial pollutant and is not commonly found in drinking water at significant levels. It is also biologically and chemically degradable (e.g., reaction with chlorine). The 1970 USEPA Community Water Supply Survey of 969 systems found an average cyanide concentration of 0.09 µg/L and a maximum of 8 µg/L.

Cyanide is readily absorbed from the lungs, GI tract, and skin. It combines with cell cytochrome and prevents oxygen transport. With chronic exposure, cyanide can be detoxified in the liver; cyanide is converted to thiocyanate. The carcinogenicity of cyanide has not been evaluated. Potassium cyanide was negative when tested for mutagenicity in bacterial systems.[9] The USEPA has not classified the carcinogenicity of cyanide (group D) because of inadequate evidence based upon inadequate carcinogenicity data in animals and humans. No MCLG or MCL has been proposed.[9,32]

## Fluoride

Fluoride occurs naturally in most soils and in many water supplies. For more than 30 years, fluoride has been added to supplies lacking sufficient natural quantities, for the purpose of reducing dental caries. Health effects of fluoride are discussed in Chap. 15.

## Hardness

Hardness is generally defined as the sum of the polyvalent cations present in water and expressed as an equivalent quantity of calcium carbonate ($CaCO_3$). The most common such cations are calcium and magnesium. Although no distinctly defined levels exist for what constitutes a hard or soft water supply, water with less than 75 mg/L $CaCO_3$ is considered to be soft and above 150 mg/L $CaCO_3$ to be hard.

An inverse relationship has been postulated between the incidence of cardiovascular disease and the amount of hardness in the water, or, conversely, a positive correlation with the degree of softness. Hypotheses outlined below suggest a protective effect from either the major or minor constituents of hard water, or, conversely, a harmful effect from elements more commonly found in soft water.[27]

Several investigators attribute a protective effect to the presence of calcium and magnesium. A moderate increase in calcium in the diet has been observed to lower levels of circulating organ cholesterol; this is speculated as a possible factor in relating water hardness and cardiovascular disease. Magnesium is theorized to protect against lipid deposits in arteries and may also have some anticoagulant properties that could protect against cardiovascular diseases by inhibiting blood clot formation.

A limited number of studies have been carried out which suggest that minor constituents often associated with hard water may exert a beneficial effect on the cardiovascular system. Candidate trace elements include vanadium, lithium, manganese, and chromium. On the other hand, other investigators suggest that certain trace metals found in higher concentrations in soft water, such as cadmium, lead, copper, and zinc, may be involved in the induction of cardiovascular disease.

Whether drinking water provides enough of these elements to have any significant impact on the pathogenesis of cardiovascular disease is uncertain. Hard water generally supplies less than 10 percent of the total dietary intake for calcium and magnesium, for example. Water provides even smaller proportions of the total intake for the various suspect trace metals. Given the level of uncertainty in this area, the USEPA currently has no national policy with respect to the hardness or softness of public water supplies. However, the USEPA strongly supports corrosion-control measures (some of which add hardness) to reduce exposure to lead.

## Lead

Lead occurs in drinking water primarily from corrosion of lead pipes and solders, especially in areas of soft water. Federal surveys have found lead in 539 of 706 supplies using groundwater with a mean concentration of 13 μg/L and a range from 5 to 182 μg/L. The same surveys found lead in 100 of 119 supplies using surface water, with a mean concentration of 14 μg/L and a range of 5 to 32.5 μg/L.[9]

Health effects of lead are generally correlated with blood test levels. Infants and young children absorb ingested lead more readily than do older children and young adults.[30] Lead exposure across a broad range of blood lead levels is associated with a continuum of pathophysiological effects, including interference with heme synthesis necessary for formation of red blood cells, anemia, kidney damage, impaired reproductive function, interference with vitamin D metabolism, impaired cognitive performance, delayed neurological and physical development, and elevations in blood pressure.[36] The USEPA has classified lead as a probable human carcinogen (group B2), because some lead compounds cause renal tumors in rats.[9]

Under the interim primary drinking water regulations, the MCL for lead is 0.05 mg/L. The USEPA has proposed setting the MCLG for lead at zero based on subtle effects at low blood lead levels, the overall USEPA goal of reducing total lead exposures, and probable carcinogenicity at very high doses.[36] An MCL of 0.005 mg/L following treatment has been proposed,[36] along with monitoring at the tap and treatment requirements if lead is too high.

## Mercury

Mercury occurs primarily as an inorganic salt in water, and as organic (methyl) mercury in sediments and fish. Federal surveys conducted on 106 supplies between 1978 and 1980 show 14 groundwater and 5 surface water systems having a mercury concentration greater than 2 μg/L.

Inorganic mercury is poorly absorbed in the GI tract, does not penetrate cells, and is not as toxic as methyl mercury. The primary target organ of inorganic mercury is the kidney. Methyl mercury targets the central nervous system (CNS), causing death and/or mental and motor dysfunctions. Fish contaminated with methyl mercury caused the well-known mercury poisonings in Japan's Minamata Bay and Niigata, where a large number of deaths and CNS disorders occurred. The carcinogenicity of mercury has not been evaluated.[9]

The USEPA has not classified the carcinogenicity of mercury (group D) because of inadequate evidence.[9] The proposed MCLG and MCL are 0.002 mg/L.[32]

## Molybdenum

Molybdenum is an essential trace element in the diet. Average daily intake of molybdenum is 0.1 to 0.46 mg/day, the bulk of which comes from food. A 1970 survey of finished waters found molybdenum in 30 percent of the samples with a mean of 85.9 $\mu$g/L and a range of 3 to 1024 $\mu$g/L. The estimated safe intake for molybdenum matches the typical intake at 0.15 to 0.5 mg/day.[27]

Acute effects of overexposure to molybdenum include damage to the liver and kidneys. Chronic exposure can result in weight loss, bone abnormalities, and male infertility. Carcinogenicity data are inadequate because the only study conducted used an inorganic pigment containing lead and chromate, both of which could skew the results. The USEPA has not classified the carcinogenicity of molybdenum (group D) because of inadequate evidence. No MCLG has been proposed.[9]

## Nickel

Nickel is common in drinking water. The USEPA's Community Water Supply Survey (CWSS) detected nickel in 86 percent of groundwater supplies and 84 percent of surface supplies tested, usually around 1 ppb.

Most ingested nickel is excreted; however, some absorption from the GI tract does occur. Acute effects of overexposure include decreased weight gain, blood and enzyme changes, and changes in organ iron content. Nickel compounds are carcinogenic via inhalation and injection in laboratory animals. The incidence of respiratory tract cancers in nickel refinery workers is significantly higher than for the population as a whole. Nickel has not, however, been shown to be carcinogenic via oral exposure. Several studies suggest that it is not carcinogenic at 5 mg/L in drinking water given to rats and mice. Nickel chloride tested negative in bacterial mutagenicity screening; however, both nickel chloride and nickel sulfate tested positive in a mutagenicity test.[9]

The USEPA has classified nickel as a probable human carcinogen (group B1) based upon inhalation exposure. This classification is based upon limited evidence of carcinogenicity in humans and sufficient evidence in animals. No MCLG has been proposed.[9]

## Nitrate and nitrite

Nitrate is one of the major ions in natural waters. The mean concentration of nitrate nitrogen in a typical surface water supply would be around 1 to 2 mg/L; however, individual wells can have significantly higher concentrations. A survey conducted in South Dakota found that out of 1000 wells, 4 percent had $NO_3$-N greater than 100 mg/L, 9 percent had greater than 50 mg/L, 17 percent had greater than 20 mg/L, and 27 percent had greater than 10 mg/L.[37] Nitrite does not typically occur in natural water at significant levels; its presence would indicate likely wastewater contamination and/or a lack of oxidizing conditions.

Nitrate in drinking water causes two adverse health effects: induction of methemoglobinemia, especially in infants, and the potential formation of carcinogenic nitrosamines. Methemoglobinemia occurs because nitrate is reduced to nitrite in saliva and in the GI tract. This occurs to a much greater degree in infants than in adults (100 versus 10 percent) because of more alkaline conditions in their upper GI tract. The nitrite then oxidizes hemoglobin to methemoglobin, which cannot act as an oxygen carrier in the blood. Anoxia and death can then occur.[26] Carcinogenic nitrosamines are formed when nitrate or nitrite are administered with nitrosatable compounds. Ingestion of nitrate or nitrite without nitrosatable compounds has no demonstrated carcinogenicity.[9]

The USEPA has not classified the carcinogenicity of nitrate and nitrite (group D) because of inadequate evidence. MCLGs and MCLs of 10 mg/L and 1 mg/L for nitrate-N and nitrite-N, respectively, have been proposed.[9,32] In addition, an MCL for total nitrate-N and nitrite-N of 10 mg/L has been proposed.[32]

## Selenium

Selenium is an essential dietary element, with most intake coming from food. The levels found in food reflect local soil conditions. For example, New Zealand has very low soil concentrations resulting in a typical human intake of 56 μg/day. In contrast, local soils in South Dakota are high in selenium and inhabitants may intake 7000 μg/day.[27]

The concentration of selenium in drinking water is typically low. NIPDWR compliance monitoring showed 150 groundwater and 6 surface water systems with a selenium concentration greater than 10 μg/L. In the 1978 USEPA CWSS and the USEPA Rural Water Survey (RWS), 42 of 329 groundwater samples had a selenium concentration greater than 5 μg/L, and 10 of 329 had a selenium concentration greater than 10 μg/L. In the same surveys, 2 of 115 surface supplies had a selenium concentration greater than 5 μg/L.[9]

Acute toxicity caused by high selenium intake is observed in laboratory animals and in animals grazing in highly seleniferous areas. Effects include renal, liver, and heart damage. Similar toxicity in humans has not been demonstrated. Persons living in areas with high selenium in the food and water demonstrate variable effects, not conclusively linked to selenium. Dermatitis, hair loss, abnormal nail formation, and psychological disturbances have all been attributed to chronically high selenium intakes. Selenium reacts in vivo with other elements, protecting against heavy metal toxicity from mercury, cadmium, silver, and thallium.[9]

Naturally occurring selenium compounds have not been shown to be carcinogenic in animals. Selenium may inhibit tumor formation.[9] Teratogenicity testing has demonstrated sensitivity of the chick embryo to selenium.[26]

The USEPA has not classified the carcinogenicity of selenium (group D) because of inadequate evidence.[9] An MCL of 0.010 mg/L exists under the NIPDWR. An MCLG and a new MCL of 0.05 mg/L have been proposed.[32]

### Silver

Trace amounts of silver are found in natural and finished water originating from natural sources and industrial wastes. It has a low solubility, 0.1 to 10 mg/L depending on the pH and chloride concentration. Silver can be used as a disinfectant, but typically is not,[26] except in some home water treatment units. The 1969 CWSS found 309 out of 677 groundwaters having a silver concentration in the range of 0.1 to 9 μg/L, and 50 out of 109 surface waters with silver concentrations of 0.1 to 4 μg/L. NIPDWR compliance monitoring found 12 groundwater supplies and 1 surface water supply with silver greater than 50 μg/L.[9]

Chronic exposure to silver results in argyria, a blue-gray discoloration of the skin and organs. This is a permanent aesthetic effect and results from 1 to 5 g of accumulation. The USEPA considers argyria to be an aesthetic effect, not an adverse health effect, because it does not impair functioning of the body or cause other physiological problems. No known cases of argyria caused by exposure to silver through drinking water have been reported.[31] Silver shows no evidence of carcinogenicity or mutagenicity. Consequently, silver was removed from the list of 83 contaminants listed in the 1986 SDWA amendments.[31] The USEPA has proposed an SMCL for silver of 0.09 mg/L.[32]

## Sodium

Sodium is a major constituent in drinking water. A survey of 2100 finished waters conducted between 1963 and 1966 by the U.S. Public Health Service found concentrations ranging from 0.4 to 1900 mg/L, with 42 percent having sodium greater than 20 mg/L and 5 percent having greater than 250 mg/L. Of a typical daily intake of 2400 mg, drinking water, at a typical concentration of 21 mg/L, contributes 1 percent.

Sodium is associated with high blood pressure and heart disease in the "at-risk" population, comprised of persons genetically predisposed to essential hypertension. In addition, certain diseases are aggravated by a high salt intake, including congestive heart failure, cirrhosis, and renal disease. Similarly harmful effects for the population as a whole have not been conclusively shown in numerous epidemiological studies.[27]

Intake from food is generally the major source of sodium. The typical intake for normal adults is 1100 to 3300 mg/day. For persons requiring restrictions on salt intakes, sodium levels are usually limited somewhere between 500 and 2000 mg/day. Some severe cases require intakes of less than 500 mg/day. In 1968, the American Heart Association recommended a drinking water concentration of 20 mg/L. Where water supplies contain more than 20 mg/L, dietary sodium restriction to less than 500 mg/day is difficult to achieve and maintain.[26]

The USEPA has not proposed an MCLG for sodium because of insufficient data showing the association between sodium in drinking water and hypertension in the general population and because of the normally minor contribution of drinking water to the total dietary intake of sodium.[9] The USEPA has suggested a guidance level of 20 mg/L for the protection of the at-risk population, as recommended by the American Heart Association.[9]

## Sulfate

Sulfate occurs in almost all natural water. The 1970 CWSS of 969 supplies found sulfate in the concentration range of less than 1 to 770 mg/L with a median concentration of 4.6 mg/L. Three percent of the samples had sulfate greater than 250 mg/L.

High concentrations of sulfate in drinking water result in transitory diarrhea. Acute diarrhea can cause dehydration, particularly in infants and young children. Persons living in areas having high sulfate concentrations in their drinking water easily adjust, with no ill effects. The USEPA is considering an MCLG and MCL of 400 mg/L sulfate to protect infants. The current SMCL for sulfate is 250 mg/L based upon aesthetic effects.

### Vanadium

Vanadium is not common in drinking water, although it may occur locally near residue piles from milling and mining operations. Source water surveys conducted between 1962 and 1967 had a 3.4 percent detection frequency with a maximum concentration of 300 μg/L and a mean of 40 μg/L. Finished waters surveyed demonstrated comparable results. Vanadium is poorly absorbed in the GI tract. Inhalation can inflame the lungs and cause some changes in blood cholesterol and enzyme levels. No evidence exists of chronic oral toxicity associated with vanadium. The USEPA removed vanadium from the list of 83 contaminants to be regulated under the 1986 SDWA amendments because available data do not justify establishment of an MCL at this time (1988).[31,38]

### Zinc

Zinc commonly occurs in source water and may be added to finished water through corrosion of metal pipes. Zinc was found in 76 percent of source waters surveyed between 1962 and 1967. The range of concentrations found was 2 to 1183 μg/L, with a mean of 64 μg/L. The same surveys found zinc in 77 percent of finished waters, with a range of 3 to 2010 μg/L.

Adverse health effects associated with zinc result more from too low an intake rather than from an excessive intake. This is not a common problem in the United States. Zinc deficiency results in growth failure, loss of taste, and hypogonadism. The adult requirement for zinc is 15 mg/day. Drinking water contributes about 3 percent of this requirement. In excess, zinc has been reported to cause muscular weakness and pain, irritability, and nausea. The level of zinc associated with these effects was 40 mg/L over a long period.

The USEPA established an SMCL of 5 mg/L for zinc based upon taste.[39] Zinc was one of the seven constituents removed from the list of 83 contaminants to be regulated under the 1986 SDWA amendments because the available data indicate that zinc in drinking water does not pose a public health risk.[9,31]

### Disinfectants and Inorganic Disinfection By-products

Chlorine has been the primary drinking water disinfectant in the United States for more than 70 years. A recent (1987) survey[40] found that about 85 percent of the U.S. utilities using surface waters and about 80 percent using groundwaters rely on chlorine. Other less common disinfectants used in drinking water treatment include chloramines, chlorine dioxide, ozone, and potassium permanganate. Other materials that can act as disinfectants include iodine, bromine, ferrate, silver, hydrogen peroxide, and ultraviolet (UV) light. Of these, iodine and bromine have been used to a very limited extent for drinking water disinfection, while the others, except UV light, are even less established.

From a health standpoint, the omnipotence of chlorine has come into question as by-products of chlorination are identified. Although judicious use of alternative disinfectants to replace or augment chlorine can provide equivalent microbiological control, questions about other disinfectants and/or the health effects of their by-products also exist.

As part of the primary drinking water regulations, the USEPA is evaluating the health effects of individual disinfectants as well as of their organic and inorganic by-products. Information on organic disinfection by-products (DBPs) is given in a subsequent section of this chapter. Information on the disinfectants themselves and their inorganic by-products is given below.

## Chlorine and chloramines

At room temperature, chlorine is a greenish-yellow poisonous gas. When added to water, however, chlorine combines with water to form hypochlorous acid that then ionizes to form hypochlorite ion. Under typical drinking water conditions, negligible chlorine gas ($Cl_2$) remains in solution. The relative amounts of hypochlorous acid and hypochlorite ion formed are dependent on pH and temperature. See Chap. 14 for a complete discussion of chlorine chemistry.

The most significant inorganic compounds that can be formed during chlorine disinfection (if ammonia is present) are chloramines. Hypochlorous acid reacts with ammonia to form monochloramine ($NH_2Cl$), dichloramine ($NHCl_2$), trichloramine or nitrogen trichloride ($NCl_3$), and other minor by-products. Monochloramine is the principal chloramine formed under usual drinking water treatment conditions and is, with increasing frequency, used as a disinfectant itself.

Chlorine reacts in water with residual organic material to produce THMs and a variety of other chlorinated and oxidized substances. Similar substances are generated in vivo by reactions between chlorine and gastric contents.

In humans, observed adverse health effects associated with monochloramine in drinking water have been limited to hemodialysis patients. Chloramines in dialysis baths cause oxidation of hemoglobin to methemoglobin and denaturation of hemoglobin. In addition, monochloramine inhibits the hexose monophosphate shunt that protects red blood cells from oxidative damage.[41] Tests conducted on healthy human volunteers to evaluate the effects of monochloramine in drinking water at doses up to 24 mg/L (short term), and 5 mg/L (for 12 weeks), showed no effects.[41]

In laboratory animals, hepatocellular effects (increased mitotic figures, cellular hypertrophy, and unusual chromatin patterns) were observed in rats and mice given monochloramine in drinking water at high doses ( > 100 mg/L) for 90 days. Lower doses produced no effects. Another study showed decreased glutathione levels in rats exposed to 1, 10, and 100 mg/L monochloramine for 6 months. The same study also showed decreases in red blood cell count and hematocrit at the 10- and 100-mg/L levels after 3 months. Overall, however, an apparent lack of dose- or time-dependent response existed for either the glutathione levels or the hematocrit parameters.[41]

Monochloramine was found to be weakly mutagenic to a strain of *Bacillus subtilis* and to *Vicia fabia* seeds. An organic concentrate treated with chloramine, however, produced only half as many revertants in the assay as did the same mixture treated with chlorine.[41] Carcinogenicity tests are under way. In one such study where water disinfected by monochloramine was concentrated via reverse osmosis and used in a mouse skin tumor initiation-promotion assay, positive findings were attributed to organic by-products, rather than to the monochloramine itself.

### Chlorine dioxide, chlorite, and chlorate

Chlorine dioxide is a green-yellow gas that decomposes readily and with explosive force to chlorine and oxygen. It is, therefore, usually manufactured on-site. In water, $ClO_2$ is soluble to 2.9 mg/L and stable in closed containers. Its primary use is as a bleach for the paper and textile industries; however, it has more recently been used in drinking water to control phenols, oxidation of iron and manganese, and for final disinfection prior to distribution.[27]

In drinking water, the predominant reaction products of chlorine dioxide are chlorite, chlorate, and chloride. Approximately 50 to 70 percent of the chlorine dioxide will initially react to form chlorite. A fraction of the chlorite will, in turn, be reduced to chloride.[42] Metabolic studies using rats indicate that this conversion also takes place in vivo.

The primary concern with using chlorine dioxide has been the toxic effects attributed to residual chlorite and chlorate. Because chlorine dioxide converts to chlorite in vivo, health effects attributed to the former may, in some degree, be caused by the latter. Nevertheless, chlorine dioxide, chlorite, and chlorate have each been evaluated separately in humans and in laboratory animals. Findings to date for each are described below.

**Chlorine dioxide.** Human volunteers given 250 mL of 40 mg/L $ClO_2$ experienced short-term (5 min) headaches, nausea, and light-headedness. In another study, subjects given increasing doses of $ClO_2$ every third day from 0.1 to 24 mg/L for a total of 16 days demonstrated no ill effects. Similarly, subjects given a dose of 5 mg/L daily for 5 weeks showed no effects.[41]

Higher doses in laboratory animals have generated abnormalities in blood, thyroid, and neurological systems. In rats, doses of 0.18 to 0.72 mg/kg did not produce methemoglobinemia, but did cause a reduction in blood glutathione concentrations. Rats and mice given 1, 10, 100, or 1000 mg/L $ClO_2$ in drinking water for up to 12 months showed an increase in catalase activity at 10, 100, and 1000 mg/L in mice and at 1000 mg/L in rats. No glutathione or methemoglobin effects were found. Another study demonstrated decreased osmotic fragility in rat erythrocytes after dosing with 10 and 100 mg/L $ClO_2$. African green monkeys given 100 mg/L $ClO_2$ (9.5 mg/kg) showed no hematological effects but did demonstrate a decrease in serum levels of thyroxine. This effect is thought to be caused by oxidation of dietary iodide in the GI tract. The oxidized iodide then binds to food or tissue and is unavailable for absorption.

In tests with mice no mutational effects were found. The carcinogenicity of $ClO_2$ has not been evaluated. Teratogenicity tests have been conducted with rats. Pups born to dams given $ClO_2$ at doses of 100 mg/L showed significant depression in serum thyroxine concentrations and a decrease in locomotor and exploratory behavior. Another group of pups whose dams were also dosed at 100 mg/L showed a significant reduction in cell numbers in their cerebella. Other tests have been less conclusive.[41]

**Chlorite.** Human volunteers given 1 L of water with increasing doses of sodium chlorite from 0.01 to 2.4 mg/L every third day for a total of 16 days showed no effects. Another group given 500 mL of water containing 5 mg/L sodium chlorite for 12 weeks showed no significant effects. A group of at-risk subjects lacking in G-6-PD (an enzyme), which renders them susceptible to oxidative stress, were given the same 12-week regimen described above. Changes in blood parameters were noted, but the small number of subjects involved made the significance of these findings questionable.[41]

Adverse effects of chlorite on the hematological systems of laboratory animals are well documented. A series of studies with rats, monkeys, and cats has shown chlorite to be a slightly less potent oxidant of hemoglobin than is nitrite, and less specific in its oxidation of cellular constituents. Subchronic doses of about 10 mg/(kg · day) (100 mg/L in drinking water for rats) for 30 to 60 days caused definitive damage to erythrocytes. In addition to oxidizing hemoglobin, chlorite was found to disrupt the red cell membrane, deplete the red cell glutathione, and induce production of hydrogen peroxide. These injuries to the red blood cells can cause an overall reduction in cell numbers, leading to hemolytic anemia.

Recent mutagenicity testing of chlorite in mouse assays was negative. The one carcinogenicity test conducted produced negative results for direct carcinogenicity and equivocal results for chlorite as a tumor promoter. In a variety of reproductive effect testing, chlorite was associated with a decrease in the growth rate of rat pups between birth and weaning, when the dams were given 100 mg/L chlorite in the drinking water during pregnancy. Some sperm aberrations were noted in males given 100 mg/L or higher doses.

**Chlorate.**   The human tests conducted with chlorine dioxide and chlorite (increasing dose over 16 days and a 5 mg/L dose for 12 weeks) were also conducted with chlorate. The increasing dose range was 0.01 to 2.4 mg/L. No clinically significant effects were noted.[41] Because of its use as a weed killer, cases of chlorate poisoning have been reported. Methemoglobinemia, hemolysis, and renal failure occurred in people consuming gram amounts. The lowest fatal dose reported was 15 g.[41]

Fewer animal tests have been conducted with chlorate than with chlorite. A study in which rats and roosters were given 10 or 100 mg/L chlorate for 4 months demonstrated decreased blood glutathione, osmotic fragility of red blood cells, and abnormalities in red blood cell morphology. No methemoglobin was detected. A subsequent study using African green monkeys given chlorate in drinking water at doses of 25 to 400 mg/L for 30 to 60 days did not result in any signs of red blood cell damage or oxidative stress. (These same doses of chlorite resulted in decreased methemoglobin.[41])

Testing of chlorate in the same mouse assays as chlorite was also negative. No carcinogenicity or reproductive effect tests have been conducted.[41]

### Bromine

Bromine has been used in swimming pools for disinfection and in cooling towers, but its use in drinking water has not been recommended. In the past, bromides have been used in drugs to control epilepsy, and they are still used in sedatives. Ethylene bromide is used as an anti-knock additive in gasoline, and methyl bromide is a soil fumigant. In water at neutral pH values, bromine exists primarily as hypobromous acid (HOBr), while below pH 6, $Br_2$, $Br_3^-$, bromine chloride, and other halide complexes occur. In the presence of ammonia or organic amines, bromamines are formed.[43] Most health effects information is on bromide because of its pharmaceutical uses.

Elemental bromine is extremely reactive and corrosive, producing irritation and burning to exposed tissues. Bromide occurs normally in blood at a range of 1.5 to 50 mg/L. Sedation occurs at a plasma concentration of about 960 mg/L, corresponding to a maintenance dose of 17 mg/(kg · day). Psychotic symptoms and neurological effects, including loss of muscular coordination, occur above approximately 1600 mg/L. GI disturbances can also occur at high doses. These effects have all been duplicated in laboratory animals. In addition, high doses in dogs were found to cause hyperplasia of the thyroid.[27] No data have been developed on the mutagenicity, carcinogenicity, or teratogenicity of bromine.

### Iodine

In water, iodine can occur as iodine ($I_2$), hypoiodous acid (HOI), iodate ($IO_3$), or iodide ($I^-$). Iodoamines do not form to any appreciable extent.[43] Iodine has been used to disinfect both drinking water and swimming pools. An iodine residual of about 1.0 mg/L is required for effective disinfection. Iodine is also used in pharmaceuticals, photographic materials, antiseptics, and catalysts. Iodine is an essential trace element, required for synthesis of the thyroid hormone. The estimated adult requirement is 80 to 150 µg/day. Deficiency results in goiter, a compensatory hyperplasia of the thyroid. Most intake of iodine is from food, especially seafood, and, in the United States, table salt that has been supplemented with potassium iodide.

Iodine is an irritant, with acute toxicity caused by irritation of the GI tract. A dose of 2 to 3 g may be fatal, although acute iodine poisoning is rare. Chronic adverse effects can resemble a sinus cold and can include skin lesions and GI tract irritation. These effects are reversible with cessation of iodine intake. Doses causing such effects may vary among populations and individuals.

### Ozone

Ozone is a very powerful oxidant. It is moderately soluble in water and is typically used at a concentration of a few milligrams per liter for drinking water disinfection. Over a thousand systems in Europe use ozone, and its use in the United States is increasing. Because it decomposes rapidly in water, it cannot be used to maintain a residual in distribution systems.

In general, ozone oxidizes inorganic elements to stable high oxidation states. The following table gives a summary of such reactions.[43,44]

| Inorganic constituent | Oxidation product |
|---|---|
| Ferrous iron ($Fe^{2+}$) | Ferric iron ($Fe^{3+}$) |
| Manganese ($Mn^{2+}$) | $Mn^{4+}$, $MnO_4^-$ |
| Bromine | Hypobromite, bromate |
| Bromamine | Nitrate and bromide |
| Nitrite | Nitrate |
| Ammonia | Nitrate |
| Cyanide | Cyanate |
| Hypochlorite | Chloride and chlorate (slow reaction with 77% chloride formed) |
| Chloramines | Chloride and nitrate |
| $ClO_2$ | Chlorate |
| Chlorite | $ClO_2$ then chlorate |

Some reactions (e.g., ozone and chloramines) occur slowly in water. The health and/or aesthetic effects of these compounds are discussed in this and other sections of this chapter.

## Particulates

Particles in water consist of finely divided solids, larger than molecules, but generally not distinguishable by the naked eye (size ranges from 1 to 150 $\mu$m) (Fig. 2.3). They include inorganic materials such as clays, silt, and asbestos, as well as living or dead organic matter such as algae, bacteria, and other microorganisms. Particles are an aesthetic concern in that their large surface area causes scattering of incident light and degrades visibility within the water. They can also be of significant health concern in that a variety of toxic materials can either comprise particulates or be adsorbed to them. These include pathogenic microorganisms, heavy metals, and chlorinated hydrocarbons. Particulates can interfere with disinfection during water treatment as microorganisms adsorbed to particulate material may be shielded against disinfectants. In addition, certain particulates can exert a disinfectant demand.[9]

## Organic Constituents

Organic constituents in water derive from three major sources: (1) the breakdown of naturally occurring organic materials, (2) domestic and

**Figure 2.3** Particles observed in a northern California utility's raw water supply. Sizes of particles indicated by marker lines of 5 to 50 µg.  (*Courtesy James M. Montgomery Consulting Engineers, Inc.*)

commercial activities, and (3) reactions that occur during water treatment and transmission. The first source predominates and is comprised of humic materials, microorganisms and their metabolites, and petroleum-based, high-molecular-weight aliphatic and aromatic hydrocarbons. These organics are typically benign, although some are nuisance constituents such as blue-green algae and its odoriferous me-

**Figure 2.4** Blue-green algae, *Oscillatoria limosa*. (*Courtesy James M. Montgomery Consulting Engineers, Inc.*)

Humics + Br$^-$ + NH$_3$ + Cl$_2$ ⟶ CO$_2$ + *New Organics:* + Smaller humics + N$_2$ + Cl$^-$

Trihalomethanes

+

Dihaloacetonitriles

+

Halogenated carboxylic acids

+

Halogenated amines

+

Halogenated phenols

+

Halogenated ketones

+

Halogenated aromatics

+

Halogenated humics

+

Aldehydes

+

Aromatics

+

Phthlatates

**Figure 2.5** Chlorination by-products. (*Source: James M. Montgomery Consulting Engineers, Inc., Water Treatment Principles and Design, John Wiley & Sons, Inc., Copyright 1985.*)

tabolite methylisoborneol (Fig. 2.4). A few of the petroleum products can have adverse health effects. In addition, humics can serve as precursors for formation of trihalomethanes and other organohalogen oxidation products during disinfection.

Organics derived from domestic and commercial activities are constituents of wastewater discharges, agricultural runoff, urban runoff, and leachate from contaminated soils. Most of the organic contaminants identified in water supplies as having adverse health concerns are part of this group. They include pesticides such as chlordane and carbofuran, solvents such as trichlorobenzene and tetrachloroethylene, metal degreasers such as trichloroethylene and trichloroethane, and plasticizers and monomers such as polychlorinated biphenols and epichlorohydrin.

Organic contaminants formed during water treatment include disinfection by-products such as THMs (e.g., chloroform) and haloacetonitriles (e.g., trichloroacetonitrile) (Fig. 2.5). Other compounds such as acrylamide are components of coagulants (e.g., polyacrylamide) that can leach out during treatment. During finished water transmission, undesirable components of pipes, coating, linings, and joint adhesives such as polynuclear aromatic hydrocarbons (PAHs) have been shown to leach into water.

The predominance of naturally occurring organics is illustrated in Fig. 2.6. Shown is the distribution of organic materials in the Mississippi River. The naturally occurring, high-molecular-weight constituents comprise approximately 80 percent of the total, while anthropogenic organics, generally weighing less than 400 g/mol, comprise about 15 percent (the shaded area). The concentrations of the naturally occurring fraction are typically in the milligram per liter range, while the concentrations of the synthetic organics are generally in the microgram or nanogram per liter range.

Composite measures of organic constituents exist. Total organic carbon (TOC) is used to track the overall organic content of water. Total organic halogen (TOX) indicates the presence of halogenated organics. Both of these methods are technologically more simple and economically more attractive than measurement of individual compounds. They are useful in general comparisons of water supplies, in identifying pollution sources, and in helping to determine when additional, more specific analyses might be required. The TOC measure has become so common that values from source and finished supplies throughout the United States can be compared on this basis. Figure 2.7 provides some illustrative values from a variety of natural waters. The TOX measure is useful because a large number of organics that are known or potential health hazards are halogenated.

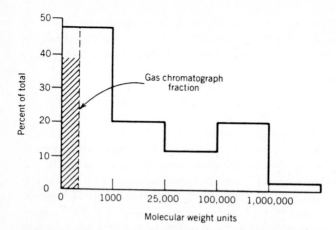

**Figure 2.6** Organic size distribution, Mississippi River water. (*Source: James M. Montgomery Consulting Engineers, Inc., Water Treatment Principles and Design, John Wiley & Sons, Inc., Copyright 1985; after C. Anderson and W. J. Maier, "The Removal of Organic Matter from Water Supplies by Ion Exchange," Report for Office of Water Research and Technology, Washington, D.C., 1977.*)

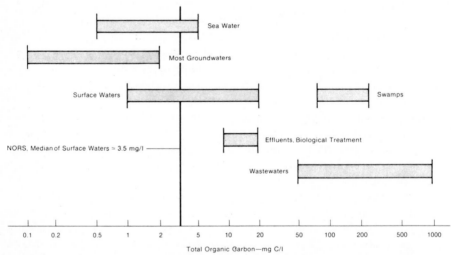

**Figure 2.7** Ranges of TOC reported for a variety of natural waters. (*Source: M. C. Kavanaugh, Coagulation for Improved Removal of Trihalomethane Precursors, Journal AWWA, vol. 70, no. 11, Nov. 1978, p. 613.*)

Health aspects of organic constituents of interest in drinking water treatment are summarized below, grouped according to chemical characteristics. Drinking water regulations and health advisories are listed in Table 2.5.

### Volatile organic chemicals (VOCs)

**Benzene.**    Benzene is produced by petroleum refining, coal tar distillation, coal processing, and coal coking. It is also used as an intermediate in the production of styrene, cyclohexane, detergents, and pesticides. Gasoline in the United States contains approximately 0.8 percent benzene by volume. Benzene is slightly soluble in water and has been detected in public drinking water supplies. In the 1976 to 1977 USEPA National Organics Monitoring Survey (NOMS), benzene was detected in 0 of 111 samples, 7 of 113 samples (mean 0.4 µg/L), and 4 of 16 samples (mean 0.95 µg/L).[27] Chronic occupational exposure to benzene has been shown to be toxic to the hematopoietic (blood-forming) system. Effects include anemia, thrombocytopenia or leukopenia, and leukemia. Benzene-induced carcinogenesis (increased tumors and leukemias) has been observed in exposed animals.[9] Changes in chromosome number and chromosome breakage have been observed in humans and laboratory animals. A single teratogenicity test showed positive results, but at doses just below lethal.[27]

The USEPA has classified benzene as a human carcinogen (group A). A drinking water concentration of 10 µg/L benzene has been calculated to represent an excess lifetime cancer risk of 1/100,000. An MCLG has been established at 0 and an MCL at 0.005 mg/L.[9,38]

**Carbon tetrachloride.**    Carbon tetrachloride is used primarily in the manufacture of chlorofluoromethane but also in grain fumigants, fire extinguishers, solvents, and cleaning agents.[26] Carbon tetrachloride is lethal when ingested at doses as low as 2 mL. Acute effects in rats include biochemical and morphological changes in the liver, including cirrhosis, necrosis, and blood enzyme changes. Mutagenicity testing using rats and bacteria was negative. Carcinogenicity bioassays produced hepatomas in mice, rats, and hamsters. No teratogenic effects have been demonstrated.[34]

The USEPA has classified carbon tetrachloride as a probable human carcinogen (group B2) based on evidence of carcinogenicity in animals. An MCLG of 0 and an MCL of 0.005 mg/L have been established.[38]

**Monochlorobenzene.**    Chlorobenzene is a solvent used in the manufacture of pesticides and dyestuffs and in cold cleaning operations. Dur-

**TABLE 2.5  USEPA Drinking Water Regulations and Health Advisories for Organic Contaminants**

| Chemical | Regulations — Status reg.* | NIPDWR, mg/L | MCLG, mg/L | MCL, mg/L | Health advisories — Status HA* | 10-kg child 1-day, mg/L | 10-kg child 10-day, mg/L | 10-kg child Longer-term, mg/L | 70-kg adult Longer-term, mg/L | 70-kg adult RfD, mg/(kg·day) | 70-kg adult DWEL, mg/L | 70-kg adult Life-time, mg/L | 70-kg adult $10^{-4}$ Cancer risk, mg/L | USEPA Cancer Group |
|---|---|---|---|---|---|---|---|---|---|---|---|---|---|---|
| Acenaphthylene | — | — | — | — | — | — | — | — | — | 0.06 | — | — | — | — |
| Acifluorfen | P | — | — | — | F | 2 | 2 | 0.1 | 0.4 | 0.013 | 0.4 | — | 0.1 | B2 |
| Acrylamide | P | — | zero | TT† | F | 1.5 | 0.3 | 0.02 | 0.07 | $2 \times 10^{-4}$ | 0.007 | — | 0.001 | B2 |
| Acrylonitrile | L | — | — | — | — | 0.02 | 0.02 | 0.001 | 0.004 | $1 \times 10^{-4}$ | 0.004 | — | 0.007 | B1 |
| Adipates (diethylhexyl) | T | — | 0.5 | 0.5 | — | — | — | — | — | 0.7 | 20 | 0.5 | — | C |
| Alachlor | P | — | zero | 0.002 | F | 0.1 | 0.1 | 0.01 | 0.01 | 0.01 | 0.4 | — | 0.04 | B2 |
| Aldicarb | P | — | 0.01 | 0.01 | F | 0.01 | 0.01 | 0.01 | 0.01 | 0.0013 | 0.04 | 0.01 | — | D |
| Aldicarb sulfone | P | — | 0.04 | 0.04 | F | 0.06 | 0.06 | 0.06 | 0.04 | 0.006 | 0.2 | 0.04 | — | D |
| Aldicarb sulfoxide | P | — | 0.01 | 0.01 | F | 0.01 | 0.01 | 0.01 | 0.04 | 0.0013 | 0.04 | 0.01 | — | D |
| Aldrin | — | — | — | — | — | $3 \times 10^{-4}$ | $3 \times 10^{-4}$ | $3 \times 10^{-4}$ | $9 \times 10^{-4}$ | $3 \times 10^{-5}$ | $9 \times 10^{-4}$ | — | $2 \times 10^{-4}$ | B2 |
| Ametryn | — | — | — | — | F | 9 | 9 | 0.9 | 3 | 0.009 | 0.3 | 0.06 | — | D |
| Ammonium sulfamate | — | — | — | — | F | 20 | 20 | 20 | 80 | 0.25 | 8 | 2 | — | D |
| Anthracene (PAH) | L | — | 0.003 | 0.003 | — | — | — | — | — | 0.3 | — | — | — | D |
| Atrazine | P | — | 0.003 | 0.003 | F | 0.1 | 0.1 | 0.05 | 0.2 | 0.005 | 0.2 | 0.003 | — | C |
| Baygon | — | — | — | — | F | 0.04 | 0.04 | 0.04 | 0.1 | 0.004 | 0.1 | 0.003 | — | C |
| Bentazon | — | — | — | — | F | 0.3 | 0.3 | 0.3 | 0.9 | 0.0025 | 0.09 | 0.02 | — | D |
| Benz(a)anthracene (PAH) | T | — | zero | $1 \times 10^{-4}$ | — | — | — | — | — | — | — | — | — | B2 |
| Benzene | F | — | zero | 0.005 | F | 0.2 | 0.2 | — | — | — | — | — | 0.1 | A |
| Benzo(a)pyrene (PAH) | T | — | zero | $2 \times 10^{-4}$ | — | — | — | — | — | — | — | — | — | B2 |
| Benzo(b)fluoranthene (PAH) | T | — | zero | $2 \times 10^{-4}$ | — | — | — | — | — | — | — | — | — | B2 |
| Benzo(g,h,i)perylene (PAH) | T | — | — | — | — | — | — | — | — | — | — | — | — | D |
| Benzo(k)fluoranthene (PAH) | T | — | zero | $2 \times 10^{-4}$ | — | — | — | — | — | — | — | — | — | B2 |
| bis-2-Chloroisopropyl ether | — | — | — | — | F | 4 | 4 | 4 | 13 | 0.04 | 1 | 0.3 | — | — |

| Compound | | | | | | | | | | | | | |
|---|---|---|---|---|---|---|---|---|---|---|---|---|---|
| Bromacil | — | — | — | — | F | 5 | 5 | 3 | 9 | 0.13 | 5 | 0.09 | C |
| Bromobenzene | — | — | — | — | D | — | — | — | — | — | — | — | — |
| Bromochloroacetonitrile | L | — | — | — | D | — | — | — | — | — | — | — | — |
| Bromochloromethane | — | — | — | — | D | — | — | — | — | — | — | — | — |
| Bromodichloromethane (THM) | L | 0.1 | — | — | D | 7 | 7 | 1 | 0.7 | 0.02 | 0.7 | 0.03 | B2 |
| Bromoform (THM) | L | 0.1 | — | — | D | 15 | 15 | 1 | 0.7 | 0.02 | 0.7 | — | B2 |
| Bromomethane | — | — | — | — | F | 0.1 | 0.1 | 0.1 | 0.5 | 0.001 | 0.05 | 0.01 | D |
| Butylate | — | — | — | — | F | 2 | 2 | 1 | 4 | 0.05 | 2 | 0.35 | D |
| Butylbenzene (n-) | — | — | — | — | D | — | — | — | — | — | — | — | — |
| Butylbenzene (sec-) | — | — | — | — | D | — | — | — | — | — | — | — | — |
| Butylbenzene (tert-) | — | — | — | — | D | — | — | — | — | — | — | — | — |
| Butylbenzylphthalate (BBP) | T | — | — | — | — | — | — | — | — | 0.2 | — | — | C |
| Carbaryl | — | — | — | — | F | 1 | 1 | 1 | 1 | 0.1 | 4 | 0.7 | D |
| Carbofuran | P | — | 0.04 | 0.04 | F | 0.05 | 0.05 | 0.05 | 0.2 | 0.005 | 0.2 | 0.04 | E |
| Carbon tetrachloride | F | — | zero | 0.005 | F | 4 | 0.2 | 0.07 | 0.3 | $7 \times 10^{-4}$ | 0.03 | 0.03 | B2 |
| Carboxin | — | — | — | — | F | 1 | 1 | 1 | 4 | 0.1 | 0.7 | — | D |
| Chloral hydrate | L | — | — | — | D | 7 | 7 | 0.2 | 0.6 | 0.002 | 0.06 | 0.06 | D |
| Chloramben | — | — | — | — | F | 3 | 3 | 0.2 | 0.5 | $5 \times 10^{-4}$ | 0.5 | 0.1 | D |
| Chlordane | P | — | zero | 0.002 | F | 0.06 | 0.06 | 0.2 | 0.5 | $4.5 \times 10^{-5}$ | 0.002 | 0.003 | B2 |
| Chlorodibromomethane (THM) | L | 0.1 | — | — | D | 7 | 7 | 1 | 0.7 | $0.065 \times 10^{-4}$ | 0.7 | 0.1 | C |
| Chloroethane | L | — | — | — | D | — | — | — | — | — | — | — | — |
| Chloroform (THM) | L | 0.1 | — | — | D | — | — | — | — | 0.01 | — | 0.6 | B2 |
| Chloromethane | L | — | — | — | D | — | — | — | — | — | — | — | — |
| Chlorophenol (2,4,6-) | L | — | — | — | D | 0.03 | 0.03 | 0.03 | 0.1 | 0.003 | 0.1 | 0.02 | B2 |
| Chlorophenol (2,4-) | L | — | — | — | D | — | — | — | — | — | — | 0.3 | D |
| Chlorophenol (2-) | L | — | — | — | D | 0.05 | 0.05 | 0.05 | 0.2 | 0.005 | 0.2 | 0.04 | D |
| Chloropicrin | L | — | — | — | — | — | — | — | — | — | — | — | — |
| Chlorothalonil | — | — | — | — | F | 0.2 | 0.2 | 0.2 | 0.5 | 0.015 | 0.5 | 0.15 | B2 |
| Chlorotoluene (o-) | L | — | — | — | F | 2 | 2 | 2 | 7 | 0.02 | 0.7 | 0.1 | D |
| Chlorotoluene (p-) | L | — | — | — | F | 2 | 2 | 2 | 7 | 0.02 | 0.7 | 0.1 | D |

TABLE 2.5 USEPA Drinking Water Regulations and Health Advisories for Organic Contaminants (*Continued*)

| Chemical | Status reg.* | Regulations NIPDWR, mg/L | MCLG, mg/L | MCL, mg/L | Status HA* | 10-kg child 1-day, mg/L | 10-day, mg/L | Longer-term, mg/L | 70-kg adult Longer-term, mg/L | RfD, mg/(kg·day) | DWEL, mg/L | Life-time, mg/L | $10^{-4}$ Cancer risk, mg/L | USEPA Cancer Group |
|---|---|---|---|---|---|---|---|---|---|---|---|---|---|---|
| Chrysene (PAH) | — | — | zero | $2 \times 10^{-4}$ | — | — | — | — | — | — | — | — | — | B2 |
| Cyanazine | L | — | — | — | F | 0.1 | 0.1 | 0.02 | 0.07 | 0.002 | 0.07 | 0.01 | — | D |
| Cyanogen chloride | L | — | — | — | D | — | — | — | — | — | — | — | — | — |
| Cymene (*p-*) | L | — | — | — | D | — | — | — | — | — | — | — | — | — |
| 2,4-D | P | 0.1 | 0.07 | 0.07 | F | 1 | 0.3 | 0.1 | 0.4 | 0.01 | 0.4 | 0.07 | — | D |
| Dacthal (DCPA) | T | — | — | — | F | 80 | 80 | 5 | 20 | 0.5 | 20 | 4 | — | D |
| Dalapon | T | — | 0.2 | 0.2 | F | 3 | 3 | 0.3 | 0.9 | 0.026 | 0.9 | 0.2 | — | D |
| Diazinon | — | — | — | — | F | 0.02 | 0.02 | 0.005 | 0.02 | $9 \times 10^{-5}$ | 0.003 | $6 \times 10^{-4}$ | — | E |
| Dibenz(a,h)-anthracene (PAH) | T | — | zero | $3 \times 10^{-4}$ | — | — | — | — | — | — | — | — | — | B2 |
| Dibromoacetonitrile | L | — | — | — | D | — | — | — | — | — | — | — | — | — |
| Dibromochloropropane (DBCP) | P | — | zero | $2 \times 10^{-4}$ | F | 0.2 | 0.05 | — | — | — | — | — | 0.003 | B2 |
| Dibromomethane | L | — | — | — | D | — | — | — | — | 0.1 | — | — | — | D |
| Dibutyl phthalate (DBP) | T | — | — | — | — | — | — | — | — | — | — | — | — | D |
| Dicamba | L | — | — | — | F | 0.3 | 0.3 | 0.3 | 1 | 0.03 | 1 | 0.2 | — | D |
| Dichloroacetaldehyde | L | — | — | — | D | — | — | — | — | — | — | — | — | — |
| Dichloroacetic acid | L | — | — | — | D | 50 | 50 | 0.5 | 2 | 0.005 | 0.2 | 0.003 | — | C |
| Dichloroacetonitrile | L | — | — | — | D | — | — | — | — | 0.008 | — | — | — | C |
| Dichlorobenzene (*p-*) | F | — | 0.075 | 0.075 | F | 10 | 10 | 10 | 40 | 0.1 | 4 | 0.075 | — | C |
| Dichlorobenzene (*o-,m*) | P | — | 0.6 | 0.6 | F | 9 | 9 | 9 | 30 | 0.089 | 3 | 0.6 | — | D |
| Dichlorodifluoromethane | — | — | — | — | F | 40 | 40 | 9 | 30 | 0.2 | 5 | 1 | — | D |
| Dichloroethane (1,1-) | L | — | — | — | D | — | — | — | — | — | — | — | — | — |
| Dichloroethane (1,2-) | F | — | zero | 0.005 | F | 0.7 | 0.7 | 0.7 | 2.6 | — | — | — | 0.04 | B2 |
| Dichloroethylene (1,1-) | F | — | 0.007 | 0.007 | F | 2 | 1 | 1 | 4 | 0.009 | 0.4 | 7 | — | C |

| Compound | | | | | | | | | | | | | | |
|---|---|---|---|---|---|---|---|---|---|---|---|---|---|---|
| Dichloroethylene (cis-1,2-) | P | — | 0.07 | 0.07 | F | 4 | 4 | 3.5 | 1 | 0.01 | 0.4 | 0.07 | — | D |
| Dichloroethylene (trans-1,2-) | P | — | 0.1 | 0.1 | F | 20 | 2 | 2 | 6 | 0.02 | 0.6 | 0.1 | — | D |
| Dichloromethane | T | — | zero | 0.005 | F | 10 | 2 | — | — | 0.06 | 2 | — | 0.5 | B2 |
| Dichloropropane (1,1-) | — | — | — | — | D | — | — | — | — | — | — | — | — | — |
| Dichloropropane (1,2-) | P | — | zero | 0.005 | F | — | 0.09 | — | — | — | — | — | 0.06 | B2 |
| Dichloropropane (1,3-) | L | — | — | — | D | — | — | — | — | — | — | — | — | — |
| Dichloropropane (2,2-) | L | — | — | — | D | — | — | — | — | — | — | — | — | — |
| Dichloropropene (1,1-) | L | — | — | — | D | — | — | — | — | — | — | — | — | B2 |
| Dichloropropene (1,3-) | L | — | — | — | F | 0.03 | 0.03 | 0.03 | 0.100 | $3 \times 10^{-4}$ | 0.01 | — | 0.02 | B2 |
| Dieldrin | L | — | — | — | F | $5 \times 10^{-4}$ | $5 \times 10^{-4}$ | $5 \times 10^{-4}$ | 0.002 | $5 \times 10^{-5}$ | 0.002 | — | $2 \times 10^{-4}$ | B2 |
| Diethylphthalate (DEP) | T | — | — | — | D | — | — | — | — | 0.8 | — | — | — | D |
| Diethylhexylphthalate (DEHP) | T | — | zero | 0.004 | D | — | — | — | — | 0.02 | — | — | 0.3 | B2 |
| Dimethrin | — | — | — | — | F | 10 | 10 | 10 | 40 | 0.3 | 10 | 2 | — | D |
| Dimethylphthalate (DMP) | L | — | — | — | — | — | — | — | — | — | — | — | — | D |
| Dinitrotoluene (2,4-) | — | — | — | — | D | — | — | — | — | — | — | — | — | — |
| Dinoseb | T | — | 0.007 | 0.007 | F | 0.3 | 0.3 | 0.01 | 0.04 | 0.001 | 0.04 | 0.007 | — | D |
| Dioxane (p-) | — | — | — | — | F | 4 | 0.4 | 0.3 | 1 | 0.03 | 1 | — | 0.7 | B2 |
| Diphenamid | — | — | — | — | F | 0.3 | 0.3 | 0.3 | 1 | 0.0022 | 1 | 0.2 | — | D |
| Diquat† | T | — | 0.02 | 0.02 | — | — | — | — | — | $4 \times 10^{-5}$ | — | — | — | D |
| Disulfoton | — | — | — | — | F | 0.01 | 0.01 | 0.003 | 0.009 | 0.001 | 0.001 | $3 \times 10^{-4}$ | — | E |
| Diuron | — | — | — | — | F | 1 | 1 | 0.3 | 0.9 | 0.002 | 0.07 | 0.01 | — | D |
| Endothall | T | — | 0.1 | 0.1 | F | 0.8 | 0.8 | 0.2 | 0.2 | 0.02 | 0.7 | 0.1 | — | D |
| Endrin | T | $2 \times 10^{-4}$ | 0.002 | 0.002 | F | 0.02 | 0.02 | 0.003 | 0.01 | 0.003 | 0.009 | 0.002 | — | D |
| Epichlorohydrin | P | — | zero | TT† | F | 0.1 | 0.1 | 0.07 | 0.07 | 0.002 | 0.07 | — | 0.4 | B2 |
| Ethylbenzene | P | — | 0.7 | 0.7 | F | 30 | 3 | 1 | 3 | 0.1 | 3 | 0.7 | — | D |
| Ethylene dibromide (EDB) | P | $5 \times 10^{-5}$ | zero | $5 \times 10^{-5}$ | F | 0.008 | 0.008 | — | — | — | — | — | $4 \times 10^{-5}$ | B2 |
| Ethylene glycol | — | — | — | — | F | 20 | 6 | 6 | 20 | 2 | 40 | 7 | — | D |
| ETU | L | — | — | — | F | 0.3 | 0.3 | 0.1 | 0.4 | $3 \times 10^{-5}$ | 0.001 | — | 0.02 | B2 |

**TABLE 2.5 USEPA Drinking Water Regulations and Health Advisories for Organic Contaminants (Continued)**

| Chemical | Regulations: Status reg.* | NIPDWR, mg/L | MCLG, mg/L | MCL, mg/L | Health advisories: Status HA* | 10-kg child: 1-day, mg/L | 10-kg child: 10-day, mg/L | 10-kg child: Longer-term, mg/L | 70-kg adult: Longer-term, mg/L | 70-kg adult: RfD, mg/(kg·day) | 70-kg adult: DWEL, mg/L | 70-kg adult: Lifetime, mg/L | 70-kg adult: $10^{-4}$ Cancer risk, mg/L | USEPA Cancer Group |
|---|---|---|---|---|---|---|---|---|---|---|---|---|---|---|
| Fenamiphos | — | — | — | — | F | 0.009 | 0.009 | 0.005 | 0.02 | $25 \times 10^{-5}$ | 0.009 | 0.002 | — | D |
| Fluometuron | — | — | — | — | F | 2 | 2 | 2 | 5 | 0.013 | 0.4 | 0.09 | — | D |
| Fluorene (PAH) | T | — | — | — | — | — | — | — | — | 0.04 | — | — | — | D |
| Fluorotrichloromethane | — | — | — | — | F | 7 | 7 | 3 | 12 | 0.3 | 10 | 2 | — | D |
| Fonofos | — | — | — | — | F | 0.02 | 0.02 | 0.02 | 0.07 | 0.002 | 0.07 | 0.01 | — | D |
| Formaldehyde | — | — | — | — | D | 10 | 5 | 5 | 20 | 0.15 | 5 | 1 | — | B1-Inhal |
| Gasoline | — | — | — | — | D | — | — | — | — | — | — | 5(benzene) | — | — |
| Glyphosphate | T | — | 0.7 | 0.7 | F | 20 | 20 | 1 | 1 | 0.1 | 4 | 0.7 | — | D |
| Heptachlor | P | — | zero | $4 \times 10^{-4}$ | F | 0.01 | 0.01 | 0.005 | 0.005 | $5 \times 10^{-4}$ | 0.02 | — | $8 \times 10^{-4}$ | B2 |
| Heptachlor epoxide | P | — | zero | $2 \times 10^{-4}$ | F | 0.01 | — | $1 \times 10^{-4}$ | $1 \times 10^{-4}$ | $13 \times 10^{-6}$ | $4 \times 10^{-4}$ | — | $4 \times 10^{-4}$ | B2 |
| Hexachlorobenzene | T | — | zero | 0.001 | F | 0.05 | 0.05 | 0.05 | 0.2 | $8 \times 10^{-4}$ | 0.03 | — | 0.002 | B2 |
| Hexachlorobutadiene | — | — | — | — | F | 0.3 | 0.3 | 0.1 | 0.4 | 0.002 | 0.07 | 0.001 | 0.05 | C |
| Hexachlorocyclopentadiene | T | — | 0.05 | 0.05 | — | — | — | — | — | 0.007 | 0.2 | — | — | D |
| Hexane (n-) | — | — | — | — | F | 10 | 4 | 4 | 10 | — | — | — | — | D |
| Hexazinone | — | — | — | — | F | 3 | 3 | 3 | 9 | 0.03 | 1 | 0.2 | — | D |
| HMX | — | — | — | — | F | 5 | 5 | 5 | 20 | 0.05 | 2 | 0.4 | — | D |
| Hypochlorite | L | — | — | — | — | — | — | — | — | — | — | — | — | — |
| Hypochlorous acid | L | — | — | — | — | — | — | — | — | — | — | — | — | — |
| Indeno(1,2,3-c,d)pyrene (PAH) | T | — | zero | $4 \times 10^{-4}$ | D | — | — | — | — | — | — | — | — | B2 |
| Isophorone | L | — | — | — | D | 15 | 15 | 15 | 15 | 0.2 | 7 | 0.1 | 0.9 | C |
| Isopropylbenzene | — | — | — | — | D | — | — | — | — | — | — | — | — | — |
| Lindane | P | 0.004 | $2 \times 10^{-4}$ | $2 \times 10^{-4}$ | F | 1 | 1 | 0.03 | 0.1 | $3 \times 10^{-4}$ | 0.01 | $2 \times 10^{-4}$ | 0.003 | C |

| Chemical | | | | | | | | | | | | | |
|---|---|---|---|---|---|---|---|---|---|---|---|---|---|
| Malathion | — | — | — | D | 0.2 | 0.2 | 0.2 | 0.8 | 0.02 | 0.8 | 0.2 | — | D |
| Maleic hydrazide | — | — | — | F | 10 | 10 | 5 | 20 | 0.5 | 20 | 4 | — | D |
| MCPA | — | — | — | F | 0.1 | 0.1 | 0.1 | 0.4 | 0.0015 | 0.053 | 0.011 | — | E |
| Methomyl | — | — | — | F | 0.3 | 0.3 | 0.3 | 0.3 | 0.025 | 0.9 | 0.2 | — | D |
| Methoxychlor | P | 0.4 | 0.4 | F | 6 | 2 | 0.5 | 2 | 0.05 | 2 | 0.4 | — | D |
| Methyl ethyl ketone | — | — | — | F | 80 | 8 | 3 | 9 | 0.025 | 0.9 | 0.2 | — | D |
| Methyl parathion | L | — | — | F | 0.3 | 0.3 | 0.03 | 0.1 | $25 \times 10^{-5}$ | 0.009 | 0.002 | — | D |
| Methyl tert butyl ether | L | — | — | D | 3 | 3 | 0.5 | 2 | 0.005 | 0.2 | 0.04 | — | D |
| Metolachlor | L | — | — | F | 2 | 2 | 2 | 5 | 0.15 | 5 | 0.1 | — | C |
| Metribuzin | — | — | — | F | 5 | 5 | 0.3 | 0.9 | 0.025 | 0.9 | 0.2 | — | D |
| Monochloroacetic acid | L | — | — | D | — | — | — | — | — | — | — | — | — |
| Monochlorobenzene | P | 0.1 | 0.1 | F | 2 | 2 | 2 | 7 | 0.02 | 0.7 | 0.1 | — | D |
| Naphthalene | — | — | — | D | 0.5 | 0.5 | 0.5 | 2 | 0.04 | 1 | 0.3 | — | D |
| Nitroguanidine | — | — | — | D | 10 | 10 | 10 | 40 | 0.1 | 4 | 0.7 | — | D |
| Nitrocellulose (non-toxic) | — | — | — | F | — | — | — | — | — | — | — | — | — |
| Oxamyl (Vydate) | T | 0.2 | 0.2 | F | 0.2 | 0.2 | 0.2 | 0.9 | 0.025 | 0.9 | 0.2 | — | E |
| Ozone by-products | L | — | — | — | — | — | — | — | — | — | — | — | — |
| Paraquat | — | — | — | F | 0.1 | 0.1 | 0.05 | 0.2 | 0.0045 | 0.2 | 0.03 | — | E |
| Pentachloroethane | — | — | — | D | — | — | — | — | — | — | — | — | — |
| Pentachlorophenol | P | 0.2 | 0.2 | F | 1 | 0.3 | 0.3 | 1 | 0.03 | 1 | 0.2 | — | D |
| Phenanthrene (PAH) | T | — | — | — | — | — | — | — | — | — | — | — | — |
| Phenol | — | — | — | D | 6 | 6 | 6 | 20 | 0.06 | 20 | 4 | — | D |
| Picloram | T | 0.5 | 0.5 | F | 20 | 20 | 0.7 | 2 | 0.07 | 2 | 0.5 | — | D |
| Polychlorinated biphenols (PCBs) | P | $5 \times 10^{-4}$ | zero | P | — | — | 0.001 | 0.004 | — | — | — | $5 \times 10^{-4}$ | B2 |
| Prometon | — | — | — | F | 0.2 | 0.2 | 0.2 | 0.5 | 0.015 | 0.5 | 0.1 | — | D |
| Pronamide | — | — | — | F | 0.8 | 0.8 | 0.8 | 3 | 0.075 | 3 | 0.05 | — | C |
| Propachlor | — | — | — | F | 0.5 | 0.5 | 0.1 | 0.5 | 0.013 | 0.5 | 0.09 | — | D |
| Propazine | — | — | — | F | 1 | 1 | 0.5 | 2 | 0.02 | 0.7 | 0.01 | — | C |
| Propham | — | — | — | D | 5 | 5 | 5 | 20 | 0.02 | 0.6 | 0.1 | — | D |
| Propylbenzene (n-) | — | — | — | D | — | — | — | — | — | — | — | — | — |

TABLE 2.5 USEPA Drinking Water Regulations and Health Advisories for Organic Contaminants (Continued)

| Chemical | Regulations | | | | | Health advisories | | | | | | | | USEPA Cancer Group |
|---|---|---|---|---|---|---|---|---|---|---|---|---|---|---|
| | | | | | | 10-kg child | | | 70-kg adult | | | | | |
| | Status reg.* | NIPDWR, mg/L | MCLG, mg/L | MCL, mg/L | Status HA* | 1-day, mg/L | 10-day, mg/L | Longer-term, mg/L | Longer-term, mg/L | RfD, mg/(kg·day) | DWEL, mg/L | Life-time, mg/L | $10^{-4}$ Cancer risk, mg/L | |
| Pyrene (PAH) | T | — | — | — | — | — | — | — | — | 0.03 | 1 | — | — | D |
| RDX | — | — | — | — | F | 0.1 | 0.1 | 0.1 | 0.4 | 0.003 | 0.1 | 0.002 | 0.03 | C |
| Simazine | T | — | 0.001 | 0.001 | F | 0.5 | 0.5 | 0.05 | 0.2 | 0.002 | 0.06 | 0.001 | — | C |
| Styrene | P | — | zero/0.1 | 0.005/0.1 | F | 20 | 2 | 2 | 7 | 0.2 | 7 | 0.1 | 0.001 | B2/C |
| 2,4,5-T | L | — | — | — | F | 0.8 | 0.8 | 0.8 | 1 | 0.01 | 0.35 | 0.07 | — | D |
| 2,3,7,8-TCDD (dioxin) | T | — | zero | $5 \times 10^{-8}$ | F | $1 \times 10^{-6}$ | $1 \times 10^{-7}$ | $1 \times 10^{-8}$ | $4 \times 10^{-8}$ | $1 \times 10^{-9}$ | $4 \times 10^{-8}$ | — | $2 \times 10^{-8}$ | B2 |
| Tebuthiuron | — | — | — | — | F | 3 | 3 | 0.7 | 2 | 0.07 | 2 | 0.5 | — | D |
| Terbacil | — | — | — | — | F | 0.3 | 0.3 | 0.3 | 0.9 | 0.013 | 0.4 | 0.09 | — | E |
| Terbufos | — | — | — | — | F | 0.005 | 0.005 | 0.001 | 0.005 | $13 \times 10^{-5}$ | 0.005 | $9 \times 10^{-4}$ | — | D |
| Tetrachloroethane (1,1,1,2-) | L | — | — | — | F | 2 | 2 | 0.9 | 3 | 0.03 | 1 | 0.07 | 0.1 | C |
| Tetrachloroethane (1,1,2,2-) | L | — | — | — | D | — | — | — | — | — | — | — | — | — |
| Tetrachloroethylene | P | — | zero | 0.005 | F | 2 | 2 | 1 | 5 | 0.01 | 0.5 | — | 0.07 | B2 |
| Toluene | P | — | 2 | 2 | F | 20 | 3 | 3 | 10 | 0.3 | 10 | 2 | — | D |
| Toxaphene | P | 0.005 | zero | 0.005 | F | 0.5 | 0.04 | — | — | 0.1 | 0.0035 | — | 0.003 | B2 |
| 2,4,5-TP | P | 0.01 | 0.05 | 0.05 | F | 0.2 | 0.2 | 0.07 | 0.3 | 0.0075 | 0.3 | 0.05 | — | D |
| Trichloroacetaldehyde | L | — | — | — | D | — | — | — | — | — | — | — | — | — |
| Trichloroacetic acid | L | — | — | — | D | 30 | 30 | 30 | 100 | 0.3 | 10 | 0.2 | — | C |
| Trichloroacetonitrile | L | — | — | — | D | — | — | — | — | — | — | — | — | — |
| Trichlorobenzene (1,2,4-) | T | — | 0.009 | 0.009 | F | 0.1 | 0.1 | 0.1 | 0.5 | 0.001 | 0.05 | 0.009 | — | D |
| Trichlorobenzene (1,3,5-) | — | — | — | — | F | 0.6 | 0.6 | 0.6 | 2 | 0.006 | 0.2 | 0.04 | — | D |

| | | | | | | | | | | | | | | |
|---|---|---|---|---|---|---|---|---|---|---|---|---|---|---|
| Trichloroethane (1,1,1-) | F | — | 0.2 | 0.2 | F | 100 | 40 | 40 | 100 | 0.09 | 1 | 0.2 | — | D |
| Trichloroethane (1,1,2-) | T | — | 0.005 | 0.003 | F | 0.6 | 0.4 | 0.4 | 1 | 0.004 | 0.1 | 0.003 | 0.06 | C |
| Trichloroethanol (2,2,2-) | L | 0.1 | — | — | — | — | — | — | — | — | — | — | — | — |
| Trichloroethylene | F | — | 0.005 | zero | F | — | — | — | — | 0.007 | 0.3 | — | 0.3 | B2 |
| Trichloropropane (1,1,1-) | — | — | — | — | D | — | — | — | — | — | — | — | — | — |
| Trichloropropane (1,2,3-) | — | — | — | zero | F | 0.6 | 0.6 | 0.6 | 2 | 0.006 | 0.2 | 0.04 | — | — |
| Trihalomethanes (total) | F | 0.1 | — | zero | — | — | — | — | — | — | — | — | — | — |
| Trifluralin | L | — | — | — | F | 0.03 | 0.03 | 0.03 | 0.1 | 0.0075 | 0.26 | 0.005 | 0.5 | C |
| Trimethylbenzene (1,2,4-) | — | — | — | — | D | — | — | — | — | — | — | — | — | — |
| Trimethylbenzene (1,3,5-) | — | — | — | — | D | — | — | — | — | — | — | — | — | — |
| Trinitroglycerol | — | — | — | — | F | 0.05 | 0.05 | 0.05 | 0.05 | — | — | 0.05 | — | — |
| Vinyl chloride | F | — | 0.002 | zero | F | 3 | 3 | 0.01 | 0.05 | — | — | — | $15 \times 10^{-4}$ | A |
| Xylenes | P | — | 10 | 10 | F | 40 | 40 | 40 | 10 | 2 | 60 | 10 | — | D |

Note: See Table 2.4 for source and footnotes.

ing chlorination of water, chlorobenzene may be formed.[26] Finished water in 9 of 10 water supplies surveyed by the USEPA contained chlorobenzene, the maximum concentration being 5.6 μg/L.[26]

Acute effects of chlorobenzene include CNS depression with high doses leading to death from respiratory failure. Target organs in chronic exposure studies are the liver, CNS, and kidneys.

The USEPA classifies chlorobenzene as having inadequate animal evidence of carcinogenicity (group D) based on a National Toxicology Program (NTP) bioassay in which an increased occurrence of neoplastic nodules of the liver in high-dose male rats was not statistically significant and carcinogenic effects were not noted in female rats or in mice of either sex. A projected upper limit excess lifetime cancer risk ($10^{-6}$) concentration in drinking water has been calculated as 2 μg/L.[34] The USEPA has proposed an MCLG and an MCL of 0.1 mg/L for monochlorobenzene.[32]

**Dichlorobenzene.**  o-Dichlorobenzene is a solvent used in the production of organic chemicals, pesticides, and dyes. m-Dichlorobenzene is also a solvent; however, little usage or health effects data are available on it. Both isomers are ubiquitous in the air and have been found in the milk of nursing mothers in several states. Dichlorobenzene appears to vaporize rapidly from surface water, and it is estimated that 99.3 percent of the population served by public drinking water supplies receives water with 0 to 0.5 μg/L.

The primary toxic effects of o-dichlorobenzene are CNS depression; blood dyscrasias (blood "poisoning"); and lung, kidney, and liver toxicity. Similar data are not available for the meta isomer, although a few short-term assays suggest similar toxicity and the short-term assessments developed for the ortho isomer are also applied to the meta. Some mutagenic activity has been observed in bacterial systems. Preliminary carcinogenicity testing indicated no carcinogenicity for o-dichlorobenzene.

The carcinogenicity of both o- and m-dichlorobenzene has not been categorized by the USEPA (group D) because of inadequate evidence. The USEPA has proposed an MCLG and an MCL of 0.6 μg/L for o-dichlorobenzene.[32] Because of insufficient data, no MCLG for the meta isomer has been proposed. The MCLG and MCL for p-dichlorobenzene are 0.075 mg/L.

**1,2-Dichloroethane (ethylene dichloride).**  1,2-Dichloroethane is used in the manufacture of vinyl chloride and tetraethyl lead, as a lead scavenger in gasoline, as an insecticidal fumigant, in tobacco flavoring, as a constituent of paint varnish and finish removers, as a metal degreaser, in soap and scouring compounds, in wetting agents, and in ore flotation. It is biologically intransigent and has a solubility of 1:120. The USEPA water supply surveys detected 1,2-dichloroethane in 11 of 83 systems (26 percent) at a mean concentration of 1 μg/L and in 26 of 80 systems (33 percent) at concentrations ranging from less than 0.2 to 6 μg/L.[26]

Human exposure to high vapor concentrations results in irritation of the eyes, nose, and throat, with continued exposure leading to CNS depression and injury to the liver, kidneys, and adrenals. Laboratory animals show the same responses to high vapor concentrations, while lower chronic doses lead to hepatic and renal damage. 1,2-Dichloroethane is weakly mutagenic in bacterial test systems. When administered by gavage to laboratory rats and mice, 1,2-dichloroethane caused carcinomas in the stomach, mammary glands, and alveoli. It has not been found to be teratogenic in mammalian laboratory animals.[27]

The USEPA has classified 1,2-dichloroethane as a probable human carcinogen (group B2) based on its positive carcinogenicity in animal ingestion studies. The MCLG for 1,2-dichloroethane is 0 and the MCL is 0.005 mg/L.[38]

**1,1-Dichloroethylene (vinylidene chloride).**   1,1-Dichloroethylene is used primarily as an intermediate in the synthesis of copolymers for food packaging films and coatings. It is also used in the synthesis of 1,1,1-trichloroethane. 1,1-Dichloroethylene has a water solubility of 0.4 g/L. It has been detected in drinking water at concentrations up to 0.1 μg/L.[26]

No definitive human health studies on 1,1-dichloroethylene have been conducted. Acute effects noted in rats include hepatotoxicity (e.g., a decrease in liver enzyme levels) and nephrotoxicity (kidney toxicity). Chronic effects in laboratory animals also target the liver. 1,1-Dichloroethylene was mutagenic in Ames testing but not in in vitro mammalian assays. Carcinogenicity test results have been equivocal. Two inhalation studies with mice and rats produced kidney adenocarcinomas and lung tumors; however, many other inhalation and feeding studies have been negative.

The USEPA has classified 1,1-dichloroethylene as a possible human carcinogen (group C), because of limited evidence of carcinogenicity in animals in the absence of human data. The final MCLG and MCL for 1,1-dichloroethylene are both 0.007 mg/L.[38]

**1,2-Dichloroethylene.**   1,2-Dichloroethylene exists in both the cis and trans forms. Their proportion depends on production conditions. Both are used, alone or in combination, as solvents and chemical intermediates. They have water solubilities of 3.5 g/L (cis) and 6.3 g/L (trans).

Their principal source in water appears to be in situ transformations from other chlorinated hydrocarbons. Both isomers have been found in water. The maximum concentration found for the cis isomer was 16 μg/L and for the trans isomer was 1.0 μg/L.

No human health effects data exist. For *trans*-1,2-dichloroethylene a mouse feeding study provided a no-observed-adverse-effect level (NOAEL) of 17 mg/(kg · day) for male mice. Some minimal kidney effects in rats have also been observed. No positive mutagenic effects were observed in *E. coli* test systems. No teratogenicity or carcinogenicity data are available.

No long-term carcinogenicity studies on *cis*- and *trans*-1,2-dichloroethylene have been carried out. The USEPA has not classified the carcinogenicity of either compound (group D) because of inadequate evidence. Based upon the results of the mouse feeding study an MCLG and an MCL of 0.1 mg/L for *trans*-1,2-dichloroethylene have been proposed.[32] An MCLG and an MCL of 0.07 mg/L for *cis*-1,2-dichloroethylene have been proposed based upon toxicity data for 1,1-dichloroethylene.[32]

**Methylene chloride (dichloromethane).** Methylene chloride is used in the manufacture of paint removers, insecticides, solvents, cleaners, pressurized spray products, and fire extinguishers. Methylene chloride is soluble at about 1 part to 50 parts of water. It is formed during chlorination.[26] A USEPA Region V survey detected methylene chloride in 8 percent of finished water supplies at a mean concentration of less than 1 μg/L, while another USEPA survey found 9 out of 10 supplies containing levels up to 1.6 μg/L.[26] However, laboratory contamination is a common confounder of methylene chloride occurrence data.

Observations on workers exposed to methylene chloride showed no increase in incidence of disease. In rats the first observable signs of toxicity occur at 0.001 mL/kg. Rats maintained on drinking water containing 0.13 g/L methylene chloride for 91 days showed no adverse effects. Very high concentration inhalation studies (10,000 ppm) showed some liver injury. Methylene chloride was mutagenic in a *Salmonella typhimurium* assay, but not in a *Drosophila* test. Limited carcinogenicity testing has been negative, as has teratogenicity testing.[27]

Methylene chloride is on the list of 83 contaminants to be regulated by the USEPA under the 1986 SDWA amendments. An MCLG and MCL will be proposed in 1990 (probably, 0 and 0.005 mg/L, respectively).

**Tetrachloroethylene [perchloroethylene (PCE)].** Tetrachloroethylene is used as a solvent, a heat transfer agent, and in the manufacture of

slightly soluble in water. It was detected in 2 of 10 finished waters surveyed at levels of 5.6 and 0.27 µg/L.[26]

Acute effects of vinyl chloride in humans include CNS dysfunction, organic disorders of the brain, skin alterations, Raynaud's syndrome, and acoosteolysis. Also found in vinyl chloride workers have been hepatic angiosarcomas. Chronic effect tests in laboratory animals have demonstrated angiosarcomas, hepatocellula carcinomas, and adenocarcinomas. Vinyl chloride is a mutagen, producing positive results in several microbial tests and in one in vitro mammalian assay. In addition, an increased number of chromosome aberrations have been noted in industrial workers.[34]

Vinyl chloride's carcinogenicity is proven in humans, as well as mice, hamsters, and rats. Angiosarcomas of the liver were the most common cancer. Also occurring were brain tumors, hepatomas, nephroblastomas, and lung tumors. Most experiments utilized inhalation exposure although some utilized gavage. In most experiments an indication of a dose response occurred. Teratogenicity tests have been negative in rats and rabbits. Inadequate data exist for humans.[34]

The USEPA has classified vinyl chloride as a human carcinogen (group A) and has calculated a $10^{-5}$ risk for a drinking water concentration of 0.01 mg/L.[9] The MCLG has been established at 0 and the MCL at 0.002 mg/L.[38]

**Paradichlorobenzene (PDB).**   Paradichlorobenzene is used in very large amounts as an insecticidal fumigant, lavatory deodorant, and for moth control. It is produced commercially by chlorination of benzene or chlorobenzene. The $m$- and $o$-isomers occur as impurities in technical grade paradichlorobenzene. PDB is soluble in water at 80 mg/L. In the USEPA's Ground Water Supply Survey (GWSS) (1982), PDB was detected in 9 of 945 supplies at concentrations up to 1.3 µg/L. Other state and national surveys have detected PDB at concentrations up to 9 µg/L.[26]

Subchronic toxicity studies conducted on laboratory animals demonstrated lesions and weight changes of the liver and kidneys. All available animal mutagenicity studies have been negative. A recently completed (1986) NTP bioassay on $p$-dichlorobenzene demonstrated treatment-related increases in the incidence of renal tubular cell adenocarcinomas and hepatocellular adenomas in male and female mice. The PDB was administered by gavage. Hepatoblastomas were also found in males. A previous long-term inhalation study revealed no treatment-related cancer increase.[47]

The USEPA has classified paradichlorobenzene as a possible human carcinogen (group C) because of limited evidence of carcinogenicity in animals. An MCLG and MCL of 0.075 mg/L have been established.[38]

## Synthetic organic contaminants (SOCs)

**Acrylamide.** Acrylamide is used as a starting material in polymers such as polyacrylamide for enhancing water and oil recovery from wells, as a flocculant in water treatment, in food processing and papermaking, as a soil conditioner, and in permanent-press fabrics. No monitoring data exist to detail its occurrence in water; however, its high solubility ($2.15 \times 10^6$ mg/L) and its use within the drinking water industry make its presence likely.[9]

Acrylamide's principal toxic effect is peripheral neuropathy (nervous system disease). Subchronic animal studies have also demonstrated atrophy of skeletal muscles in hind quarters, testicular atrophy, and weakness of the limbs. Case reports on humans indicate similar effects following exposure via the dermal, oral, or inhalation routes.

Recent evidence that acrylamide is carcinogenic in mice and rats. Acrylamide was shown to initiate skin tumors in female mice when administered orally, topically, or via intraperitoneal injection. An increase in lung adenomas was also observed in those mice receiving oral exposure. An unpublished study indicates increased tumor incidence at several sites in rats exposed via drinking water.[9] The NAS calculated an upper 95 percent confidence estimate of lifetime cancer risk of $2.1 \times 10^{-5}$, assuming consumption of 1 L/day containing 1 µg/L of acrylamide.[48]

Acrylamide was negative in the Ames mutagenicity test. Chromosome aberrations in the spermatogonia of mice were found after exposure to 5 mg/(kg · day) for 2 to 3 weeks. The USEPA has classified acrylamide as a probable carcinogen (group B2). An MCLG of 0 has been proposed.[9,32] A treatment technique that limits the amount of acrylamide used to treat drinking water has been proposed in lieu of an MCL.[32]

**Alachlor.** Alachlor is an herbicide used primarily on corn and soybeans. It is slightly soluble in water. It has been found in surface water in the midwest (from agricultural runoff) at levels up to 104 µg/L. Groundwater can also be contaminated; wells in four states sampled ranged from 0.04 to 16 µg/L.[9]

Alachlor has low acute oral toxicity. Chronic effects noted in laboratory feeding studies are hepatotoxicity and uveal degeneration syndrome (eye degenerative disease with secondary cataract formation).[9] Chronic feeding studies using mice have demonstrated alachlor-instigated lung tumors. Other studies with rats have demonstrated stomach, thyroid, and nasal turbinate tumors.

The USEPA has classified alachlor as a probable human carcinogen (group B2). An MCLG of 0 and an MCL of 0.002 mg/L have been proposed.[32]

**Aldicarb, aldicarb sulfoxide, and aldicarb sulfone.** Aldicarb is a heavily used pesticide known as Temek. It is applied to cotton primarily to control insects, mites, and nematodes. It is very soluble in water but not persistent. Aldicarb sulfoxide and aldicarb sulfone are metabolites and are the forms of aldicarb most often detected in water. (The parent compound is typically present as a very small fraction.) Aldicarb and its metabolites have been found in well waters from various states with levels up to 50 μg/L. They have also been detected in surface water.[9]

Aldicarb's acute toxicity is the highest of any of the widely used pesticides. It and its metabolites act through cholinesterase inhibition. If not fatal, their effects are transitory. The few mutagenicity tests conducted suggest that aldicarb is not a mutagen; however, not enough data exist to be certain. Aldicarb has not been shown to be a carcinogen in animals; an NCI bioassay and two additional studies showed no statistically significant increases in tumors.

The USEPA has not classified the carcinogenicity of aldicarb and its metabolites (group D). MCLGs and MCLs of 0.01 mg/L, 0.01 mg/L, and 0.04 mg/L have been proposed for aldicarb, aldicarb sulfoxide, and aldicarb sulfone, respectively.[32]

**Atrazine.** Atrazine is an herbicide and plant growth regulator used primarily on corn and soybeans. It is slightly soluble in water and has been detected in surface and groundwater. The USEPA found two groundwater systems with trace amounts of atrazine, while groundwater systems in three midwestern states tested positive with concentrations in the range of 0.8 μg/L. The USEPA also found one large surface supply with 0.1 μg/L atrazine in its finished water. The National Screen Program (NSP) for organics in drinking water, conducted by the USEPA's Office of Drinking Water, found 29 percent of the surface systems tested had atrazine in the range of 0.1 to 2.9 μg/L.[9]

Atrazine appears to have a low chronic toxicity in animals; a 2-year rat feeding study at 100 mg/L detected no adverse effects. Dogs fed atrazine showed cardiac pathologies at 150 and 1000 ppm. An NOAEL from this study was developed into a human drinking water equivalent level (DWEL) of 0.2 mg/L. At a dose of 21.4 mg/(kg · day) no significantly greater incidence of hepatomas was observed in an 80-week mouse study. A 2-year feeding study in rats did result in an increased number of mammary tumors in females.[32] Atrazine has not been shown to be mutagenic in microorganism assays.[9]

The USEPA has categorized atrazine in group C (possible human carcinogen). An MCLG and an MCL of 0.003 mg/L have been proposed.[32]

**Carbofuran.**   Carbofuran is a carbamate insecticide and nematocide used primarily on corn. It has a solubility of 700 mg/L and a half-life in soils of 30 to 60 days.[34] Carbofuran has been detected in the groundwater in five states at levels of 1 to 50 µg/L.[9]

Like the other carbamate insecticides, carbofuran is a potent cholinesterase inhibitor. When not fatal, its effects are transitory. Adverse effects found at subacute doses have been a slight decrease in rat pup survival, and testicular degeneration and aspermia in dogs. The mutagenicity of carbofuran has been tested in a number of short-term assays, the majority of which have been negative. Two assays produced equivocal results. Teratogenicity tests were negative in mice, rats, rabbits, and dogs. A 2-year feeding study to determine carcinogenicity in mice and rats produced no statistically significant increase in tumors.

The USEPA categorized carbofuran as noncarcinogenic to humans (group E).[9] An MCLG and an MCL of 0.04 mg/L have been proposed.[32]

**Chlordane.**   Chlordane is a broad-spectrum insecticide. Technical chlordane is a mixture of stereoisomers and other chlorinated analogs, including heptachlor. Its solubility is 150 to 220 µg/L. Prior to its recent ban, chlordane was the most extensively utilized insecticide for subterranean termite control. Prior to its 1977 cancellation for agricultural and home use, chlordane was also used to control soil insects and ants.

Chlordane is occasionally reported in wells near areas treated for termites. Incidents of contaminated drinking water systems from improperly controlled tank filling operations have also occurred.[9]

The primary acute and chronic effects of chlordane exposure include neurotoxicity, induction of hepatic microsomal enzyme activity, and other liver changes. Chlordane was shown to be mutagenic when tested with transformed human cells in culture. An NCI cancer study found a dose-dependent incidence of hepatocellular carcinoma in male and female mice. In rats, hepatic nodules and liver hyperplasia were observed.

The USEPA has classified chlordane as a probable human carcinogen (group B2). An upper limit excess lifetime cancer risk ($10^{-6}$) has been calculated as 0.02 µg/L. An MCLG of 0 and an MCL of 0.002 mg/L have been proposed.[9,32]

**2,4-D.**   2,4-Dichlorophenoxyacetic acid, or 2,4-D, is a systemic herbicide used to control broadleaf weeds. It is used in agriculture and for home and aquatic applications (e.g., reservoirs). It is very soluble (540 mg/L) and undergoes both chemical and biological degradation in the environment. In water, 2,4-D has a half-life of about 1 week; in soil, 2,4-D persists for about 6 weeks.[26]

2,4-D has not been detected in many drinking water systems. The National Organics Reconnaissance Survey (NORS), conducted in 1975 by the USEPA's Office of Drinking Water, found one surface system with 0.04 μg/L. The NSP also found one with 1.1 μg/L. The RWS did not find any with levels greater than the minimum quantifiable limit of 0.01 μg/L. National compliance reports show 2,4-D sometimes detected but always below the existing NIPDWR MCL of 0.1 mg/L.[9]

The short-term acute effects of 2,4-D include muscular incoordination, hindquarter paralysis, stupor, coma, and death. Four microbial systems showed no mutagenic activity, and in vivo mammalian assays have also been negative. Some increased mitotic gene conversion was noted in one test conducted with *Saccharomyces cerevisiae*. Carcinogenicity and teratogenicity data are inconclusive. No dose-dependent tumor formation has been reported; however, well-designed bioassays upon which to base definitive conclusions are lacking. The USEPA has not classified the carcinogenicity of 2,4-D (group D). The NIPDWR MCL of 0.1 mg/L is based upon an NOAEL of 8 mg/(kg · day). A more recently derived NOAEL of 1 mg/(kg · day) forms the basis for the proposed MCL of 0.07 mg/L.[9,32] An MCLG of 0.07 mg/L has also been proposed.[32]

**Dibromochloropropane (DBCP).** Dibromochloropropane is a soil fumigant used for nematode control on crops. It is moderately soluble (about 100 mg/L). Evidence suggests that it may readily leach into aquifers. Groundwater samples in five states detected DBCP, with one state reporting 62 of 92 positives. Concentrations were 0.02 to 20 μg/L. In 1977 all uses of DBCP were canceled.[9]

Acute effects in rats include impaired renal function, hepatocellular necrosis, loss of spermatogenic elements in testes, and testicular atrophy. Similar effects occur with subchronic oral exposures, while chronic exposures result in a higher incidence of toxic tubular nephropathy. Antifertility effects have also been noted in humans working with DBCP. Some positive mutagenic effects have been observed in bacterial systems, but again they may be caused by impurities, especially epichlorohydrin. No teratogenicity data are available.[30]

Carcinogenicity tests with rats and mice have been positive. The NCI found a dose-related incidence of squamous cell carcinomas of the forestomach of mice and rats exposed via gavage. Another dietary study conducted with rats showed an increased incidence of renal tube carcinomas, hepatocellular carcinomas, and stomach squamous cell carcinomas. Based on this carcinogenicity data, the USEPA has classified DBCP as a probable human carcinogen (group B2). The proposed MCLG is 0 and the proposed MCL is 0.0002 mg/L.[32]

**1,2-Dichloropropane.**   1,2-Dichloropropane is a versatile compound used as a soil fumigant, metal degreaser, a lead scavenger for antiknock fluids, and a component of dry-cleaning fluids. It is also an intermediate in the production of perchloroethylene and carbon tetrachloride.[26] The GWSS found 1,2-dichloropropane in 6 of 466 random samples with a mean concentration of 3.7 μg/L, and in 7 of 178 nonrandom samples with a mean of 3.7 μg/L. In several counties in California, 66 of 410 wells contained 1,2-dichloropropane.[9]

Toxic effects target the liver with centrilobular necrosis, liver congestion, and hepatic fatty changes. Positive mutagenicity test results were obtained using *Salmonella typhimurium* and Chinese hamster ovary cells. An NTP carcinogenicity study reported a significant increase in nonneoplastic liver lesions and mammary gland adenocarcinomas in female rats but not in males when 1,2-dichloropropane was administered via gavage. When tests were run with mice, nonneoplastic liver lesions were observed in males, and hepatocellular adenomas were observed in both males and females. The results of these tests suggest 1,2-dichloropropane is carcinogenic in mice. Results for rats are equivocal.[9]

The USEPA has classified 1,2-dichloropropane as a probable human carcinogen (group B2). An MCLG of 0 and an MCL of 0.005 mg/L[32] have been proposed.

**Dinoseb.**   Dinoseb has been used since 1945 as an herbicide and insecticide. It is used to control weeds in cereal and vegetable crops. It is slightly soluble in water (52 mg/L) but can form salts that are more soluble.[34]

Intoxication in humans results in fatigue, sweating, and psychological effects. Acute effects include prostration, convulsions, and death. Chronic effects include decreased growth and decreased organ weights of the liver, spleen, heart, lungs, and brain. Mutagenicity tests have been positive in one microbial assay and negative in two. One 18-month carcinogenicity test in mice found no increase in tumors. Teratogenic effects were found using parenteral doses but were not found using oral doses.[34] An MCLG and MCL will be proposed in 1990.

**Endrin.**   Endrin belongs to the cyclodiene group of insecticides, along with aldrin, chlordane, heptachlor, and dieldrin. It is a commercially used insecticide and rodenticide with a solubility of 0.25 mg/L. Like the other cyclodienes, endrin is persistent and accumulates in the food chain. The USEPA banned endrin in 1976 for most uses, and it is no longer produced in the United States. The NIPDWR included an MCL for endrin of 0.0002 mg/L.

Surveys conducted in the 1960s and early 1970s found widespread occurrence of cyclodienes in U.S. surface water. In a 1958 to 1965 survey of U.S. rivers, endrin was found at an average concentration of 0.008 to 0.214 μg/L. The highest concentrations were found in the lower Mississippi River basin with 46 percent of the grab samples positive for endrin. The USEPA's 10-city drinking water survey also detected endrin. More recent surveys (NSP and RWS) reported no detections of endrin, and the few detections in the federal compliance reports include no levels greater than the MCL.[9]

The cyclodienes are notable for their persistence, tendency to be stored in animal fat, and toxicity with the CNS as a target site. Human illness and death have occurred from poisoning during the manufacture, spraying, or accidental ingestion of cyclodienes.

Endrin has not been shown to be mutagenic. Carcinogenicity tests have been negative in four studies and positive in one. The positive test, however, showed no dose response, and the greatest number of tumors occurred in the lowest dose group.

The USEPA has not classified the carcinogenicity of endrin (group D). Endrin is rarely detected in drinking water and all its uses have been canceled.[9] An MCLG and MCL will be proposed in 1990.

**Epichlorohydrin (ECH).** Epichlorohydrin is a halogenated alkyl epoxide. It is used as a raw material in the manufacture of flocculants and of epoxy and phenoxy resins and as a solvent for resins, gums, cellulose, paints, and lacquers. The flocculants are sometimes used in food preparation and in potable water treatment, and the lacquers are sometimes used to coat the interiors of water tanks and pipes. Epichlorohydrin is quite soluble in water ($6.6 \times 10^4$ mg/L) and has pronounced organoleptic properties. The threshold odor concentration is 0.5 to 1.0 mg/L, while the threshold for irritation of the mouth is at 0.1 mg/L.[27]

Epichlorohydrin is rapidly absorbed and accumulates in the kidneys, liver, pancreas, spleen, and adrenals. Overexposure through inhalation in an industrial situation resulted in eye and throat irritation, nausea, headache, and dyspnea (difficulty in breathing). Within 2 days of exposure, an enlarged liver and bronchitis were reported, and 2 years later, liver damage was present.[27]

Acute and subacute exposure in laboratory animals results in CNS depression, respiratory tract irritation, weight loss, leucocytosis (increase in numbers of white blood cells), and kidney damage. Chronic exposure in laboratory animals can result in emphysema, bronchopneumonia, kidney tubule swelling, interstitial hemorrhage of the heart, and brain lesions. ECH or its metabolite alpha-chlorhydrin can produce reversible or nonreversible infertility in laboratory animals.[27]

ECH is considered a potent mutagen, having produced positive results in a number of procaryotic systems and eucaryotic cell cultures. It has also been shown to be carcinogen following oral and inhalation exposure. Laboratory rats given oral doses of epichlorohydrin developed dose dependent increases in forestomach tumors characterized as squamous and basal cell hyperplasias and in squamous cell carcinomas and papillomas. No teratogenic studies on ECH have been reported.[9]

Based on ECH's documented animal carcinogenicity, the USEPA has categorized ECH as a probable human carcinogen (group B2). A $10^{-6}$ upper lifetime cancer risk is calculated for a drinking water concentration of 4 μg/L. An MCLG of 0 has been proposed.[32] A treatment technique that will limit the amount of ECH used to treat drinking water has been proposed in lieu of an MCL.[32]

**Ethylbenzene.**  Ethylbenzene is an organic solvent used in the production of styrene. It is slightly soluble in water (1.52 mg/L) and has been detected in some groundwater supplies. The GWSS detected ethylbenzene in 3 of 466 random samples with a mean of 0.87 μg/L, and in 3 of 479 nonrandom samples with a mean of 0.78 μg/L.[9]

Ethylbenzene is readily absorbed and would be expected to accumulate in the adipose tissue. It is not very toxic in acute exposure testing. Subacute and chronic testing show some liver and kidney pathologies and some CNS disorders. Ethylbenzene does not appear to be a mutagen; however, only limited testing with *S. typhimurium* has been done. Its carcinogenicity has not been adequately tested. The NCI is currently conducting a long-term bioassay for which the data are not yet available.

The USEPA has not classified the carcinogenicity of ethylbenzene (group D).[9] An MCLG of 0.7 mg/L has been proposed based on non-carcinogenic effects.[9] The proposed MCL is 0.7 mg/L.[32]

**Ethylene dibromide (EDB).**  Ethylene dibromide is a pesticide used as a soil fumigant, a postharvest fumigant, and for termite control. It was also used as an additive in leaded gasoline. Most uses were canceled in 1984. Citrus quarantine treatment and several minor uses remain.[9] EDB is very soluble (4500 mg/L) and highly volatile. It has been found in the groundwater and finished drinking water in eight states. Isolated samples have had levels between 0.1 and 560 µg/L.

EDB is highly toxic during acute exposure. The target areas are the lungs, liver, spleen, kidney, and CNS. Chronic exposure may affect the liver, stomach, adrenal cortex, and reproductive system. The testes exhibit atrophy and antispermatogenic effects.

EDB exhibits mutagenic effects in in vitro bacterial and eucaryotic cell systems. It is a potent carcinogen in rats and mice when they are exposed via inhalation or gavage. An increase in squamous cell carcinomas of the forestomach, hemangiosarcomas of the circulatory system, hepatocellular carcinomas, and liver neoplastic nodules was observed in an NCI study with·rats and mice exposed via gavage.

The USEPA has classified EDB as a probable human carcinogen (group B2). A drinking water concentration of 0.0006 µg/L has been associated with an upper limit excess lifetime cancer risk of $10^{-6}$. An MCLG of 0 has been proposed. The proposed MCL is 0.00005 mg/L.[32]

**Heptachlor and heptachlor epoxide.**    Heptachlor and heptachlor epoxide belong to the cyclodiene group of insecticides, along with endrin and chlordane. Technical-grade heptachlor is a waxy substance containing chlordane and having a water solubility of 0.56 mg/L. It oxidizes rapidly, both photochemically and biologically, to heptachlor epoxide. Heptachlor epoxide is more stable and persistent and has a water solubility of 0.350 mg/L.[26]

In 1978 the USEPA canceled all uses of heptachlor except its use in controlling subterranean termites and for dipping roots and tops of nonfood plants. Prior to its restriction, heptachlor was widely used for control of termites, ants, and other soil insects in agricultural and landscape applications. Heptachlor has been detected in drinking water. In a 1958 to 1965 survey, heptachlor was found in surface water at concentrations of 0.0 to 0.0031 µg/L and heptachlor epoxide at less than 0.001 to 0.008 µg/L. More recently, a rural water supply survey in one state showed 62.5 percent of samples in one county and 45.5 percent in another had concentrations greater than 0.01 µg/L.[9]

Isolation of the health effects of heptachlor is difficult because of its in vivo conversion to the epoxide. Because the type of effects observed after exposure to either compound are similar, the USEPA has initially treated them together in its proposed MCLG.

Symptoms of acute heptachlor intoxication include CNS disturbances, such as tremors, convulsions, paralysis, and hypothermia. Lower doses result in microsomal enzyme induction, hyperplasia, hepatic vein thrombosis, and cirrhosis. Heptachlor epoxide was reported to cause a significant increase in hepatic carcinomas in rats receiving oral doses for 108 weeks. This early study (1959) was followed by an NCI dietary study with mice and rats using heptachlor that gave similar results for the mice but negative results for the rats.

The USEPA has classified both heptachlor and heptachlor epoxide as probable human carcinogens (group B2). The projected upper limit excess lifetime cancer risk ($10^{-6}$) occurs for heptachlor at a drinking water concentration of 0.01 μg/L, and for heptachlor epoxide at 0.0006 μg/L. MCLGs of 0 have been proposed for both heptachlor and heptachlor epoxide.[32] An MCL of 0.0004 mg/L has been proposed for heptachlor and 0.0002 mg/L for heptachlor epoxide.[32]

Lindane.    Lindane is an insecticide comprised of 99 percent gamma isomers of benzene hexachloride. It is registered for commercial and home applications and is used in shampoos for human head lice and animal lice and fleas. It is slightly soluble in water (7.8 mg/L) and persists in aerobic but not anaerobic soils. In the latter it undergoes biotransformation.[27]

Lindane has occasionally been found at low levels in water supplies. The NORS found two of eight surface water systems with 0.01 μg/L and trace concentration. The NIPDWR national compliance data show no violations of the NIPDWR MCL of 0.004 mg/L. The RWS found 1 of 71 groundwater systems with a level of greater than 0.002 μg/L (the minimum quantification limit).

The liver and kidney are the primary targets for toxic effects. Acute exposure can result in neurological and behavioral effects. Chronic exposure leads to liver hypertrophy, kidney tubular degeneration, and interstitial nephritis.

The NCI conducted lifetime carcinogenicity bioassay tests in mice. Increased hepatocellular carcinomas were observed in low-dose but not high-dose males. A separate, earlier study reported increased liver tumors in both males and females with no distinction made regarding dose. Mutagenicity tests have all been negative.

The USEPA has currently classified lindane as a possible human carcinogen (Group C) because of indefinite oncogenic data. A projected upper lifetime cancer risk ($10^{-6}$) for a drinking water concentration of 0.03 μg/L has been calculated. An MCLG and MCL of 0.0002 mg/L have been proposed.[9,32]

**Methoxychlor.**    Methoxychlor has been used as an insecticide for about 40 years. It is widely used for home and garden applications and on domestic animals, trees, and in water. It is chemically related to dichlorodiphenyltrichloroethane (DDT).

Methoxychlor is not very soluble (0.28 mg/L) and has not been detected in any national drinking water surveys. Compliance reports all give levels less than the NIPDWR MCL of 0.1 mg/L. Locally, however, methoxych or has been found in waters near high-use areas. In one county 46 percent of the rural water supplies detected methoxychlor with a mean concentration of 0.033 µg/L; another 65 percent were positive with a mean of 0.023 µg/L.[9]

Methoxychlor exposure results in a wide range of effects such as CNS disturbances, chronic nephritis, cystic tubular nephropathy, and testicular atrophy. Available evidence suggests that methoxychlor is not mutagenic or carcinogenic. A 1978 NCI bioassay with rats given methoxychlor in their diets for 78 weeks reported inconclusive results, while a number of mutagenicity studies using bacteria, yeast, *Drosophila melanogaster,* and mammalian cell cultures were all negative.[9]

The USEPA has not classified the carcinogenicity of methoxychlor (group D). An MCLG and an MCL of 0.4 mg/L[32] have been proposed.

**Polynuclear aromatic hydrocarbons (PAHs).**  Polynuclear aromatic hydrocarbons are a diverse class of compounds consisting of substituted and unsubstituted polycyclic and heterocyclic aromatic rings. They are formed as a result of incomplete combustion of organic compounds in the presence of insufficient oxygen. Although small amounts of PAHs derive from endogenous sources, most derive from human activities. Benzo(a)pyrene is the most thoroughly studied because of its ubiquity and its known carcinogenicity in laboratory animals. It is formed in the pyrolysis of naturally occurring hydrocarbons and found as a constituent of coal, coal tar, petroleum, shale, kerosene, fuel combustion products, and cigarette smoke.[30]

PAHs are extremely insoluble in water, e.g., the solubility of benzo(a)pyrene is only 10 ng/L. Their solubility can be increased by the action of detergents and other surfactants. Total PAH concentrations in surface waters have been reported to range from 0.14 to 2.5 µg/L. Because a large fraction of the PAHs are associated with particulates, water treatment can lower these concentrations to a reported range of 0.003 to 0.14 µg/L. PAHs can leach from tar or asphalt linings of distribution pipelines, resulting in elevated levels reaching the consumer. PAHs decrease in cement-lined pipes, presumably because of adsorption.[30] Benzo(a)pyrene has been reported to occur in surface waters at levels ranging from 0.0006 to 0.35 µg/L and in finished water at levels ranging from less than 0.1 to 2.1 ng/L.

Studies indicate an increased mortality from lung cancer in workers exposed to PAH-containing substances, such as coal gas, tars, soot, and coke oven emissions, and in people who smoke cigarettes; however, the PAHs in these substances have not been definitely linked with the observed lesions. In laboratory animals, subchronic and chronic doses of PAHs produce systemic toxicity, manifested by inhibition of body growth, and degeneration of the hematopoietic and lymphoid systems. PAHs appear to select for organs with proliferating cells, such as the intestinal epithelium, bone marrow, lymphoid organs, and testes. The mechanism of toxicity with these organs may be binding of PAH metabolites to DNA. Carcinogenic PAHs can also suppress the immune system, while noncarcinogenic PAHs do not have this effect. Most of the carcinogenic PAHs have also been demonstrated to be mutagenic.[30]

PAHs are on the list of 83 contaminants to be regulated by the USEPA under the 1986 SDWA amendments. An MCLH and MCL will be proposed in 1990.

**Polychlorinated biphenyls (PCBs).** Polychlorinated biphenyls are a class of colorless and stable compounds each containing a biphenyl nucleus with two or more substituent chlorine atoms. Technical PCBs are a mix often containing between 40 and 60 different chlorinated biphenyls (arochlors). They are heat-resistant and prior to their cancellation in 1976, PCBs were used in plasticizers, heat-transfer fluids, hydraulic fluids, compressor lubricants, capacitors, and transformers. They are generally insoluble in water.[26]

The NOMS (1976 to 1977) found PCBs in 6 percent of finished groundwater supplies at levels of 0.1 μg/L and in approximately 2 percent of finished surface waters at less than or equal to 1.4 μg/L. One state reported 32 of 163 groundwater supplies had up to 1.27 μg/L in finished water in 1978.[9]

PCBs have low acute toxicity. Subchronic and chronic effects are of more concern because of the bioaccumulation of PCBs. Exposure of laboratory animals to PCBs results in liver toxicity; specific effects include liver enlargement, fatty infiltration, centrilobular necrosis, and modification of prophyrin metabolism. Decreased reproductive function is also observed.

Mutagenicity testing in rat systems with various arochlors has been negative. One PCB, 4-chlorobiphenyl, tested positive in *S. typhimurium*. No human teratogenic effects are known, although PCBs can cross the placenta.

The NCI tested arochlor 1254 for carcinogenicity in Fisher rats. After 2 years' exposure to arochlor 1254 in the diet, a trend of increased lymphomas and leukemias occurred in the treated rats, but the incidence was not statistically higher than for the controls. A higher incidence of hyperplastic (nontumorous increase in cell numbers) nodules, often regarded to be preneoplastic (pretumor forming), was also observed. Another rat feeding study using arochlor 1260 did produce hepatocellular carcinomas.[27]

PCBs are categorized by the USEPA as probable human carcinogens (group B2). A $10^{-6}$ cancer risk has been calculated for a drinking water concentration of 0.0079 µg/L, and an MCLG of 0 proposed.[9,32] The proposed MCL is 0.0005 mg/L.[32]

**Pentachlorophenol (PCP).**  Pentachlorophenol is used primarily as a wood preservative but also as an herbicide, defoliant, insecticide, and fungicide. In 1984 the USEPA initiated actions to cancel all uses of PCP except wood preservation and to substantially restrict its use in that application. These actions have been challenged, and administrative proceedings are under way.[9]

Pentachlorophenol is slightly soluble in water (18 mg/L) and was detected by the NSP in two surface drinking water systems at 1.3 and 12 µg/L. None of the 12 groundwater systems tested in the NSP contained PCP at levels greater than the minimum quantification limit (1.0 µg/L).

Target areas of PCP toxicity are the liver, kidneys, and CNS. In the liver, increased liver weights and induction of hepatic enzymes are observed. In the kidney, increased weight and pigmentation occurs, and in the CNS, capillary congestion and chromatolysis of nerve cells are seen. PCP is fetotoxic and has adverse effects on reproduction. For all these toxic effects, contaminants in the makeup of PCP could be a factor.

The mutagenicity of pentachlorophenol has been tested using *S. typhimurium*, *E. coli*, and *Serratia marcescens*. All proved negative. Two oral carcinogenicity tests also generated negative results.[9] However, the NTP has recently completed work and a draft report showing dose-related increases in liver and adrenal tumors in mice.[32]

PCP has not been classified by the USEPA for carcinogenicity (group D), although the final NTP report may serve as a basis for a group B2 classification (probable human carcinogen). An MCLG and MCL of 0.2 mg/L have been proposed.[9,32]

**Simazine.**  Simazine is an herbicide applied to field crops, aquatic weeds, and algae. It is nonvolatile and has a low solubility of 5 mg/L. The NSP found 12 percent of finished drinking water from surface sources to have levels of simazine between 0.1 and 4.4 µg/L. One state reported levels from 0.18 to 0.63 µg/L in three treatment plants in 1983. The NSP also found 1 of 12 groundwater sources sampled with simazine at 1 µg/L. In California, 6 of 166 wells were found to have levels of 0.5 to 3.5 µg/L.[9]

Limited health effects data on simazine exist. It appears to have low toxicity. One 2-year rat feeding study found no difference between the test animals and controls in gross appearance or behavior. A 2-year dog feeding study identified slight thyroid hyperplasia and a slight increase in serum enzymes at 1500 ppm. No adverse effects on reproduction were observed after a three-generation rat feeding study. Mutagenicity tests conducted with four strains of *S. typhimurium* were negative. The potential carcinogenicity of simazine has not been tested. No data exist on its teratogenicity.[9]

Simazine is on the list of 83 contaminants to be regulated by the USEPA under the 1986 SDWA amendments. An MCLG and MCL will be proposed in 1990.

Styrene (vinyl benzene). Styrene is used in the manufacture of plastics, synthetic rubbers, resins, and insulation. Some resins manufactured from styrene are used in the treatment of potable water. Styrene is only slightly soluble in water and has not been detected in any national monitoring surveys.[9]

Styrene is readily absorbed and accumulates in the adipose tissue. It has a low acute toxicity. Primary effects noted include a decreased weight gain, increased liver and kidney weights, and lung congestion. Styrene was not mutagenic in *Salmonella typhimurium*. Positive mutagenicity was observed, however, using yeast, fruit flies, cultured mammalian cells, and in vivo rat and mice assays. Several carcinogenicity studies have been conducted. Results have been equivocal in most; however, an NCI assay showed a significant increase in alveolar and bronchiolar adenomas and carcinomas in male mice with a positive dose-response trend.[32]

The USEPA has classified styrene as a probable human carcinogen (group B2) based on positive results in animal studies that, however, are not definitive. Thus, at the same time, the USEPA is considering its previous classification as a possible human carcinogen (group C). To go along with the alternative carcinogenicity classifications, two MCLG, of 0 and 0.1 mg/L have been proposed.[32] Two MCLs of 0.005 and 0.1 mg/L have also been proposed.[32]

**2,3,7,8-TCDD (dioxin).** 2,3,7,8-Tetrachlorodibenzo-*p*-dioxin is not manufactured purposefully. It occurs as a contaminant of 2,4,5-trichlorophenol that is used in the production of several herbicides, such as 2,4,5-trichlorophenoxyacetic acid and silvex. It also may be formed during pyrolysis of chlorinated phenols, chlorinated benzenes, and polychlorinated diphenyl esters. Dioxin is not mobile in soils and is not expected to be found in drinking water. Thus far, it has not been detected in any water supplies, although it has been found in 32 hazardous waste sites.[9]

Dioxin is readily absorbed and accumulates in adipose tissue. It is only slowly metabolized. Noncarcinogenic effects include thymic atrophy, weight loss, and liver damage. Mutagenicity test results have been conflicting. Dioxin is, however, a potent animal carcinogen. Oral exposure leads to hepatocellular carcinomas, thyroid carcinomas, and adrenal cortical adenomas.[9]

The USEPA has classified dioxin as a probable human carcinogen (group B2). An MCLG and MCL will be proposed in 1990.

**2,4,5-TP (silvex).** 2,4,5-TP, or silvex, is an herbicide that was used to control weeds and brush on rangeland, pastures, turf, lawns, and along canals and other waterways. In 1979, the use of silvex was suspended. Subsequently all registrations were canceled.[9]

Silvex is moderately soluble in water (140 mg/L) and has been detected in water systems. Of eight surface systems sampled during the NORS, one large system was found to have 0.02 µg/L of 2,4,5-TP. In the NSP none of the 105 surface samples contained it, and in the RWS, none of the 21 surface systems contained silvex in excess of the minimum quantification limit of 0.1 µg/L. The U.S. Geological Survey found concentrations of 0.03 to 0.08 µg/L in a survey of 15 finished drinking waters from surface supplies in Florida, and concentrations of 0.04 to 0.30 µg/L in groundwater supplies from the same state.

Exposure to high doses of 2,4,5-TP causes a variety of effects, including depression, muscle weakness, and minor kidney and liver damage. Subchronic exposure produces histopathologic changes in the liver and kidneys. A single mutagenicity test using *S. typhimurium* was negative. Limited carcinogenicity testing has been done. The two existing studies reported no increase in tumors after chronic oral exposure of laboratory animals; however, neither study employed recent NCI bioassay procedures. Their findings are, therefore, regarded as inconclusive.[9]

The USEPA has not classified 2,4,5-TP with regard to carcinogenicity (group D) because of inadequate data. An NIPDWR MCL of 0.01 mg/L exists for silvex. An MCLG of 0.05 mg/L and a new MCL of 0.05 mg/L have been proposed.[32]

**Toluene (methyl benzene).** Toluene is used in lead-free gasoline; as a starting material in the production of benzene and other chemicals; and as a solvent for paints, coatings, gums, oils, and resins. It is moderately soluble in water (534.8 mg/L) and has been detected in numerous water systems. In the CWSS, toluene was measured in two groundwater systems at concentrations of 0.505 and 0.56 µg/L. Three surface supplies had 0.52, 0.72, and 1.62 µg/L. In the NSP, approximately 20 percent of surface supplies tested had toluene in their finished water. The mean concentration was 0.295 µg/L, and the high value was 1.4 µg/L. State agencies have reported levels as high as 2500 µg/L.[9]

Most health effects data come from inhalation studies. Acute and chronic health effects include CNS depression and damage to the lungs, liver, and kidneys. Mutagenicity tests with *S. typhimurium* and *E. coli* were negative. Only one long-term carcinogenicity bioassay has been conducted, again using inhalation exposure. Toluene was determined not to be carcinogenic under the conditions of the test. The NCI is currently (1985) conducting a study of exposure via inhalation and gavage.[9]

The USEPA has not classified toluene regarding carcinogenicity. An MCLG of 2.0 mg/L and MCL of 2.0 mg/L have been proposed.

**Toxaphene.** Toxaphene is a persistent, broad-spectrum insecticide. Its current registered uses are limited, but formerly it was used extensively on food and fiber crops. The solubility of toxaphene is about 0.4 mg/L. It has been detected in 27 systems tested by the USEPA; however, compliance data from the NIPDWR do not report levels in excess of the NIPDWR MCL of 0.005 mg/L.[9]

In acute exposures, toxaphene causes a variety of adverse effects to the CNS, including salivation, behavioral changes, and convulsions. The critical target organ in chronic or subchronic exposure is the liver. Toxaphene was shown to be mutagenic using *S. typhimurium*. NCI carcinogenicity tests with rats and mice have also been positive. Toxaphene in the diet caused dose-related increases in hepatocellular carcinomas in male and female mice and also caused increases in thyroid tumors in male and female rats.

The USEPA has classified toxaphene as a probable human carcinogen (group B2) and has calculated a projected upper limit excess lifetime cancer risk $(10^{-6})$ for a drinking water concentration of 0.03 μg/L. An MCLG of 0 and an MCL of 0.005 mg/L have been proposed.[9,32]

**Xylene.** Xylene occurs as three isomers (ortho, meta, and para). The three are treated as one in the USEPA health effects evaluation and regulatory procedures. Xylene is used as a component of aviation and automobile gasoline, as a solvent, and also in the synthesis of many organic chemicals, pharmaceuticals, and vitamins. It is slightly soluble in water and has been detected in federal and state surveys at levels up to 750 μg/L in groundwater and 5.2 μg/L in surface water supplies. The GWSS detected xylene in about 3 percent of groundwater tested, and the CWSS found it in about 6 percent of surface water tested.[9]

The primary toxic effects of xylene are CNS disturbances and liver damage. A long-term carcinogenicity bioassay conducted by the NTP found no increased incidence of neoplastic lesions in rats or mice of both sexes.

The USEPA has not classified xylene regarding carcinogenicity (group D). An MCLG[9] and an MCL of 10 mg/L[32] have been proposed.

## Organic Disinfection By-products (DBPs)

The use of oxidants for disinfection; taste, odor, and color removal; and for decreasing coagulant demand also produces undesirable organic by-products. Surveys conducted since the mid-1970s have determined that chloroform and other trihalomethanes, formed during drinking water chlorination, are the organic chemicals occurring the most consistently and at overall highest concentrations of any organic contaminant in treated drinking water. In addition, water chlorination can produce a variety of other compounds including haloacetic acids, halonitriles, haloaldehydes, and chlorophenols. Alternative disinfectants, chloramines, chlorine dioxide, and ozone, can also react with source water organics to yield organic by-products.[49]

Exactly which compounds are formed, their formation pathways, and health effects are not well known. Table 2.6 summarizes the organic by-products produced from the four most common disinfectants. The paucity of definitive data in Table 2.6 is largely caused by difficulties in the analysis of disinfection by-products; many are not susceptible to even highly sophisticated methods of extraction and analysis. This is particularly true for the nonvolatile compounds, which comprise the bulk of the DBPs and the polar low-molecular-weight substances.

Discussion of the health effects of compounds that either have been found or are suspected to occur in the drinking water, and for which health effects data exist, is given below.

### Trihalomethanes (THMs)

The THMs include trichloromethane, or chloroform; dibromochloromethane; dichlorobromomethane; and bromoform. Other iodine-containing, mixed halide THMs have been reported, but they are rare.[27] The frequency diagram shown in Fig. 2.8 illustrates the concentration distribution of the four common THMs in the 80-city NORS survey. Median values in that study, which are comparable to subsequent studies, were as follows (in micrograms per liter):

| | |
|---|---|
| Chloroform | 21 |
| Bromodichloromethane | 6 |
| Dibromochloromethane | 1.2 |
| Bromoform | < 0.1 |

**TABLE 2.6  Organic By-products Produced from the Four Most Common Disinfectants**

| Disinfectant | Research status | Compound | Levels detected, μg/L | |
|---|---|---|---|---|
| | | | A | B[a] |
| Chlorine | Compounds found to date comprise a small fraction of the TOC or TOX content of samples tested; many more, particularly nonvolatiles, remain to be detected. | Chloroform | 65 | 28 |
| | | Bromodichloror... | 8.7 | 7.6 |
| | | Dibromochloror... | 2.4 | 6.8 |
| | | Dichloroacetoni... | 1.0 | 2.2 |
| | | Bromochloroace... | 0.5 | 0.5 |
| | | Chloropicrin | 0.3 | 0.4 |
| | | Bromoform | < 0.5 | 0.2 |
| | | Dibromoacetoni... | < 0.3 | < 0.2 |
| | | Dichloroacetic a... | | 10–100 |
| | | Trichloroacetal... | | 10–100 |
| | | 1,1,1-Trichlorop... | | < 10 |
| | | Trichloroacetic... | | < 10 |
| | | Chloroacetic aci... | | < 10 |
| Chloramine | To the extent that chloramine hydrolyzes to form hypochlorous acid, the same products as are formed during chlorination can be expected, but at much lower concentrations. | Trihalomethane | 19[b] | |
| | | Dichloroacetic a... | | |
| | | Trichloroacetic a... | | |
| Chlorine dioxide | Does not form trihalomethanes. Very few aldehydes studies done simulating drinking water conditions. Postulated compounds include aldehydes, quinones, and epoxides.[43] | $C_2$-$C_a$ aliphatic... | | |

**TABLE 2.6    Organic By-Products Produced from the Four Most Common Disinfectants** *(Continued)*

| Disinfectant | Research status | Compounds produced | Levels detected, µg/L | |
|---|---|---|---|---|
| | | | A | B[a] |
| Ozone | Ozonation of naturally occurring organics produces compounds that can be found without treatment, because ozone mimics environmental oxidation processes.[f] Few studies of ozone reactions with synthetic organics under drinking water conditions exist. | Phthalates<br>Fatty acids<br>Carboxylic acids<br>Ketones<br>Aldehydes<br>Toluene[e] | | |

[a]Data from two recent USEPA surveys (A. Reding et al., 1986; B. Stevens et al., 1987).

[b]Data from the NORS survey. Ten out of 80 cities surveyed used chloramine disinfection. The range of THM concentrations was 1 to 81 µg/L; 19 was the average.

[c]DCA and TCA were found during laboratory chlorination of humic acids (Johnson et al., 1986, cited in Ref. 41).

[d]Produced during treatment of Ohio River water with ClO$_2$ (Stevens et al., 1978, cited in Ref. 43).

[e]All the above compounds found during ozonation of fulvic acid (Lawrence et al., cited in Glaze, 1986). A series of aldehydes $n$-hexanol, $n$-heptanol, $n$-octanol, and $n$-nonanol were detected in the waterworks in Zurich, Switzerland, after ozonation (Schalekamp, 1977).

[f]Glaze, 1986.

References for above footnotes:

Reding, R., et al.: "Measurement of Dihaloacetonitriles and Chloropiorin in Drinking Water," USEPA Office of Drinking Water, Technical Support Div., Cincinnati, Ohio, 1986.

Stevens, A. A., et al.: "Detection and Control of Chlorination By-products in Drinking Water," *Proc. Conf. Current Research in Drinking Water Treatment*, Cincinnati, Ohio, 1987.

Glaze, W. H.: "Reaction Products of Ozone: A Review," *Environmental Health Perspectives*, vol. 69, 1986, pp. 151–157.

Lawrence, J., et al.: "The Ozonation of Natural Waters: Product Identification," *Ozone: Science and Engineering*, vol. 2, 1980, pp. 55–64.

Schalekamp, M.: Experience in Switzerland with Ozone, Particularly in Connection with the Neutralization of Hygienically Undesirable Elements Present in Water. *Proc. AWWA Ann..Conf.*, Anaheim, Calif., 1977.

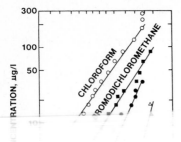

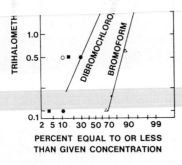

PERCENT EQUAL TO OR LESS
THAN GIVEN CONCENTRATION

**Figure 2.8** Frequency distribution of trihalomethane data (NORS). [*Reproduced with permission, from R. L. Jolley et al. (eds.), Water Chlorination: Environmental Impact and Health Effects, vol. 1, Butterworth Publishers, Stonehaven, Mass., 1983.*]

A number of epidemiological studies have indicated an association between chlorinated drinking water and increased mortality from a variety of cancers. Because confounding factors such as smoking and diet typically were not accounted for, and data were largely obtained from public records rather than from individuals, no definitive conclusions are based on these studies. The results are suggestive, however, particularly for bladder and possibly colon and rectal cancer, and further studies have been recommended.[27,41] THMs are currently regulated under the NIPDWR. An MCL of 0.10 mg/L for total THMs is based on a running annual average of quarterly samples. Health effects information for each compound is given below.

**Trichloromethane (chloroform).**   Chloroform is a volatile, slightly soluble compound formed via reaction of chlorine with various organics during drinking water treatment. It is also used as a refrigerant and aerosol propellant; as a grain fumigant; and as a general solvent for adhesives, pesticides, fats, oils, rubbers, alkaloids, and resins. Formerly it was a component of cough syrups, toothpastes, liniments, and toothache compounds; however, in 1976 the Food and Drug Administration banned its use in human drugs or cosmetic products.[26]

The USEPA surveys conducted during the mid-1970s indicated that 95 to 100 percent of finished supplies in the United States contained chloroform. The median concentration was 21 μg/L, the highest was 311 μg/L.[26]

In humans, 30 and 100 mL ingestions of chloroform resulted in severe but nonfatal poisonings. A 200-mL ingestion was fatal. Chronic toxicity tests with guinea pigs demonstrated decreased blood albumin ratios and blood catalase activity. Some deaths occurred with liver necrosis and cirrhosis, and lipoid degeneration and proliferation of interstitial cells in the myocardium. Similar tests conducted with rats demonstrated only impaired ability to develop conditioned reflexes.[27] Subsequent short-term (<90 days) and longer-term (<2 years) chronic toxicity tests conducted by a variety of researchers have demonstrated low-level adverse effects particularly on blood enzymes and on liver weights and fat content in mice, rats, and dogs.[41] Chloroform mutagenicity tests were negative using *S. typhimurium* and positive using yeast.

A number of chloroform carcinogenicity assays have been carried out, the first in 1945. Some investigators used drinking water as the vehicle for exposure; others employed corn oil, toothpaste, arachis oil, olive oil, or Emulphor (an emulsifying agent used to produce aqueous emulsions of lipophilic chemicals in water). Tests conducted in 1986 illustrated that the use of corn oil (versus Emulphor) enhanced hepatotoxic effects of chloroform; thus results from corn oil tests are less useful in the drinking water scenario.[41]

A study by Jorgenson and coworkers[50] tested chloroform in drinking water in rats and mice for 104 weeks at doses of 0 to 160 mg/kg (rats) and 0 to 263 mg/kg (mice). High-dose male rats had an increased incidence of adenomas and adenocarcinomas of the renal tubes. An earlier NCI study (using corn oil) with rats and mice found an increased incidence of renal tumors in rats after 111 weeks and a significant increase in hepatocellular carcinomas in mice after 92 or 93 weeks. (Again, the corn oil vehicle is assumed to have enhanced these observed liver effects.[41])

Using the data from the Jorgenson study, the NAS has calculated a carcinogenic risk for human exposure to chloroform in drinking water. An upper 95 percent confidence level cancer risk, assuming daily consumption of 1 L of water containing 1 μg/L chloroform, is $8.9 \times 10^{-8}$.[41] No carcinogenicity rating according to guidelines has as yet been adopted formally.

**Dibromochloromethane.** Dibromochloromethane is usually produced less frequently and in lower concentrations during water chlorination than is chloroform. Aside from its occurrence as a disinfection by-product, dibromochloromethane is used as a chemical intermediate in the manufacture of fire-extinguishing agents, aerosol propellants, refrigerants, and pesticides.

Effects from acute exposure in mice include sedation, fatty infiltration of the liver, pale kidneys, and hemorrhaging of the adrenals. No data have been developed on effects from chronic exposure. Dibromochloromethane tested positive in the *S. typhimurium* mutagenicity assay.[41]

A 2-year carcinogenicity assay was conducted by the NTP. Rats and mice were exposed to dibromochloromethane in corn oil via gavage for 104 or 105 weeks

while in females, a significant increase in adenomas occurred. Both sexes showed an increased incidence of nonneoplastic liver lesions, and the males also showed an increased incidence of nephrosis (kidney disease characterized by degeneration of the renal tubules). No testing with drinking water as the vehicle of exposure has been done.

Despite the limitations of the NTP study, a carcinogenic risk to humans exposed to dibromochloromethane in their drinking water has been calculated as $8.3 \times 10^{-7}$ (upper 95 percent confidence level, assuming consumption of 1 L/day containing 1 $\mu$g/L).[40]

**Dichlorobromomethane.** Like dibromochloromethane, dichlorobromomethane is also considered to be insoluble in water, yet has been found as a disinfection by-product.

Very few health effects data on dichlorobromomethane exist. Subacute exposure in mice leads to fatty infiltration of the liver, pale kidneys, and hemorrhaging of the adrenals, the same effects noted for dibromochloromethane.[27] Dichlorobromomethane tested positive in the *S. typhimurium* mutagenicity assay. One carcinogenicity assay using intraperitoneal injections in mice over an 8-week period (cumulative dose of 2400 mg/kg) did not yield any pulmonary tumors. No further carcinogenicity testing has been done and no carcinogenicity risk estimates have been developed.[27]

**Bromoform.** Bromoform is the least commonly formed THM disinfection by-product. Commercially, bromoform is used in pharmaceutical manufacturing; as an ingredient in fire-resistant chemicals and gauge fluid; and as a solvent for waxes, greases, and oils. It is not biodegradable in water and is soluble at 1 part to 800 parts water.[26]

Acute effects in mice and guinea pigs from exposure to bromoform via subcutaneous injection include liver and kidney histopathology. Chronic effects assessed by inhalation exposure testing in rats also included disorders in kidney and liver functions. Bromoform was weakly positive in mutagenicity testing using *Salmonella*.[26]

The carcinogenicity of bromoform has been tested via intraperitoneal injections in mice. Injections of 48 mg/kg bromoform at a rate of three injections per week for 7½ weeks resulted in a statistically significant increase in pulmonary tumors. A higher dose of 100 mg/kg did not result in pulmonary tumors. No carcinogenicity risk estimates have been developed.[27]

## Haloacids

**Dichloroacetic acid (DCA).**  Dichloroacetic acid is soluble in water and has been found in surveys designed to detect disinfection by-products in drinking water. DCA is used as a chemical intermediate, and in medicine for treatment of severe lactic acidosis that has not responded to other treatments.[26]

Adverse effects of DCA are seen in the neurological, blood, reproductive, and ocular systems. A man who received 50 mg/kg DCA for 16 weeks for lactic acidosis demonstrated reversible polyneuropathy. Laboratory rats and dogs receiving subacute doses demonstrated reduced erythrocyte, hematocrit, and hemoglobin levels; aspermatogenesis (rats) and prostate gland atrophy (dogs); eye lesions; hind limb weakness; and brain lesions. Other tests using laboratory animals have shown similar results.[41]

Positive mutagenicity test results using bacterial systems are believed to be caused by an impurity in the DCA, namely dichloroacetaldehyde. One carcinogenicity test has been conducted with DCA administered intraperitoneally to mice at a dose equivalent to 5 g/L of drinking water. Hepatocellular carcinomas developed after 61 weeks. No carcinogenicity risk estimates have been developed from these test results.[41]

**Trichloroacetic acid (TCA).**  Trichloroacetic acid has been detected in a few samples during USEPA surveys of drinking water and in laboratory tests designed to determine possible products of disinfection.[41] TCA dissolves to the extent of 1.2 kg/L of water. It is used commercially in organic synthesis, as a reagent for detection of albumin, in medicine to remove warts, in pharmacy, and in herbicides.

Few health effects data exist. One study investigated renal toxicity of TCA in mice. No other toxicity tests have been reported.[41] Two mutagenicity tests using *S. typhimurium* were negative. The carcinogenicity assay carried out for DCA also included TCA. TCA was administered intraperitoneally to mice at a dose equivalent to 5 g/L in drinking water for 61 weeks. Hepatocellular carcinomas occurred in 32 percent of the test mice, while none occurred in the controls.[41]

## Haloaldehydes

The haloacetaldehydes include chloroacetaldehyde, dichloroacetaldehyde, and trichloroacetaldehyde. Trichloroacetaldehyde is soluble in water and was found in the recent USEPA 10-city survey. It also has a long history of use as a hypnotic agent in its monohydrate form, chloral hydrate; however, it is no longer widely used because it is habit forming.

The haloacetaldehydes are precursors of anesthetic through the toxic metabolite of ethanol, possibly responsible for alcohol's liver damage, CNS toxicity, and fetal alcohol syndrome. They are also some or the intermediates found in the metabolism of the haloethylenes (e.g., vinyl chloride), compounds that have been identified as carcinogens. Little health effects information on the haloacetaldehydes themselves exists, however. The available data are summarized below for chloroacetaldehyde and trichloroacetaldehyde; none exist for dichloroacetaldehyde.

**Chloroacetaldehyde.** Various studies report that chloroacetaldehyde is irritating and corrosive to lipids and membrane structures. It also causes a decrease in liver enzyme activity. Subchronic effects observed in rats subjected to intraperitoneal injections of 2.2 or 4.5 mg/kg chloroacetaldehyde for 30 days included a decrease in red blood cell components and an increase in white blood cell components. Also noted were organ-to-body-weight ratio increases, with overall body weight decreases. Similar results were obtained in a longer-term (12 week), lower-dose (0.4 to 3.8 mg/kg) experiment, again using rats and intraperitoneal injections. This latter experiment also noted possibly premalignant necroses in the respiratory tract at the two highest doses.[41]

All chloroacetaldehydes were found to be mutagenic using a variety of *S. typhimurium* strains. Carcinogenicity tests conducted before 1980 using skin bioassays, subcutaneous injections, and intragastric feeding were all negative.[41]

**Trichloroacetaldehyde.** Chloral hydrate was administered in drinking water to mice for 90 days at a dose range approximately equivalent to 18 to 173 mg/kg. Absolute liver weights and liver-to-body-weight ratios increased in males. Liver enzyme activity was also affected. A slight decrease in humoral immune function was observed in females. A test using similar doses for 14 days found no behavioral or coordination changes in mice.

Perinatal exposure of mice to 21.3 and 204.8 mg/(kg · day) in drinking water produced no teratogenic effects. No carcinogenicity tests have been reported.[41]

## Haloketones

The haloketones include 1,1,1-trichloroacetone, 1,1,3,3-tetrachloro-acetone, and hexachloroacetone. The haloketones are postulated to occur as disinfection by-products, and surveys for their presence have started.

The haloketones are principally found as chemical intermediates in industrial processes. The only health effects information available shows four haloketones, 1,1,3-trichloroacetone, 1,1,3,3-tetrachloro-acetone, pentachloroacetone, and hexachloroacetone, to be mutagenic using various strains of *Salmonella*.[41]

## Haloacetonitriles

The haloacetonitriles include chloroacetonitrile (CAN), dichloro-acetonitrile (DCAN), trichloroacetonitrile (TCAN), bromochloro-acetonitrile (BCAN), and dibromoacetonitrile (DBAN). Three of these, DCAN, BCAN, and DBAN, have been identified in the recent USEPA surveys. Commercially, the chlorinated acetonitriles are used as insecticides and fungicides.[41]

Subchronic effects of DCAN and DBAN have been investigated using rats exposed by corn oil gavage. For DCAN, doses of 65 mg/kg for 90 days and 90 mg/kg for 14 days produced decreased weight gains and decreased weights and ratios (organ to body) of the liver, spleen, thymus, lungs, and kidneys. A decrease in blood cholesterol levels was also noted. An NOAEL of 8 mg/kg was determined. For DBAN the protocol and effects were similar. An NOAEL of 23 mg/kg was determined.[41] CAN, DCAN, and TCAN were all fetotoxic (decreased offspring weight and postnatal growth) when a dose of 55 mg/kg was administered to pregnant rats by gavage.

DCAN was shown to be mutagenic in *Salmonella*. It also induced sex-linked recessive lethals in *Drosophila* and sister chromatid exchanges in in vitro Chinese hamster ovary cells. BCAN was shown to be a mutagen using *Salmonella* and also induced sister chromatid exchange in in vitro Chinese hamster ovary cells. CAN, TCAN, and DBAN were negative in *Salmonella* but induced sister chromatid exchange in Chinese hamster ovary cells.[41]

DCAN, DBAN, and CAN applied topically initiated skin tumors in mice. In another carcinogenicity assay, DCAN, DBAN, TCAN, and BCAN were administered orally 3 times per week for 8 weeks. After 9 months, TCAN- and BCAN-dosed mice showed a significant increase in lung tumors. The number of tumors in the DCAN- and DBAN-dosed mice were not statistically significant. In another test, CAN, DCAN, and TCAN tested in a rat liver bioassay for tumor initiating activity produced fewer tumors than did the positive controls.

## Chloropicrin

Chloropicrin, also known as trichloronitromethane or nitrochloroform, is slightly soluble in water (0.17 g/100 g). In 1980 it was detected in chlorinated drinking water in the Netherlands at up to 3.0 μg/L. Subsequently, it was reported to occur in France, Japan, and the United States. Commercially, chloropicrin is used in organic synthesis, dye-

vere fibrosing peribronchitis and peribronchiolitis. An NCI study determined that 32 mg/kg body weight was the NOAEL for a study where mice were exposed by gavage to chloropicrin at 25 mg/kg for 13 weeks and 35 mg/kg for 65 weeks. Exposure to similar doses for 78 weeks resulted in reduced body weights and decreased survival.[41]

The NCI study described above detected no significant increase in tumors that were of the same sort as the controls. A low incidence of different tumors, i.e., squamous cell papillomas and carcinomas of the stomach, was, however, detected. Chloropicrin was tested for mutagenicity in a number of *Salmonella* strains, in *E. coli*, and in *Drosophila*. An indirect positive result was obtained only in one *Salmonella* strain (indirect because it required activation with a liver extract, S9). A possible weak direct mutagenic effect was observed with *E. coli*, and equivocal results were obtained with *Drosophila*.[41]

## Chlorophenols

The chlorophenols include monochlorophenols (2-chlorophenol, *o*-chlorophenol), dichlorophenols (2,4-dichlorophenol), and trichlorophenols (2,4,6-trichlorophenol). Health effects data have been developed primarily for 2,4-dichlorophenol and 2,4,6-trichlorophenol.

**2,4-Dichlorophenol.**   2,4-Dichlorophenol is used in organic synthesis. It is slightly soluble in water and has been detected in drinking water surveys at levels up to 36 μg/L.[41]

Like most of the other chlorophenols, 2,4-dichlorophenol has a low oral toxicity. A 6-month dietary study with mice fed 45 to 230 mg/kg body weight per day detected slight liver histopathology at the highest dose. An NOAEL of 100 mg/L was reported.[26] A subsequent study with mice exposed to 2,4-dichlorophenol in drinking water at 0 to 2.0 mg/mL (0 to 491 mg/kg) for 90 days found no consistent effects at any dose.[41] The only carcinogenicity data come from a skin assay where topical application with an initiator for 39 weeks promoted papillomas and carcinomas.[26]

**2,4,6-Trichlorophenol.** 2,4,6-Trichlorophenol is used as a fungicide, slimicide, bactericide, and wood preservative. Its production was discontinued in 1975 because of the high cost of removing dioxin impurities.

2,4,6-Trichlorophenol is also formed during chlorination of water or wastewater-containing phenols. In 1975, researchers in the Netherlands found various trichlorophenol isomers in surface water at levels from 0.003 to 0.1 μg/L.[30]

Two short-term feeding studies have been conducted. A 7-week study using rats and mice showed a reduction in growth rate at levels greater than 500 mg/(kg · day). A 14-day study using rats given daily oral doses of 0 to 200 mg/kg detected a minimal effect on liver detoxification capabilities.[30]

2,4,6-Trichlorophenol tested negative in mutagenicity assays using various *Salmonella* strains. Conflicting results were obtained with *Saccharomyces cerevisiae*. A carcinogenicity test in mice using skin application for 15 weeks failed to produce skin papillomas. Another test employing oral administration in mice for 18 months detected nondefinitive tumor incidence. A more recent NCI feeding study using mice and rats exposed for 105 and 107 weeks showed a dose-related decrease in body weight. Male rats developed a significant increase in lymphomas or leukemias, while both male and female mice developed hepatocellular carcinomas or adenomas. None of these tests accounted for the possible role of dioxin impurities contained in the 2,4,6-trichlorophenol.

## Radionuclides

Radionuclides are radioactive atoms that break down to release energy (radioactivity). The energy is released in one of three forms: (1) alpha radiation, consisting of large, positively charged helium nuclei; (2) beta radiation, consisting of electrons or positrons; or (3) gamma radiation, consisting of electromagnetic, wave-type energy similar to x-rays. Each of these forms reacts differently within the human body. Alpha particles travel at speeds as high as 10 million meters per second, and when ingested, these relatively massive particles can be very damaging. Beta particles travel at about the speed of light. Their smaller masses allow greater penetration but create less damage. Gamma radiation has tremendous penetrating power but has limited effect at low levels.

Radiation is generally reported in units of curies (Ci), rads, or rems. One curie equals $3.7 \times 10^{10}$ nuclear transformations per second. A common fraction is the picocurie (pCi), which equals $10^{-12}$ curies. By definition, 1 g of radium has 1 Ci of activity. By comparison, 1 g of uranium 238 has $0.36 \times 10^{-6}$ Ci. A rad quantifies the absorbed dose given to tissue or matter, such that a rad of alpha particles creates more damage than a rad of beta particles. A rem quantifies radiation in terms of its dose effect, such that equal doses expressed in rem produce the same biological effect regardless of the type of radiation involved. Thus an effective equivalent dose of 0.1 mrem/year from any

corresponds to about $10^{-?}$ excess lifetime cancer risk level. Numerically, a rem is equal to the absorbed dose in rads times a quality factor (Q) that is specific to the type of radiation in and reflects its biological effects. (Number of rems = Q × number of rads.)

The process of alpha and beta radiation leads to a different element, while gamma ray emission does not. The isotope that decays is called the parent, and the new element is called the progeny. Different isotopes decay at different rates. The half-life of an isotope is the time required for one-half of the atoms present to decay and can range from billions of years to millionths of a second. Isotopes with longer half-lives have lower activities (in terms of curies). Isotopes with very short half-lives are not significant in that they do not survive transport through drinking water distribution systems.

A final factor distinguishing different isotopes is their source. Radioactivity in water can be naturally occurring (typical) or man-made (occasional). Naturally occurring radiation derives from elements in the earth's crust or from cosmic ray bombardment in the atmosphere.

Man-made radiation comes from three general sources: nuclear fission from weapons testing, radiopharmaceuticals, and nuclear fuel processing and use. About 200 man-made radionuclides occur or potentially occur in drinking water; however, only strontium and tritium have been detected on a consistent basis. A naturally occurring decay series includes alpha emissions, while a man-made series generally lacks alpha emission (with a few exceptions such as americium 241 and plutonium 239).

Humans receive an annual dose of radiation of about 200 mrem from all sources. The USEPA estimates that drinking water contributes about 0.1 to 3 percent of a person's annual dose. Local conditions can alter this considerably, however. Some 500 to 4000 public water systems are estimated to have greater than 10,000 pCi/L of radon, a level that corresponds to an annual effective dose equivalent of 100 mrem/year.[51]

Based on occurrence in drinking water and health effects, the radionuclides of most concern are radium 226, radium 228, uranium, and radon 222. These are all naturally occurring isotopes. Radium 228 is a beta emitter, whose decay gives rise to a series of alpha-emitting daughters, while the others are all alpha emitters. Natural uranium actually combines uranium 234, uranium 235 plus uranium 238; however, uranium 238 makes up 99.27 percent of the composition.

Concentrations of these and other radionuclides (both natural and man-made) that have been detected in drinking water are described in Table 2.7. These numbers should be viewed with caution as the data sources for the values shown are limited.[52–56]

Radioactivity can cause developmental and teratogenic effects, genetic effects, and somatic effects including carcinogenesis. The carcinogenic effects of nuclear radiations (alpha, beta, and gamma) on the cell is thought to be ionization of cellular constituents leading to changes in the cellular DNA and thence to DNA-instigated cellular abnormalities. All radionuclides are considered to be carcinogens; however, their target organs differ. Radium 228 is a bone-seeker, causing bone sarcomas. Radium 226 induces head carcinomas. Radon 222 is a gas and can be inhaled during showers, washing dishes, and so forth, or ingested. A direct association between radon 222 and lung cancer has been shown. Because of its high concentration in various systems (see Table 2.7), radon-induced carcinogenicity is of great concern. Uranium is not a demonstrated carcinogen but accumulates in the bones in a similar way to radium. Thus the USEPA intends to classify it as a carcinogen. Uranium also has a demonstrated toxic effect on human kidneys leading to kidney inflammation and changes in urine composition. Under the NIPDWR, MCLs are set for gross alpha at 15 pCi/L, for radium 226 and 228 at 5 pCi/L, and for gross beta emitters at 4 mrem/year. Revised MCLs and MCLGs will be proposed for radionuclides in 1990.

## Aesthetic Quality

Aesthetic components of drinking water quality include taste and odor, turbidity, color, mineralization, hardness, and staining. These problems can originate in source water, within the treatment plant, in distribution systems, and in consumer plumbing. Many are addressed in the national interim secondary drinking water regulations (see Chap. 1). The rationale behind regulating aesthetic aspects of public water supplies is to deter consumers from seeking more pleasant but perhaps less safe sources of drinking water.

### Taste and odor

Taste problems in water derive in part from salts [total dissolved solids (TDS)] and the presence of specific metals such as iron, copper, manganese, and zinc. In general, waters with TDS less than 1200

**TABLE 2.7    Occurrence of Radionuclides in Drinking Water**

| Nuclide | Data sources* | Concentration, pCi/L |
|---|---|---|
| Uranium | USGS sampled > 34,000 surface and > 55,000 groundwaters in 1970s | Average: 2<br>Median: 0.1–0.2<br>Maximum: 600 |
| | USEPA estimate of average popula-tion weighted occurrence | 0.3–2.0 |
| | USEPA estimate of average popula-tion weighted occurrence, based on data from USEPA Office of Radia-tion survey of 2500 groundwater sys-tems and 2 east coast surveys | For groundwater systems > 1000: 0.4–1.0<br>For groundwater systems < 1000: 0.6–1.5<br>All surface: 0.1–0.5 |
| Radium 228 | USEPA estimate of average popula-tion weighted occurrence based on average Ra 228/Ra 226 activity ratio of 1 | 0.4–1.0 |
| Radon | USEPA Office of Radiation survey plus data from other surveys com-bined to give a population weighted average | Large groundwater: 240<br>Small groundwater: 780<br>All groundwater: 420 |
| | Limited survey data | Surface: 5–10 |
| | Maximum detected (private well) | 2,000,000 |
| | USEPA system estimates<br>10,000–40,000 systems<br>5000–30,000 systems<br>1000–10,000 systems<br>500–4000 systems | <br>Approximately 10<br>Approximately 100<br>Approximately 1000<br>Approximately 10,000 |
| Thorium 232 | FR (*Federal Register*) | Maximum detected: 0.1 |
| | USEPA estimated mean upper limit | 0.01 |
| Thorium 230 | FR | Maximum detected in uncontaminated water: 0.4 |
| | USEPA estimated mean upper limit | 0.04 |
| Lead 210 | FR | Average in Connecticut groundwater: 0.02 |
| | USEPA estimated occurrence range | 0.04–0.11 |
| Polonium 210 | FR | Maximum near uranium mines: 2.7 |
| | NIPDWR compliance data | 1 system >5 |
| | USEPA estimated mean based on as-sumed Pe 1210/Pb 210 activity ratio of 1.5 | 0.04–0.13 |

*Data reported from Ref. 51.

mg/L are acceptable to consumers, although levels less than about 650 mg/L are preferable.[57] Specific salts may be more significant in terms of taste, notably magnesium chloride and magnesium bicarbonate. The sulfate salts, magnesium sulfate and calcium sulfate, on the other hand, have been found to be relatively inoffensive.[57] Testing of metals in drinking water showed the following taste thresholds (in mg/L):[45]

| Zinc | 4–9 |
| Copper | 2–5 |
| Iron | 0.04–0.1 |
| Manganese | 4–30 |

Objectionable tastes and odors may also occur in water contaminated with synthetic organics and/or as a result of water treatment. Odor thresholds of some common chlorinated solvents identified in urban groundwater are shown in Table 2.8.

Many consumers object to the taste of chlorine, which has a taste threshold of about 0.2 mg/L at neutral pH.[57] An alternative disinfectant, monochloramine, has been found to have a taste threshold of 0.48 mg/L.[57] Chlorine can also react with organics to create taste and odor problems. Most notorious is the phenolic odor resulting from reactions between chlorine and phenols. Trihalomethanes have been detected by smell at 0.1 mg/L for chloroform and 0.3 mg/L for bromoform.[57] Fluoride can also cause a distinct taste above about 2.4 mg/L.

Decaying vegetation and metabolites of microbiota are probably the most universal sources of taste and odor problems in surface water. The organisms most often linked to taste and odor problems are the filamentous bacteria actinomycetes and blue-green algae, although other algal types, bacteria, fungi, and protozoans are also cited. Table 2.9 illustrates the great number of algae that have been shown to create taste and/or odor problems.

TABLE 2.8    Odor Thresholds of Chlorinated Solvents

| Solvent | Detection odor threshold, mg/L |
|---|---|
| 1,4-Dichlorobenzene | 0.0003 |
| Trichloroethylene | 0.5 |
| Tetrachloroethylene | 0.3 |
| Carbon tetrachloride | 0.2 |

SOURCE: Van Gemert, L. J., and Nettenbreijer, A. H. (eds.): *Compilation of Odour Threshold Values in Air and Water.* Natl. Inst. for Water Supply, Voorburg, Netherlands; and Centr. Inst. for Nutr. & Food Res. TNO, Zeist, Netherlands, 1977.

**TABLE 2.9  Algae-Generated Tastes and Odors**

| Algae class | Odor description | | Taste description | Tactile sensation |
| | Moderate quantities of algae | Large quantities of algae | | |
| --- | --- | --- | --- | --- |
| Cyanophyceae | | | | |
| *Anabaena* | Grassy, musty, nasturtium | Rotten, septic, medicinal | | |
| *Anabaenopsis* | — | Grassy | | |
| *Microcystis* or *Anacystis* | Grassy, musty | Rotten, septic, medicinal | Sweet | |
| *Nostoc* | Musty | Rotten, septic, medicinal | | |
| *Oscillatoria* | Grassy | Musty, spicy | | |
| *Rivularia* | Grassy | Musty | | |
| Chlorophyceae | | | | |
| *Actinastrum* | — | Grassy, musty | | |
| *Ankistrodesmus* | — | Grassy, musty | | |
| *Chara* | Garlic, skunk | Musty, garlic | | |
| *Chlamydomonas* | Musty, grassy | Fishy, septic, medicinal | Sweet | Sickly sweet, oily |
| *Chlorella* | — | Musty | | |
| *Cladophora* | — | Septic | | |
| *Closterium* | — | Grassy | | |
| *Cosmarium* | — | Grassy | | |
| *Dictyosphaerium* | Grassy, nasturtium | Fishy | | |
| *Eudorina* | — | Fishy | | |
| *Gloeocystis* | — | Rotten, medicinal | | |
| *Gonium* | — | Fishy | | |
| *Hydrodictyon* | — | Rotten, septic | | |
| *Nitella* | Grassy | Grassy, rotten | Bitter | |
| *Pandorina* | — | Fishy | | |
| *Pediastrum* | — | Grassy | | |
| *Scenedesmus* | — | Grassy | | |
| *Spirogyra* | — | Grassy | | |
| *Staurastrum* | — | Grassy | | |
| *Tribonema* | — | Fishy | | |
| *Ulothrix* | — | Grassy | | |
| *Volvox* | Fishy | Fishy | | |
| Diatoms | | | | |
| *Asterionella* | Spicy, geranium | Fishy | | |
| *Cyclotella* | Grassy, spicy, geranium | Fishy | | |
| *Diatoma* | — | Aromatic | | |
| *Fragilaria* | Grassy, spicy, geranium | Musty | | |
| *Melosira* | Grassy, spicy, geranium | Musty | — | Sickly sweet, oily |
| *Meridion* | — | Spicy | | |
| *Pleurosigma* | — | Fishy | | |

**TABLE 2.9    Algae-Generated Tastes and Odors (Continued)**

| Algae class | Odor description — Moderate quantities of algae | Odor description — Large quantities of algae | Taste description | Tactile sensation |
|---|---|---|---|---|
| Diatoms (Cont.) | | | | |
| Stephanodiscus | Grassy, spicy, geranium | Fishy | — | Sickly sweet, oily |
| Synedra | Grassy | Musty, fishy | — | Sickly sweet, oily |
| Tabellaria | Grassy, spicy, geranium | Fishy | | |
| Chrysophyceae | | | | |
| Dinobryon | Violets, fishy | Fishy | — | Sickly sweet, oily |
| Mallomonas | Violets | Fishy | | |
| Synura | Cucumber, rotten, medicinal, muskmelon | Fishy | Bitter | Dry, metallic, sickly sweet, oily |
| Uroglenopsis | Cucumber | Fishy | — | Sickly sweet, oily |
| Euglenophyceae | | | | |
| Euglena | — | Fishy | Sweet | |
| Dinophyceae | | | | |
| Ceratium | Fishy | Rotten, septic, medicinal | Bitter | |
| Glenodinium | — | Fishy | — | Sickly sweet, oily |
| Peridinium | Cucumber | Fishy | — | |
| Cryptophyceae | | | | |
| Cryptotomonas | Violets | Violets, fishy | Sweet | |

Adapted from Palmer, C. M.: 1962. *Algae in Water Supplies*. U.S. Public Health Service Pub., No. 657, U.S. Dept. HEW, Publ. Health Serv., 1962; and Seppovaara, A.: "The Effect on Fish of the Mass Development of Brackish Water Plankton," *Aqua Fennica*, 1971, pp. 118–129.

The metabolites responsible for the tastes and odors are in the process of being identified. Two highly studied metabolites of actinomycetes and blue-green algae are geosmin and methylisoborneol (MIB). These compounds are responsible for the common earthy-musty odors in water supplies and have been isolated from many genera of actinomycetes (e.g., *Actinomyces, Nocardia,* and *Streptomyces*) and of blue-green algae (e.g., *Anabaena* and *Oscillatoria*). Both geosmin and MIB can have odor threshold concentrations of less than 10 ng/L.[57] These and other compounds more recently identified are described in Table 2.10. Many other odors are as yet unidentified.

In groundwater and in some distribution systems a highly unpleasant odor of hydrogen sulfide may occur as the result of anaerobic bac-

**TABLE 2.10     Structure of Various Compounds Isolated from Odor-Causing Aquatic Organisms[45]**

| Compound | Structure | Associated organisms |
|---|---|---|
| Methylisoborneol (MIB) | | *Actinomycetes*<br>*Oscillatoria curviceps*<br>*Oscillatoria tenuis* |
| | | *Anabaena scheremetievi* |
| Mucidone | | *Actinomycetes* |
| Isobutyl mercaptan | $CH_3$<br>$CH_3$—$CHCH_2$—$SH$ | *Microcystis flos-aquae* |
| *N*-Butyl mercaptan | $CH_3(CH_2)_3$—$SH$ | *Microcystis flos-aquae*<br>*Oscillatoria chalybea* |
| Isopropyl mercaptan | $CH_3$<br>$CH_3$—$CHSH$ | *Microcystis flos-aquae* |
| Dimethyl disulfide | $CH_3$—$S$—$S$—$CH_3$ | *Microcystis flos-aquae*<br>*Oscillatoria chalybea* |
| Dimethyl sulfide | $CH_3$—$S$—$CH_3$ | *Oscillatoria chalybea*<br>*Anabaena* |
| Methyl mercaptan | $CH_3SH$ | *Microcystis flos-aquae*<br>*Oscillatoria chalybea* |

terial action on sulfates. This rotten egg odor can be detected at less than 100 ng/L. The bacteria most often responsible for hydrogen sulfide production is *Desulfovibrio desulfuricans*. Other bacterially produced sulfur compounds creating swampy and fishy tastes and odors in distribution systems include dimethylpolysulfides and methyl mercaptan.

## Turbidity and color

The appearance of water can be a significant factor in consumer satisfaction, in large part because colored or turbid water is so perceptible. Low levels of color and turbidity are also important for many industries. Typical finished water has color values ranging from 3 to 15 and turbidities from 0 to 1 NTU.

Sources of color in water can include natural metallic ions (iron and manganese), humic and fulvic acids from humus and peat materials, plankton, dissolved plant components, and industrial wastes. The added presence of turbidity increases the apparent but not true color of water. Color removal is typically achieved by the processes of coagulation, flocculation, and filtration.

Turbidity in water is caused by the presence of suspended matter such as clay, silt, finely divided organic and inorganic matter, plankton, and other microscopic organisms. Turbidity's association with health aspects of drinking water were described earlier in the section on particulates. Controlling turbidity as a component of treatment, as is dictated in the revised drinking water regulations, will ensure levels below visual detection limits.

### Mineralization

Water with high levels of salts, measured as TDS, may be less palatable to consumers, and, depending upon the specific salts present, may have a laxative effect on the transient consumer. High levels of sulfates are implicated in this latter aspect. Sulfate may also impart taste, at levels above 300 to 400 mg/L. Concentrations of chloride above 250 mg/L may give water a salty taste.

High levels of salts can also adversely affect industrial cooling operations, boiler feed, and specific processes requiring softened or demineralized water such as food and beverage industries and electronics firms. Typically, high levels of salts will force these industries to pretreat their water. High levels of chloride and sulfate can accelerate corrosion of metals in both industrial and consumer systems.

Removal of salts requires expensive treatment such as demineralization by ion exchange, electrodialysis, or reverse osmosis and is not usually done for drinking water. For water with an excessively high salt concentration, blending with lower salt supplies may ameliorate the problem.

### Hardness

Originally, the hardness of a water was understood to be a measure of the capacity of the water for precipitating soap.[58] It is this aspect of hard water that is the most perceptible to consumers. The primary components of hardness are calcium and magnesium, although the ions of other polyvalent metals such as aluminum, iron, manganese, strontium, and zinc may contribute if present in sufficient concentrations. Hardness is expressed as an equivalent quantity of calcium carbonate ($CaCO_3$). Water having less than 75 mg/L of $CaCO_3$ is generally considered soft. That having between 75 and 150 mg/L of $CaCO_3$ is said to be moderately hard. That having 150–300 mg/L of $CaCO_3$ is considered hard, and water having greater than 300 mg/L of $CaCO_3$ is classified as very hard.[59]

Calcium is of importance to industry as a component of scale. The precipitation of $CaCO_3$ scale on cast-iron and steel pipes helps inhibit corrosion, but the same precipitate in boilers and heat exchangers adversely affects heat transfer.

Staining

Staining of laundry and household fixtures can occur in water with iron, manganese, or copper in solution. In oxygenated surface water of neutral or near-neutral pH (5 to 8), typical concentrations of total iron are around 0.05 to 0.2 mg/L. In groundwater, the occurrence of iron at concentrations of 1.0 to 10 mg/L is common. Higher concentrations (up to 50 mg/L) are possible in low-bicarbonated, low-oxygen water. If water under these latter conditions is pumped out of a well, for example, red-brown ferric hydroxide will begin to precipitate (on fixtures and laundry) as soon as oxygen begins to mix with the water. Manganese is often present with iron in groundwater and may cause similar staining problems, yielding a dark brown to black staining precipitate. Excess copper in water may create blue stains.

## Summary

Health and aesthetic aspects of water quality are the driving force behind water quality regulations and water treatment practice. Because of the complexity of the studies summarized in this chapter, the reader is urged to review the cited literature for more details on any contaminant of particular interest. Because new information on waterborne disease-causing organisms and chemical contaminants is being discovered, review of literature since the publication of this chapter is recommended prior to using the information herein as the basis for decision making.

## References

1. G. F. Craun, "Introduction," in G. F. Craun (ed.), *Waterborne Diseases in the United States*, CRC Press, Boca Raton, Fla., 1986.
2. A. P. Dufour, "Diseases Caused by Water Contact," in G. F. Craun (ed.), *Waterborne Diseases in the United States*, CRC Press, Boca Raton, Fla., 1986.
3. G. F. Craun, "Surface Water Supplies and Health," *Journal AWWA*, vol. 80, no. 2, 1988.
4. G. F. Craun, "Statistics of Waterborne Outbreaks in the U.S. (1920–1980)," in G. F.

Craun (ed.), *Waterborne Diseases in the United States,* CRC Press, Boca Raton, Fla., 1986.

5. G. F. Craun, "Recent Statistics of Waterborne Disease Outbreaks (1981–1983)," in G. F. Craun (ed.), *Waterborne Diseases in the United States,* CRC Press, Boca Raton, Fla., 1986.

6. C. D. Klassen, "General Principles of Toxicology," in *Workshops on Assessment and Management of Drinking Water Contamination,* USEPA 600/M-86/026, Office of Drinking Water, Washington, D.C. (Revised March 1987).

7. J. H. Weisburger and G. M. Williams, "Carcinogen Testing: Current Problems and New Approaches," *Science,* vol. 214, October 23, 1989, p. 401.

8. D. M. Maron and B. N. Ames, "Revised Methods for the Salmonella Mutagenicity Test," *Mutation Research,* vol. 113, 1983, p. 173.

9. *Federal Register,* title 50, part 219, November 13, 1985, p. 46936.

10. M. Sobsey and B. Olson, "Microbial Agents of Waterborne Disease," in P. S. Berger and Y. Argaman (eds.), *Assessment of Microbiology and Turbidity Standards for Drinking Water,* EPA 570-9-83-001, Office of Drinking Water, Washington, D.C., July 1983.

11. T. D. Brock and K. M. Brock, *Basic Microbiology with Applications,* Prentice-Hall, Inc., Englewood Cliffs, N.J., 1978.

12. *Federal Register,* title 52, part 212, November 3, 1987, p. 42178.

13. C. L. R. Bartlett, "Potable Water as Reservoir and Means of Transmission in *Legionella,*" in C. Thornsberry et al. (eds.), *Legionella, Proc. Second Int. Symp.,* American Society for Microbiology, Washington, D.C., 1984.

14. R. Mitchell, *Water Pollution Microbiology,* Wiley, New York, 1972.

15. V. C. Rao et al., "Removal of Hepatitis A Virus and Rotavirus by Drinking Water Treatment," *Journal AWWA,* vol. 80, no. 2, February 1988, p. 59.

16. A. S. Tombes et al., "Surface Morphology of *Giardia* Cysts Removed from a Variety of Hosts," in W. Jakubowski and J. C. Hoff (eds.), *Waterborne Transmission of Giardiasis,* EPA 600/6-79-001, June 1979.

17. R. C. Rendtorff, "The Experimental Transmission of *Giardia Lamblia* among Volunteer Subjects," in W. Jakubowski and J. C. Hoff (eds.), *Waterborne Transmission of Giardiasis,* EPA 600/6-79-001, June 1979.

18. *Federal Register,* title 52, part 212, November 3, 1987, p. 42178.

19. G. S. Logsdon et al., "Alternative Filtration Method for Removal of *Giardia* Cysts and Cyst Models," *Journal AWWA,* vol. 73, no. 2, February 1981, p. 111.

20. R. Fayer and B. L. P. Ungar, "*Cryptosporidium* spp. and Cryptosporidiosis," *Microbiological Reviews,* vol. 50, no. 4, December 1986, p. 458.

21. J. B. Rose, "Occurrence and Significance of *Cryptosporidium* in Drinking Water," *Journal AWWA,* vol. 80, no. 2, February 1988, p. 53.

22. C. N. Haas et al., "Removal of New Indicators by Coagulation and Filtration," *Journal AWWA,* vol. 77, no. 2, February 1985, p. 67.

23. V. P. Olivieri, "Measurement of Microbial Quality," in P. S. Berger and Y. Argaman (eds.), *Assessment of Microbiology and Turbidity Standards for Drinking Water,* EPA 570/9-83-001, Office of Drinking Water, Washington, D.C., July 1983.

24. M. J. Allen and E. E. Geldreich, "Bacteriological Criteria for Groundwater Quality," *Groundwater,* vol. 13, 1975, p. 45.

25. E. J. Calabrese et al., *Inorganics in Drinking Water and Cardiovascular Disease,* Princeton Scientific Publishing Co., Inc., Princeton, N.J., 1985.

26. National Academy of Sciences Safe Drinking Water Committee, *Drinking Water and Health,* National Academy Press, Washington, D.C., 1977.

27. National Academy of Sciences Safe Drinking Water Committee, *Drinking Water and Health,* vol. 3, National Academy Press, Washington, D.C., 1980.

28. P. O. Ganrot, "Metabolism and Possible Health Effects of Aluminum," *Environmental Health Perspectives,* vol. 65, 1986, p. 363.

29. D. P. Perl, "Relationship of Aluminum to Alzheimer's Disease," *Environmental Health Perspectives,* vol. 63, 1985, p. 149.

30. National Academy of Sciences Safe Drinking Water Committee, *Drinking Water and Health,* vol. 4, National Academy Press, Washington, D.C., 1982.

31. *Federal Register,* title 53, part 14, January 22, 1988, p. 1892.

32. *Federal Register,* title 54, part 97, May 22, 1989, p. 2206.
33. L. J. McCabe et al., "Survey of Community Water Supplies," *Journal AWWA,* vol. 62, no. 11, November 1970, p. 670.
34. National Academy of Sciences Safe Drinking Water Committee, *Drinking Water and Health,* vol. 5, National Academy Press, Washington, D.C., 1983.
35. *Federal Register,* title 48, part 194, October 5, 1983, p. 45502.
36. *Federal Register,* title 53, part 160, August 18, 1988, p. 31516.
37. C. J. Johnson et al., "Fatal Outcome of Methemoglobinemia in an Infant," *Journal*

41. National Academy of Sciences Safe Drinking Water Committee, *Drinking Water and Health,* vol. 7, National Academy Press, Washington, D.C., 1987.
42. E. M. Aieta and J. D. Berg, "A Review of Chlorine Dioxide in Drinking Water Treatment," *Journal AWWA,* vol. 78, no. 6, June 1986, p. 62.
43. National Academy of Sciences Safe Drinking Water Committee, *Drinking Water and Health,* vol. 2, National Academy Press, Washington, D.C., 1980.
44. R. G. Rice and M. Gomez-Taylor, "Occurrence of By-products of Strong Oxidants Reacting with Drinking Water Contaminants—Scope of the Problem," *Environmental Health Perspectives,* vol. 69, November 1986, p. 31.
45. James M. Montgomery Consulting Engineers, Inc., *Water Treatment Principles and Design.* John Wiley & Sons, New York, 1985.
46. *Federal Register,* title 49, June 12, 1984, p. 24330.
47. *Federal Register,* title 52, part 74, April 17, 1987, p. 12876 .
48. National Academy of Sciences Safe Drinking Water Committee, *Drinking Water and Health,* vol. 6, National Academy Press, Washington, D.C., 1986.
49. J. R. Fowle, III, and F. C. Kopfler, "Water Disinfection: Microbes vs Molecules—An Introduction of Issues," *Environmental Health Perspectives,* vol. 69, November 1986, p. 3.
50. T. A. Jorgenson et al., "Carcinogenicity of Chloroform in Drinking Water to Male Osborne-Mendel Rats and Female B6C3F1 Mice," *Fund. Appl. Toxicol.,* vol. 5, 1985, p. 760.
51. *Federal Register,* title 51, part 189, September 30, 1986, p. 34836.
52. T. R. Horton, "Methods and Results of EPA's Study of Radon in Drinking Water," EPA SW-520/5-83-027, 1983.
53. C. R. Cothern and W. L. Lappenbusch, "Compliance Data for the Occurrence of Radium and Gross Alpha Particle Activity in Drinking Water Supplies in the U.S.," *Health Physics,* vol. 46, 1984, p. 503.
54. C. T. Hess et al., "The Occurrence of Radioactivity in Public Water Systems of the U.S.," *Health Physics,* vol. 48, 1985, p. 553.
55. J. Mitchell et al., "Gamma-ray Spectrometry for Determination of Radium-228 and Radium-226 in Natural Waters," *Analytical Chemistry,* vol. 53, 1981, p. 1885.
56. C. R. Cothern et al., "Drinking Water Contribution to Natural Background Radiation," *Health Physics,* vol. 50, January 1986, p. 33.
57. American Water Works Association Research Foundation & Lyonnaise des Eaux, *Identification and Treatment of Tastes and Odors in Drinking Water,* J. Mallevialle and I. H. Suffet (eds.), American Water Works Association, Denver, Colo., 1987.
58. American Public Health Association, American Water Works Association, Water Pollution Control Federation, *Standard Methods for the Examination of Water and Wastewater,* 16th ed., A. E. Greenberg et al. (eds.), American Public Health Association, Washington, D.C., 1985.
59. C. N. Sawyer and P. L. McCarty, *Chemistry for Sanitary Engineers,* 2d ed., McGraw-Hill, Inc., New York, 1967.

# Guide to Selection of Water Treatment Processes

**Carl L. Hamann, Jr., P.E.**

*Director for Water Technology and Innovation*
*CH2M Hill, Inc.*
*Denver, Colorado*

**J. Brock McEwen, P.E.**

*Senior Process Consultant*
*CH2M Hill, Inc.*
*Denver, Colorado*

**Anthony G. Myers, P.E.**

*Senior Process Consultant*
*CH2M Hill, Inc.*
*Milwaukee, Wisconsin*

Water treatment process selection is a complex task involving the consideration of many factors. The selection of water treatment processes to be used on a given water supply source is dictated by the need to produce acceptable finished water quality at the most attractive overall cost. The choice of a water treatment scheme depends on

- Water supply source quality
- Desired finished water quality
- Reliability of process equipment
- Operational requirements and personnel capabilities

- Flexibility in dealing with changing water quality and equipment malfunctions
- Available space for construction of treatment facilities
- Waste disposal constraints
- Capital and operating costs (including chemical availability)

flocculation may affect performance of the filtration process.

Water treatment processes can be grouped according to their general function in the water treatment scheme. However, each water source is unique, and process modifications to fit a particular treatment objective must be considered on a case-by-case basis. The purpose of this chapter is to provide a general description of water treatment processes and their effectiveness for meeting broad treatment objectives. It is intended to serve as a guide and direct the reader to more detailed information in the following chapters.

## Water Source Selection

The selection of a water supply source involves a review of the alternative sources available and their respective characteristics. Factors to consider when selecting a water supply source include

- Safe yield
- Water quality
- Collection requirements (intake structure, wells, etc.)
- Treatment requirements (including the cost and feasibility of residue disposal)
- Transmission and distribution requirements

The ability to furnish both the quantity and quality of water needed continuously, without failure, must be considered. Studies to determine water supply safe yield and water quality should be coupled with water demand projections in a water system master plan.

Water quality will affect the treatment processes selected as well as the cost of water treatment. Evaluation of alternative water sources must include not only an analysis of water treatment costs but also of the cost of water collection, transmission, and distribution. Location of the water supply source with respect to the treatment plant and point of use may significantly affect the cost of supplying water.

Although water quality is variable from source to source, surface waters have many qualities in common. Likewise, groundwater supplies have many similar characteristics.

### Surface water sources

Surface water supplies are developed from streams, rivers, lakes, or impounding reservoirs. The location and elevation of the water source may offer the advantage of gravity flow to the treatment facilities. Surface water supplies always require some type of treatment.

Streams or rivers have the characteristic of rapid changes in water quality. During heavy rains or spring runoff, changes in turbidity and other constituents require flexible and reliable treatment processes and close operator attention. In addition, rivers and streams are more susceptible to accidental spills and transport of contaminants into the treatment plant. Consequently, treatment process selection should consider the occurrence of such events.

Lakes and impounding reservoirs have the characteristic of seasonal changes in water quality, but these changes are more gradual and less dramatic in nature than those of streams and rivers. During summer months the lake may stratify into distinct layers such that warmer water stays near the surface and cooler water is trapped below with little intermixing. This condition can lead to oxygen depletion at lower depths. Under these reducing conditions, iron and manganese can be solubilized. Taste and odor problems may also increase because of release of anoxic and/or anaerobic decay products and hydrogen sulfide.

Upper lake levels are susceptible to algal blooms if carbonate, nutrient, and temperature conditions are favorable. Algal blooms can cause changes in source water turbidity, alkalinity, taste, odor, pH, and other characteristics.

Other considerations for choosing surface water supply locations include surrounding land use, vegetation, soil characteristics, and topography.

### Groundwater sources

Groundwater supplies are developed by construction of wells, galleries, and ranney collectors, or the collection of spring water. Groundwater is relatively constant in quality from season to season. However, groundwater supplies may be highly variable in quality from one well location to another. Changes in hydrogeological conditions can produce different water quality over a relatively short distance.

The cost of investigating a new groundwater supply is typically more expensive than investigating a surface water supply. Several test wells are often required to determine aquifer yield and water quality characteristics of the groundwater supply. Supplying groundwater to the treatment facility is often more expensive than supplying surface water because of increased pumping requirements.

Groundwater quality is usually superior to that of surface water with respect to bacteriological content, turbidity, and total organic concentrations. On the other hand, the mineral content (hardness, iron, manganese) of groundwater may be inferior and require additional treatment. Groundwater supplies are frequently pumped into the distribution system with minimal treatment.

evaluation.

Finally, some communities have the option of using both surface and groundwater supplies. Use of these supplies can be optimized to provide a more reliable water source that uses the best features of both supplies.

## General Capabilities of Unit Processes

This section provides an overview of the water treatment processes discussed elsewhere in this book. The objective is to present the components and general applications of unit processes and provide a "road map" to the contents of the remainder of the book. A more detailed evaluation of each unit process is presented in the referenced chapters.

### Air stripping and aeration (Chap. 5)

Air stripping and aeration are processes used in water treatment to remove dissolved gases such as carbon dioxide, hydrogen sulfide, other taste- and odor-causing compounds, and volatile organic compounds (VOC). Aeration is also used to add oxygen from air into water, to oxidize iron and manganese, and to prevent formation of reducing environments that exacerbate taste and odor problems.

Several methods of contacting air with water to strip dissolved gases and/or oxygenate water are available. The selection of the most appropriate method is generally based on contaminant removal efficiency and economics in comparison to alternative processes available for the desired contaminant removal.

Spray nozzles have been used for many years in the water treatment field. The most common application is the fountain-type aerator that sprays water under pressure into the open atmosphere. Fountain-type aerators are used in source water reservoirs to control taste and odor problems, prevent anaerobic decay of natural organic matter accumulated in the reservoir, and prevent solubilization of iron and manganese present in reservoir bottom sediments.

Cascading tray aerators have also been used in water treatment for many years. Cascading tray aerators use multistage waterfalls, generally placed at the water plant headworks, to control taste and odor problems and precipitate soluble iron and manganese.

Diffused aeration is not often used in the water treatment field but is sometimes employed in source water reservoirs to control taste and odor problems and oxidize iron and manganese. Diffused aeration involves releasing compressed air bubbles from a diffuser element located at the bottom of a water column. Diffused aeration is sometimes proposed for VOC removal but is not generally cost-effective or as efficient as packed-tower air stripping.

Packed-tower stripping is widely used in the water treatment field because of increasing concern and regulation of VOCs in drinking water. Packed-tower stripping generally consists of high-surface-area packing material supported and contained in a cylindrical shell. Water flow is normally downward through the packing material with either forced draft or induced draft upward airflow. The high surface area of the packing material provides a higher liquid-gas transfer area than the other aeration and stripping methods presented. Thus, the packed-tower stripper is more effective for removal of VOCs that are more volatile than the more traditional contaminants of concern such as carbon dioxide and hydrogen sulfide.

Off-gas treatment of contaminants stripped from the water may be necessary depending on the characteristics and quantity of the gaseous contaminants, as well as on the location of the stripping operation and prevailing air quality regulations.

### Coagulation processes: mixing, destabilization, and flocculation (Chap. 6)

Coagulation processes are commonly included in water treatment plants to promote aggregation of small particles into larger particles that can be subsequently removed by sedimentation and/or filtration. Sedimentation and filtration processes are discussed in Chaps. 7 and 8, respectively.

Coagulation processes differ from chemical precipitation in that coagulation is based on the destabilization of stable particulate suspensions in water, while chemical precipitation is based on compound solubility chemistry.

Particulate suspensions commonly removed with coagulation processes include clay- and silt-based turbidity, natural organic matter, and other associated constituents, such as microbial contaminants, toxic metals, synthetic organic chemicals, iron, and manganese. These associated contaminants often adsorb to or combine with turbidity and natural organic suspensions, thus enabling their removal by coagulation processes.

The coagulation process is comprised of three sequential steps. These are metal-salt coagulant formation, particle destabilization, particle aggregation. Coagulant formation and particle destabilization are promoted in a rapid-mixing stage where treatment chemicals are added, hydrolyzed, and dispersed throughout the particulate suspension to cause destabilization. Particle aggregation is then promoted in a flocculation stage where interparticle collisions create larger particles amenable to separation from the treated water.

Alum and iron (III) salts are the most common treatment chemicals used in water treatment, with alum probably having the most widespread use. The formation of the most effective metal salt coagulant species for particle destabilization is dependent on particulate suspension quantity and quality, chemical dose, pH, temperature, ionic strength, and reaction time. Jar tests are generally performed to select coagulation process chemical dosages.

Synthetic organic polymers are sometimes used in addition to or in place of metal salt coagulants when jar tests demonstrate acceptable performance. Synthetic organic polymers are attractive substitutes for metal salt coagulants because they generate less chemical sludge. However, substitution of synthetic organic polymers for metal salt coagulants is typically not practical because the performance of synthetic organic polymer coagulation is water-quality specific and is difficult to optimize over the broad range of source water quality conditions often encountered in municipal water treatment plants.

Synthetic organic polymers are more often used as coagulant aids in combination with metal salt primary coagulants. Synthetic organic polymers can enhance particle aggregation and produce larger, stronger, and/or denser aggregates more readily removed in downstream sedimentation and/or filtration processes. Again, jar tests are generally required to select effective coagulant aids.

Increased attention has recently been focused on the presence of residual chemical coagulants in finished waters. As a result, chemical treatments for coagulation should be selected to both optimize treatment performance and minimize the magnitude of chemical doses.

Several methods are available to provide rapid mixing when using metal salt coagulants and to disperse coagulant species throughout the water to be treated. The method of mixing inherently includes some detention time that, in turn, affects the reaction time available for formation of coagulant species. Consequently, the rapid-mixing stage is not only a mechanism for coagulant dispersal but also a factor in the character of coagulant species formed.

The most common rapid-mixing method used in water treatment is a back-mix mechanical reactor. Other rapid-mixing methods include in-line blenders, hydraulic jumps, motionless static mixers, and diffuser-injection devices.

Flocculation is typically performed in a basin baffled into three or more compartments with mechanical mixing provided in each stage to promote interparticle collisions and aggregation. In an effort to reduce the capital and operating costs associated with this traditional form of flocculation, contact flocculation performed within up-flow sludge blanket clarifiers, buoyant media roughing clarifiers, and granular media filters (in-line filters) are gaining increased application. See Chaps. 7 and 8, which address sedimentation and filtration processes, for further discussions regarding contact flocculation.

### Sedimentation and flotation (Chap. 7)

Sedimentation and flotation are solid-liquid gravity separation processes. Sedimentation processes promote gravity settling of solid particles to the bottom of the water column where accumulated solids are removed. Flotation processes introduce gas bubbles into the water that attach to solid particles and create bubble-solid agglomerates that float to the top of the water column where accumulated solids are removed. Flotation is not generally used in place of sedimentation in water treatment. However, flotation is gaining acceptance in Europe as an alternative to sedimentation for algal-laden, low-turbidity, low-alkalinity, colored waters that produce lower-density floc less amenable to sedimentation.

Sedimentation is generally used in combination with coagulation and flocculation to remove floc particles and improve subsequent filtration efficiency. Absence of sedimentation prior to filtration results in shorter filter runs, poorer filtrate quality, and dirtier filters more difficult to backwash. Sedimentation is particularly necessary for high-turbidity and highly colored waters that generate substantial solids during the coagulation and flocculation processes. Sedimentation is sometimes unnecessary prior to filtration (direct filtration) in instances where flocculation solids production is low and filtration can effectively handle the solids loading. Conditions appropriate for direct filtration are discussed in Chap. 8.

Sedimentation is also sometimes employed at the head of a water treatment plant in the form of a presedimentation basin to allow gravity settling of denser solids that do not require coagulation and flocculation for promotion of solids separation. Application of a presedimentation basin is most commonly found where surface waters have a high silt or turbidity content. In some cases coagulants may be employed ahead of presedimentation basins.

Sedimentation process efficiency is based on the effectiveness of solid-liquid separation. Sedimentation is most effective when hydraulics approach plug and laminar flow conditions so that short circuiting and turbulence are minimized. Plug inlet flow distribution and effluent collection, wind and other eddies, and density differences caused by temperature or concentration all can contribute to flow instability. Therefore, design of an effective sedimentation process is dependent on proper consideration of physical design factors such as surface loading rate, size and shape of process tankage, inlet and outlet flow arrangements, baffling, and solids removal methods.

There are many types of sedimentation tanks. A horizontal-flow rectangular tank is the traditional design. Multistory horizontal-flow rectangular tanks reduce land requirements in comparison to the traditional rectangular basins but also increase the complexity of structural requirements and mechanical sludge removal. Circular radial-flow tanks and a variety of proprietary designs including variations of integral premixing, solids recirculation, and sludge scraping are available. The circular solids-contact clarifier is commonly used for softening applications in water treatment. Inclined plates and tube settlers increase the available area for solids settling, decrease the solids settling depth, promote flow stability, and decrease land requirements in comparison to more traditional rectangular sedimentation tanks. Tube settlers have found widest application in retrofits of existing horizontal-flow rectangular and circular tanks for increasing water flow-through rates and delaying physical expansions. Plate settlers are not readily amenable to retrofits and are gaining greater acceptance in the United States as an alternative to conventional rectangular basins. Equal flow distribution to inclined plates and tube settlers is essential to obtain good performance. Floc blanket clarifiers promote contact flocculation and strain rising solids within a suspended floc blanket to achieve solids separation. Continually drawing excess floc off the top of the blanket maintains a constant blanket depth. Floc blanket clarifiers do not require separate upstream flocculation basins as do traditional rectangular sedimentation tanks. In addition, contact flocculation within the floc blanket precludes the necessity of submerged mechanical components to impart flocculation energy. However, floc blanket clarifiers are more sensitive to chemical treatment, rapid change in source water quality, and flow-rate changes than conventional flocculation-sedimentation systems.

Many options exist for separating solids from water prior to filtration. Project-specific conditions and criteria will dictate choice.

## Filtration (Chap. 8)

Filtration is the process relied on in most water treatment facilities for the removal of suspended particulate matter. Common particulates removed in water treatment filtration are clay and silt, colloidal and precipitated natural organic matter, metal salt precipitates from coagulation, lime softening precipitates, iron and manganese precipitates, and microorganisms. Suspended particulate removal not only furnishes a high-clarity, aesthetically pleasing finished water but also removes particulates that can both increase disinfectant demand and shield microorganisms from disinfection.

Filtration of surface waters is mandated by the USEPA surface water treatment rule (SWTR) to provide protection against the passage of pathogens into the finished water supply. The SWTR credits the filtration process with a specified removal of enteroviruses and *Giardia lamblia*. Consequently, filtration is considered an integral part of the overall water disinfection process. See Chap. 14 for a detailed discussion of disinfection.

Granular media filters are the most common type of filter used in potable water treatment. Granular media filters can either be contained in vessels open or closed to the atmosphere with gravity flow or pressurized passage through the filter media, respectively.

The granular media type, size, gradation, shape, and depth determine pore volume, pore size, and pore tortuosity that affect solids holding capacity, head loss characteristics, filtrate quality, and backwash flow requirements. The most common materials used in granular media filters are sand, crushed anthracite coal, garnet, ilmenite, and granular activated carbon (GAC). Sand, dual media, and mixed media (more than two media) filters are the most common configurations in the United States. However, deep-bed single medium filters using coarse sand, anthracite coal, or GAC are gaining acceptance particularly in direct filtration applications.

The rate of filtration directly impacts overall filter performance. Traditional rapid sand filters are generally operated at a rate of 2 gallons per minute per square foot ($gpm/ft^2$). Dual- and mixed-media filters provide greater solids holding capacity and improved depth removal, which is less subject to surface binding, and thus are commonly operated at rates of up to 5 or 6 $gpm/ft^2$. In addition, the advent of synthetic organic polymers as coagulants and filter aids allows formation of stronger floc particles more resistant to the greater shear forces exhibited at higher filtration rates. Thus, filtrate quality can be maintained in spite of the higher rates.

Efficient backwashing of granular media filters is necessary to maintain optimum filtered water quality, as well as to prevent gradual deterioration of the media and its eventual replacement. The backwash method most commonly used is an up-flow water wash with fluidization of the media. Deep filter beds with deep solids penetration combined with the use of synthetic organic polymer treatment aids increases the forces necessary to wash accumulated solids from the media. Auxiliary surface water wash or air scour prior to or during up-flow fluidization are often used to enhance solids removal under these more demanding conditions. Filters having the capability to be continuously backwashed are also an option for consideration.

Adequate pretreatment prior to filtration is essential to successful filter performance. Pretreatment will affect the mass solids loading to the filter, the solids particle size distribution, the particulate shear resistance, and particle net charge that in turn affect the ability of the granular media to retain solids.

Direct filtration omits sedimentation basins prior to filtration, and in-line filtration omits both flocculation and sedimentation basins prior to filtration. In both processes, the pretreatment goal is to form a pinpoint-sized floc that is filterable. Filterability is enhanced as pin floc undergo contact flocculation within the granular media pore spaces producing slightly larger floc particles more readily retained in the media. Direct or in-line filtration is appropriate for waters where the source water turbidity and chemical coagulant dosage requirements are relatively low. Direct and in-line filtration are not appropriate for waters high in turbidity and/or color. In addition, the reduced detention time inherent with these processes provides less time for seasonal taste and odor control and for operator response to source water quality changes.

Slow sand filters operate at low filtration rates and rely on cake filtration at the surface of the filter for particulate straining. As the surface cake develops during the filtration cycle, the cake itself assumes the dominant role in filtration rather than the granular media. Coagulation and sedimentation are generally not provided prior to slow sand filtration; however, the source water quality must have relatively low turbidity for the slow sand filter to produce acceptable finished water and reasonable filter run lengths. Slow sand filters are relatively inexpensive to construct and operate and are attractive for small installations where source water quality is favorable. Slow sand filters have also been found particularly effective for removal of *Giardia* from low-turbidity water, such as in treatment of snowmelt waters in resort facilities.

Precoat filtration is widely used in filtration of water for swimming pools and has also been used for potable water treatment when direct or in-line filtration of surface water is applicable, or when removal of low concentrations of iron and manganese precipitants from groundwater is practiced. Precoat filters use a thin layer of very fine material such as diatomaceous earth deposited on a permeable, rigid support structure. The precoat filter physically strains solids from the water. When the surface cake builds to a thickness that imparts a head loss impractical for further filtration, the accumulated solids and diatomaceous earth are washed from the support septum, the support structure is coated with a new layer of diatomaceous earth, and a new filter cycle is initiated. Like slow sand filtration, precoat filtration is very effective for removal of *Giardia lamblia* cysts and chemical coagulation pretreatment is unnecessary. As a result, precoat filtration is attractive for small water treatment facilities with operators unskilled in chemical treatment. Like slow sand filtration, precoat filtration is not appropriate for water with high turbidity and/or water that requires pretreatment for algae, color, and taste and odor control.

Monitoring of turbidity and head loss from each filter is an important process control factor for maintaining optimum filter performance. Turbidity and head loss monitoring enables operators to terminate filter cycles prior to turbidity breakthrough. Turbidity and head loss monitoring also enable operators to observe and control pretreatment, filtration rate, and filter to waste periods for production of the highest-quality finished water.

## Ion exchange and inorganic adsorption
## (Chap. 9)

Ion exchange with synthetic resins or adsorption onto activated alumina are generally considered in circumstances where mineral quality of the source water supply necessitates treatment capability beyond that possible with more conventional water treatment processes.

The largest application of ion exchange in water treatment is for softening. Smaller water utilities sometimes prefer ion exchange softening over lime softening because of lower capital costs and ease of automation. Cation or anion exchange or inorganic adsorption may be appropriate for water sources with contaminant ions of toxic or radioactive substances such as barium, arsenic, chromium, fluoride, nitrate, radium, and uranium.

Activated alumina adsorption is finding application in fluoride and arsenate removal from source water. However, ion exchangers and inorganic adsorption units are relatively costly to construct and operate for removal of one particular contaminant. Their application is generally more feasible for point-of-use devices. Ion exchangers are more effective for stable water quality sources such as groundwater supplies as opposed to variable quality surface water supplies.

An ion exchanger is typically comprised of a bed packed with ion exchange resin beads presaturated with an exchangeable ion or with

activated alumina granules possessing exchangeable surface hydroxides. Source water is continuously passed through the packed bed in
either an up-flow or down-flow mode until appearance of an unwanted
contaminant is detected at undesirable levels in the bed effluent. At
this stage the ion exchange media is reactivated with a regenerant solution and rinsed with water in preparation for another treatment cy-

Factors to consider in the application of ion exchange and inorganic
adsorption to municipal water treatment include pretreatment to mitigate fouling of exchange media caused by suspended solids, precipitates, and biological growth; special construction materials to handle
corrosive salt and regenerant solutions; and spent regenerant solution
and spent media disposal.

The feasibility and application of ion exchange resins and inorganic
adsorbent will likely increase as new, stricter inorganic contaminant
drinking water standards are promulgated. Radium, nitrate, fluoride,
arsenic, and barium are treatment challenges that may induce wider
application of ion exchange and inorganic adsorption particularly for
small community water treatment plants.

### Chemical precipitation (Chap. 10)

Chemical precipitation processes are commonly used in water treatment for water softening and iron and manganese removal. Chemical
precipitation is also effective for heavy metals and radionuclide removal when these contaminants are present in a water source. Chemical precipitation can also provide some dissolved organic chemical removal and reduce viruses and bacteria as well.

Lime precipitation is the most common form of chemical precipitation used for softening and removal of heavy metals, radionuclides,
dissolved organics, and viruses. Although coagulation with metal
salts involves chemical precipitation, extensive coverage of this topic
is given in Chap. 6. Oxidation of reduced iron and manganese to form
iron and manganese precipitates for subsequent filtration is the most
common method of iron and manganese removal practiced in the
United States. Potassium permanganate, molecular oxygen, and chlorine are the most common oxidants used to precipitate soluble iron
and manganese.

Lime, lime-soda ash, or caustic soda softening are commonly performed in either a conventional rapid-mix, flocculation, and sedimentation process train or a solids-contact softener that combines the
rapid-mix, flocculation, and sedimentation processes into a single
unit. Recirculation of precipitant is often employed in softening to provide precipitation nuclei to enhance the rate of precipitation and in-

crease particle collisions and agglomeration in the flocculation step. The solids-contact softener is the most common system applied for water softening.

The selection of lime, lime-soda ash, or caustic soda softening is based on cost, total dissolved solids criteria, sludge production, carbonate and noncarbonate hardness, and chemical stability. Water containing little or no noncarbonate hardness can be softened with lime alone. However, water with high noncarbonate hardness may require both lime and soda ash to achieve the desired finished water hardness. Softening with lime or lime-soda ash is generally less expensive than caustic softening. Caustic soda softening increases the total dissolved solids of the treated water, while lime and lime-soda ash softening often decreases total dissolved solids. Caustic soda softening produces less sludge than lime and lime-soda ash softening. Caustic soda does not deteriorate during storage, while hydrated lime may adsorb carbon dioxide and water in storage and quick lime may slake in storage causing feeding problems. The final selection is generally based on cost, water quality, and owner and operator preference.

Softened water has high causticity and scale-formation potential; hence, recarbonation is employed to reduce pH and mitigate scaling of downstream processes and pipelines. On-site combustion generation of carbon dioxide or liquid carbon dioxide is the most common source of carbon dioxide for recarbonation. Recarbonation can be applied in either a single-stage or two-stage process. In the single-stage process, the pH of the softened water is lowered after the sedimentation step. The two-stage process is most often employed in situations where excess lime is applied to elevate the pH to 11.0 or higher for optimal removal of magnesium, heavy metals, radionuclides, dissolved organics, and virus inactivation. In the first stage of the two-stage process, the pH is lowered to the minimum solubility of calcium carbonate to precipitate excess lime in a secondary clarification basin. Following secondary clarification, second-stage recarbonation is employed to lower the pH further and mitigate downstream scaling. In many situations, single-stage recarbonation is used because the additional capital costs of a second clarification basin and second recarbonation unit are eliminated. However, two-stage recarbonation affords greater operational flexibility, lower chemical costs for a given goal, and generally improves product water quality.

Selection of the lime, lime-soda ash, or caustic soda chemical precipitation process must adequately address the disposal of generated sludges. Many methods are available to concentrate and dewater sludges for increasing ease of handling and disposal. These methods include gravity thickening, dewatering lagoons, sand drying beds, centrifugation, vacuum filtration, pressure filtration, belt filtration, sludge pelletization, recalcination, and land application. These methods may be used separately or in combination to achieve a specific management objective.

Formerly water treatment plants discharged waste sludges into rivers or lakes; however, this practice is changing and in many cases is prohibited. Ultimate disposal of lime or caustic sludges now includes options such as discharge to sanitary sewers, landfills, drying lagoons, and land application.

Although softening is effective for iron and manganese removal, it is too costly to implement for only this purpose. More often potassium permanganate, chlorine, or aeration are used to precipitate soluble iron and manganese. For treatment of surface water sources, the oxidant is generally applied prior to coagulant addition. Groundwater applications sometimes combine potassium permanganate oxidation with manganese-greensand filtration to provide continual regeneration of manganic dioxide deposits on the greensand and to prevent pink-water filter breakthrough.

Excess lime or caustic soda addition to raise the pH to 11.0 or higher has been found effective for toxic heavy metal and radium precipitation and removal. High pH treatment also promotes precipitation and/or coprecipitation of soluble organic contaminants. However, coagulation with metal salts is generally more effective than lime softening for removal of natural organic matter. High pH treatment is also highly effective for virus inactivation.

## Membrane processes (Chap. 11)

Membrane processes include many different alternatives, such as reverse osmosis (RO), electrodialysis (ED), electrodialysis reversal (EDR), ultrafiltration (UF), and nanofiltration (NF). The RO and ED/EDR processes are actively used in the municipal water treatment field primarily for desalting or brackish water conversion. UF and NF are emerging as viable potable water treatment processes for removing particulates, color, trihalomethane (THM) precursors, and some inorganics (hardness). The common component shared among the various membrane processes is a membrane able to reject or select passage of certain dissolved species based on compound size, shape, and/or charge.

Membrane processes are usually considered in circumstances such as desalination, brackish water conversion, and for removal of specific ions that are difficult to remove with other processes. Membrane processes are also frequently evaluated for wastewater reuse applications to provide softening and removal of organics, radionuclides, heavy metals, bacteria, and viruses. ED/EDR systems are appropriate for brackish water conversion but do not provide barriers to other nonionic dissolved species. RO is appropriate for desalting seawater and brackish water and also provides an effective barrier to other dissolved organic and inorganic contaminants as well as bacteria and viruses. UF and NF provide a similar but less effective contaminant barrier than RO because of larger membrane pores.

The treatment performance capabilities and limitations of the RO, NF, and UF systems are dependent on several factors. The specific quality of the feed water and the product water quality desired influence the type of system selected. Generally, the more contaminated the feed water and the higher the desired product water quality, the greater the likelihood of membrane fouling caused by particulate matter, scaling, and biofouling.

Provision for adequate pretreatment to control feed-water quality is essential for the RO, UF, and NF processes optimum membrane performance. Suspended solids must be removed prior to membrane treatment to mitigate plugging. Chemical addition to control pH is necessary to protect membrane materials from decomposition and to mitigate membrane scaling from solids precipitation. Antiscalants are generally necessary to further reduce membrane fouling caused by solids precipitation. Biocides compatible with membrane materials are often added to control biological growths on membranes that cause plugging and/or membrane decomposition.

Proper membrane selection and system configuration for RO, UF, and NF are critical. The physical configuration of the membrane system contributes to the overall system performance. Multiple membrane stages provide greater product water recovery and less brine generation when all other process performance factors remain constant. The operating pressure of RO, UF, and NF also affect performance. Higher feed-water salinity requires greater operating pressure to provide a given product water quality and also increases brine production for a given membrane system configuration.

ED/EDR systems are often considered as an alternative to RO for brackish water conversion. In the operation of an ED/EDR system, selective cation and anion membranes are combined with a dc electric field to demineralize or deionize water. EDR is an ED process in which the polarity of the dc electric field is reversed periodically to reverse the direction of ion movement and provide automatic flushing of scale-forming materials from membrane surfaces. The ED/EDR process performance is dependent on feed-water salinity and the quantity and quality of product water desired. Once these factors are known, the size of the unit can be determined and the membrane stack can be designed with an appropriate combination of hydraulic and electrical staging to achieve the desired desalting. ED/EDR systems, unlike the

RO process, are not effective for removal of dissolved organic and microbiological contaminants. Like the RO, NF, and UF processes, the ED/EDR cation and anion membranes are subject to scaling caused by solids precipitation that may require antiscalant or acid addition and eventually require periodic cleaning.

availability of alternative water supply sources requiring less sophisticated treatment.

### Chemical oxidation (Chap. 12)

Chemical oxidants are used in water treatment for many purposes including

- Biological growth control in pipelines and basins
- Color removal
- Taste and odor control
- Reduction of specific organic compounds
- Flocculation aids
- Iron and manganese oxidation and subsequent precipitation and removal
- Disinfection

The most common oxidants used in water treatment are chlorine, chloramines, ozone, chlorine dioxide, and potassium permanganate. Other processes that generate a hydroxyl radical are under intensive study. Chlorine is the most common oxidant used in water treatment, but application of alternative oxidants is increasing because of concerns regarding the health effects of chlorinated by-products produced when chlorine reacts with naturally occurring organic compounds present in some source waters.

Chloramines are produced from the reaction of ammonia with aqueous chlorine. Chloramines are weak oxidants and are not effective for color removal or iron and manganese precipitation. Chloramines are most commonly used in water treatment as an alternative disinfectant to chlorine so as to reduce corrosivity and/or formation of chlorinated by-products. Chloramine disinfection is not as strong as chlorine, but chloramines have been successfully used for years by several major municipal water utilities.

Ozone is an extremely strong oxidizing agent and has several potential water treatment applications. Ozone is often used in Western Europe but until recently was not seriously considered as an alternative to chlorine in the United States because of its relative high cost. However, with the increasing concern and regulation regarding chlorinated by-products, use of ozone is gaining greater acceptance in the United States. Although ozone is more expensive than chlorine, the higher costs associated with the use of ozone may be offset by iron and manganese precipitation, taste- and odor-control efficacy, color removal potential, and flocculant aid potential.

Chlorine dioxide is a powerful oxidant prepared on site by the reaction of chlorine and sodium chlorite under low pH conditions. Chlorine dioxide is capable of oxidizing iron and manganese, color removal, taste and odor control, and disinfection. However, the USEPA has recommended that the total residual level of chlorine dioxide and its by-products, chlorite and chlorate, be limited to 1.0 mg/L (1988). Consequently, this limits application of chlorine dioxide in water treatment to cases where demand is low.

Potassium permanganate has been used widely in the water treatment industry for precipitation of iron and manganese. Potassium permanganate is also used for taste and odor control and color removal; however, relatively high doses are generally required to oxidize difficult taste and odor problems or effectively treat highly colored waters. In addition, potassium permanganate doses must be carefully controlled to prevent development of pink water.

Chemical oxidation performance is influenced by pH, temperature, oxidant dosage, reaction time, and the presence of substances that interfere with the desired redox reaction. The pH often affects the form of the oxidizing agent that in turn affects the rate of oxidation. In general, rates of chemical oxidation increase with increasing temperature. Most oxidation reaction rates are increased with increasing oxidant dosage. Sufficient reaction time must be provided to allow the chemical oxidation reactions to proceed to an extent providing treatment benefits. Pretreatment and/or adjustment of the aforementioned factors may be necessary to allow desired chemical oxidation reactions to proceed when substances are present in the water that compete for the oxidizing agent.

The use of chemical oxidation will depend on its economy compared to alternative methods of contaminant removal, concern regarding health effects attributable to oxidant by-products, and overall benefits provided by the chemical oxidation.

### Adsorption of organic compounds
### (Chap. 13)

Powdered or granular activated carbon (PAC or GAC) adsorption is generally used for the removal of dissolved organics, color, and taste-

and odor-causing compounds in municipal water treatment. Generally, high-molecular-weight, nonpolar compounds are adsorbed more effectively than low-molecular-weight, polar compounds.

PAC is commonly used for controlling seasonal taste and odor problems experienced in surface water supplies. Common points of PAC

time for contaminant adsorption, minimal interference from other treatment chemicals, and ability to prevent degradation of finished water quality caused by PAC fines. The PAC type and dose are selected by adding the PAC and measuring the contaminant removal. Typical PAC doses range from 1 to 50 mg/L. The effectiveness of PAC is also water-source specific and is difficult to predict without performing bench, pilot, or full-scale testing.

Although GAC treatment is more expensive than PAC treatment, GAC is far more effective than PAC for removing a wide range of organics including taste- and odor-causing compounds, total organic carbon (TOC), THM formation potential (THMFP), VOCs, and synthetic organic contaminants (SOC). The performance of GAC is influenced by the location of the process within the overall treatment train because this determines how much pretreatment precedes GAC. Pretreatment can reduce the organic load on the GAC process, remove suspended solids that may interfere with the adsorption process and/or cause hydraulic plugging, and change the adsorbability or biodegradability of the organic compounds entering the GAC process.

The contact time and application rate, with a given feed-water quality and desired finished water quality, determine the size of the GAC contactor and the activated carbon usage rate, which in turn control the capital and operating costs for the GAC process. As contact time increases, contactor size and capital costs increase, but activated carbon usage rate and operating cost decrease. Therefore, an optimum contact time exists that minimizes the total cost of the GAC process. The application rate also affects GAC contactor head loss characteristics and associated hydraulic design and costs.

Biological activity on GAC allows biological oxidation of some organic compounds, thus reducing activated carbon usage rates. However, biological activity also increases microbial concentrations in GAC effluent; thus disinfection must be provided downstream of GAC treatment.

VOCs are reversibly adsorbed and desorbed back into the bulk water flow when they are displaced by competing compounds or when a decrease in influent concentration shifts the adsorption-desorption equilibrium. (This is generally true for all organics.) These compounds may control activated carbon usage rates if GAC is the primary barrier to VOC contamination.

The GAC adsorption process is relatively expensive, and thermal reactivation of the spent activated carbon may become cost-effective depending on the facility capacity. GAC reactivation can be performed on site or contracted to an off-site facility depending on economics. In some small plants, disposal and total replacement of the GAC is most economical.

### Disinfection (Chap. 14)

Disinfection is provided in water treatment as one of the multiple barriers to assure the production of a microbiologically safe finished water quality. Disinfectants may also be added to the water treatment process to control nuisance biological growths such as algal growth in open basins and channels.

Disinfectants include free and combined chlorine, chlorine dioxide, ozone, ultraviolet (UV) irradiation, and other miscellaneous disinfectants, such as potassium permanganate, heat, or extremes in pH. Free chlorine is the most common disinfectant used in the United States. However, increasing regulation of chlorination by-products is increasing the application of alternative disinfectants.

Although free-chlorine disinfection produces trace quantities of harmful by-products, free chlorine is an excellent biocide and provides a persistent residual to maintain the microbiological safety of the finished water as it passes through the distribution system.

The advent of THM regulations and the spector of further chlorinated by-products regulation has increased the application of combined chlorine disinfection in the form of chloramines. Chloramines are less reactive with natural organic matter than free chlorine and hence reduce chlorinated by-product formation. Chloramines are less powerful biocides than free chlorine but are stable and provide a persistent residual for distribution system protection.

Chlorine dioxide is a more powerful biocide than free chlorine but does not persist as long as chlorine. However, the use of chlorine dioxide disinfection is limited because residuals greater than 0.4 to 0.5 mg/L cause adverse taste and odor problems. In addition, chlorine dioxide and its by-products, chlorite and chlorate, have been found to cause harmful effects in animals; hence, total residual concentrations should be less than 1 mg/L to minimize harmful effects.

Ozone is the most powerful disinfectant of those used in water treatment. However, ozone is highly unstable in water and lacks persistent residual. As a result, application of additional secondary disinfectants such as chlorine or chloramine is generally necessary to protect water in the distribution system. Ozone is not commonly used in the United States but has been used successfully in western Europe for years. Ozone disinfection will likely increase in the United States as standards for chlorinated by-products become more stringent.

UV radiation is a good biocide but like ozone provides no persistent re~~~~~~~~~~~~~~~~~~~~~~~~~~~~~~~~~~~~~~~~~~~~~~~~~~~~~~~tical for smaller capacity plants because of the capital and operating expense necessary to ensure adequate water contact with lamp surfaces ~~~large-capacity plants. An effective cleaning program must be ~~~~~~~~~ to ensure that biological and/or chemical foulants do not block UV transmission into the water. Also like ozone, UV radiation requires a secondary disinfectant to provide residual for distribution protection.

The performance of the disinfection process is dependent on the type of disinfectant, disinfectant dosage, pH, temperature, presence of interfering substances, feed-water microbiological quality, and contact time. The stronger the disinfectant selected, the more quickly the disinfection process proceeds. Increasing disinfectant dosage will increase the rate of disinfection under given conditions but may exacerbate formation of harmful by-products depending on the type of disinfectant. The pH may affect the form and in turn the efficiency of the disinfectant depending on the type of disinfectant. Increasing temperature, other factors remaining constant, increases the rate of disinfection. Pretreatment is essential to remove suspended solids and/or reducing agents that compete with the microorganisms for disinfectant. The concentration and type of microorganisms in the influent to the disinfection process may influence the selection of disinfectant type. The CT value (disinfectant concentration times the contact time) provided must be adequate for microbial kill under the most difficult conditions foreseen.

Control of the primary disinfectant application and maintenance of disinfectant residual are important components of the process design to optimize performance and minimize costs. Applying disinfectant at a dose greater than required to effect microbial kill and/or maintain residual is a needless cost and may exacerbate formation of harmful products depending on disinfectant type. As a result, disinfectant application, microbial monitoring, and residual measurement should be combined to achieve disinfection goals at the least cost.

## Development of Treatment Process Schemes for Basic Water Quality Problems

Two general categories of water supply sources are surface water and groundwater. Traditional treatment processes for dealing with water quality problems from both sources will be presented here. Many variations and options are available to treat water other than those presented in this chapter. However, discussing the fundamentals of water treatment process schemes will lay the groundwork for more detailed discussions in later chapters.

### Surface water

Basic surface water quality problems are defined as those associated with particulate content, color, taste, odor, and microbiological content. A conventional process scheme for surface water sources having seasonally high concentrations of particulates, color, taste, odor, and acceptable hardness is shown in Fig. 3.1. The basic treatment processes include coagulation, flocculation, and sedimentation prior to filtration. Some treatment plants include two stages of coagulation, flocculation, and sedimentation that provide more flexibility and often improve water quality. Source water screening is commonly included to remove larger materials from the water. A primary oxidant or disinfectant is used to control bacteria content, algal growth, taste, and odors. Chlorine has commonly been used in the past, but concern over the health effects of halogenated by-products has increased the use of ozone, chlorine dioxide, and potassium permanganate. Iron and aluminum salts are commonly used as coagulants, although polymers used along with or in place of metal salts are being used more frequently. Addition of either iron or aluminum salts depresses the pH, so lime, soda ash, or caustic soda are added to raise pH and reduce corrosiveness. Alternate points of pH adjustment will depend on the treatment objective. More than one application point provides addi-

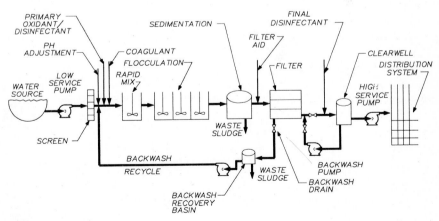

NOTE:
Chemical application points may be different than shown above. This is one potential alternative.

**Figure 3.1**   Conventional treatment, surface water.

tional flexibility. After sedimentation, a polymer or metal salt filter aid may be added to improve filter performance. A filter aid can also be added to the backwash water to reduce initial turbidity breakthrough. After filtration, a final disinfectant (usually chlorine or chloramines) is added to lower the microbiological content in the distribution system. A common practice in conventional water treatment is filter backwash water recycling. This can be an acceptable water conservation and waste stream reduction process.

Modifications to the conventional surface water treatment plant are necessary to deal with certain water quality problems. For example, water with high taste and odor content can be treated with potassium permanganate, PAC, or ozone added to the source water. The presence of SOCs or intense taste and odor compounds may require GAC either as a postfilter adsorber or as a replacement to conventional sand-anthracite filter media.

Surface water with low particulate content but high color could also be treated with a process similar to that shown in Fig. 3.1. Water with consistently low particulates, color, taste, odor, and acceptable hardness can generally be treated without the sedimentation process and have the coagulation and flocculation processes followed by filtration. This treatment process scheme is known as direct filtration. Traditional flocculation basins may be replaced by a contact basin to aid flocculation in this mode. In some situations, flocculation may be eliminated and the coagulant added with proper mixing just before the filter. This treatment process scheme is known as in-line filtration. Both direct and in-line filtration process schematics are shown in Fig. 3.2.

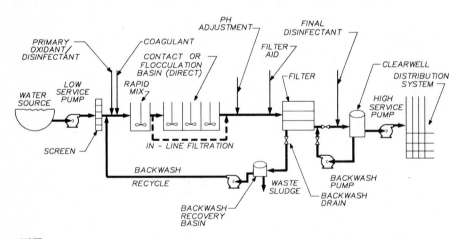

NOTE:
Chemical application points may be different than shown above. This is one potential alternative.

**Figure 3.2**  Direct and in-line filtration treatment, surface water.

Surface or groundwater sources high in hardness can be treated by a lime softening process discussed in the following section and shown in Fig. 3.3. If the surface water has a high particulate content, such as river water sources, presedimentation basins (with or without coagulant addition) can be provided before lime addition.

### Groundwater

Basic groundwater quality problems are typically associated with high hardness, iron, and manganese. Process schemes for sources high in hardness may include lime treatment for calcium and magnesium reduction, and soda ash addition if significant noncarbonate hardness is present in the supply. Such a treatment process is shown in Fig. 3.3. In excess lime softening, lime is mixed with the water supply and a flocculant aid (such as an anionic polymer) may be added to enhance the agglomeration of precipitates that are formed.

The precipitated calcium carbonate and magnesium hydroxide are removed in a settling basin. Solids-contact clarifiers combine rapid mix, flocculation, and settling into one treatment basin. Because of capital cost and space reductions, they are usually an attractive alternative in softening groundwater. In addition, solids-contact clarifiers may reduce deposition and scaling problems in pipes and channels between unit processes. Recirculating some of the lime sludge to the rapid-mix stage improves precipitation and particle agglomeration. Settled water has a high pH (10.6 to 11.0), and the pH must be reduced by addition of carbon dioxide. Figure 3.3 shows a two-stage recarbonation system where initial recarbonation is used to neutralize excess lime resulting in formation of additional calcium carbonate precipitate. Soda ash may also be added after initial recarbonation if noncarbonate hardness is high. Solids formed in these reactions are

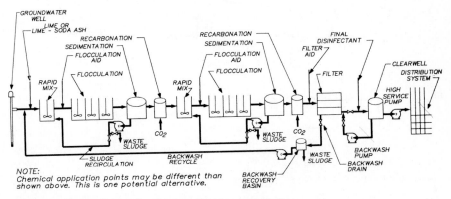

NOTE:
Chemical application points may be different than shown above. This is one potential alternative.

**Figure 3.3**   Excess lime softening treatment, groundwater.

removed by secondary mixing, flocculation, and sedimentation facilities. After secondary settling, carbon dioxide is added again to further reduce the pH and form a stable water. This is followed by filtration and final disinfection.

Although two-stage recarbonation is more effective in hardness removal and controlling the stability of the softened water, a less expensive, single-stage recarbonation process is sometimes used following secondary sediment. Aeration is sometimes used before lime softening to remove carbon dioxide from the source water. Because lime reacts with carbon dioxide, reducing the carbon dioxide concentration reduces the amount of lime required. An economic analysis of the cost of aeration versus the savings from reduced lime dosage should be performed before aeration is included in the process scheme.

Split-treatment lime softening is a variation of the softening process and is shown in Fig. 3.4. This process should be considered where the magnesium content of the supply is moderate to low, and taste, odor, and color are not excessive. Excess lime is added to a portion of the total water flow to reduce calcium and magnesium concentrations. A portion of the source water is mixed with the softened water after primary sedimentation to neutralize the excess lime and remove some calcium from the source water split stream. The precipitated solids are removed by secondary flocculation and sedimentation; then the water is recarbonated and filtered. Split-treatment softening can provide more flexibility in obtaining desired hardness levels. Chemical costs are reduced because only a portion of the flow is treated with excess lime, and the source water split neutralizes excess lime to reduce or eliminate carbon dioxide requirements. In addition, lime sludge production often decreases.

Ion exchange softening may be an attractive alternative in some circumstances. It is applicable for water low in particulates, organics,

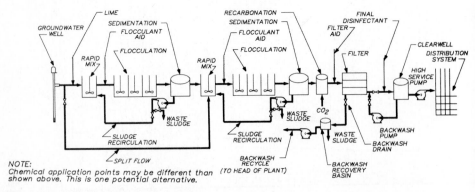

NOTE:
Chemical application points may be different than shown above. This is one potential alternative.

**Figure 3.4**   Split-flow lime softening treatment, groundwater.

iron, and manganese. Ion exchange offers advantages over lime softening for water with variable hardness concentrations and high
noncarbonate hardness content. A typical ion exchange softening process is shown in Fig. 3.5. Pretreatment to remove iron and manganese,
if they are present in the source water, should precede ion exchange.
High organic content can also foul certain ion exchange resins.

The most common ion exchange softening resin is a sodium cation
exchange (zeolite) resin that exchanges sodium for divalent cations.
After the resin has reached its capacity for hardness removal, it is
backwashed, regenerated with a sodium chloride solution, and rinsed
with finished water. This places the resin back in the sodium form so
that it can resume softening. A portion of the source water is typically
bypassed around the softening vessel and blended with softened water. This provides calcium ions to help stabilize the finished water.

Anion exchange resins are used in water treatment with equipment
similar to that shown in Fig. 3.5. Anions such as nitrates and sulfates,
along with other compounds, are removed with this process.

Process schemes for groundwater sources high in iron and manganese, but with acceptable hardness, generally include a means of oxidizing and removing the precipitated iron and manganese compounds.
Lime treatment is effective for iron and manganese removal, but it is
not economical unless softening is also required. A general iron and
manganese removal process is shown in Fig. 3.6. In water with high

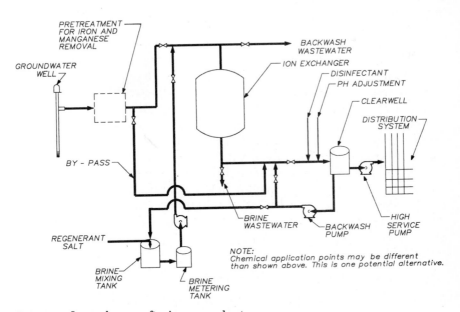

**Figure 3.5**  Ion exchange softening, groundwater.

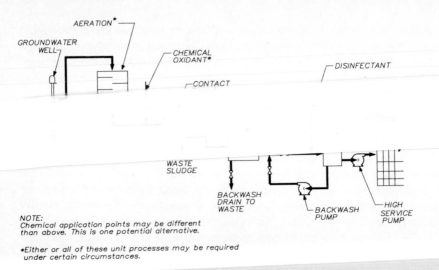

**Figure 3.6**   Iron and manganese treatment, groundwater.

iron but low manganese concentration, only aeration and filtration may be required. The rate of iron precipitation is highly dependent upon pH, and oxidation occurs rapidly at a pH of 8 or higher. At lower pH, more contact time may be required before filtration. Chemical oxidants may be used in place of or in addition to aeration. Chlorine is a common oxidant for iron removal, but ozone, chlorine dioxide, and potassium permanganate can be used for both iron and manganese removal.

Manganese oxidizes slowly with aeration at pH values below 9.5. Therefore, aeration is not an effective manganese removal process. Potassium permanganate oxidation is generally favored for waters with high manganese content. Other chemical oxidants can also be used. When potassium permanganate is used, coatings of manganese dioxide form on the grains of filter media, which assist in iron and manganese removal. Manganese-coated greensand is sometimes used instead of sand or anthracite as a filter media. The ion exchange properties of manganese-coated greensand make it an effective media for iron and manganese removal. In some cases, pressure filters are used instead of gravity filters to eliminate repumping requirements. The presence of organics can impede precipitation of iron and manganese.

## Contaminant Removal Effectiveness

The level of contaminant removal with treatment processes described herein will be generally discussed. The objective is to direct the reader to water treatment processes that have the potential for removing a contaminant of concern. Many exceptions and special conditions exist under which a water treatment process will effectively remove a contaminant. Further information can be obtained from the chapters covering that particular process.

A list of contaminants, water treatment processes, and a range of removal efficiencies is contained in Table 3.1. Contaminants are grouped by primary contaminants—those with enforceable standards that adversely affect health; secondary contaminants—those with nonenforceable standards but are of concern for aesthetic reasons; and proposed contaminants—those with enforceable standards planned for the future. Removal efficiencies are defined by the following rating system:

| Rating | Percent removal |
| --- | --- |
| Excellent | 90–100 |
| Good | 60–90 |
| Fair | 20–60 |
| Poor | 0–20 |

Some contaminants have a wide range of removal efficiencies for a particular water treatment process. This may be caused by a variety of conditions under which removal occurs, or the contaminant group may include many species that have different removal efficiencies (i.e., SOCs, pesticides).

A particular treatment process may not be effective in and of itself for removal of a contaminant. However, many of the processes are commonly used together to achieve effective removal. For example, aeration and lime softening are usually followed by sedimentation and/or filtration.

## Summary

The purpose of this chapter is to provide an overview of treatment processes and applications. It is intended to serve as a guide and direct the reader to the following chapters which expand on each treatment process and provide more detailed information.

Certain water quality concerns do not have an individual chapter dedicated to them. The problem may be addressed in several chapters. For example, as noted in Table 3.1, iron and manganese removal is addressed in Chaps. 5, 9, 10, and 12. Likewise, taste and odor is covered in Chaps. 5, 12, and 13; whereas THMs and disinfection by-products are covered in Chaps. 12 and 14.

**TABLE 3.1  General Effectiveness of Water Treatment Processes for Contaminant Removal[1-45]**

| Contaminant categories | Aeration and stripping (Chap. 5) | Coagulation processes, sedimentation, filtration (Chaps. 6,7,8) | Lime softening (Chap. 10) | Ion exchange — Anion (Chap. 9) | Ion exchange — Cation (Chap. 9) | Membrane — Reverse osmosis (Chap. 11) | Membrane — Ultrafiltration (Chap. 11) | Membrane — Electrodialysis (Chap. 11) | Chemical oxidation, disinfection (Chaps. 12,14) | Adsorption — GAC (Chap. 13) | Adsorption — Activated alumina (Chap. 9) |
|---|---|---|---|---|---|---|---|---|---|---|---|
| **A. Primary contaminants** | | | | | | | | | | | |
| **1. Microbial and turbidity** | | | | | | | | | | | |
| Total coliforms | P | G–E | G–E | P | P | E | E | — | E | P | P–F |
| *Giardia lamblia* | P | G–E | G–E | P | P | E | E | — | E | P | P–F |
| Viruses | P | G–E | G–E | P | P | E | E | — | E | | P–F |
| *Legionella* | P | G–E | G–E | P | P | E | E | — | E | | P–F |
| Turbidity | P | E | G | F | F | E | E | — | P | | P–F |
| **2. Inorganics** | | | | | | | | | | | |
| Arsenic ( + 3) | P | F–G | F–G | G–E | P | F–G | — | F–G | P | F | G–E |
| Arsenic ( + 5) | P | G–E | G–E | G–E | P | G–E | — | G–E | P | F | E |
| Asbestos | P | G–E | — | — | — | — | — | — | P | | — |
| Barium | P | P–F | G–E | P | E | E | — | G–E | P | | P |
| Cadmium | P | G–E | E | P | E | E | — | E | P | | P |
| Chromium ( + 3) | P | G–E | G–E | P | E | E | — | E | F | | P |
| Chromium ( + 6) | P | P | P | E | P | G–E | — | G–E | P | | P |
| Cyanide | P | — | — | — | — | G | — | G | E | | — |
| Fluoride | P | F–G | P–F | P–F | P | E | — | G | P | F | E |
| Lead | P | E | E | P | F–G | E | — | E | P | F | P |
| Mercury (inorganic) | P | F–G | F–G | P | F–G | F–G | — | F–G | P | | P |
| Nickel | P | F–G | E | P | E | E | — | E | P | | P |
| Nitrate | P | P | P | G–E | P | G | — | G | P | | P |
| Nitrite | F | P–F | P | G–E | P | G | — | G | G–E | | P |
| Radium (226 and 228) | P | P | G–E | P | E | G–E | — | G–E | P | P–F | P–F |
| Selenium ( + 6) | P | P | P | G–E | P | E | — | E | P | G–E | G–E |
| Selenium ( + 4) | P | F–G | F | G–E | P | E | — | E | P | G–E | G–E |

184

| | | | | | | | | | | | | |
|---|---|---|---|---|---|---|---|---|---|---|---|---|
| **3. Organics** | | | | | | | | | | | | |
| VOCs | G–E | P | P–F | P | P | F–E | F–E | F–E | P–G | F–E | P–G | P |
| SOCs | P–F | P–G | P–F | P | P | F–E | F–E | F–E | P–G | F–E | P–E | P–G |
| Pesticides | P–F | P–G | P–F | P | P | F–E | F–E | F–E | P–G | G–E | G–E | P–G |
| THMs | G–E | P | P | P | P | F–G | F–G | F–G | P–G | F–E | P–F | P |
| THM precursors | P | F–G | P–F | F–G | — | G–E | F–E | G–E | F–G | F–E | P–F | P–F |
| **B. Secondary contaminants** | | | | | | | | | | | | |
| Hardness | P | P | E | P | E | E | G–E | E | P | P | P | P |
| Iron | F–G | F–E | E | P | G–E | G–E | G | G–E | G–E | P | P | P |
| Manganese | P–F | F–E | E | P | G–E | G–E | G | G–E | F–E | E | P | P |
| Color | P | F–G | F–G | P–G | — | — | — | — | F–E | G–E | G–E | G |
| Taste and odor | F–E | P–F | P–F | P–G | — | — | — | — | P | G–E | G–E | P–F |
| Total dissolved solids | P | P | P | P | P | G–E | P–F | G–E | P | P | P | P |
| Chloride | P | P | P | F–G | P | G–E | P | G–E | P–F | P | P | — |
| Copper | P | G | G–E | P | F–G | E | — | E | P | F–G | P | — |
| Sulfate | P | P | P | G–E | P | E | P | E | P | F | — | G–E |
| Zinc | P | F–G | G–E | P | G–E | E | — | E | G–E | P | F–G | — |
| TOC | F | P–F | G | — | G–E | G | G–E | P–G | P | F–G | P | — |
| Carbon dioxide | G–E | P–F | E | P | P | P | P | P | F–E | P | P | P |
| Hydrogen sulfide | F–E | P | F–G | P | P | P | P | P | P | P | P | P |
| Methane | G–E | P–E | P | P | P | P | P | P | P | P | P | P |
| **C. Proposed contaminants** | | | | | | | | | | | | |
| VOCs | G–E | P | P–F | P | P | F–E | F–E | F–E | P–G | F–E | P–G | P |
| SOCs | P–F | P–G | P–F | P | P | F–E | F–E | F–E | P–G | F–E | P–E | P–G |
| Disinfection by-products | — | P–E | P–F | P–F | — | P | F–G | F–G | F–G | F–E | P–G | — |
| Radon | G–E | P | P | P | P | P | P | P | P | E | P–F | P |
| Uranium | P | G–E | E | E | G–E | E | — | — | P | F | P–F | G–E |
| Aluminum | P | F | F–G | P | G–E | E | — | — | P | — | — | — |
| Silver | F–G | G–E | P | G | — | E | — | — | F–G | P–F | — | — |

P—poor (0 to 20 percent removal); F—fair (20 to 60 percent removal); G—good (60 to 90 percent removal); E—excellent (90 to 100 percent removal); "—"—not applicable/insufficient data

Note: Costs and local conditions may alter a processes applicability.

# References

1. F. B. DeWalle et al., *Removal of Giardia Lamblia Cysts by Drinking Water Plants*, USEPA, EPA-600/S2-84-069, Municipal Environmental Research Laboratory, Cincinnati, Ohio, May 1984.

moval of *Giardia* Cysts," *Journal AWWA*, vol. 77, no. 2, February 1985, p. 61.

5. P. Payment et al., "Elimination of Viruses and Indicator Bacteria at Each Step of Treatment During Preparation of Drinking Water at Seven Water Treatment Plants," *Applied Environ. Microbiol.*, vol. 49, no. 6, June 1985, pp. 1418–1428.
6. O. M. Aly and S.D. Faust, "Removal of 2,4 Dichlorophenoxyacetic Acid Derivatives from Natural Waters," *Journal AWWA*, vol. 57, no. 2, February 1986, p. 221.
7. C. A. Buescher et al., "Chemical Oxidation of Selected Organic Pesticides," *Journal Water Pollution Control Fed.*, vol. 36, no. 8, August 1964.
8. M. D. Cummins, *Removal of Ethylene Dibromide from Contaminated Ground Water by Packed Column Air Stripping*, USEPA, Office of Drinking Water, Technical Support Division, August 1984.
9. V. H. Edwards and P. F. Schubert, "Removal of 2,4-D and Other Persistent Organic Molecules from Water Supplies by Reverse Osmosis," *Journal AWWA*, vol. 66, no. 10, October 1974, p. 610.
10. Environmental Science and Engineering, *Study of Effectiveness of Activated Carbon Technology for Removal of Specific Materials from Organic Chemical Processes*, EPA Rep. No. 68032610, 1981.
11. R. E. Hansen, "Experiences With Removing Organics from Water," *Public Works*, October 1977.
12. E. Hinden et al., "Organic Compounds Removed by Reverse Osmosis," *Water Sewage Works*, 1968.
13. G. Lettinga et al., "The Use of Flocculated Powdered Activated Carbon in Water Treatment," *Prog. Water Tech.*, vol. 10, pp. 537–554, 1978.
14. B. Morgeli, "The Removal of Pesticides from Drinking Water," *Sulzer Tech. Review*, vol. 54, no. 2, February 1972.
15. R. L. Mumford and J. L. Schnoor, "Air Stripping of Volatile Organics in Water," *Proc. AWWA Annual Conf.*, Miami Beach, Fla., 1982.
16. G. G. Robeck et al., "Effectiveness of Water Treatment Process in Pesticide Removal," *Journal AWWA*, vol. 57, no. 2, February 1965, pp. 181–199.
17. J. E. Singley et al., "Use of Powdered Activated Carbon for Removal of Specific Organic Compounds," Sem. Proc.: "Controlling Organics in Drinking Water," *Proc. AWWA Annual Conf.*, San Francisco, Calif., 1979.
18. T. J. Sorg and O. T. Love, "Reverse Osmosis Treatment to Control Inorganic and Volatile Organic Contamination," Sem. Proc.: "Experiences with Ground Water Contamination," *Proc. AWWA Annual Conf.*, Dallas, Texas, 1984.
19. W. L. Brinck et al., *Determination of Radium Removal Efficiencies in Water-Treatment Processes*, USEPA, ORP/TAD-76-5, 1976.
20. D. Clifford, "Processes for Removal of Inorganic Contaminants from Drinking Water," *Water/Engineering and Management*, July 1982, P-R31.
21. J. F. Ferguson and M. Q. Anderson, "Chemical Forms of Arsenic in Water Supplies and Their Removal," in A. J. Rubin (ed.), *Chemistry of Water Supply Treatment and Distribution*, Ann Arbor Science, Ann Arbor, Mich., 1974.
22. D. H. Furukawa, "Removal of Heavy Metals from Water Using Reverse Osmosis," in *Traces of Heavy Metals in Water Removal Processes and Monitoring*, USEPA, EPA-902/9-74-001, November 1973, pp. 179–188.
23. M. R. Huxstep, *Inorganic Contaminant Removal from Potable Water by Reverse Osmosis*, USEPA, 1982.

24. G. S. Logsdon and J. M. Symons, "Mercury Removal by Conventional Water Treatment Techniques," *Journal AWWA*, vol. 71, no. 9, September 1979, pp. 454–466.
25. G. S. Logsdon, *Water Filtration for Asbestos Fiber Removal*, USEPA, EPA-600/2-79-206, 1979.
26. G. S. Logsdon et al., "Removal of Heavy Metals by Conventional Treatment," in *Proc. 16th Water Quality Conference: Trace Metals in Water Supplies: Occurrence, Significance and Control*, University of Illinois, Champaign-Urbana, 1974.
27. *Manual of Treatment Techniques for Meeting the Interim Primary Drinking Water Regulations Revised*, USEPA, EPA-600/8-77-005, 1978.
28. W. A. McRae and E. J. Parsi, "Removal of Trace Heavy Metal Ions from Water by Electrodialysis," in J. E. Sabadell (ed.), "Traces of Heavy Metals in Water Removal Processes and Monitoring," *Proc. Symp. conducted by the Center for Environmental Studies and the Water Resources Program*, Princeton University, USEPA, EPA-902/9-74-001, 1974.
29. F. O. Mixon, *Removal of Toxic Metals from Water by Reverse Osmosis*, U.S. Department of Interior, Office of Saline Water, R & D Prog. Rep. No. 889, 1973.
30. F. Rubel, Jr., and R. D. Woosley, "The Removal of Excess Fluoride from Drinking Water by Activated Alumina," *Journal AWWA*, vol. 71, no. 1, January 1979, pp. 45–49.
31. E. A. Sigworth and S. B. Smith, "Adsorption of Inorganic Compounds by Activated Carbon," *Journal AWWA*, vol. 64, no. 6, June 1972, p. 386.
32. F. W. Sollo, Jr., et al., *Fluoride Removal from Potable Water Supplies*, Illinois State Water Survey, Urbana, Ill., Res. Rep. No. 136, September 1978.
33. T. J. Sorg et al., "Treatment Technology to Meet the Interim Primary Drinking Water Regulations for Inorganics," part 3, *Journal AWWA*, vol. 70, no. 12, December 1978, pp. 680–691.
34. T. J. Sorg and G. S. Logsdon, "Treatment Technology to Meet the Interim Primary Drinking Water Regulations for Inorganics," part 5, *Journal AWWA*, vol. 72, no. 7, July 1980, p. 411.
35. R. R. Trussell et al., *Selenium Removal from Ground Water Using Activated Alumina*, USEPA, EPA-600/2-80-153, 1980.
36. G. M. Zemansky, "Removal of Trace Metals during Conventional Water Treatments," *Journal AWWA*, vol. 66, no. 10, October 1974, p. 606.
37. D. C. Argo, "Use of Lime Clarification and Reverse Osmosis in Water Treatment," *Journal WPCF*, vol. 56, October 1984, p. 1238.
38. C. A. Blanck, "Trihalomethane Reduction in Operating Water Treatment Plants," *Journal AWWA*, vol. 71, no. 9, September 1979, p. 525.
39. L. L. Harms and R. W. Looyenga, "Chlorination Adjustments to Reduce Chloroform Formation," *Journal AWWA*, vol. 69, no. 5, May 1977, p. 258.
40. D. E. Johnson and S. J. Randtke, "Removing Nonvolatile Organic Chlorine and Its Precursors by Coagulation and Softening," *Journal AWWA*, vol. 75, no. 5, May 1983, p. 249.
41. S. W. Maloney et al., "Ozone-GAC Following Conventional U.S. Drinking Water Treatment," *Journal AWWA*, vol. 77, no. 2, February 1985, p. 62.
42. D. A. Reckhow and P. C. Singer, "The Removal of Organic Halide Precursors by Preozonation and Alum Coagulation," *Journal AWWA*, vol. 76, no. 4, April 1984, p. 156.
43. J. M. Symons et al., *Treatment Techniques for Controlling Trihalomethanes in Drinking Water*, USEPA, EPA-600/2-81-156, Cincinnati, Ohio, 1981.
44. J. B. Weil, "Aeration and Powdered Activated Carbon Adsorption for the Removal of Trihalomethanes from Drinking Water," Master of Engineering Thesis, University of Louisville, Louisville, Ky., December 1979.
45. P. R. Wood and J. DeMarco, "Effectiveness of Various Adsorbents in Removing Organic Compounds from Water. Part I: Removing Purgeable Halogenated Organics," in M. J. McGuire and I. H. Suffet (eds.), *Activated Carbon Adsorption of Organics from Aqueous Phase*, vol. II, Ann Arbor Science, Ann Arbor, Mi., 1980.

# 4

# Source Water
# Quality Management

## Robert H. Reinert, P.E.

*Vice President*
*Malcolm Pirnie, Inc.*
*Paramus, New Jersey*

## John A. Hroncich

*Director of Marketing*
*Laboratory Resources, Inc.*
*Westwood, New Jersey*

The decline in quality of drinking water sources became a concern when population growth and industrial development produced a concentration of society's wastes that imperiled public health. The need to both protect public health and minimize the cost of public water supply leads to the practice of source water quality management.

Source water quality management is defined as the science and practice of protecting surface water and groundwater resources used for production of drinking water, both present and future. The public must be provided water of the highest quality for drinking and general municipal uses. The control of source water quality both facilitates the economical production of safe drinking water and enhances its value.[1] Source water quality management provides a means to determine impacts on drinking water sources resulting from both natural factors and human activities, to estimate the immediate and long-term effects of such impacts, and to prevent the occurrence of public water supply system problems that are difficult or expensive to correct.

The importance of source water quality management is evident as competition grows for the world's finite water resources. A new emphasis on source control as an effective management technique has occurred as a result of society's interest in and concern with health, aesthetic, and recreational aspects of water resources. Public water supply systems, faced with the possibility of having to install water treatment facilities to meet increasingly stringent standards required by regulatory agencies, are recognizing that effective source water quality programs can be an important factor in directing program costs for water treatment.

Source water quality management is the first step in ensuring an adequate supply of safe drinking water and should be an integral element in every water utility operation. This chapter is intended mainly for those sources of supplies for which management programs may be expected to produce real and discernible benefits, i.e., the so-called protected or highly managed watersheds and aquifers. The principles of source water quality management, however, apply to all potable water sources, including large river basins and regional aquifer systems.

## Water Source Characteristics

Water supply sources may be surface waters or groundwaters. The surface water and groundwater resources of an area typically are closely related and are interconnected by the hydrologic cycle. Because surface water and groundwater are treated differently under federal regulations, knowing the difference is important.

### Hydrologic cycle

The hydrologic cycle is the constant movement of water above, on, and below the earth's surface (Fig. 4.1).[2] Although the hydrologic cycle has neither a beginning nor an end, discussing its principal features by starting with evaporation from vegetation, the oceans, and exposed moist surfaces on the land is convenient. This moisture forms clouds that return the water to the land surface or oceans in the form of precipitation.

Precipitation occurs in several forms, including rain, snow, and hail, but only rain is considered here. Initially, rain wets vegetation and other surfaces; it then begins to infiltrate into the ground. When and if the rate of precipitation exceeds the rate of infiltration, overland flow occurs. Initial infiltration replaces soil moisture; excess infiltration percolates slowly, moving downward and laterally to sites of groundwater discharge, such as springs on hillsides, or seeps in the bottoms of streams and lakes or beneath the ocean. Water reaching

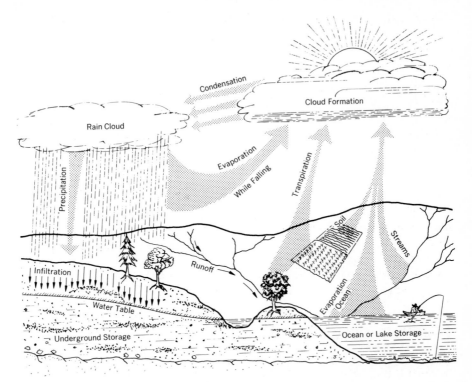

**Figure 4.1** Hydrologic cycle.

streams, both by overland flow and from groundwater discharge, moves to the oceans, where it is again evaporated to continue the cycle.

The concept of the hydrologic cycle is important in the development of a source water quality management strategy. Contaminants can be introduced into water from various sources throughout the hydrologic cycle. Contaminants may be diluted, concentrated, or carried through the cycle with the water. Source water quality management seeks to minimize contaminant input to water resources used for sources of drinking water.

### Surface water sources

Surface water is the term used to describe water on the land surface. It may be running, such as in streams, rivers, and brooks, or quiescent, such as in lakes, reservoirs, impoundments, and ponds. Surface water is produced by runoff of precipitation and by groundwater seepage. For regulatory purposes, surface water is defined as all water open to the atmosphere and subject to surface runoff.[3]

After surface water has been produced, it follows the path of least resistance. A series of brooks, creeks, streams, and rivers carries water from an area of land surface that slopes down toward one primary water course. This drainage area is known as a watershed or drainage basin. Fig. 4.2 is example of a basin surrounded by a ridge of high ground, called the watershed divide, that separates one drainage area from another.

Surface water quality is highly influenced by the point within the watershed where water is diverted for treatment. The quality of streams, rivers, and brooks will vary according to seasonal flow and

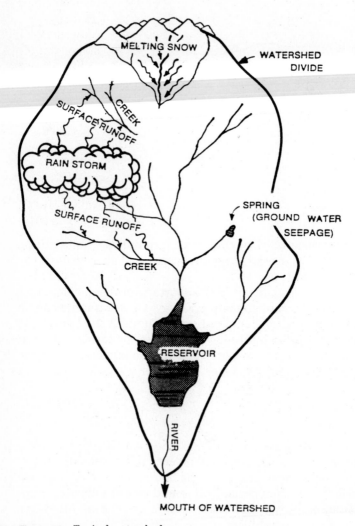

Figure 4.2  Typical watershed.

may change significantly because of precipitation and accidental spills. Lakes, reservoirs, impoundments, and ponds typically have less sediment than rivers but are subject to greater impacts from microbiological activity than river sources.

Quiescent water bodies, whether natural or human-made, are living ecosystems. Each is unique and changes in character from year to year. In addition, water bodies, including source water lakes and reservoirs, age over a relatively long period of time as a result of natural processes.[4] This aging process, generalized in Fig. 4.3, is caused by microbiological activity that is directly related to the nutrient levels in the water body and can be accelerated by human activity. The upper line in Fig. 4.3 generalizes the aging of a lake, reservoir, impoundment, or pond through natural processes. The lower line generalizes the influence of human activity. The middle line represents the quality that should be achieved when source water quality management techniques are applied. The impact of microbiological activity is discussed further in a later section.

### Groundwater sources

All water beneath the land surface is referred to as underground, or subsurface, water, although an exact definition for regulatory purposes is undecided at this time (1989). Underground water occurs in two different zones. One zone is immediately below the land surface in most areas, contains both water and air, and is referred to as the un-

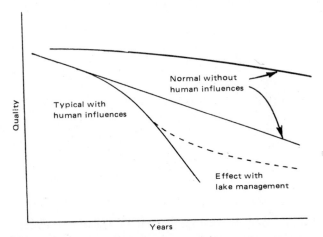

**Figure 4.3** Reservoir quality degradation as a function of time. (*Source: R. S. Geney, "Reservoir Management for Water Quality Improvement," Sem. Proc.: "Water Quality Management in Reservoirs," Proc. AWWA Annual Conf., Orlando, Fla., 1988.*)

saturated zone. The unsaturated zone is almost invariably underlain by a zone in which all interconnected openings are full of water, referred to as the saturated zone. Water in the saturated ~~zone is it undergroound~~ ... that is available to supply wells and springs ... it in the same water ~~level~~ ... Recharge ... groundwater is correctly approaching recharge of the saturated zone occurs by ... infiltration of water from the ... surface through the unsaturated zone.

... that underlie the earth's surface ... be classified either as aquifers or as confining beds. An aquifer is a rock unit that will yield water in a usable quantity to a well or spring. In geologic usage, "rock" includes unconsolidated sediments. A confining bed is a rock unit having very low hydraulic conductivity that restricts the movement of groundwater either into or out of adjacent aquifers.

Groundwater occurs in aquifers under two different conditions. Where water only partly fills an aquifer, the upper surface of the saturated zone is free to rise and decline. The water in this type of aquifer is said to be unconfined, and the aquifer is referred to as an unconfined aquifer. Unconfined aquifers are also widely referred to as water-table aquifers. Wells that open to unconfined aquifers are referred to as water-table wells. The water level in these wells indicates the position of the water table in the surrounding aquifer.

Where water completely fills an aquifer that is overlain by a confining bed, the water in the aquifer is said to be confined. Wells in such aquifers are referred to as pressure wells or artesian wells. The water level in artesian wells stands at some height above the top of the aquifer but not necessarily above the land surface. If the water level in an artesian well stands above the land surface, the well is a flowing artesian well. The water level in tightly cased wells open to a confined aquifer stands at the level of the potentiometric surface of the aquifer.

Some groundwater sources may be subject to contamination from surface waters. Springs, infiltration galleries, shallow wells, and other collectors in subsurface aquifers may be hydraulically connected to nearby surface water sources, depending on local geology. For regulatory purposes, these sources are referred to as groundwaters[3] "under the direct influence of surface waters."

## Factors Influencing Source Water Quality

Both natural and human factors influence the quality of a water source. Management of the quality of a source water must begin with the identification of those factors involved that individually or jointly affect the quality of the water source.

The degree of impact of operative factors varies depending on the type and characteristics of the source involved. Surface water, for example, is generally, but not exclusively, more vulnerable to human contamination than groundwater because of its direct exposure to human activity. Vulnerability itself will vary with size and type, for example, river versus lake or impoundment. A surface supply can restore itself more rapidly than a ground supply through seasonal and short-term runoff cycles. The slow rates of movement of groundwater are such that restoration and replenishment may take decades. Variations in weather patterns, i.e., prolonged drought or flood conditions, more directly affect surface water than groundwater. Quality effects will be different for lakes and impoundments than for free-flowing rivers, and different for sand and gravel than for rock aquifers. Oil spills may be isolated or captured in the case of surface water, but once present in an aquifer, they are very difficult to remove. A volatile chemical may remain in a groundwater supply, but that same chemical can volatilize out of a surface supply. A source water management program must take into account the nature of the supply as well as the active factors that influence water quality. Factors influencing source water quality are classified as either natural or human as shown in Table 4.1.

### Natural factors

Natural factors cannot be readily controlled and may have a significant impact on the quality of a source water. Factors considered here include climate, watershed characteristics, geology, microbiological growth, fire, saltwater intrusion, and density or thermal stratification.

**Climate.**   The primary effect on water quality caused by climatic conditions is that of precipitation. Wet climates or periods of heavy precipitation result in high rates of runoff or flood conditions that can cause resuspension of sediment and increases in turbidity, color, metals, or other contaminants. The flushing effect of heavy precipitation on watersheds can introduce an accumulation of organic compounds through runoff, with deleterious effects on water quality. Under dry conditions or prolonged drought conditions, lower rates of runoff can cause stagnation, thereby increasing the likelihood of microbiological activity and algal growth. Reduced flows under dry conditions exacerbate the impact of point source discharges because of reduction in the dilution effect and in the assimilative capacity of the water body. Temperature is also an important climatological factor affecting biological activity rates, oxygen saturation, and mass transfer coefficients.

TABLE 4.1    Factors Influencing Source Water Quality

Natural Factors

~~~~~~~~ intrusion
Density (thermal) stratification
Human Factors
  Point:
    Wastewater discharges
    Industrial discharges
    Hazardous waste facilities
    Mine drainage
    Spills and releases
  Nonpoint:
    Agricultural runoff
    Livestock
    Urban runoff
    Land development
    Landfills
    Erosion
    Atmospheric deposition
    Recreational activities

**Watershed characteristics.**    The various natural characteristics of a watershed can have a significant effect on water quality. Topography, for example, affects flow rates. Steep slopes may result in erosion of topsoil or stream banks, thereby introducing debris, sediment, and nutrients that can increase algae, color, and turbidity. Residence time in lakes and reservoirs is also a function of topography and affects water quality through its effect on sedimentation and biological activity. Vegetative cover can affect water quality in a number of ways. Decomposition of vegetative matter can cause color and is a source of fulvic and humic compounds (natural organic matter) associated with the formation of disinfection by-products. As a protective mechanism, vegetative cover provides a natural filtering action of runoff for nonpoint source pollutants and can provide a buffer to human activities. The specific vegetation will affect the types and amounts of nutrients and natural organic compounds introduced to the water supply.

Wildlife found in the watershed area primarily affect water quality

from a microbiological stand          biquitous Canada goose, with its high.
water supply professionals
deer, muskrat, and other warn.
*Giardia lamblia,* a protozoan patho
risk for the insufficiently treated supply.

**Geology.**  Local geology directly impacts the quality of both surface
and groundwater. Groundwater with high hardness, for example, derives its calcium and magnesium mineral content from the subsurface
geologic formation through which it travels. The presence of
radionuclides in groundwater, as another example, results from local
geology. Radon, a natural contaminant of increasing concern to health
authorities, is a decay product of uranium and is found in groundwater throughout the northeast and in some other parts of the country.[5]

A dramatic example of the effect of geology on water quality occurred in the eruption of Mt. Saint Helens in 1980. The vast amount of
ash released was ultimately deposited on earth and resulted in substantial increases in suspended matter in surface water supplies.[6]

Soils play a significant role in water quality, often acting as buffering agents for acidic precipitation. If soil buffering capacity is limited
or nonexistent, the higher acidity of runoff can adversely affect biological activity in lakes and reservoirs and may create treatment
problems.

**Microbiological growth (nutrients).**  The age or state of a body of water
depends on nutrient levels and microbiological activity. The natural
life cycle of a water body involves three stages known as trophic levels: oligotrophic (low nutrients, minimal microbiological activity),
mesotrophic (moderate nutrients, moderate microbiological activity),
and eutrophic (high nutrients, high microbiological activity). Water
quality effects associated with eutrophication include depleted oxygen
levels, increased microbiological activity, high turbidity and color,
and the formation of trihalomethane (THM) precursors.[7] Resultant
operational problems may occur, including interference with the
treatment process, filter clogging, and taste and odor problems.[8]

The most common indicator of eutrophication is the presence of algae. Algae is a ubiquitous plant organism found in the aquatic environment. Current estimates indicate that over 21,000 species of algae
exist, exhibiting a diversity with respect to size, organization, physi-
ology, biochemistry, and reproduction. Algae are found in fresh,
brackish, or highly saline waters and may be attached to rocks or suspended in the water column. Algae have been shown to exist in both
hot and cold temperature extremes. They utilize a wide range of or-

ganic and inorganic substances and can tolerate a wide range of pH values.[9] The algae commonly found in water supply sources are the blue-green algae that inhabit freshwater environments.[10]

Sudden and dramatic increases in algal populations occurring during a bloom can cause aesthetic and other water quality problems. In a typical situation, low water temperature and shortened periods of light during winter months cause a decrease in photosynthetic activity. The available nutrients remain unused and accumulate. As days lengthen and temperatures increase, photosynthetic activity accelerates and algal blooms occur. The increased microbiological activity continues until nutrients are exhausted and organism populations decrease.[11] A series of algal blooms may occur during the summer.

A National Eutrophication Survey (NES)[12] was conducted by the USEPA from 1972 to 1975. In the study, a total of 250 lakes in the eastern and southeastern United States were sampled for algae. A total of 117 species and 25 genera of freshwater algae were found responsible for various water problems.[12] The report concluded that no algal species can be considered as an indicator of eutrophic conditions. Although the NES identified a number of species associated with certain water quality problems, sampling and analysis for these species did not indicate their presence in all cases and under all circumstances.[12]

Oxygen deficiency caused by microbiological activity, if carried to completion, creates a chemical-reducing environment that permits solubilization of minerals and nutrients normally bound in bottom sediment. Iron and manganese, for example, can be reduced and solubilized, causing water quality problems.

Two important chemical reactions that affect reservoir water quality occur under anaerobic conditions:[4]

$$6e^- + 8H^+ + SO_4^{2-} \rightarrow H_2S + 2H_2O + 2OH^- \qquad (4.1)$$

$$2FePO_4 + 3H_2S \rightarrow 2HPO_4^{2-} + 2FeS + S + 4H^+ \qquad (4.2)$$

Equation (4.1) represents the reduction of sulfate to form hydrogen sulfide. Hydrogen sulfide is very toxic to biological organisms and can cause taste and odor problems. Hydrogen sulfide can react with iron phosphate to form iron sulfide, releasing phosphate [Eq. (4.2)].

Internal recycling of nutrients in a water body occurs through reactions such as Eqs. (4.1) and (4.2). As a result, reservoir water quality improvement programs must address both the inflow of nutrients from external watershed sources and the internal recycling of nutrients within the reservoir.

**Fire.**  Although brush and forest fires on a watershed can occur as a result of human activity, fire is considered a natural factor because nature is often the cause through a combination of drought and lightning. Where large expanses of brush and forest are destroyed by fire, adverse effects on water quality can result. The destruction of brush and forest eliminates their function as natural filters and increases the likelihood of erosion because of increased runoff rates. Additionally, reduced vegetative uptake of moisture through evapotranspiration contributes to increased runoff. Ash and charred wood can leach nitrates, and burnt wood may contribute to increases in phenol concentration that, in combination with chlorine, can create taste and odor problems. Fire-fighting activities also can create adverse water supply effects. Heavy streams or loads of water applied to the fire contribute to sediment and organic loading in the water supply, and any chemicals used may have direct adverse effects. On the positive side, forest fires are a natural means to rejuvenate forests with new and young vegetation.

**Saltwater intrusion.**    Saltwater intrusion can occur in both surface and groundwater supplies located in coastal regions or along upstream ends of tidal estuaries. In a typical river situation, the saltwater-freshwater interface moves upstream during low flow or drought periods. In the case of groundwater supplies, overpumping of an aquifer hydraulically connected to salt- or brackish water can cause the saltwater "wedge" to move and intrude on the aquifer. Both situations can result in an increase in salinity and will aggravate any THM production. They are best addressed by adjustments in the operation of the water supply involved.

**Density (thermal) stratification.**    Stratification of surface water reservoirs can have a significant impact on source water quality. Figure 4.4 shows a typical oxygen-temperature relationship during summer and winter seasons.[4] In shallow reservoirs (about 20 ft deep or less), summer temperatures and oxygen levels can vary according to the amount of wind mixing. During extended still periods, the surface temperature rises relative to the bottom temperature; stratification can develop because of density differences, and an oxygen deficiency can occur at the sediment-water interface, causing the problems described earlier. In winter, temperature levels are typically uniform and oxygen levels adequate. Ice cover, however, can affect the natural aeration caused by wind mixing leading to anaerobic conditions.

In deep reservoirs (about 20 ft deep or more), a distinct temperature stratification develops that effectively separates the interaction of the upper water, referred to as the epilimnion, from the lower water, referred to as the hypolimnion. The temperature transition zone between the epilimnion and hypolimnion is known as the thermocline, or metalimnion.

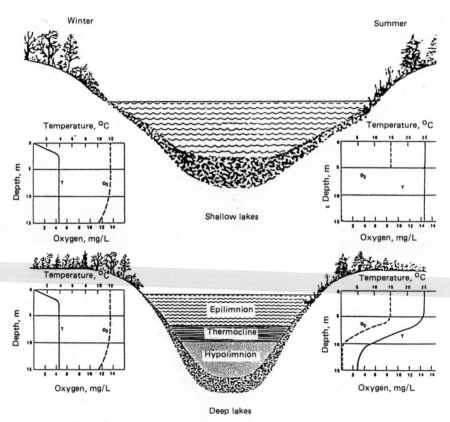

**Figure 4.4** Typical reservoir oxygen-temperature relationships during summer and winter seasons. (*Source: R. S. Geney, "Reservoir Management for Water Quality Improvement," Sem. Proc.: "Water Quality Management in Reservoirs," Proc. AWWA Annual Conf., Orlando, Fla., 1988.*)

Thermally stratified reservoirs contain warm-water fish species, such as small-mouth bass, in the upper levels and cold-water fish species, such as trout, in the lower levels. Assuming sufficient oxygen is available, a balanced ecosystem represents a healthy body of water and that is instrumental in maintaining water quality.

Stratification of deep reservoirs can cause significant water quality problems. In summer, the epilimnion will contain adequate oxygen levels because of wind mixing and photosynthesis, but the bottom stratum will be isolated by the thermocline. Without oxygen replacement, the hypolimnion will experience oxygen deficiency; the more degraded the water body, the greater the oxygen deficiency during summer months with 0 mg/L common. In winter, stratification may not cause oxygen deficiency in the hypolimnion, except in severely degraded water bodies, because the solubility of oxygen is higher at colder water temperatures and microbiological activity is less during the winter than during the summer. Deep portions of a thermally stratified reservoir are unaffected by wind action.

The intensity of thermal stratification in reservoirs is measured by calculating a parameter known as "stability." Stability is defined as the work that must be done or energy needed to lift the entire weight of a body of water the vertical distance between the center of gravity when the body of water is in a given state of stratification and the center of gravity when the water body is isothermal. Stability also may be thought of as the minimum energy required to completely mix a stratified body of water. By calculating stability changes, the effectiveness of destratification measures can be evaluated.[13]

Either because a reservoir is weakly stratified and the energy input from the wind and other natural forces is greater than the stability, or because stability declines naturally in the fall as the water column approaches an isothermally warm condition, the water body mixes or "overturns," often relatively suddenly. These reservoirs can experience particularly poor water quality when the anoxic, lower water is mixed throughout the water column. When turnover of a lake or reservoir occurs, nutrients and anoxic water move from the lake bottom to the surface, feeding algae and degrading water quality.

### Point source human factors

Human factors influencing source water quality are usually categorized as two distinct types: point or nonpoint (Table 4.1). Point sources are those sources of contamination characterized by a single or discrete conveyance, such as the terminus of a pipe or collection of pipes. Nonpoint sources, in contrast, involve large and diffuse sources of contamination such as agricultural runoff. Five types of point sources of contamination are discussed below.

**Wastewater discharges.**  The impact of wastewater discharges on a receiving water depends on the watershed, in-stream quality, and other discharges and withdrawals. The design and operation of the wastewater treatment facility are key to assessing and maintaining consistent control of the receiving source water quality. Any imbalance or interruptions to the treatment system, including flood flows in combined sewers and plant process failures, can result in violation of discharge quality standards and adverse effects on the receiving source water body. Sewer pipe failures and other catastrophic failures can cause a substantial release of untreated wastewater to the water source, resulting in a high degree of bacterial contamination, depletion of dissolved oxygen, suffocation of fish, and introduction of excessive amounts of organic and inorganic contaminants. Frequent discharges of untreated wastewater with high quantities of phosphorus and nitrogen contribute to accelerated rates of algal activity and eutrophication. The collection system also plays a role in water quality protection. Older systems or those that are not designed and maintained to prevent excessive leakage can contribute pollution throughout the collection grid. Collection systems installed over sensitive aquifers require careful attention to design, construction, and operation.

**Industrial discharges.**  Industrial activities can affect the quality of a source water supply through the release of contaminants by air, water, or soil. An industrial facility may introduce contaminants directly to a water body through a regulated discharge or accidentally as in the case of a spill. Treatment ponds and lagoons or storage ponds and lagoons can impact groundwater through subsurface percolation. The possible release or spillage of product exists in the case of storage tanks. Underground storage tanks and piping are of particular concern, especially if secondary containment and monitoring systems are not employed. Stack emissions can introduce contamination locally, regionally, or at great distances. Both surface and groundwater can be affected by airborne contamination. Some industrial operations involve the extraction of oil or other minerals by pumping large quantities of water into the subsurface formation and can cause quality effects on the groundwater involved. Use of injection wells as a means of waste disposal creates an obvious concern with respect to aquifers.

**Hazardous waste facilities.**  Hazardous waste is defined by the Federal Resource Conservation and Recovery Act (RCRA)[14] as any waste material that is flammable, corrosive, reactive, or toxic. Hazardous waste facilities may include treatment, storage, and disposal facilities. The operation of a hazardous waste facility within a water supply watershed or aquifer requires extensive precautions to prevent accidental or intentional release of hazardous contamination to the potable water source. Inactive hazardous waste facilities that have been either abandoned or for which operation has been suspended or terminated may require cleanup. All these facilities should be closely monitored for any emission or release into the surrounding environment. The Federal Comprehensive Environmental Response and Compensation Liability Act (CERCLA, or Superfund),[15] reauthorized and amended in 1986 by the Superfund Amendments and Reauthorization Act (SARA),[16,17] regulates such sites and provides funding for cleanup activities.

**Mine drainage.** Mining operations that extract or excavate minerals from surface and subsurface areas create conditions that often affect water quality. Disturbance of surface topography in conjunction with high rates of runoff can cause erosion and resuspension of sediments, turbidity, color, and other quality concerns. Drainage from mining activities may cause a change in acidity of the receiving water, thereby affecting water chemistry and resuspension of iron, manganese, and other contaminants.

**Spills and releases.** Spills and accidental or intentional releases can occur in conjunction with many human activities and in a variety of ways. Watersheds traversed by major highways and rail freight lines are subject to transportation accidents and chemical spills. Many spills and releases occur in conjunction with petroleum product storage installations. A spectacular example of an oil spill occurred over the New Year's holiday in January of 1988, when one million gallons of diesel fuel from a ruptured storage tank flowed into Pennsylvania's Monongahela River.[18] This catastrophic spill created a supply disaster for numerous water utilities up to 800 miles (mi) downstream in the Monongahela and Ohio river valleys and required emergency responses for over a month.

Leaks and spills from underground tanks and piping are of special concern because the release may remain undetected for long periods of time. Spills of oils and other low-density products generally are easier to clean up because of the floatability of the material; emulsifying action because of stream turbulence creates more difficult situations. Soluble materials are generally more difficult to deal with and may require neutralization, oxidation, precipitation, or adsorption. In some situations compounds may be allowed to dissipate through the natural processes of assimilation and degradation. In most cases, however, long-term implications of the spill and its effect on water must be known.

**Nonpoint source human factors**

Because of the larger areas involved and their diffuse nature, nonpoint sources of contamination are more complex and difficult to control than point sources. Although initial efforts to improve the quality of the nation's water resources were focused on point sources of pollution, attention is currently shifting to understanding and determining ways to control nonpoint sources.

**Agricultural runoff.** Application of pesticides, herbicides, and fertilizers are key agricultural activities that affect the quality of groundwater and surface water.[19,20] Certain pesticides and herbicides are banned by the government because of their toxicity to humans or their adverse effect on the environment; others are or will be controlled through regulation. Pesticides and herbicides usually decompose and break down as they perform their intended function, but to varying degrees and over widely different time frames depending on the chemical involved. Low-level residues are often found where complete decay does not occur. The application of fertilizers to enhance crop production is of major concern to water suppliers because dissolved nutrients in the runoff accelerate eutrophication of the receiving water body. Nitrogen, in the form of nitrates, is a contaminant found in groundwater underlying agricultural areas. Erosion caused by improper tilling techniques is yet another agricultural activity that can adversely affect water quality through increased sediment load, color, and turbidity.

**Livestock.** The presence of livestock on watersheds and over aquifers has an obvious and direct effect on bacterial contamination. Feedlots have been shown to contribute nitrates to wells and fecal coliform bacteria to surface supplies. Uncontrolled or overgrazing eliminates the vegetative cover that prevents erosion, thereby increasing the sediment loading in surface water and creating attendant quality problems.

**Urban runoff.** Runoff from highways, city streets, and commercial areas contains various contaminants that are washed from the surface and carried to the receiving body. Contaminants include petroleum products, metals such as cadmium and lead from automobiles, salt and other de-icing compounds, and silt and sediment from land erosion and wear on road and sidewalk surfaces. Bacterial contamination from human and animal sources is also often present. High-density urban populations centralize the impacts of human activities on relatively small areas, concentrating the effects of contamination. The initial flushing activity during storm events typically creates a contamination peak, with diminishing effects as pollutants are washed away.

**Land development.** The development of previously undisturbed or rural land involves several effects that influence water quality. Development decreases the buffer zones of natural vegetative cover that filter runoff. Development also increases runoff coefficients that may cause erosion and increases in sediment loading. Land development increases activities previously not present causing less water to percolate into groundwater. Human activities that affect water quality, including the application of chemicals, storage of fuel, and disposal of waste, increase with development. Development taking place on rezoned lands is of particular concern from the standpoint of any previous activities that may affect water quality, e.g., development over an abandoned landfill.

**Landfills.** The greatest water quality concern posed by a landfill is with the leachate. Whether introduced through direct precipitation or groundwater movement, water percolating through the diverse material contained in a landfill has an obvious and potentially serious effect on the source water involved. Quality factors depend upon the type, amount, and spatial extent of fill material disposed of on the site. Leachate can be controlled by capping the landfill with low-permeability cover to reduce introduction and percolation of precipitation, and by installation of leachate collection systems to retain, treat, and/or remove the polluted water. In all cases, effective monitoring systems are required.

**Erosion.** Although a characteristic effect of a number of natural and human factors that influence source water quality, the diversity and areal extent of erosion throughout a watershed qualifies it for separate status as a nonpoint source. Erosion of the soil causes soil particles and nutrients to enter surface water runoff; the increased sediment loading creates turbidity, color, and eutrophication. Resuspension of the soil particles during storm events creates a cyclic buildup of sediment at quiescent locations. This buildup may adversely affect both the quality and quantity of a source. Storage capacities of lakes and ponds are reduced by the volume of the sediment, and recharge to aquifers from streams and river bottoms can also be adversely affected.

**Atmospheric deposition.** Airborne contamination, transported great distances through the atmosphere and deposited on the earth as precipitation, is a nonpoint source of increasing prominence and concern. The popular term "acid rain" is commonly used to describe atmospheric deposition in general. Understanding of the phenomenon and its effects, although incomplete, confirms it as an important source water quality factor. Sulfur released into the atmosphere by industrial stack emissions is converted by natural chemical processes into acid, thus increasing the acidity of precipitation. Acid rain has been shown to cause increasing acidification of lakes and reservoirs, in some cases adversely affecting fish life and other biological organisms. The buffering effect present in the soils of certain regions may ameliorate the affect of acidic runoff, but such buffering capacity may not be without limit. Possible leaching of metals from soils may also create water quality problems. Industrial stack emissions are also sources of other organic and inorganic contaminants that can adversely affect surface and groundwater quality. Considerable concern exists with incinerators and resource recovery plants that may release low levels of dioxin because of incomplete combustion of refuse or refuse-derived fuel. Automobile emissions and even residential woodburning stoves also contribute significantly to the problems of airborne contamination.

**Recreational activities.** Recreational activities, such as swimming, boating, and camping in water supply reservoirs and watersheds, can impact surface water quality. The extent and significance of the impacts of recreational activities on water quality, however, have not been clearly defined and are the subject of continuing debate.[21]

Several studies have been conducted to evaluate the water quality impacts from recreational use of surface water sources. A 1964 study found that highest levels of a bacteriological indicator organism in Forrest Lake, Mo., occurred in zones of greatest recreational activity, but bacteriological counts were not evident at the intake structure.[22] Comparisons of bacteriological densities in two reservoirs near Hartford, Conn., found that coliform counts in the reservoir that allowed fishing, boating, and swimming were about 10 times greater than those in a nearby reservoir with a similar watershed but with no recreational use.[23] Total coliform counts increased in two Springfield, Mass., water supply reservoirs after they were opened to fishing.[24] Other studies, however, did not find significant impacts of recreational uses on source water quality.[25,26]

The dual use of backcountry areas for recreation and domestic water supply has also been the focus of research. A study of remote streams in California's Sierra Nevada mountains suggests that the intensity of human use may play a significant role in *Giardia* contamination.[27] A study in Montana reported that opening a previously closed municipal watershed to recreation and logging coincided with an unexpected decrease in bacteriological counts in the stream, thought to be caused by the displacement of large animal herds outside the watershed by human activity.[28] Further research is needed to quantify the impact of dispersed recreation on water quality.[29]

## Regulatory Protection of Source Waters

Regulatory protection of public water supply sources is partially provided through federal environmental programs and state and local laws and ordinances. Water suppliers must be aware of federal, state, and local laws and take advantage of the protection they can afford whenever possible.

### Federal environmental programs

In recent years, protection of surface and groundwater throughout the United States has received considerable impetus through a series of environmental laws enacted by Congress. Although each of these laws is designed to address a separate and distinct environmental purpose, a linkage exists among all relative to drinking water quality. Maintaining the quality of drinking water sources is an explicit or implicit purpose of each of these laws, and standards for drinking water quality inherently constitute a basis for measuring the degree of protection required. Federal environmental laws are usually implemented through individual state programs.

The Safe Drinking Water Act (SDWA), originally passed by Congress in 1974 and most recently amended in 1986,[30,31] contains source protection provisions. The original act included a provision to protect aquifers that are sole sources of drinking water supply and established a program to control the underground injection of contaminants. The 1986 amendments provide national guidance to states for the creation of local wellhead protection programs designed to protect groundwater quality.

The Clean Water Act (CWA)[31] was the first of the series of national environmental laws, and one that directly regulates the introduction of contaminants into the nation's surface and groundwater. Passed by Congress in 1972 with the objective of rendering water "swimmable and fishable," the CWA established a national pollutant discharge elimination system (NPDES). Sections 201 and 208 of the CWA protect water quality through regional planning programs for point and nonpoint sources of contamination. The NPDES permits, setting numeric limits for specific quality parameters, are required by the CWA for discrete discharges.

Other national environmental laws provide direct and/or indirect protection of source waters. The RCRA[14] regulates creation, transportation, and disposal of hazardous wastes and contains groundwater provisions. The Federal Insecticide, Fungicide, and Rodenticide Act (FIFRA)[32] regulates the availability and application of pesticides and herbicides. CERCLA[15] and SARA[16] regulate inactive or abandoned hazardous waste sites and provide funding for cleanup activities. The Toxic Substances Control Act (TSCA)[33] regulates new chemical products entering the marketplace, and the Clean Air Act (CAA)[34] regulates atmospheric contaminants, indirectly protecting water supplies from airborne introduction of contaminants.

## State and local programs

Regulatory protection of public water supply sources is more directly provided through state and local laws and ordinances. In addition to the implementation of federal laws and regulations, individual state programs exist that provide source protection through water codes, sanitary regulations, regulations of inland wetland areas, and other means of watercourse and aquifer protection. At the local level, municipal ordinances can provide significant protection through land use controls that regulate development activities on key watershed areas. The individual water supply utility can best integrate these protective mechanisms into its own source water quality management program by working cooperatively with state and local authorities to support the initiating legislation, to implement the regulations, and to provide effective enforcement to mutual advantage. Such participation by the water utility should be directed toward the adoption of practical laws and regulations that provide tangible benefits in terms of enhanced protection of source waters.

Comprehensive protection has been provided in some states through the adoption of statewide water quality standards and criteria. These state programs generally establish quality standards for surface and groundwater and may include goals, best-use determinations, and a classification system. In Connecticut, for example, the general statutes require the commissioner of environmental protection to adopt standards for water quality for all state waters.[35] Water quality standards (WQS) have been adopted both for surface water as required by section 303 of the federal CWA and also for groundwater, recognizing the interrelationship between surface and groundwater and the competing uses of water resources for drinking and for wastewater assimilation. Water quality classifications based on the WQS establish designated uses for surface and groundwater and identify criteria supporting those uses.

The state of New Jersey, as another example, has adopted regulatory protection programs for both surface and groundwater that are based on the best use of the resource. Surface waters are classified according to whether they have value for ecological, water supply, or other purposes. Numerical standards for a variety of parameters that drive other state regulatory programs are established for a given stream or water body. A similar approach is used for groundwater with the addition of antidegradation and degradation policies that are applied to specific aquifers depending on environmental sensitivity and water supply considerations. In some cases, a measured amount of a contaminant may be allowed, provided that the numerical standard for the aquifer is not exceeded. The individual public water supplier can advantageously employ such statewide water quality standards and criteria as part of an effective source water quality management program by working to ensure recognition of drinking water requirements in their adoption and implementation.

Governmental action at the state or regional level may be employed to address specific source water emergencies. In the 1960s, the Delaware River Valley experienced one of the worst droughts in recorded history, causing flows in the Delaware River to drop to extremely low levels. The reduced river flows caused movement of saltwater upstream threatening the water supplies of Camden, N.J., and Philadelphia, Pa. The Camden water supply is obtained from wells that tap an aquifer system recharged largely by the Delaware River estuary. The city of Philadelphia draws source water directly from the river. Recognizing that a potentially serious threat existed, the Delaware River Basin Commission acted to implement a salinity control policy. Elements of that policy included a revised sodium standard, guidelines for drought warnings and emergency declarations, a withdrawal formula to reduce water withdrawn by upstream users, regulation of releases from upstream reservoirs used for storage, and management of future water demand through conservation and water loss control. Additionally, other groundwater withdrawal limits were established and measures adopted to protect perennial streams. The water supplies of both Philadelphia and Camden were, thus, maintained throughout the multiyear drought, in large part because of this action by the commission.[36]

Localized public programs are also available to public water supply utilities for the enhancement of source water protection. A basic grassroots approach can be directed toward public education, both in the schools and at large. Progressive managers have recognized a new and expanded role for the water utility in education and other community-sponsored activities and events directed toward protection of water quality at the source. Soil conservation service activities on the local level can be effective in terms of educational benefits and actual protection achieved and may involve participation by the water utility. Public education and awareness can be promoted through a speakers program involving presentations on source protection at meetings of local organizations and other such events. Municipal sponsorship of household hazardous waste pickups helps to avoid indiscriminate disposal of harmful contaminants, and the participation or cooperation of water suppliers can be beneficial both in terms of enhanced protection and improved community relations.

## Utility Source Water Quality Management Programs

Water utilities can directly assess and influence source water quality through operational activities and procedures. The fundamental objective of such operational activities is to maintain or enhance water quality by controlling or eliminating sources of contamination.[37] Operational and management aspects of source water quality management are discussed in this section, while specific control techniques are discussed in the next section.

### Sanitary surveys

A sanitary survey is an on-site review of the water source, facilities, equipment, operation, and maintenance of a public water system for the purpose of evaluating the adequacy of such source, facilities, equipment, operation, and maintenance for producing and distributing drinking water.[37] A sanitary survey of the source of supply, conducted on a systematic basis, is essential to source water quality management and may be conducted by the utility alone or in cooperation with local or state authorities.[3] The survey should be designed to identify all areas of concern within the supply system, including the entire contributory watershed or aquifer area, and should be conducted by a qualified sanitary engineer or watershed inspector. The survey should assess the potential for contamination of the supply through an inventory of all significant installations, activities, and other possible sources of contamination, and of pollutants of concern and their avenues of movement. Water samples should be collected at appropriate locations and times to guide the assessment and establish a baseline for measurement purposes. For maximum effectiveness, sanitary surveys must be conducted periodically, usually annually, and supplemented by systematic follow-up.

The Hackensack Water Company (HWC), Hackensack, N.J., conducted sanitary surveys of certain tributaries and streams to evaluate the present condition of a body of water based on the existing activities. In one particular case, the town of Alpine, N.J., has no wastewater disposal system other than septic tanks. Because of recharge to the subsurface and because the watershed streams were influent streams, concern over bacteriological quality was raised. Samples were collected along the stream and at a location before the stream enters and exits the town. Inspectors walked the stream in search of any environmental hazards. The final report indicated that the septic system effluent was not entering the tributary stream, thus reassuring the citizens.

### Monitoring

Regular monitoring of source water quality is the next most basic element in the water supplier's source water quality management program. Monitoring should include sufficient parameters to indicate all quality concerns and should be conducted at appropriate locations throughout the source of supply. The monitoring program should include protocols for frequency of sampling and methodology of analysis and should be designed to establish baseline data to indicate both short- and long-term trends. Such monitoring can serve as a trigger mechanism to detect the occurrence of problems at their earliest stages.

Continuous monitoring at key locations within river basins, employed by water utilities with the flexibility and resources to handle supply emergencies, can provide a valuable early warning in the case of chemical spills or other critical situations. An example of an effective, cooperative monitoring system occurs in Missouri, where 14 water utilities have formed the Missouri River Public Water Supplies Association and created a voluntary monitoring system for the Missouri River. This association maintains a data collection and exchange program that includes both long-term water quality trends as well as short-term event alerts. Recently the program was expanded to include organic monitoring over an 800-mi reach of the river, both as a "spills" alert mechanism and to establish a database for contamination by synthetic organic contaminants (SOC).

In addition to their triggering function, ongoing monitoring programs can be helpful in the analysis and solution of operating problems. Historically, taste and odor problems in drinking water have not been approached from a watershed perspective. Taste and odor in drinking water can occur from a number of sources including inorganic and organic contaminants, biological sources, and anthropogenic origins (dispersed urban pollution, wastewater effluents, pesticides, and herbicides), as well as from treatment and distribution facilities.[38] Specific organic and inorganic compounds and biological organisms contribute taste and odor to both surface and groundwater supplies. Because such taste and odor problems are caused by natural and human-made activities in the watershed, monitoring at strategic locations to detect the problem at the source and guide its solution is possible.

### Watershed control

Increasingly, utility operational programs must take into account recreational impacts on source water quality. Diminished open space, population growth, and increased leisure time have created public pressure for fishing, boating, swimming, hunting, and camping on water supplies. The AWWA policy statement pertaining to recreational uses of water supply reservoirs emphasizes the need to protect the source, recommending that body contact recreation be prohibited and that any recreational activity be conducted under very strict supervision by the utility.[39] As mentioned previously, studies done to date indicate that recreational impacts on water supply sources usually are minimal and localized in nature.

Water supply systems with highly protected or managed surface sources commonly occur in the northwestern and northeastern parts of the country. Seattle, Wash., and Portland, Ore., in the northwest and New York City and Boston, Mass., in the northeast are prominent examples of large systems with high-quality surface sources for which disinfection treatment without filtration has been sufficient to date. Seattle currently owns approximately 75 percent of its two watersheds and exercises a high level of control over the remainder through agreements with other owners. New York City has an extensive ownership-control and management program for its high-quality Delaware and Catskill systems.

Bridgeport Hydraulic Company (BHC) in Connecticut and the HWC are examples of utilities with highly managed surface supplies. A direct benefit of BHC's source water management program is that complete treatment until this time has been required at only one of three surface sources, although complete treatment for all three is planned for the future. The comprehensive program employed at HWC, in conjunction with treatment, enables that company to provide high-quality water in a cost-effective manner. Table 4.2 lists selected water utilities with highly protected, highly managed surface sources, listing basic data that describe sources of supply and showing the various source water management program elements employed.

### Storm-water management

Storm-water management is an example of engineering techniques that are useful in enhancing or maintaining source water quality. In general, storm-water management is used to control runoff carrying pollutants from nonpoint sources and may involve a variety of techniques including erosion control measures, detention ponds, porous pavements, and curbing. Reduction of runoff rates can reduce erosion and minimize sedimentation problems. Detention basins and controlled releases can reduce pollutant loadings on receiving waters and thereby on treatment facilities. Storm-water management can also be useful in supplementing natural recharge of aquifers.

### Emergency response procedures

Emergency response capability enables the water supply utility (often in conjunction with local or state personnel) to contain and clean up

TABLE 4.2    Elements of Source Water Management Programs for Selected Surface Systems with Highly Protected Managed Watersheds

| | Seattle, Wash. | Portland, Ore. | New York City, N.Y. | Boston, Mass. | BHC, Conn. | HWC, N.J. |
|---|---|---|---|---|---|---|
| Type of supply | Reservoir, river | Reservoir, river | Reservoir | Reservoir, river | Reservoir | Reservoir |
| Safe yield, mgd | 170 | 108 | 1290 | 300 | 66 | 125 |
| Filtered (yes/no) | No | No | No | No | (Partial) | Yes |
| Watershed area, mi$^2$ | 162 | 106 | 1928 | 474 | 96 | 113 |
| Program elements | | | | | | |
| Watershed owned or controlled, mi$^2$ | 156 | 106 | 138(+)* | 180 | 26 | 10 |
| % | 100 | 100 | 7(+) | 38 | 27 | 9 |
| In-situ treatment | | | | | | |
| Aeration | — | — | — | — | X | — |
| Algicide or herbicide | — | — | X | X | X | X |
| Wildlife control | X | (Partial) | — | (Partial) | (Partial) | — |
| Forest management | X | X | X | X | X | — |
| Emergency response | X | X | X | X | X | X |
| Sanitary survey | X | — | X | X | X | X |
| Source monitoring | X | X | X | X | X | X |
| Watershed inspection | X | X | X | X | X | X |
| Security patrol | X | X | X | X | X | X |
| Fencing | — | (Partial) | — | (Partial) | — | X |
| Public education | X | X | — | X | X | X |

*Approximately one-half of watershed dedicated to forest preserves.

spills and may prevent major incidents of contamination of the source water supply. Local participation can be extremely important; established lines of communication between the utility and police and fire departments will assure a timely response. Local health officials and sanitarians should be familiar with water supply concerns and with the watershed areas involved. Residents on the watershed may be employed as stream watchers to detect problems and notify the utility.

Watershed surveillance by utility personnel is of great advantage to the water supplier. Surveillance and maintenance of water supplies and watershed areas requires an effective mapping system, one that shows all important elements of the supply system including watershed and property lines. Critical sites should be shown, such as areas particularly exposed to contamination. Good-quality maps containing all necessary information help to provide an expeditious response to a contamination incident. Regular inspections of the watershed by the utility are also important to the detection and monitoring of existing and potential sources of contamination. Regular sampling and analysis of source waters at key locations should be conducted to determine trends and detect problems. Watershed and source of supply surveillance can be greatly enhanced through a security patrol that may include utility personnel who are highly trained and qualified as police officers. In addition to preventing vandalism or sabotage, a security patrol can help to detect spills and other emergencies and also provide a beneficial "presence" on the watershed for the water supplier.

Protection of water quality on the watershed requires a rapid emergency response capability that can only occur with regional and local cooperation. The HWC, for example, has a watershed located in a region where land use varies from urban to open space. The company maintains a patrol force, made up of constables with arrest powers, who monitor the watershed on a 24-h/day basis. Each member has direct access to the company's sanitary engineers in the environmental management department. In the event of a spill, the sanitary engineer, in turn, has access to local and state regulatory agencies that will respond to a serious event where water quality is impacted. Additionally, the company works closely with local police who alert company personnel of any serious occurrences.[40]

Source water management practices are employed to a considerably varying degree depending on the nature of the supply, geographic and demographic factors, and utility and regulatory attitudes. Management programs for large river basin supplies typically emphasize monitoring and emergency response capability, as contrasted with highly managed surface sources that may employ and derive real and discernible benefit from all elements of a comprehensive management program.

### Groundwater sources

Management of groundwater sources is an extremely complex process and warrants special mention. Physical characteristics of aquifers vary, and threats to quality are site-specific. The sole-source aquifer system in Long Island, for example, is created in the terminal moraine of two glaciers and includes three major aquifers consisting of sand and clay deposits ranging from 200 to 1400 ft in total thickness. Widespread potential for contamination exists because of a population in the millions and extensive industrial and commercial development with the attendant presence of industrial solvents, wastewater discharges, underground storage tanks, and significant agricultural activity.[41] This contrasts sharply with highly localized aquifers in the glacial buried valleys of small rivers and streams where light residential development in primary recharge areas may constitute the only threat to quality. In other situations, regional or interstate aquifers are used for public water supply. In all cases, however, a logical, comprehensive source management program, including appropriate response capability, is highly desirable.

An example of such a comprehensive groundwater source management program can be found in the aquifer protection plan recently implemented by the BHC for 15 separate groundwater supplies in 11 different Connecticut communities.[42] The plan delineates critical areas surrounding each supply, determines the potential for contamination, identifies required monitoring wells, and recommends aquifer-specific response and protection programs. Essential elements of the plan include

- Analysis of well-field hydrogeology
- Determination of yields and zones of influence
- Identification of potential for contamination

  Sources

  Action levels
- Monitoring wells for early warning
- Preparation of monitoring and emergency response plans
- Local institutional approaches for groundwater protection

The effectiveness of a groundwater source management program depends on the degree to which the potential for contamination is accurately identified and the practicality of the response, remediation, and protection measures that are developed. All residential and commercial development and industrial and agricultural activities within the well field zone of influence and upstream of the general direction of groundwater flow should be inventoried and inspected, and monitoring systems designed to detect and control contaminants that may be introduced to the groundwater.

Institutional approaches to emergency response and preventative measures ideally should provide emergency coordination between the water supplier and appropriate community groups including state and local emergency response teams. Examples of other institutional approaches include creation of aquifer protection districts with special permits for significant uses and prohibition of incompatible criteria, and local educational programs designed to gain public support and participation.

Dade County in Florida (Miami area) offers an example of an institutional approach to the protection of the Biscayne aquifer.[43] Protection of the aquifer's eastern recharge area is provided through the

East Everglades Management Plan, which employs a combination of best management practices for development, land use controls, and land acquisitions. Wellhead protection is achieved by means of the Dade County Wellfield Protection Program that consists of five parts, including water management, water and waste treatment, public awareness and involvement, environmental regulations and enforcement, and land use policy.

## Modeling

Mathematical modeling of source water systems can be a useful management tool for qualitative as well as the traditional quantitative purposes. Such models may be categorized as mechanistic or empirical. Mechanistic models are mathematical descriptions of fundamental scientific and engineering principles, whereas empirical models are developed from an analysis of data fitted to theory, and can be created reflecting site-specific circumstances. A number of other criteria are used to describe or classify mathematical models including static vs. dynamic to indicate time-dependency, and deterministic vs. stochastic based on use of either expected values or variability and probability functions for the parameters.[44] However classified, the usefulness of mathematical models has been enhanced by the increased availability of computers and their improving flexibility of use.

Modeling of water quality in supply systems can be valuable as a tool to predict quality-related future events, such as changes in source water quality occasioned by natural or human-created events. Models can develop qualitative data in conjunction with quantitative dimensioning, as in the movement of groundwater in an aquifer and resulting quality changes. Models may be developed for all types of supply systems, including rivers, impoundments, and groundwater. Models to assess the quality of river or lake receiving water are used to predict a number of parameters, including dissolved oxygen, dissolved solids, bacteria, nutrients, pesticides, metals, and algal growth. Models have been developed for erosion, toxicity, and the eutrophication[45] processes, including lake trophic state evaluation. Although models are useful, in all cases confirming the validity of the model is necessary by testing field data against that generated by the model.

An excellent example of surface water modeling involved 34 water bodies or parts thereof that were included in the Organization of Economic Cooperation and Development (OECD) eutrophication study.[46] In that study, plotting the mean depth to hydraulic residence time vs. the phosphorus loading for water bodies with different trophic states yielded an empirical model (Fig. 4.5) that separated the three trophic

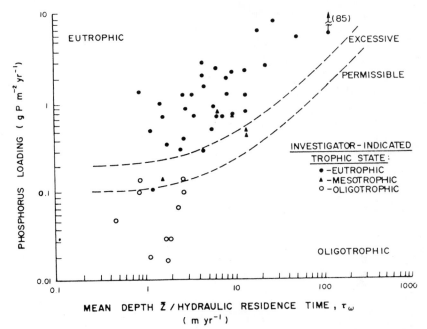

**Figure 4.5** U.S. OECD data applied to Vollenweider P loading mean depth–hydraulic residence time relationship. *(Source: Reprinted with permission from R. A. Jones and G. F. Lee, "Recent Advances in Assessing Impact on Phosphorus Loads of Eutrophication-Related Water Quality," Water Research, vol. 16, 1982, p. 503. Copyright 1982, Pergamon Press.)*

states. Using typical phosphorus export coefficients (Table 4.3) estimates can be made of phosphorus loading in a water body with a given "mean depth to hydraulic residence time" and the expected trophic state of the water body can be predicted.[46]

Another example of empirical modeling is presented in Fig. 4.6*a*, *b*, and *c*. In this case, the normalized annual areal phosphorus loading is plotted vs. mean summer chlorophyll-a concentration, mean summer secchi depth, and mean summer hypolimnetic oxygen depletion rate. Equations for the lines of best fit in these plots yield models by which

**TABLE 4.3   Watershed Total Phosphorus Export Coefficients**

| Watershed land use | Total P export coefficient, $g/(m^2)(yr)$ |
|---|---|
| Urban | 0.1 |
| Rural and agriculture | 0.05 |
| Forest | 0.01 |
| Rainfall and dry fallout | 0.02 |

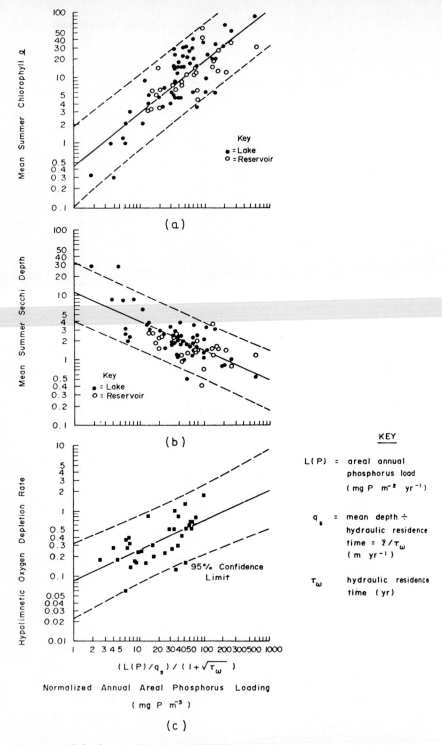

**Figure 4.6** P load eutrophication-related water quality response relationships for U.S. bodies of water. (*Source: Reprinted with permission from R. A. Jones and G. F. Lee, "Recent Advances in Assessing Impact on Phosphorus Loads of Eutrophication-Related Water Quality," Water Research, vol. 16, 1982, p. 503. Copyright 1982, Pergamon Press.*)

the independent variable can be predicted after calculating the normalized annual areal phosphorus loading.[46]

Calculations using the above models can determine whether a hypolimnion is likely to go septic during a summer stratification cycle or be used to estimate the duration that hypolimnion water could be used from a reservoir prior to possible water quality deterioration from the onset of reducing conditions. If the time to anaerobic conditions is very short, artificial destratification to avoid such problems could be employed.

Similarly, models exist for groundwater supplies that simulate hydraulic conductivities, storativities, limits, and sources of recharge. Such models are not only useful for determining aquifer yields but can also be utilized for qualitative purposes, including prediction of movement of contaminant plumes, identification of zones of influence with respect to the introduction of contaminants, and as a basis for the design of remediation programs.[45]

As an example of a mechanistic model for groundwater, with the assumptions of a one-dimensional, horizontal, single-phase flow in a saturated, unconsolidated, homogeneous medium, the contaminant transport equation is

$$\frac{-u\ \partial C}{\partial x} + \frac{D\ \partial^2 C}{\partial x^2} - \frac{\rho_b}{\epsilon}\frac{\partial S}{\partial t} + \frac{\partial C}{\partial t}\text{rn} = \frac{\partial C}{\partial t} \qquad (4.3)$$

where: $u$ = average fluid velocity
  $C$ = contaminant concentration
  $x$ = distance in direction of flow
  $D$ = dispersion coefficient
  $\rho_b$ = bulk density of soil
  $\epsilon$ = soil void ratio
  $S$ = mass of contaminant sorbed per unit dry mass of soil
  $t$ = time
  rn = reaction to degradation[45]

For a situation where degradation is zero, the first term $(-u\ \partial C/\partial x)$ relates to contaminant movement with the moving groundwater, the second term $(D\partial^2 C/\partial x^2)$ to contaminant movement by dispersion (turbulent mixing + molecular diffusion), and the third term $(\rho_b/\epsilon)(\partial S/\partial t)$ to retardation by sorption. The sum of these three effects results in the change in concentration at a point down-gradient with time $(\partial C/\partial t)$.[45] Neglecting dispersion and assuming that $dS/dC = K_d$ (the distribution coefficient), Eq. (4.3) can be rewritten as

$$\frac{-u\,\partial C}{\partial x} = \left(1 + \frac{\rho_b K_d}{\epsilon}\right)\frac{\partial C}{\partial t} \tag{4.4}$$

The term $(1 + \rho_b K_d/\epsilon)$ is known as $t_r$, the relative residence time or retardation factor. A $t_r$ of 5, for example, means that because of sorption on the soil, the contaminant front would take 5 times longer to reach a certain point than the groundwater itself.[45]

In any given situation, the bulk density of a soil and its void ratio are usually known, with $K_d$ the only unknown for calculation of the retardation factor. Studies have shown the $K_d$ can be approximated by knowing the organic content of the soil $f_{oc}$ and the octanol-to-water coefficient $K_{ow}$ (a measure of sorbability) of the contaminant. These are related as follows[46]:

$$(K_d)_i = 6.3 \times 10^{-7} f_{oc} (K_{ow})_i \tag{4.5}$$

Calculation of retardation factors using known data and Eq. (4.5) allows prediction of when the front of a contaminant will reach a certain point, e.g., a source well for a water supply, based on soil properties. Furthermore, if more than one contaminant is present in the contaminant plume, a comparison can be made as to which one will reach a given location first.

Another term in Eq. (4.3) relates to dispersion, the spreading of the concentration front. Some material moves faster and some moves more slowly as the groundwater passes through the pores of the soil, causing a spreading of the contaminant front. Dispersion tendency is measured by the Peclet number, $P_e = ux/D$. If $P_e$ is greater than 1000, dispersion can be neglected; if $P_e$ is less than 5, complete mixing occurs. Figure 4.7 presents contaminant front patterns at some downgradient point for four different Peclet numbers.[45] In groundwater hydrology $D$ is often assumed to equal $\alpha u$, where $\alpha$ is expressed in length units, commonly taken as 10 m. The Peclet number $P_e$ then becomes $ux/\alpha u$, or $x/\alpha$ or $x/10$. Thus, $P_e$ increases and the dispersion effect de-

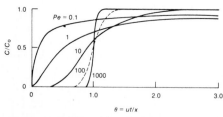

**Figure 4.7**   Effect of dispersion for $P_e = 0.1$, 1, 10, 100, and 1000. (Source: P. V. Roberts et al., "Movement of Organic Contaminants in Groundwater: Implications for Water Supply," Journal AWWA, vol. 74, no. 8, August 1982, p. 408.)

creases the further the contaminant moves through the ground.[45] Figure 4.8 presents the form of a contaminant front after 10 days of movement ($x = 50$ m, $P_e = 50/10 = 5$) and after 1000 days of movement ($x = 5000$ m, $P_e = 5000/10 = 500$). Note that the front takes 25 days to pass the 10-day point ($25/10 = 2.5$) when the $P_e$ is 5, and 400 days to pass the 1000-day point ($400/1000 = 0.4$) when the $P_e$ is 500. Thus, dispersion is less when the Peclet number is higher.[45]

## Source Water Quality Control Techniques

A variety of control techniques to improve the quality of water sources is available. The technique employed depends on the contaminant or problem involved and may be designed specifically to control nutrient concentrations, plant biomass, or other physical or chemical factors. Examples of techniques commonly employed include reservoir destratification, application of herbicides and algicides such as copper sulfate, and blending of different surface waters. Mechanical harvesting, biological controls, and sediment covers are also useful for controlling aquatic plants and algae under certain circumstances.

### Natural control measures

So-called "natural" control measures offer a means to maintain or enhance water quality, especially on relatively protected watersheds. Buffer zones adjacent to streams, reservoirs, or aquifers can be created through utility ownership or usage restrictions, thereby providing for filtration of runoff through natural vegetation. Buffer zones also increase protection by preventing development that may introduce pollutants, cause erosion problems, or create unauthorized access and other problems of trespass. No standards for the creation of buffer zones around water sources exist, but common sense should prevail and the required buffer areas should be determined taking into account exposure to contamination and the degree of treatment pro-

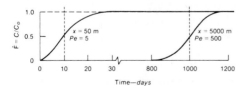

**Figure 4.8** Propagation of a nonsorbing solute. Dispersivity $a = 10$ m, average velocity $u = 5$ m/day, $u_r = 1.0$. (*Source: P. V. Roberts et al., "Movement of Organic Contaminants in Groundwater: Implications for Water Supply," Journal AWWA, vol. 74, no. 8, August 1982, p. 408.*)

vided. A common and obvious way to create buffer zones is through direct ownership by the water supplier. Easements, open space, or limited-use development also may be employed. Where land is owned by the water supplier, property lines and boundaries should be carefully marked and maintained; fencing also may be installed if necessary.

Local governments can be encouraged to set aside land areas important to water supplies as open space or protected areas. Development can be limited by federal, state, or local government action in environmentally sensitive areas; an example of this is the Pine Barrens area in New Jersey, which, in addition to its social value as open space, also provides protection to surface water and the underlying aquifer. Preservation of wetlands maintains the natural filtration and cleansing provided by these critical areas and has the added benefit of maintaining their storm-water catchment potential.

Management of the flora and fauna occurring adjacent to water sources and on watersheds can also be an important factor in optimizing source water quality. Wildlife, including geese, sea gulls, deer, and beaver may contribute significantly to bacterial loadings and eutrophication of reservoirs, necessitating control of their population and movement. *Giardia lamblia* is known to be transmitted by beaver, muskrat, deer, and other warm-blooded animals. Other pathogenic organisms, including the protozoan *Cryptosporidium,* may be distributed on watersheds by warm-blooded animals.

Burgeoning populations of the Canada goose occur throughout the east, and the bacterial contribution resulting from their feces in and around reservoirs can be significant. Means to control wildlife populations may include hunting, trapping, or physically removing certain species from the watershed.

Proper silvicultural activities in forested areas in conjunction with an appropriate forest management plan can also produce qualitative benefits for source water, especially on highly protected watersheds. Deciduous trees can be replaced by coniferous varieties around bodies of water to minimize introduction of leaves that contribute THM precursors. The integrity of the forested area can be maintained by careful planning and harvesting. Sale of timber also generates income, but control of access roads and logging activities is necessary to avoid erosion and other detrimental effects.

### Algal (nutrient) control

Algal control is perhaps the most common in situ treatment for surface water, principally by application of copper sulfate pentahydrate and other chelated copper compounds. Potassium permanganate also has been found to be an effective chemical for algal control in some cases. Although information is limited at this point, those few utilities that use potassium permanganate do find it effective.[47]

Copper sulfate application methods and dosages will vary depending on the specific reservoir being treated; caution is required because copper sulfate addition can have a detrimental effect on fish and other biological life.[47] Application methods include dissolving the copper sulfate crystals using porous bags pulled by a boat and using specially designed boats with either an application hopper that feeds copper sulfate crystals directly to the surface of the body of water, or with a spray pump through which dissolved copper sulfate is sprayed along the reservoir water surface. Timing of application is important. Thus, regular monitoring of algal numbers or chlorophyll-a to determine when algal blooms begin will allow a utility to apply the algicide before a high-standing crop develops.

Another basic method of algal control involves reducing the amount of nutrients available for algal growth and metabolism by controlling their inflow to the body of water.[48] Nutrient control can best be accomplished through a comprehensive materials monitoring program in which nutrient inputs are identified for reduction or elimination.[47] Means of reducing nutrient inflow include wastewater treatment facilities designed to remove phosphorus and nitrogen, watershed management programs that may involve the introduction of high-quality water from upstream reservoirs, forestry or vegetation management, the adoption of ordinances, and educational programs for entire watershed areas. In some cases, such as at the Wahnbach reservoir near Bonn, West Germany, treatment of the inflowing stream to the reservoir may be necessary.[49]

In situ treatment involving the application of alum sludge to inhibit phosphorus release from lake bottom sediments has been proven successful in studies conducted in Lake Eola, Fla. The study found that phosphorus release was inhibited by the use of alum sludge; calcium carbonate sludge did not prove successful. Alum sludge applied to the surface at a 200-mg/L concentration results in significant reductions of chlorophyll-a, dissolved orthophosphate, total phosphorus, total organic carbon (TOC), and turbidity. Also, results showed removal of cadmium, zinc, copper, iron, lead, and nickel greater than 90 percent. Research has shown that during anaerobic periods, algal blooms are minimized by reducing internal phosphorus recycling. Alum sludge application can also reduce costs for disposal and future coagulant use.[50]

### Destratification and aeration

Destratification of reservoirs and lakes can be effectively used to combat both turnover and anaerobic conditions and in some cases for algal control. Although the water volumes are high, the energy of stability is low, the volume of hypolimnitic water is relatively small, and the mixing device need only be placed at the deepest part of the reservoir.

**Mixing.**  Mixing techniques to accomplish destratification include diffused air aeration and mechanical or hydraulic mixing. Aeration is a process in which air is injected over wide areas of the reservoir to promote mixing. In mechanical or hydraulic mixing, water is pumped from one level of the reservoir to another area of different density.[51] Both aeration and mechanical or hydraulic mixing cause a circulation and a mixing action that results in a homogeneous water column and helps to remove anaerobic conditions in the bottom waters. Mechanical or hydraulic mixing systems generally have been found to be more costly to operate and, therefore, are less widely used than diffused-air systems. Mixing, however, also causes biological and microbiological changes to occur that should be carefully monitored, but it generally has been found to be an effective technique in most applications.[51] Timing is important, and destratification should be started early in the warming season.

The effectiveness of reservoir aeration was demonstrated in a study in the Occoquan Reservoir in Virginia.[52] The following benefits were found:

1. Elimination of anaerobic conditions and the resultant generation of hydrogen sulfide.

2. Reduction in the amount of chlorine required to oxidize organic materials.

3. Reduction in the amount of PAC required to combat taste and odor problems.

4. Uniform temperature and quality of water at the raw water intake.

5. Some manganese removal through oxidation.

6. Ability to draw from the bottom intake because of improved quality.

7. No evidence of any prolific algal growths after the aeration system was placed in service and algicide was used.

Destratification at Lake Casitas in California was successfully employed for algal control as part of a comprehensive program that also included watershed management, chemical treatment, and multilevel releases.[47] A low-energy mechanical destratifier was used in Lake Eureka, Ill., resulting in improved water quality that allowed reinstatement of the lake as a water supply source.[53]

**Hypolimnetic aeration.**    Hypolimnetic aeration is a technique that increases the dissolved oxygen content at the hypolimnion while maintaining thermal stratification. It is accomplished by injecting very small bubbles of air or oxygen at the hypolimnion or by spraying the hypolimnetic water into the air and returning it to the hypolimnion. This technique can be effective in maintaining oxygen levels at the critical sediment-water interface, suppressing release of iron, manganese, phosphorus, nitrogen, and metals.[4,54]

### Remediation of groundwater contamination

A systematic, site-specific response to the occurrence of groundwater contamination should include provision to hydraulically isolate the contamination, treat the affected supply, and/or secure an alternative supply. The following chapters in this book present details on treatment techniques that are effective in remediating contaminated groundwater. In some cases, in situ treatment can be utilized to improve groundwater quality or control contamination. Barrier wells, for example, can be used to divert a contaminant plume away from production wells.[55]

A fairly recent development for in situ treatment of groundwater is the use of microorganisms for removal of organic contaminants.[56] In situ microbiological treatment must be conducted under carefully controlled conditions, and a proper mix of nutrients, microorganisms, and physical factors are needed to attain optimal contaminant removal.

### Summary

Source water quality management is the first step in ensuring an adequate supply of safe drinking water. By knowing the condition and characteristics of its water sources, and the factors that influence source water quality, water utilities can influence that quality through operational activities and procedures. Sanitary surveys, monitoring, watershed control, storm-water management, and emergency response procedures are essential elements of an ongoing source water quality management plan. When needed, source water quality control techniques should be implemented to correct specific problems, such as chemical addition to control algal growths or hypolimnetic aeration to correct oxygen deficiency in a reservoir. By effectively managing water supply sources, water utilities will be able to ensure that the highest-quality source water is provided and lessen the burden on subsequent treatment processes.

## References

1. "AWWA Policy Statement on Quality of Water Supply Sources," *1989–90 Officers and Committee Directory,* AWWA, Denver, Colo., 1989.
2. AWWA Manual M21, *Groundwater,* AWWA, Denver, Colo., 1989.
3. USEPA, "Surface Water Treatment Regulations," *Federal Register,* title 54, part 124, June 29, 1989, p. 27486.
4. R. S. Geney, "Reservoir Management for Water Quality Improvement," Sem. Proc.: "Water Quality Management in Reservoirs," *Proc. AWWA Annual Conf.,* Orlando, Fla., June 1988.
5. J. P. Longtin, "Radon, Radium, and Uranium Occurrence in Drinking Water from Groundwater Sources," Sem. Proc.: "Radionuclides in Drinking Water," *Proc. AWWA Annual Conf.,* Kansas City, Mo., June 1987.
6. E. Hindin, "Treatment of Mount St. Helens Volcanic Ash Suspensions by Plain Sedimentation, Coagulation, and Flocculation," *Journal AWWA,* vol. 73, no. 3, March 1981, p. 160.
7. W. W. Walker, "Significance of Eutrophication in Water Supply Reservoirs," *Journal AWWA,* vol. 75, no. 1, January 1983, p. 38.
8. R. S. Gupta and M. S. Destora, "Algae Pollutants and Potable Water," in R. B. Pojasek (ed.), *Drinking Water Quality Enhancement Through Source Protection,* Ann Arbor Science, Ann Arbor, Mich., 1977, p. 431.
9. C. J. Aexoponlons and H. C. Bold, *Algae and Fungi,* MacMillan, New York, 1971.
10. American Public Health Association, American Water Works Association, and Water Pollution Control Federation, *Standard Methods for the Examination of Water and Wastewater,* 16th ed., Published by American Public Health Association, Washington, D.C., 1985.
11. E. P. Odum, *Fundamentals of Ecology,* Saunders, Philadelphia, Pa., 1971.
12. W. D. Taylor et. al., "Phytoplankton Water Quality Relationships in U.S. Lakes. Part VIII: Algae Associated with or Responsible for Water Quality Problems," USEPA Environmental Monitoring Systems Laboratory, NTIS, EPA-600/3-80-100, December 1980.
13. J. M. Symons and G. G. Robeck, "Calculation Technique for Destratification Efficiency, in J. M. Symons (ed.), *Water Quality Behavior in Reservoirs,* Bureau of Water Hygiene, U.S. Public Health Service Publ. No. 1930, Cincinnati, Ohio, 1969.
14. "Resource Conservation and Recovery Act of 1976," PL 100-582, U.S. Congress, Washington, D.C., November 1, 1988.
15. "Comprehensive Environmental Response, Compensation, and Liability Act of 1980," PL 100-647, U.S. Congress, Washington, D.C., November 10, 1988.
16. "Emergency Planning and Community Right to Know Act of 1986," PL 99-499, U.S. Congress, Washington, D.C., October 17, 1986.
17. J. Haley and B. Hanson, "Superfund's Role in Protecting Ground Water and Providing Safe Drinking Water," Sem. Proc.: "Impact of Hazardous Waste Sites on Water Utilities," *Proc. AWWA Annual Conf.,* Orlando, Fla., June 1988.
18. R. M. Clark et al., "The Great Ohio River Oil Spill of 1988," Sem. Proc.: "Impact of Hazardous Waste Sites on Water Utilities," *Proc. AWWA Annual Conf.,* Orlando, Fla., June 1988.
19. D. M. Fairchild, "A National Assessment of Ground Water Contamination from Pesticides and Fertilizers," in M. Fairchild (ed.), *Ground Water Quality and Agricultural Practices,* Lewis Publishers, Chelsea, Mich., 1987.
20. L. W. Cantor et al., *Ground Water Quality Protection,* Lewis Publishers, Chelsea, Mich., 1987.
21. AWWA Journal Roundtable Discussion, "Debating Recreational Use," *Journal AWWA,* vol. 79, no. 12, December 1987, p. 10.

22. D. A. Roseberry, "Relationship of Recreational Use to Bacterial Densities in Forrest Lake," *Journal AWWA,* vol. 56, no. 6, June 1964, p. 43.
23. A. J. Minkus, "Recreational Use of Reservoirs," *Journal New England Water Works Association,* vol. 79, 1965, p. 32.
24. P. C. Karalekas and J. P. Lynch, "Recreational Activities of Springfield, Mass., Water Reservoirs Past and Present," *Journal New England Water Works Association,* vol. 79, 1965, p. 18.
25. H. S. Peavy and C. E. Matney, "The Effects of Recreation on Water Quality and Treatability," in R. B. Pojasek (ed.), *Drinking Water Enhancement Through Source Protection,* Ann Arbor Science, Ann Arbor, Mich., 1977.
26. R. D. Lee et al., "Watershed Human Use Level and Water Quality," *Journal AWWA,* vol. 61, no. 5, 1970, p. 412.
27. T. J. Suk et al., "The Relation Between Human Presence and Occurrence of *Giardia* Cysts in Streams in the Sierra Nevada, California," *Journal Freshwater Ecology,* vol. 41, no. 1, 1987, p. 71.
28. D. G. Stuart et al., "Effects of Multiple Use on Water Quality of High Mountain Watersheds: Bacteriological Investigations of Mountain Streams," *Applied Microbiol.,* vol. 22, no. 6, 1971, p. 1048.
29. H. H. Christensen et al., "Human Use in a Dispersed Recreation Area and Its Effect on Water Quality," *Proc. Conf. Recreation Impacts on Wildlands,* Seattle, Wash., 1979.
30. "Safe Drinking Water Act," PL 100-572, U.S. Congress, Washington, D.C., October 31, 1988.
31. "Federal Water Pollution Control Act," as amended by the Clean Water Act of 1977, PL 100-688, U.S. Congress, Washington, D.C., November 18, 1988.
32. "Federal Insecticide, Fungicide and Rodenticide Act," PL 100-532, 102 Stat. 2654, U.S. Congress, Washington, D.C., October 25, 1988.
33. "Toxic Substance Control Act," PL 100-368, U.S. Congress, Washington, D.C., July 18, 1988.
34. "Clean Air Act," PL 98-213, U.S. Congress, Washington, D.C., December 8, 1983.
35. *Connecticut General Statutes,* section 22a-426.
36. J. P. Featherstone et al., "Opportunities for Conjunctive Use of Ground and Surface Water in the Delaware River Basin," *Proc. NWWA Eastern Regional Conf. Conjunctive Use of Ground and Surface Water in Delaware River Basin,* National Water Well Association, 1983.
37. P. C. Karalekas, "Sanitary Survey Techniques," *Proc. AWWA 1986 Water Quality Technology Conf.,* Portland, Ore., November 20, 1986.
38. AWWA Research Foundation and Lyonnaise Des Eaux, *Identification and Treatment of Tastes and Odors in Drinking Water,* Published by AWWA, Denver, Colo., 1987.
39. "AWWA Policy Statement on Recreational Use of Domestic Water Supply Reservoirs," AWWA, Denver, Colo., 1987.
40. J. A. Hroncich, "Surface Water Supply Protection," *Proc. AWWA Annual Conf.,* Kansas City, Mo., June 1987.
41. F. V. Padar, "Management of Long Island Aquifer Contamination by Organics Chemicals," New York Water Pollution Control Association Conference Paper, January 19, 1981.
42. YWC, Inc., "Aquifer Protection Plan," Bridgeport Hydraulic Company, Bridgeport, Conn., 1988.
43. D. Yodel, "Protection of Wellfields and Recharge Areas in Dade County, Florida."
44. K. H. Rickhow and S. C. Chapra, *Engineering Approaches for Lake Management,* vol. I, Butterworth, Woburn, Mass., 1983.
45. P. V. Roberts et al., "Movement of Organic Contaminants in Groundwater: Implications for Water Supply," *Journal AWWA,* vol. 74, no. 8, August 1982, p. 408.
46. R. A. Jones and G. F. Lee, "Recent Advances in Assessing Impact of Phosphorus Loads on Eutrophication-Related Water Quality," *Water Research,* vol. 16, 1982, p. 503.
47. AWWA Research Foundation, *Current Methodology for the Control of Algae in Surface Reservoirs,* Published by AWWA, Denver, Colo., 1987.

48. R. K. Raman, "Controlling Algae in Water Supply Impoundments," *Journal AWWA,* vol. 77, no. 8, August 1985, p. 41.
49. H. Bernhardt and H. Schell, "Energy-input-controlled Direct Filtration to Control Progressive Eutrophication," *Journal AWWA,* vol. 74, no. 5, May 1982, p. 261.
50. H. H. Harper et al., "Reuse of Water Treatment Sludges for Improvement of Reservoir Water Quality," *Proc. AWWA Annual Conf.,* Las Vegas, Nev., June 1983.
51. B. Henerson-Sellers and H. R. Markland, *Decaying Lakes: The Origins and Control of Cultural Eutrophication,* John Wiley & Sons, New York, 1987.
52. F. F. Eunpu, "Control of Reservoir Eutrophication," *Journal AWWA,* vol. 65, no. 4, April 1973, p. 268.
53. R. K. Raman and B. R. Arbuckle, "Long-Term Destratification in an Illinois Lake," *Journal AWWA,* vol. 81, no. 6, June 1989, p. 66.
54. H. Bernhardt, "Aeration of Wahnbach Reservoir Without Changing the Temperature Profile," *Journal AWWA,* vol. 59, no. 8, August 1967, p. 943.
55. G. H. Emrich and W. W. Beck, "Methods of Containing Contaminated Ground Water," Sem. Proc.: "Organic Chemical Contaminants in Groundwater: Transport and Removal," *Proc. AWWA Annual Conf.,* St. Louis, Mo., June 1981.
56. J. M. Thomas and C. H. Ward, "In situ biorestoration of organic contaminants in the subsurface," *Environ. Science Tech.,* vol. 23, no. 7, July 1989, p. 760.

# Air Stripping and Aeration

## David A. Cornwell, Ph.D., P.E.

*President*
*Environmental Engineering & Technology, Inc.*
*Newport News, Virginia*

The beneficial effect of bringing air into contact with water has been recognized since the days of the Roman aqueducts when the sides were intentionally left rough in order to increase the turbulence and increase the purifying effect of air. As a unit process for water treatment, aeration has been used more to improve the palatability of water or to improve treatment economics rather than to improve the potability of water. Historical applications of aeration have included the removal of hydrogen sulfide, which causes tastes and odors, or the removal of carbon dioxide, which can reduce the lime demand in lime softening treatment. More recently aeration has been utilized to remove trace volatile organic contaminants (VOC) from water. These VOCs are believed to cause adverse health effects. In most water plant applications, air is brought into contact with water in order to remove a substance from water, a process referred to as desorption or stripping. To a more limited extent, aeration has also been used to transfer a substance from air or a gas phase into water, a process called gas absorption. Examples of this include the addition of oxygen from air into water in order to oxidize iron or manganese or the addition of ozone to water for oxidation. The overall process of aeration can, therefore, be defined as the process of bringing water into contact with air in order to expedite the transfer of a gas between the two phases. The term "gas" is used here to refer to those substances that technically are gases at standard temperature and pressure as well as those compounds that technically are volatile liquids. In either case, the transfer itself is of the gas form of the compound.

In developing systems to accomplish gas transfer, the total amount of transfer that is possible and the rate at which that transfer is accomplished are of primary concern. In order for transfer to take place, a concentration gradient must exist between the liquid and air phase. This concentration gradient of the two-phase system is indicative of the departure from equilibrium that exists between the air and water. If equilibrium is established, then no gradient exists and hence a net transfer will not take place. Therefore, it is necessary to consider both the rate of transfer and the equilibrium phenomenon in order to describe gas transfer.

## Theory of Gas Transfer

### Equilibrium

In considering the amount of gas or the rate at which a gas can be transferred in and out of water, consideration must be given to which gases are soluble in water and what factors affect their solubility. Two basic mechanisms affect solubility. First, some gases undergo a chemical reaction with water, examples being hydrogen sulfide, ammonia, chlorine, and carbon dioxide. Secondly, there are gases that do not chemically react with water, such as oxygen, methane, and chloroform. In the case when a reaction does not take place, the mechanism of solubility is simply that all systems tend to maximum disorder. Thermodynamically this is referred to as an increase in the entropy of the system. The diffusion of and hence solubility of oxygen through water is an increase in the entropy of the system as compared to the two components (oxygen and water) existing separately. This force of increasing entropy is opposed by the attraction of the water molecules to themselves. Therefore, for a gas to increase its solubility in water, it must overcome the forces that attract water together.[1]

Many models have been proposed to account for the properties of water. Water has generally been concluded to be structured as a tetrahedron, bound together by the forces of hydrogen bonding. Figure 5.1a shows such an arrangement. The dashed lines represent the hydrogen bond that is the force of attraction between the water molecules. A constant interaction of bond formation and bond breaking occurs, producing short-lived clusters of water molecule groups, such as shown in Fig. 5.1b.

For a gas to solubilize, it must be more attracted to water than water is to itself. Consider two very similar compounds but with differing solubilities. Ethylene ($C_2H_4$) has approximately twice the solubility as

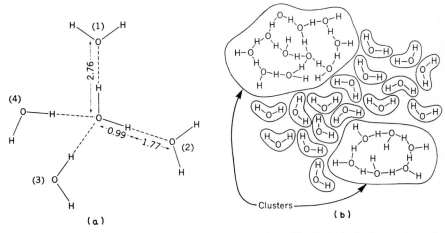

**Figure 5.1**  (a) Water molecule bonding distance and angles. Tetrahedral structure of fully coordinated water. Molecules 1 and 2 as well as the central $H_2O$ molecule lie entirely in the plane of the paper. Molecule 3 lies above this plane, molecule 4 below it, so oxygens 1, 2, 3, and 4 lie at the corners of a regular tetrahedron. Distances are in angstroms. (*Source: From J. T. Edsall and J. Wyman, Biophysical Chemistry, vol. 1, p. 31, Copyright 1958. Reproduced by permission of Academic Press, Orlando, Fla.*) (b) Clustering effect of water molecules. Two "flickering clusters" separated by unbonded water molecules. The molecules in the interior of the clusters are tetracoordinated but not drawn as such in this two-dimensional diagram. (*Source: G. Nemethy and H. A. Scheraga, Journal Chem. Phys., vol. 36, 1962, p. 3387. American Institute of Physics.*)

ethane ($C_2H_6$). The difference is due to the structures of the two hydrocarbons and their ability to break the hydrogen bonding between the water molecules. Figure 5.2a shows the structure of ethane, which is a saturated hydrocarbon with all the valence electrons utilized in localized single bonds. Ethylene is unsaturated as shown in Fig. 5.2b and includes a delocalized double bond. The double bond has an excess of electrons and hence tends to attract the positive end of a water dipole. This produces a dipole-induced attraction between water and ethylene resulting in greater solubility of ethylene in water.

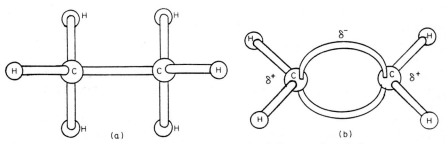

**Figure 5.2**  (a) Orbital diagram for ethane; (b) orbital diagram for ethylene.

Oxygen and nitrogen can be similarly compared. Both have delocalized electrons and so are slightly polarized in water. Oxygen is even more soluble than nitrogen because oxygen contains two unpaired electrons that are more susceptible to attraction by water molecules.

When gas undergoes a chemical reaction with water, termed hydrolysis, gas solubility is completely dependent upon the degree to which that reaction takes place. In general, when gas reacts with water, ionic species are formed. The ionic species can be considered completely soluble for practical applications in water treatment. In aeration, reactions that are reversible, that is, the ionic species can reassociate to the gas form, are of primary concern. Consider the reaction of hydrogen sulfide with water.

$$H_2S + H_2O = H_3O^+ + HS^- \tag{5.1}$$

$$HS^- + H_2O = H_3O^+ + S^{2-} \tag{5.2}$$

The amount of the total sulfide species in water is dependent upon the pH. As base is added, the hydrogen ion concentration is decreased, the reaction is shifted to the right, and the sum of the sulfide present is increased. In stripping the hydrogen sulfide from the water, only that portion of the total sulfide that is hydrogen sulfide ($H_2S$) can be stripped.

With this introduction to gas solubility, a method is needed to describe the degree of solubility. Dalton's law of partial pressures states that when a container is filled with a mixture of $N$ gases, each gas exerts the same pressure as it would if it were the only gas in the vessel. The pressure of each component of the gas mixture will depend solely on the number of moles of that gas present. These individual pressures are called partial pressures, their sum being equal to the total pressure:

$$P_{\text{total}} = P_1 + P_2 + P_3 + \cdots P_N \tag{5.3}$$

where $P_1$ is proportional to $n_1$ and $n_1$ is the number of moles of gas 1 present. Therefore, the ratio of the partial pressure of a gas to the total pressure is equal to the ratio of the number of moles of that component to the total number of moles:

$$\frac{P_1}{P_{\text{total}}} = \frac{n_1}{n_1 + n_2 + n_3 + \cdots + n_N} = p_1 \quad \text{li} \quad \frac{gas}{gas} \tag{5.4}$$

The ratio is called the mole fraction and is designated by the symbol $p$.

If air containing a specific gas, for example, oxygen, is placed in a closed vessel together with oxygen-free water, oxygen will begin to transfer into the water. A portion of the oxygen molecules in the water will transfer back to the air phase. Eventually the number of molecules leaving the air phase and entering the water will equal the number of molecules leaving the water and entering the air. The effective concentration in the water phase is constant, indicating that equilibrium has been reached. If now more oxygen is injected into the air, thereby increasing the mole fraction of oxygen in air, a net positive transfer will again take place into the water until a new, higher equilibrium concentration is reached in the water. The concentration of gas at equilibrium in the water phase, expressed as the mole fraction, is designated $c$.

If several data points were collected via the above procedure, a distribution curve as shown in Fig. 5.3 could be drawn. Distribution curves are specific for each gas and are only accurate for one temperature and total pressure. In water treatment, the atmospheric pressure is generally of concern, and therefore only temperature is considered a variable. The following principles are common for most gas-water systems applicable in water treatment:

1. At equilibrium, no net transfer of gas occurs between the phases.

2. If the system is not at equilibrium, gas transfer will occur in order to reach equilibrium. The time to reach equilibrium may be instantaneous or very long.

3. For each gas, temperature, and total pressure, a set of equilibrium conditions exists that can be shown graphically in the form of a distribution curve.

4. Generally, as the temperature increases, the solubility decreases; and as the total pressure increases, the solubility increases.

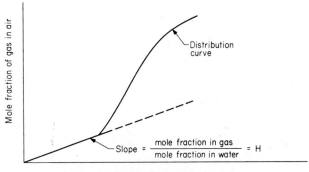

**Figure 5.3** Equilibrium distribution of a gas between air and liquid phase at constant temperature.

A special region of the distribution curves exists at dilute concentrations and is characterized by a straight line, shown in Fig. 5.3. This straight-line portion can be mathematically expressed as a linear equation passing through zero and is known as Henry's law:

$$p = \frac{Hc}{P_T} \qquad (5.5)$$

where $c$ = the mole fraction of gas in water
   $p$ = mole fraction of gas in air
   $H$ = proportionality constant, known as Henry's constant (the slope of the straight-line portion of the distribution curve)
   $P_T$ = total pressure atm

For water treatment, $P_T$ is usually 1 atmosphere (atm). Various units are used by different investigators for the concentrations in the two phases and, therefore, the units for Henry's constant vary. Because of units, care must be taken in using the relationship, especially when obtaining constants from different sources. Below is a discussion of the primary methods of reporting Henry's law.

Probably the most common method of expressing Henry's law is with units of $c$ and $p$ as mole fractions:

$$p = \frac{Hc}{P_T} \qquad (5.5)$$

where $p$ = mol gas/mol air
   $c$ = mol gas/mol water

   $H$ = atm, actually = $\dfrac{\text{atm (mol gas/mol air)}}{\text{mol gas/mol water}}$
   $P_T$ = atm, usually = 1

Recall that according to Dalton's law, a mole of gas per mole of air is the same as the partial pressure of the gas or is also the same as the volume of gas per volume of air. A useful conversion factor when calculating $c$ is that 1 L of water contains 55.6 mol of water:

$$\frac{1000 \text{ g/L}}{18 \text{ g/mol}} = 55.6 \text{ mol/L}$$

Another method of reporting Henry's law is to utilize concentration units. In this case, the total pressure $P_T$ is commonly defined as 1, and hence it is left off of the equation and atm is dropped from the units of $H$. In this case, any set of mass per volume or mole per volume units can be used (as long as $p$ and $c$ are the same), and hence it is often referred to as the dimensionless or unitless Henry's law constant:

$$p = H_u c \qquad (5.6)$$

where $p$ = concentration units, e.g., $kg/m^3$, mol/L, mg/L
$\quad H_u$ = unitless
$\quad c$ = same concentration units used for $p$

At 1 atm pressure and 0°C, 22.412 L of air is 1 mol of air. At other temperatures, 1 mol of air is $0.082T$ L [where $T$ = temperature in kelvin (K)] of air. The following conversion between $H$ and $H_u$ can be made:

$$H_u = \left[ H \frac{\text{atm (mol gas/mol air)}}{\text{mol gas/mol water}} \right] \left( \frac{\text{mol air}}{0.082T \text{ L air}} \right) \left( \frac{\text{L water}}{55.6 \text{ mol}} \right)$$

$$= \frac{H}{4.56T} \text{ or } H_u = H \times 7.49 \times 10^{-4} \text{ at } 20°C \qquad (5.7)$$

Another method for reporting Henry's constant is to use mixed units for $p$ and $c$. This is very common because units of partial pressure in the air phase and concentration units in the water phase tend to be used. Different variations are available. Two are shown below:

$$p = \frac{H_m c}{P_T} \qquad (5.8)$$

where $p$ = mol gas/mol air (partial pressure)
$\quad c$ = mol gas/$m^3$ water
$\quad H_m$ = atm × $m^3$ water/mol gas

$$= \left[ H \frac{\text{atm(mol gas/mol air)}}{\text{mol gas/mol water}} \right] \left( \frac{m^3 \text{ water}}{55,600 \text{ mol}} \right)$$

$$= \frac{H}{55,600}$$

Finally, milligram per liter units for $c$ may be used. This is very useful in water treatment:

$$p = \frac{H_D c}{P_T} \qquad (5.9)$$

where $p$ = mol gas/mol air (partial pressure)
$\quad c$ = mg/L
$\quad H_D$ = (atm)(L)/mg
$$H_D = \frac{H_m}{MW} = \frac{H}{55,600 \text{ MW}}$$
$\quad$ MW = molecular weight of gas of interest

Values for Henry's constant shown in Table 5.1 are calculated at 20°C. In general, increasing temperature will decrease solubility in the water phase. The change in Henry's law constant for temperature can be determined from the following relationship:

$$\log H = \frac{-\Delta H}{R_c T} + J \tag{5.10}$$

where $H$ = atm

$\Delta H$ = heat absorbed in the evaporation of 1 mol of gas from solution, kilocalorie/kmol, (kcal/kmol)

$R_c$ = gas constant, 1.987 kcal/kmol

$T$ = temperature, K

$J$ = empirical constant

Table 5.2 shows values for $\Delta H$ and $J$. Note that this relationship is unit-dependent and is only valid for the units shown. To convert between units for different temperatures, $H$ should first be adjusted for temperature and then the desired unit conversion made by the methods presented for Eqs. (5.6) to (5.9).

## Mass transfer

The degree to which the gas-water system deviates from equilibrium provides the driving force for diffusion. Consider a situation where a gas is diffusing from air into water. For this to occur, a concentration gradient in the direction of transfer in each phase must exist. This can be shown graphically in terms of the distance away from the air-water interface, as shown in Fig. 5.4. The concentration of gas in the bulk of

**TABLE 5.1    Henry's Law Constants at 20°C**

| | $H^*$ atm | $H_u{}^\dagger$ dimensionless | $H_D{}^\dagger$ atm · L/mg | $H_m{}^\dagger$ (atm)(m³)/mol |
|---|---|---|---|---|
| Oxygen | $4.3 \times 10^4$ | $3.21 \times 10$ | $2.42 \times 10^{-2}$ | $7.73 \times 10^{-1}$ |
| Methane | $3.8 \times 10^4$ | $2.84 \times 10$ | $9.71 \times 10^{-2}$ | $6.38 \times 10^{-1}$ |
| Carbon dioxide | $1.51 \times 10^2$ | $1.13 \times 10^{-1}$ | $6.17 \times 10^{-5}$ | $2.72 \times 10^{-3}$ |
| Hydrogen sulfide | $5.15 \times 10^2$ | $3.84 \times 10^{-1}$ | $2.72 \times 10^{-4}$ | $9.26 \times 10^{-3}$ |
| Vinyl chloride | $3.55 \times 10^5$ | $2.65 \times 10^2$ | $1.02 \times 10^{-1}$ | $6.38$ |
| Carbon tetrachloride | $1.29 \times 10^3$ | $9.63 \times 10^{-1}$ | $1.51 \times 10^{-4}$ | $2.32 \times 10^{-2}$ |
| Trichloroethylene | $5.5 \times 10^2$ | $4.1 \times 10^{-1}$ | $7.46 \times 10^{-5}$ | $9.89 \times 10^{-3}$ |
| Benzene | $2.4 \times 10^2$ | $1.8 \times 10^{-1}$ | $5.52 \times 10^{-5}$ | $4.31 \times 10^{-3}$ |
| Chloroform | $1.7 \times 10^2$ | $1.27 \times 10^{-1}$ | $2.55 \times 10^{-5}$ | $3.06 \times 10^{-3}$ |
| Bromoform | $3.5 \times 10$ | $2.61 \times 10^{-2}$ | $2.40 \times 10^{-6}$ | $6.29 \times 10^{-4}$ |
| Ozone | $5.0 \times 10^3$ | $3.71$ | $1.87 \times 10^{-3}$ | $8.99 \times 10^{-2}$ |

*$H$ values from Ref. 2.

†$H_u$, $H_D$, and $H_m$ calculated via Eqs. (5.7) to (5.9).

**TABLE 5.2    Temperature Correction Factors[2] for $H$ (in atm)**

|  | $\Delta H \times 10^{-3}$ | $J$ |
|---|---|---|
| Oxygen | 1.45 | 7.11 |
| Methane | 1.54 | 7.22 |
| Carbon dioxide | 2.07 | 6.73 |
| Hydrogen sulfide | 1.85 | 5.88 |
| Carbon tetrachloride | 4.05 | 10.06 |
| Trichloroethylene | 3.41 | 8.59 |
| Benzene | 3.68 | 8.68 |
| Chloroform | 4.00 | 9.10 |

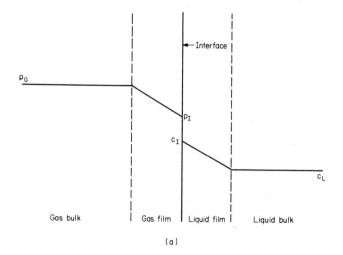

(a)

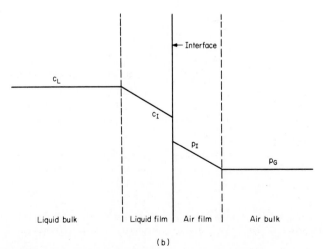

(b)

**Figure 5.4** (a) Two-resistance concept, gas absorption; (b) Two-resistance concept, stripping.

the air phase $p_G$ falls to a lower value at the interface, $p_I$; hence a driving force is set up from the bulk to the interface. Similarly, in the liquid phase, the concentration decreases from the interface, $c_I$ to the bulk liquid, $c_L$; hence a driving force is set up into the liquid from the interface. The bulk concentrations in the gas and liquid phase are clearly not in equilibrium with each other or transfer would not occur. At the same time, these bulk concentrations cannot be used directly as the driving force because they are not "equal" at equilibrium but differ as described by the equilibrium distribution curves. To overcome this problem, Lewis and Whitman[3] assumed that the only diffusional resistances are those in the two phases, and no resistance exists at the interface itself. As a result, interfacial concentrations are at equilibrium and can be described by the distribution curves, or in dilute regions, by Henry's law, $p_I = Hc_I$ in Fig. 5.4a and b.

Rates of mass transfer are described by a term called flux. Flux is the mass transfer per time through a given area and is a function of the driving force. The driving force in the air is the difference between the bulk concentration and the interface concentration. For steady-state mass transfer, the rate at which the gas reaches the interface from the air phase must equal the rate at which the gas leaves the interface and enters the water phase. The flux relationship for each phase is Fick's law and is written as

$$F = \frac{dW}{dt\,A} = k_G(p_G - p_I) = k_L(c_I - c_L) \tag{5.11}$$

where $F$ = flux
$W$ = mass transferred
$A$ = area through which transfer takes place
$t$ = time
$c$ = water-phase mass concentration; subscript $L$ designates bulk concentration and $I$ designates interface concentration shown in Fig. 5.4
$p$ = air-phase concentration; subscript $G$ designates bulk concentration and $I$ designates interface concentration
$k$ = transfer coefficients, in this case called interface transfer coefficient because interface concentrations are used to compute the driving force

From the above, computing a change in mass with time is fairly straightforward, if $p_I$ and $p_G$ or $c_I$ and $c_L$ are known. Unfortunately, the analytical techniques are not available to measure the concentration at the interface ($p_I$ or $c_I$); therefore, only an overall effect, which is computed in terms of bulk concentration, can be determined.

Consider the situation shown in Fig. 5.5. At time $t$, the bulk concentration in air $p_G$ and the bulk concentration in water $c_L$ are not in equilibrium with each other. On the other hand, the interface concentrations $p_L$ and $c_I$ are in equilibrium with each other. Considering the water phase, the previous equilibrium distribution curve discussion indicates that $c_L$ has an associated concentration in the air phase with which it would be in equilibrium, shown on Fig. 5.5 as $p^*$. The driving force in the air phase is then described as the difference between the actual concentration at time $t$ in the air, $p_G$, and the concentration that should exist if the gas were in equilibrium with the water-phase concentration $c_L$:

$$F = K_G(p_G - p^*) \tag{5.12}$$

and similarly,

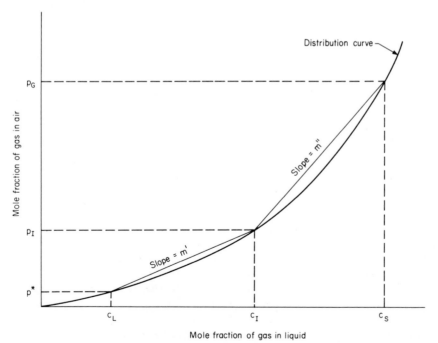

**Figure 5.5** Concentration differences of gas in air and water phases.

$$F = K_L(c_s - c_L) \tag{5.13}$$

where $K$ is the mass transfer coefficient, in this case called the overall mass transfer coefficient because the driving force is based on the bulk and equilibrium concentrations; and $c_s$ is the water-phase concentration in equilibrium with $p_G$.

Figure 5.5 also shows the distribution curve approximated by two chords with slopes of $m'$ and $m''$. By the geometry of the figure,

$$p_G - p^* = (p_G - p_I) + (p_G - p_I) + m'(c_I - c_L) \tag{5.14}$$

Substituting for the concentration differences given in Eqs. (5.11) to (5.13) results in

$$\frac{F}{K_G} = \frac{F}{k_G} + \frac{m'F}{k_L} \tag{5.15}$$

or

$$\frac{1}{K_G} = \frac{1}{k_G} + \frac{m'}{k_L} \tag{5.16}$$

similarly,

$$\frac{1}{K_L} = \frac{1}{m''k_G} + \frac{1}{k_L} \tag{5.17}$$

If dilute conditions exist, then Henry's law applies, the equilibrium distribution curve is a straight line, and

$$m' = m'' = H \tag{5.18}$$

which results in

$$\frac{1}{K_G} = \frac{1}{k_G} + \frac{H}{k_L} \tag{5.19}$$

$$\frac{1}{K_L} = \frac{1}{Hk_G} + \frac{1}{k_L} \tag{5.20}$$

If Henry's constant is large such as for highly volatile compounds, then Eq. (5.20) shows that the liquid-phase resistance controls the diffusion rate. Compounds of concern in water treatment tend to be liquid-phase controlled or a combination of liquid- and gas-phase controlled. Again consistent units must be used with the above expressions so that $H$, $K_L$, $K_G$, $k_L$, and $k_G$ are consistent.

The flux equation for the liquid phase using concentration in units of mg/L is expressed as

$$F = \frac{dc}{dt}\frac{V}{A} = K_L (c_t - c_s) \qquad (5.21)$$

or

$$\frac{dc}{dt} = K_L \frac{A}{V}(c_t - c_s) \qquad (5.22)$$

where $K_L$ = overall transfer coefficient, cm/h
   $A$ = interfacial area through which transfer occurs, cm$^2$
   $V$ = volume containing the interfacial area, cm$^3$
   $c_t$ = concentration in liquid bulk phase at time $t$, mg/L
   $c_s$ = concentration in equilibrium with gas at time $t$ in mg/L, as given by

$$p_t = H_D c_s \qquad (5.23)$$

Often the term $A/V$ is replaced by $a$, the specific interfacial area. Further, $a$ can be combined with $K_L$, resulting in a constant that is sometimes empirically derived, $K_L a$, with units of time$^{-1}$. In order to find changes of concentration with time, the values of $K_L$ and $a$ or $K_L a$ must be known. The specific interfacial area and hence $K_L a$ is dependent upon water quality factors and the type of equipment used to accomplish the gas-liquid contact. Therefore, methods to calculate mass transfer coefficients will be covered within the unit operations section.

## Unit Operations

### Packed towers

The use of packed towers for gas-liquid contacting equipment is fairly recent to the water treatment industry. In the past, stripping practices have dealt with contaminants that are highly volatile such as hydrogen sulfide, and relatively simple equipment was sufficient to provide the needed gas-liquid contact. The class of chemicals known as VOCs and their associated subset trihalomethanes (THM) are, however, less volatile and have required the use of more sophisticated contacting equipment for efficient removal.

A typical packed column consists of a cylindrical shell containing a support plate for the packing material. Packing material is usually individual pieces randomly dumped into the column, sometimes referred to as dumped packing. Figure 5.6 shows examples of typical commercially available packing shapes that are made of ceramic, stainless steel, and various plastic materials. Because of weight and cost considerations, plastic is usually used in water treatment operations. Fixed packing is also available. Fixed packing comes in prefabricated

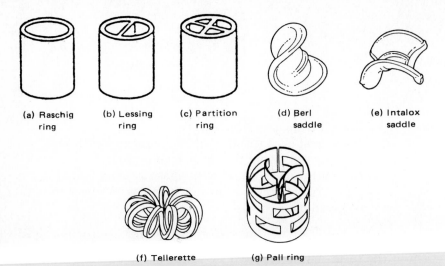

(a) Raschig    (b) Lessing    (c) Partition    (d) Berl    (e) Intalox
ring          ring           ring         saddle      saddle

(f) Tellerette        (g) Pall ring

**Figure 5.6**   Example packing materials for air stripping towers. (*Source: R. H. Perry and C. H. Chilton, Chemical Engineers Handbook, McGraw-Hill Publishers, New York, 1973, reprinted with permission.*)

sheets that are carefully placed in the tower. Although more expensive, higher transfer efficiency is claimed. Towers are normally operated using a countercurrent flow pattern with water falling down through the tower and air passing upward. The towers can also be designed for cocurrent or cross-current operation.

Mathematical expressions describing the stripping process are used to predict the removal that a tower will achieve. This requires the use of mass balance expressions across the tower and the kinetic equations previously presented. Figure 5.7 schematically shows a tower cross section with an air velocity $G$ passing upward containing influent $i$ and effluent $e$ concentrations of gas $p$. The downward water velocity $L$ contains $c$ amounts of gas. If no chemical reaction takes place, then a simple material balance would show[4]

Gas lost by water = gas gained by air

If the gas concentration is assumed to be very dilute so that the volumes of air and water present are not affected by the small transfer of gas, then the following is true:

$$L \, dc = G \, dp \tag{5.24}$$

where $L$ is the liquid velocity and $G$ is the gas velocity.

The liquid and air velocities can be expressed using several different units, and care must be exercised to assure consistency in the specific application. Velocity can be expressed as m/s, specific loading

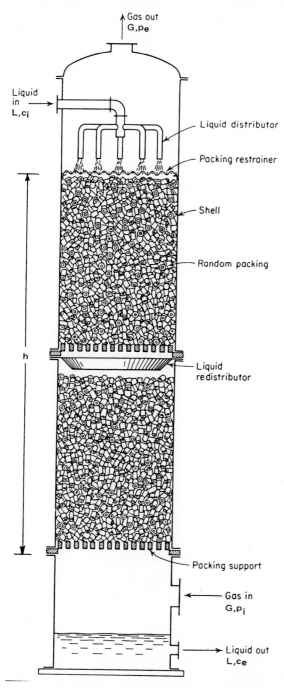

**Figure 5.7** Typical air stripping tower. (*Source: R. D. Treybal, Mass Transfer Operations, McGraw-Hill Publishers, New York, 1968, reprinted with permission.*)

rate $m^3/(m^2)(s)$, mass loading rate $kg/(m^2)(s)$, or molar loading rate $mol/(m^2)(s)$. Conversion between mass loading rate and velocity is accomplished using the density of water $r_L$, or of air $r_G$, which are found in Table 5.3. Conversion between molar loading rate and velocity can be done by the molar concentration of water (55.6 mol/L) or of air (1 mol air = $0.082T$ L of air).

The kinetic equation, Eq. (5.22), stated that at any point in the tower

$$\frac{dc}{dt} = K_L a(c_t - c_s) \tag{5.25}$$

or

$$dt = \frac{dc}{K_L a(c_t - c_s)} \tag{5.26}$$

and multiplying each side by $L$, the liquid velocity, yields

$$L\,dt = dh = \frac{L\,dc}{K_L a(c_t - c_s)} \tag{5.27}$$

where $dh$ is the differential height.

The term $(c_t - c_s)$ is the driving force DF of the reaction and is constantly changing with depth because $c_t$ is changing, resulting in the following:

$$h = \frac{L}{K_L a} \int_{c_1}^{c_2} \frac{dc}{d(c_t - c_s)} = \frac{L}{K_L a} \int_{c_1}^{c_2} \int_{DF} \frac{dc}{d(\text{DF})} \tag{5.28}$$

The integral of DF is the same as the log mean of the exit and entrance driving forces $DF_{lm}$, so the expression can be reduced to

$$h = \frac{L}{K_L a} \frac{(c_i - c_e)}{DF_{lm}} \tag{5.29}$$

TABLE 5.3    Properties of Air and Water

| | Water | | | Air | | |
|---|---|---|---|---|---|---|
| | 10°C | 20°C | 30°C | 10°C | 20°C | 30°C |
| Viscosity $u$, kg/(m)(s) | 0.00131 | 0.0010 | 0.0008 | $1.75 \times 10^{-5}$ | $1.82 \times 10^{-5}$ | $1.84 \times 10^{-5}$ |
| Density $r$, kg/m³ | 1000 | 998 | 996 | 1.25 | 1.20 | 1.16 |
| Surface tension $S$, n/m | 0.074 | 0.073 | 0.071 | | | |

where $c_i$ is the concentration of gas in the water at the influent to the tower, and $c_e$ is the concentration of gas in the water at the exit to the tower.

Again care must be exercised so that the units are consistent. The log mean of the driving force is given by

$$DF_{lm} = \frac{DF_e - DF_i}{\ln (DF_e/DF_i)} \qquad (5.30)$$

**Example Problem 1**    Chloroform at a concentration of 119 µg/L is to be reduced to 11.9 µg/L by an air stripping tower. What is the required height for the following conditions?

| | |
|---|---|
| $L$ | 73 m$^3$ water/(m$^2$ tower cross section)[hour (h)] |
| $G$ | 2200 m$^3$ air/(m$^2$ tower cross section)(h); (30:1 air-to-water volume ratio) |
| $T$ | 20°C = 293 K |
| $H_D$ | 2.55 × 10$^{-5}$ atm L/mg (Table 5.1) |
| $K_La$ | 30 h$^{-1}$ (given value) |

**Solution**    Assume $p_i = 0$ (that is, no chloroform exists in the air entering the tower) and calculate $p_e$ by the material balance equation, Eq. (5.24).

Because the molecular weight of chloroform is 119, $C_i = 1 \times 10^{-3}$ mol/m$^3$. Therefore,

$$L\, dc = G\, dp \qquad \text{[Eq. (5.24)]}$$

$$73(1 \times 10^{-3} - 1 \times 10^{-4}) = 2200(p_e - 0)$$

$$p_e = 2.99 \times 10^{-5} \frac{\text{mol gas}}{\text{m}^3 \text{ air}}$$

Because $0.082T$ L of air are in each mole of air,

$$p_e = 2.99 \times 10^{-5} \frac{\text{mol gas}}{\text{m}^3 \text{ air}} \frac{0.082(293) \times 10^{-3} \text{ m}^3 \text{ air}}{\text{mol air}}$$

$$= 7.2 \times 10^{-7} \frac{\text{mol gas}}{\text{mol air}}$$

The driving forces are then calculated.

| | Concentration in air $p$, mol gas/mol air | Concentration in water $c$, mg/L | $c_s = p/H_D$, mg/L | $DF = c - c_s$ |
|---|---|---|---|---|
| Exit (top) | 7.2 × 10$^{-7}$ | 0.119 | 0.0282 | 0.0908 |
| Entrance (bottom) | 0 | 0.0119 | 0 | 0.0119 |

The log mean of the driving force is found by Eq. (5.30)

$$DF_{lm} = \frac{0.0908 - 0.0119}{\ln \left( \dfrac{0.0908}{0.0119} \right)} = 0.039$$

and from Eq. (5.29)

$$h = \frac{73}{30} \frac{(0.119 - 0.0119)}{0.039} = 6.7 \ m$$

The above calculation procedures can be used in a variety of applications. For example, the entrance air concentration of the gas to be removed need not be zero. Henry's law can be nonlinear (concentrated solutions) and the procedures will still apply. Assuming that the solutions are dilute, that Henry's law applies, and that the solute concentration in the entrance air is zero, then the equations can be presented in a more simplified format. This was first developed by Colburn[5] in 1941, further refined by Treybal,[6] and reported on in the water supply field by Kavanaugh (see Ref. 2). In the simplification, the overall tower height $h$ (referred to as $Z$ in some references) is defined as the product of the height of one transfer unit (HTU) times the number of transfer units (NTU) required for the desired removal:

$$h = (\text{HTU}) \, (\text{NTU}) \tag{5.31}$$

Equation (5.31) is the same as Eq. (5.28) broken down into two defined components. The height of a transfer unit is the constant portion of Eq. (5.28):

$$\text{HTU} = \frac{L}{K_L a} \tag{5.32}$$

The number of transfer units is the integral portion of Eq. (5.28). If the simplifying assumptions stated above are applied, then the integral is

$$\text{NTU} = \frac{R}{R - 1} \ln \frac{(c_i/c_e)(R - 1) + 1}{R} \tag{5.33}$$

where $R$ is called the stripping factor defined as

$$R = \frac{H_u G}{L} \tag{5.34}$$

Any set of units can be used for $H$, $G$, and $L$ as long as $R$ is unitless. One of the factors that makes these relationships useful is that Treybal[6] has graphed Eq. (5.33) so that knowing the stripping factor and desired removal, NTU can be found quickly. Such a graph is shown in Fig. 5.8.

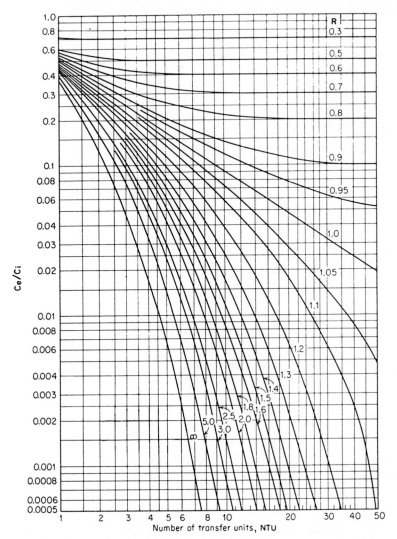

**Figure 5.8** Number of transfer units for absorbers or strippers with constant absorption or stripping factor. (*Source: R. D. Treybal, Mass Transfer Operations, McGraw-Hill Publishers, New York, 1968, reprinted with permission.*)

**Example Problem 2** Solve example problem 1 using the graph of Fig. 5.8.

| | |
|---|---|
| $c_i$ | 119 µg/L |
| $c_e$ | 11.9 µg/L |
| $L$ | 73 m³/(m²)(h) |
| $G$ | 2200 m³/(m²)(h) |
| $H_u$ | 0.127 (Table 5.1) |
| $K_L a$ | 30 h⁻¹ |

**Solution** From Eq. (5.32),

$$\text{HTU} = \frac{L}{K_L a} = \frac{73}{30} = 2.43 \text{ m}$$

From Eq. (5.34),

$$R = \frac{H_u G}{L} = \frac{0.127(2200)}{73} = 3.83$$

and the desired ratio of effluent to influent concentration is 0.10. Interpolating on Fig. 5.8 using $R = 3.8$ and $c_e/c_i = 0.1$, NTU is approximately 2.8. The total tower height is found from Eq. (5.31):

$$h = (\text{HTU})(\text{NTU})$$

$$= (2.43)(2.8) = 6.8 \text{ m}$$

The above calculations assume that the mass transfer coefficient $K_L a$ is known. Unfortunately, that is not usually the case. Various investigators have attempted to correlate mass transfer coefficients to physical and chemical properties of the gas of interest and physical properties of the packing material. Primarily two such methods have been utilized in the water treatment field. First are correlations by Sherwood and Holloway[7] that showed the liquid-phase mass transfer coefficient was dependent on the liquid velocity but independent of the gas velocity. Working with oxygen, hydrogen, and carbon dioxide and using Raschig rings and Berl saddles the following relationship was found:

$$K_L a = D_L m \left(\frac{L}{u_L}\right)^{1-n} \left(\frac{u_L}{r_L D_L}\right)^{0.5} \tag{5.35}$$

where $m$ and $n$ are constants reported by the investigator for the following units only, as $m$ and $n$ have units:

| | |
|---|---|
| $K_L a$ | $\text{h}^{-1}$ |
| $D_L$ | Diffusion coefficient for gas of interest in water, $\text{ft}^2/\text{h}$ |
| $L$ | Liquid velocity, lb water/(h)(ft$^2$) |
| $u_L$ | Liquid viscosity, lb/(ft)(h) |
| $r_L$ | Liquid density, lb/ft$^3$ |

The diffusion coefficient can be estimated from the Wilke-Change correlation[1]:

$$D_L = B\frac{T}{u_L} \tag{5.36}$$

where $D_L$ = diffusion coefficient of gas in water, $m^2/s$
$\quad\quad T$ = absolute temperature, K
$\quad\quad u_L$ = liquid viscosity, kg/(m)(s)
$\quad\quad B$ = conversion constant

For more information on $B$, or to find $B$ for other gases, consult Ref. 8. Values for $B$ are reported in Table 5.4.

A second model for predicting $K_L a$ that is reportedly useful in the water treatment field[9] was developed by Onda et al.[10] This model allows individual computation of $k_G$, $k_L$, and $a$. Its advantage is that it accounts for the gas-phase resistance and the liquid-phase resistance. The equations can be used with any consistent set of units, but the following are given as useful.

For liquid-phase resistance,

$$k_L \left( \frac{r_L}{u_L g} \right)^{1/3} = 0.0051 \left( \frac{L}{a_w u_L} \right)^{2/3} \left( \frac{u_L}{r_L D_L} \right)^{-0.5} (a_t d_p)^{0.4} \quad\quad (5.37)$$

and

$$\frac{a_w}{a_t} = 1 - \exp\left[ -1.45 \left( \frac{s_c}{s} \right)^{0.7} \left( \frac{L}{a_t u_L} \right)^{0.1} \left( \frac{L^2 a_t}{p_L^2 g} \right)^{-0.05} \left( \frac{L^2}{r_L s a_t} \right)^{0.2} \right]$$

$$(5.38)$$

**TABLE 5.4    Computation of Molecular Diffusion Coefficients**

| Compound | Formula | $B \times 10^{15}$ | $D_L{}^*$ at 20°C $\times 10^{10}$, $m^2/s$ | $D_G{}^\dagger$ at 20°C $\times 10^5$, $m^2/s$ |
|---|---|---|---|---|
| Vinyl chloride | $C_2H_3Cl$ | 3.85 | 11.3 | |
| Methane | $CH_4$ | 6.18 | 18.1 | |
| Carbon tetrachloride | $CCl_4$ | 2.76 | 8.08 | 0.81 |
| Tetrachloroethylene | $C_2Cl_4$ | 2.57 | 7.52 | 0.77 |
| Trichloroethylene | $CCHCl_3$ | 2.86 | 8.37 | 0.85 |
| Chloromethane | $CCH_3Cl$ | 4.49 | 13.1 | |
| 1,1,1-Trichloroethane | $C_2H_3Cl_3$ | 2.75 | 8.04 | 0.77 |
| Benzene | $C_6H_6$ | 3.04 | 8.91 | |
| Chloroform | $CHCl_3$ | 3.12 | 9.15 | 0.87 |
| 1,2-Dichloroethane | $C_2H_4Cl_2$ | 3.10 | 9.08 | |
| Bromoform | $CHBr_3$ | 2.99 | 8.75 | 0.75 |
| Carbon dioxide | $CO_2$ | — | 19.6 (25°C) | 1.38 |
| Hydrogen sulfide | $H_2S$ | — | 16.1 (25°C) | |
| Oxygen | $O_2$ | — | 20.3 | 2.19 |

$^*D_L$ from Ref. 2.
$^\dagger D_G$ from Refs. 8 and 9.

For gas-phase resistance,

$$\frac{k_G}{a_t D_G} = 5.23 \left(\frac{G}{a_t u_G}\right)^{0.7} \left(\frac{u_G}{r_G D_G}\right)^{1/3} (a_t d_p)^{-2.0} \qquad (5.39)$$

The overall mass transfer coefficient according to two-phase resistance theory is then found by multiplying Eq. (5.20) by $1/a$:

$$\frac{1}{K_L a} = \frac{1}{H_u k_G a} + \frac{1}{k_L a} \qquad (5.40)$$

Example units for the Onda correlations are

| | |
|---|---|
| $K_L a$ | Overall mass transfer, $s^{-1}$ |
| $k_L$ | Liquid-phase mass transfer, m/s |
| $k_G$ | Gas-phase mass transfer, m/s |
| $r_L$ | Liquid density, kg/m$^3$ |
| $r_G$ | Gas density, kg/m$^3$ |
| $u_L$ | Liquid viscosity, kg/(m)(s) |
| $u_G$ | Gas viscosity, kg/(m)(s) |
| $g$ | Gravity, m/s$^2$ |
| $L$ | Liquid velocity, kg/(m$^2$)(s) |
| $G$ | Gas velocity, kg/(m$^2$)(s) |
| $a_w$ | Wetted packing area, m$^2$/m$^3$ (used as equal to $a$) |
| $a_t$ | Total packing area, m$^2$/m$^3$ |
| $D_L$ | Liquid diffusivity, m$^2$/s |
| $D_G$ | Gas diffusivity, m$^2$/s |
| $d_p$ | Nominal packing diameter, m |
| $H_u$ | Henry's constant, unitless |
| $s_c$ | Critical surface tension of packing material, N/m |
| $s$ | Liquid surface tension, N/m = 0.073 at 20°C |

Selected values for $D_G$ and $D_L$ are shown in Table 5.4. Values for the critical surface tension of various packing materials are given in Table 5.5.

Neither the Onda nor the Sherwood-Holloway equations can be relied on to predict performance for a site-specific situation unless pilot plant verification is available or data are available from similar water quality conditions. Once one of the equations has been shown to apply to a site-specific application, it can be very useful for selecting packing height for various parameters such as temperature, liquid loading rate, and air-to-water ratio. For slightly soluble gases, the Onda equa-

**TABLE 5.5    Critical Surface Tension of Packing Materials[8]**

| Packing material | $s_c \times 10^3$, N/m |
|---|---|
| Carbon | 56 |
| Ceramic | 61 |
| Glass | 73 |
| Paraffin | 20 |
| Polyethylene | 33 |
| Polyvinyl chloride | 40 |
| Steel | 75 |

tion is more theoretically correct because it accounts for both liquid-and gas-phase resistance. Its disadvantage is that it does not have empirical constants that can be easily adjusted for site-specific situations. Therefore, either the model fits the data and can be used or it does not. On the other hand, the Sherwood-Holloway equation is semiempirical and can often be adjusted to "fit the data." Therefore, even though the equation may not be theoretically correct, if it describes the data throughout the design range, it can be very useful.

Figure 5.9 shows a graphical solution to adjusting the Sherwood-Holloway equation for a set of 1,1,1-trichloroethane removal data.[11] The Sherwood-Holloway equation is adjusted by finding the constants $m$ and $n$ for Eq. (5.35). Taking the logarithm of both sides of Eq. (5.35) results in the following:

$$\log (K_L a) - 0.5 \log \frac{D_L u_L}{r_L} = (1 - n) \log \frac{L}{u_L} + \log m \qquad (5.41)$$

Thus, a plot of the left-hand term against $\log (L/u_L)$ has a slope of $1 - n$ and $\log m$ as the intercept. Table 5.6 shows values for $m$ and $n$ as found in the original research and as found for actual water applications.

Figure 5.10 shows a comparison of actual data from Table 5.7 on chloroform removal to that predicted by the Onda equation. Also in-

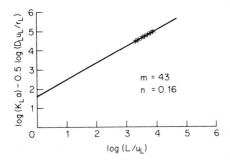

**Figure 5.9** Use of Sherwood-Holloway equation to approximate 1,1,1-trichloroethane removal using Tripak packing material.

**TABLE 5.6  Sherwood-Holloway Equation Coefficients for Determination of Mass Transfer Coefficient $K_L a$**

| Packing | Size | | Empirical constants | |
|---|---|---|---|---|
| | mm | in | $m$ | $n$ |
| Raschig rings[a] | 50 | 2.0 | 80 | 0.22 |
| Raschig rings[a] | 38 | 1.5 | 90 | 0.22 |
| Raschig rings[a] | 25 | 1.0 | 100 | 0.22 |
| Berl saddles[a] | 38 | 1.5 | 160 | 0.28 |
| Berl saddles[a] | 25 | 1.0 | 170 | 0.28 |
| InTalox saddles[b] | 25 | 1.0 | 63 | 0.28 |
| Berl saddle[c] | 25 | 1.0 | 12 | 0.25 |
| Glitsch saddle[c] | 25 | 1.0 | 27 | 0.22 |
| Tripak[d] | 50 | 2.0 | 41 | 0.14 |
| Tripak[e] | 50 | 2.0 | 66 | 0.28 |
| Tripak[f] | 50 | 2.0 | 43 | 0.16 |

[a]From Sherwood and Holloway[7] based on desorption of hydrogen, oxygen, and carbon dioxide.
[b]Based on desorption of THMs (all four groups).[11]
[c]Based on desorption of chloroform.[11]
[d]Based on desorption of 1,1,1-trichloroethane.[11]
[e]Based on desorption of trichloroethylene.[1]
[f]Based on desorption of 1,1-dichloroethane.[11]

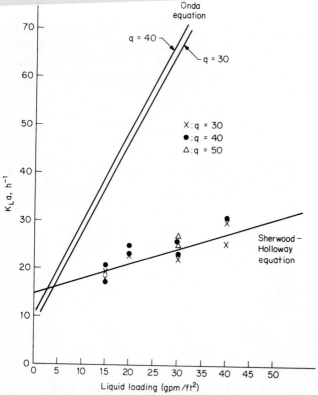

**Figure 5.10**  Comparison of chloroform removal (from Table 5.7) using Tripak media at 70°F water temperature to that predicted by the Onda equation and an adjusted Sherwood-Holloway equation.

**TABLE 5.7    Chloroform Removal Data from City of Chesapeake, Virginia, Pilot Tower**

| Tower height, ft | $L$, gpm/ft$^2$ | $K_L a$, h$^{-1}$ |
|:---:|:---:|:---:|
| Air-to-water ratio = 30:1 | | |
| 9.7 | 15 | 17.9 |
| 9.7 | 20 | 22.7 |
| 30 | 40 | 25.6 |
| 30 | 30 | 22.43 |
| 30 | 40 | 30.0 |
| Air-to-water ratio = 40:1 | | |
| 9.7 | 15 | 20.8 |
| 9.7 | 20 | 23.2 |
| 9.7 | 15 | 17.5 |
| 30 | 30 | 25.8 |
| 30 | 40 | 30.8 |
| 30 | 30 | 23.3 |
| Air-to-water ratio = 50:1 | | |
| 30 | 30 | 18.15 |
| 30 | 30 | 21.4 |
| 30 | 30 | 34.9 |
| 30 | 30 | 27.6 |

cluded is a plot of the Sherwood-Holloway equation adjusted in the above manner to fit the data using $m = 1175$ and $n = 0.67$. The Sherwood-Holloway equation is independent of air-to-water ratio, and therefore it does not account for a difference in $K_L a$ because of a change in air-to-water ratio. On the other hand, the equation was adjusted to reasonably fit the data. In this example, the theoretical Onda equation tended to overestimate the mass transfer coefficient at higher liquid loading rates, and therefore a design based on that equation would have resulted in too short of an air stripping tower.

### Diffused aeration

Diffused aeration is a term used in the environmental field to generally refer to the process of bringing air bubbles in contact with a volume of water. Air is compressed to the differential head required and then released at near the bottom elevation of the water volume through bubble diffusers. The purpose of the diffusers is to distribute the air uniformly through the water cross section and to produce the desired air bubble size. Diffused aeration has not found widespread application in the water treatment field. It has been used for aerating source water supply reservoirs and has found some use in VOC removal. Diffused aeration has a much higher power cost for VOC removal than do air stripping towers. For this reason, it is generally only considered when the diffused aeration process can take place in existing tanks, such as clearwells, thereby eliminating much of the capital cost associated with towers. If only seasonal aeration is needed, such as for summer THM removal, economics may also favor the use of diffused aeration in an existing clearwell.

The basic rate equation as developed for the two-resistance (film) theory, Eq. (5.22), also applies to diffused aeration:

$$\frac{dc}{dt} = -K_L a \, (c_s - c_t) \tag{5.42}$$

In the case of oxygen transfer, the oxygen content of the bubble is often assumed not to change as it rises through the tanks. This is fairly reasonable because the oxygen content of air is about 21 percent and only a small amount of this is transferred to the water. Because the partial pressure of oxygen in the air bubble is constant, $c_s$ is constant and Eq. (5.42) can be solved by simple integration.

In the removal of compounds from water, such as VOCs or $H_2S$ by diffused aeration, $c_s$ is, however, not constant. The partial pressure of the gas of concern generally starts at 0 and increases as the bubble rises. Therefore, $c_s$ is constantly changing. The relationship can be solved by setting the change in concentration in the air equal to the change in concentration in the water which from the two-resistance theory must be true. The total mass of gas loss from the water volume must equal the total mass transfer into the air volume. On the air side, the instantaneous change in gas concentration is[12]

| Concentration rate of change in the bubble | differential reactor volume | gas volume / water volume | = |
|---|---|---|---|
| $\dfrac{dp}{dt}$ | $A \, dz$ | $\dfrac{Q_G t_B}{V}$ | (5.43) |

where $Q_G$ = volumetric rate of gas flow
$t_B$ = detention time of bubble in reactor (bubble rise time)
$V$ = reactor volume

The instantaneous change of gas concentration in the air phase equals on the water side the instantaneous change of gas concentration in the differential reactor volume:

$$-K_L a(c_s - c_t)A \, dz \tag{5.44}$$

Substituting $p = H_u c_s$ in Eq. (5.43) and setting Eqs. (5.43) and (5.44) equal to each other results in

$$\frac{Q_G t_B}{V} \frac{H_u dc_s}{dt} = -K_L a(c_s - c_t) \tag{5.45}$$

Equation (5.45) is integrated over the range of the inlet and exit $c_s$ values. In most cases, $c_s$ at the inlet is zero because the inlet air does not contain any of the gas to be removed. $c_s$ at the exit $(c_{s(e)})$ can be found by

$$p_e = H_u c_{s(e)} \tag{5.46}$$

where

$$p_e = \frac{c_e - c_i}{q} \tag{5.47}$$

and $q$ is the volumetric air-to-water flow ratio. By combining Eqs. (5.46) and (5.47)

$$c_{s(e)} = \frac{c_e - c_i}{qH_u} \tag{5.48}$$

Equation (5.45) is integrated to result in the following for gas leaving water and entering the air.[13]

$$F = Q_G H_u c_e \left(1 - \exp\frac{-K_L a\, V}{H_u Q_G}\right) \tag{5.49}$$

where $F$ = mass transfer rate (mass transferred per time)
$Q_G$ = volumetric flow rate of air (e.g., ft$^3$/min, m$^3$/s)
$V$ = reactor volume

If the liquid volume in the tank is considered completely mixed and the air rises as a plug flow, the resulting expression is[12]

$$\frac{c_e}{c_i} = \frac{1}{1 + qH_u[1 - \exp(-K_L a V/H_u Q_G)]} \tag{5.50}$$

where $q$ is the air-to-water feed ratio $Q_G/Q_L$.

Until the term $K_L a V/H_u Q_G$ is less than about 4, the exponent term becomes essentially zero. When the exponent term is zero, the air and water have reached an equilibrium condition and the driving force has dropped to zero at some point within the tank. Therefore, the tank volume is not being fully used and the air-to-water ratio could be increased to obtain more removal. On the other hand, if a new tank were being designed, its volume could be made smaller without affecting the removal.

**Example Problem 3** Evaluate the same example considered for the packed tower where a 25.3-ft-high tower was found to provide 90 percent removal of chloroform at the stated conditions. Assume removal is to take place in a clearwell with a 1-h detention time at a 1-mgd flow (92 ft$^3$/min) with the following given:

$c_i$     119 μg/L
$K_La$   30 h$^{-1}$
$q$      30
$H_u$    0.127

**Solution** First find $Q_G = 30Q_L = 2760$ ft$^3$/min.
Then the term raised to the $e$ power of Eq. (5.50) is

$$\frac{K_La(V)}{H_uQ_G} = \frac{30(92)(60)}{0.127(2760)(60)} = 7.9$$

and therefore from Eq. (5.50)

$$c_e = \frac{c_i}{1 + qH_u[1 - \exp(-K_LaV/H_uQ_G)]}$$

$$= \frac{119}{1 + 30(0.127)[1 - \exp(-7.9)]}$$

$$= \frac{119}{1 + 3.81(1 - 0.0004)}$$

$$= 24.7 \text{ μg/L or 79 percent removal}$$

Note that it would be possible to reduce the volume of this tank by about a factor of 2 (assuming $K_La$ did not change as discussed in the next section) without significantly increasing the effluent concentration. If the volume were reduced by a factor of 10, however, and $K_La$ remained unchanged,

$$c_e = \frac{c_i}{1 + qH_u\{1 - \exp[-30(92)(6)/0.127(2760)(60)]\}}$$

$$= 38.7 \text{ μg/L or 67.5\% removal}$$

In the original 1-h detention time tank, the required $q$ to achieve the desired 90 percent removal can be determined:

$$\frac{c_e}{c_i} = 0.1 = \frac{1}{1 + q(0.127)}$$

$$q = 71$$

Therefore, the gas flow must be 6532 ft$^3$/s. Note that providing this much air may be impractical.

In the above diffused aeration example, $K_L a$ was assumed to be 30 $h^{-1}$. In reality, this value must either be experimentally determined or developed from generalized empirical equations. Matter-Muller et al.[13] have presented methods for experimentally finding $K_L a$ values in batch-diffused air reactors.

Calderbrook and Moo-Young[14] have presented an empirical correlation for free rise of single air bubbles or bubble swarms rising through liquids. For bubble diameters less than 1 mm,

$$\frac{K_L}{D_L} = \frac{2.0}{d} + 0.31\left[\frac{(r_L - r_G)g}{u_L D_L}\right]^{1/3} \tag{5.51}$$

and for bubble diameters greater than 2.5 mm,

$$\frac{K_L}{D_L} = 0.42\left(\frac{u_L}{r_L D_L}\right)^{1/2}\left[\frac{r_L(r_L - r_G)g}{u_L^2}\right]^{1/3} \tag{5.52}$$

where $d$ = bubble diameter, cm
$\quad D_L$ = diffusivity of gas in liquid, $cm^2/s$ (Table 5.4, after changing units)
$\quad u_L$ = liquid viscosity, $g/(cm)(s)$
$\quad r_L$ = liquid density, $g/cm^3$
$\quad r_G$ = air density, $g/cm^3$
$\quad K_L$ = cm/s

Their data indicate that for $d$ between 1 and 2.5 mm, linear extrapolation should be made between the two equations for the given bubble diameter. Any consistent set of units may be used, with an example as shown.

The specific interfacial area must be found for the air bubble diameter of interest. In the case of freely rising air bubbles through water, the specific interfacial area can be theoretically calculated based on the bubble diameter ($d$). This is found as

$$a = \frac{(\text{number of bubbles})(\text{area per bubble})}{\text{tank volume}}$$

The number of bubbles in the tank is the amount of air in the tank divided by the volume of one bubble:

$$\text{Number of bubbles} = \frac{Q_G t}{\pi d^3/6} \tag{5.53}$$

where $t$ is the detention time of air in the tank. Therefore,

$$a = \frac{Q_G t}{V\pi d^3/6}(\pi d^2) \tag{5.54}$$

$$a = \frac{Q_G t}{V} \frac{6}{d} \tag{5.55}$$

An alternative to the above expression can be found by recalling $q = Q_G/Q_L$ and $T_d = V/Q_L$:

$$a = q\frac{t}{T_d} \frac{6}{d} \tag{5.56}$$

The quantity $qt/T_d$ (or $Q_G t/V$) is called the gas holdup. In the absence of data for the gas holdup, the term is found by setting the bubble detention time equal to the bubble rise time integrated over the height of the tanks:

$$t = t_B = \int \frac{d_z}{v} = \frac{h}{v} \tag{5.57}$$

The rise velocity of the air bubble $v$ is estimated by the terminal rise velocity for isolated bubbles in the same manner that the velocity of a falling particle is calculated:

$$V = \left[\left(\frac{4g}{3C_D}\right)\left(\frac{r_L - r_g}{r_L}\right)d\right]^{1/2} \tag{5.58}$$

The coefficient of drag $C_D$ is a function of Reynolds number $R_n$.

$$C_D = \begin{cases} 24/R_n & \text{for } R_n < 1 \\ 18.5/R_n^{0.66} & \text{for } 1 < R_n < 1000 \\ 0.44 & \text{for } R_n > 1000 \end{cases}$$

$$R_n = \frac{v\, dr_L}{u_L}$$

For laminar flow, $R_n < 1$ which corresponds to $d < 0.013$ cm, the equation reduces to Stokes law:

$$v = \frac{g}{18u_L}(r_L - r_G)d^2 \tag{5.59}$$

and for turbulent flow, $R_n > 1000$ which corresponds to $d > 0.31$ cm,

$$v = 1.82\left[\frac{(r_L - r_G)dg}{r_L}\right]^{1/2} \tag{5.60}$$

Fine bubble diffusers produce bubbles with diameters in the range of 0.2 cm, so they fall in the transition zone:

$$v = [0.072g(r_L - r_G)d^{1.6}r_L{}^{-0.4}u_L{}^{-0.6}]^{0.714} \qquad (5.61)$$

The density of air for the above equation is approximately 0.0012 g/cm$^3$ (Table 5.3).

**Example Problem 4** Predict the overall mass transfer coefficient at 20°C for chloroform in a 15-ft-deep clearwell of 1 h detention time using diffused aerators producing 0.2-cm-diameter bubbles. As in the previous example, use $q = 30$, $Q_L = 92$ ft$^3$/min, and $Q_G = 2760$ ft$^3$/min.

**Solution** First find the rise velocity of the bubble. Because $0.013 < d < 0.31$, the transition range applies:

$$v = [0.072(981)(0.2)^{1.6}(0.998 - 0.0012)(0.998)^{-0.4}(0.01)^{-0.6}]^{0.714}$$

$$= 23.9 \text{ cm/s}$$

The bubble residence time, assumed equal to the free rise time, is [Eq. (5.57)]

$$t = \frac{h}{v} = \frac{15(30.48)}{23.9} = 19.1 \text{ s}$$

The interfacial area is found from Eq. (5.56):

$$a = q\frac{t}{T_d}\frac{6}{d} = \frac{30(19.1/60)}{60}\frac{6}{0.2}$$

$$= 4.8 \text{ cm}^{-1}$$

In order to find $K_L$, interpolation should be made between Eqs. (5.51) and (5.52). The diffusivity at 20°C from Table 5.4 is $9.15 \times 10^{-6}$ cm$^2$/s.

First estimate of $K_L$:

$$\frac{K_L}{D_L} = \frac{2.0}{d} + 0.31\left[\frac{(r_L - r_G)g}{u_L D_L}\right]^{1/3}$$

$$\frac{K_L}{9.15 \times 10^{-6}} = \frac{2.0}{0.2} + 0.31\left[\frac{(0.998 - 0.0012)(981)}{(0.01)(9.15 \times 10^{-6})}\right]^{1/3}$$

$$K_L = 0.006 \text{ cm/s}$$

Second estimate of $K_L$:

$$\frac{K_L}{D_L} = 0.42\left(\frac{u_L}{r_L D_L}\right)^{1/2}\left[\frac{r_L(r_L - r_G)g}{u_L{}^2}\right]^{1/3}$$

$$\frac{K_L}{9.15 \times 10^{-6}} = 0.42\left[\frac{0.01}{(0.998)(9.15 \times 10^{-6})}\right]^{1/2}\left[\frac{(0.998)(0.998 - 0.0012)981}{(0.01)^2}\right]^{1/3}$$

$$K_L = 0.026 \text{ cm/s}$$

So the interpolated $K_L$ is about 0.016 cm/s, and the resulting $K_L a$ is

$$K_L a = 0.016(4.8) = 0.077 \text{ s}^{-1}$$

Substitution of the interfacial area term into the integrated mass transfer expression of Eq. (5.50) and replacement of $t$ with $h/V$ results in

$$\frac{c_e}{c_i} = \frac{1}{1 + qH_u \left[1 - \exp\left(-K_L h 6 / v H_u d\right)\right]} \tag{5.62}$$

Theoretically for fine bubble diffusers, the bubble diameters are in the range of 0.2 cm. In this case, for nearly any practical basin size the system reaches equilibrium and removal is given simply by

$$\frac{c_e}{c_i} = \frac{1}{1 + qH_u} \tag{5.63}$$

In practice, fine bubble diffusers do not reach the ideal case.

Coarse bubble diffusers may typically have bubble diameters in the range of 2.5 cm. In this case, equilibrium is not reached in practically sized basins. Figure 5.11 shows an example plot of percent removal of chloroform for coarse bubble diffusers ($d = 2.5$ cm) as a function of tank depth and air-to-water ratio based on Eq. (5.62). At about 15 ft deep the coarse bubble diffuser reaches equilibrium and additional height does not help removal. Note that the coarse bubble diffuser asymptotically reaches the removal to be expected from a fine bubble diffuser for the same air-to-water ratio. In both cases, the equations

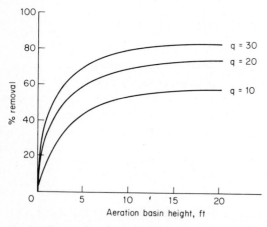

**Figure 5.11** Theoretical chloroform removal using coarse bubble diffused aeration.

predict that removal is independent of liquid detention time, as has been shown in experimental data by Bilello and Singley.[15]

### Spray nozzles and tray aerators

Spray nozzles and tray aerators are air-water contact devices that have been used for many years in the water treatment field. Applications include the addition of oxygen to water for the purpose of iron and manganese oxidation, the removal of carbon dioxide from water, and the removal of hydrogen sulfide from water. To a lesser extent, these contactors have been used for removal of taste- and odor-causing organic compounds from water.

Pressurized spray nozzles used in the water industry are either hollow cone nozzles or full cone nozzles. Full cone nozzles deliver a uniform area coverage in a round, square, or rectangular pattern (usually round is used). Hollow cone nozzles provide a rim of spray concentrated in the circumference and nearly void of spray in the center. Generally, hollow cone nozzles will produce smaller droplets but require more pressure per flow through the nozzle. Spray nozzles in the water treatment field have generally been designed as fountain-type aerators, spraying into the open atmosphere, often into a raw water reservoir. Mass transfer rate again depends upon $K_L$ and $a$. Manufacturers will occasionally have mass transfer data available for oxygen but seldom for other compounds of interest in water treatment.

Calderbrook and Moo-Young[14] have reported that their correlation for $K_L$ for air bubbles [Eqs. (5.51) and (5.52)] applies equally as well for water drops. Using their relationships, and for an open atmosphere spray fountain, the specific interfacial area is

$$a = \frac{6}{d} \tag{5.64}$$

The diameter of water droplets produced is a major controlling factor in the mass transfer rate. Manufacturers have data available on the droplet diameter for their different nozzle designs. Nozzles that produce drops with diameters from 2 to 10,000 $\mu$m can be selected. Droplet size data are reported in various ways, four of which are

*Arithmetic mean:*   The arithmetic mean is a simple weighted average based on the diameter of all the individual droplets in the spray sample.

*Surface mean:*   The surface mean is the diameter of a droplet whose surface area, if multiplied by the total number of droplets, will equal the total surface of all the particles in the spray sample.

*Volume mean:*   Volume mean is the diameter of a droplet whose volume, if multiplied by the number of droplets, will equal the total volume of the sample.

*Sauter mean:*   The sauter mean, also correctly called "volume surface mean," is the diameter of a droplet whose ratio of volume to surface area is equal to that of the entire spray sample.

The sauter mean diameter should be used in mass transfer calculations because it is based on a volume to surface area as is the interfacial area term.

In an open atmosphere situation $c_s$ remains essentially constant because of an infinite air-to-water ratio. Mass transfer is therefore given by the integrated Lewis-Whitman equation:

$$c_e - c_i = (c_s - c_i)[1 - \exp(-K_L at)] \tag{5.65}$$

In this case, $t$ is the time of contact between the water drops and the atmosphere:

$$t = \frac{2v_d \sin \alpha}{g} \tag{5.66}$$

where $\alpha$ is the angle of spray measured from the horizontal, and $v_d$ is the exit velocity of the drop from the nozzle:

$$v_d = C_v (2gh)^{1/2} \tag{5.67}$$

The velocity coefficient of the nozzle $C_v$ varies greatly (from 0.4 to 0.95), so that velocity or $C_v$ must be obtained directly from the manufacturer.

**Example Problem 5**   Find the transfer of chloroform from the previous example for a nozzle with a spray angle of 45° and an operating pressure of 40 pounds per square inch (psi). The sauter mean diameter is obtained from the manufacturer as 0.1 cm and $C_v = 0.45$.

**Solution**   From the last example, $K_L$ is 0.006 cm/s. From Eq. (5.64):

$$a = \frac{6}{d} = \frac{6}{0.1} = 60 \text{ cm}^{-1}$$

and from Eq. (5.67):

$$v_d = C_v(2gh)^{1/2}$$

$$= 0.45 \, [2(981)(40)(70.3)]^{1/2} = 1057 \text{ cm/s}$$

[Note: psi (70.3) = cm of water head.]

The time of contact is found from Eq. (5.66):

$$t = \frac{2v_d \sin \alpha}{g}$$

Therefore, from Eq. (5.65):

$$= \frac{2(1057)(0.71)}{981} = 1.52 \text{ s}$$

$$c_e - c_i = (c_s - c_i)\,[1 - \exp\,(-K_L at)]$$

$$c_e - 119 = (0 - 119)\{1 - \exp\,[-0.006(60)(1.52)]\}$$

$$c_e = 68.8 \ \mu g/L = 42 \text{ percent removal}$$

For a very high expenditure of energy (pressurizing all the water to 40 psi) only 42 percent removal is achieved. If free carbon dioxide was removed in the above system ($D_L = 2 \times 10^{-5}$), $K_L$ would be calculated as 0.01 resulting in an overall removal of 60 percent. This is in line with full-scale systems that remove 50 to 70 percent of the carbon dioxide. The tradeoffs between pressure, flow, and droplet diameter should all be considered in selecting the optimal nozzle.

Cascading tray aerators, such as that shown in Fig. 5.12, have also been used to aerate source waters. Mass transfer takes place during surface aeration within the tray and in falling laminar sheets. Mass transfer data on these types of aerators are not available in the literature. Design must be based on experiences of similar utilities and manufacturers' recommendations.

The applications of spray nozzles and tray aerators are essentially limited to $CO_2$ and $H_2S$ removal and oxygen addition where efficiencies of 50 to 80 percent are acceptable (although some systems are op-

**Figure 5.12**   Tray aerator. (*Courtesy of Crom Corporation.*)

erated for VOC removal). Both $CO_2$ and $H_2S$ are reacting gases. Stripping can only remove that portion of the gas that is in the gaseous state. Hydrogen sulfide reacts with water as follows:

[Eq.(5.1)]    $H_2S + H_2O = H_3O^+ + HS^-$        $K_1 = 1 \times 10^{-7}(25°C)$

[Eq.(5.2)]    $HS^- + H_2O = H_3O^+ + S^{2-}$        $K_2 = 1 \times 10^{-15}(25°C)$

The amount of total sulfide present as $H_2S$ can be calculated by

$$\%H_2S = 100\frac{[H^+]^2}{[H^+]^2 + K_1[H^+] + K_1K_2} \tag{5.68}$$

For example, at pH 8 only 10 percent of the total sulfide present is $H_2S$. Therefore, the most that can be removed by aeration is 10 percent if the aerator is 100 percent efficient. At a pH of approximately 7, 50 percent of the $H_2S$ is removable. In using the equilibrium equations, the amount of sulfide present as $H_2S$ at the pH prior to stripping should be found. The percent $H_2S$ remaining after stripping can be calculated and added to the original sulfide present as $HS^-$ and $S^{2-}$ to find the total sulfide remaining. Any reformation of $HS^-$ and $S^{2-}$ to $H_2S$ during stripping is negligible because the pH is increasing during removal thereby preventing back-dissociation. In actual practice, removal of $H_2S$ may not be as efficient as predicted above because most water high in $H_2S$ also contains $CO_2$. Stripping for $H_2S$ removal will also remove $CO_2$. The removal of $CO_2$ will tend to increase the pH and cause some dissociation of $H_2S$ to $HS^-$.

Some of the remaining sulfide may be oxidized by oxygen added to the water during the stripping process:

$$2H_2S + O_2 = 2H_2O + 2S(s) \tag{5.69}$$

This sulfur may be further oxidized to sulfate, however, and then reduced back to sulfide in the distribution system.

Another application for the transfer of oxygen into water by relatively simple contact devices is for the oxidation of iron and manganese. In the presence of dissolved oxygen, soluble $Fe^{2+}$ and $Mn^{2+}$ should be oxidized to their respective insoluble state:

$$2Fe^{2+} + \tfrac{1}{2}O_2 + 5H_2O \rightarrow 2Fe(OH)_3(s) + 4H^+ \tag{5.70}$$

$$Mn^{2+} + \tfrac{1}{2}O_2 + H_2O \rightarrow MnO_2(s) + 2H^+ \tag{5.71}$$

Theoretically, for the oxidation of iron, 1 mg of oxygen will oxidize

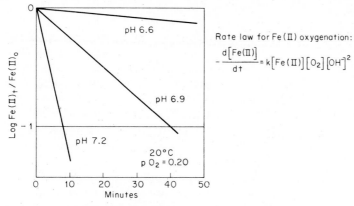

**Figure 5.13** Oxygenation of Fe(II).

7 mg of iron, and for manganese, 1 mg of oxygen will oxidize 1.5 mg of manganese. Therefore, the oxygen demands and correspondingly the oxygen gas transfer requirements are very small. The equipment need not be sophisticated to accomplish the necessary oxygen transfer.

Unfortunately, the rate of reaction between the oxygen and iron or manganese is slow (discussed in Chap. 10). Figure 5.13 shows the reaction time for iron for different pH values. In practice, reaction times tend to be even slower than indicated. Typical holding times following aeration for iron removal are 1 h.

Figure 5.14 shows the same relationship for manganese. The reaction is very slow, and in fact impractical, below pH 9.5. Generally a stronger oxidant than oxygen is required for manganese removal.

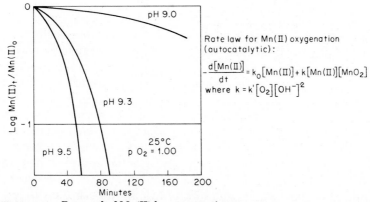

**Figure 5.14** Removal of Mn(II) by oxygenation.

## Off-Gas Treatment

Generally, the most cost-effective method for VOC removal is by air stripping. Concern has been expressed, however, about the resulting VOC air pollution from the off-gas associated with air stripping operations. When VOC removal is via a natural draft aerator or spray nozzle, the volume of air and, hence, the VOC concentration in the air is essentially impossible to calculate. Only the total mass of VOC discharged per time can be determined. For air stripping towers the air flow rate is known and the VOC concentration in the off-gas can be calculated from the material balance equation, Eq. (5.24).

One method for removing VOCs from off-gas is by adsorption of the VOCs in off-gas onto GAC. Figure 5.15 shows a treatment scheme to remove VOCs from air stripping off-gas using a gas-phase GAC adsorption bed followed by steam regeneration of the exhausted GAC, reported on in detail by Crittenden et al.[16] Crittenden et al. found the gas-phase adsorption of VOC to be a much more efficient process than direct water-phase adsorption, as long as the gas phase was heated to remove humidity. The carbon usage rate for gas-phase adsorption is less than half that of water-phase adsorption. For most VOCs the critical mass transfer zone length (discussed in Chap. 13) is less than 30 cm and cross-sectional loading rates can be very high.

Regenerating carbon on site appears to be more economical than carbon replacement, although more work is needed in this area. The steam requirement is in the range of 15 to 20 kg steam/kg GAC. Crittenden et al. found a decrease in adsorption capacity, however, to 60 percent of original over three regeneration cycles.[16] It was speculated that this was caused by a buildup of some compounds and that a higher steam temperature would alleviate the problem.

## Emerging Technology

Air stripping for the removal of VOCs is a well-established technology in the water treatment field. New areas, however, appear to primarily center on the treatment of the off-gases as discussed in the previous section, and in new, efficient technologies to accomplish air-liquid contact. For example, Roberts and Levy[17] have presented information on the energy advantages of using mechanical surface aerators in existing basins rather than placing diffused aerators in the basins or building packed towers. They concluded that mechanical surface aerators can compete very well with packed towers based on energy consumption. Therefore, the economics of construction of a basin, if necessary, and aerators, versus a column and blowers would determine the lowest cost alternative. If off-gas treatment is necessary, the surface aer-

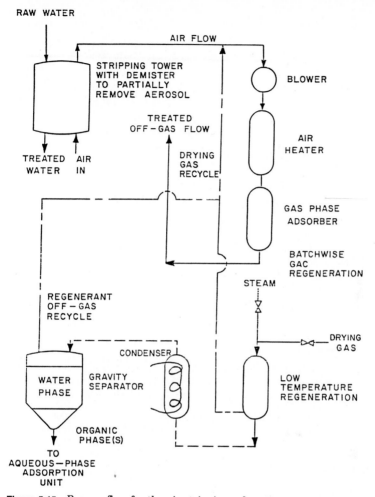

RAW WATER

AIR FLOW

STRIPPING TOWER
WITH DEMISTER
TO PARTIALLY
REMOVE AEROSOL

BLOWER

TREATED
OFF-GAS FLOW

AIR
HEATER

TREATED   AIR
WATER    IN

DRYING
GAS
RECYCLE

GAS PHASE
ADSORBER

BATCHWISE
GAC
REGENERATION

REGENERANT
OFF-GAS
RECYCLE

STEAM

DRYING
GAS

CONDENSER

WATER
PHASE

GRAVITY
SEPARATOR

LOW
TEMPERATURE
REGENERATION

ORGANIC
PHASE(S)

TO
AQUEOUS-PHASE
ADSORPTION
UNIT

**Figure 5.15**   Process flow for the air stripping solvent recovery process.

ation system would have more difficulty achieving the collection of the
off-gas for subsequent treatment.

Research is being conducted on the use of cascade air stripping.[18] In
a cascade air stripping system the countercurrent packed tower is divided into two or more stages. For example, in a two-stage system,
halfway up the tower the air injected at the bottom of the tower would
be wasted and fresh air would be added to the upper half of the tower.
Because mass transfer, as shown in Eq. (5.28), is a function of the
driving force, the addition of fresh air should increase the driving
force and reduce the overall tower height requirements. This process
may prove particularly useful for removal of less volatile compounds.

Methods are also being developed for in-well aeration of groundwater containing VOCs.[19] In-well aeration has been evaluated using airlift pumps. In this situation air is introduced into the well through an open pipe called the eductor. Because aerated water in a pipe is less dense than the surrounding water, water flows up and out the eductor. This accomplishes both pumping and aeration for VOC removal. The water leaving the well is placed in a holding tank for air-water separation prior to being pumped to the distribution system. In-well aeration can also be accomplished by placing a diffuser in the well and using a standard submersible pump. Ideally, the pump is located below the diffuser, both of which are below the screened opening of the well in order to create a countercurrent air-water flow pattern.

## References

1. J. E. Singley, personal communication.
2. J. M. Montgomery, Consulting Engineers, Inc., *Water Treatment Principles and Design*, Wiley-Interscience, New York, 1985.
3. W. K. Lewis and W. G. Whitman, "Principals of Gas Absorption," *Industrial Eng. Chem.*, vol. 16, 1924, p. 1215.
4. O. Levenspiel, *Chemical Reaction Engineering*, John Wiley & Sons, New York, 1972.
5. A. P. Colburn, *Industrial Eng. Chem.*, vol. 33, 1941, p. 111.
6. R. D. Treybal, *Mass-Transfer Operations*, McGraw-Hill, New York, 1968.
7. T. K. Sherwood and A. L. Holloway, "Performance of Packed Towers—Liquid Film Data for Several Packing," *Trans. AIChE*, vol. 36, 1940, p. 39.
8. R. H. Perry and C. H. Chilton, *Chemical Engineer's Handbook*, McGraw-Hill, New York, 1973.
9. P. V. Roberts et al., "Evaluating Two-Resistance Models for Air Stripping of Volatile Organic Contaminants in a Countercurrent Packed Column," *Environ. Science Tech.*, vol. 19, 1985, p. 164.
10. K. Onda et al., *Journal Chem. Eng. Japan*, vol. 1, 1968, p. 56.
11. M. Bishop and D. Cornwell, "Data Evaluation and Design Procedures for Air Stripping of Trihalomethane and Other Volatile Organic Compounds," *Proc. AWWA Annual Conf.*, Dallas, Texas, June 1984.
12. J. E. Bailey and D. F. Ollis, *Biochemical Engineering Fundamentals*, McGraw-Hill, New York, 1977.
13. C. Matter-Muller et al., "Transfer of Volatile Substances from Water to the Atmosphere," *Water Research*, vol. 15, 1981, p. 1271.
14. P. H. Calderbrook and M. B. Moo-Young, "The Continuous Phase Heat and Mass Transfer Properties of Dispersions," *Chem. Eng. Science*, vol. 16, 1961, p. 39.
15. L. J. Bilello and J. E. Singley, "Removal of Trihalomethanes by Packed-Column and Diffused Aeration," *Journal AWWA*, vol. 78, no. 2, February 1986, p. 62.
16. J. E. Crittenden et al., *An Evaluation of the Technical and Economic Feasibility of the Air Stripping Solvent Recovery Process*, AWWA Research Foundation, Denver, Colo., 1987.
17. P. V. Roberts and J. A. Levy, "Energy Requirements for Air Stripping Trihalomethanes," *Journal AWWA*, vol. 77, no. 4, April 1985, p. 138.
18. N. Nirmalakhandan et al., "Enhanced Removals of Semivolatile Organic Contaminants from Water by Cascade Air Stripping," *Proc. AWWA Annual Conf.*, Orlando, Fla., June 1988.
19. J. A. Coyle et al., *Control of Volatile Organic Contaminants in Groundwater by In-Well Aeration*, USEPA, EPA-600/S2-88/020, April 1988.

# 6

# Coagulation Processes: Destabilization, Mixing, and Flocculation

**Appiah Amirtharajah, Ph.D., P.E.**

*Professor, School of Civil Engineering*
*Georgia Institute of Technology*
*Atlanta, Georgia*

**Charles R. O'Melia, Ph.D., P.E.**

*Professor, Department of Geography & Environmental Engineering*
*The Johns Hopkins University*
*Baltimore, Maryland*

Coagulation is a process for combining small particles into larger aggregates. It is an essential component of accepted water treatment practice in which coagulation, sedimentation, and filtration processes are combined in series to remove particles from water. This useful perspective is, however, incomplete in two important ways. First, many pollutants of concern to human health are particles themselves (e.g., pathogenic organisms) or are associated with solid particles (e.g., certain toxic metals and SOCs), so coagulation for particle aggregation is an important component of treatment for the removal of many health-related pollutants. Second, humic substances react with most coagulants. This has long been the basis for the removal of color by aluminum and iron (III) salts. These organic substances are present in all surface waters and in many groundwaters. They complex metals and may also associate with some SOCs. Humic substances have been shown to be precursors in the formation of THMs when chlorine is used for disinfection and oxidation. They adsorb on natural particles and act as stabilizing agents in surface water.

For these reasons, considerable attention has come to be directed at the removal of humic substances by coagulation in water treatment, even when color removal is not a primary objective.

Particles in natural water vary widely in origin, concentration, and size. Some are constituents of land-based or atmospheric sources (e.g., clays, silts, other terrestrial detritus, pathogenic organisms, asbestos fibers), and some are produced by chemical and biological processes within the water source (e.g., algae; precipitates of $CaCO_3$, $FeOOH$, $MnO_2$; aquatic organic detritus). Particle size may vary by several orders of magnitude, from a few tens of nanometers (e.g., viruses) to a few hundred micrometers (e.g., zooplankton). All can be effectively removed by properly designed and operated coagulation, sedimentation, and filtration facilities. A comparison of the size spectrum of waterborne particles and filter pores is shown in Fig. 6.1.

Humic substances are the major organic constituent of unpolluted waters. They are derived from soil and are also produced within natural water and sediments by chemical and biological processes such as the decomposition of vegetation. Humic substances are anionic polyelectrolytes of low to moderate molecular weight; their charge is primarily caused by carboxyl and phenolic groups; they have both aromatic and aliphatic components and can be surface active; they are refractive and can persist for centuries or longer. Humic substances are defined operationally by the methods used to extract them from water or soil. Typically they are divided into the more soluble fulvic acids (FAs) and the less soluble humic acids (HAs), with FAs predominating in most waters.

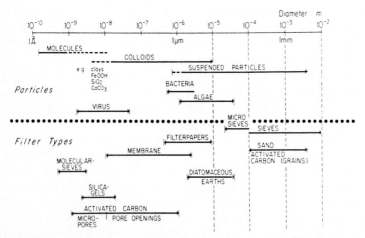

**Figure 6.1**  Size spectrum of waterborne particles and filter pores. (*From Stumm and Morgan.*[1])

The process of coagulation as practiced in water treatment can be considered as three separate and sequential steps: coagulant formation, particle destabilization, and interparticle collisions. Coagulant formation and particle destabilization occur in rapid-mixing tanks; interparticle collisions occur predominantly in flocculation tanks. When alum $[Al_2(SO_4)_3 \cdot 14.3H_2O]$ is used in coagulation, the actual chemical species operative in the process are formed during and after the alum is mixed with the water to be treated; $Al^{3+}$ and $SO_4^{2-}$, the ions in alum, are not directly involved in the coagulation process. Coagulants are sometimes formed prior to their addition to rapid-mixing facilities. Examples include the use for several decades of activated silica, more recent applications of synthetic organic polymers, and recent introductions of preformed iron (III) and aluminum species [polyaluminum chloride (PACl) and polyiron chloride (PICl)]. Coagulant reactions with humic substances and with solid particles (particle destabilization) can be fast and occur primarily in rapid-mixing tanks. Interparticle contacts necessary for aggregation are brought about by fluid flow and slow mixing; these reactions are designed to occur in flocculation tanks.

Coagulation has been a necessary component of rapid-filtration plants in the United States since the 1880s. Rapid-mix, flocculation, sedimentation, and filtration facilities have been used in series. Alum and iron (III) salts have been employed as chemicals, with alum probably having the most widespread use. In the 1930s, Baylis introduced activated silica as a "coagulant aid." This material, formed on site in separate facilities, is an anionic polymer or a small negatively charged colloid. Synthetic organic polymers were introduced in the 1960s, with cationic polymers having the greatest use. These have led to changes in pretreatment and filtration practice, including multimedia filters, high-rate filtration, direct filtration (rapid mixing, flocculation, and filtration, but no sedimentation), and in-line filtration (rapid mixing and filtration only). The development of new chemicals, advances in filter design, and changing water quality standards and goals have stimulated substantial diversity in the design and operation of coagulation processes, and such change can be expected to continue into the future.

## Definitions

Coagulation is a complex process. Many reactants and reactions are possible, and several are involved in any specific application. Often some terminology is defined to simplify and clarify discussion; at times the definitions have seemed to complicate communication. Terms such as coagulation and flocculation have different meanings to different people. Although they are used many times in this chapter, no unique correct and universal definitions for such terms exist. The usage given in this chapter to a few common terms is described in the following definitions.

Coagulation is considered to encompass all reactions, mechanisms, and results in the overall process of particle aggregation within a water being treated, including in situ coagulant formation (where applicable), chemical particle destabilization, and physical interparticle contacts. The coagulation process is usually accomplished with two types of tanks arranged in series: rapid-mixing tanks for coagulant addition, in situ coagulant formation, and particle destabilization; and flocculation tanks to promote interparticle contacts. The physical process of producing contacts is termed flocculation, and it occurs primarily in flocculation tanks.

These definitions of coagulation and flocculation are derived from engineering practice, exemplified by the work on the design of flocculation tanks by Camp.[2] Other definitions for these words are in use. LaMer[3] considered only the chemical mechanisms in particle destabilization and used the terms coagulation and flocculation to distinguish between two of them. LaMer defined destabilization by simple salts such as NaCl (indifferent electrolytes) as coagulation. Destabilization of particles by adsorption of large organic polymers and the subsequent formation of particle-polymer-particle bridges was termed flocculation. Distinguishing, as LaMer did, among various modes of particle destabilization is important, and this is done subsequently in this chapter. LaMer's distinctions are incorporated, but his terminology is not.

## Stability of Colloids

Some colloids are stable indefinitely, and some are not. Colloids that are stable indefinitely are energetically or thermodynamically stable; they have been termed reversible colloids. Examples include ordered structures from soap and detergent molecules called micelles, proteins, starches, large polymers, and some humic substances. Colloids that are not stable indefinitely coagulate; they have been termed irreversible. Examples of these thermodynamically unstable particles include clays, metal oxides, and microorganisms, i.e., virtually all solid particles present in natural water.

Some irreversible colloids coagulate slowly, and others coagulate rapidly. The terms stable and unstable are most often applied to irreversible colloids and hence have a kinetic meaning and not a thermodynamic or energetic one. A kinetically stable colloid is an irreversible or thermodynamically unstable suspension that coagulates at a very slow rate. A kinetically unstable colloid is an irreversible colloid that coagulates rapidly. In water treatment, the coagulation process is used to increase the rate or kinetics at which particles aggregate, i.e., to transform a stable suspension into an unstable one. Particles that may have been in lake water for months or years are aggregated in an hour or less. The design and operation of a coagulation process requires changing and controlling the kinetics of particle aggregation. Physics and chemistry are involved; coagulation is a physicochemical process.

Important forces occurring between particles include van der Waals forces, electrostatic forces caused by adsorbed molecules, and hydrodynamic forces. Attractive van der Waals forces and hydrodynamic forces are ubiquitous. Forces from adsorbed polymers operate when these molecules are present; they are frequently added in water treatment plants. Natural organic matter (primarily humic substances) adsorbs on colloidal particles and has been recognized as contributing to the stability of particles in many natural waters. Stability arising from polymer adsorption has been termed steric stabilization.

### Origins of colloidal stability

For purposes of discussion and analysis, the division of colloidal stability into two cases is useful: (1) electrostatic stabilization and (2) steric stabilization. Each case will be further subdivided into two aspects: (1) consideration of the structure of a single solid-liquid interface, and (2) evaluation of the forces between two such interfaces as they come into close proximity.

**Electrostatic stabilization.** Most particles in water have electrically charged surfaces, and the sign of the charge is usually negative.[4,5] Three important processes for producing this charge are considered in the following discussion.

First, surface groups on the solid may react with water and accept or donate protons. For an oxide surface such as silica with surface silanol groups ($\equiv$SiOH),

$$\equiv SiOH_2^+ \rightleftharpoons \equiv SiOH + H^+ \tag{6.1a}$$

$$\equiv SiOH \rightleftharpoons \equiv SiO^- + H^+ \tag{6.1b}$$

An organic surface can contain carboxyl and amino groups and react as follows:

$$\equiv SiOH_2^+ \rightleftharpoons \; \equiv SiOH + H^+$$

$$\equiv SiOH \rightleftharpoons \; \equiv SiO^- + H^+$$

$$\text{(6.2}a\text{)}$$

$$\text{(6.2}b\text{)}$$

In these reactions the surface charge on a solid particle depends upon the concentration of protons ($[H^+]$) or the pH ($= -\log [H^+]$) in the solution. As pH increases (i.e., $[H^+]$ decreases), Eqs. (6.1) and (6.2) shift to the right and the surface charge becomes increasingly negative. Silica is negatively charged in water with a pH above 2; proteins contain a mixture of carboxyl and amino groups and usually have a negative charge at a pH above about 4.

Second, surface groups can react in water with solutes other than protons. Again, using silica as a representative oxide,

$$\equiv SiOH + Ca^{2+} \rightleftharpoons \equiv SiOCa^+ + H^+ \tag{6.3}$$

$$\equiv SiOH + HPO_4^{2-} \rightleftharpoons \equiv SiOPO_3H^- + OH^- \tag{6.4}$$

These surface complex formation reactions involve specific chemical reactions between chemical groups on the solid surface (e.g., silanol groups) and adsorbable solutes (e.g., phosphate ions). Surface charge is again related to solution chemistry.

Third, a surface charge may arise because of imperfections within the structure of the particle; this is called isomorphic replacement. It is responsible for a substantial part of the charge on many clay minerals. Clays are layered structures. Typically, sheets of silica tetrahedra are cross-linked with sheets of alumina octahedra. The silica tetrahedra have an average composition of $SiO_2$ and may be depicted as follows:

(No net charge)

If an Al atom is substituted for an Si atom during the formation of this lattice, a negatively charged framework results:

(Net charge −1)

Similarly, a divalent cation such as Mg or Fe (II) may substitute for an Al atom in the aluminum oxide octahedral network, also producing a negative charge. The sign and magnitude of the charge produced by such isomorphic replacements are independent of the characteristics of the aqueous phase after the crystal is formed.

A colloidal suspension does not have a net electrical charge, and the primary charge on the particle must be counterbalanced in the system. Figure 6.2 shows schematically a negatively charged colloidal particle with a cloud of ions (diffuse layer) around the particle. Because the particle is negatively charged, an excess of ions of opposite charge (positive) accumulate in the interfacial region. Ions of opposite charge accumulating in the interfacial region together with the primary charge form an electrical double layer. The diffuse layer results from electrostatic attraction of ions of opposite charge to the particle (counterions), electrostatic repulsion of ions of the same charge as the particle (similions), and thermal or molecular diffusion that acts against the concentration gradients produced by the electrostatic effects. The formation of diffuse layers is shown in Figs. 6.2 and 6.3a.

When an electrical potential is applied across a suspension of negatively charged particles, they move toward the positive electrode. The potential that causes the motion of the particles is associated with the plane of shear of fluid around the particles and is called its zeta potential or its electrokinetic potential. The exact location of this slipping plane has been discussed for decades. Lyklema[6] locates the slipping plane at the outer border of the Stern layer as shown in Fig. 6.2. Zeta potential and other parameters such as streaming current measurements are indirect measurements of the charge on particles.

Because of the primary charge, an electrostatic potential (voltage) exists between the surface of the particle and the bulk of the solution. This electric potential can be pictured as the electric pressure that must be applied to bring a unit charge having the same sign as the primary charge from the bulk of the solution to a given distance from

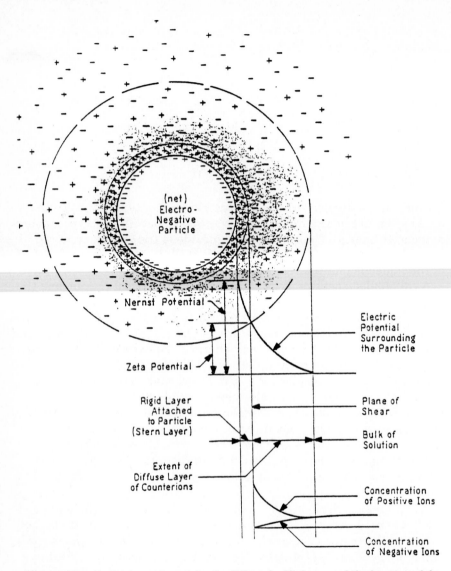

**Figure 6.2** Negatively charged particle, the diffuse double layer and the location of the zeta potential.

the particle surface, shown schematically in Fig. 6.3*b*. The potential has a maximum value at the particle surface and decreases with distance from the surface. This decrease is affected by the characteristics of the diffuse layer, and, thus, by the type and concentration of ions in the bulk solution.

Mathematical treatment of the double layer shown schematically in Figs. 6.2 and 6.3 was developed independently by Gouy and Chapman.

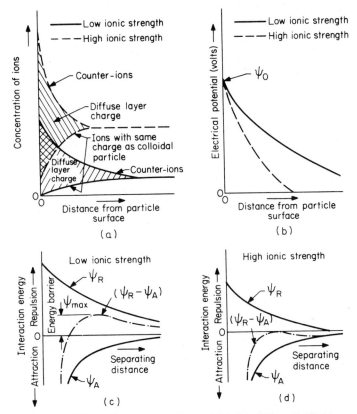

**Figure 6.3** Schematic representations of (a) the diffuse double layer; (b) the diffuse layer potential; and (c and d) two cases of particle-particle interaction energies in electrostatically stabilized colloidal systems.

The result is termed the Gouy-Chapman model; it is well presented by Verwey and Overbeek.[7] Electrostatic or coulombic attraction and diffusion are the interacting processes responsible for the formation of the diffuse layer. In these models of the double layer, ions are treated as point charges and have no other physical or chemical characteristics.

When two similar colloidal particles approach each other, their diffuse layers begin to overlap and to interact. This electrostatic interaction between the particles results in a repulsive force between them. A repulsive potential energy $\Psi_R$ is produced that increases as the distance separating the particles decreases. Such repulsive interactions are illustrated in Fig. 6.3c and d.

Certain attractive forces exist between all types of particles, no matter how dissimilar they may be. These attractive forces, termed van der Waals forces, are involved in the coagulation and filtration of many suspensions. They arise from interactions among dipoles, either permanent or induced, in the atoms comprising the colloidal particles and water. The van der Waals force between two particles decreases with increasing separating distance between them. An attractive potential energy $\Psi_A$ results from these attractive forces, and this energy also decreases as the separation increases. Schematic curves of the van der Waals attractive potential energy of interaction are shown in Fig. 6.3c and d. Unlike electrostatic repulsive forces, the van der Waals forces are essentially independent of the composition of the solution; they depend upon the kind and number of atoms in the colloidal particles and the solvent (water).

Summation of $\Psi_R$ and $\Psi_A$ yields the net interaction energy between two colloidal particles. This sum with the proper signs is $\Psi_R - \Psi_A$ and is shown schematically as a function of separating distance in Fig. 6.3c and d. When electrostatic repulsion dominates during particle-particle interactions, the suspension is said to be electrostatically stabilized and to undergo "slow" coagulation. Electrostatic stabilization is fundamental to the current understanding of coagulation in water treatment. For additional insight into electrostatic stabilization, the imaginative and extensive mathematical treatment given by Verwey and Overbeek[7] and an excellent summary written by Lyklema[6] should be consulted. Valuable summaries of the methods, mathematics, and meanings of models for electrostatic stabilization are also contained in texts by Stumm and Morgan[8] and by Morel.[9]

Mathematically sophisticated and physically elegant, the Gouy-Chapman model provides a good qualitative picture of the origins of electrostatic stabilization. It does not provide a quantitative, predictive tool for such important characteristics as coagulant requirements and coagulation rates, and for this reason the mathematics of the model are omitted from this review. Its principal drawback lies in the characterization of all ions as point charges. This allows for physical description of electrostatics and omits description of chemical interactions. The sodium ion ($Na^+$), the dodecylamine ion ($C_{12}H_{25}NH_3^+$), the proton ($H^+$), and the dihydroxyaluminum ion [$Al\,(OH)_2^+$] have identical charges but are not identical coagulants.

**Steric stabilization.**    Steric stabilization can result from the adsorption of polymers at solid-water interfaces. Large polymers can form adsorbed segments on a solid surface with loops and tails extending into solution,[6] as illustrated in Fig. 6.4. Adsorbed polymers can be either stabilizing or destabilizing, depending on the relative amounts of polymer and solid particles, the affinities of the polymer for the solid and for water, electrolyte type and concentration, and other factors. A stabilizing polymer may contain two types of groups, one of which has

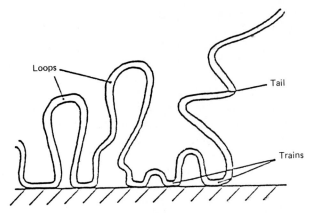

**Figure 6.4** Illustration of adsorbed polymer configuration with loops, trains, and tails. (*From Lyklema.*[6])

a high affinity for the solid surface and a second, hydrophilic group that is left "dangling" in the water.[10] The configuration of such an interfacial region is difficult to characterize either theoretically or experimentally. This, in turn, prevents quantitative formulation of the interaction forces between two such interfacial regions during a particle-particle encounter in coagulation. Some useful qualitative descriptions can, however, be made.

Gregory[10] summarized two processes that can produce a repulsion when two polymer-coated surfaces interact at close distances. These are illustrated in Fig. 6.5. First, the absorbed layers can each be compressed by the collision, reducing the volume available for the adsorbed molecules. This reduction in volume restricts the movement of the polymers (a reduction in entropy) and causes a repulsion between the particles. Second, and more frequently, the adsorbed layers may interpenetrate on collision, increasing the concentration of polymer segments in the mixed region. If the extended polymer segments are strongly hydrophilic, they can prefer the solvent to reaction with

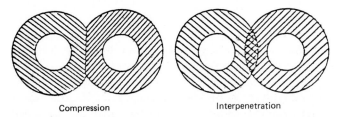

Compression                Interpenetration

**Figure 6.5** Two possible repulsive interactions of adsorbed polymer layers in sterically stabilized colloidal systems. (*From Gregory.*[10])

other polymer segments. An overlap or mixing then leads to repulsion. These two processes are separate from and in addition to the effects of polymer adsorption on the charge of the particles and the van der Waals interaction between particles. Charged polymers (polyelectrolytes) can alter particle charge; organic polymers can also reduce the van der Waals attractive interaction energy.

Steric stabilization is widely used in the manufacture of industrial colloids such as paints and waxes. Natural organic materials such as humic substances are ubiquitous in water supplies. They are anionic polyelectrolytes, absorb at interfaces, can be surface active, and may contribute to particle stability by steric effects.

## Destabilization

Different chemical coagulants can bring about the destabilization of suspensions in different ways. Four distinct methods or mechanisms are presented here: (1) compression of the double layer, (2) adsorption to produce charge neutralization, (3) enmeshment in a precipitate, and (4) adsorption to permit interparticle bridging.

### Double-layer compression

Simple salts such as sodium chloride are sometimes called "indifferent electrolytes." This means that the ions produced in solution ($Na^+$ and $Cl^-$ in this case) act as point charges and have no chemical characteristics such as hydrolysis and adsorption reactions in coagulation. The electrostatic model of Derjaguin, Landau, Verwey, and Overbeek (DLVO)[7] provides a useful description of coagulation by such salts. Results of the use of such coagulants were summarized by Hardy in 1900 in the Schulze-Hardy rule, which states that the destabilization of a colloid by an indifferent electrolyte is brought about by ions of opposite charge to the colloid (counterions) and that the coagulation effectiveness of the ions increases markedly with ion charge. For example, the concentrations of $Na^+$, $Ca^{2+}$, and $Al^{3+}$ required to destabilize a negatively charged colloid are observed to vary approximately in the ratio of $1:10^{-2}:10^{-3}$.

The interactions between such indifferent coagulants and colloidal particles are purely electrostatic; ions of similar charge to the primary charge of the colloid are repelled and counterions are attracted. This is illustrated in Fig. 6.3a. Destabilization by counterions occurs by compression of the diffuse layer surrounding the particles. High concentrations of electrolyte in solution (i.e., high ionic strength or high total dissolved solids) produce correspondingly high concentrations of counterions in the diffuse layer. The volume of the diffuse layer necessary to maintain electroneutrality is lowered, and the effective thickness of the diffuse layer is reduced (Fig. 6.3a and b). The range of the repulsive interaction between similar colloidal particles decreases (Fig. 6.3c and d), the attractive van der Waals interaction can dominate at all separations (Fig. 6.3d), the activation energy barrier can disappear, and electrostatic stabilization can be eliminated. A good example of destabilization by double-layer compression occurs when particles in freshwater of rivers with low ionic strength mix with high ionic strength seawater. This causes the particles to be destabilized by double-layer compression and accumulation of particles results in the formation of deltas at river mouths.

In the coagulation of surface water, the effectiveness of a coagulant is often determined using batch or jar tests, in which the dosage of coagulant is varied and residual turbidity is measured after appropriate periods of flocculation and sedimentation. Schematic curves of residual turbidity as a function of coagulant dosage for a natural water treated with $Na^+$, $Ca^{2+}$ and $Al^{3+}$ are presented in Fig. 6.6a. The curves are based on the empirical Schulze-Hardy rule and the theoretical DLVO model. They provide a good understanding of some simple

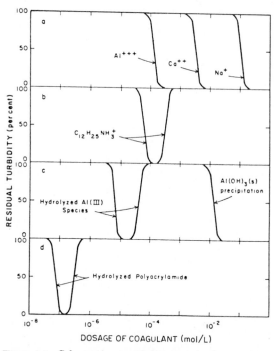

**Figure 6.6**  Schematic coagulation curves (jar test results) for four types of coagulants.

electrostatic phenomena in colloidal stability and, to a lesser extent, in coagulation. They do not, however, describe processes that are important in freshwater and in water treatment practice. The phenomena described by the theoretical DLVO model and the empirical Schulze-Hardy rule are of limited interest to engineers and chemists concerned with coagulation in natural water and in treatment systems. Coagulants are generally not "indifferent electrolytes"; they undergo many and more important reactions in addition to electrostatic ones. The simple $Al^{3+}$ species often noted in the Schulze-Hardy rule occurs at trivial concentrations when aluminum salts are added as coagulants in water treatment. Electrostatics can explain much about particle stability, but to understand coagulation in water treatment, understanding other destabilization mechanisms is necessary.

### Adsorption and charge neutralization

Considering the energy involved in a simple electrostatic interaction between a colloidal particle and a coagulant ion is instructive. The electrochemical energy may be represented by $zF\psi_0$ where $z$ is the charge of the coagulant ion [e.g., $+1$ for $Al(OH)^{2+}$], $F$ is the Faraday constant, and $\psi_0$ is the potential difference between the particle surface and the bulk of the solution. For a monovalent counterion and a particle for which a potential difference of 100 millivolts (mV) exists, the attractive electrostatic energy involved is only 9.6 kilojoules/mol (kJ/mol). Because covalent chemical bonds have bond energies of 200 to 400 kJ/mol, and even hydrogen bond energies are on the order of 20 kJ/mol, many coagulant-colloid interactions can overshadow electrostatic effects in the destabilization of colloids. The relative values of these energies indicate the importance of chemical effects in comparison to electrostatic effects.

The ability of a coagulant to destabilize a colloidal dispersion is actually a composite of coagulant-colloid, coagulant-solvent, and colloid-solvent interactions. For example, one type of coagulant, long chain organic amines such as dodecylamine $(C_{12}H_{25}NH_3^+)$, is a surface-active substance; thus it accumulates at interfaces. These ions are effectively squeezed out of the water at an available interface because of the lack of interaction between the water molecules and the $CH_2$ groups in their aliphatic chains. In contrast, sodium and calcium ions interact strongly with water molecules and are not surface-active.

Schematic jar test results in which $C_{12}H_{25}NH_3^+$ is used as a coagulant are shown in Fig. 6.6b, adapted from the work of Tamamushi and Tamaki.[11] Two important differences are apparent in comparing the singly charged coagulant species $Na^+$ and $C_{12}H_{25}NH_3^+$ in Fig. 6.6a and b. First, sodium ions are effective as coagulants only at concentrations above about $10^{-1}\,M$, while the organic amine with identical charge produces destabilization at concentrations as low as $6 \times 10^{-5}\,M$. Second, overdosing a suspension with $Na^+$ is not possible, while restabilization occurs at coagulant dosages above about $4 \times 10^{-4}\,M$ with $C_{12}H_{25}NH_3^+$. This restabilization is also accompanied by a charge reversal; that is, the net charge on the particles is reversed from negative to positive by an adsorption of an excess of counterions. If electrostatic interactions were the primary force for destabilization, such an adsorption of an excess of counterions to produce charge reversal and restabilization would not be possible.

For the results presented in Fig. 6.6b, probably coagulant-solvent interactions (in this case, more precisely, the lack of coagulant-solvent interactions between dodecylamine ions and water) are primarily responsible for the adsorption of the coagulant at the particle-water interface. For coagulants such as hydrolyzed aluminum ions and cationic polyelectrolytes commonly used in water treatment, specific adsorption is common, but in this case is generally caused by chemical interactions between coagulant and colloid. Because adsorption is dominant, jar test results can resemble Fig. 6.6b and c rather than Fig. 6.6a. Overdosing and charge reversal are possible with iron(III) and aluminum salts and common when organic polymers are used. The chemical species of inorganic salts that cause overdosing are discussed subsequently.

### Enmeshment in a precipitate

When a metal salt such as $Al_2(SO_4)_3$ or $FeCl_3$ is added to water in concentrations sufficiently high to cause precipitation of a metal hydroxide [e.g., $Al(OH)_3(s)$ or $Fe(OH)_3(s)$], colloidal particles can be enmeshed in these precipitates as they are formed and also collide with them afterward. This has been termed "sweep-floc" coagulation by Packham.[12] The process, a combination of destabilization and transport, is extensively used in the treatment of waters of variable turbidity and dissolved organic carbon (DOC) in conjunction with rapid-mixing, flocculation, sedimentation, and filtration facilities in series. For water with low turbidity and color, solids are added to the water to be treated by precipitation of aluminum hydroxide [$Al(OH)_3(s)$] in order to improve physically the kinetics of flocculation and thereby produce settleable aggregates in conventional flocculation tanks. The $Al(OH)_3(s)$ curve in Fig. 6.6c shows sweep-floc conditions. As used in water treatment, polymeric organic coagulants do not form voluminous precipitates and do not enhance physical flocculation kinetics, and so they have no application as sole coagulants in the treatment of water having low turbidity and color by conventional treatment with rapid mixing, flocculation, sedimentation, and filtration. However, polymeric organic coagulants have been used successfully as sole coagulants in direct and in-line filtration applications.

### Adsorption and interparticle bridging

Synthetic organic polymers came into use in water treatment over 2 decades ago. Cationic, nonionic, and anionic polymers have been found to be effective. LaMer and coworkers[13] and others[14,15] have developed a bridging theory that provides a useful description of the ability of polymers with high molecular weights to destabilize colloidal dispersions; a more recent summary has been written by Gregory.[10]

Destabilization by bridging occurs when segments of a polymer chain absorb on more than one particle, thereby linking the particles together. When a polymer molecule comes into contact with a colloidal particle, some of the reactive groups on the polymer adsorb at the particle surface, leaving other portions of the molecule extending into the solution (Fig. 6.4). If a second particle with some vacant adsorption sites contacts these extended loops and tails, attachment can occur, as illustrated in Fig. 6.7a adapted from Gregory.[10] A particle-polymer-particle aggregate is formed in which the polymer serves as a bridge. Effective bridging requires that absorbed polymers extend far enough from the particle surface to attach to other particles and also that some free surface is available for adsorption of the extended segments. If excess polymer is added and adsorbed, the particles are restabilized by surface saturation and can be sterically stabilized as depicted in Fig. 6.7b. The destabilization and restabilization of a negatively charged dispersion by an anionic polyelectrolyte is also shown in Fig. 6.6d.

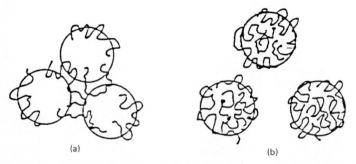

(a)                                      (b)

**Figure 6.7** (a) Aggregation and (b) restabilization of colloidal particles by adsorbed polymers. (*From Gregory.*[10])

## Aqueous chemistry of Si, Al, and Fe(III)

**Silica.** In 1937, Baylis[16] reported the use of sodium silicate modified by acidification ("activated silica") as a coagulant aid to improve the treatability of water from Lake Michigan for Chicago when alum was used as the primary coagulant during cold winter months. Some 30 years later, Stumm and coworkers[17] provided an explanation for the process. Many factors can be important, and both thermodynamic and kinetic considerations require attention. The following is a summary, adapted from Stumm et al.[17] and O'Melia.[18]

Beginning with thermodynamics is useful; a solubility diagram for amorphous silica is presented in Fig. 6.8. Silicate equilibria used in constructing this diagram are listed in Table 6.1. Below a pH value of approximately 9, the solubility of amorphous $SiO_2$ is constant at $2 \times 10^{-3}$ $M$ (120 mg/L as $SiO_2$). The principal soluble species is a monomer, $Si(OH)_4$. At pHs above 9, the solubility of amorphous $SiO_2$ is increased by the formation of anionic monomeric and polymeric species. Commercial concentrated silicate solutions are in this pH range and contain stable anionic polymeric species such as $Si_4O_6(OH)_6^{2-}$.

In preparing activated silica, a concentrated sodium silicate solution is acidified so that it is oversaturated with respect to the precipitation of amorphous silica (Fig. 6.8, point $A$). The precipitation begins with polymerization reactions in which monomers condense to form Si—O—Si bonds.[17] These polysilicates react further by cross-linking and aggregation to form negatively charged silica sols. Continued reaction can produce a silica gel. These reactions can be terminated or reversed by diluting the acidified solution near or just out of the in-

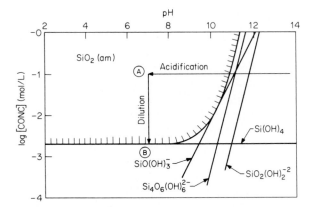

**Figure 6.8** Solubility of $SiO_2$(am). Thermodynamic data in Table 6.1. (*C. R. O'Melia, "Polymeric Inorganic Flocculants," in Brij M. Moudgil and P. Somasundaran, eds., Flocculation, Sedimentation, and Consolidation, Engineering Foundation, New York, 1985.*)

**TABLE 6.1    Silicate,[*] Iron(III),[†] and Aluminum[‡] Equilibria**

| Reaction | $\log K(25°C)$ |
|---|---|
| 1. $SiO_2(am) + 2H_2O = Si(OH)_4$ | $-2.7$ |
| 2. $Si(OH)_4 = SiO(OH)_3^- + H^+$ | $-9.46$ |
| 3. $SiO(OH)_3^- = SiO_2(OH)_2^{2-} + H^+$ | $-12.56$ |
| 4. $4Si(OH)_4 = Si_4O_6(OH)_6^{2-} + 2H^+ + 4H_2O$ | $-12.57$ |
| 5. $Fe^{3+} + H_2O = FeOH^{2+} + H^+$ | $-2.2$ |
| 6. $FeOH^{2+} + H_2O = Fe(OH)_2^+ + H^+$ | $-3.5$ |
| 7. $Fe(OH)_2^+ + H_2O = Fe(OH)_3 + H^+$ | $-6$ |
| 8. $Fe(OH)_3 + H_2O = Fe(OH)_4^- + H^+$ | $-10$ |
| 9. $2Fe^{3+} + 2H_2O = Fe_2(OH)_2^{4+} + 2H^+$ | $-2.9$ |
| 10. $3Fe^{3+} + 4H_2O = Fe_3(OH)_4^{5+} + 4H^+$ | $-6.3$ |
| 11. $Fe(OH)_3(am) = Fe^{3+} + 3OH^-$ | $-38.7$ (estimated) |
| 12. $\alpha - FeO(OH) + H_2O = Fe^{3+} + 3OH^-$ | $-41.7$ |
| 13. $Al^{3+} + H_2O = AlOH^{2+} + H^+$ | $-4.97$ |
| 14. $AlOH^{2+} + H_2O = Al(OH)_2 + H^+$ | $-4.3$ |
| 15. $Al(OH)_2^+ + H_2O = Al(OH)_3 + H^+$ | $-5.7$ |
| 16. $Al(OH)_3 + H_2O = Al(OH)_4^- + H^+$ | $-8.0$ |
| 17. $2Al^{3+} + 2H_2O = Al_2(OH)_2^{4+} + 2H^+$ | $-7.7$ |
| 18. $3Al^{3+} + 4H_2O = Al_3(OH)_4^{5+} + 4H^+$ | $-13.94$ |
| 19. $13Al^{3+} + 28H_2O = Al_{13}O_4(OH)_{24}^{7+} + 32H^+$ | $-98.73$ |
| 20. $Al(OH)_3(am) = Al^{3+} + 3OH^-$ | $-31.5$ |
| 21. $Al(OH)_3(c) = Al^{3+} + 3OH^-$ | $-33.5$ |

*From Stumm et al.[17]
†From Flynn.[19]
‡From Baes and Mesmer.[20]

solubility limit (Fig. 6.8, point $B$). A wide variety of activated silicas containing species ranging from anionic polysilicates to colloidal $SiO_2$ precipitates can be prepared by varying the concentration and basicity of the initial stock solution, the method and extent of acidification, the aging time, the extent of dilution, and the temperature. These activated silicas, while thermodynamically unstable, can remain active as coagulants for periods up to a few weeks. These same principles hold for partial neutralization of $AlCl_3$ solutions and preparation of PACl. The PACl solutions that have been developed are useful for treatment of cold, soft, turbid waters. These "polymeric inorganic coagulants" are discussed in a subsequent section.

**Aluminum.**    A solubility diagram for aluminum in water is presented in Fig. 6.9$a$ using thermodynamic data from Table 6.1. A crystalline aluminum oxide [gibbsite, $Al(OH)_3(c)$, $\log K = -33.5$, i.e., the solubility product is $10^{-33.5}$] is considered as the solid phase. At alkaline

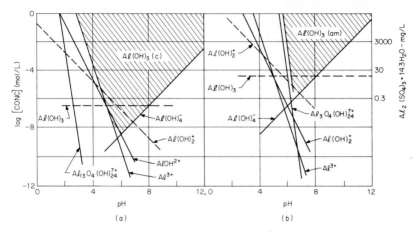

**Figure 6.9** Solubility of aluminum at equilibrium with (a) gibbsite and (b) amorphous Al $(OH)_3$. Thermodynamic data from Table 6.1.

pH values (pH > 8), the principal soluble species present at equilibrium is the monomeric anion $Al(OH)_4^-$. At lower values of pH (pH < 6), the dominant soluble species at equilibrium with gibbsite are cationic monomers such as $Al^{3+}$ and $Al(OH)^{2+}$. In this pH range, however, aluminum speciation in water is kinetically controlled and depends on many factors.

Freshly precipitated solids formed by addition of an aluminum salt to water are normally amorphous [$Al(OH)_3$(am), log $K \approx -31.5$] and more soluble than indicated by gibbsite equilibria. A solubility diagram for aluminum considering the formation of $Al(OH)_3$(am) is shown in Fig. 6.9b. Following Baes and Mesmer,[20] aluminum hydrolysis is evaluated considering three polymeric species [$Al_2(OH)_2^{4+}$, $Al_3(OH)_4^{5+}$, and $Al_{13}O_4(OH)_{24}^{7+}$] and five monomers [$Al^{3+}$, $AlOH^{2+}$, $Al(OH)_2^+$, $Al(OH)_3$, and $Al(OH)_4^-$] in equilibrium with freshly precipitated $Al(OH)_3$(am). Lines for the $Al(OH)_2^+$ and $Al(OH)_3$ species are dashed to indicate that their concentrations are uncertain because of questionable thermodynamic data. Aluminum is least soluble at a pH of about 6.2. A $10^{-4}$ M solution of Al [corresponding to a dosage of alum, $Al_2(SO_4)_3 \cdot 14.3H_2O$, of 30 mg/L] can precipitate $Al(OH)_3$(am) in the pH range of 5.8 to 8. Below a pH of 5.7, freshly precipitated $Al(OH)_3$(am) can be quite soluble and polymeric species such as $Al_{13}O_4(OH)_{24}^{7+}$ can predominate. The differences in aluminum solubility and speciation between Fig. 6.9a and b are substantial and important in water treatment practice. The figures indicate that the domain for formation of the amorphous precipitate is smaller than that for formation of crystalline species. The figures also indicate that for higher values of the solubility product for $Al(OH)_3$(am) the domain for precipitation is moved upward to higher alum doses. Aluminum chemistry at neutral and acidic pH values is generally governed by the kinetics of reactions with water (hydrolysis), with particles, and with natural organic matter; equilibrium diagrams such as Fig. 6.9 are useful but incomplete, because they do not incorporate the rates or kinetics of formation of the hydrolysis species. A procedure to incorporate kinetic boundaries in equilibrium diagrams based on supersaturation levels of aluminum and iron is indicated later in this chapter. As with silica, "polymeric" aluminum species such as PACl can range from small polyelectrolytes to colloidal precipitates, depending on the methods used in their preparation.

When commercial alum (molecular weight = 600 with 14.3 molecules of water) is added to water, the series of hydrolysis reactions shown in Table 6.1 (see nos. 13 to 20) occur as intermediate reactions prior to precipitation of $Al(OH)_3(s)$. These reactions illustrate that alum hydrolysis produces hydrogen ions ($H^+$) and hence acts like an acid and often reduces the pH of the water. The pH reduction would depend on the alkalinity of the water. For every 1 mg/L of alum that reacts to produce a precipitate of aluminum hydroxide, 0.5 mg/L of natural alkalinity expressed as $CaCO_3$ is consumed. Thus, for example, when an alum dose of 30 mg/L is used, the alkalinity required for reaction is 15 mg/L and often when treating low alkalinity waters, lime or caustic soda is added to provide the necessary alkalinity and to keep the pH conditions in the range where aluminum hydroxide precipitate may be formed and optimum coagulation occurs. Alum used for water treatment is available in lump, ground, or powdered forms as well as a concentrated solution. It is readily soluble in water and easily applied as a solution or as a dry material.

**Iron.**    Leprince et al.[21] reported that partially neutralized "polymeric" $FeCl_3$ solutions were superior to untreated solutions in removing turbidity from synthetic suspensions at low temperatures. Just as kinetically stable anionic polymeric silicates have been formed by acidification of basic sodium silicate solutions, cationic polymeric and colloidal Fe(III) and Al species have been formed by neutralization of selected salts of these metals.

A solubility diagram for Fe(III) in water is presented in Fig. 6.10 using thermodynamic data in Table 6.1. Two solubility limits are depicted. The lower boundary is for a solution at equilibrium with $\alpha$-FeOOH (goethite, log $K$ = $-41.7$). The upper boundary is for a solution at equilibrium with freshly precipitated amorphous $Fe(OH)_3(am)$

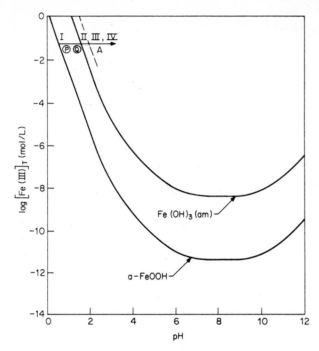

**Figure 6.10** Solubility of Fe(III). Thermodynamic data from Table 6.1. Curve denotes base titration of $6.26 \times 10^{-2}$ $M$ Fe(III) solution. Regions I, II, III, and IV are discussed in text.

($\log K = -38.7$). At high pH (pH > 10), the principal soluble species present at equilibrium with either solid is the monomeric anion $Fe(OH)_4^-$. At lower pH levels (pH < 6), the dominant soluble species are cationic monomers such as $Fe^{3+}$ and $Fe(OH)_2^+$. Iron(III) is least soluble at a pH of about 8. A $10^{-4}$ $M$ solution of Fe(III) can precipitate $Fe(OH)_3$(am) above a pH of about 3. Comparing hydrolysis equilibria in Table 6.1 and solubilities as illustrated in Figs. 6.9 and 6.10, iron(III) is a stronger acid and less soluble than aluminum. As with aluminum, the chemistry of Fe(III) in water treatment is governed by the kinetics of reactions with water and pollutants.

The reactions of iron(III) salts prior to the formation of $Fe(OH)_3$(am) are shown in Table 6.1. Iron salts are also acid salts and consume alkalinity during precipitation as a hydroxide. In operational practice, iron(III) is typically added in the form of ferric sulfate or ferric chloride. Ferric sulfate is available as a commercial coagulant in the form of an anhydrous granular material that is readily soluble in water. Ferric chloride is available as a liquid and crystal and in anhydrous form. The liquid and crystal forms are very corrosive and must be handled in rubber-lined tanks or in glass. The anhydrous form can be handled in steel only so long as moisture is excluded.

### Polymeric inorganic coagulants

A recent development in water treatment has been the preparation and use of "polymeric" Fe(III) and Al species as coagulants.[21,22] The following discussion, while focused on polymeric iron(III) preparations, also serves as a basis for considering aluminum preparations.

In preparing polymeric ferric species, concentrated solutions of iron(III) salts are partially neutralized by the addition of base as indicated by arrow $A$ in Fig. 6.10. A typical neutralization (titration) adapted from the work of van der Woude and de Bruyn[23] is presented in Fig. 6.11. Solution pH is plotted as a function of base added as represented by $r$, the number of OH ligands added per Fe(III) atom. Addition of base increases $r$ and usually, but not always, raises the pH. In the case considered here, a solution containing $6.25 \times 10^{-2}$ $M$ $Fe(NO_3)_3$ {log $[Fe(III)]_T = -1.2$} is titrated with sodium hydroxide. This is also shown by arrow $A$ in Fig. 6.10. The initial pH is 0.7, and the initial solution is slightly oversaturated with respect to the precipitation of goethite (Fig. 6.10, point $P$). Iron(III) solutions prepared by the addition of a ferric salt to water are normally oversaturated in this way; the addition of acid is required to prepare solutions that are actually thermodynamically stable.

The titration may be divided into four regions (after van der Woude

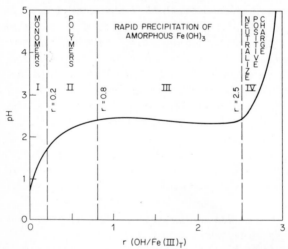

**Figure 6.11** Typical base titration of $Fe(NO_3)_3$ solution, $[Fe(III)]_T = 6.25 \times 10^{-2}$ $M$. (*Adapted from van der Woude and de Bruyn.*[23])

and de Bruyn[23]). In region I, Figs. 6.10 and 6.11 with $0 \leq r \leq 0.2$, the principal Fe(III) species are $Fe^{3+}$, $FeOH^{2+}$, $Fe(OH)_2^+$, and $Fe_2(OH)_2^{4+}$. Although the solution becomes increasingly oversaturated with respect to goethite as $r$ is increased, a precipitate is not observed. If base addition is stopped, the solution composition (including pH) does not change with time. Polymers larger than the dimer [$Fe_2(OH)_2^{4+}$] are not detected.

In region II, corresponding to $0.2 < r \leq 0.8$, polymers larger than the dimer are formed. Metastable solutions containing cationic polymeric Fe(III) species may be prepared. The lower boundary of region II ($r = 0.2$) corresponds well with the solubility limit of $Fe(OH)_3(am)$ in Fig. 6.10, point $Q$. As with activated silica, thermodynamically unstable polymeric species are formed when the system becomes oversaturated with respect to the formation of $Fe(OH)_3(am)$, and eventually a precipitate is formed in this region. The upper boundary, observed by van der Woude and de Bruyn[23] at $r = 0.8$, indicates the onset of rapid precipitation and the formation of a colloidal amorphous ferric hydroxide phase. The width of region II and particularly the location of the upper boundary have particular significance in the possible preparation of polymeric Fe(III) species for practical applications. The conditions of preparation and reaction kinetics are important. Polymerization and precipitation kinetics are affected by the concentration of the iron solution $[Fe(III)]_T$; the type, concentration, and method of base addition; ionic strength; temperature; and specific anion effects.[23,24]

In region III, for $0.8 < r < 2.5$, van der Woude and de Bruyn[23] report rapid precipitation of $Fe(OH)_3(am)$. At $r = 2.5$, precipitation is nearly complete and corresponds to a pH of approximately 2.4. Additional base ($r > 2.5$, region IV) may only neutralize the positive charge on the precipitate.

Results reported by Schneider[24] are in qualitative agreement with these concepts. Polynuclear oxyhydroxocomplexes [$Fe_pO_r(OH)_s^{3p-(2r+s)}$] were observed in highly supersaturated solutions, with precipitation kinetics very much dependent on foreign nuclei, method of base addition, and specific anion effects. Leprince et al.[21] report the existence of polymeric species of Fe(III) up to $r = 2.5$ in partially neutralized $FeCl_3$ solutions, indicating that the upper boundary of region II is not fixed at $r = 0.8$. In a situation analogous to the case of silica, a wide variety of "polymeric" iron (III) preparations ranging from small cationic polymers to colloidal amorphous iron(III) precipitates can be obtained, depending upon conditions used in their preparation.

Partially neutralized solutions of aluminum salts have been used with some success in water treatment. Bottero et al.[22] reported that partially neutralized aluminum chloride solutions were effective coagulants for clay suspensions. Polyaluminum chloride (PACl) is used extensively in Japan for the treatment of cold, soft, turbid waters; it is also used in potable water treatment in France and Germany, and its use in the United States may increase in the future.[25] The preparation and use of these materials have many similarities with the Si and Fe(III) coagulants previously described. As with Si and Fe(III), "polymeric" aluminum species may range from small polymers to colloidal precipitates, depending on many factors in their preparation. Several polymeric aluminum preparations are commercially available with a range of values of $r$; some contain additives such as sulfate and cationic organic polyelectrolytes. Some French investigators[21,22] advocate on-site preparations of these coagulants by partial neutralization or by heating, permitting a specific tailoring of the inorganic "polymeric" coagulant to the water to be treated.

## Organic polymers

A polymer is a chain of small subunits or monomers. Some synthetic polymers contain only one monomer; others contain two or three different types of subunits. The total number of subunits in a synthetic polymer can be varied, producing materials with different molecular weights and, if the subunits are charged, with different charge densities. Polymer chains may be linear or branched to varying degrees.

If a monomeric unit in a polymer contains ionizable groups (e.g., carboxyl, amino, or sulfonic groups), the polymer is termed a polyelectrolyte. Depending upon the type of ionizable groups on the monomeric units, a polyelectrolyte may be cationic, anionic, or ampholytic (containing both positive and negative groups, such as proteins). Polymers without ionizable groups are termed nonionic. Examples of these materials are presented in Table 6.2 adapted from Gregory.[10]

"As a rule, any polymer adsorbs on any surface."[7] Many polymer segments can be in contact with a surface simultaneously (Figs. 6.4 and 6.7a). Even a low bonding energy for each segment makes their collective affinity for a surface very high, and their adsorption can appear to be virtually irreversible. In many cases this adsorption can be quite specific. Bonds can be formed between particular functional groups on the polymer molecule and specific sites on a solid surface. Other polymer properties that affect performance are molecular weight, charge density, and degree of branching. Solution characteristics are also important; in particular, pH can affect the charge on a polyelectrolyte as well as the solid particles to be aggregated. For example, polyacrylic acid and hydrolyzed polyacrylamide (a copolymer of acrylic acid and acrylamide) are uncharged at pH levels below about 4

Some Synthetic Ploymers, [1]

| Nonionic | Anionic | Cationic |
|---|---|---|
| $\left[\begin{array}{c} -CH-CH_2- \\ \phantom{-}CONH_2 \end{array}\right]_n$ | $\left[\begin{array}{c} -CH-CH_2- \\ \phantom{-}COO^-\ Na^+ \end{array}\right]_n$ | $\left[-CH_2-CH_2-NH_2-\right]_n$ |
| Polyacrylamide<br>M.W. $\approx 10^6$ | Sodium<br>polyacrylate<br>M.W. $\approx 10^6$ | Polyethylene imine<br>M.W. = 600 to 100,000 |
| $\left[\begin{array}{c} -CH-CH_2- \\ \phantom{-}OH \end{array}\right]_n$ | $\left[-CH-CH_2-\right]_n$ <br> (benzene ring) $SO_3^-\ Na^+$ | $\left[\begin{array}{c} \phantom{xx}CH_2 \\ -CH \phantom{xxx} CH-CH_2- \\ \phantom{-}CH_2 \phantom{xxxx} CH_2 \\ \phantom{xxx}\overset{+}{N}Cl^- \\ \phantom{xx}CH_3 \phantom{xx} CH_3 \end{array}\right]_n$ |
| Polyvinyl alcohol | Sodium polystyrene<br>sulfonate | Polydiallyldimethyl-<br>ammonium chloride<br>("Cat-Floc")<br>M.W. = 10,000 to 100,000 |
| $\left[-CH_2-CH_2-O\right]_n$ <br> Polyethylene oxide | | $\left[-CH-CH_2-\right]_n$ <br> (pyridinium ring) $\overset{+}{N}$ <br> $H\,Br^-$ <br> Polyvinylpyridinium<br>bromide |

[1] From Gregory [10]

and negatively charged above that pH because of the acid-base chemistry of the carboxyl group ($-COOH + H_2O = -COO^- + H_3O^+$). Polydiallyldimethylammonium chloride (Poly-DADMAC) has a constant positive charge below a pH of about 10 because of the acidity of the quaternary nitrogen atoms in the molecules (Table 6.2).

Positively charged (cationic) polyelectrolytes are the most common type used in water treatment. They can function as destabilizing agents by bridge formation, charge neutralization, or both. A practical consequence of the ability of cationic polymers to adsorb on negative colloids by both electrostatic and chemical interactions and to neutralize the primary negative charge on natural particles is that these chemicals do not need a large molecular weight to be effective in destabilization. Most cationic polymers used in water treatment applications have molecular weights less than a million. Their effectiveness is also less affected by solution conditions (pH, ionic strength, hardness) than other polymers.

Most particles in water are negatively charged, and the electrostatic repulsion between them depends upon the ionic strength of the solution (Fig. 6.2). In dilute solutions, the diffuse layer surrounding the charged particles can be extensive and prevent the particles from contacting each other. Anionic and nonionic polymers function as destabilizing chemicals for such negatively charged particles by forming bridges over the gap caused by this repulsive interaction (Fig. 6.7a). Because of the thick diffuse layer in dilute solutions, only high-molecular-weight anionic and nonionic polymers should be effective bridges. In saline waters, lower-molecular-weight polymers can be effective.[26] Limited data suggest that the minimum molecular weight necessary for nonionic and anionic polymers to be effective is on the order of one million, with larger sizes used for most of these polymers. With anionic polymers, a critical electrolyte concentration is needed to promote adsorption on negatively charged particles. Ions such as $Ca^{2+}$ promote adsorption by forming complexes with anionic groups on the polymer chain and negative sites on the particle surface. Critical $Ca^{2+}$ concentrations for adsorption of anionic polymers are on the order of 1 mM (100 mg/L as $CaCO_3$), corresponding to moderately hard waters.[26]

## Coagulation diagrams

Coagulation diagrams are generalizations that are useful tools for predicting the chemical conditions under which coagulation occurs. The diagrams have been used to define the coagulant dosage and pH conditions for (1) turbidity removal, (2) color removal, (3) direct filtration, and (4) selection of rapid-mixing devices.[27–29] The generalized framework of the diagrams associates specific chemical conditions with different mechanisms of coagulation, and hence the diagrams are also valuable guides to planning, analyzing, and interpreting jar tests and pilot plant studies.

### Aluminum coagulation diagrams

**Turbidity removal.**  The diagrams (e.g., Fig. 6.12) for alum were developed by a superimposition of the chemical conditions where coagulation occurs on the thermodynamic stability diagram for $Al(OH)_3(s)$ (e.g., Fig. 6.9a and b). The coagulation diagrams typically apply for temperatures of 15 to 25°C and should be used with caution for cold water. For the development of the alum coagulation diagram,[27,28] the major

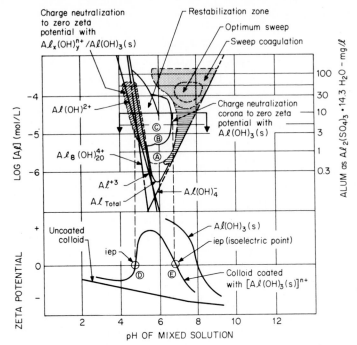

**Figure 6.12**  The alum coagulation diagram and its relationship to zeta potential.

coagulation results of several researchers[30–36] were plotted on a single diagram of log [Al] in moles per liter versus the pH of the mixed solution, and superimposed on the thermodynamic solubility diagram that formed the framework for the coagulation results. On the basis of these and other data, Amirtharajah and Mills[27] developed the alum coagulation diagram shown in Fig. 6.12 for turbidity removal.

Coagulation occurs when soluble hydrolysis species (e.g., $AlOH^{2+}$) or solid aluminum hydroxide (which can itself be charged because of surface complexes) interact with colloidal particles. The interaction between the colloid and aluminum hydroxide and the relationship of the zeta potential to the coagulation diagram are shown in the lower portion of Fig. 6.12. The isoelectric point (iep) for aluminum hydroxide is in the pH range of 7.0 to 9.0, depending on the ions in solution, often the anions. The data shown in Fig. 6.12 were based on the assumption that aluminum hydroxide has an iep of 8.0. The interaction of the positively charged aluminum hydroxide with negatively charged colloids produces two points of zero zeta potential at pH values of 4.8 and 6.8 at points $D$ and $E$ of Fig. 6.12. Favorable coagulation can be expected at both these conditions of pH. Between these two values the coated colloid is restabilized because of excess adsorption of positively charged species (see Fig. 6.6c). The corona region around the restabilization zone has a zero zeta potential and predicts the chemical conditions that are suitable for direct filtration (see Chap. 8).

Stoichiometry in coagulation at a fixed pH (for example pH = 6.0) implies a linear relationship between coagulant dose and colloid concentration and will result in a series of lines in Fig. 6.12 (e.g., A, B, C) that will define the lower horizontal boundary of the restabilization zone for increasing colloid concentration. The series of lines will be related directly with the required alum dosage for coagulation. The destabilization that causes an upper bound to exist on the restabilization zone has been thought to occur by two mechanisms. The positively charged restabilized particles can be destabilized by the negatively charged sulfate ($SO_4^{2-}$) counterions; alternatively coagulation can result from formation of a sulfate complex or enmeshment in a precipitate.[32,37] The kinetics of precipitation may be catalyzed by outer-sphere complexes of the form $Al(H_2O)SO_4^+$. A schematic diagram presented by Letterman and Vanderbrook[38] illustrating the effects of sulfate ($SO_4^{2-}$) ion on destabilizing a suspension already restabilized by an excess coating of alum hydrolysis species is shown in Fig. 6.13.

As the alum dosage is increased to nearly 30 mg/L ($10^{-4}$ $M$) with a pH of 7.0 to 8.0, the precipitation of solid aluminum hydroxide tends to occur to a greater extent and the sweep coagulation mechanism tends to dominate. On the basis of literature evidence and empirical

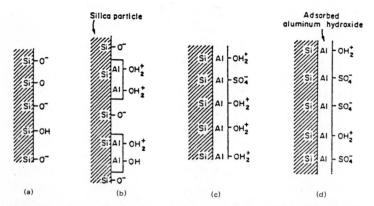

**Figure 6.13** Schematic diagram of alum-treated particles. (a) Negatively charged particle; (b) particle destabilized by charge neutralization; (c) particle restabilized by excess alum hydrolyses species; (d) destabilization by adsorption of sulfite ions. (After Letterman and Vanderbrook.[38])

results, an approximate region of optimum sweep coagulation is defined in Fig. 6.12.

Three important constraints in the use of the alum coagulation diagram need to be emphasized. First, the restabilization zone shown in Fig. 6.12 is a function of the surface area of the colloid. The zone is shown for alum dosages of 2 to 30 mg/L and pH ranges of 5.0 to 6.8. These typical domains are applicable for waters with low colloid concentrations that are generally used as sources for potable water supply. A significantly higher colloid surface area would shift the lower bound of the restabilization zone to higher alum dosages that may suppress the restabilization zone. Second, high concentrations of anions such as phosphate ($PO_4^{3-}$), silicate [$SiO(OH)_3^-$], or sulfate ($SO_4^{2-}$) may cause suppression of charge reversal and restabilization. Background sulfate ion concentrations in raw water greater than 10 to 14 mg/L as sulfate tend to eliminate charge reversal of the colloid.[35,38] See Fig. 6.13. Third, as discussed previously, a significant concentration of natural organics in the form of humic substances could control the alum dosages required for coagulation and would alter the zones of coagulation shown on the diagram.

**Color removal.** When color in the form of humic substances is present, defining the chemical conditions for effective removal on the alum coagulation diagram is also possible.[29] Regions where effective color removal occurs for two initial concentrations of humic acids are shown in Fig. 6.14. At low concentrations of color (initial color = 100 CU or TOC = 4 mg/L), two distinct regions of color removal are defined. These regions may be associated with two removal mechanisms. In the pH range of 6 to 8, conditions for rapid formation of amorphous solid $Al(OH)_3$(am) exist and removal probably occurs by adsorption of the humic acid on the precipitate of $Al(OH)_3$(am). In the pH range of 4 to 5.5, the humic acid can be assumed to be precipitated by a charge neutralization mechanism through soluble or incipiently solid phase aluminum hydrolysis species possibly leading to an aluminum-humate precipitate through specific chemical interactions. At higher initial concentrations of color (450 CU) the two zones of removal merge to form a single region of color removal (Fig. 6.14). These broad generalizations on the mechanisms of color removal by alum have been suggested and validated by Dempsey et al.[25,39] and others.[29] The color removal domains shown on Fig. 6.14 apply to high-molecular-weight (> 50,000) humic acids (Aldrich). For natural humic substances, the lower-molecular-weight fractions (< 10,000) are not removed with the same degree of high efficiency by coagulation. Therefore, other processes such as adsorption must be considered for removal of these low-molecular-weight fractions. The results in Fig. 6.14 illustrate that efficient color removal is favored by a lower pH than that

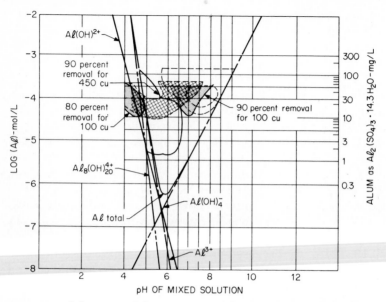

**Figure 6.14**  Color removal domains on the alum coagulation diagram for initial humic acid of 4 mg/L as TOC (100 CU) and 20 mg/L as TOC (450 CU) with zero turbidity.

for optimum turbidity removal. For example, compare the optimum sweep zone of Fig. 6.12 with the color removal zones of Fig. 6.14. Because of the importance of the natural organic matter (NOM) in a number of treatment processes, NOM is included in a following section.

**Direct filtration.**  Chemical pretreatment leading to particle destabilization is the single most important parameter controlling the effectiveness of filtration for particle removal. Therefore, the alum coagulation diagram is a suitable means by which the chemical conditions for direct filtration may be predicted. Particles in the corona region around the restabilization zone are destabilized with a zero zeta potential (see Fig. 6.12). Hence, chemically treated waters with these values of alum dosages and pH conditions will filter quite well.[40]

Effluent turbidity at various depths of an in-line filtration (rapid-mixing filtration) pilot plant with an alum dosage of 5 mg/L and variable pH conditions is shown in Fig. 6.15. Also shown in the figure is a section of the alum coagulation diagram at an alum dose of 5 mg/L. The data in this figure show that the best effluent is produced at the location of the corona region. Similar results were obtained at other alum dosages of 8 and 15 mg/L. This concept is synthesized in Fig. 6.16 with data from several sources in the literature and defines a recommended region for direct filtration of water for turbidity removal. Direct filtration as a unit process is discussed in Chap. 8.

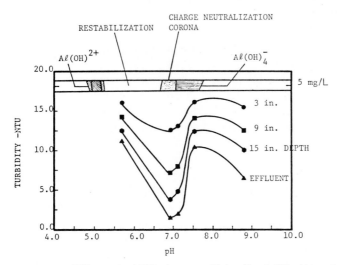

**Figure 6.15** Effluent turbidity versus pH in direct filtration at alum dose of 5 mg/L. Section of coagulation diagram from Fig. 6.12 is also shown at top of figure.

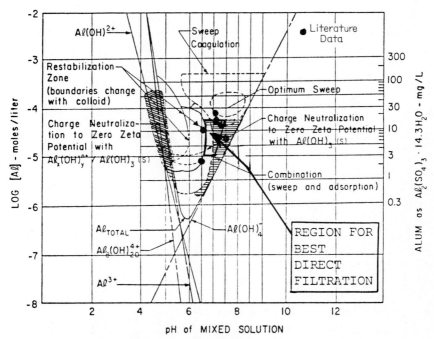

**Figure 6.16** The domain for "best" direct filtration on the alum coagulation diagram. Other studies from the literature are also shown.

### Iron coagulation diagrams

**Turbidity removal.**  The basic concepts and general conclusions indicated for alum are transferable to iron coagulants. Johnson and Amirtharajah[28] have shown the particular conditions of ferric chloride dosage and pH under which effective coagulation occurs for turbidity removal. The iron coagulation diagram is illustrated in Fig. 6.17. The original coagulation diagrams were derived from thermodynamic equations that defined the formation of amorphous ferric hydroxide. A kinetic boundary with a supersaturated concentration of iron 10 times higher than that required for the formation of the amorphous phase and boundaries for the formation of crystalline phases such as goethite ($\alpha$-FeOOH) are also included on the diagram as shown. These boundaries may enable a prediction of the speciation of polymeric forms of iron in the coagulant solutions and a characterization of their effectiveness and mode of action[41] as discussed previously with reference to Fig. 6.10.

**Color removal.**  Recent research has extended the iron coagulation diagram for color removal (Aldrich humic acid). Color removal domains on the iron coagulation diagram for two ranges of initial color, 100 CU (humic acid = 4 mg/L) and 500 CU (humic acid = 20 mg/L), are shown in Fig. 6.18. With data of this type, prediction of the chemical conditions for effective color (or DOC) removal using ferric chloride is possible. The removal efficiencies for natural humic substances will be less than that indicated in Fig. 6.18, even though the pH-dosage domains of removal will be similar.

### Coagulation of natural organic matter

Humic substances are ubiquitous in surface water supplies and are also present in many groundwaters. They are of natural origin, being

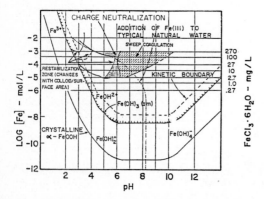

**Figure 6.17**  The iron coagulation diagram.

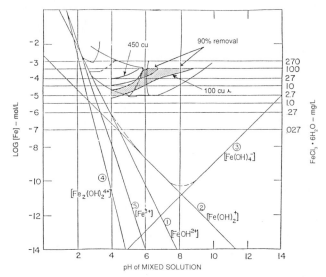

**Figure 6.18** Color removal domains on the iron coagulation diagram for initial humic acid of 4 mg/L (100 CU) and 20 mg/L (450 CU) with zero turbidity.

derived from soil and decayed wood and vegetation or produced within surface water and sediments by biological processes. They can cause several problems in water treatment plants including (1) the formation of THMs and other halogenated organics during disinfection, (2) the presence of color in water, (3) utilization of the adsorption capacity of activated carbon beds, (4) exertion of chlorine and other oxidant demands, and (5) possible transport of organic and inorganic pollutants as humic complexes through the water treatment facility.

Some excellent investigations of the properties of color and of the coagulation of colored water were conducted by Black, Packham, and their coworkers.[42–45] These studies provide much of the base for our present understanding of the coagulation of color and humic substances, and their work is useful in assessing present problems and new developments in this area.

The studies of Black and Packham[42–45] were prompted by a need to remove color from water supplies. As a result, they studied water containing high concentrations of color and humic substances. Correspondingly, the water contained high concentrations of DOC. Although not reported, the DOCs of these waters probably ranged upward from 25 mg/L. These investigators reported a direct stoichiometry between the concentration of color in the water to be treated and the concentration of iron(III) or aluminum required to accomplish effective color removal. They observed that coagulation of color was best at acidic pH values, about 4.5 to 5.5 with iron(III) and 5 to 6 with alum. They proposed that color removal was accomplished by precipitation of the color-causing humic substances by formation of Al- or Fe-humate precipitates. Present interest in the removal of DOC from water supplies extends downward to concentrations of DOC an order of magnitude lower, and this difference in concentrations can lead to differences in treatment practice as discussed subsequently. Nevertheless, the occurrence of a stoichiometry between the color in the water and the dosage of aluminum or ferric salt required for color removal reported in these early investigations is a consistent observation in more recent studies, as is the importance of pH in the removal process.

Several investigators[44,46–48] have demonstrated that humic materials can be precipitated by synthetic organic cationic polyelectrolytes used alone as coagulants. In many of these cases, the composite solids formed from humic materials and a synthetic organic polyelectrolyte were not settleable but, when the proper polymer dosage was applied, were always filterable. All these investigations demonstrated a stoichiometry between the concentration of humic material treated and the dosage of cationic polyelectrolyte required for effective coagulation and subsequent sedimentation or filtration. Good removals occurred at or near zero electrophoretic mobilities of the precipitates, with poor removals occurring at lower and higher polyelectrolyte dosages. Overdosing produced restabilization of the precipitates by forming positively charged particles. In an investigation where the charge on the untreated humic materials was measured,[46] an equivalence of charge between the anionic humic substances and the optimum dosage of cationic organic polyelectrolyte was noted. These experiments were carried out over a pH range from 5 to 8. Because the anionic charge on humic substances becomes less negative as the pH is decreased while the positive charge on many synthetic organic polyelectrolytes increases with decreasing pH, polymer requirements can decrease with decreasing pH.

Edzwald et al.[49] have demonstrated that cationic polyelectrolytes can be used in the direct filtration of waters containing humic substances, producing small destabilized particles that are filterable. In contrast, coagulant is required for particle growth in flocculation when sedimentation is used for removal. The DOC concentration below which contact opportunities in flocculation limit the production of settleable precipitates in the removal of DOC by synthetic organic polyelectrolytes has not been established. Very limited experience suggests that it is on the order of 2 mg/L or less.

Treatment of water containing such low DOC concentrations and low turbidities by organic polymers will require direct or in-line filtration; higher DOC concentrations can probably be treated either by conventional rapid-mix–flocculation–sedimentation filtration systems or by in-line or direct filtration. At high DOC concentrations, flocculation and sedimentation are required prior to filtration in order to prevent excessive head loss and short filter runs. This upper limit in DOC concentration has not been established; Edzwald et al.[49] have suggested a limit of 5 mg DOC per liter, based in part on costs associated with the high polymer dosages required to react stoichiometrically with high DOC concentrations. These investigators also report that synthetic organic cationic polyelectrolytes can be expected to remove about 40 percent of the natural TOC and THM precursors in many surface water supplies.

As discussed previously, humic substances can be removed from solution by iron(III) and aluminum salts by two mechanisms. Both involve a stoichiometry between DOC concentration and coagulation dose, and both are very pH dependent. Semmens and Field[50] indicated that the removal of natural organic substances from a water supply high in DOC (10 to 16 mg DOC per liter) by alum at pH values of 6.0 and higher was accomplished by adsorption of the organics on precipitated $Al(OH)_3(s)$, and considered that removal at a pH of 5.0 or lower occurred primarily by precipitation of soluble organic molecules by soluble aluminum species that were probably cationic aluminum polymers. In the pH range from 5.0 to 6.0, both processes were reported to occur simultaneously. The dosage of alum required to remove organic substances from this water varied with pH, and was lowest at pH 5.0.

These results reported by Semmens and Field[50] are qualitatively consistent with other recent investigations[25,39,51] that focused on waters lower in DOC. For the treatment of humic substances with aluminum salts, Dempsey and coworkers[25,39] indicate that at a pH less than 5 the removal of humic substances occurs by direct precipitation with monomers and small polymers of aluminum. At a pH greater than 7, removal occurs by adsorption of humic substances on a precipitate of $Al(OH)_3(s)$. Both processes can occur in the pH range from 5 to 7, depending on such factors as the concentration of DOC, the dosage of aluminum, and the type of aluminum coagulant (e.g., alum or PACl). These factors are illustrated by the 20 percent removal domains of fulvic acid and turbidity shown in Fig. 6.19. At low DOC concentrations (and correspondingly low aluminum dosages because of stoichiometry), the formation of humic-aluminum precipitates can be favored over sorption of humic substances on precipitated $Al(OH)_3(s)$. The result is that in the treatment of source water high in TOC, the precipitation of aluminum hydroxide is kinetically favored over the formation of humic-aluminum precipitates, coagulant dosages are higher, direct filtration is eliminated as a treatment option, and sludge production is increased.[52]

When ferric salts are used as coagulants for removal of DOC, simi-

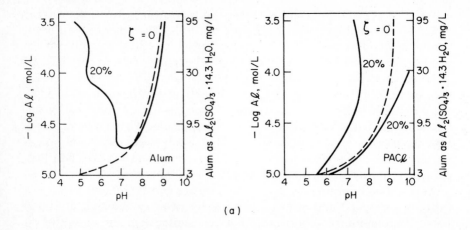

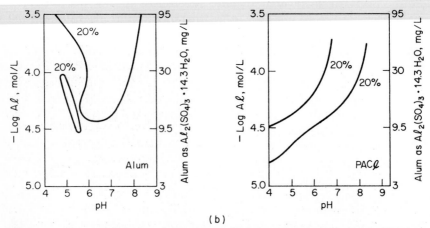

**Figure 6.19** Removal domains of 20 percent of fulvic acid with alum and PACl. Dotted lines are for zeta potential $\zeta = 0$. (*a*) Suspensions were 12.5 mg kaolinite/L and 0.17 mg FA/L; (*b*) suspensions were 0 mg kaolinite/L and 3.5 mg FA/L. (*After Dempsey et al.*[25])

lar statements can be made about the mechanisms of removal and the factors involved in establishing their occurrence, except that the specific pH limits are somewhat lower. Sinsabaugh et al.[53] showed that molecular size was the most important characteristic affecting DOC removability with ferric sulfate. Secondary factors were the charge and solubility of the organic compounds. For the reservoir water studied, the upper limit for TOC removal was 70 percent. The residual 30 percent TOC were characterized as low-molecular-weight ($< 1000$), nonpolar, neutral compounds of low to intermediate solubility. These compounds were the dominant precursors for THM and TOX formations from coagulated waters.

Edzwald and coworkers have reported approximately 60 percent removal of TOC and THM precursors from surface water by direct filtration[49] with alum and in conventional coagulation and sedimentation.[52] Although these results do not apply to all waters, they do provide a useful estimate of the effectiveness of aluminum and iron(III) salts in the treatment of natural organic substances in surface water.

### Factors affecting coagulation

The complexity of the coagulation process is evident as all the following factors affect it in diverse ways: (1) coagulant dosage; (2) pH; (3) colloid concentration, often measured by turbidity; (4) TOC and DOC (or color); (5) anions or cations in solution; (6) mixing effects; (7) electrophoretic mobility or zeta potential; and (8) temperature. The major effects of several of these variables such as coagulant dosage, pH, colloid concentration, color, and zeta potential were discussed in the prior sections and are made evident by analyzing the coagulation diagrams (Figs. 6.12, 6.14, 6.17, and 6.18). The effects of mixing are discussed in a subsequent section. The influence that ions in the solution phase and temperature have on the coagulation process is highlighted in this section.

**Ions in solution.** Specific anion effects such as the influence of sulfate ($SO_4^{2-}$) on suppressing charge reversal and on accelerating the kinetics of precipitation of aluminum hydroxide have been noted previously (Fig. 6.13). In addition, divalent cations ($Ca^{2+}$, $Mg^{2+}$, etc.) have a pronounced effect on coagulation of natural colloidal particles associated with humic substances. Black et al.[54] suggested three mechanisms by which calcium ions affect anionic polymer-clay systems: (1) divalent cations compress the double layer of a negative colloid, (2) the ions reduce the repulsive forces between the polymer and clay particles and enhance adsorption of the polymer on the clay particle, and (3) the range of the repulsive barrier between adsorbed anionic polymers is reduced by the calcium ions. Strong evidence indicates that divalent cations are generally necessary for anionic polymers to flocculate negative colloids.

O'Melia and coworkers[51,55] have recently shown that calcium ($Ca^{2+}$) exerts a destabilizing influence on natural colloidal particles that are stabilized by fulvic acids. The stability of these particles may be measured in terms of a collision efficiency factor $\alpha$, the fraction of the total number of collisions that are successful in producing aggregates by various transport processes [see later sections and Eq. (6.28)]. For water with a fulvic acid concentration of 5 mg/L as DOC, $\alpha$ increased from 0.035 to 0.2 as calcium was increased from 0 to $10^{-3}$ $M$. Hundt and O'Melia[51] found that calcium, at concentrations ranging from $5 \times 10^{-4}$ to $2 \times 10^{-3}$ $M$ and over a wide range of pH values, enhances removal of fulvic acid.

**Temperature effects.** Temperature plays an important role in turbidity removal by metal-ion coagulants; decreased removal occurs at lower temperatures. Few studies[21,56] have been made on coagulation of very cold water (0 to 4°C). Morris and Knocke[56] have recently shown that temperature had a great impact on coagulation with alum; although effects were lower with ferric chloride, they were still significant. In some typical tests, residual turbidities increased from 0.5 to 0.8 NTU at 25 and 5°C, to 2.4 NTU at 1°C with alum as a coagulant at dosages greater than 10 mg/L. The effect of decreased efficiency of coagulation of cold water has been substantiated with considerable plant-scale and pilot-scale empirical data. Some alleviation of these effects can be obtained by changing coagulants from alum to ferric chloride[56] or polymeric iron chloride[21] or by switching coagulation mechanisms from sweep coagulation to adsorption-charge neutralization by adding solids such as bentonite clay. The basic causes for the decreased efficiency of coagulation at low temperatures is still a controversial topic. Some of the reasons suggested for the decreased efficiency caused by cold temperatures are (1) increased viscosity and its effect on sedimentation, (2) change in the structure to smaller aggregates at lower temperature, and (3) decreases in the rates of hydrolysis and precipitation.

### Coagulation control strategies

Many variables affect the destabilization process and the physical and chemical properties of the agglomerated particles that are produced. The optimum dosages of coagulants for a specific water are easily and effectively determined using the simple jar test (the jar test simulates the processes of coagulation, flocculation, and sedimentation in a laboratory batch test). The effectiveness of destabilization after coagulant addition and sedimentation in jar tests may be determined from residual turbidity or other parameters such as (1) zeta potential, (2) streaming current, (3) particle counting, and (4) pilot filter effluent quality. This section presents the jar test and describes some of the parameters used for coagulation control in water treatment plant operations and design.

**Jar tests.** The jar test is universally recognized as the most valuable and most commonly used tool for coagulation control.[57–60] Singley[57] has recently summarized the application of the jar test for optimizing existing plant operations and for developing design criteria for new plant designs or for plant expansion. The jar test may be used for the following: (1) coagulant selection, (2) dosage selection, (3) coagulant aid selection and its dosage selection, (4) determination of optimum pH, (5) determination of point of addition of pH adjustment chemicals and coagulant aids, (6) optimization of mixing energy and time for rapid mixing and slow mixing (see following sections), and (7) determination of dilution of coagulant and other similar measurements.

Laboratory-scale jar testing apparatus are manufactured by a number of suppliers. They consist of several standard jars and a mixing device with standard mixer paddles. Recent innovations in jar tests[57,59] use a rectangular jar (11.5 × 11.5 × 22 cm deep) of clear plastic with a sampling tap 10 cm below the water level. The sampling tap enables characterization of surface loading rates and sedimentation process efficiency. Additional details for performing jar tests are given in the text by Hudson.[60]

A typical jar test for optimizing the dosage of coagulant includes the following steps: (1) While rapidly mixing the water, add different dosages of coagulants to the six jars that contain the same source water. Multiple syringes can be used to inject the coagulant quickly into each jar near the impeller. (2) Continue to rapidly mix the coagulants for 0.5 to 1.0 min at the maximum mixing intensity possible [commonly 100 revolutions per minute (rpm)]. (3) Slowly mix the suspensions at 25 to 35 rpm for 15 to 20 min. (4) Allow the floc to settle for 30 to 45 min without stirring. (5) Measure the turbidity of the settled water by pipetting samples of the settled water from just below the surface of water in the jars or, alternatively, by drawing samples from a port located at a fixed distance below the surface. Also, samples can be taken from the port at various times to obtain a settling velocity versus turbidity curve.[60] Typical results of the residual turbidity as a function of coagulant dose for alum and polymer may be as shown in Fig. 6.6c and d. The lowest residual turbidity corresponds to the optimum coagulant dose. The rotational speed of the stirrer in revolutions per minute can be calibrated to the average velocity gradient ($\overline{G}$ value) in the jar, and hence the jar test may be used to evaluate different mixing energies for rapid mixing and slow mixing.[60]

In addition to residual turbidity in jar tests other parameters such as zeta potential, streaming current, and particle size analysis may be used to supplement the data from jar test as strategies for coagulation control.

**Zeta potential.** The zeta potential is the potential at the plane of shear within the double layer and is shown in Fig. 6.2. It is associated with the mobility of charged particles, termed electrophoretic mobility. The zeta potential depends on the potential at the surface of the particle (the Nerst potential) and the thickness of the double layer (see Fig. 6.2). It can be measured by a zeta meter, and its value determines the extent of the electrostatic forces of repulsion between charged particles. For the measurement, the motion of a representative number of particles is timed over a fixed distance under an applied electric field.

The zeta potentials of particles in natural water are typically from $-20$ to $-40$ mV. Suspensions that are well destabilized by the charge neutralization mechanism have zeta potentials close to zero. For plant operation, the zeta potentials of coagulated particles after the rapid-mixing or slow-mixing operation in jar tests are measured. For conventional treatment, the zeta potential after the rapid mixing is preferred because large floc are difficult to measure because of sedimentation. Hence, the optimum coagulant dosage is determined when the zeta potential of the charge-neutralized particles is close to zero. The zeta potential is, however, not a reliable predictor of optimum sweep coagulation. The relationship between zeta potential and charge neutralization is also shown in the coagulation diagram for alum (Fig. 6.12). Note that the zeta potential measures the charge characteristics of particles but does not indicate whether sufficient floc volume is available for flocculation.

**Streaming current detectors.** An alternate instrument for measuring the charge on particles is the streaming current detector (SCD). This device is a cylinder containing a piston with a reciprocating motion and an annular space between piston and cylinder. A sample of the water containing the charged particles is drawn into and out of the annular space as the piston moves up and down and generates an alternating current at electrodes attached to the ends of the cylinder. This alternating current is called a streaming current, which is directly proportional to the charge on the particles. A quantitative theory explaining the generation of streaming current is not available, but a qualitative rationale has been suggested. Charged particles in the water sample are presumed to be temporarily immobilized on the piston and cylinder surfaces. As the water containing the mobile counterions in the double layer is forced past the immobile particles, the streaming current is generated.

Comparisons between zeta potential and SCD data indicate that a strong correlation exists between these measurements and that either one is suitable for determination of charge neutralization. However, the optimum SCD reading varies with source water pH so that changes in pH require different SCD goal readings. The potential advantage of the SCD is that it is capable of continuous flow monitoring for coagulation control, and feedback loop control systems may be designed for automatically adjusting the coagulant dosage for effective destabilization, if the pH remains constant.

**Residual aluminum control.** One of the reasons for using zeta potential or SCD measurements is to avoid overdosing the coagulant and thus to minimize the residual aluminum in the finished water. In the past, alum has been used extensively and successfully as a coagulant in water treatment. In recent times, questions have been raised about the health implications of elevated levels of aluminum in Alzheimer's disease and in dialysis patients. These questions have yet to be answered conclusively. In addition, postprecipitation of aluminum residuals in the distribution system has been suggested as causing significant decreases in carrying capacity. Several state, national, and international agencies have suggested as a guideline a limit of 50 μg Al per liter (0.55 mg/L as alum).

A recent survey by Letterman and Driscoll[61] of over 100 water utilities indicated that nearly 60 percent of the reporting utilities had total aluminum concentrations higher than the guideline level of 50 μg Al per liter. The survey results indicated that the high total Al concentrations in the treated water were associated with (1) high source water total Al concentration, (2) high treated water turbidity, (3) the use of atomic absorption spectrophotometry to measure Al, and (4) aluminum as a contaminant from the lime used for postfiltration pH adjustment.

The most common strategy to reduce residual aluminum levels in treated water is the adjustment of pH to 6.0, close to the minimum solubility (see Fig. 6.12). Whether pH control can significantly affect total residual aluminum levels has been questioned,[61] because changes in pH seem to affect speciation of the soluble Al but not the total Al in the treated water. Other effective strategies for reduction of total aluminum residuals are (1) use of alternative coagulants such as iron, (2) reduction of alum dosage by alum-polymer combinations, and (3) effective removal of particulate matter [may be $Al(OH)_3(s)$] during filtration.

**Example Problem 1.**    Figure 6.20a shows a schematic representation of typical jar test data using aluminum or iron salts at constant pH (see Stumm and O'Melia[62] and O'Melia[63]). (a) Analyze and interpret the jar test data on the alum coagulation diagram. (b) What would be the typical pH values for alum that would produce the data shown in Fig. 6.20a? (c) On the basis of the given data and suggestions for additional tests, recommend whether direct filtration or conventional treatment would be effective treatment trains.

**Solution**    (a) Figure 6.20b shows the charge neutralization, restabilization, and sweep coagulation regions corresponding to the schematic data in Fig. 6.20a. These are zones 2, 3, and 4 in Fig. 6.20a. The different colloid concentrations $S_1$, $S_2$, $S_3$, and $S_4$ would produce a series of lines as shown on the coagulation diagram. If the turbidity or colloid concentration is low ($S_1$), then the alum dose for charge neutralization is so low (say < 0.3 mg/L) that no coagulation occurs (because of insufficient particle contacts and particle mass) until sweep coagulation conditions are reached with alum doses greater than 30 mg/L (the high doses producing a large mass of particles). With a colloid concentration of $S_3$, coagulation will occur over a charge neutralization region (say alum doses of 2 to 5 mg/L), then restabilization (alum doses of 6 to 30 mg/L), and at higher alum doses ( > 30 mg/L) sweep coagulation will occur. With a high colloid concentration $S_4$, the charge neutralization and sweep coagulation boundaries become one at alum doses greater than about 20 mg/L. The alum doses corresponding to charge neutralization for colloid surface areas $S_2$, $S_3$, and $S_4$ will be directly correlated ("stoichiometric").

(b) From the alum coagulation diagram analysis, the pH and alkalinity of the source water had to be such that with alum addition the pH was in the range of 5.5 to 6.5. Only these pH values will give the results shown in Fig. 6.20a.

(c) For water with the low colloid concentrations $S_1$ and $S_2$, it would be feasible to add sufficient bentonite to increase particles in source water and add alum and polymer to destabilize the particles and use direct or inline filtration. Some examples of these types of waters are the cold clear waters in the northwest that are treated with this type of process train. For the water with the colloid concentration $S_3$ an optimum treatment train would be direct filtration with alum doses of 2 to 5 mg/L. This would need to be confirmed with paper filtration or pilot filtration tests. Water with the characteristics shown for $S_4$ is unlikely to be used in water treatment applications because of the very high colloid concentrations. If such water is used for water treatment, conventional treatment would be required because the alum dosage for coagulation is 20 mg/L or greater.

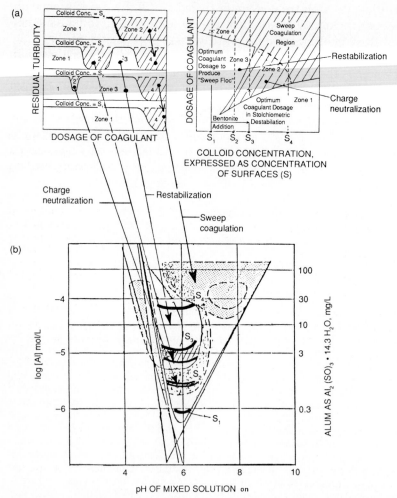

**Figure 6.20** Data and solution to Example 1. The analysis and interpretation of jar test data for coagulation with aluminum salts.

## Particle Transport Processes

Destabilized colloids must be brought into contact with one another for aggregation to occur. Three physical processes transport particles such that interparticle collisions occur. These contacts lead to the formation of particle aggregates. The transport processes are

1. *Brownian diffusion (perikinetic flocculation):*   A random motion of the particles caused by their continuous bombardment by surrounding water molecules. The driving force for this transport mechanism is the thermal energy of the fluid. The parameter quantifying this driving force is $kT$, the product of Boltzmann's constant $k$ and the absolute temperature $T$.

2. *Fluid shear (orthokinetic flocculation):*   Causes velocity differences or gradients in either laminar or turbulent fluid fields. The particles follow the motion of the suspending fluid resulting in interparticle contacts. The fluid velocity gradient $G$ is directly related to the energy dissipated per unit mass of the fluid. $G$ can be visualized as the change of velocity with distance in the flow field $dv/dz$ and thus has units of reciprocal time.

3. *Differential settling:*   Produces vertical transport of particles resulting in collisions. The driving force for differential settling is gravity, and the parameter controlling the mechanism is the settling velocity of the particle.

Current understanding of particle transport in coagulation processes derives from the work of Smoluchowski.[64] The following is a summary of the major results from more recent presentations[65,66] of the work.

### Brownian diffusion

The motion of a colloidal particle in a fluid is governed approximately by Stokes' law, which gives the drag force $F$ acting on the particle, in a direction opposite to the velocity.

$$F = -6\pi\mu Rv \tag{6.5}$$

where  $\mu$ = dynamic viscosity
       $R$ = radius of particle
       $v$ = velocity

When the particle falls vertically at a steady velocity after the initial period of acceleration, by Newton's laws,

$$0 = F_g - 6\pi\mu Rv \tag{6.6}$$

where $F_g$ is the gravitational force. If $\psi_g$ is a gravitational potential energy, then the force experienced at a distance $r$ from an assumed datum is given by $-d\psi_g/dr$ in a one-dimensional treatment. The negative sign is used because the force is in a direction opposite to increasing $r$. Hence, Eq. (6.6) can also be written as

$$0 = -\frac{d\psi_g}{dr} - 6\pi\mu Rv \tag{6.7}$$

For brownian diffusion, the random motion of particles is caused by the thermal energy of the fluid that is related to its chemical potential. The chemical potential is the partial molar Gibbs free energy of a substance, and a difference in chemical potential may be regarded as the tendency of a substance to diffuse from one place to another. A solute dissolved in a solvent has a chemical potential $\mu_i$ given (for ideal solutions) by

$$\mu_i = \mu_{i0} + kT \ln n \tag{6.8}$$

where $\mu_{i0}$, the standard state chemical potential, is a function of temperature and $n$ is the concentration of the solute. The equation for brownian diffusion is obtained in an analogous manner to Eq. (6.7) by substituting $\mu_i$ for $\psi_g$. From Eqs. (6.7) and (6.8),

$$0 = -\frac{d}{dr}(\mu_{i0} + kT \ln n) - 6\pi\mu Rv \tag{6.9}$$

Because $\mu_{i0}$ and $kT$ are not functions of $r$,

$$v = -\frac{kT}{6\pi\mu R}\frac{d}{dr}(\ln n)$$

$$= -\frac{kT}{6\pi\mu Rn}\frac{dn}{dr} \tag{6.10}$$

where $dn/dr$ is the rate of change of concentration with distance. The total flux of particles $J_j$ is given by the product velocity times concentration, $vn$, so that

$$J_j = -\frac{kT}{6\pi\mu R}\frac{dn}{dr} \tag{6.11}$$

According to Fick's law of diffusion with $D^\infty$ as the diffusion coefficient for isolated particles,

$$J_j = -D^\alpha \frac{dn}{dr} \tag{6.12}$$

Combining Eqs. (6.11) and (6.12) gives the Stokes-Einstein equation,

$$D^\alpha = \frac{kT}{6\pi\mu R} \tag{6.13}$$

Consider a colloidal suspension with one particle functioning as a central $j$ particle fixed in space. Because of brownian motion, particles around the central particle collide with it and become one with it (it functions as a sink). Hence a reduced concentration of particles around the central particle will occur and diffusion causes a steady stream of particles toward the central particle. Under steady-state conditions, the concentration at any position in space is independent of time. Assuming that the steady-state flux of particles toward the central particle is equal to the rate of capture by the central particle, this collision rate $N$ is the surface area of a sphere of radius $r$ times the particle flux $J_j$:

$$N = -4\pi r^2 J_j = \text{constant} \tag{6.14}$$

From Eqs. (6.12) and (6.14),

$$N = 4\pi r^2 D^\alpha \frac{dn}{dr} \tag{6.15}$$

Using the boundary conditions that $n = n_\infty$ at $r = \infty$ and $n = 0$ at $r = 2R$ (the radius of collision between two particles),

$$\int_0^{n_x} dn = \frac{N}{4\pi D^\alpha} \int_{2R}^{\infty} \frac{dr}{r^2}$$

After integration,

$$n_\alpha = \frac{N}{8\pi D^\alpha R}$$

Hence,

$$N = 8\pi n_x D^\alpha R \tag{6.16}$$

Equation (6.16) indicates that the rate of collisions in brownian diffusion depends on the steady-state number concentrations of particles outside the collision sphere of radius $2R$. The above result can be extended to a system containing a particle concentration of $n_i$ diffusing particles of radius $R_i$ and $n_j$ central particles per unit volume of radius $R_j$. Because the central particle can also diffuse, $D^\infty = D_i + D_j$ and $2R = R_i + R_j$. This is a critical assumption in the theory when interactions are considered in a rigorous analysis. Using this assumption,

$$N = 4\pi n_i n_j (D_i + D_j)(R_i + R_j) \tag{6.17}$$

Using Eq. (6.13), the corresponding equation for $i$ and $j$ particles is

$$N = \frac{2kT}{3\mu}\left(\frac{1}{R_i} + \frac{1}{R_j}\right)(R_i + R_j)n_i n_j \tag{6.18}$$

Equation (6.18) can be expressed in terms of diameters of particles $d_i$ and $d_j$ as

$$(N_{ij})_p = \frac{2kT(d_i + d_j)^2}{3\mu d_i d_j} n_i n_j \tag{6.19}$$

where $(N_{ij})_p$ is the rate of binary perikinetic collisions. Analysis of Eq. (6.18) shows that (1) the collision rate increases with increasing temperature, (2) the collision rate is a minimum for equally sized particles, and (3) the rate is independent of particle size for a suspension of equal sizes. The above transport theory of brownian collisions in the absence of an electrostatic potential barrier to coagulation is sometimes referred to as rapid coagulation.

For a monosized suspension $R_i = R_j$ and $n_i = n_j = n$. The value of $N$ in Eq. (6.18) must also be reduced by a factor of one-half to avoid counting each collision twice. Therefore,

$$N = \frac{4kT}{3\mu}n^2 \tag{6.20}$$

Because each collision results in a reduction in particle concentration by one, the rate of collision leads to the kinetics of the process $dn/dt$ that is second order.

$$\frac{dn}{dt} = -\frac{4kT}{3\mu}n^2 \tag{6.21}$$

Integrating Eq. (6.21) and rearranging,

$$n = \frac{n_0}{[1 + (4kTtn_0/3\mu)]} \tag{6.22}$$

where $n_0$ and $n$ are the concentrations of particles at times $t = 0$ and $t = t$. Application of Eq. (6.22) with appropriate values of $k$, $T$, and $\mu$ indicates that a water containing 10,000 viruses per milliliter and no other colloidal particles would require about 200 days before the concentration was reduced by one-half.[63] In water treatment, the addition of a few tens of milligrams per liter of coagulant increases the concentration of particles formed by nucleation and precipitation by several orders of magnitude, and hence the process of perikinetic flocculation is accelerated considerably.

### The electrostatic correction

Smoluchowski's equation (6.19) assumes that every collision leads to the formation of aggregates and that the particles are completely destabilized. If only a fraction $\alpha$ of the collisions lead to permanent aggregation, then Eq. (6.19) can be modified as

$$(N_{ij})_p = \frac{2}{3}\,\alpha\,\frac{kT(d_i + d_j)^2}{\mu d_i d_j}\,n_i n_j \tag{6.23}$$

When $\alpha = 1.0$, the rate of collisions is given by Eq. (6.19) for rapid coagulation. One of the factors causing $\alpha$ to be less than 1.0 is double-layer repulsion. The electrostatic barrier to collisions reduces the number of collisions that are effective in aggregation.

If $\psi$ is the net electrostatic potential energy, then $d\psi/dr$ is the force experienced by a particle because of electrostatic effects at a radial distance $r$ from the central $j$ particle. With the electrostatic correction, Eq. (6.9) can be modified by including $d\psi/dr$ as an additional term,

$$0 = -\frac{d\mu_i}{dr} - \frac{d\psi}{dr} - 6\pi\mu R\upsilon \tag{6.24}$$

With the extra term caused by electrostatic effects, Eq. (6.11) will be

$$J_j = -\frac{1}{6\pi\mu R}\left(kT\frac{dn}{dr} + \frac{d\psi}{dr}\right) \tag{6.25}$$

Hence,

$$(N_{ij})_p = -4\pi r^2 J_j = \frac{2r^2}{3\mu R}\left(kT\frac{dn}{dr} + \frac{d\psi}{dr}\right) \tag{6.26}$$

Integrating Eq. (6.26) with boundary conditions $n = n_0$ at $r = \infty$ and $n = 0$ at $r = 2R$, the equation corresponding to Eq. (6.16) with the electrostatic correction is

$$(N_{ij})_p = \frac{16\pi n_\infty D^\alpha R}{2R\displaystyle\int_{2R}^{\infty}(1/r^2)e^{(\psi/kT)}\,dr} \tag{6.27}$$

The above theory was developed by Fuchs,[67] and the denominator in Eq. (6.27) is called Fuchs' stability factor $W$ for slow coagulation. W can be expressed in terms of the current terminology for collision efficiency factor $\alpha$ as its reciprocal.

$$\alpha = \frac{1}{2R \int_{2R}^{\infty} (1/r^2)e^{\psi/kT}dr} \tag{6.28}$$

Using the DLVO theory (see destabilization by double-layer compression) with $\psi$ as a function of distance $r$, calculating $\alpha$ including electrostatic repulsion and van der Waals attraction forces is possible. If $\psi_{max}$ is the energy barrier in the DLVO theory, then $\alpha$ can be approximated by

$$\alpha \approx 2R\kappa e^{-(\psi_{max}/kT)} \tag{6.29}$$

in which $\kappa$ is the reciprocal thickness of the double layer.

This equation states that the collision efficiency is reduced significantly as the energy barrier (see $\psi_{max}$, Fig. 6.3c) increases. $\psi_{max}$ and $1/\kappa$ can be calculated from equations given in standard references.[1,7,9] For a suspension of ionic strength $I = 10^{-3}$ $M$ NaCl at 25°C, $\psi_{max} = 5kT$, and $d_i = 0.1$ $\mu$m. $1/\kappa$ is calculated as 9 nm and, from Eq. (6.29), $\alpha = 0.075$. Thus, even a small energy barrier of $5kT$ reduces the collision efficiency factor from 1.0 to 0.075. With this $\alpha$ correction, the period of 200 days for 50 percent reduction of viruses indicated by Eq. (6.22) will be extended to 2667 days. Modification of the above equations for binary encounters in brownian coagulation between particles of different sizes and additional details are given by Spielman[68,69] and Valioulis and List.[70]

### The hydrodynamic retardation correction

Modern analyses of particle encounters by brownian diffusion and fluid shear recognize that as particles approach each other, hydrodynamic forces tend to slow down and inhibit actual contact. These viscous forces arise because liquid must be drained between the two particles for contact to occur, and hence the average velocity of the draining liquid increases as particles approach each other. This is in contradiction with the requirement of zero velocity or no-slip condition needed on the surfaces of the contacting particles as a boundary condition from considering water as a continuum. The theoretical consequence of this analysis is that collisions are prevented altogether, unless a rapidly increasing attractive force like van der Waals attraction is simultaneously included in the computations for the collision process.

**Perikinetic encounters with corrections**

In Smoluchowski's equation for perikinetic motion the effect of hydro-dynamic forces is neglected, and as a consequence the relative diffusivity between two particles is assumed to be $D^\infty = D_i + D_j$. This assumption holds for $h/R_i$ values greater than 20 where $h$ is the sep-aration gap between the particles. At lower values of separation dis-tance the relative diffusivity between particles is a function of the sep-aration distance and the relative diffusivity $D = D(r)$. The inclusion of the hydrodynamic retardation correction using $D(r)$ in Eq. (6.27) gives

$$(N_{ij})_p = \frac{16\pi n_\infty D^\infty R}{2R \int\limits_{2R}^{\infty} \frac{D^\infty}{D(r)} \frac{1}{r^2} e^{(\psi/kT)} dr} \tag{6.30}$$

Numerical solutions for this correction are presented in several publications.[68-70]

Smoluchowski's perikinetic model ignores double-layer repulsion, van der Waals attractive forces, and viscous hydrodynamic interac-tions. Under conditions of double-layer compression or charge neutral-ization where particles are destabilized, the viscous hydrodynamic re-tardation is nullified by the attractive van der Waals forces. This explains the historical success of the Smoluchowski model ($\alpha = 1.0$) in describing rapid coagulation results. In general, for a system under slow coagulation with a significant double-layer repulsion barrier, the electrostatic correction is much larger than the hydrodynamic effect. In contrast, for double layers that are thin compared with particle di-mensions, the hydrodynamic correction is important and would reduce the value of $\alpha$ by a factor on the order of 10.

Recently Valioulis and List[70] have generated numerical computa-tions of collision efficiencies ($\alpha$ values) of particles in brownian diffu-sion including the effects of double-layer forces at constant surface charge, hydrodynamic forces, and van der Waals forces. Their results have been incorporated into a single diagram in Fig. 6.21. From this figure, the collision efficiency can be determined for varying values of the Hamaker constant $A$ in van der Waals attraction, the ionic strength $I$, and different ratios of radii of particles ($R_i/R_j$). Sudden transitions from no coagulation ($\alpha = 0$) to rapid coagulation are indi-cated by the single, almost-vertical lines for all particle radii ratios. This illustrates the importance of particle destabilization for effective collision efficiency. The figure also shows that if the particles are de-stabilized (i.e., are to the right of the vertical lines), then increases in collision efficiency factors are principally dependent on increasing the ratios of $R_i/R_j$.

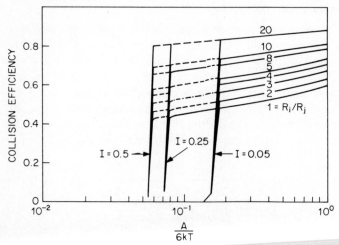

**Figure 6.21** Collision efficiencies in brownian coagulation. Variations for Hamaker's constant $A$, particle radii $R_i/R_j$, and ionic strength $I$.

### Fluid shear

In the second transport process, collisions between particles occur because of bulk motion of the fluid in laminar shear. As for brownian diffusion, consider a $j$ particle at the center of a coordinate system and the flow of liquid into a sphere of influence of radius $R_{ij} = R_i + R_j$ as in Fig. 6.22. The velocity of the fluid at a height $z$ is given by $z(dv/dz)$, as the velocity gradient is uniform. Assume that the particles follow streamlines and consider the rate of fluid flow $dq$ into the disc of thickness $dz$ (or area $dA$)

$$dq = z\left(\frac{dv}{dz}\right)dA = z\left(\frac{dv}{dz}\right)2(R_{ij}^2 - z^2)^{1/2}\,dz \qquad (6.31)$$

For $n_i$ particles of size $i$ per unit volume, the number of collisions $N_i$ with the central $j$ particle in unit time is

$$N_i = \int n_i\,dq = \int_0^{R_{ij}} 2n_i z\left(\frac{dv}{dz}\right)2(R_{ij}^2 - z^2)^{1/2}\,dz$$

$$= \frac{4}{3}n_i\,(R_{ij})^3\,\frac{dv}{dz} \qquad (6.32)$$

If the concentration of central particles is $n_j$, then the total collision rate $(N_{ij})_0$ is

$$(N_{ij})_0 = \frac{4}{3}\,n_i n_j R_{ij}^3\,\frac{dv}{dz}$$

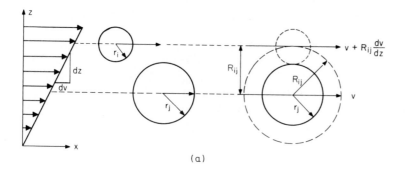

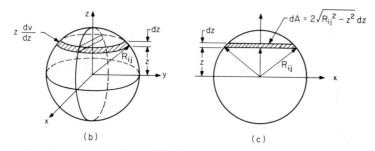

**Figure 6.22**  Schematic for Smoluchowski's theory of orthokinetic flocculation (*a*) velocity gradient and collision sphere; (*b*) flow-through collision sphere; (*c*) areal projection in direction of flow.

$$= \frac{4}{3} n_i n_j (R_i + R_j)^3 \frac{dv}{dz} \qquad (6.33)$$

Equation (6.33) is also expressed in terms of diameters of particles $d_i$ and $d_j$ and mean velocity gradient $dv/dz = \overline{G}$ as

$$(N_{ij})_0 = \frac{1}{6} n_i n_j (d_i + d_j)^3 \overline{G} \qquad (6.34)$$

Equation (6.34) and the corresponding equation for brownian diffusion [Eq. (6.19)] were developed by Smoluchowski in 1917.[64] These equations have been derived here to show their physical and mathematical bases and, more importantly, the assumptions inherent in their development so that current and more rigorous theories with the electrostatic and hydrodynamic corrections can be clearly explained.

### Orthokinetic encounters with corrections

In transport processes caused by fluid shear, effects that are similar to those described for perikinetic encounters also occur. The corrections to Smoluchowski's equation (6.34) can be listed as (1) external forces, namely, van der Waals forces, double-layer interactions, and gravity,

(2) hydrodynamic retardation, and (3) curvature of streamlines around particles.

Van de Ven and Mason[71] have included hydrodynamic interactions and external force fields in the analysis of binary encounters in orthokinetic flocculation. When double-layer repulsion is negligible, the effect of hydrodynamic retardation and van der Waals attraction gives the following empirical expression from numerical analysis.

$$\alpha \approx 0.9\left(\frac{A}{36\pi\mu\overline{G}R^3}\right)^{0.18} \tag{6.35}$$

in which $A$ is the Hamaker constant (approximately $10^{-20}$ J). For particles of radius $R = 5$ $\mu$m at 20°C and $\overline{G} = 20$ s$^{-1}$, Eq. (6.35) gives $\alpha = 0.14$. According to Eq. (6.35), hydrodynamic effects causing reductions in effective collisions are significant during orthokinetic encounters even if the particles are destabilized. In contrast to this analysis, Spielman[68] suggests that if the interacting particles are considered porous (i.e., the particles are loosely bound aggregates through which fluid flow can occur), then the Smoluchowski theory that neglects these corrections may describe destabilized flocculating systems better than analyses that incorporate them. All the above discussion applies to the case of equally sized spheres, i.e., to homocoagulation. Adler[72] has recently developed rigorous theory for heterocoagulation (i.e., unequally sized spheres) including the effects of external forces. Adler's results indicate that in most cases homocoagulation is favored over heterocoagulation. This result is in contradiction with Smoluchowski's equation that predicts a minimum in coagulation with equal-sized particles when both perikinetic and orthokinetic processes are considered together. Which theory is more applicable to practical systems is still controversial and cannot be accurately predicted at this time; it may depend on the characteristics of the particles such as their surface irregularities and their porosity.

### Turbulence and transport processes

The rigorous theories of orthokinetic transport described above have assumed laminar flow conditions. In practical treatment units for rapid mixing during particle destabilization and slow mixing during particle aggregation, the flow fields are turbulent. The applicability of the above theories to turbulent flow fields needs to be considered. Some basic concepts in turbulence are summarized subsequently in the section describing initial mixing.

Saffman and Turner[73] considered orthokinetic flocculation under two assumptions: (1) the turbulent flow field was homogeneous and isotropic, which is an idealization with the turbulence independent of position and direction, and (2) the particles were small compared to the Kolmogoroff microscale, Fig. 6.23. This microscale approximately defines size of the eddies where the energy is dissipated primarily by viscous effects. Saffman and Turner's[73] analysis was similar to that described previously for brownian motion and considered the diffusional flux of particles to a central particle, with the modification that the kinetic energy of the small-scale turbulent motion was the driving force rather than the thermal energy of the fluid. The collision frequency was derived as

$$(N_{ij})_0 = 0.162 \, n_i n_j (d_i + d_j)^3 \left(\frac{\epsilon}{\nu}\right)^{1/2}$$

$$= \frac{1}{6.18} \, n_i n_j (d_i + d_j)^3 \left(\frac{\epsilon}{\nu}\right)^{1/2} \tag{6.36}$$

in which $\epsilon$ is the rate of energy dissipation per unit mass of fluid and $\nu = \mu/\rho$ is the kinematic viscosity. Note that $(\epsilon/\nu)^{1/2}$ is equal to the mean velocity gradient $\overline{G}$. The fact that the functional form of the equation for orthokinetic collisions in turbulent fields [Eq. (6.36)] is the same as that derived by Smoluchowski for viscous conditions [Eq. (6.34)], except for the numerical constant being $1/6.18$ rather than $1/6$, is remarkable. This implies that the conceptual framework for the collision process is fundamentally similar in both laminar and turbulent conditions and also that quantitative treatments from different theoretical bases give results which are consistent. In the balance of the chapter, Eq. (6.34) will be used for both laminar and turbulent fields.

Delichatsios and Probstein[74] developed a kinetic model for flocculation in isotropic turbulent flow for both conditions where the radius of the collision sphere $(d_1 + d_2)/2$ is smaller than or greater than the Kolmogoroff microscale. For the viscous subrange the equation was similar in form to Eq. (6.36) with a constant of 0.051 instead of 0.162. For the collision sphere being larger than the microscale (i.e., the inertial subrange), the expression was

$$(N_{ij})_0 = 0.427 n_i n_j (d_i + d_j)^{7/3} \epsilon^{1/3} \tag{6.37}$$

The principal interest in this equation is the absence of viscosity in the power dissipation term. Thus for particles larger than the microscale, the collision frequency is independent of viscosity by Eq. (6.37).

### Differential settling

Camp[75] derived an expression for particle transport whereby faster-settling particles of diameter $d_i$ overtake those that settle more slowly with diameter $d_j$ and coalesce with them. In quiescent settling, one particle of diameter $d_i$ with settling velocity $v_i$ will contact in unit time all particles of diameter $d_j$ and velocity $v_j$ in a cylindrical volume given by $(\pi/4)(d_i + d_j)^2(v_i - v_j)$. If $n_i$ and $n_j$ are the number concentrations of $d_i$ and $d_j$ particles, the total number of contacts per unit time per unit volume by differential settling, $(n_{ij})_d$, is given by

$$(N_{ij})_d = \frac{\pi}{4}(d_i + d_j)^2(v_i - v_j)n_i n_j \tag{6.38}$$

In Eq. (6.5), substituting the weight of the particle for the drag force $F$ gives

$$\frac{\pi d^3}{6}(s - 1)\rho g = 6\pi\mu R v = 3\pi\mu d v$$

in which $s$ is the specific gravity of the particle and $d$ is the diameter of the particle. Rearranging with $v = \mu/\rho$, the Stokes' settling velocity is

$$v = \frac{d^2}{18}\frac{(s - 1)g}{\nu} \tag{6.39}$$

Using the Stokes' velocities for $v_i$ and $v_j$ in Eq. (6.38),

$$(N_{ij})_d = \frac{\pi g(s - 1)}{72\nu}(d_i + d_j)^3(d_i - d_j)n_i n_j \tag{6.40}$$

Relationships [Eqs. (6.19), (6.34), and (6.40)] for the transport processes of brownian diffusion, fluid shear, and differential settling are used subsequently for analyzing particle aggregation during flocculation.

## Initial Mixing

The initial mixing unit is a reactor that is usually designed to provide encounters between molecules and colloidal particles in the source water and the coagulant species, whether these species are produced within the mixing unit itself (e.g., from alum) or externally (e.g., organic cationic polyelectrolytes). These encounters are controlled by the hydrodynamic parameters, geometry, and molecular properties of the water to be treated and the kinetics of the coagulation reactions that occur in initial mixers (also called rapid-mixing units). The rapid dispersion of chemicals throughout the water for these encounters is accomplished by creating a high intensity of turbulence in the water. An understanding of the processes occurring in initial mixers can only be obtained by an analysis at the microscopic level of the mechanisms of coagulation coupled with the characteristics of the turbulent fluid field in the mixer. Proper design of the rapid-mixing unit is essential to good water treatment. If this unit is improperly selected, the coagulation-flocculation-sedimentation-filtration treatment train will be ineffective.

## Turbulence and mixing

Hinze[76] defined turbulence as an irregular condition of flow in which the various quantities show a random variation with time and space coordinates so that statistically distinct average values can be discerned. Figure 6.23 shows a schematic diagram of a turbulent field. An instantaneous velocity $U$ at a point in a turbulent field can be represented by a time average value $\overline{U}$ and a superimposed fluctuating velocity $u$ as shown in Fig. 6.24. Hence,

$$U = \overline{U} + u \qquad (6.41)$$

The mean velocity of the fluctuation $u$ is necessarily zero; however, the root mean squared (rms) velocity of fluctuations $(\overline{u^2})^{1/2}$ represents the variance of the velocity and is defined as the intensity of turbulence $u'$.

$$u' = (\overline{u^2})^{1/2} \qquad (6.42)$$

The intensity of turbulence is sometimes expressed as a fraction of the mean flow velocity, $u'/\overline{U}$. Consider the turbulent field as random motion with a family of eddies. For nonperiodic flows (i.e., flows that do

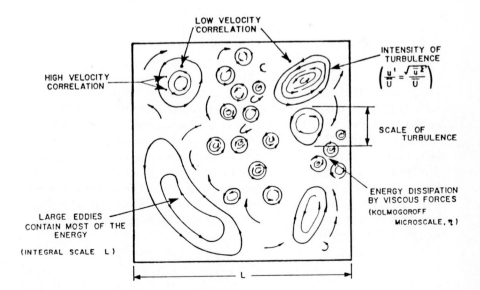

**Figure 6.23**  Schematic diagram of turbulent field.

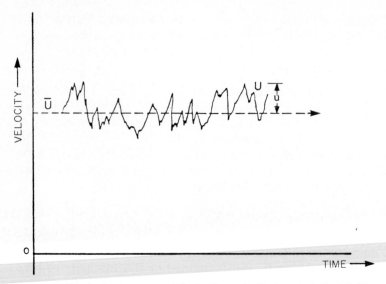

**Figure 6.24**  The instantaneous and fluctuating velocity in a turbulent field.

not have a cyclic variation), a high degree of correlation exists between velocities at distances that are small compared to the eddy diameter while little correlation could be expected at points far apart in relation to the eddy size. The length scale of turbulence can be crudely interpreted as the average size of turbulent eddies or the size of the packet of fluid within which high correlations of fluid velocity exist. The concepts of the intensity of turbulence, the scale of turbulence, and the correlation functions are illustrated schematically in Fig. 6.23. As shown in Fig. 6.23, very large vortices or large eddies arise from the interaction of the mean flow with the boundary. These eddies have a macroscale and carry a large fraction of the turbulent energy of the system. The inertial forces in the system transfer the energy via cascades from the largest eddies to the smallest eddies where they are dissipated by viscous effects into heat. This cascade of energy destroys the original turbulence characteristics of the macroscale eddies that were related to system geometry, and the smallest eddies (microscale) are in a state of universal equilibrium. For high intensities of turbulence, the universal equilibrium range is subdivided into a low eddy size, the viscous dissipation subrange, and a larger eddy size, the inertial convection subrange. These subranges are divided by the Kolmogoroff microscale $\eta$ given by

$$\eta = \left(\frac{\nu^3}{\epsilon}\right)^{1/4} \tag{6.43}$$

The above concepts of turbulence are related to the characteristics of mixing. Mixing could be defined as the commingling of two or more ingredients that retain their separate existence. The micromixed state is ultimately caused by molecular diffusion. The random motions of turbulence break up unmixed clumps of pure components reducing them in scale and increasing their surface area so that diffusion can then cause mixing to the molecular level, which is termed micromixing.

Danckwerts[77] defined the scale of segregation $L_s$ and the intensity of segregation $I_s$ in an analogous fashion to Eqs. (6.41) and (6.42). The instantaneous concentration $C$ at a point can be defined as equal to a mean value $\overline{C}$ and a fluctuating value $c$.

$$C = \overline{C} + c$$

The rms concentration fluctuation is $c' = (\overline{c})^{1/2}$, and the intensity of segregation is defined as a ratio of the mean square concentration fluctuations yielding

$$I_s = \frac{(c')^2}{(\overline{c_0}')^2} \tag{6.44}$$

where $c_0'$ is the initial mean square concentration fluctuation. The intensity of segregation defined the "goodness of mixing" and at the unmixed condition $I_s = 1$, while at the micromixed condition $I_s = 0$.

This analogous quantitative treatment of turbulence and mixing enables use of the turbulence equations to gain insights into the effects of mixing on the chemical reactions that control coagulation. For example, in a fast, irreversible reaction the extent of reaction is controlled by the extent of the mixing process. The degree of conversion for such reactions is increased when one of the reactants is in excess. Several studies in the literature[78] indicate that the formation of hydroxyaluminum species for charge neutralization is a fast, irreversible reaction. In terms of the rates of initial mixing and the kinetics of the alum hydrolysis reactions, empirical data show[78] that increased ratios of alkalinity to alum significantly increase the extent of reaction at given conditions of mixing.

### Initial mixing and mechanisms of coagulation

As noted earlier, coagulation of turbidity in water treatment by iron(III) and aluminum salts occurs predominantly by two mechanisms: (1) adsorption of hydrolysis species on the colloid causing charge neutralization and (2) sweep coagulation where interactions occur with the precipitating hydroxide. Using alum as an example, Fig. 6.25 summarizes in schematic form the predominant mechanisms. The reactions that precede charge neutralization with alum are extremely fast and occur

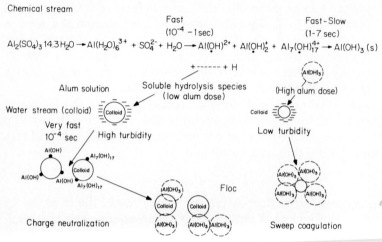

**Figure 6.25**    Reaction schematics of coagulation.

within microseconds without formation of Al(III) hydrolysis polymers and within 1 s if polymers are formed.[63,79] The formation of aluminum hydroxide precipitate before sweep coagulation is slower and occurs in the range of 1 to 7 s.[80] An analysis of the two modes of coagulation implies that for charge neutralization, rapid dispersion (less than 0.1 s) of the coagulants in the source water stream is imperative so that the hydrolysis products that develop in 0.01 to 1 s will cause destabilization of the colloid. In contrast, for sweep coagulation wherein the hydroxide formation is in the range of 1 to 7 s, extremely short dispersion times and high intensities of mixing are not as crucial as in charge neutralization. In recent times attempts have been made[27,81] to distinguish between the requirements for initial mixing on the basis of the major mode of coagulation. These will be discussed in the next two sections.

**Turbulent rapid mixing for charge neutralization with inorganic coagulants.** For colloidal particles to be destabilized by the charge neutralization mechanism of coagulation with inorganic salts, transport or collisions between the colloids and the incipiently forming products of the hydrolysis reactions must occur. This mode of coagulation at low dosages of chemicals often produces small destabilized pinpoint floc that are ideal for direct or in-line filtration. Amirtharajah and Trusler[81] have applied this concept of transport during the destabilization step to develop a theory for analyzing the required turbulent energy for initial mixing. Figure 6.26 shows schematically the framework for development of the model. The model assumes that eddies of the dimensions of the Kolmogoroff microscale of turbulence ($\eta$) interact with colloidal particles (diameter $d_i$) and cause their destabiliza-

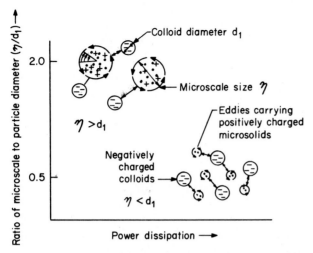

Colloid diameter $d_1$

Microscale size $\eta$

$\eta > d_1$

Eddies carrying positively charged microsolids

Negatively charged colloids

$\eta < d_1$

Ratio of microscale to particle diameter $(\eta/d_i)$ →

Power dissipation →

**Figure 6.26** Schematic diagram of destabilization caused by coagulants carried by microscale eddies. (*Appiah Amirtharajah and S. L. Trusler, "Destabilization of Particles by Turbulent Rapid Mixing," ASCE Jour. Envir. Engrg., Vol. 112, p. 1085, 1986.*)

tion. The theory used the turbulence transport equations [Eqs. (6.36) and (6.37)] for the viscous subrange ($\eta > d_i$) and the inertial subrange ($\eta < d_i$) and gave the simple result that for particle destabilization in the charge neutralization mode under all conditions of the universal equilibrium range of turbulence, rapid mixing should avoid the range of energy dissipation where $\eta = 1.33d_i$ to $2.00d_i$. In other words, rapid mixing should avoid the mixing conditions that would produce a Kolmogoroff microscale of the same order as the size of the particles to be destabilized. Amirtharajah and Trusler[81] performed limited experiments in a backmix reactor and found reasonable correspondence with the above theory. For a suspension with 3-μm colloidal particles, the theory indicated a minimum destabilization rate in the range of average velocity gradients $\overline{G}$ of 1500 to 3500 s⁻¹, which is to be avoided for effective destabilization. The results provide a theoretical rationale for the two types of rapid-mixing units used in practice, which are backmix reactors having mean $\overline{G}$ values of 700 to 1000 s⁻¹ and in-line blenders having $\overline{G}$ values of 3000 to 5000 s⁻¹. An important conclusion from the theory and experiments was that particle destabilization for charge neutralization with inorganic coagulants seems to be controlled by the turbulence around the impeller zone of a backmix-type rapid-mixing device.

**Rapid mixing for sweep coagulation.** In sweep coagulation, physical interaction occurs between the voluminous precipitates formed (iron or aluminum hydroxide) and the source water colloids. In typical water treatment practice under sweep coagulation conditions, the water is supersaturated by 3 to 4 orders of magnitude and the hydroxide is precipitated very quickly. Under these circumstances, the chemical conditions for rapid precipitation and subsequent flocculation of particles are significantly more important than transport interactions between the colloid and the hydrolysis products during destabilization. Thus, only the chemical aspects of the destabilization step and the transport aspects of flocculation are important (see schematics in Fig. 6.25). Amirtharajah and Mills[27] showed that when sweep coagulation is dominant, the results are indifferent to rapid-mixing energy inputs. Figure 6.27 shows that rapid mixing with $\overline{G}$ values from 300 to 16,000 $s^{-1}$ in jar tests produced the same settled water turbidities after 30 min. The chemical conditions for these tests were under optimum sweep coagulation conditions defined by the coagulation diagram for alum with an alum dosage of 30 mg/L and a pH of 7.8 (Fig. 6.12).

**Rapid mixing with polymers.** The mechanisms of coagulation with organic polymers are charge neutralization and interparticle bridging. Because the competing reactions of adsorption onto the colloids and precipitation as a hydroxide that occur simultaneously for the inor-

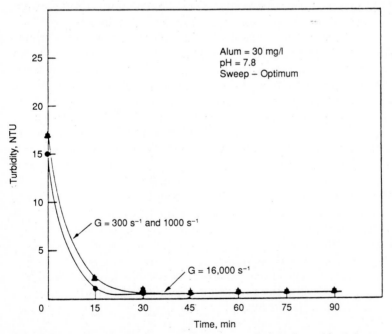

**Figure 6.27**  Response of sweep coagulation to variations in rapid mixing.

ganic metal coagulants do not occur for organic polymers, it may be assumed that high intensities of mixing are not imperative. The evidence in the literature[82–85] supports this and suggests that interparticle bridging and possible breakup of aggregated floc caused by high intensities of turbulence are important and controlling mechanisms. Leu and Ghosh[85] recently concluded that the initial rapid-mixing period is critical for flocculation with polymers used as sole coagulants.

In comparison to the inorganic coagulants, rapid-mixing criteria for low-molecular-weight cationic polymers (used as primary coagulants) are mean velocity gradients of 400 to 800 s$^{-1}$. The corresponding mixing times suggested by one study[84] are 60 s at a $\overline{G}$ of 400 s$^{-1}$ decreasing to 30 s at a $\overline{G}$ of 800 s$^{-1}$, such that $\overline{G}t$ values are in the range of 15,000 to 30,000. In contrast to these low $\overline{G}t$ values, Leu and Ghosh[85] found that optimum $\overline{G}t$ values were $2 \times 10^5$ for several polymers that were tested. The rapid-mixing and $\overline{G}t$ values noted are with polymers used alone, and the periods include the entire flocculation process. In the recent study,[85] all polyelectrolytes tested (molecular weights from $4.8 \times 10^4$ to $1.1 \times 10^7$) produced flocs that ruptured at $\overline{G}$ values greater than 350 s$^{-1}$. The $\overline{G}t$ values corresponding to the onset of floc breakup ranged from 2.8 to $4.5 \times 10^5$, higher values being associated with polymers of higher molecular weight. The above results confirm that rapid mixing with polymers is based on the flocculation stage with floc breakup as the important limiting constraint.

Tomi and Bagster[83] rationalized similar results by suggesting that during the initial stages of flocculation, aggregates grow rapidly in size reaching a critical maximum size ($d_{max}$). This maximum size is inversely related to the energy input during mixing ($d_{max} \propto \epsilon^{-0.5}$). At the critical size, rupture occurs followed by a gradual floc degradation leading to a diminution in size. This erosion mechanism is a function of time ($d_{max} \propto t^{-0.2 \text{ to} -0.5}$). These concepts confirm that both time and intensity of mixing are important parameters for polymer coagulation and $\overline{G}t$ is a valid parameter.

### Initial mixing units

Several alternative initial mixing devices are currently available for use in water treatment plants: (1) backmix reactors, (2) in-line blenders, (3) hydraulic jumps, (4) diffusers and injection devices, and (5) motionless static mixers. A comparative evaluation of these devices for use in design is described elsewhere.[86] The following is a brief description of these units, their characteristics, and the particular circumstances where they would be appropriately used. Some typical rapid-mixing units are shown in Fig. 6.28.

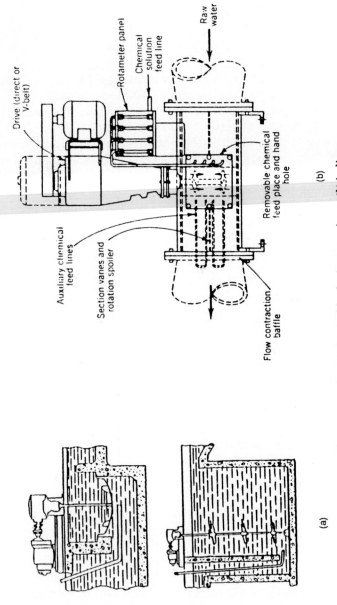

Drive (direct or V-belt)

Rotameter panel

Chemical solution feed line

Raw water

Auxiliary chemical feed lines

Section vanes and rotation spoiler

Removable chemical feed place and hand hole

Flow contraction baffle

(b)

(a)

**Figure 6.28** Rapid mixers used in practice. (*a*) Mechanical backmix type mixers; (*b*) in-line blenders.

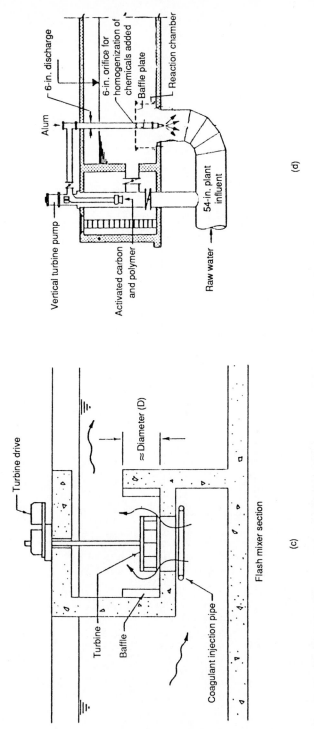

**Figure 6.28** Rapid mixers used in practice. (*Continued*) (*c*) Radial turbine flash mixer; (*d*) pump injection mixer.

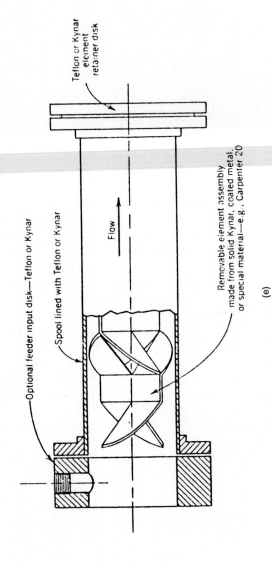

Teflon or Kynar element retainer disk

Flow

Removable element assembly made from solid Kynar, coated metal, or special material—e.g., Carpenter 20

Optional feeder input disk—Teflon or Kynar

Spool lined with Teflon or Kynar

(e)

**Figure 6.28** Rapid mixers used in practice. (*Continued*) (*e*) Static (motionless) mixer.

**Backmix mechanical reactors.**   The most common form of rapid mixer in water treatment processes is a backmix reactor shown in Fig. 6.28a. The velocity gradient or $\bar{G}$-value concept of Camp and Stein[87] is the traditional design approach used to determine the mechanical power needed for mixing. The expression for $\bar{G}$ is

$$\bar{G} = \left(\frac{P}{\mu V}\right)^{1/2} = \left(\frac{\epsilon}{\nu}\right)^{1/2} \tag{6.45}$$

where $\bar{G}$ = mean velocity gradient
   $P$ = power dissipated
   $V$ = volume of reactor

Although the $\bar{G}$ value is a somewhat simplistic parameter for the design of rapid mixers, the preceding analytical sections show that collisions between the colloids and the incipient hydrolysis products play a significant role in charge-neutralization destabilization, and collisions between particles of hydroxide play a major role in sweep coagulation. The $\bar{G}$ value is an excellent single parameter related to collision frequency based on Smoluchowski's theories; therefore, it is not surprising that it has historically played such a dominant role in both initial mixing and flocculation for over 40 years.

Energy dissipation in a stirred tank will vary with location in the tank because of inhomogeneities in the intensities of turbulence. Cutter[88] found that a stirred tank could be partitioned into three zones: (1) maximum turbulence intensity near the mixer blades, (2) an impeller stream zone, and (3) a bulk zone. The energy dissipation in the three zones and the corresponding $\bar{G}$ values in the zones as a factor of the "fictitious mean velocity gradient" $\bar{G}$ are shown schematically in Fig. 6.29. As stated above particle destabilization probably occurs in the region close to the impeller.

Gemmell[89] and the guidelines in *Water Treatment Plant Design*[90] suggest that backmix reactors be designed for 10- to 30-s contact time with mean velocity gradients of 700 to 1000 $s^{-1}$. The water horsepower (hp) input for these contact times ranges from 0.9 to 1.2 hp per million gallons per day (mgd). Letterman et al.[80] and Camp[91] suggest a rapid-mixing time of 1 to 2 min at these same velocity gradients to optimize the overall coagulation process. Providing a detention time of 1 to 2 min increases the cost of the rapid-mixing unit significantly because of the increased volume of the structure and the increased

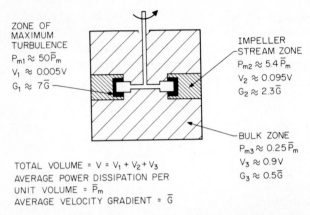

ZONE OF
MAXIMUM
TURBULENCE
$P_{m1} \approx 50\bar{P}_m$
$V_1 \approx 0.005V$
$G_1 \approx 7\bar{G}$

IMPELLER
STREAM ZONE
$P_{m2} \approx 5.4\bar{P}_m$
$V_2 \approx 0.095V$
$G_2 \approx 2.3\bar{G}$

BULK ZONE
$P_{m3} \approx 0.25\bar{P}_m$
$V_3 \approx 0.9V$
$G_3 \approx 0.5\bar{G}$

TOTAL VOLUME = $V = V_1 + V_2 + V_3$
AVERAGE POWER DISSIPATION PER
UNIT VOLUME = $\bar{P}_m$
AVERAGE VELOCITY GRADIENT = $\bar{G}$

**Figure 6.29**  Partitioned energy dissipation in a stirred tank.

power requirements for the motor. These long mixing times assist the flocculation step for sweep coagulation, which has been shown[27,78] to be indifferent to rapid-mixing conditions. Hence the least-cost design appropriate for both mechanisms of coagulation is probably an in-line blender followed by a well-designed flocculation unit. Some flexibility in operation will be lost because of omission of the backmix reactor, which is amenable to variation of $\bar{G}$ values by changing rotation speeds.

**In-line blenders.**   These manufactured devices have been a result of the concept that the hydrolysis species that cause destabilization by charge neutralization need to interact within 0.01 to 1 s with the colloid.[63,92] The value of $\bar{G}$ developed in these devices is estimated at 3000 to 5000 s$^{-1}$ with contact times of 0.5 to 1.0 s. A typical unit is shown in Fig. 6.28b. Typical power input for these devices is a water horsepower of 0.5 hp/mgd of flow. Kawamura[93] favors the use of in-line blenders because they provide good instantaneous mixing with little short circuiting and costs can be reduced by omitting a conventional rapid-mixing facility. Head losses through the unit are typically 1 to 3 ft resulting in an additional power requirement of 0.16 to 0.51 water hp/mgd of flow. The conceptual insights gained over the last few years[27,78,81] also tend to support the use of in-line blenders for most initial mixing units in water treatment.

Monk and Willis[94] illustrate (Fig. 6.28c) the design of an in-line radial turbine flash mixer used in a 600-mgd direct filtration plant. Each rapid mixer treated a flow of 50 mgd and the power requirement was 0.5 hp/mgd with a mixing time of 0.25 s.

**Hydraulic jumps.** Because many water treatment plants have a Parshall flume at the front end of a treatment train to monitor the flow rate, incorporating a hydraulic jump immediately downstream of the flume by an abrupt drop is simple and economical. The coagulants are introduced upstream of the standing wave, and typical residence times are 2 s with $\overline{G}$ values of 800 s$^{-1}$. The advantage of the hydraulic jump is the absence of any mechanical equipment; however, flexibility in operation is limited because the operator cannot modify or change the mixing intensities, and also a head loss (1 ft or greater) occurs at the jump.

**Diffusers and injection devices.** Several injection and diffuser devices have been tested[86,95,96] at the pilot scale and shown to be superior to mechanical backmix-type mixers. The coagulants are fed through orifices in a series of tubes or through venturi arrangements. These mixing devices have significant theoretical advantages that merit their use in practice. Possible disadvantages that may be limiting constraints for practical use are clogging of the orifices and inflexibility in operation, because turbulence cannot be varied.

Kawamura[93] describes the design of an injection nozzle-type initial mixing unit shown in Fig. 6.28d. The velocity gradients generated in these devices are 700 to 1000 s$^{-1}$ with mixing times on the order of 0.5 to 1.0 s.

**Motionless static mixers.** In recent times static in-line mixers have been developed. These produce turbulence and mixing from the fixed sloping vanes within the mixer. They have been effective and economical in several installations, but head loss is significant, resulting in water horsepower requirements of 0.5 to 1.0 hp/mgd. A typical unit is shown in Fig. 6.28e. A major constraint on the use of this mixer is that mixing efficiency is directly related to the flow rate through the mixer.

## Flocculation

The second stage in the overall process of coagulation is flocculation. In this stage physical processes transform smaller particles into larger aggregates or flocs. The rate of aggregation in this transformation is determined by the rate at which interparticle collisions occur. As the aggregates grow in size, hydrodynamic shearing forces can cause flocs to break up. The processes of aggregation and breakup can occur simultaneously leading to a steady-state distribution of floc sizes. Collisions between suspended particles occur by three distinct transport processes: brownian diffusion, fluid shear, and differential sedimentation. When applied to flocculation, the first two processes are often called perikinetic and orthokinetic flocculation. The quantitative characterization of these processes has been detailed in an earlier section on particle transport processes. The collision frequencies $N_{ij}$ between particles of size $d_i$ and $d_j$ of concentrations $n_i$ and $n_j$ given by Eqs. (6.19), (6.34), and (6.40) are repeated here.

Perikinetic flocculation:

$$(N_{ij})_p = \frac{2}{3} \frac{kT}{\mu} \frac{(d_i + d_j)^2}{d_i d_j} n_i n_j \tag{6.19}$$

Orthokinetic flocculation:

$$(N_{ij})_o = \frac{1}{6} (d_i + d_j)^3 \overline{G} n_i n_j \tag{6.34}$$

Differential settling:

$$(N_{ij})_d = \frac{\pi g (s - 1)}{72 v} (d_i + d_j)^3 (d_i - d_j) n_i n_j \tag{6.40}$$

These expressions for the transport mechanisms can be combined into a single equation as formulated by Friedlander.[97]

$$N_{ij} = k(i,j) n_i n_j \tag{6.46}$$

in which $k(i,j)$ is a collision frequency function similar to a "bimolecular" rate constant. This function has three separate descriptions for the three mechanisms of flocculation as shown in Table 6.3. O'Melia[98] compared the collision frequency functions for the three transport mechanisms as shown in Fig. 6.30. In Fig. 6.30a calculations are shown for the collisions of particles of size $d_j$ from 0.01 to 1000 μm

**TABLE 6.3    Agglomeration Kinetics of Colloidal Suspensions**

| Transport mechanism | Rate constant for heterodisperse suspensions | Rate constant* for $d_i = d_j = d$ |
|---|---|---|
| Brownian diffusion | $k_p = \dfrac{2}{3} \dfrac{kT}{\mu} \dfrac{(d_i + d_j)^2}{d_i d_j}$ | $k_p = \dfrac{4kT}{3\mu}$ |
| Laminar shear | $k_o = \dfrac{1}{6} (d_i + d_j)^3 \overline{G}$ | $k_o = \dfrac{2}{3} d^3 \overline{G}$ |
| Differential settling | $k_d = \dfrac{\pi g(s - 1)}{72 v} (d_i + d_j)^3 (d_i - d_j)$ | $k_d = 0$ |

*When $d_i = d_j = d$, the number of collisions has to be multiplied by one-half since the $i$ and $j$ particles are indistinguishable.
Source: From Stumm and Morgan.[1]

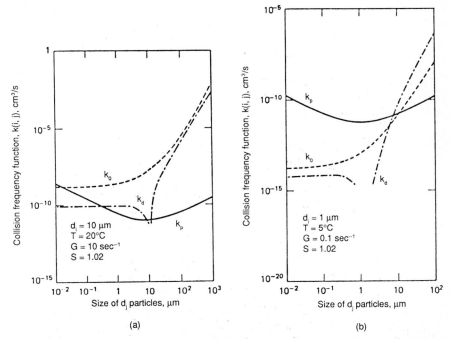

**Figure 6.30**    Effects of particle size on collision frequency function for flocculation tanks. (*C. R. O'Melia, "Coagulation in Waterwater Treatment" in K. J. Ives, ed., The Scientific Basis of Flocculation. Klewar Academic Publishers, The Netherlands, 1978.*)

with particles of size $d_i$ equal to 10 μm under a $\overline{G}$ value of 10 s$^{-1}$. A minimum in contact efficiency exists for a range of $d_j$ values from 0.1 to 10 μm, and the rate constant for shear dominates for all particle sizes $d_j$. In contrast, Fig. 6.30$b$ shows that when $d_i = 1$ μm and $\overline{G}$ is reduced to 0.1 s$^{-1}$, the rate constant for brownian diffusion dominates by several orders of magnitude when $d_j$ is less than 1 μm. The larger sizes for $d_j$ (>10 μm), however, have higher rate constants for shear and differential settling. A minimum in rate constant exists around $d_i \approx d_j \approx 1$ μm. These results suggest that heterogeneity may enhance coagulation. In contrast, when the analysis includes the corrections for hydrodynamic retardation, external forces, and curvature of streamlines for solid spheres in orthokinetic collisions, Adler[72] indicates that in most cases collisions between equal-sized spheres are favored over heterocoagulation.

All the above models for flocculation are based on collisions between particles of two sizes. In real systems, particles and flocs formed have a distribution of sizes, and hence the two particle collision models must be summed over a distribution of sizes. Lawler et al.,[99] following the formulation of Friedlander,[97] used the discrete form of the

Smoluchowski[64] equation to develop the generalized model with the collision frequency function $k(i,j)$ as

$$\frac{dn_m}{dt} = \frac{1}{2}\alpha \sum_{i+j=m} k(i,j)n_i n_j - \alpha n_m \sum_{i=1}^{\infty} k(i,j)n_i \qquad (6.47)$$

in which $dn_m/dt$ is the rate of change of number concentration of particles of size $d_m$, and $\alpha$ is the collision efficiency factor. In Eq. (6.47), the first term expresses the rate of formation of particles of size $d_m$ by the collision of two smaller particles of size $d_i$ and $d_j$ with particle volume being conserved (the coalescent sphere assumption). The second term expresses the rate of loss of particles of size $d_m$ caused by collisions with other particles to form aggregates of size larger than $d_m$. The expressions for $k(i,j)$ for the three transport mechanisms are considered additive, and hence Eq. (6.47) can be solved numerically. The above model does not include floc breakup. With the availability of personal computers and particle size analyzers, the use of the above model enables the determination of particle size distributions through the flocculation and sedimentation operations. Lawler et al.[99] modeled the expected changes in particle size distributions in the flocculation process as part of an integral design model of a water treatment plant. Recently Lawler and Wilkes[100] have tested the above model with measurements of particle sizes in a softening plant. The model was calibrated with one set of data to determine $\alpha$, chosen as a fitting parameter, and another set of data was used for model verification. Some of their results are shown in Fig. 6.31. The particle size distribution function plotted as the ordinate is developed in the next section. The model predictions agree with the experimental results for the large particles ($d_p > 8$ μm) but not for the smaller diameters. Lawler and Wilkes[100] concluded that the model predictions tend to fit the measured results for either the small or large particles but not for both. They suggested that the most likely area for improvement in the model was in the collision frequency functions for fluid shear and differential sedimentation. Figure 6.31 also illustrates the use of measurements of continuous particle size distributions and their representation in graphical form for purposes of analysis.

### Particle size distribution functions

The similarity and fundamental nature of physicochemical processes such as flocculation and sedimentation that control particle sizes in natural environmental systems (lakes and streams) and in engineered systems (flocculation and sedimentation tanks) result in particle size distributions having an equation of the same type in source and treated waters. Particle size distributions are often used to present

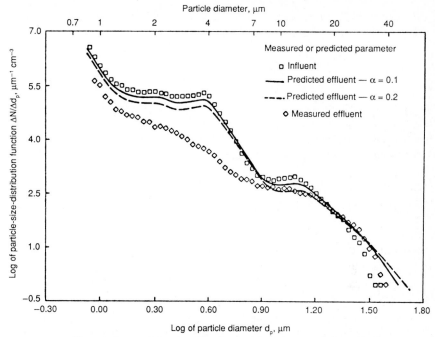

**Figure 6.31** Change in particle size distribution function, model predictions versus measured results.

data on flocculation effectiveness. Data on these distributions can be collected using particle size analyzers. These data are usefully analyzed with particle size distribution functions.

Friedlander[97] defined particle size distribution functions for diameter, surface area, and volume in the same form. Additional details on the use of these functions are given by O'Melia.[98] For number concentration,

$$\frac{dN}{d(d_p)} = \dot{n}(d_p) \tag{6.48}$$

where $dN$ = number of particles per unit fluid volume in size range $d_p$ to $d_p + d(d_p)$, number per cm$^3$
$d_p$ = diameter of particles, μm
$\dot{n}(d_p)$ = particle size distribution function, number per (cm$^3$) (μm)

The function $\dot{n}(d_p)$ is a defined analytical representation of the slope obtained from experimental measurements of the cumulative size distribution of particles. These measurements are made with particle counting instruments. Figure 6.32a shows the interrelationship between $N$ and $\dot{n}(d_p)$. Similar particle size distribution functions for

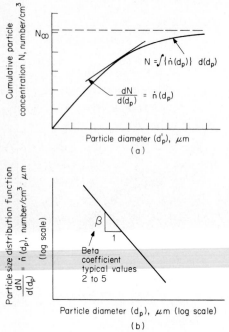

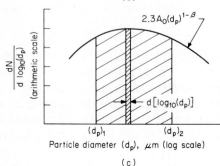

Figure 6.32 Particle size distributions: (a) cumulative function; (b) size distribution function; (c) number distribution function.

surface area $\dot{n}\,(s_p)$ and volume $\dot{n}\,(V_p)$ are defined as

$$\frac{dN}{d(s_p)} = \dot{n}\,(s_p) \tag{6.49}$$

$$\frac{dN}{d(V_p)} = \dot{n}\,(V_p) \tag{6.50}$$

where $s_p$ is the surface area of a particle in square micrometers and $V_p$ is the volume of a particle in cubic micrometers. The number concentration of particles $N_{1-2}$ between sizes $(d_p)_1$ and $(d_p)_2$ is given by

$$\int_{(d_p)1}^{(d_p)2} dN$$

Multiplying and dividing $dN$ by $d[\log (d_p)]$ gives

$$N_{1-2} = \int_{(d_p)1}^{(d_p)2} dN = \int_{(d_p)1}^{(d_p)2} \frac{dN}{d[\log (d_p)]} \, d[\log (d_p)] \qquad (6.51)$$

Equation (6.51) shows that integration or area under the curve $dN/d[\log(d_p)]$ between the limits gives the number concentration of particles between sizes $(d_p)_1$ and $(d_p)_2$. This is shown in Fig. 6.32c.

Because $2.3 \log (d_p) = \ln (d_p)$, by differentiation using the chain rule,

$$\frac{dN}{d[\log (d_p)]} = 2.3 \, (d_p) \, \dot{n} \, (d_p) \qquad (6.52)$$

Similar functions for volume and surface area distributions are given as[98]

$$\frac{dV_p}{d[\log (d_p)]} = \frac{2.3\pi}{6} \, (d_p)^4 \, \dot{n} \, (d_p) \qquad (6.53)$$

$$\frac{ds}{d[\log (d_p)]} = 2.3\pi \, (d_p)^3 \, \dot{n} \, (d_p) \qquad (6.54)$$

These theoretical formulations have a significant meaning because a considerable body of empirical data indicates[9,66] that particle size distributions in natural and engineered aerosols and aquasols follow a power law function. Thus Eq. (6.48) may be represented as,

$$\frac{dN}{d(d_p)} = \dot{n} \, (d_p) = A_0(d_p)^{-\beta} \qquad (6.55)$$

where $A_0$ is the coefficient related to the total concentration of particulate matter and $\beta$ is the coefficient that characterizes the size distribution. Equation (6.55) is an important empirical result, but some theoretical justification[101] can be given for the existence of the relationship. Figure 6.32b shows a plot of Eq. (6.55) in the straight-line form.

$$\log [\dot{n} \, (d_p)] = \log A_0 - \beta \log (d_p) \qquad (6.56)$$

In aquatic systems $\beta$ has values from 2 to 5 with much of the data[9,66] being represented by $\beta = 4$ in the particle size range $d_p \approx 1$ to $100$ $\mu$m. Substituting Eq. (6.55) in Eqs. (6.52) to (6.54) shows that the number contribution of particles over an elemental size range is proportional to $(d_p)^{1-\beta}$, whereas the volume contribution is $(d_p)^{4-\beta}$ and the surface area is $(d_p)^{3-\beta}$. Friedlander[97] showed by dimensional analysis that coagulating aerosol particles would evolve under certain circumstances into a steady-state distribution of sizes given by power law expressions with fixed $\beta$ coefficients. The underlying idea was inspired by Kolmogoroff's equilibrium theory for turbulence discussed previously. Friedlander[97] conceptualized the evolution of a steady-state size distribution by a balance between (1) the production and coagulation of primary particles and (2) the removal because of sedimentation. The analysis assumes that the size distribution is in local equilibrium, maintained by a flux of particles through the size distribution, and *that collisions between particles of similar size* determine the shape of the particle size distributions.

Hunt[101] has extended the above ideas to aquasols. In order to complete a dimensional analysis, Hunt assumed that in a particle size range where only one coagulation mechanism dominates, $\dot{n}(d_p)$ is a function of particle size $d_p$, one of the parameters for each coagulation mechanism $[k' = kT/\mu, \overline{G}, S = g(s - 1)/v]$, a parameter for removal of particles from the system by sedimentation $[g(s - 1)/v]$, and the flow rate of solids through the system $E$ (volume of solids per volume of suspension per time). Hence for the range of sizes where orthokinetic flocculation dominates,

$$\dot{n}(d_p) = f(d_p, \overline{G}, E) \tag{6.57}$$

With dimensional analysis the particle size distribution function for orthokinetic flocculation was derived as

$$\dot{n}(d_p) = A_{SH} \left(\frac{E}{\overline{G}}\right)^{1/2} (d_p)^{-4} \tag{6.58}$$

where $A_{SH}$ is a dimensionless constant. Comparing Eqs. (6.55) and (6.58), it is seen that the beta coefficient for shear is theoretically derived to be 4.0. Similar analyses for perikinetic flocculation and differential settling gave values of $\beta = 2.5$ and $4.5$, respectively. Figure 6.33 illustrates in a schematic form typical beta coefficients, diameter of particles, and the corresponding dominant coagulation mechanism. Recent research[102] suggests that these concepts are valid only for velocity gradient flocculation and are generally not applicable to perikinetic flocculation and differential sedimentation because of the strong dependence of these mechanisms on hydrodynamic and van der Waals forces and the absolute and relative sizes of interacting particles.

The above theories have important applications. The fact that $\beta$

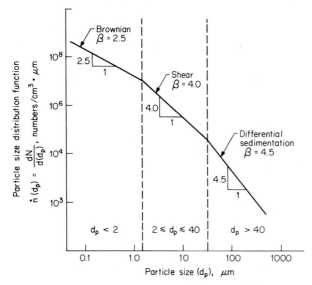

**Figure 6.33** Beta coefficients and major coagulation mechanisms for different ranges of particle sizes.

equals 4.0 for orthokinetic flocculation implies that the particle size distributions of source water from lakes would have a $\beta$ value of 4.0 in contrast to water from rivers. The particle size distribution of flocs from flocculators can be characterized by Eq. (6.58), and hence the removals in the sedimentation process may be predicted. If the $\beta$ value for a particular water is measured, using Eq. (6.56) and data from particle size analyzers, then the dominant transport process that influences the particles in that water may be determined from a comparison with $\beta$ values indicated in Fig. 6.33. Many recent papers[85,100] on flocculation use particle size distributions to determine the efficiency of alternative flocculants.

### Aggregation and breakup

Interparticle collisions cause aggregation and formation of flocs, and increased mixing with increased velocity gradients accelerates this process. However, if the agitation is too vigorous, then the turbulent shear forces developed will cause flocs to break up. Therefore, the analysis for the overall kinetics of flocculation needs to combine the processes of aggregation and breakup for a realistic description of the process, as well as for a determination of the distribution of floc sizes. In all practical applications of flocculation, the hydrodynamic regime is turbulent. Therefore, an understanding of some aspects of turbulence as summarized previously is necessary for describing certain aspects of floc breakup.

Consider the process of aggregation in orthokinetic flocculation. Modifying Eq. (6.34) with a collision efficiency factor $\alpha$ for a monosized suspension of primary particles with $d_i = d_j = d$ and $n_i = n_j = n$ gives at any instant,

$$(N_{ii})_0 = \frac{\alpha}{6} (2d)^3 \overline{G} n^2 \tag{6.59}$$

The floc volume fraction $\phi$ is defined as the volume of floc per unit volume of suspension and for spherical flocs may be represented as

$$\phi = \frac{\pi d^3}{6} n \tag{6.60}$$

Because each effective collision results in the loss of two particles, $(N_{ii})_0 = -2 \, dn/dt$. Combining this result with Eqs. (6.59) and (6.60) gives the kinetics of orthokinetic aggregation as a pseudo first-order process,

$$\frac{dn}{dt} = -\frac{4\alpha}{\pi} \phi \overline{G} n = -K \phi \overline{G} n \tag{6.61}$$

in which $K$ is the empirical coefficient dependent on system chemistry and mixing physics. Applying Eq. (6.61) in a mass balance equation for a continuous stirred tank reactor (CSTR) of volume $V$ with a flow rate $Q$ gives

$$V \frac{dn}{dt} = Q(n^0 - n) - (K \phi \overline{G} n) V \tag{6.62}$$

At steady state $dn/dt = 0$; in addition using the detention time $t_d = V/Q$ for the CSTR and $n = n^0$ at $t = 0$ gives

$$\frac{n}{n^0} = \frac{1}{1 + K \phi \overline{G} t_d} \tag{6.63}$$

As derived by Harris et al.,[103] Eq. (6.63) can be applied sequentially for a series of $m$ identical continuous stirred tank reactors to give

$$\frac{n_1^m}{n_1^0} = \left\{ \frac{1}{1 + K \phi \overline{G} (t_d/m)} \right\}^m \tag{6.64}$$

where $n_1^m$ = number of *primary* particles in effluent from $m$th reactor
$n_1^0$ = number of primary particles in influent to system
$t_d$ = overall residence time of $m$ reactors

At the present time, only limited progress has been made for a quantitative understanding of the floc disaggregation or breakup process, in comparison to particle aggregation. In dilute agitated suspensions floc breakup is caused by the interaction of individual flocs with fluid forces. A floc is an aggregate of primary particles bound together by intermolecular forces and encompassing a substantial fraction of fluid within its framework. Kaufman and coworkers[104,105] made a significant effort toward characterizing floc breakup from a microscopic viewpoint. The principal mechanisms of disaggregation or floc breakup generally accepted[68,104,105] are (1) surface erosion of primary particles from the floc and (2) fracture of the floc to form smaller aggregates. Intuitively, the simultaneous processes of aggregation and breakup will result in a stable floc size. Parker et al.[105] in their analysis of floc breakup by each of the above mechanisms defined a stable floc size. For the surface erosion model they equated the surface shearing stresses to the shear strength of the surface layers, while for the filament fracture mode the stress in a filamentous strand was equated to its tensile strength. Hence, a stable floc size $d_s$ was derived as

$$d_s = C_s \overline{G}^{-l} \tag{6.65}$$

in which $C_s$ is the coefficient related to floc strength and $l$ is the coefficient dependent on breakup mode and size regime of eddies that cause disruption. For the erosion of flocs larger than the Kolmogoroff microscale $\eta$ [Eq. (6.43)], $l = 2$ and for flocs smaller than $\eta$, $l$ was equal to 1. For the fracture mechanism, in both fluid regimes $l = 0.5$. Some experimental investigations[104,106] have reported the value of the exponent $l$ to be 1.0, i.e., the size of the largest floc is inversely proportional to $\overline{G}$. Results from Lagvankar and Gemmell's study[106] on size-density variation of flocs from an iron salt are shown in Fig. 6.34. A

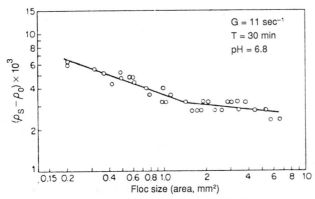

**Figure 6.34** Variation of iron floc density with floc size. Coagulant dose = $1.6 \times 10^{-4}$ $M$.

change in slope occurs at a floc size of 1.5 mm$^2$ in projected area, and flocs have an increasing density with decreasing size. At sizes larger than 1.5 mm$^2$ the floc is a loose aggregate with little variation in density. These researchers suggested that the higher-velocity gradients cause fragments of flocs to break off and that these fragments have a higher density corresponding to the size of the fragment rather than the size of the parent floc.

In a series of papers, Tambo and coworkers[107–110] have reported on extensive studies of flocculation. Their theoretical development suggested that Eq. (6.65) with numerical values for $l$ was $d_s \propto \overline{G}^{-(0.76\ \text{to}\ 0.66)}$ for flocs much smaller than the Kolmogoroff microscale $\eta$ and $d_s \propto \overline{G}^{-(1.0\ \text{to}\ 0.8)}$ when the flocs were larger than $\eta$. These values of $l$ suggest that floc breakup is occurring by both mechanisms. Results that summarize some of their significant experimental data are shown in Fig. 6.35. All the data points were on the lines shown in the figure. The alum to clay ratio (ALT) values denote the ratio of alum in milligrams per liter as aluminum (0.5 to 1000) to kaolinite clay in milligrams per liter (20 to 1000). Also shown on the figure is the microscale $\eta$. The experimental value for the slopes of the lines ($l$ values) in Fig. 6.35 is approximately 0.7, validating the theory for the viscous dissipation subrange. The size of the flocs, however, is of the same order as the Kolmogoroff microscale, contradicting the initial assumption of the theory ($d_s \ll \eta$). Tambo and Hozumi[108] suggested on the basis of experiments that the theory for the viscous subrange may be extended

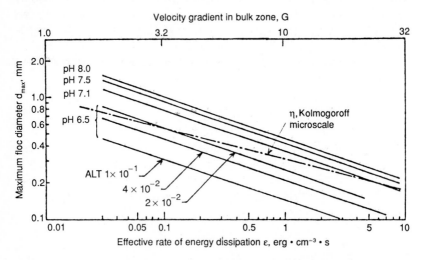

**Figure 6.35** Maximum floc diameter versus effective rate of energy dissipation in the bulk zone for clay-aluminum flocs. (*Reprinted with permission from Water Research, vol. 13, p. 429; Norohito Tambo and Y. Watanabe, "Physical Aspect of Flocculation Process. I. Fundamental Treatise," Copyright 1979, Pergamon Press.*)

to the microscale range ($d_s \approx \eta$). Amirtharajah and Trusler[81] made a similar extension of their theory for particle destabilization by turbulent rapid mixing. The velocity gradients $G$ and effective rate of energy dissipation per unit volume $\epsilon$ shown in Fig. 6.35 were computed as 15 percent of average energy dissipation $\bar{\epsilon}$ in the entire reactor. The flocculation in the reactor was assumed to occur in a bulk zone similar to that shown in Fig. 6.29. As shown in Fig. 6.35, the maximum floc diameter of clay aluminum flocs is increased with increasing pH over the range of pH from 7.0 to 8.0 at a given velocity gradient. In addition, at these pH values the maximum size of floc was independent of the ratio of alum to clay (ALT ratio from $1 \times 10^{-1}$ to $2 \times 10^{-2}$). These results correspond to the sweep coagulation domain shown in Fig. 6.12 and may be rationalized by the smaller positive zeta potential of $Al(OH)_3(s)$ as pH increases from 7.0 to 8.0. The binding forces caused by attractive van der Waals forces between aluminum hydroxide particles may be increased because of the lower repulsive barrier caused by a lower positive zeta potential, and this contributes to the larger stable size of flocs composed mainly of $Al(OH)_3(s)$. In this analysis the iep (see Fig. 6.12) for $Al(OH)_3(s)$ is assumed to be greater than a pH of 8.0 and only the characteristics of $Al(OH)_3(s)$ (i.e., not the clay) control flocculation. At a pH of 6.5, the maximum size of flocs is smaller at a given $\overline{G}$ value and is also dependent on ALT ratios. These data coincide with the charge-neutralization corona region (Figs. 6.12 and 6.16), and because the flocs are smaller in size with high density, they have characteristics that are suitable for direct filtration. The effect of varying $\overline{G}$ to produce flocs of different sizes and densities has important practical consequences and is discussed subsequently under tapered flocculation.

Parker et al.[105] expressed the rate of production of primary particles by breakup from larger particles as

$$\frac{dn_1}{dt} = K'_B \overline{G}^l \tag{6.66}$$

where $K'_B$ is the floc breakup coefficient and the exponent $l$ has values of 4 for unstable flocs greater than the microscale $\eta$ and equals 2 for flocs smaller than $\eta$. Modifying Eq. (6.61) for aggregation of primary particles $n_1$ and defining an aggregation constant $K_A = K\phi$ gives

$$\frac{dn_1}{dt} = -K_A \overline{G} n_1 \tag{6.67}$$

Equations (6.66) and (6.67) in combination give the kinetic expression for the overall process of flocculation with aggregation (first-order kinetics) and breakup (zero-order kinetics) as

$$\frac{dn_1}{dt} = -K_A\overline{G}n_1 + K'_B G^l \tag{6.68}$$

Argaman and Kaufman[104] used $l = 2$ and applied Eq. (6.68) to a flocculator system considered as a series of CSTRs as illustrated in Eq. (6.62) and developed the following:

$$\frac{n_1{}^m}{n_1{}^0} = \frac{1 + K_B\overline{G}^2 \dfrac{t_d}{m} \displaystyle\sum_{i=0}^{i=m-1} \left(1 + K_A\overline{G}\,\dfrac{t_d}{m}\right)^i}{\left(1 + K_A\overline{G}\,\dfrac{t_d}{m}\right)^m} \tag{6.69}$$

where $m$ is the number of cells or CSTRs in series and $K'_B = K_B n_1{}^0$. When $m = 1$, Eq. (6.69) simplifies to

$$\frac{n_1{}^1}{n_1{}^0} = \frac{1 + K_B\overline{G}^2 t_d}{1 + K_A\overline{G} t_d} \tag{6.70}$$

The coefficients $K_A$ and $K_B$ can be determined empirically in laboratory and pilot-scale tests. In the above development $\phi$ has been included in the coefficient $K_A$, and $n_1{}^0$ is included in the coefficient $K'_B$. Typical values for kaolin clay-alum systems and alum coagulation of natural particles are $K_A = 1.8$ to $4.5 \times 10^{-5}$ and $K_B = 0.8$ to $1.0 \times 10^{-7}$ s.

Two important conclusions can be drawn from the work of Argaman and Kaufman[104] (1) a minimum time exists below which no additional flocculation occurs, whatever the value of $\overline{G}$ (i.e., aggregation and breakup balance each other) and (2) compartmentalization significantly reduces the overall detention time required for the same degree of treatment. Figure 6.36 presents Eqs. (6.69) and (6.70) and also illustrates these conclusions. Compartmentalization is similar to a series of CSTRs, and hence the effluent from each cell is the influent to the following cell. Thus, progressively, conversion is taking place at lower concentrations of primary particles. This improves the overall conversion of primary particles into flocs, as compared to a single cell with the same detention time. Mathematically, this is illustrated by Eqs. (6.69) and (6.70). Graphically, it is illustrated in Fig. 6.36, when for example at a total detention time of 1000 s and a $\overline{G}$ value of 40 s$^{-1}$ (point $A$), the fractional ratio of primary particles remaining is approximately 0.40 for a single-cell system ($m = 1$) and equal to 0.26 for a four-cell system ($m = 4$). Amirtharajah[111] has presented examples of design of flocculators using these performance curves.

Kao and Mason[112] and Quigley and Spielman[113] have made direct visual observations on the breakup of assemblages of spheres and

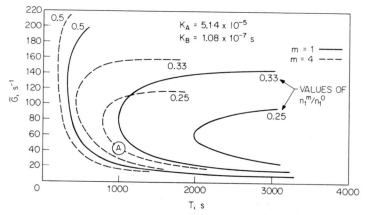

**Figure 6.36** Flocculator performance curves with aggregation and breakup. (*Y. Argaman and W. J. Kaufman, "Turbulence and Floccula-tion," ASCE Jour. Sanit., Engrg. Div., vol. 96, p. 223, 1970.*)

compacted chemical aggregates in laminar flows. They confirmed the simultaneous occurrence of the two breakup modes caused by erosion and fragmentation. In contrast to the assumptions of Kaufman's group that postulated erosion over the entire surface of the flocs, the actual erosion of primary particles was found to occur along the principal axes of extension of a floc deforming in a field of flow. Ferric hydroxide flocs underwent gross splitting into a few daughter fragments while simultaneously exhibiting continuous streaming caused by erosion of microparticles from the extremities of the fragments.[113] The erosion mode tends to reduce the size of the parent floc continuously.

Spielman and coresearchers[68,114,115] have attempted to incorporate models with the two modes of breakup into Smoluchowski's description of floc aggregation given by Eq. (6.47) to produce a generalized floc size population balance equation. The detailed floc size population balance equations may be found in the references noted. In their analyses, breakage functions for ferric hydroxide-clay floc systems by the two mechanisms had the following functional forms: (1) the floc splitting frequency varied as the 0.6 power of the parent floc volume, the first power of the turbulence shear rate, and produced two to three daughter fragments and (2) the erosion rate coefficient was independent of the turbulence shear rate.

Letterman[116] has summarized the overall effect of mixing intensity on the rate of orthokinetic flocculation as shown in Fig. 6.37. The flocculation of hypothetical particle size distributions in a batch system as a function of time and $\bar{G}$ values is illustrated by a series of curves. A high $\bar{G}$ value results in a relatively rapid ($t_0$ to $t_2$) disappearance of primary particles $d$ and the rapid formation of relatively small flocs

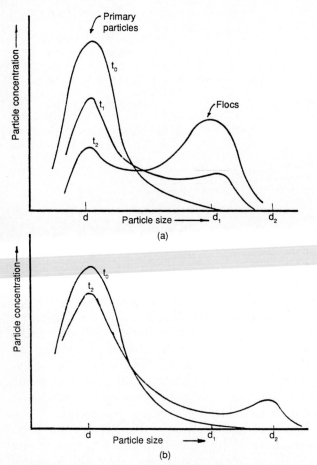

**Figure 6.37**  Hypothetical particle size distributions for flocculation in a batch system. (a) High $\overline{G}$ value; (b) low $\overline{G}$ value.
(*After Letterman.*[116])

(size $d_1$, in Fig. 6.37$a$). As shown in Fig. 6.37$b$, a lower $\overline{G}$ value causes
the rate of floc formation ($t_0$ to $t_2$) to be slower, but the size of flocs
formed $d_2$ is larger. These characteristics of floc formation are one reason for tapered flocculation, with higher $\overline{G}$ values in the first compartment and progressively smaller $\overline{G}$ values in the third and fourth cells
of a four-cell flocculator. The higher $\overline{G}$ values cause a rapid transformation of the primary particles into higher density flocs, and the
lower $\overline{G}$ value cells cause the buildup of progressively larger size flocs
for better settling. Compartmentalization and tapering improve the
process of flocculation significantly.

**Example Problem 2.**   The rates of formation and breakup of flocs in terms of
number concentration are given by the expressions $K_A \overline{G} n_1$ and $K_B \overline{G}^2 n_1{}^0$, where

$n_1$ and $n_1{}^0$ are the number concentrations of primary particles at times $t$ and $0$ and $K_A$ and $K_B$ are coefficients.

(a) (i) Derive a differential equation for the nonsteady-state variation of $n_1$ in a CSTR of detention time $t_d$ using the continuity equation. (ii) Simplify the equation for steady-state conditions, and determine the ratio of primary particles in influent to effluent. (iii) Develop an equation for the variation of $n_1$ in a plug flow reactor at steady state with a detention time $t_d$.

(b) A flocculator designed to treat 5 mgd is 60 ft long, 20 ft wide, and 10 ft deep. The total useful power input into the flocculator is 0.5 hp. Using the equations developed above compare the percentage removal of primary particles in a CSTR and plug flow configuration of the flocculator. What would be the flocculation in a real reactor? Use a temperature of 8°C and $K_A = 4.5 \times 10^{-5}$ and $K_B = 1.0 \times 10^{-7}$ s.

**Solution.**   (a) (i) Descriptive mass balance equation is

$$\text{Accumulation} = \text{inflow} - \text{outflow} \pm \text{utilization or generation}$$

$$V\frac{dn_1}{dt} = Qn_1{}^0 - Qn_1 + V(-K_A\overline{G}n_1 + K_B\overline{G}^2n_1{}^0)$$

[similar to Eq. (6.62) with breakup term included].

$$\text{Detention time} \quad t_d = \frac{V}{Q}$$

Therefore,

$$\frac{dn_1}{dt} = \frac{1}{t_d}(n_1{}^0 - n_1) - K_A\overline{G}n_1 + K_B\overline{G}^2n_1{}^0$$

$$\frac{dn_1}{dt} = n_1\left(-\frac{1}{t_d} - K_A\overline{G}\right) + \left(\frac{n_1{}^0}{t_d} + K_B\overline{G}^2n_1{}^0\right)$$

(ii) For steady-state conditions $dn_1/dt = 0$. Therefore,

$$0 = \frac{1}{t_d}(n_1{}^0 - n_1) - K_A\overline{G}n_1 + K_B\overline{G}^2n_1{}^0$$

Rearranging,

$$n_1(1 + K_A\overline{G}t_d) = n_1{}^0(1 + K_B\overline{G}^2t_d)$$

Therefore,

$$\frac{n_1}{n_1{}^0} = \frac{1 + K_B\overline{G}^2t_d}{1 + K_A\overline{G}t_d} \tag{6.70}$$

(iii) The mathematical solution for a plug flow reactor at steady state is identical to that of a batch reactor with time $t$ equal to detention time $t_d$. Hence,

$$\frac{dn_1}{dt} = -K_A\overline{G}n_1 + K_B\overline{G}^2n_1{}^0$$

Separating variables and integrating,

$$\int_{n_1{}^0}^{n_1} \frac{dn_1}{-K_A\overline{G}n_1 + K_B\overline{G}^2n_1{}^0} = \int_0^{t_d} dt$$

$$-\frac{1}{K_A\overline{G}} \ln\left(-K_A\overline{G}n_1 + K_B\overline{G}^2n_1{}^0\right)\Big]_{n_1{}^0}^{n_1} = t\Big]_0^{t_d}$$

$$-\frac{1}{K_A\overline{G}} \ln \frac{-K_A\overline{G}n_1 + K_B\overline{G}^2n_1{}^0}{-K_A\overline{G}n_1{}^0 + K_B\overline{G}^2n_1{}^0} = t_d$$

$$\frac{K_B\overline{G}^2n_1{}^0 - K_A\overline{G}n_1}{K_B\overline{G}^2n_1{}^0 - K_A\overline{G}n_1{}^0} = e^{-K_A\overline{G}t_d}$$

(b)

$$Q = 5 \text{ mgd} = 5 \times 1.547 = 7.74 \text{ ft}^3/\text{s}$$

Volume of flocculator $V = 60 \times 20 \times 10 = 12,000 \text{ ft}^3$

Detention time $t_d = \dfrac{V}{Q} = \dfrac{12,000}{7.74} = 1550 \text{ s}$

Power dissipated $P = 0.5 \text{ hp} = 0.5 \times 550 = 275 \text{ (ft)(lb)/s}$

Viscosity of water $\mu$ at 8°C (from standard tables) $= 2.9 \times 10^{-5} \text{ (lb)(s)/ft}^2$
Using Eq. (6.45),

$$\overline{G} = \left(\frac{P}{\mu V}\right)^{1/2} = \left(\frac{275}{2.9 \times 10^{-5} \times 12,000}\right)^{1/2} = 28.1 \text{ s}^{-1}$$

For the CSTR,

$$\frac{n_1}{n_1{}^0} = \frac{1 + K_B\overline{G}^2 t_d}{1 + K_A\overline{G}t_d}$$

$$= \frac{1 + 1.0 \times 10^{-7} \times (28.1)^2 \times 1550}{1 + 4.5 \times 10^{-5} \times (28.1) \times 1550}$$

$$= 0.379$$

Hence, there are 37.9 percent primary particles remaining after flocculation, or removal efficiency is 62.1 percent. For the plug flow reactor

$$\frac{K_B\overline{G}^2n_1{}^0 - K_A\overline{G}n_1}{K_B\overline{G}^2n_1{}^0 - K_A\overline{G}n_1{}^0} = e^{-K_A\overline{G}t_d}$$

$$\frac{1.0 \times 10^{-7} \times (28.1)^2 \times n_1{}^0 - 4.5 \times 10^{-5} \times 28.1 \times n_1}{1.0 \times 10^{-7} \times (28.1)^2 \times n_1{}^0 - 4.5 \times 10^{-5} \times 28.1 \times n_1{}^0}$$

$$= \exp\left[-4.5 \times 10^{-5} \times (28.1) \times 1550\right]$$

$$\frac{n_1}{n_1{}^0} = 0.195$$

Hence, there are 19.5 percent primary particles remaining after flocculation, or removal efficiency is 80.5 percent. The flocculation in a real reactor would be in between that of a CSTR and a plug flow reactor. In addition, in a real reactor three or four cells will be used in a flocculator and the velocity gradient would be tapered.

## Sludge blanket flocculation

The flocculation zones of up-flow sludge blanket or solids contact clarifiers (Fig. 6.38) function as fluidized beds with hydrodynamical similarities and dissimilarities to a horizontal flow flocculator (Fig. 6.39a, b, and d). The fluid-particle system in a sludge blanket functions like a turbulent field[111] similar to a horizontal flow unit, but a dissimilarity arises from the high concentration of solids.

Hudson,[117] Ives,[118] and Tambo and Hozumi[108] have analyzed flocculation in sludge blanket zones by starting from Smoluchowski's equation for orthokinetic flocculation and using simplifying assumptions. Consider Eq. (6.34):

$$(N_{ij})_o = \frac{1}{6}(d_i + d_j)^3 \overline{G} n_i n_j$$

In a sludge blanket, let the $n_j$ particles be flocs ($d_j \approx 1000$ to $2000$ μm) and the incoming primary particles be $n_1$ ($= n_i$) particles ($d_i = d_1 \approx 1$ to $10$ μm). Because $d_1 \ll d_j$, then $(d_1 + d_j) \approx d_j$. Therefore, the collisions between the primary particles and the flocs in the blanket are given by

$$(N_{ij})_o = \frac{1}{6} d_j^3 \overline{G} n_1 n_j \tag{6.71}$$

The number of collisions per unit time is equal to the rate of disappearance of $n_1$ particles multiplied by a collision efficiency factor $\alpha$, and therefore

$$-\frac{dn_1}{dt} = \frac{\alpha}{6} d_j^3 \overline{G} n_1 n_j \tag{6.72}$$

Using Eq. (6.60) for representing the floc volume in the blanket $n_j d_j^3/6 = \phi/\pi$ and substituting in Eq. (6.72) gives

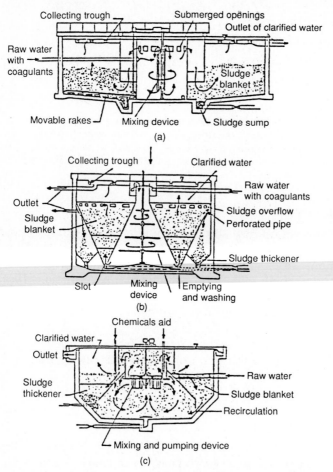

**Figure 6.38** Up-flow solids contact clarifiers. (a) Mechanically agitated bed clarifier; (b) hydraulically fluidized bed clarifier; (c) sludge circulation clarifier. (*Appiah Amirtharajah, "Design of Flocculation Systems," in R. L. Sanks, ed., Water Treatment Plant Design for the Practicing Engineer, Ann Arbor Science, Mich., 1978.*)

$$-\frac{dn_1}{dt} = \frac{\alpha\,\overline{G}\phi}{\pi}n_1 \tag{6.73}$$

The above equation is a first-order kinetic expression that can be integrated for a batch reactor by separating the variables, giving the effluent concentration of primary particles $n_1{}^{l}$ as

$$\frac{n_1{}^{1}}{n_1{}^{0}} = \exp\left(-\frac{\alpha\,\overline{G}\phi t_L}{\pi}\right) \tag{6.74}$$

where $t_L$ is the liquid detention time. The variables affecting floccula-

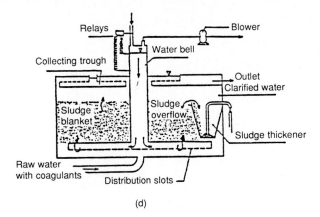

(d)

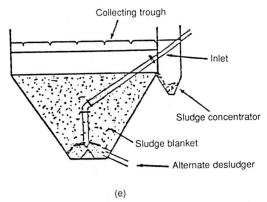

(e)

**Figure 6.38** Up-flow solids contact clarifiers. (*Continued*)
(*d*) Unsteady discharge clarifier (Pulsator) and (*e*) hopper
bottom, hydraulically fluidized clarifier.

tion in a sludge blanket may be represented by $\overline{G}\phi t_L$ as in Eq. (6.74).
The magnitude of $\overline{G}\phi t_L$ has been suggested[111] to be approximately 100
for design of sludge blanket clarifiers.

An alternate approach for analysis of these units, developed by
Bond[119,120] and Tesarik[121], considers the sludge blanket as a fluidized
bed and uses its hydrodynamic characteristics. Using hindered settling theory and the continuity equation, Bond[119] showed that

$$v_s = v_p(1 - f\phi^{2/3}) \tag{6.75}$$

where $v_s$ = settling velocity of whole suspension
$v_p$ = settling velocity of individual particles
$f$ = shape factor = 2.78 for ferric and alum floc

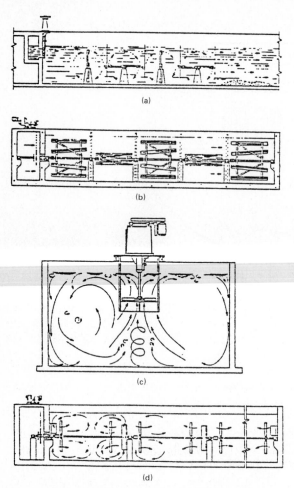

**Figure 6.39** Flocculators with mechanical mixing. (a) Paddle type; (b) reel type; (c) axial flow type with vertical axis; (d) axial flow propellor type. (*Appiah Amirtharajah, "Design of Flocculation Systems," in R. L. Sanks, ed., Water Treatment Design for the Practicing Engineer, Ann Arbor Science, Mich., 1978.*)

Bond[120] used the above equation for analysis and design of sludge blanket units by establishing an applicable value for $v_p$ based on typical values of temperature, coagulants used, and, if possible, pilot-scale tests. For design, the nominal up-flow velocity $v_u$ at the slurry separation line (which is the same as $v_s$) was equated to $0.5v_p$. For flocculation without the use of polymers, typical values for $v_p$ at 15°C were (1) floc for color removal with alum = 0.18 ft/min, (2) floc for turbidity removal with alum = 0.30 ft/min, and (3) softening floc = 0.30 ft/min.

Tesarik[121] used the following Richardson and Zaki equation for analyzing these clarifiers.

$$\frac{v_u}{v_p} = \epsilon_0^n = (1 - \phi)^n \tag{6.76}$$

where $v_u$ = nominal up-flow velocity
$\epsilon_0$ = porosity
$n$ = coefficient = 4 for inorganic hydrous oxide flocs

Application of the above expressions for design of sludge blanket clarifiers is given in more detail by Amirtharajah.[111]

The terminology that differentiates among solids contact, sludge blanket, and slurry recirculation clarifiers is not universally accepted nor well defined. Some characteristics are common to several units, and often the clarifiers have proprietary names. In general, a clarifier that utilizes a zone where previously formed and settling floc interacts with incoming particles is termed a solids contact clarifier. In this sense, all the clarifiers shown in Fig. 6.38a to e are solids contact units. If a significant recirculation of slurry or sludge between the mixing-flocculation zone and the clarification zone is used with an absence of a well-defined sludge blanket, these would be slurry recirculation units. A good example is Fig. 6.38c. In contrast, well-defined stable blankets with little mixing are formed in sludge blanket clarifiers. Good examples are Fig. 6.38d and e.

The sludge blanket unit is more sensitive to changes in flow rate and temperature. In addition, the source water solids and coagulant doses must be adequate to build a stable blanket on startup. Activated silica and other coagulant aids are often necessary for these clarifiers. Slurry recirculation solids contact units are the superior unit for softening processes. The presence of preformed calcium carbonate and magnesium hydroxide greatly accelerates the precipitation reactions. In addition, because of the slurry being recycled these units are relatively insensitive to rapid changes in source water quality. Because of the high recirculation rates, these solids contact clarifiers may also require more than the usual power for the impeller drives.

## Flocculation units

The physical unit in which the process of flocculation occurs is a mixing tank in which slow stirring promotes floc formation and conditioning. The mixing may be done mechanically or hydraulically. Three types of mechanical mixing devices are commonly used for flocculators: (1) paddle- or reel-type devices, (2) turbines, and (3) axial flow propellers. The shaft that carries the mixing device for all these types of flocculators may be installed horizontally or vertically. Some typical flocculation units are shown in Fig. 6.39.

Paddle- or reel-type devices (Fig. 6.39a and b) rotate at low speeds (2 to 15 rpm), and their tip speed is limited to 1 to 2 ft/s (0.3 to 0.7 m/s). Bench-scale studies[104] have used a stator and rotor device similar to reel-type units and found the stator-rotor configuration superior to a turbine. Plate turbine flocculators are flat-bladed devices connected to a disc or radius arm. The plane of the flat blades is in the plane of the rotating shaft having speeds of 10 to 15 rpm. Several researchers[104,111,122] have found this unit the least effective for flocculation. Peripheral velocities of the plates need to be limited to 2 ft/s for weak floc and 4 ft/s for strong floc. The third type of device is an axial flow propeller with pitched blades (Fig. 6.39c and d) inclined at 35° to the plane perpendicular to the axis of rotation. The units may be installed vertically (Fig. 6.39c) or horizontally (Fig. 6.39d) and have flow-straightening cross vanes behind or ahead of the propeller. They are high-energy flocculation devices operating at 150 to 1500 rpm and up to velocity gradients equal to 90 s$^{-1}$.

Equations (6.63) and (6.70) show that the dimensionless product $\overline{G}t_d$ is a parameter that represents the degree of flocculation. Camp[2] analyzed the designs of several existing treatment plants and found that satisfactory flocculation occurred at $\overline{G}$ values ranging from 20 to 60 s$^{-1}$ and $\overline{G}t_d$ values from $2 \times 10^4$ to $2 \times 10^5$. Most textbooks recommend $\overline{G}t_d$ values of $10^4$ to $10^5$. The large effect that particle concentration has on flocculation is not reflected in the above parameter. Equations (6.34) and (6.61) show that particle number concentration or floc volume fraction is directly related to the rate of flocculation. Several researchers have suggested[111,118] that the parameter group $\overline{G}\phi t_d \approx 100$ be used as the criterion for design of solids contact clarifiers. The value of $\overline{G}\phi t_d$ for design is approximately 10 to 30 for horizontal flow flocculators. Andreu-Villegas and Letterman[123] have indicated that optimum values exist for the velocity gradient and the flocculation time. Their empirical results can be combined[111] to give the following equation for optimum values of $\overline{G}$ for an alum floc–kaolin clay system functioning under the sweep floc mechanism.

$$(\overline{G}^*)^{2.8}Ct_d = 4 \cdot 4 \times 10^6 \tag{6.77}$$

where $\overline{G}^*$ = optimum mean velocity gradient, s$^{-1}$
$\quad\quad t_d$ = flocculation time, min
$\quad\quad C$ = alum concentration, mg/L (ranges from 10 to 50)

The prior discussion on densities of floc and breakup (Figs. 6.34, and 6.36, and 6.37) indicate that improvement and economy in flocculation may be obtained by using compartmentalization (commonly three or four cells in a flocculator) and tapering the velocity gradients from a $\overline{G}$ value of 40 to 60 s$^{-1}$ in the first cell down to 15 to 25 s$^{-1}$ in the last cell. The higher $\overline{G}$ values in the first cell enhance the production of high-density flocs quickly, and a minimum $\overline{G}$ value of 10 to 15 s$^{-1}$ is needed in the last cell to prevent the settling of the flocs in the flocculator.

For horizontal flow flocculators mixing by hydraulic means is possible. Commonly, vertical baffles are arranged for over and under or around the end flow patterns. The baffles cause hydraulic head loss or energy dissipation, which are associated with velocity gradients in the liquid stream. Typical designs use 0.8 to 1.3 ft/s for the horizontal velocity of flow. These units are useful in circumstances that seek to limit the mechanical equipment used in the design of flocculators.

A very important consideration in design of flocculators is to be certain that the hydraulic connections between the flocculators and sedimentation tanks are designed with $\overline{G}$ values less than that of the last cell of the flocculator. Otherwise the higher $\overline{G}$ values will tend to break up the flocs before they reach the sedimentation unit.

In contrast to the above flocculators, solids contact units combine the process of flocculation and up-flow clarification in a single structure. Hence significant savings in costs may be achieved. Several design variants are possible. Figure 6.38 illustrates several types of solids contact clarifiers. Some are the sludge blanket type, while others are slurry recirculation clarifiers. As an understanding of the processes occurring in a sludge blanket improves, this type of unit will be used more frequently. In fact, some of the recent innovative designs using suspended media clarifiers are using the principles of fluidized bed flocculation for flocculation and sedimentation in a single unit.

## Coagulation in Softening

The process of removing hardness is called softening. Hardness is principally caused by calcium and magnesium ions that exist as soluble chemical species [for example, $Ca(HCO_3)_2$ and $Mg(HCO_3)_2$] in hard groundwater. In lime–soda ash softening these soluble chemical species are converted into insoluble precipitates of calcium carbonate [$CaCO_3(s)$] and magnesium hydroxide [$Mg(OH)_2(s)$]. Thus, lime–soda ash softening should be analyzed as precipitation phenomena. The focus of this section is (1) to pinpoint some of the similarities and dissimilarities between coagulation with inorganic salts and organic polymers as contrasted with softening reactions, (2) summarize the factors that need to be considered when coagulants such as alum are used in conjunction with softening, and (3) discuss the removal of humic substances by softening.

Randtke[124] has indicated that the most significant difference between lime–soda ash softening and metal salt coagulation is that calcium carbonate, unlike iron and aluminum hydroxides, is precipitated as a dense crystalline solid (calcite) having a very low surface area (approximately 1 to 5 $m^2/g$) that is about two orders of magnitude lower than iron hydroxide. In addition, calcite [$CaCO_3(s)$] crystals are negatively charged, while aluminum and iron hydroxides are generally formed at pH values less than their isoelectric points and hence are positively charged (see Fig. 6.12). In contrast to calcite, magnesium precipitates as an amorphous hydroxide with large surface area at pH values greater than 11. Under most precipitation conditions $Mg(OH)_2(s)$ is also positively charged. These characteristics of the precipitates indicate that for effective calcium removal, clean crystals of $CaCO_3(s)$ are needed to act as nuclei to accelerate the kinetics of formation and precipitation of calcite. This is often achieved by using solids contact clarifiers as treatment units for softening in which sludge is recirculated. A water softening plant with calcium and magnesium removal functions as an effective coagulation unit because of the coagulating capacity of the magnesium hydroxide precipitate. The mechanisms tend to approximate sweep coagulation.

Randtke and coworkers[124–126] have studied the removal of humic substances by lime softening. They found that fulvic acids were removed by coprecipitation with rapid adsorption of the fulvic acid onto calcite crystals followed by retardation of crystal growth and changes in crystal morphology. Calcite was a poor adsorbent, and its surface repelled most humic substances. The partial removal of humic substances was caused by the affinity of certain functional groups (particularly carboxyl acids) for the calcium ions on the crystal surfaces. The removal of humic substances by softening was increased by increasing the pH, increasing the amount of solids precipitated, decreasing the TOC concentration, precipitating magnesium hydroxide, and delaying the addition of soda ash to create a calcium-rich environment.

Coagulants such as metal salts and organic polymeric coagulants are often used to agglomerate the fine calcium and magnesium precipitates to enhance their settleability. Typical dosages used are 10 mg/L or less of alum and iron salts. Black and Christman[127] investigated a number of coagulants and coagulant aids in lime–soda ash softening and found that activated silica was effective when added prior to lime additions. Organic polymers have not been extensively used in softening plants. The use of polymers for coagulation of softening sludges must be determined on an individual basis after laboratory-scale and plant-scale evaluations.

## Emerging Technology

Because of the increasingly stringent drinking water regulations and the need to adequately remove toxic inorganic and organic compounds at very low concentrations, several technologies are currently emerging to cater to these conditions. Two technologies that are becoming increasingly common are the use of (1) ozone and other oxidants, alone or in combination, and (2) membrane processes, especially reverse osmosis. Ozone has been shown to improve the process of coagulation and enhance the removal of organics. The exact mechanisms that cause this to occur are still controversial. Reverse osmosis is being considered as an alternative process for removal of low concentrations of organic and inorganic toxics. The next decade will see additional developments in emerging technologies, as well as improvements in the processes of mixing, coagulation, and flocculation.

## References

1. W. Stumm and J. J. Morgan, *Aquatic Chemistry,* 2nd ed., John Wiley & Sons, New York, 1981.
2. T. R. Camp, "Flocculation and Flocculation Basins," *ASCE Trans.,* vol. 120, 1955, pp. 1–16.
3. V. K. LaMer, "Coagulation Symposium Introduction," *Journal Colloid Science,* vol. 19, 1964, p. 291.
4. R. A. Niehof and G. I. Loeb, "The Surface Charge of Particulate Matter in Sea Water," *Limnology Oceanography,* vol. 17, 1972, p. 7.
5. K. A. Hunter and P. S. Liss, "The Surface Charge of Suspended Particles in Estuarine and Coastal Waters," *Nature,* vol. 282, 1979, p. 823.
6. J. Lyklema, "Surface Chemistry of Colloids in Connection with Stability," in K. J. Ives (ed.), *The Scientific Basis of Flocculation,* Sijthoff and Noordhoff, The Netherlands, 1978.
7. E. J. W. Verwey and J. Th. G. Overbeek, *Theory of the Stability of Lyophobic Colloids,* Elsevier, Amsterdam, 1948.
8. W. Stumm and J. J. Morgan, "Chemical Aspects of Coagulation," *Journal AWWA,* vol. 54, no. 8, 1962, p. 971.
9. F. M. M. Morel, *Principles of Aquatic Chemistry,* Wiley-Interscience, New York, 1983.
10. J. Gregory, "Effects of Polymers on Colloid Stability," in K. J. Ives (ed.), *The Scientific Basis of Flocculation,* Sijthoff and Noordhoff, The Netherlands, 1978.
11. B. Tamamushi and K. Tamaki, "The Action of Long-chain Cations on Negative Silver Iodide Sol," *Kolloid Zeitschrift,* vol. 163, 1959, p. 122.
12. R. F. Packham, "Some Studies of the Coagulation of Dispersed Clays with Hydrolyzing Salts," *Journal Colloid Interface Science,* vol. 20, 1965, p. 81.
13. V. K. LaMer and T. W. Healy, "Adsorption-Flocculation Reactions of Macromolecules at the Solid-Liquid Interface," *Reviews Pure Applied Chemistry,* vol. 13, 1963, p. 112.
14. A. S. Michaels, "Aggregation of Suspensions by Polyelectrolytes," *Industrial Engineering Chemistry,* vol. 46, 1954, p. 1485.
15. R. A. Ruehrwein and D. W. Ward, "Mechanism of Clay Coagulation by Polyelectrolytes," *Soil Science,* vol. 73, 1952, p. 485.
16. J. R. Baylis, "Silicates as Aids to Coagulation," *Journal AWWA,* vol. 29, no. 9, 1937, p. 1355.
17. W. Stumm et al., "Formulation of Polysilicates as Determined by Coagulation Effects," *Environ. Science Tech.,* vol. 1, 1967, p. 221.
18. C. R. O'Melia, "Polymeric Inorganic Flocculants," in M. Moudgil and P. Somasundaran (eds.), *Flocculation, Sedimentation, and Consolidation,* Engineering Foundation, New York, 1985.
19. C. M. Flynn, Jr., "Hydrolysis of Inorganic Iron (III) Salts," *Chemical Reviews,* vol. 84, 1984, p. 31.

20. C. F. Baes and R. E. Mesmer, *The Hydrolysis of Cations,* John Wiley & Sons, New York, 1976.
21. A. Leprince et al., "Polymerized Iron Chloride: An Improved Inorganic Coagulant," *Journal AWWA,* vol. 76, no. 10, 1984, p. 93.
22. J. Y. Bottero et al., "Studies of Hydrolyzed Aluminum Chloride Solutions. Part 1: Nature of Aluminum Species and Composition of Aqueous Solutions," *Journal Physical Chemistry,* vol. 84, 1980, p. 2933.
23. J. H. A. van der Woude and P. L. de Bruyn, "Formation of Colloidal Dispersions from Supersaturated Iron (III) Nitrate Solutions. Part 1: Precipitation of Amorphous Iron Hydroxide," *Colloids Surfaces,* vol. 8, 1983, p. 55.
24. W. Schneider, "Hydrolysis of Iron (III)—Chaotic Olation versus Nucleation," *Comments Inorganic Chemistry,* vol. 3, 1984, p. 205.
25. B. A. Dempsey et al., "Polyaluminum Chloride and Alum Coagulation of Clay-Fulvic Acid Suspensions," *Journal AWWA,* vol. 77, no. 3, 1985, p. 74.
26. J. Gregory, "The Action of Polymeric Flocculants," in B. M. Moudgil and P. Somasundaran (eds.), *Flocculation, Sedimentation, and Consolidation,* Engineering Foundation, New York, 1985.
27. A. Amirtharajah and K. M. Mills, "Rapid-Mix Design for Mechanisms of Alum Coagulation," *Journal AWWA,* vol. 74, no. 4, 1982, p. 210.
28. P. N. Johnson and A. Amirtharajah, "Ferric Chloride and Alum as Single and Dual Coagulants," *Journal AWWA,* vol. 75, no. 5, 1983, p. 232.
29. G. A. Edwards and A. Amirtharajah, "Removing Color Caused by Humic Acids," *Journal AWWA,* vol. 77, no. 3, 1985, p. 50.
30. E. Matijevic, "Colloid Stability and Complex Chemistry," *Journal Colloid Interface Science,* vol. 43, 1973, p. 217.
31. G. P. Hanna and A. J. Rubin, "Effect of Sulfate and Other Ions in Coagulation with Aluminum (III)," *Journal AWWA,* vol. 62, no. 5, 1970, p. 315.
32. A. J. Rubin and T. W. Kovac, "Effect of Aluminum (III) Hydrolysis on Alum Coagulation," in A. J. Rubin (ed.), *Chemistry of Water Supply, Treatment, and Distribution,* Ann Arbor Science, Ann Arbor, Mich., 1974.
33. R. F. Packham, "The Coagulation Process. I: Effect of pH and the Nature of the Turbidity; II. Effect of pH on the Precipitation of Aluminum Hydroxide," *Journal Applied Chemistry,* vol. 12, 1962, p. 556.
34. E. Matijevic and L. J. Stryker, "Coagulation and Reversal of Charge of Lyophobic Colloids by Hydrolyzed Metal Ions. III: Aluminum Sulfate," *Journal Colloid Interface Science,* vol. 22, 1966, p. 68.
35. P. L. Hayden and A. J. Rubin, "Systematic Investigation of the Hydrolysis and Precipitation of Aluminum (III)," in A. J. Rubin (ed.), *Aqueous Environmental Chemistry of Metal,* Ann Arbor Science, Ann Arbor, Mich., 1974.
36. N. J. McCooke and J. R. West, "The Coagulation of a Kaolinite Suspension with Aluminum Sulfate," *Water Research,* vol. 12, 1978, p. 793.
37. H. de Hek et al., "Hydrolysis-Precipitation Studies of Aluminum (III) Solutions. Part 3: The Role of the Sulfate Ion," *Journal Colloid Interface Science,* vol. 64, 1978, p. 72.
38. R. D. Letterman and S. G. Vanderbrook, "Effect of Solution Chemistry on Coagulation with Hydrolyzed Al (III): Significance of Sulfate Ion and pH," *Water Research,* vol. 17, 1983, p. 195.
39. B. A. Dempsey et al., "The Coagulation of Humic Substances by Means of Aluminum Salts," *Journal AWWA,* vol. 76, no. 4, 1984, p. 141.
40. A. Amirtharajah, "Some Theoretical and Conceptual Views of Filtration," *Journal AWWA,* vol. 80, no. 12, December 1988, p. 36.
41. C. R. O'Melia et al., "Chemical Aspects of Coagulation III," *Proc. AWWA Annual Conf.,* Washington, D.C., June 23–27, 1985.
42. R. F. Packham, "Studies of Organic Colour in Natural Water," *Proc. Water Treatment and Examination,* vol. 13, 1964, p. 316.
43. E. S. Hall and R. F. Packham, "Coagulation of Organic Color with Hydrolyzing Coagulants," *Journal AWWA,* vol. 57, no. 9, 1965, p. 1149.
44. A. P. Black and D. G. Willems, "Electrophoretic Studies of Coagulation for the Removal of Organic Color," *Journal AWWA,* vol. 53, 1961, p. 589.

45. A. P. Black and R. F. Christman, "Characteristics of Colored Surface Waters," *Journal AWWA,* vol. 55, no. 6, 1963, p. 753.
46. N. Narkis and M. Rebhun, "Stoichiometric Relationship Between Humic and Fulvic Acids and Flocculants," *Journal AWWA,* vol. 69, no. 6, 1977, p. 325.
47. J. K. Edzwald et al., "Polymer Coagulation of Humic Acid Water," *ASCE Journal Environ. Engineering Division,* vol. 103, no. EE6, 1977, p. 989.
48. H. T. Glaser and J. K. Edzwald, "Coagulation and Direct Filtration of Humic Substances with Polyethylenimine," *Environ. Science Tech.,* vol. 13, 1979, p. 299.
49. J. K. Edzwald et al., "Organics, Polymers, and Performance in Direct Filtration," *ASCE Journal Environ. Engineering,* vol. 113, no. 1, 1987, p. 167.
50. M. J. Semmens and T. K. Field, "Coagulation: Experiences in Organics Removal," *Journal AWWA,* vol. 72, no. 8, 1980, p. 476.
51. T. R. Hundt and C. R. O'Melia, "Aluminum-Fulvic Acid Interactions: Mechanisms and Applications," *Journal AWWA,* vol. 80, no. 4, 1987, p. 176.
52. R. E. Hubel and J. K. Edzwald, "Removing Trihalomethane Precursors by Coagulation," *Journal AWWA,* vol. 79, no. 7, 1987, p. 98.
53. R. L. Sinsabaugh III et al., "Removal of Dissolved Organic Carbon by Coagulation with Iron Sulfate," *Journal AWWA,* vol. 78, no. 5, 1986, p. 74.
54. A. P. Black et al., "Destabilization of Dilute Clay Suspensions with Labeled Polymers," *Journal AWWA,* vol. 57, no. 12, 1965, p. 1547.
55. C. R. O'Melia, "Particle-Particle Interactions," in W. Stumm (ed.), *Aquatic Surface Chemistry,* Wiley-Interscience, New York, 1987.
56. J. K. Morris and W. R. Knocke, "Temperature Effects on the Use of Metal Ion Coagulants for Water Treatment," *Journal AWWA,* vol. 76, no. 3, 1984, p. 74.
57. J. E. Singley, "Coagulation Control Using Jar Tests," Sem. Proc.: "Coagulation and Filtration: Back to the Basics," *Proc. AWWA Annual Conf.,* St. Louis, Mo., 1981.
58. A. P. Black et al., "Review of the Jar Test," *Journal AWWA,* vol. 49, no. 11, 1957, p. 1414.
59. H. E. Hudson, Jr., and E. G. Wagner, "Conduct and Uses of Jar Tests," *Journal AWWA,* vol. 73, no. 4, 1981, p. 218.
60. H. E. Hudson, Jr., *Water Clarification Processes Practical Design and Evaluation,* Van Nostrand Reinhold, New York, 1981.
61. R. D. Letterman and C. T. Driscoll, "Survey of Residual Aluminum in Finished Water," *Journal AWWA,* vol. 80, no. 4, 1988, p. 154.
62. W. Stumm and C. R. O'Melia, "Stoichiometry of Coagulation," *Journal AWWA,* vol. 60, no. 5, 1968, p. 514.
63. C. R. O'Melia, "Coagulation & Flocculation," in W. J. Weber, Jr. (ed.), *Physicochemical Processes for Water Quality Control,* Wiley-Interscience, New York, 1972.
64. M. Smoluchowski, "Versuch einer mathematischen Theorie der Koagulationskinetic kolloider Losunger," *Zeitschrift Physicalische Chemie,* vol. 92, 1917, p. 129.
65. J. Th. G. Overbeek, "Kinetics of Flocculation," in H. R. Kruyt (ed.), *Colloid Science 1, Irreversible Systems,* Elsevier, Amsterdam, 1952.
66. C. R. O'Melia, "Aquasols: The Behavior of Small Particles in Aquatic Systems," *Environ. Science Techn.,* vol. 14, 1980, p. 1052.
67. N. Fuchs, "Uber die Stabilitat und Aufladung der Aerosole," *Zeitschrift Physik,* vol. 89, 1934, p. 736.
68. L. A. Spielman, "Hydrodynamic Aspects of Flocculation," in K. J. Ives (ed.), *The Scientific Basis of Flocculation,* Sijthoff and Noordhoff, The Netherlands, 1978.
69. L. A. Spielman, "Viscous Interactions in Brownian Coagulation," *Journal Colloid Interface Science,* vol. 33, 1970, p. 562.
70. I. R. Valioulis and E. J. List, "Collision Efficiencies of Diffusing Spherical Particles: Hydrodynamic, van der Waals and Electrostatic Forces," *Advances Colloid Interface Science,* vol. 20, 1984, p. I.
71. T. G. M. van de Ven and S. G. Mason, "The Microrheology of Colloidal Suspensions. VII: Orthokinetic Doublet Formation of Spheres. *Colloid Polymer Science,* vol. 255, 1977, p. 468.
72. P. M. Adler, "Heterocoagulation in Shear Flow," *Journal Colloid Interface Science,* vol. 83, 1981, p. 106.

73. P. G. Saffman and J. S. Turner, "On the Collision of Drops in Turbulent Clouds," *Journal Fluid Mech.*, vol. 1, 1956, p. 16.
74. M. A. Delichatsios and R. F. Probstein, "Coagulation in Turbulent Flow: Theory and Experiment," *Journal Colloid Interface Science*, vol. 51, 1975, p. 394.
75. T. R. Camp, "Sedimentation and the Design of Settling Tanks," *ASCE Trans.*, vol. 111, 1946, p. 895.
76. J. O. Hinze, *Turbulence*, McGraw-Hill, New York, 1959.
77. P. V. Danckwerts, "The Effect of Incomplete Mixing on Homogeneous Reactions," *Chemical Engineering Science*, vol. 8, 1958, p. 93.
78. A. Amirtharajah, "Initial Mixing," Sem. Proc: "Coagulation and Filtration: Back to the Basics," *Proc. AWWA Annual Conf.*, St. Louis, Mo., 1981.
79. H. H. Hahn and W. Stumm, "Kinetics of Coagulation with Hydrolyzed Al(III)," *Journal Colloid Interface Science*, vol. 28, 1968, p. 134.
80. R. D. Letterman et al., "Influence of Rapid-Mix Parameters on Flocculation," *Journal AWWA*, vol. 65, no. 11, 1973, p. 716.
81. A. Amirtharajah and S. L. Trusler, "Destabilization of Particles by Turbulent Rapid Mixing," *ASCE Journal Environ. Engineering*, vol. 112, 1986, p. 1085.
82. A. Amirtharajah and S. Kawamura, "System Design for Polymer Use," Sem. Proc: "Use of Organic Polyelectrolytes in Water Treatment," *Proc. AWWA Annual Conf.*, Las Vegas, Nev., 1983.
83. D. T. Tomi and D. F. Bagster, "The Behavior of Aggregates in Stirred Vessels," *Trans. Inst. Chemical Engineers*, vol. 56, 1978, p. 1.
84. R. O. Keys and R. Hogg, "Mixing Problems in Polymer Flocculation," *Water–1979*, AIChE Symp. Series No. 190, vol. 75, 1978, p. 63.
85. R-J. Leu and M. M. Ghosh, "Polyelectrolyte Characteristics and Flocculation," *Journal AWWA*, vol. 80, no. 4, 1988, p. 159.
86. A. Amirtharajah, "Design of Rapid Mix Units," in R. L. Sanks (ed.), *Water Treatment Plant Design for the Practicing Engineer*, Ann Arbor Science, Ann Arbor, Mich., 1978.
87. T. R. Camp and P. C. Stein, "Velocity Gradients and Internal Work in Fluid Motion," *Journal Boston Soc. Civil Engineers*, vol. 30, 1943, p. 219.
88. L. A. Cutter, "Flow and Turbulence in a Stirred Tank," *American Inst. Chemical Engineers Journal*, vol. 12, 1966, p. 35.
89. R. S. Gemmell, "Mixing and Sedimentation," in AWWA, Inc., *Water Quality and Treatment, 3rd ed.*, McGraw-Hill, New York, 1971.
90. ASCE, AWWA, and CSSE, *Water Treatment Plant Design*, AWWA, New York, 1969.
91. T. R. Camp, "Floc Volume Concentration," *Journal AWWA*, vol. 60, no. 6, 1968, p. 656.
92. H. E. Hudson, Jr., and J. P. Wolfner, "Design of Mixing and Flocculation Basins," *Journal AWWA*, vol. 59, no. 10, 1967, p. 1257.
93. S. Kawamura, "Considerations on Improving Flocculation," *Journal AWWA*, vol. 68, no. 6, 1976, p. 328.
94. R. D. G. Monk, and J. F. Willis, "Designing Water Treatment Facilities," *Journal AWWA*, vol. 79, no. 2, 1987, p. 45.
95. R. J. Stenquist and W. Kaufman, "Initial Mixing in Coagulation Processes," SERL Report No. 72-2, College of Engineering, University of California at Berkeley, 1972.
96. L. Vrale and R. M. Jordan, "Rapid Mixing in Water Treatment," *Journal AWWA*, vol. 63, no. 1, 1971, p. 52.
97. S. K. Friedlander, *Smoke, Dust, and Haze*, Wiley-Interscience, New York, 1977.
98. C. R. O'Melia, "Coagulation in Wastewater Treatment," in K. J. Ives (ed.), *The Scientific Basis of Flocculation*, Sijthoff and Noordhoff, The Netherlands, 1978.
99. D. F. Lawler et al., "Integral Water Treatment Plant Design: From Particle Size to Plant Performance," in M. C. Kavanaugh and J. O. Leckie (eds.), *Particulates in Water Characterization, Fate, Effects, and Removal*, Advances in Chemistry Series No. 189, American Chemical Society, Washington, D. C., 1980.
100. D. F. Lawler and D. R. Wilkes, "Flocculation Model Testing: Particle Sizes in a Softening Plant," *Journal AWWA*, vol. 76, no. 7, 1984, p. 90.

101. J. R. Hunt, "Self-Similar Particle-Size Distributions During Coagulation: Theory and Experimental Verification," *Journal Fluid Mechanics,* vol. 122, 1982, p. 169.
102. I. A. Valioulis et al., "Monte Carlo Simulation of Coagulation in Discrete Particle Size Distributions. Part. Interparticle Forces and the Quasi-Stationary Equilibrium Hypothesis," *Journal Fluid Mechanics,* vol. 143, 1984, p. 397.
103. H. S. Harris et al., "Orthokinetic Flocculation in Water Treatment," *ASCE Journal Sanitary Engineering Div.,* vol. 92, 1966, p. 95.
104. Y. Argaman and W. J. Kaufman, "Turbulence and Flocculation," *ASCE Journal Sanitary Engineering Div.,* vol. 96, 1970, p. 223.
105. D. S. Parker et al., "Floc Breakup in Turbulent Flocculation Processes," *ASCE Journal Sanitary Engineering Div.,* vol. 98, 1972, p. 79.
106. A. L. Lagvankar and R. S. Gemmell, "A Size-Density Relationship for Flocs," *Journal AWWA,* vol. 60, no. 9, 1968, p. 1040.
107. N. Tambo and Y. Watanabe, "Physical Characteristics of Flocs. I: The Floc Density Function and Alum Floc," *Water Research,* vol. 13, 1979, p. 409.
108. N. Tambo and H. Hozumi, "Physical Characteristics of Flocs. II: Strength of Floc," *Water Research,* vol. 13, 1970, p. 421.
109. N. Tambo and Y. Watanabe, "Physical Aspect of Flocculation Process. I: Fundamental Treatise," *Water Research,* vol. 13, 1979, p. 429.
110. N. Tambo and Y. Watanabe, "Physical Aspect of Flocculation Process. II: Contact Flocculation," *Water Research,* vol. 13, 1979, p. 441.
111. A. Amirtharajah, "Design of Flocculation Systems," in R. L. Sanks (ed.), *Water Treatment Plant Design for the Practicing Engineer,* Ann Arbor Science, Ann Arbor, Mich., 1978.
112. S. V. Kao and S. G. Mason, "Dispersion of Particles by Shear," *Nature,* vol. 253, 1975, p. 619.
113. J. E. Quigley and L. A. Spielman, *Strength Properties of Liquid Borne Flocs,* Project No. A-036-DEL, Water Resource Center, University of Delaware, Newark, 1977.
114. J. D. Pandya and L. A. Spielman, "Floc Breakage in Agitated Suspensions: Theory and Data Processing Strategy," *Journal Colloid Interface Science,* vol. 90, 1982, p. 517.
115. C. F. Lu, "Kinetics of Floc Breakage and Aggregation in Agitated Liquid Suspension," unpublished Ph.D. dissertation, University of Delaware, Newark, 1982.
116. R. D. Letterman, "Flocculation," Sem. Proc.: "Coagulation and Filtration: Back to the Basics," *Proc. AWWA Annual Conf.,* St. Louis, Mo., 1981.
117. H. E. Hudson, "Physical Aspects of Flocculation," *Journal AWWA,* vol. 57, no. 7, 1965, p. 885.
118. K. J. Ives, "Theory of Operation of Sludge Blanket Clarifiers," *Proc. Inst. Civil Engineers,* vol. 39, 1968, p. 243.
119. A. W. Bond, "Behavior of Suspensions," *ASCE Journal Sanitary Engineering Div.,* vol. 86, 1960, p. 57.
120. A. W. Bond, "Upflow Solids Contact Basin," *ASCE Journal Sanitary Engineering Div.,* vol. 87, 1961, p. 73.
121. I. Tesarik, "Flow in Sludge Blanket Clarifiers," *ASCE Journal Sanitary Engineering Div.,* vol. 93, 1967, p. 105.
122. J. D. Walker, "High Energy Flocculation," *Journal AWWA,* vol. 60, no. 11, 1968, p. 1271.
123. R. Andreu-Villegas and R. D. Letterman, "Optimizing Flocculator Power Input," *ASCE Journal Environ. Engineering Div.,* vol. 102, 1976, p. 251.
124. S. J. Randtke, "Organic Contaminant Removal by Coagulation and Related Process Combinations," *Journal AWWA,* vol. 80, no. 5, 1988, p. 40.
125. S. J. Randtke et al., "Removing Soluble Organic Contaminants by Lime-Soda Softening," *Journal AWWA,* vol. 74, no. 4, 1982, p. 192.
126. M. Y. Liao and S. J. Randtke, "Removing Fulvic Acid by Lime Softening," *Journal AWWA,* vol. 77, no. 8, 1985, p. 78.
127. A. P. Black and R. F. Christman, "Electrophoretic Studies of Sludge Particles Produced in Lime-Soda Softening," *Journal AWWA,* vol. 53, no. 6, 1961, p. 737.

# Sedimentation
# and Flotation

## Ross Gregory

*WRc Swindon*
*Blagrove, Swindon*
*Wiltshire*
*England*

## Thomas F. Zabel

*WRc Medmenham*
*Medmenham, Oxfordshire*
*England*

Sedimentation and flotation are solid-liquid separation processes used in water treatment to lower the solids concentration, or load, on granular filters. As a result, filters can be operated more easily and cost effectively to produce acceptable-quality filtered water. Many sedimentation and flotation processes exist, and each has advantages and disadvantages. The most appropriate process for a particular application will depend on the water to be treated as well as local circumstances and requirements.

## History of Sedimentation

### Early history

Sedimentation for the improvement of water quality has been practiced, if unwittingly, since the day people collected and stored water in jars and other containers. Water stored undisturbed and then poured or ladled out with little agitation will improve in quality, and this technique is used to this day.

As societies developed, reservoirs and storage tanks were constructed. Although constructed for strategic purposes, reservoirs and storage tanks did improve water quality. Various examples can be cited that predate the Christian era. Ancient surface water impounding tanks of Aden were possibly constructed as early as 600 B.C. and rainwater cisterns of ancient Carthage about 150 B.C.[1] The castellae and piscinae of the Roman aqueduct system performed the function of settling tanks, even though they were not originally intended for that purpose.

### Modern sedimentation

The art of sedimentation progressed little until the industrial age and its increased need for water. Storage reservoirs developed into settling reservoirs. Perhaps the largest reservoirs constructed for this purpose were in the United States at Cincinnati, Ohio, where two excavated reservoirs held approximately 392 mil gal (1480 ML) and were designed to be operated by a fill-and-draw method, though they never were used in this way.[1] The development of settling basins led to the construction of rectangular masonry settling tanks that ensured more even flow distribution and easier sludge removal. With the introduction of coagulation and its production of voluminous sludge, mechanical sludge removal was introduced.

Attempts to make rectangular tanks more cost effective led to the construction of multilayer tanks. Very large diameter, 200-ft (60-m), circular tanks also were constructed at an early stage in the development of modern water treatment. Other industries, such as wastewater treatment, mineral processing, sugar refining, and water softening, required forms of sedimentation with specific characteristics, and various designs of sedimentation tanks particular to certain industries were developed. Subsequently, wider applications of successful industrial designs were sought. Out of this emerged circular radial-flow tanks and a variety of proprietary designs of solids-contact units with mechanical equipment for premixing and recirculation.

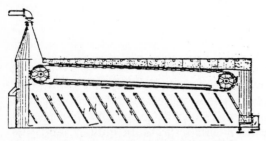

Figure 7.1   Early patent for inclined settling.[2]

The inclined plate settler also has industrial origins[2] (Fig. 7.1), although the theory of inclined settling dates back to experiments using blood in the 1920s and 1930s.[3,4] Closely spaced inclined plate systems for water treatment have their origins in Sweden in the 1950s, resulting from a search for high-rate treatment processes compact enough to be economically housed against winter weather. Inclined tube systems were spawned in the United States in the 1960s.

### Floc-blanket sedimentation

The floc-blanket process for water treatment emerged from India in about 1932 as the pyramidal Candy sedimentation tank (Fig. 7.2). A tank of similar shape was used by Imhoff in 1906 for wastewater treatment.[5] The Spaulding precipitator soon followed in 1935 (Fig. 7.3). Other designs that were mainly solids-contact clarifiers rather than true floc-blanket tanks were also introduced.

The Candy tank can be expensive to construct because of its large sloping sides, so less costly structures for accommodating floc blankets were conceived. The aim was to decrease the hopper component of tanks as much as possible yet provide good flow distribution, to produce a stable floc blanket. Development from 1945 progressed from tanks with multiple hoppers or troughs to the present flat-bottom tanks. Efficient flow distribution in flat-bottom tanks is achieved with either candelabra or lateral inlet distribution systems (Figs. 7.4 and 7.5).

A recent innovation is the inclusion of widely spaced inclined plates in the floc-blanket region (Fig. 7.5). Other developments that also have led to increased surface loadings include the use of polyelectrolytes, ballasting of floc with disposable or recycled solids, and improvements in blanket-level control. The principal centers for these developments have been the United Kingdom, France, and Hungary.

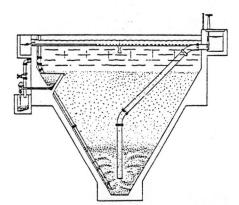

**Figure 7.2** The pyramidal Candy floc-blanket tank (by PCI).

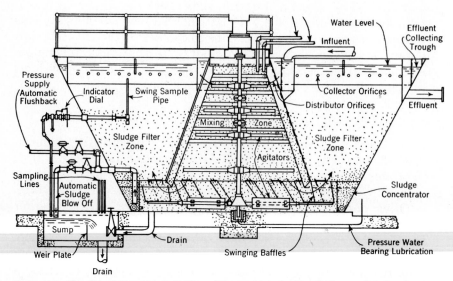

**Figure 7.3** The Spaulding Precipitator solids-contact clarifier.[6] (*Source: H. O. Hartung, Committee Report: Capacity and Loadings of Suspended Solids Contact Units, Journal AWWA, vol. 43, no. 4, April 1951, p. 263.*)

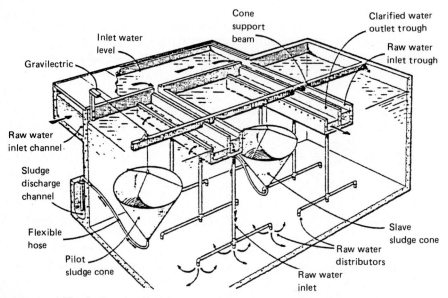

**Figure 7.4** The flat-bottom clarifier with candelabra flow distribution (by PCI).

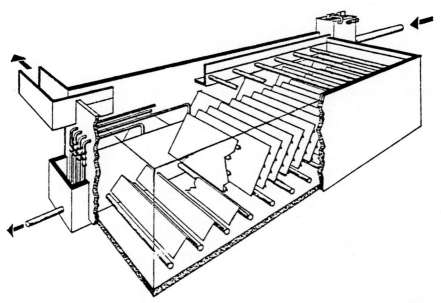

**Figure 7.5**   The Superpulsator flat-bottom clarifier with lateral-flow distribution. (*Courtesy of Infilco Degremont, Inc., Richmond, Va.*)

## Sedimentation Theory

Particle-fluid separation processes of interest to water engineers and scientists do not lend themselves easily to accurate theoretical analysis. This is because the particles are involved in flocculation and attrition and are not regular in shape, density, or size. The theory of ideal systems is, however, a useful guide to interpreting observed behavior in more complex cases.

The various regimes in settling of particles are commonly referred to as types 1 to 4. The general term "settling" is used to describe all types of particles falling through a liquid under the force of gravity. The term "sedimentation" is used to describe settling phenomena in which the particles or aggregates are suspended by hydrodynamic forces only. When particles or aggregates rest on one another, the term "subsidence" is used. The following definitions of the settling regimes are commonly used in the United States and are compatible with a comprehensive analysis of hindered settling and flux theory:

*Type 1:*   Settling or sedimentation of discrete particles in low concentration, with flocculation and other interparticle effects being negligible.

*Type 2:*   Settling or sedimentation of particles in low concentration but with coalescence or flocculation. As coalescence occurs, particle masses increase and particles settle more rapidly.

*Type 3:*   Hindered or zone settling or sedimentation in which particle concentration causes interparticle effects, which might include flocculation, to the extent that the rate of settling is a function of solids concentration. Zones of different concentrations may develop from segregation of particles with different settling velocities. Two regimes exist, a and b, with the concentration being less than and greater than the maximum flux, respectively. In the latter case the concentration has reached the point at which most particles make permanent physical contact with adjacent particles and form a loose structure. As the height of this zone develops, this structure tends to form layers of different concentrations until a state of compression is reached in the bottom layer.

*Type 4:*   Compression settling or subsidence under the layers of zone settling. The rate of compression is dependent on time and on the force caused by the weight of solids above.

## Settling of discrete particles (type 1)

**Terminal settling velocity.**   When the concentration of particles is small, each particle settles discretely, as if it is alone, unhindered by the presence of other particles. The velocity of a single particle, starting from rest, settling under gravity in a liquid will increase where the density of the particle is greater than the density of the liquid.

Acceleration continues until the resistance to flow through the liquid, or drag, equals the effective weight of the particle. Thereafter the settling velocity remains essentially constant. This velocity is called the *terminal settling velocity*. The terminal settling velocity depends on various factors relating to the particle and the liquid.

For most theoretical and practical computations of settling velocities, the shape of the particles is assumed to be spherical. The size of particles that are not spherical can be expressed in terms of a sphere of equivalent volume.

The general equation for the terminal settling velocity of a single particle is derived by equating the forces upon the particle. These forces are the drag $f_d$, buoyancy $f_b$, and an external source such as gravity $f_g$. Hence

$$f_d = f_g - f_b \tag{7.1}$$

The drag force on a particle traveling in a resistant fluid is[7]

$$f_d = \frac{C_D U_t^2 \rho A}{2} \tag{7.2}$$

where $C_D$ = drag coefficient
$\quad\ U_t$ = settling velocity
$\quad\ \rho$ = mass density of liquid
$\quad\ A$ = projected area of particle in direction of flow

Any consistent dimensionally homogeneous units may be used in Eq. (7.2) and all subsequent rational equations.

At constant settling velocity,

$$f_g - f_b = Vg(\rho_s - \rho) \tag{7.3}$$

where $V$ is the effective volume of the particle and $\rho_s$ is the density of the particle. When Eqs. (7.2) and (7.3) are substituted in Eq. (7.1),

$$\frac{C_D U_t^2 \rho A}{2} = Vg(\rho_s - \rho) \tag{7.4}$$

$$U_t = \left[\frac{2g(\rho_s - \rho)V}{C_D \rho A}\right]^{1/2}$$

When the particle is solid and spherical,

$$U_t = \left[\frac{4g(\rho_s - \rho)d}{3C_D \rho}\right]^{1/2} \tag{7.5}$$

where $d$ is the diameter of the sphere.

The value of $U_t$ is the difference in velocity between the particle and the liquid and is independent of the horizontal or vertical movement of the liquid. Therefore, the relationship also applies to a dense stationary particle with liquid flowing upward past it or a buoyant particle with liquid flowing downward.

Calculation of $U_t$ for a given system is difficult because of variation in the drag coefficient $C_D$. The drag coefficient depends on the flow regime and the nature of the flow around the particle and is related to the Reynolds number Re, based on particle diameter, as illustrated in Fig. 7.6, where

$$\text{Re} = \frac{\rho U_t d}{\mu} \tag{7.6}$$

where $\mu$ is the absolute (dynamic) liquid viscosity and $U_t$ is the velocity of particle relative to liquid.

The value of $C_D$ decreases as the value of Re increases, but at a rate depending on the value of Re, such that for spheres only:

**Region _a_: $10^{-4} <$ Re $< 0.2$.** In this region of small Re value, the laminar-flow region, the equation of the relationship approximates to

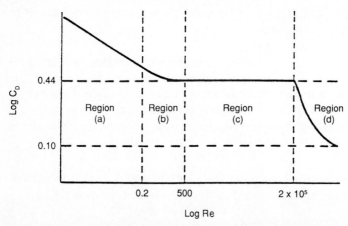

**Figure 7.6**  Variation of drag coefficient $C_D$ with Reynolds number Re for single-particle sedimentation.[8]

$$C_D = \frac{24}{Re} \tag{7.7}$$

This, substituted in Eq. (7.1), gives Stokes' equation for laminar-flow conditions:

$$U_t = \frac{g(\rho_s - \rho)d^2}{18\mu} \tag{7.8}$$

**Region b: 0.2 < Re < 500–1000.**  This transition zone is the most difficult to represent, and various proposals have been made. Perhaps the most recognized representation of this zone for spheres is that promoted by Fair, Geyer, and Okun:[9]

$$C_D = \frac{24}{Re} + \frac{3}{Re^{1/2}} + 0.34 \tag{7.9}$$

For many particles found in natural waters, the density and diameter are such to produce Re values within this region.

**Region c: 500–1000 < Re < 2 × 10⁵.**  In this region of turbulent flow, the value of $C_D$ remains approximately constant at 0.44. Substitution in Eq. (7.5) results in Newton's equation:

$$U_t = 1.74\left[\frac{(\rho_s - \rho)gd}{\rho}\right]^{1/2} \tag{7.10}$$

**Region d: 2 × 10⁵ < Re.**  The drag force decreases considerably with the development of turbulent flow at the surface of the particle, called *boundary-layer turbulence*, such that the value of $C_D$ becomes equal to

0.10. This region is unlikely to be encountered in sedimentation in water treatment.

**Effect of particle shape.** Equation (7.4) shows how particle shape affects velocity. The effect of a nonspherical shape is to increase the value of $C_D$ at a given value of Re.

As a result, the settling velocity of a nonspherical particle is less than that of a sphere having the same volume and density. Sometimes a simple shape factor $\Phi$ is determined, for example, in Eq. (7.7)

$$C_D = \frac{24\Phi}{Re} \qquad (7.11)$$

Typical values found for $\Phi$ for rigid particles are[10]

| | |
|---|---|
| Sand | 2.0 |
| Coal | 2.25 |
| Gypsum | 4.0 |
| Graphite flakes | 22 |

A shape factor value is difficult to determine for floc because the size and shape of floc particles are interlinked with the mechanics of their formation and disruption in any set of flow conditions. Floc particles with their fluffy structure are, however, likely to have relatively large values.

Further details on the settling behavior of spheres and nonspherical particles can be found in standard texts.[11]

**Settlement in tanks.** In an ideal upflow settling tank, the particles retained are those whose settling velocity exceeds the liquid upflow velocity.

$$U_t \geq \frac{Q}{A} \qquad (7.12)$$

where $Q$ is the inlet flow rate to tank and $A$ is the cross-sectional area of tank.

In a horizontal-flow rectangular tank, the settling of a particle has both vertical and horizontal components, as shown in Fig. 7.7:

$$l = \frac{tQ}{HW} \qquad (7.13)$$

where $l$ = horizontal distance traveled
$\quad t$ = time of travel
$\quad H$ = depth of liquid
$\quad W$ = width of tank

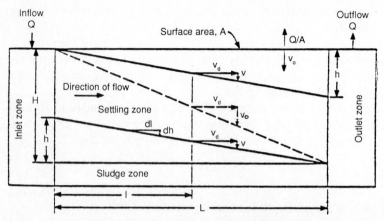

**Figure 7.7**   Horizontal and vertical components of settling velocity.[9]

and for the vertical distance traveled $h$

$$h = U_t t \qquad (7.14)$$

Hence, the settling time for a particle that has entered the tank at a given level $h$ is

$$t = \frac{h}{U_t} \qquad (7.15)$$

Substitution of this in Eq. (7.13) gives the length of tank required for settlement to occur under ideal flow conditions:

$$l = \frac{hQ}{U_t HW} \qquad (7.16)$$

or

$$U_t = \frac{hQ}{HlW} \qquad (7.16a)$$

If all particles with a settling velocity of $U_t$ are allowed to settle, then $h$ equals $H$ and $l$ equals $L$, and consequently

$$U_t = \frac{Q}{LW} \qquad (7.17)$$

This special case then defines the surface-loading or overflow rate of the ideal tank $U_t^*$:

$$U_t^* = \frac{Q}{L^*W} \tag{7.18}$$

$$U_t^* = \frac{Q}{A^*} \tag{7.18a}$$

where $L^*$ is the length of tank over which settlement ideally takes place and $A^*$ is the plan area of tank, with horizontal flow, over which the settlement ideally takes place.

All particles with a settling velocity greater than $U_t^*$ are removed. Particles with a settling velocity less than $U_t^*$ are removed in proportion to the ratio $U_t:U_t^*$.

Particles with a settling velocity $U'_t$ less than $U_t^*$ need a tank of length $L'$ greater than $L^*$ for total settlement such that

$$\frac{U_t'}{U_t^*} = \frac{L^*}{L'} \tag{7.19}$$

This ratio defines the proportion of particles with a settling velocity of $U_t'$ that settle in a length $L^*$. Equation (7.18a) states that the settling efficiency depends on the area available for settling. The same result applies to circular tanks.

Equation (7.18a) shows that the settling efficiency for the ideal condition is independent of depth $H$ and dependent on only the tank area. This principle is sometimes referred to as *Hazen's law*. In contrast, retention time $T$ is dependent on water depth $H$, as given by

$$T = \frac{AH}{Q} \tag{7.20}$$

In reality, however, depth is important because it can affect flow stability if it is large and scouring if it is small.

**Predicting settling efficiency (types 1 and 2).**  In a typical suspension a large range of particle sizes exists. To determine the efficiency of removal for a given settling time, it is necessary to consider the entire range of settling velocities present in the system. This can be accomplished either by use of sieve analyses and hydrometer tests combined with the appropriate relationship for settling velocity or by use of a settling column. When the suspension is coalescent or flocculant, as in type 2 settling, then the latter must be used. With either method a settling-velocity analysis curve is constructed from the data (Fig. 7.8).[12]

Equation (7.19) defined the proportion of particles with settling velocity $U_t$ smaller than $U_t^*$, which will be removed. If $x_t$ is the proportion of particles having settling velocities less than or equal to $U_t^*$, the

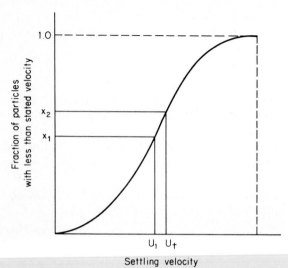

**Figure 7.8** Settling-velocity analysis curve for discrete particles.[12,13]

total proportion of particles that could be removed in settling is defined by[14]

$$F_t = (1 - x^*_t) + \int_0^x \frac{U_t}{U_t^*}\,dx_t \qquad (7.21)$$

This can be solved by using Fig. 7.8.

In the settling column test, the procedure is to place the sample in a column 6 to 8 ft (1.8 to 2.4 m) tall with taps along its length for removal of samples.[15] Over about a 2-h test period, samples are taken about every 20 min.

The suspended-solids concentration is determined in each sample and expressed as a percentage difference, removal $R$, of the original concentration. These results are charted against time and depth, and curves of equal percentage of removal are drawn. Figure 7.9 is an example of flocculant type 2 settling, with an increase in settling velocity as settlement progresses. An effective settling rate for the quiescent conditions of the column can be defined as the ratio of the effective depth divided by the time required to obtain a given percentage of removal.

For Fig. 7.9, any combination of depth $h$ and time $t$ on one of the isopercentage lines will establish a settlement velocity $U_t^*$:

$$U_t^* = \frac{h_t^*}{t^*} \qquad (7.22)$$

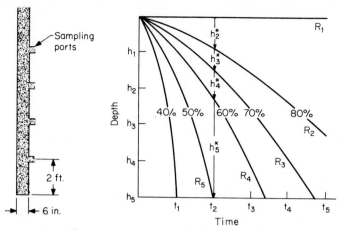

**Figure 7.9** Flow-through curve analysis, after Hazen.[12] (*Source: Metcalf and Eddy Engineers, Wastewater Engineering, 2d ed., McGraw-Hill, New York, 1979.*)

Thus, all particles with a settling velocity equal to or greater than $U_t^*$ will be removed, which means all particles represented by the particular isopercentage line selected. Particles with a velocity $U_t$ less than $U_t^*$ will be removed in the proportion $U_t/U_t^*$.

For the same time $t^*$, this would be the same as taking the depth ratios, $h_t/h_t^*$, for the same reason as in Eq. (7.19). Thus with reference to Fig. 7.9,[12,15] if $t^*$ is $t_2$, then the proportion of particles settled to level $h_5$ with velocity $U_2^*$, is $R_5$, which is 50 percent. In the same time, the particles representing the 60 percent removal $R_4$ have settled to level $h_4^*$ and so forth. The average depth settled for particles in the 50 to 60 percent range is $(h_5 + h_4^*)/2$. (Note that $h_5^* = h_5$ and $h_1^* = 0$.) Thus the average percent removal in this range is

$$R_5^* = \frac{(h_5 + h_4^*)(60 - 50)}{2h_5} \tag{7.23}$$

Similarly for the 60 to 70 percent range:

$$R_4^* = \frac{(h_4^* + h_3^*)(70 - 60)}{2h_5}$$

Thus the total removal is the sum of these:

$$\sum R = \left[\frac{(h_5 + h_4^*)}{2h_5} + \frac{(h_4^* + h_3^*)}{2h_5} + \cdots\right]\Delta R \tag{7.23a}$$

where $\Delta R$ is the constant percentage difference between lines of isopercentage. Equation (7.23) can be rearranged to give:

$$\sum R = \left[\frac{1}{2} + \frac{h_4^*}{h_5} + \frac{h_3^*}{h_5} + \frac{h_2^*}{h_5}\right]\Delta R \qquad (7.23b)$$

In practice, in the design of a settling tank to achieve comparable removal the settling rate from the test should be multiplied by a factor of 0.65 and the detention time should be multiplied by a factor of 1.75 to 2.0.[12]

### Hindered settling (types 3a and 3b)

**Particle interaction.**   Particle flow conditions are modified at high concentrations. Individual particle behavior is influenced, or hindered, by the presence of other particles, and the flow characteristics of the bulk suspension can be affected. With increased particle concentration the free area between particles is reduced, causing greater interparticle fluid velocities and alteration of flow patterns around particles. Consequently, the settling velocity of a suspension is generally less than that of a discrete particle.

When particles in a suspension are not uniform in size, shape, or density, individual particles will have different settling velocities. Particles with a settling velocity less than that of the suspension increase the effective viscosity. Smaller particles tend to be dragged down by the motion of larger particles. Flocculation may increase the effective particle size when particles are close together.

**Solids flux.**   The extent to which each factor influences the settling velocity of the suspension $U_s$ also depends on the particle concentration in the suspension $C$. The product of velocity and concentration is flux $\phi$, the mass rate of settling:

$$\phi = U_s C \qquad (7.24)$$

The relationship between $\phi$ and $C$ is shown in Fig. 7.10 and is complex because $U_s$ is affected by the concentration. The relationship can be divided into four regions.

**Region a, type 1 and type 2 settling.**   Unhindered settling occurs such that the flux increases in proportion to the concentration. A suspension of particles with different settling velocities has a diffuse interface with the clear liquid above.

**Region b, type 3a settling.**   With an increase in concentration, hindered-flow settling increasingly takes effect, and ultimately a maximum value of flux is reached. At about maximum flux the diffuse interface

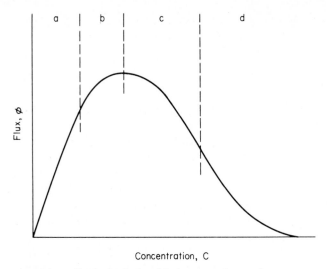

**Figure 7.10** Typical relationship between flux and concentration for batch settlement.

of the suspension becomes distinct with the clear liquid above when all particles become part of the suspension and settle with the same velocity.

**Region c, type 3b settling.** Further increase in concentration reduces flux because of the reduction in settling velocity. In this region the suspension settles homogeneously.

**Region d, type 4 settling.** Associated with the point of inflection in the flux-concentration curve, the concentration reaches the point where thickening can be seen to occur, leading ultimately to compression settling.

**Equations for hindered settling.** The behavior of suspensions in regions $b$ and $c$ has attracted considerable theoretical and empirical analysis and is most important in floc-blanket clarification. The simplest and most viable relationship is represented by the general equation[16]

$$U_s = U_0 \exp(-qC) \tag{7.25}$$

where $q$ = constant representative of suspension
$U_0$ = settling velocity of suspension for concentration extrapolated to zero
$C$ = concentration of suspension

Other empirical relationships have been proposed. The most widely accepted and tested relationship was initially developed for particles larger than 0.1-mm diameter in rigid particle fluidized systems. This relationship has been shown to be applicable to settling and is known as the *Richardson and Zaki equation*[11]:

$$U_s = U_t E^n \tag{7.26}$$

where $E$ is the porosity of the suspension and $n$ is the power value dependent on the Reynolds number of the particle.

For rigid particles this equation is valid for porosity from that at minimum fluidization velocity, about 0.6, to about 0.95. The Reynolds number Re controls the value of $n$.[11] For a suspension with uniformly sized spherical particles, $n$ equals 4.8 when Re is less than 0.2. As the value of Re increases, $n$ decreases until Re is greater than 500 when $n$ equals 2.4. A standard concentration measurement for floc particle suspensions is a problem because of variations in particle size, shape, and other factors. A simple settlement test is the easiest method.[16] The ½-h settled-solids volume is the volume occupied by the settled suspension in a graduated cylinder measured after 30 min, and it is expressed as a fraction of the total volume of the whole sample. Because such a test as this is only a relative measurement, correction factors must be included[16]:

$$U_s = U_t k_1 (1 - k_2 C^*)^r \tag{7.27}$$

where $k_1, k_2$ = constants representative of system
   $C^*$ = apparent solids volumetric concentration
   $r$ = power value dependent on system

Equation (7.26) can be substituted in Eq. (7.24) for flux with $1 - E$ substituted for $C$[11]:

$$\phi = U_t E^n (1 - E) \tag{7.28}$$

Differentiating this equation with respect to $E$ gives

$$\frac{d\phi}{dE} = U_t n E^{n-1} - U_t (n + 1) E^n \tag{7.29}$$

The flux $\phi$ has a maximum value when $d\phi/dE$ equals zero and $E$ equals $E^+$ (the porosity at maximum flux). Hence, dividing Eq. (7.29) by $U_t E^{n-1}$ and equating to zero, we get

$$0 = n - (n + 1) E^+ \tag{7.30}$$

or

$$n = \frac{E^+}{1 - E^+} \tag{7.30a}$$

This means that $E^+$, or the concentration at maximum flux $C^+$, is an important parameter in describing the settling rates of suspensions. In the case of rigid uniform spheres, if $n$ ranges from 2.4 to 4.6, the maximum flux should occur at a volumetric concentration between 0.29 and 0.18. In practice, this is the range of values found for suspensions of aluminum and iron flocs when concentration is measured as the ½-h settled volume.[16]

If Eq. (7.29) is differentiated also, then

$$\frac{d^2\phi}{dE^2} = U_t \left[ n(n-1)E^{n-2} - (n+1)nE^{n-1} \right] \tag{7.31}$$

and when $d^2\phi/dE^2 = 0$ for real values of $E$, a point of inflection will exist, given by

$$0 = n - 1 - (n+1)E \tag{7.32}$$

such that

$$E = \frac{n-1}{n+1} \tag{7.32a}$$

For rigid uniform spheres if $n$ ranges from 2.4 to 4.6, the point of inflection occurs at a concentration between 0.59 (corresponding to a packed bed) and 0.35.

The point of inflection is associated with the transition from type 3 to type 4 settling. Type 3 and type 4 settling in the context of thickening and further utilization of flux theory are considered later.

**Prediction of settling rate.** The hindered settling rate can be predicted for suspensions of rigid and uniform spheres by using Eqs. (7.5) and (7.26). For suspensions of nonuniform particles and flocculant systems, however, practical tests have to be made of the settling rate. This is done most simply by using a settling column; a 1-L measuring cylinder is usually adequate. The procedure is to fill the cylinder to the top measuring mark with the sample and to record at frequent intervals the level of the interface between the suspension and the clear-water zone. The interface is likely to be distinct enough for this purpose only if the concentration of the sample is greater than that at maximum flux. The results are plotted to produce the typical settling curve (Fig. 7.11). The slope of the curve over the constant-settling-rate period is the estimate of the type 3 settling rate for quiescent conditions. If the concentration of the sample was greater than that at in-

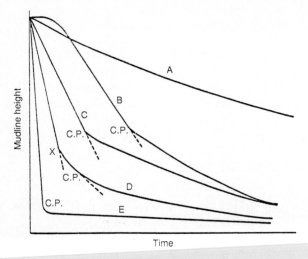

A    Concentrated, flocculated or unflocculated pulp
B    Intermediate flocculated or unflocculated pulp
C    As (B) but showing no induction period
D    Dilute, flocculated pulp
E    Dilute, unflocculated pulp
C.P. Compression point

**Figure 7.11**   Typical batch settling curves.[8]

flection in the mass flux curve, transition from regions $c$ to $d$ in Fig. 7.10, then a period of constant settling rate and the compression point (CP) will not be found as represented by line $A$.

The compression point signifies the point at which all the suspension has passed into the type 4 settling or compression regime. Up to that time a zone of solids in the compression regime has been accumulating at the bottom of the suspension with its upper interface moving upward. Thus the compression point is where that interface reaches the top of the settling suspension.

## Fluidization

When liquid is passed up through a uniform stationary bed of particles at a low flow rate, the flow behavior is similar to that observed when the flow is down through the bed. When the upward flow of liquid is great enough to cause a drag force on particles equal to the apparent weight (actual weight less buoyancy) of the particles, the particles rearrange to offer less resistance to flow and bed expansion occurs. This process continues as the liquid velocity is increased until the bed has assumed the least stable form of packing. If the upward liquid velocity is increased further, individual particles separate from one another and become freely supported in the liquid.

The bed is then said to be *fluidized*. For rigid and generally uniform particles, such as with filter sand, about 10 percent bed expansion occurs before fluidization commences. The less uniform the size and density of the particles, the less distinct is the point of fluidization. A fluidized bed is characterized by regular expansion of the bed as the liquid velocity increases from the minimum fluidization velocity until particles are in unhindered suspension, type 1 settling.

Fluidization is hydrodynamically similar to hindered, or zone type 3, settling. In a fluidized bed, particles undergo no net movement and are maintained in suspension by the upward flow of the liquid. In hindered settling, particles move downward, and in the simple case of batch settling no net flow of liquid occurs. The Richardson and Zaki equation, Eq. (7.26), has been found to be applicable to both fluidization and hindered settling,[11] as have other relationships.

In water treatment, floc-blanket clarification is more a fluidized bed rather than a hindered settling process. Extensive floc-blanket data [16] with $C^*$ determined as the ½-h settled-solids volume, such that $C^+$ tended to be in the range 0.16 to 0.20, allowed Eq. (7.27) to be simplified to

$$U_s = U_0(1 - 2.5C^*) \tag{7.33}$$

The data that enabled this simplification were obtained with alum coagulation of a high-alkalinity organic-rich river water, but the value for $k_2$ of 2.5 should hold for other types of water producing similar-quality floc. The value of $k_2$ can be estimated as the ratio of the concentration at the compression point to the ½-h settled concentration. Values predicted for $U_0$ by Eq. (7.33) are less, about one-half to one-third, than those likely to be estimated by Stokes' equation, assuming spherical particles.[16]

The theory of hindered settling and fluidization of particles of mixed sizes and different densities is more complex and is still being developed. In some situations two or more phases can occur at a given velocity, each phase with a different concentration. This has been observed with floc blankets to the extent that an early but temporary deterioration in performance occurs with an increase in upflow.[17,18] An increase in upflow leads to intermixing of the phases with a further increase in upflow limited by the characteristics of the combined phase. The theory has been developed to the extent that it explains and predicts the occurrence of intermixing and segregation, for example, in multimedia filter beds during and after backwash.[19,20]

**Example Problem 1**  Predict the maximum flux conditions for floc-blanket sedimentation.

**Solution** For a floc blanket that can be operated over a range of upflow rates, collect samples of blanket at different upflow rates. For these samples measure the ½-h settled volume. The results might be as follows:

| Upflow rate, m/h | 1.6 | 1.95 | 2.5 | 3.05 | 3.65 | 4.2 | 4.7 | 5.15 |
|---|---|---|---|---|---|---|---|---|
| Half-hour floc volume, % | 31 | 29 | 25 | 22 | 19 | 16 | 13 | 10 |
| Blanket flux = upflow rate × ½-h floc volume, % m/h | 49.6 | 56.6 | 62.5 | 67.1 | 69.4 | 67.2 | 61.1 | 51.5 |

These results predict that maximum flux occurs at an upflow rate of 3.65 m/h. If flux is plotted against upflow rate and against ½-h floc volume, then the maximum flux will be found to occur at 3.44 m/h for a ½-h floc volume of 20 percent, as shown in Fig. 7.12.

The above results can be fitted to Eq. (7.33):

$$U_s = U_0(1 - 2.5C^*)$$

$$3.44 = U_0(1 - 2.5 \times 0.2)$$

$$U_0 = \frac{3.44}{0.5} = 6.87 \ m/h$$

This means that at maximum flux the theoretical terminal settling velocity of the blanket is 6.87 m/h. The maximum rate at which a floc blanket can be operated at, assuming a stable tank, is about 70 percent of this rate, which is 4.8 m/h.

### Inclined (tube and plate) settling

The efficiency of discrete particle settling in horizontal liquid flow depends on the area available for settling. Hence, efficiency can be improved by increasing the area. Some tanks have multiple floors to achieve this. A successful alternative has been the development of lightweight structures with closely spaced inclined surfaces.

Inclined settling systems (Fig. 7.13) are constructed to be used in one of three ways with respect to the direction of liquid flow relative to the direction of particle settlement: countercurrent, cocurrent, and cross-flow. Comprehensive theoretical analyses of the various flow geometries have been made by Yao.[21] Flow conditions in the channels between the inclined surfaces must be laminar. In practice, the Reynolds number Re must be less than 800 when calculated from the mean velocity between and parallel to the inclined surfaces $U_\theta$ and hydraulic diameter of the channel $d_H$:

$$d_H = \frac{4A_H}{P} \tag{7.34}$$

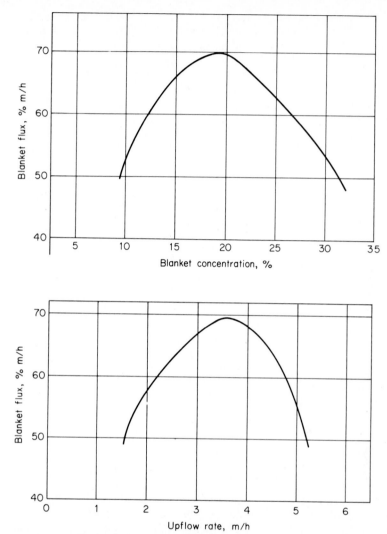

**Figure 7.12**  Relationship between blanket flux, blanket concentration, and upflow rate for Example Problem 1.

where $A_H$ is the cross-sectional area of channel to liquid flow and $P$ is the perimeter of $A_H$, such that Eq. (7.6) becomes

$$\text{Re} = \frac{\rho U_0 d_H}{\mu} \qquad (7.35)$$

**Countercurrent settling.**  The time $T$ taken for a particle to settle the vertical distance between two parallel inclined surfaces is

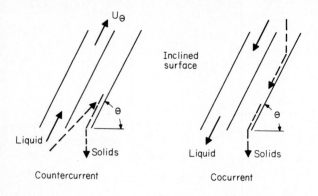

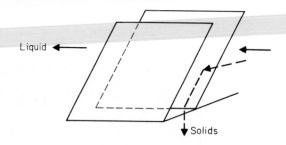

Cross-flow

**Figure 7.13** Basic flow geometries for inclined settling systems.

$$T = \frac{w}{U_t \cos \theta} \tag{7.36}$$

where $w$ is the perpendicular spacing between surfaces and $\theta$ is the angle of surface inclination. The length of surface $L_p$ needed to accommodate this time, when the liquid velocity between the surfaces is $U_\theta$, is

$$L_p = \frac{w(U_\theta - U_t \sin \theta)}{U_t \cos \theta} \tag{7.37}$$

By rearranging this equation, all particles with a settling velocity $U_t$ and greater are removed if

$$U_t \geq \frac{U_\theta w}{L_p \cos \theta + w \sin \theta} \tag{7.37a}$$

When many plates or tubes are used,

$$U_\theta = \frac{Q}{Nwb} \tag{7.38}$$

where $N$ is the number of channels made by $N + 1$ plates or tubes and $b$ is the dimension of the surface at right angles to $w$ and $Q$.

**Cocurrent settling.**   In cocurrent settling, the time taken for a particle to settle the vertical distance between two surfaces is the same as for countercurrent settling. The length of surface needed, however, has to be based on downward, not upward, liquid flow:

$$L_p = \frac{w(U_\theta + U_t \sin \theta)}{U_t \cos \theta} \tag{7.39}$$

Consequently, the condition for removal of particles is given by

$$U_t \geq \frac{U_\theta w}{L_p \cos \theta - w \sin \theta} \tag{7.39a}$$

**Cross-flow settling.**   The time for a particle to settle the vertical distance between two surfaces is again given by Eq. (7.36). The liquid flow is horizontal and does not interact with the vertical settling velocity of a particle. Hence

$$L_p = \frac{U_\theta w}{U_t \cos \theta} \tag{7.40}$$

and

$$U_t \geq \frac{U_\theta w}{L_p \cos \theta} \tag{7.40a}$$

**Other flow geometries.**   The above three analyses apply only to parallel-surface systems. Yao[21] suggested a parameter $S_c$ defined as

$$S_c = U_t \frac{(\sin \theta + L^* \cos \theta)}{U_\theta} \tag{7.41}$$

where $L^* = L/w$ is the relative length of the settler.

When $U_t^*$ is the special case that all particles with velocity $U_t$ or greater are removed, then for parallel surfaces $S_c$ equals 1.0. But the value for circular tubes is $\frac{4}{3}$ and for square conduits is $\frac{11}{8}$.[22] Identical values of $S_c$ for different systems may not mean identical behavior.

The design overflow rate is also defined by $U_t^*$ in Eq. (7.38), and Yao has shown by integration of the differential equation for a particle trajectory that the overflow rate for an inclined settler is given by

$$U_t^* = \frac{k_3 K U_\theta}{L^*} \qquad (7.42)$$

where $k_3$ is a constant equal to $8.64 \times 10^2$ m$^3$/(day)(m$^2$) and

$$K = \frac{S_c L^*}{\sin \theta + L^* \cos \theta} \qquad (7.43)$$

For given values of overflow rate and surface spacing and when $\theta = 0$, Eq. (7.42) becomes

$$\frac{S_c}{L} = \text{constant}$$

This indicates that the larger the value of $S_c$, the longer the surface length needs to be to achieve the required theoretical performance. In practice, compromises must be made between theory and the hydrodynamic problems of flow distribution and stability that each different geometry poses.

**Example Problem 2**   A tank has been fitted with 6.6-ft (2.0-m) square inclined plates spaced 2 in (50 mm) apart. The angle of inclination of the plates can be altered from 5 to 85°. The inlet to and outlet from the tank can be fitted in any way so that the tank can be used for countercurrent, cocurrent, or cross-flow sedimentation. If no allowances need to be made for hydraulic problems of flow distribution and so on, then which is the best arrangement to use?

**Solution**   Equation (7.37a) for countercurrent flow is

$$U_t \geq \frac{U_\theta w}{L_p \cos \theta + w \sin \theta}$$

Equation (7.39a) for cocurrent flow is

$$U_t \geq \frac{U_\theta w}{L_p \cos \theta - w \sin \theta}$$

And Eq. (7.40a) for cross-flow sedimentation is

$$U_t \geq \frac{U_\theta w}{L_p \cos \theta}$$

As an example, the calculation for countercurrent flow at 85° is

$$\frac{U_t}{U_\theta} = \frac{50}{2000 \cos 85 + 50 \sin 85} = \frac{50}{174.3 + 49.8} = 0.223$$

The smallest value of $U_t$ is required. Thus for the range:

| Angle θ | 5 | 15 | 30 | 45 | 60 | 75 | 85 |
|---|---|---|---|---|---|---|---|
| Countercurrent | | | | | | | |
| $U_t/U_\theta$ | 0.025 | 0.026 | 0.028 | 0.035 | 0.048 | 0.088 | 0.223 |
| Cocurrent $U_t/U_\theta$ | 0.025 | 0.026 | 0.029 | 0.036 | 0.052 | 0.106 | 0.402 |
| Cross-flow $U_t/U_\theta$ | 0.025 | 0.026 | 0.029 | 0.035 | 0.050 | 0.096 | 0.287 |

From the above, little difference exists between the three settling arrangements for an angle of less than 60°. For angles greater than 60° countercurrent flow allows settlement of particles with the smallest settling velocity.

### Flocculation

In addition to the interaction that can occur between rigid particles in hindered settling and fluidization, aggregation and attrition can occur with chemical floc. Aggregation occurs by one particle overtaking another and the two touching, and by velocity gradients created by currents and nonhomogeneous flow, namely flocculation (see Chap. 6).

The theory of flocculation detailed in Chap. 6 recognizes the role of the velocity gradient $G$ and time $t$ as well as the particle volumetric concentration $C$. For dilute suspensions optimum flocculation conditions are generally considered only in terms of $G$ and $t$:

$$Gt = \text{constant} \tag{7.44}$$

In floc blankets, the value of $G$ is usually less than in flocculators. At the "blanket clear-water" interface, the value of $G$ is about 5/s although at the bottom of the blanket in a pyramidal tank the value will be about 50/s.[16] In a flat-bottom tank with a blanket 6.5 ft (2 m) deep and an upflow velocity of 1.6 gpm/ft² (4 m/h), $t$ equals 1800 s and hence $Gt$ has a value of about 20,000. This tends to be less than is usually considered necessary for flocculation prior to inclined settling or dissolved-air flotation.

In concentrated suspensions, such as with hindered settling, the greater particle concentration $C$ contributes to flocculation by enhancing the probability of particle collisions and increasing the velocity gradient that can be expressed in terms of the head loss across the suspension.

Consequently optimum flocculation conditions for concentrated suspensions may be better represented by[9,23,24]

$$GtC = \text{constant} \tag{7.45}$$

The value of the constant at maximum flux is likely to be about 4000, when $C$ is measured as the fractional volume occupied by floc, with little benefit to be gained from a larger value.[16,24]

### Floc-blanket clarification

A simple floc-blanket tank has a vertical parallel-walled upper section with a flat or hopper-shaped base. Water that has undergone suitable chemical coagulation is fed downward into the base. The resultant expanding upward flow allows flocculation to occur, and large floc particles remain in suspension within the tank. Particles in suspension accumulate slowly at first, but increase because of enhanced flocculation and other effects to a maximum accumulation rate limited by particle characteristics and the upflow velocity of the water. When this limiting rate has been reached, a floc blanket can be said to exist.

As floc particles accumulate, the volume occupied by the suspension, known as the *floc blanket,* increases and its upper surface rises. The floc-blanket surface level is controlled by removing solids from the blanket so as to keep a zone of clear water or supernatant liquid between the blanket and the decanting troughs, launders, or weirs.

A floc blanket is thus a fluidized bed of floc particles, even though the process is regarded in water engineering as a form of sedimentation, namely, hindered settling. True hindered settling generally exists in the upper section of sludge hoppers used for removing accumulated floc for blanket-level control. Thickening takes place in the lower section of the sludge hoppers. Thus excess floc removed from the floc blanket becomes a waste stream and may be thickened to form sludge.

**Mechanism of clarification.**   Sedimentation, entrainment, and particle elutriation occur above and at the surface of a blanket. The mechanism of clarification within a floc blanket is more complex, however, and involves flocculation, entrapment, and sedimentation. In practice, the mean retention time of the water within a blanket is in excess of the requirements for floc growth to control the efficiency of the process.

A process of physical entrapment by flocculation and agglomeration, akin to surface capture in deep-bed filtration, occurs throughout a floc blanket. Probably the most important process is mechanical entrapment and straining in which rising small particles cannot pass through the voids between larger particles that comprise the bulk of the blanket. The efficiency of entrapment is affected by the spacing of the larger suspended floc particles, which in turn is related to floc quality and water velocity.

**Performance prediction.**  Within a floc blanket the relationship between the floc concentration and upflow velocity of the water is represented by Eqs. (7.25) to (7.27) for hindered settling and fluidization. Because of complexity, a simple theory for predicting solids removal efficiency has not been established, although attempts have been made.[25,26]  The relationships between settled water quality and blanket-floc concentration, upflow velocity, and flux (Fig. 7.14) are of practical importance, however.[16]

The relationships in Fig. 7.14 show that settled water quality deteriorates rapidly (point A) as floc concentration (point B) is decreased below the concentration at maximum flux (point C). Conversely, little improvement in settled water quality is likely to be gained by increasing floc concentration to be greater than at maximum flux (to the left

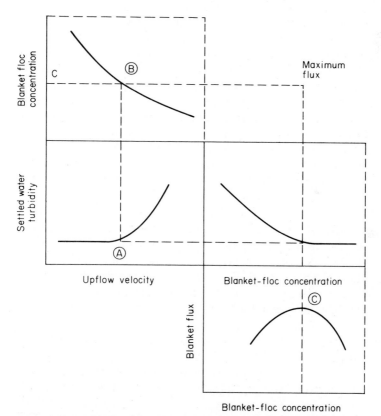

**Figure 7.14**  Typical relationships between settled water quality and blanket concentration, upflow velocity and blanket flux.[16] (*Source: R. Gregory, Floc Blanket Clarification TR111, Water Research Centre, Medmenham, U.K., 1979.*)

of points $A$ and $B$). The reason is that at concentrations greater than those at maximum flux, interparticle distances are small enough for entrapment to dominate the clarification process.

As concentration decreases below that at maximum flux, particle distances increase, especially between the larger particles, and their motion will become more intense. Thus smaller particles will be less easily entrapped and will more readily pass out of the floc blanket. Consequently the maximum flux condition represents possible optimum operating and design conditions. Maximum flux conditions and performance depend on various factors described later.

**Effects of upflow velocity.**   The surface loading for floc-blanket sedimentation is expressed as the upflow velocity or overflow rate. For some floc-blanket systems the performance curve can reflect a "temporary" or premature deterioration in water quality (Fig. 7.15). This deterioration is associated with segregation of particles, or zoning, in the blanket at low surface loading[16] because of the wide range in particle settling velocities. This also can occur with the use of powdered carbon[18] and in precipitation softening using iron coagulation.[17] As surface loading is increased, remixing occurs at the peak of the "temporary" deterioration as the lower-lying particles are brought into greater expansion.

For floc-blanket sedimentation, the "corner" of the performance curve (point $A$ in Fig. 7.14) is associated with the point of maximum flux. The limit to upflow velocity for reliable operation has been expressed by some[16,27,28] in simple terms of terminal velocity. Bond[27] noted that the blanket surface remained clearly defined up to a velocity of about one-half of the terminal settling rate $U_0$, that slight boiling occurred above $0.55U_0$, and that clarification deteriorated noticeably at about $0.65U_0$. Tambo et al.[28] found that a floc blanket is stable for velocities less than about $0.7U_0$ and very unstable at velocities greater than $0.8U_0$. Gregory[16] found that the velocity at maximum flux was about $0.5U_0$, as given by Eq. (7.25), when the maximum flux ½-h settled-solids volume is 20 percent. As the upflow velocity increased beyond this, the blanket surface became more diffuse to the extent that a blanket was very difficult to sustain for a velocity greater than $0.75U_0$. Hence, the best guideline for optimum operation is to use the velocity that creates a blanket concentration at which the blanket surface becomes diffuse.

The ½-h settled-solids volume at maximum flux has been found to be in the range of 16 to 20 percent for alum coagulation.[16] When polyelectrolyte is used as a flocculant aid or an iron coagulant, the value tends to be greater, at 25 to 30 percent. The actual value de-

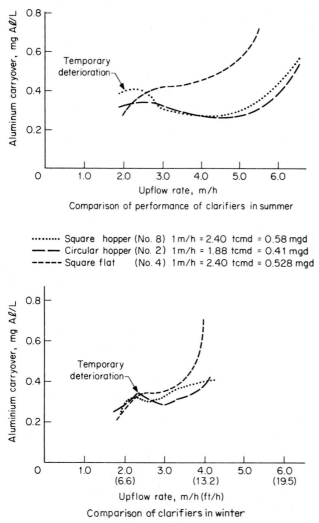

**Figure 7.15** Floc-blanket performance curves showing "temporary" deterioration in settled water quality.[18]

pends on the quality of the water as well as the choice of coagulation chemistry.

## Thickening (types 3 and 4)

With sedimentation in water treatment, the solids concentration in the inflow is usually relatively low, and the main objective is to clarify the water as efficiently as possible. Generally the sludge is removed without much attention being paid to the efficiency of its thickening.

The underflow of sludge might then be pumped to a thickener, where the main objective is to concentrate it efficiently.

The solids-handling capacity of a continuous thickener is controlled by the concentration of solids in the underflow. The liquid-handling capacity depends on the required clarity of the overflow. Thickener throughput, however, is likely to be limited by underflow rather than overflow. Traditionally the sizing of a thickener with regard to underflow is based on flux theory.

A comprehensive review of thickening theories by Pearse[8] concluded that the flux theories that have evolved, although elucidating the thickening process to a greater extent, offer little or no improvement in design prediction over the original method of Coe and Clevenger.[29] The theories most commonly used are those by Coe and Clevenger and by Talmage and Fitch.[30] Both methods have inherent limitations, but both are workable if used in conjunction with empirical safety factors.

Generally, the Coe and Clevenger method tends to underestimate the area requirement while that of Talmage and Fitch tends to overestimate it. Because of this and its experimental simplicity, the Talmage and Fitch method is often preferred.[8]

Where the underflow concentration is of critical importance, the batch test is not considered satisfactory when the required underflow concentrations are low (< 25 percent by volume), compression times are long (> 24 h), no definite compression point exists, and strong channeling is absent in the compression regime. In these cases especially, no substitute exists for semicontinuous or continuous pilot-scale testing.

**Coe and Clevenger method.** For steady-state operation in continuous thickening, Coe and Clevenger derived the equation for the flux of solids moving toward the underflow at any level:

$$\phi = \frac{U}{1/C - 1/C_u} \tag{7.46}$$

where $C_u$ is the solids concentration in underflow and $U$ is the solids settling rate at concentration $C$.

The settling rate is determined for a range of concentrations using batch tests. For a given value of $C_u$ the corresponding flux curve can be drawn (Fig. 7.16). This curve usually has a minimum at some critical value of concentration $C_c$. If the feed concentration $C_f$ is less than $C_c$ and the underflow concentration $C_u$ is greater, then the maximum solids flux that can pass to the underflow will be $\phi_c$, corresponding to the critical concentration.

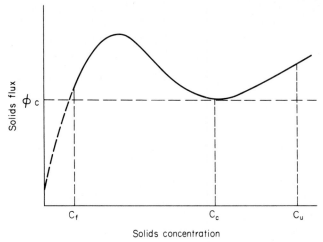

**Figure 7.16** Coe and Clevenger's flux curve for continuous thickening.[18]

If the feed concentration is greater than $C_c$, or the feed flux is greater than $\phi_c$, then a critical zone of concentration $C_c$ will form in the thickener and solids cannot pass through it as fast as they are being supplied. Excess solids add to this zone, increasing its height, until it fills the thickener and overflow occurs. This is avoided by using a thickener of sufficient area that the flux for the given feed rate does not exceed $\phi_c$.

Conversely, if the feed flux becomes less than $\phi_c$, solids pass through the critical zone faster than they are replenished and the zone disappears. Therefore, true zone settling (type 3) imposes an area demand on thickener design, but no particular depth is required.

**Talmage and Fitch method.**   The method of Talmage and Fitch draws on the theory by Kynch.[31] Kynch assumed, like Coe and Clevenger, that the solids settling rate is solely a function of the solids concentration. Consequently, the rate of upward propagation of a zone of constant concentration is also constant. Talmage and Fitch utilized this principle of constancy to determine from one batch test, rather than a series, the zone settling limit on flux at any underflow solids concentration.

With reference to Fig. 7.17, if the initial concentration and height of a suspension, the mud line, are $C_0$ and $H_0$, respectively, then the total weight of solids in the suspension is $AC_0H_0$, where $A$ is the cross-sectional area of the vessel. The compression point CP occurs when the "rate-controlling" concentration reaches the surface of the suspension, the mud line. When this has happened, all the solids must have

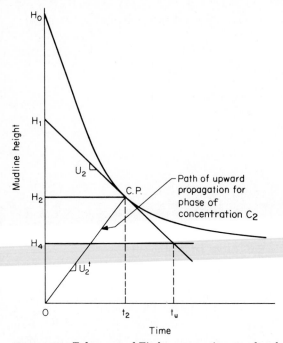

**Figure 7.17**  Talmage and Fitch construction on a batch settling curve.[8]

passed into the rate-controlling phase, as it was passing upward from the bottom of the vessel. When the concentration of the mud line is $C_2$ at time $t_2$ and the upward propagation rate of this concentration is $U_2^+$, then equating total mass leads to

$$AC_0H_0 = AC_2t_2(U_2^+ + U_2) \tag{7.47}$$

If $H_2$ is the height of the mud line at time $t_2$, then from the constancy of upward velocity

$$U_2^+ = \frac{H_2}{t_2} \tag{7.48}$$

Substituting this in Eq. (7.47) gives

$$C_2 = \frac{C_0H_0}{H_2 + U_2t_2} \tag{7.49}$$

But $U_2$ is the slope of the tangent to the curve at $(H_2, t_2)$, and if the intercept of this tangent on the $H$ axis is $H_1$, then from trigonometry

$$H_1 = H_2 + U_2t_2 \tag{7.50}$$

This means that $H_1$ would be the height of the mud line if the concentration $C_2$ were uniform throughout the suspension. Assuming that the solids-handling capacity of a thickener is controlled by the zone settling regime and is identified by the compression point, the solids at the mud line are settling at a rate of

$$U_2 = \frac{H_1 - H_2}{t_2} \qquad (7.50a)$$

When this is substituted into Eq. (7.49),

$$C_2 = \frac{C_0 H_0}{H_1} \qquad (7.51)$$

Furthermore, from trigonometry the time required to attain the underflow concentration is given by

$$t_u = \frac{(H_1 - H_u)t_2}{H_1 - H_2} \qquad (7.52)$$

Because the total quantity of solids present per unit area equals $C_0 H_0$ and it will take time $t_u$ for this amount to reach the underflow zone, the flux for concentration $C_u$ in the underflow is given by

$$\phi_u = \frac{C_0 H_0}{t_u} \qquad (7.53)$$

If the required underflow concentration lies above the compression point on the batch settling curve, then the tangent construction is not necessary. The value of $t_u$ is the intersection on the batch settling curve for the corresponding value of $H_u$.

**Sludges produced with coagulants and flocculants.** Theories of thickening explain the behavior of suspensions of rigid particles such as mining wastes, river silts, and softening sludges in gravity thickeners when they are not treated with flocculants and the thickeners do not contain rotating equipment to assist solids discharge.

Batch settling tests, although they can be used for rigid particle suspensions, are invalid for flocculated suspensions.[32,33] One reason is that the settling rate after dosing the polymer as a flocculant is affected by the laboratory experimental technique. Therefore, modeling of thickener performance must be done with a correctly designed pilot plant.

The theory is effectively limited to explaining and predicting the performance of conical thickeners, thickeners without rotating equipment to assist solids discharge.[34] A rotating rake in a continuous thickener exerts forces that are not accounted for in the theoretical models, and this enables greater solids concentrations to be achieved than are predicted from theory and laboratory tests.[32,33]

In the United Kingdom, the results from pilot-scale and full-size continuous raked thickening of coagulant sludges have been collated with sludge characteristics to produce a data base. This is used to predict thickener performance to within acceptable limits. This recently developed approach[33] has proved to be a rapid and inexpensive way of producing a design, and it has been adopted for more than 45 designs in the United Kingdom and elsewhere, including the United States.

The typical design has an agitated equalizing tank to receive presettled filter backwash and clarifier sludges. Alternatively the balancing tank can be large enough to receive wash water directly. The equalizing tank allows sludge to be fed to the thickener at a constant rate, to be dosed accurately with polymer, and to be flocculated efficiently.

If the thickened sludge is pressed, then a storage tank will be required to balance the pulsating demand of the press with the semicontinuous output from the thickener. By this approach thickened aluminum and iron sludges are typically 8 to 10 percent by weight solids, and concentrations as great as 20 percent are sometimes produced. Thickener surface loading is generally 37 gal/(ft$^2$)(h) [1.5 m$^3$/(m$^2$)(h)], and therefore the final plant is compact and is relatively diminutive alongside the traditional sludge treatment plant.

### Currents and residence time distribution in sedimentation tanks

**Ideal flow conditions.**   The simplest flow condition is plug flow when all liquid advances with equal velocity. Conditions only approximate this when turbulence is small and uniform throughout the liquid. In laminar, nonturbulent flow conditions, a uniform velocity gradient exists, with velocity zero at the wall and maximum at the center of the channel through which the liquid flows, and therefore plug flow cannot exist.

Major departures from plug- and laminar-flow conditions in sedimentation tanks are associated with currents caused by poor flow distribution and collection, wind, rising bubbles, and density differences caused by temperature or concentration.

Currents caused by these factors result in short-circuiting of flow and bulk mixing and reduce the performance of the process predicted by ideal theory. The extent of departure from the ideal can be assessed by residence time distribution analysis with the help of tracer studies.

**Tracer tests.** In simplest form, residence time distribution analysis is carried out by injecting a tracer into the liquid entering the process and monitoring the concentration of the tracer in the liquid leaving the process. The results are plotted to produce a flow-through curve, and the likely outcome when the tracer is introduced as a slug is illustrated in Fig. 7.18. Alternative tracer test methods and the ways in which results can be analyzed are described in standard texts.[35] The performance indices commonly used in analyses of flow-through curves for sedimentation and other tanks are listed in Table 7.1.[36–38] A more analytical approach is to produce what is called the $F(t)$ curve.[38–40] Here $F(t)$ represents the fraction of total tracer added that has arrived at the sampling point and is usually plotted against $t/T$, in which $t$ is the time from injection of tracer and $T$ is the theoretical residence time. Consequently, mathematical modeling predicts:

$$1 - F(t) = \left[\exp\frac{-1}{(1-p)(1-m)}\right]\left(\frac{t}{T} - p^{1-m}\right) \tag{7.54}$$

where $p$ = fraction of active flow volume acting as plug flow
$1 - p$ = fraction of active flow volume acting as mixed flow
$m$ = fraction of total basin volume that is dead space

Rebhun and Argaman[38] compared use of the $F(t)$ curve with the use of the flow-through curve and concluded that comparable results are

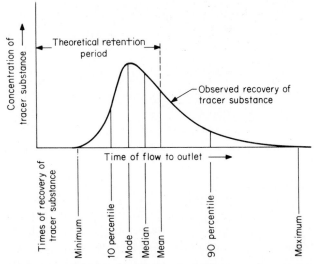

**Figure 7.18** Typical plot of flow-through curve for a tracer dosed as a slug.[9]

**TABLE 7.1    Performance Indices Commonly Used in Analyses of Flow-Through Curves.**

| Indices | Definition |
|---|---|
| $T$ | Theoretical retention time |
| $t_i$ | Time interval for initial indication of tracer in effluent |
| $t_p$ | Time to reach peak concentration (mode time) |
| $t_g$ | Time to reach centroid of curve |
| $t_{10}, t_{50}, t_{90}$ | Time of 10, 50, and 90 percent of tracer to have appeared in effluent ($t_{50}$ = mean time) |
| $t_{90}/t_{10}$ | Morril dispersion index, indicates degree of mixing; as value increases, degree of mixing increases |

Ideally, the following indices approach values of 1.0 under perfect plug-flow conditions:

| | |
|---|---|
| $t_i/T$ | Index of short-circuiting |
| $t_p/T$ | Index of modal retention time |
| $t_g/T$ | Index of average retention time |
| $t_{50}/T$ | Index of mean retention time. |

produced. Plotting the $F(t)$ curve provides, however, a quantitative measure, and its results have a clear physical meaning.

The chemical engineering dispersion index $d$, as it applies to tracer studies, was introduced by Thirumurthi.[14,35] The dispersion index $d$ is calculated from the variance of the dye dispersion curve, such that ideal plug-flow conditions are indicated when the value of $d$ approaches zero. Marske and Boyle[36] compared the dispersion index with other indices, as listed in Table 7.1. They found that the dispersion index has the strongest statistical probability to accurately describe the hydraulic performance of a contact basin. Of the conventional parameters, the Morril index $t_{90}/t_{10}$ is the best approximation of $d$.

Hazen[41] proposed that for horizontal-flow sedimentation tanks, with reference to Fig. 7.18, the number of hypothetical compartments in series within a tank $N$ is given by

$$N = \frac{t_{50}}{t_{50} - t_p} \tag{7.55}$$

This assumes that a tank consists of $N$ hypothetical compartments arranged in series. Consequently, plug flow is approached as the value of $N$ becomes very large. The proportion of particles $F_t$ with a given settling velocity $U_t$ that is removed is also dependent on $N$, as illustrated in Fig. 7.19.[42] The data in equation form produce

$$F_t = 1 - \left(1 + \frac{U_t A}{NQ}\right)^{-N} \tag{7.56}$$

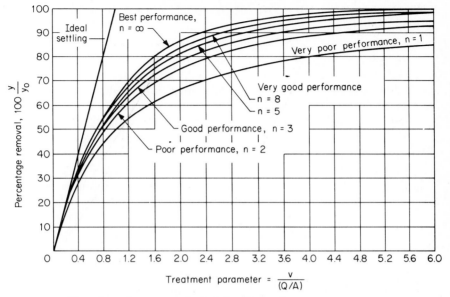

**Figure 7.19** Flow-through curve analysis (after Hazen).[9]

Hazen's theory (Fig. 7.19) predicts poorer removal efficiencies than expected from ideal conditions. For example, when $t$ equals $Q/A$ and $N$ equals infinity, only 63 percent of particles with settling velocity equal to $U_t$ are removed compared with 100 percent in the ideal case. Poorer removals occur for all settling velocities, and complete removal of particles is not achieved at any settling velocity, although at the highest settling velocity the quantity not removed might be negligibly small.

## Types of Sedimentation Tanks

### Horizontal-flow tanks

**Rectangular tanks.** With rectangular horizontal-flow tanks, water to be settled flows in at one end, and treated water flows out at the other end. The inlet flow arrangement must provide good flow distribution to maximize the opportunity for particles to settle. If flocculation has been carried out to maximize particle size, then the flow distribution system must be as nondestructive to the floc as possible, by minimizing head loss between the distribution channel and the main body of the tank. A certain amount of head loss is necessary, however, to achieve the flow distribution. Therefore, for new tanks, attaching the final stage of flocculation to the head of the sedimentation tank in a manner to assist flow distribution may be better and easier.

The length and cross-sectional shape of the tank must not encourage the development of counterproductive circulatory flow patterns and scour. Outlet flow arrangements also must ensure appropriate flow patterns. The principal differences between tanks relate to inlet and outlet arrangements; length, width, and depth ratios; and the method of sludge removal. For horizontal-flow tanks with a small length-width ratio, the end effects dominate efficiency. Inlet and outlet flow distribution substantially affects overall flow patterns and residence time distribution. When the depth is greater than the width, the length-depth ratio is more important than the length-width ratio.

A length-width ratio of 20 or more ideally is needed to approach plug flow[36,43] and maximum horizontal-flow, and presumably inclined flow, sedimentation efficiency, as shown in Fig. 7.20, by determination of the chemical engineering dispersion index. Such a high-value ratio may not be economically acceptable, and a lower ratio, possibly as low as about 5, may give acceptable efficiency if the flow distribution is good. The length-width ratio can be increased by installing longitudinal baffles or division walls.

Increasing the length-width ratio also has the effect of increasing the value of the *Froude number*

$$\mathrm{Fr} = \frac{2U^2}{d_H g} \tag{7.57}$$

The value of Fr increases because of the increase in velocity $U$ and decrease in hydraulic diameter $d_H$. Camp[13] has shown that the increase in value of Fr is associated with improvement in the flow stability.

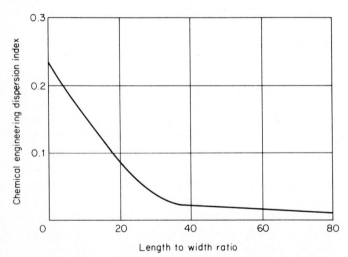

**Figure 7.20**   Effect of length-width ratio on dispersion index.[36]

Poor flow distribution may produce currents or high flow velocities near the bottom of a tank. This may cause scour, or resuspension of particles from the layer of settled sludge. Scour may cause transportation of solids along the bottom of the tank to the outlet end. An adequate tank depth can help to limit scour, and consequently depths less than 8 ft (2.4 m) are rarely encountered.[44] To avoid scour, the ratio of length to depth or surface area to cross-sectional area must be kept less than 18.[5]

Early tanks were desludged by simple manual and hydraulic methods. To avoid interruption in operation and to reduce the workforce, mechanically aided sludge removal methods were introduced. Most commonly, mechanical scrapers push the sludge to a hopper that is usually at the inlet end of the tank. Periodically, the hopper is emptied hydraulically.

If sludge is not removed regularly from horizontal-flow tanks, allowance must be made to the tank depth for sludge accumulation so that the sedimentation efficiency remains unaffected. Sludge can be allowed to accumulate until the settled water quality starts to be impaired. The tank floor should slope toward the inlet, because the bulk of solids generally settles closer to the inlet end.

The frequency of sludge removal depends on the rate of sludge accumulation. This can be estimated by mass balance calculations. Sometimes the decomposition of organic matter in the sludge necessitates more frequent sludge removal. Decomposition can be controlled by prechlorination, if this practice is acceptable. Otherwise decomposition may produce rising gas bubbles that disturb settled sludge and flow patterns and can cause taste and odor problems.

Frequent sludge removal is best carried out with mechanical sludge scrapers that sweep the sludge to a hopper at the inlet end of the tank. Frequent removal results in easy maintenance of tank volumetric efficiency and better output efficiency with continuous operation.

**Multistory tanks.** Multistory, or tray, tanks are a result of recognizing the importance of the settling area to the settling efficiency. Two basic flow arrangements are possible with multistory tanks. The trays may be coupled in parallel with flow divided between them (Fig. 7.21) or coupled in series with flow passing from one to the next. A few of the latter reverse-flow tanks exist in the United States with two levels.

The Little Falls water filtration plant of the Passaic Valley Water Commission, in Clifton, N.J., uses two-pass tanks constructed on top of two-pass tanks. Coagulated water enters the bottom pass and returns on the level above. Clarified water is removed by using submerged launders. Sludge collectors move in the direction of the flow, scraping settled material to sludge hoppers at the far end of the first

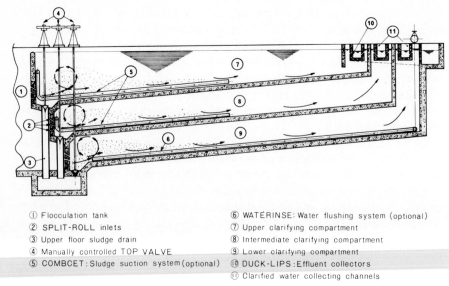

① Flocculation tank
② SPLIT-ROLL inlets
③ Upper floor sludge drain
④ Manually controlled TOP VALVE
⑤ COMBCET: Sludge suction system (optional)

⑥ WATERINSE: Water flushing system (optional)
⑦ Upper clarifying compartment
⑧ Intermediate clarifying compartment
⑨ Lower clarifying compartment
⑩ DUCK-LIPS: Effluent collectors
⑪ Clarified water collecting channels

**Figure 7.21**  Multistory horizontal tank with parallel flow on three levels. (*Courtesy of OTV, Paris, France, and Kubota Construction Co., Ltd., Tokyo, Japan.*)

pass. Each collector flight is trapped at the effluent end on the return pass so that collected material drops down into the path of the influent to the bottom pass.

Multistory tanks are attractive where the land value is high. Difficulties with these tanks include a limited width of construction for unsupported floors, flow distribution, sludge removal, and maintenance of submerged machinery. Successful installations in the United States show that these difficulties can be overcome in a satisfactory manner.

**Circular tanks.**    Circular tank flow is usually from a central feed well radially outward to peripheral weirs (Fig. 7.22). The tank floor is usually slightly conical to a central sludge well. The floor is swept by a sludge scraper that directs the sludge toward the central well.

Radial outward flow is theoretically attractive because of the progressively decreasing velocity. The circumference allows a substantial outlet weir length and hence a relatively low weir loading. Weirs must be adjustable, or installed with great accuracy, to avoid differential flow around a tank. Circular tanks are convenient for constructing in either steel or concrete, although they might be less efficient in the use of land than rectangular tanks. Sludge removal problems tend to be minimal.

The settling efficiency might be less than expected because of the problem of achieving good flow distribution from a central point to a

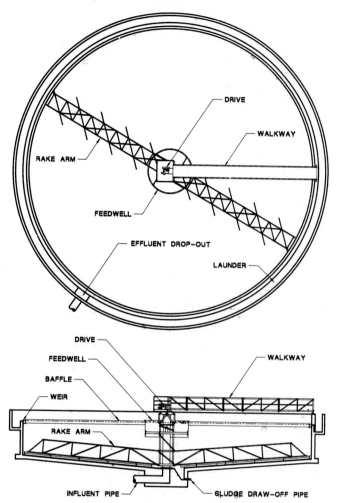

**Figure 7.22**   Circular radial-flow clarifier (by Eimco). (*Courtesy of Eimco Process Equipment Co., Salt Lake City, Utah.*)

large area. The principal differences between circular tanks are associated with floor profile and sludge-scraping equipment.

### Inclined (plate and tube) settlers

Inclined (plate and tube) settlers can be constructed with individual or prefabricated modules of plates or tubes of any suitable material. Advantages of prefabricated modules include efficient use of material, accuracy of separation distances, lightweight construction, and structural rigidity. Inclined surfaces may be contained within a suitably shaped tank for countercurrent, cocurrent, or cross-flow sedimentation. Adequate flocculation is a prerequisite for inclined settling if coagulation is carried out. The tank containing the settler system also can incorporate the flocculation stage and preliminary sludge thickening (Fig. 7.23).

The angle of inclination depends on the application, the need for self-cleaning, and the flow characteristic of the sludge on the inclined surface. If the angle of inclination of surfaces is great enough, generally greater than 50 to 60°,[22] self-cleaning of surfaces occurs. When the angle of inclination is small, the output of the settler must be interrupted periodically for cleaning. This is because the small distance between inclined surfaces allows little space for sludge accumulation. An angle as little as 7° is used when sludge removal is achieved by periodic backflushing, possibly in conjunction with filter backwashing. The typical separation distance between inclined surfaces

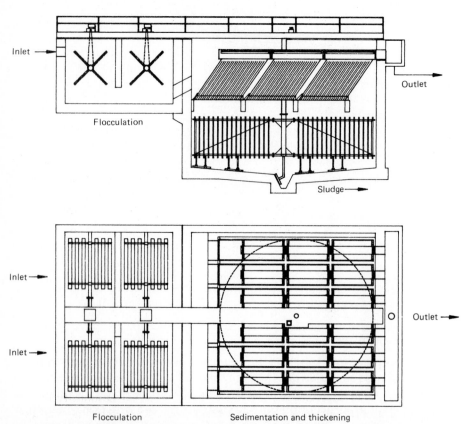

**Figure 7.23** Inclined plate settler with preflocculation and combined thickening. (*Courtesy of Zimprol/Passavant Inc.*)

for unhindered settling is 2 in (50 mm) with an inclined length of 3 to 6 ft (1 to 2 m).

The main challenge in inclined settler development has been to obtain settling efficiencies close to theoretical. Considerable attention must be given to providing equal flow distribution to each channel, producing good flow distribution within each channel, and collecting settled sludge while preventing its resuspension.

With inclined settlers, the velocity along the axis of the channels defines the flow regime. In practice, efficiency is usually examined in terms of either the surface loading based on the plan area occupied by the settling system, upflow velocity, or the loading based on the total area available for settlement.

**Countercurrent settling.** In countercurrent inclined settlers, the suspension is fed to the lower end and flows up the channels formed by the inclined surfaces (Fig. 7.13). Solids settle onto the lower surface in each channel. If the angle of inclination is great enough, the solids move down the surface counter to the flow of the liquid; otherwise, periodic interruption of flow is necessary for cleaning.

Tube settlers are used mostly in the countercurrent settling mode. Tube modules have been constructed in various formats (Fig. 7.24), including square tubes between vertical sheets, alternating inclination between adjacent vertical sandwiches, chevron-shaped tubes between vertical sheets, and hexagonal tubes.

Countercurrent modular systems are suitable for installation in existing horizontal-flow tanks and some solids-contact clarifiers to achieve upgrading and uprating. Closely spaced inclined surface systems are not cost-effective in floc-blanket clarifiers, although widely spaced [1-ft (0.3-m)], inclined plates are. Tube modules may aid uprating by acting in part as baffles that improve flow uniformity.

**Cocurrent settling.** In cocurrent settling the suspension is fed to the upper end of the inclined surfaces with flow down through the channels (Fig. 7.13). Settled solids on the lower surface move down the surface in the same direction as the liquid above. Special attention must be given to collecting settled liquid from the lower end of the upper surface of each channel to prevent resuspension of settled solids.

**Cross-flow settling.** In cross-flow settling the suspension is fed to flow horizontally between the inclined surfaces as settled solids move downward (Fig. 7.13). Resuspension of settled solids is usually less of a problem than in countercurrent and cocurrent settling. This might not be true in some systems in which the direction of inclination alternates (Fig. 7.25).[45] Alternating the inclination can allow efficient

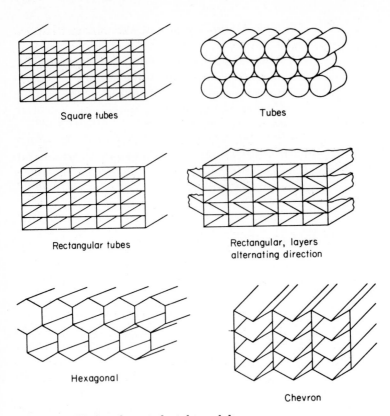

Square tubes

Tubes

Rectangular tubes

Rectangular, layers
alternating direction

Hexagonal

Chevron

**Figure 7.24**   Various formats for tube modules.

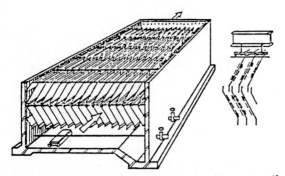

**Figure 7.25**   Alternating cross-flow lamella settler.[45]
[*Source: C. Gomella, Clarification avant filtration: ses
progres recents (Rapport General 1), Intl. Water Supply
Assoc. Intl. Conf., 1974.*]

use of tank volume and results in rigidity of modular construction. Development and application of cross-flow systems have occurred mainly in Japan.

### Solids-contact clarifiers

Solids-contact clarifiers are generally circular and contain equipment for mixing, flow circulation, and sludge scraping. A wide variety of such tanks exist, and most are of proprietary design. Hartung[6] has presented a review of eight different designs. Contact clarifiers can be grouped as premix, like the example in Fig. 7.3, or premix-recirculation types (Fig. 7.26).

In the simpler premix system, water is fed into a central preliminary mixing zone that is mechanically agitated. This premix zone is contained within a shroud that acts as the inner wall of the outer annular settling zone. Chemicals can be dosed into the premix zone. Water flows from the premix zone under the shroud to the base of the settling zone.

In the premix-recirculation system, water is drawn out of the top of the premix zone and fed to the middle of the settling zone. The recirculation rate can exceed the actual flow of untreated water to the tank such that the excess flow in the settling zone is drawn downward and under the shroud back into the premix zone. This movement recirculates solids that can assist flocculation in the premix zone.

Mechanical equipment associated with solids-contact clarifiers must

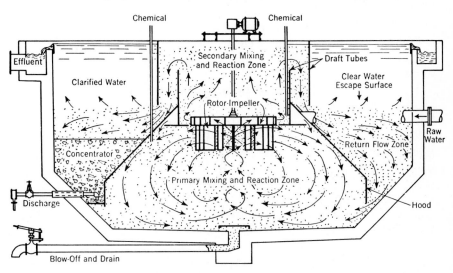

Slurry Pool Indicated by Shaded Areas

**Figure 7.26**   The Accelator solids-contact clarifier.[6]

be adjusted or tuned to the throughput. Excessive stirring motion in the premix zone can be counterproductive, while too little stirring can result in poor radial-flow distribution under the shroud as well as poor chemical mixing and flocculation. In solids-contact units, sludge settles to the tank floor and is removed with mechanical equipment. Clarifiers with recirculation to keep solids in suspension allow excess solids to accumulate in sludge pockets or concentrators, as illustrated in Fig. 7.26. Appropriate operation of these pockets contributes to controlling the concentration of solids in suspension, which affects the sedimentation efficiency of the clarifier.

### Floc-blanket clarifiers

Both types of solids-contact tanks discussed above can be considered as floc-blanket clarifiers if stable and distinct floc blankets can be established and easily maintained in the settling zone. Only a few designs of solids-contact clarifiers have been developed with this intention. Usually, the volume and concentration of solids in circulation in contact units are not likely to be as great as those needed to maintain a blanket.

**Hopper-bottom tanks.** The first purposely designed floc-blanket tanks had a single hopper bottom, square or circular in cross section. In these units coagulant-dosed water is fed down into the apex of the hopper. The hopper shape assists with even flow distribution from a single point inlet to a large upflow area. The expanding upward flow allows floc growth to occur, large particles to remain in suspension, and a floc blanket to form. The pressure loss through the floc blanket, although relatively small, also contributes to creating homogeneous upward flow.

A single hopper, conical or pyramidal, occupies only 33 percent of available space. In addition, it is expensive to construct, and its size is limited by constructional constraints. Consequently, alternative forms of hopper tanks have been developed to overcome these drawbacks yet retain the hydraulic advantage of hoppers. These include tanks with multiple hoppers, a wedge or trough, a circular wedge (premix type of clarifier, Fig. 7.3), and multiple troughs.

As a floc blanket increases in depth, the settled water quality improves, but with diminishing return (Fig. 7.27).[46] The blanket depth defines the quantity of solids in suspension. If "effective depth" is defined as the total volume of blanket divided by the area of its upper interface with supernatant, then the effective depth of a hopper is roughly one-third the actual depth. As a result, flat-bottom tanks have an actual depth that is much less than that of hopper tanks with

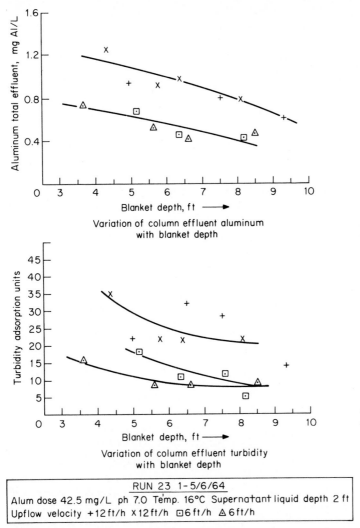

**Figure 7.27** Change in settled water quality as depth of floc blanket increases.[46]

the same effective depth. Effective depths of blankets are typically in the range 8 to 10 ft (2.5 to 3 m).

The quantity of solids in suspension, defined by blanket depth and concentration, affects the efficiency because of the effect on flocculation. In addition, the head loss assists flow distribution, ensuring a more stable blanket and thereby greater blanket concentration.

A very stable floc blanket can be operated with little depth of supernatant liquid, less than 1 ft (0.3 m), without significant carryover

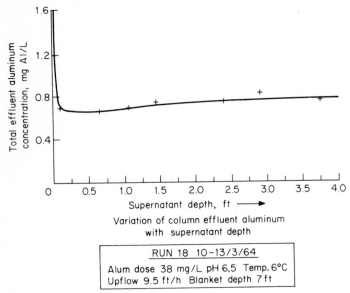

**Figure 7.28** Change in settled water quality as depth of supernatant increases above a floc blanket.[46]

(Fig. 7.28).[46] In practice, the feasibility of this approach depends on weir spacing and likely disturbance by wind. Generally, blankets operated in a manner likely to produce a blanket with a diffuse surface, have a tendency to exhibit some unstable boiling, and have poor level control methods. Thus, a supernatant liquid depth of at least 3 ft (1 m) may be necessary to minimize carryover, especially after increases in upflow velocity. Supernatant liquid depths of 6 ft (2 m) are commonly provided, but this can be unnecessarily large with good blanket-level control.

Control of the blanket surface is easily achieved by using a slurry weir or sludge hopper with sills set at an appropriately high level. Sludge hoppers can be emptied frequently by automatically timed valves. One proprietary suspended sludge hopper system utilizes a strain gauge to initiate drainage of sludge from the suspended canvas cones.

Sludge hoppers, pockets, and cones must be sized to allow efficient removal of sludge. They must be large enough to allow in situ preliminary thickening, even when the sludge removal rate has to be high. Such high rates might occur following a substantial increase in the flow-through rate during conditions of high chemical doses.

**Example Problem 3** What area will be needed for removing floc to control the blanket level in a floc-blanket clarifier, given the conditions for Example Problem 1?

**Solution** Assume that the aluminum coagulant dose to the water is 3.2 mg Al/L. The aluminum concentration in the floc blanket at maximum flux is 110 mg Al/L when the blanket floc volume concentration is 20 percent. The concentration of aluminum in the blanket is proportional to the floc volume.

From Example Problem 1 the conditions are as follows:

| Upflow rate, m/h | 1.6 | 1.95 | 2.5 | 3.05 | 3.65 | 4.2 | 4.7 | 5.15 |
|---|---|---|---|---|---|---|---|---|
| ½-h floc volume, % | 31 | 29 | 25 | 22 | 19 | 16 | 13 | 10 |

The proportion of the floc-blanket tank area needed for removing floc is determined by using a mass balance:

Al dose × total volumetric flow rate to tank = blanket Al concentration × volumetric settlement rate into removal area

But

Total volumetric flow rate to tank = upflow rate × total tank upflow area

and

Volumetric settlement rate into removal area = upflow rate × area for removal

Thus

Area for removal × blanket Al concentration = total tank upflow area × Al dose

This means that the proportion of tank area needed for removing floc is the ratio of aluminum concentration in the blanket to that being dosed. For example, at the upflow rate of 1.95 m/h

$$\text{Concentration of Al in blanket} = 110 \times \frac{29}{20}$$

Therefore

$$\text{Proportion of area needed} = \frac{3.2}{110} \times \frac{20}{29} \times \frac{1}{100\%} = 2.0\%$$

Hence

| Upflow rate, m/h | 1.6 | 1.95 | 2.5 | 3.05 | 3.65 | 4.2 | 4.7 | 5.15 |
|---|---|---|---|---|---|---|---|---|
| Proportion of area needed, % | 1.9 | 2.0 | 2.3 | 2.6 | 3.1 | 3.6 | 4.5 | 5.8 |

For a different aluminum dose, the area will need to be accordingly proportionally greater or less. In practice, a greater area will be required to cope with the short-term need to remove excess floc at a high rate to prevent the blanket from reaching the launders when the upflow rate is increased quickly.

**Flat-bottom tanks.** The pursuit of simplicity and least cost in the development of floc-blanket tanks has led to the abandonment of hoppers and the use of flat-bottom tanks. In these, good flow distribution is achieved by using either multiple downward inverted-candelabra feed pipes or laterals across the floor (Figs. 7.4 and 7.5).

An inverted-candelabra system can ensure good distribution for a wide range of flows but may obstruct installation of inclined settling systems. The opposite can apply to a lateral distribution system. In the proprietary Pulsator design, reliability of flow distribution is ensured by periodically pulsing the flow to the laterals. Pulsing has been found useful in other applications of fluidization and therefore might improve the functioning of the floc blanket.

**Inclined settling with floc blankets.**   Tube modules with normal spacing between inclined surfaces of 2 in (50 mm) have not been found cost-effective in floc-blanket tanks.[16] With the blanket surface below tube modules, settled water quality is no better than from a stable and efficient tank without modules.

With the blanket surface within the modules, floc concentration in the blanket increases by about 50 percent, but no commensurate improvement in settled water quality occurs. The failure of closely spaced inclined surfaces with hindered settling relates to the proximity of the surfaces and a circulatory motion at the blanket surface that counteracts the entrapment mechanism of the blanket.[16]

The problem with closely spaced surfaces diminishes with more widely spaced inclined surfaces. Optimum spacing is about 1 ft (0.3 m), but no optimization studies are known to have been published. Large (2.9-m) plates, however, have been shown to be preferable to shorter (1.5-m) plates.[47] The combined action of suppressing currents and inclined settling results in about a 50 percent greater throughput than a good floc blanket without inclined surfaces. The proprietary Superpulsator tank is the Pulsator design with widely spaced inclined surfaces.

**Ballasted floc systems.**   Floc produced by coagulating clay-bearing water generally settles faster than floc produced by coagulating water containing little mineral turbidity. Consequently, mineral turbidity added purposely to increase floc density can be useful. Bentonite is the usual choice, and fly ash has been considered in eastern Europe. The advantages of ballasting floc also arise with powdered activated carbon, dosed for taste and odor control, and when precipitation softening is carried out in association with iron coagulation. Sometimes fine sand has been used as the ballasting agent.

A process based on fine sand ballasting was developed in Hungary in the 1950s and 1960s. In this process, sand is recycled for economy (Fig. 7.29). The process has found favor in France and is sometimes known by the proprietary name of Cyclofloc[48] or Simtafier.[49] Recovered sand is conditioned with polyelectrolyte and added to the untreated water before the

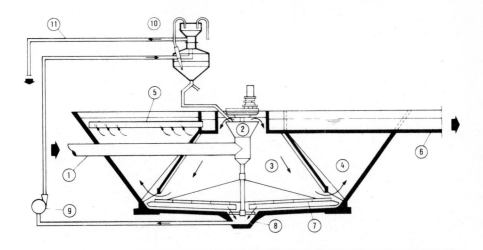

**1** Sewage inlet
**2** Bell-shaped inlet work
**3** Reaction zone
**4** Clarification zone
**5** Recovery trough
**6** Clarified water outlet

**7** Sludge scraper
**8** Sludge and microsand collection pump
**9** Recycling pump
**10** Sludge and microsand hopper
**11** Sludge discharge

**Figure 7.29**   The Cyclofloc clarification system. (*Courtesy of OTV, Paris, France, and Kubota Construction Co., Tokyo, Japan.*)

metal-ion coagulant is added. A second polyelectrolyte might be used as a flocculant aid prior to floc-blanket settling. Sand is recovered by pumping sludge through small hydrocyclones. The Fluorapide system combines sand ballasting with inclined settling.[48]

Powdered magnetite, recovered by magnetic drum filters, also has been suggested as a ballasting agent. A process from Australia,[50] with the proprietary name Sirofloc, is based on recycling magnetite, but the magnetite is chemically conditioned with sodium hydroxide such that a metal-ion coagulant is not needed. Sirofloc is efficient enough for subsequent filtration to be almost unnecessary. High-rate filtration (20 m/h) may be needed to remove residual magnetite and manganese that Sirofloc cannot normally remove. Appraisal and further development of the Sirofloc process are being pursued in the United Kingdom.[51]

## Other Factors Influencing Sedimentation Efficiency

### Flow-through rate

**Surface loading.**   The surface loading of a sedimentation tank is expressed as the flow rate per unit of surface area of that part of the tank in which sedimentation is meant to happen. The performance of all sedimentation processes is influenced by surface loading. Settled water quality deteriorates when surface loading is increased (Fig. 7.15). The reasons for this deterioration are various, but (particularly for discrete sedimentation processes) the deterioration is dependent on the simultaneous effect on prior flocculation by the change in flow rate.

The point at which the performance curve (Fig. 7.15) crosses the limit of acceptable quality defines the maximum reliable surface loading. Cost effectiveness requires the highest surface loadings possible; however, the efficiency will then be more sensitive to major variations in surface loading and water quality.

The reliable surface loading for a given type of sedimentation process depends on a wide range of factors. For example, when a highly colored, low-alkalinity water is treated in winter by floc-blanket sedimentation, the loading might be only 3 ft/h (1 m/h), whereas when a minerally turbid, high-alkalinity water is treated in summer with a flocculant aid, the loading could be greater than 23 ft/h (7 m/h). (*Note:* 1 ft/h is equivalent to 0.125 gpm/ft$^2$, hence 23 ft/h is equivalent to 2.9 gpm/ft$^2$.)

**Residence time.** The theoretical mean residence time of a process is the volume of the process from the point of dosing or end of the previous process through to the point at which the separation efficiency is measured at the outlet of the process, divided by the flow-through rate. For horizontal-flow sedimentation, the volume of the entire tank is important in assessing the effect on sedimentation efficiency. For inclined settlers the volume within the inclined surfaces, and for floc blankets the volume of the blanket itself, is most important. Depth can be used to reflect the volume of a floc blanket.

The mean residence time of a process is effectively the length divided by the velocity. Thus, the mean residence time reflects the velocity or overflow rate and will relate to sedimentation efficiency accordingly.

Flow-through curves (Fig. 7.18) are a measure of the distribution in residence time. These can be analyzed to produce estimates of efficiency of flow distribution and volumetric utilization of a tank (Fig. 7.19). Consequently the extent to which sedimentation efficiency might be improved by improving flow conditions can be estimated. Clements and Khattab[52] have shown with model studies that sedimentation efficiency is correlated with the proportion of plug flow.

### Size and shape of tank

**Size and number.** The number of tanks can affect the flexibility of plant operation. When operation is close to the limit of acceptable quality, isolation of one tank is likely to impose a reduction in total plant production. No fewer than two tanks should be used for reliability.

The size of tanks might be limited by constructional constraints, and consequently this will dictate the minimum number of tanks. While factors that affect sedimentation efficiency, such as loading, velocity, and various dimension ratios, can be maintained with any size tank, performance of extra-large tanks may become unacceptable and may limit tank size. Tanks with large surface area will be more vulnerable to environmental effects, such as wind-induced circulation.

When considering the number of tanks, the performance of a floc blanket is dependent on the upflow velocity, blanket-floc concentration, effective blanket depth, and dosing and delay time conditions. Thus, provided that the size and shape of floc-blanket tanks do not affect these factors, any such size or shape might not be expected to affect the sedimentation efficiency. The efficiency is affected, however, because differences occur between tanks in their hydrodynamic and hydraulic conditions for flocculation and flow distribution.

The length-width ratio is not relevant to floc-blanket sedimentation, although baffles have been shown to be useful.[17] With inclined settlers the length of the flow path affects the sedimentation efficiency.[22] When lengths are short, end effects may limit efficiency. Additionally, problems of flow distribution may mean that inclined plate settlers with narrow plates are more efficient than those with wide plates.

**Depth.**   Depth does not affect Hazen's law, which says that the settling efficiency is dependent on the tank area and is independent of the depth [Eq. (7.18)], but in practice the depth is interrelated with width in influencing the sedimentation efficiency of horizontal and inclined settlers. A minimum depth may be needed to limit scour, as mentioned previously. Depth also defines the spacing between surfaces and, therefore, number of layers and hence settler efficiency.

### Flow arrangements

**Inlet and outlet.**   The purpose of an inlet is to distribute incoming water uniformly over the cross section of a tank over a wide range of flow rates. The outlet arrangement is as important as the inlet in ensuring good flow distribution. Numerous investigations[14,42] have shown how inlets and outlets in horizontal-flow tanks influence the residence time distribution and settling efficiency. Tests with model tanks[53,54] showed that symmetry is desirable and that interruptions (such as by support walls and piers) can disturb uniformity. Complex arrangements can be as good as simple ones, and a uniformly fed submerged weir can give good results without a baffle.

For all types of sedimentation, low flow velocities in approach channels or pipe work that allow premature settlement must be avoided. Conversely, disruption of floc at high flows must also be avoided.

In floc-blanket tanks, the injection velocity from the inlet pipes governs the input energy, which can affect blanket stability and "boiling." Boiling causes direct carryover and a reduction in blanket concentration, resulting in deterioration of settled water quality.

Outlet weir or launder length and loading affect the outlet flow distribution. Therefore, they must be kept clean and unobstructed, but the effectiveness of long launders is questionable.[55]

**Baffling.** Baffles are useful in horizontal-flow tanks at the inlet and outlet to assist flow distribution, longitudinally to increase the length-width ratio, and as vanes to assist changes in horizontal-flow direction. Kawamura[54] did numerous model tests with diffuser walls and arrived at a number of design guides relating to the position of walls and the free area in the walls.

In hopper-bottom floc-blanket tanks, baffles might help inlet flow distribution,[56,57] but the centering of the inlet should be checked first.[16] A matrix of vertical baffles in the floc blanket is an alternative method for improving tank performance if flow distribution is poor.[17]

Vertical baffles have been installed in some solids-contact clarifiers, so that the action of the central mixer does not lead to centrifugal movement in the settling zone. Centrifugal movement will lift the floc up the outer side of the settling zone. A proprietary tube module baffle system[58] has been used to improve flow distribution in radial-flow tanks. Baffles at the water surface can help to counteract wind disturbance.

### Particulate and water quality

**Seasonal water quality.** Sedimentation efficiency varies seasonally in association with seasonal changes in temperature, alkalinity, phosphate concentration, and similar parameters, as well as the nature of color and turbidity being coagulated.

Temperature affects efficiency by influencing the rate of chemical reactions, the viscosity of water, and hence the particle settling velocity. Temperature can also be a surrogate for change in other parameters that occur on a similar seasonal basis. Changes in alkalinity, color, turbidity, and orthophosphate concentration affect coagulation reactions and the properties and rate of settling of resulting floc particles.

With floc-blanket sedimentation, for example, the reliable upflow velocity in the summer can be more than twice that in the winter (Fig.

7.15).[16,18] This is important when plant size or reliable plant output is defined.

**Example Problem 4**   What will be the floc-blanket sedimentation upflow rate at maximum flux for different temperatures, given the conditions defined in Example Problem 1?

**Solution**   The Stokes settling velocity of particles in discrete-particle settling is inversely proportional to the liquid viscosity, Eq. (7.8):

$$U_{temp1} = U_{temp2} \left( \frac{\mu_{temp2}}{\mu_{temp1}} \right)$$

The viscosity of water for different temperatures is as follows:

| Temperature, °C | 5 | 10 | 15 | 20 |
|---|---|---|---|---|
| Viscosity, cP | 1.52 | 1.31 | 1.14 | 1.01 |

If the results given in Example Problem 1 were obtained at the temperature of 10°C, then assuming no change in water chemistry and other factors that might affect the floc, the maximum flux upflow rate at 20°C can be estimated, for example as

$$U_{20°C} = \frac{3.44 \times 1.31}{1.01} = 4.46 \text{ m/h}$$

Hence for the above temperature range:

Maximum flux upflow rate, m/h    2.96    3.44    3.95    4.46

This means that the upflow rate for reliable operation in the summer could be more than 1.5 times that in the winter for this temperature range.

**Coagulation.**   The particle settling velocity is affected by various particle characteristics, principally size, shape, and density. The choice of coagulation chemistry and application efficiency affect floc particle characteristics. The choice of coagulant should be dictated by chemical considerations (see Chap. 6). Optimum chemical doses for sedimentation can be predicted reliably by jar tests. Jar tests also reliably predict color removal but not metal-ion and turbidity removal because of the difference in hydrodynamic conditions in the batch jar test and the continuous sedimentation process.

Adequate control of dosing and mixing of the chemicals should be provided. If a change in coagulant dose or pH improves the sedimentation efficiency, the current chemical doses may be incorrect or mixing or flocculation may be inadequate. The difference in efficiency between identical tanks supposedly receiving the same water can be due to differences in chemical doses if made separately to each tank, poor chemical mixing before flow splitting, or different flows through the tanks.

**Flocculation.**    The efficiency of horizontal- and inclined-flow sedimentation is dependent on prior flocculation if coagulation is carried out. The efficiency of solids-contact units is dependent on the quality of preliminary mixing. Mixing assists with chemical dispersion, flocculation, solids resuspension, and flow distribution. Because these can conflict with each other, mixing rates should be adjusted when changes are made to the flow-through rate. This is rarely done, however, because it is usually too difficult.

The extent of flocculation is dependent on the velocity gradient and time (see Chap. 6). The time delay between chemical dosing and inlet to a floc blanket, to allow preliminary flocculation in connecting pipe work and channels, affects the floc-blanket sedimentation efficiency (Fig. 7.30).[59,60] The time delay between dosing the coagulant and dosing a polyelectrolyte, and that between dosing the polyelectrolyte and the inlet to the tank, also affects the efficiency (Fig. 7.31). The reasons for needing such a delay time are not entirely understood, but

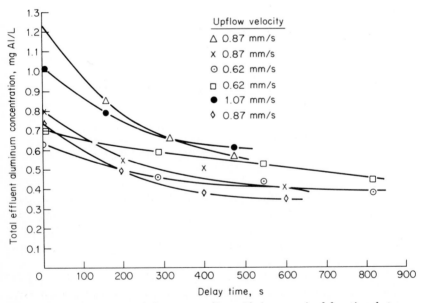

**Figure 7.30**  Change in settled water quality with increase in delay time between chemical dosing and inlet to a floc blanket.[60] (*Source: N. P. Yadav and J. T. West, The Effect of Delay Time on Floc Blanket Efficiency, TR9, Water Research Centre, Medmenham, U.K., 1975.*)

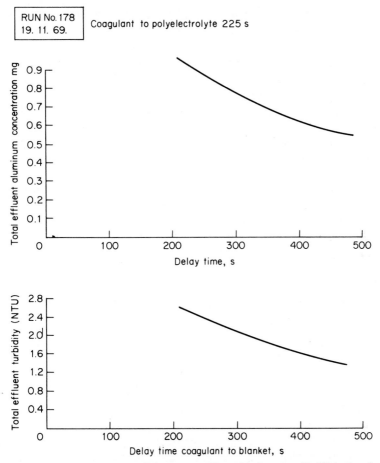

**Figure 7.31** Change in settled water quality with increase in delay time between dosing of polyelectrolyte and inlet to a floc blanket.[60] (*Source: N. P. Yadav and J. T. West, The Effect of Delay Time on Floc Blanket Efficiency, TR9, Water Research Centre, Medmenham, U.K., 1975.*)

relate to the efficiency of mixing, the rate of hydrolysis and precipitation, and the rate of formation of primary and subsequent floc particles. Useful delay times are likely to be in the range ½ to 8 min, similar to those needed for direct filtration or high-rate flocculation for inclined settling or dissolved-air flotation.

Delay time is not a substitute for initial rapid chemical dispersion. Many operators have learned that simple improvements to flow conditions and dispersion arrangements at the point of dosing can produce substantial improvements in chemical utilization and sedimentation efficiency.

**Polyelectrolytes and other agents.**  Flocculant aids improve the sedimentation efficiency by increasing floc strength and size. Jar tests can be used to select suitable polyelectrolytes and to determine a trial dosing range. Optimal doses must be carefully determined by full-plant tests. An excessive dose producing good settled water may be detrimental to filters. If the apparent optimal dose on the plant is much greater than that indicated by jar tests, then mixing and flocculation conditions must be checked.

Dosing of flocculant aids increases particle settling velocities in floc-blanket sedimentation, resulting in an increase in the blanket-floc concentration. A diminishing return also exists in increasing the dose of flocculant aid and is associated with the blanket-floc concentration, increasing with an increase in flocculant dose and becoming greater than at maximum flux (Fig. 7.32).[16] Thus the polyelectrolyte dose need be only enough to ensure that the floc concentration is greater than at maximum flux and that the blanket surface is distinct and not diffuse, and certainly not so great as to take the floc concentration into the thickening regime. Additives that increase floc density and settling velocity, such as bentonite, have an effect on sedimentation similar to flocculant aids.

### Climate and density currents

Wind can induce undesirable circulation in horizontal-flow tanks, with cross-flow winds reducing sedimentation efficiency the most.[53] Therefore, wind effects should be minimized by constructing tanks to align with prevailing winds. Strong winds can disrupt floc-blanket stability. This can be counteracted by covering tanks with roofs, placing floating covers between the launders, constructing windbreaks around each tank, aligning the launders across the prevailing wind, installing scum boards or baffles at the water surface across the prevailing wind, or installing fully submerged baffles in the supernatant liquid. Wind combined with low temperatures can cause severe ice formation and affect outlet flow. Total enclosure may be needed to prevent ice formation.

High solar radiation can cause rapid diurnal changes in water temperature and density. Rapid density changes because of temperature, solids concentration, or salinity can induce density currents that can cause severe short-circuiting in horizontal tanks[61] and inclined settlers.[62] Hudson[61] has reviewed methods of minimizing or preventing induced currents. He considered the methods to fall into several categories: (1) use of a surface weir or launder takeoff over a large part of the settling basin, (2) improved inlet arrangement, (3) schemes

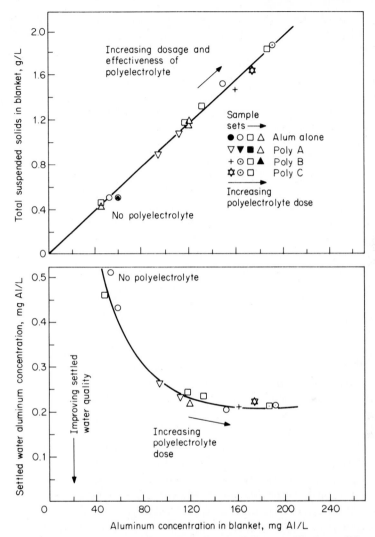

**Figure 7.32** Effect of polyelectrolyte type and dose on blanket solids concentration and settled water quality.[16] [*Source: R. Gregory, Floc Blanket Clarification, TR111, Water Research Centre, Medmenham, U.K., 1979.*]

to provide increased basin drag and friction, and (4) slurry recirculation. Substantial diurnal variation in temperature also can result in the release of dissolved gases, especially in biologically active waters. The resulting rising columns of bubbles can disrupt floc blankets.

## Flotation

### History of flotation

Over 2000 years ago, the ancient Greeks used a flotation process to separate the desired minerals from the gangue, the waste material.[63] Crushed ore was dusted onto a water surface, and mineral particles were retained at the surface by surface tension while the gangue settled. In 1860 Haynes patented a process in which oil was used for the separation of the mineral from the gangue.[64] The mineral floated with the oil when the mixture was stirred in water.

In 1905 Salman, Picard, and Ballot developed the "froth flotation" process by agitating finely divided ore in water with entrained air. A small amount of oil was added sufficient to bestow good floatability to the sulfide grains.[64] The air bubbles together with the desired mineral collected as foam at the surface while the gangue settled. The first froth flotation machine was developed by T. Hoover in 1910,[64] and except for size, it was not much different from the machines used today. Later, in 1914, Callow introduced air bubbles through submerged porous diffusers. This process is called "foam flotation."[64] The two processes, froth and foam flotation, are generally known as dispersed-air flotation and are at present widely used in the mineral industry.

Elmore suggested in 1904 the use of electrolysis to produce gas bubbles for flotation. This process, although not commercially used at that time, has been developed into electrolytic flotation.[65] Elmore also invented the dissolved-air (vacuum) flotation process whereby air bubbles are produced by applying a vacuum to the liquid, which releases the air in the form of minute bubbles.[64]

The original patent for the dissolved-air pressure flotation process was issued in 1924 to Peterson and Sveen for the recovery of fibers and white water in the paper industry.[66] In pressure dissolved-air flotation, the air bubbles are produced by releasing the pressure of a water stream saturated with air above atmospheric pressure.

Initially dissolved-air flotation was used mainly in applications where the material to be removed, such as fat, oil, fibers, and grease, had a specific gravity less than that of water. In the late 1960s, however, the process also became acceptable for wastewater and potable water treatment applications.

Dissolved-air flotation has been applied extensively for wastewater sludge thickening. In water treatment it has become accepted as an alternative to sedimentation, particularly in the Scandinavian countries and the United Kingdom, where more than 50 plants are in operation or under construction. Flotation is employed mainly for the treatment of nutrient-rich reservoir water that may contain heavy algae blooms and for low-turbidity, low-alkalinity, colored water.[67]

These types of water are difficult to treat by sedimentation, because the floc produced by chemical treatment has a low settling velocity.

## Types of flotation processes

Flotation can be described as a gravity separation process in which gas bubbles attach to solid particles to cause the apparent density of the bubble-solid agglomerates to be less than that of the water, thereby allowing the agglomerate to float to the surface. Different methods of producing gas bubbles give rise to different types of flotation processes: electrolytic flotation, dispersed-air flotation, and dissolved-air flotation.[66]

**Electrolytic flotation.** The basis of electrolytic flotation or electroflotation is the generation of bubbles of hydrogen and oxygen in a dilute aqueous solution by passing a direct current between two electrodes.[68] The bubble size generated in electroflotation is very small, and the surface loading is therefore restricted to less than 13.3 ft/h (4 m/h). The application of electrolytic flotation has been restricted mainly to sludge thickening and small wastewater treatment plants in the range 50,000 to 100,000 gpd (10 to 20 m$^3$/h). The process has been reported to be suitable for very small water treatment installations.[69]

**Dispersed-air flotation.** Two different dispersed-air flotation systems are used, foam flotation and froth flotation.[70] Dispersed-air flotation is unsuitable for water treatment because the bubble size tends to be large and either high turbulence or undesirable chemicals are required to produce the air bubbles required for flotation.

**Dissolved-air flotation.** In dissolved-air flotation, bubbles are produced by the reduction in pressure of a water stream saturated with air. The three main types of dissolved-air flotation are vacuum flotation,[67] microflotation,[71] and pressure flotation.[67] Of these three, pressure flotation is currently the most widely used. In pressure flotation, air is dissolved in water under pressure. Three basic pressure dissolved-air flotation processes can be used: full-flow, split-flow, and recycle-flow pressure flotation.[67] For water treatment applications requiring the removal of fragile floc, recycle-flow pressure flotation is the most appropriate system (Fig. 7.33). In this process the whole influent flows either initially through the flocculation tank or directly to the flotation tank if separate flocculation is not required. Part of the clarified effluent is recycled, pressurized, and saturated with air. The pressurized recycle water is introduced to the flotation tank through a pres-

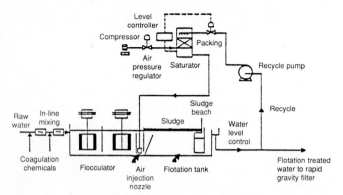

**Figure 7.33**  Schematic diagram of a flotation plant for potable water treatment.[67] (*Source: T. F. Zabel and J. D. Melbourne, "Flotation," in W. M. Lewis, ed., Developments in Water Treatment, vol. 1, Elsevier, Essex, England, 1980.*)

sure release device and is mixed with the flocculated water. In the pressure release device, the pressure is reduced to atmospheric pressure, releasing the air in the form of fine bubbles (20 to 100 μm in diameter). The air bubbles attach themselves to the flocs, and the agglomerates float to the surface. The float is removed from the surface, and clarified water is taken from the bottom of the flotation tank.

### Theory of dissolved-air flotation

To achieve efficient clarification by dissolved-air flotation, impurities present in the water must be coagulated and flocculated effectively prior to the introduction of the microbubbles to form the bubble-floc agglomerates.

**Mechanism of flotation.**  Floatable bubble-floc agglomerates might form by any of three distinct mechanisms: entrapment of bubbles within a condensing network of floc particles, growth of bubbles from nuclei within the floc, and attachment of bubbles to floc during collision. Work by Kitchener and Gouchin[72] has shown that all three mechanisms can occur but that the principal mechanism in dissolved-air flotation for potable water treatment is the attachment mechanism. Their work also indicates that the organic content of surface waters is usually high enough to render the floc surface sufficiently hydrophobic for bubble attachment. Only in very pure organic "free" systems will the flotation efficiency be reduced.

**Solubility of air in water.**  Over the temperature and pressure ranges used in dissolved-air flotation (0 to 30°C and 200 to 800 kPa), both nitrogen and oxygen obey Henry's law:

$$p = (H_e)(x) \tag{7.58}$$

where $p$ = partial pressure of gas in gas phase
$x$ = mole fraction of gas in liquid phase
$H_e$ = Henry's law constant

In a continuous saturation system, the gas phase above the water does not have the same composition as air, because oxygen is more soluble in water than nitrogen. In order that the same quantities of oxygen and nitrogen leave the saturator in the pressurized water as enter it in the compressed air, the nitrogen content of the gas in the saturator will rise, creating a nitrogen-rich atmosphere. This results in a reduction of about 9 percent in the mass of gas that can be dissolved. In assessing the performance of a continuously operating saturation system, the 100 percent saturation level should be taken as that achievable assuming a nitrogen-rich atmosphere (Fig. 7.34).[73]

**Effect of bubble size.** To achieve maximum agglomeration between gas and solid phases, the gas bubbles must rise under laminar-flow conditions. This avoids shedding of floc such as can occur in the turbulent regime. The maximum bubble diameter for laminar flow is 130 μm.[74] For bubble sizes less than 130 μm, Stokes' law, as applicable to sedimentation [Eq. (7.8)], can be used to calculate the rise rate:

$$U_t = \frac{g(\rho - \rho_g)d^2}{18\mu} \tag{7.59}$$

where $\rho_g$ is the density of the gas bubble.

The maximum bubble diameter for laminar flow can be calculated from Eq. (7.59) by assuming that the limiting value of the Reynolds number for laminar flow is Re = 1. The relationship between the rise rate and bubble diameter for single bubbles is given in Fig. 7.35.[74]

The specific gravity of floc produced in water treatment is very similar to that of water, and only very small air bubbles are required to float floc to the surface. The smaller the air bubbles that can be produced, the larger the number of bubbles produced per unit volume of gas released. The presence of large numbers of bubbles increases the chance of bubble-floc attachment in very dilute floc suspensions that are typical in water treatment applications. The smaller the bubble size, the slower the rise rate of the bubble (Fig. 7.35). Consequently, a larger flotation tank is required to allow bubbles to reach the surface (Fig. 7.36). In practice, the bubble size produced in dissolved-air flotation ranges from 10 to 120 μm with a mean size of approximately 40 μm.[73]

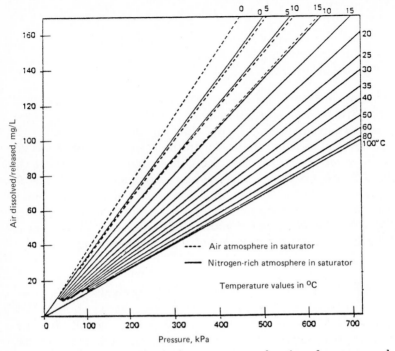

**Figure 7.34** Mass of gas dissolved in water as a function of pressure and temperature.[73]

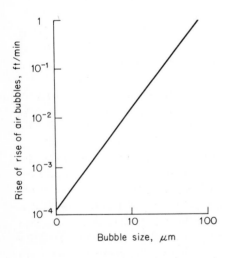

**Figure 7.35** Rate of rise of air bubbles in tapwater as a function of bubble size.[74]

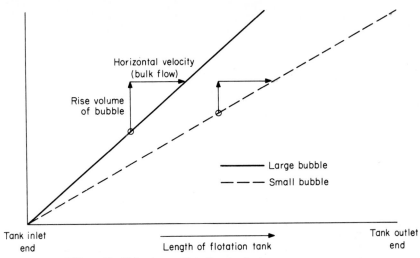

**Figure 7.36** Effect of bubble size on flotation tank size.

### Types of flotation tanks

**Circular tanks.** Circular tanks are used mainly in small flotation plants treating wastewater or for sludge thickening applications that require no preflocculation prior to flotation (Fig. 7.37).

For potable water treatment, a preflocculation stage is required prior to flotation to flocculate the impurities present in water to larger agglomerates suitable for removal by flotation. For circular tanks the transfer of the flocculated water to the flotation tank without breaking up the flocs creates problems. In larger plants the flocculated water must be introduced close to the bottom of the center section of the flotation tank to achieve even distribution. As a result, most large flotation plants for water treatment use rectangular tanks.

**Rectangular tanks.** Rectangular flotation tanks offer advantages in terms of scale-up, simple design, easy introduction of flocculated water, easy float removal, and a relatively small area requirement. Flotation tanks are currently designed with a depth of approximately 5 ft (1.5 m) and overflow rates of 26 to 40 ft/h (8 to 12 m/h) depending on the type of water treated. Tanks are equipped at the inlet with an inclined baffle (60° to the horizontal) to direct the bubble-floc agglomerates toward the surface and to reduce the velocity of the incoming water, to ensure minimum disturbance of the float layer accumulating on the water surface (Fig. 7.33). The gap between the top of the baffle and the water surface should be designed to achieve a horizontal water ve-

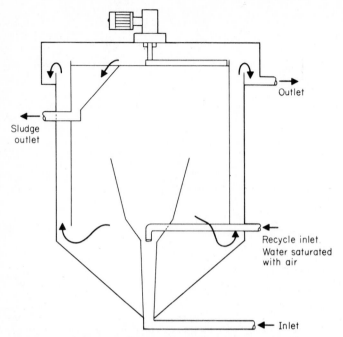

**Figure 7.37**   Schematic diagram of a circular flotation tank.

locity similar to the velocity in the top section of the inclined baffle area.

The maximum tank size is determined by hydraulic conditions and the design of the sludge removal system. Tanks with surface areas in excess of 860 ft$^2$ (80 m$^2$) are in operation. The nominal retention time in the flotation tank is between 5 and 15 min, depending on loading (upflow rate) and tank depth.[75] The flotation tank must be covered because both rain and wind can cause breakup of floated solids and because freezing of the float can cause problems. Treated water should be withdrawn, preferably over a full-width weir, to maintain uniform hydraulic conditions and to minimize changes in water level due to variations in flow through the plant.

**Combined flotation and filtration.**   The combination of flotation and filtration was pioneered in Sweden (Fig. 7.38),[67] but experience with this process has also been reported from other countries.[76,77] A rapid gravity sand or anthracite-sand filter is incorporated in the lower section of the flotation tank. This arrangement has the advantage of providing an extremely compact plant. The flotation rate of the plant is, however, limited by the filtration rate that can be achieved.

The tank depth of a flotation-filtration plant tends to be deeper, ap-

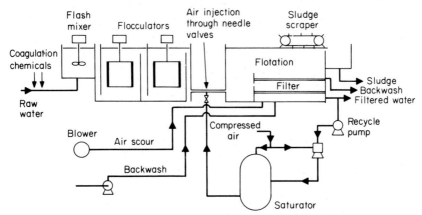

**Figure 7.38** Schematic diagram of a combined flotation-filtration plant.

proximately 8.3 ft (2.5 m), to accommodate the filter bed and underdrain system, compared with 5 ft (1.5 m) for a separate flotation unit. In addition, the flow to the plant and any coagulant dosing have to be stopped periodically to facilitate cleaning of the filter that is backwashed in the normal way by air scour and water wash. The compactness of this system makes it particularly suitable for package plants. The reported performance of the package plants in terms of treated water quality is comparable with standard flotation plants followed by rapid gravity filtration.[67]

### Air saturation systems

The air saturation system accounts for approximately 50 percent of the power cost of the flotation process. As a result, optimization of the recycle system design is important in minimizing operating costs.

Various methods are employed for dissolving air under pressure in the recycle stream. These include sparging the air into the water in a pressure vessel, or saturator, trickling the water over a packed bed, spraying the water into an unpacked saturator, entraining the air with ejectors, and injecting the air into the suction pipe of the recycle pump (Fig. 7.39).[67]

Introducing the air to the recycled water either on the suction side of the recycle pump or through eductors before entering the saturation vessel leads to substantially higher pumping costs compared with using a separate compressed-air supply.

Saturation levels of between 60 and 80 percent achieved by introducing the air on the suction side of the pump could be increased to 90 percent by providing a turbine mixer in the saturation vessel.[78]

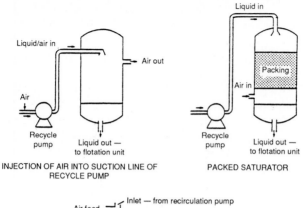

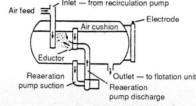

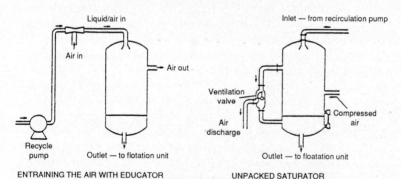

**Figure 7.39**  Different air saturation systems.[67] (*Source: T. F. Zabel and J. D. Melbourne, "Flotation," in W. M. Lewis, ed., Developments in Water Treatment, vol. 1, Elsevier, Essex, England, 1980.*)

Tests have shown that the packed saturator system has the lowest ratio of operating cost to saturation level achieved.[79,80] A possible disadvantage of using packings is the danger of blockage caused by biological growth. This problem has not been observed so far in plants treating drinking water.

Extensive research has been carried out on optimizing the design of packed saturators.[80] It has been shown that the saturator can be operated over the range of 36 to 274 ft/h [300 to 2000 $m^3/(m^2)(day)$] hy-

draulic loading without any decrease in saturation efficiency. A packing depth of 2.6 ft (0.8 m) of 1-in (25-mm) polypropylene rings is
sufficient to achieve 100 percent saturation (Fig. 7.40).[80] This packing
depth, however, was substantially higher than the 1-ft (0.3-m) figure
reported elsewhere.[78]

In terms of operating costs, an unpacked saturator must be operated
at a pressure of about 30 psi (200 kPa) above that of a packed saturator in order to supply the same amount of air to the flotation tank.[81]
Water saturated with air under pressure is corrosive. If mild steel is
used for construction of the saturation vessel, a corrosion-resistant
lining should be provided. Plastic pipes should be used for connecting
pipe work between the saturator and flotation tank.

### Factors influencing dissolved-air flotation efficiency

**Coagulation.**   To achieve efficient clarification by flotation, the coagulation pH and coagulant dose must be optimized. Optimum chemical
doses can be determined by using standard jar test apparatus (see
Chap. 6). No significant differences in primary chemical requirements
for flotation and sedimentation have been found; however, flotation
does not require the addition of polyelectrolyte as a flocculant aid. Flotation is susceptible to over- or underdosing because of the short residence time in the plant (approximately 30 min).

For best treated water quality, chemicals must be thoroughly and
rapidly mixed with the raw water. Some form of in-line mixing is preferable to the use of a flash mixer. If both a coagulant and a pH adjustment chemical are required, good mixing of the first chemical with the
source water must be completed before the second chemical is added.

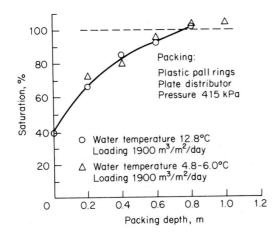

**Figure 7.40** Effect of packing depth on air dissolution.[80] (*Source: A. J. Rees, D. J. Rodman, and T. F. Zabel, Evaluation of Dissolved-Air Flotation Saturator Performance, TR143, Water Research Centre, Medmenham, U.K., 1980.*)

This is particularly important in treating soft, colored water. The order of chemical addition is not important.[82]

**Flocculation.**  Before coagulated impurities can be removed successfully by flotation, flocculation into larger agglomerates (floc) is required. The flocculation time, degree of agitation, and means of providing agitation affect flotation performance. Flocculation conditions that are different from those for sedimentation are required to produce floc suitable for flotation.

**Flocculation time.**  The flocculator usually consists of a tank subdivided by partial baffles into two equal-size compartments in series, each agitated by a slow-moving paddle. Additional compartments can be used, but this is usually not cost-effective. The flocculation time required differs according to the type of raw water being treated. For algae-laden water a flocculation time of 15 min is sufficient, whereas for soft, colored water a flocculation time of 20 min is required.

Table 7.2 shows the effect of increasing the flocculation time on treated water quality, highlighting the importance of the flocculation stage for efficient operation of the flotation process.

**Degree of agitation.**  Besides the flocculation time, the degree of agitation is very important for efficient flocculation. Agitation is usually provided by a slow-moving four-blade gate paddle in each flocculator compartment, although different paddle designs are being used. To avoid excess shear, which prevents adequate floc growth, the tip speed of the paddles should not exceed 1.6 ft/s (0.5 m/s).

The degree of agitation can be expressed by the "mean velocity gradient" $G$ (see Chap. 6). Tests have indicated that the optimum mean velocity gradient for flotation is about 70 s$^{-1}$ independent of the type of surface water treated.[83] This compares with an optimum $G$ value for horizontal sedimentation of between 10 and 50 s$^{-1}$.

**Hydraulic flocculation.**  An alternative approach to mechanical flocculation is the use of hydraulic flocculation in which the energy required

**TABLE 7.2    Effect of Increasing the Flocculation Time on Flotation-Treated Water Quality (Colored Low-Alkalinity Water)[67]**

| Flocculation time, min | Turbidity, NTU | Residual coagulant, mg Fe/L |
|---|---|---|
| 12 | 1.2 | 0.83 |
| 16 | 0.94 | 0.56 |

for flocculation is provided by the water flowing through the flocculator, which can be a baffled tank. Tests have shown that one-half the flocculation time, with a higher $G$ value ($150$ s$^{-1}$), was required for hydraulic flocculation compared with mechanical flocculation ($G = 70$ s$^{-1}$) (Fig. 7.41).[84] The difference in the $G$ value required is probably due to the more uniform velocity distribution in the hydraulic flocculators, thus avoiding excess shear and floc breakup.

The product of the mean velocity gradient and the flocculation time, $Gt$, is often used to express the flocculation conditions required. A $Gt$ value of between 40,000 and 60,000 is required for efficient flotation regardless of whether hydraulic or mechanical flocculation is employed.

**Quantity of air required for flotation.**   The quantity of air supplied to the flotation tank can be varied by altering the saturator pressure or the recycle rate or both. If a fixed orifice is used for controlling the recycling, an increase in saturator pressure is associated with a small increase in recycle rate. Thus, different nozzle sizes require different combinations of flow and pressure to deliver the same amount of air.

Experiments varying the recycle rate by using different nozzle sizes and saturator pressures have shown that treated water quality is dependent on only the total amount of air supplied, not the pressure and recycle rate employed (Fig. 7.42).[83]

The quantity of air required for treatment of surface water depends on only the total quantity of water treated and is independent of the suspended solids present, unless the suspended solids concentration is very high ($> 1000$ mg/L). The air-solids ratio required for surface water treatment is approximately 380 mL of air per gram of solids for a

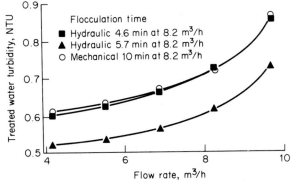

**Figure 7.41**   Comparison of hydraulic and mechanical flocculation for flotation.[84]

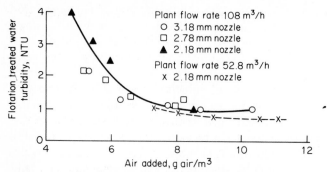

**Figure 7.42** Effect of quantity of air added and nozzle size on flotation-treated water quality.[84] (*Source: A. J. Rees, D. J. Rodman, and T. F. Zabel, Water Clarification by Flotation, 5., TR114, Water Research Centre, Medmenham, U.K., 1979.*)

solids concentration in raw water of 20 mg/L. That is much higher than for activated sludge thickening, where 15 to 30 mL of air per gram of solids is required.[85] The large air-solids ratio required for surface water treatment of low turbidity is probably necessary to ensure adequate collisions between floc particles and air bubbles, to facilitate attachment before separation.

With a packed saturator, an operating pressure of between 50 and 60 psi (350 and 420 kPa) and a recycle rate of between 7 and 8 percent corresponding to about 8 to 10 g air/m$^3$ water treated were found adequate for optimum performance.[83]

**Example Problem 5**   Calculate the recycle rate required for dissolved-air flotation, assuming

- 8 g air/m$^3$ water treated is required for effective flotation.
- Operating pressure is set at 60 psig.
- 100 percent saturation is achieved in the saturator.
- The saturator has no air bleed; therefore, a nitrogen-rich atmosphere exists in the saturator.
- The water temperature is 20°C.

**Solution**   Convert the operating pressure to kilopascals.

$$1 \text{ psig} = 6.8948 \times 10^3 \text{ N/m}^2 \text{ (Pascals, Pa)}$$

$$60 \text{ psig} = 60 \times 6.8948 \times 10^3 \text{ N/m}^2 \text{ (Pa)} = 413.7 \text{ kPa}$$

Determine the amount of air dissolved at the operating pressure. Use Fig. 7.34 and find 413.7 kPa on the horizontal axis. Go vertically upward to the 20°C (nitrogen-rich atmosphere) line, and read from the vertical axis the corresponding mass of air dissolved per liter of water, 95 mg/L.

Calculate the percentage of recycling required. If 8 g air/m$^3$ water treated is required for effective flotation and 95 mg/L can be dissolved at the operating pressure and temperature selected, then the recycling required is 8 ÷ 95 = 8.42 percent.

Calculate the recycle rate required. For a flotation plant treating 100 m³/h (440 gpm) the recycle rate required would be 8.42 m³/h (37 gpm).

**Air-release devices.** Different types of pressure-release devices are employed ranging from proprietary nozzles and needle valves to simple gate valves. To achieve effective air release, the pressure should be reduced suddenly and highly turbulent conditions should exist in the device. The velocity of the recycle stream leaving the pressure reduction device should be low enough to prevent floc breakup. As water passes through the pressure-release devices at high velocities and as air is released, erosion and cavitation can occur, and so the devices should be made from stainless steel. For larger plants a number of these devices, usually at a spacing of approximately 1 ft (0.3 m), are used to obtain good mixing and distribution between the flocculated water and the air bubbles. Good mixing of the recycle stream containing the released air bubbles with the flocculated water stream is also essential to facilitate contact between bubbles and floc.

Releasing the pressure of the recycle stream close to the point of injection into the flocculated water is important to minimize coalescence of air bubbles, which could result in a loss of bubbles available for flotation.

An air injection nozzle has been patented and developed[86] that consists of two orifice plates, to reduce the pressure and to create turbulence, and a shroud section, to decrease the velocity of the stream of recycled water before it is mixed with the flocculated water. The size of the first orifice plate, which is the smaller, controls the amount of recycled water added to the flotation tank. More than 95 percent of the bubbles produced by the nozzle were in the size range of 10 to 120 μm, with a mean size of about 40 μm. A comparison of the proprietary nozzle and a needle valve showed that the nozzle produced smaller air bubbles; however, both devices achieved similar flotation-treated water quality.[73]

**Float removal.** The sludge that accumulates on the flotation tank surface, called *float*, can be removed either continuously or intermittently by flooding or mechanical scraping. Flooding involves raising the water level in the flotation tank sufficiently by closing the treated water outlet or lowering the outlet weir to allow the float and water to flow into the float collection trough. The flooding method has the advantages of low equipment costs and minimal effect on the treated water quality, but at the expense of high water wastage (up to 2 percent of plant throughput) and very low sludge solids concentration (less than 0.2 percent). Therefore, with this float removal method, one advantage of flotation—the production of a sludge with a high solids concentration—is lost.

**Float removal system.**    The most widely used mechanical float removal devices are of two types: (1) part- or full-length scrapers usually with rubber blades that travel over the tank surface and push the float over the beach into the collection channel and (2) beach scrapers that consist of a number of rubber blades rotating over the beach.

As float is removed from the beach, float from the remainder of the flotation tank surface flows toward the beach. The beach scrapers, especially if operated continuously, have the advantage of reducing the danger of float breakup during the removal process because the float is minimally disturbed. Beach scrapers are also of simpler construction compared with full-length scrapers. They have the disadvantage, however, of producing relatively thin sludges (1 to 3 percent), because thicker sludge would not flow toward the beach. If a part- or full-length scraper is used, selecting the correct frequency of float removal and travel speed of the scraper is important to minimize deterioration in treated water quality because of float breakup.

**Effect of air-solids ratio on float.**    In water treatment, the aim is to produce water of good quality, and the thickness of the float is of secondary importance. In general, the thicker the float, which means the longer the float is allowed to accumulate on the tank surface, the more severe will be the deterioration in treated water quality during the float removal process. The variation in air-solids ratio, produced by varying the amount of air added to the system, has no influence on the float concentration produced.

**Influence of source water on float characteristics.**    The characteristics of float obtained from the treatment of different types of source water vary considerably. Therefore, the most appropriate float removal system must be selected for the source water being treated. For example, experience has shown that for cases in which a low-alkalinity, highly colored water was treated, the float started to break up after only 30 min of accumulation. The most appropriate device for such an application is a full-length scraper operating at a removal frequency and blade spacing that do not allow the float to remain on the flotation tank surface for longer than 30 min. The optimum scraper speed for this application, in terms of treated water quality and float solids concentration, is operated at 0.028 ft/s (30 m/h), producing a sludge of 1 percent solids concentration.

The float stability is independent of the primary coagulant used, and the addition of flocculant aid has no influence on the float stability. Conversely, float produced from turbid river water or stored algae-laden waters is very stable. Accumulation of float for more than 24 h does not result in float breakup or deterioration in treated water quality. Beach scrapers and part- and full-length scrapers have been used successfully for these applications, producing solids concentrations in excess of 3 percent with little deterioration in treated water quality, provided the float is not allowed to accumulate for too long. These floats were suitable for filter pressing, producing cake solids concentration between 16 and 23 percent without polyelectrolyte addition.

For optimum operation in terms of treated water quality and float solids concentration, beach scrapers should be operated continuously, the water level in the flotation tank should be adjusted close to the lower edge of the beach, and a thin, continuous float layer of about 0.4 in (10 mm) should be maintained on the surface of the flotation tank. This operation produces a float concentration, depending on the source water treated, of 1 to 3 percent with minimum deterioration of treated water quality.

Equipment costs for float removal systems are significant and can be as high as 10 to 20 percent of the total plant cost. Selection of the most appropriate and cost-effective removal system for a particular application is important.

### Performance of dissolved-air flotation plants

Extensive studies on both pilot- and full-scale plants have been conducted on the performance of the dissolved-air flotation process in treating different types of source water. The types of source water investigated include colored (low-alkalinity) water, mineral-bearing (high-alkalinity) water, and algae-bearing water. Detailed performance data have been given elsewhere.[67,76,83,87–89]

**Treatment of lowland mineral-bearing (high-alkalinity) river water.** Under optimum operating conditions, flotation reduced source water turbidities of up to 100 NTU to levels rarely exceeding 3 NTU at a design flow rate of 39.6 ft/h (12 m/h). When the source water turbidity exceeded 60 NTU, the treated water quality was improved significantly by reducing the flow rate through the plant by about 10 to 20 percent. Color was reduced from as much as 70 color units (CU) to less than 5 CU, and residual coagulant concentrations before filtration were in the range 0.25 to 0.75 mg Al/L.

A floc-blanket sedimentation plant operated at an upflow rate of 6.5 ft/h (2 m/h) produced similar treated water quality to that of the flotation plant during low-turbidity periods but better quality (by 1 to 2 NTU) when the source water turbidity was greater than 100 NTU. Selection of the correct coagulant dosage and coagulation pH was critical during flood conditions. Because of the short residence time in the flotation plant, changes in source water quality had to be followed closely to maintain optimum coagulation conditions.

Although directly abstracted river water can be treated successfully by flotation, sedimentation tends to be the more appropriate treatment process for this application, especially if the source water turbidity varies rapidly and high-turbidity peaks (> 100 NTU) occur.

**Treatment of colored (low-alkalinity) stored water.** Table 7.3 shows a comparison of the water quality achieved by flotation, sedimentation, and filtration treating a colored (low-alkalinity) stored water. The flotation plant was operated at 39.6 ft/h (12 m/h) upflow rate.

The floc-blanket sedimentation plant could only be operated at less than 3.24 ft/h (1 m/h), however, even with the addition of polyelectrolyte. The floc produced by coagulation of this water is very light and has low settling velocities. The quality of the water treated by the two processes was quite similar. Initially the residual coagulant concentration of the sedimentation-treated water was usually lower by about 0.2 mg Fe/L. By increasing the flocculation time from 12 to 16 min, however, the residual coagulant concentration in the flotation-treated water was reduced to that of the sedimentation-treated water.

Another advantage of flotation was that the plant consistently produced good treated water quality even at temperatures below 39°F (4°C), which occurred frequently during the winter months. At these

**TABLE 7.3    Comparison of Qualities Achieved with Flotation,† Sedimentation,‡ and Filtration (Colored Low-Alkalinity Water) Following Iron Coagulation[66]**

| Type of water | Turbidity, NTU | Dose, mg Fe/L | Color, Hazen deg | pH | Iron, mg/L | Manganese, mg/L | Aluminum, mg/L |
|---|---|---|---|---|---|---|---|
| Source | 3.2 | | 45 | 6.2 | 0.70 | 0.11 | 0.23 |
| Flotation-treated§ | 0.72 | 8.5 | 2 | 4.8 | 0.58 | 0.16 | 0.01 |
| Flotation-filtered | 0.19 | | | 9.0 | 0.01 | 0.02 | 0.01 |
| Sedimentation-treated | 0.50 | 6.0¶ | 0 | 5.05 | 0.36 | 0.14 | 0.10 |
| Sedimentation-filtered | 0.29 | | 0 | 10.5 | 0.01 | <0.02 | 0.10 |

†Upflow rate 39 ft/h (12 m/h).
‡Upflow rate 3.25 ft/h (1 m/h).
§Improved flotation-treated water quality similar to that achieved with sedimentation was obtained by increasing the flocculation time from 12 to 16 min.
¶Plus 0.8 mg polyelectrolyte/L required to maintain the floc blanket.

low temperatures the floc blanket in the sedimentation tank tended to become unstable, resulting in a deterioration in treated water quality.

**Treatment of algae-bearing (high-alkalinity) stored water.**  Recent emphasis on source water storage in water resources management has led to the construction of source water storage reservoirs. Severe algae problems have been experienced in some storage reservoirs containing nutrient-rich water, which in turn has led to problems in existing sedimentation treatment plants operating at less than 6.5 ft/h (2 m/h). Algae tend to float and are therefore difficult to remove by sedimentation.

Table 7.4 shows the removal efficiency of flotation and sedimentation for different algal species. At times the algae counts in the flotation-treated water were lower than those in the sedimentation and filtered water (Fig. 7.43). Efficient flocculation is essential for effective algae removal. Only 10 to 20 percent algae removal was obtained when the flotation plant was operated without coagulation.

Figure 7.44 shows a comparison of algae removal rates achieved by three different coagulants at their optimum pH for minimum coagulant residuals. Aluminum sulfate gave the best removal. For polyaluminum chloride (PAC), an equivalent dosage in terms of aluminum was required to achieve an algae removal similar to that of aluminum sulfate. The poorest treated water quality was obtained with chlorinated ferrous sulfate. Tests have shown that algae removal is improved by lowering the pH. The poorer algae removal achieved with chlorinated ferrous sulfate might have been a result of the higher coagulation pH required for minimum coagulant residual in the treated water (pH 8.3 to 8.7 for ferrous sulfate, compared with pH 6.8 to 7.2 for aluminum sulfate).

**TABLE 7.4    Comparison of Algae Removal Efficiency of Flotation† and Sedimentation‡ Using Chlorinated Ferrous Sulfate as Coagulant[67]**

| Alga type | Source water, cells/mL | Sedimentation-treated water, cells/mL§ | Flotation-treated water, cells/mL§ |
|---|---|---|---|
| *Aphanizomenon* | 179,000 | 23,000 | 2,800 |
| *Microcystis*¶ | 102,000 | 24,000 | 2,000 |
| *Stephanodiscus* | 53,000 | 21,900 | 9,100 |
| *Chlorella* | 23,000 | 3,600 | 2,200 |

†Upflow rate 39 ft/h (12 m/h).
‡Upflow rate 6.5 ft/h (2 m/h).
§Before filtration.
¶Aluminum sulfate was used as the coagulant.

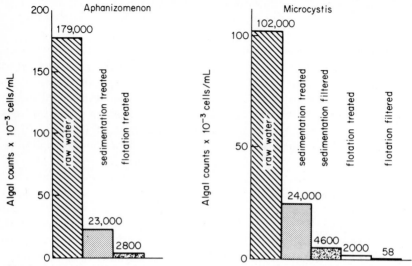

**Figure 7.43** Removal of algae by flotation [upflow rate, 39 ft/h (12 m/h)], floc-blanket sedimentation [upflow rate, 6.5 ft/h (2 m/h)], and filtration.[83]

**Filtration of flotation-treated water.** Tests comparing the performance of rapid gravity sand filters fed with water treated by flotation and by floc-blanket clarification showed that both types of water have similar turbidities and residual coagulant concentrations. The presence of air bubbles in flotation-treated water has no influence on filter performance.[83]

## Factors Influencing the Choice of Process

A comparison of the various types of sedimentation and dissolved-air flotation is summarized in Table 7.5. The factors influencing choice—those principally relating to cost, source water quality, compactness of plant, rapid start-up, and sludge removal—are discussed below.

### Typical applications

Solids separation prior to filtration is needed when the solids concentration is too great for filtration to cope with alone. When the coagulant dose required exceeds about 2 mg Al/L (equivalent to 25 mg alum/L) or 2 mg Fe/L pretreatment will be needed.

The concentration of solids in a floc blanket using alum coagulation is equivalent to about 100 mg Al/L or 200 mg Fe/L if iron coagulant is used. So that blanket-level control requirements remain reasonable, floc-blanket sedimentation becomes inappropriate for coagulant doses

| Coagulant | | Raw water algal count cells/m, microcystis | Raw water turbidity NTU |
|---|---|---|---|
| Aluminum sulphate | O | 40,000 | 3.5 – 4.1 |
| Aluminum sulphate | ● | 120,000 | 5.5 – 5.9 |
| PAC2 | ■ | 120,000 | 5.5 – 5.9 |
| Chlorinated ferrous sulphate | ▲ | 48,000 | 3.6 |

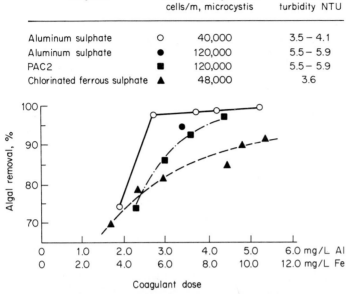

**Figure 7.44** Comparison of effectiveness of three coagulants on algae removal by flotation.[83] (*Source: A. J. Rees, D. J. Rodman, and T. F. Zabel, Water Clarification by Flotation, 5., TR114, Water Research Centre, Medmenham, U.K., 1979.*)

greater than about 7.5 mg Al/L or 15 mg Fe/L. Then other types of sedimentation will be more appropriate.

Floc-blanket sedimentation can be used with partial lime softening in conjunction with iron coagulation. For simple precipitation softening, a solids contact clarifier, which is not too complex and is easily maintained, is likely to be the best choice. Consideration should be given to using pellet reactors, however, especially when softening groundwater. When the raw water bears heavy silts, horizontal and inclined settlers are most appropriate. They will, however, need good arrangements for discharging the sediment and robust sludge-scraping equipment. When silt concentrations are very high, sedimentation prior to coagulation may be appropriate. When silt concentrations are low but algae pose a problem and for colored (low-alkalinity) water, flotation is likely to be the best choice for pretreatment. Flotation might also be a better choice for other reasons, as mentioned below.

The cost comparison of different sedimentation processes and the comparison of sedimentation with flotation must include both capital and operating costs. This is necessary because of the substantial differences in the distribution of costs in the various processes, as sum-

**TABLE 7.5  Comparison of Sedimentation and Dissolved-Air Flotation Processes**

| | Horizontal flow (rectangular) | Radial flow (circular) | Upflow, U.S.-type clarifier | Upflow floc | Floc blanket (widely spaced lates) | Inclined plates (unhindered) | Dissolved-air flotation |
|---|---|---|---|---|---|---|---|
| Sedimentation regime | Unhindered | Unhindered | Unhindered or hindered | Hindered | Hindered | Unhindered | Unhindered (hindered by air) |
| Theoretical removal of particles with $U_t < U_t^*$ | Partial depending on entrance position | Partial depending on entrance position | Partial depending on whether blanket formed | No removal | No removal | Partial depending on entrance position | Depends on entrance position |
| Appropriate for heavy silts and lime precipitate as well as light floc | Yes | Yes | Yes if bottom scraped | Flat tanks limited to light floc, hopper tanks can take wider range of settling velocities | | Yes | No |
| Appropriate for eutrophic (algal) waters | No | No | No | No | No | No | Yes |
| Ease of start-up and on/off operation | Easy | Easy | Easy for unhindered | Slow—may take several days to form a new blanket | | Quick but some skill needed | Quick and easy |
| Skill level of operation | Low | Low | Medium | Medium | Medium | High | High |
| Retention time | High | High | Medium | Medium-low | Low | Low | Low |
| Coagulation and flocculation effectiveness | Little difference if adequately designed and properly operated | | | | | | |
| Capital cost | High-medium | High-medium | Medium | Medium | Medium-low | Medium-low | Medium-low |
| Operating cost | Medium | Medium-low | Medium-low | Low | Low | Medium | Medium |

marized in Table 7.5. Floc-blanket sedimentation has a relatively high capital cost but low operating cost. In contrast, flotation has a relatively low capital cost but a high operating cost because of its energy consumption.

The chemical cost of basic coagulation should be the same for the various processes. While no advantage occurs from using polyelectrolyte as a flocculant aid with flotation, its use can be very cost-effective for any form of sedimentation because loading rates can be increased. High loading rates and hence lower capital costs are also possible with ballasted floc, but additional operating costs of the ballasting agent, conditioning chemicals, and energy for agent recovery and recirculation are incurred.

The relatively high energy cost of flotation, although still possibly less than the cost of chemicals or labor, is incurred largely for the dissolution of the air in the recycling to produce the microbubbles. Most of the remaining energy cost is required for flocculation. The cost of flocculation for discrete particle settling systems, horizontal flow, and inclined sedimentation is basically the same.

The large differences between processes in fixed and variable costs means that dissolved-air flotation is especially attractive for plants with low utilization, less than full flow, or infrequent use, because of its relatively lower capital cost and the reduced variable costs. Other factors besides cost can affect the choice of process, however. Such factors include rate of start-up, operational flexibility, ease of removing algae or color, need for coping with high mineral loadings, and production of the best treated water quality.

Dissolved-air flotation is likely to be less expensive than floc-blanket sedimentation, for a fully utilized plant, only if the sedimentation process cannot be operated at rates greater than about 2 to 3 m/h (6 to 9 ft/h).[90] This rule is easy to derive because the rating of flotation is relatively constant and independent of the type of water to be treated. As the expected plant utilization decreases, the dissolved-air flotation becomes relatively cheaper. Inclined plate sedimentation should have a similar economic advantage while the other types of unhindered sedimentation are likely to be better for full-utilization applications.

Sometimes the solids concentration in the water, after initial chemical treatment, is low enough to be within the capacity of filtration to operate without prior sedimentation or flotation and still achieve the desired objective. Just as each form of filtration has a limited capacity to remove solids, so does each form of sedimentation and flotation. This limited ability to remove solids efficiently may limit the surface loading and therefore the cost-effectiveness of the process.

Within the normal recommended application of flotation, solids loading and air requirements are independent of each other. Greater solids concentration may require a greater recycle rate or pressure to achieve efficient flotation, however, resulting in a greater operating cost. The greater the solids concentration, the more likely it is that horizontal-flow sedimentation will be selected. The extreme situation is equivalent to thickening, and then presedimentation before coagulation may be justified.

For certain types of source water, flotation does produce better water for filtration and therefore better filtrate quality than sedimentation. This may be because of better use of flocculation (this is applicable also to inclined settling) and the mechanism of flotation that makes it especially effective for algae removal. The aeration in flotation also may be attractive because it might increase dissolved oxygen concentration or even cause some desorption or stripping of contaminants.

### Compactness

Inclined settling and dissolved-air flotation are regarded as "high-rate" processes because of the relatively high surface loadings possible. This label also implies they are compact processes and so occupy less area.

That is important where land is at a premium. These processes are also likely to need shallower and smaller tanks, which is important in coping with preparing foundations in difficult sites. Both of these points are relevant where a plant has to be housed in cold climates or for other environmental or strategic reasons. The compactness of the flotation-filtration plant (flotation carried out over the filter) has made it especially attractive for package plants.

### Rapid start-up

A floc blanket can take several days to form when you are starting with an empty tank, especially when the solids concentration is low. In contrast "high-rate" inclined settlers and dissolved-air flotation can produce good-quality water within 45 min from start-up. Such rapid start-up is useful where daily continuous operation is not needed. The disadvantage of high-rate processes is that they are sensitive to failure of chemical dosing and flocculation. The advantage of slower processes is that they respond much more slowly to changes in inlet water quality. Rapid start-up plants are suitable for unstaffed sites using automatic shutdown and call-out alarms as a means of avoiding relying on difficult automatic quality control.

### Sludge removal

The ability to accumulate sludge, prior to removal, requires space, and this depends on the process:

1. In horizontal-flow sedimentation, the whole plan area of the floor of the tank is available for sludge accumulation.

2. In dissolved-air flotation, the whole plan area of the surface of the flotation tank is available, but the thickness of the floated sludge and its ease of removal are more important.

3. In solids-contact clarifiers with scrapers, the whole plan area of the floor is available.

4. Floc-blanket clarifiers have only a small area for excess floc removal as sludge. However, when it is designed well, this will be adequate to ensure some thickening before discharge.

5. Some inclined settler systems are combined with thickeners.

Some systems, such as dissolved-air flotation and those with combined thickeners, can produce sludge concentrated enough for dewatering without additional thickening (see Chap. 16). While systems that produce the more concentrated sludges are likely to incur greater operating costs for the scraping and pumping, this is likely to be more than offset by a saving in subsequent sludge treatment processes.

## Emerging Technology

Developments in sedimentation processes and dissolved-air flotation as currently known have probably reached the point of diminishing returns. Application lags knowledge, including making the plants that already exist work better. This will be done in part by paying more attention to the chemical engineering of reagent dosing, mixing, and flocculation rather than the sedimentation itself. More attention also is likely to be paid to selecting the best chemistry and process for a particular application and to the design of the chosen process with regard to materials of construction, control equipment, and compactness.

Ballasted-floc systems, particularly those using recycled ballast, do not appear to have received as much attention in the past as they might deserve. This might change in the future. An interesting process of this type, currently under development, is the Sirofloc process, which recycles magnetite. One claim for this process is that it can be so efficient that subsequent filtration may not be necessary.

Sometimes filtration has been used as an alternative to sedimentation. Various types of filtration have been used for this purpose including upflow filtration, continuous or moving-bed filters, and buoyant-media filters (see Chap. 8 on filtration). Some of these options will find more frequent application.

A wide variety of options has emerged for treatment prior to final filtration. Individual circumstances will dictate the choice. Furthermore, no single process will likely take a major share of new applications because the costs of alternatives are generally the same, unless substantial technical advantages exist.

## References

1. J. W. Ellms, *Water Purification*, McGraw-Hill, New York, (1928).
2. W. L. Barham, J. L. Matherne, and A. G. Keller, *Clarification, Sedimentation and Thickening Equipment—A Patent Review*, Bulletin no. 54, Engineering Experimental Station, Louisiana State University, Baton Rouge, La., 1956.
3. H. Nakamura and K. Kuroda, "La Cause de l'acceleration de la Vitesse de Sedimentation des Suspensions dans Les Recipients Inclines," *Keijo J. Med.* (Jpn), vol. 8, 1937, p. 265.
4. K. Kinosita, "Sedimentation in Tilted Vessels," *J. Colloid. Sci.*, vol. 4, 1949, p. 525.
5. K. H. Kalbskopf, "European Practices in Sedimentation," in *Water Quality Improvement by Physical and Chemical Processes*, Water Resources Symposium no. 3 (E. F. Gloyna and W. W. Eckenfelder, eds.), University of Texas Press, Austin, 1970.
6. H. O. Hartung, "Committee Report: Capacity and Loadings of Suspended Solids Contact Units." *J. AWWA*, vol. 43, no. 4, April 1951, p. 263.
7. L. Prandtl and O. G. Tietjens, *Applied Hydro and Aeromechanics*, Dover, New York, 1957.
8. M. J. Pearse, *Gravity Thickening Theories*, LR 261(MP), Warren Spring Laboratory, Stevenage, United Kingdom, 1977.
9. G. M. Fair, J. C. Geyer, and D. A. Okun, *Elements of Water Supply and Waste Water Disposal*, 2d ed., Wiley, New York, 1971.
10. *Water Treatment Handbook*, 5th ed., Degremont, France, 1979.
11. J. M. Coulson and J. F. Richardson, *Chemical Engineering*, vol. 2, 3d ed., Pergamon Press, Oxford, United Kingdom, 1978.
12. Metcalf and Eddy Engineers, *Wastewater Engineering*, 2d ed., McGraw-Hill, New York, 1979.
13. T. R. Camp, "A Study of the Rational Design of Settling Tanks," *Sewage Wks. J.*, vol. 8, 1936, p. 742.
14. D. A. Thirumurthi, "Breakthrough in the Tracer Studies of Sedimentation Tanks," *J. Water Pollut. Control Fed.*, vol. 41, no. 11, pt. 2, 1969, p. R405.
15. A. E. Zanoni and M. W. Blomquist, "Column Settling Tests for Flocculant Suspensions," *Proc. Am. Soc. Civil. Eng.*, vol. 101(EE3), 1975, p. 309.
16. R. Gregory, *Floc Blanket Clarification*, TR111, Water Research Centre, Medmenham, United Kingdom, 1979.
17. R. Gregory and M. Hyde, *The Effects of Baffles in Floc Blanket Clarifiers*, TR7, Water Research Centre, Medmenham, United Kingdom, 1975.
18. G. H. Setterfield, "Water Treatment Trials at Burham," *Effluent Water Treat. J.*, vol. 23, 1983, p. 18.
19. V. S. Patwardhan and Tien Chi, "Sedimentation and Liquid Fluidization of Solid Particles of Different Sizes and Densities," *Chem. Eng. Sci.*, vol. 40, 1985, p. 1051.
20. N. Epstein and B. P. LeClair, "Liquid Fluidization of Binary Particle Mixtures, II," *Chem. Eng. Sci.*, vol. 40, 1985, p. 1517.
21. K. M. Yao, "Theoretical Study of High-Rate Sedimentation," *J. Water Pollut. Control Fed.*, vol. 42, 1970, p. 218.
22. K. M. Yao, "Design of High-Rate Settlers," *Proc. Am. Soc. Civil Eng.*, vol. 99(EE5), 1973, p. 621.
23. K. J. Ives, "Theory of Operation of Sludge Blanket Clarifiers," *Proc. Inst. Civil Eng.*, United Kingdom, vol. 39, 1968, p. 243.
24. J. Vostrcil, The Effect of Organic Flocculants on Water Treatment and Decontamination of Water by Floc Blanket. Prace a Studie, Sesit 129, Water Research Institute, Prague, Czechoslovakia, 1971.

25. G. Cretu, "Contribution to the Theory of Water Treatment Using a Sludge Blanket," *Hydrotechnia Gospodarirea Apelor, Meterologia (Romania)*, vol. 13, 1968, p. 634.

26. T. Shogo, "Slurry-Blanket Type Suspended Solids Contact Clarifiers: Part 5," *Kogyo Yoshui (Japan)*, vol. 153, 1971, p. 19.

27. A. W. Bond, "Water-Solids Separation in an Upflow," *Inst. Eng. Aust., Civil Eng. Trans.*, vol. 7, 1965, p. 141.

28. N. Tambo et al., "Behaviour of Floc Blankets in an Upflow Clarifier," *J. Jpn. Water Wks. Assoc.*, vol. 44, 1969, p. 7.

29. H. S. Coe and G. H. Clevenger, "Methods for Determining the Capacities of Slime Settling Tanks," *Trans. Am. Inst. Min. Eng.*, vol. 60, 1916, p. 356.

30. W. P. Talmage and E. B. Fitch, "Determining Thickener Unit Areas," *Ind. Eng. Chem.*, vol. 47, 1955, p. 38.

31. C. J. Kynch, "A Theory of Sedimentation," *Trans. Faraday Soc. (U.K.)*, vol. 48, 1952, p. 166.

32. J. H. Warden, "Polymer Treatment of Waterworks Sludges," *The Chem. Eng. (U.K.)*, vol. 387, 1982, p. 460.

33. J. H. Warden, *Sludge Treatment Plant for Waterworks*, TR189, Water Research Centre, Stevenage, United Kingdom, 1983.

34. D. C. Dixon, *Progress in Filtration and Separation* (R. J. Wakeman, ed.), Elsevier, New York, 1979.

35. O. Levenspiel, *Chemical Reaction Engineering*, Wiley, New York, 1962.

36. D. M. Marske and J. D. Boyle, "Chlorine Contact Chamber Design—A Field Evaluation," *Water Sewage Wks.*, vol. 120, 1973, p. 70.

37. F. L. Hart and S. K. Gupta, "Hydraulic Analysis of Model Treatment Units," *Proc. Am. Soc. Civil Eng.*, vol. 104(EE4), 1979, p. 785.

38. M. Rebhun and Y. Argaman, "Evaluation of Hydraulic Efficiency of Sedimentation Basins," *Proc. Am. Soc. Civil Eng.*, vol. 91(SA5), 1965, p. 37.

39. D. Wolf and W. Resnick, "Residence Time Distribution in Real Systems," *Ind. Eng. Chem. Fund.*, vol. 2, 1963, p. 287.

40. H. E. Hudson, "Residence Times in Pretreatment," *J. AWWA*, vol. 67, no. 1, January 1975.

41. A. Hazen, "On Sedimentation," *Trans. Am. Soc. Civil Eng.*, vol. 53, 1904, p. 63.

42. G. M. Fair and J. C. Geyer, *Water Supply and Wastewater Disposal*, Wiley, New York, 1954.

43. M. J. Hamlin and A. H. Abdul Wahab, "Settling Characteristics of Sewage in Density Currents," *Water Res. (U.K.)*, vol. 4, 1970, p. 609.

44. R. S. Gemmell, "Mixing and Sedimentation," in *Water Quality and Treatment*, 3d ed., McGraw-Hill, New York, 1971.

45. C. Gomella, "Clarification avant Filtration, Ses Progrès Recents," (Rapport General 1), International Water Supply Association, International Conference, 1974.

46. D. G. Miller et al., "Floc Blanket Clarification, 1," *Water and Water Eng.*, vol. 70, 1966, p. 240.

47. J. J. Casey, K. O'Donnel, and P. J. Purcell, "Uprating Sludge Blanket Clarifiers Using Inclined Plates," *Aqua*, vol. 2, 1984, p. 91.

48. J. Sibony, "Clarification with Microsand Seeding—A State of the Art," *Water Res. (U.K.)*, vol. 15, 1981, p. 1281.

49. J. A. Webster et al., "Aspects of Design for Efficient Plant Operation for Water Treatment Processes Using Coagulants." Paper presented to Scottish Section, Institution Water Engineers and Scientists, United Kingdom, 1977.

50. D. R. Dixon, "Colour and Turbidity Removal with Reusable Magnetite Particles, VII," *Water Res. (U.K.)*, vol. 18, 1984, p. 529.

51. R. Gregory, R. J. Maloney, and M. Stokley, "Water Treatment Using Magnetite: A Study of a Sirofloc Pilot Plant, *J. Inst. Water & Environ. Manage.*, vol. 2, no. 5, 1988, p. 532.

52. M. S. Clements and A. F. M. Khattab, "Research into Time Ratio in Radial Flow Sedimentation Tanks," *Proc. Inst. Civil Eng. (U.K.)*, vol. 40, 1968, p. 471.

53. G. A. Price and M. S. Clemments, "Some Lessons from Model and Full-Scale Tests

in Rectangular Sedimentation Tanks," *Water Pollut. Control (U.K.)*, vol. 73, 1974, p. 102.

54. S. Kawamura, "Hydraulic Scale-Model Simulation of the Sedimentation Process," *J. AWWA,* vol. 73, no. 7, 1981, p. 372.

55. S. Kawamura and J. Lang, "Re-evaluation of Launders in Rectangular Sedimentation Basins," *J. Water Pollut. Control Fed.,* vol. 58, 1986, p. 1124.

56. P. E. Hale, "Floc Blanket Clarification of Water," doctoral dissertation, London University, United Kingdom, 1971.

57. B. W. Gould, "Low Cost Clarifier Improvement," *Aust. Civil Eng. Constn.,* vol. 8, 1967, p. 49.

58. Envirotech Corp., *Eimco Modular Energy Dissipating Clarifier Feedwells,* Form no. MED 121-10-72-3M, Envirotech Corporation, Brisbane, Calif.

59. D. G. Miller and J. T. West, "Pilot Plant Studies of Floc Blanket Clarification," *J. AWWA,* vol. 60, no. 2, February 1968, p.154.

60. N. P. Yadav and J. T. West, *The Effect of Delay Time on Floc Blanket Efficiency,* TR9, Water Research Centre, Medmenham, United Kingdom, 1975.

61. H. E. Hudson, "Density Considerations in Sedimentation," in *Water Clarification Processes: Practical Design and Evaluation,* Van Nostrand Reinhold, New York, 1981.

62. M. J. D. White et al., "Increasing the Capacity of Sedimentation Tanks by Means of Sloping Plates," Paper presented to Institute Water Pollution Control, East Midlands Branch, United Kingdom, November 1974.

63. A. M. Gaudin, *Flotation,* 2d ed., McGraw-Hill, New York, 1957.

64. J. A. Kitchener, "The Froth Flotation Process: Past, Present and Future—In Brief," in *The Scientific Basis of Flotation* (K. J. Ives, ed.), NATO ASI Series, Martinus Nijhoff Publishers, The Hague, Netherlands, 1984.

65. J. Bratby, "Dissolved-Air Flotation in Water and Waste Treatment," doctoral dissertation, University of Cape Town, Cape Town, South Africa, 1976.

66. H. Lundgren, "Theory and Practice of Dissolved-Air Flotation," *J. Filtrat. Separat.,* vol. 13, no. 1, 1976, p. 24.

67. T. F. Zabel and J. D. Melbourne, "Flotation," in *Developments in Water Treatment, vol. 1* (W. M. Lewis, ed.). Applied Science Publishers Ltd., London, 1980.

68. F. Barrett, "Electroflotation—Development and Application," *Water Pollut. Control,* vol. 74, 1975, p. 59.

69. M. Krofta and L. K. Wang, "Development of Innovative Flotation-Filtration Systems for Water Treatment, Part C: An Electro Flotation Plant for Single Families and Institutions," *Proc. Water Reuse Symposium III: Future of Water Reuse,* San Diego, 1984.

70. H. L. Sherfold, "Flotation in Mineral Processing," in *The Scientific Basis of Flotation* (K. J. Ives, ed.), NATO ASI Series, Martinus Nijhoff Publishers, The Hague, Netherlands, 1984.

71. M. L. Hemming, W. R. T. Cottrell, and S. Oldfelt, "Experience in the Treatment of Domestic Sewage by the Micro-Flotation Process," *Papers and Proceedings of the Water Research Centre Conference on Flotation for Water and Waste Treatment* (J. D. Melbourne and T. F. Zabel, eds.), Water Research Centre, Medmenham, United Kingdom, 1977.

72. J. A. Kitchener and R. J. Gouchin, "The Mechanism of Dissolved-Air Flotation for Potable Water: Basic Analysis and a Proposal," *Water Res. (U.K.),* vol. 15, 1981, p. 585.

73. A. J. Rees, D. J. Rodman, and T. F. Zabel, "Dissolved-Air Flotation for Solid-Liquid Separation," *J. Separat. Process Tech.,* vol. 2, 1979, p. 1.

74. M. T. Turner, "The Use of Dissolved Air Flotation for the Thickening of Waste Activated Sludge," *Effluent Water Treat. J.,* vol. 15, May 1975, pp. 243–251.

75. T. F. Zabel, "The advantages of Dissolved-Air Flotation for Water Treatment," *J. AWWA* vol. 77, no. 5, May 1985, p. 42.

76. M. Krofta and L. K. Wang, "Development of Innovative Flotation-Filtration Systems for Water Treatment. Part A: First Full-Scale Sand Float Plant in U.S.," *Proceedings Water Reuse Symposium III: Future of Water Reuse,* San Diego, 1984.

77. L. R. Van Vuuren, F. J. de Wet, and F. J. Cillie, Dissolved-Air Flotation-Filtration

Studies in South Africa, AWWA Research Foundation Research News, Denver, Colo., 1984, pp. 10–12.

78. J. Bratby and C. V. R. Marais, *Solid-Liquid Separation Equipment Scale-Up* (D. B. Purchas, ed.), chap. 5, Uplands Press Ltd., London, 1977.

79. E. R. Vrablik, "Fundamental Principles of Dissolved-Air Flotation of Industrial Wastes," *Proc. of the 14th Industrial Waste Conference,* Purdue University, Lafayette, Ind., 1959.

80. A. J. Rees, D. J. Rodman, and T. F. Zabel, *Evaluation of Dissolved-Air Flotation Saturator Performance,* TR143, Water Research Centre, Medmenham, United Kingdom, 1980.

81. T. F. Zabel and R. A. Hyde, "Factors Influencing Dissolved-Air Flotation as Applied to Water Clarification," *Papers and Proceedings of Water Research Centre Conference on Flotation for Water and Waste·Treatment* (J. D. Melbourne and T. F. Zabel, eds.), Water Research Centre, Medmenham, United Kingdom, 1977.

82. T. F. Zabel, "Flotation in Water Treatment," in *The Scientific Basis of Flotation* (K. J. Ives, ed.), NATO ASI Series, Martinus Nijhoff Publishers, The Hague, Netherlands, 1984.

83. A. J. Rees, D. J. Rodman, and T. F. Zabel, *Water Clarification by Flotation—5,* TR114, Water Research Centre, Medmenham, United Kingdom, 1979.

84. D. J. Rodman, "Investigation into Hydraulic Flocculation with Special Emphasis on Algal Removal," master's thesis, Water Research Centre, Stevenage, United Kingdom, 1982.

85. J. L. Maddock, "Research Experience in the Thickening of Activated Sludge by Dissolved-Air Flotation," *Papers and Proceedings of the Conference on Flotation for Water and Waste Treatment* (J. D. Melbourne and T. F. Zabel, eds.), Water Research Centre, Medmenham, United Kingdom, 1977.

86. Brit. Pat. Spec. nos. 1.444.026 and 1.444.027.

87. M. Krofta and L. K. Wang, "Application of Dissolved Air Flotation to the Lennox, Massachusetts, Water Supply: Water Purification by Flotation," *J. New Engl. Water Wks. Assoc.,* vol. 99, no. 3, 1985, p. 249.

88. J. A. Drajo, "Clarification of Sacramento Delta Water in a Large Scale Dissolved Air Flotation Pilot Plant," *Proc. AWWA Annual Conference,* Dallas, 1984.

89. P. D. Wilkinson, P. M. Bolas, and M. F. Adkins, British Experience with Flotation Process at Bewl Bridge Treatment Works, AWWA Research Foundation Water Quality Research News, Denver, Colo., 1981.

90. R. Gregory, "A Cost Comparison between Dissolved Air Flotation and Alternative Clarification Processes," *Papers and Proceedings of the Conference on Flotation for Water and Waste Treatment* (J. D. Melbourne and T. F. Zabel, eds.), Water Research Centre, Medmenham, United Kingdom, 1977.

# Filtration

## John L. Cleasby, Ph.D., P.E.

*Professor*
*Department of Civil and Construction Engineering*
*Iowa State University*
*Ames, Iowa*

Filtration processes discussed in this chapter are used primarily to remove suspended particulate material from water and are one of the unit operations used in the production of potable water. Particulates removed may be those in the source water or those generated in treatment processes. Examples of particulates include clay and silt particles, microorganisms, colloidal and precipitated humic substances and other organic particulates from decay of vegetation, precipitates of aluminum or iron used in coagulation, calcium carbonate and magnesium hydroxide precipitates from lime softening, and iron and manganese precipitates.

## Types of Filters

A number of different types of filters are used in potable water filtration, and they may be described in various classification schemes. The filters used for public water supplies, discussed here, are comprised of porous granular material.

One physical classification scheme is based on the type of granular medium used. Granular-bed filters commonly use a substantial depth of sand or anthracite coal or combinations thereof. A typical granular-bed filter is shown in Fig. 8.1. In contrast, precoat filters use a thin layer of very fine medium such as diatomaceous earth that is disposed of after each filter cycle. A typical precoat filter is shown in Fig. 8.2 with a circular flat plate septum that supports the precoat.

Filters may also be described by the hydraulic arrangement pro-

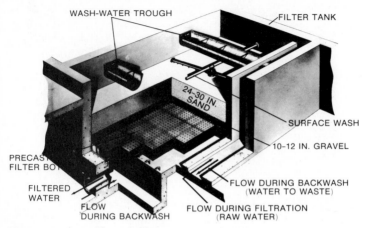

**Figure 8.1**  A rapid sand filtration system. (*Courtesy of F. B. Leopold Company.*)

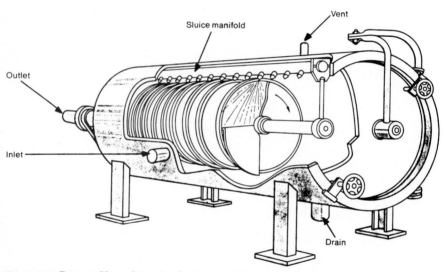

**Figure 8.2**  Precoat filter of rotating leaf type (sluice type during backwash). (*Courtesy of Manville Sales Corporation.*)

vided to pass water through the medium. Gravity filters are open to the atmosphere, and flow through the medium is achieved by gravity, such as shown in Fig. 8.1. In pressure filters the filter medium is contained in a pressure vessel. Water is delivered to the vessel under pressure and leaves the vessel at slightly reduced pressure.

Filters may also be described by the rate of filtration, that is, the flow rate per unit area. Granular-bed filters can be operated at various rates; for example, rapid sand filters operate at higher rates, but slow sand filters operate at a much lower rate that favors surface removal on the top of the sand bed.

Finally, filtration can be classified as *depth filtration* if the solids are removed within the granular material or *cake filtration* if the solids are removed on the entering face of the granular material. Rapid granular-bed filters are of the former type, while precoat and slow sand filters are of the latter type.

Thus, a filter can be fully described by an appropriate choice of descriptive adjectives. For example, a rapid gravity dual-media filter would describe a deep bed comprised of two media, usually anthracite coal over sand, operated at high enough rates to encourage depth removal of particulates within the bed and operated by gravity in an open tank.

### Dominant mechanisms, performance, and applications

Cake filtration is the physical removal by straining at the surface. In addition, for the slow sand filter the surface cake of accumulated particulates includes a variety of living and dead micro- and macro-organisms. The biological metabolism of the organisms causes some alteration in the chemical composition of the water, and the development of this dirty skin, or "schmutzdecke," enhances particulate removal as well. As the filter cake develops, the cake itself assumes a dominant role in filtration. Because of this, filtrate quality improves as the filter run progresses, and deterioration of the filtered water quality is normally not observed at the end of the filter cycle. Because the mechanism of cake filtration is largely physical straining, chemical pretreatments such as coagulation and sedimentation are not generally provided. To obtain reasonable filter cycles, however, the source water must be of quite good quality, as will be defined later.

In contrast, depth filtration involves a variety of complex mechanisms to achieve particulate removal. Particles to be removed are generally much smaller than the size of the interstices between filter grains. Transport mechanisms are needed to carry the small particles into contact with the surface of the individual filter grains, and then attachment mechanisms hold the particles to the surfaces. These mechanisms are discussed in more detail later.

Chemical pretreatment is essential to depth filtration. It serves to flocculate the colloid-size particulates into larger particles, which enhances their removal in the sedimentation tank ahead of the filter and/or enhances the transport mechanisms in filtration. In addition, chemical treatment enhances the attachment forces retaining the particles in the filter. The burden of removal in depth filtration moves progressively deeper and deeper into the bed, and if it is operated long enough, deterioration of the filtrate may be observed.

The provision of pretreatment makes the depth filtration process more versatile in meeting a variety of source water conditions. With appropriate coagulation, flocculation, sedimentation, and depth filtration, source water of high turbidity or color can be treated successfully. Better-quality source water may be treated by coagulation, flocculation, and depth filtration, a process referred to as *direct filtration*.

## Filter Media

### Types of media

The common types of medium used in granular bed filters are silica sand, anthracite coal, and garnet or ilmenite. These may be used alone or in dual- or triple-media combinations. Garnet and ilmenite are naturally occurring, high-density minerals and are described further in the next paragraph. Other types of media have also been used in some cases. For example, granular activated carbon (GAC) has been used for taste and odor reduction in granular beds that serve for both filtration and adsorption, that is, filter-adsorbers.[1] GAC is also being used after filtration for adsorption of organic compounds.

"Garnet" is somewhat of a generic term that refers to several different minerals, mostly almandite, andradite, and grossularite, which are silicates of iron, aluminum, and calcium mixtures. Ilmenite is an iron titanium ore, which invariably is associated with hematite and magnetite, both iron oxides. Garnet specific gravities range from 3.6 to 4.2, and those of ilmenite from 4.2 to 4.6. Developmental testing for triple-media filters was done with garnet at a specific gravity of about 4.1. Lower-density garnet has been used in some filters, although the desirability of such use has not been proved.

Precoat filters use diatomaceous earth or perlite as a filter medium. Diatomaceous earth (DE or diatomite) is composed of the fossilized skeletons of microscopic diatoms that grow in freshwater or marine water. Deposits of this material from ancient oceans are mined and then processed by flux calcining, milling, and air classification into various size grades for assorted filtration applications. The grades used in potable water filtration have a mean pore size of the cake from about 5 to 17 $\mu$m.

A less common medium for precoat filtration is perlite, which comes from glassy volcanic rock. It is a siliceous rock containing 2 to 3 percent water. When heated, the rock expands to form a mass of glass bubbles. It is crushed, calcined, milled, and classified into several grades. Some of the particulates may remain as small glassy spheres that may float and are ineffective as filter media.

### Important granular medium properties

A number of properties of filter media are important in affecting filtration performance and in defining the media. These properties include size, shape, density, hardness, and porosity.

**Grain size and size distribution.**   Grain size has an important effect on the filtration efficiency and on backwashing requirements for the medium. It is determined by sieve analysis using the American Society for Testing and Materials (ASTM) Standard Test C136-84a, Sieve Analysis of Fine and Coarse Aggregates.[2] A log-probability plot of a typical sieve analysis is presented in Fig. 8.3. Sieve analysis of most filter materials plots in nearly a linear manner on log-probability paper.

In the United States, the medium is described by the *effective size* (ES) and the *uniformity coefficient* (UC). The ES is that size for which 10 percent of the grains are smaller by weight. It is read from the sieve analysis curve at the 10 percent passing point on the curve, and it is often abbreviated by $d_{10}$. The UC is a measure of the size range of

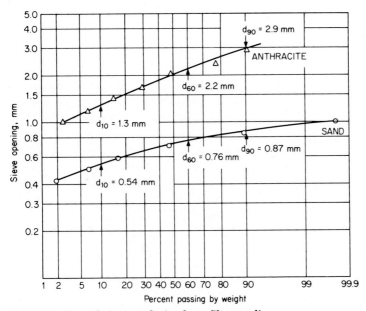

**Figure 8.3**  Typical sieve analysis of two filter media.

the medium. It is the ratio of the $d_{60}/d_{10}$ sizes read from the sieve analysis curve, with $d_{60}$ being the size for which 60 percent of the grains are smaller by weight.

In some countries, the lower and upper size ranges of the medium are specified with some maximum percentage allowance above and below the specified sizes. For example, if the allowance were 10 percent at each end, the sand shown in Fig. 8.3 would be called a 0.54- to 0.87-mm sand.

Values of $d_{10}$, $d_{60}$, and $d_{90}$ can be read from an actual sieve analysis curve such as shown in Fig. 8.3. If such a curve is not available and if a linear log-probability plot is assumed, the values can be interrelated by the following equation:

$$d_{90} = d_{10}(10^{1.67 \log \text{UC}}) \tag{8.1}$$

This relationship is useful because the $d_{90}$ size is recommended for calculation of the required backwash rate for a filter medium.

**Grain shape and roundness.** The shape and roundness of the filter grains are important because they affect the backwash flow requirements for the medium, the fixed-bed porosity, the head loss for flow through the medium, the filtration efficiency, and the ease of sieving.

Different measures of grain shape have evolved in the geological and chemical engineering literature, leading to considerable confusion in terminology. In the geological literature, the shape of a grain is its form, entirely independent of whether the edges or corners are sharp or round. Geologists define sphericity by the cube root of the ratio of the volume of a grain to the volume of its circumscribing sphere.[3]

In contrast, the chemical engineering literature defines the sphericity $\psi$ as the ratio of the surface area of an equal-volume sphere (diameter of $d_{eq}$) to the surface of the grain.[4] This is influenced by both the shape and the roundness of the grain.[4] The equivalent diameter can be determined by counting and weighing a representative sample of grains (about 100) retained between adjacent sieves. Then, by using the previously measured density of the grains (explained below), the volume per grain and equivalent spherical diameter can be calculated. In the absence of such data, the mean size for any fraction observed from the sieve analysis plot (e.g., Fig. 8.3) can be used as an acceptable approximation.

The chemical engineering definition is used in the following discussion. The sphericity of the filter medium by this definition can be determined indirectly by measuring pressure drop for flow of water or air through a bed of uniform-size grains. Bed porosity must be measured first. The Kozeny or Ergun equation for flow through porous media (presented later) is used to calculate $\psi$ after determination of all other parameters of the equation.[5]

**Grain density or specific gravity.** Grain density, the mass per unit grain volume, is important because it affects the backwash flow requirements for the medium. Grains of higher density but of the same diameter require higher wash rates to achieve fluidization. Therefore, greater hydraulic shear forces exist during backwashing, and the washing is more effectively accomplished.

Grain density is determined from the specific gravity following ASTM Standard Test C128-84, Specific Gravity and Absorption of Fine Aggregate.[2] This ASTM test uses a displacement technique to determine the specific gravity.

Specific gravity is the ratio of the mass of a body to the mass of an equal volume of water at a specific temperature. ASTM C128-84 specifies a temperature for the test of 23°C, and three alternative tests are detailed. The procedure for "bulk specific gravity, saturated surface dry" would be best from a theoretical standpoint for fluidization calculations. Starting with a reproducible saturated surface dry condition is difficult, however. Therefore, the "apparent specific gravity" that starts with an oven-dry sample is more reproducible and is an acceptable alternative for fluidization calculations. For porous materials such as anthracite coal or GAC, the sample should be soaked to fill the pores with water before final measurements are made.

**Grain hardness.** The hardness of filter grains is important to the durability of the grains during long-term service as a filter medium. Hardness is usually described by the Mohs hardness number, which is a scale of hardness based on the ability of various minerals to be scratched by another harder object. A sequence of minerals of specified hardness is listed.[6]

The two materials of known Mohs hardness that can and cannot scratch the filter medium are used to estimate the hardness of the medium. This is a rather crude test and is difficult to apply to small filter grains. For anthracite, applying the test to the uncrushed material is better.

Of the filter media listed earlier, only anthracite coal and GAC have low hardness worthy of concern. Silica sand, garnet, and ilmenite are very hard, and their hardness need not be of concern. A minimum Mohs hardness of 2.7 or 3 is often specified for anthracite coal filter medium, although measuring fractional values closer than 0.5 is doubtful.

The Mohs hardness of GAC is generally not specified, even though it is well recognized that GAC is more friable than anthracite. Rather, two standard mechanical abrasion tests are presented in the American Water Works Association (AWWA) Standard for Granular Activated Carbon (Standard B604-74)[7] to evaluate the abrasion resistance of GAC. In spite of its greater friability, the reduction in grain size of GAC due to backwashing and regeneration operations is not reported to be a serious problem.[1]

**Fixed-bed porosity.** Fixed-bed porosity is the ratio of void volume to total bed volume, expressed as a decimal fraction or a percentage. It is important because it affects the backwash flow required, the fixed-bed head loss, and the solids-holding capacity of the medium. Fixed-bed porosity is affected by the grain sphericity; angular grains (i.e., lower sphericity) have higher fixed-bed porosity.[5]

Fixed-bed porosity is determined by placing a sample of known mass and density in a transparent tube of known internal diameter. The depth of medium in the tube is used to calculate the bed volume. The grain volume is the total mass of medium in the column divided by the density. The void volume is thus the bed volume minus the grain volume. The fixed-bed porosity is substantially affected by the extent of compaction of the medium placed in the column. The loose-bed porosity can be measured in a column of water. If the bed is agitated by inversion and then allowed to settle freely in the water with no compaction, the highest porosity will be obtained, that is, the loose-bed porosity. It may be as much as 5 percent greater than porosities measured after gentle compaction of the bed. Materials of lower sphericity show greater change in porosity between the loose-bed and compacted-bed conditions.

### Sieve analysis considerations

The standard procedure for conducting sieve analysis of a filter medium is detailed in ASTM Standard Test C136-84a.[2] This standard does not specify a sieving time or a mechanical apparatus for shaking the nest of sieves. Rather, it specifies that sieving should be continued "for a sufficient period and in such manner that, after completion, not more than 1 weight percent of the residue on any individual sieve will pass that sieve during 1 minute of hand sieving," conducted in a manner described in the ASTM standard. With softer materials such as anthracite coal or GAC, in attempting to meet the 1 percent passing test, abrasion of the material may occur.

In sieving hard materials such as sand, beginning with a 100-g sample on 8-in (20-cm) sieves, and using a ro tap type of sieving machine, requiring three sieving periods of 5 min each to satisfy the 1 percent passing test is common. With some other sieving machines, the ASTM requirement will not be achieved even after three 5-min periods of sieving.

In sieving anthracite coal, the sample should be reduced to 50 g because of its lower density. The ro tap machine should be used, and the time fixed at 5 min. This will not meet the 1 percent passing test, but prolonged sieving may cause continued degradation of the anthracite, yielding a more erroneous result.

Because of sieving and sampling difficulties and because of the tolerance allowed in the manufacture of the sieves themselves (ASTM Standard Test E11-81, Wire-Cloth Sieves for Testing Purposes),[2] when filter media are specified, a reasonable tolerance should be allowed in the size. Otherwise, producers of filter material may not be able to meet the specification or a premium price will be charged. A tolerance of plus or minus 10 percent is suggested. For example, if a sand of 0.5-mm ES is desired, the specification should read 0.45- to 0.55-mm ES. If an anthracite coal of 1.0-mm ES is desired, the specification should read 0.9- to 1.1-mm ES.

### Typical properties of filter media

With the prior understanding of the importance of various filter medium properties, Table 8.1 illustrates typical measured values for some properties. The large difference in grain densities evident in Table 8.1 allows the construction of dual- and triple-media filters, with coarse grains of low-density material on top and finer grains of higher-density material beneath. Alluvial sands have the highest sphericity, and crushed materials such as anthracite, ilmenite, and some garnet have lower sphericity. Some anthracites contain an excessive amount of platey grains, resulting in lower sphericity. The loose-bed porosity is inversely related to the sphericity; that is, the lower the sphericity, the higher the loose-bed porosity. An approximate empirical relationship between sphericity and loose-bed porosity

TABLE 8.1    Typical Properties of Common Filter Media for Granular-Bed Filters[5,8,9]

| | Silica sand | Anthracite coal | Granular activated carbon | Garnet | Ilmenite |
|---|---|---|---|---|---|
| Grain density, $\rho_s$, g/cm$^3$ | 2.65 | 1.45–1.73 | 1.3–1.5† | 3.6–4.2 | 4.2–4.6 |
| Loose-bed porosity $\epsilon_0$ | 0.42–0.47 | 0.56–0.60 | 0.50 | 0.45–0.55 | ‡ |
| Sphericity $\psi$ | 0.7–0.8 | 0.46–0.60 | 0.75 | 0.60 | ‡ |

†For virgin carbon, pores filled with water, density increases when organics are adsorbed.
‡Not available.

was used in developing a predictive model for fluidization, to be presented later.[5]

## Hydraulics of Flow through Porous Media

### Head loss for fixed-bed flow

The head loss (i.e., pressure drop) that occurs when clean water flows through a bed of clean filter medium can be calculated from well-known equations. The flow through a clean filter of ordinary grain size (that is, 0.5 to 1.0 mm) at ordinary filtration velocities (2 to 5 gpm/ft$^2$, or 4.9 to 12.2 m/h) would be in the laminar range of flow depicted by the Kozeny equation[10] that is dimensionally homogenous (i.e., any consistent units may be used that are dimensionally homogenous[†]):

$$\frac{h}{L} = \frac{k\mu}{\rho g} \frac{(1 - \epsilon)^2}{\epsilon^3} \left(\frac{a}{v}\right)^2 V \tag{8.2}$$

where $h$ = head loss in depth of bed $L$
  $g$ = acceleration of gravity
  $\epsilon$ = porosity
  $a/v$ = grain surface area per unit of grain volume = specific surface $S_v$ = $6/d$ for spheres and $6/(\psi d_{eq})$ for irregular grains
  $d_{eq}$ = grain diameter of sphere of equal volume
  $V$ = superficial velocity above bed = flow rate/bed area
  $\mu$ = absolute viscosity of fluid
  $\rho$ = mass density of fluid
  $k$ = dimensionless Kozeny constant commonly found close to 5 under most filtration conditions.[10]

The Kozeny equation is generally acceptable for most filtration calculations because the Reynolds number Re based on superficial velocity is usually less than 3 under these conditions, and Camp[11] has reported strictly laminar flow up to Re of about 6:

$$\text{Re} = d_{eq} \frac{V\rho}{\mu} \tag{8.3}$$

The Kozeny equation can be derived from the fundamental Darcy-Weisbach equation for flow through circular pipes

$$h = f \frac{LU^2}{D(2g)} \tag{8.4}$$

---

†Units will not be presented for all dimensionally homogenous equations in this chapter.

where $f$ = friction factor, a function of pipe Reynolds number
   $D$ = pipe diameter
   $U$ = mean flow velocity in pipe

The derivation is achieved by considering flow through porous media analogous to flow through a group of capillary tubes of hydraulic radius $r$.[10] The hydraulic radius is approximated by the ratio of the volume of water in the interstices per unit bed volume divided by the grain surface area per unit bed volume. If $N$ is the number of grains per unit bed volume, $v$ is the volume per grain, and $a$ is the surface area per grain, then the bed volume = $Nv/(1 - \epsilon)$, the interstitial volume = $Nv\epsilon/(1 - \epsilon)$, and the surface area per unit bed volume is $Na$, leading to $r = \epsilon v/[1 - \epsilon)a]$. The following additional substitutions are made: $D = 4r$, $U$ = interstitial velocity = $V/\epsilon$, $f = 64/Re'$ for laminar flow, and $Re' = 4(V/\epsilon)r\rho/\mu$ is the Reynolds number based on interstitial velocity.

For larger filter media or higher velocities used in some applications or for velocities approaching fluidization (as in backwashing considerations), the flow may be in the transitional flow regime where the Kozeny equation is no longer adequate. Therefore, the Ergun equation,[12] Eq. (8.5), should be used because it is adequate for the full range of laminar, transitional, and turbulent flow through packed beds (Re from 1 to 2000). The Ergun equation includes a second term for turbulent head loss.

$$\frac{h}{L} = \frac{4.17\mu}{\rho g} \frac{(1 - \epsilon)^2}{\epsilon^3} \left(\frac{a}{v}\right)^2 V + k_2 \frac{1 - \epsilon}{\epsilon^3} \left(\frac{a}{v}\right) \frac{V^2}{g} \tag{8.5}$$

Note that the first term of the Ergun equation is the viscous energy loss that is proportional to $V$. The second term is the kinetic energy loss that is proportional to $V^2$. Comparing the Ergun and Kozeny equations, we see that the first term of the Ergun equation (viscous energy loss) is identical to that in the Kozeny equation except for the numerical constant. The value of the constant $k_2$ was originally reported to be 0.29 for solids of known specific surface.[12] In a later paper, however, Ergun reported a $k_2$ value of 0.48 for crushed porous solids,[13] a value supported by later unpublished studies at Iowa State University. The second term in the equation becomes dominant at higher flow velocities because it is a square function of $V$. The Kozeny equation, however, is more convenient to use and is quite acceptable up to Re = 6.

As is evident from the equation, the head loss for a clean bed depends on the flow rate, grain size, porosity, sphericity, and water viscosity. As filtration progresses and solids are deposited within the void spaces of the medium, the porosity decreases and sphericity is altered. Head loss is very dependent on porosity, and reduction in porosity causes the head loss to increase.

The ability to calculate head loss through a clean fixed bed is important in filter design because provision for this head must be made in the head loss provided in the plant. In addition, of course, head must be provided in the plant design for the increase in loss caused by clogging during the filter cycle. The clogging head loss is usually based on prior experience for similar water and treatment schemes or on pilot studies.

**Example Problem 1**  Calculate the head loss for the 3-ft-deep (0.91-m) bed of the filter sand shown in Fig. 8.3 at a filtration rate of 6 gpm/ft$^2$ (14.6 m/r) and a water temperature of 20°C, using a grain sphericity of 0.75 and a porosity of 0.42, estimated from Table 8.1.

**Solution**  Because the sand covers a range of sizes and will be stratified during backwashing, divide the bed into five equal segments and use the middle sieve opening size for the diameter term in the solution.

$$\text{Kozeny equation} \qquad \frac{h}{L} = \frac{k\mu}{\rho g} \frac{(1 - \epsilon)^2}{\epsilon^3} \left(\frac{a}{v}\right)^2 V \qquad \text{cm} \cdot \text{g} \cdot \text{s} \qquad (8.2)$$

where

$$a/v = \frac{6}{\psi d} = \frac{6}{0.75 d}$$

$\mu/\rho = v = 0.01003$ cm$^2$/s at 20°C

$g = 981$ cm/s$^2$

$k = $ Kozeny's constant, typically 5 for filter media

$V = 6$ gpm/ft$^2 = 0.408$ cm/s

$$h/L = 5 \cdot \frac{0.01003}{981} \cdot \frac{(1 - 0.42)^2}{0.42^3} \frac{6^2}{0.75^2 d^2} 0.408 = \frac{0.00606}{d^2}$$

From Fig. 8.3, select middiameters and calculate h/L for each.

| Size | Middiameter, cm | $\dfrac{h}{L}$ | Layer depth, ft | h, ft |
|------|------|------|------|------|
| $d_{10}$ | 0.054 | 2.08 | 0.6 | 1.25 |
| $d_{30}$ | 0.066 | 1.39 | 0.6 | 0.83 |
| $d_{50}$ | 0.073 | 1.14 | 0.6 | 0.68 |
| $d_{70}$ | 0.080 | 0.95 | 0.6 | 0.57 |
| $d_{90}$ | 0.087 | 0.80 | 0.6 | 0.48 |
|  |  | 1.27  average |  | 3.81 |

Alternatively, because each layer was the same depth, the average h/L can be used to calculate the head loss for a 3-ft (0.91-m) bed of sand, $L = 3$ ft (0.91 m), average $h/L = 1.27$, $h = 1.27 \times 3 = 3.81$ ft (1.16 m) of water.

## Head loss for a fluidized bed

The U.S. practice of filter backwashing for many years has been above the minimum fluidization velocity of the filter medium. Therefore, some fluidization fundamentals are essential to proper understanding of this backwashing practice.

Fluidization can best be described as the upward flow of a fluid (gas or liquid) through a granular bed at sufficient velocity to suspend the grains in the fluid. During upward flow, the energy loss (pressure drop) across the fixed bed will be a linear function of the flow rate at low superficial velocities when flow is laminar. For coarser or heavier grains, the energy loss may become an exponential function at higher flow rates if the Re enters the transitional regime, Re > 6. As the flow rate is increased further, the resistance equals the gravitational force and the particles become suspended in the fluid. Any further increase in flow rate causes the bed to expand and accommodate to the increased flow while effectively maintaining a constant pressure drop (equal to the buoyant weight of the medium). Two typical curves for real filter media fluidized by water are shown in Fig. 8.4.

The pressure drop $\Delta p$ after fluidization is equal to the buoyant weight of the grains and can be calculated from

$$\Delta p = h\rho g = L(\rho_s - \rho)g(1 - \epsilon) \tag{8.6}$$

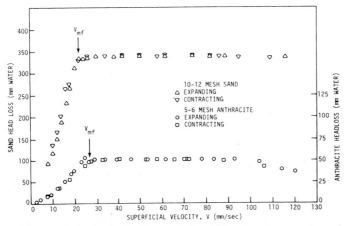

**Figure 8.4**   Head loss versus superficial velocity for 10-12 mesh sand at 25°C, $L_0$ = 37.9 cm, $\epsilon_0$ = 0.446, and for 5-6 mesh anthracite at 25°C, $L_0$ = 19.8 cm, $\epsilon_0$ = 0.581. (*Source: J. L. Cleasby and K. S. Fan, "Predicting Fluidization and Expansion of Filter Media," J. Environ. Eng. Div. ASCE, vol. 107, no. 3, p. 455. Copyright 1981. American Society of Civil Engineers.*)

in which $\rho_s$ is the mass density of the grains and the other terms are as defined before.

### Point of incipient fluidization

The point of incipient fluidization, or minimum fluidizing velocity $V_{mf}$, is the superficial fluid velocity required for the onset of fluidization. It can be defined by the intersection of the fixed-bed and fluidized-bed head loss curves, the points labeled $V_{mf}$ on Fig. 8.4.

The calculation of minimum fluidization velocity is important in determining minimum backwash flow rate requirements. The rational approach to the calculation is based on the fixed-bed head loss being equal to the constant head loss of the fluidized bed at the point of incipient fluidization. Thus, the Ergun equation [Eq. (8.5)] can be equated to the constant-head-loss equation [Eq. (8.6)] and solved for the velocity, that is, $V_{mf}$. The accuracy of the result is very dependent on using realistic values for sphericity $\psi$ and fixed-bed porosity $\epsilon$. Such data for actual media may not be available, making the calculation difficult. By substituting an approximate relation between $\psi$ and $\epsilon_{mf}$ into the aforementioned equation (Ergun equation = constant-head-loss equation), Wen and Yu[14] were able to eliminate both $\psi$ and $\epsilon_{mf}$ from the calculation of $V_{mf}$. The resulting equation is

$$V_{mf} = \frac{\mu}{\rho d_{eq}} (33.7^2 + 0.0408\text{Ga})^{0.5} - \frac{33.7\mu}{\rho d_{eq}} \tag{8.7}$$

where Ga is the Galileo number

$$\text{Ga} = d_{eq}^3 \frac{\rho(\rho_s - \rho)g}{\mu^2} \tag{8.8}$$

For a bed containing a gradation in particle sizes, the minimum fluidization velocity is not the same for all particles. Smaller grains become fluidized at a lower superficial velocity than larger grains do. Therefore, a gradual change from the fixed bed to the totally fluidized state occurs. In applying Eq. (8.7) to a real bed with grains graded in size, calculating $V_{mf}$ for the coarser grains in the bed is necessary to ensure that the entire bed is fluidized. The $d_{90}$ sieve size would be a practical diameter in this calculation. Generally, $d_{eq}$ is not conveniently available, and the $d_{90}$ diameter from the sieve analysis may be used as an acceptable approximation.

Furthermore, the minimum backwash rate selected must be higher than $V_{mf}$ for the $d_{90}$ sieve size to allow free movement of these coarse grains during backwashing. A backwash rate equal to $1.3V_{mf}$ is suggested to ensure adequate movement of the grains.[5]

**Example Problem 2** Calculate the minimum fluidization velocity for the anthracite shown in Fig. 8.3 at a water temperature of 20°C, using an anthracite density of 1.6 g/cm³, estimated from Table 8.1.

**Solution** Calculate the fluidization velocity $V_{mf}$ for the $d_{90}$ size of the anthracite as suggested in the text. The $d_{90}$ size from Fig. 8.3 = 0.29 cm. From Eq. (8.7) by Wen and Yu,

$$V_{mf} = \frac{\mu}{\rho d} (33.7^2 + 0.0408 \text{ Ga})^{0.5} - \frac{33.7\mu}{\rho d}$$

The Galileo number from Eq. (8.8) is

$$\text{Ga} = \frac{d^3 \rho (\rho_s - \rho) g}{\mu^2}$$

Solving in cm · g · s unit gives

$\mu = 0.01002$ g/(cm)(s)
$\rho = 0.998$ g/cm³
$\mu/\rho = \nu = 0.01003$ cm²/s
$g = 981$ cm/s²
$d = 0.29$ cm

$$\text{Ga} = \frac{0.29^3 \cdot 0.998(1.6 - 0.998)981}{0.01002^2} = 143,024 \quad \text{dimensionless}$$

$$V_{mf} = \frac{0.01003}{0.998 \times 0.29} (33.7^2 + 0.0408 \times 143,024)^{0.5} - \frac{33.7(0.01003)}{0.998 \times 0.29}$$
$$= 1.72 \text{ cm/s}$$

If this medium were to be backwashed with full-bed fluidization, the recommended backwash rate is 1.3 $V_{mf}$, as suggested in text:

$$\text{Backwash rate} = 1.3(1.72) = 2.23 \text{ cm/s}$$

$$2.23 \frac{\text{cm}^3}{\text{cm}^2\text{s}} \cdot \frac{L}{1000 \text{ cm}^3} \frac{1}{3.785} \frac{\text{gal}}{L} 929 \frac{\text{cm}^2}{\text{ft}^2} \frac{60 \text{ s}}{\text{min}} = 32.8 \text{ gpm/ft}^2$$

This backwash rate is higher than normal because of the very coarse anthracite grain size and the rather warm water.

## Rapid Filtration

### General description

Rapid filtration, formerly known as "rapid sand filtration," usually consists of passage of pretreated water through a granular bed at rates from 2 to 10 gpm/ft² (5 to 25 m/h). Flow is usually downward through the bed although some use of upflow filters is reported in Latin America, Russia, and the Netherlands. Both gravity and pressure filters are used although some restrictions are imposed against the use of pressure filters on polluted source water.[15]

During operation, solids are removed from the water and accumulate within the voids and on the top surface of the filter medium. This clogging results in a gradual increase in head loss across the filter if the flow rate is to be sustained. The head loss may approach the maximum head loss provided in the plant, sometimes called the *available head loss*. After a period of operation, the rapid filter is cleaned by backwashing with an upward flow of water. The operating time between backwashes is referred to as a *filter cycle* or a *filter run*. The head loss at the end of the filter run is called the *terminal head loss*.

The need for backwash is indicated by one of the following three criteria, whichever occurs first:

1. The head loss across the filter increases to the available limit or to a lower established limit, usually 8 to 10 ft (2.4 to 3.0 m) of water.

2. The filtrate begins to deteriorate in quality or reaches some set upper limit.

3. Some maximum time limit (usually 3 or 4 days) has been reached.

Typical filter cycles range from about 12 to 96 h, although some plants operate with longer cycles. Setting an upper time limit for the cycle is desirable because of concern about bacterial growth in the filter and concern that compaction of the solids removed in the filter will make backwashing difficult.

Pretreatment of surface water by chemical coagulation is essential to achieve efficient removal of particulates in rapid filters. In addition, filter aid polymers may be added to the water just ahead of filtration to strengthen the attachment of the particles to the filter media. Groundwater treated for iron and manganese removal by oxidation, precipitation, and filtration generally does not need other chemical pretreatment.

### Filter media for rapid filters

Common filter materials used in rapid filters are sand, crushed anthracite coal, GAC, and garnet or ilmenite. Typical filter media configurations are shown in Fig. 8.5. The most commonly used of these are the conventional sand and dual-media filters, but a substantial number of triple-media (mixed-media) filters have been installed in the United States. GAC replaces sand or anthracite in filter-adsorbers. It can be used alone or in dual- or triple-media configurations. The first three configurations in Fig. 8.5 are backwashed with full fluidization of the bed. Fluidization results in stratification of the finer grains of each medium near the top of that layer of medium.

The single-medium filter using coarse sand or anthracite coal (4 in Fig. 8.5) differs from the conventional sand filter in two ways. First, because the medium is coarser, a deeper bed is required to achieve

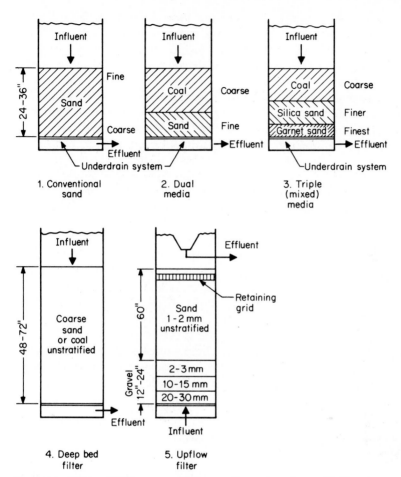

**Figure 8.5** Schematic diagrams of filter configurations for rapid filtration. Media 1, 2, and 3 are washed with fluidization whereas 4 and 5 are washed with air plus water without fluidization.

comparable removal of particulates. Second, because excessive wash rates are required to fluidize the coarse medium, it is washed without fluidization by the concurrent upflow of air and water. The air-water wash causes mixing of the medium, and little or no stratification by size occurs.

The upflow filter is used in some wastewater filtration plants and in a few potable water treatment plants in other countries. It may include a restraining grid to resist uplift, as shown in Fig. 8.5, or it may be operated with a deeper sand layer and to a limited terminal head loss so that the mass of the sand itself acts to resist uplift. The upflow filter is backwashed with air and water together during part of the backwash cycle. Wide adoption of the upflow filter is doubtful unless it is followed by a downflow filter. The reason for this is the public health concern about the potential for contamination of the filtered water caused by both the dirty backwash water and the filtered water exiting above the filter medium.

Typical grain sizes used in rapid filters are presented in Table 8.2 for various potable water applications. The UC of the filter medium is usually specified to be less than 1.65 or 1.7. For coarser filter media sizes that are to be backwashed with fluidization, however, requiring even a lower UC is beneficial, because this will minimize the $d_{90}$ size and thereby reduce the required backwash flow rate. The lower the specified UC, however, the more costly the filter medium because a greater portion of the raw material falls outside the specified size range. Therefore, the lowest practical UC is about 1.5. Anthracite coal that will meet this UC is available commercially.

In addition to the configurations of filter medium shown in Fig. 8.5, other proprietary media are being used in some applications. For example, a buoyant crushed plastic medium is being used in an upflow mode as a contact flocculator and roughing filter ahead of a downflow triple-media bed.[16]

Several manufacturers are marketing traveling backwash filters in which the filter is divided into cells. The cells are washed in sequence by a movable backwash delivery and collection system. The most common medium for this type of filter is a shallow layer of fine sand, usually about 12 in (30 cm) deep.[17]

TABLE 8.2    Typical Grain Sizes for Different Applications

| | Effective size, mm | Total depth, m |
|---|---|---|
| A. *Common U.S. Practice after Coagulation and Settling* | | |
| 1. Sand alone | 0.45–0.55 | 0.6–0.7 |
| 2. Dual media | 0.9–1.1 | 0.6–0.9 |
| Add anthracite (0.1 to 0.7 of bed) | | |
| 3. Triple media | 0.2–0.3 | 0.7–1.0 |
| Add garnet (0.1 m) | | |
| B. *U.S. Practice for Direct Filtration* | | |
| Practice not well established. With seasonal diatom blooms, use coarser top size. Dual-media coal, 1.5-mm ES | | |
| C. *U.S. Practice for Fe and Mn Filtration* | | |
| 1. Dual media similar to A-2 above | | |
| 2. Single medium | <0.8 | 0.6–0.9 |
| D. *Coarse Single-Medium Filters Washed with Air and Water Simultaneously* | | |
| 1. For coagulated and settled water | 0.9–1.0 | 0.9–1.2 |
| 2. For direct filtration | 1.4–1.6 | 1–2 |
| 3. For Fe and Mn removal | 1–2 | 1.5–3 |

૬੫

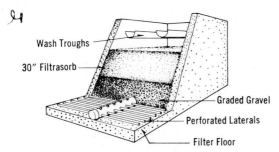

**Figure 8.6**  Rapid gravity filter with manifold and lateral underdrain system. (*After C. P. Hoover, Water Supply & Treatment, National Lime Assoc.*)

### Underdrain and support gravel

The underdrain system serves to support the filter medium, collect filtered water (in downflow filters), and distribute backwash water and air scour if air is used. A wide variety of underdrain systems is in use, but the systems can be grouped into three major types.

The first and oldest type is the manifold-lateral system in which perforated pipe laterals are located at frequent intervals along a manifold (Fig. 8.6). Perforations in the laterals are ¼ to ½ in (6 to 13 mm), located on 3- to 12-in (8- to 30-cm) spacing.

The second is the fabricated self-supporting underdrain system that is grouted to the filter floor, as in Fig. 8.1. One example of this type is a vitrified clay block underdrain now marketed by several companies. Recently, a similar plastic block underdrain has been marketed that is capable of delivering either backwash air or water or air and water simultaneously. Top openings on these types of underdrains are about ¼ in (6 mm).

The third type of underdrain is the false-floor underdrain with nozzles (Fig. 8.7). A false-floor slab, or a steel plate in pressure filters, is located 1 to 2 ft (0.3 to 0.6 m) above the bottom of the filter, thus providing an underdrain plenum below the false floor. Nozzles to collect the filtrate and distribute the backwash water are located at 5- to 8-in (13- to 20-cm) centers. Nozzles may have coarse openings of about ¼ in (6 mm), or they may have very fine openings, sufficiently small to retain the filter medium. The nozzles may be equipped with a stem protruding about 6 to 9 in (15 to 23 cm) into the underdrain plenum. One or two small holes or slots in the stem distribute air, either alone or in combination with water.

Most filters use underdrain systems with openings larger than the filter medium to be supported. As a result, using several layers of graded gravel between the underdrain openings and the filter medium is necessary to prevent the medium from leaking downward into the underdrain system. The conventional gravel system begins with

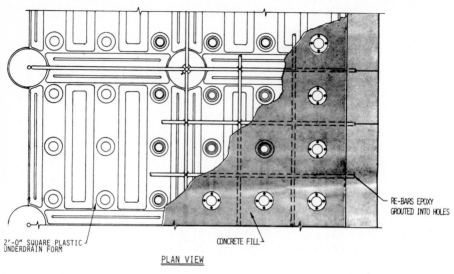

PLAN VIEW

RE-BARS EPOXY GROUTED INTO HOLES

2'-0" SQUARE PLASTIC UNDERDRAIN FORM

CONCRETE FILL

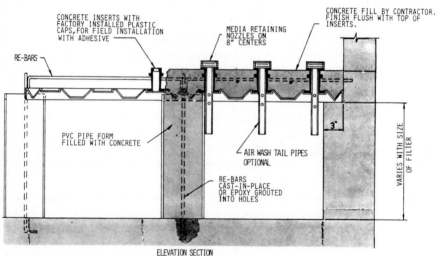

CONCRETE INSERTS WITH FACTORY INSTALLED PLASTIC CAPS, FOR FIELD INSTALLATION WITH ADHESIVE

RE-BARS

MEDIA RETAINING NOZZLES ON 8" CENTERS

CONCRETE FILL BY CONTRACTOR. FINISH FLUSH WITH TOP OF INSERTS.

PVC PIPE FORM FILLED WITH CONCRETE

AIR WASH TAIL PIPES OPTIONAL

RE-BARS CAST-IN-PLACE OR EPOXY GROUTED INTO HOLES

3"

VARIES WITH SIZE OF FILTER

ELEVATION SECTION

**Figure 8.7**  Nozzle underdrain system consisting of a monolithic, cast-in-place, concrete slab on a permanent plastic underdrain form with nozzles capable of air and water distribution. (*Multicrete II, Courtesy of General Filter Co., Ames, Iowa.*)

coarse-sized gravel at the bottom with progressively finer-sized gravel layers above up to the filter medium. These are typical rules of thumb for the gradations of adjacent layers[18]:

1. Each gravel layer should be as uniform as possible, preferably retained between sieve openings that are apart by the ratio of $\sqrt{2}$. (In the United States, a sieve size ratio of 2 is commonly accepted.[19]

2. The bottom-layer fine size should be 2 to 3 times the orifice diameter of the underdrain system.

3. The top-layer fine size should be 4 to 4.5 times the ES of the media to be retained.

4. From layer to layer, the fine size of the coarser layer should be less than or equal to 2 times the fine size of the adjacent finer layer.

5. Each layer should be at least 2 in (5 cm) thick or 3 times the coarse size of the layer, whichever is greater.

When air scour is to be delivered through the supporting gravel, great danger exists that the gravel will be disrupted during the backwash cycle, especially if air and water are used simultaneously. Two solutions are being used. One is to use a nozzle underdrain with openings small enough that gravel is not required. The other is to use a double reverse-graded gravel system graded from coarse on the bottom to fine in the middle and back up to coarse on the top. This concept, originally used by Baylis to solve gravel migration problems in large filters in Chicago,[20] is now being used in air and water backwash filters supplied by several companies.

### Rates of filtration

The traditional rates for rapid filtration were 2 gpm/ft$^2$ (5 m/h), traced to the historic study of G. W. Fuller at Louisville in the late 1890s.[21] Many studies in the 1950s and 1960s, however, led to the wide use of rates up to 4 gpm/ft$^2$ (10 m/h), even with conventional sand medium and conventional pretreatment schemes using only aluminum or iron salts for coagulation.

Increasing the filtration rate shortens the filter cycle roughly inversely with the rate. This problem was minimized by the use of dual- or triple-media filters that have become increasingly common in "high-rate" plants since the 1950s. These filters provide for better penetration of solids into the anthracite coal layer, which has a larger grain size than the traditional sand filter, and thus better utilization of the medium for solids storage during the filter cycle. The advent of synthetic organic polymers to assist both coagulation and filtration has permitted higher rates without deterioration of filtrate quality. Thus, the common high-rate filter plant today may use filtration rates up to 6 gpm/ft$^2$ (15 m/h) and usually is equipped with dual- or triple-media filters and with filter aid polymer feed systems. Acceptance of high rates varies among state regulatory agencies, however, and plant- or pilot-scale demonstrations may be required by some state agencies prior to acceptance.

Since about 1970, a few reports have been made of radically higher filtration rates in some plants, for example, 10 gpm/ft$^2$ (24 m/h) at Contra Costa, Calif.,[22] and 13.5 gpm/ft$^2$ (33 m/h) for the new Los Angeles direct filtration plant. The latter filtration rate was selected and approved by the state after more than 5 years of extensive pilot-scale studies.[23,24] Both cases involve treatment of high-quality surface water in large plants with well-qualified management and operation. Thus, filtration rates of these magnitudes probably will not become common. With poorer source water and in smaller plants with less careful surveillance or operation, lower filtration rates will continue to be the prudent choice.

### Use of GAC in rapid filtration

GAC is being used in filter-adsorbers that serve for both filtration and organic compound adsorption. The principal application to date is for taste and odor removal where full-scale experience shows successful removal for periods from 1 to 5 years before the GAC must be regenerated.[1] High concentrations of competing organic compounds, however, can reduce this duration. Most existing GAC filter-adsorbers are retrofitted rapid filters where GAC has replaced part of or all the sand in rapid sand filters or anthracite in dual- or triple-media filters. Several filter-adsorbers, however, have been initially constructed with GAC.

GAC used in retrofitted filters is typically 15 to 30 in (0.38 to 0.76 m) of 12 × 40 mesh GAC (ES, 0.55 to 0.65 mm) or 8 × 30 mesh GAC (ES, 0.80 to 1.0 mm) placed over several inches of sand (ES, 0.35 to 0.60 mm) in a dual-media configuration. These GAC materials have a higher UC (≤ 2.4) than is traditionally used for filter materials (≤ 1.6). The high UC can contribute to more fine grains in the upper layers and shorter filter cycles.

The use of retrofitted filters as filter-adsorbers for removal of less strongly adsorbed compounds, such as trihalomethanes and volatile organic compounds, is limited. The short empty-bed contact times typical of such filter-adsorbers (about 9 min) would necessitate frequent regeneration or replacement of the GAC.

Some filter-adsorbers designed for GAC initially use up to 48 in (1.2 m) of coarser GAC (ES, 1.3 mm) with lower uniformity coefficients (1.4). This provision should result in longer filter cycles and longer periods between GAC regeneration.

GAC is being used successfully on both single-medium and dual-media filter-adsorbers. It is as effective in its filtration function as conventional filter media, either sand or dual media, provided an appropriate medium size has been selected.[1] Particulates removed in the filter-adsorber, however, do impede the adsorption function somewhat compared to postfilter-adsorbers with the same contact time (see Chap. 13).

The layer of fine sand in the dual-media filter-adsorber may be essential where very low-turbidity filtered water is the goal. The use of sand and GAC in dual-media beds, however, causes difficulties in the regeneration of the GAC. The removal of the GAC from the filter-adsorber free of sand, or the separation of the sand after removal, is difficult. Sand causes difficulties in regeneration furnaces.

Bacteria proliferate in GAC beds unless the influent water contains chlorine.[1] This bacterial activity has both beneficial and detrimental effects. It benefits the water by reducing the organic content and thereby reducing microbial activity in the distribution system. Bacterial growth can shorten filter cycles, however, and disinfection is required after the filter-adsorber. Carbon fine grains have been detected in effluents from filter-adsorbers. Bacteria attached to these fine particles are difficult to kill with chlorine.

GAC is softer than anthracite, but the attrition of GAC media in full-scale plants has not been excessive. GAC losses from 1 to 6 percent per year have been reported,[1] which is not higher than typical losses of anthracite. A modest reduction in grain size of GAC has been measured at some plants. GAC has a lower density than anthracite, posing some concerns about backwashing, which are discussed later.

## Rapid Filter Performance

### General pattern of effluent quality

The quality of filtered water is poorer at the beginning of the filtration cycle and may also deteriorate near the end of the cycle if the cycle is prolonged for a sufficient time. In 1962, Cleasby and Baumann[25] presented detailed data on the filtered water quality early in the filter run, observed during filtration of precipitated iron (Fig. 8.8). The filtrate was observed to deteriorate for a few minutes and then improve over about 30 min before reaching the best level of the entire run. Thereafter, deterioration of filtrate quality occurred at the highest filtration rate of 6 gpm/ft$^2$ (15 m/h) (Fig. 8.9). Deterioration of filtrate later in the filter run, commonly called *breakthrough,* did not occur, however, at 2 and 4 gpm/ft$^2$ (5 and 10 m/h) or in a constant-head-loss run where the filtration rate declined from 6 gpm/ft$^2$ (15 m/h) at the beginning of the run to about 4 gpm/ft$^2$ (10 m/h) at the end of the run (Fig. 8.9).

The initial peak occurred at approximately the theoretical detention time of the filter plus the detention time of the effluent appurtenances to the point of turbidity monitoring. Amirtharajah and Wetstein have

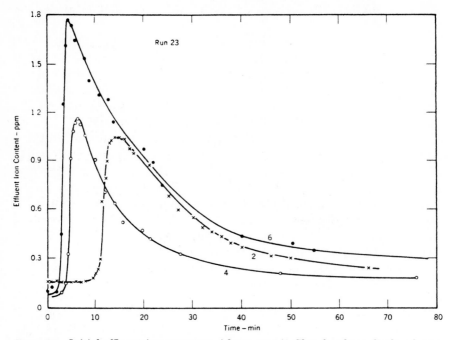

**Figure 8.8**    Initial effluent improvement. After water in filter has been displaced, turbidity drops to minimum. Numbers by curves indicate filtration rate in gallons per minute per square foot. (*Source: J. L. Cleasby and E. R. Baumann, "Selection of Sand Filtration Rates," J. AWWA, vol. 54, no. 5, May 1962, p. 579.*)

used these data and their own data to formulate a conceptual model for the initial degradation and improvement period.[26] A two-peak model was proposed. One peak results from the residual backwash water remaining in the filter at the end of the backwash. The second peak results from solids released from the filter grains as they collide on contraction of the bed at the end of backwashing.

The initial water quality degradation period also has been demonstrated in studies using *Giardia* cysts.[27] *Giardia muris* was used as a model for the human pathogen, *G. lamblia*. *Giardia muris* was spiked into a low-turbidity surface water, coagulated with alum alone or alum and cationic polymer, flocculated, and filtered through granular media filters. Initial cyst concentrations in the filtrate were from 10 to 25 times higher than those following the initial improvement period, even though the turbidity improved less than 0.1 NTU during the period.

In conventional water treatment practice, turbidity passage during the initial degradation and improvement period is small when averaged over the entire filter run and compared to the total turbidity

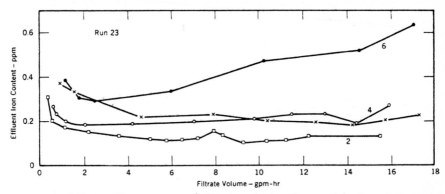

Figure 8.9 Effluent iron content and filtrate volume. Observations for curves were made after initial improvement period. Numbers by curves indicate filtration rate. Curve along X marks indicates constant pressure. Rate of this run started at 6 gpm/ft² (15 m/h) and decreased to 4 gpm/ft² (10 m/h) as head loss increased. (*Source: J. L. Cleasby and E. R. Baumann, "Selection of Sand Filtration Rates," J. AWWA, vol. 54, no. 5, May 1962, p. 579.*)

passed during the run. Therefore, little attention has been paid to the impact of the initial period on the average filtrate quality. The early practice of filtering to waste at the beginning of the filter run, to eliminate some of the turbidity carried through into the finished water, has been largely abandoned.

Elimination of the "filter-to-waste period" may not be acceptable where giardiasis is of concern because of the low infective dose for *Giardia* transmission[28] and its high resistance to chlorine disinfection.[29]

Similar observations of water quality during the filter cycle have been presented by Robeck et al.[30] during surface water treatment, shown in Figs. 8.10 and 8.11. During strong floc conditions, the filtrate quality was nearly constant after the initial improvement period (Fig. 8.10). During weak floc conditions, however, the terminal turbidity breakthrough was rather abrupt (Fig. 8.11) and occurred at low head loss. Figures 8.10 and 8.11 represent the two most typical filtrate-quality patterns observed. In some cases, however, the filtrate quality may improve throughout the run, and no terminal breakthrough is observed.[31]

From these illustrations, providing continuous monitoring of the filtrate of each filter to detect the onset of terminal breakthrough is desirable. The cycle could then be terminated at the onset of breakthrough, even if the head loss has not reached the normal maximum available at the plant. This is especially important if the passage of *Giardia* cysts, viruses, or asbestos fibers are of concern because increases in turbidity indicate simultaneous increases in other particu-

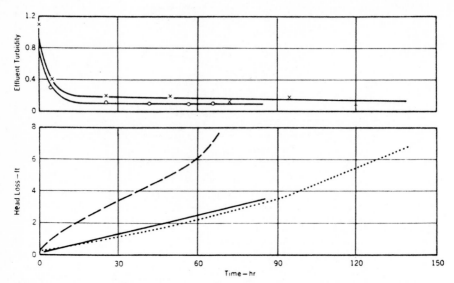

**Figure 8.10** Effect of filter media on length of run with strong floc (summer conditions). The data shown were obtained under the following operating conditions: source-water turbidity, 80 units; settling tank effluent turbidity, 2 units; alum dose, 75 mg/L; filtration rate, 2 gpm/ft$^2$ (5 m/h). In the upper graph, the curve determined by the X points is for the medium consisting of both coal and sand; and the curve determined by the circular points is for both coal alone and sand alone. In the lower graph, the dashed curve is for sand; the solid, coal; and the dotted, coal and sand combined. (*Source: G. G. Robeck, K. A. Dostal, and R. L. Woodward, "Studies of Modification in Water Filtration," J. AWWA, vol. 56, no. 2, Feb. 1964, p. 198.*)

lates, often of larger relative magnitude than the turbidity increases. For example, Figs. 8.12 and 8.13 indicate increases in viruses and asbestos fibers coinciding with breakthrough in turbidity.[30,32] Notice that the fiber count increased much more than the turbidity in Fig. 8.13. Similar evidence in Fig. 8.14 shows the change in *Giardia* cyst concentration during a turbidity breakthrough and following the backwash operation.[33] Notice that a 3-fold change in turbidity corresponded to a 30- to 40-fold change in cyst concentration.

Large numbers of very small particles can exist in filtered water with turbidity less than 1 NTU. For example, Logsdon and Symons[34] found poor correlation between numbers of asbestos fibers and turbidity in both source and filtered water from Lake Superior at Duluth, Minn. The amphibole fibers had diameters of 0.1 to 0.4 μm and, by definition, a length-width ratio of at least 3:1. The chrysotile fibers most frequently had diameters of about 0.05 μm and were counted in lengths from 0.3 to 1.3 μm. Even though these small fibers cannot be detected by turbidity measurements, the amphibole fiber count was observed to be usually less than the detection limit of the analytical

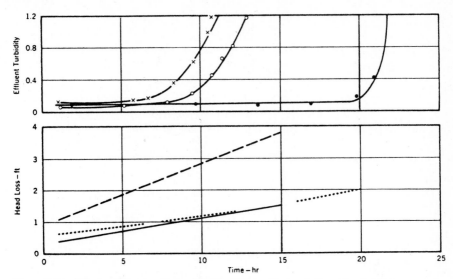

**Figure 8.11** Effect of filter media on length of run with weak floc. The data shown were obtained under the following operating conditions: source-water turbidity, approximately 20 units; alum dose, 100 mg/L; activated carbon, 2 mg/L; filtration rate, 2 gpm/ft$^2$ (5 m/h); settling tank effluent turbidity, 15 units. In the upper graph, the curve determined by the X points is for coal, that by open circles, sand; and that by solid circles, coal and sand. In the lower graph, the dashed curve is for sand; the dotted, coal and sand; and the solid, coal. (*Source: G. G. Robeck, K. A. Dostal and R. L. Woodward, "Studies of Modification in Water Filtration," J. AWWA, vol. 56, no. 2, Feb. 1964, p. 198.*)

method used when the filtered water turbidity was less than 0.1 NTU. This led to the conclusions that alum at 12 to 20 mg/L and nonionic polymer at 0.05 mg/L be used in pretreatment, that triple-media filters should be used to achieve a finished water turbidity of not greater than 0.1 NTU, that in-line turbidimeters should be used for continuous monitoring of source and finished water turbidity, and that periodic electron microscopic analyses for asbestos fibers should be conducted to check the effectiveness of the treatment process.

O'Connor et al.[35] have found large numbers of small bacterial cells in finished waters with turbidities less than 1 NTU. The method involves direct counting of bacteria on a membrane filter by using an ultraviolet microscope after staining with the fluorochrome of acridine orange [acridine orange direct count (AODC) method]. Removal of bacteria (as measured by the AODC method) in municipal treatment plants was about 1 order of magnitude in the winter and about 2 or 3 orders of magnitude in the summer. Finished waters still contained from $10^6$ to $10^9$ cells per liter, even though they generally had turbidities less than 1 NTU. These cells can include dead bacteria, autotrophic bacteria, fastidious heterotrophs, and anaerobic bacteria

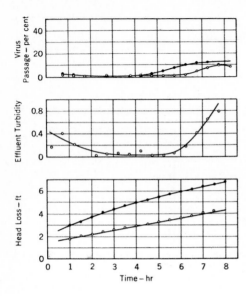

**Figure 8.12** Virus and floc break-through at high rate on coarse sand or coal. The curves with solid circles represent sand filters. Those with open circles, coal plus sand. The filtration rate was 6 gpm/ft² (15 m/h), with blended water containing an alum dose of 10 mg/L. The virus load was 8400 PFU/mL; turbidity load, 10 JTU. (*Source: G. G. Robeck, K. A. Dostal, and R. L. Woodward, "Studies of Modification in Water Filtration," J. AWWA, vol. 56, no. 2, Feb. 1964, p. 198.*)

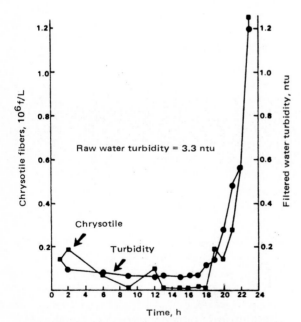

**Figure 8.13** Finished water chrysotile counts and turbidity versus time run 120, Seattle pilot plant. (*Source: G. S. Logsdon, J. M. Symons, and T. J. Sorg, "Monitoring Water Filters for Asbestos Removal," J. Environ. Eng. Div. ASCE, vol. 107, no. 6, p. 1297. Copyright 1981. American Society of Civil Engineers.*)

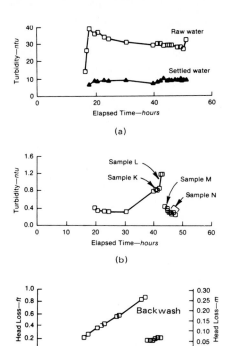

**Figure 8.14** Filter performance before and after a backwash operation during conventional treatment using alum coagulation of a surface water. Turbidity breakthrough at low head loss is reflected in higher cyst concentrations, which were reduced after the backwash. With continuous feed of *G. muris* at 11,400 cysts/L, the effluent cyst concentrations per liter were $K = 440$, $L = 240$, $M = 8.7$, and $N = 14.5$. (*a*) Turbidity of raw and settled water (test series 4); (*b*) turbidity of effluent from anthracite filter (test series 4); (*c*) head loss in anthracite filter (test series 4). (*Source: G. S. Logsdon, V. C. Thurman, E. S. Frindt, and J. G. Stoecker, "Evaluating Sedimentation and Various Filter Media for Removal of Giardia Cysts," J. AWWA, vol. 77, no. 2, February 1985, p. 61.*)

that will not be detected in the heterotrophic plate count procedure. These observations led the authors to conclude that turbidity is not a good indicator of bacterial cell count. The reductions in AODC count (1 to 3 orders of magnitude) are, however, typical of performance for turbidity reduction in conventional plants and direct filtration plants using rapid filtration.

## Mechanisms of filtration

The mechanisms involved in removal of suspended solids during rapid filtration are very complex. Many workers have discussed the various factors that may play an important role in removal. A notable example of such discussions is by O'Melia and Stumm.[36] The dominant mechanisms depend on the physical and chemical characteristics of the suspension and the medium, the rate of filtration, and the chemical characteristics of the water.

During rapid granular-bed filtration, particle removal is primarily within the filter bed, referred to as *depth filtration*. The efficiency of depth removal depends on a number of mechanisms. Some solids may be removed by the simple mechanical process of interstitial straining. Removal of other solids, particularly the smaller solids, depends on two types of mechanisms. First, a transport mechanism must bring the small particle from the bulk of the fluid within the interstices close to the surfaces of the grains. Transport mechanisms may include gravitational settling, diffusion, interception, and hydrodynamics that are affected by such physical characteristics as size of the filter medium, filtration rate, fluid temperature, and the density, size, and shape of the suspended particles.

Second, as particles approach the surface of a grain, or previously deposited solids on the grain, short-range surface forces begin to influence particle movement. If the particles have been destabilized sufficiently that electrostatic repulsive forces are minimized, collision of the particle and the grain surface (or previous deposits) is possible and attachment can occur because of short-range Van der Waal's forces. The collision and attachment are comparable to the coagulation of destabilized particles. Figure 8.15 is a scanning electron micrograph of small particles (5 to 10 μm) retained on three filter sand grains about 0.5 to 1.0 mm.

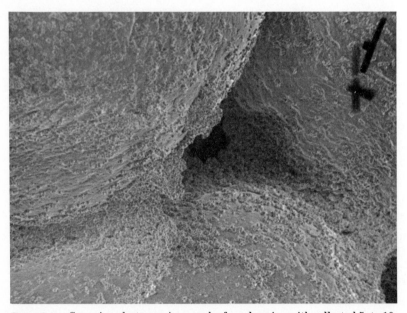

**Figure 8.15**  Scanning electron micrograph of sand grains with collected 5- to 10-μm silicate particles. Influent suspension was 200 mg/L, coagulated with cationic polymer. Photograph taken after 3 h of filtration at depth of 46 cm, sand grains about 1 mm in size. (*Courtesy of Prof. C. S. Oulman, Iowa State University.*)

## Mathematical descriptions of filtration

Many researchers have tried to describe rapid granular-bed filtration in mathematical terms. A comprehensive summary of such efforts has been presented by Herzig et al.[37] As yet, no universally acceptable model exists, and the available models are not yet used in filter design or operation. For that reason, the subject of mathematical models is covered only briefly in this chapter, to indicate the nature of the models.

**Transport models.**  Particle removal depends on a transport step and an attachment step, as discussed earlier. Some researchers have concentrated on developing models for the various possible transport mechanisms that can bring the small particles close to the grain surface so that collision and capture are possible. Flow through the filter is usually laminar, and the small particles are carried along the flow stream lines unless a transport mechanism causes particle transport across the stream lines. The particles to be removed are usually considerably smaller than the grains of the filter media. For example, typical particles range from about 0.1 to 10 $\mu$m, while typical filter grains may range from about 500 to 2000 $\mu$m.

Yao et al.[38] have presented the following models to describe the three transport mechanisms commonly considered most important in particle transport, namely, sedimentation, interception, and diffusion. These models, which are extensions of work done in air filtration, are useful in indicating the various parameters that are important in the transport of particles of different size and density in granular bed filters of different grain size, water velocity, and temperature. Each mechanism is expressed in terms of an ideal single spherical collector efficiency $\eta$, which is the ratio of the number of successful collisions to the total number of potential collisions in the cross-sectional area of the collector.

The larger or heavier particles can be transported by sedimentation. The single collector efficiency for sedimentation transport $\eta_s$ is the ratio of the Stokes' settling velocity $V_s$ to the filtration rate (i.e., approach velocity of the flow $V$):

$$\eta_s = \frac{V_s}{V} = \frac{(\rho_s - \rho)gd_p^2}{18\mu V} \tag{8.9}$$

where $d_p$ is the particle diameter and other terms are as defined earlier.

If the stream line carrying a particle passes within $d_p/2$ of the collector surface, the particle can be removed by interception. The single collector model for interception is

$$\eta_I = \frac{3}{2}\left(\frac{d_p}{d_c}\right)^2 \tag{8.10}$$

where $d_c$ is the collector diameter.

Very small colloid-size particles, less than about 1 μm, will be moved in a random pattern away from their stream lines by brownian diffusion. The single collector model for diffusive transport $\eta_D$, which incorporates Einstein's equation for the diffusion coefficient of suspended particles, is

$$\eta_D = 0.9\left(\frac{KT}{\mu d_p d_c V}\right)^{2/3} \tag{8.11}$$

where $K$ is Boltzman's constant, $T$ is absolute temperature, and the other terms are as defined earlier.

These models predict that particle collection will be hindered by higher filtration rates ($V^{-1}$ or $V^{-2/3}$), by larger grain sizes ($d_c^{-2}$ or $d_c^{-2/3}$), or by colder water of higher viscosity ($\mu^{-1}$ or $\mu^{-2/3}$). Capture is also hindered by smaller and lighter particles ($d_p^2$ and $\rho_s - \rho$) unless they are small enough for diffusive transport ($d_p^{-2/3}$). These observations based on theoretical models are in general agreement with actual filtration observations and intuition. Higher rates of filtration, larger grain sizes, and colder water yield poorer filtration. Because real suspensions contain a range of particles, all three mechanisms may be effective in removing some of the particles, so the impact of particle size on efficiency is less easily observed.

The above models have been used to predict expected removal efficiency for particles of various sizes.[38] The predicted removal is poorest for particles about 1 μm in size and better for larger size (by gravity and interception) and better for smaller size (by diffusion). Particles are presumed fully destabilized in pretreatment so that attachment will be possible if the particles approach a grain surface. Experimental studies have provided qualitative support to the prediction obtained with the transport models.[38]

**Mass balance and empirical kinetic models.**   Other investigators have developed models for filtration that do not attempt to consider the mechanisms of particle transport or attachment. Rather they combine a mass balance equation and a rational kinetic equation to predict the filtrate quality and head loss as a function of time of filtration and depth of filter medium.

The mass balance equation merely states that the mass or volume of the particulates removed from suspension in the filter must result in an equal mass or volume of accumulated solids in the pores. Mathematically,

$$V\left(\frac{\delta C}{\delta x}\right)_t + \left(\frac{\delta(\sigma + \epsilon C)}{\delta t}\right)_x = 0 \qquad (8.12)$$

where $x$ = depth
$\quad t$ = time
$\quad \sigma$ = specific deposit (volume of deposited solids/filter volume)
$\quad \epsilon$ = porosity
$\quad C$ = concentration of particulates in suspension (volume of particles/suspension liquid volume)

The $\sigma$ and $C$ terms can also be expressed in mass/volume units. In most cases, the change of accumulated solids in suspension in the pores $\epsilon C$ is trivial compared to the change in the deposited solids, so the equation is often simplified by omitting the $\epsilon C$ term.

Various rational kinetic equations have been proposed, but they all require the use of empirical coefficients. Two examples illustrate the evolution of such equations. In 1967, Heertges and Lerk[39] proposed the following:

$$\frac{\delta \sigma}{\delta t} = k_1(\epsilon_0 - \sigma)VC \qquad (8.13)$$

where $\epsilon_0$ = clean bed porosity and $k_1$ = empirical attachment coefficient. This equation states that the rate of particle deposition inside a filter lamina at a given time is proportional to the particle flux $VC$ and the volume remaining available for deposition $\epsilon_0 - \sigma$.

Adin and Rebhun[40] proposed a kinetic equation in 1977 that included a second term to allow for detachment of already deposited particles:

$$\frac{\delta \sigma}{\delta t} = k_1 VC(F - \sigma) - k_2 \sigma J \qquad (8.14)$$

where $k_2$ = empirical detachment coefficient
$\quad F$ = theoretical filter capacity (i.e., amount of deposit that would clog pores completely)
$\quad J$ = hydraulic gradient

The first term is almost identical to the previous expression, but a new filter capacity term $F$ is proposed rather than the clean bed porosity $\epsilon_0$. The second term states that the probability of detachment depends on the product of the amount of material deposited already $\sigma$ and the hydraulic gradient $J$.

The mass balance equation and one selected kinetic equation must be solved simultaneously to find $C$ and $\sigma$ as a function of the time of filtration in the cycle and depth in the filter. Various numerical or analytical solutions have been utilized to solve the equations.[39–42] One analytical solution[42] was developed based on the close similarity between the Langmuir equation for adsorption and the filtration kinetic equation.

The use of these rational models depends on the collection of pilot- or full-scale filtration data to generate the appropriate attachment and, in some cases, detachment coefficients for the particular suspension being filtered. This need, plus the complex nature of the solutions, has limited the use of such equations in routine filter design. Nevertheless, the use of such models with the appropriate empirical constants is acceptable in predicting filtrate quality, specific deposit, and head loss as a function of time and depth. Further work on modeling will no doubt continue in the future.

The two kinetic equations presented [Eqs. (8.13) and (8.14)] draw attention to a long-standing question in filtration research, namely, whether detachment of deposited solids occurs in a filter operating at a constant filter rate or new influent particles merely bypass previously clogged layers. Recent evidence on this issue indicates that detachment does occur as a result of impingement of newly arriving particles.[43] This evidence was collected by using an industrial endoscope inserted into the filter to magnify, observe, and record the deposition process on videotape. The observations were made during filtering of kaolin clay on sand filters but without added coagulant. Whether such detachment occurs when optimum coagulant addition is employed has not yet been demonstrated. Although this issue may not be fully resolved yet, the following description of filter behavior is generally accepted.

The particles removed in the filter are held in equilibrium with the hydraulic shearing forces that tend to tear them away and wash them deeper into, or through, the filter. As deposits build up, the velocities through the more nearly clogged upper layers of the filter increase, and these layers become less effective in removal. The burden of removal passes deeper and deeper into the filter.[44–46] Ultimately, inadequate clean bed depth is available to provide the desired effluent quality, and the filter run must be terminated.

If the filtration rate on a filter that contains deposited solids is suddenly increased, the hydraulic shearing forces also suddenly increase. This disturbs the equilibrium existing between the deposited solids and the hydraulic shearing forces, and some solids will be dislodged to pass out with the filtrate. This aspect of filtration is discussed in detail later.

## Effect of filtration variables on performance

Many laboratory and plant-scale studies have been conducted to compare the removal efficiency and head loss development of filters operated with different filtration variables.[47–56] A few examples of this type of research are presented in the following pages. The size of the filter medium affects the performance in two conflicting ways. Smaller grain size improves particulate removal, but also accelerates head loss development and may shorten the filter runs if the run length is determined by reaching terminal head loss. Conversely, larger grain size causes somewhat poorer particulate removal but lowers the rate of head loss development.

Laboratory and pilot-scale studies under controlled conditions have evaluated the effects of filtration rate, grain size, and viscosity.[25,47,48] Ives and Sholji[47] determined empirically that the filter coefficient $\lambda$ in a first-order removal model ($\delta C/\delta L = \lambda C$) was inversely proportional to the filtration rate, grain size, and square of the viscosity. Comparison of these observations with Eqs. (8.9), (8.10), and (8.11) shows that the filtration rate exponent agrees with Eq. (8.9), and that the grain size and viscosity exponents do not agree precisely with the three equations, but are qualitatively in agreement.

In the 1950s and 1960s, many plant-scale studies were conducted comparing filter performance at different filtration rates. These studies were generally conducted as utilities were considering uprating existing plants or building new plants with filtration rates higher than the traditional 2 gpm/ft$^2$ (5 m/h). The following examples predated the use of filter aid polymers.

George W. Fuller[21] is commonly credited with establishing a rate of filtration of 2 gpm/ft$^2$ (5 m/h) for chemically pretreated surface water that was considered practically inviolable for the first half of the twentieth century in the United States. Fuller observed that with properly pretreated water, however, higher rates gave practically the same water quality. Of equal importance, Fuller acknowledged that without adequate chemical pretreatment, no assurance of acceptable water existed even at filtration rates of 2 gpm/ft$^2$ (5 m/h). Some of the studies of higher rates are discussed in the following paragraphs.

Brown[49] compared performances at filtration rates of 2, 3, and 4 gpm/ft$^2$ (5, 7, and 10 m/h) on full-scale filters treating water that received conventional alum coagulation and sedimentation. The difference in effluent turbidity and bacterial content was considered insignificant (Table 8.3). Other plant-scale trials in a conventional plant

**TABLE 8.3    Summary of Full-Scale Filtration Results at Three Filtration Rates†[49]**

| | Filter no. | | |
|---|---|---|---|
| | 12<br>(2 gpm/ft$^2$)<br>(4.9 m/h) | 13<br>(3 gpm/ft$^2$)<br>(7.3 m/h) | 14<br>(4 gpm/ft$^2$)<br>(9.8 m/h) |
| Item | | | |
| Length of run, h | 135.2 | 116.7 | 81.3 |
| Wash water, % | 1.21 | 0.89 | 0.99 |
| Turbidity, ppm | 0.34 | 0.38 | 0.43 |
| Bacteria, colonies/mL | 0.32 | 0.42 | 0.36 |
| Coliform organisms | Negative | Negative | Negative |

†The total amount of water passing through the individual filters during the 3-year test period is not known, because they were unmetered. It may be assumed, however, that the quantities were proportional to the rates, since the accuracy of the standard venturi controllers was checked before and during the tests. The filters were operated continuously during the trial period, except when being backwashed.

using alum coagulation of a surface water demonstrated significantly higher effluent turbidity at 7.5 gpm/ft$^2$ (18.3 m/h) than at lower rates.[50]

Hudson[51] presented data from the Chicago Experimental Filtration Plant, and Baylis[52] reported on 7 years of testing on full-scale filters at Chicago. Performance at several rates from 2 to 5 gpm/ft$^2$ (5 to 12 m/h) were compared, and Baylis concluded from these studies that 5 gpm/ft$^2$ (12 m/h) did not degrade effluent quality, especially with regard to bacterial content. Prechlorination was in use at the time of the full-scale testing, however, and this may have overshadowed any detrimental effects on bacteria removal because of filtration alone. The results of solids removed on cotton plug filters, shown in Table 8.4, show somewhat poorer quality at the higher rates, but Baylis did not consider the differences to be significant. Pretreatment at Chicago in-

**TABLE 8.4    Full-Scale Filtrate Quality Data on Several Filtration Rates as Indicated by Solids Captured on Cotton Plug Filters[52]**

| | | Filtration rate, gpm/ft$^2$ (m/h) | | | |
|---|---|---|---|---|---|
| | | 2 (4.9) | 4 (9.8) | 4.5 (11.0) | 5 (12.2) |
| Year | Average of all filters | Ash, ppm‡ | | | |
| 1949 | 0.047 | | 0.055 | 0.059 | 0.066 |
| 1950 | 0.037 | 0.028 | 0.048 | 0.045 | 0.060 |
| 1951 | 0.042 | 0.039 | 0.067 | 0.064 | 0.084 |
| 1952 | 0.077 | 0.039 | 0.058 | 0.084 | 0.084 |
| 1953 | 0.060 | 0.057 | 0.067 | 0.089 | 0.087 |
| 1954 | 0.058 | 0.056 | 0.073 | 0.090 | 0.082 |
| Avg. | 0.054 | 0.044 | 0.059 | 0.072 | 0.077 |

‡Ash remaining after ignition of cotton plug filter.

volved alum coagulation of Lake Michigan water with activated silica used during times of weak flocculation. Without the activated silica, Baylis acknowledged that floc would pass through the filters even at 2 gpm/ft$^2$ (5 m/h).

The conclusion from the above literature is that higher filtration rates do result in somewhat poorer filtrate, as both theory and intuition would predict. Filtration rates up to 4 gpm/ft$^2$ (10 m/h), however, are reasonable with conventional pretreatment and without the use of filter aid polymers to increase filtration efficiency. If higher filtration rates above 4 gpm/ft$^2$ (10 m/h) are planned, however, filter aid polymers will be required to maintain filtrate quality, as illustrated by studies in which filter aid polymer was used.[22,30,53,55]

Robeck et al.[30] compared performance of pilot filters at 2 to 6 gpm/ft$^2$ (5 to 15 m/h) filtering alum coagulated surface water through single-medium and dual-media filters. They concluded that with proper coagulation ahead of the filters, the effluent turbidity, coliform bacteria, polio virus, and powdered carbon removal were as good at 6 gpm/ft$^2$ (15 m/h) as at 4 or 2 gpm/ft$^2$ (10 or 5 m/h); see Table 8.5. Pretreatment included activated silica when necessary to aid flocculation and a polyelectrolyte as a filter aid (referred to as a *coagulant aid* in the original article).

Conley and Pitman[53] showed the detrimental effect of high filtration rates up to 15 gpm/ft$^2$ (37 m/h) in the treatment of Columbia River water by alum coagulation followed by short-detention flocculation and sedimentation before filtration. A proper dose of nonionic polymer added to the water as it entered the filters, however, resulted in the same filtrate quality from 2 to 35 gpm/ft$^2$ (5 to 85 m/h). *Note:* The turbidity unit being reported in Conley's studies was later acknowledged[54] to be equivalent to about 50 Jackson turbidity units (JTU).

The use of unusually high filtration rates was reported at the Contra Costa County Water District plant in California. By precoating the dual-media filters with a small dose of polymer during the backwash operation, Harris reported successful operation at 10 gpm/ft$^2$ (24 m/h).[22] Harris also reported that the initial period of poorer water quality was eliminated by this precoating operation. The Contra Costa County Water District plant is now authorized by the state of California to operate at 10 gpm/ft$^2$ (24 m/h).

These observations and many similar studies have led to fairly common use of design rates above 2 gpm/ft$^2$ (5 m/h). For example, a 1975 survey identified over 200 plants operating at rates in excess of 3 gpm/ft$^2$ (7 m/h).[56] Although rates above 4 gpm/ft$^2$ (10 m/h) are being used in some plants, flocculant aids or filter aid polymers should be available. In evaluating the effectiveness of treatment at higher fil-

TABLE 8.5 Effect of Filtration Rate on Removal of Coliform Bacteria and Turbidity by Conventional Pretreatment and Double-Layered Filters[30]

| | Plant influent | | | | | | | Filter effluent | |
|---|---|---|---|---|---|---|---|---|---|
| No. of samples | Turbidity, JTU | Coliform bacteria count, no./ML | Alum dose, ppm | Coagulant aid dose, ppm | Settling tank effluent turbidity, JTU | Filtration rate, gpm/ft² | Length of run/h | Turbidity, JTU | Coliform bacteria removal, % |
| 5 | 430 | 19.4 | 100 | 10 | 6 | 2 | >30‡ | 0.07 | 99.8 |
| | | | | | | 4 | 28 | 0.05 | 99.8 |
| 4 | 36 | >120 | 75 | 20 | 2 | 2 | >53‡ | 0.10 | >98.3 |
| | | | | | | 4 | 29§ | 0.09 | >98.3 |
| 5 | 36 | >120 | 75 | 10 | 4 | 2 | 78§ | 0.12 | >99.1 |
| | | | | | | 4 | 30§ | 0.06 | >99.2 |
| 3 | 435 | 25 | 75 | 20 | 5 | 2 | >28‡ | 0.03 | 99.2 |
| | | | | | | 6 | >6¶ | 0.03 | 99.1 |
| 4 | 360 | 25 | 100 | 10 | 5 | 2 | >28‡ | 0.05 | 99.8 |
| | | | | | | 6 | 20 | 0.05 | 99.7 |

†Run stopped when effluent turbidity exceeded 0.5 JTU, except as noted.
‡Run stopped before 8-ft head loss or turbidity breakthrough.
§Run stopped because of 8-ft head loss; no breakthrough.
¶Run stopped because of mechanical problems.

tration rates, a goal turbidity below 0.5 NTU is justified. This idea is based on a growing body of evidence that turbidity is a weak indicator of many particulates in water, as discussed earlier.

### The importance of adequate pretreatment

Adequate and continuous pretreatment is absolutely essential to producing good filtrate. Any interruption of good pretreatment causes almost immediate deterioration of filtrate quality. Fuller's famous work[21] recognized this, as evidenced by this quotation from his 1898 report: "In all cases experience showed that for successful filtration the coagulation of the water as it enters the filter must be practically complete."

Robeck et al.[30] had a good illustration of the effect of a loss of adequate pretreatment from a study at Gaffney, S.C. (Fig. 8.16). Two filters at 2 and 5 gpm/ft$^2$ (5 and 12 m/h) were operating in parallel and producing about equal filtrate turbidity. When the source water suddenly worsened and coagulation was thereby upset, however, the effluent of both filters deteriorated sharply. One other paper demonstrated similar findings in a direct filtration pilot study, as shown in Fig. 8.17.[24] In this case, the coagulant feed was discontinued for 30 min, and filtrate turbidity rose sharply until the chemical feed was resumed.

Thus, all precautions must be taken to ensure that the chemical dosage is adequate and that the feed is reliably maintained. This is even more critical in direct filtration because of the short detention time ahead of the filters.

### Detrimental effect of sudden filtration rate increases

If the rate of filtration is suddenly increased on a dirty filter, the equilibrium is disturbed that exists between the attachment forces holding the solids in the filter and the hydraulic shearing forces tending to dislodge those solids. The result is a temporary flushing of solids deeper into the filter and into the filtrate.

Evidence of this phenomenon presented in 1963[57] showed that the amount of material flushed through the filter was greater for sudden rate increases than for gradual changes (Figs. 8.18 and 8.19). The amount of material released was greater for large increases than for small increases, but the amount was not affected by the duration of the maximum imposed rate. Different types of suspended solids encountered at different water plants exhibited different sensitivities to

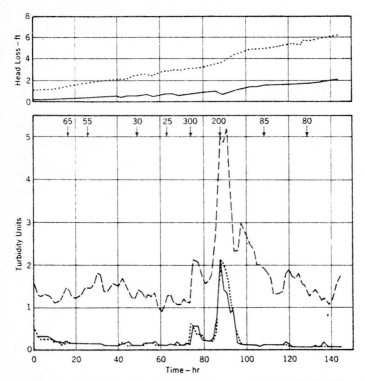

**Figure 8.16**  Turbidity values at filtration rates of 2 and 5 gpm/ft² (5 and 12 m/h). In the upper graph, the dotted curve is for 5 gpm/ft² (12 m/h); the solid, 2 gpm/ft² (5 m/h). In the lower graph, the numbers at the top are the raw-water turbidity values at the time shown during the run; the dashed curve is for filter influent; the dotted, for filter effluent at 2 gpm/ft² (5 m/h); and the solid, filter effluent at 5 gpm/ft² (12 m/h). (*Source: G. G. Robeck, K. A. Dostal, and R. L. Woodward, "Studies of Modification in Water Filtration," J. AWWA, vol. 56, no. 2, February 1964, p. 198.*)

rate increases. These observations are important in many filtration decisions.

The detrimental impact of sudden rate increases on dirty filters also is evident in the work of Tuepker and Buescher[58] and in the studies of DiBernardo and Cleasby.[59] Filtration rate increases on dirty filters should be avoided or made gradually (over 10 min).

Studies by Logsdon et al.[27] showed similar effects when the filtration rate was suddenly increased from 4.5 to 11 gpm/ft² (11 to 27 m/h). Turbidity in the effluent rose sharply and then rapidly declined. *Giardia muris* cyst concentrations followed the turbidity trends. "A four-fold increase in turbidity was accompanied by a twenty-five-fold increase in the cyst concentration in the filtered water." The same

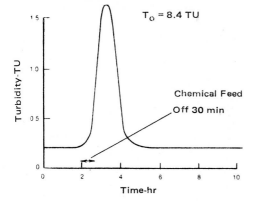

**Figure 8.17** Influence of loss of chemical feed on filter performance. Direct filtration; no flocculation; 15 m/h (6 gpm/ft$^2$); influent turbidity, 8.4 TU; chemical feed, 2 mg/L polymer and 2 mg/L alum (shut down for 30 min 2 h into operation). (*Source: R. R. Trussell, A. R. Trussell, J. S. Lang, and C. H. Tate, "Recent Development in Filtration System Design," J. AWWA, vol. 72, no. 12, December 1980, p. 705.*)

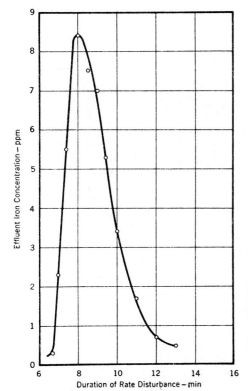

**Figure 8.18** Effect of typical rate disturbance on effluent quality. The iron concentration in the filter effluent builds up rapidly after the disturbance has been initiated. The curve represents run 9a, which had an instantaneous rate change from 2 to 2.5 gpm/ft$^2$ (5 to 6 m/h). (*Source: J. L. Cleasby, M. M. Williamson, and E. R. Baumann, "Effect of Rate Changes on Filtered Water Quality," J. AWWA, vol. 55, no. 7, July 1963, p. 869.*)

study demonstrated the detrimental effect of loss of coagulant feed and of extending the filter run into the period of terminal breakthrough. In both instances, large increases in cyst concentration were observed in the effluent.

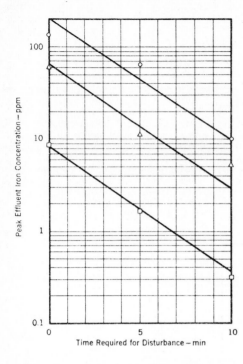

**Figure 8.19** Peak concentration versus disturbance time. The curves indicate a first-order relationship between peak concentration and time required to make the disturbance. At a base rate of 2 gpm/ft$^2$ (5 m/h), circles indicate 100 percent increase; triangles, 50 percent increase; and squares, 25 percent increase. (*Source: J. L. Cleasby, M. M. Williamson, and E. R. Baumann, "Effect of Rate Changes on Filtered Water Quality," J. AWWA, vol. 55, no. 7, July 1963, p. 869.*)

Similar disturbances sometimes occur when a dirty filter is started after brief idle periods, as evident in Fig. 8.20. Filters with automatic rate controllers sometimes exceed the target rate on start-up and disturb previously deposited solids.

Thus, systems that provide no effluent rate manipulation are attractive.[59,60] In a gravity filter without any effluent controller, imposing a sudden increase in filtration rate on the filter is impossible for the operator.

### The detrimental effects of negative head

Some filter arrangements and operating practices can result in pressures below atmospheric pressure (i.e., negative head) in the filter medium during a filter cycle. This can occur in gravity filters when the total head loss down to any point in the filter medium exceeds the static head (i.e., water depth) down to that point. Negative head is more likely to occur if gravity filters are operated with low submergence of the medium and if the filter effluent exits to the clear well below the filter medium.

Negative head is undesirable because dissolved gases in the influent water may be released in the zone of negative pressure, causing gas bubbles to accumulate between backwashing (called *air binding*). Gas accumulations cause more rapid head loss development and

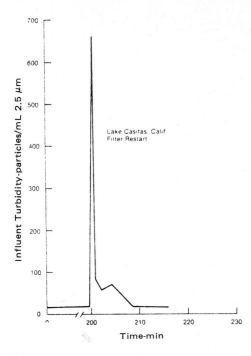

**Figure 8.20** Influence of stop-start on filter performance. Dual media; 15 m/h (6 gpm/ft²); influent particulates ( ≥ 2.5 µm), 2000 particulates/mL; polymer, 1 mg/L; alum, 4 mg/L. (*Source: R. R. Trussell, A. R. Trussell, J. S. Lang, and C. H. Tate, "Recent Development in Filtration System Design," J. AWWA, vol. 72, no. 12, December 1980, p. 705.*)

poorer filtrate quality because of the acceleration of velocity through the voids. Negative head can be completely avoided either by terminating the filter run before the total head loss reaches the submergence depth of the medium or by causing the effluent to exit at or above the surface of the filter medium.

### Head loss development during a filter run

The rate of head loss increase during the filter run is roughly proportional to the solids captured by the filter. Assuming essentially complete capture of incoming solids, the head loss will develop in proportion to the filtration rate $V$ and the influent suspended-solids concentration $C_0$. The rate of head loss development is reduced if the solids capture occurs over a greater depth of the medium, rather than in a thin upper layer of the medium. A coarser grain size encourages greater penetration of solids into the bed and thus reduces the rate of head loss development per unit mass of solids captured.

The most common head loss pattern encountered in rapid filtration is linear with respect to volume of filtrate, or nearly linear, as shown in Fig. 8.21. These data were collected during filtering of groundwater containing precipitated iron. This linear head loss would be typical for most alum or iron coagulated waters.

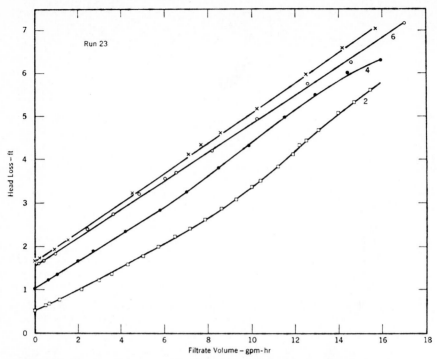

**Figure 8.21** Total head loss and filtrate volume. No optimum-rate tendency appears. Uncontrolled run (along X marks) started at 6 gpm/ft$^2$ (15 m/h) and was allowed to decrease in rate as head loss increased. The numerals adjacent to the graphs are constant rate of filtration in gpm/ft$^2$. (*Source: J. L. Cleasby and E. R. Baumann, "Selection of Sand Filtration Rates," J. AWWA, vol. 54, no. 5, May 1962, p. 579.*)

Another head loss pattern, illustrated in Fig. 8.22, is observed less commonly. This exponential pattern at lower filtration rates is caused by partial capture of solids in a surface cake and/or by very heavy solids removal within a thin layer of the medium at the top. The data in Fig. 8.22 were collected during filtration of a lime-softened water on a conventional rapid sand filter. Increasing the filtration rate reduced the exponential tendency by encouraging greater penetration of solids into the medium, and thus greater production to a given head loss was achieved (Fig. 8.22). The quality of filtrate must also be observed, however, if such filtration rate increases are being considered.

In the absence of plant- or pilot-scale data to assist in predicting head loss development, experience-derived values of mass of solids captured per unit filter area per unit head loss increase are sometimes used for prediction. This is an admittedly simplistic concept. The values used depend on the density of the solids, the ES of the medium where flow enters the filter, and the filtration rate. Typical values

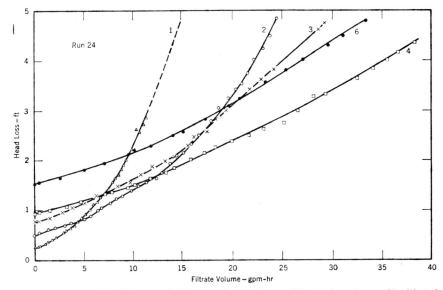

**Figure 8.22** Total head loss and filtrate volume. Ames filter influent, equally diluted with tapwater, was filtered. The numerals adjacent to graphs are constant rates of filtration in gpm/ft². (*Source: J. L. Cleasby and E. R. Baumann, "Selection of Sand Filtration Rates," J. AWWA, vol. 54, no. 5, May 1962, p. 579.*)

range from 0.035 to 0.35 lb/(ft²)(ft) [555 to 5550 g/(m²)(m)],[61] more commonly less than 0.1 lb/(ft²)(ft) [1580 g/(m²)(m)] for flocculent solids using typical potable water filter media and filtration rates. This approach can be applied only where a near-linear head loss pattern is expected.

### Acceptable run length and production per run

While the head loss increases at a faster rate per hour at higher filtration rates, the production per unit time also increases, so that production per filter run is not diminished substantially and production per day is increased substantially. Therefore, the importance commonly attached to the length of filter run can be misleading. Rather than emphasizing the effect of filtration rate on length of run, one should emphasize the effect of filtration rate on filtrate quality, net plant production, and production efficiency.

Figure 8.23 illustrates the effect of unit filter run volume on net water production at three filtration rates.[24] Unit filter run volume is the actual throughput of a filter during one filter run; it can also be called *gross production per filter run*. The figure is based on the assumption that 100 gal/ft² (4 m³/m²) is used for each backwash operation, and 30 min of downtime is required per backwash. Similar figures can be con-

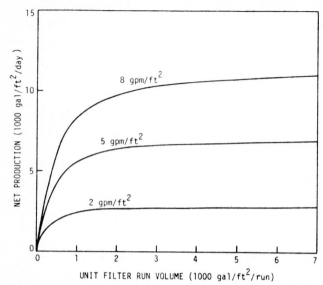

**Figure 8.23** Gross production per square foot per run versus net production, assuming 30 min downtime per backwash and 100 gal/ft² per backwash. (*Source: R. R. Trussell, A. R. Trussell, J. S. Lang, and C. H. Tate, "Recent Development in Filtration System Design," J. AWWA, vol. 72, no. 12, December 1980, p. 705.*)

structed with other assumptions. The following example illustrates the preparation of the figure.

- Assume 4 cycles/day per filter.
- Run time per cycle = 360 min/cycle – 30 min per wash = 330 min/cycle.
- Assume the desired maximum filtration rate = 5 gpm/ft² (12 m/h).
- Unit filter run volume is 5 × 330 = 1650 gal/(ft²)(cycle) [67 m³/(m²)(cycle)].
- Gross volume filtered/(ft²)(day) = 1650 × 4 cycles/day = 6600 gpd/ft² [270 m³/(m²)(day)].
- Net water production/(ft²)(day) = 6600 – backwash volume = 6600 – 400 = 6200 gpd/ft² [250 m³/(m²)(day)].
- Production efficiency expressed as percent of gross volume filtered, is (6200/6600)100 = 94 percent.

In general, allowing for backwash volumes of 100 to 200 gal/ft² (4 to 8 m³/m²) per wash and a downtime of 30 min per wash, the production efficiency will remain high if the unit filter run volume exceeds about

5000 gal/ft$^2$ (200 m$^3$/m$^2$) per run. The calculations presented above are the same whether the backwash water is recovered by recycling it through the plant or is not recovered. Above 5000 gal/ft$^2$ (200 m$^3$/m$^2$) unit filter run volume, the net production increases almost linearly with the rate. Thus, the key issue in considering the impact of high filtration rates on production is not run length, but rather the gross production per run and the filtrate quality. A unit filter run volume of 5000 gal/ft$^2$ (200 m$^3$/m$^2$) means a run length of 2500 min (42 h) at 2 gpm/ft$^2$ (5 m/h) or 1000 min (16.7 h) at 5 gpm/ft$^2$ (12 m/h). So a 42-h run at 2 gpm/ft$^2$ (5 m/h) is no better than a 16.7-h run at 5 gpm/ft$^2$ (12 m/h) because both would yield 98 percent production efficiency.

In summary, high rate filtration does have a definite effect on head loss development and run length, which must be anticipated. The benefits of greater net production per square foot per day at higher filtration rates, however, more than justify that consideration. The most rational way to view the impact of higher filtration rates is by evaluating the impact of unit filter run volume, unit backwash volume, and backwash downtime upon the net production per square foot per day and upon the production efficiency.

### Benefits of dual-media filters over single-medium filters

Early research on rapid sand filters demonstrated that most of the solids were removed in the top few inches of the sand, and thus the full bed depth was not well utilized. This occurs because the smaller sand grains rise near the surface during backwashing and the larger grains sink toward the bottom. A more ideal downflow bed would have a reverse gradation with larger grains on top and finer grains on the bottom. The dual-media filter bed, consisting of a layer of coarser anthracite coal on top of a layer of finer silica sand, was developed to encourage better penetration of solids and thus better bed utilization. It was the first of a series of media developments trying to approach the ideal reverse gradation with flow in the direction from coarser grains to finer grains, to achieve longer filter cycles and hopefully better filtrate quality.

The use of dual-media filters is now widespread in the United States. It is not a new idea. Camp[62] reported using dual media for swimming pool filters beginning about 1940 and later in municipal treatment plant filters.

Baylis[63] described early work in the mid-1930s at the Chicago Experimental Filtration Plant. They experimented with a 2-in (5-cm) layer of crushed quartz over a layer of silica sand. Later they used a

3-in (7.5-cm) layer of 1.5-mm ES anthracite over a layer of 0.5-mm ES sand. This was reported to greatly reduce the rate of head loss development in treatment of Lake Michigan water.

The benefit of dual media in reducing the rate of head loss development, and thus lengthening the filter run, is well proved by a number of studies that are summarized below. The benefit in producing better filtrate quality is less well demonstrated, however.

Conley and Pitman's work[53,54] on Columbia River water in Washington stimulated the increased use of dual media and influenced the choices of grain sizes and depths of the two layers. After comparing several sizes and depths, Conley[54] concluded that 24 in (60 cm) of 0.9-mm ES anthracite and 6 in (15 cm) of 0.43-mm ES sand were the best combination. They did not report parallel experiments with sand and dual media, so the comparison of performance is difficult. But under difficult treatment conditions when breakthrough was observed, head loss of 2 to 2.5 ft/h (0.6 to 0.76 m/h) was reported for sand filters at 6 gpm/ft$^2$ (15 m/h) while various dual-media combinations experienced 0.7 to 1.2 ft/h (0.2 to 0.36 m/h). (See Tables 8.6 and 8.7.) Again, the turbidity values in Table 8.7 must be multiplied by 50 to obtain typical JTU values, as discussed before. Note that the turbidity during these difficult treatment conditions was about the same for the single-medium and dual-media filters and that even under "optimum" dosages of alum and filter aid polymer, end-of-run breakthrough did occur in some cases with both media.

Based on their experiences, Conley and Pitman[64] concluded that the alum dosage should be adjusted to achieve low levels of uncoagulated matter in the filtrate (low turbidity) early in the filter run (after 1 h) and that the filter aid polymer should be adjusted to the minimum level required to prevent terminal breakthrough of alum floc near the end of the run. The dosage of polymer should not be higher than necessary, however, to prevent excessive head loss development.

Some of the research showing the benefits of dual media over a sin-

**TABLE 8.6    Performance of Sand Filters under Breakthrough Conditions with Optimum Treatment†[53]**

| Sand size U.S. sieve | Effluent turbidity, ppm | | Head loss, ft/h (m/h) | Filter conditioner feed, ppm |
|---|---|---|---|---|
|  | Start of run | End of run |  |  |
| 20–30 | 0.005 | 0.1‡ | 2.5 (0.76) | 0.06 |
| 30–40 | 0.004 | 0.005 | 2.0 (0.61) | 0.03 |
| 40–50 | 0.003 | 0.003 | 2.0 (0.61) | 0.02 |
| 50–60 | 0.002 | 0.002 | 2.5 (0.76) | 0.01 |

†A filtration rate of 6 gpm/ft$^2$ (15 m/h) was used.
‡Passing-retained sizes are shown.

**TABLE 8.7**  Performance of Filters under Breakthrough Conditions with Optimum Treatment†[53]

| Filter | Sand | | Anthracite | | Alum feed, ppm | Filter conditioner feed, ppm | Head loss, ft/h (m/h) | Filtrate turbidity, ppm | |
|---|---|---|---|---|---|---|---|---|---|
| | Size U.S. sieve‡ | Depth, in (mm) | Size U.S. sieve‡ | Depth, in (mm) | | | | Start of run | End of run |
| 1 | 20–30 | 8 (203) | 10–40 | 22 (560) | 12 | 0.06 | 1.2 (0.36) | 0.005 | >0.1 |
| 2 | 30–40 | 6 (152) | 10–40 | 24 (610) | 10 | 0.03 | 1.0 (0.30) | 0.004 | 0.005 |
| 3 | 30–40 | 8 (203) | 6–18 | 22 (560) | 10 | 0.05 | 0.7 (0.21) | 0.004 | 0.005 |
| 4 | 30–50 | 6 (152) | 10–30 | 24 (610) | 9 | 0.03 | 0.9 (0.27) | 0.003 | 0.004 |
| 5 | | § | 10–40 | 30 (760) | 18 | 0.06 | 1.1 (0.33) | 0.006 | >0.1 |

Note: 1 Hanford turbidity unit ≈ 50 JTU.
†Filtration rate, 6 gpm/ft² (15 m/h).
‡Passing-retained sizes are shown.
§No sand layer in this column.

gle sand medium was reported by Robeck et al.[30] They compared three filter media during filtration of alum coagulated surface water. These comparisons were made by running filters in parallel so that the benefits of dual media are more conclusively demonstrated. The data in Fig. 8.10 show results under easy summer treatment conditions with strong floc. Note that the dual-media head loss development rate is about one-half that of the sand medium, but the effluent turbidity is almost the same (the dual-media effluent turbidity is slightly inferior). No terminal breakthrough of turbidity occurred.

On the other hand, with weak floc conditions (Fig. 8.11), all three filters experienced terminal turbidity breakthrough, but the dual-media filter breakthrough was much delayed compared to the two single-medium filters. Again, the head loss of the dual-media filter was less than one-half that of the sand filter. They also demonstrated the benefit of using filter aid polymer in retarding terminal breakthrough, as shown in Fig. 8.24.

Tuepker and Buescher[58] reported plant-scale experience that dem-

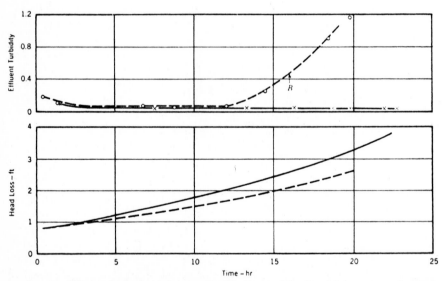

**Figure 8.24** Effect of polyelectrolyte on length of run. The data shown were obtained under the following operating conditions: raw-water turbidity, 10 units; alum dose, 75 mg/L; filtration rate 2 gpm/ft$^2$ (5 m/h); settling tank effluent turbidity, 8 units; and activated carbon, 2 mg/L. In both graphs, the dashed curve is for the filter influent with no polyelectrolyte added; and the solid curve is for 0.08 mg/L polyelectrolyte added. In the upper graph, point $B$ shows the time of filter breakthrough, 16 h. The length of run with polyelectrolyte added was more than 22 h. (*Source: G. G. Robeck, K. A. Dostal, and R. L. Woodward, "Studies of Modification in Water Filtration," J. AWWA, vol. 56, no. 2, February 1964, p. 198.*)

onstrated similar benefits of dual media over a sand medium in providing longer filter runs and 3 times the water production per unit head loss. Unfortunately, the filters were operated at different filtration rates so that filtrate quality comparisons are not conclusive.

The evidence presented above clearly demonstrates the lower head loss expected for a dual-media filter compared to a conventional sand filter. The data would indicate that for a typical dual-media filter with anthracite ES that is about double the sand ES, the head loss development rate should be about one-half the rate of the sand filter. In some cases, it may even be less than one-half. One strong benefit that was demonstrated is a delayed time of breakthrough for the dual-media filter compared to the sand filter.

### Comparison of triple-media versus dual-media filters

The benefits gained by the use of dual-media filters led to the development of the triple-media filter in which an even finer layer of high-density media (garnet or ilmenite) is added as a bottom layer.[65,66] The bottom layer of finer material would be expected to improve the filtrate quality in some cases, especially at higher filtration rates. The evidence to support this expectation is not conclusive.

This type of filter is sometimes referred to as a *mixed-media filter* because the sizes and uniformity coefficients of the three layers are selected to encourage substantial intermixing between adjacent layers. This is done to come closer to the presumed ideal configuration of "coarse to fine" filtration.

The head loss development of a filter is dominated by the grain size of the upper layer. Thus, if the upper grain sizes of a triple- and a dual-media filter are the same, the head loss increase during the run will be nearly identical for the two filters. The initial clean bed head loss will be higher for the triple-media filter, however. Depending on the depth and grain size of the fine garnet or ilmenite layer added to the triple media filter, the initial clean bed head loss may be substantially higher. Thus, for a plant with a particular total head loss, the head loss that remains available to develop during the run is reduced, which may shorten the run compared to the run obtained with a dual-media filter.[67]

Some early plant-scale comparisons of triple media with sand filters were inconclusive because of the research design.[68–70] In some cases, the sand filter was operated at 2 gpm/ft$^2$ (5 m/h) and the triple-media filter at higher rates.[68,70] If the filtrate was comparable, the triple-media filter was found to be better. Both filters should be operated at the same rates in parallel, and both filtrate and head loss should be compared. In other cases, filter aid polymer was used for the triple-media filter and not for the sand filter.[69]

Some later comparisons of triple-media versus dual-media filters have reported mixed-media filters to be superior. On Lake Superior water at Duluth, a mixed-media (triple-media) filter was superior to a dual-media filter in amphibole fiber removal.[34,71] Out of 32 filtrate samples 29 were below or near detection level for the dual-media filter, as were 18 out of 18 for the mixed-media filter. The mixed-media filter was recommended for the plant. The mixed-media filter was also reported to be superior in resisting the detrimental effects of flow disturbances.[72]

Other later studies did not find mixed-media filters to be superior to dual-media ones.[73,74] For example, Kirmeyer[73] reported pilot studies for the Seattle water supply in which two mixed-media and two dual-media filters were compared. No difference in filtrate quality was observed in either turbidity or asbestos fiber content with filtration rates of 5.5 to 10 gpm/ft$^2$ (13 to 24 m/h). Some differences in production per unit head loss were observed to favor dual-media filters at lower rates and mixed-media filters at higher rates.

A number of laboratory studies have compared triple-media and single-medium filters[75,76] or single-medium or triple-media filters arranged in normal or reverse-graded manner.[77,78] All these studies have shown clearly the head loss benefit gained by filtering in the direction of coarse grains to fine grains. Three of the studies also clearly showed benefits to the filtrate quality.[75–77] Surprisingly, no good laboratory filtration studies comparing dual media and triple media have been found.

From the foregoing references, the question of whether triple media are superior to dual media in producing better filtrate quality has not yet been fully answered. Theory would predict that it should be superior in filtrate quality, but studies have not yet shown this consistently.

## Direct Filtration

### Process description, advantages, and disadvantages

Direct filtration is a surface water treatment process that includes coagulant addition, rapid mixing, flocculation, and filtration. In some cases, the flocculation tank is omitted, and the process is referred to as *direct in-line filtration* with flocculation occurring in the filter itself.

The use of direct filtration for good-quality surface water is increasing because it offers several advantages over conventional treatment for such water. Capital costs are lower because no sedimentation tank is required. Lower coagulant dosages are generally used in direct filtration with the goal of forming a pinpoint-sized floc that is filterable, rather than a large, settleable floc. Therefore, direct filtration results in lower chemical costs compared to conventional treatment and lower sludge production, resulting in lower costs for sludge treatment and disposal. Direct filtration also results in lower operation and maintenance costs because the sedimentation tank (and sometimes the flocculation tank) need not be powered or maintained.

Several disadvantages to direct filtration exist, however. It cannot handle waters high in turbidity and/or color. Less response time for the operator to respond to changes in source water quality is available, and there is less detention time for seasonal taste and odor control efforts.

Automatic control instrumentation is very important in a direct filtration plant. Source and finished water quality monitoring are needed to alert the operator to changes, and a fail-safe plant shutdown is desirable in the event that the finished water does not meet treatment goals. If taste and odor episodes are expected, a pretreatment basin may be needed to provide the needed contact time with powdered activated carbon or oxidizing chemicals.[64]

### Appropriate source water for direct filtration

The desire to use direct filtration has prompted a number of efforts to define acceptable source water for the process.[79–83] The key issues in determining appropriate water for direct filtration are the type and dosage of coagulants needed to achieve desired filtrate quality as demonstrated by pilot- or full-scale observations. If alum is the primary coagulant, Hutchison[81] suggested that 12 mg/L of alum on a continuous basis would yield 16- to 20-h filter runs at 5 gpm/ft$^2$ (12 m/h). This was considered sufficiently long to limit backwash water volume to 4 percent of the product. By the partial substitution of polymer for alum, water with up to 25 color units (CU) was considered a suitable candidate. Hutchison also suggested that diatom levels from 1000 to 2000 areal standard units per milliliter (asu/mL) required use of coarser anthracite and more frequent use of polymer to prevent breakthrough. One areal standard unit is $20 \times 20$ μm or 400 (μm)$^2$. Anthracite with 1.5-mm ES could handle diatoms of 2500 asu/mL and produce 12-h runs at 5 gpm/ft$^2$ (12 m/h).

Wagner and Hudson[82] suggested the use of jar tests and filtration through standard laboratory filter paper as a screening technique to select appropriate water for direct filtration. Water requiring more than 15 mg/L of alum (as filter alum) to produce acceptable filtrate quality was a doubtful candidate for direct filtration. Water requiring only 6 to 7 mg/L of alum and a small dose of polymer was considered a favorable candidate. If the results of this bench-scale screening test are favorable, then pilot plant tests are needed to determine full-scale design parameters.

An American Water Works Association (AWWA) committee report[79] defined water meeting the following criteria as a "perfect candidate" for direct filtration:

| | |
|---|---|
| Color | < 40 CU |
| Turbidity | < 5 NTU |
| Algae | < 2000 asu/mL |
| Iron | < 0.3 mg/L |
| Manganese | < 0.05 mg/L |

Cleasby et al.[83] considered the AWWA committee guidelines acceptable except for turbidity, which they considered too low. During low algae seasons, they suggested turbidity limits of 12 NTU when alum is used alone or 16 NTU when cationic polymer is used alone. During high algae seasons, the suggested limits were 7 NTU with alum alone and 11 NTU with cationic polymer alone. Best performance was achieved during low algae seasons with alum dosages between 5 and 10 mg/L. During high algae seasons, dosages up to 20 mg/L were attempted with only moderate success and with runs as short as 12 h at 3 gpm/ft$^2$ (7 m/h).

### Chemical pretreatment for direct filtration

The selection of the coagulant dosage for direct filtration is best determined by full-length filter cycles or pilot- or full-scale filters. Jar tests are often misleading because the goal in direct filtration is to form small pinpoint floc that is barely visible but will filter effectively. Thus, the usual criteria used to judge jar test observations such as formation of a large floc or a clear supernatant liquor, after settling, are not appropriate for direct filtration. The use of the jar test and filter paper filtration technique[82] is somewhat better, but it provides no information on terminal breakthrough behavior or head loss generation. Substitution of a miniature granular filter for the filter paper[84] suffers from similar weaknesses. Kreissl et al.[85] demonstrated that a clean filter could be operated at very high rates to simulate the hydraulic shear that would exist in the more clogged layers of the filter late in the filter run and thus could predict whether the chemical dosage would be successful through the full length of the filter run. In spite of these various techniques of using bench or pilot tests to select chemical dosages, the most common approach involves confirmation of the selected dosage with pilot-scale or full-scale filter runs of full duration at the design filtration rate.

When alum or iron salts alone are used, the optimum dosage of coagulant for direct filtration is the lowest dosage that will achieve filtrate quality goals. An optimum dosage is normally not observed based on filtrate quality alone. Rather, as the dosage is increased, the filtrate gets marginally better, but head loss develops at a higher rate and early breakthrough is encouraged.[83] One advantage of alum or iron coagulants is that the dosage is less sensitive to source water quality.

When cationic polymer alone is used as a primary coagulant, a distinct optimum dose may be found that produces the best-quality filtrate. In pilot studies, one method of detecting whether a particular dosage is above or below the optimum is to shut off the polymer feed for a few minutes. If the filtrate improves momentarily, the prior dose is excessive. If the filtrate deteriorates immediately, the prior dose was at or below optimum.[83]

A number of studies have reported successful use of alum and cationic polymers simultaneously. Typical dosages are 2 to 10 mg/L of alum plus about 0.2 to 2 mg/L of cationic polymer.[83,86–88] These dosages are drawn from pilot- and plant-scale experience.

The impact of flocculation on direct filtration has been the subject of several studies that have been summarized by Cleasby et al.[83] These studies can assist in deciding whether flocculation should be provided in direct filtration. By comparing operation with and without flocculation, the provision of flocculation was found to improve filtrate quality before breakthrough, shorten the initial improvement period of the filter cycle, and reduce the rate of head loss development, but result in earlier breakthrough. Even though the head loss is reduced, the run length may be shortened because of early breakthrough. Current practice is to include a short flocculation period in direct filtration plants, typically about 10 min of high-energy flocculation at a velocity gradient $G$ up to $100 \ s^{-1}$, although the range of flocculation provisions in existing plants has been very great, from no flocculation up to 60 min of flocculation.[89]

In-line filtration is appropriate for source water that is consistently very low in turbidity and color. A long flocculation time is required for this water because of poor collision opportunity. Therefore, this water is more economically treated by coagulant addition and an extended rapid mix of 3 to 5 min to achieve only the initial stages of particle aggregation before filtration. Few in-line plants are being built because flocculation adds flexibility to the plant operation, and a flocculation bypass can be provided for contingencies.

### Filter details for direct filtration

Two summaries of direct filtration[79,89] indicated filtration rates of full-scale operating plants between 1 and 6 gpm/ft$^2$ (2.4 and 15 m/h). Direct filtration pilot studies at three locations showed, however, that effluent turbidity was nearly constant over the range from 2 to 12 gpm/ft$^2$ (5 to 29 m/h) and in one case as high as 18 gpm/ft$^2$ (44 m/h).[24] In these studies both alum and cationic polymer were used for chemical pretreatment. Dual media and deep beds of coarse sand were compared.

An unusual direct filtration plant is operated by the Los Angeles Department of Water and Power to treat Los Angeles aqueduct water. The plant design was based on extensive pilot studies that led to the choice of deep beds of anthracite (6 ft, 1.8 m) with an ES of 1.5 mm and UC <1.4. The design filtration rate is 13.5 gpm/ft$^2$ (33 m/h). Pretreatment includes ozone to assist coagulation with ferric chloride and cationic polymer, and flocculation is provided before filtration. The filters are backwashed by air alone first at 4 ft$^3$/(min)(ft$^2$) (1.2 m/min), followed by air plus water at 10 gpm/ft$^2$ (24 m/h). The air is terminated before overflow occurs to prevent medium loss, and the water wash is continued alone during overflow.[23] Some of the pilot studies were reported by Trussell et al.[24] Certainly, this facility represents a marked departure from current direct filtration practice.

## Flow Control in Filtration

### Why control is needed

The need for some method of filter flow control has been recognized since the early days of potable water filtration. This is reflected in the quote from the 1917 textbook by Ellms: "The rate of filtration should be carefully controlled and rapid fluctuations in rate made impossible."[90] If no attempt is made to control flow, the filters may not share the total plant flow in a reasonably equal manner or sudden changes in flow rate may occur, both of which could cause filtrate quality to suffer.

Filter control systems currently in use can be divided into two categories: mechanical control systems and nonmechanical systems that achieve control by the inherent hydraulics of the operating filters. Mechanical control devices have been most common. Some mechanical systems have been less than satisfactory because of inherent faults in design or excessive problems of maintenance to keep them operational. For that reason, a number of studies comparing alternative systems have been published.[59–60,91–98] Results of these studies form the basis of the following discussion of the various filter control systems.

### Rate control for gravity filters

Filter control alternatives for gravity filters can be separated into four main options; see Fig. 8.25.[98] The first and third systems are mechanical control systems. The other two systems depend on the hydraulics of the filter for control. The rate patterns and water-level patterns of the four systems during constant total plant flow rate are shown in Fig. 8.26. The first three systems provide constant-rate operation, and the fourth system provides declining-rate operation; see Fig. 8.26. If total flow through the plant is varied to meet varying demands of the community, then none of the systems will be truly a constant-rate or declining-rate system. Variable plant flow is a rather common situation, and filtration rates must then go up or down to meet changes in the imposed plant load.

The short-term increases in rate and water level in Fig. 8.26 are the result of removing one of the four filters from service for backwashing, with the other three filters remaining in service picking up the load of the filter removed from service. These peaks can be avoided by placing a clean filter on-line as the dirty filter is taken off-line for backwashing. After backwashing, the clean filter is held in reserve until the next backwash is required.

The special features of each of the four systems are described briefly in the following paragraphs. This is taken largely from an AWWA report on rate control alternatives.[98] After these brief descriptions, the common features of the system are described.

**Variable-controlled constant-rate system (Figs. 8.25a and 8.26a).**  This system uses a rate-of-flow controller (consisting of a flow sensor, controller, and control valve) that regulates flow through each filter based on plant loading. The controller compares the measured variable (flow rate or depth in the inlet channel) with the desired value and drives the final control element (valve) to maintain the desired value. The control system maintains equal flow from each filter if the total plant flow rate is changed.

**Inlet-split, rising head system (Figs. 8.25b, 8.26b).**  In this system, each filter receives an equal (or nearly equal) portion of the total flow. This is achieved by splitting the flow by means of an inlet weir box or ori-

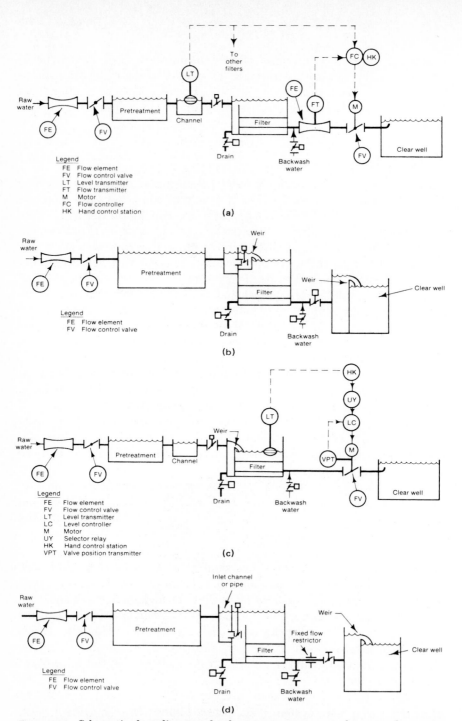

**Figure 8.25** Schematic plant diagrams for the most common control systems for gravity filters showing typical control elements provided. (*a*) Variable-controlled constant-rate system; (*b*) inlet split-rising heads; (*c*) inlet split-flow control from filter water level; (*d*) variable declining-rate system. (*Source: Committee Report, "Comparison of Alternative Systems for Controlling Flow Through Filters," J. AWWA, vol. 76, no. 1, January 1984, p. 91.*)

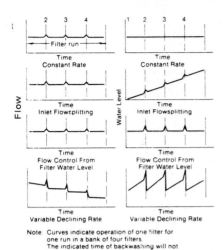

**Figure 8.26** Pattern of flow and filter water levels with constant total plant flow. (*Source: Committee Report, "Comparison of Alternative Systems for Controlling Flow Through Filters," J. AWWA, vol. 76, no. 1, January 1984, p. 91.*)

fice on each filter inlet above the maximum water level of the filter. The filter effluent discharges to the clear well at a level above a surface of the filter medium.

As solids accumulate in the filter medium, the water level rises in the filter box to provide the head required to drive the flow through the filter medium. The water level in each filter box is different and depends on the extent to which the filter medium is clogged. When the water level in a filter box reaches the maximum, that filter must be backwashed. This system requires no instrumentation for flow rate and head loss measurement on individual filters.

**Inlet split-flow control from filter water level (Figs. 8.25c, 8.26c).** This control system is a variable-controlled constant-rate system without the flow element on the outlet of each filter. The flow of filtered water is controlled by modulating a butterfly valve on the filter effluent pipe to maintain a constant water level in the filter box. Some system is required to ensure that the flow is split equally among the operating filters such as an inlet flow-splitting weir on each filter. This system does not require flow measurement instrumentation on individual filters.

**Variable declining-rate system (Figs. 8.25d, 8.26d).** In this system, flow enters the filter below the normal water level in each filter and discharges to the clear well above the level of the filter medium. Because the inlet to the filters is below the normal water level in each filter, all filters connected by a common inlet channel or pipe operate at approximately the same water level and thus have the same available head.

Therefore, the cleanest filter operates at the highest filtration rate, and the dirtiest filter operates at the lowest filtration rate.

As solids accumulate in the filter media, the water level rises in all connected filter boxes to provide the head required to drive the flow through the filter media. When the level reaches some upper desired limit, the dirtiest filter is backwashed. The filtration rate declines in a stepwise fashion. As each clean (backwashed) filter is returned to service, it assumes the highest rate of flow and all the other filters step down to lower rates of filtration. The dirtiest filter assumes the lowest rate of flow until it is backwashed.

The process requires no instrumentation for flow rate or head loss measurement on individual filters. The rate of flow through each filter is unknown, however, unless flowmetering is provided.

Less total head is needed across the filter because the rate declines. Head that was consumed in turbulent head losses in the underdrains and piping during the early high-rate part of the cycle decreases with the square of the filtration rate and becomes available for clogging head loss later in the run.

As the rate of flow declines in a dirty filter, the hydraulic shear stresses on the solids deposited in the filter medium near the end of a filter run are decreased compared with those in the constant-rate filtration process. Therefore, the declining-rate filter has less tendency for terminal breakthrough than the constant-rate filter.

**Common elements of filter control systems.**    Some common elements of the various rate control systems should be recognized. The total head provided to operate the gravity filter is the vertical distance from the water level (hydraulic grade line) in the filter inlet conduit to the water level at the downstream control point. The downstream control point is the overflow weir to the clear well for the two nonmechanical systems and is the upturned elbow or clear well level (whichever is higher) for the two mechanical systems.

If the downstream control point is located above the top surface of the medium, no possibility of having pressures below atmospheric pressure (i.e., negative head) exists anywhere within the filter medium or underdrain system because the static water level is above the surface of the medium. Also no possibility exists of accidentally allowing the filter water level to drop below the top surface of the medium (i.e., dewatering the filter). These advantages are not obtained without cost, however, because all the head needed to operate the filter must be placed above the downstream control level, and this means a deeper filter box.

The three constant-rate systems have the common characteristic that the entire operating head provided, which is essentially constant, is consumed throughout the filter run. The head that is not utilized by the dirty filter medium is wasted either in the control valve of the mechanical systems or in free fall into the filter box of the inlet split-rising head system. The declining-rate system has a lower total head requirement in the plant hydraulic profile because the total head is reallocated as the filter cycle proceeds, as described earlier.

The two nonmechanical systems have the inherent advantage that sudden changes in the filtration rate cannot be imposed on the filter. If the total plant flow is increased or a filter is removed from service, the other filters can pick up the load only by changing the water level to generate the head needed to accommodate the increased flow. This changing of water level takes time, so that the rate change occurs slowly and smoothly without mechanical devices.

Slow, smooth rate changes are also possible in properly designed mechanical control systems. Good mechanical filter control systems should have the following attributes. The rate of change of the control valve position should be proportional to the divergence of the measured variable (flow rate or depth) from the desired value. By making the valve changes slowly, some water-level variation is allowed, to avoid abrupt changes in rate. The drive for the control valve should be an electric motor rather than a pneumatic or hydraulic valve drive because the latter two drives are prone to sticking, overshooting, and hunting problems as they age. No industry standards for filter control systems are available, and so some systems do not have all the attributes listed above.

**Choice of appropriate control system.**   With the foregoing information at hand, it is possible to make some generalizations about the four systems that may affect the choice of system to meet the desired goals and the utilities' needs. Each system has a place, and a particular system should be used only when conditions are appropriate.

If full plant automation is desired, mechanical systems will be favored, although nonmechanical systems can be partially automated. If minimizing mechanical equipment is desired, nonmechanical systems will be favored. This will be especially important when equipment and future repair parts must be imported to a developing country that has scarce foreign exchange. In small nonautomated plants with four or fewer filters and with unskilled operators, the inlet split-rising head system is ideal because it is so simple to understand. In larger, non-automated plants, a variable declining-rate system may be favored because it saves head and produces better filtrate.

### Rate control for pressure filters

Rate control is just as important for pressure filters as for gravity filters because the same filtration mechanisms are functioning in both cases. Fewer options are available for pressure filters, however, because the pressure filter operates full of water under pressure. As a result, options involving a change in water level are not available, and influent gravity flow splitting is not available. Therefore, the benefits of slow rate changes by allowing water levels to change are not available to pressure filters.

The usual arrangement for a bank of pressure filters is based on the assumption that flow through the system will self-equalize. The total filter area of the filter bank is sized appropriately to meet local regulatory requirements based on feed pump capacity. The system is designed symmetrically so that flow is distributed equally to each filter in the bank. If one filter is passing more flow than it should, that filter is assumed to clog more quickly than the other filters, and the flow will be reduced because of the clogging resistance. Thus, the system is considered self-equalizing during operation. Individual filter flow controllers are not provided.

## Backwashing of Rapid Filters

Effective backwashing of rapid filters is essential to long-term successful service of filters. The goal of the backwashing operation should be to keep the filter acceptably clean, so that no progressive evidence of the development of dirty-filter problems, such as mud balls and filter cracks, occurs.

### Alternative methods of backwashing

The backwashing system is the most frequent cause of filter failure. Therefore, the selection of the backwashing system and the proper design, construction, and operation of that system are key elements in the success of the treatment plant. A number of alternative systems of backwashing are currently in use.[99]

**Upflow wash with full fluidization.**  The traditional backwash system in the United States uses an upflow water wash with full-bed fluidization. Backwash water is introduced into the bottom of the bed through the underdrain system. It should be turned on gradually over at least a 30-s interval to avoid disturbing the gravel layers or subjecting the underdrain to sudden momentary pressure increases. The filter medium gradually assumes a fluidized state as the backwash flow rate is increased and the bed expands. Typical backwash rates are 15 to 20 gpm/ft$^2$ (37 to 49 m/h) for the sizes of the media used in U.S. practice, and the resulting bed expansion is 15 to 30 percent (see the fluidization calculation procedures presented earlier). The backwash flow is continued with full fluidization until the waste wash water is reasonably clear; a turbidity of about 10 NTU is sufficient. Then the supply valve is shut off. Shutoff is not as crucial as the opening because no danger to underdrain or gravel exists. A slow shutoff will result in a greater degree of restratification.

According to Baylis, backwash by water fluidization alone is a weak washing method that does not solve all dirty-filter problems.[100] The reason for that weakness, discussed by Amirtharajah in a later paper,[101] is attributed to a lack of any abrasion occurring between the grains in a fluidized bed. For that reason, backwashing is usually assisted by an auxiliary scour system such as surface wash or air scour. Surface wash and air scour systems improve the backwashing operation by dissipating more energy in the bed during backwashing. This energy is consumed in abrasion between grains and in higher hydraulic shear forces in the bed.

**Surface wash plus fluidized-bed backwash.**    Surface wash has been extensively and successfully used to improve the effectiveness of fluidized-bed backwashing. Surface wash systems inject jets of water from orifices located about 1 to 2 in (2.5 to 5 cm) above the fixed-bed surface. Surface wash jets are operated for 1 to 2 min before the upflow wash and usually are continued during most of the upflow wash, during which time they are immersed in the fluidized filter medium. Surface wash is terminated 2 or 3 min before the end of the upflow wash.

Surface wash is accomplished either with a grid of fixed pipes placed above the granular medium or with rotary water distribution arms, containing orifices or nozzles that supply high-pressure jets of water. Orifice sizes are typically $3/32$- to $1/8$-in (2- to 3-mm) diameter and are directed downward 15 to 45° below the horizontal. Operating pressures are typically 50 to 75 psig (350 to 520 kPa).

The fixed-nozzle system consists of horizontal header pipes suspended from the wash water troughs and provided with vertical pipes closed by caps containing the orifices. The vertical pipes are on a center-to-center distance of about 2 to 3 ft (0.6 to 0.9 m) with the orifices about 2 in (5 cm) above the media. They discharge at 2 to 4 gpm/ft$^2$ (5 to 10 m/h).

In the rotary system, the filter is swept by a revolving pipe suspended at its center and is provided with nozzles at about 4- to 8-in (10- to 20-cm) spacing. These discharge at 0.5 to 1 gpm/ft$^2$ (1.2 to 2.4 m/h).

The use of surface wash has a number of advantages and disadvantages.[99] The advantages include the following:

1. Surface wash is relatively simple because only a source of high-pressure water is needed in conjunction with a system of distribution nozzles.

2. It is accessible for maintenance and repair because it is located above the media surface of the fixed bed.

Some disadvantages of surface wash systems are as follows:

1. Rotary-type washers sometimes stick in one position temporarily and do not rotate as intended.
2. If mud balls (to be discussed later) do form in the bed and reach sufficient size and density, they can sink into the fluidized bed and no longer come under the action of the surface wash jets.
3. Fixed-nozzle surface wash systems obstruct convenient access to the filter surface for maintenance and repair.

**Air scour–assisted backwash.** Air scour systems supply air to the full filter area from orifices located under the filter medium. Air scour is also used to improve the effectiveness of the backwashing operation. A substantial danger of losing the medium exists, however, if air is used during overflow, so the system must be properly designed and operated to avoid such loss. For fine sand, dual-media, or multimedia filters, air scour is applied before the water backwash with fluidization, and it may be continued at the beginning of the water backwash prior to overflow. For coarse sand filters (> 0.9 mm ES), air scour may be used simultaneously with a water backwash at subfluidization velocities during backwash overflow and is continued in this manner during most of the backwash period. The simultaneous air-plus-water backwash provides an additional benefit to the cleaning action because the interstitial water velocities are substantially increased because of the presence of the rising air bubbles.

In the case when air scour is used before the water backwash, the filter water level is first lowered at least 6 in (15 cm) below the washwater overflow level. When the air scour is started, the water level will rise because of the volume occupied by the air, but the level should remain below the lip of the backwash troughs to prevent loss of filter medium. The air scour period is typically 2 to 5 min. The air scour is terminated, and the water backwash starts to slowly expel the air from the bed before overflow begins. The water backwash is then continued alone with full-bed fluidization until the wash water is reasonably clear.

Air scour may also be used for a portion of the water backwash prior to overflow by draining the filter to nearly the surface of the filter medium. Air alone is usually applied first, and then a low rate of water backwash is added below the rate for full bed fluidization. This combined air-water backwash is a very effective backwash method, but it can be continued only until the water level is about 6 in (15 cm) below the wash-water overflow. At that level, the airflow must be terminated so that all air escapes from the bed before overflow commences. The water backwash is then continued alone with full-bed fluidization until the wash water is reasonably clear. When air scour is applied to dual-media or multimedia filters, the period of fluidization is essential to restratify the layers after the air scour period.

When air scour and water backwash are used simultaneously for coarse sand filters, the water rate is well below the fluidization velocity of the sand. After about 10 min of simultaneous air-plus-water backwash, the airflow is terminated, and the water continues to expel some of the air from the bed and to flush the remaining dirt from the water above the filter medium. During the terminal water backwash period, the flow rate may be increased but remains below the full-bed fluidization velocity. Contrary to expectations, this method of washing is very effective, even though the bed is never fluidized. A slow transport of the grains occurs, caused by the simultaneous air-plus-water flow that causes abrasion between the grains. This abrasion plus high interstitial water velocities results in an effective backwash.

Serious danger exists that the filter sand may be lost if this backwash method is not applied as follows. To prevent the loss of sand, the water and airflow rates are varied appropriately for the size of the sand, and the vertical distance from the sand surface to the wash-water overflow should be at least 1.6 ft (0.5 m). Backwash troughs are not generally used, and the dirty wash water exits over a horizontal concrete wall. The top edge of the horizontal wall is sloped 45° downward toward the filter bed so that any sand grains that fall on the sloping wall during backwashing will roll back into the filter bed.[102] Alternatively, if backwash troughs are used, specially shaped baffles can be located around each trough to prevent the loss of sand or other filter media.[103]

Air scour may be introduced to the filter either through a pipe system that is completely separate from the backwash water system or through the use of a common system of nozzles (strainers) that distribute both the air and water, either sequentially or simultaneously. In either method of distribution, if the air is introduced below graded gravel supporting the filter medium, there is concern over the movement of the finer gravel by the air, especially by air and water used concurrently, by intention or by accident. This concern has led to the use of media-retaining strainers (Fig. 8.7) in some filters which eliminates the need for graded support gravel in the filter. These strainers may clog with time, however, causing decreased backwash flow capability or, possibly, structural failure of the underdrain system.

When air scour is first used alone, followed by water backwash for typical fine filter sands of about 0.5-mm ES, the following flow rates have been used. In Britain, airflow rates of 1 to 2 scfm/ft² (18 to 36 m/h) followed by water alone at 5 to 8 gpm/ft² (12 to 20 m/h) are traditional. In the United States, for dual-media filters with about 1.0-mm ES anthracite over 0.5-mm ES sand, typical airflow rates are 3 to 5 scfm/ft² (55 to 91 m/h). This is followed by water backwash to achieve full-bed fluidization with typical rates of 15 to 20 gpm/ft² (37 to 49 m/h) depending on the water temperature.

When air scour and water backwash are used simultaneously for somewhat coarser sands of about 1.0-mm ES, airflow rates of 2 to 4 scfm/ft² (37 to 73 m/h) and simultaneous water flow rates of 6 gpm/ft² (15 m/h) represent typical practice. For coarser sands of 2.0-mm ES, airflow rates are increased to 6 to 8 scfm/ft² (110 to 150 m/h) with simultaneous water flow rates of 6 to 8 gpm/ft² (15 to 20 m/h).

Hewitt and Amirtharajah[104] did an experimental study of the particular combinations of air and subfluidization water flow that caused the formation and collapse of air pockets within the bed, a condition they called "collapse-pulsing." This condition was presumed to create the best abrasion between the grains and the optimum condition for air-plus-subfluidization-water backwashing. An empirical equation relating airflow rate, fluidization velocity, and backwash water flow rate was presented. In a companion paper,[105] Amirtharajah developed a theoretical equation for collapse-pulsing, using concepts from soil mechanics and porous-media hydraulics. The development involved equating the air pressure within the bubble to the soil stresses in an active Rankine state plus the pore-water pressures. The resulting equation was

$$0.45Q_a{}^2 + 100\left(\frac{V}{V_{mf}}\right) = 41.9 \qquad (8.15)$$

in which $Q_a$ is the airflow rate in standard cubic feet per minute per square foot and $V/V_{mf}$ is the ratio of superficial water velocity to minimum fluidization velocity based on the $d_{60}$ grain size of the medium.

The empirical and the theoretical equations give almost the same results over the range of airflow rates from 2 through 6 scfm/ft² (37 to 110 m/h). For a given airflow rate, both equations predict somewhat higher water flow rates than the typical values stated previously; the current-practice values should be used until future research reconciles the difference.

Some of the advantages of air scour auxiliary, in contrast with surface-wash systems, are the following:[99]

1. Air scour auxiliary covers the full area of rectangular filters and is adaptable to any filter dimensions.

2. It agitates the entire filter depth. Therefore, it can agitate the in-

terfaces in dual-media and multimedia beds and can reach mud balls that have sunk deep into the filter.

Some disadvantages of the air scour auxiliary are:

1. A separate air blower and piping system is needed.
2. There is the potential for loss of media, especially if air and water are used simultaneously.
3. A greater possibility of moving the supporting gravel exists if air is delivered through the gravel concurrently with water. Special gravel designs are required, as described earlier.

**Relative effectiveness of backwashing methods.** The effectiveness of the backwashing methods described previously has been evaluated in various studies.[99] The most effective backwash is achieved by simultaneous air scour and subfluidization water backwash. The use of air scour prior to water fluidization backwash is about equal in effectiveness to the use of surface wash to assist the backwash. The use of water backwash alone to fluidize the bed is the weakest washing system.

**Backwash troughs and wash water required**

The volume of wash water required to wash a filter will depend on the depth of the filter medium and the depth to which solids have penetrated the medium. The larger these depths, the larger the water volume needed to flush the dirt out of the medium and into the space above the medium. The larger the vertical distance from the surface of the medium to the wash-water overflow level, the larger the volume needed to wash the dirt out of the filter box.

The typical U.S. washing system is achieved with full fluidization, and backwash troughs are usually provided. They are located to limit the horizontal travel of dirty wash water to about 3 ft (0.9 m), and the top edge of the troughs is usually 2.5 to 3 ft (0.75 to 0.9 m) above the fixed-bed surface. This keeps the bottom of the troughs above the expanded medium during backwashing.

The flow pattern from the surface of the expanded medium to the trough edge is nearly vertical and plug flow in appearance. Transport of solids in this vertical path is quite effective with little observed sedimentation of heavy solids. Furthermore, the close spacing of the troughs minimizes the size of the dead areas along the walls and between the troughs that delays the washout of dirty water. The volume required to wash a typical filter of this type is about 100 to 150 gal/ft$^2$ (4 to 6 m$^3$/m$^2$).

In contrast, a European type of filter is washed with air and water simultaneously and has no backwash troughs. The water is allowed to travel to an overflow wall and directly into a wastewater gullet. The horizontal travel distance is usually limited to about 13 ft (4 m), and the filters are longer and narrower to limit this travel distance. As mentioned previously, the vertical distance from the fixed-bed surface to the overflow is at least 1.6 ft (0.5 m), to minimize loss of filter media.

The water volume above the medium of a European filter is almost completely mixed by the violent action of the air and water flowing simultaneously. From a theoretical analysis of a completely mixed tank, washout will occur with the same pattern whether it is withdrawn from a single outlet or multiple outlets. Resedimentation is not a problem during the horizontal transport because of the violent air-plus-water action. The total wash-water volume required is about the same as for the U.S. type of filter.

In filters with backwash troughs and washed with full-bed fluidization, the expanded media surface should be lower than the bottom of the troughs. If the expanded media rise up between the troughs, the effective vertical-flow velocity is increased and the danger of carrying the filter medium into the troughs is increased.

The wash-water volume required per wash to clean the filter could be reduced by decreasing the vertical distance from the fixed-bed surface up to the top edge of the troughs. The temptation to do this must be resisted, however, because the danger of loss of filter medium would be increased. Loss of filter medium is greater when anthracite is used. In this case, the vertical distance to the trough edges should be increased above the traditional 2.5 to 3 ft (0.75 to 0.9 m), perhaps to 3.5 to 4 ft (1.1 to 1.2 m).

### Expansion of filter medium during backwashing

When upflow wash with full fluidization is employed, the filter bed expands about 15 to 30 percent above its fixed-bed depth. The degree of expansion is affected by many variables associated with the filter medium and the water. Filter medium variables include the size and size gradation as well as the grain shape and density. Water variables include viscosity and density. The ability to predict expansion is important, for example, in determining whether the expanded medium will rise too high above the bottom of the troughs.

The following model for predicting expanded-bed porosity during backwashing was developed by extending a Reynolds-number-vs.-porosity function that was previously used for fixed beds into the expanded-bed region.[8] The modified Reynolds number $Re_1$ uses interstitial velocity $V/\epsilon$ for the characteristic velocity and a term approximating the mean hydraulic radius of the flow channel, $\epsilon/[S_v(1 - \epsilon)]$ for the characteristic length:

$$\text{Re}_1 = \frac{V}{\epsilon} \frac{\epsilon}{S_v(1 - \epsilon)} \frac{\rho}{\mu} = \frac{V\rho}{S_v(1 - \epsilon)\mu} \tag{8.16}$$

where $S_v$ = specific surface of the grains ($6/d$ for spheres and $6/\psi d_{eq}$ for nonspheres). The porosity function used previously to correlate fixed-bed pressure drop data to $\text{Re}_1$ was modified by combining it with the constant-head-loss equation for a fluidized bed [Eq. (8.6)], resulting in a new dimensionless porosity function for fluidized beds denoted as $A1$.

$$A1 = \frac{\epsilon^3}{(1 - \epsilon)^2} \frac{\rho(\rho_s - \rho)g}{S_v{}^3\mu^2} \tag{8.17}$$

By using the data for many different sizes and types of filter media, $\log A1$ was correlated with $\log \text{Re}_1$ by using a stepwise regression analysis, and the result was the following expansion correlation:[8]

$$\log A1 = 0.56543 + 1.09348 \log \text{Re}_1 + 0.17971(\log \text{Re}_1)^2$$
$$- 0.00392(\log \text{Re}_1)^4 - 1.5(\log \psi)^2 \quad (8.18)$$

The equation can be used to predict the expanded porosity $\epsilon$ of a filter medium of any uniform size (that is, $S_v$ = specific surface) at any desired backwash rate $V$. Because both $A1$ and $\text{Re}_1$ are functions of $\epsilon$, the solution is found by trial and error and is best solved by computer. A computer program for the solution has been presented.[8] When the equation is applied to a real filter medium with size gradation, the bed must be divided into several segments of approximately uniform size according to the sieve analysis data, and the expanded porosity of each size segment must be calculated. The expanded depth of each segment can then be calculated from the following equation, which is based on the total grain volume remaining the same as the bed expands:[5]

$$\frac{l}{l_0} = \frac{l - \epsilon_0}{l - \epsilon} \tag{8.19}$$

in which $l/l_0$ is the ratio of expanded-bed depth $l$ to fixed-bed depth $l_0$, $\epsilon$ is the expanded porosity, and $\epsilon_0$ is the fixed loose-bed porosity. Typical values of $\epsilon_0$ are given in Table 8.1.

**Example Problem 3**  Calculate the expansion for the $d_{50}$ size anthracite shown in Fig. 8.3 at 20°C and at the backwash rate of 32.8 gpm/ft$^2$ (2.23 cm/s) calculated in Example Problem 2. Use a sphericity of 0.55 and a density of 1.6, estimated from Table 8.1.

Solution

$$Re_1 = \frac{V\rho}{S_v(1 - \epsilon)\mu} \qquad (8.16)$$

where  $V$ = 32.8 gpm/ft$^2$ = 2.23 cm/s, as before

$\rho$ = 0.998 g/cm$^3$

$\mu$ = 0.01002 g/(cm)(s)

$\mu/\rho = \nu$ = 0.01003 cm$^2$/s

$d_{50}$ = 0.20 cm

$S_v$ = $6/\psi d$ = $6/0.55 \cdot 0.20$ = 54.54 cm$^{-1}$

$Re^1 = \dfrac{2.23}{54.54(1 - \epsilon)0.01003} = \dfrac{4.076}{1 - \epsilon}$ (dimensionless)

$\epsilon$ = desired expanded porosity

$$A1 = \frac{\epsilon^3}{(1 - \epsilon)^2} \frac{\rho(\rho_s - \rho)g}{S_v^3\mu^2} \qquad (8.17)$$

$$= \frac{\epsilon^3}{(1 - \epsilon)^2} \frac{0.998(1.6 - 0.998)981}{(54.54)^3(0.01002)^2} = \frac{36.184\epsilon^3}{(1 - \epsilon)^2} \quad \text{(dimensionless)}$$

Insert these $Re_1$ and $A1$ values into Eq. (8.18):

$$\log\left[36.184\frac{\epsilon^3}{(1 - \epsilon)^2}\right] = 0.56543 + 1.09348 \log\frac{4.076}{1 - \epsilon} + 0.17971\left(\log\frac{4.076}{1 - \epsilon}\right)^2$$

$$- 0.00392 \,(\log\frac{4.076}{1 - \epsilon})^4 - 1.5 \,(\log 0.55)^2 \qquad (8.18)$$

Note that the only unknown in the above equation is $\epsilon$, but it appears on both sides of the equation. A trial-and-error solution is necessary. A computer program in reference 8 avoids a tedious manual solution.

Solution of Eq. (8.18) yields $\epsilon$ = 0.616. The expansion of this size anthracite can be calculated from Eq. (8.19), assuming an initial porosity of 0.56 from Table 8.1.

$$\frac{l}{l_0} = \frac{1 - \epsilon_0}{1 - \epsilon}$$

$$\frac{l}{l_0} = \frac{1 - 0.56}{1 - 0.616} = 1.15 \qquad (8.19)$$

Therefore, this midsize material would be expanded by 15 percent. To get the expansion of the entire bed, the bed must be divided into about five sizes and the expanded depth of each layer calculated and totaled.

## Size stratification and intermixing during backwashing

The related phenomena of stratification and intermixing are important issues in filter construction and backwashing. These are discussed below.

**Stratification and skimming.**  In the case of a single-medium filter such as a sand filter, during backwash with fluidization the grains tend to stratify by size with the finer grains on top and the coarser grains on the bottom. The tendency to stratify at a given backwash rate (above fluidization velocity) is driven by bulk density differences between the fluidized grains of different sizes. Smaller grains expand more and have a lower bulk density (grains plus fluidizing water) and thus rise to the top of the bed. The concept of bulk density is discussed further in the next section on the intermixing of adjacent layers. This stratification tendency is upset to a varying extent by nonuniform upflow of the backwash water that creates localized regions of above-average upflow velocity. Larger grains are transported upward rapidly into the upper bed, while in adjacent regions the sand is moving downward, carrying finer grains down into the bed. These regions of excessive upflow are referred to as *sand boils,* or *jet action* in water filtration literature and *gulfstreaming* in some fluidization literature.

The tendency to stratify is used beneficially during construction of a filter to remove unwanted fine grains from the bed. The filter is washed above fluidization velocity, and the fine grains accumulate at the upper surface. After the backwash has been completed, the fine grains are skimmed from the surface to avoid leaving a blinding layer of fines on top.

Better stratification is achieved at lower upflow rates, just barely above the minimum fluidization velocity of the bed. So in preparing for skimming, first the bed should be fully fluidized, and then the upflow wash rate should be slowly reduced over several minutes to bring to the surface as many of the fine grains as possible. The fluidization and skimming process may be repeated two or three times for maximum effectiveness.

In dual- or triple-media filter beds, skimming each layer as it is completed is common. The same concepts and procedures described above are equally appropriate in this case. If the very fine grains are not removed from the lower layers, they may rise high in the next upper layer, partially negating the desired benefit of the coarse upper layer.[83]

**Intermixing of adjacent layers.**    Intermixing will tend to occur between adjacent layers of dual- and triple-media filters. For example, the upper, finer sand grains of a dual-media bed move up into the lower, coarser grains of the anthracite bed that lies above.

The tendency to intermix between two adjacent filter media can be estimated by comparing the bulk density of the two media calculated independently at any particular backwash flow rate and temperature. The *bulk density* is the mixed density of the grains and fluidizing water, calculated as follows:[9]

$$\rho_b = (1 - \epsilon)\rho_s + \rho\epsilon \qquad (8.20)$$

in which $\rho_b$ is the bulk density. The first term on the right-hand side is the density contributed by the solid fraction, and the second term is the density contributed by the water fraction.

For example, to predict the tendency of the top sand of 0.5-mm size to mix with the bottom anthracite of 2-mm size in a dual-media bed, first calculate the porosity of each of these media at any desired backwash rate above the fluidization velocity, using Eq. (8.18). Then calculate the bulk density of each layer, using Eq. (8.20). If the bulk densities are nearly the same, intermixing can be expected.

The tendency to intermix increases with the backwash flow rate because the bulk densities tend to converge at higher flow rates. An example of this for ordinary silica sand and garnet sand is illustrated in Fig. 8.27. The data in the figure represent four curves for silica sands of different uniform size. The smallest silica sand curve (triangles) decreases in bulk density the most rapidly as the backwash rate is increased, because it expands the most rapidly. All the bulk density curves would approach the same asymptote of 1.0 g/cm$^3$ as they reached the carry-out velocity of the grains (i.e., approximately the unhindered settling velocity). The fine garnet sand of 0.27-mm size (circles) would tend to intermix with the coarser silica sand of 0.78-mm size (squares) because the bulk densities converge closely within the usual backwash rates of 20 gpm/ft$^2$ (49 m/h). None of the other materials would tend to intermix at this backwash rate. Intermixing actually begins before the bulk densities become equal because of uneven flow distribution and mixing and circulation patterns that exist in the fluidized layers. Mixing was observed to occur when the bulk densities converged to within 3 to 8 lb/ft$^3$ (50 to 130 kg/m$^3$).[9]

The bulk density model would predict that inversion of layers should occur at very high wash rates, and such inversion was observed experimentally.[9] Wash rates were, however, far higher than rates used in practice.

A figure similar to Fig. 8.27 for sand and anthracite was presented by Cleasby and Woods.[9] While the same shape is evident, the tendency for sand and anthracite to intermix during backwashing is much less than for silica and garnet sand in the usual backwash flow range. This is because the bulk density differences are greater for the usual sand and anthracite sizes used in current practice.

### Backwash of GAC filter-adsorbers

The properties of GAC are sufficiently different from those of conventional anthracite or sand media that some special precautions are required related to backwashing of GAC filter-adsorbers.[1] The lower density of GAC means that lower backwash rates are required to flu-

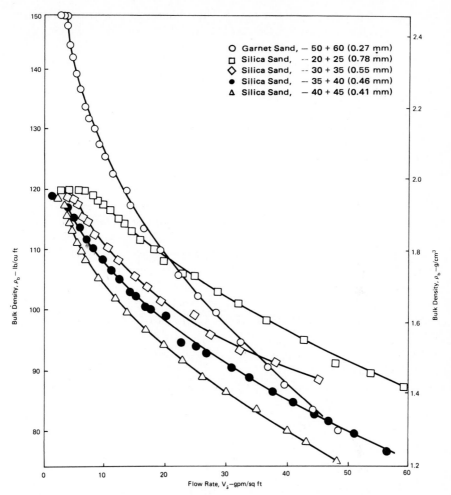

**Figure 8.27** Bulk density versus flow rate for garnet sand and silica sand. (*Source: J. L. Cleasby and C. W. Woods, "Intermixing of Dual- and Multi-Media Granular Filters," J. AWWA, vol. 67, no. 4, April 1975, p. 197.*)

idize the common size gradations currently used. Lower wash rates mean lower hydraulic shear and less effective upflow wash. Mud balls have been a common problem in GAC filter-adsorbers. Therefore, surface wash or auxiliary air scour is essential for GAC backwashing.

The higher UC of typical GAC size gradations results in a greater percentage of expansion if the full GAC bed is fluidized as well as greater potential for loss of GAC media into the backwash troughs. In retrofitting existing filters into GAC filter-adsorbers, raising the backwash troughs may be necessary to reduce this loss potential.

If sand is used in GAC filter-adsorbers in a dual-media configuration, the sand size gradation should be selected to be compatible with the GAC in its fluidization characteristics. The sand should reach full-bed fluidization at about the same backwash rate as the GAC. In retrofitting existing filters, the existing sand may not meet this requirement, and new sand may be required.

GAC is abrasive and corrosive to many metals. Therefore, all metals in contact with GAC should be resistant to abrasion and corrosion. Metals can be coated with corrosion-resistant substances.

Special precautions should be taken in preparing a new GAC filter-adsorber for service. The filter box should be disinfected before the GAC is installed. After GAC installation, the filter box must be submerged and soaked for at least 1 day to allow water to penetrate the pores of the GAC. Then backwashing must be done at a very low rate initially, to be sure that all air is out of the bed to avoid loss of GAC. When all air is removed, the backwash rate can be increased to wash out undesirable fine grains from the bed. Release of additional air and fine grains may be observed during the first week of normal backwashing operation.

### Backwash water recovery

Recovery of dirty backwash water is becoming increasingly common today for a number of reasons. It represents a rather large volume of water with low solids content. Typically, the volume is 1 to 5 percent of total plant production. Therefore, its recovery represents a savings in water resources and in the chemicals that were expended to treat it initially. Discharge of water plant wastes is usually prohibited under current pollution regulations, although some agencies are relaxing requirements for economic reasons and in recognition of the minimal environmental impact of these wastes.[106]

Most regulatory agencies require proper handling of wastes for new treatment plants and for plants being enlarged or upgraded, however. Further discussion of this subject is presented in Chap. 16.

One consideration that may discourage the use of backwash water recovery and recycling for source water containing *Giardia* cysts is the possibility that recycling may increase the cyst concentrations in the filtered water. This possibility was mentioned in a 1987 AWWA committee report on microbial considerations[107] but was not supported with any evidence or references.

### Problems in Rapid Filters

#### Dirty filter media and mineral deposits

Dirty-filter problems result from inadequate backwashing of the filter including the absence or improper operation of auxiliary scour systems. Typical manifestations of inadequate backwashing include filter cracks and mud balls. Inadequate cleaning leaves a thin layer of compressible matter around each grain of the medium. As the pressure drop across the filter medium increases during the subsequent filter run, the grains are squeezed together and cracks form in the surface of the medium, usually along the walls first. In severe cases, cracks may develop at other locations as well.

The heavier deposits of solids near the top surface of the medium break into pieces during backwash, resulting in spherical accretions referred to as mud balls. Mud balls are composed of the filter grains and the solids removed by the filter, and they range from pea size to 1 to 2 in (2.5 to 5 cm) or more. Mud ball problems are accentuated by the use of polymers as coagulant aids or as filter aids that form stronger attachments between the filter grains and the removed solids. If mud balls are small enough and of low enough density, they float on the surface of the fluidized media. If they are larger or heavier, they may sink into the filter, to the bottom, or to the coal-sand interface in dual-media filters. Subsurface accumulations can lead to solidified inactive regions of the filter bed that are not remedied in normal backwashing. This, in turn, increases the filtration rate through the remaining active portions of the bed with potential detriment to the filtrate quality and shorter filter cycles.

Surface mud balls can be removed manually by (1) ladling them from the surface with a large strainer while the backwash water is running at a low rate, (2) breaking them up with rakes, or (3) breaking them up with a high-pressure hose jet. Subsurface mud balls and agglomerations can be reduced by (1) probing the bed with lances delivering jets of high-pressure water; (2) pumping the medium through an ejector; (3) washing the filter and allowing it to stand for up to 2 days, aided by chemical additions described below; or (4) digging out the hard spots, removing, cleaning, or replacing the medium.

Mineral deposits can develop on the grains of the filter medium, causing them to change in size, shape, and density. Calcium carbonate deposits are common in many lime–soda ash softening plants if recarbonation is not sufficient to deliver stable water to the filters. Such deposits may also occur in surface water plants if lime is added ahead of filtration for corrosion control. Calcium carbonate deposition can be minimized by adding a low dosage of polyphosphate ahead of the filters. Iron oxides and manganese oxides are common deposits on the filter medium of iron and manganese removal plants. Aluminum oxide or iron oxide deposits can occur in surface water plants that use alum or iron salts for coagulation.

All these mineral deposits are subject to later leaching into the filtered water, if the pH is changed in a direction to increase solubility. Aluminum is particularly noteworthy in this regard because its minimum solubility occurs at a pH of 6 and its solubility increases rapidly above and below that pH (see Chap. 6).

The increased grain size caused by mineral deposits results in an increase in the bed depth and can impair filtration efficiency. The increased size and the altered shape and density of the grains can impair the adequacy of backwashing. Thus, cleaning the deposits from the medium or replacing the medium may become necessary.

Some lime–soda ash softening plants accept some calcium carbonate deposition intentionally, to gain additional softening and reduce carbonation expense. Periodic replacement of the filter medium is accepted as a favorable economic alternative.

Dirty-filter problems can usually be corrected by proper use of the auxiliary scour devices such as surface washers or air scour. If auxiliary scour is not available, the use of a high-pressure hose to agitate the surface of the medium prior to backwash can be helpful.

Where auxiliary scour is not successful, use of chemicals may be necessary in some cases to attempt to clean the filter medium in place. Various chemicals have been used, including chlorine, copper sulfate, acids, and alkalies.[108] Chlorine may be used where the material to be removed includes living and dead organisms or their metabolites. Copper sulfate is effective in killing algae growing on filter walls or medium. Carrying a chlorine residual in the filter influent helps control microorganisms in the filter. Acidifying the water to a pH of about 4 aids in dissolving deposits of the oxides of iron, aluminum, and manganese as well as calcium carbonate deposits. Sulfuric acid and hydrochloric acid have been used, but care must be taken to prevent high local concentrations to prevent damage to concrete filter walls. Caustic soda has been used for alum deposits and for organic deposits. It is used at the rate of 1 to 3 lb/ft$^2$ (5 to 15 kg/m$^2$). Chemical solutions are generally left in contact with the sand for 1 to 2 days, and then the filter is thoroughly washed and returned to service.

If a filter can be removed from service for an extended period, calcium carbonate deposits can be removed by feeding, at a low flow rate, a water acidified with carbon dioxide. The flow rate is selected so that the water leaves the filter with the $CO_2$ fully consumed at a pH of about 8.

### Gravel movement during backwashing

Many filters have layers of graded gravel to support the filter medium. The conventional gravel layers are graded from coarse on the bottom to fine at the top, according to appropriate size guidelines presented earlier. With this conventional gravel arrangement, some cases have been reported of mounding of the gravel caused by lateral movement of the fine gravel. In severe cases, the finest gravel layer may be completely removed from some areas of the filter bed. This can lead to leakage of the filter medium into the underdrain system and ultimately to the need to clean the underdrain and to rebuild the filter gravel and medium. Mechanisms contributing to gravel movement have been exhaustively discussed by Baylis,[20,109] and those discussions related to fluidized-bed backwashing are summarized here.

As backwashing begins, sand grains do not move apart quickly and uniformly throughout the bed. Time is required for the sand to equilibrate at its expanded spacing in the upward flow of wash water. If the backwash is turned on suddenly, it lifts the sand bed bodily above the gravel layer, forming an open space between the sand and gravel. The sand bed then breaks at one or more points, as shown in Fig. 8.28, causing sand boils and subsequent upsetting of the supporting gravel layers. This then requires frequent rebuilding of the gravel section.

Opening the backwash valve slowly is essential. The time from start to full backwash flow should be at least 30 s and perhaps longer and should be restricted by devices built into the plant. The most destructive filter-washing blunder is to turn the backwash on quickly when the bed has been drained. Gross gravel disturbance results, and rupture of the filter underdrain system has resulted in many cases, particularly with false-bottom-type underdrains.

The upward flow from the gravel is never completely uniform. As a result, in some parts of the expanded sand bed, the water and sand travel upward at rates higher than the average backwash velocity, and at other places the sand actually travels downward. Even with the most careful gravel selection and placement, the gravel is disturbed in some degree by a natural tendency toward channeling. This leads to jet action at the sand-gravel junction, as illustrated in Fig.

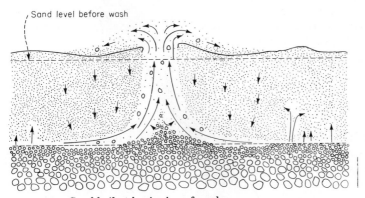

**Figure 8.28**  Sand boil at beginning of wash.

8.29. With a normal backwash rate, the average upward velocity of the water through the gravel should be less than 0.1 ft/s (0.03 m/s). The velocities of jets have been observed to be more than 10 times the average upward velocity through the area occupied by the sand. In some cases the jet velocity is so great that it will move small gravel particles several diameters larger than the sand grains adjacent to the gravel. The problem of gravel movement is greater at wash rates above 15 gpm/ft$^2$ (37 m/h).

The importance of sand in generating the gravel movement is illustrated by the observation that an upward flow through a graded gravel bed at backwash rates exceeding 25 gpm/ft$^2$ (61 m/h) will not move any of the gravel particles unless sand is present on top of the gravel.[109] The presence of the sand greatly increases the mobility of the gravel. Therefore, gravel movement studies cannot be conducted in the absence of the fine medium.

Jet action is primarily a problem in backwashing with full-bed fluidization. Gravel movement can also occur when air scour and water backwash are used simultaneously. The provisions being made to avoid gravel movement are the use of double reverse graded gravel or the use of media-retaining nozzle underdrains that require no gravel, as discussed earlier.

To determine whether gravel movement has occurred, the position of the upper gravel surface can be determined. Three methods are used: (1) A ¼-in (0.6-cm) metal rod can be pushed gently down to the

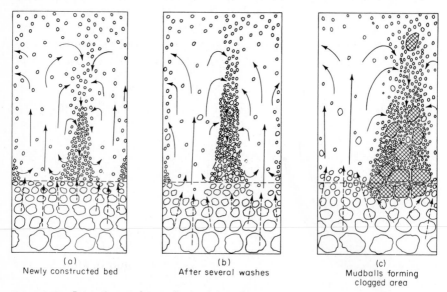

|     (a)     |     (b)     |     (c)     |
| Newly constructed bed | After several washes | Mudballs forming clogged area |

Figure 8.29   Jet action at the sand-gravel interface.

gravel surface while the filter is at rest. (2) A probe with a flat plate on the end is used during the fluidized-bed backwash. (3) Drain the filter to the gravel, insert a thin-walled pipe to the gravel, and remove the media from the pipe until the top gravel is detected. Each method has advantages and disadvantages. The first is quick, but distinguishing the fine gravel from the bottom filter medium is difficult. The second method works well if the bed is fully fluidized to the bottom; but if not, ensuring that of the gravel is being reached by the flat plate is difficult. A flat plate of about 4-in (10-cm) diameter prevents penetrating the fine gravel. This method also requires a long fluidized backwash period. The third method is the best but also the most time-consuming and labor-intensive.

Boards can be placed on the lips of the backwash troughs to gain access to filter media. The boards can serve as a convenient reference level from which to measure. Numerous measurements made on a gridiron pattern reveal the contour of the top gravel surface. Annual probing is suggested to determine the extent and progression of gravel movement, if it is occurring.

### Underdrain failures

The filter underdrain is a vital component of the rapid filter. It collects the filtered water, supports the filter gravel (if used) and filter medium, and distributes backwash water.

The most serious failures of rapid filters are usually related to problems with the underdrain. Proper design and construction of the underdrain is essential to a successful filter installation. Any underdrain system, if designed improperly, can lead to uneven distribution of the backwash water. This, in turn, can lead to problems such as dirty filter areas and gravel migration, if gravel is used to support the medium.

False-floor underdrains with nozzles have failed by several modes. If the nozzles have fine openings to retain the fine filter medium, the openings may become clogged with debris from the underdrain or backwash water or with mineral deposits. If the clogging becomes excessive, the difference in pressure from below the floor to above the floor (i.e., the pressure drop) causes uplift during backwashing that may become excessive. The problem can be repaired by removing the filter medium and cleaning the nozzles. If it is not corrected, excessive pressure drop can result in an uplift structural failure of the false floor. A simple piezometer tube can be installed to observe the underdrain plenum pressure during backwashing and to alert the operator to the degree of clogging of the nozzles.

Some modern plastic nozzles are not strong enough to accept the abuse of construction installation. They may be weakened during construction, and the nozzle heads may break off during future backwashing. The result is leakage of the medium into the underdrain plenum.

The manifold-lateral type of underdrain is not subject to uplift failure. It can fail to function properly, however, if the distribution orifices are reduced in size because of mineral deposits or enlarged in size because of corrosion or erosion.

Vitrified clay block underdrains have proved quite successful when properly installed. The manufacturers' recommendations for installation and grouting must be followed explicitly to effect a good installation. Careless work can lead to grout in the water-carrying channels. This can lead to maldistribution of wash water and, in severe cases, to excessive backwash pressures and uplift failure.

Some early plastic block underdrains with nozzles, which are no longer on the market, failed by repeated flexure of the top deck of the block during successive filtration and backwash cycles. Clogging of the nozzles accentuated the flexure stresses.

## Pressure Filters

Pressure filters are sometimes used for rapid filtration. The filter medium is contained in a steel pressure vessel. The water to be filtered enters the filter under pressure and leaves at slightly reduced pressure because of the head loss encountered in the filter medium, underdrain, and piping connections.

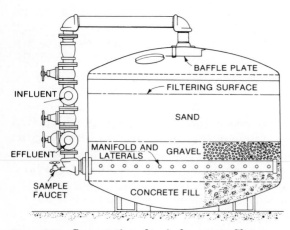

**Figure 8.30**   Cross section of typical pressure filter.

### Description of pressure filters

The pressure vessel may be a cylindrical tank with a vertical axis, such as shown in Fig. 8.30, or a horizontal-axis cylindrical tank. The horizontal cylindrical configuration has the disadvantage that the width of the filter medium is not constant from top to bottom, with it usually being wider at the top. This leaves dead areas along the walls that do not receive adequate fluidization during backwashing and therefore may not be washed effectively.

In some horizontal-axis cylindrical tanks, the filter is divided into multiple (four or five) cells by vertical bulkheads. Cost advantages are gained by this configuration because only a single pressure vessel is needed. One filter cell can be backwashed by the production of the other cells that remain in service. This requires a filtration rate sufficiently high that the total production of the operating cells is enough to fluidize the medium of the single cell being backwashed.

### Comparison of pressure and gravity filtration

While the outward appearances of a pressure filter and a gravity filter are quite different, the filtration process is the same. The same particle capture mechanisms are functioning in both. The same filter medium, the same filtration rates, and the same terminal head loss should be utilized if comparable filtrate quality is desired.

The use of higher filtration rates and high terminal head loss in a pressure filter is tempting because the influent is under pressure and more potential head loss is available. This temptation should be resisted, however, unless no detriment to the filtrate quality can be demonstrated on a case-by-case basis.

One advantage gained by pressure filtration is that water enters and leaves the filter under pressure, and so no negative pressure can ever exist in the filter medium. The potential problems associated with negative pressure discussed earlier are thus avoided.

### Operation of pressure filters

Because of the similarities between pressure and gravity filters, the operating principles are identical. For example, appropriate pretreatment is equally important to pressure and gravity filters; filtrate quality patterns will be the same; the impact of sudden rate increases will be just as detrimental; and the importance of proper backwashing is equally important.

The operation of a pressure filter is similar in most respects to that of a gravity filter. Proper backwashing of a pressure filter is more difficult, however, because the filter medium is not conveniently visible to the operator during the backwash operation. Making the visual observations that can be conveniently made for gravity filters is difficult or impossible with pressure filters. These include:

1. The presence of filter cracks before the backwash or mud balls after the backwash

2. The uniformity of backwash water distribution

3. The uniformity of rate of cleanup of the wash water over the full filter areas

4. The proper functioning of the auxiliary scour device such as surface wash or air scour

5. The elevation and appearance of the top surface of the filter medium after the backwash

6. Whether the medium is fully fluidized during the backwash (if that is the intended washing method) and the extent of expansion of the filter medium

7. The extent of loss of the filter medium

Because of these difficulties, filter failures have resulted in outbreaks of waterborne disease, partially attributed to the poor condition of the filter medium in pressure filters.[28,110] Prior concerns about the reliability of pressure filters and other concerns have caused some state regulatory agencies to exclude the use of pressure filters in the treatment of polluted source water and lime-softened water.[15]

### Applications of pressure filters

Pressure filters tend to be used in small water systems. Many pressure filters are employed in industrial water and wastewater filtration applications. They are also widely used in swimming pool filtration.

Pressure filters have one potential advantage over gravity filters. The filtrate, which is under pressure, can be delivered to the point of use without repumping. In some treatment plants, source water can be pumped from the source through the treatment plant and directly to the point of use by the source water pumps.

Some groundwater containing iron can be treated by pressure aeration and/or chemical oxidation and then filtered directly on pressure filters. This approach has received considerable application for small communities. Not all iron-bearing water responds successfully to this method of treatment, however, and prior pilot testing is usually required.

## Slow Sand Filters

### Description and history

The slow sand filter is a sand filter operated at very low filtration rates without the use of coagulation in pretreatment. The sand used is somewhat smaller than that for a rapid filter, and this, plus the low filtration rate, results in the solids being removed almost entirely in a thin layer on the top of the sand bed. This layer, composed of dirt and living and dead micro- and macroorganisms from the water (i.e., the schmutzdecke, or dirty skin), becomes the dominant filter medium as the filter cycle progresses. When the head loss becomes excessive, the filter is cleaned by draining it below the sand surface and physically removing the dirty skin along with 0.5 to 2 in (13 to 50 mm) of sand. Typical cycle lengths may vary from 1 to 6 months depending on the source water quality and the filtration rate.

Removal mechanisms in a slow sand filter are both physical and biological. Living organisms in the filter cause reductions in organic constituent concentrations and chemical transformations, such as the oxidation of ammonia to nitrate. The effectiveness of the filter improves over the first few cycles as the microflora develop and remains high thereafter.

The slow sand filter was developed in England in the early nineteenth century, and this type of filter continues to be used successfully in England, notably on Thames River water, which serves London, and in the United States on water from some upland watersheds in New England.

Slow sand filters are not successful in treating clay-bearing river water typical of most of the United States because the clay penetrates too deeply into the filter and cannot be removed in the normal surface-scraping operation. In addition, slow sand filters are not very effective for color removal, typically achieving only 25 percent removal.

Renewed interest in the slow sand filter is due to numerous recent outbreaks of giardiasis in communities using unfiltered or poorly filtered water. The slow sand filter is a simple technology requiring no knowledge of coagulation chemistry and is quite attractive for small installations treating high-quality surface water. Because of this interest, new research efforts are under way to demonstrate the efficacy of the slow sand filter in removal of *G. lamblia*.

### Physical details

A recent literature review[83] summarized early slow sand filter practice. Sand effective sizes ranged from 0.15 to 0.40 mm (with 0.3 mm most common), sand uniformity coefficients from 1.5 to 3.6 (with 2 most common), and initial bed depths from 1.5 to 5.0 ft (0.46 to 1.52

m), with 3 ft (0.9 m) most common. The sand was supported on graded gravel 6 to 36 in (0.15 to 0.91 m) deep, with 18 to 24 in (0.45 to 0.60 m) the common range.

Filtration rates ranged from 0.016 to 0.16 gpm/ft$^2$ (0.04 to 0.40 m/h) with 0.03 to 0.05 gpm/ft$^2$ (0.07 to 0.12 m/h) most common on source water that received no prior pretreatment. Flow rates higher than 0.12 gpm/ft$^2$ (0.3 m/h) were used only following some pretreatment step, to lengthen the filter cycle, such as sedimentation or plain rapid filtration without coagulants.

The available head loss for operating the filter ranged from 2.5 to 14 ft (0.76 to 4.3 m) but was most commonly from 3 to 5 ft (0.9 to 1.5 m). A survey of slow sand filtration plants in operation in the United States in 1984 indicated that the sand descriptions and filtration rates fell within the ranges noted above.[111] An experimental filter that was intentionally equipped with an unsieved local sand with an ES of 0.18 mm and a UC of 4.4 performed satisfactorily, indicating the possibility of reducing costs by using less rigid specifications.[111]

### Mechanisms of filtration and performance

Removal of particulates by a slow sand filter occurs dominantly in the dirty skin (schmutzdecke) that develops on the surface. An initial improvement period (or ripening period) occurs at the beginning of each cycle after the schmutzdecke has been removed.[83,112,113] The initial improvement period was observed to vary from 6 h to 2 weeks,[112] although most improvement periods were less than 2 days.[83,112] A filtering to waste period of 2 days is recommended where cysts of G. lamblia are of concern.[83–113] Also a distinct improvement in the performance of the filter occurs as it matures over several cycles, as indicated by coliform bacteria removal, particulate removal, and Giardia cyst removal.[83,113–115]

Early slow sand filter studies at the turn of the century at several locations indicated removal of total bacteria usually greater than 97 percent after the ripening period of the cycle[113] and often above 99 percent. A summary of the early literature reported complete removal of Salmonella typhi in an 1890 study by the Massachusetts Board of Health.[114] More recent work in England, Germany, and the Netherlands and by the World Health Organization (WHO) was also cited as demonstrating the efficiency of slow sand filters in the removal of bacteria, viruses, and organic and inorganic pollutants.[114]

Bellamy et al.[114] found that a new sand bed was able to remove 85 percent of source water coliform bacteria and 98 percent of Giardia cysts. As the sand bed matured biologically, coliform bacteria removal exceeded 99 percent and Giardia removal was virtually 100 percent.

In a midwestern pilot study, after the first four filter cycles, the average turbidity reduction was 97.8 percent or better, cyst-sized particle removal (7 to 12 μm) was 96.8 percent or better, total particle removal (1 to 60 μm) was 98.1 percent or better, total coliform bacteria removal was 99.4 percent or better, and removal of chlorophyll a was 95 percent or better.[83,113] Coliform bacteria removal got progressively better as the bed matured over several filter cycles, a behavior that was also observed in another pilot study.[115]

Turbidity removal generally parallels removal of other particulates and is generally excellent with the filtrate turbidity at full-scale plants usually below 0.5 NTU, with one exception.[112] Low-percentage turbidity removal occurred in one pilot study of lake water containing a very fine clay from mountain runoff where removal averaged from 27 to 39 percent on source water with a turbidity of 3 to 9 NTU.[114,116]

Because the slow sand filter is partially a biological process, prechlorination might be expected to be detrimental to performance. This detriment was not evident in a comparison of three full-scale plants with prechlorination to four plants without prechlorination.[112] In one pilot-scale study, a slow sand filter that received superchlorinated water with influent about 6 to 12 mg/L free chlorine and with free chlorine in the filtrate performed better than a filter without prechlorination in bacteria removal and cycle length.[117]

### Cleaning of slow sand filters

Slow sand filters are cleaned by removing the schmutzdecke along with a small amount of sand depth, an operation known as *scraping*. The sand that is removed is usually cleaned hydraulically and stockpiled for later replacement in the filter. The scraping operation can be repeated several times until the bed depth has decreased to about 16 to 20 in (0.4 to 0.5 m), at which time the depth should be replenished, referred to as *resanding the filter*.

The sand in small plants is usually skimmed by hand, using broad shovels. Scraping and resanding are labor-intensive operations, and consideration should be given to reducing the manual labor involved. A 1985 survey[112] of seven full-scale plants indicated from 4 to 42 labor hours per 1000 ft² (92 m²) of filter area for one scraping. Some of the plants were using inefficient transport systems. By the use of efficient motorized buggies or hydraulic transport and by limiting removal depth to 1 in (2.5 cm), 5 labor hours per 1000 ft² (5.4 labor hours per 100 m²) were considered adequate.[112] Resanding with 6 to 12 in (15 to 30 cm) of sand requires approximately 50 labor hours per 1000 ft² (54 labor hours per 100 m²).

One small plant reported use of asphalt rakes to scrape the dirty surface into windows that were then shoveled into buckets for transport.[118] By using this technique, only about 0.2 in (5 mm) was scraped off the filter. This was accomplished by two workers in about 30 min for an 825-ft$^2$ (77-m$^2$) filter, about 1 labor hour per 1000 ft$^2$ (92 m$^2$).

Scraping operations for the large slow sand filters serving London are conducted mechanically without manual shoveling. A specially designed track-driven machine picks up the sand and delivers it to a belt conveyer to a waiting front-end loader. The loader carries the sand up a temporary ramp and out of the filter to the filter washing and stockpiling area. Thus labor is limited mainly to machine drivers.

The longer a filter is drained for the scraping operation, the longer will be the initial improvement period during the subsequent run.[112] Therefore, cleaning should be done quickly, and the filter returned to service immediately after cleaning.

Some plants try to lengthen the time between scraping by using one to five rakings of the surface between each scraping. The run time gained with each raking diminishes, and when the scraping is finally required, up to 6 in (15 cm) of sand must be removed to reach clean sand.[83]

### Appropriate water for slow sand filters

Early references suggested that appropriate water should have less than 30 ppm turbidity and 20 ppm color.[83] The units for these measures have changed since those early reports; however, translating those values to current units is difficult. Because color removal on slow sand filters is poor (about 25 percent), the source water must have almost acceptable color.

A 1984 survey of 27 existing full-scale plants in the United States found 74 percent using lakes or reservoirs as their source water, 22 percent using rivers or streams, and 4 percent using groundwater. The mean source water turbidity was about 2 NTU with a peak of about 10 NTU.[111] Mean cycle lengths varied from 42 days in the spring to 60 days in the winter. Wide variations were reported with two reports of cycles up to 1 year.

A survey of seven New York plants, including three of those in the prior reference, revealed source water with generally less than 3 NTU turbidity. One plant with 6 to 11 NTU was not meeting the existing (1984) 1 NTU finished water turbidity MCL.[112] Average cycle lengths varied from 1 to 6 months.

Waters with periodic algal blooms may have short filter cycles during such periods. Some measure of plankton should be included in source water guidelines. A 1984 report[83] suggested that an acceptable source water should have turbidity less than 5 NTU and chlorophyll-a less than 5 mg/m$^3$ ($\mu$g/L).

In northern climates where freezing will occur on uncovered filters, the cycle length is crucial. If the cycle terminates in midwinter, the layer of ice will prevent a normal scraping operation. Therefore, if any possibility of cycles shorter than the period of ice cover on the filter exists, the filter must be covered.

### Precoat Filtration

Precoat filters use a thin layer of very fine material such as diatomaceous earth as a filter medium. In precoat filtration, the water to be filtered is passed through a uniform layer of the filter medium that has been deposited (precoated) on a septum, a permeable material that supports the filter medium. The septum is supported by a rigid structure termed a *filter element.* As the water passes through the filter medium and septum, most of the suspended particles are captured and removed. The majority of particles removed by the filter are trapped at the surface of the filter medium layer, with some being trapped within the layer. As the filter cycle proceeds, additional filter medium, called *body feed,* is regularly metered into the influent water flow in proportion to the solids being removed. Without the regular addition of body feed, the head loss across the precoat layer would increase rapidly. Instead, the dirt particles intermingle with the body feed particles so that the permeability of the cake is maintained as the thickness of the cake gradually increases. By maintaining cake permeability in this way, the length of the filter cycle is extended.

Ultimately, a gradually increasing pressure drop through the filter system reaches a point where continued filtration is impractical. The forward filtration process is stopped, the filter medium and collected dirt are washed off the septum, a new precoat of filter medium is applied, and filtration continues. A typical flow schematic is shown in Fig. 8.31.

The basic function performed by all water filters is to remove particulate matter from the water. Precoat filters accomplish this by

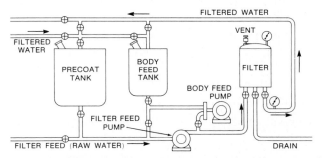

**Figure 8.31** Typical precoat filtration system. (*Courtesy of Manville Filtration & Minerals Division.*)

physically straining the solids out of the water. The thickness of the initial layer of precoat filter medium is normally $\frac{1}{16}$ to $\frac{1}{8}$ in (1.5 to 3 mm), and the water passageways through this layer are so small and numerous that even very fine particles are retained.

Five or six grades of filter medium (sometimes called *filter aid*) are commonly used. They offer a range of performance with respect to clarity and flow characteristics. With an appropriate selection from these grades, particles as small as 1 μm can be removed by the precoat filter cake. This includes most surface water impurities. Where colloidal matter or other finely dispersed particles are present, however, filtration alone may not be adequate to reduce turbidity to the required MCL.[27]

### Applications and performance

Precoat filters are widely used in industrial filtration applications and in swimming pool filtration. They have also been used in municipal potable water treatment, primarily in the direct in-line filtration of high-quality surface water (turbidity 10 NTU or less and acceptable color) and in the filtration of iron and manganese from groundwater after appropriate pretreatment to precipitate these contaminants.

Since 1949, more than 150 potable water treatment plants utilizing precoat filtration with diatomaceous earth or other filter media have been designed, constructed, and operated. About 90 percent are surface water supplies, and 10 percent are groundwater supplies. The largest existing plant is the 20-mgd San Gabriel, Calif., plant. In 1985, New York City started a 3-mgd pilot program to fully evaluate ozonation followed by precoat filtration process for the proposed 240-mgd Jerome Reservoir plant.[119]

Where the source water and other conditions are suitable, precoat filtration can offer a number of benefits to the user, including the following:

1. Capital cost savings may be possible because of smaller land and plant building requirements.

2. Treatment costs may be 40 to 60 percent less than for conventional coagulation, sedimentation, and granular media filtration when filterable solids are low,[119–121] although sedimentation would not usually be needed for such high-quality source water.

3. The process is entirely a physical and mechanical operation and does not require operator expertise in water chemistry relating to coagulation.

4. The waste filter medium is easily dewatered, and in some cases it may be reclaimed for other uses, including soil conditioning and land reclamation.

5. Acceptable finished water clarity is achieved as soon as precoating is complete and filtration starts. A filter-to-waste period is generally not necessary to bring the turbidity of the finished water within acceptable limits.

6. Terminal turbidity breakthrough is not generally observed because it is dominantly a surface filtration process.

The disadvantages of precoat filtration are as follows:

1. There is the continued cost of the filter medium usually discarded at the end of each filter cycle.

2. It is less cost-effective for water that requires pretreatment for algae, color, or taste and odor problems. Water containing only larger plankton such as diatoms can sometimes be treated economically by microstraining prior to the precoat filtration.

3. Proper design, construction, and operation are absolutely essential to prevent the dropping or cracking of the filter cake during operation that might result in failure to remove the target particulates.

4. Mechanical devices require maintenance and can fail.

Precoat filtration has been shown to remove virtually 100 percent of *Giardia* cysts over a broad range of operating conditions typical of potable water filtration.[27,122] The need for proper operation and maintenance is emphasized, however, and a precoat rate of 0.2 lb/ft$^2$ (1 kg/m$^2$) is recommended.

The capability to remove cysts and the fact that chemical coagulation is not required make precoat filtration attractive to very small communities facing filtration because of their concern about *G. lamblia* cysts in source water. Operators skilled in coagulation are less likely to be available in small installations, but operators with mechanical skills are required.

The removal of small particulates such as bacteria and viruses is dependent upon the grade of filter aid used and other operating conditions.[122] Capture of smaller particles such as bacteria, viruses, and asbestos fibers can be enhanced by using aluminum or iron coagulants or cationic polymers to coat the filter medium.[122-125]

### Filter element and septum

The filter element is a vital part of a precoat filter because it supports the septum and the filter medium. A good-quality filter element should have the following characteristics:

1. It should have enough structural strength to provide a firm support for the filter media cake.

2. It should have adequate drainage area inside the element, so that the filtered water can easily exit from the element.

3. The septum should be properly constructed to provide clear openings of proper size, so that the filter media form strong, stable "bridges" over the openings. The material of the septum should be capable of maintaining the integrity of the weave pattern, to prevent distortion of opening size or shape with continued use, and should be fabricated of corrosion-resistant construction materials for long life.

Filter elements may be either flat or tubular (Fig. 8.32). Flat elements, often referred to as *leaves,* may be rectangular or round. Tubular elements are available in several different cross-sectional shapes, but are generally round. Tubular elements are always oriented vertically in vertical tanks (Fig. 8.33). Flat leaf elements used in water treatment are oriented vertically in either vertical or horizontal tanks.

The internal construction of filter elements can vary widely, depending on the intended service. Elements intended for water service will usually incorporate a drainage member to provide strength and rigidity along with free drainage area (Fig. 8.32). Overlaying the drainage member is the septum material, consisting of either tightly woven stainless-steel wire mesh or a synthetic cloth bag, usually made from monofilament polypropylene weave.

The precoat layer forms on, and is supported by, the septum. For this to occur, the clear openings in the septum must be small enough for the filter medium particles to form and maintain stable bridges across the openings. Generally, a clear opening of 0.005 in (about 125 μm) or less in one direction is desirable.

The septum must be firmly supported so that it does not yield, flex, or become distorted as the differential pressure drop increases during the filter cycle. If the septum yields or gives as the pressure increases, the filter medium bridges may slowly break down and a small amount of filter medium may enter the finished water and adversely affect clarity.

### Filter vessel

Two basic types of precoat filter vessels that contain multiple filter elements and septa are available: the vacuum filter and the pressure filter.

**Vacuum filters.** The vessel containing vacuum filter elements and their septa is an open tank at atmospheric pressure. A filter discharge

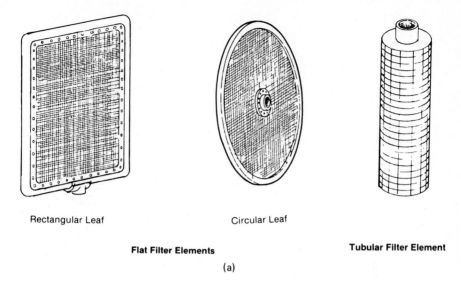

Rectangular Leaf                    Circular Leaf

**Flat Filter Elements**                              **Tubular Filter Element**

(a)

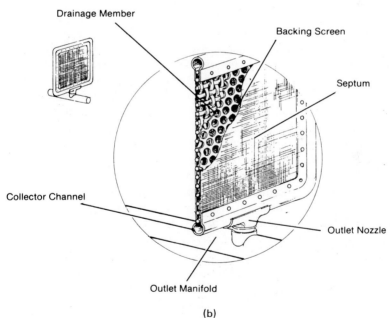

Drainage Member

Backing Screen

Septum

Collector Channel

Outlet Nozzle

Outlet Manifold

(b)

**Figure 8.32**   Types of filter elements for precoat filters and construction details of a flat leaf element.

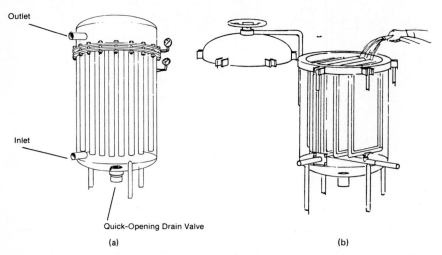

Outlet

Inlet

Quick-Opening Drain Valve

(a)                                                                                    (b)

**Figure 8.33**   Typical pressure tubular filter and leaf filter for precoat filtration. (*a*) Pressure tubular filter; (*b*) vertical tank pressure leaf filter.

pump or a vacuum discharge leg downstream of the filter creates a suction. This suction enables the available atmospheric pressure to move water through the precoat and filter medium cake as the cake builds up. The open filter tanks permit easy observation of the condition of the septa and elements and the general condition of the filter during filtration and cleaning.

The maximum available differential pressure across the vacuum filter is limited to the net positive suction head of the filter pump or the vacuum leg. This limitation influences the length of filter cycles.

The effect of any entrained air or dissolved gases coming out of solution because of the decrease in pressure must also be considered with vacuum filters. Gas will have an adverse effect on the filter cake, tending to disrupt the integrity of the filter medium on the septum.

**Pressure filters.**   A filter feed pump or influent gravity flow produces higher than atmospheric pressure on the inlet (upstream) side of a pressure filter, forcing liquid through the filter medium cake (Figs. 8.32 and 8.33). Large pressure drops across the filter are theoretically possible (limited only by the strength of the filter shell and the filter elements and septa), but the maximum economic differential pressure drop is generally limited to 30 to 40 psi (206 to 276 kPa). Typically, a higher pressure drop across a pressure filter will yield longer cycles than with vacuum filters, which in turn will yield greater suspended solids removal per pound of filter medium. Increased pumping costs for differential pressures much over 30 to 40 psi (206 to 276 kPa), however, usually offset savings in filter medium costs.

### Filter media

The filter medium is the basic means of removing contaminants from water. The medium must capture and hold back the turbidity through the filter cycle. At the same time it must be porous, and remain so, to permit water to pass through the cake. The filter tank, septum, and other equipment support and manipulate the filter medium as it performs its function. Selection of a medium appropriate to the quality of the influent and the desired quality of the effluent is an important step in economical precoat filtration.

Diatomaceous earth and perlite are both used as filter media for precoat filtration of water. Both materials are available in various grades, to allow creation of a filter cake with the desired pore size. Filter media are graded according to the flow rate and the clarity of the finished product. In general, the finer grades (those producing small cake pore diameters) will provide higher clarity, but with lower flow rates and higher head loss through the filter.

Pilot testing can determine two or three grades that will produce a desired clarity, simplifying the task of optimizing a final selection when working with a full-scale plant. As a starting point, a rule of thumb is to match the cake median pore diameter to the median particle size of the material being removed. This guideline is not easily followed in typical municipal applications, and a trial-and-error approach may be necessary that uses different grades of media. Any desired clarity-versus-flow-rate compromise within overall economic considerations will then dictate whether to select a coarser or finer grade.

### Filter operation

**Precoating.**   The precoat serves two basic purposes. First, it provides initial filtering surface to trap dirt particles when the filter begins the filtration cycle; second, it protects the septum from becoming plugged with suspended solids in the source water. It also aids in separating the cake from the septum during cleaning operations.

Successful precoating requires the uniform application of filter aid to the entire surface of the clean septum. This is accomplished by recirculating a concentrated slurry of clean water and filter medium (generally 12 percent or greater) through the filter at 1.0 to 1.5 gpm/ft$^2$ (2.4 to 3.7 m/h) until most of the medium has been deposited on the septa and the recirculating water turbidity is lowered to the desired treated water quality. Precoat recirculation may be a timed cycle based on operating experience, or it may be controlled by turbidimeter output signals.

Because particles of the medium are much smaller than the clear openings in the septum, their retention and the formation of a stable precoat take place as a result of bridging. As the particles crowd together when passing through the openings, they jam and interlock, forming a bridge over the openings (Fig. 8.34). As bridges form, additional particles are caught, and the filter septum is coated.

The amount of medium used for precoating should be adequate to cover the surface of the septum. It typically varies from 0.10 to 0.20 lb/ft$^2$ (0.5 to 1.0 kg/m$^2$) of filter area. The thickness of coating will generally be ¹⁄₁₆ to ⅛ in (1.5 to 3 mm).

The precoating system (Fig. 8.31) generally involves a small precoat preparation and storage tank equipped with a mixer as well as a larger tank used for filling up the filter. This second tank is variously referred to as a *precoating tank* or a *precoat-recycle tank*. Valves and piping connecting the tank with clean water and with the filtration vessel are also required.

**Body feeding.**   The amount of body feed to be added to the source water is determined by the nature and amount of solids to be removed. Pilot testing during representative source water quality periods will generally indicate the type and range of solids that will be encountered and the amount and type of body feed needed to provide an incompressible and permeable cake.

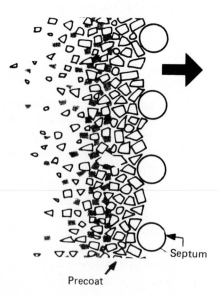

**Figure 8.34** Illustration of precoat formation with bridges of precoat between the elements of the septum material. (*Courtesy of Manville Filtration & Minerals Division.*)

Septum

Precoat

Typical body feed ratios of 1 to 10 mg/L of diatomite for each 1 mg/L of suspended solids are required, depending on the type of solids being filtered. Details about selecting the appropriate ratio are presented later. Compressible solids such as alum or iron coagulation solids require the highest ratios.

Proper control of the body feed system and rate of application is the most important factor contributing to economic operation of a precoat filtration plant. The accuracy of feed rate and continuity of feed are critical to the operation. Feeders should be designed to maintain accuracy over a wide range of feed rates so they can be closely matched to varying source water conditions. Plants may be equipped with instrumentation so that body feed rates automatically adjust to changing source water turbidity. Interruptions in body feed should be avoided, to lessen the chances of cake blinding.

Proper design of body feed metering equipment is critical for good system performance. Inadequate design in this area has been the principal cause of difficulty with precoat filtration systems in the past. Generally, body feed equipment can be classified as a dry or wet system. The wet, or slurry, feeders are the most common. The attributes of good precoat and body feed equipment have been presented in the AWWA *Precoat Filtration* manual.[119]

**Spent-cake removal.** At the end of each filter cycle, the spent filter cake is removed from the filter in preparation for the precoat that will begin the next cycle. If spent solids are not fully removed from the filter vessel, the material could be resuspended and deposited on the filter septa. The resuspended dirty material could foul the septa, although the effect would usually develop gradually so that the operator would become aware of it only after a number of cycles.

The most reliable determination of septum cleanliness requires visual observation of the bare septum, followed by observation of the uniformity and completeness of precoat. When the septa cannot be fully inspected, a higher than normal differential pressure immediately after precoating (at the start of the filtration cycle) would suggest that the septa are becoming fouled.

Techniques for removing the spent cake vary according to the different kinds of filter vessels and filter elements. The most common methods involve:

1. Sluicing the cake from the leaves with high-pressure external sprays directed on the exterior surface of the leaf

2. Reversal of flow through tubular filter elements, sometimes assisted by an air bump operation in which air under pressure is released suddenly to increase the momentary backflow

3. Draining the tank under differential air pressure, drying the cake, and then vibrating the leaves to dislodge the cake

The third method is more commonly used in industries where the liquid being filtered is quite valuable and must be reclaimed, or where dry cake could be more economically handled than a slurry.

Spent cake removed from a precoat filter is a mixture of filter media and the materials removed from the influent source water. This waste matter is usually removed from the filter in slurry form. Although some systems may be able to dispose of the slurry into a sanitary sewer system, most plants must dewater the waste material and make separate provisions for the solid and liquid wastes. Methods of waste handling are covered further in Chap. 16.

### Theoretical aspects of precoat filtration

To obtain maximum economy in the design and operation of a precoat filter, it is necessary to understand the probable effect of flow rate, terminal head loss, and body feed rate on the operation of the filter. The most important of these items affecting filter operation is the prudent use of body feed during a filter run.

**Effect of concentration of body feed.**   The results of a typical series of filter runs conducted at a constant rate of filtration and to a constant terminal pressure drop, but with varying amounts of body feed, are shown in Fig. 8.35. The total volume of filtrate is plotted versus the total head loss in feet of water. The initial head loss is caused by the clean precoat layer, septum, and filter element. With no body feed or insufficient amounts of body feed, the head loss will increase more rapidly as the run continues. This is because suspended solids that are removed on the surface of the precoat layer soon form a compressible, impermeable layer of solids, which will not readily permit the passage of water.

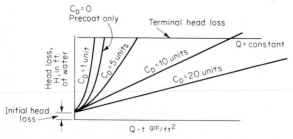

**Figure 8.35**   Effect of varying the concentration of body feed $C_D$ on the relationship of head loss to volume of filtrate for flat septa.

With increased amounts of body feed, the mixture of body feed and solids removed from the water will be less compressible and more porous and more permeable. As a result, the head loss will develop more slowly (for example, $C_D$ = 5 units in Fig. 8.35). With sufficient body feed, the cake will be incompressible, and head loss will develop in a linear manner when flat septa are used (for example, $C_D$ = 10 units in Fig. 8.35). With still higher rates of body feed, the slope of the straight-line plot will get flatter. The most economical use of body feed generally occurs at the lowest concentration that generates a linear head loss curve when flat septa are used.

The linear head loss curves shown in Fig. 8.35 for higher body feed concentrations only occur with flat septa at constant total flow rate. With tubular septa, as the body feed layer gets thicker, the area of filtration increases and the effective filtration rate decreases for the outer layers of the cake. This causes the head loss to increase more slowly as the cycle progresses.

**Effect of filtration rate.**   Consider the linear head loss curve for a flat septum at constant flow rate and adequate body feed (Fig. 8.35). The total head loss at any time is the sum of the initial head loss (caused by the precoat layer, filter element, and septum) plus the head loss of the body feed layer. The flow through the entire filter cake at the usual filtration velocities is laminar. Therefore, the following can be stated with regard to the head loss and head loss development:

1. Head loss for a particular instant (i.e., cake thickness) is proportional to the filtration rate $Q$ because of laminar-flow conditions.

2. Head loss across the body feed layer is proportional to the thickness of the layer.

3. Body feed layer thickness at constant body feed concentration and filtration rate is proportional to the volume of filtrate $V'$ = filtration rate $Q$ times running time $t$.

4. Thus, the head loss in body feed layer at any instant is proportional to $Q^2 t$.

Using the above concepts, the rate of head loss development at two different filtration rates $Q_1$ and $Q_2$ can be compared:

$$h_1 \propto Q_1^2 t_1 \quad \text{and} \quad h_2 \propto Q_2^2 t_2$$

For the case of the same terminal head loss $h_1 = h_2$, dividing these two expressions gives

$$\frac{t_1}{t_2} \simeq \frac{Q_2{}^2}{Q_1{}^2} \tag{8.21}$$

Similarly,

$$V_1' = Q_1 t_1 \quad \text{and} \quad V_2' = Q_2 t_2 \tag{8.22}$$

Dividing these two equations and inserting the previous relation for $t_1/t_2$, we get

$$\frac{V_1'}{V_2'} \simeq \frac{Q_2}{Q_1} \tag{8.23}$$

From these expressions, for a given terminal head loss, the run length is inversely proportional to the square of the filtration rate, and the filtrate volume is inversely proportional to the filtration rate. These relationships are approximate because the impact of the precoat head loss has been ignored. Precoat head loss is usually trivial, however, compared to the body feed head loss. Because of the relations illustrated by Eqs. (8.21) and (8.23), the most economical operation of precoat filters usually occurs at fairly low filtration rates, with 1 gpm/ft$^2$ (2.45 m/h) being most common but with 0.5 to 2 gpm/ft$^2$ (1.2 to 5 m/h) used in some plants.

**Mathematical model for precoat filtration.** A rigorous mathematical model has been presented for head loss development for a precoat filter operating at constant total flow rate. The model described the head loss for either flat or tubular septa operating with adequate body feed to achieve an incompressible cake.[126,127] Empirical filter cake resistance indices must be determined for the precoat and the body feed layers by pilot filter operation, in order to use the model. The body feed resistance index can be correlated empirically to the ratio of source water suspended-solids concentration to the body feed concentration.[128] The model can be used in an optimization program to select the best combination of filtration rate, terminal head loss, and body feed concentration for overall economy.[129]

## Other Filters

### Low-head continuous backwash filters

The low-head continuous backwash filter has been used on some low-turbidity source waters.[17] Several manufacturers now offer this type of filter. The filter usually consists of a shallow bed of sand about 1 ft (0.3 m) deep, with an ES of 0.4 to 0.5 mm. The bed is divided into multiple compartments with the filtrate flowing to a common effluent channel. A traveling backwash system, equipped with a backwash pump and a backwash collection system, washes each cell of the filter in succession as it traverses the length of the filter. The backwash system is capable of traversing the filter intermittently or continuously.

Because the filter is washed frequently at low head loss, the solids are removed primarily near the top surface of the fine sand. This facilitates removal of the solids during the short backwash period, which lasts only about 15 s for each cell.

This filter has the advantage that it is backwashed frequently at low head loss and, therefore, requires less depth of filter box than a conventional gravity filter. This type of filter also eliminates the need for a filter gallery, a large backwash pump, and the associated large backwash piping. Thus, the capital cost is lower than that for a conventional gravity filter plant.

Some concerns with the low-head continuous backwash system deserve mention, however. The backwash is very brief and is unassisted by auxiliary scour. Therefore, if the filter is blinded by a short-term influent flow containing high solids, it may be difficult to clean the filter by the normal backwash routine. The filter is not always able to handle high-turbidity source water and meet required filtrate quality. The flow rate and filtrate quality from individual cells are not monitored. The clean cells operate at higher rates of filtration for a brief period until the entire filter wash is completed. Thus, there is concern over the passage of solids through the newly cleaned cells during this period of higher filtration rate. The impact of the initial improvement period on the filtrate quality is of concern because the backwashes occur more frequently. Therefore, a greater portion of total operating time occurs during the initial improvement period. Research is needed to demonstrate whether these concerns are valid and to define the source water and pretreatment requirements that will result in successful filtration by the low-head continuous backwashed filter.

### Cartridge filters

A variety of cartridge filters are available for various point-of-use filtration applications. They are usually pressure filters with a medium comprised of membranes, fabric, or string. The medium is supported by a filter element and housed in a pressure vessel. The cartridge is generally disposed of after a single filter cycle.

Cartridge filters are usually rated by their manufacturers as to the particle size to be retained, with the smallest being about 0.2 μm and the largest going up to about 10 μm. Smaller retained particle sizes result in lower flow rates, higher pressure requirements, and a shorter operating period before replacement.

Cartridge filters are used for a variety of industrial needs. One use in potable water production is as a prefilter ahead of the various membrane processes such as reverse osmosis or electrodialysis (see Chap. 11).

## Emerging Technology

The desire to reduce overall costs or to treat some types of water more effectively is driving the development of new technologies involving filtration. Direct filtration, as discussed earlier, can be used only for high-quality surface water. To extend the application of direct filtration to somewhat poorer water, however, two filtration systems have been developed. Both involve the concept of a first-stage coarse media bed ahead of conventional multimedia downflow filter. Coagulating chemicals are added ahead of the coarse media bed, and it serves both as a contact flocculator and a roughing filter or clarifier to prepare the water for the multimedia filter. More coagulants or filter aid polymers may be added ahead of the second-stage filter. In one case the first stage is a bed of plastic chips with density less than that of water, so they float up against a retaining screen.[16] Flow is upward through this filter, and it is backwashed periodically with air and water simultaneously. The lower density of the air-water mixture causes the plastic to sink away from the screen, so it can be washed more effectively in a semifluidized state. The second system uses a deep bed of coarse sand as the first-stage contact "clarifier" that is operated downflow, and washed periodically upflow, using air and water simultaneously.[130]

The poor settleability of algae-laden surface water after coagulation and flocculation has led to the development of proprietary systems incorporating flotation and filtration in the same tank. See Chap. 7 for further details.

## Waste Disposal

Wastes generated in the filter operations covered in this chapter include waste backwash water from rapid filters and precoat filters and dirty sand from slow sand filter scraping.

Backwash water in rapid filtration plants is often recovered and recycled through the plant, as discussed earlier. In conventional plants, it may be recycled without sedimentation, in which case the backwash solids end up in the sedimentation tank waste solids. In direct and in-line filtration plants, the solids must be separated before recycling because this is the only waste solids discharge from the plant. The separated solids may be generated in a slurry that must be handled as a waste stream.

Spent cake removed from precoat filters at the end of each filter cycle is a mixture of filter media and organic and inorganic solids removed from the source water. It is usually transported as a slurry to a solids-separation step. The water after solids separation may be recycled or wasted. The solids are quite dewaterable because of the high content of body feed and precoat media, and they may be further dewatered for land disposal.

Dirty sand removed from slow sand filters is a mixture of sand, biological organisms from the schmutzdecke, and other organic and inorganic solids captured from the source water. The dirty sand is usually cleaned hydraulically and stockpiled for later replacement in the filter. The hydraulic cleaning generates a slurry of all solids except the sand. This slurry must be handled as a waste stream. If the sand is not recovered, the dirty sand can be disposed of as a solid waste.

Methods of handling the waste streams or solids from water plants are covered further in Chap. 16.

## References

1. S. L. Graese, V. L. Snoeyink, and R. G. Lee, *GAC Filter-Adsorbers,* American Water Works Association, Research Foundation, Denver, 1987.
2. ASTM, *1985 Annual Book of ASTM Standards,* vol. 04.02, *Concrete and Mineral Aggregates,* American Society for Testing and Materials, Philadelphia, 1985.
3. W. C. Krumbein, "Measurement and Geological Significance of Shape and Roundness of Sedimentary Particles," *J. Sed. Petrol.,* vol. 11, no. 2, August 1941, p. 64.
4. W. L. McCabe and J. C. Smith, *Unit Operations of Chemical Engineering,* 3d ed., McGraw-Hill, New York, 1976.
5. J. L. Cleasby and K. S. Fan, "Predicting Fluidization and Expansion of Filter Media," *J. Environ. Eng. Div. ASCE,* vol. 107 (EE3), June 1981, p. 455.
6. J. M. Trefethen, *Geology for Engineers,* 2d ed., Van Nostrand, Princeton, N.J., 1959.
7. American Water Works Association, *AWWA Standard for Granular Activated Carbon,* Standard B604-74, Denver, 1974.
8. A. H. Dharmarajah and J. L. Cleasby, "Predicting the Expansion of Filter Media," *J. AWWA,* vol. 78, no. 12, December 1986, p. 66.
9. J. L. Cleasby and C. W. Woods, "Intermixing of Dual and Multi-Media Granular Filters," *J. AWWA,* vol. 67, no. 4, April 1975, p. 197.
10. G. M. Fair, J. C. Geyer, and D. A. Okun, *Water and Wastewater Engineering,* vol. 2, Wiley, New York, 1968.
11. T. R. Camp, "Theory of Water Filtration," *J. Sanit. Eng. Div. ASCE,* vol. 90, no. 4, 1964, p. 1.
12. S. Ergun, "Fluid Flow through Packed Columns," *Chem. Eng. Prog.,* vol. 48, no. 2, February 1952, p. 89.
13. S. Ergun, "Determination of Geometric Surface Area of Crushed Porous Solids," *Anal. Chem.,* vol. 24, 1952, p. 388.
14. C. Y. Wen and Y. H. Yu, "Mechanics of Fluidization," *Chemical Engineering Progress Symposium Series 62,* Amer. Inst. of Chem. Engrs., vol. 62, New York, 1966.
15. Great Lakes Upper Mississippi River Board of State Sanitary Engineers, *Recom-*

*mended Standards for Water Works,* Health Education Service, Albany, N.Y., 1982.

16. H. H. Benjes, C. E. Edlund, and P. T. Gilbert, "Adsorption Clarifier Applied to Low Turbidity Surface Supplies," *Proc. AWWA Annual Conference,* Washington, June 1985.

17. S. Medlar, "This Filter Cleans Itself," *Am. City,* vol. 89, no.6, June 1974, p. 63.

18. L. Huisman, *Rapid Filtration, Part 1,* Delft University of Technology, Delft, The Netherlands, 1974.

19. American Water Works Association, *AWWA Standard for Filtering Material,* Standard B100-80, Denver, 1980.

20. American Water Works Association, *Water Quality and Treatment,* 3d ed., McGraw-Hill, New York, 1971.

21. G. W. Fuller, *The Purification of the Ohio River Water at Louisville, Kentucky,* Van Nostrand, New York, 1898.

22. W. L. Harris, "High-Rate Filter Efficiency," *J. AWWA,* vol. 62, no. 8, August 1970, p. 515.

23. Los Angeles Department of Water and Power, "Project Report on Filtration of Los Angeles Aqueduct Supply," July 1979.

24. R. R. Trussell, A. R. Trussell, J. S. Lang, and C. H. Tate, "Recent Development in Filtration System Design," *J. AWWA,* vol. 72, no. 12, December 1980, p. 705.

25. J. L. Cleasby and E. R. Baumann, "Selection of Sand Filtration Rates," *J. AWWA,* vol. 54, no. 5, May 1962, p. 579.

26. A. Amirtharajah and D. P. Wetstein, "Initial Degradation of Effluent Quality during Filtration," *J. AWWA,* vol. 72, no. 9, September 1980, p. 518.

27. G. S. Logsdon, J. M. Symons, R. L. Hoye, Jr., and M. M. Arozarena, "Removal of *Giardia* Cysts and Cyst Models by Filtration," *J. AWWA,* vol. 73, no. 2, February 1981, p. 111.

28. J. C. Kirner, J. D. Littler, and L. A. Angelo, "A Waterborne Outbreak of Giardiasis in Camus, Washington," *J. AWWA,* vol. 70, no. 1, January 1978, p. 41.

29. C. P. Hibler, C. M. Hancock, L. M. Perger, J. G. Wegrzyn, and K. D. Swabby, *Inactivation of Giardia Cysts with Chlorine at 0.5°C to 5°C,* American Water Works Association, Research Foundation, Denver, 1987.

30. G. G. Robeck, K. A. Dostal, and R. L. Woodward, "Studies of Modification in Water Filtration," *J. AWWA,* vol. 56, no. 2, February 1964, p. 198.

31. J. L. Cleasby, "Approaches to a Filterability Index for Granular Filters," *J. AWWA,* vol. 61, no. 8, August 1969, p. 372.

32. G. S. Logsdon, J. M. Symons, and T. J. Sorg, "Monitoring Water Filters for Asbestos Removal," *J. Environ. Eng. Div. ASCE,* vol. 107, no. 6, December 1981, p. 1297.

33. G. S. Logsdon, V. C. Thurman, E. S. Frindt, and J. G. Stoecker, "Evaluating Sedimentation and Various Filter Media for Removal of *Giardia* Cysts," *J. AWWA,* vol. 77, no. 2, February 1985, p. 61.

34. G. S. Logsdon and J. M. Symons, "Removal of Asbestiform Fibers by Water Filtration," *J. AWWA,* vol. 69, no. 9, September 1977, p. 499.

35. J. T. O'Connor, B. J. Brazos, W. C. Ford, L. L. Dusenberg, and B. Summerford, "Chemical and Microbiological Evaluation of Drinking Water Systems in Missouri," *Proc. AWWA Water Quality Technology Conference,* American Water Works Association, Denver, 1984.

36. C. R. O'Melia and W. Stumm, "Theory of Water Filtration," *J. AWWA,* vol. 59, no. 11, November 1967, p. 1393.

37. J. P. Herzig, D. M. LeClerc, and P. LeGoff, "Flow of Suspensions through Porous Media," *Ind. Eng. Chem.,* vol. 62, no. 5, May 1970, p. 8.

38. K. M. Yao, M. T. Habibian, and C. O'Melia, "Water and Waste Water Filtration: Concepts and Applications," *Environ. Sci. Tech.,* vol. 5, no. 11, November 1971, p. 1105.

39. P. M. Heertges and C. E. Lerk, "The Functioning of Deep Filters, Part I: The Filtration of Flocculated Suspensions," *Trans. Inst. Chem. Eng.,* vol. 45, 1967, p. T138.

40. A. Adin and M. Rebhun, "A Model to Predict Concentration and Head Loss Profiles in Filtration," *J. AWWA,* vol. 69, no. 8, August 1977, p. 444.

41. C. Tien, R. M. Turian, and H. Pendse, "Simulation of the Dynamic Behavior of Deep Bed Filters," *J. Am. Inst. Chem. Eng.,* vol. 25, no. 3, May 1979, p. 385.
42. A. A. Saatci and C. S. Oulman, "The Bed Depth Service Time Design Method for Deep Bed Filtration," *J. AWWA,* vol. 72, no. 9, September 1980, p. 524.
43. G. Clough and K. J. Ives, "Deep Bed Filtration Mechanisms Observed with Fiber Optic Endoscopes and CCTV," *Proc. 4th World Filtration Congress, Part II,* Royal Flemish Society of Engineers, Antwerp, Belgium, 1986.
44. R. Eliassen, "Clogging of Rapid Sand Filters," *J. AWWA,* vol. 33, no. 5, May 1941, p. 926.
45. D. R. Stanley, "Sand Filtration Studies with Radio-Tracers," *Proc. Am. Soc. Civil. Eng.,* vol. 81, 1955, p. 592.
46. J. T. Ling, "A Study of Filtration through Uniform Sand Filters," *Proc. ASCE Sanit. Eng. Div.,* Paper 751, 1955.
47. K. J. Ives and I. Sholji, "Research on Variables Affecting Filtration," *J. Sanit. Eng. Div. ASCE,* vol. 91, no. 4, August 1965, p. 1.
48. K. Y. Hsiung and J. L. Cleasby, "Prediction of Filter Performance," *J. Sanit. Eng. Div. ASCE,* vol. 94, no.6, December 1968, p. 1043.
49. W. G. Brown, "High-Rate Filtration Experience at Durham, N.C.," *J. AWWA,* vol. 47, no. 3, March 1955, p. 243.
50. B. A. Segall and D. A. Okun, "Effect of Filtration Rate on Filtrate Quality," *J. AWWA,* vol. 58, no. 3, March 1966, p. 368.
51. H. E. Hudson, Jr., "High Quality Water Production and Viral Disease," *J. AWWA,* vol. 54, no. 10, October 1962, p. 1265.
52. J. R. Baylis, "Seven Years of High-Rate Filtration," *J. AWWA,* vol. 48, no. 5, May 1956, p. 585.
53. W. R. Conley and R. W. Pitman, "Test Program for Filter Evaluation at Hanford," *J. AWWA,* vol. 52, no. 2, February 1960, p. 205.
54. W. R. Conley, "Experiences with Anthracite Sand Filters," *J. AWWA,* vol. 53, no. 12, December 1961, p. 1473.
55. W. R. Conley and K. Y. Hsuing, "Design and Application of Multimedia Filters," *J. AWWA,* vol. 61, no. 2, February 1969, p. 97.
56. P. H. King, R. L. Johnson, C. W. Randall, and G. W. Rehberger, "High-Rate Water Treatment: The State of the Art," *J. Environ. Eng. Div. ASCE,* vol. 101, no. 4, August 1975, p. 479.
57. J. L. Cleasby, M. M. Williamson, and E. R. Baumann, "Effect of Rate Changes on Filtered Water Quality," *J. AWWA,* vol. 55, no. 7, July 1963, p. 869.
58. J. L. Tuepker and C. A. Buescher, Jr., "Operation and Maintenance of Rapid Sand Mixed-Media Filters in a Lime Softening Plant," *J. AWWA,* vol. 60, no. 12, December 1968, p. 1377.
59. L. DiBernardo and J. L. Cleasby, "Declining Rate versus Constant Rate Filtration," *J. Environ. Eng. Div. ASCE,* vol. 106, no. 6, December 1980, p. 1023.
60. J. L. Cleasby, "Filter Rate Control without Rate Controllers," *J. AWWA,* vol. 61, no. 4, April 1969, p. 181.
61. James M. Montgomery Consulting Engineers, *Water Treatment Principles and Design,* Wiley, New York, 1985.
62. T. R. Camp, "Discussion of Conley—Experiences with Anthracite Filters," *J. AWWA,* vol. 53, no. 12, December 1961, p. 1478.
63. J. R. Baylis, "Discussion of Conley and Pitman," *J. AWWA,* vol. 52, no. 2, February 1960, p. 214.
64. W. R. Conley and R. W. Pitman, "Innovations in Water Clarification," *J. AWWA,* vol. 52, no. 10, October 1960, p. 1319.
65. W. R. Conley, "Integration of the Clarification Process," *J. AWWA,* vol. 57, no. 10, October 1965, p. 1333.
66. W. R. Conley, "High Rate Filtration," *J. AWWA,* vol. 64, no. 3, March 1972, p. 203.
67. G. G. Robeck, "Discussion of Conley," *J. AWWA,* vol. 58, no. 1, January 1966, p. 94.
68. M. C. Culbreath, "Experience with a Multi-Media Filter," *J. AWWA,* vol. 59, no. 8, August 1967, p. 1014.
69. J. E. Laughlin and T. E. Duvall, "Simultaneous Plant Scale Tests of

Mixed-Media and Rapid Sand Filters," *J. AWWA,* vol. 60, no. 9, September 1968, p. 1015.

70. G. P. Westerhoff, "Experience with Higher Filtration Rates," *J. AWWA,* vol. 63, no. 6, June 1971, p. 376.

71. D. L. Peterson, F. X. Schleppenbach, and T. M. Zaudtke, "Studies of Asbestos Removal by Direct Filtration of Lake Superior Water," *J. AWWA,* vol. 72, no. 3, March 1980, p. 155.

72. G. S. Logsdon, *Water Filtration for Asbestos Fiber Removal,* EPA 600/2-79-206, Environmental Protection Agency, December 1979.

73. G. J. Kirmeyer, *Seattle Tolt Water Supply Mixed Asbestiform Removal Study,* EPA 600/2-79-125, Environmental Protection Agency, August 1979.

74. C. H. Tate and R. R. Trussell, "Use of Particle Counting in Developing Plant Design Criteria," *J. AWWA,* vol. 70, no. 12, December 1978, p. 691.

75. E. W. J. Diaper and K. J. Ives, "Filtration through Size-Graded Media," *J. Sanit. Eng. Div. ASCE,* vol. 91, no. 3, June 1965, p. 89.

76. A. E. Rimer, "Filtration through a Trimedia Filter," *J. Sanit. Eng. Div. ASCE,* vol. 94, no. 3, June 1968, p. 521.

77. R. W. Oeben, H. P. Haines, and K. J. Ives, "Comparison of Normal and Reverse Graded Filtration," *J. AWWA,* vol. 60, no. 4, April 1968, p. 429.

78. S. S. Mohanka, "Multilayer Filtration," *J. AWWA,* vol. 61, no. 10, October 1969, p. 504.

79. Committee Report, "The Status of Direct Filtration," *J. AWWA,* vol. 72, no. 7, July 1980, p. 405.

80. R. L. Culp, "Direct Filtration," *J. AWWA,* vol. 68, no. 6, July 1977, p. 375.

81. W. R. Hutchison, "High-Rate Direct Filtration," *J. AWWA,* vol. 68, no. 6, June 1976, p. 292.

82. E. G. Wagner and H. E. Hudson, "Low-Dosage, High-Rate Direct Filtration," *J. AWWA,* vol. 74, no. 5, May 1982, p. 256.

83. J. L. Cleasby, D. J. Hilmoe, C. J. Dimitracopoulos, and L. M. Diaz-Bossio, *Effective Filtration Methods for Small Water Supplies,* USEPA Cooperative Agreement CR808837-01-0, NTIS No. PB84-187-905, Environmental Protection Agency, 1984.

84. D. A. Bowers, A. E. Bowers, and D. D. Newkirk, "Development and Evaluation of a Coagulant Control Test Apparatus for Direct Filtration," *Proc. AWWA Water Quality Technology Conference,* Nashville, 1982.

85. J. F. Kreissl, G. G. Robeck, and G. A. Sommerville, "Use of Pilot Filters to Predict Optimum Chemical Feeds," *J. AWWA,* vol. 60, no. 3, March 1968, p. 299.

86. D. G. McBride, R. C. Siemak, C. H. Tate, and R. R. Trussell, "Pilot Plant Investigations for Treatment of Owens River Water," *Proc. AWWA Annual Conference,* Anaheim, Calif., June 1977.

87. C. H. Tate, J. S. Lang, and H. L. Hutchinson, "Pilot Plant Tests of Direct Filtration," *J. AWWA,* vol. 69, no. 7, July 1977, p. 379.

88. J. T. Monscvitz, D. J. Rexing, R. G. Williams, and J. Heckler, "Some Practical Experience in Direct Filtration," *J. AWWA,* vol. 70, no. 10, October 1978, p. 584.

89. R. D. Letterman and G. S. Logsdon, "Survey of Direct Filtration Practice," *Proc. AWWA Annual Conference,* New Orleans, June 1976.

90. J. W. Ellms, *Water Purification,* 1st ed., McGraw-Hill, New York, 1917.

91. J. V. Arboleda, "Hydraulic Control Systems of Constant and Declining Flow Rate in Filtration," *J. AWWA,* vol. 66, no. 2, February 1974, p. 87.

92. J. Arboleda-Valencia, "Hydraulic Behavior of Declining-Rate Filtration," *J. AWWA,* vol. 77, no. 12, December 1985, p. 67.

93. W. W. Aultman, "Valve Operating Devices and Rate-of-Flow Controllers," *J. AWWA,* vol. 51, no. 11, November 1959, p. 1467.

94. J. R. Baylis, "Variable Rate Filtration," *Pure Water,* vol. 11, May 1959, p. 5.

95. J. L. Cleasby and L. DiBernardo, "Hydraulic Considerations in Declining-Rate Filtration Plants," *J. Environ. Eng. Div. ASCE,* vol. 106, no. 6, December 1980, p. 1043.

96. J. L. Cleasby, "Declining Rate Filtration," *J. AWWA,* vol. 73, no. 9, September 1981. p. 484.

97. H. E. Hudson, Jr., "Filter Design—Declining Rate Filtration," *J. AWWA*, vol. 51, no. 11, November 1959, p. 1455.
98. Committee Report, "Comparison of Alternative Systems for Controlling Flow through Filters," *J. AWWA*, vol. 76, no. 1, January 1984, p. 91.
99. J. L. Cleasby, A. Arboleda, D. E. Burns, P. W. Prendiville, and E. S. Savage, "Backwashing of Granular Filters (AWWA Filtration Subcommittee Report)," *J. AWWA*, vol. 69, no. 2, February 1977, p. 115.
100. J. R. Baylis, "Review of Filter Design and Methods of Washing," *J. AWWA*, vol. 51, no. 11, November 1959, p. 1433.
101. A. Amirtharajah, "Optimum Backwashing of Sand Filters," *J. Environ. Eng. Div. ASCE*, vol. 104, no. 5, October 1978, p. 917.
102. G. Degremont, *Water Treatment Handbook*. Stephen Austin & Sons Ltd., Caxton Hill, Hertford, England, 1973.
103. M. F. Dehab and J. C. Young, "Unstratified-Bed Filtration of Wastewater," *J. Environ. Eng. Div. ASCE*, vol. 103, no. 1, February 1977, p. 21.
104. S. R. Hewitt and A. Amirtharajah, "Air Dynamics through Filter Media during Air Scour," *J. Environ. Eng. Div. ASCE*, vol. 110, no. 3, June 1984, p. 591.
105. A. Amirtharajah, "Fundamentals and Theory of Air Scour," *J. Environ. Eng. Div. ASCE*, vol. 110, no. 3, June 1984, p. 573.
106. A. H. Vicory and L. Weaver, "Controlling Discharges of Water Plant Wastes to the Ohio River," *J. AWWA*, vol. 76, no. 4, April 1984, p. 122.
107. Committee Report, "Microbial Considerations for Drinking Water Regulation Revisions," *J. AWWA*, vol. 79, no. 5, May 1987, p. 81.
108. H. E. Babbitt, J. J. Doland, and J. L. Cleasby, *Water Supply Engineering*, 6th ed., McGraw-Hill, New York, 1962.
109. J. R. Baylis, "Nature and Effect of Filter Backwashing," *J. AWWA*, vol. 51, no. 1, January 1959, p. 131.
110. E. C. Lippy, "Tracing a Giardiasis Outbreak at Berlin, N.H.," *J. AWWA*, vol. 70, no. 9, September 1978, p. 512.
111. L. A. Slezak and R. C. Sims, "The Application and Effectiveness of Slow Sand Filtration in the United States," *J. AWWA*, vol. 76, no. 12, December 1984, p. 38.
112. T. R. Cullen and R. D. Letterman, "The Effect of Slow Sand Filter Maintenance on Water Quality," *J. AWWA*, vol. 77, no. 12, December 1985, p. 48.
113. J. L. Cleasby, D. J. Hilmoe, and C. J. Dimitracopoulos, "Slow Sand and Direct In-Line Filtration of a Surface Water," *J. AWWA*, vol. 76, no. 12, December 1984, p. 44.
114.. W. D. Bellamy, G. P. Silverman, D. W. Hendricks, and G. S. Logsdon, "Removing *Giardia* Cysts with Slow Sand Filtration," *J. AWWA*, vol. 77, no. 2, February 1985, p. 52.
115. K. R. Fox, R. J. Miltner, G. S. Logsdon, D. L. Dicks, and L. F. Drolet, "Pilot Plant Studies of Slow-Rate Filtration." *J. AWWA*, vol. 76, no. 12, December 1984, p. 62.
116. W. D. Bellamy, D. W. Hendricks, and G. S. Logsdon, "Slow Sand Filtration: Influences of Selected Process Variables," *J. AWWA*, vol. 77, no. 12, December 1985, p. 62.
117. E. R. Baumann, T. Willrich, and D. D. Ludwig, "For Purer Water Supply, Consider Prechlorination," *Agric. Eng.*, vol. 44, no. 3, 1963, p. 138. (The above is a condensation of "Effect of Prechlorination on Filtration and Disinfection," Paper 62-208, American Society of Agricultural Engineers, June 19, 1962.)
118. T. J. Seelaus, D. W. Hendricks, and B. A. Janonis, "Design and Operation of a Slow Sand Filter," *J. AWWA*, vol. 78, no. 12, December 1986, p. 35.
119. AWWA, *Precoat Filtration*, AWWA Manual M30, Denver, 1988.
120. A. Bryant and C. Yapijakis, "Ozonation-Diatomite Filtration Removes Color and Turbidity," *Water Sewage Works*, Part 1, 124:96 (September 1977); *Part 2*, 124:94 (October 1977).
121. J. L. Ris, I. A. Cooper, and W. R. Goodard, "Pilot Testing and Predesign of Two Water Treatment Processes for Removal of *Giardia lamblia* in Palisade, Colorado," *Proc. AWWA Annual Conference*, Dallas, June 1984.
122. K. P. Lange, W. D. Bellamy, and D. W. Hendricks, "Diatomaceous Earth Filtration of *Giardia* Cysts and Other Substances," *J. AWWA*, vol. 78, no. 1, January 1986, p. 76.

123. T. S. Brown, J. F. Malina, Jr., and B. D. Moore, "Virus Removal by Diatomaceous Earth Filtration, Part 2," *J. AWWA,* vol. 66, no. 12, December 1974, p. 735.

124. E. R. Baumann, "Diatomite Filters for Removal of Asbestos Fibers," *Proc. AWWA Annual Conference,* Minneapolis, June 1975.

125. D. E. Burns, E. R. Baumann, and C. S. Oulman, "Particulate Removal on Coated Filter Media," *J. AWWA,* vol. 62, no. 2, February 1970, p. 121.

126. J. H. Dillingham, J. L. Cleasby, and E. R. Baumann, "Diatomite Filtration Equations for Various Septa," *J. Sanit. Eng. Div. ASCE,* vol. 93, no. 1, February 1967, p. 41.

127. W. J. Weber, *Physicochemical Processes for Water Quality Control,* Wiley-Interscience, New York, 1972.

128. J. H. Dillingham, J. L. Cleasby, and E. R. Baumann, "Prediction of Diatomite Filter Cake Resistance," *J. Sanit. Eng. Div. ASCE,* vol. 93, no. 1, February 1967, p. 57.

129. J. H. Dillingham, J. L. Cleasby, and E. R. Baumann, "Optimum Design and Operation of Diatomite Filtration Plants," *J. AWWA,* vol. 58, no. 6, June 1966, p. 657.

130. J. S. MacNeill, Jr., and A. MacNeill, *Feasibility Study of Alternative Technology for Small Community Water Supply, Project Summary,* EPA-600/S2-84-191, USEPA Water Engineering Research Laboratory, Cincinnati, March 1985.

# Ion Exchange
# and Inorganic Adsorption

## Dennis A. Clifford, Ph.D., P.E.

*Professor of Environmental Engineering*
*Department of Civil and Environmental Engineering*
*University of Houston*
*Houston, Texas*

Contaminant cations such as calcium, magnesium, barium, strontium, and radium and anions such as fluoride, nitrate, fulvates, humates, arsenate, selenate, chromate, and anionic complexes of uranium can be removed from water by using ion exchange or adsorption onto activated alumina. The theory and practice of these processes are the subject of this chapter.

Ion exchange with synthetic resins and adsorption onto activated alumina are water treatment processes in which a presaturant ion on the solid phase, the "adsorbent," is exchanged for an unwanted ion in the water. To accomplish the exchange reaction, a packed bed of ion-exchange resin beads or alumina granules is used. Source water is continually passed through the bed in a downflow or upflow mode until the adsorbent is exhausted, as evidenced by the appearance (breakthrough) of the unwanted contaminant at an unacceptable concentration in the bed effluent.

The most useful ion-exchange reactions are reversible. In the simplest cases, the exhausted bed is regenerated by using an excess of the presaturant ion. Ideally, no permanent structural change takes place during the exhaustion/regeneration cycle. (Resins do swell and shrink, however, and alumina is partially dissolved during regeneration.) When the reactions are reversible, the medium can be reused many times before it must be replaced because of irreversible fouling or, in the case of alumina, attrition losses. In a typical water supply application, from 300 to as many as 60,000 bed volumes[†] (BV) of contaminated water may be treated before exhaustion. Regeneration typically requires from 1 to 5 BV of regenerant followed by 2 to 20 BV of rinse water. These wastewaters generally amount to less than 2 percent of the product water; nevertheless, their ultimate disposal is a major consideration in modern design practice. Disposal of the spent medium may also pose a problem if it contains a toxic or radioactive substance such as arsenic or radium.

## Uses of Ion Exchange

By far the largest application of ion exchange to drinking water treatment is in the area of softening, i.e., the removal of calcium, magnesium, and other polyvalent cations in exchange for sodium. The ion-exchange softening process is applied for both individual home use and municipal treatment. It can be applied for whole house [point-of-entry (POE)] softening or for softening only the water that enters the hot-water heater. Radium and barium ions are preferred by the resin to calcium and magnesium ions; thus the former are also effectively removed during ion-exchange softening. Resin beds containing chloride-form anion-exchange resins can be used for nitrate, arsenate, chromate, and selenate removal, and more applications of these processes will be seen in the future. Activated alumina is being used to remove fluoride and arsenate from drinking water, particularly in water with total dissolved solids (TDS) on a point-of-use (POU), POE, and municipal scales.

The feasibility of ion exchange or alumina adsorption is determined by the water quality (particularly the TDS level), competing ions, alkalinity, contaminant concentration, and the affinity of the resin or alumina for the contaminant ion in comparison to the competing ions. The affinity sequence determines the run length, chromatographic peaking (if any), and process costs. As previously mentioned, process selection will be affected by spent regenerant and spent medium disposal requirements, particularly if "hazardous" materials are involved. Each of these requirements is dealt with in detail in the upcoming design sections for specific processes, summarized in Table 9.1.

## Past and Future of Ion Exchange

Natural zeolites, i.e., crystalline aluminosilicates, were the first ion exchangers used to soften water on a commercial scale. Later, zeolites were completely replaced with the synthetic resins because of the

---

[†]A bed volume (BV) of feedwater is a volume equal to the volume of the resin beads plus the voids between particles, that is, the empty-bed volume.

TABLE 9.1     Advantages and Disadvantages of Packed-Bed Inorganic-Contaminant Removal Processes

Ion Exchange

Advantages
- Operates on demand
- Relatively insensitive to flow variations
- Essentially zero level of effluent contaminant possible
- Large variety of specific resins available
- Beneficial selectivity reversal commonly occurs upon regeneration

Disadvantages
- Potential for chromatographic effluent peaking
- Spent regenerant must be disposed of
- Variable effluent quality with respect to background ions
- Usually not feasible at high TDS levels

Activated Alumina Adsorption

Advantages
- Operates on demand
- Insensitive to flow and TDS background
- Low effluent contaminant level possible
- Highly selective for fluoride and arsenic

Disadvantages
- Both acid and base required for regeneration
- Medium tends to dissolve, producing fine particles
- Slow adsorption kinetics
- Spent regenerant must be disposed of

latter's faster exchange rates, longer life, and higher capacity. Aside from softening, the use of ion exchange for the removal of specific contaminants from municipal water supplies has been limited. This is primarily because of the expense involved in removing what is perceived as a minimal health risk resulting from a contaminant such as fluoride, nitrate, or chromate. The production of pure and ultrapure water by ion-exchange demineralization is the largest use of ion-exchange resins on a commercial scale. This complete removal of contaminants is not necessary for drinking water treatment, however. Furthermore, treatment costs are high compared to those for the alternative membrane processes (reverse osmosis and electrodialysis) for desalting water (see Chap. 11).

Adherence to the Safe Drinking Water Act's *maximum contaminant level* (MCL) goals for *inorganic contaminants* (IOCs) will mean greater use of ion exchange and alumina for small community water treatment to remove radium, nitrate, fluoride, barium, and other IOCs. A recent AWWA survey[1] indicates that 400 communities exceed the 10 mg/L nitrate-N MCL, 400 exceed the 4.0 mg/L fluoride MCL,[2] and 200 exceed the 1.5 mg/L secondary limit on barium. Regarding radiological contaminants, an estimated 550 communities exceed the 5.0 pCi/L MCL for radium,[3] and many others may exceed the MCL goal for radon contamination when it is established. In most of these cases, new contaminant-free sources cannot be readily developed, and a treatment system will eventually be installed.

## Ion-Exchange Materials and Reactions

An ion-exchange resin consists of a cross-linked polymer matrix to which charged functional groups are attached by covalent bonding. The usual matrix is polystyrene cross-linked for structural stability with 3 to 8% divinylbenzene. The common functional groups fall into four categories: strongly acidic (e.g., sulfonate, $-SO_3^-$), weakly acidic (e.g., carboxylate, $-COO^-$), strongly basic [e.g., quaternary amine, $-N^+(CH_3)_3$], and weakly basic [e.g., tertiary amine, $-N(CH_3)_2$].

A schematic presentation of resin matrix cross-linking and functionality is shown in Fig. 9.1. The figure is a schematic three-dimensional bead (sphere) made up of many polystyrene polymer chains held together by divinylbenzene cross-linking. The negatively charged ion-exchange sites ($-SO_3^-$) or ($-COO^-$) are fixed to the resin backbone, or *matrix,* as it is called. Mobile positively charged counterions (plus charges in the figure) are associated by electrostatic attraction with each negative ion-exchange site. The resin exchange capacity is measured as the number of fixed charge sites per unit volume or weight of resin. "Functionality" is the term used to identify the chemical composition of the fixed charge site, e.g., sulfonate ($-SO_3^-$) or carboxylate ($-COO^-$). Porosity (e.g., microporous, gel, or macroporous) is the resin characterization referring to the degree of openness of the polymer structure. An actual resin bead is much tighter than on the schematic, which is shown as fairly open for illustration purposes only. Finally, the water (40 to 60 percent by weight) present in a typical resin bead is not shown. Note, however, that this resin-bound water is an extremely important characteristic of ion exchangers because it strongly influences both the resin kinetics and thermodynamics.

### Strong- and weak-acid cation exchanges

*Strong-acid cation* (SAC) exchangers operate over a very wide pH range because the sulfonate group, being strongly acidic, is ionized throughout the pH range (1 to 14). Three typical SAC exchange reactions are shown below. In Eq. (9.1), the neutral salt $CaCl_2$, representing noncarbonate hardness, is said to be "split" by the resin, and hydrogen ions are exchanged for calcium ions even though the equilibrium liquid phase is acidic because of HCl production. Equa-

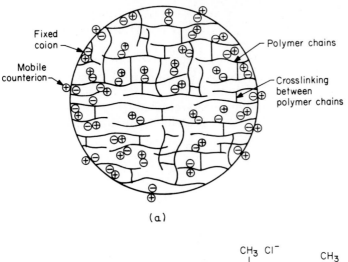

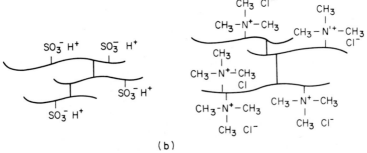

(b)

**Figure 9.1** (a) Organic cation-exchanger bead comprising polystyrene polymer cross-linked with divinylbenzene with fixed coions (minus charges) of negative charge balanced by mobile positively charged counterions (plus charges). (b) Strong-acid cation exchanger (*left*) in the hydrogen form and strong-base anion exchanger in the chloride form.

tions (9.2) and (9.3) are the standard ion-exchange softening reactions in which sodium ions are exchanged for the hardness ions $Ca^{2+}$, $Mg^{2+}$, $Fe^{2+}$, $Ba^{2+}$, $Sr^{2+}$, or $Mn^{2+}$, as either noncarbonate hardness [Eq. (9.2)] or carbonate hardness [Eq. (9.3)]. In all these reactions, R denotes the resin matrix, and the overbar indicates the solid (resin) phase.

$$\overline{2RSO_3^-H^+} + CaCl_2 \rightleftharpoons \overline{(RSO_3^-)_2Ca^{2+}} + 2HCl \qquad (9.1)$$

$$\overline{2RSO_3^-Na^+} + CaCl_2 \rightleftharpoons \overline{(RSO_3^-)_2Ca^{2+}} + 2NaCl \qquad (9.2)$$

$$\overline{2RSO_3^-Na^+} + Ca(HCO_3)_2 \rightleftharpoons \overline{(RSO_3^-)_2Ca^{2+}} + 2NaHCO_3 \qquad (9.3)$$

Regeneration of the spent resin is accomplished using an excess of concentrated (1 to 3 $M$) HCl or NaCl, and constitutes the reversal of Eqs. (9.1) through (9.3).

*Weak-acid cation* (WAC) resins can exchange ions only in the neutral to alkaline pH range because the functional group, typically carboxylate ($pK_a$ = 4.8), is not ionized at low pH. Thus, WAC resins can be used for carbonate hardness removal [Eq. (9.4)] but not noncarbonate hardness removal, as is evident in Eq. (9.5) that proceeds to the left as written:

$$2\overline{RCOOH} + Ca(HCO_3)_2 \rightarrow \overline{(RCOO^-)_2Ca^{2+}} + H_2CO_3 \qquad (9.4)$$

$$2\overline{RCOOH} + CaCl_2 \rightarrow \overline{(RCOO^-)_2Ca^{2+}} + 2HCl \qquad (9.5)$$

If Eq. (9.5) proceeded to the right, the HCl produced would be so completely ionized that it would protonate, i.e., add a hydrogen ion to the resin's weakly acidic carboxylate functional group and prevent exchange of $H^+$ ions for $Ca^{2+}$ ions. Another way of expressing the nonoccurrence of Eq. (9.5) is to say that WAC resins will not "split neutral salts," i.e., they cannot remove noncarbonate hardness. This is not the case in Eq. (9.4) in which the basic salt $Ca(HCO_3)_2$ is "split" because a very weak acid, $H_2CO_3$ ($pK_1$ = 6.3), is produced.

In summary, SAC resins split basic and neutral salts (remove carbonate and noncarbonate hardness), whereas WAC resins split only basic salts (remove only carbonate hardness). Nevertheless, as pointed out in detail later, WAC resins have some distinct advantages for softening, namely, TDS reduction, no increase in sodium, and very efficient regeneration resulting from the carboxylate's high affinity for the regenerant $H^+$ ion.

### Strong- and weak-base anion exchangers

The use of *strong-base anion* (SBA) exchange resins for nitrate removal is a recent innovation in drinking water treatment,[4,5] although they have been used in water demineralization for decades. In anion exchange reactions with SBA resins, the quaternary amine functional group $[-N(CH_3)_3^+]$ is so strongly basic that it is ionized and is therefore useful as an ion exchanger over the entire pH range of 1 to 13. This is shown in Eqs. (9.6) and (9.7) in which nitrate is removed from water by using hydroxide- or chloride-form SBA resins.

$$\overline{R_4N^+OH^-} + NaNO_3 \rightarrow \overline{R_4N^+NO_3} + NaOH \qquad (9.6)$$

$$\overline{R_4N^+Cl^-} + NaNO_3 \rightarrow \overline{R_4N^+NO_3^-} + NaCl \qquad (9.7)$$

In Eq. (9.6) the caustic (NaOH) produced is completely ionized, but the quaternary ammonium functional group has such a small affinity for $OH^-$ ions that the reaction proceeds as written. Equation (9.7) is a simple ion-exchange reaction without a pH change. Fortunately, all SBA resins have a much higher affinity for nitrate than chloride,[4] and Eq. (9.7) proceeds as written at near-neutral pH values.

*Weak-base anion* (WBA) exchange resins are useful only in the acidic pH region where the primary, secondary, or tertiary amine functional groups (Lewis bases) are protonated and thus can act as positively charged exchange sites for anions. In Eq. (9.8) chloride is, in effect, being adsorbed by the WBA resin as hydrochloric acid, and the TDS level of the solution is being reduced. In this case, a positively charged Lewis acid-base adduct ($R_3NH^+$) is formed that can act as an anion-exchange site. As long as the solution in contact with the resin remains acidic (just how acidic depends on basicity of the $R_3N$:, sometimes a pH less than or equal to 6 is adequate), ion exchange can take place as indicated in Eq. (9.9)—the exchange of chloride for nitrate by a WBA resin in acidic solution. If the solution is neutral or basic, no adsorption or exchange can take place, as indicated by Eq. (9.10). In all these reactions, R represents either the resin matrix or a functional group such as $-CH_3$ or $-C_2H_5$, and overbars represent the resin phase.

$$\overline{R_3N:} + HCl \rightarrow \overline{R_3NH^+Cl^-} \tag{9.8}$$

$$\overline{R_3NH^+Cl^-} + HNO_3 \rightleftharpoons \overline{R_3NH^+NO_3^-} + HCl \tag{9.9}$$

$$\overline{R_3N:} + NaNO_3 \rightarrow \text{no reaction} \tag{9.10}$$

Although no common uses of WBA resins are known for drinking water treatment, useful ones are possible.[4] Furthermore, activated alumina, when used for fluoride and arsenic removal, acts as if it were a weak-base anion exchanger, and the same general rules regarding pH behavior can be applied. Another advantage of weak-base resins in water supply applications is the ease with which they can be regenerated with bases. Even weak bases such as lime [$Ca(OH)_2$] can be used, and regardless of the base used, only a slight stoichiometric excess (less than 20 percent) is normally required for complete regeneration.

## Activated alumina adsorption

Packed beds of activated alumina can be used to remove fluoride, arsenic, selenium, silica, and humic materials from water. The mechanism, which is one of exchange of contaminant anions for surface hydroxides on the alumina, is generally called *adsorption,* although *ligand exchange* and *chemisorption* are more appropriate terms for the highly specific surface reactions involved.[6]

The typical activated aluminas used in water treatment are 28 × 48 mesh (0.3- to 0.6-mm-diameter) mixtures of amorphous and gamma aluminum oxide ($\gamma$-$Al_2O_3$) prepared by low-temperature (300 to 600°C) dehydration of $Al(OH)_3$. They have surface areas of 50 to 300 $m^2$/g. By using the model of a hydroxylated alumina surface subject to protonation and deprotonation, the following ligand-exchange reaction [Eq. (9.11)] can be written for fluoride adsorption in acid solution (alumina exhaustion) in which $\equiv$Al represents the alumina surface and an overbar denotes the solid phase.

$$\overline{\equiv Al - OH} + H^+ + F^- \rightarrow \overline{\equiv Al - F} + HOH \qquad (9.11)$$

The equation for fluoride desorption by hydroxide (alumina regeneration) is

$$\overline{\equiv Al - F} + OH^- \rightarrow \equiv Al - OH + F^- \qquad (9.12)$$

Another common application for alumina is arsenic removal, and reactions similar to Eqs. (9.11) and (9.12) apply for exhaustion and regeneration when $H_2AsO_4^-$ is substituted for $F^-$.

Activated alumina processes are sensitive to pH, and anions are best adsorbed below pH 8.2, a typical *zero point of charge* (ZPC) below which the alumina surface has a net positive charge and excess protons are available to fuel [Eq. (9.11)]. Above the ZPC, alumina is predominantly a cation exchanger, but its use for cation exchange is relatively rare in water treatment. An exception is encountered in the removal of radium by plain and treated activated alumina.[7,8]

Ligand exchange as indicated in Eqs. (9.11) and (9.12) occurs chemically at the surface of activated alumina. A more useful model for process design, however, is one that assumes that the adsorption of fluoride or arsenic onto alumina at the optimum pH of 5 to 6 is analogous to WBA exchange. For example, the uptake of $F^-$ or $H_2AsO_4^-$ requires a protonation of the alumina surface, and that is accomplished by preacidification with HCl or $H_2SO_4$ and reduction of the feedwater pH to approximately 6.0. The positive charge caused by excess surface protons may then be viewed as being balanced by exchanging anions, i.e., ligands such as hydroxide, flouride, and arsenate. To reverse the adsorption process and remove the adsorbed fluoride or arsenate, an excess of a strong base, e.g., NaOH, must be applied. The following series of reactions is presented as a model of the adsorption-regeneration cycle that is useful for design purposes.

The first step in the cycle is acidification in which neutral (water-washed) alumina (alumina · HOH) is treated with acid, e.g., HCl, and acidic alumina is formed as follows:

$$\overline{\text{Alumina} \cdot \text{HOH}} + \text{HCl} \rightarrow \overline{\text{alumina} \cdot \text{HCl}} + \text{HOH} \qquad (9.13)$$

When HCl-acidified alumina is contacted with fluoride ions, they strongly displace the chloride ions, providing that the alumina surface remains acidic (pH 5 to 6). This displacement of chloride by fluoride, analogous to ion exchange, is shown as

$$\overline{\text{Alumina} \cdot \text{HCl}} + \text{HF} \rightarrow \overline{\text{alumina} \cdot \text{HF}} + \text{HCl} \qquad (9.14)$$

To regenerate the fluoride-contaminated adsorbent, a dilute solution (0.25 to 0.5 $N$) of NaOH is used. Because alumina is both a cation and an anion exchanger, $Na^+$ is exchanged for $H^+$ which immediately combines with $OH^-$ to form HOH in the alkaline regenerant solution. The regeneration reaction of fluoride-spent alumina is

$$\overline{\text{Alumina} \cdot \text{HF}} + 2\text{NaOH} \rightarrow \overline{\text{alumina} \cdot \text{NaOH}} + \text{NaF} + \text{HOH} \qquad (9.15)$$

To restore the fluoride removal capacity, the basic alumina is acidified by contacting it with an excess of dilute (typically 0.5 $N$) HCl;

$$\overline{\text{Alumina} \cdot \text{NaOH}} + 2\text{HCl} \rightarrow \overline{\text{alumina} \cdot \text{HCl}} + \text{NaCl} + \text{HOH} \qquad (9.16)$$

The acidic alumina, alumina $\cdot$ HCl, is now ready for another fluoride (or arsenic or selenium) adsorption cycle, as summarized by Eq. (9.14). Alternatively, the feedwater may be acidified prior to contact with the basic alumina, thereby combining acidification and adsorption into one step, summarized by

$$\overline{\text{Alumina} \cdot \text{NaOH}} + \text{NaF} + 2\text{HCl} \rightarrow \overline{\text{alumina} \cdot \text{HF}} + 2\text{NaCl} + \text{HOH} \qquad (9.17)$$

The modeling of the alumina adsorption-regeneration cycle as being analogous to WBA exchange fails in regard to regeneration efficiency that is excellent for weak-base resins but quite poor on alumina. This is caused by the need for excess acid and base to partially overcome the poor kinetics of the semicrystalline alumina that exhibits very low solid-phase diffusion coefficients compared to resins that are well hydrated, flexible gels offering little resistance to the movement of hydrated ions. A further reason for poor regeneration efficiency on alumina is that alumina is amphoteric and reacts with (consumes) excess acid and base to produce soluble forms $[\text{Al}(\text{H}_2\text{O})_6^{3+}, \text{Al}(\text{H}_2\text{O})_2(\text{OH})_4^-]$ of aluminum. Resins are totally inert in this regard; i.e., they are not dissolved by regenerants.

### Special-purpose resins

Resins are practically without limit in their variety because polymer matrices, functional groups, capacity, and porosity are controllable during manufacture. Thus, numerous special-purpose resins have been made for water treatment applications. Bacteriostatic resins for water and resin disinfection are one example. Here long-chain quaternary amine functional groups have been added to anion resins to kill bacteria on contact with the resin surface.[9] Bacterial growth can be a major problem with anion resins used in some water supply applications because the positively charged resins tend to "absorb" the negatively charged bacteria that metabolize the adsorbed organic material—negatively charged humate and fulvate anions. Therefore, bacteriostatic anion resins have a promising future in water treatment.

The strong attraction of polyvalent humate and fulvate anions (aquatic humus) for anion resins has been used as a treatment technique for removal of these *total organic carbon* (TOC) compounds from water by using special highly porous resins. Both weak- and strong-base macroporous anion exchangers have been manufactured to remove these large anions from water. A difficulty is that the very porous resins necessary for adsorption of such large molecules tend to be structurally weak and break down easily. Regeneration can also be a problem because of the strong attraction of the aromatic portion of the anions for the aromatic resin matrix. Nevertheless, macroporous resins are being used for this purpose, and the search for better ones continues.[10]

Resins with chelating functional groups such as phosphoric acid or ethylenediaminetetracetic acid (EDTA) have been manufactured[11] that have extremely high affinities for hardness ions and troublesome metals such as $Cu^{2+}$, $Zn^{2+}$, $Cr^{3+}$, $Pb^{2+}$, and $Ni^{2+}$. These resins have a promising future in trace-metal removal and metals recovery operations, particularly with respect to decontamination of hazardous waste. The simplified structures of these resins are shown in Fig. 9.2.

## Ion-Exchange Equilibrium

### Selectivity coefficients and separation factors

Ion-exchange resins do not prefer all ions equally. This variability in preference is often expressed semiquantitatively, as a position in a selectivity sequence, or quantitatively, as a separation factor $\alpha_{ij}$ or a selectivity coefficient $K_{ij}$ for binary exchange. The selectivity, in turn, determines the run length to breakthrough for the contaminant ion; the higher the selectivity, the longer the run length. Consider, for example, Eq. (9.18), the simple exchange of $Cl^-$ for $NO_3^-$ on an anion exchanger, whose equilibrium constant is expressed in Eq. (9.19) and graphically in Fig. 9.3a:

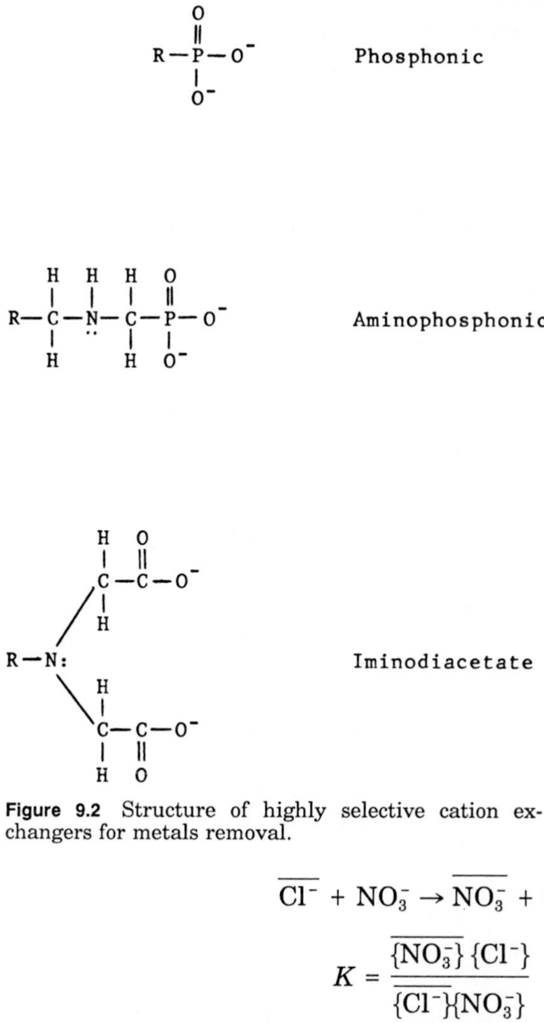

R—P—O⁻ (Phosphonic)

$$\text{R}-\overset{\overset{\displaystyle O}{\|}}{\text{P}}-\text{O}^- \qquad \text{Phosphonic}$$

Aminophosphonic

Iminodiacetate

Figure 9.2 Structure of highly selective cation exchangers for metals removal.

$$\overline{\text{Cl}^-} + \text{NO}_3^- \rightarrow \overline{\text{NO}_3^-} + \text{Cl}^- \tag{9.18}$$

$$K = \frac{\{\overline{\text{NO}_3^-}\}\{\text{Cl}^-\}}{\{\overline{\text{Cl}^-}\}\{\text{NO}_3^-\}} \tag{9.19}$$

In these equations [Eqs. (9.18) to (9.20)], overbars denote the resin phase, and the matrix designation R has been removed for simplicity; $K$ is the thermodynamic equilibrium constant, and braces denote ionic activity. Concentrations are used in practice because they are measured more easily than activities. In this case, Eq. (9.20) based on concentration, the selectivity coefficient $K_{\text{N/Cl}}$ describes the exchange. Note that $K_{\text{N/Cl}}$ includes activity coefficient terms that are functions of ionic strength and thus is not a true constant, varying with different ionic strengths.

$$K_{\text{N/Cl}} = \frac{[\overline{\text{NO}_3^-}][\text{Cl}^-]}{[\overline{\text{Cl}^-}][\text{NO}_3^-]} = \frac{q_{\text{N}}C_{\text{Cl}}}{q_{\text{Cl}}C_{\text{N}}} \tag{9.20}$$

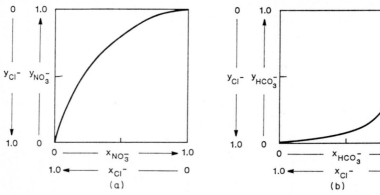

**Figure 9.3** (a) Favorable isotherm for nitrate-chloride exchange according to reaction (9.18) with constant separation factor $\alpha_{NO_3/Cl} > 1.0$. (b) Unfavorable isotherm for bicarbonate-chloride exchange with constant separation factor $\alpha_{HCO_3/Cl} < 1.0$.

where [ ] = concentration, mol/L

$q_N$ = resin-phase concentration of nitrate, equiv/L

$C_N$ = aqueous phase concentration of nitrate, equiv/L

The binary separation factor $\alpha_{N/Cl}$, used throughout chemical engineering separation practice, is a most useful description of the exchange equilibria because of its simplicity and intuitive nature:

$$\alpha_{ij} = \frac{\text{distribution of ion } i \text{ between phases}}{\text{distribution of ion } j \text{ between phases}} = \frac{y_i/x_i}{y_j/x_j} \tag{9.21}$$

For the example above

$$\alpha_{N/Cl} = \frac{y_N/x_N}{y_{Cl}/x_{Cl}} = \frac{y_N x_{Cl}}{x_N y_{Cl}} = \frac{(q_N/q)(C_{Cl}/C)}{(C_N/C)(q_{Cl}/q)} = \frac{q_N C_{Cl}}{q_{Cl} C_N} \tag{9.22}$$

where $y_i$ = equivalent fraction of ion $i$ in resin, $q_i/q$

$y_N$ = equivalent fraction of nitrate in resin, $q_N/q$

$x_i$ = equivalent fraction of ion $i$ in water, $C_N/C$

$x_N$ = equivalent fraction of nitrate in water, $C_N/C$

$q_N$ = concentration of nitrate on resin, equiv/L

$q$ = total exchange capacity of resin, equiv/L

$C_N$ = nitrate concentration of water, equiv/L

$C$ = total ionic concentration of water, equiv/L

Equations (9.20) and (9.22) show that for homovalent exchange, i.e., monovalent/monovalent and divalent/divalent exchange, the separation factor $\alpha_{ij}$ and the selectivity coefficient $K_{ij}$ are equal. This is expressed for nitrate/chloride exchange as

$$K_{N/Cl} = \alpha_{N/Cl} = \frac{q_N C_{Cl}}{C_N q_{Cl}} \qquad (9.23)$$

For exchanging ions of unequal valence, i.e., heterovalent exchange, the separation factor is not equivalent to the selectivity coefficient. Consider, for example, the case of sodium ion-exchange softening as represented by Eq. (9.24), the simplified form of Eq. (9.2):

$$2\overline{Na^+} + Ca^{2+} = \overline{Ca^{2+}} + 2Na^+ \qquad (9.24)$$

$$K_{Ca/Na} = \frac{q_{Ca} C_{Na}^2}{C_{Ca} q_{Na}^2} \qquad (9.25)$$

Using a combination of Eqs. (9.21) and (9.25), we obtain

$$\alpha_{divalent/monovalent} \text{ or } \alpha_{Ca/Na} = K_{Ca/Na} \frac{(q/C)}{(y_{Na}/x_{Na})} \qquad (9.26)$$

The implication of these equations is that the intuitive separation factor for divalent/monovalent exchange depends on the solution concentration $C$ and on the distribution ratio $y_{Na}/x_{Na}$ between the resin and water, with $q$ constant. The higher the solution concentration, the lower the divalent/monovalent separation factor; i.e., selectivity tends to reverse in favor of the monovalent ion as the ionic strength increases. This reversal of selectivity is discussed in detail below.

## Selectivity sequences

A *selectivity sequence* describes the order in which ions are preferred by a particular resin or alumina. Although special-purpose resins such as chelating resins can have unique selectivity sequences, commercially available cation and anion resins exhibit similar selectivity sequences. These are presented in Table 9.2 where the most preferred ions, i.e., those with the highest separation factors, are listed at the top of the table and the least preferred ions at the bottom. For example, the $\alpha_{Ca/Na}$ value of 1.9 means that at equal concentrations in the aqueous phase, calcium is preferred by the resin 1.9/1.0 over sodium. Weak-acid cation resins with carboxylic functional groups exhibit the same selectivity sequence as SAC resins, except that hydrogen is the most preferred cation and the magnitudes of the separation factors differs from those in Table 9.2. Similarly, WBA resins and SBA resins exhibit the same selectivity sequence, except that hydroxide is most preferred by WBA resins and the WBA separation factors differ from those in Table 9.2.

Some general rules govern selectivity sequences. In dilute solution,

**TABLE 9.2    Relative Affinities[†] of Ions for Resins**

| Strong-acid cation resins[‡] | | Strong-base anion resins[§] | |
|---|---|---|---|
| Cation $i$ | $\alpha_{i/Na^+}$[¶] | Anion $i$ | $\alpha_{i/Cl^-}$[¶] |
| $Ra^{2+}$ | 13.0 | $CrO_4^{2-}$ | 100.0 |
| $Ba^{2+}$ | 5.8 | $SeO_4^{2-}$ | 17.0 |
| $Pb^{2+}$ | 5.0 | $SO_4^{2-}$ | 9.1 |
| $Sr^{2+}$ | 4.8 | $HSO_4^-$ | 4.1 |
| $Cu^{2+}$ | 2.6 | $NO_3^-$ | 3.2 |
| $Ca^{2+}$ | 1.9 | $Br^-$ | 2.3 |
| $Zn^{2+}$ | 1.8 | $HAsO_4^{2-}$ | 1.5 |
| $Fe^{2+}$ | 1.7 | $SeO_3^{2-}$ | 1.3 |
| $Mg^{2+}$ | 1.67 | $HSO_3^{3-}$ | 1.2 |
| $K^+$ | 1.67 | $NO_2^-$ | 1.1 |
| $Mn^{2+}$ | 1.6 | $Cl^-$ | 1.0 |
| $NH_4^+$ | 1.3 | $HCO_3^-$ | 0.27 |
| $Na^+$ | 1.0 | $CH_3COO^-$ | 0.14 |
| $H^+$ | 0.67 | $F^-$ | 0.07 |

†Above values are approximate separation factors for 0.01 $N$ solution (TDS = 500 mg/L as $CaCO_3$).
‡SAC resin is polystyrene divinylbenzene matrix with sulfonate functional groups.
§SBA resin is polystyrene divinylbenzene matrix with $—N^+(CH_3)_3$ functional groups, i.e., a type 1 resin.
¶All $\alpha_{ij}$ values except for those of $Ra^{2+}$, $SeO_4^{2-}$, $HAsO_4^{2-}$, $SeO_3^{2-}$, and $CrO_4^{2-}$ are taken from the *Duolite Ion Exchange Manual*, Diamond Shamrock Chemical Company, Redwood City, CA, 1969, pp. 21, 23 (out of print). The remaining affinity values are based on research done at the University of Houston.

i.e., in the TDS range of natural water, the resin prefers ions with the highest charge and lowest degree of hydration.

From the point of view of an ion, hydrophobic ions (such as nitrate and chromate) prefer hydrophobic resins, i.e., highly cross-linked resins (macroporous) without polar functional groups, whereas hydrophilic ions such as bicarbonate and acetate prefer moderately cross-linked (gel) resins with polar functional groups. Divalent ions (such as sulfate and calcium) prefer resins with closely spaced exchange sites where their need for two charges can be satisfied.[13–15]

Activated alumina operated in the acidic to neutral pH range for anion adsorption has a selectivity sequence that differs markedly from that of anion-exchange resins. Fortunately, some of the ions, such as fluoride and arsenate, that are least preferred by resins (and therefore are not amenable to removal by resins) are highly preferred by alumina. Based on unpublished data from the University of Houston and those of other investigators,[16,17] activated alumina operated in the pH range of 5.5 to 8.5 prefers anions in the following order:

$$OH^- > H_2AsO_4^- > Si(OH)_3O^- > F^- > HSeO_3^- > SO_4^{2-} > CrO_4^{2-} \gg$$
$$HCO_3^- > Cl^- > NO_3^- > Br^- > I^- \quad (9.27)$$

Humic and fulvic acid anions are preferred to sulfate, but because of their widely differing molecular weights and structures and the different pore-size distributions of commercial aluminas, no exact position in the sequence can be assigned. Reliable separation factors for ions in the above selectivity sequence such as fluoride, arsenate, silicate, and biselenite are not available in the literature, but this is not particularly detrimental to the design effort because alumina has an extreme preference for these ions. For example, when fluoride or arsenate is removed from water, the presence of the usual competing ions, bicarbonate and chloride, is nearly irrelevant in establishing the run length to contaminant ion breakthrough.[18,19] Sulfate does, however, offer some small but measurable competition for adsorption sites. The problem with the extremely preferred ions is that they are difficult to remove from the alumina during regeneration.

## Isotherm Plots

The values of $\alpha_{ij}$ and $K_{ij}$ can be determined from a constant-temperature equilibrium plot of resin-phase concentration versus aqueous-phase concentration, i.e., the ion-exchange isotherm. Favorable and unfavorable isotherms are depicted in Fig. 9.3a and b where each curve depicts a constant separation factor, $\alpha_{NO_3/Cl}$ for Fig. 9.3a and $\alpha_{HCO_3/Cl}$ for Fig. 9.3b.

A "favorable" isotherm (convex to the $x$ axis) means that species $i$ ($NO_3^-$ in Fig. 9.3a), which is plotted on each axis, is preferred to species $j$ ($Cl^-$ in Fig. 9.3a), the hidden or exchanging species. An "unfavorable" isotherm (concave to the $x$ axis) indicates that species $i$ ($HCO_3^-$ in Fig. 9.3b) is less preferred than $j$ ($Cl^-$ in Fig. 9.3b). During column exhaustion processes, favorable isotherms result in sharp breakthroughs when $i$ is in the feed and $j$ is on the resin, whereas unfavorable isotherms lead to gradual breakthroughs under these conditions. (This is discussed in detail under the heading "Column Processes and Calculations," where Fig. 9.8 is explained.) In viewing these binary isotherms, note that

$$x_i + x_j = 1.0 \quad (9.28)$$

$$C_i + C_j = C \quad (9.29)$$

$$y_i + y_j = 1.0 \quad (9.30)$$

$$q_i + q_j = q \quad (9.31)$$

Therefore, the concentration or equivalent fraction of either ion can be directly obtained from the plot, which in Fig. 9.3a and b is a "unit" isotherm because equivalent fractions $(x_i, y_j)$ rather than concentrations have been plotted, in the range of 0.0 to 1.0. Figure 9.3a represents the favorable isotherm for nitrate-chloride exchange and Fig. 9.3b the unfavorable isotherm for chloride-bicarbonate exchange.

For nonconstant separation factors such as the divalent-monovalent $(Ca^{2+}/Na^+)$ exchange case described by Eqs. (9.24) and (9.26), a separate isotherm exists for every total solution concentration $C$. As the solution concentration or TDS level decreases, the resin exhibits a greater preference for the divalent ion, as evidenced by a progressively higher and more convex isotherm. The phenomenon can be explained by solution theory: As the solution concentration increases, the aqueous phase becomes more ordered. This results in polyvalent ion activity coefficients that are significantly less than 1.0; i.e., the tendency for polyvalent ions to escape from the water into the resin is greatly diminished, leading to a reduction in the height and convexity of the isotherm. This phenomenon of diminishing preference for higher-valence ions with increasing ionic strength $I$ of the solution has been labeled "electroselectivity" and can eventually lead to selectivity reversal, whereupon the isotherm becomes concave. This trend is shown in Fig. 9.4 where the sulfate-chloride isotherm is favorable in $0.06\ N$ solution and unfavorable in $0.6\ N$ solution.

The exact ionic strength at which electroselectivity reversal occurs is dependent on the ionic makeup of the solution and highly dependent on the resin structure[12] and its inherent affinity for polyvalent ions. Electroselectivity reversal is very beneficial to the sodium-ion-exchange softening process in that it causes the divalent hardness ions to be highly preferred in dilute solution $(I \le 0.020\ M)$ during resin exhaustion and highly nonpreferred, i.e., easily rejected, during regeneration with relatively concentrated (0.25 to 3.0 $M$) salt solution.

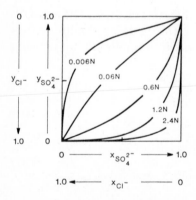

**Figure 9.4**  Electroselectivity of a typical type 1 strong-base anion-exchange resin used for divalent-monovalent $(SO_4^{2-}/Cl^-)$ anion exchange. (Based on data in Ref. 12.)

**Example Problem 1**  This example problem briefly describes the experimental technique needed to obtain isotherm data and illustrates the calculations required to construct a nitrate-chloride isotherm for a strong-base anion-exchange resin. By using the isotherm data or the plot, the individual and average separation factors $\alpha_{ij}$ can be calculated. Only minor changes are needed to apply the technique to weak-base resins or cation resins. For example, acids (HCl and $HNO_3$) rather than sodium salts would be used for equilibration of weak-base resins.

To obtain the data for this example, weighted amounts of air-dried chloride-form resin of known exchange capacity were placed into capped bottles containing 100 mL of 0.005 $N$ (5.0 meq/L) $NaNO_3$ and equilibrated by tumbling for 16 h. Following equilibration, the resins were settled, and the nitrate and chloride concentrations of the supernatant water were determined for each bottle. The nitrate and chloride equilibrium data can be found in Table 9.3. The total capacity $q$ of the resin is 3.63 meq/g. Note that the units of resin capacity used here are milliequivalents per gram rather than equivalents per liter because, for precise laboratory work, a mass rather than a volume of resin must be used.

*Outline of solution*

1. Verify that, within the expected limits of experimental error, the total concentration $C$ of the aqueous phase at equilibrium is 0.005 $N$. Large deviations from this value usually indicate that concentrated salts were absorbed in the resin and leached out during the equilibration procedure. This problem can be avoided by extensive prewashing of the resin with the same normality of salt, in this case 0.005 $N$ NaCl, as is used for equilibration.

2. Calculate the equivalent fractions $x_N$ and $x_{Cl}$ of nitrate and chloride, respectively, in the water at equilibrium.

3. Using the known total capacity of the resin $q_{Cl}$, calculate the milliequivalents of chloride remaining on the resin at equilibrium by subtracting the chloride found in the water.

4. Calculate the milliequivalents of nitrate on the resin $q_N$ by assuming that all the nitrate removed from solution is taken up by the resin.

5. Calculate the equivalent fractions $y_N$ and $y_{Cl}$ of nitrate and chloride, respectively, in the resin phase at equilibrium.

6. Calculate the separation factor $\alpha_{ij}$, which is equal to the selectivity coefficient $K_{ij}$ for homovalent exchange.

**TABLE 9.3    Example Data for Plot of Nitrate-Chloride Isotherm**

| $g$ resin 100 mL | $C_N$, meq/L | $C_{Cl}$, meq/L | $C$, meq/L | $x_N$ | $X_{Cl}$ | $y_N$ | $y_{Cl}$ | $\alpha_{ij}$ |
|---|---|---|---|---|---|---|---|---|
| 0.020 | 4.24 | 0.722 | 4.96 | 0.854 | 0.146 | 0.98 | 0.020 | 8.6 |
| 0.040 | 3.56 | 1.32 | 4.88 | 0.730 | 0.27 | 0.92 | 0.091 | 4.25 |
| 0.100 | 2.18 | 2.77 | 4.98 | 0.440 | 0.56 | 0.76 | 0.240 | 4.12 |
| 0.200 | 1.17 | 3.78 | 4.95 | 0.236 | 0.764 | 0.52 | 0.48 | 3.43 |
| 0.400 | 0.53 | 4.36 | 4.89 | 0.108 | 0.892 | 0.30 | 0.70 | 3.59 |
| 1.20 | 0.185 | 4.49 | 4.68 | 0.040 | 0.96 | 0.11 | 0.890 | 2.99 |

*Note:*  The first three columns represent experimental data. The remaining italicized columns were obtained by calculation as described in the example.

7. Repeat steps 1 through 6 for all equilibrium data points, and plot the isotherm.

**solution (The equilibrium data point for 0.200-g resin is chosen as an example.)**

1. We know that

$$C = C_N + C_{Cl} = 1.17 + 3.78 = 4.95 \text{ meq/L}$$

[This is well within the expected $\pm 5$ percent limits of experimental error; $(5.00 - 4.95)/5.00 = 1.0$ percent error.]

2.

$$x_N = \frac{C_N}{C} = \frac{1.17 \text{ meq/L}}{4.95 \text{ meq/L}} = 0.24$$

$$x_{Cl} = \frac{C_{Cl}}{C} = \frac{3.78 \text{ meq/L}}{4.95 \text{ meq/L}} = 0.76$$

*Check:* $x_N + x_{Cl} = 0.24 + 0.76 = 1.00$

3. Calculate the chloride remaining on the resin at equilibrium $q_{Cl}$:

$$q_{Cl} = q_{Cl, \text{ initial}} - \text{chloride lost to water per gram of resin}$$

$$q_{Cl, \text{ initial}} = q = 3.63 \text{ meq/g}$$

$$q_{Cl} = 3.63 \text{ meq/g} - (3.78 \text{ meq/L}) \frac{0.100 \text{ L}}{0.200 \text{ g}} = 1.74 \text{ meq/g}$$

4. Calculate the nitrate on the resin at equilibrium:

$$q_N = q_{N, \text{ initial}} + \text{nitrate lost from water per gram of resin}$$

$$q_N = 0 + [(5.00 - 1.17) \text{ meq/L}] \frac{0.100 \text{ L}}{0.200 \text{ g}} = 1.91 \text{meq/g}$$

*Check:* $q_N + q_{Cl} = 1.74 + 1.91 = 3.65 \text{ meq/g}$    (within 5 percent of 3.63)

5. Calculate the resin-phase equivalent fractions $y_N$ and $y_{Cl}$ at equilibrium.

$$y_N = \frac{1.91 \text{ meq/g}}{3.65 \text{ meq/g}} = 0.503 = 0.52$$

$$y_{Cl} = \frac{1.74 \text{ meq/g}}{3.65 \text{ meq/g}} = 0.497 = 0.48$$

6. Calculate the separation factor $\alpha_{ij}$.

$$\alpha_{ij} = \frac{y_N x_{Cl}}{x_N y_{Cl}} = \frac{0.52 \times 0.76}{0.24 \times 0.48} = 3.43$$

*Note:* Each data point will have an associated $\alpha_{ij}$ value. These $\alpha_{ij}$ values can be averaged, but it is preferable to plot the isotherm data, construct the best-fit curve, and use the curve at $x_N = 0.5$ to obtain an average $\alpha_{ij}$ value. The bad data points will be evident in the plot and can be ignored. Due to mathematical sensitivity, resin nonhomogeneity, and imprecise experimental data, the calculated $\alpha_{ij}$ values are not constant, as seen in Table 9.3. The $\alpha_{ij}$ values at the ends of the isotherm are particularly nonrepresentative.

7. Plot the isotherm of $y$ versus $x_N$. The nitrate versus chloride isotherm plot should appear similar to Fig. 9.3$a$.

## Ion-Exchange and Adsorption Kinetics

### Pure ion-exchange rates

As is usual with interphase mass transfer involving solid particles, resin kinetics are governed by liquid- and solid-phase resistances to mass transfer. The liquid-phase resistance, modeled as the stagnant thin film, can be minimized by providing turbulence around the particle, such as that resulting from fluid velocity in packed beds or mechanical mixing in batch operations. The speed of "pure" ion-exchange reactions, i.e., reactions not involving WAC resins in the RCOOH form or free-base forms of weak-base resins, can be attributed to the inherently low mass transfer resistance of the resin phase that is caused by its well-hydrated gelular nature. Resin beads typically contain 40 to 60 percent water in their boundaries, and this water can be considered as a continuous extension of the aqueous phase in the flexible polymer network. This pseudo continuous aqueous phase in conjunction with the flexibility of the resin phase can result in rapid kinetics for pure ion-exchange reactions.

### Alumina and SBA resins compared

Unlike adsorption onto *granular activated carbon* (GAC) or activated alumina, requiring on the order of hours to days to reach equilibrium, pure ion exchange using resins is a rapid process at near-ambient temperature. For example, the half time to equilibrium for adsorption of arsenate onto granular 28 × 48 mesh (0.32- to 0.65-mm-diameter) activated alumina was found to be approximately 2 days,[19] while the half time to equilibrium during the exchange of arsenate for chloride on a strong-base resin was only 5 min.[20] Similarly, the exchange of sodium for calcium on a SAC resin is essentially complete within 5 min.[21]

### Rates involving tight resin forms

In contrast, ion exchange with WAC and WBA resins can be very slow because of the tight, nonswollen nature of the acid form ($\overline{\text{RCOOH}}$) of WAC resins or free-base form, e.g., $R_3N$:, of WBA resins. In reactions involving this tight form, the average solid-phase diffusion coefficients change drastically during the course of the exchange that is often described by the progressive-shell shrinking-core model[22,23] depicted in Fig. 9.5. In these reactions, which are effectively

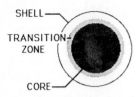

SHELL

TRANSITION
ZONE

CORE

**Figure 9.5** Progressive-shell model of ion exchange with weak resins.

neutralization reactions, either the shell or the core can be the swollen (more hydrated) portion, and a rather sharp line of demarcation exists between the tight and swollen zones.

Consider, for example, the practical case of softening with WAC resins in the $H^+$ form [Eq. (9.4)]. As the reaction proceeds, the hydrated, calcium-form shell comprising $\overline{(RCOO^-)_2Ca^{2+}}$ expands inward and replaces the shrinking, poorly hydrated core of $\overline{RCOOH}$. The entire process is reversed upon regeneration with acid, and the tight shell ($\overline{RCOOH}$) thickens as it proceeds inward and replaces the porous, disappearing core of $\overline{(RCOO^-)_2Ca^{2+}}$.

In some cases, pure ion exchange with weak resins is possible, however, and proceeds as rapidly as pure ion exchange with strong resins. For example, the "pure" exchange of sodium for calcium on a WAC resin [Eq. (9.32)] does not involve conversion of the resin RCOOH in contrast to Eq. (9.4) and would take place in a matter of minutes as with SAC resins [Eq. (9.3)].

$$2\overline{RCOO^-Na^+} + Ca(HCO_3)_2 = \overline{(RCOO^-)_2Ca^{2+}} + 2NaHCO_3 \qquad (9.32)$$

Although weak resins involving $\overline{RCOOH}$ and $\overline{R_3N}$: may require several hours to attain equilibrium in a typical batch exchange, they may still be used effectively in column processes where the contact time between the water and the resin is only 1 to 5 min. There are two reasons: (1) an overwhelming amount of unspent resin is present relative to the amount of water in the column and (2) the resin is typically exposed to the feedwater for periods in excess of 24 h before it is exhausted. Prior to exhaustion, the overwhelming ratio of resin-exchange sites present in the column to exchanging ions present in the column water nearly guarantees that an ion will be removed by the resin before the water carrying the ion exits the column. This removal takes place in the "adsorption" or "ion-exchange" or "mass transfer" zone (See Fig. 9.6) carrying the breakthrough curve of interest. For column processes the true ion-exchange equilibrium time is the time (typically more than 24 h) needed for this reaction zone to pass through the column, and not the time (typically < 3 min) needed for the water to pass through the column. This same line of reasoning applies to adsorption using GAC and activated alumina, media which typically require several days to equilibrate in batch processes. GAC and alumina columns are usually run for many days prior to regeneration because, as in ion-exchange columns, the adsorption capacity of the medium is enormous compared to the contaminants present in the feedwater.

In summary, ion exchange of small inorganic ions using strong resins is fundamentally a fast, interphase transfer process because these resins are well-hydrated gels exhibiting large solid-phase diffusion coefficients and little resistance to mass transfer. This is not the case with weak resins in the acid ($\overline{RCOOH}$) or free-base ($\overline{R_3N}$:) forms, nor is it true for alumina because these media offer considerably more solid-phase diffusion resistance. Irrespective of fast or slow batch kinetics, all these media can be effectively used in column processes for contaminant removal from water because columns exhibit enormous contaminant removal capacity and are exhausted over a period of many hours to many days. Leakage of contaminants will, however, be much more significant when media are used that exhibit relatively slow kinetics of mass transfer.

## Column Processes and Calculations

### Binary ion exchange

Ion-exchange and adsorption column operations do not result in a fixed percentage of removal of contaminant with time, which would result, for example, in a steady-state coagulation process. These column processes exhibit a variable degree of contaminant removal and gradual or sharp contaminant breakthroughs similar to (but generally much more complicated than) the breakthrough of turbidity through a granular filter. First, we consider the hypothetical case of pure binary ion exchange before proceeding to the practical drinking water treatment case of multicomponent ion exchange.

If pure calcium chloride solution is softened by continuously passing it through a bed of resin in the sodium form, ion exchange [Eq. (9.2)] immediately occurs in the uppermost differential segment of the bed (at its inlet). Here all the resin is converted to the calcium form in the moving ion-exchange zone where mass transfer between the liquid and solid phases occurs. These processes are depicted in Fig. 9.6.

The resin phase experiences a calcium wave front that progresses through the column until it reaches the outlet. Now no more sodium-form resin exists to take up calcium, and calcium "breaks through" into the effluent, as shown in Fig. 9.7. In this pure binary ion-

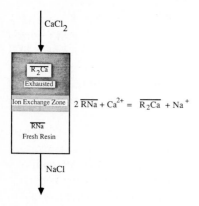

**Figure 9.6**  Resin concentration profile for binary ion exchange of sodium for calcium.

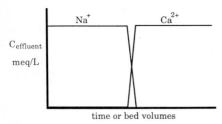

**Figure 9.7** Effluent concentration histories (breakthrough curves) for the softening reaction in Fig. 9.6.

exchange case, the effluent calcium concentration can never exceed that of the influent; this is, however, generally not true for multicomponent ion exchange, as we show later. The sharpness of the calcium breakthrough curve depends on both equilibrium (i.e., selectivity) and kinetic (i.e., mass transfer) considerations. Imperfect, i.e., noninstantaneous, interphase mass transfer of sodium and calcium always acts to reduce the sharpness of the breakthrough curve and results in a broadening of the adsorption (ion-exchange) zone. This is equivalent to saying that nonequilibrium (noninstantaneous) mass transfer produces a diffuse calcium wave and a somewhat gradual calcium breakthrough.

A breakthrough curve can be gradual even if mass transfer is instantaneous because the first consideration in determination of the shape is the resin's affinity (an equilibrium consideration) for the exchanging ions. Mass transfer is the second consideration. If the exchange isotherm is favorable, as is the case here, i.e., calcium is preferred to sodium, then a perfectly sharp (square-wave) theoretical breakthrough curve results. If the ion-exchange isotherm is unfavorable, as is the case for the reverse reaction of sodium chloride fed to a calcium-form resin, then a gradual breakthrough curve results even for instantaneous (equilibrium) mass transfer. These two basic types of breakthrough curves, sketched in Fig. 9.8, result from the solution of mass balance equations assuming instantaneous equilibrium and constant adsorbent capacity.

### Multicomponent ion exchange

The breakthrough curves encountered in water supply applications are much more complicated than those in Figs. 9.7 and 9.8. The greater complexity is caused by the multicomponent nature of the exchange reactions using natural water. Some ideal resin concentration profiles and breakthrough curves for hardness removal by ion-exchange softening and for nitrate removal by chloride-form anion exchange are sketched in Fig. 9.9a and b. The important determinants of the shapes of these breakthrough curves are (1) the feedwater composition, (2) the resin capacity, and (3) the resin's affinity for each of the ions as quantified by the separation factor $\alpha_{ij}$ or selectivity coefficient $K_{ij}$. The order of elution of ions from the resin, however, is determined solely by the selectivity sequence; two examples are shown in Table 9.2.

In carrying out the cation- or anion-exchange reactions, ions in addition to the target ion, e.g., calcium or nitrate, are removed by the resin. All the ions are concentrated, in order of preference, in bands or zones in the resin column, as shown in the resin concentration profiles of Fig. 9.9a and b. As these resin boundaries (wave fronts) move through the column, the breakthrough curves shown in the figures result. These are based on theory,[24] but have been verified in the actual breakthrough curves published by Clifford[25] and Snoeyink et al.[26]

Some useful rules can be applied to effluent histories in multicomponent ion-exchange (and adsorption) systems[24,25]

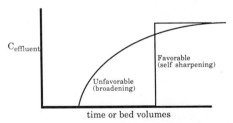

**Figure 9.8** Theoretical breakthrough curves for equilibrium ion exchange with no mass transfer limitations. An unfavorable isotherm (Fig. 9.3b) results in a broadening wave front (breakthrough), while a favorable isotherm (Fig. 9.3a) results in a self-sharpening wave front.

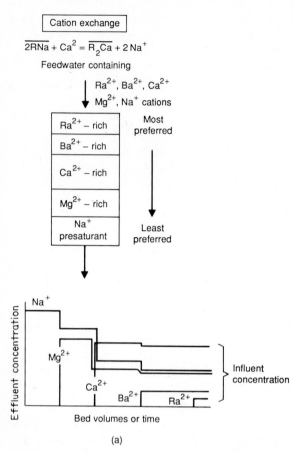

Figure 9.9 (a) Ideal resin concentration profile (*above*) and breakthrough curves (*below*) for typical softening and radium removal. Note that the column was run far beyond hardness breakthrough and slightly beyond radium breakthrough. The most preferred ion is $Ra^{2+}$ followed by $Ba^{2+} > Ca^{2+} > Mg^{2+} > Na^+$. (b) Ideal resin concentration profile (*above*) and breakthrough curves (*below*) for nitrate removal by chloride-form anion exchange with a strong-base resin. Note that the column was run far beyond nitrate breakthrough and somewhat beyond sulfate breakthrough. The most preferred ion is $SO_4^{2-}$ followed by $NO_3^-$, $Cl^- > HCO_3^-$.

■ Ions higher in the selectivity sequence than the presaturant ion tend to have long runs and sharp breakthroughs (like all those except $HCO_3^-$ in Fig. 9.9*b*); those less preferred than the presaturant ion will always have early gradual breakthroughs, as typified by $HCO_3^-$.

■ The most preferred species (radium in the case of softening and sulfate

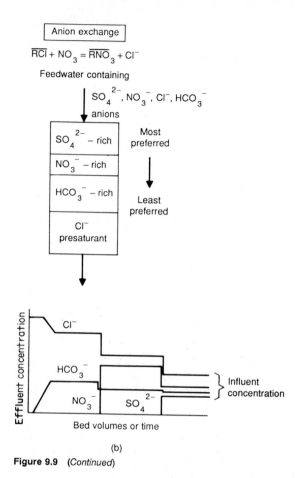

Figure 9.9 *(Continued)*

in the case of nitrate removal) are last to exit the column, and their effluent concentrations never exceed their influent concentrations.

- The species exit the column in reverse preference order, with the least preferred leaving first.

- The less preferred species will be concentrated in the column and at some time will exit the column in concentrations exceeding their influent concentrations (chromatographic peaking). This is a potentially dangerous situation depending on the toxicity of the ion in question. Good examples of chromatographic peaking, i.e., an effluent concentration greater than an influent concentration, are visible in Fig. 9.9a and b. A magnesium peak is shown in Fig. 9.9a and bicarbonate and nitrate peaks in Fig. 9.9b.

- When all the breakthrough fronts have exited the column, the entire resin bed is in equilibrium with the feedwater. When this happens, the column is exhausted, and the effluent and influent ion concentrations are equal.

- The effluent concentration of the presaturant ion ($Na^+$ in Fig. 9.9a and $Cl^-$ in Fig. 9.9b) decreases in steps as each new ion breaks through because the total ionic concentration of the water $C$ meq/L must remain constant during simple ion exchange.

The troublesome chromatographic peaking of toxic ions such as nitrate and arsenate can be eliminated only by inverting the selectivity sequence so that the undesired contaminant is the ion most preferred by the resin. Thus special-purpose resins must be designed and manufactured for specific application where chromatographic peaking is to be completely eliminated. This has been done in the case of nitrate removal and is discussed later under that heading. Peaking problems still remain with other inorganic contaminants, notably arsenic and selenium.

### Breakthrough detection and run termination

Clearly an ion-exchange column run must be stopped before a toxic contaminant is "dumped" during chromatographic peaking. Even without peaking, violation of the MCL will occur at breakthrough when the contaminant feed concentration exceeds the MCL. Detecting and preventing a high effluent concentration of contaminant depend on the sampling and analysis frequency. Generally, continuous online analysis of the contaminant, e.g., nitrate or arsenate, is too complicated, particularly in small communities where most of the inorganic contaminant problems exist.[1] On-line conductivity detection, the standard means of effluent quality determination in ion-exchange demineralization processes, is not easily applied to the detection of contaminant breakthrough in single-contaminant removal processes such as radium, barium, nitrate, or arsenate removal. This is because of the high and continuously varying conductivity of the effluents from cation or anion beds operated on typical water supplies. Conductivity should not be ruled out completely because even though the changes may be small as the various ions exit the column, a precise measurement may be possible in selected applications.

On-line pH measurement is a proven, reliable technique that can sometimes be applied as a surrogate for contaminant breakthrough. For example, pH change can be used to signal the exhaustion of a weak-acid resin ($-RCOOH$) used for carbonate hardness removal. When exhausted, the WAC resin ceases to produce acidic carbon dioxide, and the pH quickly rises to that of the feedwater. This pH increase is, however, far ahead of the barium or radium breakthrough. The pH can sometimes be used as an indicator of nitrate breakthrough and is discussed later.

The usual method of terminating an ion-exchange column run is to establish the relevant breakthrough curve by sampling and analysis and then use these data to terminate future runs based on the metered volume of throughput with an appropriate safety factor. If a breakthrough detector such as a pH or conductivity probe is applied, the sample line to the instrument can be located ahead (e.g., 6 to 12 in) of the bed outlet to provide advance warning of breakthrough.

### Typical service cycle

Ion-exchange and adsorption columns are operated on similar service cycles consisting of these steps: exhaustion, backwash, regeneration, slow rinse, fast rinse, and return to service. A simple process schematic is shown in Fig. 9.10 that includes an optional bypass for a portion of the feedwater. Bypass blending will be a common procedure for drinking water treatment applications because ion-exchange resins can usually produce a contaminant-free effluent that is purer than that required by law. Therefore, to minimize treatment costs, part of the contaminated feedwater, typically 10 to 50 percent, will be bypassed around the process and blended with the effluent to produce a

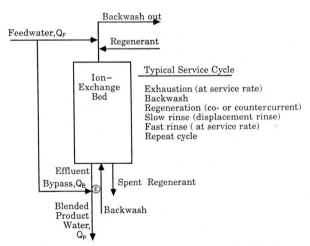

**Figure 9.10** Schematic and service cycle of ion-exchange process.

product water approaching some fraction (e.g., 70 percent) of the MCL acceptable to the regulatory agency.

### Partial regeneration and regenerant reuse

An alternative means of minimizing process costs is to use the technique of partial regeneration. This involves the use of only a fraction (e.g., 25 to 50 percent) of the regenerant required for "complete" (e.g., 90 to 100 percent) removal of the contaminant from the exhausted resin. The result is a large leakage of contaminant, on the next exhaustion run, caused by the relatively high level of contaminant remaining on the resin. Such large leakage can often be tolerated without exceeding the MCL. Partial regeneration is particularly useful in nitrate removal, as discussed in detail later. Generally, either bypass blending or partial regeneration will be used; use of both processes simultaneously is possible but creates significant process control problems.

Reuse of spent regenerant is another means of reducing costs and minimizing waste disposal requirements. The spent-regenerant solution can be considered for reuse if the following criteria are met: (1) Significant excess regenerant is in the solution, (2) the regenerant is highly preferred compared to the contaminant, and (3) some inevitable leakage of contaminant can be tolerated during the exhaustion step. Items (1) and (2) seem contradictory but do coexist in the case of fluoride removal by alumina where excess hydroxide, although highly preferred to fluoride, is required for regeneration because of kinetic considerations.

Reusing the entire spent-regenerant solution is not necessary. In the case where there is a long tail on the contaminant elution curve, the first few bed volumes of regenerant are discarded, and only the least-contaminated portions are reused. In this case a two-step roughing-polishing regeneration can be utilized. The roughing regeneration is completed with the partially contaminated spent regenerant, and the polishing step is carried out with fresh regenerant. The spent regenerant from the polishing step is then used for the next roughing regeneration.

In some unique applications, the trace contaminant can be removed from the spent regenerant by precipitation, coprecipitation, or selective adsorption, thereby enabling reuse of the regenerant solution. This technique has been used on a laboratory scale by Snoeyink et al.[26] to remove radium from regenerant calcium chloride brines by coprecipitation with barium sulfate.

The regenerant reuse techniques are relatively new to the ion-exchange field and are yet to be proved in full-scale long-term use for water supply applications. Regenerant reuse can result in some significant disadvantages: (1) increased process complexity, (2) increased contaminant leakage, (3) progressive loss of capacity caused by incomplete regeneration and fouling, and (4) the need to store and handle spent regenerants.

## Multicolumn processes

Ion-exchange or adsorption columns can be connected either in series to improve product purity or in parallel to increase throughput. A series roughing-polishing sequence is shown in Fig. 9.11. In such a process, a completely exhausted roughing column is regenerated when the partially exhausted polishing column effluent exceeds the MCL. This unregenerated polishing column becomes the new roughing column, and the old roughing column, now freshly regenerated, becomes the new polishing column. Often three columns are used. While two are in service, the third is being regenerated. A three-column system operated in this manner is referred to as a "merry-go-round system."

During normal operation of a parallel three-column system, one or two columns are in use, the third is either in standby or is being regenerated. The system is quite flexible; depending on flow requirements, one, two, or all three columns can be in service or in standby.

## Process differences: resins versus alumina

The design of a process for activated alumina exhaustion and regeneration is similar to that for ion-exchange resins but with some significant exceptions. First, contaminant leakage is inherently greater with alumina, and breakthrough curves are more gradual because alumina adsorption processes are much slower than ion exchange with strong resins. Second, effluent chromatographic peaking of the contaminant (fluoride, arsenic, or selenium) is not exhibited during alumina adsorption because these contaminants are usually the most preferred ions in the feedwater. Finally, complex, two-step base-acid

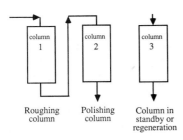

**Figure 9.11** Two-column roughing-polishing system operated in a merry-go-round fashion. After exhaustion of column 1, it will be taken out of service, and the flow sequence will be column 2 and then column 3. Following exhaustion of column 2 and regeneration of column 1, the roughing-polishing sequence will be column 3 then column 1.

regeneration is required to elute the contaminant and rinse out the excess base, in order to return the alumina to a useful form.

## Design Considerations

### Resin characteristics

Several hundred different resins are available from U.S. and European manufacturers. Of these, resins based on the polystyrene divinylbenzene matrix see widest use. Representative ranges of properties of these resins are shown in Table 9.4 for the two major categories of resins used in water treatment. Ion-exchange capacity is expressed in milliequivalents per milliliter (wet-volume capacity) because resins are purchased and installed on a volumetric basis (meq/mL $\times$ 0.0458 = kgrain $CaCO_3/ft^3$). A wet-volume capacity of 1.0 meq/mL means that the resin contains $6.023 \times 10^{20}$ exchange sites per milliliter of wet resin, including voids. The dry-weight capacity in

**TABLE 9.4    Properties of Styrene Divinylbenzyl, Gel-Type Strong-Acid and Strong-Base Resins**

| Note | Strong acid | Type 1 strong base |
|---|---|---|
| Screen size, U.S. mesh | $-16 + 50$ | $-16 + 50$ |
| Shipping weight, $lb/ft^3$ ($kg/m^3$) | 53 (850) | 44 (700) |
| Moisture content, % | 45–48 | 43–49 |
| pH range | 0–14 | 0–14 |
| Maximum operating temperature, °F (°C) | 280 (140) | $OH^-$ form 140 (60) $Cl^-$ form 212 (100) |
| Turbidity tolerance, NTU | 5 | 5 |
| Iron tolerance, mg/L as Fe | 5 | 0.1 |
| Chlorine tolerance, mg/L as $Cl_2$ | 1.0 | 0.1 |
| Backwash rate, $gpm/ft^2$ (m/h) | 5–8 (12–20) | 2–3 (4.9–7.4) |
| Backwash period, min | 5–15 | 5–20 |
| Expansion volume, % | 50 | 50–75 |
| Regenerant and concentration[†] | NaCl, 3–12% | NaCl, 1.5–12% |
| Regenerant dose, $lb/ft^3$ ($kg/m^3$) | 5–20 (80–320) | 5–20 (80–320) |
| Regenerant rate, $gpm/ft^3$(min/BV) | 0.5 (15) | 0.5 (15) |
| Rinse volume, $gal/ft^3$ (BV) | 15–35 (2–5) | 15–75 (2–10) |
| Exchange capacity[‡] kgr $CaCO_3/ft^3$ (meq/mL) | 39–41 (1.8–2.0) | 22–28 (1–1.3) |
| Operating capacity§ kgr $CaCO_3/ft^3$ (meq/mL) | 20–30 (0.9–1.4) | 12–16 (0.4–0.8) |
| Service rate, $gpm/ft^3$ (BV/h) | 1–5 (8–40) | 1–5 (8–40) |

[†]Other regenerants such as $H_2SO_4$, HCl, and $CaCl_2$ can be used for SAC resins while NaOH, KOH, and $CaCl_2$ can be used for SBA regeneration.

[‡]Kilograins of $CaCO_3$ per cubic foot are the units commonly reported in resin manufacturer literature. To convert kgr $CaCO_3/ft^3$ to meq/mL, multiply by 21.8.

§Operating capacity depends on method of regeneration, particularly on the amount of regenerant applied. See Table 9.5.

milliequivalents per gram is more precise and is sometimes used in scientific research.

The "operating capacity" is a measure of the actual performance of a resin under a defined set of conditions including, for example, feedwater composition, service rate, and degree of regeneration. The operating capacity is always less than the advertised exchange capacity because of incomplete regeneration and contaminant leakage. Some example operating capacities during softening are given in Table 9.5 where the operating capacity is seen to be a function of the amount of regenerant used.

### Bed size and flow rates

A resin bed depth of 30 in (76 cm) is usually considered the minimum, and beds as deep as 12 ft (3.67 m) are not uncommon. The *empty-bed contact time* (EBCT) chosen determines the volume of resin required and is usually in the range of 1.5 to 7.5 min. The reciprocal of EBCT is the *service flow rate* (SFR) or exhaustion rate, and its accepted range is 1 to 5 gpm/ft$^3$. These relationships are expressed as

$$\text{EBCT} = \frac{V_R}{Q_F} = \text{average fluid detention time in empty bed} \qquad (9.33)$$

$$\text{SFR} = \frac{1}{\text{EBCT}} = \frac{Q_F}{V_R} \qquad (9.34)$$

where $Q_F$ is the volumetric flow rate in gallons per minute (liters per

TABLE 9.5. Softening Capacity as a Function of Regeneration Level†

| Regeneration level | | Hardness capacity | | Regeneration efficiency | |
|---|---|---|---|---|---|
| lb NaCl / ft$^3$ resin | kg NaCl / m$^3$ resin | kg CaCO$_3$ / ft$^3$ resin | equiv CaCO$_3$ / L resin | lb NaCl / kgr CaCO$_3$ | equiv NaCl / equiv CaCO$_3$ |
| 4 | 64 | 17 | 0.78 | 0.24 | 1.40 |
| 6 | 96 | 20 | 0.92 | 0.30 | 1.78 |
| 8 | 128 | 22 | 1.00 | 0.36 | 2.19 |
| 10 | 160 | 25 | 1.15 | 0.40 | 2.38 |
| 15 | 240 | 27 | 1.24 | 0.56 | 3.30 |
| 20 | 320 | 29 | 1.33 | 0.69 | 4.11 |
| ∞ | ∞ | 45 | 2.06 | ∞ | ∞ |

†These operating capacity data are based on the performance of Amberlite IR-120 SAC resin. Other manufacturers' resins are comparable. Values given are independent of EBCT and bed depth providing the minimum criteria are met. kgr is kilograins.

minute) and $V_R$ is the resin bed volume including voids in cubic feet (cubic meters).

### Fixed-bed columns

Ion-exchange columns are usually steel pressure vessels constructed so as to provide (1) a good feed and regenerant distribution system, (2) an appropriate bed support including provision for backwash water distribution, and (3) enough free space above the resin bed to allow for expected bed expansion during backwashing. Additionally, the vessel must be lined so as to avoid corrosion problems resulting from concentrated salt solutions and, in some cases, acids and bases used for regeneration or resin cleaning. There must be minimal dead space below the resin bed where regenerants and cleaning solutions might collect and subsequently bleed into the effluent during the service cycle.

## Applications of Ion-Exchange and Adsorption Processes

### Sodium ion-exchange softening

As already mentioned, softening water by exchanging sodium for calcium and magnesium by using SAC resin [see Eq. (9.2)] is the major application of ion-exchange technology for drinking water treatment. Prior to the advent of synthetic resins, zeolites, i.e., inorganic crystalline aluminosilicate ion exchangers in the sodium form, were utilized as the exchangers. The story of one major application of ion-exchange softening at the Weymouth plant of the Metropolitan Water District of Southern California is well told by A. E. Bowers in *The Quest for Pure Water*.[27] In that application, which included 400 mgd ($1.5 \times 10^6$ m³/d) of softening capacity, softening by ion exchange eventually supplanted excess lime–soda ash softening because of better economics, fewer precipitation problems, and the requirement for a high alkalinity level in the product water to reduce corrosion. One advantage of the lime–soda ash softening process is that it reduces the TDS level of the water by removing calcium and magnesium bicarbonates as $CaCO_3(s)$ and $Mg(OH)_2(s)$. The concomitant removal of alkalinity is, however, sometimes detrimental, thus favoring ion-exchange softening that deals only with cation exchange leaving the anions intact.

As with most ion-exchange softening plants, the zeolite medium at the Weymouth plant was exchanged for resin in the early 1950s shortly after polystyrene SAC resins were introduced. The SAC softening resins used today are basically the same as these early polystyrene resins. Their main features are high chemical and physical stability, even in the presence of chlorine; uniformity in size and composition; high exchange capacity; rapid exchange kinetics; a high degree of reversibility; and long life. A historical comparison between the life of the zeolites and that of the resins indicated that zeolites could process a maximum of $1.6 \times 10^6$ gal $H_2O/ft^3$ zeolite [214,000 bed volumes (BV)] before replacement whereas the resins could process up to $20 \times 10^6$ gal $H_2O/ft^3$ resin (2,700,000 BV) before they needed replacement. The softeners designed and installed for resins at this plant in 1966 were $28 \times 56$ ft reinforced concrete basins filled to a depth of 2.5 ft, each containing 4000 $ft^3$ of resin.

The Weymouth plant utilized ion-exchange softening for over 30 years. Softening ceased in 1975 when the source water hardness was reduced by blending. At that time, the 9-year-old resin in the newest softeners was still good enough to be resold. Other interesting design features of this plant included disposal of waste brine to a wastewater treatment plant through a 20-mi-long pipe flowing at 15 $ft^3/s$ and the upflow exhaustion at 6 gpm/$ft^2$ (3.1-min EBCT in a 2.5-ft-deep bed) followed by downflow regeneration.

**Example Problem 2**  This typical design example illustrates how to establish the ion-exchange resin volume, column dimensions, and regeneration requirements.

*Design Problem*  A groundwater is to be partially softened from 275 down to 150 mg per liter of $CaCO_3$ hardness. Ion exchange has been selected instead of lime softening because of its simplicity and the ease of cycling on and off to meet the water demand. The well pumping capacity is 1.0 mgd (700 gpm), and the system must be sized to meet this maximum flow rate. The source water contains only traces of iron; therefore, potential clogging problems because of suspended solids are not significant. Note: In those applications where source water suspended solids would foul the resins, filtration pretreatment with dual- or multimedia filters will be required.

*Outline of Solution*

1. Select a resin and a regeneration level, using manufacturer's literature.
2. Calculate the allowable fraction $f_B$ of bypass source water.
3. Choose the SFR (gpm/$ft^3$ or EBCT).
4. Calculate the run length $t_H$ (h) and the bed volumes $V_F$ that can be treated prior to hardness breakthrough.
5. Calculate the volume of resin $V_R$ required.
6. Determine the minimum out-of-service time, in hours, during regeneration.
7. Choose the number of columns in the system.
8. Dimension the columns.
9. Calculate the volume and composition of wastewater.

*Calculations*

1. *Selection of resin and resin capacity.* Once the resin and its regeneration level have been specified, the ion-exchange operating capacity is fixed based on experimental data of the type found in Table 9.5. The data are for a polystyrene

SAC resin subjected to cocurrent regeneration using 10% NaCl. If a typical regeneration level of 15 lb NaCl/per cubic foot of resin is chosen, the resulting hardness capacity prior to breakthrough is 27 kgr of hardness as $CaCO_3$ per cubic foot of resin, i.e., 1.24 meq per milliliter of resin.

2. *Calculation of bypass water allowance.* Assume that the water passing through the resin has zero hardness. Actually, hardness leakage during exhaustion will be detectable, but usually it is less than 5 mg/L as $CaCO_3$. The bypass flow is calculated by writing a hardness balance at the blending point, point $E$ in Fig. 9.10, where the column effluent is blended with the source water bypass.

Mass balance on hardness at point $E$:

$$Q_B C_B + Q_F C_E = Q_P C_P \tag{9.35}$$

Balance on flow at point $E$:

$$Q_B + Q_F = Q_P \tag{9.36}$$

where    $Q_B$ = bypass flow rate
$Q_F$ = column feed and effluent flow rate
$Q_P$ = blended product flow rate, i.e., total flow rate
$C_B$ = concentration of hardness in bypass source water, 275 mg/L as $CaCO_3$
$C_E$ = concentration of hardness in column effluent, assumed to be 0 mg/L
$C_P$ = chosen concentration of hardness in blended product water, 150 mg/L as $CaCO_3$

The solution to these equations is easily obtained in terms of the fraction bypassed $f_B$:

$$f_B = \frac{Q_B}{Q_P} = \frac{C_P}{C_B} = 0.55 \tag{9.37}$$

The fraction $f_F$ that must be treated by ion exchange is

$$f_F = 1 - f_B = 0.45 \tag{9.38}$$

3. *Choosing the exhaustion flow rate.* The generally acceptable range of SFR for ion exchange is 1 to 5 gpm/ft$^3$. Choosing a value of 2.5 gpm/ft$^3$ results in an EBCT of 3.0 min and an approach velocity $v_0$ of 6.25 gpm/ft$^2$ if the resin bed is 2.5 ft deep.

$$\text{EBCT} = \frac{1 \text{ min} \cdot \text{ft}^3}{2.5 \text{ gal}} \times \frac{7.48 \text{ gal}}{1 \text{ ft}^3} = 3 \text{ min} \tag{9.39}$$

$$v_0 = \frac{\text{depth}}{\text{detention time}} = \text{SFR} \times \text{depth} \tag{9.40}$$

$$v_0 = 2.5 \text{ gpm/ft}^3 \times 2.5 \text{ ft} = 6.25 \text{ gpm/ft}^2 \tag{9.41}$$

4. *Calculation of run length.* The exhaustion time to hardness breakthrough $t_H$ and the bed volumes to hardness breakthrough $BV_H$ are calculated from a mass balance on hardness, assuming again that the resin effluent contains zero hardness. Expressed in words, this mass balance is

Equivalents of hardness removed from water during run =

equivalents of hardness accumulated on resin during run

$$Q_F C_F t_H = V_F C_F = q_H V_R \qquad (9.42)$$

where   $q_H$ = hardness capacity of resin at selected regeneration level, equiv/L $(kgr/ft^3)$

$C_F$ = feedwater hardness concentration, equiv/L

$V_R$ = volume of resin bed including voids

$Q_F t_H = V_F$, volume of water fed to column during time $t_H$

Then

$$\frac{V_F}{V_R} = BV_H = \frac{q_H}{C_F} \qquad (9.43)$$

Numerically the bed volumes to hardness breakthrough $BV_H$ following a regeneration at 15 lb NaCl/ft$^3$ is

$$BV_H = \frac{1.25 \text{ equiv CaCO}_3}{1 \text{ L resin}} \times \frac{1 \text{ L H}_2\text{0}}{275 \text{ mg CaCO}_3} \times \frac{50{,}000 \text{ mg CaCO}_3}{1 \text{ equiv CaCO}_3}$$

$$= 225 \text{ volumes of H}_2\text{O treated per volume of resin} \qquad (9.44)$$

The time to hardness breakthrough $t_H$ is related to the bed volumes to breakthrough $BV_H$ and the EBCT:

$$t_H = \text{EBCT} \times BV_H \qquad (9.45)$$

$$t_H = 3.0 \frac{\text{min}}{\text{BV}} \times 225 \text{ BV} \times \frac{1 \text{ h}}{60 \text{ min}} = 11.2 \text{ h} \qquad (9.46)$$

If the EBCT is decreased by increasing the flow rate through the bed, i.e., SFR, then the run time is proportionately shortened even though the total amount of water treated $V_F$ remains constant.

5. *Calculation of resin volume* $V_R$. The most important parameter chosen was SFR because it directly specified the necessary resin volume $V_R$ according to the following relationships, based on Eq. (9.34):

$$V_R = \frac{Q_F}{\text{SFR}} = Q_F \text{ (EBCT)} \qquad (9.47)$$

Numerically, for a column feed flow $Q_F$ of 45 percent (the amount not bypassed) of 1.0 mgd,

$$V_R = \frac{0.45 \times 10^6 \text{ gal}}{1 \text{ day}} \times 3.0 \text{ min} \times \frac{1 \text{ day}}{1440 \text{ min}} \times \frac{1 \text{ ft}^3}{7.48 \text{ gal}}$$

$$= 125 \text{ ft}^3 \qquad (9.48)$$

6. and 7. *Calculation of number of columns and minimum out-of-service time for regeneration.* For a reasonable system design, two columns are required—one in operation and one in regeneration or standby. A single-column design

with product water storage is possible but provides no margin of safety in case the column has to be serviced. Even with two columns, the out-of-service time $t_S$ for the column being regenerated should not exceed the exhaustion run time to hardness breakthrough $t_H$:

$$t_S \leq t_{BW} + t_R + t_{SR} + t_{FR} \qquad (9.49)$$

where   $t_{BW}$ = time for backwashing, 5 to 15 min
        $t_R$ = time for regeneration, 30 to 60 min
        $t_{SR}$ = time for slow rinse, 10 to 30 min
        $t_{FR}$ = time for fast rinse, 5 to 15 min

A conservative out-of-service time would be the sum of the maximum times for backwashing, regeneration, and rinsing, i.e., 2 h. This causes no problem in regard to continuous operation because the exhaustion time is more than 11 h.

8. *Calculation of column dimensions.* The resin depth $h$ has been specified earlier as 2.5 ft, thus the column height, after we allow for 100 percent resin bed expansion during backwashing, is 5.0 ft. The bed diameter $D$ is then

$$D = \sqrt{\frac{4V_R}{\pi h}} = 8\,ft \qquad (9.50)$$

The resulting ratio of resin bed depth to column diameter is 1.5:8, or 0.3:1. This is within the acceptable range of 0.2:1 to 2:1 if proper flow distribution is provided. Increasing the resin depth to 4 ft increases the column height to 8 ft and reduces its diameter to 6.3 ft.

Another alternative would be to use three columns, with two in service and one in standby. This offers a more flexible design. The resin volume of the in-service unit would be $125/2 = 62.5$ ft$^3$ each; i.e., the flow would be split between two 62.5-ft$^3$ resin beds operating in parallel. Regeneration would be staggered such that only one column would undergo regeneration at any time.

Clearly, a variety of depths and diameters are possible. Before specifying these, the designer should check with equipment manufacturers because softening units in this capacity range are available as predesigned packages.

9. *Calculation of volume and composition of wastewater.* The spent regenerant solution comprises the regenerant and the slow-rinse (displacement-rinse) volumes. These waste solutions must be accumulated for eventual disposal, as detailed in Chap. 16. The actual wastewater volume per regeneration will depend on the size of the resin bed, i.e., whether one, two, or three beds are chosen for the design. The following calculations are in terms of bed volumes, which can be converted to fluid volume once the column design has been fixed. The chosen regeneration level, 15 lb NaCl/ft$^3$ resin, is easily converted to bed volumes of regenerant required. Given that a 10% NaCl solution has a specific gravity of 1.07, the regenerant volume applied is

$$\frac{15\ \text{lb NaCl}}{1\ \text{ft}^3\ \text{resin}} \times \frac{1\ \text{lb soln}}{0.1\ \text{lb NaCl}} \times \frac{\text{ft}^3\ \text{soln}}{1.07 \times 62.4\ \text{lb soln}} = 2.25\ \text{BV} \qquad (9.51)$$

Following the salt addition, a displacement (slow) rinse of 1 to 2 BV is applied. The total spent-regenerant volume is made up of the spent regenerant (2.25 BV) and the displacement rinse (2.0 BV):

Spent regenerant = 2.25 + 2.0 = 4.25 BV

In this example, the wastewater volume amounts to approximately 1.9 percent [(4.25/225)100] of the treated water or 0.9 percent [(1.9) (0.45)] of the blended product water. Choosing 1.0 L of resin as a convenient bed size, we get 1.0 BV = 1.0 L, and the spent-regenerant volume from a 1.0-L bed is 4.25 L. For our example bed containing 125 ft$^3$ of resin, the wastewater volume is 531 ft$^3$ (4000 gal).

If the small quantity of ions in the water used to make up the regenerant solution is disregarded, the waste brine concentration, in equivalents per liter, can be calculated as follows:

$$\text{Total ionic concentration of wastewater} = \text{hardness concentration}$$
$$+ \text{ excess NaCl concentration} \quad (9.52)$$

$$\text{Wastewater hardness concentration} = \frac{\text{equiv of hardness removed}}{\text{spent-regenerant volume}}$$
$$(9.53)$$

From Table 9.5 we see that 1.24 equiv of hardness is removed from the feedwater per liter of resin at a regeneration level of 15 lb NaCl per cubic foot of resin. During steady-state operation, this is the amount of hardness removed from the resin with each regeneration. Therefore

$$\text{Waste hardness concentration} = \frac{1.24 \text{ equiv}}{4.25 \text{ L}} = 0.29 \text{ equiv/L} \quad (9.54)$$

This amounts to 14,500 mg $CaCo_3$ hardness/L, which can, if necessary, be broken down further into the separate Ca and Mg concentrations by using the known ratio of Ca to Mg in raw water.

The excess NaCl concentration is also calculated from Table 9.5 and the following relationship:

$$\text{Excess NaCl concentration} = \frac{\text{equiv NaCl applied} - \text{equiv hardness removed}}{\text{spent-regenerant volume}}$$

$$\text{Excess NaCl concentration} = \frac{1.24(3.3) - 1.24}{4.25} = 0.67 \text{ equiv NaCl/L}$$
$$(9.55)$$

This excess NaCl concentration corresponds to 39,300 mg NaCl/L. The total cation composition of the wastewater is 0.96 equiv/L, made up of 0.29 equiv/L hardness and 0.67 equiv/L sodium. These cations, in addition to chloride, are the major constituents of the wastewater. Other minor contaminant cations removed from the source water will be present in the spent regenerant. These can include $Ba^{2+}$, $Sr^{2+}$, $Ra^{2+}$, $Fe^{2+}$, $Mn^{2+}$, and others. The minor anionic contaminants in the wastewater will be bicarbonate and sulfate from the source water used for regeneration and rinsing.

**Brine disposal from softening plants.** The usual method for disposing of spent-regenerant brine is through metering into a sanitary sewer. In coastal locations, direct discharge into the ocean is a possibility. Other alternatives are properly lined evaporation ponds in arid regions or brine disposal wells in areas where such wells are permitted or already in existence. The uncontrolled discharge (batch dumping) of the spent-regenerant brine into surface water or sanitary sewers is usually not allowed because of the potential damage to biota from localized high salinity. Waste disposal is covered in Chap. 16.

### Hydrogen ion-exchange softening

Softening water without the addition of sodium is sometimes desirable. In this case, hydrogen can be exchanged for hardness ions by using either strong- or weak-base cation exchangers. Hydrogen-form strong-acid resins [Eq. (9.1)] are seldom used in this application because of the inefficiency of their regeneration and the problems of excess acid disposal. Hydrogen-form WAC exchangers are sometimes used for sodium-free softening. Only the removal of temporary hardness is possible, and this proceeds according to Eq. (9.4), resulting in partial softening, dealkalization, and TDS reduction. Regeneration is accomplished with strong acids such as HCl or $H_2SO_4$ and proceeds according to Eq. (9.5).

The partially softened, alkalinity-free column effluent must be stripped of $CO_2$ and blended with source water to yield a noncorrosive product water. Alternatively, the pH of the column effluent can be raised by adding NaOH or $Ca(OH)_2$ following $CO_2$ stripping. This approach, however, costs more and results in the addition of either sodium or hardness to the product water.

### Barium removal by ion-exchange softening

During all types of ion-exchange softening, barium is removed in preference to calcium and magnesium. This is shown graphically in Fig. 9.9a where, in theory, barium breaks through long after hardness. This is true even though barium-contaminated groundwater will always contain much higher levels of calcium and magnesium. Snoeyink et al.[26] summarized their considerable research on barium removal using hydrogen- and sodium-form SAC and WAC resins operated to barium breakthrough. They found that the main problem with using SAC resin for barium removal is the difficulty of removing barium from the exhausted resin. Barium accumulates on the resin and reduces the exchange capacity of subsequent runs if sufficient NaCl is not used for regeneration.

When WAC resins are used in the hydrogen form for barium removal, the same considerations apply as with WAC softening: Divalent cations are preferentially removed, cation removal is equivalent to alkalinity, partial desalting occurs, $CO_2$ must be stripped from the column effluent, and bypass blending will be required. The advantage of WAC resins is that barium is easily removed during regeneration with a small excess, typically 20 percent, of HCl or $H_2SO_4$. In summary, using a hydrogen-form WAC resin for barium or combined hardness and barium removal produces a better-quality product water with less wastewater volume compared to a sodium-form resin. The WAC process is, however, more complex and more expensive because of chemical costs, the need for acid-resistant construction materials, wastewater neutralization, and the need to strip $CO_2$ from the product water.

### Radium removal by ion exchange

Radium 226 and radium 228 ($^{226}$Ra and $^{228}$Ra, respectively) are natural groundwater contaminants that occur at ultratrace levels. Their current (1988) combined maximum contaminant level (MCL) is limited to 5.0 pCi/L, which corresponds to 11.1$^{226}$Ra disintegrations/(min)(L), or 0.185 Bq/L. For $^{226}$Ra alone, this corresponds to $5 \times 10^{-9}$ mg $^{226}$Ra/L. According to a recent survey,[3] more than 550 community water supplies in the United States exceed the radium MCL. Cation exchange with either sodium- or hydrogen-form resins is a very effective means of radium removal because radium is preferred to all the common cations found in water. As with hardness and barium removal, radium can be effectively removed by using sodium-form SAC or hydrogen-form WAC resins.

**Radium removal during softening.** During the normal sodium ion-exchange softening process, radium is completely removed (greater than 95 percent); thus, softening is an effective technique for meeting the radium MCL. Recently completed pilot studies on a groundwater containing 18 pCi/L total radium and 275 mg/L total hardness in Lemont, Ill.,[28,30] have resulted in the following conclusions regarding sodium ion-exchange softening for radium removal:

1. On the first exhaustion run to radium breakthrough, hardness breakthrough occurred at 300 BV while radium did not reach the MCL until 2500 BV.

2. On the second and subsequent exhaustion runs following salt regeneration at 15 lb NaCl per cubic foot of resin, radium broke through simultaneously with hardness at 300 BV. This long first run to radium breakthrough followed by shorter subsequent runs was not an anomaly but was repeated with other SAC resins.

3. When the resin bed was operated in the normal fashion, i.e., exhausted downflow and never run beyond hardness breakthrough,

no radium was removed during the first three cocurrent regenerations at 15 lb NaCl per cubic foot of resin. And five exhaustion-regeneration cycles were required to reach a steady state where radium sorption during exhaustion equaled radium desorption during regeneration.

4. Weak-acid cation resins had a lower relative affinity for radium than SAC resins did. Macroporous SAC resins had the highest affinity for radium but were more difficult to regenerate because of this high affinity.

5. Radium never broke through before hardness in any of the 80 experimental runs with five different SAC and WAC resins. Furthermore, the radium concentration of the effluent never exceeded that of the influent; i.e., chromatographic peaking of radium never occurred.

6. Radium was very difficult to remove from exhausted resins, presumably because it is a very large, poorly hydrated ion that seeks the relatively inaccessible hydrophobic regions of the resin phase. Increasing the regeneration time beyond the usual 15 to 30 min helped only slightly to remove the radium from the spent resin. In this regard, radium behaves differently from barium,[29] which was much more efficiently eluted at 30 min compared to 15 min EBCT for the regenerant.

**Calcium-form resins for radium removal.** Calcium-form SAC resins can be used for radium and barium removal[29,30] when concomitant softening is not necessary. In this process, 1 to 2 $M$ $CaCl_2$ is used as the regenerant at a level of 14.2 lb/ft$^3$, and counterflow regeneration for an extended time (60 min) is the preferred mode of operation. With counterflow regeneration, care must be taken not to mix the exhausted resin bed prior to or during regeneration. In theory, the run time to radium breakthrough is independent of the form (sodium or calcium) of the resin. This constant run length irrespective of the resin's initial condition was verified in the Lemont pilot study for the first exhaustion of a calcium-form resin that also ran for 2500 BV before radium reached the MCL.[30] The lengths of subsequent runs of calcium-regenerated resin were, however, very much a function of the regeneration conditions. The best counterflow $CaCl_2$ regeneration resulted in a typical run length of 500 BV to radium breakthrough with a continuous radium leakage prior to breakthrough of 3 pCi/L when the feed was 18 pCi/L. Cocurrent $CaCl_2$ regeneration resulted in immediate radium leakage (8 to 10 pCi/L) that continually decreased until radium breakthrough. Therefore, cocurrent $CaCl_2$ regeneration is not an acceptable way to operate the radium-removal process. When calcium-form resins are utilized for radium removal, only countercurrent regeneration can be employed. Furthermore, extreme care must be taken to keep traces of radium-contaminated resin away from the column exit during exhaustion.

**Dealing with radium-contaminated brines.**  If one is consistent with existing practices regarding the disposal of radium-contaminated brines, disposal into the local sanitary sewer should be allowed. For example, in many midwest communities where water is being softened on both a residential and a municipal scale, radium removal is also taking place and the radium-contaminated brines are being disposed of in the usual fashion, i.e., by metering into the local sanitary sewer. To determine the radium concentration in waste-softener brines, a simple calculation can be done. In a hypothetical situation in which the raw water contains as much as 20 pCi/L of radium and the softening run length is 300 BV, the 5 BV of spent regenerant contains an average of 1200 pCi/L of radium. This is a concentration factor of 60 (300/5) and is typical of softeners in general.

If necessary, a spent-regenerant brine solution could be decontaminated prior to disposal by passing it through a radium-specific adsorbent such as the Dow "complexer" or $BaSO_4$-loaded activated alumina.[8] [The Dow *radium-selective complexer* (RSC) is not really a complexer but a $BaSO_4$-impregnated SAC resin (Dow patent) that adsorbs radium onto the $BaSO_4$ even in the presence of a high concentration of competing ions such as $Ca^{2+}$, $Mg^{2+}$, and $Na^+$.] This brine decontaminant process has been tested on a small municipal scale[31] and found to work well. The RSC was loaded to a level of 2.7 nCi/cm$^3$ and was still decontaminating spent brine after 1 year of operation. Unfortunately, disposal of a radium-containing solid at a level of 2.7 nCi/cm$^3$ (2.7 × 10$^6$ pCi/L) is potentially a far more serious problem than disposal of the original radium-contaminated brine. Partially because of the disposal problem, Dow announced in December 1987 that it would no longer produce the RSC XSF43230 even though it had been proved technically very effective.

### Nitrate removal by SBA exchange

Ion exchange of chloride for nitrate is currently the simplest and lowest-cost method for removing nitrate from contaminated groundwater to be used for drinking. Only a few applications of the process (shown schematically in Fig. 9.10) now exist, and these are restricted to small community and noncommunity water supplies. As additional communities are forced into compliance, however, more applications of nitrate removal by ion exchange will be seen. This prediction is based on an AWWA survey of inorganic contaminants[1] which reported that nitrate concentration in excess of 10 mg/L $NO_3$-N MCL was the most common reason compelling the shutdown of small community water supply wells.

The anion-exchange process for nitrate removal is similar to cation-exchange softening except that (1) anions rather than cations are being exchanged, (2) nitrate is a monovalent ion whereas calcium is divalent, and (3) nitrate, unlike calcium, is not the most preferred common ion involved in the multicomponent ion-exchange process with a typical nitrate-contaminated groundwater. The latter two exceptions lead to some significant differences between softening and nitrate removal.

Chloride-form SBA exchange resins are used for nitrate removal according to Eq. (9.7). Excess NaCl at a concentration of 1.5 to 12 percent (0.25 to 2.0 $N$) is used for regeneration to produce a reversal of that reaction. The apparently simple process is not without complications, however, as detailed below.

**Effects of water quality on nitrate removal.** The source water quality and, in particular, the sulfate content influence the bed volumes that can be treated prior to nitrate breakthrough, which is shown along with that for chloride and bicarbonate in Fig. 9.12. The effect of increasing sulfate concentration on nitrate breakthrough is shown in

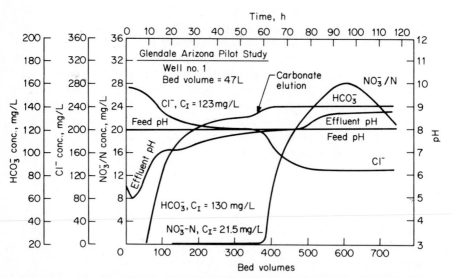

**Figure 9.12** Breakthrough curves for nitrate and other anions following complete regeneration of type 2 gel SBA resin in chloride form. $C_I$ = influent concentration.

Fig. 9.13, constructed from data obtained by spiking Glendale, Ariz., water with sodium sulfate.[32] As sulfate increased from the natural value of 43 mg/L (0.9 meq/L) to 310 mg/L (6.5 meq/L), i.e., an increase of 5.6 meq/L, the experimental run length to nitrate breakthrough decreased 55 percent from 400 to 180 BV. Similar response to increasing sulfate is expected for all commercially available type 1 and type 2 strong-base resins.

The detrimental effect of sulfate shown in Fig. 9.13 is well predicted by multicomponent chromatography calculations.[25] Such calculations can be used to estimate the effects of additional chloride and bicarbonate on nitrate breakthrough. When the chloride concentration in the Glendale water is hypothetically increased by 5.6 meq/L (200 mg/L), the calculated nitrate breakthrough occurs at 265 BV—a reduction of 34 percent. Adding 5.6 meq/L (340 mg/L) of bicarbonate only reduces the nitrate run length by 15 percent, to 340 BV. These reductions in nitrate run length are in accord with qualitative predictions based on the selectivity sequence, i.e., sulfate > nitrate > chloride > bicarbonate.

Because all commercially available SBA resins prefer sulfate to ni-

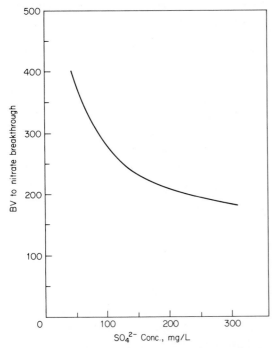

**Figure 9.13** Effect of sulfate concentration on nitrate breakthrough for source and sulfate-spiked Glendale water.

trate at the TDS levels and ionic strengths of typical groundwater, chromatographic peaking of nitrate occurs following its breakthrough. This peaking, when it does occur, means effluent nitrate concentrations exceeding the source water nitrate level. No easy way is available to calculate whether a peak will occur or what the magnitude of the expected nitrate peak will be. It depends primarily on the TDS level; the specific composition of the source water including its sulfate, nitrate, and alkalinity concentrations; and the type of SBA resin used.

As the TDS level of the source water increases, selectivity reversal can occur with the result that nitrate is preferred to sulfate. In this case, which was reported by Guter in his McFarland, Calif., nitrate-removal pilot studies,[5] sulfate broke through prior to nitrate, and no nitrate peaking occurred.

In other laboratory and pilot studies,[4,32] nitrate peaking was clearly observed with both simulated groundwater and actual groundwater. The theoretical nitrate breakthrough curve for the standard type 1 and type 2 SBA resins is depicted in Fig. 9.9b where idealized nitrate peaking is shown. Figure 9.12 depicts the pH, nitrate, and sulfate breakthrough curves observed in the Glendale nitrate-removal pilot studies.[32] In this groundwater, which is considered somewhat typical of nitrate-contaminated groundwater, the effluent nitrate concentration peaked at 1.3 times the feedwater value.

**Detecting nitrate breakthrough.**   When the usual type 1 and type 2 SBA resins are used to remove nitrate from a typical groundwater containing sulfate and having an ionic strength $I \leq 0.010$ $M$, chromatographic peaking of nitrate can occur. Even if it does not occur, the column run must still be terminated prior to the breakthrough of nitrate. If the source water concentration is reasonably constant, this termination can be initiated by a flowmeter signal when a predetermined volume of feed has passed through the column. Terminating a run at a predetermined volume of treated water cannot be generally recommended because of the variable nature of nitrate contamination in some types of groundwater. For example, in a 6-month period in McFarland,[5] the nitrate content of one well varied from 5 to 25 mg/L $NO_3$-N. Such extreme variations did not occur during the 15-month pilot study in Glendale,[32] where the nitrate content of the study well only varied from 18 to 25 mg/L $NO_3$-N.

In the Glendale pilot study, a measurable pH change, about 1.0 pH unit, accompanied the nitrate breakthrough. This pH wave is visible in Fig. 9.12 where we see that if the run had been terminated when the effluent pH equaled the feed pH, the nitrate peak would have been avoided. The observed pH increase resulted from the simultaneous elution of carbonate and nitrate. The carbonate elution was a fortunate coincidence, and it allowed pH or differential (feed-effluent) pH to be used to anticipate nitrate breakthrough. Such pH waves are not unique to the Glendale water but are expected when significant ( > 1.0 meq/L) concentrations of sulfate and bicarbonate are present and $I \leq 0.01M$. In practice, the pH detector should be located slightly upstream of the column effluent to provide a safety factor.

**Choice of resin for nitrate removal.**  Both laboratory and field studies of nitrate removal by chloride-form anion exchange have shown that no significant performance differences exist among the commercially available SBA resins. Both type 1 and type 2 polystyrene divinylbenzene SBA resins have been used successfully. Guter[5] reported on the performance of a type 1 resin in a full-scale application, and Clifford et al.[32] made extensive use of a type 2 SBA resin during a 15-month pilot-scale study in Glendale. When even potential chromatographic peaking of nitrate must be avoided, however, a special nitrate-selective resin is necessary. Guter[5] described several such special resins that were nitrate-selective with respect to sulfate based on their increased charge separation distance and hydrophobicity. The nitrate-sulfate selectivities of these resins were in accord with the predictions of Clifford and Weber,[4] who found that (1) the sulfate preference of a resin could be significantly reduced by increasing the distance between charged ion-exchange sites and (2) the nitrate preference of a resin could be improved by increasing matrix and functional group hydrophobicity.

The nitrate-sulfate selective (NSS) resins reported on by Guter have been made in experimental quantities and will become commercially available when the market for such resins justifies their manufacture. No known technical or legal impediments to their production and use exist. Basically, NSS resins are similar to the standard type 1 resins containing trimethyl amine functionality [$RN(CH_3)_3$] but with ethyl, propyl, or butyl groups substituted for the methyl groups to increase the charge separation distance and resin hydrophobicity.

**Regeneration of nitrate-laden resin.**  Regardless of the regeneration method used, all previous studies have demonstrated that sulfate is much easier to elute than nitrate. Typical sulfate and nitrate elution curves are shown in Fig. 9.14. For a complete regeneration, divalent sulfate, the most preferred ion during exhaustion, is the first to be stripped from the resin during regeneration because of the selectivity reversal that occurs in high-ionic-strength ($I > 0.25$) salt solution.

Regeneration of the nitrate-laden resin was studied extensively in Glendale by using the complete and partial regeneration techniques. For complete regeneration, i.e., the removal of more than 95 percent of

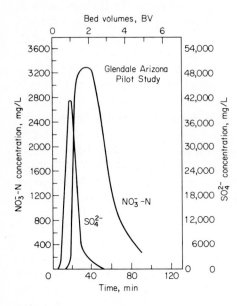

**Figure 9.14** Elution of sulfate and nitrate during complete regeneration of type 2 gel SBA resin following exhaustion run shown in Fig. 9.12. Cocurrent downflow regeneration using 12% NaCl (2 $N$) at EBCT = 19 min. Essentially 100 percent recovery of sulfate and nitrate observed.

the sorbed nitrate, the more dilute the regenerant, the more efficient it was. For example, 0.5 $N$ NaCl required 4.0 equiv of chloride per equivalent of resin, whereas 1.7 $N$ NaCl (10% NaCl) required 170 percent of this value. A further advantage of dilute regenerants is their amenability to biological denitrification of the spent brine, because the spent-regenerant denitrification rate has been found to decrease significantly with increasing NaCl concentration.[33] Presumably, the salt is somewhat toxic to the denitrifying bacteria. An obvious disadvantage of dilute regenerants is that they produce more wastewater and require longer regeneration times.

Complete regeneration is not very efficient in terms of the amount of salt required, but it does result in minimal leakage of nitrate during the subsequent exhaustion run. But because significant nitrate leakage (up to 7 mg $NO_3$-N/L) is permissible based on the MCL of 10 mg $NO_3$-N/L, complete regeneration is not required. Thus, nitrate removal is an ideal application for incomplete or partial regeneration, i.e., the removal of 50 to 60 percent of the sorbed contaminant.

Partial regeneration levels of 0.64 to 2.0 equiv per equivalent of resin (3 to 10 lb NaCl per cubic foot of resin) were studied at Glendale with the result that 1.0 equiv $Cl^-$/equiv resin was acceptable if 7.0 mg $NO_3$-N/L could be tolerated in the effluent prior to the nitrate breakthrough that occurred earlier (350 BV) than in the complete regeneration case (410 BV). Guter[5] found that regeneration levels as low as 0.64 equiv $Cl^-$ per equivalent of resin (3 lb NaCl per cubic foot of resin) were acceptable in his pilot studies in McFarland. The acceptable level must be determined for each application, and it depends primarily on the feedwater quality and the allowable nitrate leakage. Regardless of the level of salt used for partial regeneration, complete mixing of the resin bed is mandatory following regeneration. If mixing is not accomplished, excessive nitrate leakage will occur during the first 80 to 100 BV of effluent. The required mixing can be achieved mechanically or by introducing air into the backwash, but conventional backwashing, even for extended periods, will not provide adequate mixing.

In the Glendale pilot study, a comparison of the optimum complete and partial regeneration methods indicated that partial regeneration would use 37 percent less salt. Bypass blending could not be used with partial regeneration because the nitrate leakage was 7 mg NO$_3$-N/L.

**Disposal of nitrate-contaminated brine.**     Because of its eutrophication potential, nitrate-contaminated brine cannot be disposed of into rivers or lakes even if it is slowly metered into the receiving water. Furthermore, the high sodium concentration prevents disposal of spent regenerant onto land where its nitrogen content could serve as a fertilizer. To eliminate the high-sodium problem, calcium chloride could be used in place of NaCl, but this would be much more expensive and could result in the precipitation of CaCO$_3$($s$) during regeneration. Potassium chloride (KCl) is also feasible as a regenerant, but currently (1989) it is much more expensive than NaCl.

In one municipal-scale application of nitrate removal by chloride ion exchange,[5] the spent regenerant is simply metered into the sanitary sewer for subsequent biological treatment. This is probably only possible in those applications where nitrate removal is being carried out in a fraction, say less than 25 percent, of the wells in the supply system. If all the wells were being treated, the waste salt and varying salinity would surely have a negative effect on the biological treatment process.

In the Glendale studies, reuse of the spent brine was attempted but without much success. Apparently, the high affinity of the resin for the nitrate in the brine resulted in excessive nitrate on the resin following reuse. Removing the nitrate from the spent brine prior to its reuse is possible via biological denitrification. Bench and pilot-scale studies of this process have been reported by Van der Hoek et al.[33] who found that biological denitrification is feasible below about 15,000 mg NaCl/L. They also found that the denitrified spent-regenerant solution could be reused. This process, although more complex than processes using direct disposal of waste regenerants, should be considered where wastewater disposal is a major consideration.

### Fluoride removal by activated alumina

Activated alumina, a semicrystalline porous inorganic adsorbent, is an excellent medium for fluoride removal. Alumina is superior to synthetic organic anion-exchange resins because fluoride is one of the ions most preferred by alumina, whereas with resins fluoride is the least preferred of the common anions. [Compare the position of fluoride in Eq. (9.27), the activated alumina anion selectivity sequence, with Table 9.2 containing the relative affinity of anions for resins.] The usefulness of alumina as a fluoride adsorbent has been known for a long time, and municipal defluoridation of public water supplies using packed beds of activated alumina has been practiced since the 1940s.[34] More recently, POE and POU fluoride-removal systems utilizing activated alumina either have come into common use[35] or at least are being considered for individual-home treatment.[36]

The fundamentals of fluoride adsorption onto activated alumina were previously covered in the section "Activated Alumina Adsorption" where attention is drawn to Eqs. (9.11) through (9.17). The following discussion focuses on the important design considerations for municipal defluoridation. Although pH adjustment is not ordinarily performed prior to POE and POU defluoridation, the discussion may also prove useful to designers of these systems because the important factors, including pH, governing fluoride capacity are explained.

**Alumina defluoridation system design.**    A typical fluoride-removal plant utilizing activated alumina consists of two or more adsorption beds operated alternately or simultaneously. The source water pH is adjusted to 5.5 to 6.0 and passed downflow through a 3- to 5-ft-deep bed of fine (28 × 48 mesh) medium that adsorbs the fluoride. The fluoride breakthrough curve (some typical examples are shown in Fig. 9.15) is not sharp compared to the usual ion-exchange resin breakthrough curves. The effluent fluoride concentration is continuously increasing, thereby making the use of the bypass blending technique, shown in Fig. 9.10, difficult to implement. Because of the gradual and increasing leakage of fluoride, bypass blending is not ordinarily used. Instead a product water storage tank is provided to equalize the column-effluent fluoride concentration and to maximize the column run length.

A target level of fluoride, usually in the range of 1 to 3 mg/L, must be chosen for the process effluent before the operating procedures and economics can be established. The MCL for fluoride is 4 mg/L, and the secondary maximum contaminant level (SMCL) is 2 mg/L (see Chap. 15). When the source water contains 3 to 6 mg/L fluoride and a max-

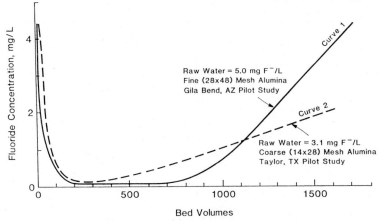

**Figure 9.15** Typical fluoride breakthrough curves for activated alumina operated at a feed pH of 5.5 and EBCT of 5 min.

imum fluoride effluent (breakthrough) concentration of 1.4 mg/L is chosen, a run length of approximately 1000 to 1300 BV can be expected. At the usual EBCT of 5 min, this corresponds to a service time of 3 to 5 days before regeneration is required. Some typical data from pilot-scale defluoridation processes are presented in Table 9.6 for source water fluoride concentrations of 2 to 5 mg/L, and some detailed design recommendations are given in Table 9.7.

Following exhaustion, the medium is backwashed and then subjected to a two-step regeneration with base followed by acid. Because many types of high-fluoride groundwater are located in hot, arid climates, the spent-regenerant brines are neutralized and sent to a lined evaporation pond for interim disposal. The ultimate disposal of high-fluoride salt residues is a problem still unsolved.

**TABLE 9.6    Fluoride Capacity of F-1 Activated Alumina Columns at pH 5.5 to 6.0—Field Results**

| Feedwater fluoride mg/L | Feedwater TDS, mg/L | Alumina mesh size† | Run length BV to 1.4 mg F⁻/L | Fluoride capacity g/m³ | Note |
|---|---|---|---|---|---|
| 2.0 | 810 | 28 × 48 | 2300 | 3700 | †‡ |
| 3.0 | 1350 | 14 × 28 | 1200 | 3000 | § |
| 5.0 | 1210 | 28 × 48 | 1150 | 4600 | ¶ |

†28 × 48 mesh is "fine" alumina, the recommended medium. 14 × 28 mesh is a "coarse" alumina that can also be used.
‡San Ysidro, N. M.; average of three pilot-scale runs, Ref. 38.
§Taylor, Tex.; average of three pilot-scale runs.
¶Gila Bend, Ariz.; typical pilot-scale run, Ref. 37.

TABLE 9.7    Process Design Criteria for Fluoride Removal by Activated
Alumina

| Parameter | Typical value or range |
|---|---|
| **Exhaustion and Backwash** | |
| Fluoride concentration | 3–6 mg/L |
| Medium[a] | Alcoa F-1 activated alumina |
| Medium size[b] | 28 × 48 mesh (0.29 to 0.59 mm) |
| Medium depth | 3–5 ft |
| Fluoride capacity | 1300–2200 grains/ft$^3$ (3000–5000 g/m$^3$) |
| Bed volumes to 1.4 mg F$^-$/L[c] | 1000–1500 |
| Exhaustion flow rate | 1.5 gpm/ft$^3$ (EBCT = 5 min) |
| Exhaustion flow velocity | 4–8 gpm/ft$^2$ |
| Backwash flow velocity | 8–9 gpm/ft$^2$ |
| Backwash time | 5–10 min using source water |
| **NaOH Regeneration** | |
| Volume of regenerant | 5 BV |
| Regenerant flow rate[d] | 0.5 gpm/ft$^3$ (EBCT = 15 min) |
| Regenerant concentration[e] | 1% NaOH (0.25 N) |
| Total regenerant contact time | 75 min |
| Displacement rinse volume | 2 BV |
| Displacement rinse rate | 0.5 gpm/ft$^3$ |
| **H$_2$SO$_4$ Neutralization** | |
| Acid concentration[f] | 2.0% (0.4 N H$_2$SO$_4$) |
| Acid volume[g] | Sufficient to neutralize bed to pH 5.5, typically 1.5 BV |
| Displacement rinse volume | 2 BV |
| Displacement rinse rate | 0.5 gpm/ft$^3$ |

[a]Mention of trade names does not imply endorsement.
[b]Coarse 14 × 28 mesh alumina has also been used successfully.
[c]Capacity and BV to fluoride breakthrough depend to some extent on the fluoride level in the source water, the sulfate level in the pH-adjusted feedwater, and the severity of regeneration.
[d]Cocurrent (downflow) or countercurrent regeneration may be utilized.
[e]Higher regenerant NaOH concentrations, e.g., 4%, have also been used successfully.
[f]Lower and higher H$_2$SO$_4$ concentrations have been used successfully. HCl can also be used, and it may be preferred to H$_2$SO$_4$ because chloride does not compete with fluoride for adsorption sites.
[g]Alternatively, the neutralization and acidification steps can be combined. In this case the initial feedwater pH is lowered to approximately 2.5, to produce a bed effluent pH slowly dropping from a high initial value ( > 13) down to a continuous effluent pH in the range of 5.5 to 6.0.

**Factors influencing the fluoride capacity of alumina.**    The fluoride capacity of alumina is very sensitive to pH, as shown in Fig. 9.15 based on laboratory equilibrium data obtained by using minicolumns of granular alumina exposed for up to 30 days to pH-adjusted fluoride solutions in deionized water. The fluoride capacities in Fig. 9.16 can be considered the "maximum attainable (equilibrium) capacities" for F-1 alumina because potentially competing anions such as sulfate, sili-

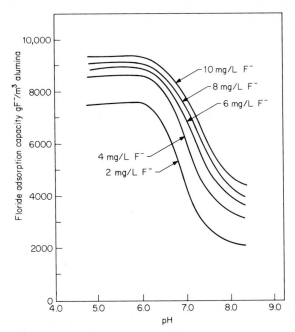

**Figure 9.16**  Effects of pH on equilibrium fluoride adsorption capacity on F-1 alumina. Distilled water solutions of NaF, pH-adjusted with $Na_2CO_3$ and $H_2SO_4$.

cate, arsenate, and selenite were minimized or entirely absent in the test solutions.[18] Pilot-scale field studies on typical fluoride-contaminated water have resulted in the observation that single columns of fine (28 × 48 mesh) alumina operated to 1.4 mg fluoride/L breakthrough and 5-min EBCT can attain approximately 50 percent of the capacities shown in Fig. 9.16.[37,38] Using coarse-mesh alumina under the same conditions yields about 37 percent of the maximum attainable (equilibrium) capacity. These percentages (37 percent for coarse medium and 50 percent for fine medium) can be used in conjunction with Fig. 9.15 to estimate the single-column fluoride capacities attainable in actual single-column installations when the natural or adjusted source water pH is known.

The inability of a pilot- or full-scale, continuously operated alumina column to closely approach fluoride equilibrium is primarily due to solid-phase mass transfer limitations. These limitations cause the fluoride breakthrough curve to be gradual and the run to be terminated long before the medium is saturated. This is a classic case in which the process economics can be improved by operating two columns in series in a roughing-polishing sequence, as depicted in Fig. 9.11. The two-column sequence will allow the roughing column to attain a near-saturation fluoride loading prior to regeneration. The obvious disadvantages of such systems are greater complexity and higher capital costs.

In POE and POU alumina systems, a closer approach to equilibrium may be expected because of the intermittent operation of the alumina bed. In these installations, the solid-phase fluoride concentration gradient has a chance to relax completely (approaching equilibrium) during the many, prolonged "off" periods experienced by the column. This results in a high concentration gradient between the liquid and the surface of the solid and much improved fluoride removal when the column is restarted. Another reason for the less-than-equilibrium fluoride capacities of actual columns is the effect of competing anions such as silicate and sulfate present in the source water. During a laboratory study[18] of the effect of competing ions on fluoride capacity at optimum pH, bicarbonate and chloride ions did not compete measurably with fluoride for adsorption sites. Sulfate did, however, cause a significant reduction in fluoride adsorption. For example, 50 mg sulfate/L lowers the fluoride capacity by about 16 percent. The maximum reduction in fluoride capacity because of sulfate is about 33 percent, and it occurs at approximately 500 mg sulfate/L. Beyond that, no further fluoride capacity reduction occurs, probably because the alumina contains a high proportion of fluoride-specific sites that are not influenced by sulfate.

Previously unpublished results of fluoride-removal pilot studies in Taylor, Tex., indicate that silicate $[Si(OH)_3O^-]$ adsorption onto alumina occurs at about pH 7 and above and can cause a serious reduction in the fluoride capacity of a column operated in the pH range of 7 to 9. This reduction is particularly detrimental during cyclic operation because silicate is only partially removed during a normal regeneration to remove fluoride. The presence of silica in the water has no observable effect on fluoride capacity for columns operated in the optimum (5.5 to 6.0) pH range.

**Regeneration of fluoride-spent alumina columns.** Prior to regeneration to elute fluoride, the exhausted column must be backwashed to remove entrained particles, break up clumps of alumina, and reclassify the medium to eliminate packing and channeling. Following a 5- to 10-min backwash, regeneration is accomplished in two steps—the first utilizing NaOH and the second utilizing $H_2SO_4$. Because fluoride leakage is such a serious problem, the most efficient regeneration is accomplished by using countercurrent flow. Nevertheless, cocurrent regeneration is more commonly used because of its simplicity. (Rubel[39] recommends a two-stage, upflow-then-downflow application of base followed by downflow acid neutralization.) A slow rinse follows both the base addition and acid addition steps to displace the regenerant and conserve chemicals. The rinsing process is completed by using a fast rinse applied at the usual exhaustion flow rate. The total amount of wastewater produced amounts to about 4 percent of the blended product water and is made up of 8-BV backwash, 5-BV NaOH, 2-BV base displacement, 1.5-BV acid, 2-BV acid displacement, and 30 to 50 BV of fast rinse water. Further details pertaining to regeneration can be found in Table 9.7.

Essentially complete (> 95 percent) removal of fluoride from the alumina is possible by using the regeneration techniques described; however, in addition to the complexity of regeneration, its efficiency is poor. For example, an alumina bed employed to remove fluoride from a typical 800 to 1200 mg TDS/L groundwater containing 4 to 6 mg fluoride/L would contain an average fluoride loading of 4600 g/m$^3$ (2000 grains F$^-$/ft$^3$). Removing this fluoride would typically require 5 BV of 1% NaOH (30 gal of 1% NaOH/ft$^3$). This equates to 19 percent base regeneration efficiency, or 5.2 hydroxide ions required for each fluoride ion removed from the alumina. When the amount of acid required (1.5 BV of 0.4 $N$ H$_2$SO$_4$) is added into the efficiency calculation, an additional 2.5 hydrogen ions are required for each fluoride ion removed and the overall efficiency drops to 13 percent, or a total of 7.7 regenerant ions required for each fluoride ion removed from the water. In spite of the complexity of the process and the inefficiency of regeneration on an ion-for-ion basis, fluoride removal by activated alumina is probably the cheapest alternative compared to other acceptable methods, such as reverse osmosis and electrodialysis.

### Arsenic removal by resins and alumina

As with other toxic inorganic contaminants, arsenic is almost exclusively a groundwater problem. Although it can exist in both organic and inorganic forms, only inorganic arsenic in the +3 or +5 valence state has been found to be significant where potable water supplies are concerned.[40]

The current (1989) MCL for total arsenic is 0.05 mg/L (see Chap. 2). Depending on the redox condition of the groundwater, either arsenite [As(III)] or arsenate [As(V)] forms will be predominant. The pH of the water is also very important in determining the arsenic speciation. The primary arsenate [As(V)] species found in groundwater in the pH range of 6 to 9 are monovalent H$_2$AsO$_4^-$ and divalent HAsO$_4^{2-}$. These anions result from the dissociation of arsenic acid (H$_3$AsO$_4$), which exhibits p$K_a$ values of 2.2, 7.0, and 11.5. Uncharged arsenious acid (H$_3$AsO$_3$) is the predominant species of trivalent arsenic found in natural water. Only at pH values above its $pK_a$ of 9.2 does the monovalent arsenite anion ($H_2AsO_3^-$) predominate.

Both the redox potential and pH are important with regard to arsenic removal from groundwater using the most economical alternatives—anion exchange and activated alumina adsorption. This is because both processes require an anionic species to effectively remove arsenic. Theoretically, at least, As(III) is not removed by either process. Arsenic(V), however, is removed by both processes. Which process to choose is primarily dependent on the sulfate and TDS levels in the source water. Because of competition by background ions for exchange sites, anion exchange is not economically attractive at high TDS (> 500 mg/L) or sulfate (> 25 mg/L) levels, whereas alumina adsorption is so specific for arsenate that it is not greatly affected by these variables.

The advantages of anion exchange are that pH adjustment is not required and that ordinary sodium chloride can be used to achieve nearly complete (85 to 100 percent) elution of arsenic from spent resin. But alumina requires both sodium hydroxide and sulfuric acid and so typically recovers only 50 to 70 percent of the adsorbed arsenic. A significant potential disadvantage of ion exchange is that chromatographic peaking of arsenic is possible depending on the sulfate level in the feedwater; i.e., the higher the sulfate, the greater the potential for arsenic peaking. This potential problem can be solved by careful monitoring of the flow or chemical quality of the effluent.

Naturally contaminated arsenic-bearing types of groundwater have been reported to have relatively high pH and high alkalinity.[41] These water types, if they are also low in sulfate, as was the case in Hanford, Calif.,[42] are very amenable to treatment by chloride-form anion exchange providing As(III) is oxidized (see below) to As(V). High-sulfate high-TDS types of water such as the water supply encountered in San Ysidro, N.M.,[38] are not amenable to ion exchange but can be effectively treated by activated alumina adsorption following oxidation of As(III) to As(V) and pH reduction to 5.5 to 6.0.

**Oxidation of As(III) to As(V).**   To achieve effective removal of arsenic from groundwater by means of columns or membranes, arsenite must be oxidized to arsenate. Figure 9.17 illustrates the importance of oxidative pretreatment ahead of activated alumina columns operated at the optimum pH of 6.0. Starting with a source water containing 100 μg As(III)/L, we see that arsenic breakthrough curve 1 reaches 50 μg As/L after only 300 BV. But curve 4, representing the same water but containing 100 μg As(V)/L, does not reach 50 μg As/L until 23,400 BV. This is an 80-fold improvement in performance due simply to convert-

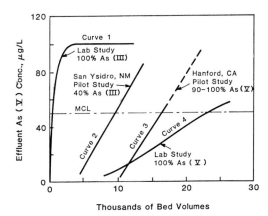

**Figure 9.17** Arsenic (total) breakthrough curves for fine (28 × 48) mesh F-1 activated alumina used in laboratory and pilot-scale columns. Feedwater pH = 6.0, arsenic (total) concentration = 88 to 100 ppb, and As(III) or As(V) percentages are indicated. EBCT = 3 to 5 min.

ing As(III) to As(V). The intermediate plots, curves 2 and 3, also illustrate the need for oxidation but are not such dramatic examples. In these curves, actual field data from pilot studies in San Ysidro and Hanford were used. Some unplanned oxidation of As(III) to As(V) within the alumina column occurred during the New Mexico study (curve 2) and resulted in better than expected performance of the alumina. The performance of the alumina on chlorinated Hanford, Calif., groundwater (curve 3) containing nearly 100 percent As(V) was somewhat poorer than expected and was probably caused by fouling of the alumina by the negatively charged black "mica" particles in the water. For properly oxidized source water in a municipal arsenic-removal application, a breakthrough curve somewhere between curves 3 and 4 would be expected. Further data on alumina column performance as a function of oxidation state and pH are presented in Table 9.8. Here, at high pH (8.6), it does not matter whether As(III) is oxidized because about the same mediocre performance is obtained, i.e., 800 to 900 BV. At relatively high pH (8.6), the adsorption or ligand-exchange capacity of alumina is severely reduced by competition from hydroxide ions. This leads to poor As(V) uptake compared to adsorption at the optimum pH of 6.0. The As(III) uptake at pH 8.6 is, however, slightly improved compared to that at pH 6.0 because of the increase in the fraction of charged $H_3AsO_2^-$ with increasing pH.

In a laboratory study[43] of the oxidation of As(III) to As(V) in the 6 to 10 pH range, As(III) was immediately converted to As(V) in the presence of 1 mg free chlorine/L. In that same study, pure oxygen could not oxidize As(III) to As(V) in 1 h, but complete oxidation of As(III) to As(V) occurred during 2 months of ambient temperature storage of synthetic groundwater. Inadvertent, unpredictable, microbially assisted oxidation of As(III) to As(V) has been observed in all the University of Houston studies with As(III).[19,38,42,43] Arsenic reduction can

**TABLE 9.8  Summary Arsenic (Total) Capacities of 28 × 48 Mesh F-1 Alumina Columns**

| Column feedwater | As(III), mg/L | As(V), mg/L | As(total), mg/L | pH | BV to 0.05 mg As/L | Arsenic (total) capacity, g/m³ | Comment |
|---|---|---|---|---|---|---|---|
| Synthetic groundwater | 100 | 0 | 100 | 6 | 300 | 20 | Pure As(III) |
| Hanford, Calif., groundwater | 80 | 10 | 90 | 6 | 700 | 60 | 90% As(III) |
| San Ysidro, N.M., groundwater[†] | 31 | 57 | 88 | 6 | 9,000 | 575 | 40% As(III) |
| Chlorinated Fallon, Nev., groundwater[‡] | 0 | 110 | 100 | 5.5 | 13,100 | 1,280 | 100% As(V) |
| Chlorinated Hanford, Calif., groundwater | 0 | 98 | 98 | 6 | 16,000 | 1,410 | 100% As(V) |
| Synthetic groundwater | 0 | 100 | 100 | 6 | 23,000 | 1,920 | 100% As(V) |
| Performance at Unadjusted pH§ | | | | | | | |
| Chlorinated Fallon, Nev., groundwater | 0 | 110 | 110 | 9 | 800 | 42 | 100% As(V) |
| Hanford, Calif., groundwater | 80 | 10 | 90 | 8.6 | 800 | 61 | 90% As(III) |
| Chlorinated Hanford, Calif., groundwater | 0 | 98 | 98 | 8.8 | 900 | 83 | 100% As(V) |

†Unplanned oxidation of As(III) to As(V) occurred on the activated alumina column, thus the run was longer than expected.

‡Measurement of As(III) was not performed. And 100% As(V) was presumed caused by the presence of free Cl₂ in the column feedwater.

§Normally POE and POU systems would be operated in this manner, i.e., without pH adjustment of the source water. Although unadjusted pH runs are much shorter than runs at pH 6, the unadjusted pH process is feasible for POE and POU applications.

occur, too. In one case involving bisulfite addition to water, As(V) was reduced to As(III). Therefore, As(III) and As(V) are readily interconvertible under appropriate conditions.

**Alumina system design for arsenic removal.**   The design of an activated alumina system for arsenic removal is basically the same as that already described for fluoride removal. Four significant differences, however, do exist: (1) Because the arsenic concentration (typically 0.06 mg/L) is so low compared to the fluoride concentration (typically 6 mg/L), a continuous arsenic run may last as long as 30 to 90 days. (2) Because of the difficulty in removing arsenic, the regenerant NaOH concentration will be 4% (1.0 $N$) as opposed to the 1% NaOH used for fluoride removal. (3) Prechlorination or an alternative form of oxidation will probably be required to ensure the presence of As(V) as opposed to As(III). (4) A spent-regenerant treatment system involving the coprecipitation and sludge disposal of arsenates with $Fe(OH)_3(s)$ or $Al(OH)_3(s)$ will probably also be required because of the greater toxicity of arsenic compared to fluoride.

**Factors influencing the arsenic capacity of alumina.**   The optimum pH for fluoride and arsenic adsorption onto alumina has not been clearly established. Based on a limited number of full-scale and pilot-scale column studies, Rubel recommends pH 5.5 for both fluoride[37] and arsenic,[44] and a combination of equilibrium isotherm studies and pilot-scale column studies indicated that the optimum pH is closer to 6.0 for both fluoride[18] and arsenic.[19] The equilibrium isotherm data for arsenic in Fig. 9.18 show the arsenic capacity declining rapidly on either side of the pH 6.0 optimum. The isotherms summarized in Fig. 9.18 were developed in synthetic background water containing approximately 1000 mg TDS/L and the competing anions of sulfate and chloride at pH 5 and 6 and sulfate, chloride, and bicarbonate at pH 7 to 9.

As with fluoride adsorption, the As(V) capacity of alumina is significantly reduced in the presence of sulfate ions but nearly unaffected by high concentrations of chloride. The influences of these ions are demonstrated in the 25°C equilibrium isotherms presented in Fig. 9.19. In actual column operation, however, the arsenic capacity obtainable is far less than the equilibrium values shown. This is evident from a comparison of the three data points near the bottom of Fig. 9.19 to the equilibrium curves. The three column capacities shown were obtained from single columns operated until the breakthrough of 0.05 mg As(V)/L. The effects of competing anions and the nonequilibrium mass transfer limitations already discussed (under the topic of fluoride removal) are reasons for the low column capacities observed. An

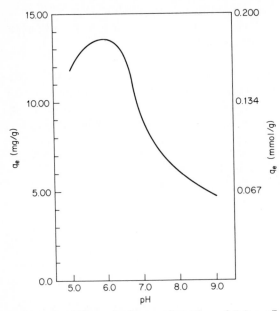

**Figure 9.18** Effect of pH on adsorption of 1.0 mg/L arsenic(V) onto F-1 activated alumina by the column equilibration method.[19]

additional reason is the possible fouling of the porous alumina by particulate and colloidal constituents such as colloidal silica and mica. The latter was identified as a problem in the Hanford study.[42] An engineering estimate of the practically achievable column capacity based on pH 6.0 operation with source water containing 0.10 mg As(V)/L is 1400 g As(V)/m$^3$ of alumina.

**Regeneration of arsenic-spent alumina.**   Arsenic is much more difficult to remove from alumina than fluoride is. For this reason, a higher concentration, 4% NaOH, should be used for the base regeneration step. The usual acid concentration, 2% H$_2$SO$_4$, can be used, however, for the acid neutralization step. Even with additional caustic, only 50 to 70 percent of the adsorbed arsenic will be eluted from the column during regeneration. Furthermore, the arsenic capacity of the alumina appears to deteriorate, and subsequent runs are progressively shorter by about 10 to 15 percent. Thus, arsenic removal by alumina is significantly different from fluoride removal. With defluoridation processes, nearly complete fluoride elution is observed, and subsequent runs are not significantly shorter. To date, only pilot studies of arsenic removal by alumina have been conducted. Because of the complexities of the process and the incomplete regeneration observed, it is not yet decided whether this process has real commercial applicability.

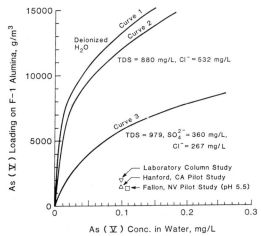

**Figure 9.19**  Equilibrium arsenic(V) adsorption onto F-1 activated alumina at pH 6.0 for deionized water and high-TDS background water.

The discussion above is detailed because arsenic is a very toxic contaminant and activated alumina is technically effective in removing it, providing the species present are forms of As(V) and the pH is reduced to 5.5 to 6.0. However, because of problems with regeneration and regenerant disposal, the engineer should consider POU or POE treatment without pH adjustment. Refer again to Table 9.8. POU systems without preoxidation or pH reduction, used intermittently, should achieve 1000-BV throughput prior to exhaustion. Exhausted medium would simply be thrown away, not regenerated. Although not verified, the spent medium would probably pass the standard extraction procedure (EP) toxicity test as a nonhazardous waste. The reason is that the arsenic loading is very low and the EP test is done at pH 5, which is near the optimum pH for arsenic adsorption onto alumina. Furthermore, arsenic-laden $Al(OH)_3(s)$ sludges from spent-regenerant treatment are known to pass the EP toxicity test, and these sludges have very similar chemistry to that of activated alumina containing adsorbed arsenic.

**Arsenic removal from spent alumina regenerants.**  During regeneration and acidification of spent alumina, enough aluminum dissolves to make precipitation of $Al(OH)_3(s)$ a feasible treatment step for the removal of arsenic from the wastewater. This is done by lowering the pH of the spent regenerant to about 6.5 by using $H_2SO_4$ or HCl.[38,44] The As(V) quantitatively coprecipitates with the resulting $As(OH)_3(s)$. Following dewatering, the arsenic-contaminated dried $Al(OH)_3(s)$ sludge easily passes the EP toxicity test. Rubel and Hathaway[44] described a dried sludge containing 1627 mg As/kg solids that yielded only 0.036 mg As/L—a value that is well below the current (1989) 5 mg/L arsenic MCL. Other observations are in accord with this result.[38]

**Ion exchange for arsenic removal.** If the source water is low in TDS (< 500 mg/L) and sulfate (< 25 mg/L), like the water in Hanford, Calif. (TDS = 213 mg/L, $SO_4$ = 5 mg/L), ion exchange may be the arsenic-removal process of choice.[42] Preoxidation to convert As(III) to As(V) is necessary, but pH adjustment is not. The chlorinated and filtered raw water is passed downflow through a 2.5- to 5-ft-deep bed of chloride-form strong-base anion-exchange resin, and the chloride-arsenate ion-exchange reaction [Eq. (9.56)] takes place in the pH range of 8 to 9. Regeneration, according to Eq. (9.57), is easily accomplished and returns the resin to the chloride form, ready for another exhaustion cycle:

$$2\overline{RCl} + HAsO_4^{2-} = \overline{R_2HAsO_4} + 2Cl^- \qquad (9.56)$$

$$\overline{R_2HAsO_4} + 2NaCl = 2\overline{RCl} + Na_2HAsO_4 \qquad (9.57)$$

**Breakthrough curves in As(V) ion exchange.** Prechlorination followed by ion exchange proved to be a very effective means of As(V) removal in Hanford. A typical set of breakthrough curves for pH, arsenic, and sulfate is shown in Fig. 9.20. Here a commercially available SBA exchange resin (Dowex 11) could effectively treat 4200 BV (16.5-day run length at a 5.6 min EBCT) of water before the arsenic level reached 0.05 mg/L. However, in spite of its very low concentration, sulfate was still the most preferred anion, and it eventually drove arsenate off the column and caused its concentration to peak at 0.135 mg As(V)/L—a value that is 160 percent of its feed concentration. Note, however, that this peak of As(V) is only a potential peak and if it did ever occur accidentally, the effluent toxicity would still be lower than the more toxic As(III) originally present in the feedwater. [Although the MCL does not distinguish between arsenic forms, As(III) is generally considered more toxic than As(V).[41]]

Unfortunately, a significant pH change did not occur simultaneously with As(V) breakthrough in Fig. 9.20. The effluent pH began at a relatively low value (5.6) and rose to the influent value of 8.7 at about 1300 BV (5 days), while As(V) did not reach 0.02 mg/L (a potential regeneration point) until about 3500 BV (14 days). In this or a similar situation, a column run could simply be terminated at a predetermined length of, say, 3500 BV, which is far short of the point at which the effluent As(V) concentration reaches its influent value of 0.085 mg/L (5200 BV or 20 days).

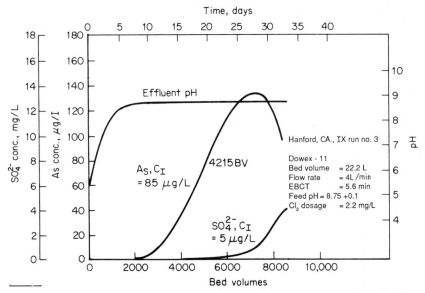

**Figure 9.20**  Typical ion-exchange exhaustion run showing arsenic, pH, and sulfate breakthrough curves from chloride-form anion-exchange column. $C_I$ = Influent concentration.

### Effect of process variables on As(V) removal by ion exchange.

Five different commercially available SBA exchange resins, including polystyrene type 1, were tested for As(V) removal performance in Hanford, Calif.[42] After the exchange capacity was taken into account, the resins differed very little in the ability to remove As(V). Dowex 11, an improved porosity polystyrene type 1 resin, and Ionac ASB-2, a polystyrene type 2 resin, seemed to have a slight edge over the others in terms of BV to arsenic breakthrough, so they were studied extensively. Resin performance deteriorated slightly following each regeneration. The bed volumes to As(V) breakthrough (0.05 mg/L) for the first three runs were 4210, 4000, and 3880 for the Dowex 11 resin. The reason for the 3 to 5 percent per cycle decrease was not established but was thought to have been caused by the mica fouling observed on the exhausted resin. Most of, but not all, the black coating was removed upon NaCl regeneration. Presumably this sort of fouling could be eliminated by multimedia filtration ahead of the anion resin.

During one run at Hanford, the As(V) concentration was deliberately increased 14-fold from 0.085 to 1.2 mg/L by spiking the feedwater. The trend predicted from equilibrium multicomponent chromatography theory[24] did occur; i.e., the run length to arsenic breakthrough did not decrease proportionately. Less than a 50 percent reduction in run length resulted; at 0.085 and 1.2 mg/L As(V), the run lengths for virgin ASB-2 resin were 4940 and 2680 BV, respectively.

With respect to EBCT, the Hanford study showed that reducing the EBCT from 5 to 1.4 min produced no significant reduction in performance. Thus, to reduce resin inventory and capital investment, shorter contact times are preferred, but higher operating costs will result because more frequent regenerations will be required.

**Regeneration of arsenic-spent resins.**   Regardless of the type of resin used in Hanford, arsenic recoveries upon downflow (cocurrent) regeneration were essentially complete. The 3 BV of 1.0 $N$ NaCl (11 lb NaCl per cubic foot of resin) was more than adequate to elute all the adsorbed arsenic. This is shown graphically in Fig. 9.21 where the arsenic is more easily eluted than even bicarbonate—a very nonpreferred ion. One reason why arsenic elutes so readily is that it is a divalent ion ($HAsO_4^{2-}$) and is thus subject to a selectivity reversal in the high-ionic-strength (> 1 $M$) environment of the regenerant solution. This ease of regeneration is a strong point in favor of ion exchange as compared to alumina for this type of source water.

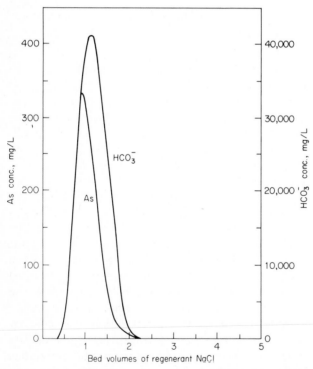

**Figure 9.21**   Elution of arsenic and bicarbonate during regeneration of a typical SBA resin with 1.0 $N$ NaCl at an EBCT of 20 min.

In confirmation of similar findings from the nitrate studies in Glendale, Ariz., in the Hanford experiments, dilute regenerants were much more efficient than concentrated ones for eluting arsenic. With dilute (0.25 $N$ NaCl) regenerant, only 1.2 equiv chloride per equivalent of resin (i.e., 1.2 $Cl^-$ ions per resin exchange site) was required for 100 percent arsenic elution whereas 2 equiv of chloride per equivalent of resin was required with 1.0 $N$ NaCl. Therefore, the 1.0 $N$ NaCl regeneration consumed 67 percent more salt than the regeneration with 0.25 $N$ NaCl. The disadvantages of dilute regenerants are a longer regeneration time and a larger volume (about 2 times larger) of more dilute wastewater to be treated for arsenic removal.

In summary, when anion exchange is appropriate for As(V) removal, the spent resin with a total exchange capacity of 1 equiv/L can be regenerated with either 2.0 BV of 1.0 $N$ NaCl (7.3 lb NaCl per cubic foot of resin) or 5 BV of 0.25 $N$ NaCl (4.6 lb NaCl per cubic foot of resin). In both cases about 2 BV of displacement rinse is required.

**Treatment of spent ion-exchange regenerant.** Arsenic(V) can be precipitated from the waste brine by using ferric iron or aluminum salts such as $FeCl_3 \cdot 6H_2O$ or $Al_2(SO_4)_3 \cdot 18H_2O$ or lime $[Ca(OH)_2]$. The stoichiometric reactions are

$$3Na_2HAsO_4 + 3H_2O + 2FeCl_3 = Fe(OH)_3(s) + Fe(H_2AsO_4)_3(s)$$

$$+ 6NaCl \quad (9.58)$$

$$2Na_2HAsO_4 + NaHCO_3 + 4Ca(OH)_2 = CaCO_3(s) + Ca_3(AsO_4)_2(s)$$

$$+ 3H_2O + 5NaOH \quad (9.59)$$

Iron or aluminum is required in greater than stoichiometric amounts in actual practice. A wastewater of pH 9.9 resulting from a 1.0 $N$ NaCl regeneration was treated to remove As(V) in Hanford, Calif. For the wastewater that contained 90 mg As(V)/L and 50,000 mg TDS/L, about 12 times the stoichiometric amount of iron (adjusted final pH 8.5) or aluminum (adjusted final pH 6.5) was required to reduce the As(V) level from 90 to less than 5 mg/L. To reduce As(V) to less than 1.5 mg/L required a 20-fold stoichiometric dosage of either metal salt. The iron floc was heavier and settled better than the alum floc. When subjected to the EP toxicity test to determine As(V) leachability from the dried sludges, both sludges passed the test with about 1.5 mg As(V)/L in the leachate. No attempts were made to reuse the decontaminated spent regenerants, but this is a possibility. If the regenerant is to be reused, consideration must be given to the anion, i.e., chloride or sulfate, chosen for the metal salt, because the anion will not be precipitated and may remain on the resin along with bicarbonate following reuse. Regenerant reuse remains an area for further study.

## Selenium removal by resins and alumina

Selenium contamination of potable water supplies at concentrations above 0.03 mg/L is not common. For example, a nationwide survey of potable water supplies conducted by the AWWA in 1982 and supplemented by U.S. Environmental Protection Agency data[1] found only 44 supplies with selenium in excess of 0.03 mg/L. All these were groundwater supplies. The current (1989) MCL for selenium is 0.01 mg/L, and the new proposed MCL is 0.05 mg/L (see Chap. 2).

When selenium is found in groundwater, it is a result of natural, not artificial, contamination. At typical groundwater pH values (7.0 to 9.5) only the anionic forms of selenious [Se(IV)] and selenic [Se(VI)] acid are found. The dissociation equations and constants for the two acids are presented in Eqs. (9.60) through (9.63).
Selenious acid dissociation equilibria:

$$H_2SeO_3 = H^+ + HSeO_3^- \qquad pK_1 = 2.55 \qquad (9.60)$$

$$HSeO_3 = H^+ + SeO_3^{2-} \qquad pK_2 = 8.15 \qquad (9.61)$$

Selenic acid dissociation equilibria:

$$H_2SeO_4 = H^+ + HSeO_4^- \qquad pK_1 = -3.0 \qquad (9.62)$$

$$HSeO_4^- = H^+ + SeO_4^{2-} \qquad pK_2 = 1.66 \qquad (9.63)$$

Under oxidizing conditions, Se(VI) will predominate, and divalent selenate ($SeO_4^{2-}$), an anion with chemical behavior similar to that of sulfate, will be found. Under reducing conditions Se(IV) will predominate, and at pH values below 8.15 the monovalent biselenite anion ($HSeO_3^-$) will be the dominant form; above pH 8.15, the divalent selenite anion ($SeO_3^{2-}$), will dominate.

Depending on the form of selenium present, either ion exchange or activated alumina adsorption is chosen for treatment. Selenite [Se(IV)] forms are best removed by alumina, and SBA exchange is the process of choice for selenate [Se(VI)]. The justification for this recommendation is based on the relative affinity values for anions on alumina [Eq. (9.27)] and anion resins (Table 9.2).

**Oxidation of selenite [Se(IV)] to selenate [Se(VI)].**  Selenate is so strongly preferred by anion resins that oxidizing selenite to selenate prior to treatment is the best approach. Although Se(IV) is considerably more difficult to oxidize than As(III), laboratory research has demonstrated that this oxidation can be readily accomplished with free chlorine. In synthetic groundwater containing sulfate, chloride, and bicarbonate at pH 8.3, the reaction is first-order in both Se(IV) and free chlorine concentrations.[45]

The selenite oxidation rate is optimum between pH 6.5 and 8.0. In this range, 60 percent of the Se(IV) is converted to Se(VI) within 5 min at a free chlorine concentration of 2 mg/L. At pH 9 only 15 percent of the Se(IV) can be converted in 5 min with 2 mg free chlorine per liter. In these studies pure oxygen was ineffective (no measurable oxidation in 1 h) while $H_2O_2$ and $KMnO_4$ were not nearly as effective as free chlorine. Finally, note that TOC was not present in the Se(IV) oxidation experiments, and it is expected to slow the oxidation rate by also reacting with the free chlorine.

**Ion exchange for selenate removal.** To date (1989), only laboratory studies with synthetic water have been done to test the possibility of using chloride-form SBA exchange for selenate removal. Even with a high TDS level (700 mg/L), high sulfate level (192 mg/L), and source water containing 0.1 mg Se(VI)/L, an acceptable run length (275 BV) to 0.03 mg Se(VI)/L was obtained.[45] As expected, selenate eluted after sulfate and was not subject to chromatographic peaking. [Chromatographic peaking of Se(IV) will occur, however, if it is present with Se(VI).] No attempts were made to regenerate the resin, but this should not be a problem because $SeO_4^{2-}$ is a divalent ion subject to selectivity reversal, as are sulfate, arsenate, and chromate, all strongly adsorbed divalent anions that are readily eluted during regeneration with NaCl.

**Activated alumina for selenite removal.** Trussell et al.[16] demonstrated that alumina adsorption is very effective for removal of Se(IV), i.e., $HSeO_3^-$, in the optimum pH range of 5 to 6. They also found, not unexpectedly, that alumina is a relatively poor medium for the removal of selenate [Se(VI)] because of strong competition from sulfate. Yaun et al.[46] also reported that alumina was good for biselenite ($HSeO_3^-$) removal and found nearly the same optimum pH range—3 to 8. Because biselenite adsorption onto alumina is approximately as strong as fluoride adsorption, complete regenerability of the spent alumina by using the typical base-acid sequence described for fluoride regeneration is expected. These laboratory studies demonstrate that the feasibility and cost of an activated alumina process for selenium removal depend on the oxidation state of the selenium. It could be an excellent process for water in which Se(IV) is the only selenium species, but it becomes less attractive as the fraction of Se(VI) increases.

In water containing both Se(IV) and Se(VI), oxidation with chlorine followed by ion exchange is more feasible than microbial reduction (or some as yet undetermined chemical reduction means) followed by activated alumina adsorption. This is especially obvious given that the product water for distribution usually has to be disinfected with chlorine.

## Chromate Removal by Anion Exchange

The two common oxidation states for chromium in natural water are Cr(III) and Cr(VI). Trivalent chromium occurs as a hydrated cation $Cr^{3+}$ at low pH, but it readily hydrolyzes to form insoluble $Cr(OH)_3(s)$ and soluble cationic and neutral hydroxide complexes such as $Cr_3(OH)_4^{5+}$, $Cr(OH)_2^+$, and $Cr(OH)_3^0$, which are significant in the 7 to 9 pH range.[47] Because of its relative insolubility under typical groundwater conditions, Cr(III) is not a significant groundwater contaminant whereas hexavalent chromium, i.e., Cr(VI) from both natural and artificial sources, is found in groundwater.[48] Chromate ($CrO_4^{2-}$), the divalent anion of chromic acid ($H_2CrO_4$; $pK_1 = 0.08$, $pK_2 = 6.5$) is the predominant form of Cr(VI) in the 7 to 9 pH range. Theoretically, anion exchange with synthetic resins is an ideal process for chromate removal because chromate is the most preferred of the common anions. (See Table 9.2.) In practice, ion exchange works well, as demonstrated by the author's previously unpublished research with chloride-form SBA exchange to remove chromate from a Scottsdale, Ariz., well. These results are summarized below. The Scottsdale well used for the field research contained 0.042 mg Cr(VI)/L present as the $CrO_4^{2-}$ anion. Other water contaminants present at the milligram/liter level were bicarbonate (244), chloride (24), sulfate (9), magnesium (28), calcium (19), and sodium (39). The pH was 7.6, TDS was 276, and silica was 32 mg/L as $SiO_2$.

### Effect of resin matrix on chromate removal

Although all anion-exchange resins strongly prefer chromate, the resin matrix exhibits a significant influence on run length to chromate breakthrough. The extreme affinity of chromate for all anion resins and the effects of resin matrix and porosity are demonstrated in Table 9.9. The longest run (32,000 BV and 98 days) was achieved with a macroporous polystyrene resin. With regard to $CrO_4^{2-}$ affinity, macroporous resins are better than the gel and improved-porosity types, and hydrophobic polystyrene resins are better than the hydrophilic polyacrylic type.

In ion-exchange processes, the resin with the highest affinity for the contaminant is not necessarily the resin of choice, after regenerability

TABLE 9.9    Effects of Resin Matrix and Porosity on Run Length to Chromate Breakthrough

| Resin | Matrix/porosity | Capacity, meq/mL | Run length to 0.010 mg Cr(VI)/L | |
|---|---|---|---|---|
| | | | BV | Days |
| IRA 900 | Polystyrene DVB†/MR‡ | 1.1 | 32,000 | 98 |
| Dowex 11 | Polystyrene DVB/iso§ | 1.2 | 20,700 | 68 |
| IRA 958 | Polyacrylic DVB/MR | 0.8 | 14,600 | 44 |

†DVB = divinylbenzene.
‡MR = macroporous.
§iso = isoporous or unique porosity gel resin designed for organic substance removal.

has been taken into account. In fact, the resin with the highest affinity and longest virgin run length is usually the hardest to regenerate. This was the case with chromate sorption onto macroporous polystyrene resin (IRA 900) observed in the Scottsdale study.

### Regenerability of chromate-spent resin

Various concentrations and amounts of salt (NaCl) along with mixtures of salt and caustic (NaOH), typical of chromate-removal processes for cooling water treatment, were experimented with to elute chromate from spent resin in Scottsdale. In the initial experiments, it was found that NaOH was not necessary to achieve a good regeneration, thus most regenerations were done with NaCl alone. Chromate was surprisingly easy to remove from the resin, as seen in Table 9.10, which compares the regenerability of three resins using 1 $N$ NaCl. The macroporous polystyrene resin, IRA 900 with the highest chromate affinity, was more difficult to regenerate than both the polystyrene gel resin (Dowex 11) and the macroporous polyacrylic resin (IRA 958).

In spite of the extreme affinity of chromate for resins, only 3 to 4 equiv of chloride per equivalent of resin (11 to 19 lb NaCl per cubic foot of resin) was required for "complete" cocurrent (downflow) regeneration. The range of maximum chromate recovery for all the 1 $N$ NaCl regenerations was 77 to 89 percent. The recovery, which was al-

TABLE 9.10    Regeneration of Resins Used for Chromate Removal

| Resin | Regenerant | lb NaCl / ft$^3$ resin | equiv chloride / equiv resin | Cr(VI) recovered, % |
|---|---|---|---|---|
| Dowex 11 | 3.7 BV 1 $N$ NaCl | 13.5 | 3.2 | 89 |
| IRA 958 | 2.6 BV 1 $N$ NaCl | 9.5 | 3.3 | 87 |
| IRA 900 | 4.0 BV 1 $N$ NaCl | 14.6 | 3.6 | 67 |

ways less than 100 percent, was attributed to Cr(VI) reduction to Cr(III) with subsequent precipitation of greenish $Cr(OH)_3(s)$, observed in the top one-fifth of the resin bed. This $Cr(OH)_3(s)$ was not completely removed during backwashing and regeneration. The seriousness of the Cr(III) precipitation problem was not extensively studied in Scottsdale. Due to the long runs, only one repeat run (with Dowex 11) was made. Following regeneration, the second run with Dowex 11 was 20,700 BV, i.e., slightly longer than the initial 19,700 BV run. Thus, no significant reduction was observed due to resin fouling with $Cr(OH)_3(s)$. This is not surprising in light of the fact that, at chromate breakthrough, only 3 percent of the resin exchange sites were in the chromate form. Although extended exhaustion/regeneration studies should be completed before the long-term stability of the process is established, these initial results are promising.

### Effect of chromate concentration

To assess the influence of chromate level on run length, the chromate concentration of the Scottsdale water was increased 4.6-fold by spiking. As predicted from *equilibrium multicomponent chromatography theory* (EMCT), the run length was not decreased proportionately.[24] With IRA 900 resin and raw water containing 0.042 mg Cr/L, the initial run length was 32,000 BV. This decreased only 32 percent when the chromate concentration was increased 360 percent to 0.195 mg/L. Actually the EMCT predicts much less of a concentration influence on breakthrough for a trace species such as chromate; but because of mass transfer limitations and nonattainment of equilibrium, more chromate leakage occurs as its concentration in the raw water increases.

### Chromate removal from spent regenerant

Chromate in the saline spent-regenerant solution can be reduced to Cr(III) and precipitated as $Cr(OH)_3(s)$ by using a reductant such as ferrous sulfate or acidic sodium sulfite. This spent-regenerant treatment process has been studied extensively on a laboratory scale.[49] The Cr(VI) reduction precipitation reactions are

$$HCrO_4^- + 3Fe^{2+} + 7H^+ \rightarrow Cr^{3+} + 3Fe^{3+} + 4H_2O \tag{9.64}$$

$$2HCrO_4^- + 3H_2SO_3 + 2H^+ \rightarrow 2Cr^{3+} + 3SO_4^{2-} + 5H_2O \tag{9.65}$$

When acidified sodium sulfite is the reductant, it is necessary to reduce the pH to below 1.5 to achieve a reasonable reaction rate. The entire process involves sulfite addition at 1.3 times the stoichiometric amount, pH reduction, 30- to 60-min reaction time, pH increase to 8.3 to precipitate $Cr(OH)_3(s)$, and sludge settling.

For Cr(VI) reduction and precipitation with the ferrous ion, the optimum pH range is 5 to 7. Outside this range, significant amounts of soluble chromium are found in the supernatant after settling. Reduction with the ferrous ion comprises ferrous sulfate addition followed by slow mixing while pH is controlled within the 5 to 7 range and then by settling.

For both reduction processes, the high salinity of the spent regenerant improves the floc formation rate and increases the floc size. When properly operated, a total chromium concentration less than 0.10 mg/L is expected in the filtered regenerant following Cr(VI) reduction and $Cr(OH)_3(s)$ precipitation. Additional research needs to be done to establish the feasibility of reusing the regenerant following chromate removal. Cost estimates comparing sodium sulfite to ferrous sulfate reduction were made for 0.1- and 4-mgd chromate-removal plants. When treatment chemicals and land disposal of sludge are taken into consideration, the ferrous sulfate reduction was significantly less costly than the sulfite reduction in spite of the fivefold greater sludge volume resulting from $Fe(OH)_3(s)$ in addition to the unavoidable $Cr(OH)_3(s)$.

### Chromate-removal process recommendations

Based on the results of the Scottsdale study, the recommended ion-exchange process for chromate removal from groundwater comprises chloride-form strong-base anion exchange with a polystyrene gel or isoporous resin. Following chromate breakthrough at about 0.01 mg/L, the spent resin should be backwashed before cocurrent regenerant with 4 BV of 1 $N$ NaCl (15 lb NaCl per cubic foot of resin) followed by the appropriate slow and fast rinses. Chromic hydroxide $[Cr(OH)_3(s)]$ should be precipitated from the spent regenerant brine by using ferrous sulfate. Consideration should be given to reuse of treated regenerant.

## Anion-Exchange Processes for Removal of Organic Substances

Naturally occurring *dissolved organic carbon* (DOC) comprising humate and fulvate anions is a common foulant for the gel porosity anion exchangers used in demineralization processes. Nevertheless, these organic substances can be removed from groundwater and surface water by using macroporous anion exchangers in the chloride form. Published research on the ion-exchange process for organic substance removal is scarce, and much remains to be done to optimize the process; nevertheless, a limited number of full-scale DOC removal processes have been installed and operated.

Kolle[50] described a DOC removal process utilizing anion exchange for treatment of a highly colored Hannover, West Germany, groundwater. The 2500-m$^3$ (15.8-mgd) full-scale process uses a macroporous polystyrene resin (Lewatit MP 500 A) in four single beds operated in parallel with EBCT = 1.2 min. Regeneration is accomplished with 2 BV of a mixture of 1.7 $N$ (10%) NaCl, and 0.5 $N$ (2%) NaOH. A striking feature of the process is the reuse of regenerant 7 times with the NaCl and NaOH concentrations adjusted after each use. The DOC is not removed from the reused regenerant and builds up to 25,000 mg/L prior to disposal, which is triggered when the sulfate concentration increases to the chloride level. This reuse method allows for a 20,000:1 ratio of product water to spent regenerant.

Typical breakthrough curves for DOC, chloride, and sulfate are shown in Fig. 9.22. Following sulfate breakthrough at 250 BV, the DOC removal remains at about 50 percent until 5000 BV—the point of run termination. Although far from complete, the 50 percent DOC removal is sufficient to increase the bacterial doubling time in the product water from 5 to 12 h and to drastically reduce the chlorine consumption with its attendant organic halide formation problems. The overall treatment efficiency was reduced only 10 percent because of fouling and resin losses over a period of 2 years of full-scale operation.

## Multiple-column anion-exchange processes for DOC removal

Brattebo et al.[51] described a similar DOC removal process designed to treat a highly colored Norwegian surface water containing 3 to 6 mg/L DOC, primarily in the > 10,000 molecular weight (MW) range. Dur-

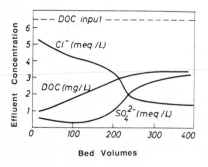

**Figure 9.22** Typical breakthrough curves for DOC, sulfate, and chloride from a macroporous SBA exchanger removing humate and fulvate anions from a Hannover, West Germany, groundwater. Column is operated to 5000 BV before regeneration. Between 400 and 5000 BV the DOC removal is approximately 50 percent. Influent sulfate = 130 mg/L.

ing their pilot studies, they verified the superiority of MP 500 A resin compared to a limited number of other polystyrene resins of macroporous (MR) and gel porosity. They also effectively reused the 1.7 $N$ NaCl/0.5 $N$ NaOH regenerant as many as 8 times by disposing a fraction of the first eluate containing most of the DOC and making up the remainder with fresh NaCl and NaOH. For water containing up to 4 mg DOC/L, they recommended the ion-exchange method implemented by using a three-column merry-go-round ion-exchange system (see Fig. 9.11) in which the resting, freshly regenerated fourth column automatically becomes the final polisher when the effluent DOC reaches a predetermined level. At this time, the lead column is taken out of service and regenerated. They also determined that the EBCT for the series was the most important design parameter, and they suggested a range of 5 to 10 min, i.e., considerably longer than used in Hannover.

The same Norwegian research group tested the multibed anion-exchange process on a small water supply using a full-scale, 3 m³/h (13 gpm) plant and reported the following conclusions: (1) NaOH is a necessary constituent of the regenerant because with NaCl alone the DOC recovery is less than 80 percent; (2) reusing up to 75 percent of the spent regenerant does not reduce the DOC adsorption capacity upon subsequent runs; (3) mass transfer of DOC is particle-diffusion-limited such that a reduction in temperature from 15 to 5°C requires a 20 percent increase in EBCT; (4) for small water supplies with relatively low (< 4 mg/L) DOC, the anion-exchange process is estimated to be less costly to build and operate than a conventional coagulation process.[52]

## Fundamentals of DOC ion exchange

In spite of the full-scale implementation of anion exchange for DOC removal, no fundamental study of the process had been undertaken prior to the work of Fu and Symons.[53] Their objectives were (1) to establish the relative importance of adsorption compared to ion exchange for removal of aquatic humic substances, (2) to determine the influence of conventional treatment on DOC removal by ion exchange, and (3) to describe the chromatographic behavior of the multicomponent DOC mixture during columnar ion exchange. Their study utilized two macroporous resins, five gel resins, and two macroporous adsorbents to remove DOC from < 1000, 1000 to 5000, 5000 to 10,000, and > 10,000 MW fractions of Lake Houston surface water. The percentages of DOC in the raw water fractions were 52, 9, 28, and 11, respectively.

Their ion-exchange isotherm experiments indicated that, within the limits of experimental error, each milliequivalent of DOC anion sorbed was balanced by an equivalent amount of chloride ion released. Therefore, the sorption mechanism for DOC removal is ion exchange with only a small percentage of the DOC in the < 1000 and > 10,000 fractions being removed by adsorption. The predominance of the ion-exchange mechanism as the means of DOC uptake on ion-exchange resins was supported by the findings with macroporous adsorbents that were, at best, capable of adsorbing only 7 percent of the DOC (in the 1000 to 5000 MW fraction).

Fu and Symons found that the resin matrix and porosity exhibit significant influences on DOC removal by ion-exchange resins. Generally, the hydrophilic polyacrylic resins are preferred for DOC removal compared to the relatively hydrophobic polystyrene resins. For all the MW fractions greater than 1000 macroporous resins performed better than the gel types; i.e., the greater the resin porosity, the better the DOC removal. At < 1000 MW, the resin porosity did not influence the DOC removal, which was, however, influenced by the resin matrix. In this low MW range, hydrophobic polystyrene resins performed somewhat better than the hydrophilic polyacrylic ones.

To establish the effect of partial DOC removal by coagulation prior to the ion-exchange process, Fu and Symons also fractionated the DOC resulting after conventional treatment. They found that 70 to 90 percent of the DOC was removed by coagulation and that the percentage of removal increases with the MW. Using this knowledge and the best resin from the isotherm tests (IRA 958, MR polyacrylic), they studied the chromatographic behavior of the < 1000 and 5000 to 10,000 MW fractions. Column study results suggested that the DOC could be divided into three categories: not removed, less preferred than sulfate, and more preferred than sulfate. About 12 percent of the DOC in the < 1000 MW fraction and 3 percent of the 5000 to 10,000 MW fraction were less preferred than sulfate. Column tests also showed that about 30 percent of the < 1000 MW fraction was not removed by the resin while only 10 percent of the 5000 to 10,000 MW fraction was not removed.

In summary, it is apparent that naturally occurring organic matter in groundwater and surface water is amenable to removal by using macroporous chloride-form strong-base anion resins of both the polystyrene and polyacrylic types. Generally the DOC that is adsorbed is preferred to the sulfate but can be eluted efficiently with a reusable mixture of NaCl and NaOH.

## Waste Disposal

To the extent possible, waste disposal considerations have been covered in the discussions of each contaminant removal process. The fo-

**TABLE 9.11  Summary of Processes for Removing Inorganic Cations**

| Contaminant and its MCL | Usual form at pH 7–8 | Removal options | Typical BV to MCL† | Pretreatment required | Typical regenerants and % recovery of sorbed contaminant±/§ | Effect of TDS on BV | Effect of hardness on BV | Notes |
|---|---|---|---|---|---|---|---|---|
| Hardness (no MCL) | $Ca^{2+}Mg^{2+}$ | Na IX softening with SAC resin | 200–700 | Iron removal | 6–12% NaCl 90–100% recovery | V. signif. reduction |  | a,b |
|  |  | $H^+$ IX softening with WAC resin | 200–500 | None | 1–2 $N$ HCl or $H_2SO_4$ | V. signif. reduction |  | c,d,e,f |
| Barium, 1.0 mg/L | $Ba^{2+}$ | Na IX softening with SAC resin | 200–700 | Iron removal | 1–12% NaCl 70 to 100% recovery | V. signif. reduction | V. signif. reduction | a–c,g,h |
| Radium 5 pCi/L | $Ra^{2+}$ | Na IX softening with SAC resin | 200–700 | Iron removal | 6–12% NaCl 0–100% recovery | V. signif. reduction | V. signif. reduction | a–c,i,j,k,l |
|  |  | Ca IX with SAC resin | 300–1500 | Iron removal | 10–15% $CaCl_2$ 50–100% recovery | Slight reduction | Slight reduction | l,m,n,o,p |
|  |  | Dow radium selective complexer | 20,000–50,000 | Iron removal | Not regenerable | Slight reduction | Slight reduction | l,o,q,r,s |
|  |  | $BaSO_4^-$ impregnated alumina | 20,000–50,000 | Iron removal | Not regenerable | Slight reduction | Slight reduction | l,o,q,r,s |
|  |  | Activated alumina | 1000–3000 | None | 2% HCl followed by 1% NaOH 70 to 100% recovery | None | Slight reduction | l,o,q,r |

†Generally, run length depends on raw-water contaminant concentration, allowable effluent concentration, competing ions, leakage, and actual resin or adsorbent used.

‡Percent recovery during the first regeneration depends on the amount of regenerant used. For steady-state exhaustion-regeneration, 100 percent of the contaminants sorbed during exhaustion is eluted during regeneration.

§Spent-regenerant disposal may be a problem if sanitary sewer disposal is not allowed.

[a]Hardness capacity depends on regeneration level (NaCl/ft³ resin).

[b]Typically, spent regenerant is disposed of in sanitary sewer.

[c]Iron removal by oxidation and filtration should be considered if total iron concentration exceeds 0.3 mg/L.

[d]Only carbonate hardness can be removed by using WAC resins.

[e]Carbon dioxide produced in the IX reaction must be removed from product water; and pH adjustment may be necessary.

[f]TDS reduction occurs as a result of removing both the hardness and the alkalinity.

[g]Barium tends to build up on the resin and break through with hardness when insufficient regenerant is used.

[h]If necessary, $BaSO_4$ can be precipitated from the spent regenerant by adding sulfate.

[i]With a virgin resin, radium breaks through long after hardness, but in cyclic operation radium and hardness eventually elute simultaneously.

[j]Even though radium accumulates on the resin, no leakage of radium occurs before hardness breakthrough.

[k]Current disposal practices allow the discharge of radium-contaminated spent regenerant to the sanitary sewer.

[l]Radon 222 is continuously generated from the radium 226 on the resin. Radon peaks can occur after idle periods.

[m]Immediate serious radium leakage occurs if extensive countercurrent regeneration is not used.

[n]$CaCl_2$ is much more expensive than NaCl as a regenerant.

[o]A process advantage is that sodium is not added to the product water.

[p]No softening is achieved in the calcium-exchange process, and magnesium is exchanged for calcium.

[q]Radon generation is more serious with the RSC and $BaSO_4^-$ alumina because of the large amount of radium on the medium.

[r]Disposal of the spent medium is a serious problem because it is considered a low-level radioactive waste.

[s]The RSC and $BaSO_4^-$ alumina may not be commercially available because of disposal problems.

**TABLE 9.12  Summary of Processes for Removing Inorganic Anions**

| Contaminant and its MCL | Usual form at pH 7–8 | Removal options | Typical BV to MCL† | Pretreatment required | Typical regenerants and % recovery of sorbed contaminant‡ | Effect of TDS on BV | Effect of $SO_4^{2-}$ on run BV | Notes |
|---|---|---|---|---|---|---|---|---|
| Fluoride 4.0 mg/L | $F^-$ | Activated alumina | 1000–2500 | pH 5.5 to 6.0 | 1% NaOH followed by 2% $H_2SO_4$ 90–100% recovery | None | Slight reduction | |
| Nitrate-N, 10 mg/L | $NO_3^-$ | Anion exchange (complete regeneration) | 300–600 | Usually none | 0.25–2.0 $N$ NaCl (1.5–12% NaCl), 90–100% recovery | Very significant reduction | Very significant reduction | a,b |
| | | Anion exchange (partial regeneration) | 200–500 | Usually none | 1.0–2.0 $N$ NaCl (6.0–12% NaCl), 50% recovery | Very significant reduction | Very significant reduction | a,c,d |
| Arsenic, 0.05 mg/L | $HAsO_4^{2-}$ As(V) | Activated alumina adsorption | 10,000–25,000 | pH 5.5 to 6.0, oxidize | 4% NaOH followed by 2% $H_2SO_4$, 70% recovery | None | Slight reduction | e,f |
| | | Anion exchange | 1000–5000 | Oxidize and prefilter to remove iron | 1.0 $N$ NaCl, > 95% recovery | Very significant reduction | Very significant reduction | a,g–i |
| Selenium, 0.01 mg/L | $HSeO_3^-$ Se(IV) | Activated alumina adsorption | 1000–2500 | pH 5.5 to 6.0 | 1% NaOH followed by 2% $H_2SO_4$ | None | Slight reduction | j |
| | $SeO_4^{2-}$ Se(VI) | Anion exchange | 300–1500 | None | 1.0 $N$ NaCl, 90–100% recovery | Very significant reduction | Very significant reduction | k |
| Chromium 0.05 mg/L | $CrO_4^{2-}$ | Anion exchange | 10,000–50,000 | None | 1.0 $N$ NaCl, 60–90% recovery | Slight reduction | Slight reduction | l,m |

†Generally run length depends on raw water contaminant concentration, allowable effluent concentration, competing ions, leakage, and the actual resin or adsorbent used.

‡Percent recovery during the first regeneration depends on the amount of regenerant used. For steady-state exhaustion regeneration, 100 percent of the contaminants sorbed during exhaustion is eluted during regeneration.

$a$Chromatographic peaking of contaminant is possible after breakthrough.

$b$No significant leakage of nitrate occurs prior to breakthrough.

$c$Continuous, significant ($>5$ mg/L) leakage of nitrate occurs following partial regeneration during all runs.

$d$Resin must be mixed mechanically following regeneration to avoid excessive early nitrate leakage.

$e$As(III) in the form of uncharged arsenious acid ($H_3AsO_3$) must be oxidized to As(V) prior to adsorption or ion exchange.

$f$Arsenic(V) can be coprecipitated from regenerant by lowering pH (to 6 to 8) to precipitate $Al(OH)_3(s)$.

$g$Some adsorption capacity may be lost following each exhaustion-regeneration cycle.

$h$Chloride-form anion exchange can be the process of choice for low-sulfate ($<50$ mg/L) and low-TDS ($<500$ mg/L) water.

$i$As(V) can be coprecipitated from spent regenerant by using ferric sulfate or alum.

$j$Se(IV) can typically be oxidized to Se(VI) by 1 to 2 mg free chlorine per liter in 30 to 60 min at pH 6.5 to 8.5.

$k$Se(IV) must be absent if ion exchange is used because it peaks before sulfate and selenate [Se(VI)] breakthrough.

$l$Cr(III) can be precipitated from the spent regenerant after reduction of Cr(VI) with ferrous sulfate or acidic sodium sulfite.

$m$Macroporous resins and polystyrene resins have a higher preference for chromate than gel and acrylic resins.

cus has been on reusing the spent regenerant to maximize the ratio of product water to spent regenerant for final disposal. Further explanations of how to deal with process residues are given in Chap. 16.

## Summary

In the first part of the chapter, the fundamentals of ion-exchange and adsorption processes were explained with the goal of demonstrating how these principles influence process design for inorganic contaminant removal. In the second part, ion-exchange and adsorption processes were described in detail for the removal of hardness, barium, radium, nitrate, fluoride, arsenic, selenium, chromate, and DOC. The selection of a process for removal of a given contaminant can be confusing because of the many variables to be considered, e.g., contaminant species, resins, adsorbents, competing ions, foulants, regenerants, and column flow patterns. Summary Tables 9.11 and 9.12 have been added to aid the reader in choosing a process. In Table 9.11 the important cation-removal process alternatives have been summarized, while in Table 9.12 anion removal is covered. The reader should refer to the discussion of fundamentals and process alternatives to answer questions arising from the tables.

## References

1. AWWA Inorganic Contaminants Committee, "An AWWA Survey of Inorganic Contaminants in Water Supplies," *J. AWWA,* vol. 77, no. 5, May 1985, p. 67.
2. U.S. Environmental Protection Agency, "National Primary Drinking Water Regulation; Fluoride," *Federal Register,* vol. 50, no. 220, Nov. 14, 1985.
3. C. R. Cothern and W. L. Lappenbusch, "Compliance Data for the Occurrence of Radium and Gross α-Particle Activity in Drinking Water Supplies in the United States," *Health Phys.,* vol. 146, no. 3, March 1984, p. 503.
4. D. A. Clifford and W. J. Weber, Jr., *Nitrate Removal from Water Supplies by Ion Exchange,* EPA-600/2-78-052, U.S. EPA, Cincinnati, 1978.
5. G. A. Guter, *Removal of Nitrate from Contaminated Water Supplies for Public Use,* EPA-600/5-81-029, U.S. EPA, Cincinnati, April 1981.
6. P. W. Schindler, "Surface Complexes at Oxide-Water Interfaces," in M. A. Anderson and A. J. Rubin (eds.), *Adsorption of Inorganics at Solid-Liquid Interfaces,* Ann Arbor Science, Ann Arbor, Mich., 1981.
7. D. Garg and D. A. Clifford, *Removal of Radium from Water by Adsorption onto Barium-Sulfate Impregnated and Plain Activated Alumina,* U.S. EPA Report on CR-813148-02 (in press).
8. D. A. Clifford, W. Vijjeswarapu, and S. Subramonian, "Evaluating Various Adsorbents and Membranes for Removing Radium from Groundwater," *J. AWWA,* vol. 80, no. 7, July 1988, p. 94.
9. G. E. Janauer, C. P. Gerba, W. C. Gihorse, M. Costello, and E. M. Heurich, "Insoluble Polymer Contact Disinfectants: An Alternative Approach to Water Disinfection," in W. J. Cooper (ed.), *Chemistry in Water Reuse,* Ann Arbor Science, Ann Arbor, Mich., 1981.
10. P. L. K. Fu and J. M. Symons, "Removal of Aquatic Organics by Anion Exchange Resins," *Proc. AWWA Annual Conf.,* Orlando, Fla., June 1988.

11. C. Calmon, "Specific Ion Exchangers," in C. Calmon and H. Gold (eds.), *Ion Exchange for Pollution Controls*, vol. 2, CRC Press, Boca Roton, Fla., 1979, chap. 15.

12. G. Boari, L. Liberti, C. Merli, and R. Passino, "Exchange Equilibria on Anion Resins," *Desalin.*, vol. 15, 1974, p. 145.

13. D. A. Clifford and W. J. Weber, Jr., "The Determinants of Divalent/Monovalent Selectivity in Anion Exchangers," *Reactive Polymers*, vol. 1, 1983, p. 77.

14. A. K. Sengupta and D. A. Clifford, "Chromate Ion-Exchange Mechanism for Cooling Water," *Ind. Eng. Chem. Fund.*, vol. 25, no. 2, 1986, pp. 249–258.

15. S. Subramonian and D. A. Clifford, "Monovalent/Divalent Selectivity and the Charge Separation Concept," *Reactive Polymers*, vol. 9, 1988, pp. 195–209.

16. R. R. Trussell, A. Trussell, and P. Kreft, *Selenium Removal from Groundwater Using Activated Alumina*, EPA-600/2-80-153, U.S. EPA, Cincinnati, August 1980.

17. G. L. Schmitt and D. J. Pietrzyk, *Anal. Chem.*, vol. 57, 1985, p. 53.

18. G. Singh and D. A. Clifford, *The Equilibrium Fluoride Capacity of Activated Alumina*, PB 81-204 075, NTIS, Springfield, Va., 1981; *Summary Report*, EPA-600/52-81-082, U.S. EPA, Cincinnati, July 1981.

19. E. R. Rosenblum and D. A. Clifford, *The Equilibrium Arsenic Capacity of Activated Alumina*, PB 84/10 527, NTIS, Springfield, Va., 1984; *Summary Report*, EPA-600/52-83-107, U.S. EPA, Cincinnati, February 1984.

20. L. L. Horng, "Reaction Mechanisms and Chromatographic Behavior of Polyprotic Acid Anions in Multicomponent Ion Exchange," Ph.D. dissertation, University of Houston, 1983.

21. Robert Kunin, *Ion Exchange Resins*, 2d ed., Robert E. Krieger Publishing Co., Huntington, N.Y., 1972.

22. F. Helfferich, "Ion Exchange Kinetics," in J. A. Marinsky (ed.), *Ion Exchange: A Series of Advances*, vol. 1, Marcel Dekker, New York, 1966.

23. F. Helfferich, "Ion-Exchange Kinetics V—Ion Exchange Accompanied by Reactions," *J. Phys. Chem.*, vol. 69, 1965, p. 1178.

24. F. Helfferich and G. Klein, *Multicomponent Chromatography: Theory of Interference*, Marcel Dekker, New York, 1970.

25. D. A. Clifford, "Multicomponent Ion Exchange Calculations for Selected Ion Separations," *Ind. Eng. Chem. Fund.*, vol. 21, May 1982, pp. 141–153.

26. V. L. Snoeyink, C. Cairns-Chambers, and J. L. Pfeffer, "Strong-Acid Ion Exchange for Removing Barium, Radium, and Hardness," *J. AWWA*, vol. 79, no. 8, August 1987, p. 66.

27. A. E. Bowers, "Ion Exchange Softening," in *The Quest for Pure Water*, vol. 2, 2d ed., AWWA, Denver, 1980.

28. D. A. Clifford, W. Vijjeswarapu, and S. Subramonian, "Removing Radium from Drinking Water Using Ion Exchange, Membranes and Specific Adsorbents: A Preliminary Report on the Lemont, Illinois, Study," *Proceedings of AWWA Seminar on Radionuclides in Drinking Water*, June 14, 1987, Kansas City, Mo., AWWA, Denver, 1987, pp. 69–82.

29. A. G. Myers, V. L. Snoeyink, and D. W. Snyder, "Removing Barium and Radium through Calcium Cation Exchange," *J. AWWA*, vol. 77, no. 5, May 1985, p. 60.

30. S. Subramonian, D. A. Clifford, and W. Vijjeswarapu, "Evaluating Ion Exchange for the Removal of Radium from Groundwater," *Journal American Water Works Association* (in press).

31. K. A. Mangelson, *Radium Removal for a Small Community Water Supply System*, PB 88-235 551/AS NTIS, Springfield, Va. (1988). *Summary Report*, EPA/600/52-88/039, USEPA, Cincinnati, Ohio (1988).

32. D. A. Clifford, C. C. Lin, L. L. Horng, and J. V. Boegel, *Nitrate Removal from Drinking Water in Glendale, Arizona*, PB 87-129 284/AS, NTIS, Springfield, Va., 1987; *Summary Report*, EPA/600/52-86/107, U.S. EPA, Cincinnati, March 1987.

33. J. P. Van der Hoek, P. J. M. Van der Ven, and A. Klapwijk, "Combined Ion Exchange/Biological Denitrification for Nitrate Removal from Ground Water under Different Process Conditions," *Water Res.*, vol. 22, no. 6 1988, pp. 679–684.

34. F. J. Maier, "Defluoridation of Municipal Water Supplies," *J. AWWA*, vol. 45, no. 8, August 1953, p. 879.

35. Water Quality Research Council, "EPA Approves Point-of-Use/Point-of-Entry Use by Public Water Systems," *Point of Use*, vol. 2, no. 3, 1987.

36. G. E. Bellen, M. Anderson, and R. A. Gottler, *Point-of-Use Treatment to Control Organic and Inorganic Contaminant in Drinking Water*, EPA/600-52-85/112, U.S. EPA, Cincinnati, January 1986.

37. F. Rubel, Jr., and R. D. Woosley, "The Removal of Excess Fluoride from Drinking Water by Activated Alumina," *J. AWWA*, vol. 71, no. 1, January 1979, p. 45.

38. D. A. Clifford and C. C. Lin, *Arsenic (III) and Arsenic(V) Removal from Drinking Water in San Ysidro, New Mexico*, U.S. EPA Report CR-813148 (in press).

39. F. Rubel, Jr., *Design Manual-Removal of Fluoride from Drinking Water Supplies by Activated Alumina*, EPA-600/2-84-134, U.S. EPA, Cincinnati, August 1984.

40. K. J. Irgolic, *Speciation of Arsenic Compounds in Water Supplies*, EPA 600/S1-82-010, Cincinnati, Ohio, November 1982.

41. National Research Council, *Drinking Water and Health*, vol. 1, National Academy of Science, Washington, 1977.

42. D. A. Clifford and C. C. Lin, "Arsenic Removal from Groundwater in Hanford, California—A Preliminary Report," University of Houston, Department of Civil/Environmental Engineering, 1986.

43. P. Frank and D. A. Clifford, *Arsenic(III) Oxidation and Removal from Drinking Water*, PB 86-158 607/AS, NTIS, Springfield, Va., 1986; *Summary Report*, EPA/600/52-86/021, U.S. EPA, Cincinnati, April 1986.

44. F. Rubel, Jr., and S. W. Hathaway, *Pilot Study for Removal of Arsenic from Drinking Water at the Fallon, Nevada, Naval Air Station*, EPA/600/52-85/094, U.S. EPA, Cincinnati, September 1985.

45. J. V. Boegel and D. A. Clifford, *Selenium Oxidation and Removal by Ion Exchange*, PB 86-171 428/AS, NTIS, Springfield, Va., 1986; *Summary Report*, EPA/600/52-86/031, U.S. EPA, Cincinnati, May 1986.

46. J. R. Yaun, M. M. Ghosh, S. M. Hornung, and R. J. Schlicher, "Adsorption of Arsenic and Selenium on Activated Alumina," *Proc. National Conference on Environmental Engineering*, American Society of Civil Engineering, New York 1983, pp. 433–441.

47. C. F. Baes and R. E. Mesmer, *The Hydrolysis of Cations*, Wiley-Interscience, New York, 1976.

48. F. N. Robertson, "Hexavalent Chromium in the Ground Water in Paradise Valley, Arizona," *Groundwater*, vol. 13, no. 6, 1975, p. 516.

49. S. Siegel and D. A. Clifford, *Removal of Chromium from Ion Exchange Regenerant Solution*, PB88-158 084/AS, NTIS, Springfield, Va., 1988; *Summary Report*, EPA/600/52-88/007, U.S. EPA, Cincinnati, April 1988.

50. W. Kolle, "Humic Acid Removal with Macroreticular Ion Exchange Resins at Hannover," in *NATO/CCMS Adsorption Techniques in Drinking Water Treatment*, CCMS 112, USEPA 570/9-84-005, U.S. EPA, Washington, September 1984.

51. H. Brattebo, H. Odegaard, and O. Halle, "Ion Exchange for Removal of Humic Acids in Water Treatment," *Water Res.*, vol. 21, no. 9, September 1987, p. 1045.

52. H. Odegaard, H. Brattebo, and O. Halle, "Removal of Humic Substances by Ion Exchange" presented at ACS Annual Meeting, Denver, April 5–10, 1987.

53. P. Fu and J. M. Symons, "Removal of Aquatic Organics by Ion Exchange Resins," *Proc. AWWA Annual Conf.*, Orlando, Fla., 1988.

# Chemical Precipitation

## Larry D. Benefield, Ph.D.

*Professor*
*Department of Civil Engineering*
*Auburn University, Alabama*

## Joe M. Morgan, Ph.D.

*Associate Professor*
*Department of Civil Engineering*
*Auburn University, Alabama*

Chemical precipitation is an effective treatment process for the removal of many contaminants. Coagulation with alum, ferric sulfate, or ferrous sulfate and lime softening both involve chemical precipitation. The removability of substances from water by precipitation depends primarily on the solubility of the various complexes formed in water. For example, heavy metals are found as cations in water, and many will take both hydroxide and carbonate solid forms. These solids have low solubility limits in water. Thus, as a result of the formation of insoluble hydroxides and carbonates, the metals will be precipitated out of solution.

Although coagulation with alum, ferric sulfate, or ferrous sulfate involves chemical precipitation, we give extensive coverage of coagulation in Chap. 6 and do not repeat that here. The discussion of the application of chemical precipitation in water treatment presented in this chapter emphasizes the reduction in the concentration of calcium and magnesium (water softening) and the reduction in the concentration of iron and manganese. Attention is also given to the removal of heavy metals, radionuclides, and organic materials.

## Fundamentals of Chemical Precipitation

Chemical precipitation is one of the most commonly used processes in water treatment. Still, experience with this process has produced a wide range of treatment efficiencies. Reasons for such variability will be explored in this chapter by considering precipitation theory and translating this to problems encountered in actual practice.

### Solubility equilibria

A chemical reaction is said to have reached equilibrium when the rate of the forward reaction is equal to the rate of the reverse reaction, so that no further net chemical change occurs. A general chemical reaction that has reached equilibrium is commonly expressed as

$$aA + bB \rightleftharpoons cC + dD \tag{10.1}$$

The equilibrium constant $K_{eq}$ for this reaction is defined as

$$K_{eq} = \frac{(C)^c(D)^d}{(A)^a(B)^b} \tag{10.2}$$

where the equilibrium activities of chemical species A, B, C, and D are denoted (A), (B), (C), and (D) and the stoichiometric coefficients are represented as $a$, $b$, $c$, and $d$, respectively. For dilute solutions, molar concentration is normally used to approximate activity of aqueous species whereas partial pressure measured in atmospheres is used for gases. By convention, the activities of solid materials (such as precipitates) and solvents (such as water) are taken as unity. Remember, however, that the equilibrium-constant expression corresponding to Eq. (10.1) must be written in terms of activities in order to describe the equilibrium in a completely rigorous manner.

The state of solubility equilibrium is a special case of Eq. (10.1) that may be attained either by formation of a precipitate from the solution phase or from partial dissolution of a solid phase. The precipitation process is observed when the concentrations of ions of a sparingly soluble compound are increased beyond a certain value. When this occurs, a solid that may settle is formed. Such a process may be described by the reaction

$$A^+ + B^- \rightleftharpoons AB(s) \tag{10.3}$$

where (s) denotes the solid form. The omission of "(s)" implies that the species is in aqueous form.

Precipitation formation is both a physical and chemical process. The physical part of the process is composed of two phases: nucleation and crystal growth. Nucleation begins with a supersaturated solution, i.e., a solution that contains a greater concentration of dissolved ions than can exist under equilibrium conditions. Under such conditions, a condensation of ions will occur, forming very small (invisible) particles.

The extent of supersaturation required for nucleation to occur varies. The process, however, can be enhanced by the presence of preformed nuclei that are introduced, for example, through the return of settled precipitate sludge back to the process.

Crystal growth follows nucleation as ions diffuse from the surrounding solution to the surfaces of the solid particles. This process continues until the condition of supersaturation has been relieved and equilibrium is established. When equilibrium is achieved, a saturated solution will have been formed. By definition, this is a solution in which undissolved solute is in equilibrium with solution.

No compound is totally insoluble. Thus, every compound can be made to form a saturated solution. Consider the following dissolution reaction occurring in an aqueous suspension of the sparingly soluble salt $AB(s)$:

$$AB(s) \rightleftharpoons AB \qquad (10.4)$$

The aqueous undissociated molecule that is formed then dissociates to give a cation and anion:

$$AB \rightleftharpoons A^+ + B^- \qquad (10.5)$$

The equilibrium-constant expressions for Eqs. (10.4) and (10.5) may be manipulated to give Eq. (10.6), where the product of the activities of the two ionic species is designed as the thermodynamic activity product $K_{ap}$:

$$K_{ap} = (A^+)(B^-) \qquad (10.6)$$

The concentration of a chemical species, not activity, is of interest in water treatment. Because dilute solutions are typically encountered, this parameter may be employed without introducing significant error into the calculations. Hence, in this chapter all relationships are written in terms of analytical concentration rather than activity. By this convention, Eq. (10.6) becomes

$$K_{sp} = [A^+][B^-] \qquad (10.7)$$

This is the classical solubility product expression for the dissolution of a slightly soluble compound where the brackets denote molar concentration. The equilibrium constant is called the *solubility product constant*. The more general form of the solubility product expression is derived from the dissolution reaction

$$A_x B_y(s) \rightleftharpoons xA^{y+} + yB^{x-} \qquad (10.8)$$

and has the form

$$K_{sp} = [A^{y+}]^x [B^{x-}]^y \qquad (10.9)$$

The value of the solubility product constant gives some indication of the solubility of a particular compound. For example, a compound that is highly insoluble will have a very small solubility product constant. Solubility product constants for solutions at or near room temperature are listed in Table 10.1.

Equation (10.9) applies to the equilibrium condition between an ion and a solid. If the actual concentrations of the ions in solution are such that the ion product $[A^{y+}]^x \cdot [B^{x-}]^y$ is less than the $K_{sp}$ value, no precipitation will occur and any quantitative information that can be derived from Eq. (10.9) will apply only where equilibrium conditions exist. Furthermore, if the actual concentrations of ions in solution are so great that the ion product is greater than the $K_{sp}$ value, precipitation will occur (assuming nucleation occurs). Still, however, no quantitative information can be derived directly from Eq. (10.9).

If an ion of a sparingly soluble salt is present in solution in a defined concentration, it can be precipitated by the other ion common to the salt, if the concentration of the second ion is increased to the point where the ion product exceeds the value of the solubility product constant. Such an influence is called the *common-ion effect*. Furthermore, precipitating two different compounds is possible if two different ions share a common third ion and if the concentration of the third ion is increased so that the solubility product constants for both sparingly soluble salts are exceeded. This type of precipitation is normally possible only when the $K_{sp}$ values of the two compounds do not differ significantly.

The common-ion effect is an example of LeChâtelier's principle, which states that if stress is applied to a system in equilibrium, the system will act to relieve the stress and restore equilibrium, but under a new set of equilibrium conditions. For example, if a salt containing cation A (e.g., AC) is added to a saturated solution of AB, then AB(s) will precipitate until the ion product $[A^+][B^-]$ has a value equal to the solubility product constant. The new equilibrium concentration of $A^+$, however, will be greater than the old equilibrium concentration, while the new equilibrium concentration of $B^-$ will be lower than the old equilibrium concentration. The following example illustrates calculations involving the common-ion effect.

**Example Problem 1**   Determine the residual magnesium concentration that exists in a saturated magnesium hydroxide solution if enough sodium hydroxide has been added to the solution to increase the equilibrium pH to 11.0.

**solution**

1. Write the appropriate chemical reaction:

TABLE 10.1    Solubility Product Constants for Solutions at or Near Room Temperature

| Substance | Formula | $K_{sp}^{\dagger}$ |
|---|---|---|
| Aluminum hydroxide | $Al(OH)_3$ | $2 \times 10^{-32}$ |
| Barium arsenate | $Ba_3(AsO_4)_2$ | $7.7 \times 10^{-51}$ |
| Barium carbonate | $BaCO_3$ | $8.1 \times 10^{-9}$ |
| Barium chromate | $BaCrO_4$ | $2.4 \times 10^{-10}$ |
| Barium fluoride | $BaF_2$ | $1.7 \times 10^{-6}$ |
| Barium iodate | $Ba(IO_3)_2 2H_2O$ | $1.5 \times 10^{-9}$ |
| Barium oxalate | $BaC_2O_4 H_2O$ | $2.3 \times 10^{-8}$ |
| Barium sulfate | $BaSO_4$ | $1.08 \times 10^{-10}$ |
| Beryllium hydroxide | $Be(OH)_2$ | $7 \times 10^{-22}$ |
| Bismuth iodide | $BiI_3$ | $8.1 \times 10^{-19}$ |
| Bismuth phosphate | $BiPO_4$ | $1.3 \times 10^{-23}$ |
| Bismuth sulfide | $Bi_2S_3$ | $1 \times 10^{-97}$ |
| Cadmium arsenate | $Cd_3(AsO_4)_2$ | $2.2 \times 10^{-33}$ |
| Cadmium hydroxide | $Cd(OH)_2$ | $5.9 \times 10^{-15}$ |
| Cadmium oxalate | $CdC_2O_4 3H_2O$ | $1.5 \times 10^{-8}$ |
| Cadmium sulfide | $CdS$ | $7.8 \times 10^{-27}$ |
| Calcium arsenate | $Ca_3(AsO_4)_2$ | $6.8 \times 10^{-19}$ |
| Calcium carbonate | $CaCO_3$ | $8.7 \times 10^{-9}$ |
| Calcium fluoride | $CaF_2$ | $4.0 \times 10^{-11}$ |
| Calcium hydroxide | $Ca(OH)_2$ | $5.5 \times 10^{-6}$ |
| Calcium iodate | $Ca(IO_3)_2 6H_2O$ | $6.4 \times 10^{-7}$ |
| Calcium oxalate | $CaC_2O_4 H_2O$ | $2.6 \times 10^{-9}$ |
| Calcium phosphate | $Ca_3(PO_4)_2$ | $2.0 \times 10^{-29}$ |
| Calcium sulfate | $CaSO_4$ | $1.9 \times 10^{-4}$ |
| Cerium(III) hydroxide | $Ce(OH)_3$ | $2 \times 10^{-20}$ |
| Cerium(III) iodate | $Ce(IO_3)_3$ | $3.2 \times 10^{-10}$ |
| Cerium(III) oxalate | $Ce_2(C_2O_4)_3 9H_2O$ | $3 \times 10^{-29}$ |
| Chromium(II) hydroxide | $Cr(OH)_2$ | $1.0 \times 10^{-17}$ |
| Chromium(III) hydroxide | $Cr(OH)_3$ | $6 \times 10^{-31}$ |
| Cobalt(II) hydroxide | $Co(OH)_2$ | $2 \times 10^{-16}$ |
| Cobalt(III) hydroxide | $Co(OH)_3$ | $1 \times 10^{-43}$ |
| Copper(II) arsenate | $Cu_3(AsO_4)_2$ | $7.6 \times 10^{-76}$ |
| Copper(I) bromide | $CuBr$ | $5.2 \times 10^{-9}$ |
| Copper(I) chloride | $CuCl$ | $1.2 \times 10^{-6}$ |
| Copper(I) iodide | $CuI$ | $5.1 \times 10^{-12}$ |
| Copper(II) iodate | $Cu(IO_3)_2$ | $7.4 \times 10^{-8}$ |
| Copper(I) sulfide | $Cu_2S$ | $2 \times 10^{-47}$ |
| Copper(II) sulfide | $CuS$ | $9 \times 10^{-36}$ |
| Copper(I) thiocyanate | $CuSCN$ | $4.8 \times 10^{-15}$ |
| Iron(III) arsenate | $FeAsO_4$ | $5.7 \times 10^{-21}$ |
| Iron(II) carbonate | $FeCO_3$ | $3.5 \times 10^{-11}$ |
| Iron(II) hydroxide | $Fe(OH)_2$ | $8 \times 10^{-16}$ |
| Iron(III) hydroxide | $Fe(OH)_3$ | $4 \times 10^{-38}$ |
| Lead arsenate | $Pb_3(AsO_4)_2$ | $4.1 \times 10^{-36}$ |
| Lead bromide | $PbBr_2$ | $3.9 \times 10^{-5}$ |
| Lead carbonate | $PbCO_3$ | $3.3 \times 10^{-14}$ |
| Lead chloride | $PbCl_2$ | $1.6 \times 10^{-5}$ |
| Lead chromate | $PbCrO_4$ | $1.8 \times 10^{-14}$ |
| Lead fluoride | $PbF_2$ | $3.7 \times 10^{-8}$ |
| Lead iodate | $Pb(IO_3)_2$ | $2.6 \times 10^{-13}$ |

**TABLE 10.1    Solubility Product Constants for Solutions at or Near Room Temperature (*Continued*)**

| Substance | Formula | $K_{sp}^{\dagger}$ |
|---|---|---|
| Lead iodide | $PbI_2$ | $7.1 \times 10^{-9}$ |
| Lead oxalate | $PbC_2O_4$ | $4.8 \times 10^{-10}$ |
| Lead sulfate | $PbSO_4$ | $1.6 \times 10^{-8}$ |
| Lead sulfide | $PbS$ | $8 \times 10^{-28}$ |
| Magnesium ammonium phosphate | $MgNH_4PO_4$ | $2.5 \times 10^{-13}$ |
| Magnesium arsenate | $Mg_3(AsO_4)_2$ | $2.1 \times 10^{-20}$ |
| Magnesium carbonate | $MgCO_3 3H_2O$ | $1 \times 10^{-5}$ |
| Magnesium fluoride | $MgF_2$ | $6.5 \times 10^{-9}$ |
| Magnesium hydroxide | $Mg(OH)_2$ | $1.2 \times 10^{-11}$ |
| Magnesium oxalate | $MgC_2O_4 2H_2O$ | $1 \times 10^{-8}$ |
| Manganese(II) hydroxide | $Mn(OH)_2$ | $1.9 \times 10^{-13}$ |
| Mercury(I) bromide | $Hg_2Br_2$ | $5.8 \times 10^{-23}$ |
| Mercury(I) chloride | $Hg_2Cl_2$ | $1.3 \times 10^{-18}$ |
| Mercury(I) iodide | $Hg_2I_2$ | $4.5 \times 10^{-29}$ |
| Mercury(I) sulfate | $Hg_2SO_4$ | $7.4 \times 10^{-7}$ |
| Mercury(II) sulfide | $HgS$ | $4 \times 10^{-53}$ |
| Mercury(I) thiocyanate | $Hg_2(SCN)_2$ | $3.0 \times 10^{-20}$ |
| Nickel arsenate | $Ni_3(AsO_4)_2$ | $3.1 \times 10^{-26}$ |
| Nickel carbonate | $NiCO_3$ | $6.6 \times 10^{-9}$ |
| Nickel hydroxide | $Ni(OH)_2$ | $6.5 \times 10^{-18}$ |
| Nickel sulfide | $NiS$ | $3 \times 10^{-19}$ |
| Silver arsenate | $Ag_3AsO_4$ | $1 \times 10^{-22}$ |
| Silver bromate | $AgBrO_3$ | $5.77 \times 10^{-5}$ |
| Silver bromide | $AgBr$ | $5.25 \times 10^{-13}$ |
| Silver carbonate | $Ag_2CO_3$ | $8.1 \times 10^{-12}$ |
| Silver chloride | $AgCl$ | $1.78 \times 10^{-10}$ |
| Silver chromate | $Ag_2CrO_4$ | $2.45 \times 10^{-12}$ |
| Silver cyanide | $Ag[Ag(CN)_2]$ | $5.0 \times 10^{-12}$ |
| Silver iodate | $AgIO_3$ | $3.02 \times 10^{-8}$ |
| Silver iodide | $AgI$ | $8.31 \times 10^{-17}$ |
| Silver oxalate | $Ag_2C_2O_4$ | $3.5 \times 10^{-11}$ |
| Silver oxide | $Ag_2O$ | $2.6 \times 10^{-8}$ |
| Silver phosphate | $Ag_3PO_4$ | $1.3 \times 10^{-20}$ |
| Silver sulfate | $Ag_2SO_4$ | $1.6 \times 10^{-5}$ |
| Silver sulfide | $Ag_2S$ | $2 \times 10^{-49}$ |
| Silver thiocyanate | $AgSCN$ | $1.00 \times 10^{-12}$ |
| Strontium carbonate | $SrCO_3$ | $1.1 \times 10^{-10}$ |
| Strontium chromate | $SrCrO_4$ | $3.6 \times 10^{-5}$ |
| Strontium fluoride | $SrF_2$ | $2.8 \times 10^{-9}$ |
| Strontium iodate | $Sr(IO_3)_2$ | $3.3 \times 10^{-7}$ |
| Strontium oxalate | $SrC_2O_4H_2O$ | $1.6 \times 10^{-7}$ |
| Strontium sulfate | $SrSO_4$ | $3.8 \times 10^{-7}$ |
| Thallium(I) bromate | $TlBrO_3$ | $8.5 \times 10^{-5}$ |
| Thallium(I) bromide | $TlBr$ | $3.4 \times 10^{-6}$ |
| Thallium(I) chloride | $TlCl$ | $1.7 \times 10^{-4}$ |
| Thallium(I) chromate | $Tl_2CrO_4$ | $9.8 \times 10^{-13}$ |
| Thallium(I) iodate | $TlIO_3$ | $3.1 \times 10^{-6}$ |
| Thallium(I) iodide | $TlI$ | $6.5 \times 10^{-8}$ |
| Thallium(I) sulfide | $Tl_2S$ | $5 \times 10^{-21}$ |
| Tin(II) sulfide | $SnS$ | $1 \times 10^{-25}$ |
| Titanium(III) hydroxide | $Ti(OH)_3$ | $1 \times 10^{-40}$ |
| Zinc arsenate | $Zn_3(AsO_4)_2$ | $1.3 \times 10^{-28}$ |

**TABLE 10.1    Solubility Product Constants for Solutions at or Near Room Temperature (*Continued*)**

| Substance | Formula | $K_{sp}$[†] |
|---|---|---|
| Zinc carbonate | $ZnCO_3$ | $1.4 \times 10^{-11}$ |
| Zinc ferrocyanide | $Zn_2Fe(CN)_6$ | $4.1 \times 10^{-16}$ |
| Zinc hydroxide | $Zn(OH)_2$ | $1.2 \times 10^{-17}$ |
| Zinc oxalate | $ZnC_2O_42H_2O$ | $2.8 \times 10^{-8}$ |
| Zinc phosphate | $Zn_3(PO_4)_2$ | $9.1 \times 10^{-33}$ |
| Zinc sulfide | $ZnS$ | $1 \times 10^{-21}$ |

†The solubility of many metals is altered by carbonate complexation. Solubility predictions without consideration for complexation can be highly inaccurate.

SOURCE: Robert B. Fischer and Dennis G. Peters, *Chemical Equilibrium*. Copyright© 1970 by Saunders College Publishing, a division of Holt, Rinehart, and Winston, Inc., reprinted by permission of the publisher.

$$Mg(OH)_2(s) \rightleftharpoons Mg^{2+} + 2OH^-$$

From Table 10.1 the solubility product constant for this reaction is $1.2 \times 10^{-11}$.

2. Determine the hydroxide ion concentration:

$$K_w = [H^+][OH^-] = 10^{-14} \text{ at } 25°C$$

Because

$$[H^+] = 10^{-pH} = 10^{-11} \text{ mol/L}$$

we know that

$$[OH^-] = \frac{10^{-14}}{10^{-11}} = 10^{-3} \text{ mol/L}$$

3. Establish the solubility product constant expression, and solve for the magnesium ion concentration:

$$K_{sp} = [Mg^{2+}][OH^-]^2$$

$$[Mg^{2+}] = \frac{1.2 \times 10^{-11}}{(10^{-3})^2}$$

$$= 1.2 \times 10^{-5} \text{ mol/L} \quad \text{or} \quad 0.29 \text{ mg/L}$$

Since hardness ion concentrations are frequently expressed as $CaCO_3$, multiply the concentration by the ratio of the equivalent weights:

$$0.29 \times \frac{50}{12.2} = 1.2 \text{ mg/L as } CaCO_3$$

## Metal removal by chemical precipitation

Consider the following equilibrium reaction involving metal (M) solubility:

$$MA_x(s) \rightleftharpoons M^{x+} + xA^- \tag{10.10}$$

$$K_{sp} = [M^{x+}][A^-]^x \tag{10.11}$$

Equation (10.11), the solubility product expression for Eq. (10.10), indicates that the equilibrium concentration (in precipitation processes this is referred to as the *residual concentration*) of the metal in solution is solely dependent on the concentration of $A^-$. When $A^-$ is the hydroxide ion, the residual metal concentration is a function of pH such that

$$\log [M^{x+}] = \log K_{sp} - x \log K_w - X\mathrm{pH} \tag{10.12}$$

This relationship is shown as line $A$ in Fig. 10.1 where $K_{sp} = 10^{-10}$, $K_w = 10^{-14}$, and $X = 2$ (assumed values). The solubility of most metal hydroxides is not accurately described by Eq. (10.12), however, because they exist in solution as a series of complexes formed with hydroxide and other ions. Each complex is in equilibrium with the solid phase, and their sum gives the total residual metal concentration. For the case of only hydroxide species and a divalent metal, the total residual metal concentration is.

$$M_T = M^{2+} + M(OH)^+ + M(OH)_2^0 + M(OH)_3^- + \cdots \tag{10.13}$$

For this situation, the total residual metal concentration is a complex function of pH, as illustrated by line $B$ in Fig. 10.1. Line $B$ shows that the lowest residual metal concentration will occur at some optimum pH value and that the residual concentration will increase when the pH is either lowered or raised from this optimum value.

Nilsson[3] computed the logarithm of the total residual metal concen-

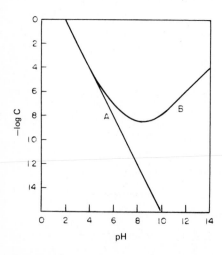

**Figure 10.1** Theoretical solubility of hypothetical metal hydroxide, with and without complex formation. $A$ = without complex formation, $B$ = with complex formation. (*Source: J. W. Patterson and R. A. Minear, "Physical-Chemical Methods of Heavy Metal Removal," in P. A. Kenkel (ed.), Heavy Metals in the Aquatic Environment, Pergamon Press, Oxford, 1975.*)

tration as a function of pH for several pure metal hydroxides; see Fig. 10.2. Bold lines show those areas where the total residual metal concentration is greater than 1 mg/L. If the rise in pH results from adding NaOH, the total residual Cr(III) and total residual Zn(II) will rise again when the pH values rise above approximately 8 and 9, respectively, because of an increase in the concentration of the negatively charged hydroxide complexes. If the rise in pH results from adding lime, then a rise in the residual concentration does not occur, because the solubilities of calcium zincate and calcium chromite are relatively low.

Numeric estimations of metal removal by precipitation as metal hydroxide should always be treated carefully because oversimplification of theoretical solubility data can lead to errors of several orders of magnitude. There are many possible reasons for such discrepancies. For example, changes in the ionic strength of a water can result in significant differences between calculated and observed residual metal concentrations when molar concentrations rather than activities are used in the computations (high ionic strength will result in a higher than predicted solubility). The presence of organic and inorganic species other than hydroxide, which are capable of forming soluble species with metal ions, will increase the total residual metal concentration. Two inorganic complexing agents that result in very high residual metal concentrations are cyanide and ammonia. Small amounts of carbonate will significantly change the solubility chemistry of some metal hydroxide precipitation systems. As a result, differences between theory and practice should be expected because in practice precipitating metal hydroxides is virtually impossible without at least some carbonate being present.

Temperature variations can explain deviations between calculated and observed values if actual process temperatures are significantly different from the value at which the equilibrium constant was evaluated. Kinetics may also be an important consideration because under

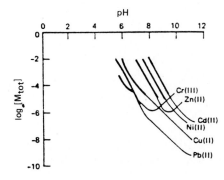

**Figure 10.2** The solubility of pure metal hydroxides as a function of pH. Heavy portions of lines show where concentrations are greater than 1 mg/L. *Note:* If NaOH is used for pH adjustment, Cr(III) and Zn(II) will exhibit amphoteric characteristics. (*Source: Reprinted with permission from Water Research, Vol. 5, R. Nilsson, "Removal of Metals by Chemical Treatment of Municipal Wastewater." Copyright 1971. Pergamon Press.*)

process conditions the reaction between the soluble and solid species may be too slow to allow equilibrium to become established within the hydraulic retention time provided. Furthermore, many solids may initially precipitate in an amorphous form but convert to a more insoluble and more stable crystalline structure after some time has passed.

Formation of precipitates other than the hydroxide may result in a total residual metal concentration lower than the calculated value. For example, the solubility of cadmium carbonate is approximately 2 orders of magnitude less than that of the hydroxide. Effects of coprecipitation on flocculating agents added to aid in settling the precipitate may also play a significant role in reducing the residual metal concentration. Nilsson[3] found that when precipitation with aluminum sulfate was employed, the actual total residual concentrations of zinc, cadmium, and nickel were much lower than the calculated values because the metals were coprecipitated with aluminum hydroxide.

In summary, the solubility behavior of most slightly soluble salts is very complex because of competing acid-base equilibria, complex ion formation, and hydrolysis. Still, many precipitation processes in water treatment can be adequately described when these reactions are ignored. This is the approach taken in this chapter. A more detailed discussion of solubility equilibria may be found in Stumm and Morgan,[4] Snoeyink and Jenkins,[5] and Benefield et al.[6]

### Carbonic acid equilibria

The pH of most natural water is generally assumed to be controlled by the carbonic acid system. The applicable equilibrium reactions are

$$CO_2 + H_2O \rightleftharpoons H_2CO_3 \rightleftharpoons H^+ + HCO_3^- \tag{10.14}$$

$$HCO_3^- \rightleftharpoons H^+ + CO_3^{2-} \tag{10.15}$$

Because only a small fraction of the total $CO_2$ dissolved in water is hydrolyzed to $H_2CO_3$, it is convenient to sum the concentrations of dissolved $CO_2$ and $H_2CO_3$ to define a new concentration term $H_2CO_3^*$. Equilibrium-constant expressions for Eqs. (10.14) and (10.15) have the forms

$$K_1 = \frac{[H^\pm][HCO_3^-]}{[H_2CO_3^*]} \tag{10.16}$$

$$K_2 = \frac{[H^\pm][CO_3^{2-}]}{[HCO_3^-]} \tag{10.17}$$

where $K_1$ and $K_2$ represent the equilibrium constants for the first and second dissociations of carbonic acid, respectively. Rossum and

Merrill[7] presented the following equations to describe the relationships between temperature and $K_1$ and $K_2$:

$$K_1 = 10^{14.8435 - 3404.71/T - 0.032786T} \qquad (10.18)$$

$$K_2 = 10^{6.498 - 2909.39/T - 0.02379T} \qquad (10.19)$$

where $T$ represents the solution temperature in kelvins (i.e., °C + 273).

The total carbonic species concentration in solution is usually represented by $C_T$ and defined in terms of a mass balance expression:

$$C_T = [H_2CO_3{}^*] + [HCO_3^-] + [CO_3^{2-}] \qquad (10.20)$$

The distribution of the various carbonic species can be established in terms of the total carbonic species concentration by defining a set of ionization fractions $\alpha$, where

$$\alpha_0 = \frac{[H_2CO_3{}^*]}{C_T} \qquad (10.21)$$

$$\alpha_1 = \frac{[HCO_3^-]}{C_T} \qquad (10.22)$$

$$\alpha_2 = \frac{[CO_3^{2-}]}{C_T} \qquad (10.23)$$

Through a series of algebraic manipulations[5]

$$\alpha_0 = \frac{1}{1 + K_1/[H^+] + K_1K_2/[H^+]^2} \qquad (10.24)$$

$$\alpha_1 = \frac{1}{[H^+]/K_1 + 1 + K_2/[H^+]} \qquad (10.25)$$

$$\alpha_2 = \frac{1}{[H^+]^2/(K_1K_2) + [H^+]/K_2 + 1} \qquad (10.26)$$

The effect of pH on the species distribution for the carbonic acid system is shown in Fig. 10.3. Because the pH of most natural waters is in the neutral range, the alkalinity (assuming that alkalinity results mainly from the carbonic acid system) is in the form of bicarbonate alkalinity.

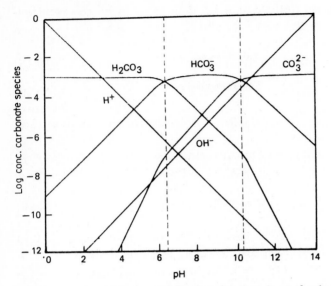

**Figure 10.3** Concentration distribution diagram for carbonic acid. (*Source: Handbook of Water Resources and Pollution Control. H. W. Gehm and J. I. Bregman, eds. Van Nostrand Reinhold Co., New York, 1976.*)

## Calcium carbonate and magnesium hydroxide equilibria

The solubility equilibrium for $CaCO_3$ is described by

$$CaCO_3(s) \rightleftharpoons Ca^{2+} + CO_3^{2-} \tag{10.27}$$

The addition of $Ca(OH)_2$ to water increases the hydroxyl ion concentration and elevates the pH which, according to Fig. 10.3, shifts the equilibrium of the carbonic acid system in favor of the carbonate ion $CO_3^{2-}$. This increases the concentration of the $CO_3^{2-}$ ion and, according to LeChâtelier's principle, shifts the equilibrium described by Eq. (10.27) to the left (common-ion effect). Such a response results in the precipitation of $CaCO_3(s)$ and a corresponding decrease in the soluble calcium concentration.

The solubility equilibrium for $Mg(OH)_2$ is described by

$$Mg(OH)_2(s) \rightleftharpoons Mg^{2+} + 2OH^- \tag{10.28}$$

According to LeChâtelier's principle, the addition of hydroxyl ions shifts the equilibrium described by Eq.(10.28) to the left (commonion effect), resulting in the precipitation of $Mg(OH)_2$ and a corresponding decrease in the soluble magnesium concentration.

The solubility product expressions for Eqs. (10.27) and (10.28) have the forms

$$K_{sp} = [Ca^{2+}][CO_3^{2-}] \tag{10.29}$$

$$K_{sp} = [Mg^{2+}][OH^-]^2 \tag{10.30}$$

The effects of temperature on the solubility product constants for calcium carbonate and magnesium hydroxide are given by the empirical equations[7-9]

$$\text{Calcium carbonate: } K_{sp} = 10^{13.870 - 3059/T - 0.04035T} \tag{10.31}$$

$$\text{Magnesium hydroxide: } K_{sp} = 10^{-0.0175t - 9.97} \tag{10.32}$$

where $T$ and $t$ are the solution temperature in kelvins and degrees Celsius, respectively. The $K_{sp}$ for calcium carbonate presented in Eq.(10.31) is based on the classical 1942 constant of Larson and Buswell.[10] A modern constant has recently been introduced by Plummer and Busenberg.[11,12]

Complex ion formation reactions that contribute to the total soluble calcium and magnesium concentrations are listed in Table 10.2. These reactions can be used to determine the effect of complex ion formation on calcium carbonate and magnesium hydroxide solubility by writing mass balance relationships for total residual calcium and total residual magnesium that consider these species. Such relationships have the form

$$[Ca]_T = [Ca^{2+}] + [CaOH^+] + [CaHCO_3^+] + [CaCO_3^0] + [CaSO_4^0] \tag{10.33}$$

which reduces to

$$[Ca]_T = \frac{K_{sp}}{\alpha_2 C_T}\left(1 + \frac{K_w K_3}{[H^+]} + K_4 \alpha_1 C_T + K_5 \alpha_2 C_T + K_6[SO_4^{2-}]\right) \tag{10.33a}$$

and

$$[Mg]_T = [Mg^{2+}] + [MgOH^+] + [MgHCO_3^+] + [MgCO_3^0] + [MgSO_4^0] \tag{10.34}$$

which reduces to

$$[Mg]_T = \frac{K_{sp}[H^+]^2}{K_w^2}\left(1 + \frac{K_2 K_7}{[H^+]} + K_8 \alpha_1 C_T + K_g \alpha_2 C_T K_{10}[SO_4^{2-}]\right) \tag{10.34a}$$

where

$$K_w = 10^{6.0486 - 4471.33/T - 0.017053T} \tag{10.35}$$

**TABLE 10.2  Complex Ion Formation Reactions of Calcium and Magnesium Ions[†]**

| Reaction | Equilibrium constant | Temperature correction $T$, K |
|---|---|---|
| **1. Calcium** | | |
| *a.* $Ca^{2+} + OH^- \rightleftharpoons CaOH^+$ | $K_3 = \dfrac{[CaOH^+]}{[Ca^{2+}][OH^-]}$ | $pK_3 = -1.299 - 260.388\dfrac{1}{T} - \dfrac{1}{298.15}$ |
| *b.* $Ca^{2+} + HCO_3^- \rightleftharpoons CaHCO_3^+$ | $K_4 = \dfrac{[CaHCO_3^+]}{[Ca^{2+}][HCO_3^-]}$ | $pK_4 = 2.95 - 0.0133T$ |
| *c.* $Ca^{2+} + CO_3^{2-} \rightleftharpoons CaCO_3^0$ | $K_5 = \dfrac{[CaCO_3^0]}{[Ca^{2+}][CO_3^{2-}]}$ | $pK_5 = 27.393 - \dfrac{4114}{T} - 0.05617T$ |
| *d.* $Ca^{2+} + SO_4^{2-} \rightleftharpoons CaSO_4^0$ | $K_6 = \dfrac{[CaSO_4^0]}{[Ca^{2+}][SO_4^{2-}]}$ | $pK_6 = \dfrac{691.70}{T}$ |
| **2. Magnesium** | | |
| *a.* $Mg^{2+} + OH^- \rightleftharpoons MgOH^+$ | $K_7 = \dfrac{[MgOH^+]}{[Mg^{2+}][OH^-]}$ | $pK_7 = -0.684 - 0.0051T$ |
| *b.* $Mg^{2+} + HCO_3^- \rightleftharpoons MgHCO_3^+$ | $K_8 = \dfrac{[MgHCO_3^+]}{[Mg^{2+}][HCO_3^-]}$ | $pK_8 = -2.319 + 0.011056T - (2.29812 \times 10^{-5})T$ |
| *c.* $Mg^{2+} + CO_3^{2-} \rightleftharpoons MgCO_3^0$ | $K_9 = \dfrac{[MgCO_3^0]}{[Mg^{2+}][CO_3^{2-}]}$ | $pK_9 = -0.991 - 0.00667T$ |
| *d.* $Mg^{2+} + SO_4^{2-} \rightleftharpoons MgSO_4^0$ | $K_{10} = \dfrac{[MgSO_4^0]}{[Mg^{2+}][SO_4^{2-}]}$ | $pK_{10} = \dfrac{707.07}{T}$ |

[†]Temperature corrections are from Truesdell and Jones.[13]

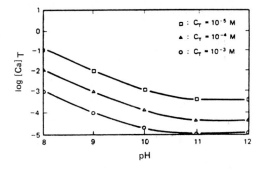

**Figure 10.4** Relationship between total soluble calcium, pH, and equilibrium total carbonic species concentration. (*Source: L. D. Benefield, J. F. Judkins, and B. L. Weand, Process Chemistry for Water and Wastewater. Copyright 1982, pp. 124, 292. Reprinted by permission of Prentice-Hall, Inc., Englewood Cliffs, New Jersey.*)

Figures 10.4 and 10.5 illustrate the effect of complex ion formation on calcium carbonate and magnesium hydroxide, respectively. For convenience, a solution temperature of 25°C and a sulfate ion concentration of zero were assumed. The results show that the equilibrium carbonic species concentration has virtually no effect on the total residual magnesium concentration (Fig. 10.5) but significantly affects the total residual calcium concentration (Fig. 10.4).

Cadena et al.[14] indicate that at 25°C the $CaCO_3^0$ species accounts for 13.5 mg/L of soluble calcium expressed as $CaCO_3$. Their work is based in part on this relationship for the variation in the dissociation constant for $CaCO_3^0$ with temperature

$$\log K = \frac{2280}{T} - 12.10 \tag{10.36}$$

where $T$ represents the temperature in kelvins. The concentration of $CaCO_3^0$ may be estimated by dividing the solubility product expression

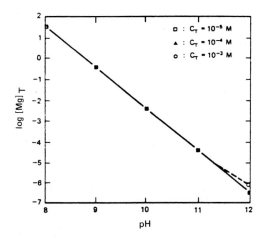

**Figure 10.5** Relationship between total soluble magnesium, pH, and final equilibrium total carbonic species concentration. (*Source: L. D. Benefield, J. F. Judkins, and B. L. Weand, Process Chemistry for Water and Wastewater. Copyright 1982, pp. 124, 292. Reprinted by permission of Prentice-Hall, Inc., Englewood Cliffs, New Jersey.*)

for calcium carbonate by the equilibrium-constant expression for $CaCO_3^0$. This gives

$$[CaCO_3^0] = \frac{K_{sp}}{K_{CaCO_3^0}} \tag{10.37}$$

A graphical representation of the variation in the $CaCO_3^0$ concentration with temperature is presented in Fig. 10.6. Trussell et al.[16] do not consider the $CaCO_3^0$ species important. These workers indicate that the concentration of $CaCO_3^0$ in a saturated solution of calcium carbonate is about 0.17 mg/L as $CaCO_3$ rather than 13.5 mg/L. Recent experimental evidence by Pisigan and Singley[17] support this. They found that the concentration of $CaCO_3^0$ is insignificant in fresh water in the pH range of 6.20 to 9.20.

For a detailed explanation of the calcium carbonate system and the ion pairs $CaHCO_3^+$ and $CaCO_3^0$, see the rigorous work of Plummer and Busenberg.[11]

### Equilibria governing iron and manganese solubilities

Iron(II) and manganese(II) are much more soluble than iron(III) and manganese(IV). Thus, iron and manganese removal usually depends on oxidation to less soluble forms. For natural water containing significant alkalinity (hence, having a pH in the range of 6.5 to 9.5), the solubilities of iron(II) and manganese(II) are controlled by their car-

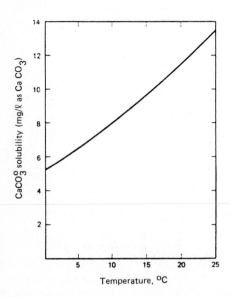

**Figure 10.6** Variation in solubility of $CaCO_3$ complex ion with temperature. (*Source: D. T. Merrill, "Chemical Conditioning for Water Softening and Corrosion Control," Proc. 5th Envir. Engr. Conf., Montana State University, June 1976.*)

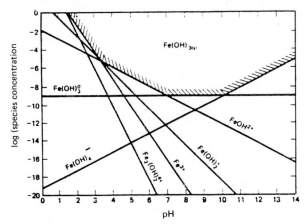

**Figure 10.7** Solubility diagram for iron(III) hydroxide. (*Source: L. D. Benefield, J. F. Judkins, and B. L. Weand, Process Chemistry for Water and Wastewater. Copyright 1982, pp. 124, 292. Reprinted by permission of Prentice-Hall, Inc., Englewood Cliffs, New Jersey.*)

bonates. For some groundwater and hypolimnetic reservoir water, however, the hydroxides or other species (e.g., sulfides) may control.

As is indicated by Fig. 10.7, iron(III) hydroxide is soluble only to a very limited extent across the entire pH range of interest in municipal water treatment. The theoretical minimum concentration of about $10^{-8}$ to $10^{-9}$ mol/L ($5.58 \times 10^{-4}$ to $5.58 \times 10^{-5}$ mg Fe/L) occurs over the range of about pH 7 to about pH 10. Thus, oxidation of relatively soluble iron(II) to the much less soluble iron(III) form and its subsequent precipitation as iron(III) hydroxide is a very effective removal technique.

The situation is similar for manganese, except that the chemistry is more complex and less well documented. The principal solubility-controlling species are thought to be manganese(IV) oxide/hydroxide (MnOOH) and manganese(IV) dioxide ($MnO_2$). From a practical viewpoint, these compounds may be considered completely insoluble in water.

For a detailed explanation of the factors affecting the solubilities of iron and manganese, see the summary work of Stumm and Morgan.[4]

## Water Softening by Chemical Precipitation

Hardness in natural water is caused by the presence of any polyvalent metallic cation. Principal cations causing hardness in water and the major anions associated with them are presented in Table 10.3. Because the most prevalent of these species are the divalent cations of calcium and magnesium, *total hardness* is typically defined as the sum

TABLE 10.3    Principal Cations Causing Hardness in Water and the Major Associated Anions[18]

| Principal cations causing hardness | Anions |
|---|---|
| $Ca^{2+}$ | $HCO_3^-$ |
| $Mg^{2+}$ | $SO_4^{2-}$ |
| $Sr^{2+}$ | $Cl^-$ |
| $Fe^{2+}$ | $NO_3^-$ |
| $Mn^{2+}$ | $SiO_3^{2-}$ |

of the concentration of these two elements and is usually expressed in terms of milligrams per liter as $CaCO_3$. In the United States there is a significant regional variation in the total hardness of both surface water and groundwater.

The hardness of water is the property that causes it to form curds (Ca or Mg oleate) when soap is used with it. Some types of water are very hard, and the consumption of soap by these types of water is commensurately high. Other adverse effects include bathtub rings, deterioration of fabrics, and in some cases stains. In recent years, many of these problems have been alleviated by the development of detergents and soaps that do not react with hardness.

Public acceptance of hardness varies from community to community, consumer sensitivity being related to the degree to which the consumer is accustomed. Because of this variation in consumer acceptance, the finished water hardness produced by different utility softening plants will range from 50 to 150 mg/L as $CaCO_3$. According to the hardness classification scale presented by Sawyer and McCarty[18] (see Table 10.4) this hardness range covers the scale from soft water to hard water.

TABLE 10.4    Hardness Classification Scale[18]

| Hardness range, mg/L as $CaCO_3$ | Hardness description |
|---|---|
| 0–75 | Soft |
| 75–150 | Moderately hard |
| 150–300 | Hard |
| > 300 | Very hard |

Hardness is classified in two ways. These classes are (with respect to the metallic ions and with respect to the anions associated with the metallic ions):

1. *Total hardness*    Total hardness represents the sum of multivalent metallic cations that are normally considered to be only calcium and magnesium. Generally, chemical analyses are performed to de-

termine the total hardness and calcium hardness in the water. Magnesium hardness is then computed as the difference between total hardness and calcium hardness.

2. *Carbonate and noncarbonate hardness* Carbonate hardness is caused by cations from the dissolution of calcium or magnesium carbonate and bicarbonate in the water. Carbonate hardness is hardness that is chemically equivalent to alkalinity, where most of the alkalinity in natural water is caused by the bicarbonate and carbonate ions. Noncarbonate hardness is caused by cations from calcium and magnesium compounds of sulfate, chloride, or silicate that are dissolved in the water. Noncarbonate hardness is equal to the total hardness minus the carbonate hardness. Thus, when the total hardness exceeds the carbonate and bicarbonate alkalinity, the hardness equivalent to the alkalinity is carbonate hardness and the amount in excess of carbonate hardness is noncarbonate hardness. When the total hardness is equal to or less than the carbonate and bicarbonate alkalinity, then the total hardness is equivalent to the carbonate hardness and the noncarbonate hardness is zero. Example Problem 2 illustrates the carbonate and noncarbonate hardness classification:

**Example Problem 2** A groundwater has the following analysis: calcium, 75 mg/L; magnesium, 40 mg/L; sodium, 10 mg/L; bicarbonate, 300 mg/L; chloride, 10 mg/L; and sulfate, 109 mg/L. Compute the total hardness, carbonate hardness, and noncarbonate hardness, all expressed as milligrams per liter of $CaCO_3$.

**solution**
1. Construct a computation table, and convert all concentrations to milligrams per liter of $CaCO_3$.
   a. The species concentration in milliequivalents per liter is calculated from the relationship

   $$[meq/L \text{ of species}] = \frac{mg/L \text{ of species}}{\text{equivalent weight of species}}$$

   b. The species concentration expressed as milligrams per liter of $CaCO_3$ is computed from

   $$[mg/L \ CaCO_3] = [mg/L \text{ of species}] \left[ \frac{50}{\text{equivalent weight of species}} \right]$$

| Chemical specie | Concentration, mg/L | Equivalent weight | Concentration, meq/L | Concentration, mg/L $CaCO_3$ |
|---|---|---|---|---|
| $Ca^{2+}$ | 75 | 20.0 | 3.7 | 187 |
| $Mg^{2+}$ | 40 | 12.2 | 3.3 | 164 |
| $Na^+$ | 10 | 23.0 | 0.4 | 22 |
| | | | 7.4 | 373 |
| $HCO_3^-$ | 300 | 61.0 | 4.9 | 246 |
| $Cl^-$ | 10 | 35.5 | 0.3 | 14 |
| $SO_4^{2-}$ | 109 | 48.0 | 2.2 | 113 |
| | | | 7.4 | 373 |

2. Draw a bar diagram of the raw water, indicating the relative proportions of the chemical species important to the softening process. Cations are placed above the anions on the diagram.

3. Calculate the hardness distribution for this water.

   Total hardness = 187 + 164 = 351 mg/L as $CaCO_3$

   Alkalinity = bicarbonate alkalinity = 246 mg/L as $CaCO_3$

   Carbonate hardness = alkalinity = 246 mg/L as $CaCO_3$

   Noncarbonate hardness = 351 − 246 = 105 mg/L as $CaCO_3$

## Process chemistry

During precipitation softening, calcium is removed from water in the form of $CaCO_3(s)$ precipitate while magnesium is removed as $Mg(OH)_2(s)$ precipitate. The concentrations of the various carbonic species and the system pH play important roles in the precipitation of these two solids.

Carbonate hardness can be removed by adding hydroxide ions and elevating the solution pH so that the bicarbonate ions are converted to carbonate form (pH above 10). Before the solution pH can be changed significantly, however, the free carbon dioxide or carbonic acid must be neutralized. The increase in carbonate concentration from the conversion of bicarbonate to carbonate causes the calcium and carbonate ion product $[Ca^{2+}][CO_3^{2-}]$ to exceed the solubility product constant for $CaCO_3(s)$, and precipitation occurs. As a result, the concentration of calcium ions, originally treated as if they were associated with the bicarbonate anions, is reduced to a low value. The remaining calcium (noncarbonate hardness), however, is not removed by a simple pH adjustment. Rather, carbonate, usually sodium carbonate (soda ash), from an external source must be added to precipitate this calcium. Carbonate and noncarbonate magnesium hardnesses are removed by increasing the hydroxide ion concentration until the magnesium and hydroxide ion product $[Mg^{2+}][OH^-]^2$ exceeds the solubility product constant for $Mg(OH)_2(s)$ and precipitation occurs.

In the lime–soda ash softening process, lime is added to provide the hydroxide ions required to elevate the pH, while sodium carbonate is added to provide an external source of carbonate ions. The least expensive form of lime is quicklime (CaO), which must be hydrated or slaked to $Ca(OH)_2$ before application. These reactions of the lime–soda ash softening process are:

$$H_2CO_3 + Ca(OH)_2 \rightarrow CaCO_3(s) + 2H_2O \qquad (10.38)$$

$$Ca^{2+} + 2HCO_3^- + Ca(OH)_2 \rightarrow 2CaCO_3(s) + 2H_2O \qquad (10.39)$$

$$Ca^{2+} + \begin{bmatrix} SO_4^{2-} \\ 2Cl^- \end{bmatrix} + Na_2CO_3 \rightarrow CaCO_3(s) + 2Na^+ + \begin{bmatrix} SO_4^{2-} \\ 2Cl^- \end{bmatrix} \qquad (10.40)$$

$$Mg^{2+} + 2HCO_3^- + 2Ca(OH)_2 \rightarrow 2CaCO_3(s) + Mg(OH)_2(s) + 2H_2O \qquad (10.41)$$

$$Mg^{2+} + \begin{bmatrix} SO_4^{2-} \\ 2Cl^- \end{bmatrix} + Ca(OH)_2 \rightarrow Mg(OH)_2(s) + Ca^+ + \begin{bmatrix} SO_4^{2-} \\ 2Cl^- \end{bmatrix} \qquad (10.42)$$

$$Ca^{2+} + \begin{bmatrix} SO_4^{2-} \\ 2Cl^- \end{bmatrix} + Na_2CO_3 \rightarrow CaCO_3(s) + 2Na^+ + \begin{bmatrix} SO_4^{2-} \\ 2Cl^- \end{bmatrix} \qquad (10.43)$$

Equation (10.38) represents the neutralization reaction between free carbon dioxide or carbonic acid and lime that must be satisfied before the pH can be increased significantly. Although no net change in water hardness occurs as a result of Eq. (10.38), this reaction must be considered because a lime demand is created. If both carbonic acid and lime are expressed in terms of calcium carbonate, stoichiometric coefficient ratios suggest that for each milligram per liter of carbonic acid (expressed as $CaCO_3$) present, 1 mg/L of lime (expressed as $CaCO_3$) will be required for neutralization.

The removal of calcium carbonate hardness is reflected in Eq. (10.39). This reaction shows that for each molecule of calcium bicarbonate present, two carbonate ions can be formed by raising the pH. One of the carbonate ions can be assumed to react with one of the calcium ions originally present as calcium bicarbonate, while the other carbonate ion can be assumed to react with the calcium ion released from the lime molecule added to elevate the pH. In both cases, calcium carbonate precipitates. If both calcium bicarbonate and lime are expressed in terms of $CaCO_3$, stoichiometric coefficient ratios show that for each milligram per liter of calcium bicarbonate (calcium carbonate hardness) present, 1 mg/L of lime (expressed as $CaCO_3$) will be required for its removal.

Equation (10.40) represents the removal of calcium noncarbonate hardness. If the calcium noncarbonate hardness is expressed in terms of $CaCO_3$, stoichiometric coefficient ratios suggest that for each milligram per liter of calcium noncarbonate hardness present, 1 mg/L of sodium carbonate (expressed as $CaCO_3$) will be required for its removal.

Equation (10.41) is somewhat similar to Eq. (10.39) in that it represents the removal of carbonate hardness, except in this case it is magnesium carbonate hardness. By elevating the pH, two carbonate ions can be formed from each magnesium bicarbonate molecule. Because no calcium is considered to be present in this reaction, enough calcium ions must be added in the form of lime to precipitate the carbonate ion as calcium carbonate before the hydroxide ion concentration can be increased to the level required for magnesium removal. The magnesium is precipitated as magnesium hydroxide. If magnesium bicarbonate and lime are expressed in terms of $CaCO_3$, stoichiometric coefficient ratios state that for each milligram per liter of magnesium carbonate hardness present, 2 mg/L of lime (expressed as $CaCO_3$) will be required for its removal.

Equation (10.42) represents the removal of magnesium noncarbonate hardness. If magnesium noncarbonate hardness and lime are expressed in terms of $CaCO_3$, stoichiometric coefficient ratios state that for each milligram per liter of magnesium noncarbonate hardness present, 1 mg/L of lime (expressed as $CaCO_3$) will be required for its removal. In this reaction, however, there is no net change in the hardness level because for every magnesium ion removed a calcium ion is added. Thus, to complete the hardness removal process, sodium carbonate must be added to precipitate this calcium. This is illustrated in Eq. (10.43), which is identical to Eq. (10.40).

Based on Eqs. (10.39) to (10.43), the chemical requirements for lime–soda ash softening can be summarized as follows if all constituents are expressed as equivalent $CaCO_3$: 1 mg/L of lime as $CaCO_3$ will be required for each milligram per liter of carbonic acid (expressed as $CaCO_3$) present; 1 mg/L of lime as $CaCO_3$ will be required for each milligram per liter of calcium carbonate hardness present; 1 mg/L of soda ash as $CaCO_3$ will be required for each milligram per liter of calcium noncarbonate hardness present; 2 mg/L of lime as $CaCO_3$ will be required for each milligram per liter of magnesium carbonate hardness present; 1 mg/L of lime as $CaCO_3$ and 1 mg/L of soda ash as $CaCO_3$ will be required for each milligram per liter of magnesium noncarbonate hardness present. To achieve removal of magnesium in the form of $Mg(OH)_2(s)$, the solution pH must be raised to a value greater than 10.5 [see Fig. 10.5 which shows the solubility of $Mg(OH)_2$ as a function of pH]. This will require a lime dosage greater than the stoichiometric requirement.

## Chemical dose calculations for lime–soda ash softening

**Calculations based on stoichiometry.** The characteristics of the source water will establish the type of treatment process necessary for softening. Four process types are listed by Humenick.[19] Each process name is derived from the type and amount of chemical added.

1. *Single-stage lime process*    The source water has high calcium, low magnesium carbonate hardness (less than 40 mg/L as $CaCO_3$). There is no noncarbonate hardness.

2. *Excess lime process*    The source water has high calcium, high magnesium carbonate hardness. There is no noncarbonate hardness. It may be a one- or two-stage process.

3. *Single-stage lime–soda ash process*    The source water has high calcium, low magnesium carbonate hardness (less than 40 mg/L as $CaCO_3$). There is some calcium noncarbonate hardness.

4. *Excess lime–soda ash process*    The source water has high calcium, high magnesium carbonate hardness and some noncarbonate hardness. It may be a one- or two-stage process.

Example Problems 3 through 6 illustrate chemical dose calculations and hardness distribution determinations for each type of process.

**Example Problem 3 (Straight Lime Softening)**    A groundwater was analyzed and found to have the following composition (all concentrations are as $CaCO_3$):

$$pH = 7.0 \qquad Alk = 260 \text{ mg/L}$$

$$Ca^{2+} = 210 \text{ mg/L} \qquad Temp = 10°C$$

$$Mg^{2+} = 15 \text{ mg/L}$$

Estimate the lime dose required to soften the water.

**solution**
1. Estimate the carbonic acid concentration.
   a. Determine the bicarbonate concentration in moles per liter by assuming that at pH = 7.0; all alkalinity is in the bicarbonate form.
   $$[HCO_3^-] = 260 \left[\frac{61}{50}\right]\left[\frac{1}{1000}\right]\left[\frac{1}{61}\right]$$

   $$= 5.2 \times 10^{-3} \text{ mol/L}$$

   b. Compute the dissociation constants for carbonic acid at 10°C from Eqs. (10.18) and (10.19).
   $$K_1 = 10^{14.8435 - 3404.71/283 - 0.032786(283)}$$

   $$= 3.47 \times 10^{-7}$$

$$K_2 = 10^{6.498 - 2909.39/283 - 0.02379(283)}$$

$$= 3.1 \times 10^{-11}$$

c. Compute $\alpha_1$ from Eq. (10.25).

$$\alpha_1 = \frac{1}{1.0 \times 10^{-7}/3.47 \times 10^{-7} + 1 + 3.1 \times 10^{-11}/1.0 \times 10^{-7}}$$

$$= 0.77$$

d. Determine the total carbonic species concentration from Eq. (10.22).

$$C_T = \frac{5.2 \times 10^{-3}}{0.77} = 6.75 \times 10^{-3} \text{ mol/L}$$

e. Compute the carbonic acid concentration from a rearrangement of Eq. (10.20) while neglecting the carbonate term because it will be insignificant at a pH of 7.0.

$$[H_2CO_3{}^*] = C_T - [HCO_3^-] = 6.75 \times 10^{-3} - 5.2 \times 10^{-3}$$

$$= 1.55 \times 10^{-3} \text{ mol/L}$$

or

$$[H_2CO_3{}^*] = 155 \text{ mg/L as CaCO}_3$$

2. Draw a bar diagram of the untreated water.

3. Establish the hardness distributed based on the measured concentrations of alkalinity, calcium, and magnesium.

Total hardness = 210 + 15 = 225 mg/L

Calcium carbonate hardness = 210 mg/L

Magnesium carbonate hardness = 15 mg/L

*Note:* Generally there is no need for magnesium removal when the concentration is less than 40 mg/L as $CaCO_3$.

4. Estimate the lime dose requirement by applying the following relationship for the straight lime process:

Lime dose for straight lime process = carbonic acid concentration + calcium carbonate hardness

$$= 155 + 210 = 365 \text{ mg/L as CaCO}_3$$

or

$$\text{Lime dose} = 365 \times \frac{37}{50} = 270 \text{ mg/L as Ca(OH)}_2$$

This calculation assumes that the lime is 100 percent pure. If the actual purity is less than 100 percent, the lime dose must be increased accordingly.

5. Estimate the hardness of the finished water. The final hardness of the water is all the $Mg^{2+}$ in the untreated water plus the practical limit of $CaCO_3$ removal. Although calcium carbonate has a finite solubility, seldom are the theoretical solubility equilibrium concentrations reached, because of factors such as insufficient detention time in the softening reactor; the interaction of $Ca^{2+}$, $CO_3^{2-}$, and $OH^-$ with soluble anionic or cationic impurities to precipitate insoluble salts in a separate phase from $CaCO_3$; and inadequate particle size for effective solids removal. For most situations the practical lower limit of calcium achievable is between 30 and 50 mg/L as $CaCO_3$. Sometimes a 5 to 10 percent excess of the stoichiometric lime is added to accelerate the precipitation reactions. In such cases the excess should be added to the lime dose established in step 4.

**Example Problem 4 (Excess Lime Softening)**   A water was analyzed and found to have the following composition, with all concentrations as $CaCO_3$:

$$pH = 7.0 \qquad \text{Alk} = 260 \text{ mg/L}$$

$$Ca^{2+} = 180 \text{ mg/L} \qquad \text{Temp} = 10°C$$

$$Mg^{2+} = 60 \text{ mg/L}$$

Estimate the lime dose required to soften the water.

**solution**

1. Estimate the carbonic acid concentration. From step 1 in Example Problem 3, the carbonic acid concentration is 155 mg/L as $CaCO_3$.

2. Draw a bar diagram of the untreated water.

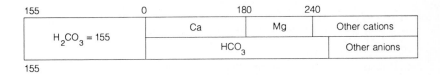

3. Establish the hardness distribution based on the measured concentrations of alkalinity, calcium, and magnesium.

$$\text{Total hardness} = 180 + 60 = 240 \text{ mg/L}$$

$$\text{Calcium carbonate hardness} = 180 \text{ mg/L}$$

$$\text{Magnesium carbonate hardness} = 60 \text{ mg/L}$$

*Note:* In determining the required chemical dose for this process, sufficient lime must be added to convert all bicarbonate alkalinity to carbonate alkalinity, to precipitate magnesium as magnesium hydroxide, and to account for the excess lime requirement.

4. Estimate the lime dose requirements by applying the following relationship for the excess lime process:

Lime dose for excess lime process = carbonic acid concentration
+ total alkalinity + magnesium hardness + 60 mg/L excess lime

$$= 155 + 260 + 60 + 60$$

$$= 535 \text{ mg/L as } CaCO_3$$

or

$$\text{Lime dose} = 535 \times \frac{37}{50} = 396 \text{ mg/L as } Ca(OH)_2$$

A high hydroxide ion concentration is required to drive the magnesium hydroxide precipitation reaction to completion. This is normally achieved when the pH is elevated above 11.0. To ensure that the required pH is established, 60 mg/L as $CaCO_3$ of excess lime is added.

5. Estimate the hardness of the finished water. See step 5 in Example Problem 3 for an explanation. Normally the practical lower limit of calcium achievable is between 30 and 50 mg/L as $CaCO_3$ while the practical limit of magnesium achievable is between 10 and 20 mg/L as $CaCO_3$ with an excess of lime of 60 mg/L as $CaCO_3$. In this case, however, the finished water calcium concentration will be slightly higher than the normal range because of the excess lime added.

**Example Problem 5 (Straight Lime–Soda Ash Process)** A water was analyzed and found to have the following composition, where all concentrations are as $CaCO_3$:

$$\text{pH} = 7.0 \qquad \text{Alk} = 260 \text{ mg/L}$$

$$Ca^{2+} = 280 \text{ mg/L} \qquad \text{Temp} = 10°C$$

$$Mg^{2+} = 10 \text{ mg/L}$$

Estimate the lime and soda ash dosage required to soften the water.

**solution**

1. Estimate the carbonic acid concentration. From step 1 in Example Problem 3, the carbonic acid concentration is 155 mg/L as $CaCO_3$.

2. Draw a bar diagram of the untreated water.

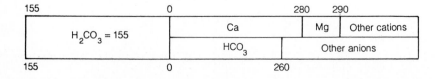

3. Establish the hardness distribution based on the measured concentrations of alkalinity, calcium, and magnesium.

  Total hardness = 280 + 10 = 290 mg/L

  Calcium carbonate hardness = 260 mg/L

  Calcium noncarbonate hardness = 280 − 260 = 20 mg/L

  Magnesium carbonate hardness = 0 mg/L

  Magnesium noncarbonate hardness = 10 mg/L

4. Estimate the lime and soda ash requirements by applying the following relationships for the straight lime–soda ash process:

  Lime dose for straight lime–soda ash process =
  
  carbonic acid concentration + calcium carbonate hardness

$$= 155 + 260$$

$$= 415 \text{ mg/L as } CaCO_3$$

or

$$\text{Lime dose} = 415 \times \frac{37}{50} =$$

$$307 \text{ mg/L as } Ca(OH)_2$$

and

Soda ash dose for straight lime–soda ash process =

  calcium noncarbonate hardness

$$= 20 \text{ mg/L as } CaCO_3$$

$$\text{Soda ash dose} = 20 \times \frac{53}{50} =$$

$$21 \text{ mg/L as } Na_2CO_3$$

5. Estimate the hardness of the finished water. See step 5 in Example Problem 3 for an explanation. The final hardness of the water is all the $Mg^{2+}$ in the untreated water plus the practical limit of calcium achievable, which is between 30 and 50 mg/L as $CaCO_3$.

**Example Problem 6 (Excess Lime–Soda Ash Process)**   A water is analyzed and found to have the following composition, where all concentrations are as $CaCO_3$:

$$pH = 7.0 \qquad Alk = 260 \text{ mg/L}$$

$$Ca^{2+} = 280 \text{ mg/L} \qquad Temp = 10°C$$

$$Mg^{2+} = 80 \text{ mg/L}$$

Estimate the lime and soda ash dosage required to soften the water.

**solution**

1. Estimate the carbonic acid concentration. From step 1 in Example Problem 3, the carbonic acid concentration is 155 mg/L as $CaCO_3$.

| 155 | | 0 | | 280 | 360 |
|---|---|---|---|---|---|
| | | | Ca | Mg | Other cations |
| $H_2CO_3 = 155$ | | | HCO$_3$ | | Other cations |

155                                    0                                260

2. Draw a bar diagram of the untreated water.

3. Establish the hardness distribution based on the measured concentrations of alkalinity, calcium, and magnesium.

   Total hardness = 280 + 80 = 360 mg/L

   Calcium carbonate hardness = 260 mg/L

   Calcium noncarbonate hardness = 280 − 260 = 20 mg/L

   Magnesium carbonate hardness = 0 mg/L

   Magnesium noncarbonate hardness = 80 mg/L

4. Estimate the lime and soda ash requirements by applying the following relationships for the excess lime–soda ash process:

   Lime dose for excess lime–soda ash process = carbonic acid concentration
              + calcium carbonate hardness + 2 magnesium carbonate hardness
   + magnesium noncarbonate hardness + 60 mg/L excess lime
   $$= 155 + 260 + (2)(0) + 80 + 60$$
   $$= 555 \text{ mg/L as } CaCO_3$$

   or

   $$\text{Lime dose} = 555 \times \frac{37}{50} = 411 \text{ mg/L as } Ca(OH)_2$$

   and

   Soda ash dose for excess lime–soda ash process

       = calcium noncarbonate hardness + magnesium noncarbonate hardness

       = 20 + 80

       = 100 mg/L as $CaCO_3$

   or

   $$\text{Soda ash dose} = 100 \times \frac{53}{50} = 106 \text{ mg/L as } Na_2CO_3$$

5. Estimate the hardness of the finished water. See step 5 in Example Problem 3 for an explanation. The practical limit of calcium achievable is between 30 and 50 mg/L as $CaCO_3$ while the practical limit of magnesium achievable is between 10 and 20 mg/L as $CaCO_3$ with an excess lime of 60 mg/L as $CaCO_3$. Although excess lime was added, no excess soda ash was added to remove these extra calcium ions.

**Calculations based on Caldwell-Lawrence diagrams.** An alternative to the stoichiometric approach is the solution of simultaneous equilibria equations to estimate the dosage of chemicals in lime–soda ash softening. A series of diagrams have been developed that make such calculations relatively easy. These diagrams are called *Caldwell-Lawrence* (C-L) *diagrams*. Only a brief discussion of the principles of these diagrams and their application is presented in this chapter. The interested reader is referred to the AWWA publication *Corrosion Control by Deposition of CaCO₃ Films*[20] for an excellent introduction to the use of C-L diagrams. A detailed discussion of the application of C-L diagrams in the solution of lime–soda ash softening problems has been presented by Merrill[21] and Benefield et al.[6]

A C-L diagram is a graphical representation of saturation equilibrium for $CaCO_3$ (Fig. 10.8). Any point on the diagram indicates the pH, soluble calcium concentration, and alkalinity required for $CaCO_3$ saturation. The coordinate system for the diagram is defined as follows:

$$\text{Ordinate} = \text{acidity} \tag{10.44}$$

$$\text{Abscissa} = C_2 = \text{Alk} - \text{Ca} \tag{10.45}$$

where acidity = acidity concentration expressed as mg/L $CaCO_3$
  Alk = alkalinity concentration as mg/L $CaCO_3$
  Ca = calcium concentration as mg/L $CaCO_3$

When C-L diagrams are employed to estimate chemical dosages for water softening, it is necessary to use both the direction format diagram and the Mg-pH nomograph located on each diagram. The general steps involved in solving water softening problems with C-L diagrams are as follows:

1. Measure the pH, alkalinity, soluble calcium concentration, and soluble magnesium concentration of the water to be treated.
2. Evaluate the equilibrium state with respect to $CaCO_3$ precipitation of the untreated water. This is done by locating the point of intersection of the measured pH and alkalinity lines. Determine the value of the calcium line that passes through that point. Compare that value to the measured calcium value. If the measured value is greater, the water is oversaturated with respect to $CaCO_3$. If the measured value is less than the value obtained from the C-L diagram, the water is undersaturated with respect to $CaCO_3$.
3. To use the direction format diagram, the water must be saturated with $CaCO_3$. Follow this procedure for establishing this point for water that is not saturated:
   a. *Source water oversaturated.* Locate the point of $CaCO_3$ satura-

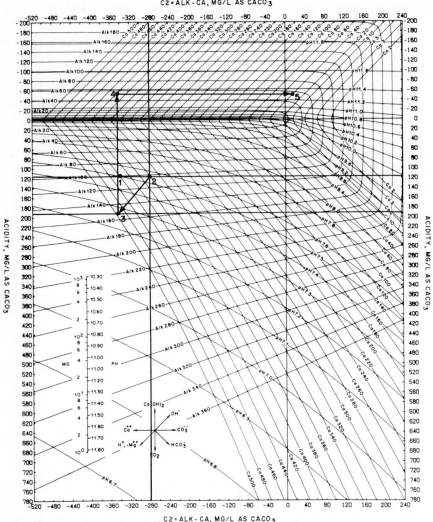

**Figure 10.8** Water-conditioning diagram for 15°C and 400 mg/L TDS. (*Source: Corrosion Control by Deposition of CaCO₃ Films, AWWA, Denver, 1978.*)

tion by allowing $CaCO_3$ to precipitate until equilibrium is established. This point is located at the point of intersection of the
horizontal line through the ordinate value given by
[acidity]$_{initial}$ and a vertical line through the abscissa value
$C_2 = [Alk]_{initial} - [Ca]_{initial}$.

*b.* *Source water undersaturated.* Locate the point of CaCO₃ saturation by allowing recycled CaCO₃ particles to dissolve until

equilibrium is established. This point is located by the same procedure followed in step 3a.

4. Establish the pH required to produce the desired residual soluble magnesium concentration. This is accomplished by simply noting the pH associated with the desired concentration on the Mg-pH nomograph.

5. On a C-L diagram, $Mg(OH)_2$ precipitation produces the same response as the addition of a strong acid. This response is indicated on the direction format diagram as down and to the left at 45°. When a C-L diagram is utilized for softening calculations, the effect of $Mg(OH)_2$ precipitation should be accounted for before the chemical dose is computed. The starting point for the chemical dose calculation is located as follows:

   a. Compute the change in the magnesium concentration as a result of $Mg(OH)_2$ precipitation:

$$\Delta Mg = [Mg]_{initial} - [Mg]_{desired} \qquad (10.46)$$

   b. From the initial saturation point construct a vector down and to the left at 45°. The magnitude of this vector should be such that the horizontal and vertical projections have a magnitude equal to Mg.

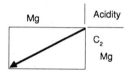

6. From the point located in step 5b, construct a vector whose head intersects the pH line established in step 4. The direction of this vector will depend on the chemical selected for the softening process. If lime is chosen, the direction format diagram shows that the vector will move vertically. But if sodium hydroxide is added, the diagram shows that the vector will move up and to the right at 45°. The required $Ca(OH)_2$ dose is given by the magnitude of the projection of the lime addition vector on the ordinate, and the required NaOH dose is given by the magnitude of either the horizontal projection on the $C_2$ axis or the vertical projection on the ordinate of the sodium hydroxide vector.

7. Evaluate the calcium line that passes through the point at the intersection of the pH line established in step 4 and the chemical addition vector established in step 6. If this line indicates that the soluble calcium concentration is too high, soda ash must be added. The required soda ash dose is determined as follows:

    *a.* Locate the calcium line representing the desired residual calcium concentration.

    *b.* Construct a vector from the point established in step 6 to intersect the desired calcium line. The direction format diagram indicates that the direction of this vector is horizontal and to the right.

    *c.* The required soda ash dose is given by the magnitude of the projection of the soda ash vector on the $C_2$ axis. The saturation state defined by the intersection of the desired calcium line and the soda ash vector describes the characteristics of the softened water.

The use of a C-L diagram for computing the required chemical dose for water softening is illustrated in Example Problem 7.

**Example Problem 7**  A source water has the following characteristics:

$$pH = 7.5 \qquad\qquad Alk = 100 \text{ mg/L as } CaCO_3$$

$$Ca^{2+} = 380 \text{ mg/L as } CaCO_3 \qquad Temp = 15°C$$

$$Mg^{2+} = 80 \text{ mg/L as } CaCO_3$$

and

$$I = 0.01\,M$$

where $I$ represents ionic strength of the water. Determine the chemical dose requirement for excess lime–soda ash softening if the finished water is to contain 40 mg/L calcium and 10 mg/L magnesium (both as $CaCO_3$).

**solution**
1. Evaluate the equilibrium state of the untreated water.
    *a.* Locate the intersection of the initial pH and initial alkalinity lines (shown as point 1 in Fig. 10.8).
    *b.* The calcium line that passes through point 1 is 440 mg/L as $CaCO_3$. Because this represents a concentration greater than 380 mg/L as $CaCO_3$, the water is undersaturated with respect to $CaCO_3$.
2. Compute the initial acidity of the untreated water.
    *a.* Construct a horizontal line through point 1, and read the acidity value at the point where that line intersects the ordinate: acidity = 117 mg/L as $CaCO_3$.
3. Compute the $C_2$ value for the untreated water:

$$C_2 = [Alk] - [Ca] = 100 - 380 = -280 \text{ mg/L as } CaCO_3$$

4. Locate the system equilibrium point at the intersection of a horizontal line through acidity = 117 and a vertical line through $C_2 = -280$ (shown as point 2 in Fig. 10.8).
5. Establish the pH required to produce the desired residual soluble magnesium concentration. This is obtained from the Mg-pH nomograph, which shows

that a pH of 11.32 is required to reduce the soluble magnesium concentration to 10 mg/L as $CaCO_3$.

6. Compute the change in the magnesium concentration as a result of $Mg(OH)_2$ precipitation:

$$\Delta Mg = [Mg]_{initial} - [Mg]_{desired}$$

$$= 80 - 10$$

$$= 70 \text{ mg/L as } CaCO_3$$

7. Construct a downward vector from point 2*b* that will account for the effects of $Mg(OH)_2$ precipitation.

   a. Draw a horizontal line through the acidity value of 117 + 70 (initial acidity + [$\Delta Mg$]) = 187 mg/L as $CaCO_3$.

   b. Beginning at point 2, construct a vector down and to the left at 45° until it intersects the horizontal line through acidity = 187 mg/L as $CaCO_3$ (shown as point 3 in Fig. 10.8).

8. Construct a vertical vector beginning at point 3 to intersect the pH = 11.32 line (shown as point 4 in Fig. 10.8). The lime dose is equal to the magnitude of the projection of this vector onto the ordinate (from point 3 up to $O$ on ordinate is 187 units, and from $O$ up to point 4 on ordinate is 50 units):

$$\text{Lime dose} = 187 + 50 = 237 \text{ mg/L as } CaCO_3$$

9. Construct a horizontal vector beginning at point 4 to intersect the Ca = 40 line (shown as point 5 in Fig. 10.8). The soda ash dose is equal to the magnitude of the projection of this vector onto the $C_2$ axis (from point 4 to $O$ on the abscissa is 350 units, and from $O$ to point 5 on the abscissa is 18 units):

$$\text{Soda ash dose} = 350 + 18 = 368 \text{ mg/L as } CaCO_3$$

*Note:* Chemical dosages computed in steps 8 and 9 are lower than those calculated by the stoichiometric approach because the C-L diagram assumes that equilibrium is achieved, which actually does not happen in real plants.

## Recarbonation

Depending on the softening process utilized (straight lime, excess lime, straight lime–soda ash, or excess lime–soda ash), the treated water will usually have a pH of 10 or greater. It is necessary to lower the pH and stabilize such water to prevent the deposition of hard carbonate scale on filter sand and distribution piping. Recarbonation is the process most commonly employed to adjust the pH. In this process, carbon dioxide ($CO_2$,) is added to the water in sufficient quantity to lower the pH to the range of 8.4 to 8.6.

When low-magnesium water is softened, no excess lime is added. After softening, the water is supersaturated with calcium carbonate and has a pH between 10.0 and 10.6. When carbon dioxide is added to this water, the carbonate ions are converted to bicarbonate ions according to the following reaction:

$$Ca^{2+} + CO_3^{2-} + CO_2 + H_2O \rightleftharpoons Ca^{2+} + 2HCO_3^- \qquad (10.47)$$

When high-magnesium water is softened, excess lime is added to raise the pH above 11 to precipitate magnesium hydroxide. For this situation enough carbon dioxide must be added to neutralize the excess hydroxide ions as well as to convert the carbonate ions to bicarbonate ions. To achieve the first requirement, i.e., to neutralize the excess hydroxide ions, carbon dioxide is added to lower the pH to between 10.0 and 10.5. In this pH range, calcium carbonate is formed as shown by Eq. (10.48), and magnesium hydroxide that did not precipitate, as well as that which did not settle, is converted to magnesium carbonate as shown by Eq. (10.49):

$$Ca^{2+} + 2OH^- + CO_2 \rightleftharpoons CaCO_3(s) + H_2O \qquad (10.48)$$

$$Mg^{2+} + 2OH^- + CO_2 \rightleftharpoons Mg^{2+} + CO_3^{2-} + H_2O \qquad (10.49)$$

Additional carbon dioxide is required to lower the pH to between 8.4 and 8.6. Here the previously formed calcium carbonate redissolves, and the carbonate ions are converted to bicarbonate ions, as described by

$$CaCO_3(s) + H_2O + CO_2 \rightleftharpoons Ca^{2+} + 2HCO_3^- \qquad (10.50)$$

$$Mg^{2+} + CO_3^{2-} + CO_2 + H_2O \rightleftharpoons Mg^{2+} + 2HCO_3^- \qquad (10.51)$$

**Process description.**    Two types of recarbonation processes are used in conjunction with the four types of softening processes previously discussed. For treatment of low-magnesium water where excess lime addition is not required, single-stage recarbonation is used. A typical plant arrangement for single-stage softening with recarbonation is shown in Fig. 10.9a. In this process, lime is mixed with the source water in a rapid-mix chamber, resulting in a pH of 10.2 to 10.5. If noncarbonate hardness removal is required, soda ash is added along with the lime. After rapid mixing, the water is slowly mixed for 40 min to 1 h, to allow the particles to agglomerate. After agglomeration the water passes to a sedimentation basin for 2 to 3 h where most of the suspended material is removed. Following sedimentation, the water, which carries some particles still in suspension, moves to the recarbonation reactor. Here carbon dioxide is added to reduce the pH to 8.5 to 9.0. Any particles remaining in suspension after recarbonation are removed during the filtration step.

For treatment of high-magnesium water where excess lime is required, two-stage recarbonation is sometimes used. A typical plant arrangement for two-stage softening with recarbonation is shown in Fig. 10.9b. In this process excess lime is added in the first stage to raise the

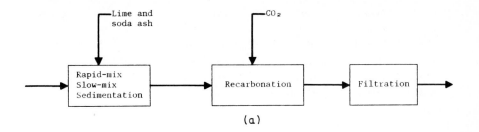

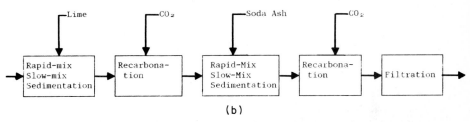

**Figure 10.9** (a) Typical plant arrangement for single-stage softening with recarbonation. (b) Typical plant arrangement for two-stage softening with recarbonation.

pH to 11.0 or higher for optimum magnesium removal. Following first-stage treatment, carbon dioxide is added to reduce the pH to 10.0 to 10.6, the optimum value for calcium carbonate precipitation. If noncarbonate hardness removal is required, soda ash is added in the second stage. During second-stage treatment, carbon dioxide is added to reduce the pH to 8.4 to 8.6. Because of the capital cost savings realized through the elimination of one set of settling basins and recarbonation units, single-stage recarbonation is usually the method of choice for high-magnesium water. Still, there are certain advantages in the use of two-stage recarbonation. These include a lower operating cost because of the requirement for lower carbon dioxide dosages and a better finished water quality. The water produced by two-stage softening and recarbonation is softer and lower in alkalinity than water produced by a single-stage softening and recarbonation process. In most situations, however, the latter advantage is not important because water hardness concentrations between 80 and 120 mg/L as $CaCO_3$ are normally acceptable for municipal use.

**Dose calculations for recarbonation.** The quantity of gas required for recarbonation varies with the quantity of water treated, the amounts of carbonate and hydroxide alkalinity in the water, and the degree to which recarbonation is to be performed. Example Problem 8 illustrates the stoichiometric approach to estimating carbon dioxide requirements.

**Example Problem 8**   Estimate the carbon dioxide requirements for the following treatment situations:

1. Example Problem 3 with single-stage recarbonation
2. Example Problem 4 with single-stage recarbonation
3. Example Problem 5 with single-stage recarbonation
4. Example Problem 6 with two-stage recarbonation

**solution**

1. For Example Problem 3. Estimate the carbon dioxide dose, using the following relationship for single-stage recarbonation for straight lime softening:

   Carbon dioxide requirement = estimated carbonate alkalinity of softened water

   where

   Estimated carbonate alkalinity of softened water = source water alkalinity
   − source water calcium hardness
   − estimated residual calcium hardness of settled softened water

   Therefore, assuming the residual calcium hardness in the settled softened water is 50 mg/L as $CaCO_3$,

   $$\text{Carbon dioxide requirement} = 260 - (210 - 50)$$

   $$= 100 \text{ mg/L as } CaCO_3$$

   or

   $$\text{Carbon dioxide requirement} = 100 \times \frac{22}{50} = 44 \text{ mg/L as } CO_2$$

2. For Example Problem 4. Estimate the carbon dioxide dose, using the following relationship for single-stage recarbonation for excess lime softening:

   Carbon dioxide requirement =
   estimated carbonate alkalinity of softened water + 2 excess lime dose +
   estimated residual magnesium hardness of settled softened water

   where

   Estimated carbonate alkalinityof softened water = source water alkalinity
   − (source water total hardness + excess lime dose
   − estimated residual total hardness of settled softened water)

   Therefore, assuming the residual calcium hardness and residual magnesium hardness in the settled softened water are 30 and 20 mg/L as $CaCO_3$, respectively, we get

   $$\text{Carbon dioxide requirement} = 260 - (240 + 60 - 50 + 2(60) + 20$$

   $$= 150 \text{ mg/L as } CaCO_3$$

   or

   $$\text{Carbon dioxide requirement} = 150 \times \frac{22}{50} = 66 \text{ mg/L as } CO_2$$

3. For Example Problem 5. Estimate the carbon dioxide dose, using the following relationship for single-stage recarbonation for straight lime–soda ash softening:

Carbon dioxide requirement = estimated carbonate alkalinity of softened water

where

Estimated carbonate alkalinity of softened water = source water alkalinity
+ soda ash dose − (source water calcium hardness
− estimated residual calcium hardness of settled softened water)

Therefore, assuming the residual calcium hardness in the settled softened water is 45 mg/L as $CaCO_3$, we have

$$\text{Carbon dioxide requirement} = 260 + 20 - (280 - 45)$$

$$= 45 \text{ mg/L as } CaCO_3$$

or

$$\text{Carbon dioxide requirement} = 45 \times \frac{22}{50} = 20 \text{ mg/L as } CO_2$$

4. For Example Problem 6. Estimate the carbon dioxide dose, using the following relationship for two-stage recarbonation for excess lime–soda ash softening:
   a. First stage:

Carbon dioxide requirement = estimated residual alkalinity of softened water

where

Estimated hydroxide alkalinity of softened water = excess lime dose
+ estimated residual magnesium hardness of settled softened water

Therefore, assuming the residual magnesium hardness in the settled softened water is 15 mg/L as $CaCO_3$, we have

$$\text{Carbon dioxide requirement} = 60 + 15$$

$$= 75 \times \frac{22}{50} = 33 \text{ mg/L as } CO_2$$

   b. Second stage:
Carbon dioxide requirement = estimated residual alkalinity of softened water

where

Estimated carbonate alkalinity of softened water = source water alkalinity
+ soda ash dose − (source water total hardness
− estimated residual total hardness of settled softened water)

Because $C_T$ is increased after first-stage recarbonation, the calcium carbonate solubility during second-stage softening is lowered. Hence, the residual calcium hardness after second-stage softening is less than that normally observed in a single-stage process. In this problem a residual calcium hardness of 30 mg/L as $CaCO_3$ is assumed. The residual magnesium hardness is the same as that assumed for first-stage treatment, or 15 mg/L as $CaCO_3$.

$$\text{Carbon dioxide requirement} = 260 + 100 - (360 - 45)$$

or

$$\text{Carbon dioxide requirement} = 45 \times \frac{22}{50} = 20 \text{ mg/L as } CO_2$$

Caldwell-Lawrence diagrams may also be used to compute the required carbon dioxide dose for both two-stage and single-stage softening. Example Problem 9 illustrates the procedure.

**Example Problem 9** Compute the chemical doses required to neutralize the softened water described by point 1 in Fig. 10.10, using two-stage recarbonation and one-stage recarbonation. Assume that the pH must be reduced to 8.7 during neutralization.

**solution**
1. Two-stage recarbonation
   a. Locate the point of minimum calcium concentration. Starting at point 1 in Fig. 10.10, construct a vector vertically down until the head intersects a horizontal line through the ordinate value $C_1 = 0$. This is the point of minimum calcium concentration, point 2 in Fig. 10.10. Note that if the vector had been extended beyond this point, it would have begun to intersect calcium lines of increasing concentration. First-stage recarbonation should be terminated at this point.
   b. Determine the Ca, alkalinity, and pH for the settled water; i.e., determine the water characteristics at point 2 in Fig. 10.10.
      The Ca and pH values can be determined directly from the C-L diagram. Distinguishing between the alkalinity lines in this region of the diagram is very difficult in many cases. For those situations alkalinity can be computed from the $C_2$ value associated with the point of minimum calcium concentration

$$C_2 = \text{Alk} - \text{Ca}$$

or

$$\text{Alk} = C_2 + \text{Ca}$$

For point 2, pH = 10.4, Ca = 10, and $C_2$ = 18 (point 3). Then

$$\text{Alk} = 18 + 10 = 28.$$

   c. Compute the $CO_2$ requirement for first-stage recarbonation. The $CO_2$ requirement is given by the magnitude of the projection of the vector onto the ordinate:

$$\text{First-stage } CO_2 \text{ dose} = 50 \text{ mg/L as } CaCO_3$$

   d. Determine the $CO_2$ requirement for second-stage recarbonation. The di-

WATER CONDITIONING DIAGRAM FOR 15 C AND 400 MG/L TOTAL DISSOLVED SOLIDS
C2 = ALK - CA, MG/L AS CACO₃

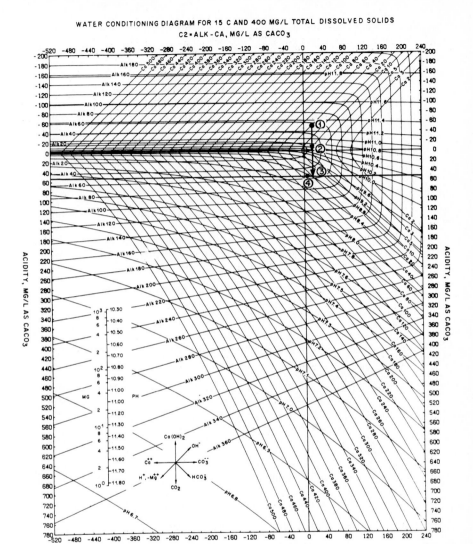

**Figure 10.10**  Example Problem 9 water-conditioning diagram for 15°C and 400 mg/L total dissolved solids. (*Source: Corrosion Control by Deposition of CaCO₃ Films, AWWA, Denver, 1978.*)

rection format diagram cannot be applied beyond point 2 because the addition of CO₂, after the CaCO₃ particles precipitated during first-stage recarbonation have been removed, will produce an undersaturated condition. To circumvent this problem, compute the acidity of the final state at a pH of 8.7, assuming CO₂ addition does not affect alkalinity as long as CaCO₃ does not precipitate. The following equation can be used:

Final acidity, mg/L as $CaCO_3$

$$= \frac{([Alk] - K_w/[H^+] + [H^+])(1 + [H^+]/K_1)}{(1 + K_2/[H^+])} + [H^+] - \frac{K_w}{[H^+]} \quad (10.52)$$

where $K_w$, $K_1$, $K_2$ = equilibrium constants for ionization of $H_2O$, $H_2CO_3{}^*$, and $HCO_3^-$, respectively

[Alk] = alkalinity of water after first-stage recarbonation, mg/L as $CaCO_3$

$[H^+]$ = hydrogen ion concentration at final state, in this case

pH = 8.7 (mg/L as $CaCO_3$)

According to this equation,

Acidity = 28 mg/L as $CaCO_3$

The $CO_2$ required to reduce the pH to 8.7 is given by the difference between the acidity of the final state (computed) and the acidity at point 2; see point 4.

Second-stage $CO_2$ dose = 28 − 0 = 28 mg/L as $CaCO_3$

   e. Calculate the total $CO_2$ requirement for first- and second-stage recarbonation:

Total $CO_2$ = 50 + 28 = 78 mg/L as $CaCO_3$ (mg/L as $CaCO_3$)

2. One-stage recarbonation
   a. Locate the point describing the limiting saturated state that can be attained by $CO_2$ addition. Starting at point 1 in Fig. 10.10, construct a vector vertically down until the head intersects calcium line 40 on the opposite side of the acidity = 0 line (shown as point 3 in Fig. 10.10). The sequence of events between points 1 and 3 can be visualized as $CaCO_3$ precipitation between points 1 and 2 and dissolution of the precipitated $CaCO_3$ between points 2 and 3. The direction format diagram cannot be used to show the reaction path past point 3 because any further $CO_2$ addition produces an undersaturated condition.
   b. Evaluate the Ca, alkalinity, and pH for the saturated conditions at point

3. These values are read directly from the C-L diagram:

Ca = 40     Alk = 58   (point 5)     pH = 8.8

   c. Compute the $CO_2$ dose requirement.
      (1) Remembering that alkalinity was assumed to not change with $CO_2$ as long as $CaCO_3$ does not precipitate, locate the final state at the intersection of the Alk = 58 and pH = 8.7 lines (shown as point 4 in Fig. 10.10).
      (2) Construct a horizontal line through point 5, and determine the acidity of the final state.

Acidity = 58 mg/L as $CaCO_3$     (point 6)

      (3) The $CO_2$ requirement is given by the change in acidity between the initial and final states (in this case between point 7 and point 6):

$CO_2$ requirement $= 50 + 58 = 108$ mg/L as $CaCO_3$ (mg/L as $CaCO_3$)

The use of C - L diagrams for computing carbon dioxide dosages does not recognize the demand for unsettled $CaCO_3$ floc. This demand may, at times, be quite high.

### Split treatment

Several types of processes fit the split treatment category.[22] This category includes processes where two or more streams are treated separately and then combined. Three such processes are split treatment with excess lime, parallel softening and coagulation, and blended lime-softened stream with another stream, either ion-exchange or reverse-osmosis treated water. In this chapter, only the split treatment with excess lime process is discussed.

For certain types of water, split treatment with excess lime softening is less costly (because of reduced chemical costs) than conventional excess lime treatment. Experience has shown that split-treatment softening should be considered when the magnesium content of the water is high and the noncarbonate hardness content is insignificant. Larson et al.[23] have suggested that the magnesium concentration of finished water not exceed 40 mg/L as $CaCO_3$ if water heater fouling is to be prevented. Because total hardness levels of 60 to 120 mg/L as $CaCO_3$ are often acceptable, the calcium concentration of finished water can be allowed to vary between 20 and 80 mg/L as $CaCO_3$.

Split-treatment excess lime softening is a treatment technique in which the source water flow is divided into two streams. One stream receives excess lime treatment for calcium carbonate hardness and magnesium hardness removal. This reduces magnesium hardness in this stream to its practical solubility limit near 10 mg/L as $CaCO_3$. The idea is to treat enough of the total water volume that when the treated and untreated streams are mixed, the magnesium concentration will be less than 40 mg/L as $CaCO_3$. Dissolved carbon dioxide in the bypassed stream is used to neutralize the hydroxide alkalinity produced by the excess lime dose. The blended water is normally allowed to settle before filtration. Because of neutralization by the free carbon dioxide contained in the untreated water, however, often recarbonation is not required. A flow diagram for a typical split-treatment excess lime softening process is presented in Fig. 10.11.

Once the desired finished water magnesium concentration is established, the bypass flow fraction can be determined from

$$X = \frac{(Mg)_e - (Mg)_t}{(Mg)_r - (Mg)_t} \qquad (10.53)$$

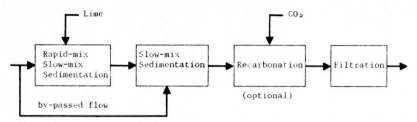

**Figure 10.11**  Flow schematic for split-treatment excess lime softening.

where $X$ = fraction of total flow to be bypassed
$(Mg)_e$ = desired magnesium concentration in plant effluent, mg/L as $CaCO_3$
$(Mg)_t$ = magnesium concentration in stream from lime treatment process, mg/L as $CaCO_3$†
$(Mg)_r$ = magnesium concentration in source water, mg/L as $CaCO_3$

The upper limit for the fraction of water to bypass can be determined by noting that the lower limit for the magnesium concentration that can be achieved in the treated water is zero. Hence, setting $(Mg)_t = 0$ and solving for $X$ give

$$(X)_{max} = \frac{(Mg)_e}{(Mg)_r} \qquad (10.54)$$

Example Problem 10 illustrates the use of C - L diagrams to determine the chemical requirements for split-treatment excess lime softening.

**Example Problem 10**  A groundwater was analyzed and found to have the following composition:

$$pH = 7.1 \qquad\qquad Alk = 300 \text{ mg/L as } CaCO_3$$

$$Ca^{2+} = 200 \text{ mg/L as } CaCO_3 \qquad Temp = 15°C$$

$$Mg^{2+} = 80 \text{ mg/L as } CaCO_3$$

Determine the lime dose required to soften this water if split-treatment excess lime softening is used. The magnesium concentration in the finished water should not exceed 40 mg/L as $CaCO_3$, and the total hardness in the effluent should not exceed 120 mg/L as $CaCO_3$.

**solution**
1. Determine the maximum fraction of water that can be bypassed from Eq. (10.54):

$$(X)_{max} = \frac{40}{80} = 0.5$$

---

†This is normally assumed to be 10 mg/L as $CaCO_3$, which is the practical solubility limit of $Mg(OH)_2(s)$.

Because this is the maximum fraction that can be bypassed and still reach the desired effluent $Mg^{2+}$ concentration, 0.4 of the flow will actually be bypassed because $Mg^{2+}$ is generally not reduced to 0 mg/L by chemical precipitation.

2. Compute the required magnesium concentration in the treated stream when 0.4 of the flow is bypassed:

$$(Mg)_2 = \frac{40 - (80)(0.4)}{1 - 0.4} = 13 \text{ mg/L as } CaCO_3$$

3. Evaluate the equilibrium state of the untreated water.
   a. Locate the intersection of the initial pH and initial alkalinity lines (shown as point 1 in Fig. 10.12).
   b. The calcium line that passes through point 1 is 370 mg/L as $CaCO_3$. Because this represents a concentration greater than 200, the water is unsaturated with respect to $CaCO_3$.
4. Compute the initial acidity of the untreated water.
   a. Construct a horizontal line through point 1, and read the acidity value where this line intersects the ordinate:

$$\text{Acidity} = 415 \text{ mg/L as } CaCO_3 \quad (\text{point } 1a)$$

5. Compute the $C_2$ value for the untreated water:

$$C_2 = [\text{Alk}] - [\text{Ca}] = 300 - 200 = 100 \text{ mg/L as } CaCO_3$$

6. Locate the system equilibrium point at the intersection of a horizontal line through acidity = 415 mg/L as $CaCO_3$ and a vertical line through $C_2$ = 100 mg/L as $CaCO_3$ (point 2 in Fig. 10.12).
7. Establish the pH required to produce the desired residual soluble magnesium concentration. This is obtained from the Mg - pH nomograph which shows that a pH of 11.25 is required to reduce the soluble magnesium concentration to 13 mg/L as $CaCO_3$.
8. Compute the change in the magnesium concentration as a result of $Mg(OH)_2$ precipitation:

$$\Delta Mg = [Mg]_{\text{initial}} - [Mg]_{\text{desired}}$$
$$= 80 - 13 = 67 \text{ mg/L as } CaCO_3$$

9. Construct a downward vector from point 2 that will account for the effects of $Mg(OH)_2$ precipitation.
   a. Draw a horizontal line through the acidity value of 415 + 67 = 482 mg/L as $CaCO_3$.
   b. Construct a vector, beginning at point 2, down and to the left at 45° until it intersects the horizontal line through acidity = 482 mg/L as $CaCO_3$ (point 3 in Fig. 10.12).
10. Construct a vertical vector, beginning at point 3, to intersect the pH = 11.25 line (point 4 in Fig. 10.12). The lime dose is equal to the magnitude of the projection of this vector (3–4) onto the acidity axis:

$$\text{Lime dose} = 482 + 43 = 525 \text{ mg/L as } CaCO_3$$

11. Combine the treated and bypassed water, and evaluate the equilibrium state of the mixture.

a. Assume that the $CaCO_3$ is infinitely soluble and compute the Ca, alkalinity, and acidity of the mixed streams.

$$[Ca]_{mix} = [Ca]_{pt\ 4}(0.6) + [Ca]_{untreated}(0.4)$$
$$= (25)\ (0.6) + (200)(0.4)$$
$$= 95\ mg/L\ as\ CaCO_3$$

$$[Alk]_{mix} = [Alk]_{pt\ 4}(0.6) + [Alk]_{untreated}(0.4)$$
$$= (50)(0.6) + (300)(0.4)$$
$$= 150\ mg/L\ as\ CaCO_3$$

$$[Acidity]_{mix} = [Acidity]_{pt\ 4}(0.6) + [Acidity]_{untreated}(0.4)$$
$$= (-43)(0.6) + (415)(0.4)$$
$$= 140\ mg/L\ as\ CaCO_3$$

b. Construct a horizontal line through the ordinate value given by an acidity of 140 mg/L as $CaCO_3$ (point 4). Locate the intersection of this line and alkalinity = 150 mg/L as $CaCO_3$ (point 5 in Fig. 10.12).

c. The calcium line that passes through point 5 is 15 mg/L as $CaCO_3$. Because this represents a concentration less than 95 mg/L as $CaCO_3$, the mixture is oversaturated with respect to $CaCO_3$.

12. Determine the final equilibrium state of the system.

a. Remove the condition of infinite solubility, and allow $CaCO_3$ to precipitate. Recall two things: Acidity remains constant during precipitation, and because equivalent amounts of alkalinity and Ca are removed, $[Alk] - [Ca] = C_2$ will remain constant during $CaCO_3$ precipitation.

b. Construct a vertical line through the $C_2$ value of $[150] - [95] = 55$ mg/L as $CaCO_3$ (point 5). Locate the intersection of this line and the horizontal line through point 5 (shown as point 6 in Fig. 10.12). The characteristics of the final equilibrium state at point 6 are pH = 8.3, alkalinity = 137 mg/L as $CaCO_3$, and Ca = 82 mg/L as $CaCO_3$. Thus, the theoretical final hardness of the water is 122 mg/L as $CaCO_3$ (Ca = 82 mg/L as $CaCO_3$ and Mg = 40 mg/L as $CaCO_3$). Particle carryover will increase this value, but it will probably still be less than 120 mg/L. A final pH of 8.3 was achieved without the need for recarbonation. Note that these values will differ from those calculated by the stoichiometric approach because the C-L diagram assumes that equilibrium is achieved, which actually does not happen in real plants.

## Softening by means of pellet reactors

A pellet softener is basically a fluidized bed of grains on which the crystallization of $CaCO_3$ takes place. One of the most commonly used pellet softener systems is the pellet reactor (see Fig. 10.13). This system was developed in 1938 by Zentner[24] in Czechoslovakia. It consists of an inverted conical tank in which the softening reactions take place in the presence of a suspended bed of fine sand 0.1 to 0.2 mm in diameter that acts as a catalyst. Source water and chemicals enter tangentially at the bottom of the cone and mix immediately. The treated water then rises through the reactor in a swirling motion. The

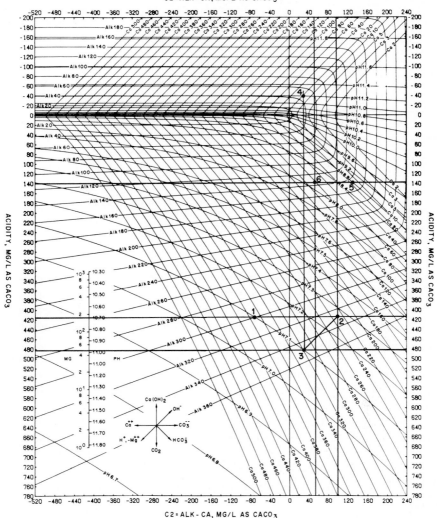

WATER CONDITIONING DIAGRAM FOR 15 C AND 400 MG/L TOTAL DISSOLVED SOLIDS
C2 = ALK - CA, MG/L AS CACO₃

ACIDITY, MG/L AS CACO₃

C2 = ALK - CA, MG/L AS CACO₃

**Figure 10.12** Example Problem 10 water-conditioning diagram for 15°C and 400 mg/L total dissolved solids. (*Source: Corrosion Control by Deposition of CaCO₃ Films, AWWA, Denver, 1978.*)

magnitude of the upward velocity is sufficient to keep the sand fluid-ized. The contact time between the treated water and the sand grains is 8 to 10 min. During this period, precipitated hardness particles at-tach to the surface of the sand grains so that the grains increase in diameter. Because a bed of large grains has a small reactive surface, some of the grains are removed regularly at the bottom of the reactor

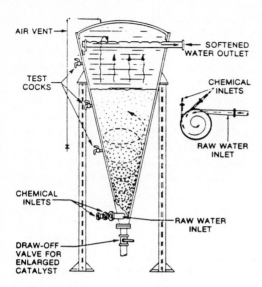

AIR VENT

SOFTENED WATER OUTLET

TEST COCKS

CHEMICAL INLETS

RAW WATER INLET

CHEMICAL INLETS

RAW WATER INLET

DRAW-OFF VALVE FOR ENLARGED CATALYST

**Figure 10.13**  Typical pellet reactor.

and replaced by smaller-diameter seeding grains. During treatment the pH should be kept low enough to prevent the precipitation of magnesium hydroxide. This material does not adhere well to the surface of the sand grains and so will pass out of the reactor, creating a high solids loading on the filter.

Advantages of the pellet reactor are its small size, low installation cost, and rapid treatment. Because removing magnesium in these systems is difficult, however, they should not be considered when the water to be treated has a high magnesium content.

Pellet reactors have been used for softening in the Netherlands for many years. Their use has resulted in a high degree of standardization so that pellet reactors are now considered an established technique in that country.[25] Over the past few years these units have also been installed at a number of locations in the United States.

## Process considerations in water softening

Although water softening is often considered a treatment process whose use is limited to groundwater, surface water supplies require softening in a number of locations. If split treatment is used with surface water, however, some considerations must be given to problems such as taste and odor in the bypass water. When surface water is being softened, preceding the softening process with coagulation may be beneficial if the water has a high turbidity, contains organic colloids (such materials have been found to inhibit the growth of $CaCO_3$ crystals), or recalcining of the sludge is practiced. Even when groundwater is softened, which is relatively free from turbidity, treatment efficiency is often increased when coagulants are added to enhance agglomeration of the colloid-size calcium and magnesium precipitates into particles that are rapidly removed by sedimentation. Small amounts of metal coagulants such as aluminum salts and iron salts are often used for this purpose. However, polymeric coagulants offer an attractive alternative because (1) a smaller volume of sludge is generated than with metal coagulants, (2) polymeric coagulants are effective over a much broader pH range than metal coagulants, and (3) sludges produced with polymeric coagulants tend to dewater more easily than sludges produced with metal coagulants.[26]

Flow diagrams for two full-scale lime softening plants in Illinois that use metal coagulants to enhance treatment efficiency are presented in Fig. 10.14. In this figure, aeration is shown in the scheme for treatment plant B. This operation is necessary to reduce the $CO_2$ content in water where it is high and thus reduce the lime requirement for carbonic acid neutralization.

The discussion of softening presented in this chapter has been limited to lime and soda ash softening. In many cases, however, caustic soda (NaOH) may be substituted for both these chemicals. Four factors must be considered in the decision of whether to use lime and soda ash or caustic soda for a particular application:

1. *Cost*   The total chemical cost will generally be less when lime and soda ash are used. Caustic soda is most competitive for either very low or very high alkalinity waters.

2. *Total dissolved solids*   A greater increase in TDS occurs in the finished water when caustic soda is used, whereas the TDS level often decreases when lime is used. Furthermore, the use of caustic soda may increase the sodium concentration to a level high enough to pose a health concern for some.

3. *Sludge production*   Generally less sludge is produced when caustic soda is used than when lime and soda ash are used. When lime and soda ash are used, sludge production increases with increasing alkalinity, while the quantity of sludge production is independent of alkalinity when caustic soda is employed for softening a water of a particular hardness.

4. *Chemical stability*   Storing and feeding caustic soda are easier than the same process for lime. Caustic soda does not deteriorate during storage, but hydrated lime may adsorb $CO_2$ and water from the air and form $CaCO_3$ while quicklime may slake in storage.

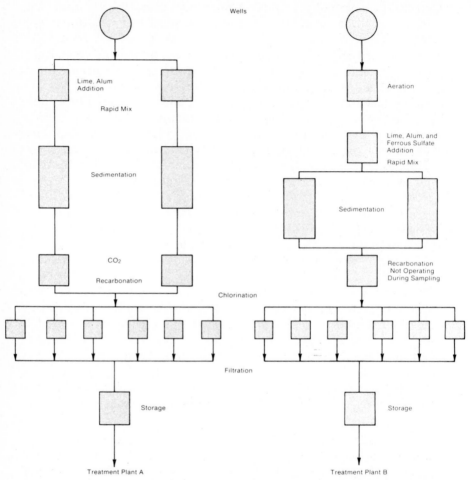

**Figure 10.14**   Flow diagram of full-scale lime softening treatment plant. (*Source: T. J. Sorg and G. S. Logsdon, "Treatment Technology to Meet the Interim Primary Drinking Water Regulations for Inorganics: Part 5," J. AWWA, vol. 72, no. 7, July 1980, p. 411.*)

In the final analysis, constraints for the use of caustic soda will probably be chemical costs and the guideline for sodium content of the finished water.

**Design considerations.**   In lime–soda ash or caustic soda softening plants, the softening process may be carried out by a unit sequence of rapid mix, flocculation, and sedimentation (see Fig. 10.15) or in a solids-contact softener where rapid mix, flocculation, and sedimentation occur in a single unit. The process begins with rapid dispersion of the chemicals into the flow stream, followed at once by violent agita-

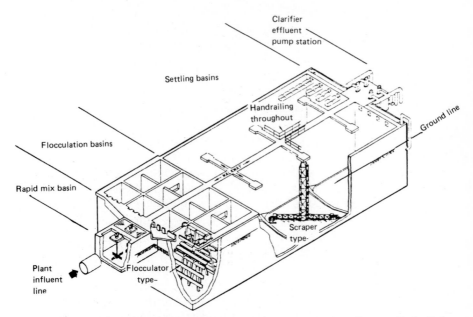

**Figure 10.15**  Chemical clarifier, Orange County Water District. (*Source: R. L. Culp, G. M. Wesner, and G. L. Culp, Handbook of Advanced Wastewater Treatment, 2d ed., Van Nostrand Reinhold, 1978.*)

tion. This is referred to as *rapid mixing* (sometimes called *initial* or *flash mixing*). Where coagulation of colloids is to occur simultaneously with softening, the initial introduction of chemicals into the water is very often a key factor in determining the amount to be used. Introduction of the destabilizing chemicals into the inlet pipeline by jet injection appears to be the most effective means of achieving rapid dispersion.[22] When the rapid-mix step is designed, remember that lime dissolves rather slowly in water. In those situations where coagulants will be added to enhance the effectiveness of the softening process, lime is added and dispersed before the addition of the coagulant.

The purpose of the flocculation step is to provide the retention time and contact opportunities needed for the chemical precipitate to grow large enough to be removed by gravity settling. During this step the water is gently mixed (sometimes called the *slow-mix step*). For large flows, rectangular basins with horizontal and vertical paddle flocculators are commonly used while smaller installations use turbine mixers or end-around flow channels. In the past, most flocculation basins consisted of single uniformly mixed tanks. Both the degree of flocculation and the subsequent clarification step, however, have been found to improve if flocculation basins are compartmentalized. Current design favors the use of basins with three or more compartments, with the velocity gradient $G$ decreasing from compartment to compartment. A minimum of three stages for conventional softening, with the $G$ value decreasing from 50 to 10 $s^{-1}$, has been recommended.[22] In groundwater lime softening, however, the recirculation of sludge is more important to the kinetics of crystal formation than compartmentalization.

The detention time in a flocculation basin is also an important parameter because it determines the amount of time that particles are exposed to the velocity gradient and thus is a measure of contact opportunity in the basin. A minimum retention time of 30 min is recommended for conventional water softening.[22] Sludge return to the head of the flocculator will reduce chemical requirements as well as provide floc nuclei for precipitate growth.[29] The estimated portion of return sludge is 10 to 25 percent of the source water flow.

Sedimentation is the next step following flocculation. Settling rates for chemical precipitates are a function of particle size and density. To ensure efficient removal of the precipitates formed in the softening process, a retention time of 1.5 to 3 h is normally provided in the sedimentation basin. Sedimentation basins are constructed in several configurations that include horizontal flow units (both rectangular and circular), inclined flow units (tube settlers mounted in rectangular or circular basins), and upflow flow units (units that incorporate chemical mixing, flocculation, and sedimentation in a single tank). The upflow clarifier, specifically the sludge-blanket clarifier, is often used in water softening operations. A schematic of this type of unit is shown in Fig. 10.16. This system combines mixing and sludge recirculation. The recirculated settled sludge provides additional particles that increase the probability of particle contact (and nuclei for crystal growth) and form a dense sludge blanket. The sludge blanket concentrates, traps, and settles out suspended particles and floc before they are discharged over the effluent weir.

Water softening produces a very unstable water of high causticity. Such water has an objectionable taste, causes filter-sand encrustation, and causes scale formation on pipes and valves. To eliminate these problems, the pH of the water is usually reduced by adding carbon dioxide. In the past $CO_2$ was produced on site by the combustion of oil or gas in either underwater or external burners. But liquid $CO_2$ is now readily available in bulk quantities. Transfer efficiencies for liquid $CO_2$ are near 100 percent, and its use significantly reduces the operation and maintenance problems normally encountered with the recarbonation process.

**Chemical feeders and mixers.**   The amounts of chemicals added to water must be carefully controlled to ensure uniform treatment. Certain chem-

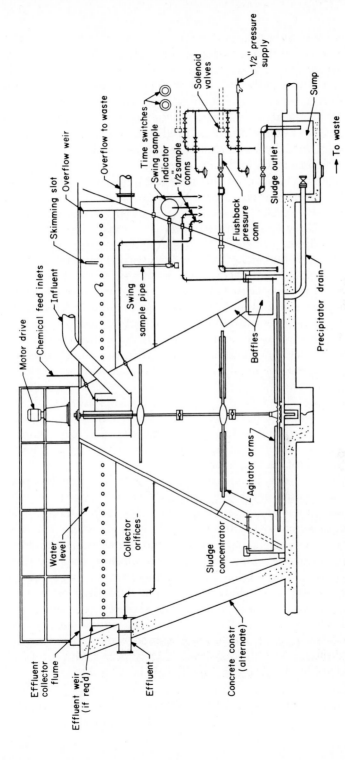

**Figure 10.16** Vertical sludge blanket clarifier. (*Source: Reproduced, with permission, from R. L. Sanks, ed., Water Treatment Plant Design, Butterworth Publishers, Stoneham, Mass., 1978.*)

icals, such as lime, soda ash, and most nonionic and anionic polymers, are available only in dry form. Because most chemicals must be in solution form prior to mixing with the water to be treated, the use of these chemicals normally requires two distinct operations: the preliminary preparation of the chemical and the feeding of the prepared chemical.

Preliminary preparation of dry chemical requires that a specific volume or weight of chemical be measured and dissolved in water. Although both dry feeders and saturators can be used for this purpose, because of their accuracy and ease of operation dry feeders are normally selected. Two types of dry feeders are available: the volumetric feeder, which meters the chemical by volume per unit time, and the gravimetric feeder, which meters the chemical by weight per unit time. Volumetric dry feeders are simpler, less expensive, and less accurate than gravimetric feeders. The types of dry feeding mechanisms, which may be controlled either volumetrically or gravimetrically, are rotating disk (suitable for feed rates less than 10 lb/h), oscillating (suitable for feed rates between 10 and 100 lb/h), rotary gate (suitable for feed rates between 200 and 500 lb/h), belt (suitable for feed rates between 500 and 20,000 lb/h), and screw (these feeders are generally volumetric and are suitable for feed rates ranging from 10 to 24,000 lb/h). Unless the savings in chemicals resulting in the greater accuracy of the gravimetric control are warranted, the smaller feeders are generally volumetric.

The dry chemical discharged from the feeder falls into a tank where it is dissolved in water to form the feed solution. This solution is fed to the source water mixing point. Where gravity delivery cannot be achieved, piston or diaphragm applicators are generally employed to deliver the feed solution under pressure.

Bulk handling of dry chemicals usually will be more economical than bag handling and batch preparation. Chemical storage facilities should be planned with sufficient capacity for at least a month of storage. Soda ash is noncorrosive and relatively safe to handle and may be stored in containers such as barrels or drums. Quicklime and hydrated lime are also noncorrosive, but care must be taken to keep the storage units airtight and watertight. In most medium to large softening plants, it is generally more economical to use quicklime. Before it can be applied, however, quicklime must be slaked. In a typical lime slaker, quicklime is combined with water in a paddle-agitated compartment to form a paste near boiling temperature. In some plants, the paste is pumped with a progressive cavity pump without dilution to avoid deposits in the line that result if it is diluted.

**Residues from lime–soda ash softening.**  The residues from water softening plants are predominantly calcium carbonate or a mixture of calcium carbonate and magnesium hydroxide. Calcium carbonate sludges are generally dense, stable, and inert materials that dry well. The sludge solids content is typically near 5 percent, although a range between 2 and 30 percent has been observed. The sludge pH is normally greater than 10.5.

Theoretically, the type of hardness removed and the chemicals used in the softening process establish the amount of sludge produced (see Table 10.5). The amount of sludge resulting from only carbonate hardness removal can be estimated from Eq. (10.55)[31]:

$$\Delta S = 86.4Q(2.0\text{Ca} + 2.6\text{Mg}) \qquad (10.55)$$

where $\Delta S$ = dry weight of sludge solids formed, kgd
$\quad Q$ = source water flow, $m^3/s$
$\quad$ Ca = calcium carbonate hardness removed, mg/L as $CaCO_3$
$\quad$ Mg = magnesium carbonate hardness removed, mg/L as $CaCO_3$

In those situations where coagulants are added to increase the efficiency of the softening process, Eq. (10.55) must be modified to account for the additional solids generated[31]:

$$\Delta S = 86.4Q(2.0\text{Ca} + 2.6\text{Mg} + 0.44\text{Al} + 1.9\text{Fe} + \text{SS} + \text{A}) \qquad (10.56)$$

where Al = alum dose as 17.1% $Al_2O_3$, mg/L
$\quad$ Fe = iron dose as Fe, mg/L
$\quad$ SS = suspended solids concentration in source water, mg/L
$\quad$ A = additional chemicals such as polymer, clay, or activated carbon, mg/L

The composition of the sludge significantly affects its dewatering characteristics. Relatively pure calcium carbonate sludge can be easily dewatered to a solids content up to 50 or 60 percent. Increasing the magnesium hydroxide content of the sludge, however, causes it to be more difficult to handle and dewater. Normally a sludge with a Ca:Mg ratio less than 2 will be difficult to dewater while a sludge with a Ca:Mg ratio greater than 5 will be fairly easy to dewater (see Fig. 10.17).[32] The presence of lime in the sludge will also have an adverse

**TABLE 10.5  Theoretical Solids Production,[31] mg Dry Solids/mg Hardness Removed as $CaCO_3$**

| Treatment chemical | Carbonate hardness | | Noncarbonate hardness | |
|---|---|---|---|---|
| | Calcium | Magnesium | Calcium | Magnesium |
| Lime and soda ash | 2.0 | 2.6 | 1.0 | 1.6 |
| Sodium hydroxide | 1.0 | 0.6 | 1.0 | 0.6 |

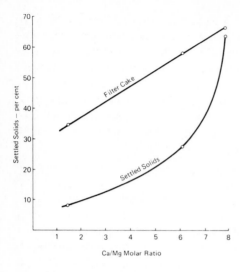

**Figure 10.17** Effect of Ca:Mg ratio on sludge solids concentration for lime sludges. (*Source: R. J. Calkins and J. T. Novak, "Characterization of Chemical Sludges," J. AWWA, vol. 65, no. 6, June 1973, p. 423.*)

effect on dewatering. Normally this is a result of poor slaking or incomplete dissolution.

Various methods are employed today to concentrate and dewater softening sludges: gravity thickening, dewatering lagoons, sand drying beds, centrifugation, vacuum filtration, pressure filtration, belt filtration, sludge pelletization, recalcination, and land application. These methods may be used separately or in combination to accomplish a specific management objective (see Chap. 16).

When the sludge is predominantly calcium carbonate, solids concentrations greater than 30 percent have been achieved in gravity thickeners at loading rates of 40 lb/(ft$^2$)(day).[31] Solids concentrations greater than 50 percent have been reported for dewatering lagoons and sand drying beds. Even higher solids concentrations can be achieved with centrifugation, vacuum filtration, and pressure filtration.

Sludge pelletization occurs during the suspended-bed (pellet reactor) softening process. The precipitated hardness is withdrawn as sandlike granules, which are much easier to dispose of than is sludge from a conventional softening process. When it leaves the reactor, the pelletized sludge is near 60 percent solids by weight. The entrained water can be readily drained to produce a residue containing greater than 90 percent solids. In the Netherlands, several methods of using the pellets have been found, including treatment of aggressive groundwater, neutralization of acid wastewater, and utilization for road construction, cement manufacturing, and metal industries.[25]

Calcium carbonate sludge from lime water softening operations can be converted to calcium oxide (quicklime) by recalcination (burning the calcium carbonate sludge in a furnace at about 1850°F). Magnesium hydroxide, clay, and other inorganic materials are not completely oxidized and remain in the quicklime as impurities and will ultimately lead to reduced lime recovery. Not only will recalcination produce lime that can be used in the softening process, but also it will dramatically reduce the volume of sludge requiring disposal and produce carbon dioxide that can be used in the recarbonation process. The basic reaction is

$$CaCO_3 \xrightarrow{\text{heat}} CaO + CO_2 \qquad (10.57)$$

The high energy demand is the greatest disadvantage of this process.

Traditionally, water treatment plant wastes have been disposed of via discharge to rivers and lakes, either directly or by way of a storm sewer. Federal law no longer allows this. Alternative methods of ultimate lime sludge disposal include discharge to sanitary sewers, landfills, and drying lagoons and land spreading in agricultural areas. Water treatment plant waste management is discussed in Chap. 16.

## Iron and Manganese Removal by Chemical Precipitation

Iron and manganese may be removed from water by various means. The most popular approach utilized in the United States involves oxidation of the more soluble iron(II) and manganese(II) sometimes encountered in natural water to relatively insoluble iron(III) and manganese(IV) and subsequent removal of the precipitates thus formed by filtration. Molecular oxygen, free available chlorine, and potassium permanganate have all been used successfully as oxidizing agents. Because unnecessary application of chlorine in drinking water treatment is now generally discouraged, the use of chlorine as an oxidant should be viewed with caution. Both iron and manganese are effectively removed via precipitation softening, but the process is generally much too costly to consider for this purpose alone. In the event that split-treatment precipitation softening is practiced and additional iron and/or manganese removal is desired, the first-stage bypass stream may be treated via oxidation.

### Fundamental process chemistry

Oxidation of iron(II) and manganese(II) by molecular oxygen, potassium permanganate, and chlorine may be described by the following chemical reactions:

$$4Fe^{2+} + O_2(g) + 10H_2O \rightleftharpoons 4Fe(OH)_3(s) + 8H^+ \tag{10.58}$$

$$3Fe^{2+} + KMnO_4 + 7H_2O \rightleftharpoons 3Fe(OH)_3(s) + MnO_2(s) + K^+ + 5H^+ \tag{10.59}$$

$$2Fe^{2+} + Cl_2 + 6H_2O \rightleftharpoons 2Fe(OH)_3(s) + 2Cl^- + 6H^+ \tag{10.60}$$

$$2Mn^{2+} + O_2(g) + 2H_2O \rightleftharpoons 2MnO_2(s) + 4H^+ \tag{10.61}$$

$$3Mn^{2+} + 2KMnO_4 + 2H_2O \rightleftharpoons 5MnO_2(s) + 2K^+ + 4H^+ \tag{10.62}$$

$$Mn^{2+} + Cl_2 + 2H_2O \rightleftharpoons MnO_2(s) + 2Cl^- + 4H^+ \tag{10.63}$$

where $(g)$ denotes the gas phase.

In each case, hydrogen ions are produced. Thus, based on these reactions alone, pH may decrease significantly in the absence of sufficient buffer capacity. The stoichiometry of each of these reactions is summarized in Table 10.6.

**TABLE 10.6    Stoichiometry of Iron and Manganese Oxidation**

| Reaction | Stoichiometric amount per mg/L iron or manganese removed, mg/L |
|---|---|
| For iron(II) oxidized with oxygen | |
| Oxygen required, as $O_2$ | 0.14 |
| Hydrogen ions produced, as $H^+$ | 0.04 |
| Alkalinity destroyed, as $CaCO_3$ | 1.80 |
| For iron(II) oxidized with permanganate | |
| Permanganate required, as $KMnO_4$ | 0.94 |
| Hydrogen ions produced, as $H^+$ | 0.03 |
| Alkalinity destroyed, as $CaCO_3$ | 1.49 |
| For iron(II) oxidized with chlorine | |
| Chlorine required, as $Cl_2$ | 0.64 |
| Hydrogen ions produced, as $H^+$ | 0.05 |
| Alkalinity destroyed, as $CaCO_3$ | 2.70 |
| For manganese(II) oxidized with oxygen | |
| Oxygen required, as $O_2$ | 0.29 |
| Hydrogen ions produced, as $H^+$ | 0.04 |
| Alkalinity destroyed, as $CaCO_3$ | 1.80 |
| For manganese(II) oxidized with permanganate | |
| Permanganate required, as $KMnO_4$ | 1.92 |
| Hydrogen ions produced, as $H^+$ | 0.02 |
| Alkalinity destroyed, as $CaCO_3$ | 1.21 |
| For manganese(II) oxidized with chlorine | |
| Chlorine required, as $Cl_2$ | 1.29 |
| Hydrogen ions produced, as $H^+$ | 0.07 |
| Alkalinity destroyed, as $CaCO_3$ | 3.64 |

In actual practice, the amount of permanganate required may be less than the theoretical stoichiometric amount. This may result from the catalytic influence of manganese dioxide on the oxidation of both iron(II) and manganese(II). The mechanism shown below has been postulated[30]:

$$2Fe^{2+} + 2MnO_2(s) + 5H_2O \rightleftharpoons 2Fe(OH)_3(s) + Mn_2O_3(s) + 4H^+$$

$$(10.64)$$

$$Mn^{2+} + MnO_2(s) + H_2O \rightleftharpoons Mn_2O_3(s) + 2H^+ \qquad (10.65)$$

Equations (10.58), (10.59), and (10.60) indicate that the actual removal of iron(II) depends on the precipitation of iron(III) (ferric) hydroxide [$Fe(OH)_3(s)$]. That is,

$$Fe^{3+} + 3OH^- \rightleftharpoons Fe(OH)_3(s) \qquad (10.66)$$

The solubility product constant $K_{sp}$ for ferric hydroxide is approximately $4.0 \times 10^{-38}$ at 25°C.[2]

On the basis of Eq. (10.66) alone, the solubility of ferric hydroxide might be concluded to consistently decrease as pH increases. This, however, is not the case. Actually the solubility of ferric hydroxide decreases with increasing pH only up to about pH 10. At this pH the hydroxide begins to dissociate to form the soluble anion $Fe(OH)_4^-$. Nevertheless, in the pH range of interest in drinking water treatment, the solubility of ferric hydroxide is quite low ($10^{-9}$ to $10^{-8}$ mol/L).

The chemistry of manganese is less straightforward than that of iron, but for practical engineering purposes manganese(II) removal may be thought of as depending on the precipitation of manganese(IV) (manganic) dioxide [$MnO_2(s)$], as shown in Eqs. (10.61), (10.62), and (10.63). Manganic dioxide is essentially insoluble over the entire pH range of interest in drinking water treatment.

Readers interested in a more detailed discussion of the process chemistry involved in the oxidation of iron and manganese are directed to Stumm and Morgan[4] and Benefield et al.[6]

## Chemical dose calculations

The stoichiometric doses of molecular oxygen and potassium permanganate required to oxidize iron(II) and manganese(II) may be calculated directly from the values in Table 10.6. Example Problem 11 illustrates the procedure.

**Example Problem 11**   What potassium permanganate dose, in milligrams per liter as $KMnO_4$, is needed to oxidize 1.3 mg/L iron(II)?

**solution**   From Table 10.6, the required dosage is

$$\text{Dose} = 1.3 \times 0.94 = 1.2 \text{ mg/L as KMnO}_4$$

For typical iron(II) and manganese(II) concentrations likely to be encountered, the calculated required stoichiometric molecular oxygen doses will be quite low. In these situations, providing conditions that are conducive to oxidation virtually ensures the presence of sufficient oxygen.

When chlorine is used as the oxidant, a dosage somewhat in excess of the theoretical amount shown in Table 10.6 may be required. One reason is that some of the chlorine may be consumed in side reactions not directly related to oxidation of iron and manganese. Furthermore, maintenance of a free-chlorine residual of about 0.4 mg/L has been suggested as necessary to ensure the effectiveness of the process.

### Process kinetics

As indicated above, the molecular oxygen, potassium permanganate, and chlorine doses required to oxidize given concentrations of iron(II) and manganese(II) may be readily estimated from stoichiometric considerations. Sizing the reactor required for a specific situation, however, requires consideration of the rate at which the oxidation process occurs. In this section, the principal factors that influence the rates at which iron(II) and manganese(II) are oxidized with molecular oxygen and potassium permanganate are considered. Much of the information available on this general topic may be attributed to the work of Stumm, Morgan, and various coworkers. For a more detailed approach and extensive bibliography, the reader is directed to their summary work[4] and to Benefield et al.[6]

**Oxidation of iron(II) with molecular oxygen.** At a pH equal to or greater than about 5, the following rate law is applicable to the oxidation of iron(II) with molecular oxygen:

$$\frac{d[\text{Fe}^{2+}]}{dt} = -k(\text{PO}_2)[\text{OH}^-]^2[\text{Fe}^{2+}] \tag{10.67}$$

where $d[\text{Fe}^{2+}]/dt$ = rate of iron(II) oxidation, mol/(L)(min)
$\quad\quad\quad k$ = reaction rate constant = $8.0(\pm 2.5) \times 10^{13}$
$\quad\quad\quad\quad$ $\text{L}^2/\text{(min)(atm)(mol)}^2$ at 20.5°C
$\quad\quad \text{PO}_2$ = partial pressure of oxygen, atm
$\quad\quad [\text{OH}^-]$ = hydroxide ion concentration, mol/L
$\quad\quad [\text{Fe}^{2+}]$ = iron(II) concentration, mol/L

That is, the oxidation may be represented as first-order with respect to iron(II) and molecular oxygen and second-order with respect to the hydroxide ion. From a practical viewpoint, however, the oxidation of iron(II) may be considered to follow pseudo first-order kinetics, as shown:

$$\frac{d[Fe^{2+}]}{dt} = -K[Fe^{2+}] \qquad (10.68)$$

where $K$ = the apparent first-order reaction rate constant = $(1.68 \times 10^{-15})/[H^+]^2$ min$^{-1}$us (assuming that the partial pressure of oxygen is 0.21 atm and that the ionization constant for water is $10^{-14}$). Equation (10.68) can be used to size reactors in most practical applications.

Equations (10.67) and (10.68) reveal that the rate of oxidation of iron(II) with molecular oxygen proceeds at an increasing rate as pH rises, other factors being held constant. In the context of normal drinking water treatment applications, the reaction may be considered to be quite slow at pH 6 and very rapid at pH 7.5 and above.

The presence of organic matter often retards the oxidation rate. This apparently results from the formation of various soluble iron(II) complexes. But the presence of catalysts such as copper and cobalt and of certain anions and organic substances that complex with iron(III) speeds the oxidation process. Other factors held constant, the oxidation rate increases by about an order of magnitude for each 15°C rise in temperature.

**Oxidation of manganese(II) with molecular oxygen.** The oxidation of manganese(II) with molecular oxygen is an autocatalytic process. One possible mechanism is shown below:

$$Mn^{2+} + O_2(g) \rightleftharpoons MnO_2(s) \qquad (10.69)$$

$$Mn^{2+} + MnO_2(s) \rightleftharpoons Mn^{2+} \cdot MnO_2(s) \qquad (10.70)$$

$$Mn^{2+} \cdot MnO_2(s) + O_2(g) \rightleftharpoons 2MnO_2(s) \qquad (10.71)$$

In relative terms, Eqs. (10.69) and (10.71) may be considered quite slow, and Eq. (10.70) may be considered to be quite rapid. This mechanism, if valid, offers a reasonable explanation of why, as noted above, a less than stoichiometric oxidant dose may be required in the "oxidation" of manganese(II) to manganese(IV). That is, removal of manganese(II) is enhanced by the presence of solid manganic dioxide and may occur within a reasonable time without complete oxidation. Equation (10.71) indicates, however, that the oxidation may eventually be completed if environmental conditions permit. This mechanism, along with Eqs. (10.64) and (10.65), also explains why the presence of manganic dioxide generally increases the apparent rate of oxidation of both iron(II) and manganese(II).

A kinetic model similar to that presented above for the oxidation of iron(II) can be developed for the oxidation of manganese(II).[4,6] Oxidation of manganese(II) with molecular oxygen is not generally practiced, however, because the pH must be raised to about 8.5 and the water must be exposed to media coated with manganic dioxide to make the process feasible.

**Oxidation of iron(II) and manganese(II) with potassium permanganate and chlorine.** Very little has been reported in the literature concerning the kinetics of oxidation of iron(II) and manganese(II) with either potassium permanganate or chlorine, because the oxidation of both metals is quite rapid at pH 7 and higher. Thus, kinetic considerations have little influence on either facility design or operation. Virtually the same situation exists for the oxidation of iron(II) with chlorine. Oxidation of manganese(II) with chlorine is similarly rapid at pH 8.5 or higher.

**Process design considerations**

The design of an iron(II) and/or manganese(II) removal process in which precipitation of iron(III) and/or manganese(IV) is involved depends to a very great degree on site-specific considerations. In the relatively rare case when the water to be treated is from a surface source, iron removal—and to a lesser extent manganese removal—can be readily integrated into the typical coagulation-flocculation-sedimentation-filtration flow scheme. Either molecular oxygen introduced via aeration or potassium permanganate can be used as the oxidant if only iron removal is desired. If manganese removal is necessary, potassium permanganate should be used.

Generally, the best approach is to apply the oxidant prior to the addition of other treatment chemicals (except when pH control is required). When small amounts of iron are being oxidized by molecular oxygen, adding more oxygen is usually unnecessary; simply providing sufficient reaction time at a suitable pH will do. In any case, the actual removal of iron and/or manganese is accomplished in the filtration step. This approach is generally more effective for iron removal than for manganese removal. Filtration is discussed in detail in Chap. 8.

When groundwater is being treated, the need for removal of other contaminants (or lack thereof) will generally dictate the best approach for iron(II) and manganese(II) removal. Often the aeration process needed to remove excess carbon dioxide and/or other gases can also provide molecular oxygen for the oxidation of iron(II) and manganese(II). Manganese(II) removal by this method will be feasible, however, only when the pH is raised to 8.5 or higher and provision for contact with manganic dioxide deposits is made. Coke tray and similar aerators may be used for this purpose. (Aeration is discussed in detail in Chap. 5.) Once again, the actual removal of iron and/or manganese is via filtration. Pressure filters are usually used.

Potassium permanganate oxidation of iron(II) and manganese(II) in groundwater is a fairly simple process. If the pH is within the range of 6 to 9, all that is required is to feed the permanganate (dry or liquid feeders may be used) and remove the precipitated iron and/or manganese via filtration. Manganese-greensand filters are especially useful for this purpose because short-term dosages of excess permanganate will be adsorbed onto the filter. Oxidation of iron(II) and manganese(II) with chlorine is similar to oxidation with permanganate.

Regardless of the oxidant used, a typical treatment system consists of chemical addition, reaction, filtration, and backwashing facilities. The following design criteria have been suggested as conservative.[19]

| | |
|---|---|
| Reaction time: | 5 to 30 min at average flow |
| Filter media: | Anthracite |
| | Effective size $\geq$ 1.5 mm |
| | Uniformity coefficient 1.2 to 1.4 |
| Filtration rate: | 5 to 10 gpm/ft$^2$ |
| Backwash rate: | 15 to 25 gpm/ft$^2$ |

## Removal of Other Contaminants by Precipitation

Although this chapter has focused on calcium, magnesium, and iron and manganese removal, the heavy metals, radionuclides, organics, and viruses may also be removed from water via some type of chemical precipitation process. Probably the most popular method of removing toxic heavy metals from water is precipitation of the metal hydroxide. This process normally involves the addition of caustic soda or lime to adjust the solution pH to the point of maximum insolubility. Figure 10.2 gives a graphical representation of the experimentally determined solubilities of several metal hydroxides of interest in water treatment. This figure also shows the amphoteric nature of certain metal hydroxides, i.e., those metal hydroxides that act as both acids and bases and will redissolve in excessively acid or alkaline solutions. Because heavy metal contamination of drinking water supplies has not been a frequent problem, few studies have been conducted that consider the removal of a specific heavy metal from drinking water by precipitation. Sorg et al.[34] have, however, discussed the application of various treatment technologies for removal of inorganic materials. Ta-

**TABLE 10.7    Effectiveness of Chemical Coagulation and Lime Softening Processes for Inorganic Contaminant Removal[34]**

| Contaminant | Method | Removal, % |
|---|---|---|
| Arsenic | | |
| $As^{3+}$ | Oxidation to $As^{5+}$ required | >90 |
| $As^{5+}$ | Ferric sulfate coagulation, pH 6–8 | >90 |
| | Alum coagulation, pH 6–7 | >90 |
| | Lime softening, pH 11 | >90 |
| Barium | Lime softening, pH 10–11 | >80 |
| Cadmium† | Ferric sulfate coagulation, pH > 8 | >90 |
| | Lime softening, pH > 8.5 | >95 |
| Chromium† | | |
| $Cr^{3+}$ | Ferric sulfate coagulation, pH 6–9 | >95 |
| | Alum coagulation, pH 7–9 | >90 |
| | Lime softening, pH > 10.5 | >95 |
| $Cr^{6+}$ | Ferrous sulfate coagulation, pH 6.5–9 (pH may have to be adjusted after coagulation to allow reduction to $Cr^{3+}$) | >95 |
| Lead† | Ferric sulfate coagulation, pH 6–9 | >95 |
| | Alum coagulation, pH 6–9 | >95 |
| | Lime softening, pH 7–8.5 | >95 |
| Mercury,† inorganic | Ferric sulfate coagulation, pH 7–8 | >60 |
| Selenium† $Se^{4+}$ | Ferric sulfate coagulation, pH 6–7 | 70–80 |
| Silver† | Ferric sulfate coagulation, pH 7–9 | 70–80 |
| | Alum coagulation, pH 6–8 | 70–80 |
| | Lime softening, pH 7–9 | 70–90 |

†No full-scale experience.

ble 10.7 summarizes the effectiveness of chemical coagulation and lime softening processes for the removal of inorganic contaminants from drinking water.

Depending on the contaminant and its concentration, either precipitation or coprecipitation or both will play a major role in removal during chemical coagulation or lime softening. In most cases, however, coprecipitation results in the removal of soluble metal ions during coagulation and lime treatment. There are four types of coprecipitation:

1. *Inclusion*    This involves the mechanical entrapment of a portion of the solution surrounding the growing particle. This type of coprecipitation is normally significant only for large crystals.

2. *Adsorption*    This type involves the attachment of an impurity onto the surface of a particle or precipitate. This type of coprecipitation is generally not important if the particle size is large when the precipitation is complete, because large particles have surface areas that are very small in proportion to the amount of precipitate they contain. Adsorption may, however, be a major means of contaminant removal if the particles are small.

3. *Occlusion*   In this form, a contaminant is entrapped in the interior of a particle of precipitate. This type of coprecipitation occurs by adsorption of the contaminant onto the surface of a growing particle, followed by further growth of the particle to enclose the adsorbed contaminant.

4. *Solid-solution formation*   In this type of occlusion, a particle of precipitate becomes contaminated with a different type of particle that precipitates under similar conditions and is formed from ions whose sizes are nearly equal to those of the original precipitate.

During coagulation, some metals will coprecipitate with either iron or aluminum hydroxide. Iron coagulants seem to perform better than aluminum coagulants, primarily because iron hydroxide is insoluble over a wider pH range and is less soluble than aluminum hydroxide. Iron coagulants form a stronger and heavier floc, and coprecipitation of iron-metal complexes appears to be a significant factor. Mukai et al.[35] found that the apparent solubility of $Cd^{2+}$, $Cu^{2+}$, and $Zn^{2+}$ could be dramatically reduced through the addition of $Fe^{3+}$. Removal efficiencies for various heavy metals for a lime-treated secondary effluent have been presented by Culp et al.[28] In all cases studied, the residual metal concentration was found to be less than 0.1 mg/L.

Because it occurs naturally and is sometimes found in drinking water, radium is a radionuclide of interest in water treatment. Radium can be effectively removed from water by lime softening. Removal efficiency is a function of pH, however, and if a high degree of treatment is required, the pH during softening should be elevated above 10.8 (see Fig. 10.18).[36]

Most natural water contains a certain amount of organic matter known as *humic substances* which, when present in high concentrations, impart a yellowish brown color to water. Rook[37] identified this material as precursors for trihalomethanes in chlorinated water. Many water supplies are treated with alum coagulation or lime softening, and although these processes are not intended to remove organic contaminants, they are generally effective in removing a significant amount of organic material. Hall and Packham[38] found that

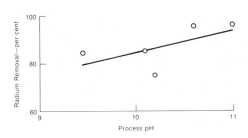

**Figure 10.18** Radium removal versus softening pH. (*Source: W. L. Brinck et al. "Radium-Removal Efficiencies in Water-Treatment Processes," J. AWWA, vol. 70, no. 1, January 1978, p. 31.*)

organic color and clay turbidity were removed by entirely different mechanisms in the coagulation process. These workers suggest that the removal of organic color with alum is a chemical process in which a partially hydrolyzed aluminum ion of empirical formula $Al(OH)_{2.5}$ interacts with ionic groups on the humic acid colloid. Such a response results in the precipitation of an insoluble humate or fulvate. Edzwald[39] found good removal of humic substances by using alum with high-molecular-weight polymers in the pH range of 4.5 to 6.5. Reductions in humic acid of 90 percent or greater were obtained at a pH of 6 with the following dosages: 10 mg/L alum, 0.5 mg/L cationic polymer, and 5 mg/L humic acid; 10 mg/L alum, 1 mg/L anionic polymer, and 5 mg/L humic acid; and 10 mg/L alum, 1 mg/L nonionic polymer, and 5 mg/L humic acid.

Soluble organic contaminants may also be removed from water by lime softening. Randtke et al.[40] discuss the removal of soluble organic contaminants from wastewater by lime precipitation. Their data indicate a chemical oxygen demand removal range between 24 and 70 percent. Johnson and Randtke[41] have presented data (see Table 10.8) that illustrates the importance of the point of chlorination on the removal of nonpurgeable organic chlorine and total organic carbon (TOC) by lime precipitation. These data suggest that prechlorination with free chlorine can have a detrimental effect on the removal of TOC by lime precipitation. Liao and Randtke[42] found that lime softening could remove a significant fraction of the fulvic acid extracted

TABLE 10.8    Removal of Nonpurgeable Organic Chlorine and Total Organic Carbon from Three Water Sources by Lime Softening[41]

| Sample | Treatment[†] | Removal of NPOCl | | Removal of TOC | |
|---|---|---|---|---|---|
| | | µg/L | % | mg/L | % |
| River water | Chlorination only | 131.5 | | 1.55 | |
| | Prechlorination, softening | 95.0 | 28 | 1.52 | 2 |
| | Softening, postchlorination | 84.8 | 36 | 1.34 | 14 |
| | Prechlorination, hydrolysis | 110.8 | 13 | | |
| Groundwater | Chlorination only | 368.5 | | 3.10 | |
| | Prechlorination softening | 253.4 | 31 | 2.48 | 20 |
| | Softening, postchlorination | 231.6 | 37 | 2.11 | 32 |
| | Prechlorination, hydrolysis | 301.4 | 18 | | |
| Secondary effluent | Chlorination only | 171.8 | | 6.41 | |
| | Prechlorination, softening | 128.1 | 25 | 5.34 | 17 |
| | Softening, postchlorination | 135.7 | 21 | 5.00 | 22 |
| | Prechlorination, hydrolysis | 134.5 | 22 | | |

†Prechlorination was at point 1, 48 h before softening. Hydrolysis was effected by softening the samples at pH 11.0 and then acidifying them to pH 2.0 after sedimentation (without solids separation), thereby dissolving the precipitated solids back into solution.

from groundwater. Conditions favoring a high removal efficiency were a high pH, a high calcium concentration, and a low carbonate concentration. The results of this study are summarized in Table 10.9. Weber and Godellah[43] have presented Figs. 10.19 and 10.20 showing the effect of alum coagulation and lime softening on TOC removal from a humic acid solution, a fulvic acid solution, and Huron River water. These data indicate that at high alum dosages more than 80 percent of the humic and fulvic acids are removed, while TOC removal from the Huron River sample exceeded 50 percent only slightly. A similar trend was noted for lime softening. Sinsabaugh et al.[44] investigated the effects of charge, solubility, and molecular size on the removal of dissolved organic carbon by ferric sulfate coagulation and settling. They found that molecular size, independent of any charge or solubility correlation, was the most significant factor. In every charge and solubility category studied, removal efficiency declined monotonically with molecular weight.

Wolf et al.[45] conducted a large-scale pilot study of virus removal by both lime and alum. For an Al:P ratio of 7:1, they observed bacterial virus removals as high as 99.845 percent for coagulation-sedimentation and 99.985 percent for coagulation-sedimentation-filtration processes. At lower alum dosages a marked decrease in virus removal occurred. They also found that treating with lime to elevate the solution pH above 11.0 resulted in excellent virus removal, but the

**TABLE 10.9    Effects of Operational Changes on the Removal of Groundwater Fulvic Acid and Water Hardness[42]**

| Process variable | TOC removed, % | Residual hardness, mg/L as $CaCO_3$ |
|---|---|---|
| pH | | |
| 9 | 28 | 25 |
| 11† | 35 | 16 |
| 12 | 44 | 12 |
| Chemical addition | | |
| Calcium-rich | 41 | 9 |
| Carbonate-rich | 29 | 7 |
| Sludge recycling | | |
| 1:1 (old:fresh) | 31 | 9 |
| 2:1 (old:fresh) | 23 | 8 |
| Additives | | |
| Ca:Mg = 6:2 | 72 | 14 |
| Ca:Mg = 6:1 | 60 | 13 |
| Ca:P = 6:0.2 | 43 | 52 |
| Ca:P = 6:0.1 | 35 | 50 |
| Two-stage chemical addition | 53 | 7 |

†Standard condition: $[CO_3^{2-}] = [CO_3^{2-}] = 6$ mM, single stage, pH = 11, no sludge recycling, no additives present, and TOC = 3 mg/L.

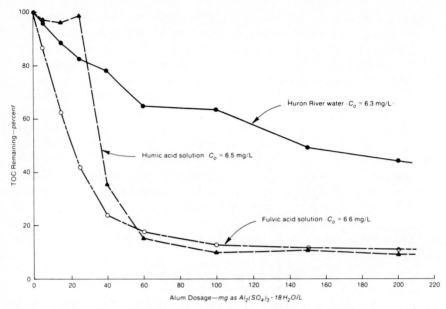

**Figure 10.19**   Removal of TOC by alum coagulation. (*Source: W. J. Weber, Jr., and A. M. Jodellah, "Removing Humic Substances by Chemical Treatment and Adsorption," J. AWWA, vol. 77, no. 4, April 1985, p. 132.*)

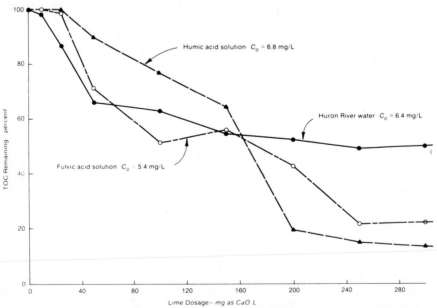

**Figure 10.20**   Removal of TOC by lime softening. (*Source: W. J. Weber, Jr., and A. M. Jodellah, "Removing Humic Substances by Chemical Treatment and Adsorption," J. AWWA, vol. 77, no. 4, April 1985, p. 132.*)

actual percentages were not quantified. Rao et al.[46] studied the influences of water softening on virus removal. They found that during calcium hardness removal at pH 9.6 rotavirus was not removed as effectively as poliovirus and hepatitis A virus. Greater than 90 percent of rotavirus, however, was removed during $Mg^{2+}$ hardness removal at pH 10.8 at 37°C. During total hardness removal at pH 11, all viruses were efficiently removed.

## References

1. R. B. Fischer and D. G. Peters, *Chemical Equilibrium,* W. B. Saunders, Philadelphia, 1970.
2. James W. Patterson and Roger A. Minear, "Physical-Chemical Methods of Heavy Metals Removal," in P. A. Krenkel (ed.,) *Heavy Metals in the Aquatic Environment,* Pergamon Press, New York, 1975.
3. R. Nilsson, "Removal of Metals by Chemical Treatment of Municipal Wastewater," *Water Res.,* vol. 5, 1971, p. 51.
4. W. Stumm and J. J. Morgan, *Aquatic Chemistry,* Wiley-Interscience, New York, 1981.
5. V. L. Snoeyink and D. Jenkins, *Water Chemistry,* Wiley, New York, 1980.
6. L. D. Benefield, J. F. Judkins, and B. L. Weand, *Process Chemistry for Water and Wastewater Treatment,* Prentice-Hall, Englewood-Cliffs, N. J., 1982.
7. J. R. Rossum and D. T. Merrill, "An Evaluation of the Calcium Carbonate Saturation Indexes," *J. AWWA,* vol. 75, no. 2, February 1983, p. 95.
8. S. D. Faust and J. G. McWhorter, "Water Chemistry," in H. W. Gehm and J. I. Bregman (eds.), *Handbook of Water Resources and Pollution Control,* Van Nostrand Reinhold, New York, 1976.
9. R. E. Lowenthal and G. V. R. Marais, *Carbonate Chemistry of Aquatic Systems: Theory and Application,* Ann Arbor Science Publishers, Ann Arbor, Mich., 1976.
10. T. E. Larson and A. M. Buswell, "Calcium Carbonate Saturation Index and Alkalinity Interpretations," *J. AWWA,* vol. 34, no. 11, November 1942, p. 1667.
11. L. N. Plummer and E. Busenberg, "The solubilities of calcite, aragonite, and vaterite in $CO_2$-$H_2O$ solutions between 0 and 90 degrees C, and an evaluation of the aqueous model for the system $CaCO_3$-$CO_2$-$H_2O$," *Geochim. Cosmochim. Acta,* vol. 46, 1982, p. 1011.
12. *Standard Methods for the Examination of Water and Wastewater,* 17th ed., American Public Health Association, Washington, 1989.
13. A. H. Truesdell and B. F. Jones, *WATES, A Computer Program for Calculating Chemical Equilibria of Natural Waters,* NTIS, U.S. Department of Commerce, Springfield, Va., PB 220464, 1973.
14. F. Cadena, W. S. Midkiff, and G. A. O'Conner, "The Calcium Carbonate Ion-Pair as a Limit to Hardness Removal," *J. AWWA,* vol. 66, no. 9, September 1974, p. 524.
15. D. T. Merrill, "Chemical Conditioning for Water Softening and Corrosion Control," *Proc. Fifth Environmental Engineers' Conference,* Montana State University, June 16–18, 1976.
16. R. R. Trussell, L. L. Russell, and J. F. Thomas, "The Langelier Index," in *Water Quality in the Distribution System,* Fifth Annual AWWA Water Quality Technology Conference, Kansas City, Mo., 1977.
17. R. A. Pisigan, and J. E. Singley, "Calculating the pH of Calcium Carbonate Saturation," *J. AWWA,* vol. 77, no. 10, October 1985, p. 83.
18. C. N. Sawyer and P. L. McCarty, *Chemistry for Sanitary Engineers,* 2d ed., McGraw-Hill, New York, 1967.
19. Michael J. Humenick, *Water and Wastewater Treatment,* Marcel Dekker, New York, 1977.
20. *Corrosion Control by Deposition of $CaCO_3$ Films,* AWWA, Denver, 1978.
21. D. T. Merrill, "Chemical Conditioning for Water Softening and Corrosion Control," in Robert L. Sanks (ed.), *Water Treatment Plant Design,* Ann Arbor Science, Ann Arbor, Mich., 1978.

22. James M. Montgomery, Consulting Engineers, *Water Treatment Principles and Design,* Wiley, New York, 1985.
23. T. E. Larson, R. W. Lane, and C. H. Neff, "Stabilization of Magnesium Hydroxide in the Solids-Contact Process," *J. AWWA,* vol. 51, no. 12, December 1959, p. 1551.
24. *Maintenance and Operation of Water Supply, Treatment, and Distribution Systems,* Department of the Air Force Regulation AFR 91-26, 1984.
25. A. Graveland, J. C. Van Dyk, P. J. deMoel, and J. H. C. M. Oomen, "Developments in Water Softening by Means of Pellet Reactors," *J. AWWA,* vol. 75, no. 12, December 1983, p. 619.
26. C. W. Reh, "Lime-Soda Softening Processes," in Robert Sanks (ed.), *Water Treatment Plant Design,* Ann Arbor Science, Ann Arbor, Mich., 1978.
27. T. J. Sorg and G. S. Logsdon, "Treatment Technology to Meet the Interim Primary Drinking Water Regulations for Inorganics: Part 5," *J. AWWA,* vol. 72, no. 7, July 1980, p. 411.
28. R. L. Culp, G. M. Wesner, and G. L. Culp, *Handbook of Advanced Wastewater Treatment,* Van Nostrand Reinhold, New York, 1978.
29. R. A. Ryder, "State of the Art in Water Treatment Design, Instrumentation and Analysis," *J. AWWA,* vol. 69, no. 11, November 1977, p. 612.
30. R. T. O'Connell, "Suspended Solids Removal," in Robert L. Sanks (ed.), *Water Treatment Plant Design,* Ann Arbor Science, Ann Arbor, Mich., 1978.
31. "Committee Report: Lime Softening Sludge Treatment and Disposal," *J. AWWA,* vol. 73, no. 11, November 1981, p. 600.
32. R. J. Calkins and J. T. Novak, "Characterization of Chemical Sludges," *J. AWWA,* vol. 65, no. 6, June 1973, p. 423.
33. R. S. Engelbrecht, J. T. O'Conner, and M. Gosh, "Significance and Removal of Iron in Water Supplies," *Fourth Annual Environmental Engineering & Water Resources Conference,* Vanderbilt University, Nashville, 1965.
34. T. J. Sorg, O. T. Love, Jr., and G. Logsdon, *Manual of Treatment Techniques for Meeting the Interim Primary Drinking Water Regulations,* U.S. EPA Report 600/8-77-005, MERL, Cincinnati, 1977.
35. S. Mukai, T. Wakamatsu, and Y. Nakahiro, "Study on the Removal of Heavy Metal Ions in Wastewater by the Precipitation-Flotation Method," *Recent Developments in Separation Science,* 67 (1979).
36. W. L. Brink, R. J. Schliekelman, D. L. Bennett, C. R. Bell, and I. M. Markwood, "Radium-Removal Efficiencies in Water-Treatment Processes," *J. AWWA,* vol. 70, no. 1, January 1978, p. 31.
37. J. J. Rook, "Haloforms in Drinking Water," *J. AWWA,* vol. 68, no. 3, March 1976, p. 168.
38. E. S. Hall and R. F. Packham, "Coagulation of Organic Color with Hydrolyzing Coagulants," *J. AWWA,* vol. 57, 1965, p. 1149.
39. J. K. Edzwald, "Coagulation of Humic Substances," *AICHE Symp. Ser.* No. 190, vol. 75, 1978, p. 54.
40. S. J. Randtke, C. E. Thiel, M. Y. Liao, and C. N. Yamaya, "Removing Soluble Organic Contaminants by Lime-Softening," *J. AWWA,* vol. 74, no. 4, April 1982, p. 192.
41. D. E. Johnson and S. J. Randtke, "Removing Nonvolatile Organic Chlorine and Its Precursors by Coagulation and Softening," *J. AWWA,* vol. 75, no. 5, May 1983, p. 249.
42. M. Y. Liao and S. J. Randtke, "Removing Fulvic Acid by Lime Softening," *J. AWWA,* vol. 77, no. 8, August 1985, p. 78.
43. W. J. Weber, Jr., and A. M. Jodellah, "Removing Humic Substances by Chemical Treatment and Adsorption," *J. AWWA,* vol. 77, no. 4, April 1985, p. 132.
44. R. L. Sinsabaugh, R. C. Hoehn, W. R. Knocke, and A. E. Linkins, "Removal of Dissolved Organic Carbon by Coagulation with Iron Sulfate," *J. AWWA,* vol. 78, no. 5, May 1986, p. 74.
45. H. W. Wolf, R. S. Safferman, A. R. Mixson, and C. E. Stringer, "Virus Inactivation during Tertiary Treatment," *J. AWWA,* vol. 66, no. 9, September 1974, p. 526.
46. V. C. Rao, J. M. Symons, A. Ling, P. Wang, T. G. Metcalf, J. C. Hoff, and J. L. Melnick, "Removal of Hepatitis A Virus and Rotavirus in Drinking Water Treatment Processes," *J. AWWA,* vol. 80, no. 2, February 1988, p. 59.

# Membrane Processes

## William J. Conlon, P.E.

*Vice President/Manager*
*Stone & Webster Water Technology*
*Services Division*
*Fort Lauderdale, Florida*

The depletion of water supplies, saltwater intrusion, and water pollution, especially by complex organic materials such as the priority pollutants, have contributed to the expanded use of advanced membrane technologies that provide superior potable water quality more efficiently than conventional treatment systems. State-of-the-art advances in membranes systems that reduce capital, operation, and maintenance costs are making membrane processes competitive with conventional unit processes.[1]

## Types of Membrane Processes

Various types and sizes of membranes exist. Figure 11.1 illustrates the membrane spectrum, showing relationships between micrometer size, molecular weight, and angstrom units. System treatment ranges are shown in Fig. 11.2, which also shows relationships to salinity or total dissolved solids.

*Reverse osmosis* (RO) is a pressure-driven process that retains virtually all ions and passes water. The pressure applied exceeds the osmotic pressure of the salt solution against a semipermeable membrane, thereby forcing pure water through the membrane and leaving salts behind.

*Electrodialysis* (ED) is a process in which ions are transferred through membranes from a less concentrated to a more concentrated solution as a result of the passage of direct electric current.[2] The flow of water is tangential to the membrane while the flow of ions is per-

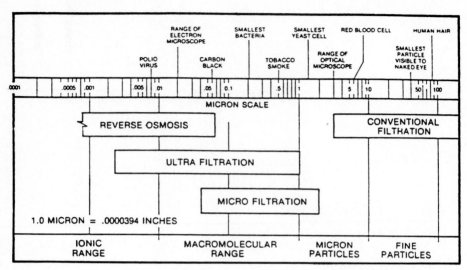

**Figure 11.1** Particle size separation comparison chart. (*Courtesy of Basic Technologies, Inc.*)

pendicular to the membrane. *Electrodialysis reversal* (EDR) is an electrodialysis process in which the polarity of the electrodes is reversed on a prescribed time cycle, thus reversing the direction of ion movement in a membrane stack.[2]

*Ultrafiltration* (UF) is a pressure-driven process for fractionating and concentrating solutions containing colloids and high-molecular-weight materials.[3] Ultrafiltration retains nonionic matter and generally passes most ionic matter depending on the *molecular weight cutoff* (MWC) of the membrane. The MWC is the membrane specification describing the nominal rejection of a known solute[4] or the MW above which most species are retained by that membrane. Molecular weight can be considered an approximate guide to molecular size, which allows membrane manufacturers to characterize the retention percentage of certain solutes, given their molecular weights. From these data, manufacturers assign a nominal MWC. These limits are not sharp cutoff limits or absolute; instead, they are dependent on particle size, shape, and charge.[5]

*Nanofiltration* (NF) or membrane softening is an emerging membrane technology.[6,7,8] The prefix "nano" means one thousand millionth, or $10^{-9}$, and refers in this case to the supposed pore size of the membrane. A nanofilter is an ultra-low-pressure membrane, passing only particles smaller than 1 nm (10 Å) yet exhibiting some characteristics of both an ultrafiltration membrane and an RO membrane. NF membranes operate generally in a higher MWC range than the

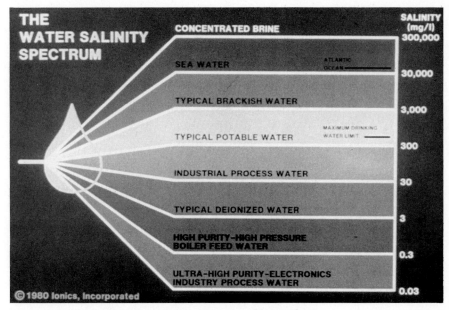

Figure 11.2  Water salinity spectrum. (*Courtesy of Ionics, Inc.*)

classic RO membrane and exhibit high rejection of divalent ions such as calcium and magnesium. NF membranes reject monovalent ions at a much lower rate than divalent ions.

Microfiltration is the separation of micrometer or submicrometer particles in a solution across a membrane material.[4] Microfiltration clarifies water and other fluids by catching suspended matter and microorganisms on the surface or inside a filter while passing dissolved substances and water.[9]

Membrane air separation processes retain membrane-impermeable gases while passing membrane-permeable gases and are not applicable at present for water separation process. For example, gas separation membranes are used to separate carbon dioxide from methane and air into nitrogen and oxygen.

Dialysis uses the differences in ion concentrations on each side of a membrane to draw off the desired ion. Dialysis retains most dissolved matter and passes microsolutes (low-molecular-weight solutes) and water.[9] Dialysis is primarily used in the medical field.

Coupled-transport membranes, which are not yet commercially available, separate one kind of dissolved ion or molecule from another via a chemical carrier that is bound within a membrane.[9] These membranes will probably be applied to the metals recovery industry.

## History and Development of Membranes

Many individuals engaged in research, manufacturing, engineering, and academia contributed to the growth and development of membrane technology. The following is a brief synopsis of the history and development of the four major types of membranes used today in potable water treatment: reverse osmosis, nanofiltration, electrodialysis, and ultrafiltration.

### Reverse osmosis

The phenomenon of osmosis was discovered in 1748[10] by Nollet, a French physicist, who found that water would diffuse spontaneously through a pig bladder membrane into alcohol.[11] Over 100 years later in 1867, Traube, a German chemist, performed experiments with artificially prepared membranes.[12]

Since the original discovery of osmosis in 1748, osmosis and RO remained undeveloped for over 200 years. Then, in the late 1950s, Reid and Breton, at the University of Florida, developed cellulose acetate (CA) RO membranes.[13] The Reid-Breton membrane, however, produced low fluxes because of excessive thickness. At about the same time at the University of California at Los Angeles, Loeb and Sourirajan developed a technique for casting porous CA membranes with high flux rates and high salt rejection, which led to the commercial manufacture of membranes.[13] The Loeb-Sourirajan membrane had 10 times the flux of the Reid-Breton membrane and had a salt removal efficiency of 95 percent.[9] In the early 1960s, Bray and Westmoreland at the Chemical Process Group of General Atomic Division developed the spiral-wound membrane.[9] Fluid Systems Division, DuPont, and Hydranautics were major driving forces in translating membrane technology from the laboratory to a unit process. Caddote developed the first thin-film composite membrane from 1972 to 1973 at North Star Research and Development Institute.[9]

Improvements have been made in almost all areas of membrane manufacturing since the early 1970s. New, improved membrane materials and casting chemicals have been developed, resulting in specialized applications for different types of membrane materials such as chlorine-resistant membranes. Improvements in module construction have improved membrane performance and durability. Advancements in manufacturing applications using computerized techniques by some manufacturers have improved quality control. Larger modules have been developed and placed on the market. Modules for low-pressure operation on brackish and softening applications are now available. All these recent developments have improved membrane economics. Research and development continues regarding membranes, modules, membrane cleaning, and rejuvenation.[6,14] Competition in the membrane market has also stimulated research and development, and there are at least 10 or more major manufacturers worldwide. There are concurrent developments in ED and UF technology.[14]

### Nanofiltration

In 1976, a Florida-based equipment manufacturer, Basic Technologies, Inc., conceived the innovative and futuristic idea of using modified RO membranes to treat water with low *total dissolved solids* (TDS) as a substitute for lime softening. Not only would the membrane reject hardness, but also it would act as a barrier to repel bacteria and virus. Further, this membrane process would remove organic-related color without generating undesirable chemical compounds such as chlorinated hydrocarbons.

In 1977, a membrane manufacturer located in California, Fluid Systems Division of UOP, formulated a special softening membrane designated ROGA 8150, with a rejection rate of approximately 47 percent for monovalent ions. This cellulose diacetate (CDA) membrane was manufactured specifically for the membrane-softening concept and for use at a plant operating at 75 percent recovery with a source water TDS of 930 mg/L. The membrane was designed to reduce the TDS just below the regulatory requirement of 500 mg/L and was the first softening membrane manufactured. Standard brackish CDA RO elements were hydrolyzed in situ with a controlled caustic solution to achieve the desired membrane properties.

### Electrodialysis

In 1940, Meyer and Strauss devised a multicompartment ED cell with ion-selective membranes. Ionics, Inc., invented the ion transfer membrane in 1948 that resulted in the development of the classical ED process, first commercially sold in 1954.[2] In the early 1970s EDR was introduced.[2]

### Ultrafiltration

In the 1930s, polymer UF membranes of various pore sizes were developed.[15] Alan Michaels at the Massachusetts Institute of Technology cast UF membranes made of polyanion and polycation mixtures and of polysulfone. The pores of the new membrane were larger than RO membrane pores and offered a way to separate proteins, small microorganisms, and colloids from solution and from each other via macromolecular separation.[9] The first UF manufacturing company was founded in 1962 by Michaels.[9]

## Reverse Osmosis

Reverse osmosis is accomplished by applying a pressure (over and above the solution osmotic pressure) to a concentrated solution, thereby forcing pure water to flow across a semipermeable membrane to the dilute side.

### Theory

Osmosis is a phenomenon that sometimes occurs when water (or another solvent) flows through a semipermeable membrane, that is, a membrane that permits the passage of a solvent but not of dissolved substances. Solutions separated by such a membrane tend to become equal in molecular concentration; thus, water will flow from a weaker to a stronger solution, with the solutions tending to have more nearly equal concentrations.[16] The flow of water across the membrane exerts a pressure called the *osmotic pressure*.

**Osmotic formula.** The osmotic pressure $\pi$ of a solution can be found from the formula

$$\pi = 1.12(T + 273)\sum \overline{mi} \qquad (11.1)$$

where $\pi$ = osmotic pressure, psi

$T$ = temperature, °C

$\Sigma \overline{mi}$ = summation of molalities of all ionic and nonionic constituents in solution[17]

The osmotic pressure of a solution increases with the concentration of the solution. A rule of thumb that works well with natural water, based on sodium chloride, is that the osmotic pressure increases by approximately 0.01 psi (0.07 kPa) for each milligram per liter.[12] Figure 11.3 is an illustration of the osmosis relationship, showing three vessels, each with a semipermeable membrane down the center. The vessel on the left demonstrates naturally occurring osmosis. The center vessel demonstrates osmotic equilibrium, and the vessel on the right shows RO.

**Water and solute diffusion.** Water flow through a semipermeable membrane can be expressed by[12]

$$F_w = A(\Delta p - \Delta \pi) \qquad (11.2)$$

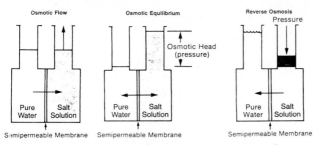

**Figure 11.3** Pictorial explanation of reverse osmosis. (*Courtesy of E.I. du Pont de Nemours & Co.*)

where $F_w$ = water flux, $g/(cm^2)(s)$
$A$ = water permeability constant, $g/(cm^2)(s)(atm)$
$\Delta p$ = pressure differential applied across membrane, atm
$\Delta \pi$ = osmotic pressure differential across membrane, atm

Salt or solute flow through a semipermeable membrane can be expressed by

$$F_s = B(C_1 - C_2) \tag{11.3}$$

where $F_s$ = salt flux, $g/(cm^2)(s)$
$B$ = salt permeability constant, cm/s
$C_1 - C_2$ = concentration gradient across membrane, $g/cm^3$

Water and solute (salt) permeability constants depend on the characteristic of the membrane used and the processing it has received.

Water flux through a membrane is dependent on the applied pressure. As feedwater pressure is increased above the solution osmotic pressure, the flow of water through the membrane increases. The flow of salt remains basically constant, however, and is proportional to the salt concentration differential. Therefore, increased quantities of product water will occur with increased applied pressure. Water flux decreases, however, as the feedwater salinity increases because the osmotic pressure increases.

The percentage of salt rejection defines the product quality:

$$\text{Salt rejection} = 100 - \text{salt passage} \tag{11.4}$$

$$\text{or} \quad = \frac{\text{feed concentration} - \text{product concentration}}{\text{feed concentration}} \times 100$$

The *rejection rate* is the ratio of solute concentration in the feed stream to the solute concentration in the product stream. Solute rejection varies with the particular membrane used, recovery, feedwater concentration, chemical valance of the ions in the solute, and other factors; for example, arsenic at a valance of $+3$ is rejected at a lesser amount than arsenic at a valance of $+5$.[18]

**Example Problem 1**    Pretreated feedwater to a brackish water RO system contains 2500 mg/L TDS. The product water analysis shows a concentration of 85 mg/L TDS. What is the percentage of salt rejection? That of salt passage?

**solution**    From Eq. (11.4),

$$\text{Salt rejection} = \frac{\text{feed concn} - \text{product concn}}{\text{feed concn}} \times 100$$

$$= \frac{2500 - 85}{2500} \times 100$$

$$= 96.6\%$$

$$\text{Salt passage} = 100 - \text{salt rejection}$$

$$= 100 - 96.6$$

$$= 3.4\%$$

Recovery represents the percentage of feed converted to product:

$$\text{Recovery} = \frac{\text{feed rate} - \text{brine rate}}{\text{feed rate}} \times 100 \qquad (11.5)$$

Generally, the higher the recovery rate, the greater the product water yield from the feedwater. Factors that determine the recovery rate are water quality and the saturation percentage of critical membrane foulants such as calcium sulfate, strontium sulfate, and barium sulfate in the concentrate.

**Example Problem 2**    Source water quality for a brackish water RO system requires that 25.0 mgd of feedwater be applied to the membranes to yield 20.0 mgd of product water. What is the plant recovery rate?

**solution**    From Eq. (11.5),

$$\text{Recovery} = \frac{\text{feed rate} - \text{brine rate}}{\text{feed rate}} \times 100$$

$$= \frac{25.0 \text{ mgd} - 5.0 \text{ mgd}}{25.0 \text{ mgd}} \times 100$$

$$= 80\%$$

The quantity of source water required for a given product water rate (also known as the *nominal plant capacity*) and recovery may be determined from

$$\text{Quantity of source water required} = \frac{\text{nominal plant capacity}}{\text{recovery percentage}} \quad (11.6)$$

**Example Problem 3**  Source water dictates that a proposed brackish water RO plant operate at a recovery rate of 75 percent. Determine the source water requirement for treatment to yield 20.0 mgd of product water.

**solution**  From Eq.(11.6),

$$\text{Quantity of source water required} = \frac{\text{nominal plant capacity}}{\text{recovery percentage}}$$

$$= \frac{20.0 \text{ mgd}}{0.75}$$

$$= 26.6 \text{ mgd}$$

The lower the MWC of the membrane, the greater the organic substance removal. Membranes with MWC values of 100 are available today. Nearly all the *synthetic organic chemicals* (SOCs) consisting of purgeables, pesticides, and acid extractable substances that the U.S. Environmental Protection Agency (EPA) has identified in the primary drinking water regulations[19] have an MW above 100. Given a membrane with an MWC of 100, we might conclude that an organic solute with an MW less than 100 would be rejected poorly and that one with an MW above 100 would be good. This should not be considered absolute inasmuch as organic acids and amines appear to follow the same pattern as inorganic acids and bases. The undissociated species are poorly rejected while the salts are well rejected.[12] This will be more significant as the maximum contaminant levels for SOCs are lowered by regulatory agencies. The EPA has funded a preliminary investigation of the effectiveness of RO membranes for removing SOCs, and funding of additional research projects in this area is likely.[20]

### Membrane materials

Membranes are manufactured from a variety of materials such as CA, CDA, cellulose triacetate, polyamide (PA), other aromatic polyamides, polyetheramides, polyetheramines, and polyetherurea. Thin-film composite (TFC) membranes may be made from a wide variety of polymers consisting of several different materials for the substrate, the thin film, and other functional layers in the membrane.

In CA membrane chemistry, the higher the acetyl content, the higher the salt rejection and the lower the water flux. The cellulosic membranes are generally cheaper and can tolerate some chlorine (generally less than 1.0 mg/L in the feed as free chlorine); however, they have several disadvantages. CA membranes are subject to biological attack and to hydrolysis, the chemical reversion of CA to cellulose by the reaction with water to form cellulose and acetic acid. The reversion occurs most rapidly at very low or high pH.[11] The rate of hydrolysis is accelerated by increased feedwater temperature or a feedwater pH above or below the optimum pH range of 5 to 6.[12] Acid pretreatment is almost always required to maintain the feed pH at optimum.

PA and TFC membranes are subject to degradation if exposed to chlorine or other oxidants; however, they are not susceptible to biological attack and resist hydrolysis. These membranes usually operate best in the range of pH 4 to 11 because they are not subject to hydrolysis there.

In time, membrane performance will change, mainly because of compaction and fouling. Membrane compaction, sometimes called "flux decline," is similar to plastic or metal "creep" under compression stress conditions. The compaction rate increases with higher applied feedwater pressures and temperatures. The majority of compaction generally occurs during the first year of operation and is irreversible.

Commercially available membranes usually operate within the temperature range of 70 to 95°F (21 to 35°C). The higher the operating temperature within this range, the greater the flux.

Membrane fouling occurs from scale, colloidal deposition, silt, metal oxides, organics, silica, and other constituents in feedwater. Pretreatment may be required to prevent premature fouling. Pretreatment includes any additional system needed to prevent damage to the membrane for a particular feedwater. A properly designed pretreatment system contains all the necessary particulate-removal filtration systems and chemical feed systems to prevent fouling and hydrolysis. A well-trained operations staff following the equipment manufacturer's recommendations will aid in reducing fouling and extending membrane life. Membrane fouling can be recognized by a decrease in productivity, increase in module $\Delta p$, and increase in salt passage. These signs, among others, indicate a need for cleaning.

Excess amounts of oxidants in the feedwater system such as ozone or permanganate warrant contact with the membrane manufacturer for advice concerning maximum allowable oxidant levels. Chlorine-resistant membranes are available for those systems containing small amounts of chlorine, or dechlorination can be achieved by using a dechlorination agent.

## Membrane configurations

Four membrane configurations are currently manufactured: spiral-wound, hollow fine fiber, tubular, and plate and frame. Of these, only the spiral-wound and hollow fine fiber types are used for municipal water treatment.

A spiral-wound membrane (Fig. 11.4) consists of two flat sheets of membrane separated by porous support or backing sheets sealed on three sides to form an envelope. The fourth side is attached with an adhesive to a hollow plastic tube that collects the product water. Typically two or more of these membrane envelopes are glued to the product water collection tube and rolled up in the form of a spiral. The spiral-wound module resembles a jelly roll when constructed and has been so nicknamed by those working in the industry. Multiples of the spiral modules are usually connected in series in a fiberglass vessel.

A hollow fine fiber module (Fig. 11.5) is a compact bundle of thousands of longitudinally aligned fibers that surround the feedwater distribution core. Each hairlike fiber is laid in the form of a U in the bundle, and both ends are encapsulated in an epoxy resin tube sheet. The bundle is encased in a cloth overwrap and screen and installed in a pressure vessel constructed of fiberglass or epoxy-coated steel.

A tubular membrane (Fig. 11.6) consists of a membrane installed inside a porous tube. The pressurized feedwater enters the inside of the membrane or membrane film and exits through the porous tube. Because of the high cost of producing the membrane, tubular units are not used in the large-volume municipal potable water treatment systems, but are used in industry. Tubular elements have a small membrane area per unit volume.

Plate and frame membranes (Fig. 11.7) are membranes sandwiched between circular or square plates. The modules resemble a plate and frame filter press. (See Chap. 16.) As with tubular membranes, this design does not lend itself to municipal potable water treatment.

### RO system components and design considerations

RO system categories are summarized in Table 11.1. A flow schematic of a typical RO system is shown in Fig. 11.8. Key components are described in the following sections.

**Feedwater supply.**   Groundwater is the preferred feedwater supply because of its general chemical stability. Early RO supply wells were constructed with steel casings and used vertical turbine pumps with steel and bronze components. Standard packing glands were an integral part of the vertical turbine unit. Through experience, corrosion of the iron and bronze system components was soon discovered to be contributing foulants to the membrane system. In addition, air entering

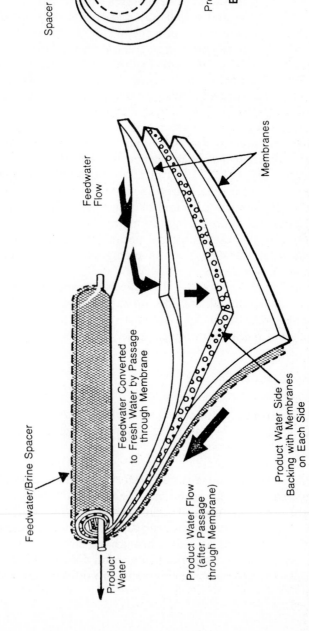

Membrane Envelope

Spacer

Product Tube

**End view**

Feedwater Flow

Membranes

Feedwater/Brine Spacer

Feedwater Converted to Fresh Water by Passage through Membrane

Product Water Side Backing with Membranes on Each Side

Product Water Flow (after Passage through Membrane)

Product Water

**Figure 11.4** Spiral wound reverse-osmosis cartridge. (*Courtesy of E.I. du Pont de Nemours & Co.*)

720

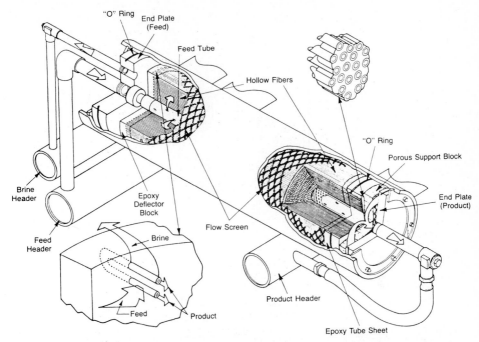

**Figure 11.5** Hollow fiber permeator construction and hookup. (*Courtesy of E.I. du Pont de Nemours & Co.*)

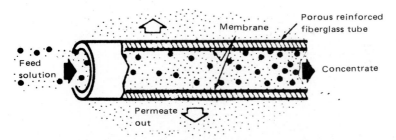

**Figure 11.6** Tubular membrane configuration. (*Courtesy of Abcor, Inc.*)

the system through pump packing glands caused oxidation of iron and sulfur compounds, causing severe fouling of membranes. Since then, only noncorrosive materials such as stainless steel, polyvinyl chloride, and fiberglass are used in groundwater supply systems. Either vertical turbine pumps with mechanical seals or submersible pumps are used, so an airtight system is virtually ensured.

Accuracy in water quality sampling and analyses is of prime importance. Generally, at least three separate analyses are recommended to ensure accurate design data. Analyses should include the critical test

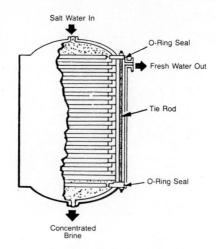

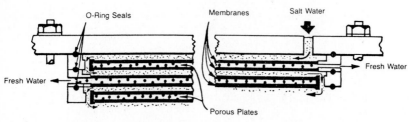

**Figure 11.7**   Plate and frame reverse osmosis. (*Courtesy of E.I. du Pont de Nemours & Co.*)

**TABLE 11.1   General Reverse-Osmosis System Categories**

| System | Transmembrane pressure operating range, psi | Transmembrane pressure operating range, kPa | Salinity TDS range, mg/L | Recovery rates, % |
|---|---|---|---|---|
| Seawater | 800–1,500 | 5,000–10,000 | 10,000–50,000 | 15–55 |
| Standard pressure | 400–650 | 3,000–4,500 | 3,500–10,000 | 50–85 |
| Low-pressure | 200–300 | 1,000–2,000 | 500–3,500 | 50–85 |
| Nanofiltration | 45–150 | 310–1,000 | Up to 500 | 75–90 |

parameters required for design of membrane systems indicated in Table 11.2.

Surface water supplies require additional pretreatment to compensate for seasonal variations in source water quality such as those occurring during wet and dry seasons. Surface water requires more stringent monitoring than groundwater because of variables that can influence pretreatment. For example, security of upstream water sup-

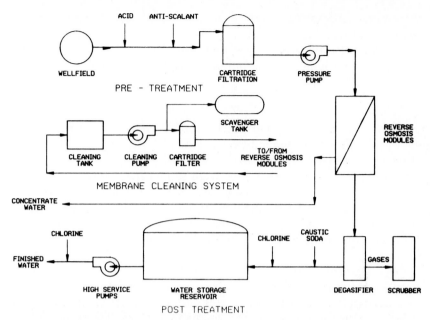

**Figure 11.8** Reverse-osmosis flow schematic. (*Courtesy of James M. Montgomery Consulting Engineers*)

**TABLE 11.2    Minimum Water Quality Data Required for RO System Design**

| | |
|---|---|
| Calcium | Carbon dioxide |
| Magnesium | Bicarbonate |
| Potassium | Sulfate |
| Manganese | Chloride |
| Sodium | Fluoride |
| Iron | Nitrate |
| Barium | Ammonia |
| Strontium | Phosphate |
| Hydroxide | Silica (soluble) |
| pH | Silica (insoluble) |
| Specific conductance | Hydrogen sulfide |
| Temperature | Silt density index |
| Chlorine residual | Suspended solids |
| Color | Turbidity (NTU) |
| Total dissolved solids | |
| Bacteriological analysis—heterotrophic plate count | |
| Total hardness (as $CaCO_3$) | |
| Total alkalinity (as $CaCO_3$) | |

NOTE: Source of feedwater should be described by stating appearance and any unusual conditions.

plies from contamination, drought, and seasonal variations can affect feedwater quality. Product water quality will vary with a changing feedwater quality. Piping, pumps, and other appurtenances should be made of the same noncorrosive materials as those used for groundwater supply.

**Pretreatment.**  A well-designed pretreatment system is an important factor in the life cycle of membranes. Turbidity and suspended solids must be removed from the feedwater before they reach the membrane. A turbidimeter or controller can be used to monitor feedwater turbidity and provide a plant shutdown feature if turbidity levels are exceeded. Suspended material such as silt and clay can become trapped in the membrane assembly, causing channeling and uneven feedwater distribution, and if there is not prompt cleaning, serious fouling may result.

The *silt density index* (SDI) is a key indicator of feedwater quality. This test or fouling index should be performed frequently on the feedwater supply.[21] The SDI indicates the amount of particulate and colloidal matter contained in the feedwater. The SDI value of the pretreated feedwater fairly accurately tracks the amount of fouling material present in the feedwater. Generally, spiral-wound membranes can tolerate a maximum SDI of 5 whereas hollow fine fiber membranes should be operated with an SDI less than 3 or 4 depending on the manufacturer.

Chemical addition is typically required to prevent membrane scaling. Sulfuric acid (66° Baumé, 93%) is usually added to the feedwater to prevent calcium carbonate ($CaCO_3$) and magnesium carbonate ($MgCO_3$) scaling of the membrane. Through the use of this pH adjustment method, brackish feedwater is acidified to yield a negative Langlier saturation index in the concentrate stream or a negative Stiff and Davis index[22] in a seawater concentrate stream. Redundant chemical feeders are often used to ensure that backup equipment is available. Two pH sensors and controllers are generally used, one in the feedwater system and the other in the product water stream. The chemical feeders are paced by ratio controllers off the feedwater flow sensor. The pH controller in the feedwater stream is the primary pH controller and is preset to give the desired pH in the feedwater stream. In the event the feedwater pH controller fails and plant shutdown does not occur, the product water pH controller will sense a pH rise and will shut down the plant, thus protecting the membrane from scaling or permanent damage. Chemical addition is accomplished prior to the cartridge prefilters and may include an in-line static mixer to ensure complete uniform mixing prior to reaching the membrane system. Precipitation of $CaCO_3$ and $MgCO_3$ can also be prevented by ion exchange (sodium cycle), but this method is typically more expensive than pH adjustment with acid.

Upon plant shutdown, membranes should be flushed with acidified feedwater or unchlorinated product water to remove concentrated brine from prolonged contact with the membranes. If the concentrate or brine reject is left standing in contact with the membrane, precipitation will eventually occur, resulting in membrane scaling. If system shutdown occurs for periods longer than a few days, the membrane elements should be disinfected or sterilized by filling them with a biocidal solution such as formaldehyde or sodium metabisulfite. Another way to place a plant on standby without using a biocide is to operate the system at least once per day for at least 30 min. Monitoring of the biological growth rate is recommended if this method is used.

Sulfate precipitation is more difficult to control than carbonates. Calcium sulfate, barium sulfate, and strontium sulfate may cause problems in brackish water systems, and calculations should be checked to ensure that solubility limits are not exceeded. Generally, sulfates are not considered a problem in seawater RO systems; however, calculations to determine solubility limits should be performed. Historically, sodium hexametaphosphate (SHMP) has been the primary sequestrant used to inhibit the rate of sulfate precipitation. SHMP has been used to control calcium sulfate at twice its saturation level in the concentrate stream. Barium and strontium are much less soluble than calcium sulfate. SHMP is injected into the feedwater stream after the acid and prior to cartridge filtration. A static mixer is also recommended to guarantee proper mixing. A predetermined rate of feed generally of 5 to 10 mg/L is used. A disadvantage of using SHMP is that it begins to hydrolyze in water to form orthophosphate, which is ineffective in controlling sulfate scale. Therefore, only a 1-day supply should be mixed for use. Calcium sulfate scale control is critical because it is considered irreversible.

Recently substitute antiscalants or long-chain polymers have become available and may ultimately replace SHMP. As an alternative to acid, SHMP or one of the new antiscalants may be added to feed, and the Langlier saturation index may be as high as +1.9. This reduces acid requirements considerably and applies to all PA membranes with good rejection at neutral pH. CA membranes still require acid addition to prevent hydrolysis. Piloting of sequestering agents should be an integral part of the selection process because one agent may work better with a certain feedwater. Flow and low-liquid-level switches should be inserted in the chemical feed system and wired into the plant instrumentation to shut down the plant in case of sequestering agent feed system failure.

Cartridge filters are necessary in both surface water and groundwater supply pretreatment. The 5-$\mu$m cartridge filters are the most common size currently used in RO plants. Cartridge vessel housings are made of stainless steel or fiberglass and are equipped with inlet-outlet pressure gauges. Cartridge filters should be changed when the pressure difference across the filters reaches 15 psi (103 kPa).

The organic content and biological activity of the feedwater must be determined prior to the system design. Organic and biological constituents are more likely to occur in a surface water supply than in a groundwater supply. Organic compounds that are immiscible in water can affect membrane performance. RO plants typically operate on feedwater containing as much as 20 mg/L of total organic carbon. Abnormally high levels of chemical oxygen demand, biochemical oxygen demand, and total organic carbon can indicate wastewater contamination that could subsequently cause a biological growth problem. Bacteria, bacterial slimes, and algae can cause fouling, product water contamination, and biological degradation of some membrane materials. Chlorine in controlled amounts can be used with certain chlorine-tolerant membranes to control bacteria and algae problems. Some membrane materials cannot tolerate any residual chlorine, and other biocides are needed to control bacteria and algae. After the contamination is eliminated or under control, the membrane can be treated with a biocide followed by cleaning with a detergent. Surface water supplies high in colloidal matter may require the addition of a cationic polymer or coagulant aid prior to conventional filtration. Aluminum sulfate and certain anionic polymers are examples of chemical additives that can cause damage and therefore should not be used. Aluminum content greater than 0.1 mg/L (as $Al^{3+}$ ion) can be a problem for the RO system.[22] The membrane manufacturer should be consulted prior to the use of any pretreatment additive to determine chemical compatibility.

Oils and greases can affect membrane performance by forming a film on the membrane surface, affecting its rejection qualities. The membrane manufacturer can recommend O-ring lubricants that will not affect membrane performance.

**Membrane blocks and arrays.** Membrane blocks or trains consist of a group of modules, typically sized for convenience. If a plant is being expanded in a modular fashion, a membrane block or train may be sized to correspond with a particular growth rate. Several blocks should be used so that one may be taken out of service for cleaning or modification. Cleaning systems are generally designed to handle a unit block. A block (Fig. 11.9) may consist of a single-stage system or a two- or three-stage array. The array staging is governed by the recovery rate, which in turn is dependent on the feedwater quality. Recovery rates of up to 55 percent for single-stage systems are achievable on brackish water. Two-stage arrays generally operate at 75 to 80 percent recovery while three-stage arrays are needed for 85 to 90 percent recovery. In a multistage array, the first-stage concentrate becomes the second-stage feedwater, and the second-stage concentrate becomes the third-stage feedwater, taking advantage of residual pressures remaining in the concentrate streams.

When the detailed water quality data, desired block or train size, conversion or recovery rate, product water design quality, minimum and maximum temperature ranges, feed pressure, and product pressure are known, a system design can be completed. A typical RO system is designed to ensure a rated water output performance over the estimated membrane lifetime. This is called the *design life* of the membrane. A design life of 3 or 5 years is common and is used so that a plant's capacity is provided after membrane compaction is essentially complete. A specific minimum product water flow and quality is

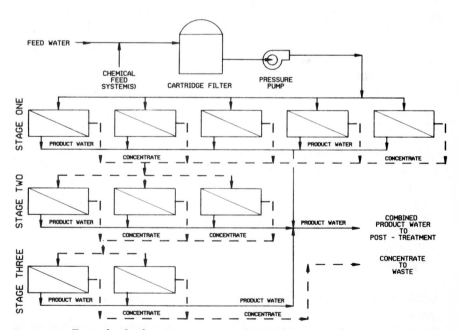

**Figure 11.9**  Example of a three-stage system using a 5-3-2 array. (*Courtesy of James M. Montgomery Consulting Engineers*)

projected for the end of that design period. The feedwater pressure is usually stated for the beginning and end of the design period.

**Instrumentation.**   A minimum level of instrumentation is required to ensure fail-safe operation and to avoid potentially catastrophic damage to the membranes. System shutdown should occur in the case of high-pressure pump, low suction pressure, low or high feedwater pH, improper feedwater flow rate, high feedwater turbidity, high feedwater pressure, high feedwater conductivity, high or low product flow rate, high product conductivity, low antiscalant level or flow, plus other conditions. System instrumentation can be as simple as "smart" relays or as sophisticated as microprocessors and computers that are used to run state-of-the-art plants. Trending analysis can be included in the computer software to give the operator early warning of potential serious problems. Computers can also be programmed to print out monthly operational reports or, via modem, can be used by the equipment manufacturer to check compliance with membrane guarantee conditions.

**Posttreatment systems.**   Degasification is needed to remove gases such as carbon dioxide and hydrogen sulfide ($H_2S$). A reduction in carbon dioxide will raise the pH and help to stabilize the product water. The removal of $H_2S$ will remove objectionable odor. If the plant is near a residential neighborhood, a scrubber may be required on the degasifier for aesthetic reasons. The product water pH must be adjusted to the pH of stabilization by using caustic soda or lime. Chlorine is added for disinfection. Because almost all organics have been removed from the product water, the chlorine demand is low and generally all chlorine added produces a free chlorine residual.

**Energy recovery.**   Until recently, energy recovery was associated with seawater RO systems with brine streams operating at as high as 850 psi (5900 kPa). Energy recovery was considered economically feasible for seawater systems, but impractical for brackish water systems. The principal energy recovery device was the Pelton wheel or turbine, which requires atmospheric discharge and during these early seawater applications had high capital and maintenance costs. In the case of seawater RO, however, energy recovery could be cost-effective because of higher brine pressures and/or greater brine stream flow rates. Other factors in favor of using the Pelton turbine are the higher efficiency of the device and its dependability.

Energy recovery is now practiced on low-pressure systems using stainless-steel submersible pumps and motors running in reverse.

These reverse-running turbines are placed in the pressurized brine or reject line. The pressurized brine or concentrate water flowing through the reject line to waste turns the submersible turbine in reverse which in turn runs the motor in reverse, thus generating electricity. The power generated is returned to the main power supply for use within the RO plant. These low-cost devices have been shown to be cost-effective in spite of their lower efficiency on brackish water systems.[23]

**Cleaning systems and membrane rejuvenation.** An integral part of membrane treatment plants is the cleaning system. Cleaning systems consist of a stainless-steel pump sized for proper flow and pressure, a fiberglass or polypropylene tank with appropriate retention capacity, a 1- to 5-$\mu$m cartridge filter, noncorrosive piping, valves, hoses, and controls. Operations personnel must be alert to system conditions that signal when cleaning is required. These are some of the signals: The salt passage increases by 15 percent or more, the module pressure drop increases by 20 percent or greater, feed pressure requirements increase by 20 percent, product flow drops or increases by 5 percent, brine flows change by 5 percent, and fouling or scaling is evident.

Cleaning solutions for chemical or biological fouling are generally prescribed by the membrane manufacturer. Specialty chemical companies now market cleaning chemicals specifically for membrane cleaning, and chemicals recommended by the membrane manufacturer can be purchased from other sources.

Safety procedures must be observed during operation of the cleaning system, especially with regard to proper ventilation and skin or body exposure to certain cleaning chemicals. Disposal of the spent cleaning chemicals needs to be done in an environmentally acceptable manner.

Depending on the degree of fouling, chemical rejuvenation of the membrane may be necessary. Each membrane manufacturer has recommended rejuvenation procedures and chemicals to restore rejection. Polyvinyl acetate–crotonic acid (PVA/CA) copolymer,[24] polyvinyl methyl ether, and tannic acid are some of the chemical rejuvenatives used.

### RO system operation and maintenance

The feedwater supply must be checked routinely for changes in water quality. Water quality parameters, both chemical and biological, should be graphed and trends charted, preferably via computer. Groundwater system maintenance consists of routine mechanical and electrical preventive measures.

Surface water supply system maintenance is more involved because of added system components such as the reservoir, intake screens, and filters for suspended solids and turbidity removal. Complete pretreatment, including sedimentation, may be required on very dirty surface water. In addition, biological monitoring and treatment are needed.

The acid injection system, consisting of bulk storage tanks, chemical feeders, ratio controllers, and injection lines, should be routinely checked for maintenance and safety-related conditions. Protective clothing, masks, and safety showers should be checked periodically and safety drills conducted. Acid quality should be verified on new shipments and certified analyses reviewed. Controllers for pH should also be checked and calibrated routinely and the pH probes cleaned.

The antiscalant chemical feed system and related instrumentation should be checked daily and maintained routinely. The chemical quality of each shipment should be reviewed and samples of each shipment kept for a set period in case operational problems develop.

Cartridge filter vessel drains should be flushed daily, and the $\Delta p$ across the filters should be checked and recorded. Upon filter changeout at 15-psi (100 kPa) $\Delta p$, proper seating of the cartridges should be ensured. Residual particulate matter inside the cartridge filter vessel should not be allowed to enter downstream piping during filter cartridge changeout. Quality control should be exercised in the purchasing of replacement cartridges. Maximum flow rates per individual cartridge should not be exceeded.

Membrane block feedwater, brine and product conductivity, flows, pH, and pressures should be logged daily. The $\Delta p$ across membrane modules in the case of hollow fine fiber units and across stages in spiral-wound systems should be checked routinely. Any unusual changes in preset membrane block operating conditions may require adjustment to the system. High-pressure pumps and motors require routine maintenance as recommended by the manufacturer.

All plant instruments should be kept in good operating condition. All instrument systems associated with fail-safe or shutdown conditions must be kept operational. Minimum spare-parts inventory recommended by the manufacturer should be kept on hand.

The posttreatment system usually consists of degasification, chlorination, and pH adjustment. The degasifier packed tower or tray aeration system requires periodic inspection and cleaning. Slime buildups occur in degasifiers and require cleaning. Degasifier blowers and motors require routine maintenance.

The pH adjustment chemical feed system will need routine daily checks and periodic maintenance. If caustic soda is used, the bulk tank, heaters, and chemical feed injection system should be scrutinized routinely. Safety precautions and chemical quality control are recommended. The chlorination system should be operated under similar guidelines as other chemical feed systems.

### RO system applications

RO can be divided into different subcategories according to salinity and operating pressure. This division is dependent on the osmotic pressure of the water and where the projected operating pressure occurs in the category of membranes available on the market. These subcategories are generally divided as shown in Table 11.1. Differences in system design and components are based on the pressure and feedwater quality.

Seawater RO is being successfully used worldwide. The installed daily capacity of brackish water RO systems exceeds 50 mil gal in Florida alone.[25] Seawater RO systems, which generally treat surface water supplies, require the following additional components: a seawater intake system, additional pretreatment such as polymer addition and conventional filtration, higher pressure-rated 317L stainless-steel piping, more costly pressure pumps and motors, and other plant equipment. Both capital and operating costs diminish with a decrease in the system pressure.

In addition to desalinization of seawater and brackish water, RO can be used where rejection of naturally occurring organics and synthetic organic contaminants is desired. The lower the MWC of a membrane, the higher the organic rejection. At present RO membranes with an MWC as low as 100 are available.

Bacteria and viruses are also effectively removed by RO, although O-ring leaks can occur when spiral-wound membranes are used. High rejection of heavy metals and other dissolved salts by RO lends itself to a variety of applications.[26] RO systems have also been used in public water supplies for radionuclide removal.[27]

### Electrodialysis and Electrodialysis Reversal[2]

ED is an electrochemical separation process in which ions are transferred through anion- and cation-selective membranes from a less concentrated to a more concentrated solution as a result of the flow of direct electric current. EDR is an ED process in which the polarity of the electrodes is reversed on a prescribed time cycle, thus reversing the direction of ion movement in a membrane stack. ED/EDR systems are used to treat brackish water for potable use or to desalt and concentrate effluents for reuse.

### Ion transport

To understand ED, the effect of a direct-current (dc) potential on an ionic solution must be understood. Imagine a rectangular tank with an electrode at each end. The tank is filled with sodium chloride (NaCl) solution (Fig. 11.10). When a dc potential is applied across the electrodes, the cations ($Na^+$) are attracted to the negative electrode, called the cathode, while the anions ($Cl^-$) are attracted to the positive electrode, called the anode. The following reduction reaction of hydrogen ion from the dissociation of water occurs at the cathode:

$$2H_2O + 2e^- \rightarrow 2OH^- + H_2 \uparrow \qquad (11.7)$$

The following oxidation reaction of hydroxide ion from the dissociation of water occurs at the anode:

$$2H_2O \rightarrow 4H^+ + O_2 \uparrow + 4e^- \qquad (11.8)$$

A reaction involving the formation of chlorine gas may also occur at the anode:

$$2Cl^- \rightarrow Cl_2 \uparrow + 2e^- \qquad (11.9)$$

ED membranes are basically cationic or anionic ion-exchange resins in sheet form. Thus they are called "cation membranes" and "anion membranes."

In the operation of an ED array or "stack" (Fig. 11.11), the dissolved ions, the selective cation and anion membranes, and the dc electric field are combined to achieve demineralization or deionization of water and concentration of removed ions in a brine stream of higher concentration. In the membrane stack, water containing dissolved ions is pumped between a cation and an anion membrane guided through a polyethylene spacer which forms a flow path. The water path is between membranes rather than through a semipermeable membrane,

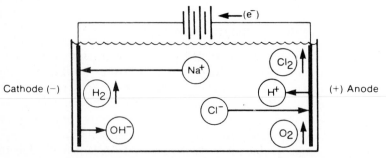

**Figure 11.10**  Effect of dc potential on an ionic solution. (*Courtesy of Ionics, Inc.*)

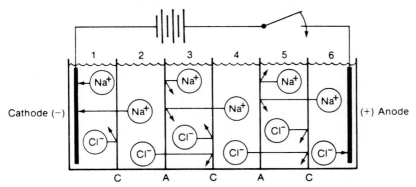

**Figure 11.11** Electrode dialysis array or "stack." (*Courtesy of Ionics, Inc.*)

as in RO. Negative ions pass through an anion membrane while positive ions are rejected. Conversely, positive ions pass through a cation membrane while negative ions are rejected. Those ions passing through either anionic or cationic membranes are collected in concentrating compartments and wind up in the brine solution. These ED membranes are electrically conductive and essentially water-impermeable under pressure. When a dc field is imposed on this stack, all the cations move toward the negative pole. If the first membrane that a cation encounters is a cation membrane, the cation will pass to the next water compartment, leaving behind a partly deionized or diluting cell. If the first membrane that the cation encounters is an anion membrane, that ion will not pass. It is, therefore, trapped in its original compartment, which becomes a concentrating compartment. As seen in Fig. 11.11, the anions undergo similar movements and restrictions in the opposite direction as they are attracted to the positive pole.

In an ED array, with alternating cation and anion membranes, each cation and anion either can move not at all or can move only one compartment before being blocked by an opposite membrane. The end result of these membrane-controlled, direct-current-induced ion movements is that one alternate set of water-containing compartments becomes demineralized and the other becomes concentrated. The flow in the ED stack, as described above, is unidirectional, unlike that in EDR.

**Polarity reversal**

Instead of the classic unidirectional ED, in EDR (Fig. 11.12) the polarity of the electrodes is reversed on a prescribed time cycle, usually every 15 to 20 min. The change in electric polarity across the mem-

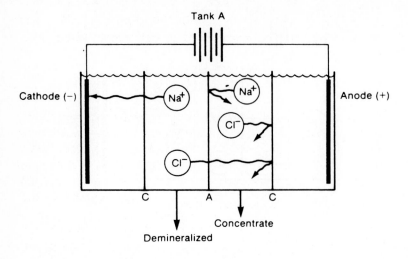

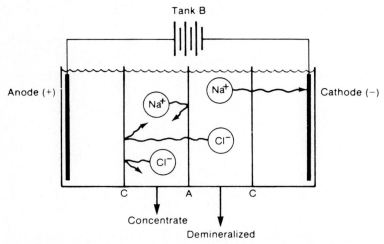

**Figure 11.12** Electrodialysis reversal process. (*Courtesy of Ionics, Inc.*)

brane stack reverses the direction of ion movement. The symmetric demineralized and concentrate flow passage automatically interchanges upon reversal. Reversal is accomplished via automatically operated valves that switch the two inlet and outlet streams so that the incoming feedwater flows into the new demineralizing compartments and the recycled concentrate stream flows into the new concentrating compartments.

Polarity reversal provides frequent, automatic flushing of scaleforming materials from the membrane surface. As a consequence of polarity reversal, EDR desalination typically requires little or no pretreatment to minimize the fouling or scaling potential of available feedwater.

## Polarization

"Current density" is defined as the amount of current carried by a unit area of membrane surface. The current density can be increased until nearly all the ions next to the membrane surface are depleted. When this occurs, the concentration polarization or limiting current density is reached. This limit is also related to the water temperature, fluid velocity in the flow path, types of ions in the solution and concentration, type of membrane, and type of flow spacer.

In ED, almost all the ions are transported *through* the membranes via electrical transport. Roughly half of the ions *approaching* the membrane surfaces from the bulk of the solution are carried by electrical transport. The rest of the ions arrive at the membrane surfaces from the stream flow by convection and diffusion. The concentration of ions in the demineralizing cell becomes depleted in the thin layer just adjacent to the membrane surfaces as ions are electrically transferred from the cell through the membranes.

At polarization, the fluid at the membrane surface is depleted of electrolyte essentially completely, thus substantially increasing the electric resistance of the membrane compartment. This abrupt increase of resistance with current density is a sign of polarization. At this point, ions resulting from the dissociation of water have concentrations comparable to those of the electrolyte at the membrane surface. A significant fraction of the current through the anion membrane is carried by hydroxide ions into the concentrate side, resulting in an increased pH and potential calcium carbonate precipitation. The pH changes in the process stream can indicate when polarization is taking place. Polarization causes several undesirable conditions, including decreased current efficiency, increased energy consumption, and reduced transfer of desired ions in the polarization region.

The limiting current density can be increased by increasing the fluid temperature, fluid velocity, or solution concentration. The maximum allowable operating current density should be 70 percent of the limiting current density in commercial ED system design, which allows for a reasonable factor of safety.

Scaling in ED systems as a result of polarization is reduced by adequate turbulence, such as occurs near the membrane surface by the tortuous path of the membrane spacers. Polarization scaling is also reduced or controlled by the addition of chemicals such as acid or antiscalant. EDR breaks up polarization scale without the need for chemicals.

Several important design considerations for ED systems must be evaluated. The ion concentration ratio should be maintained below 150 : 1 (demineralized stream versus concentrate stream); however, occasionally higher ratios are maintained (i.e., up to approximately 200 : 1; maximum is probably 250 : 1). When EDR is used without chemical addition, the concentrate stream should be held below 200% calcium sulfate saturation. With chemical addition to the concentrate stream, 350% calcium sulfate saturation is possible. The Langlier saturation index should be kept below + 2.2. Voltage applied to the membrane stack should be limited to 80 percent of that required to prevent excessive current leakage. The "limiting voltage" is defined as the voltage at which excessive electric current will travel from the electrode laterally through an adjacent membrane to the concentrate stream manifold, generating enough heat to damage some membranes and spacers in the vicinity of the electrode.

### Membrane properties

Both anion and cation transfer membranes have the appearance of translucent plastic sheets reinforced with a synthetic fiber cloth. Membrane surfaces are uniformly flat and smooth. The membranes are rectangular and are manufactured in different sizes. All membranes have the same thickness (0.5 mm) except for the heavy and thicker cation membrane used in the electrode compartment and in interstaging. In systems where additional hydraulic stages are incorporated in a single-membrane stack, one or more interstage membranes are used. This interstage cationic membrane is made twice as thick (1.0 mm) to withstand a greater hydraulic pressure than a normal membrane.

Most anion membranes are translucent white without any color tinge and are stamped with the word "anion." Positive charges are affixed to sites throughout the membrane matrix during production. The fixed positive charges are quarternary ammonium ions that repel positive ions and permit negative ions to pass through.

Cation membranes vary in color from amber-brown to white depending on the type and are stamped with the word "cation." During manufacture, negative charges are affixed to sites throughout the membrane matrix. These fixed negative charges are sulfonate groups that repel negative ions and allow positive ions to pass through.

Anion and cation membranes have several common properties. Both membranes are impermeable to water under pressure, resistant to fouling, long-lasting, resistant to change in pH from pH 1 to pH 10, semirigid, low in electric resistance, insoluble in aqueous solutions, able to operate at temperatures in excess of 115°F (46°C), and resistant to osmotic swelling.

### Energy requirements

The electric energy consumption of feedwater pumping equipment can be approximated at 2.5 kWh/1000 gal for pumping at normal system pressures of 70 to 90 psi (500 to 600 kPa). Membrane stack desalting energy approximates 2.0 kWh/1000 gal per 1000 mg/L of salts removed. Control and instrumentation energy is approximately equal to 5 percent of the total system energy consumption.

### ED/EDR system components and design considerations

Figure 11.13 depicts a flow schematic for an ED system. Water from the feedwater supply is fed into the feed pump at a pressure of 2 to 40 psi (10 to 300 kPa). The feed pump boosts the feedwater to a pressure of 70 to 90 psi (500 to 600 kPa) through a cartridge filter (typically 10 μm). Unlike RO/UF membranes, the ED product water does not pass through the membrane. As a result, provisions must be made in the pretreatment system design if unwanted colloids or organics or both are present in the feedwater. The feedwater normally requires pretreatment if it contains free chlorine over 0.5 mg/L, iron over 0.3 mg/L, hydrogen sulfide over 0.3 mg/L, turbidity above 2.0 NTU, or manganese over 0.1 mg/L.

A low-pressure brine (concentrate) pump is used to increase recovery by recycling a controlled fraction of the concentrate stream through the membrane stack. Electric equipment consists of a dc rectifier power supply and a control module that includes instrumentation. Valving includes automatic motorized valves for feedwater, stream switching (reversing of the concentrate and demineralizing streams in EDR), product water diversion, pressure regulation, and electrode stream control. The two electrode compartments present in each membrane stack are typically supplied with a separate recirculating hydraulic system called the "electrode stream."

ED/EDR is specifically designed for each application. Factors influencing the design are the quantity and quality of product water desired. The quantity determines the size of the ED unit, pumps, piping, and stack size. The quantity of salt to be removed determines the stack array.

The manner in which the membrane stack array is arranged determines the staging. By staging the stack array, sufficient membrane

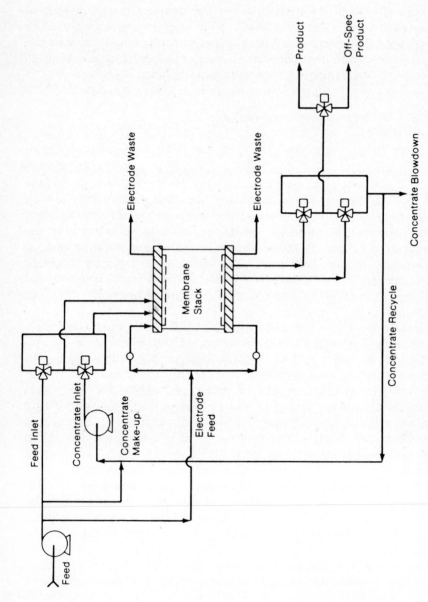

**Figure 11.13** Typical electrodialysis reversal process flow diagram. (*Courtesy of Ionics, Inc.*)

area and retention time are provided for the quantity of salt that must be removed from the demineralized stream. Two types of staging are available: hydraulic and electrical.

Hydraulic staging consists of multiple passes of water between electrodes used in the ED/EDR system to achieve greater demineralization. The maximum salt removal per hydraulic stage is 55 to 60 percent. A normal design range is 40 to 50 percent per hydraulic stage. If a higher-capacity system is desired, additional hydraulic stages may be added in series.

Electrical staging is accomplished by inserting additional pairs of electrodes into a membrane stack to optimize the dc electric system within the stack. The combination of hydraulic and electrical staging gives flexibility to the system, allowing for maximum salt removal rates while avoiding polarization and hydraulic pressure limitations.

### ED/EDR system operation and maintenance

State-of-the-art ED/EDR systems are fully automatic. The operator must record routine readings from the system instrumentation. The records can be logged by a system computer, depending on the sophistication of the instrumentation system.

Routine maintenance consists of changing cartridge filters, maintaining chemical feed systems, calibrating and maintaining instrumentation, replacing membranes occasionally, maintaining pumps and valves, cleaning membranes (depending on how clean feedwater is maintained), and occasionally replacing electrodes.

### ED/EDR system application

ED/EDR systems are used in municipal water treatment for treating brackish water to produce demineralized water that meets drinking water standards. In industry, ED/EDR is used as a roughing demineralizer for the purification of foodstuffs and for wastewater recovery.

### Ultrafiltration

UF is defined as a separation process that uses semipermeable membranes to separate macromolecules in a solution. The size and molecular shape of the solute are important factors in retention.[5,28,29] UF membranes have a much looser pore structure than RO membranes, which do not have definable pores and have only spaces between polymer fibers where a small volume of water can be taken up.[15]

The degree of rejection of solute depends on the MWC of the membrane and the molecular weight of the solute. As stated previously, the size and shape of the solute also affect the solute percentage rejected. Commercial UF membranes with sufficient flux to be applicable to municipal water treatment are available in the MWC range of 1000 to 50,000. Tighter 500-MWC membranes are available on the market and have demonstrated up to 35 to 40 percent salt rejection in pilot tests on municipal surface and groundwater supplies.[6,30] These 500-MWC membranes or lower are not UF membranes, but are nanofiltration membranes. The UF membranes in the 1000- to 50,000-MWC range do not remove dissolved salts.

Normally, driving pressures for UF systems range from 10 to 100 psi (70 to 700 kPa), enabling use of centrifugal pumps. Energy requirements vary with the system design.

### Concentration polarization

Concentration polarization is the phenomenon that causes solutes to concentrate at the membrane surface, forming a gel layer.[4] The gel layer or boundary layer will grow in thickness until eventually the hydraulic resistance reduces the flux. Concentration polarization can be reduced by increasing the turbulence and shear to the flow near the boundary surface, reducing the applied pressures, reducing the concentration of macrosolute (high-molecular-weight solutes) in the system, raising the temperature, and maximizing the solubility of macrosolutes.

### Membrane properties

Ultrafiltration membranes are commercially available in CA, polysulfone, acrylic, and other proprietary noncellulosic polymers. UF membranes have been prepared from a number of other polymers such as polycarbonate, polyvinyl chloride, polyamides, polyvinylidene fluoride, copolymers of acrylonitrile and vinyl chloride, polyacetal, polyacrylates, polyelectrolyte complexes, and cross-linked polyvinyl alcohol.[31]

UF membrane configurations are similar to RO membrane elements and include tubular, plate and frame, spiral-wound, and hollow fine fiber designs. Tubular elements are used for high-fouling applications in industry but are not applicable for large-flow municipal water treatment systems because of higher capital costs, low packing density, and increased energy requirements. UF tubular elements are similar to RO tubular elements. Plate and frame UF elements are similar to RO plate and frame devices and are used in low-flow applications in industry. Plate and frame UF membranes are not applicable for high-flow municipal water treatment system requirements. UF spiral-wound membrane elements are similar to RO spiral elements and are especially adaptable to municipal potable water applications because of their high flux capabilities and low operating pressure ranges.

Hollow fiber UF membranes differ from hollow fiber RO elements. In UF, the feed flow enters the hollow fiber core and exits through the fiber wall, whereas in RO the feedwater passes through the hollow fiber wall from the outside and exits via the hollow fiber core. At present, UF hollow fiber elements are not manufactured in large sizes, unlike their RO counterpart, and so are not as applicable to municipal water treatment systems as spiral-wound membranes.

### UF system components and design considerations

UF system components and UF flow regime are basically identical to those in RO systems (Fig. 11.14). Table 11.3 depicts the types of optional and required pretreatment that can be employed in a UF system.

UF pumping equipment is less costly and more readily available than RO pumping equipment. Standard centrifugal pumps can be

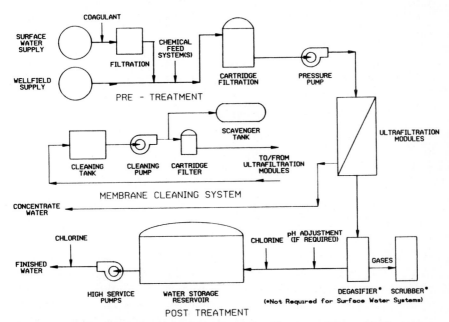

Figure 11.14  Ultrafiltration process flow schematic. (*Courtesy of James M. Montgomery Consulting Engineers.*)

TABLE 11.3  Ultrafiltration System Pretreatment Options

| Pretreatment | Optional | Required | Comments |
|---|---|---|---|
| Cartridge filtration | | X | |
| Conventional filtration | X | | If surface water supply or highly turbid supply is used |
| Acid or antiscalant | X | | If scale control is required |
| Biocide | X | | If biological growth problem develops |
| Static mixing | X | | Depending on piping layout and flow velocities |

used because of the lower UF system operating pressure and less corrosive feedwater supply (nonbrackish water).

The typical UF membrane system for municipal use is a high-recovery system with multiple low-pressure fiberglass vessels in parallel. Spiral-wound high-flux modules are now commercially available in up to 11-in (27.9-cm) diameter. Posttreatment may consist of degasification, chlorination, and stabilization of the product water, as required.

Ultrafiltration system controls typically consist of flow, pressure, temperature, turbidity, and particle counter indicators and recorders. Instruments are usually located in a central operations control panel, but some are mounted on local panels at the equipment control blocks. Instrumentation system sophistication may vary up to the latest state-of-the-art microprocessors and computers. The UF process and instrumentation system would resemble that of an RO system, including many of the fail-safe operating shutdown features.

### UF system operation and maintenance

Routine UF systems have operating parameters similar to those for an RO system, with less emphasis on pretreatment and posttreatment chemical feed systems. Maintenance duties associated with UF systems are fewer than those for brackish or seawater RO systems because of lower system operating pressures and much less aggressive feedwater quality.

### UF system applications

Several important pilot programs in Florida at municipal water treatment plants on both surface and groundwater supplies have proved UF and NF effective for removing organics, indicated by the total organic carbon, trihalomethane formation potential, and color.[30,32,33] In both pilot studies, tight low-MWC (500) NF membranes also exhibited as high as 40 percent rejection of salts. A third pilot plant project is in operation at the time of this writing (1989). UF was proven to be competitive in cost to conventional unit processes to achieve superior results and was considered as a design alternative for a large municipal system in Florida. UF technology has been used in a parallel system with RO as a potable water supply at Pheasant Walk, Fla. UF has been employed as a pretreatment process for seawater RO and has been considered as a process for removing total organic carbon, turbidity, and suspended solids from several advanced wastewater treatment plants prior to conveyance to a reuse system or injection into a groundwater supply.

## Process Performance and Selection

Given treatment process performance capabilities, a membrane process may be selected for a particular situation by considering source water quality, desired product water quality, and economics. Of the various types of membrane processes in existence, RO, NF, and ED/EDR are the only processes actively used in municipal water treatment.

RO effectively removes most organic, bacterial, and particulate matter from water. RO membranes are also effective in removing inorganic contaminants with removal rates ranging between 60 and 99 percent depending on the contaminant being removed and the membrane used.[26,34] ED/EDR effectively reduces dissolved salts from a feedwater but leaves all other dissolved nonionic matter in the product water stream.

NF membranes reject hardness, repel bacteria and virus, and remove organic-related color without generating undesirable chemical compounds such as chlorinated hydrocarbons. This process is being applied to low-TDS waters for softening and organics removal.

UF has been applied to a municipal system in the past[35] at the Pheasant Walk water treatment plant in Florida. Several ongoing pilot programs have demonstrated that UF as a viable process for removing organic contaminants from public water supplies. One such pilot program has been completed at Melbourne, Fla.[30] UF is an effective process for removing most particulate matter, microorganisms, organics, and colloids. The UF process will likely be applied more frequently in the future for the removal of organic contaminants from groundwater and surface water supplies.

## Waste Disposal

All membrane processes have waste streams. RO, NF, ED/EDR, and UF concentrate streams vary in quantity and quality and must be disposed of in accordance with federal, state, and local regulatory requirements.[36–38] Concentrate (also called *reject* or *brine*) disposal is becoming more difficult because of increasing regulatory requirements. In the past, disposal to a brackish body of water was acceptable. Today a disposal permit must be obtained by proving that there will be no detrimental effect to the receiving water.

Acceptable disposal methods include deep-well injection, ocean discharge, discharge to a brackish stream or body of water, dilution followed by spray irrigation, and retention in lined evaporation ponds in combination with other methods. Prior to discharge, concentrate streams may require additional treatment, such as degasification or aeration and chlorination. Regulatory agencies have expressed concerns regarding high levels of radionuclides, hydrogen sulfide, and other constituents. Chapter 16 addresses waste disposal in detail.

## Nanofiltration—An Emerging Technology

Water softening using low-pressure RO membranes has been accomplished in public water supplies in Florida.[8,35,39] Prototype ultra-low-pressure membranes have been successfully used in pilot studies for membrane softening applications and are now available commercially.[6,8] Softening and organics removal are accomplished on low-TDS and slightly brackish water by using these membranes. The MWC of these membranes may vary from less than 200 to 500 as opposed to tight RO membranes that have an MWC of 100. These membranes fall under the category of a new emerging technology called "nanofiltration." The difference in rejection qualities, such as higher rejection of divalent ions than monovalent ions, occurs in the manufacturing process by varying the formulation of the rejection layer bath chemicals and temperature and time in the annealing process. The same basic formulations are used as for RO membranes such as cellulosic, polyamide, and thin-film composite.

The scaling-up of spiral-wound RO and UF membranes to 11-in (27.9-cm) or 12-in (30.4-cm) elements from 8-in (20.3-cm) ones in the future may further reduce membrane process costs while approximately doubling the flow in gallons per day per element. The advantages of using these scaled-up membranes have not been proven to date in large-scale municipal or privately owned membrane process plants. Even if the advantages are proven, industry standardization on element size, as with smaller elements, is needed so that users will not be at a disadvantage when purchasing replacement elements.

## Acknowledgment

At the time this chapter was written William J. Conlon was an employee of Post, Buckley, Schuh, and Jernigan, Inc., Fort Myers, Florida.

# References

1. J. C. Hickman and W. J. Conlon, "The Economics of Low Pressure Reverse Osmosis, Membrane Processes More Economical for Potable Water Treatment than Lime Softening," *Technical Proceedings of the 12th Annual Conference,* Water Supply Improvement Association, St. Leonard, Md., May 1984.
2. L. R. Smauss, Ionics, Inc., Watertown, Mass., personal communication.
3. *Ultrafiltration for Sanitary Service,* Bulletin FD/BR-81-3, Abcor Inc., Wilmington, Mass., 1981.
4. *Ultrafiltration Handbook,* Romicon, Inc., Woburn, Mass., 1983.
5. *Understanding Ultrafiltration,* Cat. No. RP039, Millipore Corporation, Bedford, Mass., 1977.
6. W. J. Conlon, "Pilot Field Test Data for Prototype Ultra-Low Pressure Reverse Osmosis Elements," *Proc. of the Second World Congress on Desalination and Water Reuse,* vol. 3, International Desalination Association, Topsfield, Mass., November 1985.
7. P. Eriksson, *Nanofiltration Extends the Range of Membrane Filtration,* American Institute of Chemical Engineers, New York, November 1986.
8. W. J. Conlon and S. A. McClellan, "Membrane Softening, A Water Treatment Process that Has Come of Age," *J. AWWA,* vol. 81, no. 11, November 1989, p. 47.
9. T. Parrett, "Membranes—Succeeding by Separating," *Technology,* March/April 1982, pp. 20–29.
10. *How It Works, Encyclopedia of Science and Invention,* vol. 12, H. S. Stuttman, Westport, Conn., 1983, p. 1649.
11. S. S. Whipple, Dow Chemical USA, Midland, Mich., 1987, personal communication.
12. Allied Signal, "Fluid Systems," *Reverse Osmosis Principles and Applications,* October 1970, San Diego, Calif.
13. S. Sourirajan, *Reverse Osmosis,* Academic Press, New York, 1970.
14. M. E. Mattson and M. Lew, "Recent Advances in Reverse Osmosis and Electrodialysis Membrane Desalting Technology," *Desalination,* vol. 41, no. 1, Amsterdam, Netherlands, 1982.
15. H. P. Gregor and C. D. Gregor, "Synthetic Membrane Technology," *Sci. Am.,* July 1978, pp. 112–128.
16. E. B. Uvarov, D. R. Chapman, and A. Isaacs, *A Dictionary of Science,* 3d ed., Penguin Reference Books, Baltimore, 1964.
17. L. Applegate, "Membrane Separation Processes," *Chem. Eng.,* June 11, 1984, pp. 64–89.
18. T. J. Sorg and O. T. Love, Jr., "Reverse Osmosis Treatment to Control Inorganic and Volatile Organic Contamination," *Proc. AWWA Seminar: Experiences with Groundwater Contamination,* AWWA Annual Conference, Dallas, June 1984.
19. "National Primary Drinking Water Regulations; Synthetic Organic Chemicals; Monitoring for Unregulated Contaminants; Final Rule," *Fed. Reg.,* vol. 52, no. 130, July 8, 1987, p. 25690.
20. S. C. Lynch, J. K. Smith, L. C. Rando, and W. L. Yanger, *Isolation or Concentration of Organic Substances from Water—An Evaluation of Reverse Osmosis Concentration,* EPA NTIS No. PB 85-124147, January 1985.
21. D. Comstock, "Testing the Membrane Plugging Factor in Reverse Osmosis," *J. AWWA,* vol. 74, no. 9, September 1982, p. 486.
22. DuPont Company, *Permasep Engineering Manual,* Tech. Bull. 502, Wilmington, Del., December 1982.
23. W. J. Conlon and D. L. Rohe, "Energy Recovery in Low Pressure Membrane Plants," *Proc. of the First Biennial Conference,* National Water Supply Improvement Association, June 1986, St. Leonard, Md.
24. Dow Chemical USA, *DOWEX Reverse Osmosis Permeators, Guidelines for Cleaning and Salt Rejection Enhancement/Restoration,* Dow Technical Literature, Midland, Mich., June 1985.
25. G. M. Dykes, "Desalting Water in Florida," *J. AWWA,* vol. 75, no. 3, March 1983, p. 104.
26. Martin R. Huxstep and T. J. Sorg, *Reverse Osmosis Treatment to Remove Inorganic*

*Contaminants from Drinking Water,* EPA 600/S2-87/109, March 1988, Cincinnati, Ohio.

27. T. J. Sorg, R. W. Forbes, and D. S. Chambers, "Removal of Radium 226 from Sarasota County, Florida, Drinking Water by Reverse Osmosis," *J. AWWA,* vol. 72, no. 4, April 1980, p. 230.

28. R. F. Probstein, C. Calmon, and R. E. Hicks, *Separation of Organic Substances in Industrial Wastewater by Membrane Processes,* EPA 600/8-83-011, April 1983, Cincinnati, Ohio.

29. W. J. Weber and E. H. Smith, "Removing Dissolved Organic Contaminants from Water," *Environ. Sci. Tech.,* vol. 20, no. 10, October 1986, p. 970.

30. W. J. Conlon and J. D. Click, "Surface Water Treatment with Ultrafiltration," *58th Annual Technical Conference,* Florida Section AWWA, Florida Pollution Control Assn. and Florida Water & Pollution Control Assn., Orlando, Fla., November 1984.

31. H. K. Lonsdale, "The Growth of Membrane Technology, Bend Research, Inc., 1981," *J. Membrane Science,* vol. 10, 1982.

32. J. S. Taylor, D. Thompson, B. R. Snyder, J. Less, and L. Mulford, *Cost and Performance of In-Plant Trihalomethane Control Techniques,* EPA 600/S2-85/138, September 1985, Cincinnati, Ohio.

33. J. S. Taylor, D. M. Thompson, and J. K. Carswell, "Applying Membrane Processes to Groundwater Sources for Trihalomethane Precursor Control," *J. AWWA,* vol. 79, no. 8, August 1987, p. 72.

34. D. Clifford, S. Subramoniam, and T. J. Sorg, "Removing Dissolved Inorganic Contaminants from Water," *Environ. Sci. Tech.,* vol. 20, no. 11, November 1986, p. 1072.

35. G. M. Dykes and W. J. Conlon, "Use of Membrane Technology in Florida," *J. AWWA,* vol. 81, no. 11, November 1989, p. 43.

36. W. J. Conlon, "Disposal of Concentrate," *Waterworld News,* vol. 5, no. 1, January/February 1989.

37. W. J. Conlon, "Historic Development of Concentrate Regulations," *Proc. of the Disposal of Concentrates from Brackish Water Desalting Plants,* National Water Supply Improvement Association and South Florida Water Management District, Palm Beach Gardens, Fla., November 1988.

38. W. J. Conlon, "Disposal of Concentrate from Membrane Process Plants," *AWWA Seminar Proc.: Membrane Processes: Principles and Practices,* AWWA Annual Conference, Orlando, Florida, June 19, 1988.

39. W. J. Conlon, D. L. Rohe, W. T. McGivney, and S. A. McClellan, "Membrane Softening Facilities in South Florida—A Cost Update," *62nd Annual Technical Conference,* FS/AWWA, FPCA and FW and PDCOA, Fort Lauderdale, Fla., November 1988.

## Bibliography

Flinn, J. E., *Membrane Science and Technology,* Plenum Press, New York, 1970.

Nusbaum, I. and A. B. Riedinger, "Water Quality Improvement by Reverse Osmosis," Robert L. Sanks (ed.), *Water Treatment Plant Design,* Ann Arbor Science, Ann Arbor, Mich., 1979, Chap. 23.

Pohland, H. W., "Reverse Osmosis," in *Handbook of Water Purification,* McGraw-Hill, New York, 1981, Chap. 8.

Porteous, A., *Desalination Technology,* Applied Science Publishers, New York, 1983.

Scott, J., *Desalination of Seawater by Reverse Osmosis,* Noyes Data Corporation, Park Ridge, N. J., 1981.

Sourirajan, S., and T. Matsuura, *Reverse Osmosis/Ultrafiltration Process Principles,* Division of Chemistry, NRC Canada, Ottawa, Ontario, 1985.

# Chemical Oxidation

## William H. Glaze, Ph.D.

*University of North Carolina*
*Department of Environmental Science and Engineering*
*Chapel Hill, North Carolina*

In this chapter the role of oxidants in water treatment is reviewed. Specific oxidants such as chlorine, ozone, and chlorine dioxide are discussed, and some of the future trends expected in the use of chemical oxidants are outlined.

Chemical oxidants play several important roles in water treatment and may be added at several locations in the treatment process depending on the purpose of the oxidant (Fig. 12.1). Chemical oxidants are often added before water enters the treatment plant. This is usually done to control algae and other forms of biological growth that may occur in small holding basins or in pipelines leading to the plant. Oxidants are also added in the first step of treatment either as first-stage disinfectants or for a number of other purposes: control of biological growth in basins, color removal, control of tastes and odors, reduction of specific organic pollutants, precipitation of metals, and coagulant aids. In some cases oxidants are added midway through treatment to control growth on filters, to remove manganese, or to provide an extra level of disinfection. Oxidants are used as the final step in treatment most often for the purpose of disinfection.

Until recently, chlorine was the overwhelming choice when an oxidant was needed in drinking water treatment. Other oxidants such as potassium permanganate, chlorine dioxide, chloramines, ozone, and hydrogen peroxide, while useful for specialty applications, were not serious competitors of chlorine, except in western Europe where ozone is often used.[2]

In 1974 chemists discovered that chlorine reacts with natural or-

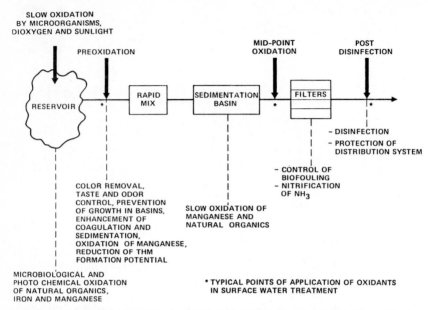

SLOW OXIDATION
BY MICROORGANISMS,
DIOXYGEN AND SUNLIGHT

PREOXIDATION

MID-POINT
OXIDATION

POST
DISINFECTION

RESERVOIR

RAPID
MIX

SEDIMENTATION
BASIN

FILTERS

– DISINFECTION
– PROTECTION OF
   DISTRIBUTION SYSTEM

– CONTROL OF
   BIOFOULING
– NITRIFICATION
   OF NH$_3$

COLOR REMOVAL,
TASTE AND ODOR
CONTROL, PREVENTION
OF GROWTH IN BASINS,
ENHANCEMENT OF
COAGULATION AND
SEDIMENTATION,
OXIDATION OF MANGANESE,
REDUCTION OF THM
FORMATION POTENTIAL

SLOW OXIDATION OF
MANGANESE AND
NATURAL ORGANICS

MICROBIOLOGICAL AND
PHOTO CHEMICAL OXIDATION
OF NATURAL ORGANICS,
IRON AND MANGANESE

* TYPICAL POINTS OF APPLICATION OF OXIDANTS
  IN SURFACE WATER TREATMENT

**Figure 12.1**  Location of points of oxidation in a typical water treatment plant using a surface water source. (*Reprinted with permission from. W. H. Glaze, "Drinking Water Treatment with Ozone," Environ. Sci. Tech., vol. 21, p. 224. Copyright 1987. American Chemical Society.*)

ganics in water to form a group of potentially carcinogenic compounds, the trihalomethanes (THMs): chloroform, bromodichloromethane, chlorodibromomethane, and bromoform.[3,4] THMs are now the subject of federal and state regulations that have caused the water treatment industry to reexamine its uses of chlorine and to consider other oxidants and disinfectants. As we discuss in greater detail, each of the alternative oxidants has advantages and disadvantages. Some oxidants such as hydrogen peroxide and chloramines are not as strong as chlorine; others such as ozone and chlorine dioxide are expensive compared to chlorine and have other properties such as instability (ozone) or by-product formation (chlorine dioxide) that may restrict their usefulness. Nonetheless, THM regulations have the effect of requiring that alternative chemical oxidants and disinfectants be considered in order to minimize THM formation. Already widespread examination of alternative oxidants is occurring, and in the future greater use of ozone, chlorine dioxide, and other chemicals is expected in drinking water plants worldwide. The use of these oxidants involves different technologies and new problems that engineers and operators will have to solve.

## Principles of Oxidation

Oxidation processes involve the exchange of electrons between chemical species so as to change the oxidation state (valence) of the species involved. Strictly speaking, oxidation processes should be referred to as *reduction-oxidation* (redox) *processes* because one species loses electrons or is oxidized while another gains electrons or is reduced.

To illustrate how oxidation processes involve the loss and gain of electrons, the oxidation of ferrous ions ($Fe^{2+}$) by hypochlorous acid is useful to consider. Hypochlorous acid is the species of aqueous chlorine that predominates at acid and neutral pH values. When hypochlorous acid is added to water containing ferrous ions, a rapid chemical reaction occurs:

$$2Fe^{2+} + HOCl + 5H_2O \rightarrow 2Fe(OH)_3 + Cl^- + 5H^+ \qquad (12.1)$$

In this process iron has been oxidized, i.e., increased in oxidation number from $+2$ to $+3$, while chlorine has been reduced, i.e., decreased in oxidation number from $+1$ to $-1$. This means that each iron atom has lost one electron, while each chlorine atom has gained two electrons. To make the total number of electrons lost equal the number gained, two iron atoms must be oxidized for each hypochlorous acid reduced, as shown in Eq. (12.1). On a weight basis this means that 52.5 mg/L of hypochlorous acid can oxidize $2 \times 55.85 = 111.7$ mg/L of divalent iron, or a ratio of approximately 0.5 mg/L of hypochlorous acid per milligram per liter of ferrous iron.[†]

## Oxidation and reduction half-reactions

An oxidation-reduction reaction such as Eq. (12.1) may be separated into a reduction half-reaction and an oxidation half-reaction, as shown in Eqs. (12.2) and (12.3), respectively:

$$HClO + H^+ + 2e^- \rightarrow Cl^- + H_2O \qquad (12.2)$$

$$2[Fe^{2+} + 3H_2O \rightarrow Fe(OH)_3 + e^- + 3H^+] \qquad (12.3)$$

$$2Fe^{2+} + HClO + 5H_2O \rightarrow 2Fe(OH)_3 + Cl^- + 5H^+ \qquad (12.1)$$

---

[†]The weight of hypochlorous acid or hypochlorite required for a reaction is often expressed in terms of chlorine equivalents. Because hypochlorous acid forms from molecular chlorine by the reaction

$$Cl_2 + H_2O \rightarrow HOCl + Cl^- + H^+$$

so 52.5 mg/L of hypochlorous acid is equivalent to 71.0 mg/L of chlorine, or from Eq. (12.1), 71.0 mg/L of chlorine can oxidize 111.7 mg/L of ferrous iron.

In each of these reactions the number of electrons lost or gained is equal to the change in oxidation state of the oxidized or reduced species. In addition, for the equations to be properly balanced, the number of atoms of each element and the net charge must be the same on both sides of the equation.

### Electrode potentials

To effect a change in oxidation state, a powerful enough oxidizing and/or reducing agent is required. The power of an oxidant or a reductant is measured by the electrode potential of the substance. For convenience, electrode potentials are written for half-reactions (Tables 12.1 and 12.2) that can be added to give complete equations, as shown above. By convention, all electrode potentials are tabulated for reduction reactions occurring under "standard conditions," i.e., where the thermodynamic activities of all substances are unity. (Activities are nearly equal to molar concentrations for components of very dilute solutes, and molar concentrations are used instead of activities in this chapter.) Under standard conditions, the electrode potential is called the "standard electrode potential" $E°$.

Values of $E°$ for half-reactions may be summed to calculate the ther-

**TABLE 12.1    Standard Electrode Potentials for Chemical Oxidants Used in Water Treatment**

| Oxidant | Reduction half-reaction | $E$, °V |
|---|---|---|
| Chlorine | $Cl_2(g) + 2e^- \rightarrow 2Cl^-$ | 1.36 |
| Hypochlorous acid | $HOCl + H^+ + 2e^- \rightarrow Cl^- + H_2O$ | 1.49 |
| Hypochlorite | $ClO^- + H_2O + 2e^- \rightarrow Cl^- + 2OH^-$ | 0.90 |
| Chloramines, basic | | |
| Monochloramine | $NH_2Cl + H_2O + 2e^- \rightarrow Cl^- + NH_3 + OH^-$ | 0.75 |
| Dichloramine | $NHCl_2 + 2H_2O + 4e^- \rightarrow 2Cl^- + NH_3 + 2OH^-$ | 0.79 |
| Chloramines, acidic | | |
| Monochloramine | $NH_3Cl^+ + H^+ + 2e^- \rightarrow Cl^- + NH_4^+$ | 1.40 |
| Dichloramine | $NHCl_2 + 3H^+ + 4e^- \rightarrow 2Cl^- + NH_4^+$ | 1.34 |
| Ozone, acidic | $O_3 + 2H^+ + 2e^- \rightarrow O_2 + H_2O$ | 2.07 |
| Ozone, basic | $O_3 + H_2O \rightarrow O_2 + 2OH^-$ | 1.24 |
| Hydrogen peroxide | | |
| Acidic | $H_2O_2 + 2H^+ + 2e^- \rightarrow 2H_2O$ | 1.78 |
| Basic | $HO_2^- + 2e^- + H_2O \rightarrow 3HO^-$ | 0.85 |
| Chlorine dioxide | $ClO_2 + 2H_2O + 5e^- \rightarrow Cl^- + 4OH^-$ | 1.71 |
| Permanganate | | |
| Acidic | $MnO_4^- + 4H^+ + 3e^- \rightarrow MnO_2 + 2H_2O$ | 1.68 |
| | $MnO_4^- + 8H^+ + 5e^- \rightarrow Mn^{2+} + 4H_2O$ | 1.49 |
| Basic | $MnO_4^- + 2H_2O + 3e^- \rightarrow MnO_2 + 4HO^-$ | 0.58 |
| Oxygen | | |
| Acidic | $O_2 + 4H^+ + 4e^- \rightarrow 2H_2O$ | 1.23 |
| Basic | $O_2 + 2H_2O + 4e^- \rightarrow 4HO^-$ | 0.40 |

**TABLE 12.2  Standard Electrode Potentials for Reactions of Interest in Water Treatment**

| Reduction half-reaction | $E°$, V |
|---|---|
| $HBrO + H^+ + 2e^- \rightarrow Br^- + H_2O$ | 1.33 |
| $MnO_2 + 4H^+ + 2e^- \rightarrow Mn^{2+} + 2H_2O$ | 1.21 |
| $ClO_2 + e^- \rightarrow ClO_2^-$ | 1.15 |
| $Fe(OH)_3 + e^- + 3H^+ \rightarrow Fe^{2+} + 3H_2O$ | 1.01 |
| $Fe^{3+} + e^- \rightarrow Fe^{2+}$ | 0.77 |
| $ClO_2^- + 2H_2O + 4e^- \rightarrow Cl^- + 4OH^-$ | 0.76 |
| $ClO_3^- + H_2O + 2e^- \rightarrow ClO_2^- + 2OH^-$ | 0.35 |
| $S(s) + 2H^+ + 2e^- \rightarrow H_2S$ | 0.14 |
| $NO_3^- + H_2O + e^- \rightarrow NO_2^- + 2HO^-$ | 0.01 |
| $NO_2^- + 2OH^- \rightarrow NO_3^- + H_2O + e^-$ | −0.01 |
| $\frac{1}{4}CO_2(g) + H^+ + e^- \rightarrow \frac{1}{24}(glucose) + \frac{1}{4}H_2O$ | −0.20 |

modynamic potential of an oxidation-reduction equation, which is related to the standard free energy change $\Delta G°$ and the equilibrium constant $K$ of the reaction by

$$\Delta G° = -nFE° = -RT \ln K \tag{12.4}$$

In Eq. (12.4), $n$ is the number of electrons transferred in the reaction [two in Eq. (12.1)], $F$ is Faraday's constant, $R$ is the gas constant, and $T$ is the absolute temperature. The larger the positive value of the potential for the chemical reaction, the larger the negative value of the free energy change and the larger the equilibrium constant of the reaction.

To illustrate, the potential of Eq. (12.1) can be calculated from the half-reaction $E°$ values. For Eq. (12.2), $E°$ is found in Table 12.1 to be 1.49 V. For Eq. (12.3) the value is obtained by reversing the sign of the $E°$ value from Table 12.2, and it is $-1.01$ V. The standard potential for the net Eq. (12.1) is the sum of these two values, or $1.49 - 1.01 = 0.48$ V. Because the potential of the reaction is positive, Eq. (12.1) should occur under standard conditions with a free energy change of $-22$ kcal:

$$\Delta G° = -(2e^-)[23 \text{ kcal/(V)}(e^-)](0.48 \text{ V}) = -22 \text{ kcal}$$

The equilibrium constant is calculated as follows:

$$\ln K = \frac{-\Delta G°}{RT} = \frac{-nFE°}{RT} = \frac{+22 \text{ kcal}}{(1.98 \times 10^{-3} \text{ kcal/deg})(298 \text{ deg})}$$

$$= +37.286$$

$$K = 1.6 \times 10^{16}$$

In other words, the reaction should proceed far to the right. At conditions other than standard, the potential of the reaction may be calculated from the Peters-Nerst equation

$$E = E° - \frac{2.303RT}{nF} \log \frac{\pi[\text{products}]^n}{\pi[\text{reactants}]^m} \tag{12.5}$$

The terms to the right of the logarithm are the products of the activities of reaction products and reactants, raised to the power of their coefficients in the balanced equation. For Eq. (12.1), the Peters-Nerst equation is

$$E = E° - \frac{0.0591}{2} \log \frac{[\text{H}^+]^5[\text{Cl}^-]}{[\text{Fe}^{2+}]^2[\text{HClO}]} \tag{12.6}$$

Suppose the following values are used for concentrations: $[\text{H}^+] = 10^{-6}\,M$, $[\text{HClO}] = 10^{-4}\,M$, $[\text{Fe}^{2+}] = 10^{-5}\,M$, and $[\text{Cl}^-] = 10^{-3}\,M$. Then the calculated value of the second term in Eq. (12.6) is 0.56 V.[†] The potential of the reaction is 1.04 V. In other words, under realistic environmental conditions the reaction is even more favorable than at theoretical standard conditions: 1.04 V > 0.48 V.

**Example Problem 1**

1. Write a balanced net reaction for the following half-reactions:

$$\text{ClO}^- + \text{H}_2\text{O} + 2e^- \rightarrow \text{Cl}^- + 2\text{OH}^-$$

$$\text{NO}_2^- + 2\text{OH}^- \rightarrow \text{NO}_3^- + \text{H}_2\text{O} + 2e^-$$

2. From the data in Tables 12.1 and 12.2, select the standard potentials for the half-reactions above. Calculate the standard potential for the net reaction.

3. Calculate the potential of the net reaction for solutions in which the concentrations of the chemical species are as follows: hypochlorite, $1.0 \times 10^{-4}\,M$; chloride, $1.5 \times 10^{-3}\,M$; nitrate, $3.9 \times 10^{-5}\,M$; nitrite, $4.5 \times 10^{-6}\,M$.

**solution**

1. Because both equations involve two electrons, the balanced chemical equation is obtained by summing the two:

$$\text{ClO}^- + \text{H}_2\text{O} + 2e^- \rightarrow \text{Cl}^- + 2\text{OH}^-$$

$$\underline{\text{NO}_2^- + 2\text{OH}^- \rightarrow \text{NO}_3^- + \text{H}_2\text{O} + 2e^-}$$

$$\text{ClO}^- + \text{NO}_2^- \rightarrow \text{Cl}^- + \text{NO}_3^-$$

2. From Tables 12.1 and 12.2 the half-reaction potentials are as follows:

---

[†]The activities of $\text{Fe(OH)}_3$ and $\text{H}_2\text{O}$ have been omitted in Eq. (12.6) according to the convention that the activities of solids and solvents will be nearly equal to 1.

$$ClO^- + H_2O + 2e^- \rightarrow Cl^- + 2OH^- \qquad 0.90 \text{ V}$$

$$NO_2^- + 2OH^- \rightarrow NO_3^- + H_2O + e^- \qquad -0.01 \text{ V}$$

$$ClO^- + NO_2^- \rightarrow Cl^- + NO_3^- \qquad E° = 0.89 \text{ V}$$

3. From Eq. (12.5) the potential of the net reaction under nonstandard conditions is given by

$$E = E° - \frac{0.0591}{2} \log \frac{[Cl^-][NO_3^-]}{[ClO^-][NO_2]}$$

$$= 0.89 - \frac{0.0591}{2} \log \frac{[1.5 \times 10^{-3}][3.9 \times 10^{-5}]}{[1.0 \times 10^{-4}][4.5 \times 10^{-6}]}$$

$$= 0.89 - 0.06 = 0.83 \text{ V}$$

**Example Problem 2**

1. Write a balanced chemical equation for the oxidation of $Mn^{2+}$ to $MnO_4^-$ by ozone under acidic conditions.

2. Calculate the standard potential, free energy, and equilibrium constant for the reaction under standard conditions.

3. Pink water is sometimes found following extensive ozonation of natural water containing reduced manganese. Is this consistent with your results? Explain.

**solution**

1. The oxidation of $Mn^{2+}$ can yield $MnO_2$ or permanganate ($MnO_4^-$). To determine whether permanganate formation is possible, the potential of the reaction from the data in Table 12.1 is calculated. The appropriate half-reactions and net reaction are

$$5[O_3 + 2H^+ + 2e^- \rightarrow O_2 + H_2O]$$

$$2[Mn^{2+} + 4H_2O \rightarrow MnO_4^- + 8H^+ + 5e^-]$$

$$5O_3 + 2Mn^{2+} + 3H_2O \rightarrow 2MnO_4^- + 6H^+ + 5O_2$$

2. From Table 12.1, the standard electrode potentials are as follows:

$$5[O_3 + 2H^+ + 2e^- \rightarrow O_2 + H_2O] \qquad 2.07 \text{ V}$$

$$2[Mn^{2+} + 4H_2O \rightarrow MnO_4^- + 8H^+ + 5e^-] \qquad -1.49 \text{ V}$$

$$5O_3 + 2Mn^{2+} + 3H_2O \rightarrow 2MnO_4^- + 6H^+ + 5O_2 \qquad 0.58 \text{ V}$$

(*Note:* The half-cell potentials for the half-reactions are *not* multiplied by 5 and 2, respectively; electrode potentials are intrinsic quantities that are independent of the amount of the material involved.) The standard free energy of the net reaction $\Delta G°$ and the equilibrium constant may be calculated from Eq. (12.4). Note that the number of electrons involved for the reaction *as written* is 10:

$$\Delta G° = -nFE° = -(10 \text{ eq})[23 \text{ kcal/(V)(eq)}](0.58 \text{ V})$$

$$= -133 \text{ kcal}$$

$$K = \exp\left(\frac{-\Delta G°}{RT}\right) = \exp\left(\frac{133 \text{ kcal}}{(1.98 \times 10^{-3} \text{ kcal/deg}) (298 \text{ deg})}\right)$$

$$= 5.2 \times 10^{97}$$

3. The reaction is highly favored thermodynamically and should proceed essentially to completion. Unless the rate of reaction is slow, formation of permanganate should be possible during ozonation of water with $Mn^{2+}$ present. The observation of pink water after ozonation of such water is consistent with this calculation.

### Limitations of using thermodynamic potentials

Sometimes the tendency is to use tables such as Table 12.1 to rank the power of oxidants. While these comparisons hold true in some cases, a reaction will not necessarily occur, in a practical sense, because it has a high negative free energy. The other factor that determines whether a chemical reaction will occur is chemical kinetics, or how fast the process may occur. A redox process may have a large equilibrium constant in theory, but a rate so slow that it does not proceed in practical terms.

This principle is illustrated by considering the oxidation of organic matter. In general, organic matter on this planet is unstable thermodynamically. All the oxidants in Table 12.1, including oxygen ($O_2$), should oxidize all the organic matter in water to carbon dioxide and water. Humic material is exceptionally stable to oxidant attack, however, illustrating that kinetics, not thermodynamics, is the determining factor.

### Chemical kinetics

Except for the simplest systems, to predict the rate of a chemical equation is not feasible from theory alone. It is possible to determine from experiments the rate of a given reaction, however, as well as how it is affected by changes in concentrations of reactants, catalysts, temperature, and other environmental variables. In this way a model of the reaction system that may be used to predict the rate of the process under other sets of conditions may be established. Such a model is usually referred to as a *rate equation*.

The rate equation for the oxidation of iron with oxygen in natural water is[5]

$$\frac{-d[\text{Fe(II)}]}{dt} = k[\text{Fe(II)}][\text{OH}^-]^2 pO_2 \tag{12.7}$$

This equation states that the rate of ferrous iron oxidation is proportional to the first power of the Fe(II) concentration, the first power of the partial pressure of oxygen in the water, and the second power of the hydroxyl ion concentration. Thus a 10-fold increase in the $\text{OH}^-$ concentration (1 pH unit) will cause a 100-fold increase in the rate of iron(II) oxidation. The proportionality constant $k$ in Eq. (12.7) is referred to as the *rate constant*.

Rate equations are often complex, indicating that the mechanism of a chemical reaction consists of more than one step. Conversely, a simple rate law does not mean that a mechanism is necessarily simple. For example, the decomposition of ozone has an extremely complex mechanism, as we discuss later, yet the rate law for overall decomposition may be quite simple.

**Example Problem 3**    Ozone reacts with phenol either as the neutral phenol molecule or as the phenolate anion. The rate expression of the primary reaction is

$$\frac{d[\text{O}_3]}{dt} = k_{\text{phenol}}[\text{phenol}][\text{O}_3] + k_{\text{phenolate}}[\text{phenolate}][\text{O}_3]$$

The rate constants are as follows:

$$k_{\text{phenol}} = 1.3 \times 10^3 \text{ L/(mol)(s)}$$

$$k_{\text{phenolate}} = 1.4 \times 10^9 \text{ L/(mol)(s)}$$

The dissociation constant for phenol to phenolate is given by

$$\text{Phenol} \rightleftharpoons \text{phenolate} + \text{H}^+$$

The dissociation constant for this reaction is

$$K_{\text{diss}} = 1.26 \times 10^{-10}$$

1. Assuming that the initial total concentration of phenol is $1.00 \times 10^{-5} M$ and that the initial concentration of ozone is $5.00 \times 10^{-4} M$, calculate the initial concentration of phenol and phenolate species at pH 2.00 and pH 10.00.
2. Calculate the value of the two terms in the kinetic expression for $d[\text{O}_3]/dt$ at pH 2 and pH 10. What fraction of the phenol is reacting as phenol and phenolate at these two pH values?

**solution**

1. The expression for the dissociation constant of phenol is

$$K_{\text{diss}} = \frac{[\text{phenolate}][\text{H}^+]}{[\text{phenol}]} = \frac{[x][1 \times 10^{-\text{pH}}]}{[\text{phenol}]_{\text{tot}} - x} = 1.26 \times 10^{-10}$$

where [phenol]$_{tot}$ is the total concentration of phenol or $1 \times 10^{-5} M$, $x$ is the concentration of phenolate, and [phenol]$_{tot} - x$ is the concentration of phenol. At pH values of 2.00 and 10.00, the concentrations of hydrogen ions are $1.00 \times 10^{-2}$ and $1.00 \times 10^{-10} M$, respectively. These values can be substituted into the expression for $K_{diss}$ to yield the following values for phenol and phenolate ion concentrations:

| | Concentration, $M$ | |
| --- | --- | --- |
| pH | Phenol | Phenolate |
| 2 | $1.00 \times 10^{-5}$ | $1.26 \times 10^{-13}$ |
| 10 | $0.44 \times 10^{-5}$ | $0.56 \times 10^{-5}$ |

2. The expression for the rate of loss of ozone is

$$\frac{d[O_3]}{dt} = k_{phenol}[phenol][O_3] + k_{phenolate}[phenolate][O_3]$$

By using the values for the initial concentrations of phenol and phenolate from the solution to part 1 and the values of the rate constants and initial concentration of ozone, the value for $d[O_3]/dt$ may be calculated. For example, at pH 2

$$\frac{d[O_3]}{dt} = [5.00 \times 10^{-4}\,M]\{[1.3 \times 10^3\,L/(mol)(s)][1.00 \times 10^{-5}\,M]\}$$
$$+ [5.00 \times 10^{-4}\,M]\{[1.4 \times 10^9\,L/(mol)(s)][1.26 \times 10^{-13}\,M]\}$$

$$\frac{d[O_3]}{dt} = [6.5 \times 10^{-6} + 8.8 \times 10^{-8}]\,M/s$$

Likewise, at pH 10.0

$$\frac{d[O_3]}{dt} = [5.00 \times 10^{-4}\,M]\{[1.3 \times 10^3\,L/(mol)(s)][0.44 \times 10^{-5}\,M]\}$$
$$+ [5.00 \times 10^{-4}\,M]\{[1.4 \times 10^9\,L/(mol)(s)][0.56 \times 10^{-5}\,M]\}$$

$$\frac{d[O_3]}{dt} = [2.9 \times 10^{-6} + 3.9]\,M/s$$

At pH 2 the majority of the reaction (99 percent) is through the phenol molecule, but the reaction is slow. At pH 10 the reverse is true: Only $7 \times 10^{-5}$ percent of the reaction goes through the phenol molecule, and the reaction is 600,000 times faster than that at pH 2.

## Catalysts

Rates of chemical reactions are often increased by substances that are not consumed in the reaction. These substances are referred to as "catalysts." Sometimes a product of a reaction is catalytic, in which case the reaction is said to be "autocatalytic," that is, it will accelerate as the reaction proceeds. This is the case for the oxidation of manganese(II) by molecular oxygen where the rate equation is[6]

$$\frac{-d[\text{Mn(II)}]}{dt} = K_0[\text{Mn(II)}] + k[\text{Mn(II)}][\text{MnO}_2] \qquad (12.8)$$

where

$$k = k_1 [\text{OH}^-]^2 p\text{O}_2$$

The second term on the right shows that the rate of oxidation of Mn(II) increases as the concentration of the manganese dioxide product increases. That is, $\text{MnO}_2$ is a catalyst for the oxidation process. One explanation is that the Mn(II) absorbs onto the solid $\text{MnO}_2$ to form a surface complex in which the Mn(II) is more susceptible to oxidation than free Mn(II).

### Effect of pH and temperature

Rates of oxidation reactions are often affected by changes in pH because oxidation agents are susceptible to changing form or mechanism under basic or acidic conditions. A common example is the case of aqueous chlorine discussed later in this chapter. Chlorine can exist in three principal forms in aqueous solutions depending on the pH: chlorine gas $\text{Cl}_2$, hypochlorite ion $\text{ClO}^-$, and hypochlorous acid HOCl. The reactivities of these forms are different, so major rate effects are observed in chlorination reactions at different pH conditions as the relative proportions of the three species change. Likewise, ozonation rates are highly dependent on pH. Above pH 8 to 9, ozone decomposes to yield the highly reactive hydroxyl radical HO; thus, rates of ozonation are often observed to change at high pH levels.[7]

Changes in temperature also have an important effect on the rates of chemical reactions. In general, the rate constant $k$ of a chemical reaction will increase with increasing temperature according to a logarithmic function:

$$\ln k = -\frac{E_a}{RT} \qquad (12.9)$$

where $E_a$ is an empirical constant, the energy of activation. For typical values of this parameter, a 10° rise in temperature will cause the reaction rate to increase twofold or threefold. The net effect of this principle applied to water treatment is that oxidation rates may be substantially retarded in colder waters.

### Pilot-scale studies

To determine the effects of temperature, pH, and other water quality variables on oxidation processes, it is generally necessary to carry out tests using the water of interest. This is best done on a pilot scale where the conditions expected in the full-scale treatment plant can be simulated and various oxidant combinations attempted.

## Oxidants Used in Water Treatment

### Chlorine

As noted earlier in this chapter, chlorine is the most common oxidant used in water treatment. Chlorine is available in liquid or gaseous form in pressurized metal tanks, as a concentrated aqueous solution (sodium hypochlorite), or a solid (calcium hypochlorite). The solution is used primarily for small treatment plants and for swimming pools. Once it has been added to water, the chemistry of all forms of chlorine is essentially the same.

When pure chlorine is added to water, the chlorine rapidly disproportionates to form hypochlorous acid (HOCl):

$$Cl_2 + H_2O \rightarrow HOCl + H^+ + Cl^- \qquad (12.10)$$

The equilibrium constant for Eq. (12.10) is $5 \times 10^{-4}$ at 25°C. This means that the concentration of $Cl_2$ in solution represents only a small fraction of the total chlorine, except at very low pH (below 1, see Fig. 12.2). For example, in water at pH 2 and $10^{-3}$ $M$ chloride concentration, a 1 mg/L chlorine solution will have only 2 percent of the chlorine as $Cl_2$.

HOCl is a weak acid ($pK_a = 7.5$) and partially dissociates to hypochlorite ion ($OCl^-$):

$$HOCl \rightleftharpoons H^+ + OCl^- \qquad (12.11)$$

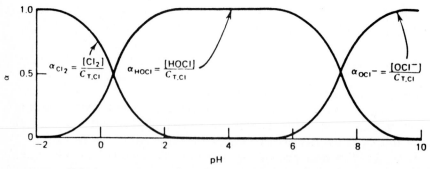

**Figure 12.2** Relative amounts of molecular chlorine, hypochlorous acid, and hypochlorite in water as a function of pH. $\alpha$ = fraction of each species, $C_{T,Cl}$ = sum of all species; $[Cl^-] = 1.0 \times 10^{-3}$ $M$. (*Source: Water Chemistry. Chap. 7 by V. L. Snoeyink and D. Jenkins. Copyright 1980. John Wiley & Sons, Inc.*)

The ration of the hypochlorous acid to hypochlorous may be calculated from

$$\log \frac{[\text{HClO}]}{[\text{ClO}^-]} = 7.5 - \text{pH} \qquad (12.12)$$

Equations (12.10), (12.11), and (12.12) shows that the species of chlorine present in water will depend on the total concentration of chlorine, the pH, and the temperature. Figure 12.2 shows a diagram of the relative amounts of the three species as a function of pH. As Fig. 12.2 shows, hypochlorous acid is the predominant species between pH 1 and pH 7, the species are at equal concentrations at a pH of 7.5 (the $pK_a$ value), and hypochlorite ion is predominant at high pH values.

Chlorine in the form of calcium hypochlorite granules dissolves in water to form the hypochlorite ion ($OCl^-$) which then reacts with water:

$$\text{OCl}^- + \text{H}_2\text{O} \rightleftharpoons \text{HOCl} + \text{OH}^- \qquad (12.13)$$

The relative percentages of HOCl and $OCl^-$ that result are determined by the pH, total chlorine concentration, and temperature, as before. In other words, no matter what form of chlorine is added to water, hypochlorite, hypochlorous acid, and molecular chlorine will be formed as described by the laws of chemical equilibrium shown in the foregoing reactions.

As shown in Table 12.1, hypochlorous acid and hypochlorite ions are both strong oxidizing agents, but HOCl should be the stronger of the two. This is true in actual practice; that is, oxidation reactions and disinfection reactions of chlorine are usually more effective at low pH values. More details on the disinfection properties of chlorine are given in Chap. 14, Disinfection.

Liquid chlorine is usually added to water by evaporating chlorine liquid to gas, metering the gas into water through an ejector, and then diffusing the solution into the flow stream at the point of application. Chlorine gas also may be added directly to the main flow or, more commonly, into a sidestream that is added to the main flow. In the latter case, a concentrated solution containing several hundred milligrams per liter of chlorine is prepared.

Solid granules of calcium hypochlorite can be added to the water directly or as a concentrated solution. Sodium hypochlorite solution is usually added directly to the main flow.

Aqueous chlorine is one of the most versatile and effective chemical oxidants. Its uses include the oxidation of manganese, color removal, control of off-tastes and odors, and a flocculent aid. Unfortunately, chlorine also forms potentially mutagenic and carcinogenic by-products, especially when it is used in the treatment of surface water.[3] From a large body of research on this subject, by-products of chlorination are formed from the chlorination of natural organic compounds present in some water sources. A variety of precursor compounds are probably involved: humic and fulvic acids (themselves a complex mixture of compounds), chlorophyll, and other components or metabolites of algae and bacteria that grow in municipal water supplies.[9]

The principal chlorination by-products of health concern are low-molecular-weight chlorinated and brominated compounds, including chloroform and other trihalomethanes, di- and trichloroacetic acids, chloropicrin, and halogenated acetonitriles.[10] A very large number of other compounds are also formed in smaller quantities, including halogenated ketones and various halogenated aliphatic compounds.[11] The group parameter *total organic halogen* (TOX) has been used to measure the combined levels of these halogenated by-products.[12] TOX levels in chlorinated water depend on the level of total organic carbon (TOC) in the raw water, but substantial variations occur depending on the source of the water and the pH at which chlorine is applied.

Bromine-substituted compounds are formed during the chlorination process as a result of the oxidation of bromide by chlorine:

$$\text{Br}^- + \text{HOCl} \rightarrow \text{Cl}^- + \text{HOBr} \tag{12.14}$$

Hypobromous acid competes with HOCl for reaction with the active sites that produce the low-molecular-weight by-products. If natural bromide levels are high, as in the case of water that is highly saline, the levels of bromine-substituted by-products produced by chlorination may exceed the levels of chlorinated compounds.[13]

In spite of these by-products, chlorine is still useful as an oxidant and disinfectant where the chemical demand is low, as in well-treated water and good-quality groundwater.

### Chloramines (combined chlorine)

Chloramines are formed by the reaction of ammonia with aqueous chlorine. This may be for the purpose of preparing chloramines to act as disinfectants, or it may occur in the process of water or wastewater chlorination when ammonia is present. The mixture that results may contain monochloramine ($\text{NH}_2\text{Cl}$), dichloramine ($\text{NHCl}_2$), and trichloramine ($\text{NCl}_3$):

$$\text{NH}_3 + \text{HOCl} \rightarrow \text{NH}_2\text{Cl} + \text{H}_2\text{O} \tag{12.15}$$

$$\text{NH}_2\text{Cl} + \text{HOCl} \rightarrow \text{NHCl} + \text{H}_2\text{O} \tag{12.16}$$

$$\text{NHCl}_2 + \text{HOCl} \rightarrow \text{NCl}_3 + \text{H}_2\text{O} \tag{12.17}$$

The relative amounts of the three species formed will depend primarily on the pH and the ratio of ammonia to chlorine. At high chlorine-ammonia ratios, Eq. (12.17) becomes important, and when free ammonia is exhausted, chlorine may oxidize the chloramines, for example, by Eq. (12.18):

$$2NH_2Cl + HOCl \rightarrow N_2(g) + 3H^+ + 3Cl^- + H_2O \qquad (12.18)$$

This equation illustrates the so-called breakpoint phenomenon where chlorine in large doses will eventually oxidize ammonia, beyond which free chlorine may become the predominant species.[14]

At a chlorine-ammonia weight ratio of 3 : 1 and a pH of 7, conditions common during chloramination of drinking water, the principal chloramine formed is monochloramine ($NH_2Cl$). Figure 12.3 shows the initial and equilibrium concentrations of mono- and dichloramine formed with equimolar aqueous solutions of chlorine and ammonia at 25°C.[8] The figure is an excellent example of how a reaction system can be controlled initially by chemical kinetics and later by thermodynamics. Initially, Eqs. (12.15) and (12.16) occur simultaneously, yielding $NH_2Cl$ and $NHCl_2$. For example, Fig. 12.3 shows that if the reaction is carried out at pH 7, the yields of $NH_2Cl$ and $NHCl_2$ will be 92 and 8 percent, respectively. If this water is allowed to stand for a day

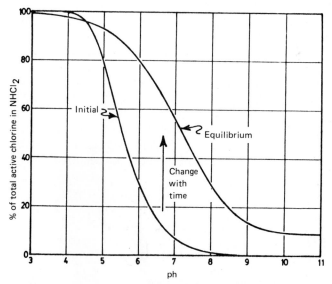

**Figure 12.3** Initial and equilibrium distributions of $NHCl_2$ and $NH_2Cl$ with equimolar aqueous chlorine and ammonia at 25°C. (*Reproduced, with permission, from R. L. Jolley et al. (eds.), "Water Chlorination," Environmental Impact and Health Effect, vol. 1, Butterworth Publishers, Stoneham, Mass., 1983.*)

or so, the monochloramine will disproportionate slowly according to

$$2NH_2Cl \rightarrow NHCl_2 + NH_3 \qquad (12.19)$$

If free chlorine is available, the ammonia will be chlorinated, disproportionation will continue, and the system will eventually come to equilibrium. Calculations show that at equilibrium the water will contain 43% $NH_2Cl$ and 57% $NHCl_2$. This illustrates that the chlorine-ammonia system is a complex one, where the relative amounts of the two principal species will change during the time that water is in a treatment plant or perhaps further as the water passes through the distribution system.

Chloramines are prepared in water treatment by the addition of ammonia to water after chlorine has already been added, by adding ammonia first, or by adding the chemicals simultaneously. To avoid the formation of trichloramine, the ratio of chlorine to ammonia in the mixing zone should be 2 : 1 to 5 : 1 on a weight basis.

Because free chlorine may be present for brief periods during chloramine formation, and because Eq. (12.5) is slowly reversible, differentiating the oxidizing effects of chloramines from hypochlorite is difficult. Laboratory studies indicate, however, that chloramines are very weak oxidants even though the oxidation potential of monochloramine is 0.75 V (Table 12.1).[15] From a practical perspective, chloramines are too weak to serve as oxidants for removal of manganese, and they do not lower THM formation potential (THMFP) or most taste and odor compounds.[16,17]

Although chloramines do not react with natural organics to yield THMs, some reaction is apparent because TOX increases when natural water is chloraminated.[18] The details of this reaction are not clear; chloramines will, however, chlorinate phenols and primary or secondary organic amines.

In general, the value of chloramines in water treatment lies not in their role as an oxidant but in that as a disinfectant. Disinfectant properties of chloramines are described in Chap. 14, Disinfection.

## Ozone

Ozone is an unstable gas that must be generated on-site.[1] The chemical equations for the formation of ozone ($O_3$) can be written in simplified form as

$$O_2 + energy \rightarrow O + O \qquad (12.20)$$

$$O + O_2 \rightarrow O_3 \qquad (12.21)$$

The energy required to dissociate oxygen in Eq. (12.20) is usually supplied by a dc or ac electric field with a peak voltage from 8000 to 20,000 V depending on the apparatus used. Under these conditions a plasma is generated that is called a "corona discharge" or a "cold plasma discharge." The yield of ozone will depend on the voltage, frequency, design of the ozonator, and type of gas used. Additional details concerning generation and feeding of ozone are provided in Chap. 14, Disinfection.

Ozone has limitations that impede its effectiveness as an oxidant in water treatment. Most important, ozone is so reactive that it will dissipate rapidly in most types of natural water, either by reacting with natural constituents such as humic material or by a spontaneous decomposition process.

Ozone decomposition is a complex chain-radical process that may be initiated by several types of water contaminants such as hydroxide ion (high pH values), natural organics, and ferrous iron.[19] The reactions shown in Eqs. (12.22) through (12.27) illustrate the scheme that occurs when hydroxide ion is the initiator.

$$HO^- + O_3 \rightarrow HO_2 + O_2^- \qquad (12.22)$$

$$HO_2 \rightleftharpoons H^+ + O_2^- \qquad (12.23)$$

$$O_2^- + O_3 \rightarrow O_2 + O_3^- \qquad (12.24)$$

$$O_3^- + H^+ \rightarrow HO_3 \qquad (12.25)$$

$$HO_3 \rightarrow O_2 + HO \qquad (12.26)$$

$$HO + O_3 \rightarrow HO_2 + O_2 \qquad (12.27)$$

These reactions constitute a chain mechanism because the $HO_2$ product of Eq. (12.27) may initiate a new chain through Eqs. (12.23) and (12.24). In pure water the chain may be very long; i.e., hundreds of ozone molecules may be decomposed by a single initiation step. In actual water, the lifetime of ozone depends on several variables including pH, temperature, TOC level, and bicarbonate level.[19]

Bicarbonate increases the lifetime of ozone by stopping the chain mechanism shown in the above equations by reacting with the hydroxyl radical intermediate HO:

$$HO + HCO_3^- \rightarrow OH^- + HCO_3 \qquad (12.28)$$

The bicarbonate radical $HCO_3$ is a relatively unreactive intermediate that cannot propagate the chain; thus, water high in bicarbonate alkalinity and low in other contaminants will retain an ozone residual for longer periods than low-alkalinity high-TOC water.[19,20]

Recently, evidence has been published that describes how the decomposition of ozone may affect the rates of oxidation processes involving ozone.[19] The hydroxyl radical HO produced in the decomposition of ozone [Eq. (12.26)] is one of the most powerful chemical agents known, capable of reacting with almost any organic compound. The rate of oxidation of an organic compound such as benzene, therefore, may increase substantially when the pH of the water is increased above 8, due to the higher rates of formation of HO through Eqs. (12.22) to (12.27). However, in water with high alkalinity the rate of oxidation may decrease as the pH is increased because of the formation of carbonate ions which are more effective scavengers of OH radicals than bicarbonate ions.[21]

Compounds that dissociate in water may react with ozone at different rates depending on the species present. For example, phenolate anion reacts much faster than phenol.[22] Hence, the rates of ozonation may increase or decrease as the pH changes, depending on the specific compound being oxidized and matrix contaminants.

Ozonation also forms by-products when used in water treatment, particularly with surface water containing natural organics. These by-products are generally polar compounds including aldehydes, aliphatic and aromatic carboxylic acids, and other polar compounds such as quinones and peroxides.[23] None of these has yet been shown to have potential health significance although little information is available on this subject. Compared to chlorine and chlorine dioxide, ozone appears to yield the smallest quantities of mutagenic by-products and under some conditions may even decrease the mutagenicity of water.[24,25]

Ozone by-products are generally more biodegradable than their precursors. Hence, water that has been ozonated may show active growth of bacteria if it is not well disinfected. This subject is discussed in Chap. 14, Disinfection.

Ozone also may form bromoform ($CHBr_3$) and other bromine-substituted by-products by reacting with bromide in natural water to form hypobromous acid (HOBr), which can react with natural organics:

$$O_3 + Br^- + H^+ \rightarrow HOBr + O_2 \qquad (12.29)$$

$$HOBr + \text{natural organics} \rightarrow CHBr_3 \text{ etc.} \qquad (12.30)$$

Field studies have shown that the yield of bromoform is not significant unless the bromide level is high, for example, in certain groundwater[26] and in seawater ozonated for control of biofouling in cooling towers.[27]

## Potassium permanganate

Potassium permanganate ($KMnO_4$) has been used as an oxidant for water treatment for decades. It may be fed into water as a solid or as a solution prepared on-site. The common point of addition is at the head of a treatment plant.

According to the thermodynamic potentials listed in Tables 12.1 and 12.2, permanganate should be capable of oxidizing virtually any organic compound and many of the inorganic pollutants of concern such as iron and divalent manganese. The equation for manganese (II) oxidation is the sum of Eqs. (12.31) and (12.32):

$$3Mn^{2+} + 6H_2O \rightarrow 3MnO_2 + 12H^+ + 6e^- \qquad E° = -1.21 \tag{12.31}$$

$$2MnO_4^- + 8H^+ + 6e^- \rightarrow 2MnO_2 + 4H_2O \qquad E° = 1.68 \tag{12.32}$$

$$3Mn^{2+} + 2MnO_4^- + 2H_2O \rightarrow 5MnO_2 + 4H^+ \qquad E° = 0.47 \tag{12.33}$$

The net reaction has a positive $E°$ value that in this case is accompanied by a favorable rate equation. The $MnO_2$ is insoluble at normal pH values and can be removed by filtration. (See below for a discussion of the precise nature of the manganese dioxide precipitate.) The manganese dioxide also removes Mn(II) ions by adsorption, thus decreasing the amount of permanganate necessary to remove a certain quantity of Mn(II). Permanganate enhances the removal of TOC, possibly by adsorption of organics on manganese dioxide flocs.[28]

Permanganate has been applied for the oxidation of cyanide, for oxidation of phenols, for taste and odor control, and for color removal. It is generally not capable of removing difficult-to-oxidize taste and odor compounds[17] or of oxidizing THM precursors[29] completely.

Permanganate oxidation yields precipitates that are sometimes difficult to handle, because they form mud balls on filters, for example. Also permanganate doses must be carefully controlled, or pink water will result.

## Chlorine dioxide

Chlorine dioxide is a powerful oxidant that is always prepared on-site[30] most economically by the reaction of chlorine and sodium chlorite [Eq. (12.34)] preferably under low-pH conditions because of improved kinetics:

$$Cl_2 + 2NaClO_2 \rightarrow 2ClO_2 + 2NaCl \tag{12.34}$$

The chlorine may be introduced to the water and then added to a solution containing hydrochloric acid and sodium chlorite or may be injected as a gas under vacuum into a stream of chlorite solution. The former method is apt to leave substantial amounts of chlorine in the solution, which may lead to the formation of chlorinated by-products.

Chlorine dioxide may also be prepared by acidification of a chlorite solution, which causes a disproportionation reaction to occur:

$$5NaClO_2 + 4HCl \rightarrow 4ClO_2 + 5NaCl + 2H_2O \qquad (12.35)$$

This method may be preferable to reaction (12.34) if the use of chlorine is not convenient, but for large-scale applications reaction (12.34) is preferred.

Chlorine dioxide is an explosive gas, but is stable in water in the absence of light and elevated temperatures. It disproportionates (chlorine valence changes from +4 on the left to +3 and +5 on the right) to form both chlorite and chlorate in a process that is base-catalyzed:

$$2ClO_2 + 2OH^- \rightarrow ClO_2^- + ClO_3^- + H_2O \qquad (12.36)$$

Because chlorite and chlorate are both strong oxidants (Table 12.2) and complexing agents with potential health impacts,[31] alkaline conditions are to be avoided when chlorine dioxide residuals are present. As noted below, chlorite is also produced in the reaction of chlorine dioxide with some natural and synthetic organics; hence, the level of chlorine dioxide used in drinking water treatment should be carefully controlled. Presently the EPA has recommended that the total residual level of the three oxidants (chlorite, chlorate, and chlorine dioxide) be no more than 1.0 mg/L.[32] Effectively this limits the use of chlorine dioxide to situations in which the demand is low.

Chlorine dioxide is capable of oxidizing iron and manganese, removing color, and lowering THM formation potential.[33] It also oxidizes many organic and sulfurous compounds that cause off-tastes and odors, although it adds a special taste to water that is objectionable to some.[17,28] No THMs are formed with chlorine dioxide treatment, and when chlorine dioxide is accompanied by chlorine resulting from the generation process, the yields of THMs are lowered.[33]

Organic compounds are oxidized by chlorine dioxide by a variety of mechanisms,[33] and in some cases chlorine substitution occurs in the products. TOX concentrations found after chlorine dioxide treatment of surface water are possibly caused by the reaction of the oxidant with phenolic constituents of natural organics. Laboratory-scale studies have shown that phenols react with chlorine dioxide to form chlorinated phenols and quinones.[33,34] The possible health effects of these by-products have not been established.

## Application of Oxidants in the Treatment Process

This section describes typical applications of oxidants and the role of the oxidant in the overall treatment process (Fig. 12.1).

### Control of iron and manganese

Iron and manganese often enter a treatment plant in soluble forms, usually as ferrous ions ($Fe^{2+}$) and manganous ions ($Mn^{2+}$) complexed to various organic and inorganic species. These metals are often found in groundwater and occasionally in surface water drawn from below the thermocline of a stratified reservoir. Manganous and ferrous ions can oxidize and precipitate in the distribution system, possibly causing fouling and discoloration problems.

Chemical oxidants are often added at the beginning of the treatment process to oxidize iron and manganese. If the addition of the oxidant is delayed, it must come before filtration. Almost all the common oxidants except chloramines will convert ferrous (+2) iron to the ferric (+3) state and manganese to the +4 state, where they precipitate as ferric hydroxide and $MnO_2$, respectively. The precise chemical composition of the precipitate will depend on the nature of the water. For example, manganese(IV) precipitate may contain $MnOOH$, or the iron may precipitate as the carbonate in water of high alkalinity or at pH values near 9.

Manganese may also be removed by the use of greensand filter medium, that is, medium containing absorbed manganese, which catalyzes the oxidation of Mn(II) by molecular oxygen. Generally speaking, manganese oxidation is slower than iron oxidation, and the former is more troublesome to handle if it is present in the source water.

Oxidation of iron and manganese with oxygen from the air is catalyzed by hydroxide ions and so may be accelerated by addition of a pH modifier such as lime or caustic soda. Strongly complexed iron and manganese are not readily oxidized with oxygen, forcing the use of a more powerful oxidant.[5,6] The most common oxidants for this purpose have been chlorine and potassium permanganate, but in Europe chlorine dioxide or ozone is also commonly used.

### Control of biological growth (biofouling) in the treatment plant

Control of biofouling is necessary whenever water is used under conditions where biological growth is possible. This includes drinking water treatment plants, cooling towers, and source water transmission lines. Control of biological growth in distribution systems is covered in Chap. 18, Microbiological Quality Control in Distribution Systems.

Treatment plants that have basins and filters open to sunlight are especially susceptible to growth of algae and other forms of microorganisms in basins and on filter media. Growth may be minimized by the use of a chemical oxidant and the maintenance of a residual level of the oxidant through the plant. Alternatively, the system may be "shocked" with occasional high doses of oxidant. Control of biofouling may also be necessary if membrane processes, such as reverse osmosis, are used in treatment. Details of this control may be found in Chap. 11, Membrane Processes.

Chlorine has been the most commonly used oxidant for control of biological growth, but other alternatives are now being investigated. Ozone and chlorine dioxide are effective substitutes for chlorine, but each has disadvantages. Ozone is so reactive toward various water contaminants that it generally does not survive very long in a treatment plant. Ozone applied as a preoxidant may dissipate by the time water has passed through the sedimentation basins, and thus ozone is not effective for control of growths there or in the filters. As mentioned previously, ozone also reacts with natural organics in water to form lower-molecular-weight organic by-products that are more biodegradable than their precursors. As a result, growth in basins and filters can actually be enhanced by preozonation. Two-stage ozonation, that is, application of ozone at the head of the plant and midway through the treatment process, may be necessary to overcome this problem.

### Color removal

Color is a property of water caused by the presence of organic and inorganic substances, usually of natural origin, which absorb visible light. The nature of these substances and the molecular basis of the color vary with the source water. In most cases, color is probably caused by natural organics (humic substances) that probably have complexed metals bound into their structures.[35] Addition of a chemical oxidant usually reduces—but may not completely remove—color. The mechanism by which this takes place is not clear. Oxidizing agents are thought to attack the chromophores (parts of the humic molecule that absorb visible light) and convert them to chemical forms that do not absorb visible light. For example, both chlorine and ozone attack carbon-carbon multiple bonds, oxidize metals, and break up chelates, all of which can contribute to absorption of visible light.

An oxidizing agent may not be capable of destroying color by itself, at least not economically. Preoxidation in combination with chemical flocculation, sedimentation, and filtration is usually the method of choice for removing color from water. Of the common oxidants, chlorine, ozone, and chlorine dioxide are most effective.

### Removal of tastes and odors

Obnoxious tastes and odors may be found in water supplies, particularly surface water supplies susceptible to growth of microorganisms such as blue-green algae. Taste and odor problems often come on suddenly and with vehemence, usually related to the growth of a microorganism in the source water. Strong tastes and odors may also be caused by microbiological or chemical processes that occur in groundwater, producing sulfur-containing compounds. Taste and odor compounds are often detectable by human senses at very low levels (nanograms per liter). Most often a given taste or odor problem is caused by a combination of chemical agents. The chemicals identified as contributors to off-tastes and odors include sulfides, saturated and unsaturated aldehydes, geosmin (trans-1.10-dimethyl-trans-9-decalol), and MIB (methylisoborneol). The last two are among the most common taste and odor compounds found in municipal water supplies. They are thought generally to be metabolites of blue-green algae and actinomycetes.

Oxidation is one common treatment method for controlling taste and odor problems, but it does not always meet with complete success.[16,17] Moreover, each of the oxidants adds new tastes and odors to the water.[36] Chlorine, potassium permanganate, and, in Europe, ozone and chlorine dioxide are common taste and odor control oxidants.

Another common method for controlling taste and odor problems is the use of powdered or *granular activated carbon* (GAC) (see Chap. 13, Adsorption of Organic Compounds). Chlorine should not be used in contact with activated carbon because the two react, decreasing the effectiveness of the chlorine and possibly forming chlorinated by-products.[37]

In Europe, chlorine dioxide and ozone are commonly used for taste and odor control, and these oxidants are also becoming more popular in North America.[37] Of the common oxidants, ozone appears to be the most effective at destroying some of the recalcitrant taste and odor compounds, particularly geosmin and methylisoborneol, but all oxidants are limited in their effectiveness of taste and odor control. In the later section new oxidant systems are discussed that may have the potential for becoming better control agents for taste and odor.

## Oxidation as an aid to flocculation

Oxidation of surface water may destabilize colloidal material, leading to improved flocculation, sedimentation, and filtration. The mechanism of this phenomenon is not clear, but several possibilities have been offered, including oxidation of organics into more polar forms, causing them to desorb from and thus destabilize clay particles; oxidation of humic material to form more polar and/or chelating groups that induce coagulation; and oxidation of metal ions to yield insoluble forms such as Fe(III).[38] One recent (1986) study (Fig. 12.4) showed that ozone is more effective as a coagulant aid than chlorine, and at least one large treatment plant using preozonation for this purpose is now operating in the United States.[39]

If preozonation improves coagulation, cost savings may result because of lower dosages of coagulant chemicals that are required. Unit processes such as flocculation and sedimentation may be eliminated or reduced in scale, also resulting in cost savings. Further research on the ozonation process may further advance its effectiveness.

## Oxidation of THM and TOX precursors

As described above, THMs and other halogenated organics are produced by the reaction of chlorine with natural organics, often called humic materials.[3,9] Both chlorine and natural organics are needed for

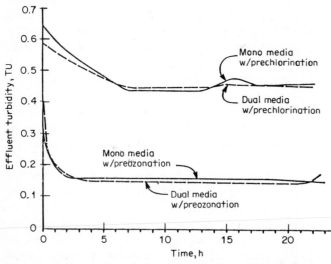

**Figure 12.4** Comparison of preozonation and prechlorination in reducing turbidity levels at Los Angeles Aqueduct pilot plant. (*Source: Ozone: Science & Engineering, vol. 8, 1986, p. 77, with permission.*)

THMs and TOX formation; thus, removing either will prevent or minimize the formation of THMs and TOX. A useful treatment strategy is to use chlorine dioxide or ozone as a preoxidant in place of chlorine to oxidize THM precursors so that chlorine could be used later in the treatment process. This results in additional benefits mentioned above. Alternatively, oxidation can be used after coagulation, sedimentation, and filtration where the oxidant demand is lower and where THM formation potential control may be equally effective with a lower dose of oxidant.

Ozone[40,41] and chlorine dioxide[30,33] are two effective oxidants to react with natural organics to decrease THM formation potential. Under certain conditions, however, ozonation can actually increase the THM formation potential of water, while under other conditions substantial reductions can occur.[42] Because of water quality variations, particularly humic composition and bicarbonate levels,[42] each case must be treated individually, and pilot-scale studies are highly recommended.

Data from a pilot-scale study in which ozone was added to a surface water after coagulation, sedimentation, and filtration is shown in Fig. 12.5.[41] Ozonation at a dose of 2.5 mg/L removed 10 to 15 percent of the THM formation potential of the water. Higher doses will achieve greater reductions but may not be cost-effective. Combined with pretreatment, ozonation removed 35 to 50 percent of the precursors.

The studies described in Refs. 41 and 42 illustrate another application of oxidants in water treatment, i.e., stimulation of biological activity in GAC and sand filters. Preozonation has been shown to enhance the degradation of natural organics and the removal of THM formation potential on GAC far beyond the point where the adsorption capacity of the GAC is exhausted.[42] More details on this process are included in Chap. 13, Adsorption of Organic Compounds.

### Oxidation of phenols and synthetic organics

Strong chemical oxidants are capable of reacting with synthetic organic chemicals present in water supplies. In principle, chemical oxidation is superior to other methods for removal of such impurities because oxidation can convert organic compounds to innocuous by-products while air stripping, GAC adsorption, and reverse osmosis transfer the problem contaminant from one phase to another. Chemical oxidation of synthetic organics is highly dependent on the nature of the organic compound, the oxidant, and other contaminants in the water. Some organic compounds are relatively easy to oxidize while others are very resistant. For example, phenolic compounds react readily with chlorine,[43] ozone,[44] and chlorine dioxide[45] (Fig. 12.6). With chlorine, objectionable tastes and odors

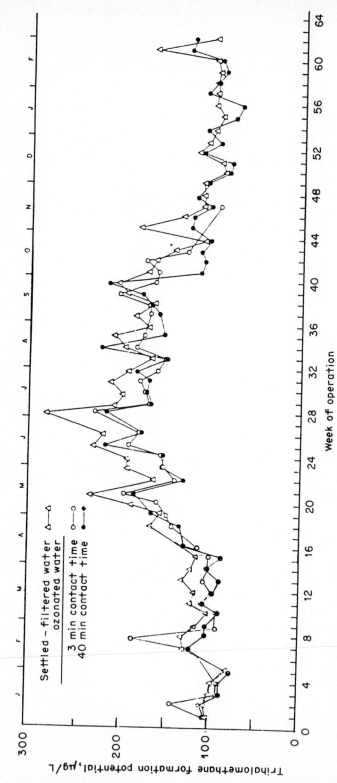

**Figure 12.5** Effect of ozonation on THMFP of alum settled-filtered water. Ozone dose was 6.3 mg/L during first 6 weeks and 2.5 mg/L thereafter. (*Source: W. H. Glaze and J. L. Wallace, "Control of Trihalomethane Precursors in Drinking Water: Granular Activated Carbon With and Without Preozonation," J. AWWA, vol. 76, no. 2, Feb. 1984, p. 68.*)

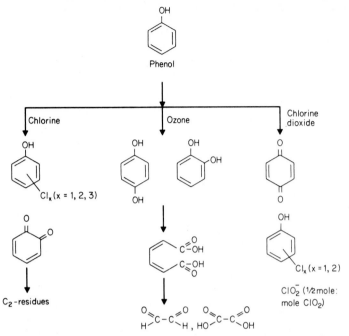

**Figure 12.6** Scheme showing principal by-products from the oxidation of phenol by chlorine,[43] ozone,[44] and chlorine dioxide.[45]

can occur because of the formation of chlorinated phenols. With ozone complete oxidation will yield small polar molecules such as glyoxalic acid, but at lower doses partially oxidized species such as hydroquinones and muconic acid are formed, as shown in Fig. 12.6. In the case of each oxidant, the yields of by-products will depend on the oxidant dose. Lower doses will favor the formation of partial oxidation products.

Ozone and the other common oxidants are theoretically capable of reacting with other synthetic organics, but they are highly selective; i.e., the rates of the processes are often too inefficient to be of practical value. This is especially true when pollutants present at micrograms per liter are competing with natural contaminants at milligrams per liter. For example, the common oxidants are usually not cost-effective for the removal of trichloroethylene and other chlorinated solvents and pesticides in polluted surface water or groundwater. New oxidation processes that show promise for these purposes are described in the next section.

**General effectiveness of water treatment
oxidants**

Tables 12.3 and 12.4 summarize the general effectiveness of the common oxidants used in water treatment. The reader must appreciate

TABLE 12.3    General Effectiveness of Water Treatment Oxidants

| Problem | Chlorine | Chloramines | Ozone | Chlorine dioxide | Potassium permanganate | Oxygen |
|---|---|---|---|---|---|---|
| Iron | E | N | E | E | E | E |
| Manganese[a] | S | N | E | E | E | N |
| Sulfide | E | N | S | S | S | E[b] |
| Taste and odor | S | N | E | E | S[c] | S[c] |
| Color | E | N | E | E | S | N |
| Flocculation aid | E | N | E | U | S[d] | N |
| THMFP | N | N | E[e] | E | S | N |
| Synthetic organics | S[f,g] | N | S[f] | S[f] | S[f] | N |
| Biological growth | E | S | N[h] | E | S | N |

Key: E = effective, S = somewhat effective, N = not effective, U = unknown.
[a]Above pH 7.
[b]By stripping.
[c]Except earthy-musty odor-causing compounds.
[d]May involve adsorption on $MnO_2$.
[e]May increase at low doses.
[f]Depending on compound.
[g]May form chlorinated by-products.
[h]Except with dual-stage ozonation.

that the generalizations do not always hold; that is, the relative effectiveness of oxidants may vary depending on the source water quality; the need for achieving multiple treatment objectives; and local conditions, for example, the cost of oxidants, environmental regulations, and so forth.

## Emerging Technology

As noted above, most oxidants discussed in this section have oxidation potentials such that they should be powerful and versatile oxidants. They do not reach their theoretical potential because they must compete with natural contaminants at much higher concentrations or because their reactions are rate-limited, that is, they are too slow. Oxygen ($O_2$) and ozone are excellent examples of such rate-limited oxidants. The oxygen-oxygen bond in $O_2$ is apparently so strong (119 kcal/mol) that much thermal energy or a catalyst such as an enzyme is necessary to break it. Once the $O_2$ bond is broken, oxidations proceed quite readily, as in biological systems. Ozone is a similar case. The $O_3$ molecule is not as stable as $O_2$ (O-O bond energy of 43 kcal/mol) but is sufficiently stable that ozone is extremely selective in its reactions.[5,22]

To overcome the kinetic barriers that limit the usefulness of chemical oxidants, researchers have devised catalysts and coreagents that

TABLE 12.4    General Advantages and Disadvantages of Water Treatment Oxidants

| Oxidant | Advantages | Disadvantages |
|---|---|---|
| Chlorine | Strong oxidant<br>Simple feeding<br>Persistent residual<br>Long history of use | Chlorinated by-products<br>Taste and odor problems possible<br>pH influences effectiveness |
| Chloramines | No THM formation<br>Persistent residual<br>Simple feeding<br>Long history of use | Weak oxidant<br>Some TOX formation<br>pH influences effectiveness<br>Taste, odor, and growth problems possible |
| Ozone | Strong oxidant<br>Usually no THM or TOX formation<br>No taste or odor problems<br>Some by-products biodegradable<br>Little pH effect<br>Coagulant aid | Short half-life<br>On-site generation required<br>Energy intensive<br>Some by-products biodegradable<br>Complex generation and feeding<br>Corrosive. |
| Chlorine dioxide | Strong oxidant<br>Relatively persistent residual<br>No THM formation<br>No pH effect | TOX formation<br>$ClO_3$ and $ClO_2$ by-products<br>On-site generation required<br>Hydrocarbon odors possible |
| Potassium permanganate | Easy to feed<br>No THM formation | Moderately strong oxidant<br>Pink $H_2O$<br>By-products unknown<br>Causes precipitation |
| Oxygen | Simple feed<br>No by-products<br>Companion stripping<br>Nontoxic | Weak oxidant<br>Corrosion and scaling |

allow the oxidants to come closer to their full potential. This section discusses some of these systems.

## Advanced oxidation processes

Ozone decomposition is initiated by a variety of water contaminants, including hydroxide ion (high pH values), humic materials, and transition-metal ions described above. Usually decomposition is caused by an electron transfer reaction to form the unstable superoxide ion $O_2^-$ or the ozonide radical ion $O_3^-$ [Eqs. (12.22) and (12.24), respectively]. This causes the onset of a series of reactions that consume more ozone and generate powerful oxidizing species, particularly the hydroxyl radical [Eq. (12.26)].

Because the HO radical is one of the most powerful oxidizing species known, when ozone decomposes, its oxidizing power is not necessarily lost if the hydroxyl radical can be efficiently utilized. New developments in oxidation chemistry to generate hydroxyl radicals by ozone decomposition and other methods are currently under study. These may yield future oxidation processes for water treatment.

Among the most promising processes are ozone decomposition initiated by ultraviolet (uv) radiation and by hydrogen peroxide. As Table 12.1 shows, hydrogen peroxide alone is a moderately powerful oxidizing agent, but it has not seen much acceptance as an oxidant for drinking water treatment.[46] Apparently the effectiveness of hydrogen peroxide as an oxidant is limited by unfavorable chemical kinetics. Hydrogen peroxide in combination with uv[47] or ozone[48] is more powerful, however, in that hydroxyl radicals are generated. Hydrogen peroxide is a weak acid, and studies have shown that it is the conjugate base $HO_2^-$ that reacts with ozone[49]

$$H_2O_2 \rightarrow HO_2^- + H^+ \qquad (12.37)$$

$$HO_2^- + O_3 \rightarrow HO_2 + O_3^- \qquad (12.38)$$

The ozonide radical ion $O_3^-$ is the precursor of the hydroxyl radical, as shown in Eqs. (12.25) and (12.26).

Ozone with uv has been shown[50] to form hydrogen peroxide in situ[46] by the following reaction:

$$O_3 + H_2O + \text{uv radiation} \rightarrow H_2O_2 + O_2 \qquad (12.39)$$

Both ozone/uv and ozone/peroxide have been shown to be more effective than ozone alone for the oxidation of natural organics (THM precursors) and synthetic organics such as tri- and tetrachloroethylene.[48,51,52] They also show promise as alternative oxidants for the destruction of refractory taste and odor compounds such as geosmin and MIB that resist other oxidants.[53]

Ozone/uv and ozone/peroxide processes suffer from two distinct disadvantages: First, they are relatively expensive compared to traditional water treatment processes. They may be competitive, however, with other processes such as GAC adsorption. Second, processes that involve radical intermediates are subject to interference from radical traps. Radical traps are substances such as bicarbonate ions[54] that react with HO radicals and therefore decrease the effectiveness of the process for the destruction of organics. Lowering of the alkalinity prior to the application of the ozone/uv or ozone/peroxide process may be necessary for water that has a high bicarbonate level. In addition to these disadvantages, the ozone/uv process involves equipment and maintenance procedures that are costly and unfamiliar to most water treatment operators.

In summary, advanced oxidation processes show promise as alternatives to traditional oxidants for the removal of substances that resist conventional removal techniques. Future research and development of these processes should determine if they can be made cost-effective for application in water treatment plants.

## Acknowledgments

At the time this chapter was written, William H. Glaze was the director of the Environmental Science and Engineering Program, School of Public Health, at the University of California, Los Angeles.

## References

1. William H. Glaze, "Drinking Water Treatment with Ozone," *Environ. Sci. Tech.*, vol. 21, 1987, p. 224.
2. G. W. Miller, R. G. Rice, C. M. Robson, R. L. Scullin, W. Kuhn, and H. Wolf, *An Assessment of Ozone and Chlorine Dioxide Technology for Treatment of Municipal Water Supplies*, EPA-600/2-78-147, EPA, Cincinnati, October 1978.
3. J. J. Rook, "Formation of Haloforms during Chlorination of Natural Waters," *Water Treat. Exam.*, vol. 23, 1974, p. 234.
4. T. A. Bellar, J. J. Lichtenberg, and R. C. Kroner, "The Occurrence of Organohalides in Chlorinated Drinking Water," *J. AWWA*, vol. 66, no. 12, December 1974, p. 703.
5. J. F. Pankow and J. J. Morgan, "Kinetics for the Aquatic Environment, 1," *Environ. Sci. Tech.*, vol. 15, 1981, p. 1155.
6. J. F. Pankow and J. J. Morgan, "Kinetics for the Aquatic Environment, 2," *Environ. Sci. Tech.*, vol. 15, 1981, p. 1306.
7. J. Hoigne and H. Bader, "Ozonation of Water: Role of Hydroxyl Radicals as Oxidizing Intermediates," *Science,* vol. 109, 1975, p. 4216.
8. V. L. Snoeyink and D. Jenkins, *Water Chemistry,* chap. 7, Wiley, New York, 1980.
9. R. L. Jolley, R. J. Bull, W. P. Davis, S. Katz, M. H. Roberts, Jr., and V. A. Jacobs, *Water Chlorination: Chemistry, Environmental Impact and Health Effects,* vol. 5, Lewis Publishers, Chelsea, Mich., 1985.
10. A. A. Stevens, L. A. Moore, C. J. Slocum, B. L. Smith, D. R. Seeger, and J. C. Ireland, "By-products of Chlorination at Ten Operating Utilities," *Water Chlorination: Chemistry, Environmental Impact and Health Effects,* vol. 6, R. L. Jolley et al., editors) Lewis Publishers, Inc., Chelsea, Mich., 1989.
11. D. L. Norwood, J. D. Johnson, R. F. Christman, and D. S. Millington, "Chlorination Products from Aquatic Humic Material at Neutral pH," in R. L. Jolley et al. (eds.) *Water Chlorination: Environmental Impact and Health Effects,* vol. 4, book 1, Ann Arbor Science, Ann Arbor, Mich., 1983.
12. W. Kuhn and H. Sontheimer, "Several Investigations on Activated Carbon for the Determination of Organic Chloro-Compounds," *Vom Wasser,* vol. 15, 1973, p. 65.
13. A. A. Stevens, R. C. Dressman, R. K. Sorrell, and H. J. Brass, "Organic Halogen Measurements: Current Uses and Future Prospects," *J. AWWA,* vol. 77, no. 4, April 1985, p. 146.
14. J. C. Morris and R. A. Isaac, "A Critical Review of Kinetic and Thermodynamic Constants for the Aqueous Chlorine-Ammonia System," in R. L. Jolley et al. (eds.), *Water Chlorination: Environmental Impact and Health Effects,* vol. 4, Ann Arbor Science, Ann Arbor, Mich., 1983.
15. E. T. Gray, Jr., D. W. Margerum, and R. P. Huffman, "Chloramine Equilibria and

the Kinetics of Disproportionation in Aqueous Solution," in R. R. Brinckman and J. M. Bellama (eds.), *Organometals and Organometalloids: Occurrence and Fate in the Environment*, American Chemical Society, Washington, 1978.

16. S. W. Krasner, S. E. Barrett, M. E. Dale, and C. J. Hwang, "Free Chlorine versus Monochloramine in Controlling Off-Tastes and -Odors in Drinking Water," *Proc. AWWA Annual Conference*, Denver, June 1986.

17. S. Lalezary, M. Pirbazari, and M. J. McGuire, Oxidation of Five Earthy-Musty Taste and Odor Compounds," *J. AWWA*, vol. 78, no. 3, March 1986, p. 62.

18. J. N. Jensen, J. J. St. Aubin, R. F. Christman, and D. J. Johnson, "Characterization of the Reaction between Monochloramine and Isolated Aquatic Fulvic Acid," in R. L. Jolley et al. (eds.), *Water Chlorination: Chemistry, Environmental Impact and Health Effects*, vol. 5, Lewis Publishers, Chelsea, Mich., 1985.

19. J. Staehelin and J. Hoigne, "Decomposition of Ozone in Water in the Presence of Organic Solutes Acting as Promoters and Inhibitors of Radical Chain Reaction," *Environ. Sci. Tech.*, vol. 19, 1985, p. 1206.

20. J. Staehelin and J. Hoigne, "Mechanism and Kinetics of Decomposition of Ozone in Water in the Presence of Organic Solutes," *Vom Wasser*, vol. 61, 1983, p. 337.

21. J. Hoigne and H. Bader, "Ozone and Hydroxyl Radical-Initiated Oxidations of Organic and Organometallic Trace Impurities in Water," in R. R. Brinckman and J. M. Bellama (eds.), *Organometals and Organometalloids: Occurrence and Fate in the Environment*, American Chemical Society, Washington, 1978.

22. J. Hoigne and H. Bader, "Rate Constants of Reactions of Ozone with Organic and Inorganic Compounds in Water. II. Dissociating Organic Compounds," *Water Res.*, vol. 17, 1983, p. 185.

23. William H. Glaze, "Reaction Products of Ozone: A Review," *Environ. Health Perspect.*, vol. 69, 1986, p. 151.

24. M. M. Bourbigot, M. C. Hascoet, Y. Levi, F. Erb, and N. Pommery, "Role of Ozone and Granular Activated Carbon in the Removal of Mutagenic Compounds," *Environ. Health Perspect.*, vol. 69, 1986, p. 159.

25. B. C. J. Zoeteman, J. Hrubec, E. de Greef, and J. J. Kool, "Mutagenic Activity Associated with By-products of Drinking Water Disinfection by Chlorine, Chlorine Dioxide, Ozone and UV Irradiation," *Environ. Health Perspect.*, vol. 46, 1982, p. 197.

26. W. J. Cooper, R. G. Zika, and M. S. Steinhauer, "Bromide-Oxidant Interactions and THM Formation: A Literature Review," *J. AWWA*, vol. 77, no. 4, April 1985, p. 116.

27. G. R. Helz, R. Y. Hsu, and R. M. Block, "Bromoform Production by Oxidative Biocides in Marine Waters," in R. G. Rice and J. A. Cotruvo (eds.), *Ozone/Chlorine Dioxide Oxidation Products of Organic Molecules*, Ozone Press International, Cleveland, Ohio, 1978.

28. M. A. Carlson, R. C. Hoehn, W. R. Knocke, and D. H. Hair, "Experiences with the Use of Chlorine Dioxide and Potassium Permanganate as Preoxidants for Trihalomethane and Manganese Control," *Proc. AWWA Annual Conf.*, Denver, June 1986.

29. P. C. Singer, J. H. Borchardt, and J. M. Colthurst, "The Effects of Permanganate Pretreatment on Trihalomethane Formation in Drinking Water," *J. AWWA*, vol. 72, no. 10, October 1980, p. 573.

30. E. M. Aieta and J. D. Berg, "A Review of Chlorine Dioxide in Drinking Water Treatment," *J. AWWA*, vol. 78, no. 6, June 1986, p. 62.

31. L. W. Condie, "Toxicological Problems Associated with Chlorine Dioxide," *J. AWWA*, vol. 78, no. 6, June 1986, p. 73.

32. J. A. Cotruvo and C. D. Vogt, "Regulatory Aspects of Disinfection," in R. L. Jolley et al. (eds.), *Water Chlorination: Chemistry, Environmental Impact and Health Effects*, vol. 5, Lewis Publishers, Chelsea, Mich., 1985.

33. A. A. Stevens, "Reaction Products of Chlorine Dioxide," *Environ. Health Perspect.* vol. 46, 1982, p. 101.

34. H. B. Amor, J. De Laat, and M. Dore, "Mode of Action of Chlorine Dioxide on Organic Compounds in an Aqueous Medium," *Water Res.*, vol. 18, 1984, p. 1545.

35. E. T. Gjessing, *Physical and Chemical Characteristics of Aquatic Humus*, Ann Arbor Science, Ann Arbor, Mich., 1975.

36. I. H. Suffet, C. Anselme, and J. Mallevialle, "Removal of Tastes and Odors by

Ozonation," *Seminar Proc.: Ozonation: Recent Advances and Research Needs,* AWWA Annual Conference, Denver, June 1986.

37. E. A. Voudrias, V. L. Snoeyink, and R. A. Larson, "Desorption of Organics Formed on Activated Carbon," *J. AWWA,* vol. 78, no. 2, February 1986, p. 82.

38. D. A. Reckhow, P. C. Singer, and R. R. Trussell, "Ozone as a Coagulant Aid," *Seminar Proc.: Ozonation: Recent Advances and Research Needs,* AWWA Annual Conference, Denver, June 1986.

39. P. W. Prendiville, "Ozonation at the 900 cfs Los Angeles Water Purification Plant," *Ozone: Sci. Eng.,* vol. 8, 1986, p. 77.

40. W. H. Glaze, G. R. Peyton, S. Lin, F. Y. Huang, and J. L. Burleson, "Destruction of Pollutants in Water with Ozone in Combination with Ultraviolet Radiation; 2, Natural Trihalomethane Precursors," *Environ. Sci. Tech.,* vol. 16, 1982, p. 454.

41. W. H. Glaze and J. L. Wallace, "Control of Trihalomethane Precursors in Drinking Water: Granular Activated Carbon with and without Preozonation," *J. AWWA,* vol. 76, no. 2, February 1984, p. 68.

42. F. A. DeGiano, "Ozone and Biodegradation in Slow Sand Filters and Granular Activated Carbon," *Seminar Proc.: Ozonation: Recent Advances and Research Needs,* AWWA Annual Conference, Denver, June 1986.

43. J. C. Morris, *Formation of Halogenated Organics by Chlorination of Water Supplies,* EPA-600/1-75-002, EPA, Washington, 1975.

44. P. C. Singer and M. D. Gurol, "Ozonation of Phenol: Mass Transfer and Reaction Kinetic Considerations," *Wasser Berlin '81,* Proc. of the Fifth Ozone World Congress, Berlin, International Ozone Association, Vienna, Va., 1981.

45. J. E. Wajon, D. H. Rosenblatt, and E. P. Burrows, "Oxidation of Phenol and Hydroquinone by Chlorine Dioxide," *Environ. Sci. Tech.,* vol. 16, 1982, p. 396.

46. H. S. Possalt and W. J. Weber, Jr., in W. J. Weber, Jr. (ed.), *Physicochemical Processes for Water Quality Control,* Wiley-Interscience, New York, 1972.

47. L. Berglind, E. Gjesing, and E. Skipperud Johansen, "Removal of Organic Matter in Water by UV and Hydrogen Peroxide," in W. Kuhn and H. Santheimer (eds.), *Oxidation Techniques in Drinking Water Treatment,* EPA-570/9-79-020, EPA, Washington, 1979.

48. J. P. Duguet, E. Brodard, B. Dussert, and J. Mallevialle, "Improvement in the Effectiveness of Ozonation of Drinking Water through the Use of Hydrogen Peroxide," *Ozone: Sci. Eng.,* vol. 7, 1985, p. 241.

49. J. Staehelin and J. Hoigne, "Decomposition of Ozone in Water: Rate of Initiation by Hydroxide and Hydrogen Peroxide," *Environ. Sci. Tech.,* vol. 16, 1982, p. 676.

50. G. R. Peyton and W. H. Glaze, "Mechanism of Photolytic Ozonation," in R. G. Zika and W. J. Cooper (eds.), *Photochemistry of Environmental Aquatic Systems,* ACS Symposium Series No. 327, American Chemical Society, Washington, 1987.

51. H. W. Prengle, Jr., C. G. Hewes, III, and C. E. Mauk, "Oxidation of Refractory Materials by Ozone with Ultraviolet Radiation," *Second International Symposium on Ozone Technology,* Ozone Press International, Jamesville, N.Y., 1976.

52. W. H. Glaze, J.-W. Kang, and E. M. Aieta, "Ozone–Hydrogen Peroxide Systems for Control of Organics in Municipal Water Supplies," *The Role of Ozone in Water and Wastewater Treatment,* Proc. Second International Conference, Edmonton, Alberta, April 28–29, 1987.

53. W. H. Glaze, J.-W. Kang, and D. Chapin, "The Chemistry of Water Treatment Processes Involving Ozone, Hydrogen Peroxide and Ultraviolet Radiation," *Ozone: Sci. Eng.,* vol. 9, 1987, p. 335.

54. B. Legube, J. P. Croue, D. A. Reckhow, and M. Dore, "Ozonation of Organic Precursors: Effects of Bicarbonate and Bromide," in R. Perry and A. E. McIntyre (eds.), *Proc. International Conference on the Role of Ozone in Water and Wastewater Treatment,* Selper Ltd., London, 1986.

# Adsorption of
# Organic Compounds

**Vernon L. Snoeyink, Ph.D.**

*Professor of Environmental Engineering*
*Department of Civil Engineering*
*University of Illinois at Urbana-Champaign*
*Urbana, Illinois*

Adsorption of a substance involves its accumulation at the interface between two phases, such as a liquid and a solid, or a gas and a solid. The molecule that accumulates, or adsorbs, at the interface is called an *adsorbate,* and the solid on which adsorption occurs is the *adsorbent.* Adsorbents of interest in water treatment include activated carbon; ion exchange resins; adsorbent resins; metal oxides, hydroxides, and carbonates; activated alumina; clays; and other solids that are suspended in or in contact with water.

Adsorption plays an important role in the modification of water quality. Activated carbon, for example, can be used to adsorb organic molecules that cause taste and odor, color, mutagenicity, and toxicity. The aluminum hydroxide and ferric hydroxide solids that form during coagulation will adsorb color-causing molecules and compounds that react with chlorine to form trihalomethanes. Adsorption of natural organic molecules on anion exchange resins may cause them to lose capacity for anions (see Chap. 9), but ion exchange resins and adsorbent resins are available that can be used for efficient removal of selected organic compounds. The calcium carbonate and magnesium hydroxide solids that form during the lime softening process also have some adsorption capacity, and pesticides adsorbed on clay particles can be removed by coagulation and filtration.

The primary focus of this chapter is the removal of organic compounds by adsorption on activated carbon because of its importance in water purification processes. A study conducted by two committees of the AWWA showed that approximately 25 percent of 645 United States utilities, including the 500 largest, used powdered activated carbon (PAC) in 1977.[1] In 1986, 29 percent of the 600 largest utilities reported using PAC.[2] While the use was predominantly for odor control, removal of nonodorous compounds also was likely. More attention now is being given to granular activated carbon (GAC) as an alternative to PAC. GAC is used in columns or beds that permit higher adsorptive capacities to be achieved and easier process control than is possible with PAC. The higher cost for GAC often can be offset by better efficiency, especially when organics must be removed on a continuous basis. GAC should be seriously considered for water supplies when odorous compounds or synthetic organic chemicals of health concern frequently are present, when a barrier is needed to prevent organic compounds from spills from entering finished water, or in some situations that require trihalomethane control. GAC has excellent adsorption capacity for many undesirable substances, and it can be removed from the columns for regeneration (sometimes called reactivation) when necessary. Although only 65 plants were using GAC in 1977,[1] principally for odor removal, the number had increased to 135 by 1986.[3] Its ability to remove many types of organic compounds probably will result in a significant increase in its usage.

This chapter also covers the use of ion exchange and adsorbent resins for the removal of organic compounds. Removal of inorganic ions by ion exchange resins and activated alumina is an adsorption process, but it is discussed in Chap. 9.

## Adsorption Theory

### Adsorption equilibrium

Adsorption of molecules can be represented as a chemical reaction

$$A + B \rightleftharpoons A \cdot B$$

where A represents the adsorbate, B the adsorbent, and $A \cdot B$ the adsorbed compounds. Adsorbates are held on the surface by various types of chemical forces such as hydrogen bonds, dipole-dipole interactions, and van der Waals forces. If the reaction is reversible, as it is for many compounds adsorbed to activated carbon, molecules continue to accumulate on the surface until the rate of the forward reaction (adsorption) equals the rate of the reverse reaction (desorption). When this condition exists, equilibrium has been reached and no further accumulation will occur.

**Isotherm equations.** One of the most important characteristics of an adsorbent is the quantity of adsorbate that it can accumulate. The constant-temperature equilibrium relationship between the quantity of adsorbate per unit of adsorbent $q_e$ and the equilibrium concentration of adsorbate in solution $C_e$ is called the *adsorption isotherm*. Several equations or models are available that describe this function,[4] but only two of the more common equations, the Freundlich and the Langmuir, are presented here.

The Freundlich equation is an empirical equation that is very useful because it accurately describes much adsorption data. This equation has the form

$$q_e = KC_e^{1/n} \tag{13.1}$$

and can be linearized as follows:

$$\log q_e = \log K + \frac{1}{n} \log C_e \tag{13.2}$$

The parameters $q_e$ (with units of mass adsorbate/mass adsorbent or moles adsorbate/mass adsorbent) and $C_e$ (with units of mass/volume or moles/volume) are the equilibrium surface and solution concentrations, respectively. The terms $K$ and $n$ are constants, and the units of $K$ are determined by the units of $q_e$ and $C_e$. Although the Freundlich equation was developed empirically, a theory of adsorption that leads to the Freundlich equation was later developed by Halsey and Taylor.[5]

The constant $K$ in the Freundlich equation is related primarily to the capacity of the adsorbent for the adsorbate, and $1/n$ is a function of the strength of adsorption. For fixed values of $C_e$ and $1/n$, the larger the value of $K$, the larger the capacity $q_e$ is. For fixed values of $K$ and $C_e$, the smaller the value of $1/n$, the stronger the adsorption bond is. As $1/n$ becomes very small, the capacity tends to be independent of $C_e$, and the isotherm plot approaches the horizontal; the value of $q_e$ then is essentially constant, and the isotherm is called *irreversible*. If the value of $1/n$ is large, the adsorption bond is weak, and the value of $q_e$ changes markedly with small changes in $C_e$.

The Freundlich equation cannot apply to all values of $C_e$, however. As $C_e$ increases, for example, $q_e$ increases [in accordance with Eq. (13.1)] only until the adsorbent approaches saturation. At saturation, $q_e$ is a constant, independent of further increases in $C_e$, and the Freundlich equation no longer applies. Also, no assurance exists that adsorption data will conform to the Freundlich equation over all concentrations less than saturation, so care must be exercised in extending the equation to concentration ranges that have not been tested.

The Langmuir equation

$$q_e = \frac{q_{max}bC_e}{1 + bC_e} \tag{13.3}$$

where $b$ and $q_{max}$ are constants and $q_e$ and $C_e$ are as defined above, has a firm theoretical basis.[6] The constant $q_{max}$ corresponds to the surface concentration at monolayer coverage and represents the maximum value of $q_e$ that can be achieved as $C_e$ is increased. The constant $b$ is related to the energy of adsorption and increases as the strength of the adsorption bond increases. The values of $q_{max}$ and $b$ can be determined from a plot of $1/q_e$ versus $1/C_e$ in accordance with a linearized form of Eq. (13.3):

$$\frac{1}{q_e} = \frac{1}{q_{max}bC_e} + \frac{1}{q_{max}} \tag{13.4}$$

$$\frac{C_e}{q_e} = \frac{1}{q_{max}b} + \frac{C_e}{q_{max}} \tag{13.5}$$

The Langmuir equation often does not describe adsorption data as accurately as the Freundlich equation. The experimentally determined values of $q_{max}$ and $b$ often are not constant over the concentration range of interest, possibly because of the heterogeneous nature of the adsorbent surface (a homogeneous surface was assumed in the model development), interaction between adsorbed molecules (all interaction was neglected in the model development), and other factors.

**Factors affecting adsorption equilibria.** Important adsorbent characteristics that affect isotherms include surface area, pore size distribution, and surface chemistry. The maximum amount of adsorption is proportional to the amount of surface area within pores that is accessible to the adsorbate. Surface areas range from a few hundred to more than 1500 $m^2/g$, but not all of the area is accessible to aqueous adsorbates. The range of pore size distributions in an arbitrary selection of GACs is shown in Fig. 13.1. A relatively large volume of micropores (pores less than 2 nm diameter, $d$)[4] generally corresponds to a large surface area and a large adsorption capacity for small molecules, whereas a large volume of macropores ($d > 50$ nm) is usually directly correlated to capacity for large molecules. Pore volume in the intermediate size range (transition pores, $2 < d < 50$ nm) is thought to be important for rapid transport of adsorbates to the small pores, although data to show this are not available. The fulvic acid isotherms in Fig. 13.2 are for the same activated carbons whose pore size distributions are shown in Fig. 13.1. Note that the activated carbons that have a relatively small volume of macropores also have a relatively low capacity for the large fulvic acid molecule. Lee et al.[7] showed that the quantity

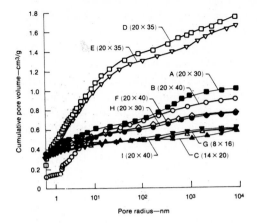

**Figure 13.1** Pore size distributions for different activated carbons. (*Source: M. C. Lee, V. L. Snoeyink, and J. C. Crittenden, "Activated Carbon Adsorption of Humic Substances," J. AWWA, vol. 73, no. 8, 1981, p. 440.*)

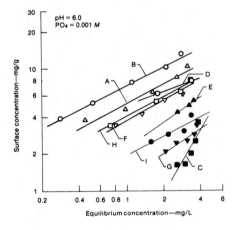

**Figure 13.2** Adsorption isotherms for peat fulvic acid. (*Source: M. C. Lee, V. L. Snoeyink, and J. C. Crittenden, "Activated Carbon Adsorption of Humic Substances," J. AWWA, vol. 73, no. 8, 1981, p. 440.*)

of humic substances of a given size that was adsorbed correlated to pore volume within pores of a given size. The relative positions of the isotherms for the activated carbons in Fig. 13.1 might be entirely different if the adsorbate were a small molecule, such as a phenol, which can enter pores much smaller than fulvic acid.

The surface chemistry of activated carbon and adsorbate properties also can affect adsorption.[8] Coughlin and Ezra[9] demonstrated that extensive oxidation of activated carbon with ammonium persulfate in aqueous solution led to large decreases in the amount of phenol, nitrobenzene, and benzenesulfonate that could be adsorbed. Kipling and coworkers[10,11] found that increased amounts of surface oxygen decreased the affinity of carbon black, an adsorbent similar to activated carbon, for benzene, and that carbon black surface covered with oxygen did not adsorb iodine. Oxidation of the activated carbon surface with aqueous chlorine also was found to increase the number of oxygen surface functional groups and, correspondingly, to decrease the adsorption capacity for phenol.[12] Thus, oxygenated surfaces do not adsorb at least simple aromatic compounds strongly.

The amount of adsorption of a molecule is a function of its affinity for water as compared to its affinity for the adsorbent. Adsorption onto GAC from water, for example, generally increases as the adsorbate's solubility decreases.[13] As a molecule becomes larger through the addition of hydrophobic groups such as $—CH_2—$, its solubility decreases and its extent of adsorption increases as long as the molecule can gain entrance to the pores. When a size increase causes the molecule to be excluded from some pores, however, adsorption capacity may decrease as solubility decreases. As molecular size increases, the rate of diffusion within the activated carbon particle decreases, especially as molecular size approaches the particle's pore diameter.

The affinity of weak organic acids or bases for activated carbon is an important function of pH. When pH is in a range where the molecule is in the neutral form, adsorption capacity is relatively high. When pH is in a range where the species is ionized, however, affinity for water is very high and activated carbon capacity accordingly is very low. Phenol that has been adsorbed on activated carbon at pH below 8, where phenol is neutral, can be desorbed if the pH is increased to 10 or above, where the molecule is anionic.[14] If adsorption occurs on resins by the ion exchange mechanism, the specific affinity of the ionic adsorbate for charged functional groups may also cause good removal.

The inorganic composition of water also can have an important effect on extent of fulvic acid adsorption as shown in Fig. 13.3. After 70 days, a small GAC column was nearly saturated with fulvic acid. Addition of $CaCl_2$ at this point resulted in a large increase in adsorbability of fulvic acid, as reflected in the reduced column effluent concentration. After 140 days, elimination of the $CaCl_2$ resulted in desorption of much of the fulvic acid.[15] Calcium ion apparently associates (complexes) with the fulvic acid anion to make it more adsorbable.[15,16] Presumably many other divalent ions can act in similar fashion, but calcium is of special interest because of its relatively high concentration. Similar effects are expected for other anionic adsorbates, but salts are not expected to have much effect on adsorption of neutral adsorbates.[17]

Inorganic substances such as iron, manganese, and calcium salts or precipitates may interfere with adsorption if they deposit on the adsorbent. Pretreatment to remove these substances, or to eliminate the supersaturation, is necessary if they are present in large amounts.

Adsorption isotherms may be determined for heterogeneous mixtures of compounds by using group parameters such as total organic

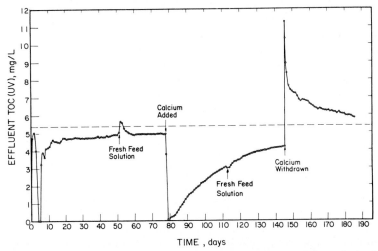

**Figure 13.3**    The effects of calcium chloride addition and withdrawal on column performance (pH = 8.3; TOC = 5.37 mg/L, peat fulvic acid; buffer = 1.0 mM NaHCO₃). (*Source: S. J. Randtke and C. P. Jespen, "Effects of Salts on Activated Carbon Adsorption of Fulvic Acids," J. AWWA, vol. 74, no. 2, 1982, p. 84.*)

carbon (TOC), dissolved organic carbon (DOC), chemical oxygen demand (COD), dissolved organic halogen (DOX), uv absorbance, and fluorescence as a measure of the total concentration of substances that is present. The mixture of compounds then is treated as a single compound in isotherm equations such as Eqs. (13.1) and (13.3). Because the compounds within a mixture can vary widely in their affinity for an adsorbent, the shape of the isotherm will depend on the relative amounts of compounds in the mixture. For example, isotherms with the shape shown in Fig. 13.4 are expected if some of the compounds are nonadsorbable, and some are more strongly adsorbable than the rest.[18] The strongly adsorbable compounds can be removed with small doses of adsorbent and yield large values of $q_e$. In contrast, the weakly adsorbable compounds can only be removed with large doses of adsorbent that yield relatively low values of $q_e$. The nonadsorbable compounds produce a vertical isotherm at low $C_e$ values. In contrast to single-solute isotherms, the isotherm for a heterogeneous mixture of compounds will be a function of initial concentration and the fraction of the mixture that is adsorbed.

The relative adsorbabilities of compounds within a mixture have an important effect on performance of adsorption columns. The nonadsorbable fraction cannot be removed regardless of the column design, whereas the strongly adsorbable fraction may cause the effluent concentration to slowly approach the influent concentration.

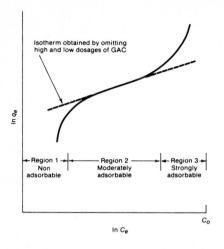

Figure 13.4 Nonlinear isotherm for a heterogeneous mixture of organic compounds. (*Source: S. J. Randtke and V. L. Snoeyink, "Evaluating GAC Adsorptive Capacity," J. AWWA, vol. 75, no. 8, 1983, p. 406.*)

**Competitive adsorption.** Competitive adsorption is important in drinking water treatment because most compounds to be adsorbed exist in solution with other adsorbable compounds. The quantity of activated carbon or other adsorbent required to remove a certain amount of a compound of interest from a mixture of adsorbable compounds is greater than if adsorption occurs without competition because part of the adsorbent's surface is utilized by the competing substances.

The extent of competition on activated carbon depends upon the strength of adsorption of the competing molecules, the concentrations of these molecules, and the type of activated carbon. Some examples illustrate the possible magnitude of the competitive effect. Jain and Snoeyink[19] showed that as $p$-bromophenol (PBP) equilibrium concentration increased from $10^{-4}$ to $10^{-3}$ $M$ (17 to 173 mg/L), the amount of $p$-nitrophenol (PNP) adsorbed at an equilibrium concentration of $3.5 \times 10^{-5}$ $M$ (~5 mg/L) decreased by about 30 percent (see Fig. 13.5). Figure 13.6 shows that the presence of 10 mg/L of a humic substance caused about 90 percent reduction in capacity for the musty-odor compound geosmin at an equilibrium concentration of 1 mg/L,[20] although Lalezary et al.[21] later used a different humic substance and found much less reduction. The adsorbability of natural organic matter in surface and groundwater varies widely, so wide variation in the extent of competition is expected.

Displacement of previously adsorbed compounds by competition can result in a column effluent concentration of a compound that is greater than the influent concentration, as shown in Fig. 13.7. A dimethylphenol (DMP) concentration about 50 percent greater than the influent resulted when dichlorophenol (DCP) was introduced to the influent of a column saturated with DMP.[22] Similar occurrences have been observed in full-scale GAC systems. Effluent concentrations in excess of influent concentrations can be prevented through

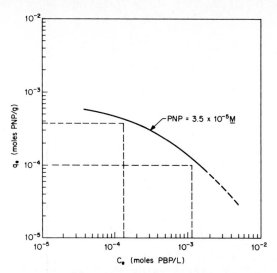

**Figure 13.5** Adsorption of *p*-nitrophenol as a function of *p*-bromophenol concentration from a solution containing neutral *p*-nitrophenol and neutral *p*-bromophenol. (*Source: J. S. Jain and V. L. Snoeyink, "Adsorption from Bisolute Systems on Active Carbon," J. Water Pollution Control Fed., vol. 45, 1973, p. 2463.*)

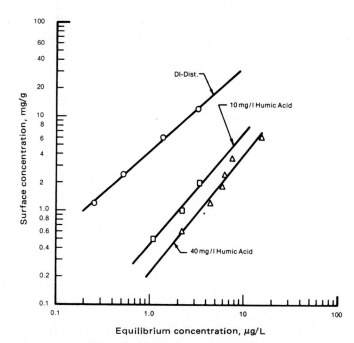

**Figure 13.6** Adsorption of geosmin. (*Source: D. R. Herzing, V. L. Snoeyink, and N. F. Wood, "Activated Carbon Adsorption of Odorous Compound 2-Methylisoborneol and Geosmin," J. AWWA, vol. 69, no. 4, 1977, p. 223.*)

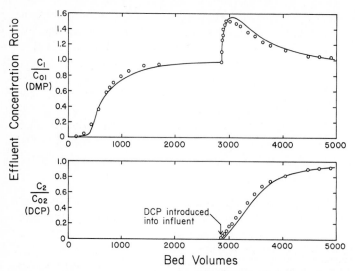

**Figure 13.7**  Breakthrough curves for sequential feed of DMP and DCP to a GAC adsorber ($C_{01}$ = 0.990 mmol/L, $C_{02}$ = 1.02 mmol/L, EBCT = 25.4 s). (*Source: W. E. Thacker, J. C. Crittenden, and V. L. Snoeyink, "Modeling of Adsorber Performance: Variable Influent Concentration and Comparison of Adsorbents," J. Water Pollution Control Fed., vol. 56, 1984, p. 243.*)

careful operation. Crittenden et al.[23] showed that the magnitude of the displacement decreased when the value of $C_{eff}/C_{inf}$ was lowered at the time the second compound was introduced. Thus, a reasonable strategy to prevent the occurrence of an undesirable compound at a concentration greater than the influent is (1) to monitor the column for that compound, and (2) to regenerate or replace the activated carbon before complete saturation occurs (i.e., before $C_{eff} = C_{inf}$).

A number of isotherm models have been used to describe competitive adsorption. A common model for describing adsorption equilibrium in multiadsorbate systems is the Langmuir model for competitive adsorption, which was first developed by Butler and Ockrent.[24] This model is based on the same assumptions as the Langmuir model for single adsorbates. Assuming, as did Langmuir, that the rate of adsorption of a species at equilibrium is equal to its rate of desorption:

$$q_{e,1} = \frac{q_{max,1} b_1 C_{e,1}}{1 + \sum_i b_i C_{e,i}}$$  (13.6)

$$q_{e,2} = \frac{q_{max,2} b_2 C_{e,2}}{1 + \sum_i b_i C_{e,i}}$$  (13.7)

where $q_{e,1}$ and $q_{e,2}$ = of solutes 1 and 2 adsorbed per unit weight, or per unit surface area, of adsorbent at equilibrium concentrations $C_{e,1}$ and $C_{e,2}$, respectively

$q_{max,1}$ and $q_{max,2}$ = maximum values of $q_{e,1}$ and $q_{e,2}$, respectively, that are obtained from single-solute isotherm analysis and that correspond to monolayer coverage of adsorbent

$b_1$ and $b_2$ = constants that are a function of the energy of adsorption of solutes 1 and 2, respectively, and that are obtained from single-solute isotherm analysis

Although this model is commonly used, it does not always give good predictions, especially for porous adsorbents. Jain and Snoeyink[19] proposed a modification to account for a fraction of the adsorption taking place without competition. This can happen if the adsorbates have a different size and only the smaller one can enter the smaller pores. Also, some of the surface functional groups may adsorb one of the compounds but not the other. Other models that can be used to describe and predict competitive effects are the Freundlich-type isotherm of Sheindorf et al.[25] and the ideal adsorbed solution theory of Radke and Prausnitz.[26] The latter has proven to be applicable to a number of situations, although it will not apply if competing species do not have equal access to all sites.

**Desorption.** Adsorption of many compounds is reversible, which means that they can desorb. Desorption may be caused by displacement by other compounds, as discussed above, or by a decrease in influent concentration. Both phenomena may occur in some situations. An analysis of desorption by Thacker et al.[27] showed that the quantity of adsorbate that can desorb in response to a decrease in influent concentration increased as (1) the diffusion coefficient of the adsorbate increased, (2) the amount of compound adsorbed increased, (3) the strength of adsorption decreased (e.g., as the Langmuir $b$ value decreased, or the Freundlich $1/n$ value increased), and (4) the activated carbon particle size decreased. Volatile organic compounds are especially susceptible to displacement because they are weakly adsorbed and diffuse rapidly. Symons,[28] for example, showed substantial desorption of chloroform and 1,2-dichloroethylene in response to a decrease in influent concentration.

### Adsorption kinetics

**Transport mechanisms.** Removal of organic compounds by physical adsorption on porous adsorbents involves a number of steps, each of which can affect the rate of removal:

1. *Bulk solution transport* Adsorbates must be transported from bulk solution to the boundary layer of water surrounding the adsorbent particle. The transport occurs by diffusion if the adsorbent is suspended in quiescent water, such as in a sedimentation basin, or by turbulent mixing, such as during turbulent flow through a packed bed of GAC, or when PAC is being mixed in a rapid mix or flocculator.

2. *Film diffusion transport* Adsorbates must be transported by molecular diffusion through the stationary layer of water (hydrodynamic boundary layer) that surrounds adsorbent particles when water is flowing past them. The distance of transport, and thus the time for this step, is determined by the rate of flow past the particle: the higher the rate of flow, the shorter the distance.

3. *Pore transport* After passing through the hydrodynamic boundary layer, adsorbates must be transported through the adsorbent's pores to available adsorption sites. Intraparticle transport may occur by molecular diffusion through the solution in the pores (pore diffusion) or by diffusion along the adsorbent surface (surface diffusion) after adsorption takes place.

4. *Adsorption* After transport to an available site, the adsorption bond is formed between the adsorbate and adsorbent. This step is very rapid for physical adsorption,[29] and as a result one of the preceding diffusion steps will control the rate at which molecules are removed from solution. If adsorption is accompanied by a chemical reaction that changes the nature of the molecule, the chemical reaction may be slower than the diffusion step and thereby control the rate of compound removal.

The transport steps occur in series, so the slowest step, called the *rate-limiting step,* will control the rate of removal. In turbulent flow reactors, a combination of film diffusion and pore diffusion very often controls the rate of removal for some types of molecules to be removed from drinking water. Film diffusion may control initially; then, after some adsorbate accumulates within the pore, pore transport may control. The mathematical models of the adsorption process, therefore, usually include both steps.

Both molecular size and adsorbent particle size have important effects on the rate of adsorption. Intraparticle diffusion coefficients, in particular, decrease as molecular size increases, and thus it takes longer to remove large-molecular-weight humic substances than low-molecular-weight phenols, for example. Adsorbent particle size is also important because it determines the time required for transport within the pore to available adsorption sites. Calculations by Randtke and Snoeyink[18] for activated carbon illustrate these points (see Table 13.1). For the low-molecular-weight dimethylphenol, nearly eight days is estimated for near-equilibrium ($C_{final} = 1.01C_e$) of 2.4-mm-

**TABLE 13.1    Time Required to Approach Equilibrium in Simulated Adsorption Experiments**

| Adsorbate | GAC mesh size | Particle diameter, mm | $C_0$,† mg/L | $t_{1.01}$‡ days |
|---|---|---|---|---|
| Dimethylphenol | 8 | 2.380 | 90.67 | 7.70 |
| (122 MW) | 30 | 0.595 | 90.67 | 0.54 |
| | 325 | 0.044 | 90.67 | 0.01 |
| Rhodamine B | 8 | 2.380 | 85.10 | 590 |
| (423 MW) | 30 | 0.595 | 85.10 | 37 |
| | 325 | 0.044 | 85.10 | 0.2 |
| Peat fulvic acid | 8 | 2.380 | 0.73 | 1840 |
| (10,000 MW) | 30 | 0.595 | 0.73 | 115 |
| | 325 | 0.044 | 0.73 | 0.7 |
| Humic acid | 8 | 2.380 | 37.93 | 5300 |
| (50,000 MW) | 30 | 0.595 | 37.93 | 333 |
| | 325 | 0.044 | 37.93 | 2.0 |

†Dosage of activated carbon = 0.25 g/L; $C_0$ selected so that $C_e$ = 0.25$C_0$ in all cases except two of the humic acid simulations.
‡$t_{1.01}$ is the time to reach a final solution concentration of 1.01$C_e$. The times given are estimates only, and will vary depending on type of activated carbon, type of humic substance, initial concentration, and other factors.
SOURCE: Randtke and Snoeyink (Ref. 18).

diameter activated carbon, but only about 15 min is required for 44-μm-diameter activated carbon. For very large-molecular-weight (approximately 50,000) humic substances, the 2.4-mm-diameter particle is expected to take much longer than a year to equilibrate, but only 2 days are required for the 44-μm-diameter particle. Calculations for 10,000 molecular weight fulvic acid showed only about 25 percent saturation of 2.4-mm-diameter particles after 40 days of contact. If the rate of adsorbate uptake is controlled by intraparticle diffusion, as it is for the humic and fulvic substances in Table 13.1, the time to reach equilibrium is directly proportional to the diameter of the particle squared.[18] Thus, the smaller the particle, the faster equilibrium is achieved in both column and complete-mix adsorption systems.

Some conclusions that can be drawn from these observations are that (1) granular carbon should be pulverized for isotherm measurement, especially when the capacity of large-molecular-weight compounds is to be determined (pulverizing does not affect the total surface available for adsorption);[18] (2) the smallest size activated carbon, consistent with other process constraints, such as head loss and loss during regeneration, should be chosen for the best kinetics; and (3) all the capacity of large activated carbon particles in a column may not be used because the time interval between activated carbon replacements is not sufficient for equilibrium to be achieved.

**Mass transfer zone and breakthrough curves for packed-bed reactors.**  The region of an adsorption column in which adsorption is taking place, the *mass transfer zone* (MTZ), is shown in Fig. 13.8*a*. The activated carbon behind the MTZ has been completely saturated with adsorbate at $C_e = C_0$, and the amount adsorbed per unit mass of GAC is $(q_e)_0$. The activated carbon in front of it has not been exposed to adsorbate, so solution concentration and adsorbed concentration are both zero. Within the MTZ, the degree of saturation with adsorbate varies from 100 percent ($q = [q_e]_0$) to zero. The length of the MTZ, $L_{MTZ}$, depends on the rate of adsorption and the solution flow rate. Anything that causes a higher rate of adsorption, such as a smaller carbon particle size, higher temperature, a larger diffusion coefficient of adsorbate, and greater strength of adsorption of adsorbate (i.e., a larger Freundlich $K$ value), will decrease the length of the MTZ. In some circumstances, $L_{MTZ}$ will be reduced sufficiently so that it can be assumed to be zero, yielding the condition shown in Fig. 13.8*b*. If $L_{MTZ}$ is negligible, analysis of the adsorption process is greatly simplified.

The *breakthrough concentration* $C_B$ for a column is defined as the maximum acceptable effluent concentration. When the effluent con-

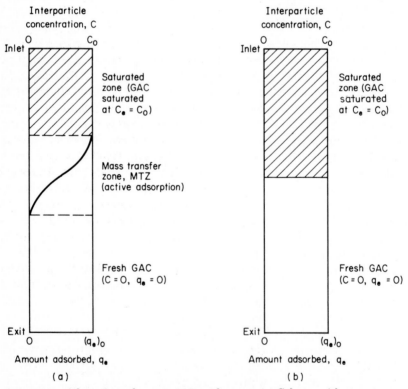

**Figure 13.8** Adsorption column mass transfer zone. (*a*) Column with mass transfer zone. (*b*) Column without mass transfer zone.

centration reaches this value, the GAC must be replaced. The *critical depth* of a column $L_{\text{critical}}$ is the depth that leads to the immediate appearance of an effluent concentration equal to $C_B$ when the column is started up. For the situation in which $C_B$ is defined as the minimum detectable concentration, the critical depth of an activated carbon column is equal to the length of the MTZ. The length of the MTZ is fixed for a given set of conditions, but $L_{\text{critical}}$ varies with $C_B$. The critical depth, the flow rate $Q$, and the area $A$ of the column can be used to calculate the minimum *empty-bed contact time* (EBCT = tank volume occupied by the activated carbon divided by the volumetric flow rate)

$$\frac{L_{\text{critical}}}{Q/A} = \text{EBCT}_{\min} \tag{13.8}$$

When $C_B$ is greater than the minimum detectable concentration, the critical depth is less than $L_{\text{MTZ}}$ and its value can be determined as shown in a later section (see Fig. 13.15 and related discussion).

The breakthrough curve is a plot of the column effluent concentration as a function of either the volume treated, the time of treatment, or the number of bed volumes (BV) treated—i.e., the volume treated divided by the volume of GAC in the contractor. The number of bed volumes is a particularly useful parameter because the data from columns of different sizes and with different flow rates are normalized. A breakthrough curve for a single, adsorbable compound is shown in Fig. 13.9. The shape of the curve is affected by the same factors that affect the length of the MTZ, in the same way. Anything that causes the rate of adsorption to increase will increase the sharpness of the curve, while increasing the flow rate will cause the curve to "spread out" over a larger volume of water treated. The breakthrough curve will be vertical if $L_{\text{MTZ}} = 0$, as shown in Fig. 13.8b. The *breakthrough capacity*, defined as the mass of adsorbate removed by the adsorber at breakthrough, and the *degree of column utilization*, defined as the mass adsorbed at breakthrough divided by the mass adsorbed at complete saturation at the influent concentration, as shown in Fig. 13.9, increase as the rate of adsorption increases.

The breakthrough curve can be used to determine the activated *carbon usage rate* (CUR)—the mass of activated carbon required per unit volume of water treated:

$$\text{CUR}\left(\frac{\text{mass}}{\text{vol}}\right) = \frac{\text{mass of activated carbon in column}}{\text{volume treated to breakthrough, } V_B} \tag{13.9}$$

Breakthrough curves are strongly affected by the presence of nonadsorbable compounds, the biodegradation of compounds in a biologically active column, slow adsorption of a fraction of the molecules

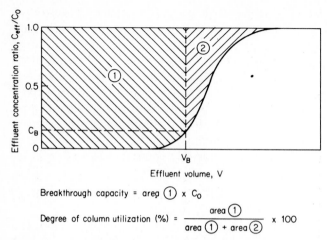

Breakthrough capacity = area ① x $C_O$

Degree of column utilization (%) = $\dfrac{\text{area ①}}{\text{area ① + area ②}}$ x 100

**Figure 13.9**   Adsorption column breakthrough curve.

present, and the critical depth of the column relative to the length of the column. Immediate leakage of adsorbable compounds occurs if the $L_{MTZ}$ is greater than the activated carbon bed depth (compare curves A and B in Fig. 13.10). Nonadsorbable compounds immediately appear in the column effluent, even when the carbon depth is greater

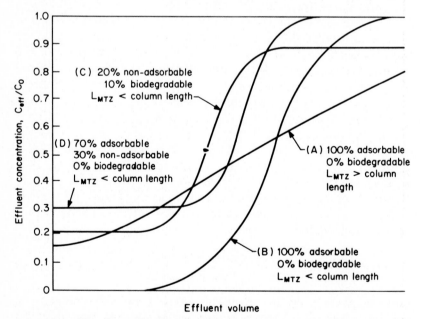

**Figure 13.10**   The effect of biodegradation and the presence of nonadsorbable compounds on breakthrough curves.

than the $L_{MTZ}$ (compare curves C and D in Fig. 13.10). Removal of adsorbable, biodegradable compounds by microbiological degradation in a column results in continual removal, even after the carbon is saturated with adsorbable compounds (see curve C in Fig. 13.10). If a fraction of compounds adsorbs slowly, the upper part of the breakthrough curve will be similar to that produced by biodegradation, but will slowly approach $C_{eff}/C_0 = 1$. Breakthrough curves shown in later sections (see Fig. 13.16 and 13.19, for example) also illustrate some of these effects.

## GAC Adsorption Systems

### Characteristics of GAC

**Physical properties.**  A wide variety of raw materials can be used to make activated carbon,[30] but the substances used for drinking water treatment carbons in the United States predominantly are wood, peat, lignite, subbituminous coal, and bituminous coal. The basic manufacturing process involves carbonization, or conversion of the raw material to a char, and activation, or oxidation to develop the internal pore structure. Carbonization, or pyrolysis, is usually done in the absence of air at temperatures less than 700°C, while activation is carried out with oxidizing gases such as steam and $CO_2$ at temperatures of 800 to 900°C. Patents describing carbonization and activation procedures are given by Yehaskel.[31]

Various characteristics of activated carbon affect its performance.* The particle shape of crushed activated carbon is irregular, but extruded activated carbons have smooth cylindrical shapes. Particle shape affects the filtration and backwash properties of GAC beds. Particle size is an important parameter because of its effect on rate of adsorption, as discussed previously. Particle size distribution refers to the relative amounts of different-size particles that are part of a given sample, or lot, of carbon, and it has an important impact on the filtration properties of GAC in GAC columns that are used both as filters to remove particles and as adsorbers (i.e., filter-adsorbers).[33] Common crushed activated carbon sizes are 12 × 40 and 8 × 30 U.S. standard mesh, which range in diameter from 1.68 to 0.42 mm and 2.38 to 0.59 mm, respectively. The uniformity coefficient is often quite large, typically about 1.9, to promote stratification during backwashing and, thus, to prevent the mixing of activated carbon particles with more

---

*Descriptions of the analytical procedures for testing activated carbon are given in ASTM standards[32] available from ASTM, 1916 Race St., Philadelphia, PA 19103, as well as in AWWA Standards B604-74 and B600-66 available from AWWA, 6666 West Quincy Ave., Denver, CO 80235.

adsorbed compounds from the top of a bed with activated carbon containing smaller amounts of adsorbed compounds from the bottom of the bed (cf. Fig. 13.8). If mixing does occur, it can lead to desorption and premature breakthrough of some compounds.[34] Usually, commercially available activated carbons have a small percentage of material, smaller than the smallest sieve and larger than the largest sieve, which significantly affects the uniformity coefficient. Extruded carbon particles all have the same diameter, but vary in length. There is no method comparable to the sieve analysis procedure to characterize the distribution of lengths, however.

The *apparent density** of activated carbon is the mass of nonstratified dry activated carbon per unit volume of activated carbon, including the volume of voids between grains. Typical values for GAC are 350 to 500 kg/m$^3$ (25 to 31 lb/ft$^3$). Distinguishing between the apparent density and the *bed density, backwashed and drained* (i.e., stratified, free of water) is important, however. The former is a characteristic of activated carbon as shipped. The latter is about 10 percent less than the apparent density and is typical of activated carbon during normal operation unless it becomes destratified during backwashing. The latter is an important parameter because it determines how much activated carbon must be purchased to fill a given size filter.

The *particle density wetted in water* is the mass of solid activated carbon plus the mass of water required to fill the internal pores per unit volume of particle. Its value for GAC typically ranges from 1300 to 1500 kg/m$^3$ (90 to 105 lb/ft$^3$), and it determines the extent of fluidization and expansion of a given size particle during backwash.

Particle hardness is important because it affects the amount of attrition during backwash, transport, and regeneration. In general, the harder the activated carbon, the less attrition there is for a given amount of friction or impact between particles. Activated carbon hardness is generally characterized by an experimentally determined hardness or abrasion number, using a test such as the ASTM ball pan hardness test that measures the resistance to particle degradation upon agitating a mixture of activated carbon and steel balls.[32] The relationship between the amount of attrition that can be expected when activated carbon is handled in a certain way and the hardness number has not been determined, however.

**Adsorption properties.**  A number of parameters are used to describe the adsorption capacity of activated carbon. The *molasses number* or

---

*This definition is based on ASTM Standard D2854-83.[32] It conflicts with ASTM Standard C128-84 for apparent specific gravity, as used in Chap. 8, which does not include the volume of interparticle voids.

*decolorizing index* is related to the ability of activated carbon to adsorb large-molecular-weight color bodies from molasses solution, and generally correlates well with the ability of the activated carbon to adsorb other large adsorbates. The *iodine number*[32] measures the amount of iodine that will adsorb under a specified set of conditions, and it generally correlates well with the surface area available for small molecules. Other numbers have been developed for specific applications, such as the *carbon tetrachloride activity,* the *methylene blue number,* and the *phenol adsorption value.* The values of these numbers give useful information about the abilities of various activated carbons to adsorb different types of organics. Isotherm data for the specific compounds to be removed in a given application, however, if available, are much better indicators of performance.

Two of the more important characteristics of an activated carbon are its pore size distribution (discussed previously) and surface area. The manufacturer provides typical data that usually include the *BET surface area.* This parameter is determined by measuring the adsorption isotherm for nitrogen gas molecules, and then analyzing the data using the *Brunauer-Emmett-Teller* (BET) isotherm equation[29] to determine the amount of nitrogen to form a complete monolayer of nitrogen molecules on the carbon surface. Multiplying the surface area occupied per nitrogen molecule ($0.162 \text{ nm}^2$ per molecule of $N_2$) by the number of molecules in the monolayer yields the BET surface area. Because nitrogen is a small molecule, it can enter pores that are unavailable to larger adsorbates. As a result, all of the BET surface area may not be available for adsorbates in drinking water.

Tabulations of single-solute isotherm constants are very useful when only rough estimates of adsorption capacity are needed to determine whether a more intensive analysis of the adsorption process is warranted. The Freundlich isotherm constants of Dobbs and Cohen,[35] as tabulated by Faust and Aly,[36] are reproduced in Table 13.2 for this purpose.* Additional values from Miltner et al.[37] have also been listed. These data can be used to judge relative adsorption efficiency. The $K$ values of isotherms that have nearly the same values of $1/n$ show the relative capacity of adsorption. For example, if a GAC column is satisfactorily removing 2-chlorophenol [$K = 51$ (mg/g) $(\text{L/mg})^{1/n}$ and $1/n = 0.41$], the removal of compounds with larger values of $K$ and approximately the same concentration will very likely be better. (An exception might occur if the organic compounds adsorb to

---

*References 35 and 37 should be consulted to determine the type of activated carbon and the experimental conditions that were used. The data in reference 35 were not determined in a way that would ensure that equilibrium was achieved for all adsorbates, but the data are suitable to show relative absorbability of compounds and to make rough estimates of activated carbon life. If precise values are needed, new isotherms should be determined, using the water to be treated.

**TABLE 13.2   Freundlich Adsorption Isotherm Constants for Toxic Organic Compounds**[†]

| Compound | $K(mg/g)(L/mg)^{1/n}$ | $1/n$ | Reference[‡] |
|---|---|---|---|
| PCB | 14,100 | 1.03 | 163 |
| Bis(2-ethylhexyl phthalate) | 11,300 | 1.5 | 35/37 |
| Heptachlor | 9,320 | 0.92 | 163 |
| Heptachlor epoxide | 2,120 | 0.75 | 163 |
| Butylbenzyl phthalate | 1,520 | 1.26 | 35/37 |
| Toxaphene | 950 | 0.74 | 163 |
| Endosulfan sulfate | 686 | 0.81 | 35/37 |
| Endrin | 666 | 0.80 | 35/37 |
| Fluoranthene | 664 | 0.61 | 35/37 |
| Aldrin | 651 | 0.92 | 35/37 |
| PCB-1232 | 630 | 0.73 | 35/37 |
| β-Endosulfan | 615 | 0.83 | 35/37 |
| Dieldrin | 606 | 0.51 | 35/37 |
| Alachlor | 479 | 0.26 | 163 |
| Hexachlorobenzene | 450 | 0.60 | 35/37 |
| Pentachlorophenol | 436 | 0.34 | 163 |
| Anthracene | 376 | 0.70 | 35/37 |
| 4-Nitrobiphenyl | 370 | 0.27 | 35/37 |
| Fluorene | 330 | 0.28 | 35/37 |
| Styrene | 327 | 0.48 | 163 |
| DDT | 322 | 0.50 | 35/37 |
| 2-Acetylaminofluorene | 318 | 0.12 | 35/37 |
| α-BHC | 303 | 0.43 | 35/37 |
| Anethole | 300 | 0.42 | 35/37 |
| 3,3-Dichlorobenzidine | 300 | 0.20 | 35/37 |
| γ-BHC (lindane) | 285 | 0.43 | 163 |
| 2-Chloronaphthalene | 280 | 0.46 | 35/37 |
| Phenylmercuric acetate | 270 | 0.44 | 35/37 |
| Carbofuran | 266 | 0.41 | 163 |
| 1,2-Dichlorobenzene | 263 | 0.38 | 163 |
| Hexachlorobutadiene | 258 | 0.45 | 35/37 |
| p-Nonylphenol | 250 | 0.37 | 35/37 |
| 4-Dimethylaminoazobenzene | 249 | 0.24 | 35/37 |
| PCB-1221 | 242 | 0.70 | 35/37 |
| DDE | 232 | 0.37 | 35/37 |
| m-Xylene | 230 | 0.75 | 163 |
| Acridine yellow | 230 | 0.12 | 35/37 |
| Dibromochloropropane (DBCP) | 224 | 0.51 | 163 |
| Benzidine dihydrochloride | 220 | 0.37 | 35/37 |
| β-BHC | 220 | 0.49 | 35/37 |
| n-Butylphthalate | 220 | 0.45 | 35/37 |
| n-Nitrosodiphenylamine | 220 | 0.37 | 35/37 |
| Silvex | 215 | 0.38 | 163 |
| Phenanthrene | 215 | 0.44 | 35/37 |
| Dimethylphenylcarbinol | 210 | 0.34 | 35/37 |
| 4-Aminobiphenyl | 200 | 0.26 | 35/37 |
| β-Naphthol | 200 | 0.26 | 35/37 |
| p-Xylene | 200 | 0.42 | 163 |
| α-Endosulfan | 194 | 0.50 | 35/37 |
| Chlordane | 190 | 0.33 | 163 |

**TABLE 13.2   Freundlich Adsorption Isotherm Constants for Toxic Organic Compounds[†] (Continued)**

| Compound | $K(mg/g)(L/mg)^{1/n}$ | $1/n$ | Reference[‡] |
|---|---|---|---|
| Acenaphthene | 190 | 0.36 | 35/37 |
| 4,4'-Methylene-bis-(2-chloroaniline) | 190 | 0.64 | 35/37 |
| Benzo[k]fluoranthene | 181 | 0.57 | 35/37 |
| Acridine orange | 180 | 0.29 | 35/37 |
| α-Naphthol | 180 | 0.32 | 35/37 |
| Ethylbenzene | 175 | 0.53 | 163 |
| o-Xylene | 174 | 0.47 | 163 |
| 4,6-Dinitro-o-cresol | 169 | 0.27 | 35/37 |
| α-Naphthylamine | 160 | 0.34 | 35/37 |
| 2,4-Dichlorophenol | 157 | 0.15 | 35/37 |
| 1,2,4-Trichlorobenzene | 157 | 0.31 | 35/37 |
| 2,4,6-Trichlorophenol | 155 | 0.40 | 35/37 |
| β-Naphthylamine | 150 | 0.30 | 35/37 |
| 2,4-Dinitrotoluene | 146 | 0.31 | 35/37 |
| 2,6-Dinitrotoluene | 145 | 0.32 | 35/37 |
| 4-Bromophenyl phenyl ether | 144 | 0.68 | 35/37 |
| p-Nitroaniline | 140 | 0.27 | 35/37 |
| 1,1-Diphenylhydrazine | 135 | 0.16 | 35/37 |
| Naphthalene | 132 | 0.42 | 35/37 |
| Aldicarb | 132 | 0.40 | 163 |
| 1-Chloro-2-nitrobenzene | 130 | 0.46 | 35/37 |
| p-Chlorometacresol | 124 | 0.16 | 35/37 |
| 1,4-Dichlorobenzene | 121 | 0.47 | 35/37 |
| Benzothiazole | 120 | 0.27 | 35/37 |
| Diphenylamine | 120 | 0.31 | 35/37 |
| Guanine | 120 | 0.40 | 35/37 |
| 1,3-Dichlorobenzene | 118 | 0.45 | 35/37 |
| Acenaphthylene | 115 | 0.37 | 35/37 |
| Methoxychlor | 115 | 0.36 | 163 |
| 4-Chlorophenyl phenyl ether | 111 | 0.26 | 35/37 |
| Diethyl phthalate | 110 | 0.27 | 35/37 |
| Chlorobenzene | 100 | 0.35 | 163 |
| Toluene | 100 | 0.45 | 163 |
| 2-Nitrophenol | 99 | 0.34 | 35/37 |
| Dimethyl phthalate | 97 | 0.41 | 35/37 |
| Hexachloroethane | 97 | 0.38 | 35/37 |
| 2,4-Dimethylphenol | 78 | 0.44 | 35/37 |
| 4-Nitrophenol | 76 | 0.25 | 35/37 |
| Acetophenone | 74 | 0.44 | 35/37 |
| 1,2,3,4-Tetrahydronaphthalene | 74 | 0.81 | 35/37 |
| Adenine | 71 | 0.38 | 35/37 |
| Dibenzo[a,h]anthracene | 69 | 0.75 | 35/37 |
| Nitrobenzene | 68 | 0.43 | 35/37 |
| 2,4-D | 67 | 0.27 | 163 |
| 3,4-Benzofluoranthene | 57 | 0.37 | 35/37 |
| 2-Chlorophenol | 51 | 0.41 | 35/37 |
| Tetrachloroethylene | 51 | 0.56 | 35/37 |
| o-Anisidine | 50 | 0.34 | 35/37 |
| 5-Bromouracil | 44 | 0.47 | 35/37 |
| Benzo[a]pyrene | 34 | 0.44 | 35/37 |

**TABLE 13.2    Freundlich Adsorption Isotherm Constants for Toxic Organic Compounds[†] (Continued)**

| Compound | $K(mg/g)(L/mg)^{1/n}$ | $1/n$ | Reference[‡] |
|---|---|---|---|
| 2,4-Dinitrophenol | 33 | 0.61 | 35/37 |
| Isophorone | 32 | 0.39 | 35/37 |
| Trichloroethylene | 28 | 0.62 | 35/37 |
| Thymine | 27 | 0.51 | 35/37 |
| 5-Chlorouracil | 25 | 0.58 | 35/37 |
| N-Nitrosodi-n-propylamine | 24 | 0.26 | 35/37 |
| Bis(2-Chloroisopropyl) ether | 24 | 0.57 | 35/37 |
| 1,2-Dibromoethene (EDB) | 22 | 0.46 | 163 |
| Phenol | 21 | 0.54 | 35/37 |
| Bromoform | 20 | 0.52 | 35/37 |
| 1,2-Dichloropropane | 19 | 0.59 | 163 |
| 1,2-trans-Dichloroethylene | 14 | 0.45 | 163 |
| cis-1,2-Dichloroethylene | 12 | 0.59 | 163 |
| Carbon tetrachloride | 11 | 0.83 | 35/37 |
| Bis(2-Chloroethyoxy) methane | 11 | 0.65 | 35/37 |
| Uracil | 11 | 0.63 | 35/37 |
| Benzo[g,h,i]perylene | 11 | 0.37 | 35/37 |
| 1,1,2,2-Tetrachloroethane | 11 | 0.37 | 35/37 |
| 1,2-Dichloropropene | 8.2 | 0.46 | 35/37 |
| Dichlorobromomethane | 7.9 | 0.61 | 35/37 |
| Cyclohexanone | 6.2 | 0.75 | 35/37 |
| 1,1,2-Trichloroethane | 5.8 | 0.60 | 35/37 |
| Trichlorofluoromethane | 5.6 | 0.24 | 35/37 |
| 5-Fluorouracil | 5.5 | 1.0 | 35/37 |
| 1,1-Dichloroethylene | 4.9 | 0.54 | 35/37 |
| Dibromochloromethane | 4.8 | 0.34 | 35/37 |
| 2-Chloroethyl vinyl ether | 3.9 | 0.80 | 35/37 |
| 1,2-Dichloroethane | 3.6 | 0.83 | 35/37 |
| Chloroform | 2.6 | 0.73 | 35/37 |
| 1,1,1-Trichloroethane | 2.5 | 0.34 | 35/37 |
| 1,1-Dichloroethane | 1.8 | 0.53 | 35/37 |
| Acrylonitrile | 1.4 | 0.51 | 35/37 |
| Methylene chloride | 1.3 | 1.16 | 35/37 |
| Acrolein | 1.2 | 0.65 | 35/37 |
| Cytosine | 1.1 | 1.6 | 35/37 |
| Benzene | 1.0 | 1.6 | 35/37 |
| Ethylenediaminetetraacetic acid | 0.86 | 1.5 | 35/37 |
| Benzoic acid | 0.76 | 1.8 | 35/37 |
| Chloroethane | 0.59 | 0.95 | 35/37 |
| N-Dimethylnitrosamine | $6.8 \times 10^{-5}$ | 6.6 | 35/37 |

[†]The isotherms are for the compounds in distilled water, with different activated carbons. The values of $K$ and $1/n$ should be used only as rough estimates of the values that will be obtained using other types of water and other activated carbon.

[‡]All values from Refs. 35 and 36 were determined by Dobbs and Cohen[35] and have been tabulated by Faust and Aly.[36]

particles that pass through the adsorber.) If the $1/n$ values are much different, however, the capacity of activated carbon for each compound of interest should be calculated at the equilibrium concentration of interest using Eq. (13.1), because the relative adsorbability will depend on the equilibrium concentration. The PAC usage rate and the use of isotherm values to estimate adsorber life will be discussed later.

### GAC contactors

GAC may be used in pressure or gravity contactors. Pressure filters enclose the GAC and can be operated over a wide range of flow rates because of the wide variations in pressure drop that can be used. These filters are sometimes cheaper to design and construct because they can be prefabricated and shipped to the site. A disadvantage is that the GAC cannot be visually observed with ease. Gravity contactors are better suited to systems when wide variations in flow rate are not desirable because of the need to remove turbidity, when large pressure drops are undesirable because of their impact on operation costs, and when visual observation is needed to monitor the condition of the GAC. For many systems the decision between pressure or gravity contactors is made on the basis of cost.

Water may be applied to GAC either upflow or downflow, and upflow columns may be either packed bed or expanded bed. Downflow columns are the most common and seem best suited for drinking water treatment. McCarty et al.[38] found that carbon fines were produced during packed bed upflow operation but not during downflow operation. Expanded bed upflow columns are suited for waters with high suspended solids concentrations when suspended solids removal is not necessary or when the solids are to be removed by a subsequent process. A higher activated carbon usage rate is expected for expanded beds, compared to packed beds with no backwashing, because mixing of the activated carbon will create a longer MTZ.

Single-stage contactors are often used for small systems, but if more than one contactor is required, lower activated carbon usage rates can be achieved by arranging them either in series or in parallel as shown in Fig. 13.11, possibly yielding a lower cost system. GAC in a single-stage contactor must be removed about the time the MTZ begins to exit the column. At this point only a portion of the activated carbon is saturated at the influent concentration, so the activated carbon usage rate may be relatively high. Alternatively, columns may be arranged in series so that the MTZ is entirely contained within the downstream columns after the lead column has been saturated with the influent concentration. When the activated carbon is replaced in the lead column, the flow is redirected so that it goes through the freshest acti-

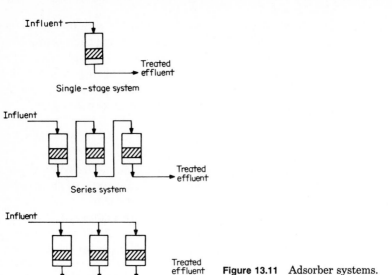

Figure 13.11   Adsorber systems.

vated carbon last. Thus, the activated carbon "moves" countercurrent to the flow of water, and lower activated carbon usage rates are achieved than with single-stage contactors. The increased cost of plumbing counters the cost benefit of reduced activated carbon usage rate, however, especially when more than two columns must be used in series.

Parallel-flow activated carbon adsorbers can also be used to decrease the activated carbon usage rate from that which is possible with a single-stage contactor.[39,40] Because the effluent from each of the units is blended, each unit can be operated until it is producing a water with an effluent concentration in excess of the treated water goal. Only the composite flow must meet the effluent quality goal. Roberts and Summers[40] showed that the fraction, or concentration, of organic matter remaining in the composite effluent, $\bar{f}$, was given by

$$\bar{f} = \frac{1}{n} \sum_{i=1}^{n} f_i \tag{13.10}$$

where $f_i$ is the fraction, or concentration, of organic matter remaining in the effluent from the $i$th adsorber, and $n$ is the number of adsorbers in parallel, each of equal capacity. Values of $f_i$ can be determined from a single breakthrough curve such as in Fig. 13.12a, assuming that replacement of GAC in each adsorber will take place at equal intervals. Given that $\theta_n$ is the number of bed volumes processed through each adsorber in the parallel system at the time of replacement, the abscissa of the break-

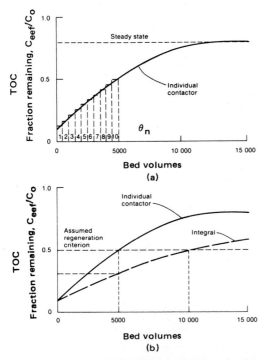

**Figure 13.12** Integral breakthrough curve for characterizing performance of multiple parallel contactor operation. (*Source: P. V. Roberts and R. S. Summers, "Performance of Granular Activated Carbon for Total Organic Carbon Removal," J. AWWA, vol. 74, no. 2, 1982, p. 113.*)

through curve for the individual contactor from 0 to $\theta_n$ is divided into $\theta_n/n$ equal increments. For a given value of $\theta_n$, the value of $f_i$ for each increment can be read from the figure, and Eq. (13.10) can be used to calculate the concentration of the blended water. For example, if 10 adsorbers are used in parallel, determination of $\bar{f}$ for several different values of $\theta_n$ using Eq. (13.10) leads to the integral curve shown in Fig. 13.12b. By operating the 10 contactors in parallel, each adsorber can process 10,000 bed volumes of throughput if the effluent TOC criterion is 50 percent of the influent, compared to 5000 bed volumes if only a single contactor were used, or if all 10 contactors were operated in parallel but were replaced at the same time.

Other flow arrangements can be used to produce lower activated carbon usage rates. The parallel-series arrangement of gravity filters used in North Holland (see Fig. 13.13) is possible. Sontheimer and Hubele[42] report that a similar arrangement using pressure filters was employed at Pforzheim, West Germany (see Fig. 13.14). Each of the

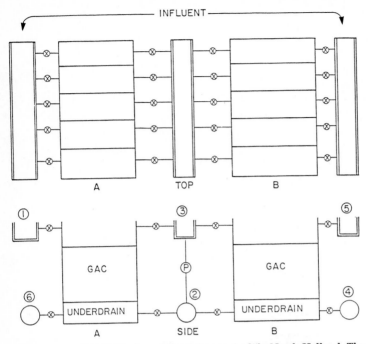

**Figure 13.13**  Parallel-series configuration at Andijk, North Holland, The Netherlands.[41] During operation, the influent in channel 1 passes through the filter cells with partially spent carbon in bank A into channel 2. It is then pumped to channel 3 and passes through the filter cells with fresh carbon in filter bank B into product water channel 4. The influent in channel 5 passes through the cells with partially spent carbon in bank B into channel 2. It is then pumped to 3, passes through the fresh carbon cells in bank A into product water channel 6. When effluent quality exceeds the desired objective, the GAC in the cell producing the lowest-quality water is replaced with fresh GAC.

two layers can be backwashed and replaced independently, and the order of flow through these layers can be reversed as shown in the figure. A 35 percent lower activated carbon usage rate for this system for removing halogenated hydrocarbons from groundwater was reported, compared to a single-stage system.

The pulsed-bed contactor can also be used to decrease carbon usage rate from that of a single contactor. The flow is applied upward through the column; the spent GAC, a fraction of the total amount present, is periodically removed from the bottom of the column, and an equal amount of fresh GAC is applied to the top. The desired effect is countered if suspended solids, or biodegradable compounds that cause extensive biofilm growth on the activated carbon, are present so

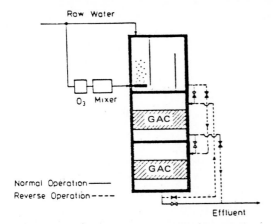

Row Water

O₃   Mixer

GAC

GAC

Normal Operation ———
Reverse Operation - - - -

Effluent

**Figure 13.14**  Process scheme for the removal of volatile halogenated hydrocarbons from groundwater at the Pforzheim water works. (*Source: H. Sontheimer and C. Hubele, "The Use of Ozone and Granular Activated Carbon in Drinking Water Treatment," in P.M Huck, and P. Toft (eds.), Treatment of Drinking Water for Organic Contaminants, Pergamon Press, New York, 1987.*)

that the bed must be backwashed. Backwashing will lead to mixing of the fresh activated carbon at the top of the bed with spent activated carbon deeper in the bed and will destroy some of the beneficial countercurrent effect. Additionally, some activated carbon fines may be produced during upflow that may require removal by a subsequent process.[38] The density of GAC particles changes very little during use if only organics are adsorbed[33] because adsorbed organics have approximately the same density as the water molecules that are displaced during adsorption. If inorganic deposits accumulate on the GAC, large changes in density may be observed, however.

GAC contactors can also be classified by their position in the treatment train. The filter-adsorber employs GAC to remove particles as well as dissolved organic compounds. These contactors may be constructed simply by removing all or a portion of the granular media from a rapid filter and replacing it with GAC. Alternatively, a new filter box and underdrain system for the GAC may be designed and constructed. Graese et al.[33] discuss these types of filters in detail. The postfilter-adsorber is preceded by a granular media filter and, thus, has as its only objective the removal of dissolved organic compounds. Backwashing of these adsorbers is usually unnecessary, but if extensive biological growth occurs they may require backwashing as often as once per week.[43,44]

## Performance of GAC Systems

### Factors affecting organic compound removal efficiency

Adsorbate and GAC properties both have important effects on adsorption that have been discussed in earlier sections. Additional factors that must be considered in the design of full-scale systems are presented here.

**GAC particle size.**  The effect of particle size on the rate of approach to equilibrium in isotherm determination was discussed previously. It has a similar effect on rate of adsorption in columns. If the rate of adsorption is controlled by intraparticle diffusion, the time to reach equilibrium with a given solution concentration in a column approximates that for a batch test. Decreasing particle size will decrease the time required to achieve equilibrium and will decrease the length of the MTZ in a column, with all other factors constant. Thus, to improve adsorption efficiency and to minimize the size of column required, the particle size selected for a contactor should be as small as possible.

The rate of head loss buildup caused by turbidity removal may limit the size of GAC that can be used in adsorbers. The smaller the GAC particle, the higher the initial head loss and rate of head loss buildup; thus, cost of energy and availability of head have an important influence on the GAC size selected for a design. Additionally, if a filter-adsorber is constructed by replacing media in an existing rapid filter, turbidity removal efficiency generally will be higher for smaller GAC than for large. If the media is too small, however, the rate of head loss buildup because of particle accumulation will be too high, and the net water production will be too small for cost-effective operation.

The commercial sizes of GAC typically are characterized by a relatively large uniformity coefficient of up to 1.9. This large coefficient causes the bed to restratify more easily after backwashing and thus helps minimize the adverse effects of backwashing on adsorption efficiency. This large uniformity coefficient also requires that a greater percentage of expansion of the adsorber be used during backwash in order to expand the bottom media.[33] Some recently designed GAC filters have used GAC with a small uniformity coefficient ($\sim 1.3$) in deep beds to improve depth removal of turbidity and to increase net water production.[45] Mixing of the media in these filters undoubtedly is more than in filters with large uniformity coefficients, so they should not be used in applications where desorption from the mixed GAC will require early GAC replacement.

Common practice is to use $12 \times 40$ U.S. standard mesh (1.68

mm × 0.42 mm) or similar activated carbon in postfilter-adsorbers because backwashing is rarely required because of particle buildup. The
same activated carbon is commonly used when 76 cm (30 in) or less is
used as the only filtration medium in filter-adsorbers. Deeper beds
that will be used to remove turbidity commonly employ 8 × 30 U.S.
standard mesh (2.38 mm × 0.60 mm) or larger activated carbon to
promote longer filter runs. The option of modifying the particle size to
obtain a better media design for a particular application is available,
but the activated carbon would have to be custom-sized for the application.

**Contact time, bed depth, hydraulic loading rate.** The most important
GAC adsorber design parameter is the contact time, most commonly
described by its EBCT:

$$\text{EBCT} = \frac{V}{Q} \tag{13.11}$$

or

$$\text{EBCT} = \frac{L_{\text{Bed}}}{Q/A} = \frac{L_{\text{Bed}}}{\text{approach velocity}} \tag{13.12}$$

where $V$ = bulk volume of carbon in contactor
$\quad Q$ = the volumetric flow rate to contactor
$\quad L_{\text{Bed}}$ = bed depth
$\quad A$ = bed area

The actual contact time is the product of the EBCT and the
interparticle porosity, and this porosity is usually about 0.4 to 0.5.

The EBCT has a significant impact on performance. For a given situation, a critical depth of GAC and a corresponding minimum EBCT
exist that must be exceeded if the adsorber is to produce any water of
acceptable quality. As the EBCT increases, the bed life (expressed in
bed volumes of product water to breakthrough) will increase until a
maximum value is reached. Correspondingly, the activated carbon usage rate will decrease to a minimum value. For example, Fig. 13.15a
shows that the operating time or service time of a column will increase with increasing depth, although the increase is not always linear with depth. These curves are commonly called *bed depth–service
time* curves, and may be used to determine the critical depth as shown
in the figure. Figure 13.15b shows that the percentage of activated
carbon in a column that is exhausted at breakthrough increases as
depth, or EBCT, increases. The mass of organic matter adsorbed per
unit mass of activated carbon increases as percent exhaustion in-

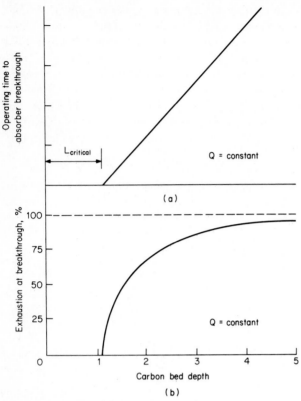

**Figure 13.15** Bed depth–service time and percent exhaustion at breakthrough versus depth.

creases, and, correspondingly, the number of bed volumes of water that can be processed before breakthrough will also increase to a maximum value.

Increasing EBCT, or bed depth at a constant hydraulic application rate, will impact annual treatment costs.[34,46,47] As contactor size increases, fixed costs increase because of the greater cost of larger systems. Operating costs decrease because of decreasing carbon usage rate and replacement frequency. Thus an optimum depth, or contact time, can be achieved.

Pilot data that show the effect of increasing EBCT on bed life have been given by Love and Eilers.[48] Data for *cis*-1,2-dichloroethylene at 7.6 m/h (3 gpm/ft$^2$) showed that as EBCT was increased from 6 to 12 min, and then to 18 min, the number of bed volumes of water processed to a breakthrough concentration of 0.1 μg/L were 4100, 7100, and 8100, respectively. The activated carbon usage rate for the 18-min

EBCT adsorber is thus about one-half the value for the 6-min EBCT adsorber. Because little change occurred in the number of bed volumes processed to breakthrough as the EBCT was increased from 12 to 18 min, little change occurred in activated carbon usage rate, indicating that the GAC was approaching complete exhaustion. Similar data were given for 1,1,1-trichloroethane and carbon tetrachloride. Collecting pilot-plant data for a range of EBCT values is important if the lowest activated carbon usage for a given application is to be determined.

Other factors must be considered when selecting a hydraulic application rate. As it increases, the thickness of the hydrodynamic boundary layer decreases, although this is not a major effect for most applications because film transport often plays a relatively small role in affecting rate of uptake of many compounds from aqueous solution. Cover and Pieroni[49] review data that show adsorbers with the same EBCT, but with different hydraulic loading rates, give essentially the same performance in terms of number of bed volumes processed to breakthrough provided the depth is considerably greater than the critical depth. Head loss will increase with increasing hydraulic rate, so energy costs must be considered. Also, if the GAC is being used to remove particles as well as dissolved organics, the effect of application rate on removal of turbidity must also be considered.

GAC EBCTs in use today range from a few minutes for some filter-adsorbers[33] to more than 4 h for the removal of high concentrations of some specific contaminants.[50] Hydraulic application rates vary from 1 to 30 m/h (0.4 to 12 gpm/ft$^2$), with a typical value being 7 to 10 m/h (3 to 4 gpm/ft$^2$).

**Backwashing.**   Backwashing of GAC filter-adsorbers is essential to remove solids and to maintain the desired hydraulic properties of the bed, and possibly to control biological growth, and is often necessary for postfilter-adsorbers. Backwashing should be minimized, however, because of its possible effect on adsorption efficiency. Mixing of the bed may take place during backwashing, and, if so, GAC with adsorbed molecules near the top of the bed will move deeper into the bed where desorption is possible. Molecules that are easily reversibly adsorbed, such as carbon tetrachloride and other volatile organic chemicals (VOCs), may be partially desorbed in this new position, leading to a spreading out of the MTZ and to early breakthrough.[34] Desorption will not occur if the molecules are irreversibly adsorbed or if they are removed by a destructive mechanism, such as biodegradation, instead of adsorption. The large uniformity coefficient of most commercial activated carbons promotes restratification after backwash, but if the underdrain system does not properly distribute the wash water or if the backwash is not carried out in a manner that aids restratification, substantial mixing of the activated carbon can occur with each backwash.[33]

**Biological activity.**    Biological activity on GAC has several beneficial aspects. Some specific compounds can be removed by biological oxidation rather than by adsorption, such as phenol,[51] the odor-causing compounds geosmin and methylisoborneol (MIB),[52,53] p-nitrophenol and salicylic acid,[54] ammonia,[55] and probably many more compounds.[56] Some evidence for biological removal of chlorinated benzenes and aromatic hydrocarbons was also found at Water Factory 21.[57] Additionally, some portion of the DOC in natural waters can be biologically oxidized on activated carbon, as shown in Fig. 13.16. Sontheimer and Hubele[42] found a small amount of biological oxidation if the water was not preozonated, but application of 1.1 mg $O_3$/mg DOC resulted in removal of 35 to 40 percent of the influent DOC by biological oxidation. Biodegradable compounds may be removed by microbes, without prior adsorption to the GAC, if a biofilm capable of degrading them was developed before they are applied. Adsorbable, biodegradable compounds may be adsorbed first if the biofilm is not

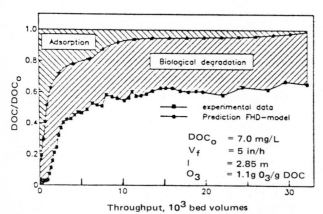

Figure 13.16   DOC removal by adsorption and biodegradation during GAC filtration of an ozonated humic acid solution. (*Source: H. Sontheimer and C. Hubele, "The Use of Ozone and Granular Activated Carbon in Drinking Water Treatment," in P.M Huck, and P. Toft, eds., Treatment of Drinking Water for Organic Contaminants, Pergamon Press, New York, 1987.*)

developed when the compounds enter the column, and then desorbed and degraded as the biofilm develops.

Preozonation does not always produce increased amounts of biological oxidation. Glaze et al.[58] studied a GAC system with a 24-min EBCT; one set of columns was preceded by ozonation at 0.5 to 0.6 mg $O_3$/mg TOC, and another parallel system was not preceded by ozone. Typical results showed about 40 to 50 percent TOC removal by GAC for the 20- to 44-week period of operation in both systems, thus indicating little effect of the ozone. About 0.6 to 0.8 mg/L of the approximately 2 mg/L TOC being removed by each column could be attributed to biological activity. Possibly no significant effect of the ozone was noted because the dose is on the lower end of the 0.5 to 1.0 mg $O_3$/mg TOC dose recommended for promoting biological oxidation,[42] or because a large fraction of the organic matter was biodegradable before it reacted with ozone.

An important observation made by Glaze et al.[58] at Shreveport was that biological activity was high during the summer months when the water temperature was in the 25° to 35°C range, but that it decreased to a relatively insignificant amount as water temperature dropped to 8 to 12°C during the winter months. Thus, biological oxidation cannot be depended upon throughout the year if low temperatures are expected.

Sontheimer and Hubele[42] give typical process parameters for an ozone-GAC system, reproduced in Table 13.3. Note, these values may not be appropriate for all designs. The 100 g DOC/m³-day biological oxidation compares well with the 75 to 150 g TOC/m³-day found by Glaze et al.[58]

An important beneficial effect of ozone-GAC systems was noted by van der Gaag et al.[59] when treating coagulated and settled Rhine River water. The mutagenicity of chlorinated effluent from an ozone-GAC system was significantly lower than the mutagenicity of chlorinated, nonozonated GAC effluent. The authors caution, however, that this response may not be the same for all waters. Sontheimer and Hubele,[42] on the other hand, reference similar results from research

TABLE 13.3  **Process Parameters for the Drinking Water Treatment by Adsorption and Biological Activity**

| | |
|---|---|
| Ozone dosage | 0.5 to 1.0 g $O_3$/g DOC |
| Biological degradation | ~ 100 g DOC/(m³)(day) |
| $O_2$ demand (for DOC oxidation) | ~ 200 g $O_2$/(m³)(day) |
| Empty bed contact time | 15 to 30 min |

SOURCE: H. Sontheimer and C. Hubele, "The Use of Ozone and Granular Activated Carbon in Drinking Water Treatment," in P. M. Huek and P. Toft (eds.), *Treatment of Drinking Water for Organic Contaminants*, Pergamon Press, New York.

in Israel. There the ozone-GAC effluent showed a much lower response in cell tissue tests, thus indicating a lower mutagenicity of that effluent. Whether the beneficial effect is caused by the ozone, by the biological treatment, or by the combination of the two must yet be determined, however.

Biologically active GAC generally does cause the concentration of microorganisms in a GAC column effluent to be higher than in the influent. A thorough analysis of this is given by Symons et al.[60] They report data from Beaver Falls, Pa., that show both coliform and standard plate counts to be higher in GAC effluent than influent when the water temperature was greater than about 10°C, even though 1 to 2 mg/L of chlorine residual was present in the influent to the bed. Apparently the GAC reduced the chlorine and allowed the bacteria to regrow. When the water temperature was below about 10°C, no regrowth was noticed. The GAC was used in filter-adsorbers that provided an EBCT of about 12 min. About 30 cm of sand was used under 60 cm of GAC. Other data from Philadelphia showed that coliform organisms such as *Citrobacter freundii, Enterobacter cloacae,* and *Klebsiella pneumonia* were found in the GAC filters. In all cases, postdisinfection produced water meeting United States Environmental Protection Agency regulations.

Even though bacteria grow readily in GAC filters, and high microorganism counts are observed in effluents from these processes, the disinfectant demand to achieve microorganism kill is much reduced by GAC filtration. An exception to this occurs if activated carbon particles penetrate the underdrain and provide a habitat for microbes that protects them from being killed by disinfectant.[61] This has been shown to occur in some filter-adsorbers. More information is needed to determine the factors affecting the production of fines and passage of carbon particles to the finished water.

Zooplankton can grow in GAC filters if the filters are biologically active. Organisms such as oligochaetes and rotatoria have been reported to increase as water was processed through GAC during the summer months at Rotterdam and North Holland,[62] probably because they use bacteria for food. Similar observations have been made in West Germany[43] and France.[44] Sontheimer[43] and Fiessinger[44] recommend backwashing with air scour once every five days or so to wash the eggs of these organisms out of the adsorber before they can hatch. In contrast, microscreening is used to remove the organisms from the GAC filtered water at North Holland.

**Control of microbial growth.** Biologically active carbon must be controlled to avoid undesirable effects. Anaerobic conditions may develop, with attendant odor problem, if the system is not kept aerobic. This may happen if large concentrations of ammonia enter the filter (1 mg/L of $NH_3$ requires about 3.8 mg/L of dissolved oxygen if it is converted to $NO_3^-$), if insufficient dissolved oxygen is in the water, or if the bed is allowed to stand idle for a period of time. Proper control should ensure that sufficient oxygen is present at all times.

Control is also possible through proper design. Figure 13.17 shows the distribution of microorganisms in a biologically active GAC filter that is treating nonchlorinated water.[63] Much larger numbers of organisms are on the activated carbon at the entrance to the bed than at greater depths. This distribution is consistent with larger amounts of adsorbed compounds in the upper level.[42] Growth in the upper part of a deep bed has the opportunity to be removed deeper within the bed when it sloughs off. Van der Kooij[62] noted that increasing EBCT by increasing the bed depth can bring about sharp reductions in numbers of organisms in GAC filtrate, but increasing EBCT by decreasing flow rate did not decrease, but sometimes increased, the plate count. Available data on the effect of backwashing on effluent organism concentration are inconclusive.[62]

Application of chlorine to GAC adsorbers does not prevent growth and increases the concentration of adsorbed chloroorganics. The potential exists for chlorine to make the activated carbon more friable and break up more easily, especially during backwash, because chlorine destroys some of the activated carbon when it is reduced.[64] More research is needed to show whether this causes attrition, however. The potential exists for formation of unique organic compounds through the catalytic action of the activated carbon surface as shown

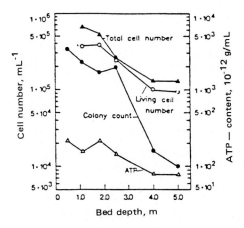

**Figure 13.17** Microbiological parameters in a GAC filter. (*Source: H. Sontheimer and C. Hubele, "The Use of Ozone and Granular Activated Carbon in Drinking Water Treatment," in P. M. Huck and P. Toft eds., Treatment of Drinking Water for Organic Contaminants, Pergamon, New York, 1987.*)

below (see Table 13.4 and related discussion). Thus, application of chlorine to GAC filters is not recommended. Rather, adequate disinfection should follow the GAC.

**Pretreatment.**    Pretreatment can have a significant impact on the performance of activated carbon systems. The concentration of organics may be lowered, the species of compounds may be changed, thus changing adsorbability and biodegradability, and the inorganic composition may be changed in a manner that affects absorbability and the tendency of the activated carbon to become fouled. The removal of natural organic matter by coagulation, sedimentation, and filtration reduces the quantity of organics that must be removed by adsorption. Lower-cost operation of GAC systems for TOC and trihalomethane precursor removal may then be possible, and reduced competition with trace organics is expected (see Chap. 10 for coverage of organic compound removal during coagulation).

Adsorption of organic acids and bases by GAC is generally affected by solution pH, so pretreatment steps that affect pH have an important effect on adsorption. In general, both the undissociated and ionized forms of an adsorbate can be adsorbed on GAC, with the undissociated form being more strongly adsorbed. The ionic form has a much higher affinity for polar water molecules and thus tends to remain in solution. Typical data are shown in Fig. 13.18 for three weak organic acids: 2,4,6-trichlorophenol, 2,4-dichlorophenol, and soil fulvic acid. Different types of humic substances may have both different capacity and capacity dependence on pH.

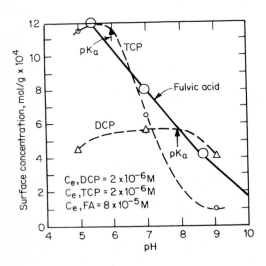

**13.18** The effect of pH on trichlorophenol (TCP), dichlorophenol (DCP), and soil fulvic acid (FA) adsorption. (TCP and DCP data were taken from Ref. 65, and FA data, expressed in moles C, from Ref. 66.)

Reactions of chlorine, or other oxidative pretreatment chemicals, such as ozone, chlorine dioxide, and permanganate, with GAC or with organic compounds in aqueous solution or on the GAC surface can alter the adsorption performance. For example, ozone can react with humic substances to produce more polar intermediates that are less adsorbable on GAC,[67] but are usually more biodegradable.[42] If TOC is more biodegradable, increased removals by microbiological activity in a GAC contactor are expected.

Chlorine-containing disinfectants ($HOCl$, $ClO_2$, or $NH_2Cl$) may react both with activated carbon and adsorbed compounds. Unusual products not characteristic of solution reactions may be formed when activated carbon is present. For example, the $HOCl$ reaction with adsorbed 2,4-dichlorophenol (2,4-DCP) will give unusual products, including a series of hydroxylated polychlorinated biphenyls (PCBs) at $HOCl$ concentrations normally encountered in drinking water treatment practice (see Table 13.4). A similar product mixture was also obtained when the GAC was first treated with $HOCl$ and then 2,4-DCP was adsorbed. The $HOCl$-activated surface caused similar compounds to form as adsorption took place. Furthermore, some of these products may desorb from the activated carbon column. Additional data are given on this effect by Voudrias et al.[68] The reaction products that have been found to date have been formed only in laboratory systems. Further research is needed to show whether they will also form in field installations in the presence of humic substances. Because such compounds might form and because their health effects are unknown, the application of chlorine-containing disinfectants to GAC adsorbers needs to be reevaluated and eliminated where possible.

Pretreatment to prevent fouling of GAC is also important. Application of water that is supersaturated with salts such as calcium carbonate will lead to blockage of the activated carbon pores and possibly to complete coverage of the particle. Iron and manganese precipitates may interfere with adsorption also. If ammonia concentration is very high, pretreatment for ammonia removal may also be necessary to prevent dissolved oxygen depletion in biologically active beds.

## Reactions of inorganic compounds with activated carbon

When activated carbon is used in water treatment, it may inadvertently contact oxidants, such as oxygen, aqueous chlorine, chlorine dioxide, and permanganate, and react with them. For example, virgin activated carbon has been shown by Prober et al.[69] to react with from

**TABLE 13.4    Reaction Products from HOCl-2,4-Dichlorophenol-GAC Reaction**

| Compound |
|---|

Source: Reprinted with permission from *Environmental Sci. Technol.*, vol. 19, 1985, p. 441, "Effects of Activated Carbon on the Reactions of Free Chlorine with Phenols," by E. A. Voudrias, R. A. Larson, and V. L. Snoeyink. Copyright 1985, American Chemical Society.

10 to 40 mg aqueous $O_2$ per gram of carbon over a time span of 1700 h. Less than 5 percent of this could be recovered as molecular oxygen, and from 10 to 50 percent was accounted for as acidic oxides on the carbon surface. For comparison, Chudyk and Snoeyink[70] found that 3 to 4 mg $O_2$ per gram of carbon had reacted over a time span of 130 h. The extent of reaction is undoubtedly affected by solution properties and the type of carbon. Activated carbon also is used purposely to destroy oxidants such as chlorine and other disinfectants. It may also adsorb other inorganic compounds such as mercury.

**Free chlorine–activated carbon reactions.**    The well-known reactions of HOCl and $OCl^-$ with activated carbon are as follows:

$$HOCl + C^* \rightarrow C^*O + H^+ + Cl^- \tag{13.13}$$

$$OCl^- + C^* \rightarrow C^*O + Cl^- \tag{13.14}$$

where $C^*$ and $C^*O$ represent the carbon surface and a surface oxide, respectively. These reactions proceed rapidly, with reaction (13.13) (pH < 7.5) being faster than reaction (13.14) (pH > 7.5).[71-73] The reactions also proceed faster at higher temperature and when small-particle-size carbon is used rather than large. For carbons made from the same raw material, the carbon with the larger pore volume gives the higher rate.[74] Typically, a column with 1.2-min EBCT, pH 7.6, 18 × 20 mesh carbon (U.S. standard) will reduce 5 mg/L free chlorine as $Cl_2$ to less than 0.5 mg/L for up to 90 days. The removals in columns with detention times equal to those used in practice will be much greater than this. The effluent concentration increases gradually with time during column operation, presumably because the buildup of surface oxides is eliminating possible reaction sites.

Reactions of free chlorine with activated carbon will result in the production of organic by-products. The total organic halogen (TOX) on the surface increases as extent of reaction increases, and some of these compounds may be found in the column effluent if the reaction proceeds for a very long time.[75]

**Combined chlorine-activated carbon reactions.** Bauer and Snoeyink[76] initially studied the reactions of monochloramine, $NH_2Cl$, with activated carbon and hypothesized that the following reactions were taking place:

$$NH_2Cl + H_2O + C^* \rightarrow NH_3(aq) + C^*O + H^+ + Cl^- \quad (13.15)$$

$$2NH_2Cl + C^*O \rightarrow N_2(g) + 2H^+ + 2Cl^- + H_2O + C^* \quad (13.16)$$

Initially, all the $NH_2Cl$ was converted to $NH_3$ and $Cl^-$ in accordance with reaction (13.15), but after a period of reaction some of the $NH_2Cl$ was converted to $N_2(g)$ and HCl in accordance with reaction (13.16). Apparently a surface oxide product of reaction (13.15) is required as a reactant in Eq. (13.16). If $NH_3$ is initially present in water that is dosed with chlorine and then contacted with carbon, 1.5 moles chlorine is required for each mole of $NH_3$ oxidized to $N_2$ at steady state, thus indicating that activated carbon is serving only as a catalyst and is not being changed over time as expected if it were a reactant. The rate of reaction was much slower than the reactions of either free chlorine or dichloramine with activated carbon.

Because activated carbon serves only as a catalyst, the effluent quality from a fixed bed of GAC receiving aqueous monochloramine does not vary with time after an acclimation period,[77,78] although some decrease in efficiency may occur if organic compounds are adsorbed. Komorita and Snoeyink[79] studied the reaction kinetics during the acclimation period and found 12 × 40 mesh carbon to reduce 2.8 mg/L $NH_2Cl$ as $Cl_2$ to less than 0.1 mg/L for 20,000 bed volumes. The life of 20 × 50 mesh GAC was much longer ( ≈ 55,000 bed volumes). Larger GAC particles performed much more poorly, however.

Dichloramine, $NHCl_2$, reacts very rapidly with activated carbon according to the following reaction:[76]

$$2NHCl_2 + H_2O + C^* \rightarrow N_2(g) + C^*O + 4H^+ + 4Cl^- \quad (13.17)$$

Kim et al.[80] found evidence of the parallel reaction on activated carbon:

$$NH_4^+ + 3NHCl_2 \rightarrow 2N_2(g) + 7H^+ + 6Cl^- \quad (13.18)$$

when excess ammonia was present. Predictions made by Kim,[81] using the model of Kim et al.,[82] showed that a 1.5-cm-deep bed of 60 × 80 mesh carbon operated at 10 m/h should reduce 40 mg/L $NHCl_2$ as $Cl_2$ to less than 10 mg/L for 400 days. The model predictions were verified only for a short time, however, and factors other than those accounted for in the model, such as adsorption of organic compounds, may prove to be of importance for long runs. The column effluent concentration gradually increases with time, possibly because of the buildup of surface oxides in Eq. (13.17).

**Chlorine dioxide, chlorite, and chlorate-activated carbon reactions.** $ClO_2$ reacts rapidly with activated carbon, but the nature of the reaction changes as pH changes. At pH 3.5, $Cl^-$, $ClO_2^-$, and $ClO_3^-$ are products found in the effluent of a column receiving only $ClO_2$, but $Cl^-$ is the predominant product. At pH 7.9, the same end products are formed, but $ClO_2^-$ is now the predominant species. The $ClO_2^-$ concentration in the column effluent is initially low, but then rapidly increases, as would be expected if the carbon initially reacted with the $ClO_2^-$ and then its capacity to react with it was used up.[83]

Separate experiments consisting of $ClO_2^-$ solutions at pH 7 applied to GAC columns showed that fresh carbon readily destroyed $ClO_2^-$, presumably by the reaction[84]

$$ClO_2^- + C^* \rightarrow C^*O_2 + Cl^- \quad (13.19)$$

The reaction capacity of the fresh carbon was saturated after 80 to 90 mg $ClO_2^-$ reacted per gram of GAC. Further study of the factors affecting the capacity of GAC for $ClO_2^-$ is needed, however. The chlorate ion was not reduced by activated carbon, but was adsorbed to a slight extent ($\approx 0.03$ mg $ClO_3^-$/g) by it, presumably by an ion exchange mechanism.[75]

**Removal of other inorganic ions.** Huang[85] has reviewed the removal of inorganic ions by activated carbon. A number of ions can be removed from water by GAC, but the capacity for many substances is quite low. The gold-cyanide complex, for example, is adsorbed onto GAC as a widely used means of recovering gold in the mining industry. Huang reviewed the data that show some removal of cadmium (+II) at higher pH, and that removal can be increased slightly by complexing it before adsorption with chelating agents. The removal of chromium involves adsorption of Cr(III) or Cr(VI), and under some conditions Cr(VI) is chemically reduced to Cr(III) by the activated carbon. Mercury adsorption is best at low pH. Capacities on the order of 0.3 to 4.1 mg/g have been observed. From 0.1 to 0.5 mmole Cu/g at equilibrium concentrations of 8 mmol/L have also been observed to adsorb. Activated carbon will also catalyze the oxidation of Fe(II) to Fe(III).

### Adsorption efficiency of full-scale systems

**Taste and odor removal.**    Many types of taste and odor problems are encountered in drinking water. Troublesome compounds may result from biological growth or industrial activities. They may be produced in the water supply, in the water treatment plant from reactions with treatment chemicals, in distribution systems, and in consumers' plumbing systems.[86] Activated carbon, both PAC and GAC, have an excellent history of success in removing taste and odor compounds from the raw water. It may also remove taste and odor precursor compounds, but this has not been well documented.

GAC sand replacement filter-adsorber systems are reported to effectively remove odor from source water typically for one to five years.[45] The life is dependent upon the intensity and frequency of appearance of taste and odor compounds, the presence of organics that compete for adsorption sites, and the concentration of these compounds that is acceptable in the treated water. Case history information is difficult to apply at different utilities because the intensity and frequency of appearance usually is not well documented, and the acceptable level of taste and odor varies from community to community. Regina, Saskatchewan, provides an interesting example of the effect of high-intensity tastes and odors on GAC performance.[87] There, 10-ft-deep postfilter-adsorbers lasted only through one taste and odor season (about five months) when the *threshold odor number* (TON) of influent musty and grassy odors was as high as 50.

Some useful observations have been made based on data from full-scale GAC systems. Background organic matter, for example, usually breaks through much earlier than the taste and odor compounds.[88,89] Further, GAC will remove even the most intense tastes and odors initially, followed by a period in which removal is partial but still good, and finally by a period in which peaks of influent taste and odor were incompletely removed and would lead to consumer complaints.[90]

The experience at Stockton East Water District showed that factors other than adsorption may complicate the removal of MIB and geosmin.[91] GAC filters were operated over an approximately two-year period when MIB and geosmin in the adsorber influent were often from 5 to 20 ng/L. Extraction of GAC taken from cores of the filter after about two years of operation showed no measurable MIB, and geosmin levels of about 0.1 to 0.2 μg/g GAC in the upper part of the filter (detection limits were 0.01 μg/g), even though other studies have shown these compounds to be strongly adsorbed.[20] These measurements are consistent with the concept that MIB and, possibly, some geosmin are removed by biodegradation, because both compounds are biodegradable.[52,53]

One hypothesis for a removal mechanism of these compounds is that adsorption capacity is needed to prevent these molecules from passing through the GAC adsorber. After some accumulate on the GAC surface, a biofilm capable of biodegrading them develops. A low solution concentration develops in the biofilm, and a reverse concentration gradient then exists in the GAC pore that allows the compounds to desorb and diffuse to the biofilm where they are biologically oxidized. The ability of GAC adsorbers to remove odor decreases over time because adsorption sites are gradually taken up by natural organics and thus are not available for MIB and geosmin. GAC regeneration or replacement is necessary when adsorption capacity has been used up by these competitors. Additional study is needed to prove this hypothesis.

Studies of the adsorption of odors of industrial origin at Nitro, W.Va., provided another example of GAC adsorbers that were spent in less than the typical one to five years.[92] A series of small-scale experiments and a full-scale test were conducted using 8 × 30 U.S. standard mesh (2.4 × 0.6 mm) GAC and EBCTs of 3.8 to 15 min. Odor-free water was produced for as long as 26 days, depending on the contact time and the TON of the applied water, which varied from 40 to 400. A finer carbon, 20 × 50 mesh (0.8 × 0.3 mm), produced an odor-free water for twice as long as the 8 × 30 mesh, thus demonstrating the importance of adsorption kinetics for this application.

GAC adsorbers with an 11-min EBCT were used to control a sulfide odor problem at the Goleta Water District in California.[93] The odor was attributed to hydrogen polysulfides, which were not removed by aeration, and which GAC was only partially effective in removing.[94] Influent TONs varied from 6 to 1000, while effluent TONs ranged from odor-free to 35.[89] These beds were in service for two years before the activated carbon was replaced.

**Total organic carbon and trihalomethane precursors.** The type and adsorbability of natural organic matter in water vary widely from location to location. Sufficient differences in adsorbability exist, so adsorption tests should be done with the water in question if the results are needed for design. Reviewing reported results is informative, though, to obtain estimates of the range of adsorbability.

Roberts and Summers[40] presented an excellent summary of full-scale plant performance for TOC removal. They evaluated removals from 47 different plants, including some wastewater reclamation plants. The ranges of design conditions are given in Table 13.5, and typical breakthrough curves are given in Fig. 13.19. The breakthrough curves reach an effluent concentration plateau about 10 to 25 percent below the influent concentration. They found that the amount of organic matter remaining in the effluent immediately after start-up generally decreased as the EBCT increased to about 20 min, and that the time to reach a steady-state effluent concentration increased as EBCT increased (see Fig. 13.20), although the scatter of the data negates this plot as an effective predictive tool. These plots can be used to show that a GAC adsorber with an EBCT of 10 min typically will take 60 to 70 days to reach a steady-state effluent concentration and that 9000 to 10,000 bed volumes of water will be processed during this time.

Other indicators of performance can also be used. For example, Graese et al.[33] summarized TOC removal data for a criterion of 50 percent removal and showed that adsorbers with EBCT values of less than 10 min had lives of less than 30 days, and that increased EBCT increased time of operation. To obtain the lowest-cost adsorption system for TOC removal, the EBCT probably must be much in excess of what is used at many existing plants.

The performance for GAC removal of *trihalomethane formation potential* (THMFP) usually parallels the removal of TOC.[95] Symons et al.[60] have summarized much of the research prior to 1981. Typical results are those from Jefferson Parish, La., where activated carbon with an EBCT of 23 min treated water with an influent THMFP of 100 to 250 μg/L. The effluent THMFP was 25 μg/L after 2000 bed volumes, 50 μg/L after 3300 bed volumes, 100 μg/L after 5000 bed vol-

**TABLE 13.5   Design Conditions for GAC Adsorbers[†]**

| Parameter | Median | Range | Typical range |
|-----------|--------|-------|---------------|
| Empty bed contact time, min | 10 | 3–34 | 5–24 |
| Depth of bed, meters | 1.0 | 0.2–8 | 0.5–4 |
| Hydraulic loading,‡ m/h | 6 | 1.9–20 | 2.6–17 |
| Influent concentration, mg/L as TOC | 3.5 | 1–16 | 2–6 |

†Data from 47 plants were analyzed.
‡m/h × 0.42 = gpm/ft$^2$.
Source: Roberts and Summers.[40]

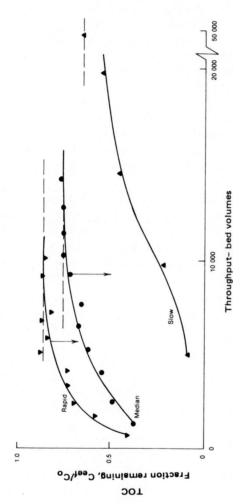

**Figure 13.19** Representative TOC breakthrough curves. *(Source: P. V. Roberts and R. S. Summers, "Performance of Granular Activated Carbon for Total Organic Carbon Removal," J. AWWA, vol. 74, no. 2, 1982, p. 113.)*

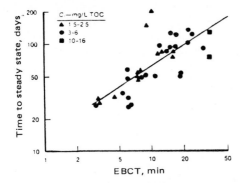

**Figure 13.20** Observed run length at which steady state is reached. (*Source: P. V. Roberts and R. S. Summers, "Performance of Granular Activated Carbon for Total Organic Carbon Removal," J. AWWA, vol. 74, no. 2, 1982, p. 113.*)

umes, and 150 µg/L after 8000 bed volumes. Removals can vary widely, depending on whether the GAC influent is preozonated (see the earlier discussion of biological activity on GAC) and on the source of the water, however.

The bromide concentration of the water has an important bearing on the removal of THMFP. Symons et al.[60] observed that formation of brominated THMs occurs more rapidly in GAC adsorber product water. Graveland et al.[96] made a similar observation and further noted that a shift to the formation of more highly brominated forms of THMs in GAC effluent. This is because of the rapid formation of bromine-substituted THMs compared to chloroform.

**Volatile organic chemical and other synthetic organic chemical removal.** Volatile organic chemicals include compounds such as tetrachloroethylene and trichloroethylene, which are adsorbed relatively strongly, and 1,1,1-trichloroethane, 1,2-dichloroethane, and chloroform, which are adsorbed relatively weakly. The isotherm data in Table 13.2 show the relative adsorbability of these compounds. GAC adsorbers can be used to remove VOCs directly from contaminated water[48,50,97] or from the off-gases from air-stripping towers.[98] The best type of system will depend on the type of VOC to be removed and the air emission standards that are in effect.

VOCs are commonly found in contaminated groundwater, and although this type of water is generally low in TOC,[95,99] large competitive effects are still noted. For example, Zimmer et al.[99] studied the adsorption of three VOCs—tetrachloroethylene, trichloroethylene, and 1,1,1-trichloroethane—from distilled water and from groundwater in field installations. The full-scale field installations were 2.5 to 3 m (8.2 to 9.8 ft) deep and had EBCTs from 12 to 15 min. Water containing 0.4 to 2 mg/L of DOC was applied at a rate of 10 to 15 m/h (4 to 6 gpm/ft). Several types of GAC were used in the field installations. As the data in Table 13.6 show, the amount of water containing 50 μg/L of tetrachloroethylene that could be processed in the field was about 8 percent of the distilled water value. The capacity for trichloroethylene and 1,1,1-trichloroethane was reduced to 22 percent and 24 percent, respectively, of the distilled water value. Apparently natural organic compounds adsorbed and occupied much of the GAC surface.

Experimental data given by Love and Eilers[48] were used by Hess[97] to predict the effect of influent concentration and type of VOC on GAC life (Fig. 13.21), and of trichloroethylene influent concentration and effluent concentration at breakthrough on GAC life (Fig. 13.22). The figures show that GAC adsorber life decreases as influent concentration increases and as the required effluent concentration decreases.

Crittenden et al.[98] developed mass transfer models to predict the removal of VOCs in air-stripper off-gas by GAC, and investigated the regeneration of this GAC with steam. They found that adsorption efficiency was highly dependent on the relative humidity of the off-gas, and that heating of the gas stream to reduce the relative humidity to 40 to 50 percent was beneficial. They found air stripping followed by GAC to purify the off-gas to be a good alternative to treatment of the water by GAC alone for several VOC treatment applications, even though steam regeneration of spent gas-phase GAC was not feasible. This advantage results because the natural organics that interfere with adsorption from the aqueous phase are not present in the stripper off-gas.

Symons et al.[60] summarized most of the data available in 1981 on trihalomethane adsorption. Consistent with the size of the Freundlich K values listed in Table 13.2, the brominated THMs are adsorbed much better than chloroform. Graese et al.[33] presented data from Cincinnati and Miami for an 80 percent chloroform removal criterion that showed that about 5000 to 6000 bed volumes of water could be processed before breakthrough. If the EBCT was less than 8 to 10 min, however, fewer bed volumes could be processed because of the length of the MTZ. The performance will change as the breakthrough criterion and the composition of the water change.

Pesticides are common *synthetic organic chemicals* (SOCs) that require removal. GAC beds were evaluated for the removal of selected pesticides spiked into river water.[100] One activated carbon column was exhausted for removal of background organic matter as measured by carbon chloroform extract and COD. Pesticide-spiked, sand-filtered river water was applied to the exhausted activated carbon column, and the effluent from it was applied to a second, fresh column. With concentrations of dieldrin as high as 4.3 μg/L, the effluent from the first column was reduced to 0.3 μg/L. Further reduction in the second column reached concentrations as low as 0.05 μg/L and often below

**TABLE 13.6  GAC Life with and without Competition from Natural Organics**

| Compound | Single-solute isotherm constants[†] | | Bed volume to saturation without competition, $C_o = 50$ µg/L[‡] | Observed capacity of full-scale adsorbers[§] | | Bed volumes to breakthrough ($\sim 5$ µg/L) $C_o = \sim 50$ µg/L |
| --- | --- | --- | --- | --- | --- | --- |
| | $K$ (mg/g) $(\text{L/mg})^{1/n}$ | $1/n$ | | $K'$ (mg/g) $(\text{L/mg})^{1/n'}$ | $1/n'$ | |
| Tetrachloroethylene | 219 | 0.42 | 620,000 | 20.4 | 0.46 | 51,000[¶] |
| Trichloroethylene | 78 | 0.46 | 197,000 | 27.4 | 0.61 | 44,000 |
| 1,1,1-Trichloroethane | 23 | 0.60 | 38,000 | 62 | 1.4 | 9,400 |

[†] For F-100 GAC, after Zimmer et al.[99]

[‡] This number is calculated under the assumption that all influent compound is adsorbed until complete saturation is reached.

[§] An "adsorber isotherm" of the form $q = K'C_o^{1/n'}$, where $C_o$ is the adsorber influent concentration, $K'$ and $1/n'$ are constants, and $q$ is the amount adsorbed (mg/g) at the time of adsorber breakthrough, was used to describe the data. These data are based on the performance of several full-scale plants in West Germany.[99]

[¶] 52,500 bed volumes can be processed in one year if the EBCT is 10 min.

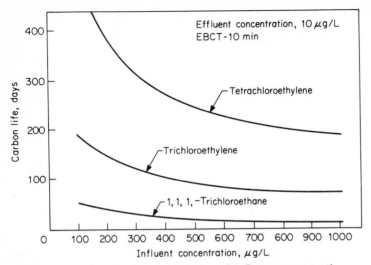

**Figure 13.21**   Effect of contaminant type and influent concentration on carbon life. (Note: 100 days = 14,400 bed volumes if the EBCT = 10 min.)   (*Source: A. F. Hess et al., "GAC Treatment Designs and Costs for Controlling Volatile Organic Compounds in Ground Water," ACS Meeting, Atlanta, Ga., 1981.*)

the detection limit of 0.01 µg/L. This occurred in spite of sudden increases in the COD of the influent river water. At Jefferson Parish, La., 18 chlorinated hydrocarbon insecticides at a combined concentration of 18 to 88 ng/L were removed to less than about 5 ng/L for one year in an adsorber with an EBCT of 20 min. Alachlor at concentrations of 13 to 593 ng/L was reduced to the limits of detectibility during the same period. Atrazine was present at 30 to 560 ng/L and was essentially 100 percent removed by the postfilter-adsorber (24-min EBCT), but concentrations of 10 to 20 ng/L were often found in the effluent of the filter-adsorber (14-min EBCT).[102] Adsorbers with an 8.5-min EBCT were found to remove more pesticide spikes than 2.8-min EBCT adsorbers,[101] as expected because of the larger column. Other analyses at Jefferson Parish showed that phthalates, *n*-alkanes, and substituted benzenes at the ng/L level were not removed by GAC.[102]

Polynuclear aromatic hydrocarbons are removed well by GAC. Tap water spiked with 50 µg/L of naphthalene was passed through a GAC column at Cincinnati, Ohio.[88] After seven months of operation at 5 m/h (2 gpm/ft$^2$), the *nonvolatile total organic carbon* (NVTOC) front for 50 percent removal had penetrated the first 51 cm (1.7 ft) of the bed, while the 50 percent removal point for naphthalene was only about 5 cm (0.2 ft) down the column. Riverbank filtration followed by activated carbon treatment reduced the concentration of polynuclear aro-

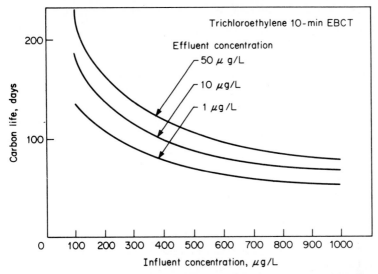

**Figure 13.22**  Effect of influent and effluent contaminant concentrations on carbon life. (Note: 100 days = 14,400 bed volumes if the EBCT = 10 min.)  (*Source: A. F. Hess et al., "GAC Treatment Designs and Costs for Controlling Volatile Organic Compounds in Ground Water," ACS Meeting Atlanta, Ga., 1981.*)

matic hydrocarbons in water by about 99 percent, whereas rapid sand filtration followed by ozonation or chlorination was not effective,[103] probably because of the absence of GAC.

**Filter-adsorber versus postfilter-adsorber performance.**  Although sand replacement filter-adsorbers have proven to be very effective in many taste and odor removal applications, with typical bed lives of 1 to 5 years, their use for removing weakly adsorbed compounds, including THMs and some VOCs, is limited.[33] EBCTs and activated carbon depths in sand replacement systems are often limited by the existing filter design. The typical EBCT of operating sand replacement systems is about 9 min at existing flow rates, but much less at design flow rates, which are generally much larger. Control of weakly adsorbed organics with such short EBCTs would require regeneration or replacement of GAC after only a few weeks or months of operation. Replacement or regeneration at a high frequency would be operationally cumbersome, and GAC loss during regeneration would be a significant operational cost. In addition, filter-adsorbers exhibit earlier contaminant breakthrough, and lower organic loadings (mass contaminant adsorbed per unit mass of GAC) than postfilter-adsorbers (see Graese et al.[33] for a review).

Reduced adsorption efficiency in filter-adsorbers may be partially caused by reduced stratification resulting from mixing that occurs during more frequent backwashing. The lower organic loadings and earlier breakthrough for filter-adsorbers may also be attributed to interference by the trapped floc. The type of coagulants and filter aids used, as well as the application of other compounds, such as calcium carbonate or manganese, to the filter, may influence filter backwashing frequency and adsorption performance. A thorough economic analysis is required to determine whether the costs associated with a higher carbon usage rate in the filter-adsorber, if any, would be justified by the reduced capital costs of the filter-adsorber.

### Turbidity removal by filter-adsorbers

A current review of the available data on the efficiency of GAC for turbidity reduction compared to sand or coal showed that GAC is as good or better.[33] Comparisons are sometimes difficult to make in side-by-side studies, even when the effective sizes of the media are the same, because GAC generally has a larger uniformity coefficient than do typical rapid filter media. When turbidity loadings were high, a more rapid rate of head loss buildup was experienced because of the larger uniformity coefficient. When the turbidity loading was low, the difference in uniformity coefficient seemed to make little difference in performance.

A sand layer is often retained in the filter below the GAC if turbidity is to be removed. The efficiency of GAC at removing turbidity indicates this is not necessary for good performance, although it may be useful if the decision is made to use a large-diameter GAC or if very low levels of turbidity are desired. Sand below the GAC is a nuisance when GAC must be removed, because the sand often mixes with the GAC. It interferes with resale of the regenerated product if the regenerated carbon is to be reused, and it may interfere with regeneration itself. Additional sand may be required after GAC removal to keep the same amount of sand under the GAC. Gravel under the GAC can cause similar problems, but gravel is necessary to prevent the GAC from entering many types of underdrains.

Nozzle underdrains can be used that do not require either sand or gravel, and they can be used to provide both air and water for backwashing. These appear to have significant advantages over the types that require gravel; however, a pressure-relief device on the wash water influent line should be provided to prevent high-pressure damage to underdrains when and if the nozzles clog.[104] Monitoring of the plenum pressure during backwashing can give advance warning of nozzle clogging problems.

If sand or another medium is to be placed under GAC, care must be taken to size this medium so that it fluidizes properly relative to the GAC. Use of a backwash rate that is 1.3 times the minimum fluidization velocity of the $d_{90}$* sand or GAC, whichever is greater, ensures that both layers will be fluidized.[105]

The backwash system must be properly designed to prevent accumulation of mudballs in the GAC. Mudballs accumulate if the GAC is not properly cleaned, and the GAC is more difficult to clean because it is lighter than sand and anthracite. The density of GAC wetted in water ranges from 1.3 to 1.5 g/cm$^3$; thus, providing sufficient energy for cleaning the bed is more difficult if water wash alone is used. Provision of good surface scour or air scour is important for good performance.

## GAC Performance Estimation

### Isotherm determination and application

Useful information can be obtained from the adsorption isotherm. It has been used in previous sections to show the variation in adsorbability of different types of organic compounds, the differences between activated carbons, and the magnitude of competitive effects. Because adsorption capacity of a GAC is a major determinant of the activated carbon's performance in columns, the isotherm can be used (1) to compare candidate activated carbons, (2) as a quality control measure of purchased activated carbons (some variation from lot to lot of a given manufacturer's activated carbon is expected), and (3) as an input function to the mathematical methods that can be used to predict performance (discussed later). Important limitations to the use of isotherms must be recognized if they are to be properly used, however. The purpose of this section is to demonstrate how isotherms can be used to estimate performance and to note several important factors in isotherm determination.

**Types of adsorbates.**   The determination procedures must take into account the adsorbate characteristics. The simplest procedure can be followed if a single nonvolatile adsorbate is present in an aqueous solution with no background organics that might compete for adsorption sites. This isotherm is not a function of initial concentration, and no precautions need be taken to avoid volatilization. The test is conducted by agitating known volumes of the solution with accurately weighed portions of GAC until equilibrium is achieved. The time required for equilibrium can be estimated by using the data in Table

---

*90 percent by weight of the medium is smaller than this size.

13.1, and can be determined by measuring solution concentration versus time. When equilibrium is reached, the data are analyzed by determining

$$q_e = \frac{(C_0 - C_e)V}{m}$$     (13.20)

where $q_e$ = amount adsorbed per unit mass of carbon
$\quad C_0$ = initial concentration
$\quad C_e$ = equilibrium concentration
$\quad V$ = solution volume
$\quad m$ = mass of carbon

The units of $q_e$ (mol/g; mass/g) depend on the units of $C$ (mol/L; mass/L), $V$ (L), and $m$ (g). The data can be plotted to determine the Freundlich parameters [plot ln $q_e$ versus ln $C_e$ according to Eq. (13.2)] or other parameters as appropriate.

Adsorption from a mixture of adsorbable organics is more complicated because of the effect of the different adsorbabilities of compounds in the mixture. Figure 13.4 shows a typical isotherm for a heterogeneous mixture of compounds. The small fraction of strongly adsorbable compounds is often overlooked when determining an isotherm because of the analytical difficulty of obtaining good data for $C_e$ close to $C_0$ (i.e., for very small doses of activated carbon). This fraction is important because it is responsible, at least in part, for the slow approach of effluent concentration to influent concentration in column operation. The effect of initial concentration on the adsorption of a fulvic acid is shown in Fig. 13.23.[18] The difference is caused primarily by the different mass concentration of strongly adsorbable compounds in each of the solutions. The same type of difference is expected if the amount of a single compound that is adsorbed in the presence of different concentrations of background organics is measured. The isotherm of the single compound will be a function of the initial concentration of background organics and the carbon dose if competition occurs between them.

Other adsorbate-related factors must be carefully controlled during isotherm determination. If the adsorbate is volatile, equilibrium must

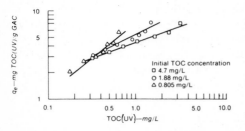

Figure 13.23   The effect of initial TOC concentration on the adsorptive capacity of carbon A (14 × 20 mesh pulverized prior to adsorption) for peat fulvic acid. (*Source:* S. J. Randtke and V. L. Snoeyink, "Evaluating GAC Adsorptive Capacity," J. AWWA, vol. 75, no. 8, 1983, p. 406.)

take place in a headspace-free vessel, and mixing of the system must take place by continual inversion of the test bottle. Control of pH is important for compounds whose isotherms are a function of pH, and other salts in solution can be important for ionic substances or for substances that compete with organic ions for adsorption sites. The reader is referred to Randtke and Snoeyink[18] for a more detailed discussion of these factors.

**Activated carbon.**    The major concerns related to activated carbon are (1) obtaining a representative sample for testing, and (2) preparing it to obtain a rapid approach to equilibrium without destroying its representative nature. The adsorption capacity of some commercial activated carbons is a function of particle size because of the manufacturing process that is used,[18] and, thus, selecting a sample for testing with the same particle size distribution as the bulk activated carbon is important. Activated carbon sizes may separate during shipment, so special equipment or techniques are required to obtain the representative sample. With equipment such as a sample reducer and a sample splitter,* representative samples from large quantities (bags) of GAC can be obtained. The sample reducer reduces sample sizes greater than 4.5 kg (10 lb) to representative samples that are smaller on a ratio of 16:1. Further reduction to a size that can be used for laboratory tests is possible using the sample splitter, or riffle. The particle size distribution and quality of the final sample are then representative of those of the original large quantity of GAC. Using such equipment to establish that a shipment of GAC meets specification is important. A simpler alternative that yields a good approximation of the characteristics of a batch is to sieve the sample and use only those particles in the dominant size range.

The time to reach equilibrium is an important function of particle size, but the adsorption capacity of a sample of activated carbon is not altered by crushing it to a smaller particle size.[18] Grinding a granular activated carbon to 325 mesh (0.044 mm) or less reduced the time to reach equilibrium to two days for a large humic acid molecule, and smaller molecules will equilibrate much faster. Avoiding significant biological activity should be possible if the sample is not seeded with microorganisms and the test time is two days or less. Filtration of the sample through 0.45-μm-pore-diameter filter paper will be necessary to remove the activated carbon before testing for the concentration of residual organics. Filter paper must be selected that does not contribute organic compounds that interfere with the quantitative analysis of the test compounds.

---

*Tyler Co., Menton, Ohio, is one supplier of this equipment.

Isotherms can be used to obtain a rough estimate of activated carbon loading and bed life. These are useful in determining the applicability of GAC. Assuming that (1) all of the GAC in an adsorber will reach equilibrium with the adsorber influent concentration, and (2) the capacity obtained by extrapolating the isotherm data to the initial concentration is a good value, the bed life, in bed volumes of water, can be calculated.* For example, if $(q_e)_0$ is the mass adsorbed (mg/g) when $C_e = C_0$, the bed life $Y$, the volume of water that can be treated per unit volume of carbon, can be calculated:

$$Y = \frac{(q_e)_0 \text{ (mg/g GAC)}}{(C_0 - C_1) \text{ mg/L}} \cdot \rho_{GAC}(g/L) \qquad (13.21)$$

where $C_e$ = equilibrium concentration
$C_0$ = influent concentration
$C_1$ = average effluent concentration for entire column run
$\rho_{GAC}$ = apparent density of GAC

$C_1$ is zero for a strongly adsorbed compound that has a sharp breakthrough curve, and is the concentration of nonadsorbable compounds when such substances are present. An estimate of the activated carbon usage rate (CUR), the rate at which activated carbon is spent, is given by

$$\text{CUR (g/L)} = \frac{(C_0 - C_1) \text{ (mg/L)}}{(q_e)_0 \text{ (mg/g)}} \qquad (13.22)$$

**Example Problem 1.**    Estimate the bed life and carbon usage rate for a GAC adsorber that is to remove 10 μg/L of bromoform from solution, given $\rho_{GAC} = 500$ g/L.

**solution.**

1. From Table 13.2, $K = 20$ (mg/g) (L/mg)$^{1/n}$ and $1/n = 0.52$ for bromoform.

2. Applying the Freundlich equation, Eq. (13.1), we have

$$(q_e)_0 = KC^{1/n}$$

$$= 20 \text{ (mg/g) (L/mg) } (0.01 \text{ mg/L})^{0.52}$$

$$= 1.82 \text{ mg/g}$$

3. Applying Eq. (13.21), assuming that $C_1 = 0$, gives

$$Y = \frac{(q_e)_0 \text{ (mg/g)}}{(C_0 - C_1) \text{ mg/L}} \cdot \rho_{GAC} \text{ (g/L)}$$

---

*These assumptions are consistent with assuming that the length of the MTZ is negligible.

$$= \left(\frac{1.82 \text{ mg/g}}{0.01 \text{ mg/L}}\right) 500 \text{ g/L}$$

$$= 91{,}000 \text{ L H}_2\text{O/L GAC} = \text{bed life}$$

4. Applying Eq. (13.22) to obtain CUR gives

$$\text{CUR} = \frac{(C_0 - C_1) \text{ mg/L}}{(q_e)_0 \text{ mg/g}}$$

$$= \frac{0.01 \text{ mg/L}}{1.82 \text{ mg/g}}$$

$$= 0.0055 \text{ g GAC/L H}_2\text{O}$$

*Note:* The bed life will be reduced and the CUR will be increased if other organics that compete for adsorption sites are in the water.

Several limitations exist on the use of isotherm data to estimate activated carbon usage rate. This approach is only valid for columns in series or for very long columns for which the assumption that all the activated carbon in the column is in equilibrium with the influent concentration is valid. Furthermore, it provides no indication of the effect of biological activity. Finally, when competition does occur, its impact in a batch test often is not the same as is experienced in a column because the molecules will separate in a column according to their strength of adsorption. GAC particles in a column then are exposed to different types of organics versus time than are particles in a batch test. Competitive effects of natural organics on trace organics may be much greater in a column for this reason.[99] Symons et al.[60] have some comparisons of calculated versus actual $CHCl_3$ breakthrough times.

### Small-scale column tests and applications

Different types of small-scale column tests have been developed to obtain data that can be used to estimate the performance of large contactors. Rosene et al.[106] developed the *high-pressure minicolumn* (HPMC) technique, which uses a high-pressure liquid chromatography column loaded to a depth of 2 to 2.5 cm (0.8 to 1 in) with about 50 mg of activated carbon (see Fig. 13.24). The activated carbon is crushed until it passes a 0.149-mm (100 mesh) sieve. This column is used with a headspace-free reservoir and a flow rate of 2 to 3 mL/min to determine the capacity of activated carbon for volatile compounds such as chloroform. The headspace-free reservoir is not required if the adsorbable molecules are not volatile.

Bilello and Beaudet[107] modified the HPMC by using only the $230 \times 325$ mesh fraction ($63 \times 44$ μm) of crushed GAC in a similar

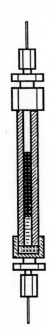

1/8" STAINLESS STEEL TUBING

SWAGELOK 1/4" TO 1/8" REDUCER

2.1mm ID, 1/4" OD STAINLESS
STEEL COLUMN

ACTIVATED CARBON

GLASS WOOL

5um STAINLESS STEEL FRIT

1/4" TO 1/16" STAINLESS
STEEL REDUCER

1/16" TEFLON TUBING

**Figure 13.24** High-pressure mini-column. (*Source: Rosene et al.*[106] *Reproduced, with permission, from M. J. McGuire and I. H. Suffet, eds., Activated Carbon Adsorption for the Aqueous Phase, vol. 1, Stoneham, Mass., Butterworth Publishers, 1980.*)

column. Initial attempts to use all crushed particles less than 200 mesh (74 μm) led to pressure drops through the column in excess of 4000 psi (kPa), but the 230 × 325 mesh fraction allowed a pressure drop of less than 100 psi (kPa) for solutions with low turbidity. Prefiltration of samples containing many particles was sometimes necessary to prevent excessive head loss. A vibrator is used to load the column to eliminate pockets of voids.

The HPMC technique has several advantages. It allows rapid determination of GAC capacity under conditions that are closer to those experienced in full-scale installations, and is especially advantageous for volatile compounds because it is easier to avoid their loss. It has the potential for being used in a way that will give good information on the extent of competition, although more research is needed to determine how these tests should be run to duplicate the competition caused by natural organics in full-scale systems. Thus, it can be used to screen potentially applicable carbons. It also allows a rapid determination of the order of elution of compounds expected from a carbon column. Its disadvantages include the large pressure drop, the need for a high-pressure pump, and the difficulty of getting good kinetic information that can be scaled up to predict accurately the performance of large adsorbers.

Other types of small columns can be used to obtain good kinetic data, which, together with equilibrium data from isotherms or HPMC tests, can be used in mathematical models to predict the performance of larger columns. For example, Liang and Weber[108] used 0.3- to 0.4-mm-diameter GAC, a column diameter of 10 mm (0.4 in), a GAC depth of 10 to 16 mm (0.4 to 0.6 in), and hydraulic application rates of 1.7 to 3.6 gpm/ft$^2$ (4.3 to 9 m/h). Their model accurately predicted the effluent profiles of columns with depths of 60 to 100 mm (2.4 to 3.9 in).

Crittenden et al.[109] developed an alternative procedure called the *rapid small-scale column test* (RSSCT) for scaling the data obtained from small columns to predict the performance of larger columns. In their first studies, they used a mathematical model to develop the relationship between the breakthrough curves of large-scale and small-scale columns. Based on the assumption that the internal diffusion coefficient is not a function of GAC particle diameter, they found that

$$\frac{\text{EBCT}_{\text{SC}}}{\text{EBCT}_{\text{LC}}} = \left[\frac{d_{\text{SC}}}{d_{\text{LC}}}\right]^2 = \frac{t_{\text{SC}}}{t_{\text{LC}}} \tag{13.23}$$

where $d$ is the diameter of the GAC particle, the subscript SC refers to the small column, and LC refers to the large column. The fraction $t_{\text{SC}}/t_{\text{LC}}$ is the time required to conduct a small-column test divided by the time necessary to conduct a large-column test. This equation is valid for various combinations of pore and surface diffusion controlled adsorption. It predicts that a column with 0.1-mm-diameter GAC can be 0.01 times as large as a column with 1-mm-diameter GAC and will produce the same breakthrough curve. Furthermore, the time required to process a given number of bed volumes of water through the small column is 0.01 times the time required to process the same number of bed volumes through the large column.

Subsequent studies by Crittenden et al.[110] showed that the internal diffusion coefficient could not always be assumed to be constant as a function of GAC particle diameter. They developed a second relationship based on the assumption that the internal diffusion coefficient varied linearly with particle diameter, consisting of

$$\frac{\text{EBCT}_{\text{SC}}}{\text{EBCT}_{\text{LC}}} = \frac{d_{\text{SC}}}{d_{\text{LC}}} = \frac{t_{\text{SC}}}{t_{\text{LC}}} \tag{13.24}$$

where each term has the same definition as in Eq. (13.23). Considerably longer tests are required if Eq. (13.24) applies instead of Eq. (13.23). For example, if the particle diameter ratio is 0.1 to 1.0, the small-scale test will require one-tenth the time of the full-scale test if its EBCT is one-tenth as large. Thus, the time savings is not as much as can be achieved if the diffusion coefficient is constant with changing GAC particle diameter. Tests (1989) have shown Eq. (13.24) to apply reasonably well to some TOC removal data, and further research may show it to be generally applicable for this.

The RSSCT technique is very promising because it has the potential to significantly reduce the time required to obtain typical performance data. It does have limitations, however. Much of the variability in effluent concentration is caused by changes in influent concentration. A test that lasts from a fraction of a day to at most a few days cannot account for the changes expected over a much longer time interval. In addition, the short-term test cannot determine the effect of biological activity in the full-scale column. More importantly, the test requires further development to establish the conditions under which it can be used reliably. At this time, this technique should not be used without verifying the accuracy of the predicted values.

### Pilot-plant testing

Pilot testing is useful in several situations. Isotherm or minicolumn tests or the experience of others may provide good reason to anticipate that activated carbon will be a technically and economically feasible solution to a water quality problem. Each water is somewhat unique with respect to day-to-day and season-to-season variation in contaminants, background organics, other water quality parameters that can affect adsorption, and the pretreatment that will be given to the water and, thus, questions remain about how activated carbon will perform on the water in question. The pilot study should show whether activated carbon is effective, should permit a determination of the best design parameters to use for the full-scale system, and should establish the best operating procedure to use. A good estimate of the cost of the full-scale system can then be made. Although pilot testing is expensive, the excessive cost of operating an improperly designed system because of insufficient information may more than justify the cost of a pilot study to obtain that information.

Pilot studies are not always necessary. Available information may be sufficient to determine whether activated carbon should be used instead of an alternative process, and whether a particular design will achieve the desired removal efficiency or the desired degree of protection from upstream spills. Use of GAC for taste and odor removal from many water supplies provides an example of this (see previous discussion of taste and odor removal performance). Also, the cost of a pilot study may not be justified by the potential savings that would be made possible by an optimized design. GAC adsorbers for removal of VOCs from small supplies thus may be designed on the basis of existing information, possibly supplemented with isotherm or minicolumn test information, with an appropriate safety factor to compensate for the lack of a pilot test.

Extensive (1988) research by the USEPA and AWWA Research Foundation (AWWARF) has the objective of developing laboratory test procedures that will give accurate projections of full-scale performance with minimal or no pilot testing. Special attention is being given to techniques for predicting the competitive effects of background organics. Thus, the circumstances for which a pilot study is appropriate may change in the near future.

**Pilot-plant design and operation.** The pilot plant should be designed with the same pretreatment that will be used in the full-scale system, and it should be run to give needed information for design and operation of the full-scale system. It should be operated at the location where the full-scale adsorber is to be constructed so that the water quality for pilot and full-scale will be the same. For designing a pilot plant, the following data are often useful:

1. A detailed chemical analysis of the water including the concentrations of specific organics, TOC, THMFP, taste and odor compounds, turbidity, and inorganics that might affect adsorption and removal by biodegradation such as dissolved oxygen, pH, hardness, and alkalinity. Preferably, these analyses will be done at different times so that the range of concentrations expected at the full-scale plant can be estimated accurately. The analyses should also show what pretreatment is necessary, if any.

2. Isotherm tests or minicolumn tests, preferably on water samples taken at different times, coupled with an analysis of the experience of others to obtain estimates of activated carbon usage rates for the different candidate activated carbons. The best activated carbon for pilot testing can then be selected. Activated carbon cost and head loss during operation should be included in this consideration. A preliminary sizing of the full-scale plant should be done to establish probable flow rate, EBCT, and GAC replacement frequency. The pilot plant then should be designed and operated to show whether these are optimum or whether they should be modified for best performance.

The pilot plant should include the necessary pretreatment processes and three or four GAC columns in series such as are shown in Fig. 13.25. Alternatively, one or two long columns with several taps can be used so that samples can be taken at different depths. The materials of construction may range from plastics such as clear acrylic and polyethylene to the more expensive stainless steel, poly-

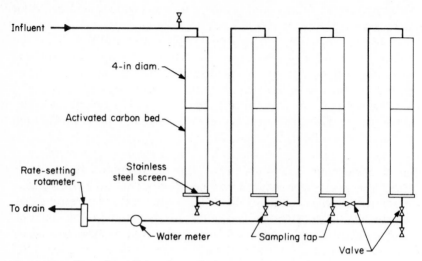

Influent

4-in diam.

Activated carbon bed

Rate-setting rotameter

Stainless steel screen

To drain

Water meter    Sampling tap    Valve

**Figure 13.25**  Downflow pilot carbon columns.

tetrafluoroethylene (PTFE), and glass. Clear acrylic or similar material can be used if the purpose of the GAC is to remove TOC or THMFP. The study of Kreft et al.[111] showed no leaching of VOCs or TOC from, or adsorption of VOCs and TOC to, clear acrylic columns and polyethylene tubing in a pilot plant. They did find that phthalates leached from transparent vinyl plastic tubing, so this material should not be used where phthalates will interfere. The possibility that some SOCs may be adversely affected by some materials of construction cannot be ruled out, however. Stainless steel, PTFE, and glass can be used if avoiding any question of possible organics added by plastic is necessary.

The total depth of the activated carbon should be sufficient to allow determination of the optimum EBCT (or optimum depth at a given flow rate). A total depth of 3 to 4 m (10 to 13 ft), greater than the depth likely to be used in the full-scale plant, should be used. Column diameter to average particle diameter should be at least 50:1.[112] When GAC is added to a column, care must be taken to eliminate the air from the GAC pores. Partially filling the pilot column with water and then slowly adding the GAC is one effective means of accomplishing this. If possible, the duration of the test should be long enough to show the impact of seasonal variations in quality of the source water. The concentrations of all substances that might have a significant effect on the primary contaminant should be monitored. The data that are obtained can be used to calculate total annual costs (capital plus operating costs) for alternative designs to determine the most economical design.

**Analysis of column data.**  The GAC effluent data collected at different EBCT values should be plotted as breakthrough curves to show the effect of EBCT on effluent quality. The effluent concentration $C_{eff}$ or the effluent concentration divided by the influent concentration, $C_{eff}/C_o$, can be plotted against time of operation, volume of water processed (see Fig. 13.26), or bed volumes of water processed (see Fig. 13.27). Because flow rate may vary with time, the latter two abscissa parameters are the most useful and most commonly used. Because most full-scale plants use time, a double abscissa is suggested with time being one independent variable. Using the number of bed volumes as the abscissa has the important advantage of normalizing the data collected at different depths and at different flow rates. The number of bed volumes processed must be calculated using only the volume of carbon above the sample port, however. As EBCT values increase, the $C_{eff}/C_o$ versus bed volume breakthrough curves eventually will superimpose as shown in Fig. 13.27; when this occurs, the CUR will be a minimum and a function only of the breakthrough criterion that is used. The CUR can be calculated from the breakthrough curve:

$$\text{CUR (g/L)} = \frac{\rho_{GAC}\ (\text{g/L})}{\text{bed volumes to breakthrough}} \qquad (13.25)$$

where $\rho_{GAC}$ is the apparent density of the GAC.

Flow rates of 5 to 12.5 m/h (2 to 5 gpm/ft$^2$) are usually used for pilot testing.[112] Within this range, the assumption can be made that at a given EBCT, activated carbon usage rate is not a function of flow rate.

The $C_{eff}/C_0$ versus bed volume curves should be carefully analyzed to determine whether the number of bed volumes to breakthrough increases to a maximum value and then begins to decrease, as EBCT increases. Arbuckle[113] observed that GAC at the end of an adsorber train adsorbed less trace compound per unit mass than GAC at the front of the adsorber train. Similar effects were observed for VOC adsorption from groundwater[99,114] and for adsorption of chlorinated organics from Rhine River water.[115] The explanation given for this phe-

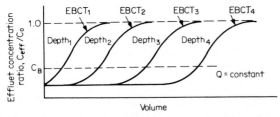

**Figure 13.26**  Breakthrough curves as a function of total volume of water processed.

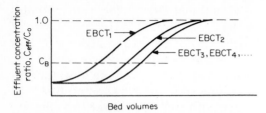

Figure 13.27   Breakthrough curves as a function of the number of bed volumes processed.

nomenon was that large molecules of background organics adsorbed first in the lower reaches of the adsorber and fouled the activated carbon so that it could not adsorb as much trace compound. The phenomenon was called the *premature exhaustion* or *preloading* effect, and resulted in higher activated carbon usage rates as EBCT is increased beyond a certain value. The implication of this phenomenon is that a single EBCT exists that will give the lowest activated carbon usage rate. If this effect does exist for an application, GAC adsorber designs and operating procedures that minimize it need to be considered.

Bed depth–service time plots (see Fig. 13.15) also are useful as an aid to interpreting breakthrough curve data. For a constant flow rate, the length of time to breakthrough (ordinate) is plotted as a function of GAC depth (abscissa). Data from plots similar to Fig. 13.26 are used, except that time of operation instead of volume is used as the abscissa. This allows a determination of the $L_{critical}$ for a certain breakthrough concentration.

**Parallel column analysis.**   One of the objectives of pilot testing should be to determine the activated carbon usage rate if GAC columns are to be used in parallel. Data from an individual GAC pilot contactor, plotted as shown in Fig. 13.27, are needed to develop the integral breakthrough curve shown in Fig. 13.12b. The integral curve is developed, assuming that (1) each contactor will be the same size, (2) each contactor will be operated for the same number of bed volumes $\theta_n$ before regeneration or replacement, and (3) only one contactor is regenerated or replaced at one time at intervals of $\theta_n/n$ bed volumes for $n$ contactors in parallel. For a given value of $n$, $\bar{f}$ versus $\theta$ is calculated by applying Eq. (13.10) for different assumed values of $\theta_n$. (The procedure is illustrated in Fig. 13.12 and related discussion for $\theta_n = 5000$ bed volumes and an $n$ of 10.) The number of bed volumes to breakthrough can then be determined from the integral curve. Separate integral curves should be calculated for each possible value of $n$. An economic analysis can then be made for each parallel arrangement to determine the best choice.

**Example Problem 2.** An ozone-GAC pilot plant was used to remove THMFP from a shallow reservoir supply after coagulation, sedimentation, and filtration. The GAC columns had a diameter of 4 in ID. Samples were taken from ports located 20 in (1), 44 in (2), and 64 in (3) from the inlet. A flow rate of 0.24 gpm gave corresponding EBCT values of 4.6, 10.1, and 14.8 min. The data (except for points at large volumes processed at ports 2 and 3) are shown in Fig. A. Calculate the GAC replacement frequency if four GAC adsorbers, each with 15 min EBCT, are to be operated in parallel. The THMFP concentration of the treated water is to be 25 $\mu$g/L. The $C_0$ for THMFP is 60 $\mu$g/L.

**solution.**

1. Plot the breakthrough curve (shown in Fig. A) as a function of the bed volumes of water treated. To convert volume of water processed to bed volumes, divide the volume processed by the volume of GAC above the port. The volume of GAC above each port is 1.09, 2.39, and 3.48 gal for ports 1, 2, and 3, respectively. For example, the point (vol = 20,000 gal, $C_{\text{eff}}/C_0 = 0.28$) for port 2 in Fig. A corresponds to

$$\text{BV} = \left(\frac{20,000}{2.39} = 8400\right) \qquad \left(\frac{C_{\text{eff}}}{C_0} = 0.28 \text{ in Fig. B}\right)$$

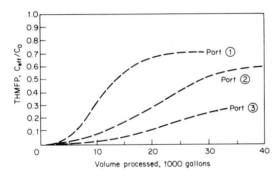

Example problem 2, Fig. A

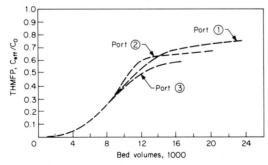

Example problem 2, Fig. B

2. Use Fig. B to obtain a good estimate of the breakthrough curve expected for a full-scale system. More emphasis is given to data from ports 2 and 3 because the depth of bed for these ports should be much greater than the critical depth. The EBCT for port 3 is also about 15 min, and should have about the same amount of biological degradation as is expected for the full-scale system. The resulting breakthrough curve is shown in **Fig. C**.

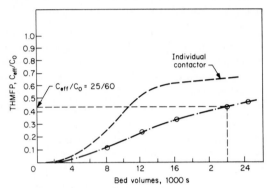

**Example problem 2, Fig. C**

3. Determine the expected breakthrough curve when four columns in parallel are blended (integral curve) by applying Eq. (13.10), as discussed earlier.

| Assumed values of $\theta_4$ | Corresponding values of $f_i$ (from Fig. C) | $f_i$ |
|---|---|---|
| 8000 BV | $f_1$ (2000 BV) = 0.02 | $\dfrac{1}{n}\sum f_i = \dfrac{0.43}{4} = 0.11$ |
| $\left(\dfrac{8000}{4} = 2000\right)$ | $f_2$ (4000 BV) = 0.05 <br> $f_3$ (6000 BV) = 0.13 | |
| | $f_4$ (8000 BV) = $\underline{0.23}$ <br> $\sum f_i = 0.43$ | |
| 12,000 BV | $f_1$ (3000) = 0.03 | $\dfrac{1.03}{4} = 0.26$ |
| $\left(\dfrac{12,000}{4} = 3000\right)$ | $f_2$ (6000) = 0.13 | |
| | $f_3$ (9000) = 0.35 <br> $f_4$ (12,000) = $\underline{0.52}$ <br> $\sum f_i = 1.03$ | |
| 16,000 BV | | 0.36 |
| 20,000 BV | | 0.48 |

4. Use the integral curve to determine the number of bed volumes that can be processed for a breakthrough concentration $C_B$ of 25 µg/L ($C_{\text{eff}}/C_0 = 25/60 = 0.42$). As shown in Fig. C, each column can be operated for

22,000 bed volumes. Given the EBCT of 15 min (or 96 BV/day), each adsorber can be operated for 22,000 BV/(96 BV/day) = 230 days before GAC replacement. One adsorber will be replaced every 58 days ( = 230/4).

**Series column analysis.**   The advantage of series operation of columns was discussed earlier. The pilot test should be carefully designed to show the reduction in activated carbon usage rate that can be achieved by such operation.

The simplest case is experienced if only a single compound with no competing adsorbates is present in the water to be treated. The breakthrough curves for each column in a two-column series are shown in Fig. 13.28a. Each column is equal in size and thus has the same EBCT. The pilot test can be terminated when the effluent concentration in the second column reaches $C_B$, because moving column 2 to position 1 and placing a fresh GAC column in position 2 should give the same breakthrough curves shown in Fig. 13.28a between $V_1$ and $V_2$. The activated carbon usage rate (g/L) is then the mass of GAC in one column (g) divided by the total volume of water processed between replacements, $V_2 - V_1$ (L). The number of bed volumes processed is $V_2 - V_1$ divided by the volume of GAC in one column.

The situation is more complex if competing organics are present with the contaminant or if a mixture of compounds is to be removed. The pilot plant must begin with fresh GAC in each position and then operated until the $C_{\text{eff}}$ in the second column reaches $C_B$. The data should be plotted as shown in Fig. 13.28a. Column 2 then must be moved to position 1, a fresh GAC column must be placed in position 2, and operation must continue until $C_{\text{eff}}$ for the second column again

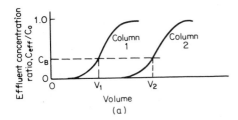

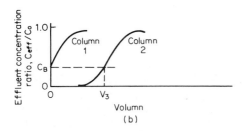

**Figure 13.28** Pilot test breakthrough curves for two columns in series. (a) Fresh GAC in each column at start of test. (b) Column 1 was in the second position until $C_B$ was reached and then moved to the first position. Fresh GAC was in column 2 at start of test.

reaches the breakthrough concentration. These data should be plotted as shown in Fig. 13.28$b$. The volume $V_3$ probably will not be the same as volume $V_2 - V_1$ because of the effects of competition. The activated carbon usage rate (g/L) for this case now becomes the mass of GAC in one column (g) divided by $V_3$ (L).

The plots in Fig. 13.28 show that column 1 is completely saturated when column 2 reaches $C_B$ (point $A$). If this does not happen, a lower activated carbon usage rate probably can be achieved with more than two columns in series. The pilot test to determine this carbon usage rate, assuming competing organics are present, must be run until the column initially in the last position rotates into the first position, followed by continued operation until this sequence of columns reaches breakthrough.

### Mathematical models

Several useful mathematical models of the adsorption process are available.[4,116] These models have led to a better understanding of the adsorption process, to determine high-priority areas for future study, and to determine potential cost-effective ways to design and operate activated carbon systems. As of this writing, the models do not permit the making of long-term, accurate predictions of trace compound removal from a mixture of adsorbable organics, however. The purpose of this section is to summarize what can be done with models and to refer the reader to other sources of information on the subject.

The *homogeneous surface diffusion model* (HSDM) and modifications of it have been widely used to predict performance of adsorption systems. Crittenden and Weber[117,118] extensively developed this model, and it has been employed to accurately predict adsorption and desorption for single-solute and bisolute systems with similar-size adsorbates.[22,27] Hand et al.[119] developed a simplified approach to solving the model. A major difficulty in extending the use of the model is the accurate description of the competitive interactions between adsorbates with different characteristics, especially size. The nature of the competition between trace compounds and natural organics also is not well understood and is difficult to predict based on laboratory tests.

Some good success with the prediction of TOC (or other group parameter) adsorption from heterogeneous mixtures of organics has been achieved. Lee et al.[7] treated humic substances as a single compound in the HSDM with some success. Sontheimer et al.[4] developed the approach of approximating solutions with many adsorbates by a hypothetical solution with only a few fictive components. The concentration and adsorption properties of the fictive components are assigned to give a composite adsorption isotherm that is the same as the isotherm of the mixture. The HSDM combined with a competitive adsorption equilibrium model can then be used to predict the performance of various types of adsorbers for adsorption of the mixture of fictive compounds. Summers and Roberts[120] developed a solution to this same problem, using the assumptions that DOC could be treated as a single compound and that it had a linear isotherm. Their model has an analytical solution, and they were able to obtain good agreement between predictions and the performance of operating systems.

Weber and coworkers[116,121] have dealt with the problem of one or more target compounds in the presence of complex mixtures of organics by incorporating the effects of the background organics in the kinetic and equilibrium constants for the target compound(s). The target compound(s) are then modeled as though only they are present in the water.

A good application of mathematical models as of 1989 is for predicting pilot-plant results and determining factors that should be tested in pilot studies. Jones and Symons,[122] for example, developed a model that permits the comparison of batch equilibrium and dynamic pilot-plant data. Pilot studies should be used to verify the predictions and to investigate aspects of performance that cannot be modeled as yet.

## PAC Adsorption Systems

### Comparison to GAC

The primary characteristic of PAC that differentiates it from GAC is its particle size. Commercially available PACs typically show 65 to 90 percent passing a number 325 mesh (44-μm) sieve. For comparison, Kruithof et al.[123] give the particle size distributions of two powdered activated carbons available in the Netherlands that show 23 to 40 mass percent smaller than 10-μm diameter and 10 to 18 percent larger than 74 μm. The particle size distribution is important because the smaller PAC particles in the absence of aluminum or ferric floc adsorb organic compounds more rapidly than large particles,[124] although studies to show the effect of particle size in the presence of floc have not yet been done. PAC is made from a wide variety of material, including wood, lignite, and coal. Its apparent density ranges from 0.36 to 0.74 g/cm$^3$ (23 to 46 lb/ft$^3$) and depends on the type of materials and the manufacturing process. Iodine number, molasses number, and phenol number are often used to characterize PAC. The AWWA standard for PAC specifies a minimum iodine number of 500, for example.[125] Note that, in general, commercial GAC products have higher iodine numbers.

The primary advantages to using PAC are the low capital invest-

ment costs and the ability to change the PAC dose as the water quality changes. The latter advantage is especially important for systems
that do not require an adsorbent for much of the year. Sontheimer[126]
noted the disadvantages of high operating costs (if high doses were required for long periods of time), the inability to regenerate, the low
TOC removal, the increased difficulty of sludge disposal, and the difficulty of completely removing the PAC particles from the water. He
further noted that GAC is used for these reasons along the lower
Rhine, where an adsorbent is constantly required. Research may show
how to overcome some of these problems, however.

### Application of PAC

PAC can be fed as a powder using dry feed machines or as a slurry
using metering pumps.[127] Common points of PAC addition in conventional plants during treatment include the plant intake, rapid mix,
and filter influent. Another point of addition that should be considered, although it is not commonly used, is a continuous flow slurry
contactor that precedes the rapid mix. The PAC can be intensely
mixed with the water, enabling rapid adsorption onto the small PAC
particles, and then incorporated into the floc in the rapid mix for subsequent removal by sedimentation and filtration. Table 13.7 summarizes some of the important advantages and disadvantages to each of
these points.

Important criteria for selecting the point of addition include

1. The provision of good mixing or good contact between the PAC and
   all the water being treated

2. Sufficient time of contact for adsorption of the contaminant

3. Minimal interference of treatment chemicals with adsorption on
   PAC

4. No degradation of finished water quality

The PAC must be added in a way that ensures its contact with all of
the flow. Addition at locations other than listed above may not
achieve this objective.

Sufficient time of contact is necessary also, and the time required is
an important function of the characteristics and concentration of the
molecule to be adsorbed.[128] In the absence of competition and floc particle interference, 15 min is sufficient time for molecules such as
dimethylphenol (molecular weight = 122, $C_0$ = 90 mg/L, activated
carbon dose = 250 mg/L) to equilibrate with a 325 mesh (44-$\mu$m diameter) particle if mixing is good. As the molecular size increases, the

TABLE 13.7   Advantages and Disadvantages of Different Points of PAC Addition

| Point of addition | Advantages | Disadvantages |
|---|---|---|
| Intake | Long contact time, good mixing | Some substances may adsorb that otherwise would be removed by coagulation, thus increasing the activated carbon usage rate. |
| Rapid mix | Good mixing during rapid mix and flocculation; reasonable contact time | Possible reduction in rate of adsorption because of interference by coagulants.<br><br>Contact time may be too short for equilibrium to be reached for some contaminants.<br><br>Some competition may occur from molecules that otherwise would be removed by coagulation. |
| Filter inlet | Efficient use of PAC | Possible loss of PAC to the clear well and distribution system. |
| Slurry contactor preceding rapid mix | Excellent mixing for the design contact time, no interference by coagulants, additional contact time possible during flocculation and sedimentation | A new basin and mixer may have to be installed.<br><br>Some competition may occur from molecules that otherwise would be removed by coagulants. |

rate of diffusion decreases. For example, rhodamine B dye (molecular weight of 422) requires approximately 5 h to come to equilibrium, a 10,000 molecular weight fulvic acid requires about 17 h, and a 50,000 molecular weight humic acid requires about two days. Commercial shipments of PAC may contain a large percentage of particles larger than 325 mesh, so longer times would be required to equilibrate these. The adsorption kinetics and equilibrium capacity depend on the type of PAC used, so these values should be taken as rough estimates only. If insufficient time is allowed for equilibration, an increased PAC dose must be used to compensate.

The effect of PAC particle size on required contact time in the absence of floc particles and competition was shown by Najm et al.[124] for a continuous stirred tank reactor. In 15 min, 500 μg/L trichlorophenol was lowered to 25 μg/L by 14-μm-diameter PAC, but only to 275 μg/L by 100-μm-diameter PAC. The composite PAC sample with an average diameter of 40 μm lowered the concentration to 180 μg/L. In no case was equilibrium achieved in 15 min, and the concentration in the test with the composite sample was still decreasing after 120 min.

When adding PAC at the rapid mix, incorporation of PAC into floc particles is one factor that may reduce rate of adsorption.[126,129] The adsorbate must diffuse through the part of the floc surrounding the PAC particle and then into the particle itself in order to be adsorbed. Gauntlett and Packham[129] conducted jar tests that showed the removal rate of chlorophenol by PAC in the absence of alum addition to be most rapid, and that addition of PAC after alum addition gave a better rate of removal than when it was added just before alum (Fig. 13.29). PAC added after the alum floc formed adhered to the outer surface of the alum floc rather than being incorporated into the floc, thus avoiding interferences. Najm et al.[179] found little reduction, however, in the rate of trichlorophenol adsorption on PAC because of the incorporation of the PAC particles into coagulant floc.

Addition at the intake has the advantage of providing extra contact time, but the disadvantage of adsorbing many compounds that otherwise would be removed by coagulation, flocculation, and sedimentation. On-site tests are recommended to determine whether one factor outweighs the other.

Addition of PAC just before the filter is advantageous because the PAC can be kept in contact with the water longer, thereby better us-

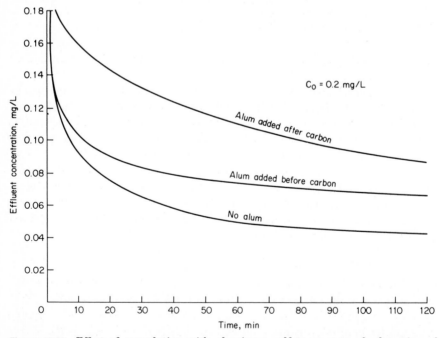

**Figure 13.29** Effect of coagulation with aluminum sulfate on rate of adsorption of monochlorophenol. (*Source: R. B. Gauntlett and R. F. Packham, "The Use of Activated Carbon in Water Treatment," Proc. Conf. on Activated Carbon in Water Treatment, Univ. Reading, Water Res. Assn., Medmenham, England, 1973.*)

ing its capacity. The average PAC residence time is equal to one-half of the time between two successive backwashings, assuming PAC is continuously added to the filter influent. The addition must be performed carefully to avoid penetration of PAC to the distribution system, however. The maximum dosage of PAC is limited by the ability of the filter to retain the PAC, and by the rate of head loss buildup in the filter, which is expected to increase as PAC dosage increases. The maximum dosage may not be sufficient to remove the amount of contaminant present. Selected polyelectrolytes may be added to retain the PAC, but careful monitoring of the practice is required.

Sontheimer[126] suggested adding a separate reactor between the sedimentation basin and the filter to increase the time of contact, and noted that this procedure would have the advantage of having eliminated competing organics to the maximum possible extent by coagulation-sedimentation. A major disadvantage is that the PAC would be removed from the water by another coagulation and filtration step.

Careful attention must be paid to the interaction of PAC with water treatment chemicals. Activated carbon is an efficient chemical reducing agent that will chemically reduce substances such as free and combined chlorine, chlorine dioxide, ozone, and permanganate, thereby increasing the demand for these substances and the cost of treatment. Reaction of activated carbon with chlorine will reduce the adsorption capacity of the activated carbon for selected compounds.[64,130] Lalezary-Craig et al.[131] found a reduction in the ability of PAC to adsorb both geosmin and MIB when PAC was applied to water containing free chlorine and monochloramine, and the effect of the monochloramine appeared to be greater than that of chlorine. Furthermore, the disinfecting ability of the oxidant will be lost.

Addition of PAC to a water that is supersaturated with $CaCO_3$ or other precipitates, or an increase of pH to cause supersaturation just after PAC is added, such as in lime softening, may lead to coating of the particle with precipitates and to a corresponding decrease in adsorption efficiency. Also, adsorption at high pH is often poorer than at low pH because many organic contaminants are weak acids that ionize at high pH.

Various techniques of applying PAC to improve the ability of PAC to adsorb slow-diffusing compounds such as trihalomethane precursors are available. For example, addition of PAC to solids contact clarifiers has the potential for improved adsorption efficiency because the carbon can be kept in contact with the water for a longer time.[132,133] Richard[104] noted that the PAC dose for detergent removal could be reduced by 25 to 40 percent if the activated carbon were added to the influent of a floc blanket clarifier instead of to a conventional system.

More research is needed to optimize this process, especially for TOC removal, and to determine coagulation procedures that can be used as a coagulant to increase adsorption kinetics, such as using polyelectrolyte alone instead of alum or ferric salts.[128,133]

In the Roberts-Haberer process,[134-136] buoyant polystyrene spheres 1 to 3 mm in diameter are coated with PAC. The spheres are held in a reactor by means of a screen, and the water to be treated is passed upflow through the media. After saturation, the PAC is removed from the beds by backwashing the media with a high flow rate. New PAC can then be applied to the beds. A primary advantage of the process is the ability to retain the PAC until its adsorption capacity is utilized, without incorporation of the PAC in floc as is necessary in most other applications. Recovery of the PAC for regeneration may be possible. In addition, the Roberts-Haberer filter may be used as a roughing filter to remove suspended solids ahead of the rapid filter, thereby reducing the solids load to the rapid filter. Careful studies are needed of the adsorption efficiency for specific organics and of the ability of the polystyrene beads to retain the PAC in order to determine whether the process will be more cost effective than GAC, however.

Jar tests are often used to determine an estimate of the PAC dose that is required to achieve the desired removal in conventional plants. Use of a mixing rate that does not allow the PAC to settle is important[128]; the chemical doses, rapid mix, and settling time should correspond as nearly as possible to those that will be encountered in the full-scale plant. The percent removal or final concentration of contaminant in the jar test can be plotted against dose to determine the required amount to be added. This dosage will only be an estimate of the full-scale plant dosage because of the many differences between treatment in a jar and in the plant. The dosage, therefore, should be modified based on full-scale performance of the PAC. Some bench-scale or pilot testing also may be necessary to determine the best polymer to prevent PAC from penetrating the rapid filter.

### PAC performance

Many reports of successful removal of taste and odor using PAC are in the literature and a few are summarized below. Twenty-five percent of the 683 United States water plants surveyed (including the 500 largest utilities) in 1976 reported that they used PAC,[1] and similar data have been reported for 1984.[2] The predominant reason for its use is taste and odor control, but it also can be used to remove nonodorous organics. PAC doses range from a few to more than 100 mg/L, but are typically less than 25 to 50 mg/L.[137] The TON is most commonly used to measure its effectiveness. Reports of the effectiveness of PAC are complicated because an odor may be caused by the combination of the effects of more than one compound, or a compound may mask the odor of another compound and the odor of the second may appear after the first compound has been adsorbed. Furthermore, odors may form after treatment through chemical or biological reactions in the distribution system,[86] thus making PAC appear ineffective. A reliable catalog of odors that can be removed well by PAC does not exist, probably because of the mentioned complicating factors. As analytical capabilities improve and the ability to qualitatively and quantitatively analyze compounds that cause odor advances, however, development of such a catalog may be possible.

PAC performance for removal of THMs and VOCs has not been very good, consistent with theoretical calculations. The minimum amount of PAC required to achieve a given removal can be calculated from the pure solute isotherm data in Table 13.2. For this calculation, the PAC is assumed to reach equilibrium with the final concentration because it moves through the treatment plant with the water being treated in most applications. The Freundlich isotherm is used to calculate the surface concentration at an equilibrium equal to the final concentration, and the amount to be removed divided by the surface concentration yields the minimum PAC dose:

$$(\text{PAC dose})_{min} \text{ (g/L)} = \frac{(C_0 - C_e) \text{ (mg/L)}}{q_e \text{ (mg/g)}} \qquad (13.26)$$

where $q_e = K C_e^{1/n}$

**Example Problem 3.**   Calculate the minimum PAC dose required to reduce chloroform concentration from 50 to 5 µg/L. The $K$ and $1/n$ values are 2.6 $(\text{mg/g})(\text{L/mg})^{1/n}$ and 0.73, respectively.

**solution.**   Assuming that the PAC will equilibrate with the final chloroform concentration (i.e., $C_e = 5$ µg/L), Eq. (13.1) yields

$$q_e = K C_e^{1/n}$$

$$= 2.6 \text{ (mg/g)}(\text{L/mg})^{0.73} (0.005 \text{ mg/L})^{0.73}$$

$$= 0.054 \text{ mg/g}$$

and Eq. (13.26) gives

$$(\text{PAC dose})_{min} = \frac{(C_0 - C_e) \text{ (mg/L)}}{q_e \text{ (mg/g)}}$$

$$= \frac{(0.05 - 0.005) \text{ (mg/L)}}{0.054 \text{ (mg/g)}}$$

$$= 0.830 \text{ g/L}$$

*Note:* The actual dose will probably be larger than 830 mg/L for the reasons given below.

The calculations for THMs and VOCs are shown in Table 13.8. These numbers will change as the type of PAC is changed, resulting in a different isotherm, as competing substances are present that lower the amount of compound that can be absorbed, and as the time allowed for adsorption or the degree of mixing does not permit equilibrium to be achieved. The data show that compounds such as trichloroethylene, tetrachloroethylene, and bromoform possibly can be reduced in concentration with reasonable PAC doses, but that reducing the concentrations of poorly adsorbing compounds such as chloroform and carbon tetrachloride requires very high doses and, therefore, would probably be uneconomical. The investigation of means of efficient contact to achieve adsorption is reasonable for compounds with the lower doses.

Consistent with these calculations, only 20 percent reduction of 70 μg/L carbon tetrachloride was achieved with 30 mg/L PAC in jar tests,[138] and similar poor removals were noted in full-scale plants.[139] Symons et al.[60] reported that more than 100 mg/L of PAC were required to remove about 50 percent of the 60 μg/L chloroform in jar tests. Also, Singley et al.[140] reported only minor removal of dichloroethane ($C_0 = 5$ μg/L) with up to 27 mg/L PAC in a full-scale plant. Removals of several other VOCs were measured, but because of the variability of influent concentrations, precise calculations of removals were difficult.[140]

The removals of compounds that adsorb more strongly than trichloroethylene (for example, compounds with Freundlich $K$ values

**TABLE 13.8   Theoretical PAC Doses Required for 90 Percent Reduction of THMs and Selected VOCs**

| | $K$, mg/g · (L/mg)$^{1/n}$ | $1/n$ | PAC dose, mg/L (100 μg/L to 10 μg/L)† | PAC dose, mg/L (10 μg/L to 1 μg/L)‡ |
|---|---|---|---|---|
| Tetrachloroethylene | 51 | 0.56 | 23 | 8 |
| Trichloroethylene | 28 | 0.62 | 55 | 23 |
| Bromoform | 20 | 0.52 | 49 | 16 |
| Carbon tetrachloride | 11 | 0.83 | 375 | 250 |
| 1,1,2,2-Tetrachloroethane | 11 | 0.37 | 45 | 10 |
| Dichlorobromomethane | 7.9 | 0.61 | 187 | 75 |
| 1,1,2-Trichloroethane | 5.8 | 0.60 | 250 | 98 |
| 1,1-Dichloroethylene | 4.9 | 0.54 | 225 | 75 |
| Dibromochloromethane | 4.8 | 0.34 | 90 | 20 |
| Chloroform | 2.6 | 0.73 | 1000 | 530 |
| 1,1,1-Trichloroethane | 2.5 | 0.34 | 173 | 38 |
| 1,2-Dichloroethane | 3.6 | 0.83 | 1140 | 770 |

†$C_0 = 100$ μg/L and $C_e = 10$ μg/L; see Eq. (13.19) for calculation procedure.
‡$C_0 = 10$ μg/L and $C_e = 1$ μg/L

greater than about 30 $(mg/g)(L/mg)^{1/n}$ in Table 13.2) are much better than the removals of some of the chlorinated solvents. Singley et al.[141] showed about 90 percent removal of nonvolatile synthetic organic chemicals with PAC doses of 10 to 15 mg/L (2 h contact time) at the Sunny Isles water treatment plant. Chlorobenzene in particular was well removed, as were several polynuclear organic compounds. The odorous compound 2,4-dichlorophenol has also been shown to adsorb well.[142,143] Organics may adsorb to particles,[144] however, and strongly adsorbing compounds may not be removed well by PAC if such particles are not removed by clarification processes. El-Dib et al.[145] found that from 15 to 65 mg/L PAC were required to reduce 1 to 4 mg/L of toluene, o-xylene, ethylbenzene, and cyclohexane to 0.1 mg/L. Benzene, which is more poorly adsorbed, required 150 mg/L for the same influent level. Pesticides and herbicides are generally quite well removed.[100,142,146]

TOC and THMFP removals by PAC in conventional plants have not been very high, possibly because removal is limited by slow kinetics of adsorption and equilibrium capacity. Typical TOC isotherms for natural waters show adsorption capacities in the range of 10 to 50 mg/g. If the PAC dose is 25 mg/L, theoretically only 0.25 mg/L TOC can be removed if the equilibrium capacity is 50 mg/g. Thus, poorly adsorbing natural organics can be a reason for poor TOC removals. In addition, many natural organic molecules are large and diffuse slowly; thus, achieving equilibrium in the contact time available is impossible. Removals that have been observed include about 50 percent removal at a PAC dose of 100 mg/L and a contact time of less than 30 min.[147] At Contra Costa, Calif., addition of up to 40 mg/L PAC (simulating both addition to the rapid mix and to settled water before filtration) showed no effect on THMFP removal.[148] Hoehn et al.[132] reported that a dose of 18 to 25 mg/L PAC to the influent of a floc blanket reactor at Newport News, Va., increased the removal of TOC and THMFP by 9 to 20 percent over that which could be achieved by coagulation alone. The TOC of the raw water averaged 6.4 mg/L, and about 40 percent removal was achieved by coagulation alone.

## Thermal Regeneration of GAC

### Regeneration principles

The regeneration process can be described as consisting of four basic steps[149]:

1. Drying, including loss of highly volatile adsorbates, at temperatures up to 200°C

2. Vaporization of volatile adsorbates and decomposition of unstable

adsorbates to form volatile fragments, at temperatures of 200 to 500°C

3. Pyrolysis of nonvolatile adsorbates and adsorbate fragments to form a carbonaceous residue or char on the activated carbon surface, at temperatures of about 500 to 700°C

4. Oxidation of the pyrolyzed residue using steam or carbon dioxide as the oxidizing agent, at temperatures above about 700°C

The drying step eliminates the 40 to 50 percent water associated with the spent GAC.[150] The proportions of adsorbates that are volatilized (steps 1 and 2) or converted to char (step 3) depend on the nature of the adsorbate and the strength of the adsorbate-activated carbon bond.[151] Thus, the type of activated carbon also plays an important role. Suzuki et al.[152] related the amount volatilized to the boiling point of the compound (the lower the boiling point, the larger the amount of volatilization) and its degree of aromaticity (the more aromatic, the smaller the amount of volatilization). Molecules such as lignin, phenolic compounds, and humic substances produced the largest amount of char. Tipnis and Harriott[153] studied a series of phenolic compounds, for example, and showed that 20 to 50 percent was converted to char. Van Vliet[149] cautions that pyrolysis should not take place above 850°C because a char with a more graphitized structure similar to that of the base-activated carbon will be formed. The stronger oxidation conditions required to remove the graphitized char will also result in extensive attack, and loss, of the activated carbon.

Oxidation to gasify the char is critical to returning the GAC as closely as possible to virgin conditions. The objective is to remove the char, with minimal loss of the activated carbon, in a manner that does not alter the structure and adsorption properties of the activated carbon. The reactions that are used are similar to those used to produce virgin activated carbon. Steam and carbon dioxide are the oxidizing agents:

$$H_2O(g) + C(s) \rightarrow CO(g) + H_2(g) \qquad (13.27)$$

$$CO_2(g) + C(s) \rightarrow 2CO(g) \qquad (13.28)$$

CO can then react with steam:

$$CO(g) + H_2O(g) \rightarrow CO_2(g) + H_2(g) \qquad (13.29)$$

Reactions (13.27) and (13.28) are endothermic and, thus, require heat in order to proceed. This property makes these reactions easy to control by controlling the heat input rate.[154] In contrast, the reaction between oxygen and carbon,

$$\frac{3}{2}O_2(g) + C(s) \rightarrow CO_2(g) + CO(g) \qquad (13.30)$$

is exothermic, or heat-producing, and is thus self-promoting and difficult to control.[149,155] Thus, oxygen is more likely to attack the base-activated carbon structure, altering the pore volume distribution and increasing activated carbon loss, and should not be used under normal regeneration conditions.

Factors affecting regenerated product quality are primarily those that affect the oxidation step. Important factors are the type of oxidizing gas, the time and temperature of oxidation, the amount and type of adsorbate on the carbon, the type and quantity of inorganic substances that accumulate on the carbon, and the type of activated carbon.

The activated carbon should be relatively homogeneous and have a lower reactivity with steam and carbon dioxide than the char deposit.[151] This difference in reactivity should make possible the selection of regeneration conditions that minimize oxidation of the activated carbon while removing the deposit. Juntgen[151] found that the reactivity of several activated carbons varied widely, depending on the raw material used to make the activated carbon. For example, activated carbon made from wood was far more reactive than activated carbon made under the same conditions from hard coal. Juhola[156] reported that activated carbons with a large volume of large pores, such as lignite-activated carbon, usually regenerate at a lower temperature than activated carbons with smaller pores. The larger pores made desorbing the adsorbate molecules easier and also allowed oxidizing gases to reach the deposits of carbonized adsorbate.

Regeneration studies indicate a trade-off between the time required for reactivation and the temperature of reactivation: the higher the temperature, the shorter the reactivation time.[156] A temperature of 925°C was near optimum for one activated carbon that was used to remove organics from secondary effluent.[156] At lower temperature the recovery of small pores that were needed for this application was not as good as at higher temperature; also, the oxidation rate was very slow at 700°C. At temperatures above 925°C, the carbon structure itself was very reactive, and activated carbon losses were expected to be high, although they were not measured. The optimum temperature for reactivation has been shown to depend on the type of activated carbon.[151] Van Vliet and Venter[157] thoroughly studied the regeneration of GAC from a wastewater reclamation plant and found optimum regeneration conditions to include a temperature and time of regeneration range of 800°C for 10 min to 850°C for 5 min. Longer times and higher temperatures resulted in higher losses and a less desirable pore structure, while shorter times and lower temperatures gave incompletely regenerated GAC.

Schuliger and MacCrum[150] showed that the type of adsorbed organic compound significantly affects the time of regeneration. Regeneration times from 5 to 125 min were required, with a typical range being 20 to 60 min for a variety of applications. The required time increased as the quantity of adsorbed organics increased.

In the application of activated carbon for purification of surface water and possibly groundwater, the need to remove adsorbed humic substances will probably control the regeneration step even though the primary purpose in applying the activated carbon is to remove trace organic compounds. Because the humic substances are the predominant organic fraction in most surface waters and they readily adsorb,[158] the quantities of adsorbed trace organics will be small by comparison to the amount of humic substance adsorbed because of their much lower concentrations. Thus, the quantity of pyrolyzed deposit will probably depend on the manner in which humic substances are converted to char. Suzuki et al.[152] found that 26 percent of one humic acid was converted to char during pyrolysis. Humic substances vary widely in molecular weight; the larger molecules will probably adsorb in the larger activated carbon pores, while the smaller molecules and most trace organics will likely adsorb in the smaller pores. Good removal of organics from all pores will then be necessary if the regenerated carbon is to function well.

Certain inorganic substances if present in the water being treated may form a precipitate on the activated carbon surface, or may accumulate on the activated carbon surface because of their association with the organic molecules that are being adsorbed. Juhola[156] noticed an increase in ash content of several percent after a few regenerations of a bituminous carbon that had been used to remove organics from municipal secondary effluent, and that recovering the small pores during regeneration was difficult. When the inorganic materials were removed with an HCl wash prior to the regeneration step, the small pore recovery was very good. Smisek and Cerny[159] indicated that oxides and carbonates of metals such as sodium, potassium, iron, and copper catalyze the reactivation step. Umehara et al.[160] found that the $Na_2SO_4$ produced by regeneration of a carbon loaded with the adsorbed detergent, sodium dodecyl benzene sulfonate, had a significant catalytic effect on the rate of oxidation of char. Thus, the optimum temperature and other regeneration conditions are likely to change as ash builds up on the carbon surface. Poggenburg et al.,[161] reporting on a study of the Rhine River water treatment in West Germany, noted no increase in ash content over three reactivations of a particular activated carbon. Thus, ash buildup after regeneration is a function of a type of water and is not to be expected in every case.

### Regeneration furnaces

Common types of furnaces are the rotary kiln, the multiple hearth, and the fluidized bed furnace, but the infrared furnace, which is electrically heated, is also receiving much attention. Diagrams of these furnaces are shown in Figs. 13.30 through 13.33. While the multiple hearth furnace is perhaps the most commonly used, the furnaces most recently installed in the United States and West Germany for drinking water treatment are fluidized bed. Scrubbers are used to remove particles from the off-gases, and afterburners, which are maintained at 750 to 1000°C by burning fuel with excess oxygen, are used to destroy organic compounds that are volatilized but not combusted to $CO_2$ during the reactivation step. A diagram of a complete regeneration system is shown in Fig. 13.34.

**Regeneration performance.** Loss of mass during regeneration is very important because of its impact on cost of using activated carbon. DeMarco et al.[164] reported total losses (transport plus loss in the furnace) of 16 to 19 percent in an experimental system at Cincinnati, while Koffskey and Lykins[165] observed 9 percent loss (7 percent in the furnace and 2 percent during transport) at Jefferson Parish, La. Proper design of the transport system is very important because of the high losses that can take place in a poorly designed system.[163] Schuliger et al.[163] note that losses are a function of the transport sys-

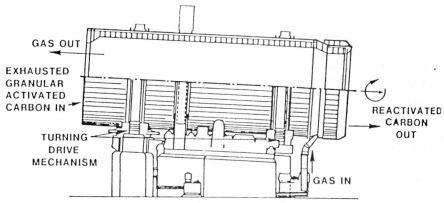

**Figure 13.30**  Cross section of rotary kiln furnace.  *(Source: O. T. Love, Jr., "Experience with Reactivation of Granular Activated Carbon," Sem. Proc., Controlling Organics in Drinking Water, AWWA Ann. Conf., San Francisco, Calif., June 1979.)*

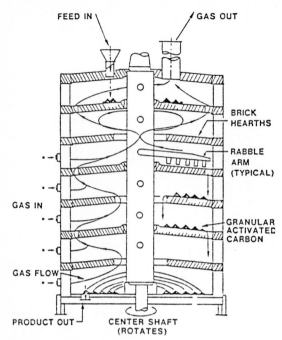

**Figure 13.31** Cross section of multihearth furnace. (*Source: O. T. Love, Jr., "Experience with Reactivation of Granular Activated Carbon," Sem. Proc., Controlling Organics in Drinking Water, AWWA Ann. Conf., San Francisco, Calif., June 1979.*)

tem design, the type of adsorbate, the loading of adsorbate on the carbon, and the type of furnace.

Along with the loss in mass may come a change in pore size distribution. Often small pores less than 2 nm in diameter are lost while larger pores are created.[166] Pore size distribution has a significant effect on both rate of uptake and capacity for a particular adsorbate.[161,167] Modifying regeneration conditions to produce the best pore size distribution for a particular adsorbate should be possible, but research to show how this can be done for drinking water treatment applications is needed.

Other observations are that the particle diameter becomes smaller during regeneration; this has the disadvantage of creating higher head losses in fixed-bed adsorbers, but has the advantage of allowing the rate of uptake of molecules to occur more rapidly. The decrease in particle diameter should be minimized as much as possible because this decrease normally correlates with loss in mass. This change in particle size, however, apparently does not affect the rate of regeneration of the particles.[156] The hardness of activated carbon will de-

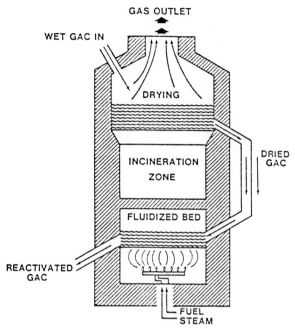

**Figure 13.32** Cross section of a two-stage fluidized bed furnace. (*Source: O. T. Love, Jr., "Experience with Reactivation of Granular Activated Carbon," Sem. Proc., Controlling Organics in Drinking Water, AWWA Ann. Conf., San Francisco, Calif., June 1979.*)

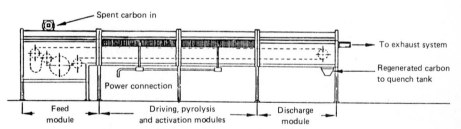

**Figure 13.33** Infrared regeneration furnace. (*Source: O. T. Love, Jr., "Experience with Reactivation of Granular Activated Carbon," Sem. Proc., Controlling Organics in Drinking Water, AWWA Ann. Conf., San Francisco, Calif., June 1979.*)

crease upon regeneration because of the destruction of part of the structure of the activated carbon. As the activated carbon becomes less hard, the losses in the furnace and during handling will increase.

Generally, density, molasses number or decolorization index, iodine number, and similar parameters are used to measure the quality of the activated carbon. An improved procedure involves the use of adsorbate molecules in the water to be treated to test the quality of the

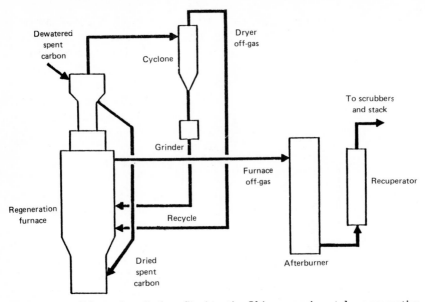

**Figure 13.34** Schematic of the Cincinnati, Ohio, experimental regeneration system. (*Source: W. G. Schuliger, G. N. Riley, and N. J. Wagner, "Thermal Reactivation of Granular Activated Carbon: A Proven Technology," AWWA Ann. Conf., Kansas City, Mo., June 1987.*)

product. Poggenburg et al.[161] used this approach to evaluate activated carbons for removal of organics found in Rhine River water. Their procedure involved extensive use of adsorption isotherms and adsorption rate tests and modeling of the performance of full-scale adsorbers. Changes in activated carbon quality produced by regeneration then can be expressed as changes in the length of time to breakthrough and the amount of material adsorbed at breakthrough for a column of a given design. Because the rate of adsorption depends on pore size distribution and adsorbate size, as well as other factors, using the molecule in the water to be treated as a test substance will yield a meaningful comparison of the activated carbon quality for the application in question.

Kornegay[46] analyzed the cost of regeneration as a function of furnace size. He showed rapidly increasing unit costs as activated carbon usage became smaller than 5000 to 6000 lb/day. Kornegay noted that on-site thermal regeneration will not be economical if usage is smaller than 500 to 1000 lb/day, while Hess et al.[97] proposed a lower economical limit of about 2000 lb/day. Transport of spent GAC to off-site thermal regeneration facilities may be appropriate for intermediate values of activated carbon usage rate. The cost of regeneration would then include transportation to and from the site, the cost of makeup activated carbon, and (presumably) profit for the owner of the furnace. This approach to regeneration is common in Europe where transportation distances are relatively short, the incidence of GAC usage for surface water is high, and the number of on-site furnaces is small. Further, assurance would have to be given that only water-treatment-plant-activated carbon would be returned to a plant. Inclusion of regenerated activated carbon that had been used for municipal or industrial waste treatment would be unacceptable.

### Regeneration by-products

Activated carbon must be regenerated in a way that does not pollute the atmosphere. For this reason, scrubbers and afterburners are used to minimize particulate and gaseous emissions, respectively, from operating facilities (see Fig. 13-34). Off-gas control is important in controlling dioxins and furans that are produced during regeneration. Lykins et al.[168] studied operating regeneration systems at Cincinnati and Jefferson Parish and found these compounds in various effluent streams, before treatment of these effluent streams, even though they were not on GAC entering the furnace. A small amount of the dioxins and furans do pass through the stack gas control equipment, but Lykins et al.[168] found the concentrations to be so low that the estimated cancer risk was minimal and negligible.

### Adsorption on Resins

Many types of synthetic resins that differ both in the matrix that supports the functional groups and in the functional groups themselves have been used for removal of organic compounds. Some of the more common matrices and functional groups are shown in Fig. 13.35. A few resins, called *adsorbent resins,* such as the styrene divinylbenzene (SDVB) resin and the phenol-formaldehyde (PF) resin, have been made without functional groups in addition to those that are part of the matrix. More typically, resins have functional groups that make it possible for substances to be taken up by ion exchange or by specific interaction with the functional group. The strong-base ion exchange resins, for example, frequently have the quaternary ammonium functional group (Fig. 13.35:A1), whereas the weak-base resins generally have an amine functional group (Fig. 13.35:A2). The strong-acid and weak-acid functional groups can remove substances from water by cation exchange; typical functional groups are the sulfonate group (Fig. 13.35:C1) and the carboxyl group (Fig. 13.35:C2). In addition, the

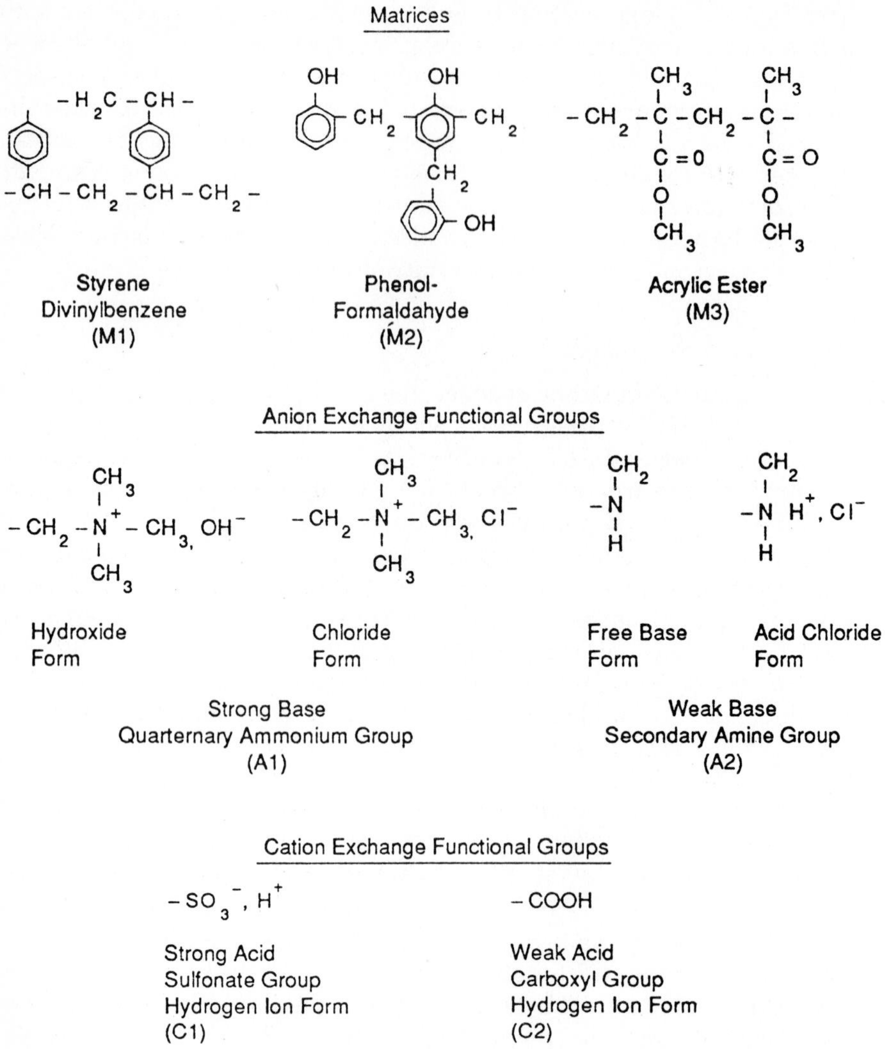

## Matrices

| Styrene Divinylbenzene (M1) | Phenol-Formaldehyde (M2) | Acrylic Ester (M3) |

## Anion Exchange Functional Groups

| Hydroxide Form | Chloride Form | Free Base Form | Acid Chloride Form |

| Strong Base Quarternary Ammonium Group (A1) | Weak Base Secondary Amine Group (A2) |

## Cation Exchange Functional Groups

$$- SO_3^-, H^+ \qquad\qquad - COOH$$

| Strong Acid Sulfonate Group Hydrogen Ion Form (C1) | Weak Acid Carboxyl Group Hydrogen Ion Form (C2) |

**Figure 13.35** Matrices and functional groups of resins commonly used for water purification. (*Source: B. R. Kim, V. L. Snoeyink, and F. M. Saunders, "Adsorption of Organic Compounds by Synthetic Resins," J. Water Pollution Control Fed., vol. 48, 1976, p. 120.*)

degree of cross-linking between the polymers that constitute the matrix can be varied and thus resins with different pore size distributions can be prepared.

Kolle[170] found good removal of TOC during groundwater treatment at Hannover, West Germany, using a macroporous strong-base SDVB resin. The resin was regenerated with a NaCl-NaOH solution (100 and 20 g/L, respectively) and the solution could be reused several times before disposal. The volume of spent regenerant used could be limited to one unit per 25,000 units of product water. Brattebo et al.[171] used the same regenerant and the same resin for a surface water in Norway and concluded that an average regenerant reuse of 78 percent would not degrade the resin. The strong-base SDVB resin also was used in England on Thames River water and resulted in much poorer removals of TOC than could be achieved with activated carbon.[172] The amount of alkylbenzene sulfonate (ABS) adsorbed during the test and the capacity for trace concentrations of dichlorophenol were also low.

Fu and Symons[173] concentrated the organics in Lake Houston, Texas, water by reverse osmosis and investigated their removal as a function of molecular weight by anion exchange. They found the removal to be predominantly by ion exchange between the organic anion and the chloride ion. They found the acrylic strong-base exchanger to have the higher capacity and, generally, to have the better performance.

The performance of weak-base resins under different conditions of adsorption differs markedly from that of the strong-base resins. One of the major differences is the pH dependence of adsorption. Adsorption of phenol on a weak-base PF resin, for example, takes place on the functional group in the free-base form which predominates at pH 5 and above; as pH decreases below pH 5, acid is adsorbed by the resin and competes with phenol, thereby decreasing phenol capacity. As pH increases above approximately 8 to 9, capacity for phenol decreases because the OH groups of the PF matrix ionize, and phenol itself ionizes. The maximum adsorption for phenol occurs near neutral pH. On the other hand, ABS is negatively charged and is taken up by the acid form of the resin (Fig. 13.35:A2) by an exchange mechanism. The capacity of weak-base resins for ABS is very good at low pH, but they do not adsorb very much on the free-base form of the resin.

An advantage of weak-base resins over strong-base resins is that the former usually can be regenerated more easily. They have a lower affinity for anionic organics, and as a result the organics can be removed by sodium hydroxide solution or salt solution more easily. Strong-acid (Fig. 13.35:C1) and weak-acid (Fig. 13.35:C2) cation exchangers have found little application for removal of organic materials from water, presumably because most of the organics in water are either negatively charged or neutral.

In recent years SDVB resins with no functional groups (Fig. 13.35:M1) have received much attention as an adsorbent. They have been used to concentrate pesticides and related compounds from raw and finished waters in a study of contamination of water supplies,[174] and they have been shown to be particularly advantageous for removal of chlorinated pesticides from an industrial wastewater.[175] Regeneration was readily accomplished with acetone or isopropanol, and the spent regenerant could be reclaimed for reuse. This resin also adsorbs phenols, although its capacity is somewhat less than that of weak-base resins.[176] The SDVB resin did not remove TOC from Thames River water to an appreciable extent.[172]

The Water Research Center in England evaluated synthetic resins in comparison to activated carbon for application in water treatment plants using coagulated and filtered Thames River water.[172] They eliminated the SDVB resins because of very low capacity, and selected three strong-base resins and one weak-base resin for additional testing. The results showed that the activated carbon removed TOC more effectively than any of the resins and that the strong-base SDVB and acrylic matrix resins performed somewhat better than the weak-base PF resin. Activated carbon removed the trace concentrations of the pesticide γ-BHC very effectively; the weak-base resin removed it nearly as well, but the strong-base resin removed very little.

Jayes and Abrams[177] reported on the results of the application of a PF weak-base resin for removing color at the Lawrence, Mass., test facility. The resin was generally quite effective for removing this material, although at times when the river water was extremely high in color a substantial amount of leakage through the column occurred. Regeneration was accomplished with 4 lb of sodium hydroxide per cubic foot. On the basis of this test a resin life in excess of 200 cycles of adsorption-regeneration was estimated.

Commercial humic acid was adsorbed very well on weak-base PF resins and strong-base resins; the capacities achieved were comparable to those for GAC.[178] The capacity of the weak-base PF resin decreased when the pH was either decreased from 8.3 to 5.5 or increased from 8.3 to 9.5. At low pH, the amine functional groups adsorb acid, thereby reducing the amount of organic material that can be adsorbed at those sites. As the pH is increased above pH 8.3, the OH functional groups on the PF resin matrix ionize, thus giving the resin a negative charge. Because of the negative charge on humic acid, which increases as pH is increased, adsorption capacity on the negatively charged resin is low. The strong-base resin did not show a great variation in capacity with changes in pH over the range of 5.5 to 9.5. The weak-base PF resin had essentially no capacity for the earthy-musty odor compound, MIB.[180] The SDVB adsorbent resin did adsorb MIB, but its capacity was lower than those of low-activity bituminous carbons.

A general conclusion concerning resins is that they are not applicable as a general adsorbent for drinking water treatment. They are more selective than activated carbon and do not meet the criterion that an adsorbent must be able to remove a wide variety of compounds. They may be applicable in specific situations that require only a particular type of contaminant be removed.

## Waste Disposal

Activated carbon waste is an important consideration in drinking water treatment. Spent GAC must be disposed of, recognizing that adsorption is a reversible process. Exposure of spent GAC to percolating rainwater, for example, can lead to leaching of adsorbed compounds and possibly to contamination of soil and groundwater. Perhaps the best solution is to thermally destroy the adsorbed compounds by thermal regeneration or by combustion of the spent GAC.

PAC waste may be subject to similar leaching problems if disposed of on land. If it is used only to adsorb natural, odorous compounds, however, this may not cause a problem. If PAC is part of waste alum or iron sludge being discharged to a receiving body of water, the black color of the sludge may cause a visual problem.

Chapter 16 deals with these issues in more detail.

## References

1. AWWA Committee, "Measurement and Control of Organic Contaminants by Utilities," *J. AWWA*, vol. 69, no. 5, 1977, p. 267.
2. *1984 Utility Operating Data*, AWWA, Denver, Colo., 1986.
3. J. L. Fisher, Calgon Carbon Corp., personal communication, Nov. 1986.
4. H. Sontheimer, J. C. Crittenden, and R. S. Summers, *Activated Carbon for Water Treatment*, 2d ed., DVGW-Forschungstelle am Engler-Bunte-Institut der Universitat Karlsruhe, Karlsruhe, West Germany, 1988.
5. G. Halsey and H. S. Taylor, *J. Chem. Phys.*, vol. 15, 1947, p. 624.
6. I. Langmuir, *J. Amer. Chem. Soc.*, vol. 40, 1918, p. 1931.
7. M. C. Lee, V. L. Snoeyink, and J. C. Crittenden, "Activated Carbon Adsorption of Humic Substances," *J. AWWA*, vol. 73, no. 8, 1981, p. 440.
8. V. L. Snoeyink and W. J. Weber, Jr., "Surface Functional Groups on Carbon and Silica," in J. F. Danielli et al. (eds.), *Progress in Surface and Membrane Science*, Pergamon, New York, 1972.
9. R. W. Coughlin and F. Ezra, "Role of Surface Acidity in the Adsorption of Organic Pollutants on the Surface of Carbon," *Environmental Sci. Technol.*, vol. 2, 1968, p. 291.
10. C. G. Gasser and J. J. Kipling, *Proc 4th Conf. on Carbon*, 1959, p. 55.
11. J. J. Kipling and P. V. Shooter, *J. Colloid Interface Sci.*, vol. 21, 1966, p. 238.
12. V. L. Snoeyink, H. T. Lai, J. H. Johnson, and J. F. Young, "Active Carbon Dechlorination and the Adsorption of Organic Compounds," in A. Rubin (ed.), *Chemistry of Water Supply, Treatment and Distribution*, Ann Arbor Science, Ann Arbor, Mich., 1974.
13. W. J. Weber, Jr., *Physicochemical Processes*, Wiley-Interscience, New York, 1972.
14. R. D. Fox, R. T. Keller, and C. J. Pinamont, *Recondition and Reuse of Organically Contaminated Waste Sodium Chloride Brines*, U.S. Environmental Protection Agency, EPA-R2-73-200, USEPA, Washington, D. C. 1973.
15. S. J. Randtke and C. P. Jepsen, "Effects of Salts on Activated Carbon Adsorption of Fulvic Acids," *J. AWWA*, vol. 74, no. 2, 1982, p. 84.
16. W. J. Weber, Jr., T. C. Voice, and A. Jodellah, "Adsorption of Humic Substances:

The Effects of Heterogeneity and System Characteristics, *J. AWWA*, vol. 75, no. 12, 1983, p. 612.

17. V. L. Snoeyink, W. J. Weber, Jr., and H. B. Mark, Jr., "Adsorption of Phenol and Nitrophenol by Active Carbon," *Environmental Sci. Technol.*, vol. 3, 1969, p. 918.

18. S. J. Randtke and V. L. Snoeyink, "Evaluating GAC Adsorptive Capacity," *J. AWWA*, vol. 75, no. 8, 1983, p. 406.

19. J. S. Jain and V. L. Snoeyink, "Adsorption from Bisolute Systems on Active Carbon," *J. Water Pollution Control Fed.*, vol. 45, 1973, p. 2463.

20. D. R. Herzing, V. L. Snoeyink, and N. F. Wood, "Activated Carbon Adsorption of Odorous Compound 2-Methylisoborneol and Geosmin," *J. AWWA*, vol. 69, no. 4, 1977, p. 223.

21. S. Lalezary, M. Pirbazari, and M. J. McGuire, "Evaluating Activated Carbons for Removing Low Concentrations of Taste- and Odor-Producing Organics," *J. AWWA*, vol. 78, no. 11, 1986, p. 76.

22. W. E. Thacker, J. C. Crittenden, and V. L. Snoeyink, "Modelling of Adsorber Performance: Variable Influent Concentration and Comparison of Adsorbents," *J. Water Pollution Control Fed.*, vol. 56, 1984, p. 243.

23. J. C. Crittenden, B. W. C. Wong, W. E. Thacker, V. L. Snoeyink, and R. L. Hinrichs, "Mathematical Model of Sequential Loading in Fixed-Bed Adsorbers," *J. Water Pollution Control Fed.*, vol. 52, 1980, p. 2780.

24. J. A. V. Butler and C. Ockrent, "Studies in Electrocapillarity. III," *J. Phys. Chem.*, vol. 34, 1930, p. 2841.

25. C. Sheindorf, M. Rebhun, and M. Sheintuch, "A Freundlich-Type Multicomponent Isotherm," *J. Colloid Interface Sci.*, vol. 79, no. 1, 1981, p. 136.

26. C. J. Radke and J. M. Prausnitz, "Thermodynamics of Multi-Solute Adsorption from Dilute Liquid Solutions," *AIChE J.*, vol. 18, 1972, p. 761.

27. W. E. Thacker, V. L. Snoeyink, and J. C. Crittenden, "Desorption of Organic Compounds during Operation of GAC Adsorption Systems," *J. AWWA*, vol. 75, no. 3, 1983, p. 144.

28. J. M. Symons, *Removal of Organic Contaminants from Drinking Water Using Techniques other than Granular Activated Carbon Alone—A Progress Report*, Drinking Water Research Division, U.S. Environmental Protection Agency, Cincinnati, Ohio, May 1972.

29. A. W. Adamson, *Physical Chemistry of Surfaces*, 4th ed., Wiley, New York, 1982.

30. J. W. Hassler, *Activated Carbon*, Chemical Publishing, New York, 1974.

31. A. Yehaskel, *Activated Carbon Manufacture and Regeneration*, Noyes Data Corporation, Park Ridge, N. J., 1978.

32. "Refractories; Carbon and Graphite Products; Activated Carbon," in *Annual Book of ASTM Standards*, vol. 15.01, American Society for Testing Materials, Philadelphia, 1988.

33. S. L. Graese, V. L. Snoeyink, and R. G. Lee, "Granular Activated Carbon Filter-Adsorber Systems," *J. AWWA*, vol. 79, no. 12, 1987, p. 64.

34. M. R. Wiesner, J. J. Rook, and F. Fiessinger, "Optimizing the Placement of GAC Filtration Units," *J. AWWA*, vol. 79, no. 12, 1987, p. 39.

35. R. A. Dobbs and J. M. Cohen, *Carbon Adsorption Isotherms for Toxic Organics*, U.S. Environmental Protection Agency, EPA-600/8-80-023, 1980.

36. S. D. Faust and O. M. Aly, *Chemistry of Water Treatment*, Ann Arbor Science, Ann Arbor, Mich., 1983.

37. R. J. Miltner, T. F. Speth, D. D. Endicott, and J. M. Reinhold, *Final Internal Report on Carbon Use Rate Data*, U.S. Environmental Protection Agency, June 1987.

38. P. L. McCarty, D. Argo, and M. Reinhard, "Operational Experience with Activated Carbon at Water Factory 21," *J. AWWA*, vol. 71, no. 11, 1979, p. 683.

39. J. J. Westrick and J. M. Cohen, "Comparative Effects of Chemical Pretreatment on Carbon Adsorption," *J. Water Pollution Control Fed.*, vol. 48, 1976, p. 323.

40. P. V. Roberts and R. S. Summers, "Performance of Granular Activated Carbon for Total Organic Carbon Removal," *J. AWWA*, vol. 74, no. 2, 1982, p. 113.

41. B. Schultink, Provincial Waterworks of North Holland, personal communication, 1982.

42. H. Sontheimer and C. Hubele, "The Use of Ozone and Granular Activated Carbon in Drinking Water Treatment," in P. M. Huck and P. Toft (eds.), *Treatment of Drinking Water for Organic Contaminants,* Pergamon, New York, 1987.

43. H. Sontheimer, personal communication, 1983.

44. F. Fiessinger, personal communication, 1983.

45. S. L. Graese, V. L. Snoeyink, and R. G. Lee, *GAC Filter Adsorbers,* American Water Works Association Research Foundation, Denver, Colo., 1987.

46. B. H. Kornegay, "Control of Synthetic Organic Chemicals by Activated Carbon— Theory, Application, and Regeneration Alternatives," *Seminar on Control of Organic Chemicals in Drinking Water,* U.S. Environmental Protection Agency, 1979.

47. M. C. Lee, J. C. Crittenden, V. L. Snoeyink, and M. Ari, "Design of Carbon Beds to Remove Humic Substances," *J. Env. Eng. Div., ASCE,* vol. 109, no. 3, 1983, p. 631.

48. O. T. Love and R. G. Eilers, "Treatment of Drinking Water Containing Trichloroethylene and Related Industrial Solvents," *J. AWWA,* vol. 74, no. 8, 1982, p. 413.

49. A. E. Cover and L. J. Pieroni, *Evaluation of the Literature on the Use of Granular Activated Carbon for Tertiary Waste Treatment,* U.S. Department of the Interior, FWPCA, Report No. TWRC-11, 1969.

50. V. L. Snoeyink, *Control Strategy-Adsorption Techniques. Occurrence and Removal of Volatile Organic Chemicals from Drinking Water,* American Water Works Association Research Foundation, AWWARF/KIWA Report, 1983.

51. W. A. Chudyk and V. L. Snoeyink, "Bioregeneration of Activated Carbon Saturated with Phenol," *Environmental Sci. Technol.,* vol. 18, 1984, p. 1.

52. J. K. G. Silvey, "Studies on Microbiotic Cycles in Surface Waters," *J. AWWA,* vol. 56, no. 1, 1964, p. 60.

53. E. Namkung and B. E. Rittman, "Removal of Taste- and Odor-Causing Compounds by Biofilms Grown on Humic Substance," *J. AWWA,* vol. 79, no. 7, 1987, p. 109.

54. J. DeLaat, F. Bouanga, and M. Dore, "Influence of Microbiological Activity in Granular Activated Carbon Filters on the Removal of Organic Compounds," in H. A. M. de Kruif and H. J. Kool (eds.), *Organic Micropollutants in Drinking Water and Health,* Elsevier, New York, 1985.

55. G. Bablon, C. Ventresque, and R. Ben Aim, "Developing a Sand-GAC Filter to Achieve High-Rate Biological Filtration," *J. AWWA,* vol. 80, no. 12, 1988, p. 47.

56. B. E. Rittmann and P. M. Huck, "Biological Treatment of Public Water Supplies," *CRC Crit. Rev. Environmental Control,* vol. 19, 1989, p. 2.

57. P. L. McCarty, D. Argo, and M. Reinhard, "Operational Experiences with Activated Carbon Adsorbers at Water Factory 21," *J. AWWA,* vol. 71, no. 11, 1979, p. 683.

58. W. H. Glaze, K. L. Dickson, D. P. Wilcox, K. R. Johansson, E. Chang, and A. W. Busch, *Evaluation of Biological Activated Carbon for Removal of Trihalomethane Precursors,* Report to the U.S. Environmental Protection Agency, 1982.

59. M. A. van der Gaag, J. C. Kruithof, and L. M. Puijker, "The Influence of Water Treatment Processes on the Presence of Organic Surrogate and Mutagenic Compounds in Water," in H. A. M. de Kruif and H. J. Kool (eds.), *Organic Micropollutants in Drinking Water and Health,* Elsevier, New York, 1985.

60. J. M. Symons, A. A. Stevens, R. M. Clark, E. E. Geldreich, O. T. Love, Jr., and J. DeMarco, *Treatment Techniques for Controlling Trihalomethanes in Drinking Water,* U.S. Environmental Protection Agency, EPA-60012-81-156, 1981.

61. M. W. LeChevallier, T. S. Hassenauer, A. D. Camper, and G. A. McFeters, "Disinfection of Bacteria Attached to Granular Activated Carbon," *Appl. Environmental Microbiol.,* vol. 48, no. 5, 1984, p. 918.

62. D. van der Kooij, *Biological Processes in Carbon Filters. Activated Carbon in Drinking Water Technology,* KIWA/AWWARF Report, Denver, Colo., 1983.

63. P. Topalian, referenced by H. Sontheimer and C. Hubele, "The Use of Ozone and Granular Activated Carbon in Drinking Water Treatment," in P. M. Huck and P. Toft (eds.), *Treatment of Drinking Water for Organic Contaminants,* Pergamon, New York, 1987.

64. V. L. Snoeyink and M. T. Suidan, "Dechlorination by Activated Carbon and Other Reducing Agents," in J. D. Johnson (ed.), *Disinfection: Water and Wastewater,* Ann Arbor Science, Ann Arbor, Mich., 1975.

65. C. J. Murin and V. L. Snoeyink, "Competitive Adsorption of 2,4-Dichlorophenol and 2,4,6-Trichlorophenol in the Nanomolar to Micromolar Concentration Range," *Environmental Sci. Technol.,* vol. 13, 1979, p. 305.

66. J. J. McCreary and V. L. Snoeyink, "Characterization and Activated Carbon Adsorption of Several Humic Substances," *Water Research,* vol. 14, 1980, p. 151.

67. A. S. C. Chen, V. L. Snoeyink, and F. Fiessinger, "Activated Alumina Adsorption of Dissolved Organic Compounds before and after Adsorption," *Environmental Sci. Technol.,* vol. 21, 1987, p. 83.

68. E. A. Voudrias, R. A. Larson, and V. L. Snoeyink, "Effects of Activated Carbon on the Reactions of Free Chlorine with Phenols," *Environmental Sci. Technol.,* vol. 19, 1985, p. 441.

69. R. Prober, J. J. Pyeha, and W. E. Kidon, "Interaction of Activated Carbon with Dissolved Oxygen," *AIChE J.,* vol. 21, 1975, p. 1200.

70. W. A. Chudyk and V. L. Snoeyink, *The Removal of Low Levels of Phenol by Activated Carbon in the Presence of Biological Activity,* University of Illinois Water Resources Center Report, No. 154, 1981.

71. M. T. Suidan, V. L. Snoeyink, and R. A. Schmitz, "Performance Predictions for the Removal of Aqueous Free Chlorine by Packed Beds of Granular Activated Carbon. Water—1976: I. Physical Chemical Wastewater Treatment," *AIChE Symp. Ser.,* vol. 73, 1976, p. 18.

72. M. T. Suidan, V. L. Snoeyink, and R. A. Schmitz, "Reduction of Aqueous HOCl with Granular Activated Carbon," *J. Environmental Eng. Div., Amer. Soc. Civil Eng.,* vol. 103, 1977, p. 677.

73. M. T. Suidan, V. L. Snoeyink, and R. A. Schmitz, "Reduction of Aqueous Free Chlorine with Granular Activated Carbon—pH and Temperature Effects," *Environmental Sci. Technol.,* vol. 11, 1977, p. 785.

74. M. T. Suidan, V. L. Snoeyink, W. E. Thacker, and D. W. Dreher, "Influence of Pore Size Distribution on the HOCl-Activated Carbon Reaction," in A. J. Rubin (ed.), *Chemistry of Wastewater Technology,* Ann Arbor Science, Ann Arbor, Mich., 1978.

75. L. M. J. Dielmann, III, *The Reaction of Aqueous Hypochlorite, Chlorite and Hypochlorous Acid with Granular Activated Carbon,* M. S. Thesis, Environmental Engineering Program, University of Illinois, Urbana, Ill., 1981.

76. R. C. Bauer and V. L. Snoeyink, "Reactions of Chloramines with Active Carbon," *J. Water Pollution Control Fed.,* vol. 45, 1973, p. 2290.

77. B. R. Kim and V. L. Snoeyink "The Monochloramine-Activated Carbon Reaction: A Mathematical Model," in I. H. Suffet and M. McGuire (eds.), *Activated Carbon Adsorption of Organics from the Aqueous Phase,* Ann Arbor Science, Ann Arbor, Mich., 1980.

78. B. R. Kim and V. L. Snoeyink, "The Monochloramine-GAC Reaction in Adsorption Systems," *J. AWWA,* vol. 72, no. 8, 1980, p. 488.

79. J. D. Komorita and V. L. Snoeyink, "Monochloramine Removal by Activated Carbon," *J. AWWA,* vol. 77, no. 1, 1985, p. 62.

80. B. R. Kim, V. L. Snoeyink, and R. A. Schmitz, "Removal of Dichloramine and Ammonia by Granular Carbon," *J. Water Pollution Control Fed.,* vol. 50, 1978, p. 122.

81. B. R. Kim, *Analysis of Batch and Packed Bed Reactor Models for the Carbon-Chloramine Reactions,* Ph.D. Thesis, Environmental Engineering Program, University of Illinois, Ill., 1977.

82. B. R. Kim, R. A. Schmitz, V. L. Snoeyink, and G. W. Tauxe, "Analysis of Models for Dichloramine Removal," *Water Research,* vol. 12, 1978, p. 317.

83. A. S. C. Chen, R. A. Larson, and V. L. Snoeyink, "Reactions of Chlorine Dioxide with Hydrocarbons: Effects of Activated Carbon," *Environmental Sci. Technol.,* vol. 16, 1982, p. 268.

84. E. A. Voudrias, L. M. J. Dielmann, III, V. L. Snoeyink, R. A. Larson, J. J. McCreary, and A. S. C. Chen, "Reactions of Chlorite with Activated Carbon," *Water Research,* vol. 17, no. 9, 1983, p. 1107.

85. C. P. Huang, "Chemical Interactions between Inorganics and Activated Carbon," in Cheremisinoff and Ellerbusch (eds.), *Carbon Adsorption Handbook,* Ann Arbor Science, Ann Arbor, Mich., 1978.

86. J. Mallevialle and I. H. Suffet, (eds.), *Identification and Treatment of Tastes and Odors in Drinking Water,* American Water Works Association Research Foundation, Denver, Colo., 1987.

87. L. Gammie and G. Giesbrecht, "Full-Scale Operation of Granular Activated Carbon Contactors at Regina/Moose Jaw, Saskatchewan," *Proc. AWWA Annual Conf.,* Denver, Colo., June 1986.

88. G. G. Robeck, *Evaluation of Activated Carbon,* Water Supply Research Laboratory, National Environmental Research Center, Cincinnati, Ohio, March 1975.

89. O. T. Love, Jr., G. G. Robeck, J. M. Symons, and R. Buelow, "Experience with Activated Carbon in the U.S.A.," *Proc. Conf. on Activated Carbon in Water Treatment,* Univ. of Reading, Water Research Association, Medmenham, England, April 1973.

90. D. B. Ford, "The Use of Granular Carbon Filtration for Taste and Odor Control," *Proc. Conf. on Activated Carbon in Water Treatment,* Univ. of Reading, Water Research Association, Medmenham, England, April 1973.

91. J. Thomas, personal communication, 1986.

92. K. A. Dostal, R. C. Pierson, D. G. Hager, and G. G. Robeck, "Carbon Bed Design Criteria Study at Nitro, West Virginia," *J. AWWA,* vol. 57, no. 5, 1965, p. 663.

93. C. H. Lawrence, "California Plant Uses Diatomite and Carbon Filters," *Water Waste Eng.,* vol. 5, 1968, p. 46.

94. J. T. Monsitz and L. D. Ainesworth, "Detection and Control of Hydrogen Polysulfide in Water," *Public Works,* vol. 101, 1970, p. 113.

95. B. W. Lykins, Jr., R. M. Clark, and J. Q. Adams, "Granular Activated Carbon for Controlling THMs," *J. AWWA,* vol. 80, no. 5, 1988, p. 85.

96. A. Graveland, J. C. Kruithof, and P. A. N. M. Nuhn, "Production of Volatile Halogenated Compounds by Chlorination after Carbon Filtration," *ACS Meeting,* Atlanta, Ga., April 1981.

97. A. F. Hess, "GAC Treatment Designs and Costs for Controlling Volatile Organic Compounds in Ground Water," *Amer. Chem. Soc. Meeting,* Atlanta, Ga., 1981.

98. J. C. Crittenden, R. D. Cortright, B. Rick, S. R. Tang, and D. Perram, "Using GAC to Remove VOCs from Air Stripper Off-Gas," *J. AWWA,* vol. 80, no. 5, 1988, p. 73.

99. G. Zimmer, H. J. Brauch, and H. Sontheimer, "Activated Carbon Adsorption of Humic Substances in the Presence of Humic Substances," *Amer. Chem. Soci. Meeting,* Denver, Colo., April 1987.

100. G. G. Robeck, K. A. Dostal, J. M. Cohen, and J. F. Kreissl, "Effectiveness of Water Treatment Processes in Pesticide Removal," *J. AWWA,* vol. 57, no. 2, 1965, p. 181.

101. James M. Montgomery, Consulting Engineers, Inc., *Pre-Ozonation/Deep Bed Filtration Pilot Plant Study,* Report to Contra Costa Water District, Sept. 1986.

102. W. E. Koffskey and N. V. Brodtmann, *Organic Contaminant Removal in Lower Mississippi Drinking Water by Granular Activated Carbon,* Report to U.S. Environmental Protection Agency, 1981.

103. J. B. Andelman, "World Health Organization, European Standard for Organic Matter in Drinking Water," in V. L. Snoeyink (ed.), *Organic Matter in Water Supplies: Occurrence, Significance, and Control, Proc. 15th Wtr. Quality Conf.,* University of Illinois Bull., Urbana, Ill. June 1973, p. 122.

104. Y. Richard, personal communication, 1986.

105. J. Cleasby and K. Fan, "Predicting Fluidization and Expansion of Filter Media," *J. Env. Eng. Div., ASCE,* vol. 107, no. EE3, 1981, p. 455.

106. M. R. Rosene, R. T. Deithorn, J. R. Lutchko, and N. J. Wagner, "High Pressure Technique for Rapid Screening of Activated Carbon for Use in Potable Water," in M. J. McGuire and I. H. Suffet (eds.), *Activated Carbon Adsorption for the Aqueous Phase,* vol. I, Ann Arbor Science, Ann Arbor, Mich., 1980.

107. L. J. Bilello and B. A. Beaudet, "Evaluation of Activated Carbon by the Dynamic Minicolumn Adsorption Technique," in M. J. McGuire and I. H. Suffet (eds.), *Treatment of Water by Granular Activated Carbon,* Amer. Chem. Soc., 1983.

108. S. Liang and W. J. Weber, Jr., "Parameter Evaluation for Modelling Multicomponent Mass Transfer in Fixed-Bed Adsorbers," *Chem. Eng. Commun.,* vol. 35, 1985, p. 49.

109. J. C. Crittenden, J. K. Berrigan, and D. W. Hand, "Design of Rapid Small-Scale Adsorption Tests for a Constant Diffusivity, *J. Water Pollution Control Fed.,* vol. 58, 1986, p. 312.

110. J. C. Crittenden, J. K. Berrigan, D. W. Hand, and B. W. Lykins, Jr. "Design of Fixed-Bed Adsorption Tests for Nonconstant Diffusivities," *J. Environmental Eng. Div., ASCE,* vol. 113, no. 2, 1987, p. 243.

111. P. Kreft, A. Trussell, J. Lang, M. Kavanaugh, and R. Trussell, "Leaching of Organics from a PVC-Polyethylene-Plexiglas Pilot Plant," *J. AWWA,* vol. 73, no. 10, 1981, p. 558.

112. B. H. Kornegay, "Determining Granular Activated Carbon Process Design Parameters," *AWWA Ann. Conf.,* Kansas City, Mo., June 1987.

113. W. B. Arbuckle, "Premature Exhaustion of Activated Carbon Columns," in J. M. McGuire and I. H. Suffet (eds.), *Activated Carbon Adsorption,* vol. 2 Ann Arbor Science, Ann Arbor, Mich., 1980.

114. G. Bauldauf and G. Zimmer, "Removal of Volatile Chlorinated Hydrocarbons by Adsorption in Water Treatment," *Vom Wasser,* vol. 66, 1986, p. 21.

115. R. S. Summers, B. Haist, J. Kohler, J. Ritz, G. Zimmer, and H. Sontheimer, "The Influence of Background Organic Matter on GAC Adsorption," *J. AWWA,* vol. 81, no. 5, 1989, p. 66.

116. W. J. Weber, Jr., and E. H. Smith, "Simulation and Design Models for Adsorption Processes," *Environmental Sci. Technol.,* vol. 21, no. 11, 1987, p. 1040.

117. J. C. Crittenden and W. J. Weber, Jr., "A Predictive Model for Design of Fixed-Bed Adsorbers: Multicomponent Model Verification," *J. Environmental Eng. Div., ASCE,* vol. 104, 1978, p. 1175.

118. J. C. Crittenden and W. J. Weber, Jr., "A Predictive Model for Design of Fixed-Bed Adsorbers: Single Component Model Verification," *J. Environmental Eng. Div., ASCE,* vol. 104, 1978, p. 433.

119. D. W. Hand, J. C. Crittenden and W. E. Thacker, "Simplified Models for Design of Fixed-Bed Adsorption Systems," *J. Environmental Eng. Div., ASCE,* vol. 110, 1984, p. 440.

120. R. S. Summers and P. V. Roberts, "Simulation of DOC Removal in Activated Carbon Beds," *J. Environmental Eng. Div., ASCE,* vol. 110, no. 1, 1984, p. 73.

121. W. J. Weber, Jr., and M. Pirbazari, "Adsorption of Toxic and Carcinogenic Compounds from Water," *J. AWWA,* vol. 74, no. 4, 1982, p. 203.

122. R. L. Jones and J. M. Symons, *Procedure for Comparing Actual Adsorption Behavior in Continuous Flow Columns with that Expected from Using Batch Equilibrium Isotherm Data,* Research Report, Univ. of Houston, Department of Civil and Environmental Engineering Houston, Texas, June 1989.

123. J. C. Kruithof, J. A. P. Meijers, H. G. M. M. Smeenk, B. J. van der Veer, and J. Roelands, "Selection of Brands of Activated Carbon for Adsorptive Properties," in *Activated Carbon in Drinking Water Technology,* American Water Works Association Research Foundation, Denver, Colo. 1983.

124. I. M. Najm, V. L. Snoeyink, M. T. Suidan, C. H. Lee, and Y. Richard, "Adsorption on Powdered Activated Carbon: Effect of Particle Size and Background Organics," *J. AWWA,* vol. 82, no. 1, 1990, p. 65.

125. American Water Works Association, *AWWA Standard for Powdered Activated Carbon,* B600-78, Denver, Colo., 1978.

126. H. Sontheimer, "The Use of Powdered Activated Carbon," *Translation of Reports of Special Problems of Water Technology,* vol. 9, *Adsorption,* U.S. Environmental Protection Agency, Report EPA-600/9-76-030, December 1976.

127. American Water Works Association. *Water Quality and Treatment,* 3d ed., McGraw-Hill, New York, 1971.

128. J. A. P. Meijers and R. C. van der Leer, "The Use of Powdered Activated Carbon in Conventional and New Techniques," *Activated Carbon in Drinking Water Technology,* American Water Works Association Research Foundation, Denver, Colo. 1983.

129. R. B. Gauntlett and R. F. Packham, "The Use of Activated Carbon in Water Treatment," *Prof. Conf. on Activated Carbon in Water Treatment*, Univ. of Reading, Water Research Association, Medmenham, England, April 1973.

130. M. J. McGuire and I. H. Suffet, "Aqueous Chlorine/Activated Carbon Interactions," *J. Environmental Eng. Div., ASCE*, vol. 110, no. 3, 1984, p. 629.

131. S. Lalezary-Craig, M. Pirbazari, M. S. Dale, T. S. Tanaka, and M. J. McGuire, "Optimizing the Removal of Geosmin and 2-Methylisoborneol by Powdered Activated Carbon," *J. AWWA*, vol. 80, no. 3, 1988, p. 73.

132. R. C. Hoehn, S. R. Lavinder, C. Hamann, Jr., E. R. Hoffman, J. McElroy, and E. G. Snyder, THM-Precursor Control with Powdered Activated Carbon in a Pulsed-Bed, Solids Contact Clarifier, *AWWA Ann. Conf.*, Kansas City, Mo., June 1987.

133. G. Lettinga, W. A Beverloo, and W. C. van Lier, "The Use of Flocculated Powdered Activated Carbon in Water Treatment," *Prog. Water Tech.*, vol. 10, 1978, p. 537.

134. K. Haberer and S. Normann, "Untersuchungen zu einer Neuartigen Pulverkohle-Filtrationstechnik fur die Wasseraufbereitung," *Vom Wasser*, vol. 49, 1979, p. 331.

135. K. Haberer and S. Normann, "Entwicklung eines Kurztalt-Filtrations Verfahrens zum Einsatz vor Pulverkohle in der Wasseraufbereitung," *Gas und Wasserfach, Wasser/Abwasser*, vol. 118, 1980, p. 393.

136. R. C. Hoehn, P. E. Johnson, B. H. Kornegay, and K. P. Rogenmuser, "A Pilot-Scale Evaluation of the Roberts-Haberer Process for Removing Trihalomethane Precursors from Surface Water with Activated Carbon," *AWWA Ann. Conf.*, Dallas, Texas, June 1984.

137. *Water Treatment Handbook*, 5th ed., Degremont, Paris, 1979.

138. U.S. Environmental Protection Agency, "The Analysis of Trihalomethanes in Finished Waters by the Purge and Trap Method," Environmental Monitoring and Support Laboratory, Cincinnati, Ohio, Sept. 1977.

139. D. R. Seeger, E. J. Slocum, and A. A. Stevens, "GC/MS Analysis of Purgeable Contaminants in Source and Finished Drinking Water," *Proc. 26th Ann. Conf. on Mass Spectrometry and Applied Topics*, St. Louis, Mo., May 1978.

140. J. E. Singley, B. A. Beaudet, and A. L. Ervin, "Use of Powdered Activated Carbon for Removal of Specific Organic Compounds," *Sem. Proc.: Controlling Organics in Drinking Water, AWWA Ann. Conf.*, San Francisco, Calif., 1979.

141. J. E. Singley, *Minimizing Trihalomethane Formation in a Softening Plant*, U.S. Environmental Protection Agency, Water Supply Res. Div., Municipal Environ. Res. Lab., Cincinnati, Ohio, 1977.

142. O. M. Aly and S. D. Faust, "Removal of 2,4-Dichlorophenoxyacetic Acid Derivatives from Natural Waters," *J. AWWA*, vol. 57, no. 2, 1965, p. 221.

143. R. H. Burttschell, A. A. Rosen, F. M. Middleton, and M. B. Ettinger, "Chlorine Derivatives of Phenol Causing Taste and Odor," *J. AWWA*, vol. 51, no. 2, 1959, p. 205.

144. AWWA Committee Report, "Organics Removal by Coagulation: A Review and Research Needs," *J. AWWA*, vol. 71, no. 10, 1979, p. 588.

145. M. A. El-Dib, A. S. Moursy, and M. I. Badawy, "Role of Adsorbents in the Removal of Soluble Aromatic Hydrocarbons from Drinking Water," *Water Research*, vol. 12, 1978, p. 1131.

146. A. M. El-Dib and O. A. Aly, "Removal of Phenylamide Pesticides from Drinking Waters. II. Adsorption on Powdered Carbon," *Water Research*, vol. 11, 1977, p. 617.

147. O. T. Love, J. K. Carswell, R. J. Miltner, and J. M. Symons, *Treatment for the Removal of Trihalomethanes in Drinking Water*. U.S. Environmental Protection Agency, Drinking Water Res. Div., Cincinnati, Ohio, 1976.

148. A. L. Lange and E. Kawczynski, "Trihalomethane Studies, Contra Costa County Water District Experience," *Proc. Water Treatment Forum VII, California-Nevada Section AWWA*, Palo Alto, Calif., April 1978.

149. B. M. van Vliet, "Regeneration Principles," *Proc. Symp. on Design and Operation of Plants for the Recovery of Gold by Activated Carbon*, South African Inst. of Mining and Metallurgy, Johannesburg, October 1985.

150. W. A. Schuliger and J. M. MacCrum, "Granular Activated Carbon Reactivation System Design and Operating Conditions," *AIChE Meeting*, Detroit, June 1973.

151. H. Juntgen, "Phenomena of Activated Carbon Regeneration," *Translation of Re-*

ports on *Special Problems of Water Technology,* vol. 9, *Adsorption,* U.S. Environmental Protection Agency, Report EPA-600/9-76-030, 1976.

152. M. Suzuki, D. M. Misic, O. Koyama, and K. Kawazoe, "Study of Thermal Regeneration of Spent Activated Carbons: Thermogravimetric Measurement of Various Single Component Organics Loaded on Activated Carbons," *Chem. Eng. Sci.,* vol. 33, 1978, p. 271.

153. P. R. Tipnis and P. Harriott, "Thermal Regeneration of Activated Carbons," *Chem. Eng. Commun.,* vol. 46, 1986, p.11.

154. A. J. Juhola, "Manufacture, Pore Structure and Application of Activated Carbons. I," *Kemia-Kemi,* No. 11, 1977.

155. A. W. Loven, "Perspectives on Carbon Regeneration," *Chem. Eng. Progr,* vol. 69, 1973, p. 56.

156. A. J. Juhola, *Optimization of the Regeneration Procedures for Granular Activated Carbon,* U.S. Environmental Protection Agency, Report No. 17020 DAO, 1970.

157. B. M. van Vliet and L. Venter, "Infrared Thermal Regeneration of Spent Activated Carbon from Water Reclamation," *Water Sci. Technol.,* vol. 17, 1984, p. 1029.

158. J. J. McCreary and V. L. Snoeyink, "Granular Activated Carbon in Water Treatment," *J. AWWA,* vol. 69, no. 8, 1977, p. 437.

159. M. Smisek and S. Cerny, *Active Carbon,* Elsevier, New York, 1970.

160. T. Umehara, P. Harriott, and J. M. Smith, "Regeneration of Activated Carbon. Part 2: Gasification Kinetics with Steam," *AIChE,* vol. 29, no. 5, 1983, p. 737.

161. W. Poggenburg, B. Fokken, B. Strack, and H. Sontheimer, "Untersuchungen zur Optimierung der Aktivkohleanwendung ber der Trinkwasseraufbereitung am Rhein Unter besonderer Berucksichtigung der Regeneration nach thermischen Verfahren," Helft 12, Veroffentlichugen des Bereichs und Lehrstuhls fur Wasserchemie, Universitat Karlsruhe, Federal Republic of Germany, 1979.

162. O. T. Love, Jr., "Experience with Reactivation of Granular Activated Carbon," *AWWA Sem. Proc., Controlling Organics in Drinking Water, AWWA Ann. Conf.,* San Francisco, Calif., June 1979.

163. W. G. Schuliger, G. N. Riley, and N. J. Wagner, "Thermal Reactivation of Granular Activated Carbon: A Proven Technology." *AWWA Ann. Conf.,* Kansas City, Mo., June 1987.

164. J. DeMarco, R. Miller, D. Davis, and C. Cole, "Experiences in Operating a Full-Scale Granular Activated Carbon System with On-Site Reactivation," in M. J. McGuire and I. H. Suffet (eds.), *Treatment of Water by Granular Activated Carbon,* American Chemical Society, Washington, D.C., 1983.

165. W. E. Koffskey and B. W. Lykins, Jr., "Experiences with Granular Activated Carbon Filtration and On-Site Reactivation at Jefferson Parish, Louisiana," *AWWA Ann. Conf.,* Kansas City, Mo., June 1987.

166. R. A. Hutchins, "Economic Factors in Granular Carbon Thermal Regeneration," *Chem. Eng. Progr.,* vol. 69, 1973, p. 48.

167. A. J. Juhola, "Manufacture, Pore Structure and Application of Activated Carbons. II," *Kemia-Kemi,* No. 12, 1977.

168. B. W. Lykins, Jr., R. M. Clark, and D. H. Cleverly, "Polychlorinated Dioxin and Furan Discharge during Carbon Reactivation," *J. Environmental Eng. Div. ASCE,* vol. 114, no. 2, 1988, p. 300.

169. B. R. Kim, V. L. Snoeyink, and F. M. Saunders, "Adsorption of Organic Compounds by Synthetic Resins," *J. Water Pollution Control Fed.,* vol. 48, 1976, p. 120.

170. W. Kolle, *Use of Macroporous Ion Exchangers for Drinking Water Purification,* Translation of Reports on Special Problems of Water Technology, (H. Sontheimer (ed.), Environmental Protection Agency, Report No. 600/9-76-030, Cincinnati, Ohio, 1976.

171. H. Brattebo, H. Odegaard, and O. Halle, "Ion Exchange for the Removal of Humic Substances in Water Treatment," *Water Research,* vol. 21, no. 9, 1987, p. 1045.

172. R. B. Gauntlett, *A Comparison between Ion-Exchange Resins and Activated Carbon for the Removal of Organics from Water,* Water Research Center Technical Report TR 10, Medmenham, England, 1975.

173. P. L. K. Fu and J. Symons, "Removal of Aquatic Organics by Anion Exchange Resins," *AWWA Ann. Conf.,* Orlando, Fla., June 1988.

174. G. A. Junk, J. J. Richard, H. J. Svec, and J. S. Fritz, "Simplified Resin Sorption for Measuring Selected Contaminants," *J. AWWA,* vol. 68, no. 4, 1976, p. 218.

175. C. Kennedy, "Treatment of Effluent from Manufacture of Chlorinated Pesticide with a Synthetic Polymeric Adsorbent, Amberlite XAD-4," *Environmental Sci. Technol.,* vol. 7, 1973, p. 138.

176. J. S. Kumagai and W. J. Kaufman, *Removal of Organic Contaminants, Phenol Sorption by Activated Carbon and Selected Macroporous Resins,* San. Engineering Research Laboratory, Report No. 68-8, Univ. of California, Berkeley, Calif., July 1968.

177. D. A. Jayes and I. M. Abrams, "A New Method of Color Removal," *New Engl. Water Works Assoc.,* vol. 82, 1968, p. 15.

178. P. H. Boening, D. D. Beckmann, and V. L. Snoeyink, "Activated Carbon vs. Resin Adsorption of Humic Substances," *J. AWWA,* vol. 72, no. 1, 1980, p. 54.

179. I. N. Najm, V. L. Snoeyink, M. T. Suidan, and Y. Richard, "Powdered Activated Carbon in Floc Blanket Reactors," *AWWA Ann. Conf.,* Los Angeles, Calif., June 1989.

180. W. A. Chudyk, V. L. Snoeyink, D. D. Beckman, and T. J. Temperly, "Activated Carbon vs. Resin Adsorption of 2-Methylisoborneol and Chloroform," *J. AWWA,* vol. 71, no. 9, 1979, p. 529.

# Disinfection

## Charles N. Haas, Ph.D.

*Professor*
*Pritzker Department of Environmental Engineering*
*Illinois Institute of Technology*
*Chicago, Illinois*

Disinfection is a process designed for the deliberate reduction of a number of pathogenic microorganisms. While other water treatment processes, such as filtration or coagulation-flocculation-sedimentation, may achieve pathogen reduction, this is not generally their primary goal. A variety of chemical or physical agents may be used to carry out disinfection. The concept of disinfection preceded the recognition of bacteria as the causative agent of disease. Averill,[1] for example, proposed chlorine disinfection of human wastes as a prophylaxis against epidemics in 1832. Chemical addition during water treatment for disinfection became accepted only after litigation on its efficacy.[2] The prophylactic benefits of water disinfection soon became apparent, particularly with respect to the reduction of typhoid and cholera.

While classic waterborne diseases have been controlled, a residual of waterborne disease remains, transmitted by newly recognized agents such as the viruses,[3,4] certain bacteria (*Campylobacter,*[5] *Yersinia,* or *Mycobacteria,*[6] for example), and protozoans (*Giardia,*[7] *Cryptosporidium*[8]). Occasional outbreaks of drinking-water-associated hepatitis have also occurred.[9] In addition, new viral agents continually are being found to be capable of waterborne transmission.

In 1989 the majority of community water supplies in the United States practice disinfection, predominantly by the use of chlorine or chlorine compounds. Hoehn et al.[10] summarized the state of practice in the late 1970s (Tables 14.1 and 14.2), at least with regard to chlorination. With increasing concern for removal of some of the more resistant pathogens, such as *Giardia,* and for the formation of chlorina-

TABLE 14.1    Distribution of Disinfectant Contact Times between Points of Application and First Customer's Tap (227 respondents)*

| Contact time, min | Percent with contact time equal to or less than that indicated |
|:---:|:---:|
| 2 | 6 |
| 10 | 24 |
| 20 | 34 |
| 30 | 43 |
| 60 | 57 |
| 90 | 58 |
| 120 | 65 |
| 180 | 69 |
| 250 | 77 |
| 420 | 85 |
| 600 | 91 |
| 780 | 95 |
| 1440 | 99 |

*Source: From Ref. 10.

TABLE 14.2    Distribution of Chlorine Residual Concentration and Respondents Reporting Free, Combined, and Total Residuals

| Chlorine concentration, mg/L | Percent of respondents reporting a residual less than or equal to the concentration indicated | | |
|:---:|:---:|:---:|:---:|
| | Free | Combined | Total |
| 0 | 4 | 3 | — |
| 0.10 | 8 | 22 | 1 |
| 0.20 | 15 | 42 | 4 |
| 0.40 | 26 | 58 | 11 |
| 0.60 | 39 | — | 20 |
| 0.80 | 51 | 73 | 27 |
| 1.00 | 66 | 81 | 43 |
| 1.30 | 71 | 86 | 54 |
| 1.70 | — | 93 | — |
| 2.00 | 91 | 98 | 81 |
| 2.50 | 95 | — | 91 |
| 2.75 | 98 | — | — |
| 3.00 | — | — | 97 |
| 3.50 | — | — | 99 |
| 4.00 | — | 100 | — |
| 4.75 | 100 | — | — |
| 5.00 | — | — | 100 |

Source: From Ref. 10.

tion by-products, other disinfectants are gaining popularity. This chapter covers the use of chlorine, as well as the major alternative agents, for the purpose of disinfection.

## History of Disinfection

### Chlorine

Chlorine gas was first prepared by Scheele in 1774 but was not regarded as a chemical element until 1808.[11] Early uses of chlorine include the use of Javelle water (chlorine gas dissolved in an alkaline potassium solution) in France for waste treatment in 1825[12] and its use as a prophylactic agent during the European cholera epidemic of 1831.[11]

Late in the nineteenth century, electrolytic generation of chlorine and hypochlorites became sufficiently competitive with the chemical oxidation synthesis routes to spur the use of chlorine compounds for disinfection. The growth of chlorine disinfection for both water and wastewater applications occurred simultaneously. In 1909, production of liquid chlorine commenced at Niagara Falls, N.Y. In 1912, Wallace and Tiernan developed and installed a direct-feed gas chlorinator at the Boonton, N.J., waterworks, and went on to develop solution feed chlorinators.[13]

The acceptance of chlorine, or chlorine compounds, as disinfectants was extremely rapid. Within two years of the first full-scale U.S. applications of chlorine, at Bubbly Creek (Chicago) and the Jersey City Water Company in 1908, chlorine was introduced as a disinfectant at New York City (Croton), Montreal, Milwaukee, Cleveland, Nashville, Baltimore, and Cincinnati, as well as at other, smaller treatment plants. Frequently, dramatic reductions in typhoid accompanied the introduction of this process.[14] By 1918, over 1000 cities, treating more than 3 bil gal/day of water, were employing chlorine as a disinfectant.[2]

Chloramination, the addition of both chlorine and ammonia, either sequentially or simultaneously, was first employed in Ottawa, Canada, and Denver, Colo., in 1917. Both of these early applications employed prereaction of the two chemicals prior to their addition to the full flow of water. Somewhat later, preammoniation (the addition of ammonia prior to chlorine) was developed. In both cases, the process was advocated for its ability to prolong the stability of residual disinfectant during distribution and for its diminished propensity to produce chlorophenolic taste and odor substances. Shortages of ammonia during World War II reduced the popularity of the chloramination process. Recent concerns for organic by-products of the chlorination, however, have increased the popularity of chloramination.[15]

### Chlorine dioxide

Chlorine dioxide was first produced from the reaction of potassium chlorate and hydrochloric acid by Davy in 1811.[16] Not until the industrial-scale preparation of sodium chlorite, from which chlorine dioxide may more readily be generated, however, did its widespread use occur.[17]

Chlorine dioxide has been used widely as a bleaching agent in pulp and paper manufacture.[17] Despite early investigations on the use of chlorine dioxide as an oxidant and disinfectant,[18] however, its ascendancy in both water and wastewater treatment has been slow. As recently as 1971,[19] it was stated that "...$ClO_2$ has never been used extensively for water disinfection."

By 1977, 84 potable water treatment plants in the United States were identified as using chlorine dioxide treatment, although only one of these relied upon it as a primary disinfectant.[16] In Europe, chlorine dioxide is used as either an oxidant or disinfectant in almost 500 potable water treatment plants.[16]

### Ozone

Ozone was discovered in 1783 by Van Marum and named by Schonbein in 1840. In 1857, the first electric discharge ozone generation device was constructed by Siemens, with the first commercial application of such devices occurring in 1893.[20]

Ozone was first applied as a potable water disinfectant in 1893 at Oudshoorn, Netherlands. In 1906, Nice, France, installed ozone as a treatment process, and this plant represents the oldest ozonation installation in continual operation.[21] In the United States, ozone was first employed for taste and odor control at the New York City Jerome Park Reservoir in 1906.[22] In 1987, five water treatment facilities in the United States were using ozone oxidation primarily for taste and odor or trihalomethane precursor removal.[23]

### Ultraviolet radiation

The biocidal effects of ultraviolet radiation (uv) have been known since it was established that short-wavelength uv was responsible for microbial decay often associated with sunlight.[24] By the early 1940s, design guidelines for uv disinfection were proposed.[25] Ultraviolet has been accepted for treating potable water on passenger ships.[26] Historically, however, it has met with little enthusiasm in public water supply applications because of the lack of a residual following application. In wastewater treatment, in contrast, about 100 plants in the United States are either using, currently designing, or constructing uv disinfection facilities.[27]

### Other disinfecting agents

A variety of other agents may be used to affect inactivation of microorganisms. These include heat, extremes in pH, metals (silver, copper), surfactants, and permanganate. Heat is useful only in emergencies, as in "boil water" orders, and is uneconomical. Alkaline pHs (during high lime softening) may provide some microbial inactivation, but are not usually sufficient as a sole disinfectant. Potassium permanganate has been reported to achieve some disinfecting effects; however, the magnitudes have not been well characterized. In this chapter, therefore, primary consideration will be given to chlorine compounds, ozone, chlorine dioxide, and uv.

## Disinfectants and Theory of Disinfection

### Basic chemistry

**Chlorine and chlorine compounds.**    Chlorine may be used as a disinfectant in the form of compressed gas under pressure or in water solutions, solutions of sodium hypochlorite, or solid calcium hypochlorite. The three forms are chemically equivalent because of the rapid equilibrium that exists between dissolved molecular gas and the dissociation products of hypochlorite compounds.

Elemental chlorine ($Cl_2$) is a dense gas that, when subject to pressures in excess of its vapor pressure, condenses into a liquid with the release of heat and with a reduction in specific volume of approximately 450-fold. Hence, commercial shipments of chlorine are made in pressurized tanks to reduce shipment volume. When chlorine is to be dispensed as a gas, supplying thermal energy to vaporize the compressed liquid chlorine is necessary.

The relative amount of chlorine present in chlorine, or hypochlorites, is expressed in terms of "available chlorine." The concentration of hypochlorite (or any other oxidizing disinfectant) may be expressed as available chlorine by determining the electrochemical equivalent amount of $Cl_2$ to that compound. By Eq.(14.1), 1 mole of elemental chlorine is capable of reacting with two electrons to form inert chloride:

$$Cl_2 + 2e^- \rightleftharpoons 2Cl^-$$

(14.1)

From reaction (14.2), 1 mole of hypochlorite ($OCl^-$) may react with two electrons to form chloride:

$$OCl^- + 2e^- + 2H^+ \rightleftharpoons Cl^- + H_2O \tag{14.2}$$

Hence, 1 mole of hypochlorite is electrochemically equivalent to 1 mole of elemental chlorine, and may be said to contain 70.91 g of available chlorine (identical to the molecular weight of $Cl_2$).

Calcium hypochlorite [$Ca(OCl)_2$] and sodium hypochlorite ($NaOCl$) contain 2 moles and 1 mole of hypochlorite per mole of chemical, respectively, and, as a result, 141.8 g and 70.91 g available chlorine per mole, respectively. The molecular weights of $Ca(OCl)_2$ and $NaOCl$ are, respectively, 143 and 74.5, so pure preparations of the two compounds contain 99.2 and 95.8 weight percent available chlorine; hence, they are effective means of supplying chlorine for disinfection purposes.

Commercially, calcium hypochlorite is available as a dry solid. In this form, it is subject to a loss in strength of approximately 0.013 percent per day.[28] Recently, calcium hypochlorite has become available in a tablet form for use in automatic feed equipment at low-flow treatment plants.

Sodium hypochlorite is available in 1 to 16 weight percent solutions. Higher concentration solutions are not practical because chemical stability rapidly diminishes with increasing strength. At ambient temperatures, the half-life of sodium hypochlorite solutions varies between 60 and 1700 days, respectively, for solutions of 18 and 3 percent available chlorine.[28,29]

When a chlorine-containing compound is added to a water containing insignificant quantities of kjeldahl nitrogen, organic material, and other chlorine-demanding substances, a rapid equilibrium is established among the various chemical species in solution. The term *free available chlorine* is used to refer to the sum of the concentrations of molecular chlorine ($Cl_2$), hypochlorous acid ($HOCl$), and hypochlorite ion ($OCl^-$), each expressed as *available chlorine*.

The dissolution of gaseous chlorine to form dissolved molecular chlorine is expressible as a phase equilibrium, and may be described by Henry's law:

$$Cl_2(g) = Cl_2(aq) : H(\text{mol/L-atm}) = \frac{[Cl_2(aq)]}{P_{Cl_2}} \tag{14.3}$$

In Eq. (14.3), quantities within brackets represent molar concentrations, $P_{Cl_2}$ is the gas phase partial pressure of chlorine in atmospheres, and $H$ is Henry's law constant, estimated from the following equation[30]:

$$H = 4.805 \times 10^{-6} \exp\left(\frac{2818.48}{T}\right) \quad \text{mol/(L)(atm)} \qquad (14.4)$$

Dissolved aqueous chlorine reacts with water to form hypochlorous acid, chloride ions, and protons as indicated by Eq. (14.5).

$$Cl_2(aq) + H_2O \rightleftharpoons H^+ + HOCl + Cl^- \qquad (14.5)$$

$$K_H = \frac{[H^+][HOCl][Cl^-]}{[Cl_2(aq)]}$$

This reaction typically reaches completion in 100 ms[31,32] and involves elementary reactions between dissolved molecular chlorine and hydroxyl ions. The extent of chlorine hydrolysis, or disproportionation (because the valence of chlorine changes from 0 on the left to $+1$ and $-1$ on the right), as described by Eq. (14.5), decreases with decreasing pH and increasing salinity; hence, the solubility of gaseous chlorine may be increased by the addition of alkali or by the use of fresh water rather than brackish.

Hypochlorous acid is a weak acid and may dissociate according to reaction (14.6):

$$HOCl \rightleftharpoons OCl^- + H^+ \qquad (14.6)$$

$$K_a = \frac{[OCl^-][H^+]}{[HOCl]}$$

The $pK_a$ of hypochlorous acid at room temperature is approximately 7.6.[33] Morris[33] has provided a correlating equation for $K_a$ as a function of temperature:

$$\ln K_a = 23.184 - 0.0583T - 6908/T \qquad (14.7)$$

In Eq. (14.7), $T$ is specified in kelvins (K = °C + 273). Figure 14.1 illustrates the effect of pH on the distribution of free chlorine between $OCl^-$ and $HOCl$.

One practical consequence of the reactions described by Eqs. (14.3) to (14.7) is that the chlorine vapor pressure over a solution depends on solution pH, decreasing as pH increases (because of the increased formation of nonvolatile hypochlorite acid). Therefore, the addition of an alkaline material such as lime or sodium bicarbonate will reduce the volatility of chlorine from accidental spills or leaks and thus minimize danger to exposed personnel.

**Example Problem 1.**   The solution produced by a gas chlorinator contains 3500 mg/L available chlorine at a pH of 3. What is the equilibrium vapor pressure of

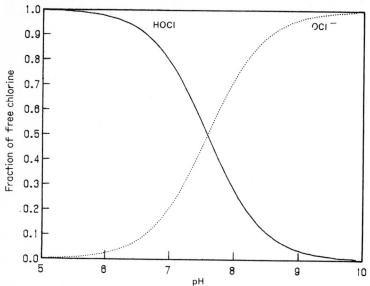

**Figure 14.1**    Effect of pH on distribution of free chlorine forms at 20°C.

this solution at 20°C? (The value of the hydrolysis constant $K_H$ is $4.5 \times 10^4$ at this temperature.)

**solution**    The pH is sufficiently low so that the dissociation of hypochlorous acid to form hypochlorite can be ignored. Therefore, a balance over chlorine species yields

$$[Cl_2] + [HOCl] = (3500 \times 10^{-3})/70$$

The factor of 70 reflects the fact that 1 mole of either dissolved chlorine or hypochlorous acid contains 70 g of available chlorine.

The hydrolysis equilibrium constant can be used to develop an additional equation:

$$4.5 \times 10^4 = \frac{[H^+][Cl^-][HOCl]}{[Cl_2]}$$

or, because the pH is given,

$$4.5 \times 10^7 = \frac{[Cl^-][HOCl]}{[Cl_2]}$$

Because chlorine gas was used to generate the dissolved free chlorine, the disproportionation reaction requires that for each mole of HOCl produced 1 mole of $Cl^-$ must have been produced. If the initial concentration of chloride (in the feedwater to the chlorinator) was minimal, then a third equation results:

$$[Cl^-] = [HOCl]$$

These three equations can be manipulated to produce a quadratic equation in the unknown $[Cl_2]^{1/2}$:

$$[Cl_2] + 6708[Cl_2]^{1/2} - 0.05 = 0$$

The single positive root is the only physically meaningful one; hence

$$[Cl_2]^{1/2} = 7.45 \times 10^{-6}$$

or

$$[Cl_2] = 5.55 \times 10^{-11}$$

Henry's law constant can be computed as 0.072 $M/(L)(atm)$, and therefore the partial pressure of chlorine gas is found:

$$P_{Cl_2} = 5.55 \times 10^{-11}/0.072 = 7.7 \times 10^{-10} \text{ atm} = 0.77 \text{ ppb}$$

The 1-h occupational no-observed-effect level is reported as 4 ppm. Therefore, this level is of no apparent health concern to the workers.

**Chlorine dioxide.** Chlorine dioxide ($ClO_2$) is a neutral compound of chlorine in the +IV oxidation state. It has a boiling point of 11°C at atmospheric pressure. The liquid is denser than water, and the gas is denser than air.[34]

Chemically, chlorine dioxide is a stable free radical that, at high concentrations, reacts violently with reducing agents. It is explosive, with the lower explosive limit in air variously reported as 10 percent[30,35] or 39 percent.[34] As a result, virtually all applications of chlorine dioxide require the synthesis of the gaseous compound in a dilute stream (either gaseous or liquid) on location as needed.

The solubility of gaseous chlorine dioxide in water may be described by Henry's law, and a fit of the available solubility data[36] results in the following relationship for Henry's law constant (in units of 1/atm):

$$\ln H = \text{mole fraction dissolved } ClO_2(aq)/P_{ClO_2} \qquad (14.8)$$

$$= 58.84621 + 47.9133/T - 11.0593 \ln T$$

Under alkaline conditions, the following disproportionation into chlorite ($ClO_2^-$) and chlorate ($ClO_3^-$) occurs[37]:

$$2ClO_2 + 2OH^- \rightleftharpoons H_2O + ClO_3^- + ClO_2^- \qquad (14.9)$$

In the absence of carbonate, which is catalytic, the reaction is governed by parallel first- and second-order kinetics.[37,38] The half-life of aqueous chlorine dioxide solutions decreases substantially with increasing concentration and with pH values above 9. Even at neutral pH values, however, in the absence of carbonate at room temperature, the half-life of chlorine dioxide solutions of 0.01, 0.001, and 0.0001 mol/L is 0.5, 4, and 14 h, respectively. Hence, the storage of stock solutions of chlorine dioxide for even a few hours is impractical.

The simple disproportionation reaction to chlorate and chlorite is insufficient to explain the decay of chlorine dioxide in water free of extraneous reductants. Equation (14.9) predicts that the molar ratio of chlorate to chlorite formed should be 1:1. Medir and Giralt,[39] however, found that the molar ratio of chlorate:chlorite:chloride:oxygen produced was 5:3:1:0.75, and that the addition of chloride enhanced the rate of decomposition and resulted in the predicted 1:1 molar ratio of chlorite:chlorate. Thus, the oxidation of chloride by chlorate, and the possible formation of intermediate free chlorine, may be of significance in the decay of chlorine dioxide in demand-free systems.[37]

The concentration of chlorine dioxide in solution is generally expressed in terms of "g/L as chlorine" by multiplying the molarity of chlorine dioxide by the number of electrons transferred per mole of chlorine dioxide reacted and then multiplying this by 35.5 g $Cl_2$ per electron mole. Conventionally, the five-electron reduction [Eq. (14.10)] is used to carry out this conversion. (Note, however, that the typical reaction of chlorine dioxide in water, being reduced to chlorite, is a one-electron reduction.)

$$ClO_2 + 5e^- + 4H^+ \rightleftharpoons Cl^- + 2H_2O \qquad (14.10)$$

Hence, 1 mole of chlorine dioxide contains 67.5 g of mass, and is equivalent to 177.5 ( = 5 × 35.5) g $Cl_2$. Therefore, 1 g of chlorine dioxide contains 2.63 g "as chlorine." In examination of any studies on chlorine dioxide, due care with regard to units of expression of disinfectant concentration is warranted.

**Ozone.**  Ozone is a colorless gas, produced from the action of electric fields on oxygen. It is highly unstable in the gas phase; in clean vessels at room temperature the half-life in air is 20 to 100 h.[40]

The solubility of ozone in water can be described by a temperature and pH-dependent Henry's law constant. The following provisional relationship ($H$ in 1/atm) has been suggested[41]:

$$H = 3.84 \times 10^7 \, [OH^-] \exp\left(\frac{-2428}{T}\right) \qquad (14.11)$$

Practical ozone generation systems have maximum gaseous ozone concentrations of about 50 g/m³; thus, the maximum practical solubility of ozone in water is about 40 mg/L.[27] Upon dissolution in water, ozone can react with water itself, with hydroxyl ions, or with dissolved chemical constituents, as well as serving as a disinfecting agent. Details of these reactions will be discussed later and in Chap. 12.

### Disinfectant demand reactions

**Chlorine.** In the presence of certain dissolved constituents in water, each of the disinfectants may react and transform to less effective chemical forms. In the case of chlorine, these principally involve reactions with ammonia and amino nitrogen compounds. In the presence of ammonium ion, free chlorine reacts in a stepwise manner to form chloramines. This process is depicted in Eqs. (14.12) to (14.14):

$$NH_4^+ + HOCl \rightleftharpoons NH_2Cl + H_2O + H^+ \qquad (14.12)$$

$$NH_2Cl + HOCl \rightleftharpoons NHCl_2 + H_2O \qquad (14.13)$$

$$NHCl_2 + HOCl \rightleftharpoons NCl_3 + H_2O \qquad (14.14)$$

These compounds, monochloramine ($NH_2Cl$), dichloramine ($NHCl_2$), and trichloramine ($NCl_3$), each contribute to the total (or combined) chlorine residual in a water. The terms *total available chlorine* or *total oxidants* refer, respectively, to the sum of free-chlorine compounds and reactive chloramines, or total oxidizing agents. Under normal conditions of water treatment, if any excess ammonia is present, at equilibrium the amount of free chlorine will be much less than 1 percent of total residual chlorine. Each chlorine atom associated with a chloramine molecule is capable of undergoing a two-electron reduction to chloride; hence, each mole of monochloramine contains 71 g available chlorine, each mole of dichloramine contains $2 \times 71 = 142$ g, and each mole of trichloramine contains $3 \times 71 = 223$ g of available chlorine. Inasmuch as the molecular weights of mono-, di-, and trichloramine are 51.6, 86, and 110.5, respectively, the chloramines contain, respectively, 1.38, 1.65, and 2.02 g available chlorine per gram. The efficiency of the various combined chlorine forms as disinfectants differs, however, and thus, the concentration of available chlorine is insufficient to characterize process performance. On an approximate basis, for example, for coliforms, the biocidal potency of $HOCl:OCl^-:NH_2Cl:NHCl_2$ is approximately 1:0.0125:0.005:0.0166, and for viruses and cysts the combined chlorine forms are even considerably less effective.[42] As reaction (14.12) indicates, the formation of monochloramine is accompanied by the loss of a proton, because chlorination reduces the affinity of the nitrogen moiety for protons.[43]

The significance of chlorine speciation on disinfection efficiency was graphically demonstrated by Weber, as shown in Fig. 14.2.[44] As the dose of chlorine is increased, the total chlorine residual (i.e., remaining in the system after 30 min) increases until a dose of approximately

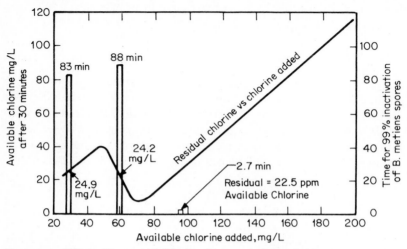

**Figure 14.2**  Effect of increased chlorine dosage on residual chlorine and germicidal efficiency; pH 7.0, 20°C, $NH_3$ 10 mg/L.  (*Adapted from G. R. Weber et al., "Effect of Ammonia on the Germicidal Efficiency of Chlorine in Neutral Solutions," J. AWWA, vol. 32, no. 11, 1940, p. 1904.*)

50 mg/L, whereupon residual chlorine decreases to a very low value and subsequently increases linearly with dose indefinitely. The "hump and dip" behavior is paralleled by the sensitivity of microorganisms to the available chlorine residual indicated by the time required for 99 percent inactivation of *Bacillus metiens* spores. At the three points indicated, the total available chlorine is approximately identical at 22 to 24 mg/L, yet a 32-fold difference in microbial sensitivity occurred.

The explanation for this behavior is the "breakpoint" reaction between free chlorine and ammonia (Fig. 14.3). At doses below the "hump" in the chlorine residual curve (zone 1), only combined chlorine is detectable. At doses between the "hump" and the "dip" in the curve, an oxidative destruction of combined residual chlorine accompanied by the loss of nitrogen occurs (zone 2).[46] One possible reaction during breakpoint is

$$2NH_3 + 3HOCl \rightleftharpoons N_2 + 3H^+ + 3Cl^- + 3H_2O \qquad (14.15)$$

This reaction also may be used as a means to remove ammonia nitrogen from water or wastewaters.[47] Finally, after the ammonia nitrogen has been completely oxidized, the residual remaining consists almost exclusively of free chlorine (zone 3). The minimum in the chlorine residual versus dose curve (in this case $Cl_2:NH_4^+$—N weight ratio of 7.6:1) is called the *breakpoint* and denotes the amount of chlorine that must be added to a water before a stable free residual can be obtained.

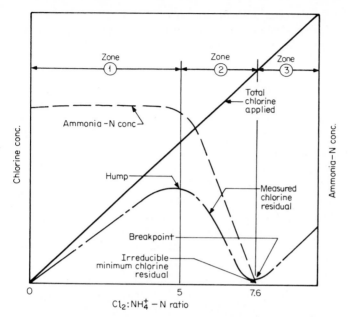

**Figure 14.3** Schematic idealization of the breakpoint curve. (*Source: Adapted from G. C. White, Disinfection of Wastewater and Water for Reuse, Van Nostrand Reinhold, New York. Copyright 1978.*)

In their investigations of the chlorination of drinking water, Griffin and Chamberlin[48,49] observed that

1. The classical "hump and dip" curve is only seen at water pHs between 6.5 and 8.5.
2. The molar ratio between chlorine and ammonia nitrogen dose at the breakpoint under ideal conditions is 2:1 corresponding to a mass dose ratio ($Cl_2:NH_4^+$—N) of 10:1.
3. In practice, mass dose ratios of 15:1 may be needed to reach breakpoint.

The breakpoint reaction may also affect the pH of a water. If sodium hypochlorite is used as the source of active chlorine, as breakpoint occurs, the pH decreases, caused by an apparent release of protons during the breakpoint process. If gaseous chlorine is used, this effect is obscured by the release of protons by hydrolysis of gaseous chlorine according to Eqs. (14.5) and (14.6).[50]

The oxidation of ammonia nitrogen by chlorine to gaseous nitrogen at the breakpoint would theoretically require 1.5 mol of chlorine ($Cl_2$) per mole of nitrogen oxidized according to reaction (14.15). The observed stoichiometric molar ratio between chlorine added and ammonia nitrogen consumed at breakpoint is typically about 2:1, suggesting that more oxidized nitrogen compounds are produced at breakpoint rather than $N_2$ gas. Experimental evidence[51] indicates that the principal addition oxidized product may be nitrate formed via reaction (14.16):

$$NH_4^+ + 4HOCl \rightleftharpoons NO_3^- + 4Cl^- + 6H^+ + H_2O \qquad (14.16)$$

Depending upon the relative amount of nitrate formed in comparison to nitrogen at breakpoint, between 1.5 and 4.0 mol of available chlorine may be required, which is consistent with the available data.

Below the breakpoint, inorganic chloramines decompose by direct reactions with several compounds. For example, monochloramine may react with bromide ions to form monobromamine.[52] If trichloramine is formed, as would be the case for applied chlorine doses in excess of that required for breakpoint, it may decompose either directly to form nitrogen gas and hypochlorous acid or by reaction with ammonia to form monochloramine and dichloramine.[53] In distilled water, the half-life of monochloramine was approximately 100 h.[54] Even in this simple circumstance, however, the decomposition products have not been completely characterized. Valentine et al.[55] have found that the decomposition of pure solutions of monochloramine produces an unidentified product that absorbs uv light at 243 nm and is capable of being oxidized or reduced.

Where the pH is also below 9.0 (so that the dissociation of ammonium ion is negligible), the amount of combined chlorine in dichloramine relative to monochloramine after reactions (14.12) and (14.13) have attained equilibrium is given by the following relationship:[50]

$$A = \frac{BZ}{1 - \sqrt{1 - BZ(2 - Z)}} - 1 \qquad (14.17)$$

In Eq. (14.17), $A$ is the ratio of available chlorine in the form of dichloramine to available chlorine in the form of monochloramine, $Z$ is the ratio of moles of chlorine (as $Cl_2$) added per mole of ammonia nitrogen present, and $B$ is defined by Eq. (14.18):

$$B = 1 - 4K_{eq}[H^+] \qquad (14.18)$$

The equilibrium constant in Eq. (14.18) refers to the direct interconversion between dichloramine and monochloramine as follows:

$$H^+ + 2NH_2Cl \rightleftharpoons NH_4^+ + NHCl_2 \qquad (14.19)$$

$$K_{eq} = \frac{[NH_4^+][NHCl_2]}{[H^+][NH_2Cl]^2}$$

At 25°C, $K_{eq}$ has a value of $6.7 \times 10^5$ L/mol.[50,56] From these relationships, determination of the equilibrium ratio of dichloramine to
monochloramine as a function of pH and applied chlorine dose ratio is
possible (assuming no dissipative reactions other than those involving
the inorganic chloramines). As pH decreases and the Cl:N dose ratio
increases, the relative amount of dichloramine also increases (Fig.
14.4). As the Cl:N molar dose ratio increases, the relative amount of
dichloramine also increases. As the Cl:N molar dose ratio increases
beyond unity, the amount of dichloramine relative to monochloramine
rapidly increases as well. For the conversion from dichloramine to
trichloramine, the equilibrium constant given at 0.5 $M$ ionic strength
and 25°C indicates that the amount of trichloramine to be found in
equilibrium with di- and monochloramine at molar dose ratios of up to
2.0 is negligible.[56] This agrees with experimental measurement of the
individual combined chlorine species as a function of approach to
breakpoint.[57]

These findings, coupled with the routine observation of the
breakpoint at molar doses at or below 2:1 (weight ratios Cl₂:N below
10:1), indicate that trichloramine is not an important species in the
breakpoint reaction. Rather, the breakpoint reaction leading to oxidation of ammonia nitrogen and reduction of combined chlorine is initiated with the formation of dichloramine.

The kinetics of formation of chloramine species have been investigated by various researchers since initial attempts by Weil and
Morris.[43] The formation of monochloramine is a first-order process in

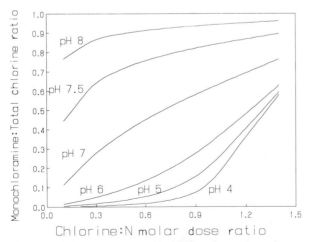

**Figure 14.4**  Effect of pH and $Cl_2:NH_4^+$ molar ratio on fractional amount of combined chlorine existing as
monochloramine at equilibrium at 20°C.

each of hypochlorous acid and uncharged ammonia. Solely through kinetic arguments, however, determining whether this, or a process involving hypochlorite ions reacting with ammonium cations, is the actual mechanism of reaction is not possible. If the neutral species are selected as the reactants, then the rate of formation of monochloramine ($r$) may be described by[58]

$$r[\text{mol/(L)(s)}] = 6.6 \times 10^8 \exp\left(\frac{-1510}{T}\right)[\text{HOCl}][\text{NH}_3] \quad (14.20)$$

Because hypochlorous acid dissociates into hypochlorite with a $pK_a$ of approximately 7.4 and ammonia is able to associate with a proton to form the ammonium cation, with the $pK_a$ for the latter of approximately 9.3, for a constant chlorine:nitrogen dose ratio, the maximum rate of monochloramine formation occurs at a pH where the product $\text{HOCl} \times \text{NH}_3$ is maximized, which is at the midpoint of the two $pK$ values or 8.4. At this optimum pH and the usual temperatures encountered in practice, the formation of monochloramine attains equilibrium in seconds to 1 min; however, at either a higher or lower pH, the speed of the reaction slows.

A number of the other reactions in the chlorine-ammonia system may be kinetically limited. Recently, these have been reviewed, and Table 14.3 is a compilation of the known reaction kinetics involving chlorine, ammonia, and intermediate species.

The reaction of $\text{NH}_2\text{Cl}$ with HOCl to form $\text{NHCl}_2$ is catalyzed by a number of acidic species that may be present in water.[59] Possibly, a number of the other reactions in Table 14.3 can also be catalyzed in a similar manner; however, insufficient data is available to evaluate this possibility.

When free chlorine is contacted with a water-containing ammonia, the initial velocity of monochloramine formation is substantially greater than the velocity of the subsequent formation of dichloramine.

TABLE 14.3    Summary of Chlorine Reaction Kinetics

| Reaction | Forward rate expression | Reverse rate expression |
|---|---|---|
| $\text{NH}_3 + \text{HOCl} \rightleftharpoons \text{NH}_2\text{Cl} + \text{H}_2\text{O}$ | $6.6 \times 10^8 \exp(-1510/T)$ | $1.38 \times 10^8 \exp(-8800/T)$ |
| $\text{NH}_2\text{Cl} + \text{HOCl} \rightleftharpoons \text{NHCl}_2$ | $3 \times 10^5 \exp(-2010/T)$ | $7.6 \times 10^{-7}$ L/(m)(s)† |
| $\text{NHCl}_2 + \text{HOCl} \rightleftharpoons \text{NCl}_3$ | $2.0 \times 10^5 \exp(-3420/T)$ | $5.1 \times 10^3 \exp(-5530/T)$ |
| $2\text{NH}_2\text{Cl} \rightleftharpoons \text{NHCl}_2 + \text{NH}_3$ | $80 \exp(-2160/T)$ | $24.0$ L/(m)(s)† |

Notes: Rates are in units of L/mol-s.
Concentrations are in mol/L.
Reactions are elementary and water is at unit activity.
$T$ is absolute temperature (kelvins).
†This is the rate constant at 25°C.
SOURCE: From Ref. 58.

Hence, relative to equilibrium levels, an initial accumulation of monochloramine will occur, if large dose ratios are employed, until the dichloramine formation process can be driven.[60]

The kinetic evolution of the chlorine-ammonia speciation process in batch systems is described by a series of coupled ordinary differential equations. While these are highly nonlinear, various authors have applied numerical integration techniques for their solution and, below the breakpoint, have found reasonable concordance between model predictions and experimental measurements.[51,59,61,62]

The breakpoint process involves a complex series of elementary reactions, of which Eqs. (14.15) and (14.16) are the net results. Saunier and Selleck[51] proposed that hydroxylamine ($NH_2OH$) and NOH may be intermediates in this reaction. Sufficient evaluation of their proposed kinetic scheme for the breakpoint process, however, has not yet been achieved to justify its use for design applications.

A wide variety of reactions of chlorine with other constituents in a water can occur. These may lead to a loss of disinfectant from the system, or its transformation to a less effective species.

Morris[63] has determined that organic amines could react with free chlorine to form organic monochloramines. The rate laws for these reactions follow patterns similar to the inorganic monochloramine formation process, except that the rate constants are generally less. In addition, the rate constants for this process correlate to the relative basicity of the amine reactant. Organic chloramines may also be formed by the direct reaction between monochloramine and the organic amine, and this is apparently the most significant mechanism of organic $N$-chloramine formation at higher concentrations such as might exist at the point of application of chlorine to a water.[64] Pure solutions of amino acids and some proteins also display breakpoint curves of identical shape to those of ammonia solutions.[65,66]

Free chlorine can react with organic constituents to produce chlorinated organic by-products. Murphy et al.[67] noted that phenols, amines, aldehydes, ketones, and pyrrole groups are readily susceptible to chlorination. Granstrom and Lee[38] found that phenol could be chlorinated by free chlorine to form chlorophenols of various degrees of substitution. The kinetics of this process depend on both phenolate ions and hypochlorous acid. If excess ammonia was present, however, the formation of chlorophenols was substantially inhibited.

DeLaat et al.[68] have determined that polyhydric phenols are substantially more reactive than simple ketones in the production of chloroform, and that the rates of these processes are first order in each of the phenol concentration and the free chlorine concentration. Significantly, the reactivity of these compounds was observed to be greater than the reactivity of ammonia with hypochlorous acid. Therefore, even if subbreakpoint chlorination is practiced, some chloroform may be formed rapidly prior to the conversion of free to combined chlorine. Chapter 12 presents additional discussion of the formation of trihalomethanes and other disinfection by-products that can arise from reactions with naturally occurring dissolved humic substances.

The reactivity of the chlorine species with compounds responsible for taste and odor depends on the predominant form of chlorine present. In field tests, Krasner et al.[69] determined that free chlorine, but not combined chlorine, could remove tastes and odors associated with organic sulfur compounds.

The rates of reaction between free-chlorine residuals and other inorganic compounds likely to be present in water are summarized in Table 14.4.[70] These reactions are generally first order in the oxidizing agent (hypochlorous acid or hypochlorite anion) and in the reducing agent.

Nitrites present in partially nitrified waters react with free chlorine via a complex, pH-dependent mechanism.[71] While combined chlorine residuals were generally thought to be unreactive with nitrite, Valentine[72] has found that the rate of decay of monochloramine in the presence of nitrite was far greater than would be predicted based on reaction of the equilibrium free chlorine, implicating a direct reaction between $NH_2Cl$ and $NO_2^- - N$.

The rate of exertion of chlorine demand in the complex milieu of water has been the subject of numerous studies. The most systematic work has been that of Taras,[46] who chlorinated pure solutions of various organic compounds and found that chlorine demand kinetics could be described by Eq. (14.21):

**TABLE 14.4  Summary of Kinetics of HOCl and OCl⁻ Reduction by Miscellaneous Reducing Agents after Wojtowicz**

| Oxidizing agent | Reducing agent | Oxidation product | Log $k$, L/m-s, 25°C |
|---|---|---|---|
| $OCl^-$ | $IO^-$ | $IO_4^-$ | $-5.04$ |
| $OCl^-$ | $OCl^-$ | $ClO_2^-$ | $-7.63$ |
| $OCl^-$ | $ClO_2^-$ | $ClO_3^-$ | $-5.48$ |
| $OCl^-$ | $SO_3$ | $SO_4^{2-}$ | $3.93$ |
| $HOCl$ | $NO_2^-$ | $NO_3^-$ | $0.82$ |
| $HOCl$ | $HCOO^-$ | $H_2CO$ | $-1.38$ |
| $HOCl$ | $Br^-$ | $BrO^-$ | $3.47$ |
| $HOCl$ | $OCN^-$ | $HCO_3^-, N_2$ | $-0.55$ |
| $HOCl$ | $HC_2O_4^-$ | $CO_2$ | $1.20$ |
| $HOCl$ | $I^-$ | $IO^-$ | $8.52$ |

Source: From Ref. 70.

$$D = kt^n \qquad (14.21)$$

where $t$ = time, h
$\quad D$ = chlorine demand
$\quad k, n$ = empirical constants

In subsequent work, Feben and Taras[73] found that chlorine demand exertion of waters blended with wastewater could be correlated to Eq. (14.21), with the value of $n$ correlated to the 1-h chlorine demand.

More recently, Haas and Karra[74] developed Eq. (14.22) to describe chlorine demand exertion kinetics.

$$D = C_0[1 - Xe^{-k_1 t} - (1 - X)e^{-k_2 t}] \qquad (14.22)$$

In Eq. (14.22), $X$ is an empirical constant, typically 0.4 to 0.6, and $k_1$ and $k_2$ and rate constants, typically 1.0/min and 0.003/min, respectively, and $C_0$ is the chlorine dose in milligrams per liter.

**Example Problem 2.** A water supply is to be postammoniated. If the water has a pH of 7.0, a free-chlorine residual of 1.0 mg/L, and a temperature of 25°C, how much ammonia should be added such that the ratio of dichloramine to monochloramine is 0.1? (Assume that, upon the addition of ammonia, none of the residual dissipates.)

**solution**   From Eqs. (14.17) and (14.18), the following is determined:

$$B = 1 - 4K_{eq}(10^{-7}) = 1 - 4(6.7 \times 10^5)(10^{-7})$$

$$= 0.732$$

From Eq. (14.16), noting that the problem condition specifies $A = 0.1$, the following equation is to be solved:

$$0.1 = \frac{0.732Z}{1 - [1 - 0.732(2 - Z)Z]^{1/2}} - 1$$

This can be rearranged into a quadratic equation

$$-0.289Z^2 + 0.134Z = 0$$

The single nonzero root gives $Z = 0.463$, which is molar ratio of chlorine (as $Cl_2$) to ammonia nitrogen. Because chlorine has a molecular weight of 70, 1 mg/L of free chlorine has a molarity of $1.43 \times 10^{-5}$. Therefore, $3.09 \times 10^{-5}$ molarity of ammonia is required, or (multiplying by the atomic weight of nitrogen, 14) a concentration of 0.43 mg/L as N of ammonia must be added.

**Dechlorination.**   When the chlorine residual in a treated water must be lowered prior to distribution, the chlorinated water can be dosed with a substance that reacts with, or accelerates the rate of decomposition of, the residual chlorine. Compounds that may perform this function include thiosulfate, hydrogen peroxide, ammonia, sulfite-bisulfite-sulfur dioxide, and activated carbon; however, only the latter two materials have been widely used for this purpose in water treatment.[75]

**Chlorine Dioxide.** The reaction of chlorine dioxide with material present in waters containing chlorine dioxide demand appears to be less significant than in the case of chlorine. Rather, the dominant causes of loss of chlorine dioxide during disinfection may be the direct reactions with water and interconversions to chlorite and chloride, as outlined above [Eqs. (14.9), (14.10)]. At milligram per liter concentrations, ammonia nitrogen, peptone, urea, and glucose have insignificant chlorine dioxide demand in 1 h.[76,77]

Masschelein[78] concluded that only the following organic-$ClO_2$ reactions are of significance to water applications:

1. Oxidation of tertiary amines to secondary amines and aldehydes
2. Oxidation of ketones, aldehydes, and, to a lesser extent, alcohols to acids
3. Oxidation of phenols
4. Oxidation of sulfhydryl-containing amino acids

Wajon et al.[79] found a reaction stoichiometry of 2 mol of chlorine dioxide consumed per mole of phenol (or hydroquinone) consumed. Products formed included chlorophenols, aliphatic organic acids, benzoquinone, and (in the case of phenol) hydroquinone. The mechanism appeared to include the possible formation of hypochlorous acid as an intermediate that would chlorinate, and the rate of this process was found to be base-catalyzed and first order in each of the reactants.

In general, chlorine dioxide itself has been found to produce fewer organic by-products with naturally occurring dissolved organic material, although some nonpurgeable organic halogenated compounds are formed.[80] In practice, however, chlorine dioxide may be generated in a manner in which chlorine is present as an impurity. Therefore, the reactions of such a stream may also include those discussed above regarding chlorine reactions.

**Ozone.** Upon addition to water, ozone reacts with hydroxide ions to form hydroxyl radicals and organic radicals. These radicals cause increased decomposition of ozone and are responsible for nonselective (compared to the direct ozone reaction) oxidation of a variety of organic materials. Carbonate, and possibly other ions, may act as radical scavengers and slow this process.[81,82]

Gurol and Singer[83] determined ozone decomposition kinetics in various aqueous solutions to be second order in ozone concentration and base-promoted. Some systematic difference between various buffer systems employed does occur, with borate giving higher decomposition rates than phosphate and with phosphate at higher ionic strength giving lower decomposition rates than phosphate at lower ionic strength (1 $M$ versus 0.1 $M$). This effect was suggested as being caused by phosphate being a radical scavenger (and radical decomposition being important at higher pH values).

As a result of these decomposition processes, the half-life of ozone in water, even in the absence of other reactive constituents, is quite short, on the order of seconds to minutes. Water chemistry may exert a strong influence on the rate and extent of ozone demand in a given application. Reactions of ozone in aqueous solution are discussed further in Chap. 12.

Bromide reacts with ozone under aqueous conditions typical of drinking water disinfection. Products of the reaction may be hypobromous acid, hypobromite, and bromate. Higher concentrations of bromide can reduce the rate of ozone decomposition. Under alkaline conditions, this may be influenced by trace metal catalysts and organic sinks for radicals and oxidized bromine species.[84]

Ozone will react with cyanides at a very fast rate. The mechanism involves reaction of the cyanide ion (to form unknown products), and the process is inhibited by iron complexes but catalyzed by copper complexes of cyanide.[85]

**Ultraviolet interactions.**  With uv disinfection systems, the equivalent of "demand" results from dissolved and suspended materials, such as proteins, humic material, and iron compounds, that may absorb the radiation and thus shield microorganisms. Huff et al.[26] found that intensity monitoring within the reactor itself could be used to correct for such effects.

One particular problem unique to physical systems such as uv is the need to ensure complete mixing in the transverse direction to ensure that all microorganisms may come equally close to the uv source. Cortelyou et al.[86] analyzed this effect for batch uv reactors, and the analysis was extended to flow-through reactors by Haas and Sakellaropoulous.[87] This phenomenon results in the desirability to achieve turbulent flow conditions in a uv reactor.

## Assessment of Microbial Quality

The microbial quality of a source water, or the efficacy of a treatment system for removing microorganisms, can be assessed either by direct monitoring of pathogens or by the use of an indicator system. Because pathogens are a highly diverse group, generally requiring a highly specialized (and often insensitive and expensive) analytical technique for each pathogen, the use of indicator organisms is a more popular technique.

An indicator group of organisms can be used either to assess source water contamination or degree of treatment; however, often the same indicator group is used to assess both properties. This poses severe constraints on the group of indicator organisms chosen. Bonde[88] has proposed that an ideal indicator must

1. Be present whenever the pathogens concerned are present

2. Be present only when the presence of pathogens is an imminent danger (i.e., they must not be able to proliferate to any greater extent in the aqueous environment)

3. Occur in much greater numbers than pathogens

4. Be more resistant to disinfectants and to the aqueous environment than pathogens

5. Grow readily on relatively simple media

6. Yield characteristic and simple reactions enabling, as far as possible, an unambiguous identification of the group

7. Be randomly distributed in the sample to be examined, or it should be possible to obtain a uniform distribution by simple homogenization procedures

8. Grow widely independent of other organisms present when inoculated in artificial media (i.e., the indicator bacteria should not be seriously inhibited in their growth by the presence of other bacteria)

The use of coliforms as indicator organisms stems from the pioneering work of Phelps.[89] The basic rationale was that coliforms and enteric bacterial pathogens originate from a common source, namely human fecal contamination. Subsequent work by Butterfield and others[90–93] confirmed that these organisms were at least as resistant to free or combined chlorine as enteric bacterial pathogens.

The coliform group is a heterogenous conglomerate of microorganisms, including forms native to mammalian gastrointestinal tracts, as well as a number of exclusively soil forms. The common *fermentation tube* (FT) and *membrane filter* (MF) procedures are subtly different in the organisms they enumerate. Classically, coliforms have been defined as "Gram-negative, non-sporeforming bacteria which (sic) ferment lactose at 35–37°C, with the production of acid and gas."[94] The FT procedure, however, ignores anaerogenic and lactose-negative coliforms, and the MF procedure ignores non-lactose-fermenting strains.[95]

Furthermore, interferences can selectively reduce coliforms as measured by one method or the other. Allen et al.,[96] for example, found that high concentrations (> 500 to 1000/mL) of *standard plate count* (SPC) organisms appeared to reduce the recovery of coliforms by the MF technique when compared to the FT technique.

The fecal coliform group of organisms is that subset of coliforms that are capable of growing at elevated temperature (44.5°C). The original rationale for development of this test was to provide a more selective indicator group, excluding mesophillic coliforms indigenous, primarily, to soils. Total coliforms, however, continue to be the basic U.S. microbiological standard because their absence ensures the absence of fecal coliforms, which is a conservative standard.

While coliforms, either fecal or total, may be reasonably good indicators of fecal contamination of a water supply, as early as 1922[97] reservations were expressed about the relative resistance of coliforms to chlorine vis-à-vis pathogenic bacteria, and the resulting adequacy of the coliform test as an indicator of disinfection efficiency. In more recent work, coliforms have been found to be more sensitive to disinfection by one or more forms of chlorine than various human enteric viruses[98,99] and the protozoan pathogens *Naegleria*[100] and *Giardia.*[101] In addition, viruses[102] and protozoan cysts[103] have been found to be more resistant to $ClO_2$ inactivation than coliforms. Farooq[104] has determined that coliforms are more resistant to ozone than viruses. Rice and Hoff[105] found that *Giardia lamblia* cysts survived exposure to uv doses sufficient to effect over 99.99 percent inactivation of *Escherichia coli*. In full-scale water treatment plants practicing conventional treatment, and meeting turbidity and coliform standards in the presence of free residual chlorine, isolating human enteric viruses has been possible.[106]

As a result of the problems with the coliform group of organisms, workers have investigated alternative indicator systems with greater resistance to disinfectants than coliforms. Among the most successful of these are the acid-fast bacteria and yeasts studied by Engelbrecht and associates.[107–113] In addition, work using endotoxins,[114] *Clostridia,*[115] and bacteriophage[116] has been carried out. In addition, to some degree, *heterotrophic plate count* (HPC) organisms may provide a conservative indicator of treatment efficiency. Despite these studies, however, in U.S. practice, no alternative to the total coliform group of organisms has found widespread application.

## Disinfection Kinetics

The information needed for the design of a disinfection system includes knowledge of the rate of inactivation of the target, or indicator, organism(s) by the disinfectant. In particular, the effect of disinfectant

concentration on the rate of this process will determine the most efficient combination of contact time (i.e., basin volume at a given design
flow rate) and dose to employ.

The major precepts of disinfection kinetics were enunciated by
Chick,[117] who recognized the close similarity of microbial inactivation
by chemical disinfectants to chemical reactions. Disinfection is analogous to a bimolecular chemical reaction, with the reactants being the
microorganism and the disinfectant, and can be characterized by a
rate law as are chemical reactions:

$$r = -kN \qquad (14.23)$$

In Eq. (14.23), $r$ is the inactivation rate [(organisms killed/(volume)
time)] and $N$ is the concentration of viable organisms. In a batch system, this results in an exponential decay in organisms, because the
rate of inactivation equals $dN/dt$, assuming that the rate constant ($k$)
is actually constant.

Watson[118] proposed Eq. (14.24) to relate the rate constant of inactivation $k$ to the disinfectant concentration $C$:

$$k = k'C^n \qquad (14.24)$$

In Eq. (14.24), $n$ is termed the coefficient of dilution, and $k$ is presumed independent of disinfectant concentration and, by Eq. (14.23),
microorganism concentration.

From the Chick-Watson law, when $C$, $n$, and $k$ are constant (i.e., no
demand, constant concentration), the above rate law may be integrated so that in a batch system the following relationship arises:

$$\ln \frac{N}{N_o} = -k'C^n t \qquad (14.25)$$

In Eq. (14.25), $N$ and $N_o$ are, respectively, the concentrations of viable
microorganisms at time $t$ and time 0. When disinfectant composition
changes with time, or when a configuration other than a batch (or
plug-flow) system is used, the appropriate rate laws characterizing
disinfectant transformation[119] along with the applicable mass bal-
ances must be used to obtain the relationship between microbial inactivation and concentration and time.

Inactivation of microorganisms in batch experiments, even when
disinfectant concentration is kept constant, does not always follow the
exponential decay pattern predicted by Eq. (14.25). Indeed, two common types of deviations are noted (Fig. 14.5). In addition to the linear
Chick's law decay, the presence of "shoulders" or time lags until the
onset of disinfection are often observed (curve 2 in Fig. 14.5). Also,
some microorganisms and disinfectants exhibit a "tailing" wherein

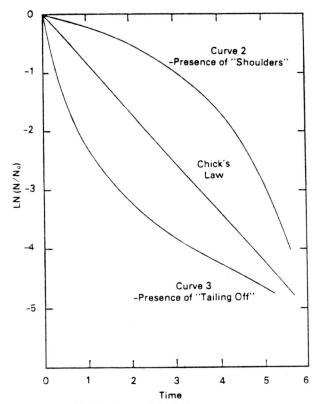

Figure 14.5    Chick's law and deviations.

the rate of inactivation progressively decreases (curve 3 in Fig. 14.5). In some cases, a combination of both of these behaviors is seen.

Even if deviations from Chick-Watson behavior are observed, plotting combinations of disinfectant concentration and time to produce a fixed percent inactivation is generally possible. Such plots tend to follow the relationship $C^n t$ = constant, where the constant is a function of the organism, pH, temperature, form of disinfectant, and extent of inactivation. Such plots are linear on a log-log scale (Fig. 14.6). If $n > 1$, a proportionate change in disinfectant concentration produces a greater effect than a proportionate change in time. In many cases,[120] the Chick-Watson law $n$ value is close to 1.0, and hence a fixed value of the product of concentration and time ("$Ct$ product") results in a fixed degree of inactivation (at a given temperature, pH, etc.).

In the chemical disinfection of a water, the concentration of disinfectant may change with time, and, particularly during the initial moments of contact, the chemical form(s) of halogens such as chlorine undergo rapid transformations from the free to the combined forms.

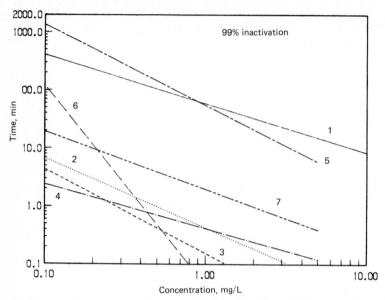

**Figure 14.6**  Concentration-time relationships for 99 percent inactivation of various microorganisms by various disinfectants. (1) *Giardia lamblia;* free chlorine, 5°C, pH 6 (*Hoff and Akin, Ref. 207*). (2) *E. coli;* free chlorine, 2 to 5°C, pH 8.5 (*Haas and Karra, Ref. 119*). (3) *E. coli;* free chlorine, 20 to 25°C, pH 8.5 (*Haas and Karra, Ref. 119*). (4) Poliovirus 1 (Mahoney); free chlorine, 2°C, pH 6 (*Haas and Karra, Ref. 119*). (5) *E. coli;* combined chlorine, 3 to 5°C, pH 7 (*Haas and Karra, Ref. 119*). (6) Poliovirus 1 (Mahoney); ozone, 20°C, pH 7.2 (*Roy et al., Ref. 135*). (7) *Giardia muris;* ozone, 5°C, pH 7 (*Wickramanayke et al., Ref. 167*).

Because $C$ would thus not be a single-valued constant, disinfection results obtained in batch systems typically exhibit "tailing," the degree of which may depend on the demand and the concentration of reactive constituents (such as ammonia) in the system.[121] Determination of the disinfectant residual (and its chemical forms) is more critical than the disinfectant dose in these systems. According to Heukelekian and Smith,[122] "...the control of chlorination cannot be based on the dosage of chlorine, because...a constant dosage of chlorine [does not] produce a constant number of coliform organisms per ml (sic)...."

In the chlorine system, for example, by knowing the rate laws for inactivation by individual separate species and the dynamics of chlorine species interconversions, as described previously, one can formulate an overall model for chlorine inactivation.[119] In this computation, the individual rates are usually assumed to be additive,[123] although this assumption has not yet been experimentally verified.

The presence of "shoulders" in inactivation curves is often seen in organisms that form clumps. This means that more than one cell must be inactivated to get inactivation of a colony or plaque-forming unit.

For example, Rubin et al.[100] found that cysts of *Naegleria gruberii* in demand-free water showed shoulder-type inactivation to free chlorine. Similarly, when cells of *E. coli* were agglutinated, they displayed shoulder-type inactivation, which was absent in unagglutinated cultures.[124] Severin et al.[125] found shoulder-type inactivation curves in the case of *E. coli* with preformed chloramines (i.e., solutions of ammonia and chlorine prereacted to form combined chlorine prior to addition of microorganisms), *Candida parapsilosis* (a yeast organism proposed as a possible disinfection-resistant indicator) with both preformed chloramines, and free chlorine and poliovirus with iodine.

Shoulder inactivation curves may be explained by either a multitarget model,[126] by a series event model,[125] or by a diffusional model.[127] Tailing inactivation curves may be explained either by a vitalistic hypothesis, in which individuals in a population are nonidentical and their inherent resistance is distributed in a permanent (time-independent) manner, or by a mechanistic concept.[128] In the latter case, four particular mechanisms have been advanced, leading to tailing:

1. Conversion to resistant form during inactivation (hardening)

2. Existence of genetic variants in subpopulation

3. Protection of a subpopulation, or variations in received dose of disinfectant

4. Clumping of a subpopulation

The hardening process and resultant tailing has received wide attention, following discoveries of apparent hardening in the formaldehyde inactivation of poliovirus prepared for the Salk vaccine. Gard[129] has proposed an empirical rate law for this behavior, which has been used by Selleck et al.[130] in the analysis of wastewater chlorination kinetics. Tailing behavior has been found for viral and coliform inactivation by ozone[131] and for coliform inactivation by free chlorine.[121,132]

Hom[133] has developed a flexible, but highly empirical, kinetic formulation based on modifying Eqs. (14.23) and (14.24) to the following form:

$$r = -k'Nt^mC^n \qquad (14.26)$$

Depending on the value of $m$, both shoulders and tailing may be depicted by Eq. (14.26). In early work, Fair et al.[123] used a model of the form of (14.26) with $m = 2$ to analyze *E. coli* inactivation by free and combined chlorine.

Another class of models can be obtained by assuming that inactivation is other than first order in surviving microbial concentrations.

Depending on the order chosen, either tailing or shoulders can be produced. For example, Roy et al.,[134,135] using continuously stirred tank reactor studies on inactivation of poliovirus 1 with ozone in demand-free systems, developed the following rate law:

$$r = -kCN^{0.69} \qquad (14.27)$$

A similar model was used by Benarde et al.[136] to analyze *E. coli* inactivation by chlorine dioxide.

Disinfection, like all other rate processes, is temperature-dependent. This dependency may be quantified by the Ahrrenius relationship:

$$k = k_0 \exp(-E/RT) \qquad (14.28)$$

where *k* = rate constant characterizing reaction (such as the Chick-Watson *k'* value)
$T$ = absolute temperature
$R$ = ideal gas constant
$k_0$ = frequency factor
$E$ = activation energy, energy/mole

The value of $E$ is always positive, and as it increases the effect of temperature becomes more pronounced. The values of $E$ and $k_0$ may be determined from rates of inactivation obtained as a function of temperature. As $E$ increases, the effect of temperature on the rate increases. For example, an $E$ of 10 kcal/mol doubles the rate between 10 and 20°C. In contrast, activation energies for breaking hydrogen bonds are 3 to 7 kcal/mol.[137] Activation energies less than this range suggest physical (e.g., transport) limitations rather than chemical reactions.

In general, microbial inactivation kinetics have been determined in batch systems. Real contactors, however, are continuous and may have nonideal flow patterns with "back mixing" or short-circuiting. Although discussing means to extrapolate from kinetics obtained under ideal hydraulic conditions to less ideal contractors is beyond the scope of this chapter, this material is well discussed elsewhere.[27,138,139]

In the application of kinetic models to the analysis of uv disinfection processes, the "concentration" of disinfectant is replaced by the incident light intensity (in units of energy per unit area). This may be determined either by direct measurement in the actual disinfection reactor or by modeling of physical aspects of light transmission.[20,27]

## Mode of Action of Disinfectants

### Chlorine

Since Nissen (1890),[140] free chlorine at low pH has been known to be more biocidal than free chlorine at high pH. Holwerda[141] proposed that hypochlorous acid was the specific agent responsible for inactivation and, thus, the pH effect. Fair et al.[123] determined that the pH dependency of free-chlorine potency correlated quantitatively with the dissociation constant of hypochlorous acid. More recently, Chang[142] determined that, at low pH, the association of chlorine with cysts of *Entamoeba hystolytica* was greater than at high pH. Friberg[143,144] and Haas and Engelbrecht,[145,146] using radioactive free chlorine, found similar results applied with respect to bacteria, and that the microbial binding of chlorine could be described by typical chemical isotherms. With respect to viruses, this association of chlorine parallels the biocidal efficacy of hypochlorous acid, hypochlorite, and monochloramine.[147-149]

Once taken into the environment of the living organism, chlorine may enter into a number of reactions, with critical components causing inactivation. In bacteria, respiratory, transport, and nucleic acid activity are all adversely affected.[145,146,150,151] In bacteriophage f2, the mode of inactivation appears to be disruption of the viral nucleic acid.[149] With poliovirus, however, the protein coat, and not the nucleic acid, appears to be the critical site for inactivation by free chlorine.[152,153]

The rate of inactivation of bacteria by monochloramine is greater than could be attributed to the equilibrium-free chlorine present in solution. This argues strongly for a direct inactivation reaction of combined chlorine.[74] Although organic chloramines are generally measured as combined or total chlorine by conventional methods, they are of substantially lower effectiveness as disinfectants than are inorganic chloramines.[154,155]

The activation energy for inactivation by chlorine has been computed by several researchers (Table 14.5). In general, the less active

**TABLE 14.5  Activation Energies for Chlorine Inactivation of Various Microorganisms**

| Chemical species | Microorganism | pH | Kinetic law† | E, kcal/mol | Ref. |
|---|---|---|---|---|---|
| Free chlorine | *E. coli* | 7.0 | CW | 8.2 | 123 |
| | | 8.5 | CW | 6.4 | 123 |
| | | 9.8 | CW | 12.0 | 123 |
| | | 10.7 | CW | 15.0 | 123 |
| | Poliovirus 1 | 6.0 | DB | 17.8 | 127 |
| Chloramines | *E. coli* | 7.0 | CW | 12.0 | 123 |
| | | 7.0 | SE | 18.0 | 125 |
| | | 8.5 | CW | 14.0 | 123 |
| | | 9.5 | CW | 20.0 | 123 |
| | *C. parapsilosis* | 7.0 | SE | 9.8 | 125 |

†CW = Chick-Watson, SE = series event, DB = diffusional model (binding reaction)

forms of disinfectant have higher activation energies than the more active disinfectants (compare free chlorine as a function of pH, or free versus combined chlorine, for example).

Surprisingly, in 1972, Scarpino et al.[156] reported that viruses were more sensitive to free chlorine at high pH than at low pH. Other authors have confirmed these findings with viruses and with bacteria. Hypochlorite can form neutral ion pairs with sodium, potassium, and lithium (particularly at ionic strengths approaching 0.1 $M$), however, and these ion pairs can increase disinfection efficiency by free chlorine at high pH.[157] Whether such effects can be important at lower ionic strengths has yet to be determined.

### Chlorine dioxide

pH has much less effect on inactivation efficiency by chlorine dioxide than for chlorine, although the results are more inconsistent. Benarde et al.,[158] working with *E. coli,* and Scarpino et al.,[159] working with poliovirus 1, found that the degree of inactivation by chlorine dioxide increases as pH increases. For amoebic cysts, however, as pH increases, the efficiency of inactivation by chlorine decreases.[160] The physiological mode of inactivation of bacteria by chlorine dioxide has been attributed to a disruption of protein synthesis.[136] In the case of viruses, chlorine dioxide preferentially inactivated capsid functions rather than nucleic acids.[161,162]

Benarde et al.[136] computed the activation energy for the inactivation of *E. coli* by chlorine dioxide at pH 6.5 as 12 kcal/mol. An identical number was computed for the disinfection of poliovirus 1 by chlorine dioxide at pH 7.[159] These are similar to the activation energies for chlorine inactivation (Table 14.5).

### Ozone

Understanding of the mode of inactivation of microorganisms by ozone remains hindered by difficulties in measuring low concentrations of dissolved ozone. The effect of pH on ozone inactivation of microorganisms appears to be predominantly associated with changing the stability of residual ozone, although additional work is needed. Farooq[104] found little effect of pH on the ability of dissolved ozone residuals to inactivate acid-fast bacteria. Roy[163] found a slight diminution of the virucidal efficacy of ozone residuals as pH decreased; however, Vaughn (cited in Hoff[120]) noted the opposite effect.

In actual ozone reactors, the potential exists for inactivation to occur by direct physical contact between microorganisms and the ozone-rich gas bubbles, as well as by action of dissolved ozone and its reaction products. Farooq et al.[164] found that direct contact was more important than the activity of dissolved ozone, although both mechanisms yield some degree of inactivation. Corroboration of this finding was made by Dahi and Lund,[165] who found that inactivation of bacteria by gaseous ozone was enhanced in the presence of ultrasound, which, presumably, increased mass transfer into the dissolved phase.

Bacterial cells lacking certain deoxyribonucleic acid polymerase gene activity were found to be more sensitive to inactivation by ozone than were wild-type strains, strongly implicating physicochemical damage to deoxyribonucleic acid as a mechanism of inactivation by ozone.[166] For poliovirus, the primary mode of inactivation by ozone also appears to be nucleic acid damage.[134,135]

Activation energies for ozone inactivation of *Giardia* cysts were reported by Wickramanayake et al.[167] For the two cysts examined, at pH 7, the activation energies were 9.7 and 16.7 kcal/mol. For poliovirus type 1, Roy et al.[134,135] estimated an activation energy of 3.6 kcal/mol at pH 7.2. If the latter value is correct, its low value suggests that ozone inactivates virus by a diffusional, rather than a reaction-controlled, process.

### Ultraviolet light

The mode of inactivation of microorganisms by ultraviolet radiation is quite well characterized. Specific deleterious changes in nucleic acid arise upon exposure to uv radiation.[168] These may be repaired by light-activated or dark repair enzymes in vegetative microorganisms. The phenomenon of photoreactivation of uv disinfected microorganisms has been demonstrated in municipal effluents.[169] The operation of these repair processes in microorganisms discharged to actual distribution systems or receiving waters is not clear, however.

Severin et al.[170] have shown that the series event model for inactivation describes the kinetics of uv disinfection quite well. Kinetic parameters for inactivation were found to be as follows:

|  | Activation energy, kcal/mol |
|---|---|
| *E. coli* | 0.554 |
| *Candida parapsilosis* | 0.562 |
| f2 virus | 1.023 |

Activation energies for uv inactivation using this model were lower than for chemical disinfection (indicating the relative insensitivity to temperature).[170] These are considerably less than the activation energies for chemical disinfectants in accord with the apparent (purely physical) mechanism of uv inactivation. This also indicates, as a practical matter, that the effect of temperature on uv performance is much less than for chemical agents such as ozone or chlorine.

The pH dependency of uv inactivation has not been characterized in controlled systems. Because the mechanism of uv inactivation appears to be purely physical, however, pH would not be expected to dramatically alter the efficiency of uv disinfection.

### Influence of physical and physiological conditions on inactivation

The physiological state of microorganisms, especially vegetative bacteria, may influence their susceptibility to disinfectants. Milbauer and Grossowicz[171] found that coliforms grown under minimal conditions were more resistant than cells grown under enriched conditions. Similarly, Berg et al.[172] found that chemostat-grown cells produced at high growth rates were more sensitive to disinfectants than were cells harvested from low growth rates. A simple subculturing of aquatic strains of *Flavobacterium* has been found to increase sensitivity to chlorine disinfection.[155]

Postexposure conditions can also influence apparent microbial response to disinfectants.[173] In recent work, three New England water treatment plants and distribution systems were sampled for total coliforms using both standard MF techniques and media to recover sublethally injured organisms (m-T7 agar medium). From 8 to 38 times as many coliforms were recovered on the m-T7 agar medium than the standard medium.[174] This sublethal injury does not adversely affect pathogenicity to mice.[175]

With viruses, the phenomenon of multiplicity reactivation can occur when individual viruses, inactivated by different specific events, are combined in a single host cell to produce an infectious unit. This has been demonstrated to occur in enteric viruses inactivated by chlorine.[176]

The survivors from disinfection can exhibit inheritable increased resistance to subsequent exposure. This was first demonstrated for poliovirus exposed to chlorine.[177] Demonstration in bacteria, however, has not been consistent.[132,178] The apparent increase in microbial resistance by clumping has already been discussed.

Microorganisms can also be partially protected against the action of disinfectants by adsorption to or enmeshment in nonviable solid particles present in a water. Stagg et al.[179] and Hejkal et al.[180] found that fecal material protected poliovirus against inactivation by combined chlorine. Boardman and Sproul[181] found that kaolinite, alum flocs, and lime sludge each were found to increase the resistance of poliovirus 1 to free chlorine in demand-free systems. For chlorine dioxide, bentonite turbidity protects poliovirus against the action of chlorine dioxide.[182] Although there was some protection from ozone, inactivation was afforded to coliform bacteria and viruses by fecal matter and by cell debris, at ozone doses usually employed in disinfection, and achieving more than 99.99 percent inactivation within 30 s was still possible.[183,184]

**Example Problem 3.**  A certain water supply has operational problems due to high levels of HPC organisms. To maintain adequate system water quality, a decision has been made to keep the concentration of HPC organisms below 10/mL at the entry point to the distribution system (i.e., following disinfection). Disinfection using free residual chlorine is practiced. As part of the laboratory investigation to develop design criteria for this system, the inactivation of the HPC organisms is determined in batch reactors (beakers). The pH and temperature are held constant at the expected final water conditions. Using water with an initial HPC of 1000/mL, the following data are taken:

| $Cl_2$ residual, mg/L | Contact time, min | HPC remaining, No./mL |
|---|---|---|
| 0.5 | 10 | 40 |
| 0.5 | 20 | 4 |
| 0.5 | 30 | 1 |
| 1 | 5 | 35 |
| 1 | 10 | 4 |
| 1 | 15 | 1 |
| 1.5 | 2 | 98 |
| 1.5 | 5 | 10 |
| 1.5 | 10 | 1 |

From this information, determine the best fit using the Hom inactivation model, and compute the necessary chlorine residual that will achieve HPC $<10$/mL (from an initial concentration of 1000/mL) at a contact time of 10 min?

**solution**    The best approach to this problem would be the use of a maximum likelihood technique to fit the Hom model to the data. Regression methods, however, can be used in two different ways. In a batch system, the Hom model becomes

$$\ln \frac{N}{N_0} = -kC^n t^m$$

This can be rearranged as

$$\ln\left(-\ln \frac{N}{N_0}\right) = \ln(k) + n \ln(C) + m \ln(t)$$

A multiple linear regression using $\ln[-\ln(N/N_0)]$ as the dependent variable and $\ln(C)$ and $\ln(t)$ as the dependent variables produces an intercept [equal to ln $(k)$] and slopes equal to $n$ and $m$. This computation can be handled on common spreadsheet programs, as well as statistical packages. Transformation of the data given produces the following values of the dependent and independent variables:

| ln $[-\ln (N/N_0)]$ | ln $(t)$ | ln $(C)$ |
|---|---|---|
| 1.169032 | 2.302585 | -0.69314 |
| 1.708642 | 2.995732 | -0.69314 |
| 1.932644 | 3.401197 | -0.69314 |
| 1.209678 | 1.609437 | 0 |
| 1.708642 | 2.302585 | 0 |
| 1.932644 | 2.708050 | 0 |
| 0.842768 | 0.693147 | 0.405465 |
| 1.527179 | 1.609437 | 0.405465 |
| 1.932644 | 2.302585 | 0.405465 |

The result is

$k = 1.111331$

$n = 0.683281$

$m = 0.695641$

Correlation coefficient $= 0.9973$

In the absence of programs to compute multiple linear regression, a two-step linear regression can be used to fit the data. The first step is to compute at each individual concentration the regression line to fit the equation

$$\ln\left(-\ln\frac{N}{N_0}\right) = \text{YINT} + m \ln (t)$$

The term YINT is the $y$-intercept of the regression line. From values of YINT versus ln($C$), both $k$ and $m$ can be computed via a regression of the form

$$\text{YINT} = \ln (k) + n \ln (C)$$

From the first set of regression equations, the following intermediate results are obtained:

| $C$ | slope($m$) | YINT |
|---|---|---|
| 0.5 | 0.704053 | -0.43820 |
| 1 | 0.664724 | 0.150146 |
| 1.5 | 0.680825 | 0.389090 |

The average slope is accepted as the average $m$ value ($m = 0.683201$). Any systematic or extreme variations in $m$ as a function of concentration would be accepted as lack of fit of the Hom model to the data.

From the second regression [of YINT versus ln ($C$)], the slope is $n$

($n = 0.763346$) and the intercept is 0.106870. Therefore, the value of $k$ is 1.112799.

The differences between the two techniques reflect the different manner in which the two procedures weight errors in the data.

If the results of the multiple linear regression are used, the final estimation equation becomes

$$\ln (N/N_0) = -1.111331 C^{0.683281} t^{0.695641}$$

Inserting the known information gives the following results:

$$\ln (10/1000) = -1.111331 C^{0.683281} t^{0.695641}$$

$$C^{0.683281} = 0.8334$$

$$C = 0.766 \text{ mg/L}$$

Hence, if a 10-min contact time is accepted as a worst-case condition, and assuming good contactor hydraulics, the maximum chlorine residual required to achieve the design inactivation is 0.766 mg/L. From this information and the chlorine demand of the water, the capacity of a chlorine feed system can be computed.

## Disinfectant residuals

One factor important in evaluating the relative merits of alternative disinfectants is their ability to maintain microbial quality in a water distribution system. With respect to chlorine, free-chlorine residuals may serve to protect the distribution system against regrowth, or at least to serve as a sentinel for the presence of contamination.[185] Other studies have noted, however, the lack of correlation between distribution system water quality and the form or concentration of chlorine residual.[109] Similarly, LeChevallier[186] has reported that microbial slimes grown in tap water may be more sensitive to inactivation by combined chlorine than to free chlorine transported by the overlying water. Regardless of the disinfectant chosen, the water distribution system can never be regarded as biologically sterile. As shown by Means et al.,[187,188] shifts in the dominant form of disinfectant (e.g., from free chlorine to monochloramine) can result in shifts in the taxonomic distribution of microorganisms that inhabit the distribution system.

The ability of chlorine dioxide residuals to maintain distribution system microbial water quality has not been studied. With respect to both ozone and uv, their absence of a residual necessitates the addition of a second disinfectant if a residual in the distribution system is desired.

## Application of Technologies

### Chlorine

**Sources of chemicals.**  Chlorine may be obtained for disinfection in three forms or be generated on site. For very small water treatment plants, solid calcium hypochlorite [$Ca(OCl)_2$] can be applied as a dry powder or in proprietary "tablet" type dispensers. It is relatively more expensive than the other chemical forms, and, particularly in hard waters, the use of calcium hypochlorite can lead to scale formation.

Generally, on a per unit mass basis of active chlorine, the least expensive form of chlorine at large use rates is liquified gas. The use of liquified chlorine gas carries with it certain risks associated with accidental leakage of the gas. As a result, a number of utilities have elected to use the somewhat more expensive sodium hypochlorite, $NaOCl$, as a source of disinfectant.

Upon addition to a water, chlorine gas will reduce the pH and alkalinity, while sodium hypochlorite will raise the pH and alkalinity. In a poorly buffered water using chlorine gas, the addition of a pH control agent may be necessary to control the distribution system water stability. Use of sodium hypochlorite, on the other hand, produces high pH's where chlorine ($OCl^-$) is less effective.

Chlorine and hypochlorites have been produced from the electrolysis of brines and saline solutions since the early days of the twentieth century.[189] This remains an attractive option for remote treatment plants near a cheap source of brine. The basic principle is the use of a direct-current electrical field to effect the oxidation of chloride ion with the simultaneous and physically separated reduction of water to gaseous hydrogen.

In actual practice, operating electrolytic chlorine generating units at voltages as high as 3.85 V is necessary to provide reasonable rates of generation. At these overvoltages, however, additional oxidations, such as formation of chlorate, ohmic heating, and incomplete separation of hydrogen from oxidized products with subsequent dissipative reaction, combine to produce system inefficiencies. For typical electrolytic generating units, current efficiencies of 97 percent may be obtained along with energy efficiencies of 58 percent.[30] These efficiencies are related to the physical configuration of the electrolysis cells, brine concentration, and desired degree of conversion to available chlorine.[190,191]

**Source water chlorination.**  Source water chlorination, or *prechlorination,* is designed to minimize operational problems associated with biological slime formation on filters, pipes, and tanks, and to release potential taste and odor problems from such slimes. In addition, source water chlorination can be used for the oxidation of hydrogen sulfide or reduced iron and manganese. Probably the most common point of addition of chlorine for source water chlorination is the rapid-mix basin (where coagulant is added).

Because of present concerns for minimizing the formation of chlorine-substituted by-products, the use of source water chlorination is being supplanted by the use of other chemical oxidants (e.g., ozone, permanganate ion) for the control of biological fouling, odor, or reduced iron or manganese.

**Postchlorination.**   Postchlorination, or *terminal disinfection,* is the primary application for microbial reduction. Addition of chlorine either immediately before the clear well or immediately before the sand filter is most common. In the latter case (sometimes referred to as *within-plant chlorination*), the filter itself serves, in effect, as a contact chamber for disinfection.

In general, the use of specific contact chambers subsequent to the addition of chlorine to a water has been uncommon. Instead, the clear well, or finished water reservoir, serves the dual function of providing contact to ensure adequate time for microbial inactivation prior to distribution. The distribution system itself, from the entry point until the first consumer's tap, provides additional contact time.

The hydraulic characteristics of most finished water reservoirs, however, are not compatible with the ideal characteristics of chlorine contact chambers. The latter are most desirably plug flow, while the former usually have a large degree of dispersion.

**Superchlorination-dechlorination.**   In the process of superchlorination-dechlorination, which has generally been employed when treating a poor-quality water (with high ammonia nitrogen concentrations, or perhaps with severe taste and odor problems), chlorine is added beyond the breakpoint. This oxidizes the ammonia nitrogen present [Eqs. (14.15), and (14.16)]. Generally, the residual chlorine obtained at this point is higher than may be desired for distribution. The chlorine residual may be decreased by the application of a reducing (dechlorinating) agent (sulfur compounds or activated carbon).

A modern application of chlorination-dechlorination in water treatment may be the judicious application where both high degrees of microbial inactivation as well as low organic by-product formation are desired. Holding water with free chlorine for a period (sufficient to ensure disinfection, but not so long as to produce substantial by-products) then partially (or completely) dechlorinating the water to minimize the production of organic by-products may be possible.

**Chloramination.** Chloramination, the simultaneous application of chlorine and ammonia, or the application of ammonia prior to the application of chlorine (preammoniation), resulting in a stable combined residual, has been a long-standing practice at many utilities (Table 14.6). In current practice, chloramination has been gaining popularity as a means to minimize organic by-product formation.

For example, Jefferson Parish, La., has been using simultaneous addition of ammonia and chlorine for a disinfection process for over 20 years.[193] To further reduce trihalomethane formation, the point of disinfection addition was changed to immediately upstream of the filters (rather than upstream of clarifiers) to permit removal of precursors prior to disinfection. This change maintained satisfactory distribution system water quality, using a chloramine residual of 1.6 mg/L as $Cl_2$ exiting the filters.

A survey of utilities (1984) using combined chlorine indicated that[192]

1. Seventy percent use anhydrous ammonia, 20 percent aqua ammonia, and 10 percent ammonium sulfate.

2. Most utilities use 3:1 to 4:1 chlorine to ammonia feed ratios by weight. Excess ammonia is generally used to make monochloramine predominant; however, some use higher ratios to form more effective dichloramine.

3. No clear consensus exists as to the most preferable point of ammonia application (i.e., source water or postammoniation).

**TABLE 14.6   Utilities with Long Experience of Chloramine Use**

| City | Years of chloramine use (as of April 1984) |
|------|--------------------------------------------|
| Denver | 70 |
| Portland, Ore. | 60 |
| St. Louis | 50 |
| Boston | 40 |
| Indianapolis | 30 |
| Minneapolis | 30 |
| Dallas | 25 |
| Kansas City, Mo. | 20 |
| Milwaukee | 20 |
| Jefferson Parish, La. | 20 |
| Philadelphia | 15 |
| Houston | 2 |
| Miami, Fla. | 2 |
| Orleans Parish, La. | 2 |
| San Diego | 2 |

SOURCE: From Ref. 192.

Preformed chloramine residuals (i.e., chloramines allowed to react for sufficient time to stabilize prior to their use as a disinfectant) are more effective bactericides (*E. coli*, demand-free batch tests) at pH 6 than at pH 8 and at high $Cl_2$:N weight ratios (5:1) rather than at low weight ratios (down to 2:1), presumably because of the greater concentrations of the more biocidal dichloramine. Concurrent addition of ammonia and chlorine is as effective as preammoniation (and at pH 6 is nearly as effective as free residual chlorination). Both concurrent addition of chlorine and ammonia and preammoniation are more effective than addition of preformed chloramines, except at pH 8, where all three modes behave in a similar manner.[194]

In pilot-plant studies, Means et al.[188] found that concurrent and sequential (chlorine at rapid mix and ammonia at end of flocculation tank) methods gave better performance at removing HPC bacteria than preammoniation (but poorer than free chlorination). Concurrent addition gave about as low a total trihalomethane level as preammoniation.

In the presence of concentrations of organic nitrogen similar to ammonia, preformed chloramines may give better performance than dynamically formed chloramines from preammoniation because of the favorable competition for chlorine by many organic N compounds (and their low biocidal potency) relative to inorganic nitrogen.[155]

A major concern with chloramination arises during the transition from free chlorination to chloramination.[187,188] Before-and-after data on distribution system water quality was collected at the Metropolitan Water District of Southern California when the distribution system was changed from a free-chlorine residual to a combined (monochloramine) chlorine residual. No effect occurred on coliform counts; however, the plate counts increased dramatically. In one of the reservoirs, a precipitous drop in chlorine residual occurred following the switchover, associated with nitrification in the reservoir and growth of microorganisms. This was postulated to occur as a reaction between nitrites and monochloramine.

Hospitals and kidney dialysis centers must be alerted to switchovers from free residual chlorination to chloramination. Birrell et al.[195] and Eaton et al.[196] reported cases of chloramine-induced hemolytic anemia in such centers, when appropriate treatment of dialysis water was not provided.

## Chlorine dioxide

**Generation.** Generation of chlorine dioxide on a continuous basis is necessary for its use as a disinfectant. Although a few European potable water treatment plants have been reported to use the acid-chlorite generation process,[16] the most common synthesis route for disinfectant $ClO_2$ generation is the chlorine-chlorite process.

Theoretically, chlorine dioxide may be produced by either the oxidation of a lower-valence compound or reduction of a more oxidized compound of chlorine. Chlorites ($ClO_2^-$) or chlorous acid ($HClO_2$) may be oxidized by chlorine or persulfate ($S_2O_8^{2-}$) to chlorine dioxide, or may undergo autooxidation (disproportionation) to chlorine dioxide in solutions acidified with either mineral or organic acids. Chlorates ($ClO_3^-$) may be reduced by use of chlorides, sulfuric acid, sulfur dioxide, or oxalic acid or by electrochemical means to form chlorine dioxide.[35,78] For practical purposes in water treatment, chlorine dioxide is generated exclusively from chlorite inasmuch as the reductive processes using chlorate as a starting material are capital intensive and competitive only at larger capacities.[35,78]

In the acid-chlorite process, sodium chlorite and hydrochloric acid react according to Eq. (14.29):

$$5NaClO_2 + 4HCl \rightleftharpoons 4ClO_2 + 5NaCl + 2H_2O \qquad (14.29)$$

The resulting chlorine dioxide may be evolved as a gas or removed in solution. Mechanistically, this process occurs by a series of coupled reactions, some of which may involve the in situ formation of chlorine, the catalysis by chloride, and the oxidation of chlorite by chlorine.[34,35,37,78] In addition, the yield of the reaction as well as the rate of the process are improved by low pH values in which both gaseous chlorine and chlorous acid formation are favored. Under these favorable conditions, the reaction proceeds for several minutes; however, to achieve these conditions, excess hydrochloric acid is required.

During the acid-chlorite reaction, the following side reactions result in chlorine production:

$$5ClO_2^- + 5H^+ \rightleftharpoons 3ClO_3^- + Cl_2 + 3H^+ + H_2O \qquad (14.30)$$

$$4ClO_2^- + 4H^+ \rightleftharpoons 2Cl_2 + 3O_2 + 2H_2O \qquad (14.31)$$

$$4HClO_2 \rightleftharpoons 2ClO_2 + HClO_3 + HCl + H_2O \qquad (14.32)$$

If the stoichiometric requirements of the reactants are met, then close to 100 percent of the conversion of chlorite may occur, and a final pH below 0.5 will result.[35,78]

Alternatively, chlorine dioxide may be produced by the oxidation of chlorite with chlorine gas according to reaction (14.33):

$$2NaClO_2 + Cl_2 \rightleftharpoons NaCl + 2ClO_2 \qquad (14.33)$$

As in the previous case, low pHs accelerate the rate of this process, as does excess amounts of chlorine gas. If chlorine gas is used in stoichiometric excess, however, the resultant product may contain a mixture of unconsumed chlorine as well as chlorine dioxide.

The rate of the direct reaction between dissolved $Cl_2$ and chlorite has been measured,[32] with a forward second-order rate constant given by

$$k_f = 1.31 \times 10^{11} \exp\left(-\frac{4800}{T}\right) \quad \text{L/(m)(s)} \qquad (14.34)$$

In the chlorine-chlorite process, sodium chlorite is supplied as a solid powder or a concentrated solution. A solution of chlorine gas in water is produced by a chlorinator-ejector system of design similar to that used in chlorination. The chlorine-water solution and a solution of sodium chlorite are simultaneously fed into a reactor vessel packed with Raschig rings (or other similar materials) to promote mixing.[16] From Eq. (14.33), 1 mole of chlorine is required for 2 moles of sodium chlorite, or 0.78 part of $Cl_2$ per part of $NaClO_2$ by weight. For this reaction to proceed to completion, however, reducing the pH below that provided by the typical chlorine-water solution produced by an ejector, generally with hydrochloric acid, is necessary. In the absence of acid, at 1:1 feed ratios by weight, only 60 percent of the chlorite typically reacts.[16]

To provide greater yields, several options exist. First, producing chlorine-water solutions in excess of 3500 mg/L using pressurized injection of gas is possible. In this case, however, an excess of unreacted chlorine will occur in the product solution, and the resultant disinfectant will consist of a mixture of chlorine and chlorine dioxide. The second option consists of acid addition to the chlorine and chlorite solution. For example, a 0.1 $M$ HCl/mol chloride addition enabled the production of a disinfectant solution of 95 percent purity in terms of chlorine dioxide and achieved a 90 percent conversion of chlorite to chlorine dioxide, the remainder being undetermined.[197] A third process, developed by CIFEC (Paris, France), involves recirculation of the chlorinator ejector discharge water back to the ejector inlet to produce a strong (5 to 6 kg/m$^3$) chlorine solution, typically at pH below 3.0, and, in this manner, to increase the efficiency of chlorite conversion.[16,45] This last option is capable of producing 95 to 99 percent pure solutions of chlorine dioxide.[45] The intricacies of chlorine dioxide reactions and by-products necessitate the careful process monitoring during operation of the generator.[198]

**Example Problem 4.** A water utility has chlorination capacity of 1000 tons/day (454 kg/day) and is considering a switchover to chlorine dioxide to be generated using the chlorine-chlorite process. If the existing chlorination equipment is to be used, what is the maximum production capacity of chlorine dioxide, and how much sodium chlorite must be used under these conditions? Assume ideal stoichiometry and no excess chlorite or chlorine requirements.

**solution.**     From Eq. (14.33), 2 moles of sodium chlorite ($NaClO_2$) react with 1 mole of chlorine to produce 2 moles of chlorine dioxide. Because chlorine has a molecular weight of 70, the current chlorinators have a capacity of 454,000/70 = 6486 mol/day of chlorine. Therefore, 12,971 mol/day of sodium chlorite are required, and the result would be an equal number of moles of chlorine dioxide. The molecular weights are

Chlorine dioxide  35 + 2(16) = 67

Sodium chlorite  23 + 35 + 2(16) = 89

Therefore, the sodium chlorite required is 12,971(89) = $1.15 \times 10^6$ g/day (2540 lb/day). The chlorine dioxide produced would be 12,971(67) = $0.87 \times 10^6$ g/day (1914 lb/day).

**Application of chlorine dioxide.**  The use of chlorine dioxide is limited by two factors. First, the maximum residual that does not cause adverse taste and odor problems is 0.4 to 0.5 mg/L as $ClO_2$.[35] Second, the chlorite produced by reduction of chlorine dioxide as demand is exerted has been found to cause certain types of anemia and other physiological effects in animals, and, therefore, the maximum $ClO_2 + ClO_2^- + ClO_3^-$ residual should be 1 mg/L to minimize this effect.[199]

Initial reports[200] suggested that chlorine dioxide residuals have moderate stability during distribution. Whether the analytical methods used in that study could adequately differentiate chlorine dioxide from its reaction products, however, is not clear.

Insofar as chlorine dioxide is produced free of chlorine (in the acid-chlorite process or in "optimized" chlorine-chlorite processes), the reactions with organic material to produce chlorinated by-products appear less significant than with chlorine.[201–203]

## Ozonation

The use of ozone in water treatment in the United States has been confined to source water oxidation.[23] Additional information on the use of ozone as an oxidant is provided in Chap. 12.

**Generation.**  Because of its instability, ozone is produced by gas-phase electrolytic oxidation of oxygen either by using very dry air or pure oxygen. The ozone-enriched gaseous phase is then contacted with the water to be treated in a bubble contactor (either diffused air or turbine mixed) or in a countercurrent tower contactor. Because of the cost of ozone, it is highly desirable to maximize the efficiency of ozone transfer from gas to liquid.

Most ozone generators used in water treatment use one of two designs.[23] The most common for large plants is a bank of glass tube generators as shown in Fig. 14.7. Small plants may use this type of generator (on a smaller scale) or a plate-type generator in which the ozone is generated between ceramic plates. Cooling the tubes or plates increases the efficiency of ozone production. Ozone generators may use pure oxygen, oxygen-enriched air, or air as the feed gas. If air is used, the most economical operation gives a product stream that contains about 2 percent ozone by weight. Enhancement of the amount of oxygen in the stream increases the economical yield of ozone; for example, pure oxygen can generate a stream containing 5 to 7 percent ozone economically. In any case, the gas stream must have a very low dew point ($-50°C$) and must be free of organic vapors. Figure 14.8 shows a flow diagram for a plant that utilizes oxygen-enriched air.

Ozone generator designs that utilize different electrode configurations are also available, such as the surface discharge models. Also, ozone may be generated by irradiation of air with high-energy ultraviolet radiation (with wavelengths less than 200 nm). Photochemical generators are not yet capable of producing as much ozone as plasma generators, but may be particularly useful for small-scale applications such as swimming pool disinfection.

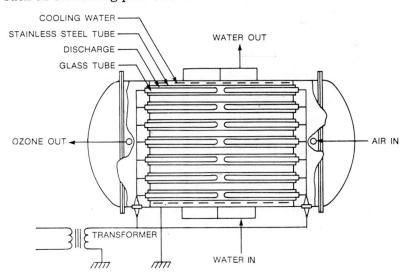

**Figure 14.7**  Large-scale, tube-type generator for production of ozone from air or oxygen by cold plasma discharge.

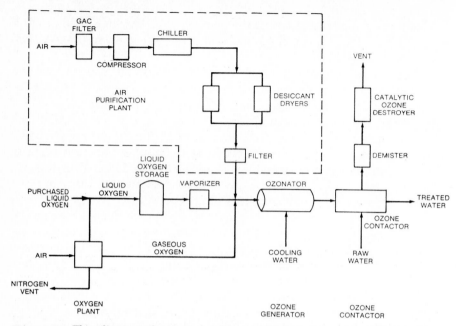

**Figure 14.8**  Flow diagrams for air and oxygen purification for ozone production from oxygen-enriched air. The air purification unit may be omitted when pure oxygen is used or it may be used without oxygen enrichment.

**Ozone contactors.** After generation, ozone is piped to a contactor where the ozone is transferred into the water. The most common type of contactor is the countercurrent sparged tank with diffuser (Fig. 14.9). In this reactor ozone-containing gas forms small bubbles as it is passed through a porous stone at the bottom of the tank. As the bubbles rise through the tank, ozone is transferred from the gas phase into water according to the rate equation

$$\text{Rate of transfer } [\text{mol}/(\text{m}^3)(\text{s})] = K_1 a(C^* - C) \qquad (14.35)$$

Here $C$ ($\text{mol/m}^3$) is the prevailing concentration of ozone in the liquid, $C^*$ is the concentration at saturation, and $K_1 a$ is an overall transfer coefficient ($\text{s}^{-1}$). The value of $C^*$ depends on the percentage of ozone in the gas and may be calculated from the equation

$$C^* = \frac{P_{\text{gas}}}{H} \qquad (14.36)$$

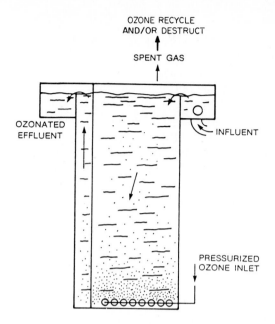

OZONE RECYCLE
AND/OR DESTRUCT

SPENT GAS

OZONATED
EFFLUENT

INFLUENT

PRESSURIZED
OZONE INLET

**Figure 14.9** Diagram of typical countercurrent sparged column for transfer of ozone from gas to liquid phase.

where $P_{gas}$ is the partial pressure of ozone in the gas phase (atm) and $H$ is the Henry's law constant for ozone (0.082 atm-m$^3$/mol at 25°C). At a partial pressure of ozone of 0.02 atm (corresponding to 3 weight percent of ozone in air), Eq. (14.36) predicts that the concentration of ozone in water at saturation will be 0.24 mol/m$^3$ (12 mg/L). More details on theory and application of liquid to gas transfer may be found in Chap. 5.

Ozone contactors such as that shown in Fig. 14.7 may be sized to transfer ozone on a small or a very large scale. Alternately, one may use other designs such as sparged stirred tank reactors and venturi injectors. If injected under a large head of water pressure, ozone may be transferred at higher rates because the saturation concentration of ozone, the $C^*$ term in Eq. (14.35), is higher.

Ozone contactors must have systems for the collection of ozone off-gas. Ozone is toxic and must be kept within Occupational Safety and Health Administration allowable limits within the treatment plant and surrounding areas.[204] In some regions of the United States ozone discharge from the treatment plants may be regulated. As a consequence of these requirements, thermal and catalytic ozone destroyers are routinely used in plants employing ozonation.

Ozone in water is so highly corrosive that only certain materials of construction may be used in plants that generate and utilize ozone for water treatment. Metal in contact with ozone should be 304 stainless steel, and gasket materials should be some form of inert polymer, such as a fluorocarbon. Concrete is a typical material for construction of basins, but joints must be caulked with an inert material.

Unfortunately, the characteristics that promote efficient gas-liquid mass transfer, particularly the desirability of intense agitation, lead to hydraulic characteristics that decrease disinfection efficiency (short-circuiting). Therefore, laboratory results on ozone inactivation of microorganisms are poor predictors of field performance, unless detailed aspects of field-scale hydraulic features are considered.

In one pilot study of GAC-treated surface water (with no other pretreatment),[205] "complete viral inactivation" (more than seven logs) was found using a contactor with 57 s of contact and an ozone dosage of 1.45 mg/L (with a transferred dose, i.e., amount of ozone transferred to the water, of 1.13 mg/L). Satisfactory coliform results (< 2/100 mL) were obtained at a dose of 1.29 mg/L (i.e., transferred dose of 1.04 mg/L).

The contact time required for ozone inactivation of microorganisms at commonly employed ozone doses (several mg/L) is quite short (seconds to several minutes) rather than the characteristically longer disinfection times commonly used for chlorine or chloramines. The lack of a persistent residual following ozone treatment has generally necessitated application of an additional terminal disinfection process such as chloramination.

### Ultraviolet radiation

Ultraviolet radiation can be effectively produced by the use of mercury vapor (or, more recently, antimony vapor) lamps. While, to date, limited experience with uv disinfection in water treatment has accumulated, a substantial operational database associated with uv disinfection of wastewater effluents exists.[20,26]

Ultraviolet has been employed in a variety of physical configurations. In two of these, the uv lamp(s) are surrounded by quartz sheaths, and the jacketed lamps are immersed in the flowing water. The flow may be in a closed or open vessel and may be parallel or perpendicular to the lamp axes. In the third configuration, the water flows through Teflon tubes (which are relatively transparent to uv radiation) surrounded by uv lamps.

For design purposes, ensuring that turbulence occurs (to allow all elements of fluid to come sufficiently close to the lamp surfaces) and yet minimizing the degree of transverse mixing (short-circuiting) is necessary. The latter must explicitly be considered in design calculations.[27]

The contact times for uv disinfection systems can be quite short, generally under 1 min. Therefore, the space required for uv disinfection units is relatively small. Because no residual is created, some additional terminal disinfection process would usually be required. The potential for uv reactions to produce organic by-products is minor because the intensities required for uv disinfection are less than those needed to cause photochemical effects. Operationally, employing an effective cleaning program to periodically remove biological and chemical fouling materials from lamp jacket or Teflon tube surfaces is essential.[20,27]

### Use of multiple disinfectants

In view of the reexamination of disinfection practices that has occurred over the past 15 years, a number of case studies of multiple disinfectants have been described.[54,206] These might include a relatively reactive primary disinfectant (such as $O_3$ or $ClO_2$) followed by a secondary disinfectant that is used to maintain residual in the distribution system.

The Louisville Water Company, for example, conducted a plant-scale examination of a process in which $ClO_2$ was added between the coagulation basin effluent and the softening basin influent along with excess ammonia to convert the free chlorine to combined chlorine (predominantly monochloramine).[206] Some additional chlorine was added after filtration to provide a monochloramine residual in the distribution system. Although some regrowth in the filters occurred, the distribution system bacterial quality was not altered, and the trihalomethane concentration was reduced from 30 µg/L (free chlorination) to less than 5 µg/L.

### Relative Comparisons

Table 14.7 summarizes important aspects and technical advantages and disadvantages of the major technologies. The dominant disinfection technology in the United States is presently free residual chlorination. Chloramination, chlorination with partial dechlorination, and chlorine dioxide treatment are viable alternative primary disinfectants. In addition, ozone, or possibly uv, when supplemented with a chemical that can produce a lasting residual in the distribution system, have application.

**TABLE 14.7    Applicability of Alternative Disinfection Techniques**

| Consideration | $Cl_2$[†] | $Cl_2/deCl_2$ | $O_3$ | $ClO_2$ | UV |
|---|---|---|---|---|---|
| Size of plant | All sizes | All sizes | Medium to large | Small to medium | Small to medium |
| Equipment reliability | Good | Fair to good | Fair to good | Good | Fair to good |
| Relative complexity of technology | Simple to moderate | Moderate | Complex | Moderate | Simple to moderate |
| Safety concerns | Yes | Yes | Moderate | Yes | Minimal |
| Bactericidal | Good | Good | Good | Good | Good |
| Virucidal | 1[‡] | 1[‡] | Good | Good | Good |
| By-products of possible health concern | 2[‡] | 2[‡] | 3[‡] | Yes | No |
| Persistent residual | Long | None | None | Moderate | None |
| Contact time | Moderate | Moderate | Short | Moderate | Short |
| Reacts with ammonia | Yes | Yes | No | No | No |
| pH dependent | Yes | Yes | Slight | Slight | No |
| Process control | Well developed | Well developed | Developing | Developing | Developing |

†Includes chloramination.

‡1 = moderate for free residual chlorination; poor for combined residual chlorination; 2 = fewer by-products with combined residual chlorination, 3 = health significance of by-products is unresolved at present.

# References

1. C. Averill, Facts Regarding the Disinfecting Powers of Chlorine. Letter to Hon. J. I. Degraff, Mayor of the City of Schenectady, SS Riggs Printer, Schenectady, N.Y., 1832.
2. J. Race, *Chlorination of Water,* Wiley, New York, 1918.
3. J. L. Melnick et al., "Viruses in Water," *Bull. World Health Organization,* vol. 56, 1978, p. 499.
4. J. W. Mosley, "Transmission of Viral Diseases by Drinking Water," in G. Berg (ed.), *Transmission of Viruses by the Water Route,* Wiley, New York, 1966.
5. S. R. Palmer et al., "Waterborne Outbreak of *Campylobacter* Gastroenteritis," *Lancet,* 1983, pp. 287–290.
6. E. E. Geldreich, "Water-borne Pathogens," in R. Mitchell (ed.), *Water Pollution Microbiology,* Wiley-Interscience, New York, 1971.
7. D. Juranek, "Waterborne Giardiasis," *Waterborne Transmission of Giardiasis,* U.S. Environmental Protection Agency, EPA 600/9-79-001, 1979.
8. J. B. Rose, "Detection of *Cryptosporidium* in Drinking Water Supplies," *Proc. AWWA Ann. Conf.,* Kansas City, Mo., June 1987.
9. M. L. Rosenberg et al., "The Risk of Acquiring Hepatitis from Sewage-Contaminated Water," *Amer. J. Epidemiology,* vol. 112, 1980, p. 17.
10. R. C. Hoehn et al., "Committee Report—Disinfection, Water Quality Control and Safety Practices of the Water Utility Industry in 1978 in the United States," *J. AWWA,* vol. 75, no. 1, 1983, p. 51.
11. L. R. Belohlav and E. T. McBee, "Discovery and Early Work," in J. S. Sconce (ed.),

*Chlorine: Its Manufacture, Properties and Uses,* ACS Monograph No. 154, Van Nostrand Reinhold, New York, 1962.

12. J. C. Baker, "Use of Chlorine in the Treatment of Sewage," *Surveyor,* vol. 69, 1926, p. 241.

13. H. A. Faber, "How Modern Chlorination Started. The Story of the Solution Feed Process as It Began Forty Years Ago," *Water and Sewage Works,* vol. 99, 1952, p. 45.

14. A. G. Hooker, *Chloride of Lime in Sanitation,* Wiley, New York, 1913.

15. R. L. Wolfe, N. R. Ward, and B. H. Olson, "Inorganic Chloramines as Drinking Water Disinfectants: A Review," *J. AWWA,* vol. 76, no. 5, 1984, p. 74.

16. G. W. Miller et al., *An Assessment of Ozone and Chlorine Dioxide for Treatment of Municipal Water Supplies,* U.S. Environmental Protection Agency, EPA 600/8-78-018, 1978.

17. W. H. Rapson, "From Laboratory Curiosity to Heavy Chemical," *Chem. Can.,* vol. 18, no. 1, 1966, p. 2531.

18. R. N. Aston and J. F. Synan, "Chlorine Dioxide as a Bactericide in Waterworks Operation," *J. New Engl. Water Works Assoc.,* vol. 62, 1948, p. 80.

19. J. C. Morris, "Chlorination and Disinfection—State of the Art," *J. AWWA,* vol. 63, no. 12, 1971, p. 769.

20. WPCF, *Wastewater Disinfection,* Manual of Practice FD-10, Water Pollution Control Federation, Alexandria, Va., 1986.

21. R. G. Rice, C. M. Robson, G. W. Miller, and A. G. Hill, "Uses of Ozone in Drinking Water Treatment," *J. AWWA,* vol. 73, no. 1, 1981, p. 44.

22. AWWA, *Water Quality & Treatment,* 3d ed., McGraw-Hill, New York, 1971.

23. W. H. Glaze, "Drinking Water Treatment with Ozone," *Environmental Sci. Technol.,* vol. 21, 1987, p. 224.

24. A. Downes and T. Blount, "Research on the Effect of Light upon Bacteria and Other Organisms," *Proc. Roy. Soc. (London),* vol. 26, 1877, p. 488.

25. M. Luckeish et al., "Germicidal Energy," *General Electric Rev.* 1944.

26. C. B. Huff et al., "Study of Ultraviolet Disinfection of Water and Factors in Treatment Efficiency," *U.S. Public Health Reports,* vol. 80, 1965, p. 695.

27. U.S. Environmental Protection Agency, *Design Manual: Municipal Wastewater Disinfection,* EPA/625/1-86/021, 1986.

28. E. J. Laubusch, "Sulfur Dioxide," *Public Works,* vol. 94, no. 8, 1963, p. 117.

29. R. J. Baker, "Characteristics of Chlorine Compounds," *J. Water Pollution Control Fed.,* vol. 41, 1969, p. 482.

30. A. J. Downs and C. J. Adams, *The Chemistry of Chlorine, Bromine, Iodine and Astatine,* Pergamon, Oxford, 1973.

31. J. C. Morris, "The Mechanism of the Hydrolysis of Chlorine," *J. Amer. Chem. Soc.,* vol. 68, 1946, p. 1692.

32. E. M. Aieta and P. V. Roberts, "The Chemistry of Oxo-Chlorine Compounds Relevant to Chlorine Dioxide Generation," in R. L. Jolley et al. (eds.), *Water Chlorination: Environmental Impact and Health Effects,* vol. 5, Lewis, Chelsea, Mich., 1985.

33. J. C. Morris, "The Acid Ionization Constant of HOCl from 5°C to 35°C," *J. Phys. Chem.,* vol. 70, 1966, p. 3798.

34. M. G. Noack and R. L. Doerr, "Chlorine Dioxide, Chlorous Acid and Chlorites," H. F. Mark et al. (eds.), *Kirk-Othmer Encyclopedia of Chemical Technology,* vol. 5, 3d ed., John Wiley, New York, 1979.

35. W. J. Masschelein, "Use of Chlorine Dioxide for the Treatment of Drinking Water," *Oxidation Techniques in Drinking Water Treatment, Drinking Water Pilot Project Report IIA,* U.S. Environmental Protection Agency, EPA 570/9-79-020, 1979.

36. R. Battino, *Chlorine Dioxide Solubilities,* Pergamon, New York, 1982, p. 454.

37. G. Gordon et al., "The Chemistry of Chlorine Dioxide," *Progr. in Inorganic Chem.,* vol. 15, 1972, p. 201.

38. M. L. Granstrom and G. F. Lee, "Rates and Mechanisms of Reactions Involving Oxychloro Compounds," *Public Works,* vol. 88, no. 12, 1957, p. 902.

39. M. Medir and F. Giralt, "Stability of Chlorine Dioxide in Aqueous Solution," *Water Research,* vol. 16, 1982, p. 1379.

40. T. C. Manley and S. J. Niegowski, *Kirk-Othmer Encyclopedia of Chemical Technology,* vol. 14, 2d ed., 1967, p. 410.
41. J. A. Roth, *Ozone Solubilities,* Pergamon, New York, 1982, p. 474.
42. S. L. Chang, "Modern Concept of Disinfection," *ASCE J. Sanitary Eng. Div.,* vol. 97, 1971, p. 689.
43. I. Weil and J. C. Morris, "Equilibrium Studies on *N*-Chloro Compounds," *J. Amer. Chem. Soc.,* vol. 71, 1949, p. 3.
44. G. R. Weber et al., "Effect of Ammonia on the Germicidal Efficiency of Chlorine in Neutral Solutions," *J. AWWA,* vol. 32, no. 11, 1940, p. 1904.
45. G. C. White, *Disinfection of Wastewater and Water for Reuse,* Van Nostrand Reinhold, New York, 1978.
46. M. J. Taras, "Preliminary Studies on the Chlorine Demand of Specific Chemical Compounds," *J. AWWA,* vol. 42, no. 5, 1950, p. 462.
47. T. A. Pressley, F. B. Dolloff, and S. G. Roan, "Ammonia Nitrogen Removal by Breakpoint Chlorination," *Environmental Sci. Technol.,* vol. 6, 1972, p. 622.
48. A. E. Griffin and N. S. Chamberlin, "Some Chemical Aspects of Breakpoint Chlorination," *J. New Engl. Water Works Assoc.,* vol. 55, 1941, p. 3.
49. A. E. Griffin and N. S. Chamberlin, "Relation of Ammonia Nitrogen to Breakpoint Chlorination," *Amer. J. Public Health,* vol. 31, 1941, p. 803.
50. J. E. McKee et al., "Chemical and Colicidal Effects of Halogens in Sewage," *J. Water Pollution Control Fed.* vol. 32, 1960, p. 795.
51. B. M. Saunier and R. E. Selleck, "The Kinetics of Breakpoint Chlorination in Continuous Flow Systems," *J. AWWA,* vol. 71, no. 3, 1979, p. 164.
52. T. W. Trofe et al., "Kinetics of Monochloramine Decomposition in the Presence of Bromine," *Environmental Sci. Technol.* vol. 14, 1980, p. 544.
53. J. L. S. Saguinsin and J. C. Morris, "The Chemistry of Aqueous Nitrogen Trichloride," in J. D. Johnson (ed.), *Disinfection: Water and Wastewater* Ann Arbor Science, Ann Arbor, Mich., 1975.
54. R. N. Kinman and R. F. Layton, "New Method for Water Disinfection," *J. AWWA,* vol. 68, no. 6, 1976, p. 298.
55. R. L. Valentine et al., "A Spectrophotometric Study of the Formation of an Unidentified Monochloramine Decomposition Product," *Water Research,* vol. 20, no. 8, 1986, p. 1067.
56. E. T. Gray et al., "Chloramine Equilibria and the Kinetics of Disproportionation in Aqueous Solution," *Organometals and Organometalloids, Occurrence and Fate in the Environment,* ACS Symp. Ser. No. 82, American Chemical Society, Washington, D.C., 1978.
57. G. C. White, *Handbook of Chlorination,* Van Nostrand Reinhold, New York, 1972.
58. J. C. Morris and R. A. Isaac, "A Critical Review of Kinetics and Thermodynamic Constants for the Aqueous Chlorine," in R. L. Jolley et al. (eds.), *Water Chlorination: Environmental Impact and Health Effects,* vol. 5, Lewis, Chelsea, Mich., 1985.
59. R. L. Valentine and C. T. Jafvert, "General Acid Catalysis of Monochloramine Disproportionation," *Environmental Sci. Technol.,* vol. 22. no. 6, 1988, p. 691.
60. A. T. Palin, *Chemistry and Control of Modern Chlorination,* LaMotte Chemical Products Co., Chestertown, Md., 1983.
61. W. R. Haag and M. H. Lietzke, "A Kinetic Model for Predicting the Concentration of Active Halogen Species in Chlorinated Saline Cooling Waters," in R. L. Jolley et al. (eds.), *Water Chlorination: Environmental Impact and Health Effects,* Butterworths, Stoneham, Mass., 1980.
62. R. A. Isaac et al., "Subbreakpoint Modeling of the $HOBr-NH_3-OrgN$ Reactions," in R. L. Jolley et al. (eds.), *Water Chlorination: Environmental Impact and Health Effects,* vol. 5, Lewis, Chelsea, Mich., 1985.
63. J. C. Morris, "Kinetics of Reactions between Aqueous Chlorine and Nitrogen Compounds," in S. D. Faust (ed.), *Principles and Application of Water Chemistry,* Wiley, New York, 1967.
64. R. A. Isaac and J. C. Morris, "Rates of Transfer of Active Chlorine between Nitrogenous Substances," in R. L. Jolley et al. (eds.), *Water Chlorination: Environmental Impact and Health Effects,* Butterworth, Stoneham, Mass., 1980.
65. N. C. Wright, "The Action of Hypochlorites on Amino Acids and Proteins. The Effect of Acidity & Alkalinity," *Biochem. J.,* vol. 30, 1936, p. 1661.

66. R. W. R. Baker, "Studies on the Reaction between Sodium Hypochlorite and Proteins. I. Physiocochemical Study of the Course of the Reaction," *Biochem. J.* vol. 41, 1947, p. 337.
67. K. L. Murphy et al., "Effect of Chlorination Practice on Soluble Organics," *Water Research,* vol. 9, 1975, p. 389.
68. J. DeLaat et al., "Chlorination de Composes Organics: Demand on Chlore et Reactivite vis à de la Furmation des Trihalomethanes: Incidence de L'Azote Ammoniacal," *Water Research,* vol. 16, 1982, p. 1437.
69. S. W. Krasner et al., "Free Chlorine Versus Monochloramine in Controlling Off Tastes and Odors in Drinking Water," *Proc. AWWA An. Conf.,* Denver, Colo., June 1986.
70. J. A. Wojtowicz, "Chlorine Monoxide, Hypochlorous Acid and Hypochlorites," in H. F. Mark et al. (eds.), *Kirk-Othmer Encyclopedia of Chemical Technology,* vol. 5, 3d ed., Wiley, New York, 1979.
71. J. M. Cachaza, "Kinetics of Oxidation of Nitrite by Hypochlorite in Aqueous Basic Solution," *Can. J. Chem.,* vol. 54, 1976, p. 3401.
72. R. L. Valentine, "Disappearance of Monochloramine in the Presence of Nitrite," in R. L. Jolley et al. (eds.), *Water Chlorination: Environmental Impact and Health Effects,* vol. 5, Lewis, Chelsea, Mich., 1985.
73. D. Feben and M. J. Taras, "Chlorine Demand Constants of Detroit's Water Supply, *J. AWWA,* vol. 42, no. 5, 1950, p. 453.
74. C. N. Haas and S. B. Karra, "Kinetics of Microbial Inactivation by Chlorine. I. Review of Results in Demand-Free Systems," *Water Research,* vol. 18, 1984, p. 1443.
75. V. L. Snoeyink and M. T. Suidan, "Dechlorination by Activated Carbon and Other Dechlorinating Agents," in J. D. Johnson (ed.), *Disinfection: Water and Wastewater,* Ann Arbor Science, Ann Arbor, Mich., 1975.
76. R. S. Ingols and G. M. Ridenous, "Chemical Properties of Chlorine Dioxide in Water Treatment," *J. AWWA,* vol. 40, no. 11, 1948, p. 1207.
77. C. Sikorowska, "Influence of Pollutions on Chlorine Dioxide Demand of Water, *Gaz. Woda. Tech. Sanit.,* vol. 35, no. 12, 1961, p. 4645.
78. W. J. Masschelein, *Chlorine Dioxide,* Ann Arbor Science, Ann Arbor, Mich., 1979.
79. J. E. Wajon et al., "Oxidation of Phenol and Hydroquinone by Chlorine Dioxide," *Environmental Sci. Technol.,* vol. 16, 1982, p. 396.
80. C. Rav-Acha, "The Reactions of Chlorine Dioxide with Aquatic Organic Materials and Their Health Effects," *Water Research,* vol. 18, 1984, p. 1329.
81. J. Hoigne and H. Bader, "Ozonation of Water: Role of Hydroxyl Radicals as Oxidizing Intermediates," *Science,* vol. 190, 1975, p. 782.
82. J. Hoigne and H. Bader, "The Role of Hydroxyl Radical Reactions in Ozonation Processes in Aqueous Solution," *Water Research,* vol. 10, 1976, p. 377.
83. M. D. Gurol and P. C. Singer, "Kinetics of Ozone Decomposition: A Dynamic Approach," *Environmental Sci. Technol.,* vol. 16, no. 7, 1982, p. 377.
84. W. J. Cooper, R. G. Zika, and M. S. Steinhauer, "The Effect of Bromide Ion in Water Treatment. II. A Literature Review of Ozone and Bromide Interactions and the Formation of Organic Bromine Compounds," *Ozone Sci. Eng.,* vol. 7, 1985, p. 313.
85. M. D. Gurol et al., "Oxidation of Cyanides in Industrial Wastewaters by Ozone," *Environmental Progr.,* vol. 4, no. 1, 1985, p. 46.
86. J. R. Cortelyou et al., "The Effects of Ultraviolet Irradiation on Large Populations of Certain Waterborne Bacteria in Motion," *Appl. Microbiol.,* vol. 2, 1954, p. 227.
87. C. N. Haas and G. P. Sakellaropoulous, "Rational Analysis of Ultra-Voilet Disinfection Reactors," *Proc. ASCE Environmental Engineering Speciality Conf.,* 1979.
88. G. J. Bonde, "Bacteriological Methods for Estimation of Water Pollution," *Health Lab. Sci.,* vol. 3, no. 2, 1966, p. 124.
89. E. B. Phelps, "The Disinfection of Sewage and Sewage Filter Effluents," *USGS Water Supply Paper 229,* 1909.
90. C. T. Butterfield et al., "Influence of pH and Temperature on the Survival of Coliforms and Enteric Pathogens When Exposed to Free Chlorine," *U.S. Pub. Health Rep.,* vol. 58, 1943, p. 1837.
91. C. T., Butterfield et al., "Influence of pH and Temperature on the Survival of Coliforms and Enteric Pathogens When Exposed to Chlorine," *U.S. Pub. Health Rep.,* vol. 61, 1946, p. 157.

92. E. Wattie and C. T. Butterfield, "Relative Resistance of *Escherichia coli* and *Eberthella typhosa* to Chlorine and Chloramines," *U.S. Publ. Health Rep.*, vol. 59, 1944, p. 1661.

93. P. W. Kabler, "Relative Resistance of Coliform Organisms and Enteric Pathogens in the Disinfection of Water with Chlorine," *J. AWWA*, vol. 43, no. 7, 1951, p. 553.

94. APHA, AWWA, WPCF, *Standard Methods for the Examination of Water and Wastewater*, A. E. Greenberg, L. S. Clesceri, and R. R. Trussell (eds.), American Public Health Association, Washington, D.C., 1989.

95. J. A. Clark and J. E. Pagel, "Pollution Indicator Bacteria Associated with Municipal Raw and Drinking Water Supplies," *Canad. J. Microbiol.*, vol. 23, 1977, p. 465.

96. M. J. Allen et al., "The Impact of Excessive Bacterial Populations in Coliform Methodology," *1977 Ann. Meeting of the American Society for Microbiology*, 1977.

97. Anon. editorial, "Is Chlorination Effective against All Waterborne Disease." *J. Amer. Med. Assoc.*, vol. 78, 1922, p. 283.

98. S. Kelly and W. W. Sanderson, "The Effect of Chlorine in Water on Enteric Viruses," *Amer. J. Pub. Health*, vol. 48, 1958, p. 1323.

99. W. O. K. Grabow et al., "Inactivation of Hepatitis a Virus, Other Enteric Viruses and Indicator Organisms in Water by Chlorination," *Water Sci. Technol.*, vol. 17, 1985, p. 657.

100. A. J. Rubin et al., "Disinfection of Amoebic Cysts in Water with Free Chlorine," *J. Water Pollution Control Fed.*, vol. 55, no. 9, 1983, p. 1174.

101. E. L. Jarrol et al., "Effect of Chlorine on *Giardia lamblia* Cyst Viability," *Appl. Environmental Microbiol.*, vol. 41, 1981, p. 483.

102. P. V. Scarpino et al., "Effect of Particulates on Disinfection of Enteroviruses and Coliform Bacteria in Water by Chlorine Dioxide," *Proc. AWWA Water Qual. Tech. Conf.*, Kansas City, Mo., December 1977.

103. J. G. Leahy, *Inactivation of Giardia muris Cysts by Chlorine and Chlorine Dioxide*, M.S. Thesis, Ohio State University, 1985.

104. S. Farooq, *Kinetics of Inactivation of Yeasts and Acid-Fast Organisms with Ozone*, Ph.D. Thesis, University of Illinois at Urbana-Champaign, 1976.

105. E. W. Rice and J. C. Hoff, "Inactivation of *Giardia lamblia* Cysts by Ultraviolet Irradiation," *Appl. Environmental Microbiol.*, vol. 42, no. 3, 1981, p. 546.

106. J. B. Rose et al., "Isolating Viruses from Finished Water," *J. AWWA*, vol. 78, no. 1, 1986, p. 56.

107. R. S. Engelbrecht et al., *New Microbial Indicators of Disinfection Efficiency*, U.S. Environmental Protection Agency, EPA 600/2–77/052, 1977.

108. R. S. Engelbrecht et al., *Acid-Fast Bacteria and Yeasts as Indicators of Disinfection Efficiency*, U.S. Environmental Protection Agency, EPA 600/2–79/091, 1979.

109. C. N. Haas et al., "Microbial Alterations in Water Distribution Systems and Their Relationship to Physical-Chemical Characteristics," *J. AWWA*, vol. 75, no. 9, 1983, p. 475.

110. C. N. Haas et al., "The Ecology of Acid-Fast Organisms in Water Supply, Treatment, and Distribution Systems," *J. AWWA*, vol. 75, no. 3, 1983, p. 139.

111. C. N. Haas and S. B. Karra, "Kinetics of Wastewater Chlorine Demand Exertion," *J. Water Pollution Control Fed.*, vol. 56, no. 2, 1984, p. 170.

112. C. N. Haas et al., "Removal of New Indicators by Coagulation-Flocculation and Sand Filtration," *J. AWWA*, vol. 77, no. 2, 1985, p. 67.

113. C. N. Haas et al., "Field Observations on the Occurrence of New Indicators of Disinfection Efficiency," *Water Research*, vol. 19, 1985, p. 323.

114. C. N. Haas et al., "The Utility of Endotoxins as a Surrogate Indicator in Potable Water Microbiology," *Water Research*, vol. 17, 1983, p. 803.

115. V. J. Cabelli, "*Clostridium perfringens* as a Water Quality Indicator," in A. W. Hoadley and B. J. Dutka (eds.), *Bacterial Indicators/Health Hazards Associated with Water*, ASTM, Philadelphia, Pa., 1977.

116. W. O. K. Grabow, "The Virology of Waste Water Treatment," *Water Research*, vol. 2, 1968, p. 675.

117. H. Chick, "An Investigation of the Laws of Disinfection," *J. Hygiene*, vol. 8, 1908, p. 92.

118. H. E. Watson, "A Note on the Variation of the Rate of Disinfection with Change in the Concentration of the Disinfectant," *J. Hygiene,* vol. 8, 1908, p. 536.

119. C. N. Haas and S. B. Karra, "Kinetics of Microbial Inactivation by Chlorine. II. Kinetics in the Presence of Chlorine Demand," *Water Research,* vol. 18, 1984, p. 1451.

120. J. C. Hoff, *Inactivation of Microbial Agents by Chemical Disinfectants,* U.S. Environmental Protection Agency, EPA/600/2-86/067, 1986.

121. V. P. Olivieri, et al., "Inactivation of Virus in Sewage," *ASCE J. Sanit. Eng. Div.,* vol. 97, no. 5, 1971, p. 661.

122. H. Heukelekian and M. B. Smith, "Disinfection of Sewage with Chlorine," *Sewage Ind. Wastes,* vol. 22, 1950, p. 1509.

123. G. M. Fair et al., "The Behavior of Chlorine as a Water Disinfectant," *J. AWWA,* vol. 40, no. 10, 1948, p. 1051.

124. S. Carlson et al., "Water Disinfection by Means of Chlorine: Killing of Aggregate Bacteria," *Zbl. Bakteriol. Hygiene. I. Abt. Orig. B.,* vol. 161, 1975, p. 233.

125. B. F. Severin et al., "Series Event Kinetic Model for Chemical Disinfection," *J. Environmental Eng.,* vol. 110, no. 2, 1984, p. 430.

126. C. W. Hiatt, "Kinetics of the Inactivation of Viruses," *Bacteriol. Rev.,* vol. 28, 1964, p. 150.

127. C. N. Haas, "A Mechanistic Kinetic Model for Chlorine Disinfection," *Environmental Sci. Technol.,* vol. 14, no. 3, 1980, p. 339.

128. O. Cerf, "Tailing of Survival Curves of Bacterial Spores," *J. Appl. Bacteriol.,* vol. 42, 1977, p. 1.

129. S. Gard, "Theoretical Considerations in the Inactivation of Viruses by Chemical Means," *Ann. NY Acad. Sci.,* vol. 83, 1960, p. 638.

130. R. E. Selleck, H. Collins, and G. C. White, "Kinetics of Bacterial Deactivation with Chlorine," *ASCE J. Environmental Eng. Div.,* vol. 104, 1978, p. 1197.

131. E. Katzenelson, B. Kletter, and H. I. Shuval, "Inactivation Kinetics of Viruses and Bacteria in Water by Use of Ozone," *J. AWWA,* vol. 66, no. 12, 1974, p. 725.

132. C. N. Haas and E. C. Morrison, "Repeated Exposure of *Escherichia coli* to Free Chlorine: Producation of Strains Possessing Altered Sensitivity," *Water, Air, Soil Pollution,* vol. 16, 1981, p. 233.

133. L. W. Hom, "Kinetics of Chlorine Disinfection in an Ecosystem," *ASCE J. Sanitary Eng. Div.,* vol. 98, no. 1, 1972, p. 183.

134. D. Roy et al., "Kinetics of Enteroviral Inactivation by Ozone," *ASCE J. Environmental Eng. Div.,* vol. 107, no. 5, 1981, p. 887.

135. D. Roy et al., "Mechanism for Enteroviral Inactivation by Ozone," *Appl. Environmental Microbiol.,* vol. 41, no. 3, 1981, p. 718.

136. M. A. Benarde et al., "Kinetics and Mechanism of Bacterial Disinfection by Chlorine Dioxide," *Appl. Microbiol.,* vol. 15, no. 2, 1967, p. 257.

137. J. E. Bailey and D. F. Ollis, *Biochemical Engineering Fundamentals,* 2d ed., McGraw-Hill, New York, 1986.

138. R. R. Trussell and J. Chao, "Rational Design of Chlorine Contact Facilities," *J. Water Pollution Control Fed.,* vol. 49, no. 4, 1977, p. 659.

139. C. N. Haas, "Micromixing and Dispersion in Chlorine Contactors," *Environmental Technol. Lett.,* vol. 9, 1988, p. 35.

140. F. Nissen, *Zeitschr. F. Hygiene,* vol. 8, 1890, p. 62.

141. K. Holwerda, *Mededeelingen van den Dienst der Volksgezondheid in Ned-Indie,* vol. 17, 1928, p. 251.

142. S. L. Chang, "Destruction of Micro-Organisms," *J. AWWA,* vol. 36, no. 11, 1944, p. 1192.

143. L. Friberg and E. Hammarstrom, "The Action of Free Available Chlorine on Bacteria and Bacterial Viruses," *Acta Patholol. Microbiol. Scand.,* vol. 38, 1956, p. 127.

144. L. Friberg, "Further Qualitative Studies on the Reaction of Chlorine with Bacteria in Water Disinfection," *Acta Patholol. Microbiol. Scand.,* vol. 40, 1957, p. 67.

145. C. N. Haas and R. S. Engelbrecht, "Physiological Alterations of Vegetative Microorganisms Resulting from Aqueous Chlorination," *J. Water Pollution Control Fed.,* vol. 52, 1980, p. 1976.

146. C. N. Haas and R. S. Engelbrecht, "Chlorine Dynamics during Inactivation of Coliforms, Acid-Fast Bacteria and Yeasts," *Water Research,* vol. 14, 1980, p. 1749.

147. V. P. Olivieri, et al., Reaction of Chlorine and Chloramines with Nucleic Acids under Disinfection Conditions, in R. L. Jolley et al. (eds.), *Water Chlorination: Environmental Impact and Health Effects,* Butterworths, Stoneham, Mass., 1980.

148. W. H. Dennis et al., "The Reaction of Nucleotides with Aqueous Hypochlorous Acid," *Water Research,* vol. 13, 1979, p. 357.

149. W. H. Dennis et al., "Mechanism of Disinfection: Incorporation of Cl-36 into f2 Virus," *Water Research,* vol. 13, 1979, p. 363.

150. C. Venkobachar et al., "Mechanism of Disinfection," *Water Research,* vol. 9, 1975, p. 119.

151. C. Venkobachar et al., "Mechanism of Disinfection: Effect of Chlorine on Cell Membrane Functions," *Water Research,* vol. 11, 1977, p. 727.

152. K. M. Tenno, R. S. Fujioka and P. C. Loh, "The Mechanism of Poliovirus Inactivation by Hypochlorous Acid," in R. L. Jolley et al. (eds.), *Water Chlorination: Environmental Impact and Health Effects,* Butterworths, Stoneham, Mass., 1980.

153. R. S. Fujioka, K. M. Tenno and P. C. Loh, "Mechanism of Chloramine Inactivation of Poliovirus: A Concern for Regulators," in R. L. Jolley et al. (eds.), *Water Chlorination: Environmental Impact and Health Effects,* Lewis, Chelsea, Mich., 1985.

154. T. H. Feng, "Behavior of Organic Chloramines in Disinfection," *J. Water Pollution Control Fed.,* vol. 38, 1966, p. 614.

155. R. L. Wolfe and B. H. Olson, "Inability of Laboratory Models to Accurately Predict Field Performance of Disinfectants," in R. L. Jolley et al. (eds.), *Water Chlorination: Environmental Impact and Health Effects,* Lewis, Chelsea, Mich., 1985.

156. P. V. Scarpino et al., "A Comparative Study of the Inactivation of Viruses in Water by Chlorine," *Water Research,* vol. 6, 1972, p. 959.

157. C. N. Haas et al., "Alteration of Chemical and Disinfectant Properties of Hypochlorite by Sodium, Potassium and Lithium," *Environmental Sci. Technol.,* vol. 20, 1986, p. 822.

158. M. A. Benarde et al., "Efficiency of Chlorine Dioxide as a Bactericide," *Appl. Microbiol.,* vol. 13, no. 5, 1965, p. 776.

159. P. V. Scarpino et al., *Effect of Particulates on Disinfection of Enteroviruses and Coliform Bacteria in Water by Chlorine Dioxide,* U.S. Environmental Protection Agency, EPA-600/2-79-054, 1979.

160. Y. S. R. Chen, O. J. Sproul, and A. J. Rubin, "Inactivation of *Naegleria gruberi* Cysts by Chlorine Dioxide," *Water Research,* vol. 19, no. 6, 1985, p. 783.

161. C. I. Noss, W. H. Dennis, and V. P. Olivieri, "Reactivity of Chlorine Dioxide with Nucleic Acids and Proteins," in R. L. Jolley et al. (eds.), *Water Chlorination: Environmental Impact and Health Effects,* Lewis, Chelsea, Mich., 1985.

162. V. P. Olivieri et al., "Mode of Action of Chlorine Dioxide on Selected Viruses," in R. L. Jolley et al. (eds.), *Water Chlorination: Environmental Impact and Health Effects,* Lewis, Chelsea, Mich., 1985.

163. D. Roy, *Inactivation of Enteroviruses by Ozone,* Ph.D. Thesis, University of Illinois at Urbana-Champaign, 1979.

164. S. Farooq et al., "The Effect of Ozone Bubbles on Disinfection," *Progr. Water Technol.,* vol. 9, no. 2, 1977, p. 233.

165. E. Dahi and E. Lund, "Steady State Disinfection of Water by Ozone and Sonozone," *Ozone Sci. Eng.,* 1980, p. 13.

166. C. Hamelin and Y. S. Chung, "Role of the POL, REC, and DNA Gene Products in the Repair of Lesions Produced in *E. coli* DNA by Ozone," *Studia (Berlin),* vol. 68, 1978, p. 229.

167. G. B. Wickramanayake et al., "Effects of Ozone and Storage Temperature on *Giardia* Cysts," *J. AWWA,* vol. 77, no. 8, 1985, p. 74.

168. J. Jagger, *Introduction to Research in Ultraviolet Photobiology,* Prentice-Hall, Englewood Cliffs, N.J., 1967.

169. O. K. Scheible and C. D. Bassell, *Ultraviolet Disinfection of a Secondary Wastewater Treatment Plant Effluent,* U.S. Environmental Protection Agency, EPA-600/2-81-152, 1981.

170. B. F. Severin et al., "Effects of Temperature on Ultraviolet Light Disinfection," *Environmental Sci. Technol.*, vol. 17, 1983, p. 717.

171. R. Milbauer and N. Grossowicz, "Effect of Growth Conditions on Chlorine Sensitivity of *Escherichia coli*," *Appl. Microbiol.*, vol. 7, 1959, p. 71.

172. J. D. Berg et al., "Disinfection Resistance of *Legionella pneumophila Escherichia coli* Grown in Continuous and Batch Culture," in R. L. Jolley et al. (eds.), *Water Chlorination: Environmental Impact and Health Effects*, Lewis, Chelsea, Mich., 1985.

173. R. Milbauer and N. Grosswicz, "Reactivation of Chlorine-Inactivated *Escherichia coli*," *Appl. Microbiol.*, vol. 7, 1959, p. 67.

174. G. A. McFeters, J. S. Kippin, and M. W. LeChevallier, "Injured Coliforms in Drinking Water," *Appl. Environmental Microbiol.*, vol. 51, no. 1, 1986, p. 1.

175. A. Singh, R. Yeager, and G. A. McFeters, "Assessment of *in vivo* Revival, Growth, and Pathogenicity of *Escherichia coli* Strains after Copper and Chlorine Induced Injury," *Appl. Environmental Microbiol.*, vol. 52, no. 4, 1986, p. 832.

176. D. C. Young and D. G. Sharp, "Partial Reactivation of Chlorine Treated Enterovirus," *Appl. Environmental Microbiol.*, vol. 37, no. 4, 1979, p. 766.

177. R. C. Bates, S. M. Sutherland, and P. T. B. Shaffer, "Development of Resistant Poliovirus by Repetitive Sublethal Exposure to Chlorine," in R. L. Jolley et al. (eds.), *Water Chlorination: Environmental Impact and Health Effects*, Butterworths, Stoneham, Mass., 1978.

178. C. Leyval et al., *Environmental Technol. Lett.*, vol. 5, no. 8, 1984, p. 359.

179. C. H. Stagg et al., "Chlorination of Solids Associated Coliphages," *Progr. Water Technol.*, vol. 10, no. 12, 1978, p. 3817.

180. T. W. Hejkal et al., "Survival of Poliovirus within Organic Solids during Chlorination," *Appl. Environmental Microbiol.*, vol. 38, 1979, p. 114.

181. G. D. Boardman and O. J. Sproul, "Adsorption as a Protective Mechanism for Poliovirus," *Proc. AWWA Ann. Conf.*, Anaheim, Calif., June 1977.

182. F. A. O. Brigano et al., "Effect of Particulates on Inactivation of Enteroviruses in Water by Chlorine Dioxide," *78th Ann. Meeting American Society for Microbiology*, 1978.

183. D. M. Howser et al., "Ozone Inactivation of Cell and Fecal Associated Viruses and Bacteria," *51st Ann. Conf. of Water Pollution Control Federation*, 1978.

184. O. J. Sproul et al., "Effect of Particulate Matter on Virus Inactivation by Ozone," *Proc. AWWA Ann. Conf.*, Atlantic City, N.J., June 1978.

185. M. C. Snead et al., *Benefits of Maintaining a Chlorine Residual in Water Supply Systems*, U.S. Environmental Protection Agency, EPA-600/2-80-010, 1980.

186. M. LeChevallier, "Destruction of Bacterial Biofilms," *6th Conf. Water Chlorination: Environmental Impact and Health Effects*, 1987.

187. E. G. Means, K. N. Scott, M. L. Lee, and R. L. Wolfe, "Bacteriological Impact of a Changeover from Chlorine to Chloramine Disinfection in a Water Distribution System," *Proc. AWWA Ann. Conf.*, Denver, Colo., June 1986.

188. E. G. Means, T. S. Tanaka, D. J. Otsuka, and M. J. McGuire, "Effect of Chlorine and Ammonia Application Points on Bactericidal Efficiency," *J. AWWA*, vol. 78, no. 1, 1986, p. 62.

189. S. Rideal, "Application of Electrolytic Chlorine to Sewage Purification and Deodorization in the Dry Chlorine Process," *Trans. Faraday Soc.*, vol. 4, 1908, p. 179.

190. J. E. Bennett, "On-Site Generation of Hypochlorite Solutions by Electrolysis of Seawater," *AIChE Symp. Ser.*, vol. 74, no. 178, 1978, p. 265.

191. S. A. Michalek and F. B. Leitz, "On Site Generation of Hypochlorite," *J. Water Pollution Control Fed.*, vol. 44, no. 9, 1972, p. 1697.

192. R. Trussell and P. Kreft, "Engineering Considerations of Chloramine Application," *AWWA Sem. Proc.: Chloramination for THM Control: Principles and Practices, AWWA Ann. Conf.*, Dallas, Texas, 1984.

193. N. V. Brodtmann and P. J. Russo, "The Use of Chloramine for Reduction of Trihalomethanes and Disinfection of Drinking Water," *J. Amer. Water Works Assoc.*, vol. 71, no. 1, 1979, p. 40.

194. N. R. Ward, R. L. Wolfe, and B. H. Olson, "Disinfectant Activity of Inorganic

Chloramines with Pure Culture Bacteria: Effect of pH, Application Technique and Chlorine to Nitrogen Ratio," *Appl. Environmental Microbiol.*, vol. 48, 1984, p. 508.

195. D. J. Birrell et al., "An Epidemic of Chloramine Induced Anemia in a Hemodialysis Unit," *Medical J. Australia*, vol. 2, 1978, p. 288.

196. J. W. Eaton et al., "Chlorinated Urban Water. A Cause of Dialysis Induced Hemolytic Anemia." *Science*, vol. 181, 1973, p. 463.

197. R. W. Jordan, "Improved Method Generates More Chlorine Dioxide," *Water Sewage Works*, vol. 127, no. 10, 1980, p. 44.

198. W. C. Lauer et al., "Experience with Chlorine Dioxide at Denver's Reuse Plant," *J. AWWA*, vol. 78, no. 6, 1986, p. 79.

199. J. A. Cotruvo and C. D. Vogt, "Regulatory Aspects of Disinfection," in R. L. Jolley et al. (eds.), *Water Chlorination: Environmental Impact and Health Effects*, Lewis, Chelsea, Mich., 1985.

200. H. W. Augenstein, "Use of Chlorine Dioxide to Disinfect Water," *J. AWWA*, vol. 66, no. 12, 1974, p. 716.

201. B. W. Lykins and M. H. Griese, "Using Chlorine Dioxide for Trihalomethane Control," *J. AWWA*, vol. 78, no. 6, 1986, p. 88.

202. E. M. Aieta and J. Berg, "A Review of Chlorine Dioxide in Drinking Water Treatment," *J. AWWA*, vol. 78, no. 6, 1986, p. 62.

203. A. A. Stevens et al., "Organic Halogen Measurements: Current Uses and Future Prospects," *J. AWWA*, vol. 77, vol. 4, 1985, p. 146.

204. *Federal Register*, 19CFR 1910.1000, July 1, 1985.

205. R. A. Morin et al. "Ozone Disinfection Pilot Plant Studies at Laconia, New Hampshire," *J. New Engl. Water Works Assoc.*, September, 1975, p. 206.

206. S. A. Hubbs, M. Goers, and J. Siria, "Plant-Scale Examination and Control of a Chlorine Dioxide-Chloramination Process at the Louisville Water Company," in R. L. Jolley et al. (eds.), *Water Chlorination: Environmental Impact and Health Effects*, Butterworths, Stoneham, Mass., 1980.

207. J. C. Hoff and E. W. Akin, "Microbial Resistance to Disinfectants: Mechanisms and Significance," *Environ. Health Perspect.*, vol. 69, pp. 7–13, 1986.

# Water Fluoridation

## Thomas G. Reeves, P.E.

*National Fluoridation Engineer*
*U.S. Public Health Service*
*Centers for Disease Control*
*Atlanta, Georgia*

Fluoridation of public water supplies has been practiced since 1945. Few public health measures have been accorded greater clinical and laboratory research, epidemiologic study, clinical trials, and public attention than has water fluoridation. This chapter will present an overview of the history and public health and engineering aspects of fluoridation.

Fluoridation is the deliberate adjustment of the fluoride concentration of a public water supply in accordance with scientific and medical guidelines. Fluoride, a natural trace element, is present in small but widely varying amounts in practically all soils, water supplies, plants, and animals, and is a normal constituent of all diets.[1] The highest concentrations in mammals are found in bones and teeth. Virtually all public water supplies in the United States contain at least trace amounts of fluoride from natural sources.

## History

The discovery of the relationship between fluoride in drinking water and dental health has an interesting and intriguing history. The series of studies that led to a demonstration that fluoridated water had caries-inhibitory properties was one of the most extensive programs carried out in the epidemiology of chronic disease. It began in 1901, when a U.S. Public Health Service (USPHS) physician stationed in Naples, Italy, wrote that black teeth observed in emigrants from a nearby region were popularly believed to have been caused by using water charged with volcanic fumes. It was later determined that the water supply contained an extremely high amount of fluoride and that everyone drinking it was afflicted with discolored (or "mottled") teeth, a condition referred to as *dental fluorosis.*

In its mildest form, dental fluorosis is characterized by very slight, opaque, whitish areas on some posterior teeth. As the defect becomes more severe, discoloring is more widespread, and changes in color range from shades of gray to black. In the most severe cases, gross calcification defects occur, resulting in pitting of the enamel. In some of the latter cases, teeth are subject to such severe attrition that they wear down to the gum line, and complete dentures must be obtained.

In 1916 Dr. Frederick S. McKay, a practicing dentist, reported that many of his patients in Colorado Springs, Colo., had this defect.[2] After further study, McKay later concluded that the condition was caused by an undetermined substance in the drinking water. McKay recommended that the water supply of Oakley, Idaho, be changed because of the high incidence of such dental defects among the children there. The supply was changed in 1925 to a nearby spring that had been used by a few other children whose teeth were not discolored.

Several years elapsed before the cause of dental fluorosis was discovered. Almost simultaneously, two different groups of scientists working independently with different tools and methods identified the causal agent. A. W. Petrey, a chemist who was head of the testing division of the laboratory at the Aluminum Company of America at Pittsburgh, noticed the calcium fluoride band in a spectroscopic examination for aluminum in a water sample from Bauxite, Ark. The chief chemist of these laboratories, H. V. Churchill, reported in 1931 that similar examinations of water samples from areas where dental fluorosis was endemic invariably showed the presence of fluoride.[3]

At almost the same time, Drs. H. V. Smith, M. C. Smith, and E. M. Lantz at the University of Arizona reported the cause of mottling by duplicating the condition in rats by feeding concentrated naturally fluoridated water and comparing the results with the mottling observed when a diet high in fluorides was used.[3] The strikingly similar appearance of the mottling uncovered a long-standing mystery.

On April 10, 1931, an abstract of Churchill's report appeared in *Industrial and Engineering Chemistry.* This was the first printed notice of the possible relationship between fluoride and dental fluorosis. The September issue of the *Journal of the American Water Works Association* (AWWA) carried Churchill's complete paper.[4] Churchill found that endemic regions of mottling had waters containing 2 mg/L or more fluoride, while those areas without mottling had water supplies with less than 1.0 mg/L. This division of fluoride waters was confirmed by the Smiths in Arizona, who reported that water sources from nonendemic mottling areas contained less than 0.72 mg/L fluoride.[5]

Soon after this, Dr. Clinton T. Messner, head of the USPHS, assigned the dentist Dr. H. Trendley Dean to do research on dental fluorosis. Dean began carefully prepared and executed epidemiologic investigations. He confirmed that many localities have water supplies containing fluoride. Areas with the largest number of such supplies containing the highest levels of fluorides include those states running from North Dakota to Texas, those along the Mexican border, and Illinois, Indiana, Ohio, and Virginia. Similar supplies are found in the British West Indies, China, Holland, Italy, Mexico, North Africa, South America, Spain, and India.

Through observation of thousands of children in communities with varying fluoride levels, Dean established what he termed a *mottled enamel index*—a numerical method for measuring the severity of fluorosis.[6] Using this index, he established the fluoride level below which the use of such water contributed no significant discoloration. This level in the latitude of Chicago was about 1.0 mg/L.

Many investigators, including McKay, observed during the 1920s that less decay occurred in children whose teeth were afflicted with mottling. In order to confirm this, Dean examined 7257 children in 21 cities with water supplies containing varying fluoride levels. Results of this study, some of which are shown in Fig. 15.1, revealed a remarkable relationship between waterborne fluorides and fluorosis and caries incidence.[7] Three conclusions were drawn from Dean's study:

1. When the fluoride concentration exceeds about 1.5 mg/L, any further increase does not significantly decrease the decayed, missing, and filled (DMF) tooth incidence, but does increase the occurrence and severity of mottling.

2. At a fluoride concentration of about 1.0 mg/L, the optimum occurs—maximum reduction in caries with no aesthetically significant mottling. At this level DMF tooth rates were reduced by 60 percent among the 12- to 14-year-old children.

3. At fluoride concentrations below 1.0 mg/L, some benefits occur, but caries reduction is not as great and decreases as the fluoride level decreases until, as zero fluoride is approached, no observable improvement occurs.

Studies on fluoride were interrupted by World War II, but in 1945 and 1947, four classic studies were initiated with the intent to demonstrate conclusively the benefits of adding fluoride to community drinking water.[8–11] Fluoridation began in January 1945 in Grand Rapids, Mich.; in May 1945 in Newburgh, N.Y.; in June 1945 in Brantford, Ont.; and in February 1947 in Evanston, Ill. When compared to a nonfluoridated "control city" a 50 to 65 percent reduction in

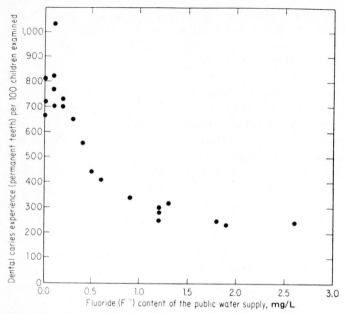

**Figure 15.1** Relation between the amount of dental caries (permanent teeth) observed in 7257 selected 12- to 14-year-old white school children of 21 cities of four states and the fluoride content of public water supply. (*Source: J. F. McClure, "Ingestion of Fluoride and Dental Caries—Quantitative Relations Based on Food and Water Requirements of Children 1–12 Years Old," Am. J. Diseases Children, vol. 66, 1943, p. 362.*)

dental caries was found in the fluoridated cities without evidence of any adverse effects. These initial studies established fluoridation as a practical and effective public health measure that would prevent dental caries.

Once the safety and effectiveness of fluoridation had been established, engineering aspects needed to be developed before community water fluoridation could be implemented. In the 1950s and 1960s, Franz J. Maier, a sanitary engineer, and Ervin Bellack, a chemist, both with the USPHS, made major contributions to the engineering aspects of water fluoridation. Maier and Bellack helped determine which chemicals were the most practical to use in water fluoridation, the best mechanical equipment to use, and the best process controls. Bellack also contributed to major advances in fluoride testing. In 1963, Maier published the first comprehensive book on the technical aspects of fluoridation,[12] and, in 1972, Bellack, then with the U.S. Environmental Protection Agency (USEPA), published an engineering manual[13] that has only recently been replaced.

Over the past 40 years, fluoride and fluoridation have been the subject of numerous studies undertaken by the USPHS, state health departments, and nongovernmental research organizations. Since 1970, over 3700 such studies have been conducted.[14] These studies have overwhelmingly supported the beneficial effect of water fluoridation.

Recently, studies in the United Kingdom, and in several states by the Robert Wood Johnson Foundation, have reaffirmed the safety and benefits of fluoridation.[15,16] In a 1983 United Kingdom study, dental caries experience of 5-, 12-, and 15-year-olds from Anglesby (fluoridated) was compared with nonfluoridated Afron. The study found an approximate 50 percent reduction in DMF teeth in Anglesby verses Afron.[15] The Robert Wood Johnson Foundation conducted a 4-year study (1977–1981) to assess whether dental caries in children could be reduced or eliminated by various combinations of preventive measures, including water fluoridation. The study concluded that fluoridation is an important, extremely effective, yet inexpensive preventive measure.[16]

Concerns have recently been expressed that increases in the prevalence of fluorosis are occurring in communities with negligible and optimal water-fluoride concentrations, because of increased total fluoride consumption from various sources.[17] Studies from the University of Minnesota and the National Institute of Dental Research indicate that no increase in fluoride consumption from foods and beverages over the last 40 years has occurred and preliminary investigations indicate that no increase in the prevalence of fluorosis in either fluoridated or nonfluoridated communities can be demonstrated.[18,19]

## Present Status of Fluoridation

The Centers for Disease Control (CDC) estimated that approximately 130 million Americans, or about 60.5 percent of those served by public water supplies, consumed fluoridated water daily as of January 1, 1988.[20] Some 9.0 million of these people are served by naturally fluoridated supplies. Approximately 70 percent of all cities with populations of 100,000 or more have fluoridated water. More than 22 states, the District of Columbia, and Puerto Rico provide fluoridated water to over one-half of their population. As of 1985, eight states require fluoridation, at least for cities above a minimum population. The increase in the U.S. population served by fluoridated drinking water systems is shown in Fig. 15.2.[21]

In 1981, approximately 38 countries reported community water fluoridation benefiting approximately 208 million people. The United States, Ireland, Canada, Brazil, Australia, Venezuela, and the United Soviet Socialist Republic have large populations consuming fluoridated water. The city-states of Hong Kong and Singapore are totally fluoridated.

Considerable progress has been made toward achieving community

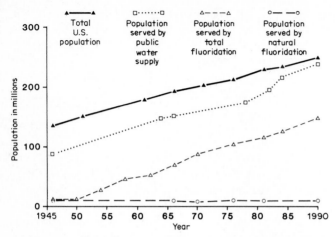

**Figure 15.2** Fluoridation growth in the United States, 1945–1984. (*Source: Centers for Disease Control, "Dental Caries and Community Water Fluoridation Trends," MMWR, vol. 34, no. 6, 1985 p. 77.*)

fluoridation in Central and South America, especially in Brazil. Brazil requires fluoridation for all communities with populations over 50,000. The Pan American Health Organization (a branch of the World Health Organization) has been very active in the promotion of fluoridation in Latin America.

Although community water fluoridation has been shown to be safe and the most cost-effective method to prevent dental caries, a small percentage of the population continues to oppose its introduction into community water systems. When fluoridation is being considered for adoption by a community, persons opposed to fluoridation often attempt to refute the benefits, safety, and efficacy of this effective public health measure. Charges against fluoridation and the corresponding truths have been discussed elsewhere.[22] Assistance in responding to false charges against fluoridation may be obtained from the Dental Disease Prevention Activity of the CDC, Atlanta, Ga. The National Institute of Dental Research, Bethesda, Md., a branch of the National Institutes of Health, is another source of information concerning fluoridation.

## Causes of Dental Caries

Tooth decay is a complex process, and all factors involved are not entirely understood. It is usually characterized by loss of tooth structure (enamel, dentin, and cementum) as a result of destruction of these tissues by acids. Evidence indicates that acids are produced by the action of oral bacteria and enzymes on sugars and carbohydrates entering the mouth. This takes place beneath the plaque, an invisible film composed of gummy masses of microorganisms that adhere to the teeth. Oral bacteria are capable of converting some of the simpler sugars into acids, and the bacteria and enzymes acting in combination are capable of converting carbohydrates and more complex sugars into acids. The production of acids is a result of the natural existence of bacteria and enzymes in the mouth.

Nearly everyone is attacked by dental caries, the most prevalent chronic disease of humans.[23] It is truly universal. Until water fluoridation became widespread, almost 98 out of 100 Americans experienced some tooth decay by the time they reached adulthood. The highest tooth decay activity is found in school children. Tooth decay begins in early childhood, reaches a peak in adolescence, and diminishes during adulthood.[24]

### Dental Benefits of Fluoride in Drinking Water

When water containing fluoride is consumed, some fluoride (about 50 percent) is retained by fluids in the mouth and is incorporated onto teeth by surface uptake (topical effect). The rest (about 50 percent) enters the stomach where it is rapidly adsorbed by diffusion through the stomach walls and intestine. Fluoride enters the blood plasma and is rapidly distributed throughout the body, including teeth (systemic effect). Because of the systemic effect, the fluoride ion is able to pass freely through all cell walls and is available to all organs and tissues of the body. Distributed in this fashion, the fluoride ion is available to all skeletal structures of the body in which it may be retained and stored in proportions that generally increase with age and intake.

Bones, teeth, and other parts of the skeleton tend to attract and retain fluoride. Soft tissues do not retain fluoride. Fluoride is a "bone seeker," with about 96 percent of the fluoride found in the body deposited in the skeleton. Because teeth are part of the skeletal system, incorporation of fluoride in teeth is basically similar to that in other bones. It is most rapid during the time of the child's formation and growth. Erupted teeth differ from other parts of the skeleton in that once they are formed, with the exception of the dentin (inner part of the tooth) and the root, cellular activity virtually ceases. As a result, very little change occurs in the fluoride level in teeth after they are formed. Children must drink the proper amount of fluoridated water during early development of permanent teeth, preferably before they start school, in order to realize full benefits.

High levels of fluoride in drinking water have been found to cause adverse health effects. As a result, the USEPA has established regulatory limits on the fluoride content of drinking water. Based on a detailed review of health effects studies on fluoride,[25,26] the USEPA set a maximum contaminant level of 4 mg/L in water systems to prevent crippling skeletal fluorosis.[27] A secondary level of 2 mg/L was established by USEPA to protect against objectionable dental fluorosis.[27] These limits, to be reviewed by USEPA every three years, or as new health data become available, primarily impact systems that have naturally high fluoride levels.

The relationship between dental caries, dental fluorosis, and fluoride level is shown in Fig. 15.3.[28] The beneficial effect of optimally fluoridated water ingested during the years of tooth development has been amply demonstrated. At the optimal concentration in potable water, fluoride will reduce dental caries from 20 to 40 percent among children who ingest this water from birth.[29] Evidence that water fluoridation is effective in preventing caries has been repeatedly demonstrated, starting with the the initial community trials in the United States and Canada in the 1940s. In recent years, however, the relative impact of water fluoridation appears to have diminished as other sources of fluoride supplementation (toothpastes, food, etc.) have increased. Continuation of benefits into adult life is inevitable. Stronger teeth result in fewer caries, which require fewer and less extensive fillings, fewer extractions, and fewer artificial teeth. Fluoridated water helps prevent cavities on exposed roots as a result of receding gums in adults who develop periodontal disease. Early evidence indicated that higher levels of fluoride would strengthen bones of older

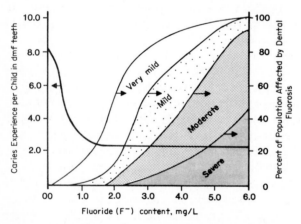

Figure 15.3   Dental caries and dental fluorosis in relation to fluoride in public water supplies.   (*Source: J. M. Dunning, Principles of Dental Public Health, Harvard University Press, Cambridge, Mass., 1962.*)

people, thereby reducing the incidence of bone fractures.[30,31] This now appears not to be true.

Various studies in fluoridated communities over the last 30 years have shown a dramatic increase in the number of teenagers that are completely caries-free. Teenagers without lifetime exposure realize benefits from fluoridation, and benefits increase for those with lifetime exposure. Conservative estimates indicate that 20 percent of the teenagers in a fluoridated community will be caries-free.[32] This is about six times as many as are caries-free in a fluoride-deficient community. Fluoride makes teeth more resistant to bacterial acids and inhibits the growth of certain kinds of bacteria that produce acids.[33] In addition, fluoride appears to actually aid in the remineralization of teeth.[34,35]

As a result of the positive effects of fluoride on teeth, fluoridation can substantially reduce costs associated with restorative dentistry. For every dollar spent (1980) on water fluoridation, a potential $50 in dental bills may be saved.[36] In 1985, the CDC estimated the cost of fluoridation to be about the cost of a candy bar per person per year.

### Optimal Fluoride Levels

The optimal fluoride level in water is the level that produces the greatest protection against caries with the least risk of fluorosis. Initially, this figure was obtained after examining the teeth of thousands of children living in various places with differing fluoride levels. It was not based on any direct or accurate knowledge of how much water children drank at various times and at different places. Early in such investigations, variation in the optimal figure was observed, depending on the local air temperature, which had a direct bearing on the amount of water children at different ages consumed. Studies in California and Arizona, where temperatures are considerably above the average of other parts of the United States, showed a definitely lower optimal fluoride level. This was demonstrated by observing dental fluorosis prevalence in places with various natural fluoride levels in their public water supplies and by estimating the actual quantities of water ingested by children of various age groups and weights.

The optimal fluoride level for a water system is usually established by the appropriate state regulatory agency. Optimal fluoride concentrations and control ranges recommended by the USPHS and CDC may be used as guidelines if state limits have not been established. These levels, shown in Table 15.1,[37] are based on the annual average of the maximum daily air temperature in the particular school or community.

Many water supplies contain fluorides naturally. For these systems,

TABLE 15.1    Optimal Fluoride Levels Recommended by the U.S. Public Health Service, Centers for Disease Control

| Annual average of maximum daily air temperatures;*°F | Recommended fluoride concentration | | Recommended control community systems | | Range school systems, mg/L | |
|---|---|---|---|---|---|---|
| | Community, mg/L | School,† mg/L | 0.1 Below | 0.5 Above | 20% Low | 20% High |
| 40.0–53.7 | 1.2 | 5.4 | 1.1 | 1.7 | 4.3 | 6.5 |
| 53.8–58.3 | 1.1 | 5.0 | 1.0 | 1.6 | 4.0 | 6.0 |
| 58.4–63.8 | 1.0 | 4.5 | 0.9 | 1.5 | 3.6 | 5.4 |
| 63.9–70.6 | 0.9 | 4.1 | 0.8 | 1.4 | 3.3 | 4.9 |
| 70.7–79.2 | 0.8 | 3.6 | 0.7 | 1.3 | 2.9 | 4.3 |
| 79.3–90.5 | 0.7 | 3.2 | 0.6 | 1.2 | 2.6 | 3.8 |

*Based on temperature data obtained for a minimum of 5 years.
†Based on 4.5 times the optimal fluoride level for communities.
SOURCE: From T. G. Reeves, Ref. 37.

the question of the practicability of supplementing natural fluoride with enough fluoride to bring the concentration up to the optimal level must be addressed. Usually, addition of the small amount of fluoride needed to reach the optimal level can be shown to be economically justified based on the resulting benefits to the community.

## Fluoride Chemicals and Chemistry

Fluorine, a gaseous halogen, is the thirteenth most abundant element found in the earth's crust. It is a pale yellow noxious gas that is highly reactive. It is the most electronegative of all elements and cannot be oxidized to a positive state. Fluorine is not found in a free state in nature; it is always found in combination with chemical radicals or other elements as fluoride compounds.

Fluoride can be found in a solid form in fluoride-containing minerals such as fluorspar, cryolite, and apatite. Fluorspar is a mineral containing from 30 to 98 percent calcium fluoride ($CaF_2$). Cryolite ($Na_3AlF_6$) is a compound of aluminum, sodium, and fluoride. Apatite [$Ca_{10}(PO_4, CO_3)_6(F, Cl, OH)_2$] is a deposit of a mixture of calcium compounds that include calcium phosphates, calcium fluorides, and calcium carbonates. Trace amounts of sulfates are usually present as impurities. Apatite contains from 3 to 7 percent fluoride and is the main source of fluoride used in water fluoridation at the present time. Apatite is also the raw material for phosphate fertilizers. Cryolite is not a major source of fluoride in the United States.

Fluoride is widely distributed in the lithosphere and hydrosphere. Because of the dissolving power of water and movement of water in the hydrologic cycle, fluoride is found naturally in all waters. High concentrations of fluoride are not common in surface water, but may occur in groundwater, hot springs, and geothermal fluids.

Fluoride forms compounds with every element except helium, neon, and argon. Polyvalent cations such as aluminum, iron, silicon, and magnesium form stable complexes with the fluoride ion. The extent to which complex formation takes place depends on several factors, including the complex stability constant, pH, and the concentrations of fluoride and the complexing species.[38]

Sodium fluoride, sodium silicofluoride, and hydrofluosilicic acid are the three most commonly used fluoride chemicals in the United States. Standards for these chemicals are published by AWWA for use by the water industry.[39–41] All chemicals used for fluoridation should be comparable in quality to the requirements of these standards.

From time to time, shortages of fluoride chemicals have occurred. Generally, most "shortages" are not shortages at the manufacturer's plant, but a temporary shortage at the local distributor level. Local shortages are usually eliminated quickly. In the past, shortages at the manufacturing level, especially of hydrofluosilicic acid and sodium silicofluoride, have occurred.

Hydrofluosilicic acid and sodium silicofluoride (and most sodium fluoride) are by-products of phosphoric acid manufacture, the main ingredient of phosphate fertilizer. Sales of fertilizer will have a direct effect on the volume of fluoride chemicals produced. In the past, slow sales of fertilizer have resulted in a temporary shortage of these two chemicals. Shortages have been relatively mild because the number of fluoridated communities was smaller and lower volumes of sodium silicofluoride and hydrofluosilicic acid were needed than at the present time. Shortages occurred in 1955 to 1956, in the summer of 1969, in the spring and summer of 1974, in the summer of 1982, and in the early part of 1986, this last one being the most severe.

During production of fluoride chemicals, trace amounts of impurities may be introduced into the chemical, especially arsenic, lead, and/or zinc. Normally, impurities are at levels far below that which would necessitate the establishment of maximum impurity limits.[42]

### Sodium fluoride

Sodium fluoride (NaF) is a white, odorless material available either as a powder or in the form of crystals of various sizes. It has a molecular weight of 42.00, a specific gravity of 2.79, and a practically constant solubility of 4.0 g/100 mL (4 percent) in water at temperatures generally encountered in water treatment practice. When added to water, sodium fluoride dissociates into sodium and fluoride ions:

$$NaF \rightleftharpoons Na^+ + F^- \tag{15.1}$$

The pH of a sodium fluoride solution varies with the type and amount of impurities present. Solutions prepared from common grades of sodium fluoride have a pH near neutrality (approximately 7.6). Sodium fluoride is available in purities ranging from 97 to over 98 percent, with impurities consisting of water, free acid or alkali, sodium silicofluoride, sulfites, and iron, plus traces of other substances. Approximately 8.6 kg (19 lb) of sodium fluoride will add 1 mg/L of fluoride to 1.0 mil gal (3.8 ML) of water.

### Sodium silicofluoride

Sodium silicofluoride ($Na_2SiF_6$) is a white, odorless crystalline material with a molecular weight of 188.06 and a specific gravity of 2.679. Its solubility varies from 0.44 g/100 mL of water at 0°C to 2.45 g/100 mL at 100°C.

When sodium silicofluoride is dissolved in water, virtually 100 percent dissociation occurs rapidly:

$$Na_2SiF_6 \rightleftharpoons 2Na^+ + SiF_6^{2-} \tag{15.2}$$

Silicofluoride ions ($SiF_6^{2-}$) may react in two ways. The most common is hydrolysis of $SiF_6^{2-}$, releasing fluoride ions and silica ($SiO_2$):

$$SiF_6^{2-} + 2H_2O \rightleftharpoons 4H^+ + 6F^- + SiO_2 \tag{15.3}$$

Silica, the main ingredient in glass, is very insoluble in water. Alternatively, $SiF_6^{2-}$ dissociates very slowly, releasing fluoride ions and silicon tetrafluoride ($SiF_4$):

$$SiF_6^{2-} \rightleftharpoons 2F^- + SiF_4 \tag{15.4}$$

Silicon tetrafluoride is a gas that will easily volatilize out of water when present in high concentrations. It also reacts quickly with water to form silicic acid or silica:

$$SiF_4 + 3H_2O \rightleftharpoons 4HF + H_2SiO_3 \tag{15.5}$$

$$SiF_4 + 2H_2O \rightleftharpoons 4HF + SiO_2 \tag{15.6}$$

Solutions are acidic, with saturated solutions usually exhibiting a pH of between 3 and 4 (approximately 3.6). Sodium silicofluoride is available in purities of 98 percent or higher. Principal impurities are water, chlorides, and silica. Approximately 6.3 kg (14 lb) of sodium silicofluoride will add 1 mg/L of fluoride to 1.0 mil gal (3.8 ML).

## Hydrofluosilicic acid

Hydrofluosilicic acid, also known as hexafluosilicic, silicofluoric, or fluosilicic acid ($H_2SiF_6$), has a molecular weight of 144.08 and is available commercially as a 20 to 35 percent aqueous solution. It is a straw-colored, transparent, fuming, corrosive liquid having a pungent odor and an irritating action on the skin. Solutions of 20 to 35 percent hydrofluosilicic acid have a low pH (1.2), and, at a concentration of 1 mg/L in poorly buffered potable waters, a slight depression of pH can occur.

Hydrofluosilicic acid dissociates in solution virtually 100 percent. Its chemistry is very similar to $Na_2SiF_6$:

$$H_2SiF_6 \rightleftharpoons 2HF + SiF_4 \tag{15.7}$$

$$SiF_4 + 2H_2O \rightleftharpoons 4HF + SiO_2 \tag{15.8}$$

$$SiF_4 + 3H_2O \rightleftharpoons 4HF + H_2SiO_3 \tag{15.9}$$

Hydrofluosilicic acid should be handled with great care because of its low pH and the fact that it will cause a "delayed burn" on skin tissue. Hydrofluosilicic acid (23 percent) will freeze at approximately 4°F ($-15.5$°C). Approximately 20.8 kg (46 lb) of 23 percent acid are required to add 1 mg/L of fluoride to 1.0 mil gal (3.8 ML).

Hydrofluoric acid and silicon tetrafluoride are common impurities in hydrofluosilicic acid that result from production processes. Hydrofluoric acid is an extremely corrosive material. Its presence in hydrofluosilicic acid, whether from intentional addition (i.e., "fortified" acid) or from normal production processes, demands careful handling. Hydrofluosilicic acid fumes are lighter than air and will rise.

## Other fluoride chemicals

Ammonium silicofluoride, magnesium silicofluoride, potassium fluoride, hydrofluoric acid, and calcium fluoride (fluorspar) are being, or have been, used for water fluoridation. Each material has properties that make it desirable in a specific application, but each also has undesirable characteristics. None of these chemicals has widespread application in the United States. Calcium fluoride, however, is widely used in South America.

## Fluoride Feed Systems

Three methods of feeding fluoride are common in community water supply systems:

1. Dry chemical feeder with a dry fluoride compound

2. Chemical solution feeder with a liquid fluoride compound or with a prepared solution of a dry chemical

3. Fluoride saturator

The first two methods are also commonly used to feed other water treatment chemicals. The saturator is a unique method for feeding fluoride.

Selection of the best fluoridation system for a situation must be based on several factors, including population served or water usage rate, chemical availability, cost, and operating personnel available. Although many options will be possible, some general limitations are imposed by the size and type of facility. In general, very large systems will use the first two methods, whereas smaller systems will use either an acid feeder or the saturator.

Manuals describing considerations and alternatives involved in selecting an optimal fluoridation system are available.[22,37] Factors important in the selection, installation, and operation of a fluoride feed system are the type of equipment used, the fluoride injection point, safety, and waste disposal.

### Type of equipment

Fluoride chemicals are added to water as liquids, but they may be measured in either liquid or solid form. Solid chemicals must be dissolved into solution before feeding. This is usually accomplished by using a dry chemical feeder that delivers a predetermined quantity of chemical in a given time interval. Two types of dry feeders exist. Each has a different method of controlling the rate of delivery. A volumetric dry feeder delivers a measured *volume* of dry chemical per unit of time. A gravimetric dry feeder delivers a measured *weight* of chemical per unit of time.

Many water treatment plants that treat surface water will utilize dry feeders to add other treatment chemicals and will use dry feeders for fluorides to maintain consistency with other equipment. Dry feeders are used almost exclusively to feed sodium silicofluoride because of the high cost of sodium fluoride.

The saturator feed system is unique to fluoridation and is based on the principle that a saturated fluoride solution will result if water is allowed to trickle through a bed containing a large amount of sodium fluoride. A small pump is used to feed the saturated solution into the water being treated. Although saturated solutions of sodium fluoride can be manually prepared, automatic feed systems are available.

### Fluoride injection point

Ideally, the fluoride injection point should be at a location through which all water to be treated passes. In a treatment plant, this could be a channel where other water treatment chemicals are added, a main coming from the filters, or the clearwell. If a combination of facilities exist, such as a treatment plant for surface water plus supplemental wells, a point where all water from all sources passes must be selected. If no common point exists, a separate fluoride feeding installation is needed for each facility.

Another consideration in selecting the fluoride injection point is the possibility of fluoride losses in filters. Whenever possible, fluoride should be added after filtration to avoid substantial losses that can occur, particularly with heavy alum doses or when magnesium is present and the lime–soda ash softening process is being used. A fluoride loss of up to 30 percent can result if the alum dosage rate is 100 mg/L.[13] In some situations, addition of fluoride before filtration may be necessary, such as when the clearwell is inaccessible.

When other chemicals are being fed, the question of chemical compatibility must be considered. The fluoride injection point should be as far away as possible from the injection point of chemicals that contain calcium, in order to minimize loss of fluoride by local precipitation. For example, if lime is being added to the main leading from the filters for pH control, fluoride can be added to the same main at another point or at the clearwell. If lime is added to the clearwell, fluoride should be added to the opposite side. If injection point separation is not possible, an in-line mixer must be used to prevent local precipitation of calcium fluoride and to ensure that the added fluoride dissolves.

In a single-well system, the well pump discharge can be used as the fluoride injection point. If more than one well pump is used, the line leading to the distribution system can be used as the injection point. In a surface water treatment plant or softening plant, the ideal location of the fluoride injection point is in the line from the filters to the clearwell. This location provides for maximum mixing. Sometimes the clearwell is located directly below the filters, and discharging chemicals directly to the clearwell is difficult. In this situation the fluoride injection point must be at another location, such as in the main line to the distribution system or before the filters.

### Safety considerations

Fluoride levels to which a water plant operator may be exposed can be much higher than the optimal level. Proper handling of fluoride chemicals is necessary to prevent overexposure. Dusts are a particular problem when sodium fluoride and sodium silicofluoride are used. Operators must be aware of the hazards involved in the feeding of fluoride chemicals and should always follow accepted safety practices. Manuals describing operational hazards and safety practices for fluoride chemical feed systems are available.[22,37,43]

## Waste disposal

An important consideration in the design and operation of a fluoridation system is the disposal of empty fluoride chemical containers. The temptation to reuse fiber drums is difficult to overcome because the drums are convenient and sturdy. Paper bags are dusty and could cause a hazard if they are burned, and empty acid drums could contain enough acid to cause contamination. The best practice is to rinse all empty containers, including paper bags, with water. After all traces of fluoride are removed, bags and drums should be disposed of according to the requirements of the state's environmental protection program solid waste division.

## School Fluoridation

Children living in areas not served by community water supplies are often unable to benefit from drinking water that has been optimally fluoridated. In 1985, about 17 million people (7 percent of the U.S. population) lived in areas that lacked central water systems. Because community fluoridation is not feasible for these areas, other ways of preventing dental caries must be developed if the natural fluoride level in the water supply is inadequate. Fluoridation of water supplies of rural schools has been implemented in many areas and is particularly appealing because it reaches sizable numbers of children with minimal demands on personnel, equipment, and funds. As of June 1987, 459 schools in 12 states provide fluoridated water to approximately 150,700 students.[44] School fluoridation is another way to provide fluoridated water, but it is not considered an alternative to community water fluoridation because of its limitations.

The most obvious limitation of school water fluoridation is that children are approximately 6 years old before they begin attending school, whereas maximum dental benefits occur when fluoridated water is consumed from birth. Data obtained from communities with controlled fluoridation, however, indicate that children who are 6 years of age or older at the time fluoridation is initiated do derive dental benefits. These findings are not surprising, considering that, at age 6, a significant amount of calcification will occur in later-erupting permanent teeth. In addition, considerable fluoride uptake occurs between the completion of permanent tooth calcification and eruption. Evidence also indicates that the topical action of fluoridated water will confer some caries inhibition to erupted teeth.

A second factor limiting the effectiveness of school water fluoridation is that exposure to fluoridated water in a school is intermittent. Children attend school only 5 days a week for only part of the day and only part of the year. One concern is the generally short water lines found in a school system. The impact of an overfeed would be more serious than in a community system, because a slug of fluoride only needs to travel several feet rather than many yards or even miles before it might be consumed.

Several studies have been conducted to determine the optimal fluoride level for school fluoridation systems. Studies at Pike County, Ky., and Elk Lake, Pa., found a 35 to 40 percent reduction in dental cavities in school children consuming water at 4.5 times the recommended optimal level for community fluoridation.[45,46] Results of a 12-year study at a school near Seagrove, N.C., showed that little additional benefits resulted when the fluoride level was increased to 7 times the recommended optimal level for communities.[47] As a result, 4.5 times the optimal level recommended for community systems has been established as the recommended level for school fluoridation systems.

When school fluoridation is considered, the issue of safety must always be addressed. Full-time exposure to fluoride levels as low as twice the optimum can cause some degree of dental fluorosis. Yet, early epidemiologic studies have found that children consuming fluoride-free water at home were uniformly free of any objectionable signs of dental fluorosis when water with a natural fluoride at a level of 6 mg/L was consumed at school. Because fluorosis is a development disturbance that is produced only at the initial stage of enamel formation, teeth of school-age children may be too advanced to be adversely affected by higher levels of fluorides. Other epidemiologic findings support this evidence.[45–48]

## Alternatives to Water Fluoridation

Alternative means of providing the benefits of fluoride besides the fluoridation of municipal water supply systems are available. Municipal water fluoridation, however, is the most cost-effective means available for reducing the incidence of caries in a community. This conclusion is based on the mass of evidence demonstrating the efficacy of the measure and on the most current information on costs of implementing fluoridation. Alternative methods should only be considered in situations where municipal water fluoridation is not possible.

In general, five alternatives to water fluoridation exist that use either the topical or systemic method:

| Topical fluoride methods | Systemic fluoride methods |
| --- | --- |
| 1. Gels | 1. Tablets |
| 2. Mouth rinses | 2. Drops |
| 3. Dentifrices | |

Topical fluoride methods can be used in conjunction with water fluoridation (optimally fluoridated water in community or school water systems or naturally fluoridated water). Systemic fluoride methods are sufficient alone in preventing tooth decay, and other methods should not be used in conjunction with them. The cost and effectiveness of alternatives to municipal water fluoridation are shown in Table 15.2.[49]

Topical fluoridation application may be professionally or self-applied. Table 15.3 ranks these methods according to use and shows the percent reduction in DMF teeth.[50] Only those fluoride-containing toothpastes that are accepted by the Council of Dental Therapeutics of the American Dental Association ensure effectiveness.

Systemic fluoride supplements require daily administration and may be tablets or drops. The amount of fluoride in tablets or drops depends on the fluoride concentration in the water and on the age of the child. Fluoride supplements may be combined with vitamins. Recommendations of the American Dental Association and the American Academy of Pediatrics on fluoride supplements are shown in Table 15.4.[51]

## Emerging Technology

A venturi-type fluoride feed system using a saturator and sodium fluoride has been developed by the Indian Health Service, which is a branch of the USPHS.[52] A study of 10 venturi saturators has been

**TABLE 15.2    Comparison of Effectiveness of Different Types of Fluoride Applications**

| Procedure | Cavities prevented per $100,000 spent | Cost/cavity prevented |
|---|---|---|
| 1. Water fluoridation | | |
| a. Municipal | 500,000 | $ 0.20 |
| b. School | 111,100 | 0.90 |
| 2. Topical Fluorides | | |
| a. Supervised application of paste or rinse in school | 55,555 | 1.82 |
| b. Professional application of topical fluoride | 25,600 | 3.90 |
| 3. Systemic fluorides | | |
| a. Supervised distribution of fluoride tablets in school | 16,542 | 6.06 |
| b. Individually prescribed fluoride tablets or drops | 10,000 | 10.00 |

SOURCE: From C. Gish, Ref. 49.

TABLE 15.3    Effectiveness of Topical Fluoride Applications

| Application | Frequency | Cavity reduction, % |
|---|---|---|
| 1. Professionally applied | | 35 |
| 2. Self-applied | | |
| a. 0.2% NaF rinse | Weekly | 25 |
| b. Supervised brushing with 9.0% stannous fluoride | 2/year | 25 |
| c. Toothbrushing at home with 0.1% fluoride dentifrice | Daily | 20 |

SOURCE: From S. B. Heifetz, Ref. 50.

TABLE 15.4    Daily Supplemental Fluoride Dosage Schedule

| Age, years | Concentration of fluoride in water, mg F/day | | |
|---|---|---|---|
| | Less than 0.3 mg/L | 0.3 mg/L to 0.7 mg/L | Greater than 0.7 mg/L |
| Birth to 2 | 0.25 | 0.0 | 0.0 |
| 2–3 | 0.50 | 0.25 | 0.0 |
| 3–13 | 1.00 | 0.50 | 0.0 |

*2.2 mg of sodium fluoride contains 1 mg of fluoride (F).
SOURCE: From American Dental Association, Ref. 5.

funded by CDC with the state of Minnesota to evaluate the performance characteristics of these units.

Recent shortages of fluoride chemicals have stimulated research to find an alternative chemical that is readily available and not subject to shortages. Calcium fluoride is readily available in fluorspar rock and is prevalent throughout the United States, especially in southern Illinois. Calcium fluoride is also used as the feed chemical for many water systems in Brazil and other South American countries. A three-year demonstration project funded by CDC to determine the feasibility of using calcium fluoride to fluoridate community water supply systems is being conducted by the Ohio State University Water Resources Foundation. Results have not looked promising.

# References

1. H.C. Hodges and F.A. Smith, in J.H. Simons (ed.), *Fluorine Chemistry,* Vol. 4, Academic Press, New York, 1965.
2. F.S. McClure, *Water Fluoridation, the Search and the Victory,* National Institutes of Health, Bethesda, Md., 1970.
3. D.R. McNeil, *The Fight for Fluoridation,* Oxford University Press, New York, New York, 1957.
4. H.V. Churchill, "The Occurrence of Fluorides on Some Waters of the United States," *J. AWWA,* Vol. 23, No. 9, 1931, p. 1399.

5. M.C. Smith, E.M. Lantz, and H.V. Smith, "The Cause of Mottled Enamel, a Defect of Human Teeth," *University of Arizona Agricultural Experiment Station Bulletin,* No. 32, 1931.

6. H.T. Dean, "Chronic Endemic Dental Fluorosis (Mottled Enamel)," *JAMA,* Vol. 107, 1936, p. 1269.

7. F.J. McClure, "Ingestion of Fluoride and Dental Caries—Quantitative Relations Based on Food and Water Requirements of Children 1–12 Years Old," *Am. J. Diseases Children,* Vol. 66, 1943, p. 362.

8. H.T. Dean F.A. Arnold, J. Phillip, and J.W. Knutson, *Studies on Mass Control of Dental Caries through Fluoridation of Public Water Supply,* Public Health Reports 65, Grand Rapids-Muskegon, Mich., 1950.

9. D.B. Ast, D.J. Smith, B. Wachs, H.C. Hodges, H.E. Hilleboe, E.R. Schesinger, H.C. Chase, K.T. Cantwell, and D.E. Overton, "Newburgh-Kingston Caries-Fluorine Study: Final Report," *J. Am. Dental Assoc.,* Vol. 52, 1956, p. 290.

10. H.K. Brown and M. Poplove, "Brantford-Sarnia-Statford, Fluoridation Caries Study, Final Survey, 1963," *J. Can. Dental Assoc.,* Vol. 31, No. 8, 1965, p. 505.

11. I.N. Hill, J.R. Blayney, and W. Wolf, "Evanston Fluoridation Study—Twelve Years Later," *Dental Progr.,* vol. 1, 1961, p. 95.

12. Franz J. Maier, *Manual of Water Fluoridation Practice,* McGraw-Hill, New York, 1963.

13. E. Bellack, *Fluoridation Engineering Manual,* U.S. Environmental Protection Agency, Washington, D.C., 1972; reprinted September 1984.

14. *Michigan Department of Public Health Policy Statement on Fluoridation of Community Water Supplies and Synopsis of Fundamentals of Relation of Fluorides and Fluoridation to Public Health,* Michigan Department of Public Health, 1979.

15. D. Jackson, P.M.C. James, and F.D. Thomas, "Fluoridation in Anglesby 1983: A Clinical Study of Dental Caries," *Br. Dental J.,* vol. 158, 1985, p. 45.

16. *FL 130 National Prevention Dentistry Demonstration Program Reaffirms Benefits of Community Water Fluoridation,* United States Department of Health and Human Services, Public Health Service, Centers for Disease Control, Atlanta, Ga., 1985.

17. D. Leverett, "Prevalence of Dental Fluorosis in Fluoridated and Nonfluoridated Communities—A Preliminary Investigation," *J. Pub. Health Dentist.,* vol. 46, 1986, p. 4.

18. L. Singer, and R.H. Ophaug, "Fluoride Intake of Humans" in J.L. Shupe, H.B. Peterson, and N.C. Leone (eds.), *Fluorides, Effects on Vegetation, Animals and Humans.* Paragon Press, Salt Lake City, Utah, 1983.

19. W.S. Driscoll, H.S. Horowitz, R.J. Meyer, S.B., Heifex, A. Kingman, and E.R. Zimmerman, "Prevalence of Dental Caries and Dental Fluorosis in Areas with Negligible, Optimal, and Above-Optimal Fluoride Concentrations in Drinking Water," *J. Am. Dental Assoc.,* vol. 113, 1986, p. 29.

20. *Fluoridation Census 1985,* United States Department of Health and Human Services, Public Health Service, Centers for Disease Control, Atlanta, Ga., 1988.

21. "Dental Caries and Community Water Fluoridation Trends—U.S.," *Morbidity and Mortality Weekly Report,* vol. 34, no. 6, 1985, p. 77.

22. *Water Fluoridation Principles and Practices,* 2d ed., AWWA Manual M4, AWWA, Denver, Colo., 1984.

23. D.F. Striffler, W.O. Young, and B.A. Burt, *Dentistry, Dental Practice and the Community,* 3d ed., Saunders, Philadelphia, 1983.

24. *The Prevalence of Dental Caries in United States Children, 1979–1980. The National Dental Caries Prevalence Survey,* National Institute of Dental Research, National Caries Program, National Institute of Health, Bethesda, Md., December 1981.

25. United States Environmental Protection Agency, *Final Draft for the Drinking Water Criteria Document on Fluoride,* Criteria and Standards Division, Office of Drinking Water, Washington, D.C., 1985.

26. "National Primary Drinking Water Regulations; Fluoride; Final Rule and Proposed Rule," *Federal Register,* vol. 50, 1985, p. 47142.

27. "National Primary and Secondary Drinking Water Regulations; Fluoride; Final Rule." *Federal Register,* vol. 51, April 2, 1986, p. 11396.

28. J.M. Dunning, *Principles of Dental Public Health,* Harvard University Press, Cambridge, Mass., 1962.
29. E. Newbrun, "Effectiveness of Water Fluoridation," *J. Public Health Dentistry,* vol. 49, no. 5, special issue, 1989.
30. J. Jowsey, L.B. Riggs, P.J. Kelly, and D.L. Hoffman, "Effect of Combined Therapy with Sodium Fluoride, Vitamin D, and Calcium in Osteoporosis," *Am. J. Med.,* vol. 53, 1972, p. 43.
31. L.B. Riggs, E. Seeman, S.F. Hodgson, D.R. Taves, and W.M. O'Fallon, "Effects of the Fluoride/Calcium Regimen of Vertebral Fracture Occurrence in Postmenopausal Osteoporosis," *N. Engl. J. Med.,* vol. 306, No. 8, 1982, p. 446.
32. *FL-98 Caries-Free Teenagers Increase with Fluoridation,* United States Department of Health and Human Services, Public Health Service, Centers For Disease Control, Atlanta, Ga., 1978.
33. Paul H. Keyes, "Present and Future Measures for Dental Caries Control," *J. Am. Dental Assoc.,* vol. 79, 1969, p. 1395.
34. Leon M. Silverstone, "The Significance of Remineralization in Caries Prevention," *J. Can. Dental Assoc.,* vol. 50, No 2, 1984, p. 157.
35. J.R. Mellberg and D.E. Mallon, "Acceleration of Remineralization, in vitro, by Sodium Monofluorophosphate and Sodium Fluoride," *J. Dental Res.,* vol. 63, No. 9, 1984, p. 1130.
36. C.W. Gish, "The Dollar and Cents of Prevention," *J. Indiana Dental Assoc.,* vol. 58, 1979.
37. Thomas G. Reeves (ed.), *Water Fluoridation—A Training Course Manual for Engineers and Technicians,* United States Department of Health and Human Services, Public Health Service, Centers for Disease Control, Centers for Disease Control, Atlanta, Ga., 1986.
38. Bert A. Eichenberger and Kenneth Y. Chen, "Origin and Nature of Selected Inorganic Constituents in Natural Waters," in Roger A. Minear and Lawrence H. Keith (ed.), *Water Analysis, Vol. 1: Inorganic Species, Part 1,* Academic Press, New York, 1982.
39. *AWWA B701, Standard for Sodium Fluoride,* AWWA, Denver, Colo.
40. *AWWA B702, Standard for Sodium Silicofluoride,* AWWA, Denver, Colo.
41. *AWWA B703, Standard for Hydrofluosilicic Acid,* AWWA, Denver, Colo.
42. National Academy of Sciences, Committee on Water Treatment Chemicals, *Water Chemicals Codex,* National Academy Press, Washington, D.C., 1982.
43. *Safety Practice for Water Utilities,* AWWA Manual M3 AWWA, Denver, Colo., 1983.
44. Center for Disease Control, Dental Disease Prevention Activity Files, July 1987.
45. H.S. Horowitz, S.B. Heifetz, F.E. Law, and W.S. Driscoll, School Fluoridation Studies in Elk Lake, Pennsylvania and Pike County, Kentucky—Results after 8 Years," *Am. J. Pub. Health,* vol. 58, No. 12, 1968, p. 2240.
46. H.S. Horowitz, S.B. Heifetz, and F.E. Law, "Effect of School Water Fluoridation on Dental Caries: Final Results in Elk Lake, PA, after 12 Years," *J. Am. Dental Assoc.,* vol. 84, 1972, p. 832.
47. H.S. Horowitz, S.B. Heifetz, and J.A. Brunelle, "Effect of School Water Fluoridation on Dental Caries: Results in Seagrove, NC, after 12 Years," *J. Am. Dental Assoc.,* vol. 106, 1983, p. 334.
48. G.A. Kempf and F.S. McKay, "Mottled Enamel in a Segregated Population," *Pub. Health Rep.,* vol. 45, 1930, p. 2923.
49. Charles Gish, "Relative Efficiency of Methods of Caries Prevention in Dental Public Health," *Proc. Workshop on Preventive Methods in Dental Public Health,* University of Michigan, Ann Arbor, Mich., June 1978.
50. S.B. Heifetz, "Cost-Effectiveness of Topically Applied Fluorides," *Proc. Workshop on Preventative Methods in Dental Public Health,* University of Michigan, Ann Arbor, Mich., June 1978.
51. American Dental Association Council on Dental Therapeutics, *Accepted Therapeutics,* 39th ed., American Dental Association, Chicago, Ill., 1982.
52. J. Leo, *The Venturi Fluoridator,* Indian Health Service, United States Public Health Service, August 1981.

# Water Treatment Plant Waste Management

## Peter W. Doe, P.E.

*Consultant*
*Havens and Emerson, Inc.*
*Saddle Brook, New Jersey*

Earlier chapters indicate that the preparation of a new supply of potable water involves source development, chemical analysis to determine water quality and treatment requirements, treatment plant design and construction, and system operation to distribute potable water to the consumer. The amount of contamination to be removed from the source water dictates the degree and type of treatment required and, consequently, the amount of residue that will remain after treatment. The characteristics and quantities of residues generated by the treatment process depend on the water source and the chemicals and processes used.

A residue is a remnant—that is, something remaining after another part has been taken away. In this case, potable water is removed, leaving undesirable constituents behind. Referring to all residues as wastes is inappropriate because it implies that no beneficial use can be made of them. This is not necessarily the case. For example, in the softening process, calcium hardness is removed in the form of calcium carbonate, and, if little or no magnesium is present, this can be recalcined to form new lime. Obviously, referring to this residue as a waste is incorrect. Remembering that the quantity of residue depends on the type and amount of contamination that must be removed is important.

For many years the primary place to deposit residue was in dumps or the nearest river.[1] The development of means to dispose of residues has progressed in a very intermittent manner. Emphasis on environ-

TABLE 16.1    Methods of Residue Disposal from Water Treatment Plants, Reported in 1971

| Disposal method | Percent of plants |
| --- | --- |
| To streams or lakes | 92.4 |
| To storm sewers or surface drains | 3.5 |
| To sanitary sewers | 0.3 |
| To sludge beds | 3.1 |
| To city reservoirs, irrigation ditches, dry creeks, impounding basins | 0.8 |

SOURCE: From AWWA.[1]

mental issues caused a big improvement in techniques for dewatering sludges, starting in the 1950s and 1960s. Construction of major waste disposal plants followed. New issues are now arising, particularly the difficulty of dewatered sludge disposal. This and new exotic wastes from new means of water treatment such as air stripping and granular activated carbon beds will necessitate a review of the whole spectrum of residue disposal. A new era in the treatment and disposal of wastes is just beginning.

Looking at the situation as reported in the last edition of this book (1971),[1] which is summarized in Table 16.1, is interesting. Even at that time, the major portion of residue was returned to streams and lakes. Since then, the situation has changed completely. Legislation, intended to clean up the environment, has affected the whole concept of residue treatment, ranging from collection through treatment. The benefits of source selection and process selection discussed later are unavailable except for new plants.

This chapter will discuss the philosophy of disposal for all residues ranging from sludges to liquid wastes, such as brine waste, whether their disposal was thought of at the time of design or whether it has become a necessity caused by subsequent legislation. The chapter also describes the means available to the water industry to reduce the amount of residue for disposal, commencing with plant design to avoid the production of the residual waste in the first place, or at least to try to reduce the quantity to negligible proportions and, if that is not completely possible, to prepare the material in such a way that acceptable disposal or reuse is possible.

*Management* is the key word. Three questions must be answered: What must be removed? Where will it be disposed? What treatment is necessary to prepare it for that disposal? At all times, the goal of waste management is to dispose of residue in an economical and environmentally acceptable manner.

## Environmental Considerations

Before describing legislation and regulations governing waste discharge, examining both the reasons for their presence and the rationale behind regulation of discharges is appropriate.

Sedimentation is the earliest known form of water purification, and this and subsequent filtration formed the first major efforts to provide potable water for the public. Rainwater systems in Venice in the fifth century A.D. had filters, but no large-scale filters were attempted in the United States until 1870. The first use of coagulation before filtration was recorded in 1885. In 1960, there were 19,236 facilities, of which 6149 were using a treatment process capable of producing a sludge.[2] In other words, the problem of waste disposal did not arrive until the twentieth century.

Wastes must go somewhere. Historically, the most widely used methods were lagooning, to settle and dry the solids, or direct discharge to a river. The total volume of sludge to be disposed of rose only slowly with the increased use of sedimentation; dirty backwash water disposal became a problem only when the filtration process became an established part of the overall water treatment process. Even as late as 1953, the means of disposal of filter washings indicated that over 92 percent went to a stream or lake.[3] The environmental movement in the 1960s caused an intense review of the effect of wastes on the environment. This led to legislation, which is discussed in the next section.

Many in the water industry feel that solids, coming as they do from a river and consisting primarily of inert materials such as fine sand, silt, and clay, are innocuous by virtue of being a concentration of the materials already present in the raw water. When remixed with the receiving water, the impact of water treatment plant wastes is sometimes undetectable except for the presence of chemicals added during the treatment process. Under certain circumstances, however, returning concentrated wastes to a river or lake is unsightly and undesirable. The normal situation is for waste discharge permits to limit discharges to 30 mg/L suspended solids for the wastes. Just as the quantity of waste from water treatment plants is increasing, the quantity of land available for disposal of wastes is decreasing, and a firm and environmentally acceptable policy is needed now. In addition, as treatment processes become more complex, so will wastes and so will the potential effect of their discharge on the environment.

Aluminum toxicity to aquatic organisms is perhaps the major concern regarding the effects of alum sludge discharge. Much of the research related to alum sludge discharge has been done outside the water supply field. A recent report[4] summarized current research on the environmental effects of alum sludge discharge. Little research has been conducted on the toxic components in alum sludge and their effects on receiving streams.[4,5] Even the research on identification of toxic components in alum sludges is sparse.[6-8] Significant research must be completed before alum sludge discharge practices can be supported or eliminated.

## Regulatory Aspects of Waste Discharges

On October 18, 1972, the United States Congress, overriding a presidential veto, enacted PL 92-500, the Federal Water Pollution Control Act Amendments of 1972. Responding largely to the public demand for a better environment and cleaner water, in particular, the new law ended 2 years of intensive negotiation and compromise. It expanded the federal role in water pollution control, authorized a high level of federal funding for the construction of public-owned waste treatment works, and opened new avenues for public participation. Its objective was to reach, whenever attainable, a water quality that provides for the protection and propagation of fish, shellfish, and wildlife by July 1, 1983. Two new phrases were coined—the "best practicable control" and the "best available technology" that is economically achievable. Wastes recognized in the act were mainly from industry, water pollution control plants, and toxic substances. Surprisingly, waste residues from water treatment plants were included as industrial wastes under Section 303(e)(3)(G), which states that "the Administrator shall approve...controls over the disposition of all residual waste from any water treatment processing...."

This caused dissent from owners of water treatment plants, on the principle that they had only taken out of the river or lake those impurities that nature had put there and were simply returning them again to the same stream. To some extent this argument has been upheld, but elsewhere the states that have either adopted federal standards or issued more rigorous ones themselves have interpreted the water treatment process as an industry and subject to waste discharge regulations established under this act. As an example, in the Commonwealth of Virginia, 104 water treatment plants were issued waste discharge permits, and of this number 66 provided treatment. The remaining 38 had schedules to provide treatment by July 1, 1984.[9]

The next enactment in water pollution control legislation was the Clean Water Act of 1977 (PL 95-217) that was signed into law on December 27, 1977, and which significantly amended certain provisions of PL 92-500. The law itself is complicated, and, apart from noting the necessity of compliance, it will not be expanded on here. The reader is referred to the Bureau of National Affairs Report[10] to study the act and its amendments in detail. The act, among other things, called for establishment of effluent limitation guidelines for specific industries, including the water supply industry.

After passage of the legislation, the U.S. Environmental Protection Agency (USEPA) was faced with the necessity to develop effluent guidelines. A preliminary draft report for the waterworks industry had been released in 1974. The draft report received very unfavorable reviews by the water supply industry because its findings were based on a severe simplification of the water industry. The report considered only two treatment processes and assumed only two raw water qualities. From these general assumptions exemplary treatment processes and costs were derived, and the impact of PL 92-500 on the water industry was theorized.

This approach was criticized by the industry for a variety of reasons:

1. The quality of raw water treated by water utilities varies widely.

2. Many types of water treatment processes are currently in use.

3. Costs of treating water treatment wastes and sludge outweigh the possible environmental benefits.

Public Law 92-500 endeavored to assist owners of publicly operated treatment works in construction of works to reduce pollution by awarding grants for the purpose. Because water treatment was considered an industry, federal funding to meet the requirements of the act was not included. Consequently, the owners of water treatment plants have been reluctant to spend large sums of money in the reduction of pollution.

Since the release of that 1974 study, USEPA has found need to re-evaluate its findings, but the effort is at a temporary halt. At this writing, no federal effluent guidelines for the disposal of waste from water treatment plants exist, and each state is currently issuing waste discharge permits that tend to be tailor-made to that state's situation. The lack of federal guidelines creates an ongoing problem, and USEPA regional offices and states tend to base their permits on previous water quality standards. USEPA, although reportedly working on a guidance document, gives the matter low priority. Federal action, if and when it does occur, could have a significant impact on how states apply clean water rules to water treatment plant waste disposal practices.

Regulations governing water treatment plant waste discharges vary from state to state and within a single state, depending on the nature of the receiving stream. In Missouri, for example, rules for the high-flow, high-solids content Missouri and Mississippi rivers are different from those for other streams of the state. Solids removed during water treatment are allowed to be returned to the Missouri River along with chemicals precipitated during treatment. For streams other than these two, suspended solids limits prevent direct discharge of water treatment wastes. St. Louis County Water Company plants on the Missouri River discharge wastes directly to the stream.[5] On the other hand, plants on the Meramec River, which is a tributary of the Missouri River, impound waste for liquid-solid separation. The clear supernatant liquid is discharged to the stream. Solids are disposed of on company-owned land under a state solid waste permit.

Water treatment plants are becoming affected by additional environmental programs. These include PCB transformer rules under the Toxic Substances Control Act,[11] underground tank notification rules under Resource Conservation and Recovery Act 1984 Amendments,[12] waste oil disposal practices under Resource Conservation and Recovery Act hazardous waste rules,[12] and volatile organic chemical emission rules under the Clean Air Act.[13] Future environmental regulations will likely influence water treatment plant waste disposal practices.

## Sources of Water Treatment Plant Wastes

Two major sources of residue exist: the material in the raw water itself, whether it is soluble or insoluble, and the chemicals used to remove this material. Previous chapters have illustrated typical source water and the types of contaminants found. Each treatment process creates a specific residue. Wastes from the treatment processes described in this book are listed in a general fashion in Table 16.2. The total residue for treatment and disposal at a facility will be a combination of wastes from one or more processes.

One type of residue allows easy disposal and will not be considered

TABLE 16.2    Sources of Water Treatment Plant Wastes

| Treatment process | Residue | Chapter reference |
| --- | --- | --- |
| Presedimentation | Solids | 4 |
| Air stripping/aeration | Off-gases | 5 |
| Coagulation, flocculation, sedimentation | Slurry† | 6,7 |
| Filtration | Liquid/slurry | 8 |
| Ion exchange/inorganic adsorption processes | Liquid/brine† | 9 |
| Chemical precipitation (softening) | Slurry† | 10 |
| Membrane processes | Liquid/brine† | 11 |
| Chemical oxidation | None† | 12 |
| Adsorption processes for organic compounds | Slurry/solid† | 13 |
| Disinfection | None | 14 |
| Water fluoridation | None | 15 |
| Corrosion/deposition control | None | 17 |

†Solid wastes associated with chemical feed systems will be generated.

further. This is trash that is removed where inlet screens are used, particularly on river supplies such as the River Seine, Paris, and the Passaic Valley Water Commission plant at Little Falls on the Passaic River, New Jersey. It consists mainly of floating river debris such as branches, leaves, and so forth, and is disposed of on site or at a sanitary landfill.

## Residue Characteristics

Characteristics of water treatment plant residues are location- and season-specific. The subject is best illustrated in a general fashion by Table 16.3. Residues can be organic or inorganic, liquid or solid. Characteristics of particular wastes are covered in more detail later.

Wide physical and chemical changes in a waste can occur even when treating the same source water supply using the same chemicals, because of seasonal factors. As an extreme example, the 1960s drought at the Stocks filtration plant, Yorkshire, England (direct filtration), reduced the level of the reservoir drastically and caused a large increase in the organic material that had to be removed from the raw water. Filter runs in the pressure filters decreased to a maximum of 5 or 6 h. The backwash water that normally settled easily in two holding basins fitted with stirrers (designed based on earlier research[14]) took on a greenish tinge because of the very high organic matter. This would not settle, and instead of the customary 2.5 to 3 percent settled solids produced, only 0.5 percent could be achieved even in the new type of stirring thickener being tested at the time. The drought finally ended, and rain washed the silt and manganese deposits from the exposed sides of the reservoir bottom into the main stream. Filter runs suddenly increased, and the stirred sludge settled to an incredible 11 percent solids. The change in the type of solids was accentuated by the sludge becoming dark brown to almost black.

With river supplies in particular, variations in source water quality cause significant changes in the physical characteristics of residues. Certain other physical properties may be present with an alum sludge. For example, the thixotropic nature (behaves as a thick viscous liquid when stirred, but sets like a jelly if agitation ceases) causes complications in the design and operation of sludge pumping facilities and, particularly, force mains.

Anaerobic conditions in sedimentation basins impact sludge characteristics and treated water quality. Manganese and iron associated with sludge may be solubilized if anaerobic conditions develop in the sludge blanket.[15] Anaerobic conditions may develop rapidly in manually cleaned basins, even in winter conditions. Mechanically cleaned basins will usually remain aerobic if they are cleaned regularly.

**TABLE 16.3  Physical Characteristics of Wastes**

| Process for treatment | Chemical used | Residue produced | Physical appearance | | |
|---|---|---|---|---|---|
| | | | Color | Settlement | Description |
| Softening | | | | | |
| Precipitation reactor | Lime and soda | $CaCO_3$<br>$CaCO_3 + Mg(OH)_2$ | White†<br>White | Good<br>Fair | Settled solids like toothpaste |
| Pellet reactor | Lime | $CaCO_3$ | White† | Instantly | Hard crystalline spheres 1 to 2 mm diameter |
| Coagulation | | | | | |
| Alum | Aluminum sulfate | $Al(OH)_3$ | Brown | Very poor | Thick chocolate consistency, viscous, thixotropic |
| Iron salts | Ferrous or ferric salts | $Fe(OH)_3$ | Red/brown | Poor | When dried it forms hard lumps |
| Polymer | Polymer | Polymer | N/A | Poor | Depends on polymer |
| Aeration | Air to precipitate iron and Mn | Insoluble ferric and Mn compounds | Red/brown | Good | Highly colored solids; will settle, dry, and crack |
| Softening by ion exchange | Resins | Acid or brine from regeneration | Clear | N/A | Liquid; high total dissolved solids |
| Filtration | | | | | |
| Backwash water | — | — | Brown | Poor to fair | Agglomeration of fine floc removed from surface of filter |
| Diatomaceous earth | — | Spent media | White to tan | Good | Mainly spent media with particulate matter removed from water |

†Tan/brown if iron is present.

## Quantities of Water Treatment Residues

The quantity of residue produced in the form of solids is calculated relatively easily by using a solids balance and appropriate chemical equations. The first step is to review all the solids that may appear as residue. This is summarized for some of the more common items in Table 16.4. An investigation of this nature will result in a list of all solids likely to be found in the residue. The next step is to calculate the residue that results from a given quantity of the chemical. For this the chemical equation concerned and the molecular weights (MW) of the substances concerned are used. As an example,

$$Al_2(SO_4)_3 \cdot 14H_2O + 3Ca(HCO_3)_2$$
$$\rightleftharpoons \underset{\text{Solid}}{2Al(OH)_3} + \underset{\text{Soluble}}{3CaSO_4} + \underset{\text{Gas}}{6CO_2} + \underset{\text{Water}}{14H_2O} \qquad (16.1)$$

where 594g (1 lb) of solids are found on the left side of the equation and 156g (0.26 lb) are found on the right side.

Repeating for other chemicals listed in Table 16.4, the pounds of solids produced by 1 lb of chemical can be calculated. Table 16.5 gives some examples. The anticipated dose of each required chemical will be determined by jar tests for a new source or by log book records in order that minimum, maximum, and average doses can be determined. Fi-

**TABLE 16.4    Likely Residues from Treatment Processes**

| Item/chemical | Comes from | Appears in residue as | Solid |
|---|---|---|---|
| Dissolved solids | Raw water | Dissolved solids | Only if precipitated |
| Suspended (silt) solids | Raw water | Silt—no change | Yes |
| Organic matter | Raw water | Probably no change | Yes |
| Aluminum salts | Chemical coagulation | Aluminum hydroxide | Yes |
| Iron salts | Chemical coagulation | Ferric hydroxide | Yes |
| Polymer | Chemical treatment | No change | Yes |
| Lime | Chemical treatment and pH correction | Either calcium carbonate or, if using lime water, impurities only | Yes |
| Powdered activated carbon | Taste and odor control | Powdered activated carbon | Yes |
| Chlorine, ozone | Disinfection | In solution | No |

TABLE 16.5    Theoretical Solids Production

| Item/chemical | Pounds of solids produced per pound of chemical |
|---|---|
| Suspended solids (silt) | 1.0 |
| Organic matter | 1.0 |
| Aluminum sulfate | 0.26 as $Al(OH)_3$† |
| Ferric chloride | 0.66 as $Fe(OH)_3$† |
| Polymer | 1.0 |
| Lime | Allow 0.10 as insoluble fraction |
| Powdered activated carbon | 1.0 |

†Depending on the hydrated form of the alum or ferric chloride.

nally, the total residual solids production can be calculated by using the estimated design daily finished water output. Typical results are given in Table 16.6.

The quantity of water in which waste solids are contained depends primarily on the ability of the operator to wash the filter or remove sludge from a sedimentation basin, using the minimum amount of water. Thus, both the quantity of the waste and the amount of dry solids it contains must be reported. For example, in a case in Virginia the sedimentation basins were permitted to fill to 25 to 50 percent capacity with sludge at 2 percent solids or more before discharging to a nearby stream. Because of the clear water above the sludge and because water from high-pressure hoses used to wash the basin was also mixed with it, the average solids concentration of the discharge to the stream was less than 0.5 percent. The situation was rectified by installation of sludge-collecting mechanisms.

Large treatment facilities can generate astronomical quantities of waste. For example, Chicago pumps just over 1000 mgd (4000 ML/d) from Lake Michigan. Of the total pumpage, 361,225 mil gal (1,367,000 ML) in 1 year was pumped directly from the lake, and 5252 mil gal (19,880 ML) was recirculated filtered wash water (1.5 percent). Filter backwash water is recycled to intake basins, but the basin sediment cannot be recycled or returned to the lake and must be disposed of by

TABLE 16.6    Typical Impact of Source Water on Residue Production

| Source water | Average range of residue (lb dry solids per mil gal) |
|---|---|
| Good-quality reservoir water | 100–150 |
| Fair-quality reservoir water | 150–250 |
| Average river water | 200–300 |
| Poor-quality reservoir water | 250–350 |
| Poor-quality river water | 350–450 |

way of the metropolitan sanitary district sewer system to their processing facilities. During January 1985, the Jardine and South water filtration plants totaled an average of 140,530 lb (63,744 kg) of sediment per day or 140 lb/mil gal (17 kg/ML).[16]

Some degree of control can be exercised over the quantity of residue created:

1. Residue quantity depends on the degree of contaminant removal. For example, the degree of softening carried out on hard waters will affect the quantity of sludge produced.

2. Processes used to remove contaminants influence residues produced. As an example, the choice between ion exchange or lime addition to remove hardness will influence the character and quantity of waste produced.

3. The physical form of the waste can sometimes be controlled. For example, hardness reduction by lime softening can be achieved by using either a precipitative process, which produces a fine calcium carbonate sludge, or the pellet reactor process, which produces crystalline residue in the form of small-diameter calcium carbonate pellets.

In the initial stages of investigation into the treatment of a new water source, careful review of the degree of treatment required and the treatment options is important. Several alternatives are often available, and a decision must be made as to which process train is the most cost-effective overall. Consideration must be given to the sum of the cost of the treatment of the water itself and the cost of disposal of the residues created. The impact of source selection, process selection, and the choice of treatment chemicals is discussed in more detail in the next section.

## Reduction of Quantity of Residues

Several measures can be implemented to minimize or reduce the quantity of wastes generated from a treatment system. Measures include those that may be taken at the initial stages of a project and those possible later during system operation.

### Selection of raw water source

In the early planning of any project, a decision must be made as to the source water and, if a river or underground aquifer, the location of the actual point of diversion. A number of analyses must be made of the source water in order to arrive at the most economic means of design. The cleaner the source water, the less the residue that must be removed by the treatment process. In European countries, the tendency is to use headwaters where water quality is highest and to convey the water to a treatment plant located between intake and the city to be served. A choice must be made between having a source water transmission main or a treated water transmission main or part of each. In larger countries, such as the United States, this option is usually not feasible.

Having selected a water source, the next task is to determine the desired or required treated water quality and estimate the amount of contaminants or material to be removed. Contaminants removed will appear as the major part of the residue from the treatment plant. For example, in the development of a well field in northern England, over 50 boreholes and 40 wells were drilled, indicating variations in total hardness from 155 to 760 mg/L as $CaCO_3$. The combined yield of the wells was over 25 mgd (95 ML/d). After the maximum capacity of the well field was tested and compared to the potential demand required by the distribution system, a decision was made as to which wells would be connected and, in particular, which would give the maximum yield for the removal of the minimum quantity of contaminants. Operation of this well field was possible so that, under normal circumstances, a minimum of softening is necessary and, hence, a minimum quantity of residue is produced. If during future pumping the yield of the chosen wells declines, then other, previously drilled, wells would be brought into operation, although the quantity of residue would be greater. Another important factor in deciding which wells to use was the generally low ratio of magnesium to calcium in the hardness of the water.

Concerning reservoirs, the cleaner the reservoir water is, the fewer contaminants that need to be removed and, hence, the smaller is the quantity of residues obtained. In the development of the Lake Sidney Lanier source for Gwinnett County, Georgia, an early analysis of Lake Lanier water characterized it as excellent and revealed minimal contaminants. The analysis is given in Table 16.7. The water was of such good quality that direct filtration was considered in the earliest design stage. After discussions with the state of Georgia, a plant was designed that could operate in either a direct or conventional filtration mode and, consequently, save on construction and operating costs.

### Choice of process

Two broad classifications of potable water treatment exist: coagulation, sedimentation, and filtration for surface water, and softening or iron and manganese removal for groundwater, although river water is

TABLE 16.7    Lake Lanier Water Analysis at Depth between 20 and 25 ft

|  | Maximum | Minimum | Average |
|---|---|---|---|
| pH | 7.8 | 6.6 | 6.9 |
| Alkalinity, total (mg/L) | 12 | 8 | 9 |
| Turbidity (JTU) | 9.0 | 1.5 | 2.3 |
| Conductance (mho/cm) | 45 | 15 | 31 |
| Iron (mg/L) | 0.09 | 0.04 | 0.06 |
| Manganese (mg/L) | 0.14 | 0 | 0.05 |
| Dissolved solids (mg/L) | 66 | 11 | 13 |
| Color, APHA | 20 | 2 | 7 |

often softened in the Midwest. Previous chapters discussed other, less used, treatment methods. At this stage of design, experience is necessary to select the treatment process scheme. Although a certain process produces potable water inexpensively, the disposal of residues from that process may cost many times more than the process itself. A significant proportion of the total plant cost is often associated with residue disposal.

Surface water treatment normally revolves around production of a sturdy floc that settles in a sedimentation basin, the removal of the finer solids then being accomplished by filtration. Jar tests should be carried out to indicate which coagulant will not only produce an acceptable water but will also give the maximum quantity of residue. Polymers are now partially replacing the trivalent metal salts, such as aluminum sulfate, in coagulating suitable water in preparation for direct filtration and, by virtue of their molecular structure, will produce smaller quantities of residue. Polymers used as coagulant aids may reduce the quantity of residue from coagulation if their use results in reduction of the primary coagulant dosage.

The question is one of process selection, recognizing that the choice of process has a direct bearing on the quantity of residue. This point will be illustrated by two examples. In relation to softening, for water with a suitable magnesium-calcium ratio, the choice often lies between a pellet reactor or a conventional suspended precipitate reactor. Each process variation produces different types of sludge. The pellet reactor process (see Chap. 10) differs from other lime-softening processes in that it is seeded with fine sand or crushed pellets to provide a nucleus. Lime is introduced as a slurry or saturated solution at the bottom of a tall contact reactor. The calcium carbonate crystallizes on the nuclei. This process is attractive because the volume of waste is reduced from a considerable quantity of calcium carbonate slurry for a conventional process to a smaller quantity of crystalline beads that are bled off from time to time. The pellets or beads are almost like small ball bearings. No dewatering is necessary, because any water removed from the reactor during the debeading process simply runs away from the pile of waste solid beads. Another advantage of the process is that if the impurities in the pellets are low they can be used as a soil conditioner to lower soil acidity.

Another example relates to removal of iron and manganese from certain aquifers. The conventional method is to pump groundwater to the surface for treatment by aeration to oxidize the iron and manganese, remove the precipitate by sedimentation and filtration, and convey the sludge from time to time to lagoons and ultimate disposal. An alternative to the conventional method is in situ treatment. Introduced in Europe a decade ago, in situ treatment involves changing conditions around a well so that iron and manganese are oxidized and removed from water in the ground. By utilizing a sufficiently large treatment zone, oxidized iron and manganese can be retained in the soil during the useful life of the production well, thus eliminating waste residue and the need for sludge-handling facilities. The system is applicable particularly in glacial and fluvial aquifers, but soil conditions, hydrogeology, and water chemistry can be limiting.[17] Operating costs are claimed to be lower than for other removal technologies because building needs are moderate and only a portion of the produced water actually passes through treatment equipment. Reduced well maintenance, no sludge production, and no chemical requirements are the claimed advantages.

### Choice of chemicals

Once the general method of treatment has been established, treatment chemicals must be selected and, apart from their efficiency in the water treatment process, the residues created by these chemicals must be determined. Although a certain method of treatment using a specific chemical is cost-effective, the process may have an overwhelming effect on the cost of treating its waste. For example, at the small town of Ljungskile, Sweden, a small lake was selected as the source of future water supply. This water had low turbidity but high color. Ozone was considered as an oxidant, it being particularly effective on water with low turbidity and high color.

Alternatively, powdered activated carbon could also be used for color removal, but (see Table 16.8) it can form a high percentage of the total waste of the plant. The owner was unwilling to tolerate any waste from the treatment plant either for deposit on the land or for disposal in the nearby creek. Ozone was therefore chosen as the

TABLE 16.8    Analyses of Residue from Plants Using Powdered Activated Carbon and Alum

| Component | Fraction of total residue solids, % | |
|---|---|---|
| | Plant A | Plant B |
| Aluminum hydroxide | 34.0 | 24.1 |
| Powdered activated carbon | 24.0 | 27.1 |
| Suspended solids in raw water | 42.0 | 48.8 |
| Total | 100.0 | 100.0 |

method of treatment for color removal in order to minimize the quantity of waste produced to negligible proportions.

### Recovery of chemicals

In addition to minimizing residue quantity by choice of process, the quantity may also be reduced by recovering some of the constituents of the waste. For example, some coagulants, such as aluminum, can be recovered. Recovery of polymers, however, is not feasible.

Chemical recovery is achieved by performing chemical reactions with the sludge, thereby altering sludge characteristics. When chemical recovery is considered, the rhetorical question must always be asked: why change the chemical characteristics of the sludge by the recovery of certain chemicals when so much effort has been spent trying to eliminate this waste from the source water?

The cost of recovery is an important consideration and must be carefully evaluated when considering chemical recovery. A comparison must be made between the cost of chemical recovery plus the cost of makeup chemical and the cost of new chemical without recovery plus the cost of disposing of a portion of the sludge.

The use of coagulant recovery systems leads to an interesting ethical discussion as to whether a water utility must only supply water or whether it should compete with the chemical industry in order to supply water in the most economical manner. The latter case would certainly allow for the consideration of operations necessary for coagulant recovery. Each case must be evaluated on its own merits.

**Recalcination of lime-softening residues.**    Before discussing typical lime recovery plants, we review the chemistry of the lime-softening reaction. Lime softening proceeds as described by three equations (summarized from Chap. 10):

Free $CO_2$:

$$Ca(OH)_2 + CO_2 \rightleftharpoons CaCO_3 + H_2O \qquad (16.2)$$

Magnesium hardness:

$$2Ca(OH)_2 + Mg(HCO_3)_2 \rightleftharpoons 2CaCO_3 + Mg(OH)_2 + 2H_2O$$
$$(16.3)$$

and the most important equation of the three:

$$Ca(OH)_2 + Ca(HCO_3)_2 \rightleftharpoons 2CaCO_3 + 2H_2O \qquad (16.4)$$

74 tons      162 tons      200 tons      36 tons

Recalcining of the $CaCO_3$ can be carried out in accordance with the equation

$$CaCO_3 \overset{\Delta}{\rightleftharpoons} CaO + CO_2 \qquad (16.5)$$

200 tons      112 tons      88 tons

This is then slaked in accordance with the equation

$$CaO + H_2O \rightleftharpoons Ca(OH)_2 \qquad (16.6)$$

112 tons      36 tons      148 tons

If the small proportion of $CO_2$ and $Mg(HCO_3)_2$ is ignored, twice as much lime is produced than was used for the softening process, because part of the recovered lime has been produced from the calcium hardness of the water; that is, 148 tons (134,000 kg) of $Ca(OH)_2$ are produced from only 74 tons (67,000 kg) of lime. This, of course, assumes 100 percent recovery, whereas only 90 to 95 percent is obtained in practice. The principle still applies that an excess of lime is produced, however, and, unless a market is found for this excess lime, calcining only sufficient lime for use at the softening plant is economical but a partial sludge disposal problem will remain.

Calcining can take place in two types of kilns: horizontal and vertical. The horizontal kiln is rotary and takes up a lot of land: it is not aesthetic, but it is comparatively foolproof. The vertical kiln can be easily enclosed in a building, however, and requires very little space, but being nonrotary, difficulties are often experienced in producing the lime in the form of pellets. Recalcining is carried out at several locations in the United States, including Dayton, Ohio; Miami, Fla.; San Diego, Calif.; and Lansing, Mich.

**Reclamation of iron and aluminum.**   The cost of iron salts is usually low because they are a by-product of other industries. As a result, the recovery of iron, although feasible, has not been pursued. Aluminum sulfate, however, is generally more costly than iron, and recovery processes have been investigated for many years. Some of the earliest work was carried out by Isaacs and Vahidi[18] in England and by Roberts and Roddy[19] in Tampa, Fla., using sulfuric acid. The Tampa experiments were important in that they led from the laboratory stage to the construction of a pilot plant and, for a short time, full-scale use of reclaimed alum at the Tampa water treatment plant. Later work at the Tampa plant was carried out by Cornwell[20] in a very interesting project involving liquid-ion exchange. This process was found to be successful in practice.[21]

One alum recovery system involving the reaction between sulfuric acid and aluminum hydroxide in the sludge was attempted at the Daer Water Board treatment plant works near Hamilton, Scotland. This 25-mgd (95-ML/d) plant utilizes sedimentation tanks and rapid gravity filters. The method of recovering the coagulant has been described by Webster.[22] A pilot plant was operated for nearly a year to ensure that no buildup of color or chemicals occurred in the reconstituted coagulant. Water with high iron concentrations produces a buildup of iron in the recovered coagulant. In this case no problem occurred, and the dose utilized was 6 mg/L of fresh alum plus 14 mg/L of recovered alum. A 70 percent recovery was achieved by adding acid to the sludge after mechanical freezing and thawing.

Full-scale testing of alum recovery used in connection with a sand bed sludge dewatering process was recently completed in Durham, N. C.[23] Sulfuric acid acidification was used, resulting in a 27 to 53 percent reduction in solids production on a dry-weight basis. The overall recovery in the amount of dilute liquid alum was 64 to 79 percent, based on recovery of both supernatant and bed drainage. The reuse of only recovered alum as a coagulant in the water plant showed acceptable performance based on final water quality standards. A 40 percent reduction in sludge treatment, transport, and disposal costs and a 28 percent reduction in alum costs are expected once full-scale operation is underway.[23]

### Recirculation and reuse of filter backwash waters

As mentioned earlier, 90 percent of source water suspended solids will be removed by sedimentation under normal circumstances, leaving the remaining 10 percent of lighter floc to be passed forward to the surface of the filters (see Chap. 8). During filter backwashing, the bed is agitated and dirty water is removed to settling basins. Presently, many treatment plants return backwash water to a river or an adjacent water course. This practice, however, cannot continue, and other means must be found for disposal. Solids captured by filters are fine and have been carried over from sedimentation basins. Because they were not removed by sedimentation the first time, they may not settle any better when backwashed from the filters. Storage of solids in the filter, however, often causes some solids to agglomerate, and thus they are easier to settle when backwashed from the filter. A growing tendency is for backwash water to be returned to the head of the treatment plant where it can mix and settle with the heavier floc in the sedimentation basins. Equalization tanks must be used in order to avoid a shock load to the plant. Particles in the wash water might also serve as nuclei for flocculation of the source water. Although wash water normally accounts for 1 to 5 percent of the total plant output, reusing this water is often necessary in cases where the state has reduced the diversion permits from the source concerned. Disposal of residues from filter backwashing is further discussed later.

Chemical characteristics of wash water must be carefully investigated before deciding to recirculate backwash waters. Several authors, notably Cyril Gomella (Paris, France), have questioned the wisdom of mixing dirty backwash water with the source water, because so much time and energy was spent in separating the impurities in the first place. In a discussion with the author, Gomella asked, "How do we know what harmful ingredients there are which have been removed from the raw water? Why consider returning these?" Discovery of new contaminants may well prompt questions as to whether contaminants are currently being taken out of the water by good fortune, about which no knowledge exists at this moment. The possibility of recycling contaminants to the source water must be evaluated when recirculation and reuse of filter backwash water are considered.

### Processes for Concentration and Treatment of Residues

Water treatment plant waste management involves five stages (Fig. 16.1).

*Stage 1:*    Residues are separated from the treated water in the various treatment processes utilized at the facility.
*Stage 2:*    Residue streams are collected and combined, with or without intermediate thickening, in the collection stage.
*Stage 3:*    Residues may be concentrated before further treatment.
*Stage 4:*    Residues may undergo further treatment.
*Stage 5:*    Residues are disposed.

Processes for residue concentration and treatment (stages 2 and 3) are discussed in this section.

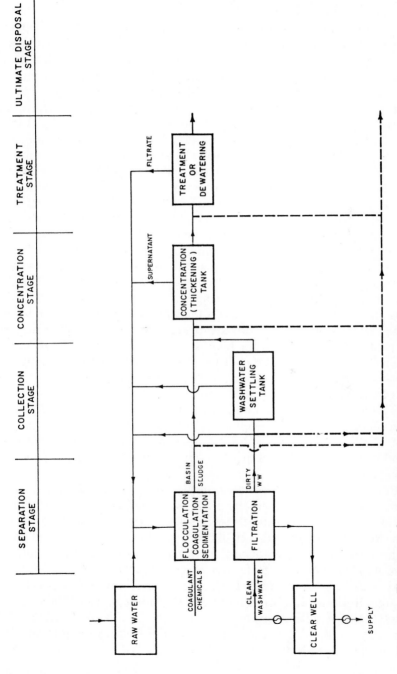

**Figure 16.1**  Five stages of water treatment plant waste management.

## Lagoons

The time-honored method of sludge disposal at many plants in many countries is simple lagooning. Where ample land is available without causing public nuisance and in favorable climatic conditions, disposal by lagooning on at least a short-term basis is usually the economic solution. The forces of nature are used to dewater and dry the sludge in one of two ways, depending on the climate: (1) in warm arid climates sludge is dried by the sun; (2) in cold climates with long periods of freezing, the drying process is aided by natural freezing and thawing.

In warm areas of the country, the sludge surface dries and cracks into a "giant's causeway" type of surface. The sludge dries to a depth of 18 to 24 in (0.46 to 0.61 m) and can then be picked up by a front-end loader for disposal in a landfill. In areas of high temperature and low humidity, such as Arizona and California, this system will produce an almost dry sludge in a relatively short time. Drying time is site-specific and depends primarily on the prevailing climatic conditions.

Further north, the adoption of natural freezing and thawing turns the sludge into a fine material resembling coffee grains.[24,25] Unfortunately, such a system can only be used in areas where a considerable portion of the winter has temperatures below freezing and a prolonged period of summer well above freezing.

Although simple lagooning is inexpensive initially, mechanical equipment must eventually be used to remove the solids for transport to adjacent land or a sanitary landfill. Simple lagooning is undoubtedly one of the least expensive alternatives for solids disposal and can be used both for alum and lime-softening sludges. For example, lagoons are used for the disposal of softening sludge at St. Petersburg, Fla. This large plant treats well water by aeration, to remove hydrogen sulfide, and then lime softening with polyelectrolyte as a coagulant aid. Sludge is piped into large lagoons. Ample swampland was available adjacent to the plant, and when a lagoon is filled a new one is constructed.

Alum sludge does not dewater very well in temperate areas. The Elizabethtown Water Company in New Jersey has been lagooning alum sludge at the Raritan-Millstone water treatment plant for many years. Recently 38,000 yd$^3$ (29,000 m$^3$) of sludge was removed from one lagoon that had been drying for 7 years. Average sludge concentrations were found to vary from 1 percent concentration at 1-ft (0.3-m) depth to 21 percent at 9 ft (3 m). The overall average was 7.5 percent. The anticipated design output of the plant at year 2005 is 120 mgd (450 ML/d), with average solids production of 54,000 lb/d (25,000 kg/d)—that is, 450 lb/mgd (55 kg/ML), which requires significant lagoon or drying bed area.

A cost comparison between lagooning and mechanical dewatering, recently carried out at Elizabethtown, indicates that rising disposal and labor costs have increased the cost of lagooning to the range of mechanical dewatering and disposal. The decision whether to invest in sludge dewatering facilities or to continue with the present practice of lagoon drying and trucking to a decreasing number of industrial landfills has not yet been made.[26]

In summary, under certain circumstances lagooning is a satisfactory and economic sludge disposal method. In view of the cost of land, a careful distinction between lagooning and dumping must be made. Some iron or alum sludges, for example, never thicken beyond 8 or 9 percent solids even after many years of lagooning, after which the lagoon ceases to be a means of dewatering and begins to function as a dump.

### Drying beds

An improvement in lagoon performance can be made by providing underdrains at the base of the lagoon. Economics are an important factor, but provision of underdrains or wedge wire may accelerate the drying process. In effect, the lagoon is operated as a filter. The head available is small, but filtration does occur in addition to evaporation. In some cases, coarse sand is used at the bottom of the lagoon, and collecting drains convey filtrate to a nearby water course. Removal of dried sludge may be a little easier than with a normal lagoon, but the cost of removal will still be considerable.

Vacuum-assisted drying beds are designed to dewater sludge in shallow beds. The floor of each bed comprises rigid porous plates, and when a vacuum is applied below these plates, filtrate is withdrawn from the sludge. Laboratory-scale testing of this process on alum sludge samples showed that the system could effectively dewater polymer-conditioned sludge.[27] Cake solids of 13 to 14 percent were achieved during the tests using a 22-h cycle time.

Vacuum-assisted drying beds are attractive because of their relatively low capital cost and energy requirements. The process may not, however, produce a cake of sufficient dryness for final disposal in many public sanitary landfills.

### Gravity thickening

After removal from a sedimentation basin, most sludges can be further thickened. Thickening can be economically attractive because a reduced sludge volume results, allowing for smaller dewatering unit processes if further treatment is desired. Some dewatering systems will perform more efficiently with higher feed solids concentrations. Thickening tanks can also serve as equalization basins to provide a uniform feed to downstream dewatering processes. Although several types of thickeners exist, gravity thickeners are used almost exclusively in water treatment.

Gravity sludge thickeners are usually circular settling basins with a scraper mechanism in the bottom (Fig. 16.2), or else they are equipped with sludge hoppers. Thickeners may be operated on a continuous-flow basis or a batch basis. For continuous-flow thickeners, sludge usually enters the thickener near the center of the basin and is distributed radially. Settled water exits the thickener over a peripheral weir or trough, and thickened sludge is drawn off the basin floor. For tanks equipped with a scraper mechanism, the scraper is located at the thickener bottom and rotates slowly. This movement directs the sludge to the draw-off pipe near the bottom center of the basin. The slow rotation of the scraper mechanism also prevents bridging of the sludge solids. The basin floor is usually sloped to the center to facilitate collection of the thickened sludge.

Batch fill and draw-thickening tanks are often equipped with bottom hoppers. Sludge flows into the tank, usually from a batch removal of sludge from the sedimentation basin, until the thickening tank is full. After quiescent settling, a telescoping decant pipe is used to remove supernatant liquid. The decant pipe may be continually lowered as solids settle until the maximum desired supernatant liquid-solids concentration is reached, or until the sludge will not thicken further.

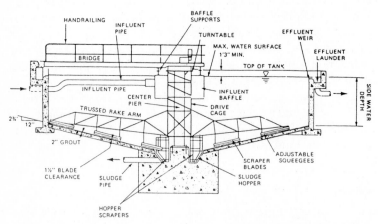

Figure 16.2 Circular gravity thickener. (*Source: P. Innocenti, "Techniques for Handling Water Treatment Sludge," Opflow, vol. 14, no. 2, 1988, p. 1.*)

Thickened sludge is pumped out of the bottom hoppers to further treatment or disposal.

Design of batch or continuous-flow thickeners is usually based on previous experience of similar full-scale installations or on laboratory settling tests.[28] Small pilot-scale thickener tests are very difficult to operate, and the results obtained from them are not always representative of full-scale units.

### Vacuum filtration

Many types of vacuum filters exist. Each is subject to the same limitation; that is, the maximum theoretical pressure differential that can be applied is atmospheric, 14.7 psi (103 kPa). In practice, a differential pressure of about 10 psi (70 kPa) is achieved.

The equipment itself consists of a horizontal cylindrical drum that rotates partially submerged in a vat of sludge that, in order to assist dewatering, is usually conditioned by either a coagulant or a body feed such as fly ash (Fig. 16.3). The drum surface is covered by a filtering medium that is fine enough to retain a thin cake of sludge solids as it is formed. The filtering medium usually consists of a fabric mesh. The drum surface is divided into sections around its circumference. Each section is sealed from its adjacent section and the ends of the drum. A vacuum is applied to the appropriate zone and subsequently to each section of the drum. From 10 to 40 percent of the drum surface is submerged in a vat containing the sludge slurry. The submerged area is the cake-forming zone. When the vacuum is applied to this zone, it causes filtrate to pass through, leaving a cake formed on the cloth. The next zone, the cake-drying zone, represents from 40 to 60 percent of the drum surface. In this zone, moisture is removed from the cake

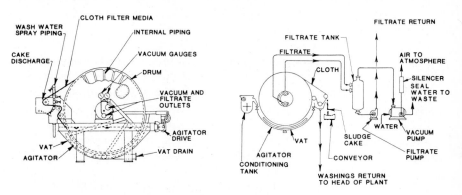

**Figure 16.3** Typical vacuum filter schematic. (*Source: P. Innocenti, "Techniques for Handling Water Treatment Sludge," Opflow, vol. 14, no. 2, 1988, p. 1.*)

under vacuum. The zone terminates at the point where the vacuum is shut off. Finally, the sludge cake enters the cake discharge zone where it is removed from the medium. This is accomplished by the filter cloth belt leaving the drum surface and passing over a small-diameter discharge roll that facilitates cake discharge. No vacuum is applied to this zone.

Vacuum filters are usually not efficient for an alum type of sludge unless precoating is used. Even with wastewater treatment sludges, the cake solids concentration does not usually exceed 10 to 12 percent solids, with 20 percent as the absolute upper limit. Vacuum filters are, therefore, unlikely to be applied to water treatment sludges in circumstances where a cake of 25 to 30 percent solids must be produced for landfill disposal.

An ideal application of vacuum filtration for water treatment residues is for dewatering lime-softening sludge. For example, the treatment plant in Boca Raton, Fla., a conventional lime-softening plant that reduces hardness from 240 to about 70 mg/L as $CaCO_3$, uses this method. Lime sludge is piped from precipitators to a thickening tank and from there to a 6-ft (2-m) diameter vacuum filter. A 55 to 60 percent cake solid is achieved. The final product is sufficiently stable for use in road construction or for subbase stabilization of new public roads. It is also in demand by local residents for improving the moisture-retaining qualities of sandy topsoil indigenous to the area.

### Belt filters

A new development now in competition with vacuum filtration is the belt filter (Fig. 16.4). This technology has only been available in the last decade. Applicable primarily for wastewater raw and digested sludges, certain water treatment sludges may nevertheless be sufficiently adaptable for the use of a belt filter.

Belt filters utilize endless permeable belts to compress and shear the sludge to remove water. A three-stage press is most often selected (Fig. 16.5). In the first stage after conditioning, sludge is drained by gravity to a nonfluid consistency. In the second stage, pressure is applied and gradually increased. In the third stage, filtrate is removed to discharge ports. In the last stage, the cake is sheared and further dewatered. The sludge must be conditioned in such a way that it is capable of being retained on the top belt. This is usually accomplished by using a polymer.

Belt filters are designed so that sludge is sufficiently dewatered to leave a cake (even if it is a wet one) at the top of the first stage. Some manufacturers use a small vacuum on the underside to achieve further dewatering. Sludge from the first stage is directed into the pres-

**Figure 16.4**  Belt filter.  (*Courtesy of Ashbrook-Simon-Hartley, Houston, Texas.*)

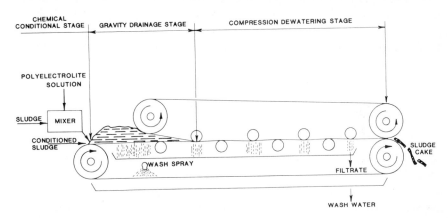

**Figure 16.5**  Belt filter press schematic.  (*Source: P. Innocenti, "Techniques for Handling Water Treatment Sludge," Opflow, vol. 14, no. 2, 1988, p. 1.*)

sure zone, where a wedge and decreasing-diameter rollers are used to gradually increase the pressure of both belts on the sludge, which then more effectively remove water. A limit exists, however, because too much pressure on a poorly conditioned sludge might produce extrusion through the porous belt.

In the final stage, shear is applied to the cake by running the belts in a serpentine path around very small diameter rollers. The differential speed of the two belts opens up the cake and allows water to be released through channels to the final output where doctor blades lift the cake off the belt onto a conveyor or into a hopper.

Many belt filter units are available. A relatively recent unit utilizes a series of dams or plows applied to the top gravity section to, in effect, turn over the sludge cake and increase dewatering. Few full-scale applications of this new system have been made to water treatment sludge.

### Filter presses

The filter press is another process option available for dewatering sludge for final land disposal. Early filter presses were frequently used in Europe for dewatering thin slurries, such as china clays and wastewater sludges. Their practical use for water treatment residues is relatively new, dating from around 1965. Experiments commenced in England in 1956, but were disappointing until the advent of the use of polymers as conditioners. The first known use in the United States was at the Atlanta waterworks and at the Little Falls treatment plant of the Passaic Valley Water Commission.[29]

At the commencement of a filtering cycle, sludge is forced into contact with the cloth, which retains the solid matter while passing the liquid filtrate. Very quickly the cloth becomes coated with a cake of sludge solids, and all future filtering occurs through this cake, which increases in depth as succeeding layers build up. The type of cloth does not affect the rate of filtration after the first few minutes, and it can be ignored from a theoretical point of view. Filtration may be described by the Carman-Kozeny and D'Arcy equations (see Chap. 8).[28,30,41]

Filter presses (Fig. 16.6) are very heavy, cumbersome pieces of equipment demanding costly foundations and relatively large buildings. Apart from minor refinements, for several decades filter press design changed little until the advent of the diaphragm filter a few years ago. The original design of the plate and frame filter (Fig. 16.7) consisted of a series of frames into which sludge is passed under high pressure, up to 225 psi (1570 kPa), in order to dewater the sludge against the outer cloth-covered plate. The depth of the cake was consequently fixed, being governed by the distance between filter plates. Different sizes of plates were manufactured to give cakes of, for example, ¾, 1, or 1 ½ in (19, 25, or 38 mm) depth. Such filters have been used to dewater sludge in an acceptable manner for many years, with sludge cakes being sufficiently dry for sanitary landfill disposal. In some cases, such as Kenosha, Wis., where the raw sludge is a mixture of water treatment sludge and sewage sludge, the cake can be used as

Figure 16.6 Conventional filter press. (*Courtesy of Zimpro/Passavant, Rothschild, Wis.*)

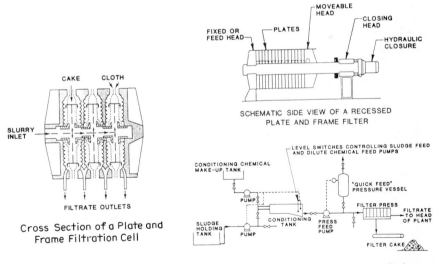

Figure 16.7 Plate and frame filter press schematic. (*Source: P. Innocenti, "Techniques for Handling Water Treatment Sludge", Opflow, vol. 14, no. 2, 1988, p. 1.*)

a cover for the landfill. Improvements in design have occurred over the years.

A considerable change in design resulted with the introduction in the last few years of diaphragm filters (Fig. 16.8). The advantage of this system is that the thickness of the cake is infinitely variable within the limits of the machine dimensions. Sludge is filtered through a cloth for a fixed period of time, perhaps 20 min, at which stage the sludge supply is cut off and water or compressed air is applied behind an expandable diaphragm that further squeezes water out of the sludge. The cake is dislodged by shaking or by rotating the cloth, depending on the manufacturer's design, and falls into a hopper for disposal. Hanging cakes, where the cake refuses to leave the cloth, an unhappy feature of the older plate and frame press, is consequently eliminated. Diaphragm presses also have the advantage that conditioning of the sludge, although desirable in some circumstances, is not always necessary. As a result, although output is somewhat reduced, a considerable amount of capital and operations costs are eliminated. A diaphragm press is currently installed at the Moores Bridges water treatment plant in Norfolk, Va., and is designed for a peak output of 138,000 lb (62,600 kg) of solids per week.[31]

**Figure 16.8**  Diaphragm filter press.  (*Courtesy of Envirex, Inc., Waukesha, Wis.*)

### Freezing and thawing

The most satisfactory end product from a gelatinous sludge for coagulant recovery or dewatering for landfilling is produced by the freezing and thawing process. This can be accomplished by mechanical means, if energy is relatively inexpensive, or naturally, where climatic conditions are acceptable. The change in the physical appearance of the sludge from a sticky thick liquid to a clear supernatant liquid with a small quantity of solids resembling coffee grains can only be described as spectacular. The cleanliness of the process and the ease of transporting the grains to landfill are among the major advantages.

The advantages of mechanically freezing wastewater sludge was first discovered by Clements, Stephenson, and Regan in London over 40 years ago.[32] The principle was later used by R. H. Burns for freezing radioactive waste at Harwell, England.[33] The process was adapted in the 1950s for water treatment alum sludge produced at the Stocks filtration plant, where the design and operation of a mechanical freezing plant for sludge dewatering was demonstrated.[24] The freezing principle has been tried successfully elsewhere, notably in Tokyo, Japan, Daer, Scotland, and at a pilot plant on the River Danube near Munich, West Germany.

The mechanism of the freezing process is postulated to be as follows. The collodial nature of the sludge is broken down because, when freezing commences, crystals of pure ice form. As they increase in size, thus dewatering the sludge, they ultimately coalesce and imprison particles of the more concentrated sludge. These particles are subjected to tremendous pressure from the ice and then converted into small granules that, during thawing, fall to the bottom of the tank. Aspects of the freezing mechanism are discussed further by Logsdon and Edgerley.[34] The principle of volume reduction and easy settling of solids after thawing is inescapable, but the cost of the energy required for mechanical freezing, unfortunately, is considerable.

The natural freezing system was first used at Copenhagen, where volume reduction occurred in several of the 16 lagoons when they froze during the winter and then thawed in the spring. The method can be used in areas where a long winter freeze is anticipated. This situation exists at the Trap Falls treatment plant of the Bridgeport Hydraulic Company (Connecticut).[35] At Trap Falls, the design is for sludge to be transported from the plant to three lagoons. The first lagoon is filled during the summer. Sludge is allowed to settle and freeze during the winter. Upon thawing the following spring, sludge volume has been reduced to such an extent that sludge removal by a front-end loader is possible (Fig. 16.9). The sludge is then deposited on adjacent grassland. This initially was an experimental process in the design of the plant, but it has proved so successful that three additional lagoons have been constructed.

### Centrifugation

Centrifuges are used to thicken some sludges, but their use is limited because the output from a centrifuge is in liquid form, albeit a concentrated liquid. If, for example, preparing sludge for disposal on a sanitary landfill where the concentration must not be less than 20 or 25 percent solids is necessary, a centrifuge is unlikely to be able to achieve this. Furthermore, energy requirements for centrifuging are very high. Centrifuges are primarily classification machines, rather than a dewatering method which concentrates or thickens sludge for subsequent treatment. Because of their limited use in water treatment, centrifuges will not be discussed in detail, but references are available.[28,30]

## Treatment of Specific Waste Streams

Previous sections have discussed sources, quantities, and characteristics of water treatment plant wastes and unit processes for concentration and treatment of residues. This section reviews treatment of specific wastes and waste streams.

### Residue from presedimentation

One impurity in untreated water is suspended solids caused by mud and disturbance of bottom sediments in a reservoir or the suspended solids carried by a river. Suspended solids are sometimes removed by presedimentation in order to reduce the load on a conventional coagulation-flocculation-sedimentation process. Source water suspended solids are far more easily removed before they become part of the treatment plant residue. Presedimentation is employed at a number of plants in the United States, including Omaha, Neb., and in London, England, as a means of reducing the silt content of River Thames water.

Source water solids can be present in the form of turbidity and suspended matter that can be measured quantitatively in a laboratory. An approximate relationship between suspended solids and turbidity often exists for each individual source water, and determining this relationship is sometimes helpful. Suspended solids are often the major constituent of the total residue.

(a)

(b)

Figure 16.9   Natural freezing. (a) Drying sludge (right) and frozen and thawed (left);
(b) thawed material removed and placed on land.

Presedimentation may be accomplished by using sedimentation basins devoted to this purpose or by using impounding reservoirs. Solids deposited naturally in impounding reservoirs must be dredged out from time to time.

### Residues from chemical coagulation

Suspended solids, both organic and inorganic, that do not settle in presedimentation basins (where provided) are removed by coagulant addition, flocculation, and sedimentation. Metal salt coagulants react in various ways (see Chap. 6), ultimately resulting in removal of particulate matter in the form of a hydroxide sludge. The material settles to the bottom of the sedimentation tank, where it collects and is removed either manually or mechanically, depending on the plant design. The chemical characteristics of residues from the coagulation-flocculation-sedimentation process are determined primarily by the coagulant used.

Sludge characteristics may be affected by choices made in the operation of the coagulation process. Thickening and dewatering characteristics of aluminum and ferric hydroxide sludges respond to changes in a variety of operation variables. Studies by Knocke et al.[36] found that increases in the rate and extent of dewatering occurred when the coagulation pH was reduced, the coagulant to influent turbidity ratio was reduced, or the coagulant mechanism was adjusted from enmeshment to adsorption-charge neutralization. Changes observed in sludge characteristics were a direct result of changes in the size and relative water content of the floc.[36]

**Coagulation with aluminum salts.**  Aluminum sulfate (alum) is by far the most widely used coagulant. The trivalent aluminum ions react with hydroxyl ions to form $Al(OH)_3$, as described by Eq. (16.1). The hydroxide sludge formed is one of the most difficult sludges to dewater. All hydroxide sludges are thixotropic, which causes problems in sludge force mains. The sludge requires comparatively extensive processing to concentrate and dewater.

**Coagulation with iron salts.**  Iron salts produce a sludge that is almost as difficult as alum to dewater. Ferrous and ferric sulfate and ferric chloride are common coagulants. In practice, ferrous ions are oxidized and form ferric hydroxide. The end result from any of these coagulants is a residue high in ferric hydroxide that is difficult to dewater.

**Coagulant aids.**  Sometimes the floc produced is very light, and attaching a chemical of high molecular weight is helpful. Clays such as bentonite have been used as weighting agents, and improvement also results when activated silica is used. Polymers of high molecular weight are currently the most popular coagulant aid in settling. The advantage of improved settling is, however, offset to some extent by the necessity of treating a small additional quantity of sludge because of the additional chemical added, which, in addition to increasing the overall weight of solids slightly, certainly influences the physical characteristics of the sludge itself. More details on the use of coagulant aids are included in Chap. 6.

**Residues from iron and manganese removal.** One of the traditional methods of treating groundwater containing iron and manganese is by oxidation using aeration or by chemical oxidation followed by sedimentation to remove the hydrated ferric and manganic oxides. The sludge is comparable to alum or iron sludges and is difficult to dewater (see Chap. 10).

**Residues from precipitative softening.** The hardness of virtually all water is caused by soluble calcium and magnesium found in the source water because of the dissolution of various earth minerals. The degree of softening required to render a water acceptable is a matter of judgment and affects the quantity and characteristics of softening sludge produced.

Chapter 10 indicates that the residues from softening consist of $CaCO_3$ and $Mg(OH)_2$. Calcium carbonate, when precipitated in a softening reactor, forms very fine particles that are difficult to settle but relatively easy to dewater. Magnesium hydroxide, however, is difficult to dewater, and a decision is often made in view of this to add only sufficient lime to precipitate calcium carbonate (see Chap. 10). The quantity of $Mg(OH)_2$ present in the sludge has a considerable effect on the physical characteristics of the residue. Most plants usually remove some magnesium along with $CaCO_3$, but in small quantities magnesium has a relatively small influence. The major residue produced by softening with lime by precipitation is a white precipitate of calcium carbonate. Calcium carbonate can, of course, be used to recover quicklime by *recalcination*.

As with other wastes, the usual practice is to dispose of this waste in lagoons. In some locations, multiple lagoons may be used, so when one fills to overflowing a second lagoon is placed in service. This disposal practice is likely to continue for many years to come.

At Boca Raton, Fla., precipitated calcium carbonate is concentrated in a gravity-thickening tank and then dewatered by a vacuum filter. The cake from the vacuum filter can be used for road foundation material, sold, or given away to local residents to improve garden soil texture.[37]

Another method of softening using lime is in the pellet reactor process. The end product is a small pellet 0.04 to 0.06 in (1 to 1.5 mm) in diameter that is very hard and almost pure white crystalline calcium carbonate, in some cases colored by iron. This is of value agriculturally in regulating the pH of acid soils, and at the Northwest Water Authority plant at Broughton, England, it is sold to local farmers.

### Residues from filter backwashing

Direct discharge of filter backwash wastes is usually prohibited under current pollution regulations, although some agencies are relaxing requirements for economic reasons and in recognition of the minimal environmental impact of these wastes.[38] Most agencies, however, require proper handling of backwash wastes for all new treatment plants and for plants being enlarged or upgraded.

If the quantity of water loss can be tolerated, then allowing dirty filter backwash water to settle in separate gravity thickeners or holding tanks and decanting of the supernatant liquid to a water course may be desirable. Settled solids from the backwash water should then be added to the solids from sedimentation for further processing and disposal. Processing of filter backwash water in this manner will eliminate the possibility of recycling contaminants to the source water.

Despite concerns expressed earlier, recovery of dirty backwash water is becoming increasingly common today for several reasons. Backwash water represents a rather large volume of water with low solids content. Typically, the volume is 1 to 5 percent of total plant production. Recovery, therefore, represents a savings in water resource and in chemicals that were expended to treat it initially.

The common approach to backwash water recovery is to discharge the wash water to a holding tank from which it is pumped back to the inlet of the treatment plant at an equalized flow rate. Another alternative when using reservoir water is to pump the water back into the reservoir if the solids are low and the reservoir is large. The solids are comparatively negligible and settle readily in a large reservoir and water is recovered.

For conventional plants, solids separation before return is not common, and some holding tanks are mixed to keep solids in suspension. Unmixed tanks may slope steeply to a bottom withdrawal in order to return solids without mixing. Such tanks are dewatered after each backwash. Some regulatory agencies may require settling of solids and return of the supernatant liquid. In this case, provisions for removing the settled solids from the holding tank, such as sludge scrapers, are needed.

In direct filtration plants, because solids are only captured in the filtration step, backwash recovery must include a solids separation step with supernatant liquor returned to the plant. This may be accomplished in several ways: (1) in a separate small treatment plant providing coagulation, flocculation, and sedimentation; (2) in lagoons; and (3) in the raw water reservoir.

The recycle of the backwash water and its content of solids directly back to the plant causes some concerns. For example, if *Giardia lamblia* cysts are present in the solids, the recycle will lead to greater cyst concentrations in the filter influent water and a higher risk of cyst passage through the filters. Similar concerns have been expressed for viruses. Concerns over microbial pathogens are accentuated in plants that have discontinued chlorination ahead of filtration to minimize trihalomethane production. Other problems may also be magnified by backwash recovery, including taste and odor problems and trihalomethane production.

In spite of these expressed concerns, the practice of backwash water recovery is quite common. No quantitative demonstrations to support the concerns have been presented. In specific plants, if a problem is demonstrated, wasting the supernatant liquor after settling the backwash water may become necessary. Lagoons are typically used for disposal of this waste stream. Supernatant liquid from lagoons will usually meet water pollution control standards, and the accumulated sludge is periodically piped to drying beds where it is allowed to dry and is eventually removed from the site to a landfill.

### Residues from precoat filtration

Diatomaceous earth is a mineral formed by minute marine organisms. It is an excellent filter medium for waters that have low suspended solids and is used particularly in small water treatment plants. The process is described in Chap. 8. The residue from backwashing is easily settled or dewatered and can be conveniently landfilled.

### Residues from ion exchange processes

Ion exchange processes (Chap. 9) generate brine solutions that are liquids with a high total dissolved solids concentration. Cation units exchange or remove from the inlet raw water positive ions such as calcium, magnesium, sodium, potassium, iron, and manganese by exchange with a resin-available ion such as sodium or hydrogen. When the resin has depleted all of its available ion, it must be regenerated by passing a concentrated solution of the required available ion through the resin.

Brines, generally sodium chloride, when passed through the exhausted resin bed, will attach a sodium ion to the resin as the available ion, releasing the cations exchanged, such as calcium and magnesium (hardness), into the concentrated chloride solution where they will then pass out to waste as a brine. The brine must then be diluted or discharged to an appropriate treatment facility.

Other cation resins are regenerated by the use of a concentrated hydrogen ion solution (acid) such as sulfuric acid or hydrochloric acid. Because the hydrogen ion is very active, all positive ions are replaced. Effluent of the unit then is generally weakly acidic and must be neutralized or blended before use. This effluent may be blended with raw water, or the effluent of a sodium exchanger, or adjusted with a basic chemical (i.e., $Na_2CO_3$, NaOH, etc.), or passed through an anion exchanger, as indicated later.[39]

Anion units are usually regenerated with a basic material, though in the dealkalizer application they are regenerated with sodium chloride. The anion units are classified either as weakly basic (regenerated with sodium carbonate), which will remove the strong anions (chloride, sulfate, and nitrate), or as strongly basic (regenerated with sodium hydroxide), which will remove most anions (chloride, sulfate, nitrate, bicarbonate, and silica).

Regenerants from the anion exchanger contain the anions removed plus sodium carbonate from the weakly basic anion exchanger, or sodium hydroxide from the strongly basic anion exchanger. The latter, because of its high pH, is more critical to neutralize prior to discharge. These wastes can be neutralized with an acid or by combination with the regeneration wastes of a hydrogen cation exchanger with the proper stoichiometric balance.[39]

Waste from the regeneration of ion exchange units is usually high in chlorides and other salts. In coastal areas, ocean disposal is practiced. In areas remote from the sea, a drain tile field similar to that used for septic tanks is adequate but may not be acceptable to regulatory agencies. Older plants usually discharge to a river or sanitary sewer. An interesting case of disposal at sea is at Sarasota, Fla., where an 8-mgd (30-ML/d) plant softens water from 36 wells. Seawater for regeneration is pumped to the system through an 8-in (200-mm) diameter pipe, and spent wash water is returned to the sea by means of a short dike and culvert. The quantity of wastewater in this case is rather high, approximately 20 percent, but the simplicity of operation more than makes up for the additional cost of this water.

### Residues from membrane processes

In membrane processes (Chap. 11), reverse osmosis, ultrafiltration, and electrodialysis, the inlet flow is split into two streams; one is product water with reduced ion concentrations, and the second is reject water containing the increased concentration of the inlet raw ions plus any treatment chemicals. These wastes are generally classified as a brine solution. Previously, discharge to the sea was acceptable to regulatory agencies, but now solar evaporation ponds or deep injection well disposal is sometimes necessary.

### Residues resulting from intermittent treatment measures

Residues are sometimes generated when short-term or intermittent treatment measures are implemented. The type and characteristics of the residues produced and the optimal disposal method will be determined by the type of treatment. Short-term or intermittent treatment is usually implemented to correct water-quality problems caused by seasonal factors or unusual circumstances.

Periodically surface water supplies exhibit strong tastes and odors that are often the result of seasonal algal growth and decay. One of the accepted practices for reduction of taste and odor thus caused is the application of activated carbon. Although water can be passed through a bed of granular activated carbon, a convenient periodic measure is to use powdered activated carbon. Dosed into the raw water or at alternate points, powdered carbon ultimately appears in the sludge, usually from the sedimentation basin, backwash water, or both. It is yet another constituent of the total residue of the plant. The amount can be considerable, and Table 16.8 indicates the extent in the waste residue from two water treatment plants in Norfolk, Va.

An inert filler material such as activated carbon in such quantities will have an important effect on the physical characteristics of the sludge. Also, changes in activated carbon dose will have a direct effect on the dewatering properties of the sludge itself.

### Ultimate Disposal of Residues

Three major disposal areas for residues exist:

1. The source from whence it came
2. The nearest wastewater pollution control plant
3. A sanitary landfill or other means of land disposal

The choice of disposal area will dictate the type of treatment required and the degree of concentration needed to prepare it for that method of disposal (Fig. 16.10). Sludge concentration prior to disposal in a river is unnecessary. For land disposal, however, such as at a sanitary land-

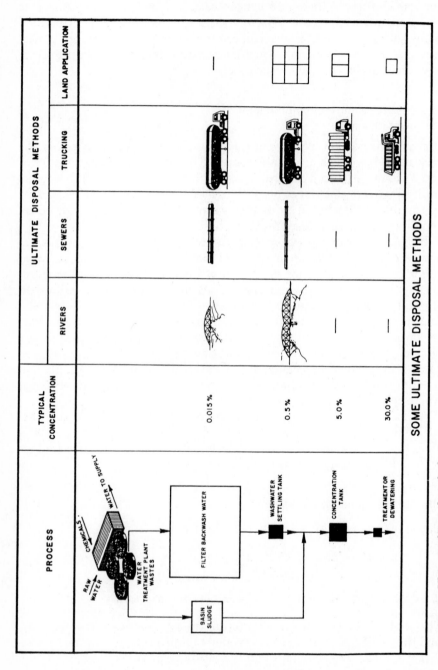

Figure 16.10  Ultimate disposal methods.

fill, having the residue as dry as possible is important to prevent problems at the landfill caused by residue instability.

The problem, therefore, is to increase the sludge concentration to perhaps 5 percent so that city sewers can handle it more easily, or 25 to 30 percent for a sanitary landfill. Such techniques for concentration were discussed earlier. The process of getting the residue in a form acceptable for final disposal is usually expensive, and carrying out site or laboratory tests on one or more alternatives before design is wise. To this end, having at least one year's records of the source water is customary, particularly noting suspended solids and turbidity, together with records of the chemical coagulant used in the treatment process. A mass balance can be calculated from such information. Naturally, the more historical data available, the better. An examination of the records for perhaps 10 years would highlight peak years as well as peak months. This is very important in order that ample capacity or storage be provided for periods of high residue production. This also gives the opportunity to carry out laboratory or pilot tests on the types of dewatering equipment under consideration and the chemicals required, if any, for treatment.

### Landfills

Landfills may be on public land, such as a city sanitary landfill, or on private land. Landfill operators commonly require a concentration of at least 15 percent solids by weight, sometimes as high as 30 percent, before acceptance of the residue. The minimum concentration required depends on the operator or local sanitary landfill regulations. Disposal is usually a matter of negotiation between the water treatment plant owner and the state, county, or sanitary landfill owner.

For alum sludges, which are the most common residue from water treatment plants in the United States, a good hard cake has a concentration of at least 25 percent solids. At the lower concentrations the cake is sloppy to handle and may revert to a liquid if wetted again. Sludge disposal as a top layer on a landfill prior to topsoiling is a useful disposal method.

### Disposal into sewers

Water treatment sludge is one of many wastes generated in a community. Central treatment of water plant sludges at the local wastewater pollution control plant is sometimes practiced. This method is used in Philadelphia, Pa., and at several other plants in the United States where the pollution control plant and the water works are under the same management. As long as the flow can be evened out by using equalizing tanks, there is usually no problem in treating the material at the wastewater pollution control plant.

Sometimes sufficient free alum is available to assist as a coagulant at the waste treatment plant. On the negative side, suspended solids in the water treatment sludge will accentuate the problem of solids disposal at the wastewater plant. For example, the presence of silts and clays may be abrasive to centrifuges and other equipment used for sludge concentration. If incineration is practiced, additional inert solids will increase the cost to heat the sludge, because only organics will burn off during incineration. The system is nevertheless most satisfactory for a utility that is responsible for both water and wastewater treatment, but in all cases careful investigation is needed, because success will vary with the type of sludge and type of treatment used at the wastewater treatment plant.

No physical connection must exist between the water treatment plant and sanitary sewers. Special manholes with an air break must be constructed at the point where water treatment sludge enters the sanitary sewer system. Cleasby[40] warns that

1. Discharge of lime-softening sludges to a sanitary sewer should be avoided because of the significant increased load on the sludge-handling system at the waste treatment plant. This is especially true if the sludge is treated by anaerobic digestion.

2. Discharge of sludges from surface water clarification plants to a sanitary sewer will also increase the sludge load at the waste treatment plant, but not generally as seriously as in the case of lime sludge disposal.

3. Because water plant sludges are largely inorganic and nonputrescible, they can be better disposed of by storage in properly designed sludge storage lagoons, particularly when land is readily available.

## Emerging Technology

The disposal of residues from water treatment plants has progressed considerably in the last 10 years to the stage where "bolt on" treatment and disposal methods for existing plants are now available. The traditional lime-softening residues, alum sludges, and the like will soon be joined by new, exotic types of wastes that may well require innovative means of treatment and disposal. New types of treatment will produce new types of wastes. The advent of new methods of analysis for water have detected hitherto unsuspected impurities in raw water, which may be toxic or carcinogenic, and that should be removed. Trihalomethanes are a good example. More and more compounds are being found in raw water that should not be present in safe potable water. Not only must these be removed, but the residues that new treatment processes will produce must be identified and properly handled.

## References

1. American Water Works Association, *Water Quality and Treatment,* McGraw-Hill, New York, 1971.
2. *Survey Report,* U.S. Public Health Service, 1963.
3. John B. Dean, "Disposal of Wastes from Filter Plants and Coagulation Basins," *J. AWWA,* vol. 45, no. 11, 1953, p. 1229.
4. Sludge Disposal Committee, "Committee Report: Research Needs for Alum Sludge Discharge," *J. AWWA,* vol. 79, no. 6, 1987, p. 99.
5. T. Gloriod, personal communication, March 1986.
6. M. H. Robert and R. J. Diaz, "Assessing the Effects of Alum Sludge Discharges into Coastal Streams," *Proc. AWWA Seminar on Recent Advances in Sludge Treatment and Disposal,* AWWA, Denver, Colo., 1985.
7. D. S. Lamb and G. C. Bailey, "Acute and Chronic Effects of Alum to Midge Larva," *Bull. Env. Contam. Toxicol.* vol. 27, 1981, p. 59.
8. C. T. Driscoll et al., "Effect of Aluminum Speciation on Fish in Dilute Acidified Water," *Nature,* vol. 284, 1980, p. 161.
9. Memorandum, Virginia State Water Control Board, June 1983.
10. Bureau of National Affairs, Inc. *Environment Report,* 71.5101, January 1985.
11. Public law 94-469, approved October 11, 1976; as amended by Public Law 98-620, approved November 8, 1984. See Bureau of National Affairs, Inc., *Chemical Regulation Reporter,* reference file 1, p. 91:0101.
12. Public law 94-580, approved October 21, 1976; as amended by Public Law 98-616, approved November 8, 1984. See Bureau of National Affairs, Inc., *Chemical Regulation Reporter,* reference file 2, p. 91:0301.
13. Public Law 84-159, approved July 14, 1955; amended by Public Law 98-213, approved December 8, 1983. See Bureau of National Affairs, Inc., *U.S. Environmental Laws,* Washington, D.C., 1986.
14. P. W. Doe, "The Treatment and Disposal of Washwater Sludge," *J. Inst. Water Eng. Scient.,* vol. 12, no. 6, 1958.
15. Robert C. Hoehn, John T. Novak, and William E. Cumbie, "Effects of Storage and Preoxidation on Sludge and Water Quality," *J. AWWA,* vol. 79, no. 6, 1987, p. 89.
16. James J. Costello, personal communication, 1985.
17. B. D. Rundell and S. J. Randtke, "In-situ Groundwater Treatment for Iron and Manganese: Fundamental, Practical and Economic Considerations," *Proc. AWWA Ann. Conf.,* Kansas City, Mo., June 1987.
18. P. C. G. Isaacs and I. Vahidi, "The Recovery of Alum Sludge," *Proc. Soc. Water Treat. Exam.* vol. 10, 1961, p. 91.
19. J. Moran Roberts and Charles P. Roddy, "The Recovery and Reuse of Alum Sludge at Tampa," *J. AWWA,* vol. 52, no. 7, 1960, p. 857.
20. David A. Cornwell, *Ninety Percent Alum Recovery Without Sludge Filtration,* American Water Works Association Research Foundation, Denver, Colo., 1980.
21. D. A. Cornwell, G. C. Cline, J. M. Przybyla, and D. Tippin, "Demonstration Testing of Alum Recovery by Liquid Ion Exchange," *J. AWWA,* vol. 73, no. 6, 1981, p. 326.
22. J. A. Webster, "Operational and Experimental Experience at Daer Water Treatment Works with Special Reference to the Use of Activated Silica and the Recovery of Alum Sludge," *J. Inst. Water Eng.,* vol. 20, no. 3, 1966, p. 167.

23. M. M. Bishop et. al., "Testing of Alum Recovery for Solids Reduction and Reuse," *J. AWWA*, vol. 79, no. 6, 1987, p. 76.
24. P. W. Doe, D. Benn, and L. R. Bays, "The Disposal of Washwater Sludge by Freezing," *J. Inst. Water Eng. Scient.*, vol. 19, no. 4, 1965.
25. Gary S. Logsdon and E. Edgerley, Jr., "Sludge Dewatering by Freezing," *J. AWWA*, vol. 63, no. 11, 1971, p. 734.
26. R. B. Palasits, personal communication, 1986.
27. A. R. White, "Alum Recovery—An Aid to the Disposal of Water Plant Solids," *ASCE Spring Convention*, 1984.
28. David A. Cornwell et. al., *Handbook of Practice: Water Treatment Plant Waste Management*, American Water Works Association Research Foundation, Denver, Colo., 1987.
29. Peter W. Doe and Wendell R. Inhoffer, "Design of Wash-Water and Alum-Sludge Disposal Facilities," *J. AWWA*, vol. 65, no. 6, 1973, p. 404.
30. Linvil G. Rich, *Unit Operations of Sanitary Engineering*, Wiley, New York, 1961.
31. Peter W. Doe, E. Blair, and P. E. Malmrose, "Water Treatment Plant Waste Disposal Study for City of Norfolk, Virginia," *Annual Meeting AWWA Virginia Section*, Norfolk, Va., October 1983.
32. G. S. Clements, R. J. Stephenson, and C. J. Regan, "Sludge De-Watering by Freezing with Added Chemicals," *J. Inst. Sewage Purification*, Part 4, 1950, p. 318.
33. R. H. Burns et. al., "Present Practices in the Treatment of Liquid Wastes at the Atomic Energy Research Establishment, Harwell," *Proc. Symp. Practices in the Treatment of Low- and Intermediate-Level Radioactive Wastes*, International Atomic Energy Agency, Vienna, 1966.
34. G. S. Logsdon and E. Edgerley, Jr., "Sludge Dewatering by Freezing," *J. AWWA*, vol. 63, no. 11, 1971, p. 734.
35. Peter W. Doe et. al., "Design, Construction, and Treatment of the Trap Falls Water Treatment Plant," *Proc. AWWA Ann. Conf.*, Miami, Fla., May 1982.
36. William R. Knocke, Jeff R. Hamon, and Betsy E. Dulin, "Effects of Coagulation on Sludge Thickening and Dewatering," *J. AWWA*, vol. 79, no. 6, 1987, p. 89.
37. P. W. Doe, *A Report on the Disposal of Sludge from Water Treatment Plants*, British Water Works Association, 1967.
38. A. H. Vicory and L. Weaver, "Controlling Discharges of Water Plant Wastes to the Ohio River," *J. AWWA*, vol. 76, no. 4, 1984, p. 122.
39. E. J. Connelley, personal communication, 1985.
40. J. Cleasby, "Should Water Plant Sludges Be Dumped into Sanitary Sewers?" *IWPCA Meeting*, 1969.
41. T. A. Wolfe, P. W. Doe, and P. E. Malmrose, "Theory and Design of Mechanical Dewatering of Water Plant Sludges," *AWWA/WPCF Joint Residuals Management Conference*, San Diego, Calif., 1989.

# Internal Corrosion and Deposition Control

## Michael R. Schock

*Research Chemist*
*U.S. Environmental Protection Agency*
*Drinking Water Research Division*
*Cincinnati, Ohio*

Corrosion is one of the most important problems in the water utility industry. It can affect public health, public acceptance of a water supply, and the cost of providing safe water. Many times the problem is not given the attention it needs until expensive changes or repairs are required.

Two potentially toxic metals (lead and cadmium) that may occur in tap water arise almost entirely because of leaching caused by corrosion. Three other metals, usually present because of corrosion, cause staining of fixtures, or metallic taste, or both. These are copper (blue stains and metallic taste), iron (red-brown stains and metallic taste), and zinc (metallic taste).

The corrosion products in the distribution system can also shield bacteria, yeasts, and other microorganisms (see Chap. 18). In a corroded environment, these organisms can reproduce and cause many problems, such as bad tastes, odors, and slimes. Such organisms can also cause additional corrosion.

Corrosion-caused problems that add to the cost of water include

1. Increased pumping costs, caused by corrosion products

2. Loss of water and water pressure, caused by leaks

3. Water damage to the dwelling that would require that pipes and fittings be replaced

4. Replacing hot water heaters

5. Responding to customer complaints of "colored water," "stains," or "bad taste," which is expensive both in terms of money and public relations

*Corrosion* is the deterioration of a substance or its properties because of a reaction with its environment. In the waterworks industry, the "substance" that deteriorates may be a metal pipe or fixture, the cement in a pipe lining, or an asbestos-cement (A-C) pipe. For internal corrosion, the "environment" of concern is water.

All waters are corrosive to some degree. A water's corrosive tendency will depend on its physical and chemical characteristics. Also, the nature of the material with which the water comes in contact is important. For example, water corrosive to galvanized iron pipe may be relatively noncorrosive to copper pipe in the same system. Corrosion inhibitors added to the water may protect a particular material, but may have no effect on, or even be detrimental to, other materials.

Physical and chemical actions between pipe material and water may cause corrosion. An example of a physical action is the erosion or wearing away of a pipe elbow because of excess flow velocity in the pipe. An example of a chemical action is the oxidation or rusting of an iron pipe. Biological growths in a distribution system (Chap. 18) can also cause corrosion by providing a suitable environment in which physical and chemical actions can occur. The actual mechanisms of corrosion in a water distribution system are usually a complex and interrelated combination of these physical, chemical, and biological systems. They depend greatly on the materials themselves and the chemical properties of the water. The purpose of this chapter is to provide an introduction to the concepts involved in corrosion and deposition phenomena in potable waters. Each material has a body of literature devoted to it. For the detailed consideration of the forms of corrosion of each metal or piping materials and specific corrosion inhibition practices to be employed, comprehensive text[1] and water treatment journal articles must be consulted.

Table 17.1, modified slightly from the original source,[2] briefly relates various types of materials to corrosion resistance and the potential contaminants added to the water. In general plumbing, the more inert, nonmetallic pipe materials, such as concrete, asbestos cement, and plastics, are more corrosion-resistant.

## Corrosion, Passivation, and Immunity

### Electrochemical reactions

Almost all mineral salts dissolve in water to some extent, from insignificant traces to gross concentrations exceeding that of salt in seawa-

TABLE 17.1    Corrosion Properties of Materials Frequently Used in Water Distribution Systems

| Plumbing material | Corrosion resistance | Associated potential primary contaminants |
| --- | --- | --- |
| Copper | Good overall corrosion resistance; subject to corrosive attack from high velocities, soft water, chlorine, dissolved oxygen, low pH, and high inorganic carbon levels (alkalinities). | Copper and possibly iron, zinc, tin, arsenic, cadmium, and lead from associated pipes and solder. |
| Lead | Corrodes in soft water with pH < 8, and in hard waters with high inorganic carbon levels (alkalinities). | Lead. |
| Mild steel | Subject to uniform corrosion; affected primarily by high dissolved oxygen levels. | Iron, resulting in turbidity and red water complaints. |
| Cast or ductile iron (unlined) | Can be subject to surface erosion by aggressive waters. | Iron, resulting in turbidity and red water complaints. |
| Galvanized iron | Subject to galvanic corrosion of zinc by aggressive waters; corrosion is accelerated by contact with copper materials; corrosion is accelerated at higher temperatures as in hot-water systems; corrosion is affected by the workmanship of the pipe and galvanized coating. | Zinc and iron; cadmium and lead (impurities in galvanizing process). |
| Asbestos-cement | Good corrosion resistance; immune to electrolysis; aggressive waters can leach calcium from cement; polyphosphate sequestering agents can deplete the calcium and substantially soften the pipe. | Asbestos fibers; increase in pH, calcium. |
| Plastic | Resistant to corrosion. | |
| Brass | Good overall resistance; different types of brass respond differently to water chemistry; subject to dezincification by waters of pH > 8.3 with high ratio of chloride to carbonate hardness. Conditions causing mechanical failure may not directly correspond to those promoting contaminant leaching. | Lead, copper, zinc. |

SOURCE: From AWWA, Ref. 2.

ter. When they become dissolved, these salts separate into two types of ions, namely anions and cations, which have opposite electric charges and are kept apart by water itself. These ions are responsible for the ability of "water" to conduct an electric current. Pure water has relatively few ions. Only 0.0001 percent of water separates into hydrogen cations and hydroxide anions. Therefore, pure water, free from mineral salts, has very little capacity for carrying electric current, making its electric conductivity extremely low.

When minerals are dissolved in water, however, the resultant ions provide the necessary conductivity to permit corrosion currents to flow, with negatively charged anions moving to the anode and positively charged cations moving to the cathode. Their accumulation at the respective electrodes is limited by other reactions that take place at these points.

For corrosion of any type to occur, the presence of all the components of an electrochemical cell is required. These are an anode, a cathode, a connection between the anode and cathode for electron transport, and an electrolyte solution that will conduct ions between the anode and cathode. The anode and cathode are sites on the metal that have a difference in potential between them, because metals are not completely homogeneous. If any component is absent, a corrosion cell does not exist and corrosion will not occur. Oxidation and dissolution of the metal takes place at the anode. The electrons generated by the anodic reaction migrate to the cathode, where they are discharged to a suitable electron acceptor, such as oxygen. The positive ions generated at the anode will tend to migrate to the cathode, and the negative ions generated at the cathode will tend to migrate to the anode, in response to the concentration gradients and to maintain an electrically neutral solution.

At the phase boundary of a metal in an electrolyte, a potential difference exists between the solution and the metal surface. This potential is the result of the tendency of a metal to go to the equilibrium state with the electrolyte. This oxidation reaction, representing a loss of electrons by the metal, can be written as

$$Me \rightleftharpoons Me^{z+} + ze^- \tag{17.1}$$

This equation indicates that the metal corrodes, or dissolves, as the reaction goes to the right. This reaction will proceed until the metal is in equilibrium with the electrolyte containing ions of this metal. The current that results from the oxidation of the metal is called the *anodic current*. In the reverse reaction, the metal ions are reduced by combination with electrons. The current resulting from the reduction (the reaction going to the left) is called the *cathodic current*. At equilibrium, the forward reaction proceeds at the same rate as the reverse reaction, and the anodic current is equal to the cathodic current. Thus, no net corrosion occurs at equilibrium.

The velocity of an electrochemical reaction, unlike that of a normal chemical reaction, is strongly influenced by the potential itself. Figure

17.1 shows schematically the dependency of the electric current, which is equal to the velocity of the reaction, on the electrode potential. The solid curve represents the sum of the current produced by the oxidation and reduction half-reactions. The net current is zero when the rate of reduction is equal to the rate of oxidation, and the system is at equilibrium. Figure 17.2 gives a qualitative picture of the charge distribution in the phase boundary of such a metal electrode.

Corrosion results from the flow of electric current between electrodes (anodic and cathodic areas) on the metal surface. These areas may be microscopic and in very close proximity, causing general uniform corrosion and, often, "red water." Alternatively, they may be large and somewhat remote from one another, causing pitting, with or without tuberculation. Electrode areas may be induced by various conditions, some because of the characteristics of the metal and some because of the character of the water at the boundary surface. Especially significant are variations in the composition of the metal or the water from point to point on the contact surface. Impurities in the metal, sediment accumulations, adherent bacterial slimes, and accumulations of the products of corrosion are all related either directly or indirectly to the development of electrode areas for corrosion circuits.

In almost all forms of pipe corrosion, the metal goes into solution at the anode areas. Because a movement of electrons occurs on solution of the

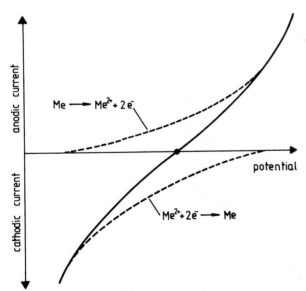

**Figure 17.1**   Potential versus current for an electrochemical reaction. (*Source: Internal Corrosion of Water Distribution Systems, American Water Works Association Research Foundation, Denver, Colo., 1985.*)

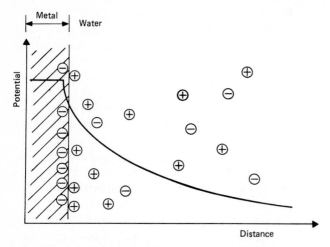

**Figure 17.2** Potential distribution at an electrode surface. (*Source: Internal Corrosion of Water Distribution Systems, American Water Works Association Research Foundation, Denver, Colo. 1985.*)

metal, the metal develops an electric potential. Electrons liberated at these areas flow through the metal to the cathode areas where they become involved in another chemical reaction, and the metal develops another electric potential. Control of corrosion by water treatment methods aims at retarding either or both of the primary electrode reactions. Figure 17.3 shows an example of corrosion reactions taking place on a pipe surface with proximate anodic and cathodic areas.[3]

If a metal is protected by *immunity,* it is thermodynamically stable and is therefore incorrodible.[4] Frequently, this region of electrochemical behavior is only possible when water itself is not chemically stable, so it is only encountered in potable water systems when the consumption of externally supplied energy (*cathodic protection*) occurs.

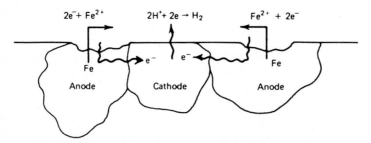

**Figure 17.3** Corrosion of iron in acid solution. (*Source: Water Chemistry, V. L. Snoeyink and D. Jenkins. Copyright © 1980, John Wiley & Sons, Inc.*)

*Passivation* occurs when the metal is not stable, but it becomes protected by a stable film. The protection can be perfect or (more usually) imperfect, depending on whether the film effectively shields the metal from contact with the solution.[4] True passivation films must satisfy several requirements to effectively limit corrosion. They must be electrically conductive, mechanically stable (neither flaking or cracking), and continuous.

The formation of scales, such as CaCO$_3$(s) and iron carbonates on corroding iron or steel, are normally thicker and have higher porosity than the passivating films and, thus, are not directly comparable. Similar reactions occur on other metals. They do not decrease the corrosion rate as much as oxide films do, and the corrosion current-potential relationship for passivating films does not occur for such scales. Further, scale formation on steel reduces the corrosion rate by decreasing the rate of oxygen transport to the metal surface, thus decreasing the rate of the cathodic reaction, whereas passivating films on steel cause an anodic-controlled corrosion reaction.

## The Nernst equation

Corrosion of metals in water will take place if the corrosion products are at a lower free-energy level than the metal itself. The difference in free-energy levels is the driving force of the reaction. A metal may go into solution as an ion, or it may react in water with another element or molecule to form a complex, an ion pair, or an insoluble compound. The free-energy difference under such conditions depends on the electrochemical potential, which, in turn, is a function of the type of metal and the solid- and aqueous-phase reaction products. Electrons (electricity) will then flow from certain areas of a metal surface to other areas through the metal.

The equilibrium potential of a single electrode can be calculated by using the Nernst equation for the reaction

$$E_{me/me^{z+}} = E^{\circ}_{me/me^{z+}} - \frac{RT}{zF} \ln \{Me^{z+}\} \qquad (17.2)$$

where $E_{me/me^{z+}}$ = potential (volts)
$E^{\circ}_{me/me^{z+}}$ = standard potential (volts), a constant that can be obtained from tables of standard reduction potentials
$\{\ \}$ = activity of the ion $Me^{z+}$
$R$ = ideal gas constant (about 0.001987 kcal/deg · mol)
$T$ = absolute temperature (K)
$F$ = Faraday constant (23.060 kcal/V)
$z$ = number of electrons transferred in the reaction
$me/me^{z+}$ = subscript indicates reaction written as a reduction

The Nernst equation written as Eq. (17.2) is for a single electrode, as-suming that the electrode is coupled with the normal hydrogen elec-trode at which the reaction

$$2H^+ + 2e^- \rightarrow H_2(g) \qquad E° = 0.00 \qquad (17.3)$$

takes place, and reactants and products are assumed by thermody-namic convention to equal $1$.[5,6] Further, the convention $\Delta G = -zFE$ is used, where $\Delta G$ is the free-energy change for the complete reaction (kcal/mol).[6,7] In this convention, standard "half-cell" potentials are tabulated with the reactions written as reductions.

The general form of the Nernst equation, for the balanced net reac-tion of two half-cells, each of the form

$$ox + ze^- \rightleftharpoons red \qquad (17.4)$$

where ox and red indicate oxidized and reduced species, respectively, is[6,7]

$$E_{red/ox} = E°_{red/ox} - \frac{RT}{zF} \ln \frac{\{red\}}{\{ox\}} \qquad (17.5)$$

The subscript red/ox indicates overall cell potentials for the total bal-anced reaction.

At 25°C, and with the conversion to base 10 logarithms for conve-nience of calculation, Eq. (17.5) can be rewritten as

$$E_{red/ox} = E°_{red/ox} - \frac{0.0591}{z} \log Q \qquad (17.6)$$

where $Q$ is the reaction quotient ($\{red\}/\{ox\}$). At equilibrium, no elec-trochemical current is generated, and the oxidants and reductants are at their equilibrium activities. Thus, the reaction quotient $Q$ becomes equal to the equilibrium constant $K$ for the overall reaction.[7]

In drinking water systems, the oxidation half-cell reaction of a metal, such as iron, zinc, copper, or lead, is coupled with the reduction of some oxidizing agent, such as dissolved oxygen or chlorine species. Example half-cell reactions are

$$O_2 + 2H_2O + 4e^- \rightleftharpoons 4OH^- \qquad (17.7)$$

$$HOCl + H^+ + e^- \rightleftharpoons (1/2)Cl_2(aq) + H_2O \qquad (17.8)$$

$$(1/2)Cl_2(aq) + e^- \rightleftharpoons Cl^- \qquad (17.9)$$

Equations (17.8) and (17.9) can be combined to yield the net reac-tion

$$HOCl + H^+ + 2e^- \rightleftharpoons Cl^- + H_2O \qquad (17.10)$$

which represents a significant oxidizing half-cell reaction for metals in drinking water.

If Eq. (17.7) is written in the Nernst form [Eq. (17.5)] and the ionic strength of the water is low enough that it can be assumed to have unit activity ($\{H_2O\} = 1$), then

$$E_{O_2/OH^-} = E°_{O_2/OH^-} - \frac{0.0591}{4} \log \frac{\{OH^-\}^4}{\{O_2\}} \qquad (17.11)$$

The dependence of pH of the oxidation potential of the dissolved oxygen reaction is clearly shown, since the $[OH^-]$ raised to the fourth power is in the numerator. Similarly, the oxidation potential of the hypochlorous acid reaction is directly related to pH:

$$E_{HOCl/Cl^-} = E°_{HOCl/Cl^-} - \frac{0.0591}{2} \log \frac{\{Cl^-\}}{\{HOCl\}\{H^+\}} \qquad (17.12)$$

because the $[H^+]$ is in the denominator. By thermodynamic definition, the corrosion (and, hence, dissolution of metals from plumbing materials) can only occur if the overall cell potential exceeds the equilibrium cell potential.[4,6,7]

Reactions such as Eqs. (17.7) and (17.10) can be combined with metal oxidation half-cells [Eq. (17.1)] to show overall corrosion reactions likely to occur in drinking water. Examples are

$$2Fe(metal) + O_2 + 2H_2O \rightleftharpoons 2Fe^{2+} + 4OH^- \qquad (17.13)$$

$$2Pb(metal) + O_2 + 2H_2O \rightleftharpoons 2Pb^{2+} + 4OH^- \qquad (17.14)$$

$$Pb(metal) + HOCl + H^+ \rightleftharpoons Pb^{2+} + Cl^- + H_2O \qquad (17.15)$$

$$Fe(metal) + OCl^- + 2H^+ \rightleftharpoons Fe^{2+} + Cl^- + H_2O \qquad (17.16)$$

The overall Nernst expression for Eq. (17.16), at 25°C, is

$$E_{Fe^{2+}/OCl^-} = E°_{Fe^{2+}/OCl^-} - \frac{0.0591}{2} \log \frac{\{Fe^{2+}\}\{Cl^-\}}{\{OCl^-\}\{H^+\}^2} \qquad (17.17)$$

Note that in Eq. (17.17) the activities are for the free aqueous species, not the total concentrations. The overall potential of the reaction, therefore, will mostly depend upon the ionic strength, the temperature, the hydrolysis of the metal, the presence of complexing agents for the metal, and the limitation of the free-metal ion by direct and indirect solubility limitations, such as the formation of corrosion product solids. The reactions can also be stifled by a barrier to the diffusion of oxidants to the surfaces of the materials, where "fresh" metal is available to oxidize. An example of the difference the chemical environment can make on $E°$ is shown by the standard reduction potential for the half-cell[7]

$$Fe^{3+} + e^- \rightleftharpoons Fe^{2+} \qquad (17.18)$$

being 0.770 V in 1 $M$ HCl, but 0.43 V in 1 $M$ $H_3PO_4$.

Electrochemical cells can also develop through reactions where both the oxidant and reductant are solids immersed in an electrolyte solution. Such reactions form the basis behind galvanic corrosion, where different metals in piping are connected by a conductive joint. The driving force for current flow is actually the differences in responses of each metal of the couple to its aqueous chemical environment (the potable water). This topic will be considered in more detail later, along with other types of corrosion.

## Corrosion kinetics

A three-step process is involved in governing the rate of corrosion of pipe: (1) transport of dissolved reactants to the metal surface; (2) electron transfer at the surface; and (3) transport of dissolved products from the reaction site.[8] When either or both of the transport steps are the slowest, *rate-limiting step,* the corrosion reaction is said to be under *transport control.* When the transfer of electrons at the metal surface is rate-limiting, the reaction is said to be under *activation control.* The formation of solid natural protective scales that inhibit transport are often an important factor in transport control. This section presents an overview of the concepts involved in the rates of corrosion reactions in potable water systems. Numerous reference articles and texts exist that present a detailed development of the theories that are the basis for many direct electrochemical rate-measuring techniques.

For corrosion to occur, the system must be in disequilibrium. A thermodynamic force drives a spontaneous change away from the original metallic form toward more energetically favored species. Typically, a net flow of electrons from anodic sites to cathodic sites occurs as metal spontaneously oxidizes and goes into solution. The system is out of thermodynamic equilibrium. The current acts to bring the metal and solution closer to an equilibrium state.

Corrosion is often described, as was done previously, in terms of numerous tiny galvanic cells on the surface of the corroding metal. Such localized anodes and cathodes as those described are not fixed on the surface, but are statistically distributed on the exposed metal over space and time. The electrochemical potential of the surface is determined by the mixed contributions to potential of both the cathodic and anodic reactions, averaged over time and over the surface area. Both the individual anodic and cathodic half-reactions are reversible and occur in both directions at the same time. When the "electrode" is at its equilibrium, the rates of reaction in both the cathodic and anodic half-cells are equal.

The exchange current density represents the equilibrium rate of the forward and backward directions of the half-reaction (in units of amperes per square centimeter, although it is not actually a chemical reaction rate). The exchange current density is equal to the equilibrium rate of oxidation (or reduction) in moles per square centimeter per second, times a conversion factor of $nF$ (number of electrons transferred in the reaction multiplied by the Faraday constant).[8]

The exchange current density, therefore, is a measure of the rate of oxidation and reduction of species for a half-reaction when those rates are at the appropriate equilibrium electrode potential for the half-reaction, and it depends on the particular half-reaction and on the nature of the metal surface.

If the actual electrode potential differs from the equilibrium electrode potential (given by the Nernst equation), then the rates of the forward and backward directions of the half-reaction will also differ. The system will then be out of equilibrium.

In most potable water distribution systems, the rate of corrosion is subject to limitation by a transport step for the motion of dissolved species to or from the surface sites. The phenomenon of rate limitation by transport is referred to in corrosion literature as *concentration polarization*. This suggests that the polarization, or difference in potential, in a corroding metal system is related to concentration differences between the surface and the bulk solution.[8] Where surface scales are present, the rate of transport of reacting ions through the scale may be the rate-limiting step in the corrosion process.

Given the thermodynamic basis for corrosion described above, and the body of knowledge about kinetic factors that affect the rate of corrosion of metals, several properties of the water passing through a pipe or device that influence the rate of corrosion can be identified. Some of the water-related properties are (1) concentration of dissolved oxygen, (2) pH, (3) temperature, (4) water velocity, (5) concentration and type of chlorine residual, (6) chloride and sulfate ion concentration, and (7) concentration of dissolved inorganic carbon and calcium.

These properties interrelate, and their effect depends on the plumbing material as well as the overall water quality. References specific to the type of plumbing situation (pipe, soldered joint, galvanic connection, faucet, or flow-control device) and the material of interest should be consulted for the most appropriate information on corrosion rate control. Some generalizations will be considered later.

## Solubility diagrams

In order to display solubility relationships in a relatively simple two-dimensional manner, the total solubility of some constituent can be plotted as a function of some master variable.[6,7,9] Other solution parameters can affect solubility, such as the ionic strength, or the concentration of dissolved species (ligands) that can form coordination compounds or complexes with the metal. The solubility diagram will not necessarily give an answer to the minimization of the rate of corrosion, because that is a kinetic parameter dependent on the relative rates of the oxidation, dissolution, diffusion, and precipitation reactions. It also cannot predict the ability of a pipe coating produced to adhere to the pipe surface or the permeability of a coating to oxidants from the water solution or pitting agents. It does give important information for estimating the ability of a water quality to provide attainment of maximum contaminant levels (MCLs) for drinking water, the potential for precipitating passivation films on pipe surfaces, or the deposition of other solids important in water treatment, such as calcium carbonate, octacalcium phosphate, aluminum or ferric hydroxide, and so forth.

In order to construct this type of diagram, an aqueous mass balance equation must be written for the metal (or other constituent of interest), with the total solubility $(S_t)$ as the unknown. The mass balance expression should include the concentration of the uncomplexed species (free-metal ion), along with all complexes to be included in the model. An example of such a mass balance expression is given in Table 17.2. The solubility constant expression for each solid of interest is rearranged and solved for the free-species concentration and substituted into the mass balance equation. A diagram is then constructed for each solid, and the curves are superimposed. The points of minimum solubility are then connected, giving the final diagram. This procedure is discussed in more detail by Schock[10] for lead(II) and by Snoeyink and Jenkins[7] for iron(II).

Because these diagrams are inherently two-dimensional (solubility on the $y$-axis, pH on the $x$-axis), and additional variables that have an important impact on solubility usually exist (such as temperature, ionic strength, carbonate concentration or alkalinity, orthophosphate concentration, etc.), these diagrams must display solubility and species concentration lines with these other variables fixed. For instance, Fig. 17.4 displays a solubility diagram for lead in the $Pb-H_2O-CO_2$ system, showing the important aqueous species and the stability domains of two lead solids. In this figure, the dissolved inorganic carbon (DIC) concentration was fixed at 0.00025 mol/L (3 mg C/L), with a temperature of 25°C and an assumed ionic strength of 0.005. Figure

**TABLE 17.2    Example Mass-Balance and Solubility Constant Expressions for the Construction of a Solubility Diagram for Lead(II) at 25°C[†]**

| Simple $H_2O$-Pb-$CO_2$ System |
| --- |
| Total soluble Pb(II), $S_t = S_{t,CO_2}$ |

$S_{t,CO_2} = [Pb^{2+}] + [PbOH^+] + [Pb(OH)_2^0] + [Pb(OH)_3^-]$
$\quad + [Pb(OH)_4^{2-}] + 2[Pb_2(OH)^{3+}] + 3[Pb_3(OH)_4^{2+}]$
$\quad + 4[Pb_4(OH)_4^{4+}] + 6[Pb_6(OH)_8^{4+}] + [PbHCO_3^+]$
$\quad + [PbCO_3^0] + [Pb(CO_3)_2^{2-}]$

| $H_2O$-Pb-$CO_2$-$PO_4$ System |
| --- |

$S_t = S_{t,CO_2} + S_{t,PO_4}$
$S_{t,PO_4} = [PbHPO_4^0] + [PbH_2PO_4^+]$

| $H_2O$-Pb-$CO_2$-$PO_4$-$SO_4$ System |
| --- |

$S_t = S_{t,CO_2} + S_{t,PO_4} + S_{t,SO_4}$
$S_{t,SO_4} = [PbSO_4^0] + [Pb(SO_4)_2^{2-}]$

[†]Subscripts indicate the system to which the total solubility contribution applies.

17.5 shows the same basic system, but with a total DIC concentration of 0.0025 mol/L (30 mg C/L).

In both diagrams, the implicit assumption was also made that the solution redox potential was not high enough to cause the formation of Pb(IV) aqueous or solid species. Additionally, when metal solubility becomes high, care must be taken to ensure that computational constraints on the system, such as the fixed ionic strength and the total ligand concentration (mass balance), are not violated by the presence of high concentrations of dissolved metal species.

The solubility diagrams in Figs. 17.4 and 17.5 illustrate three noteworthy features of lead chemistry in drinking water. First, as the DIC level is increased, the domain of lead carbonate stability is extended to a higher pH limit (pH $\approx$ 8.5 instead of pH $\approx$ 7), and the solubility of lead is decreased within the lead carbonate stability domain. Second, the aqueous carbonate complexes compete successfully for lead primarily at the expense of Pb(OH)$_2^0$ and PbOH$^+$. The third significant effect shown is the enhancement of lead solubility in the region where basic lead carbonate is the controlling solid, brought about by the carbonate complexation. Note that if the same complex formation constants were used, but a less-soluble constant was used for PbCO$_3(s)$ solubility, the simple carbonate solid would be predicted to be stable over a wider pH range. Similar information can be obtained through the careful construction and study of solubility diagrams for other metals, such as copper and zinc.

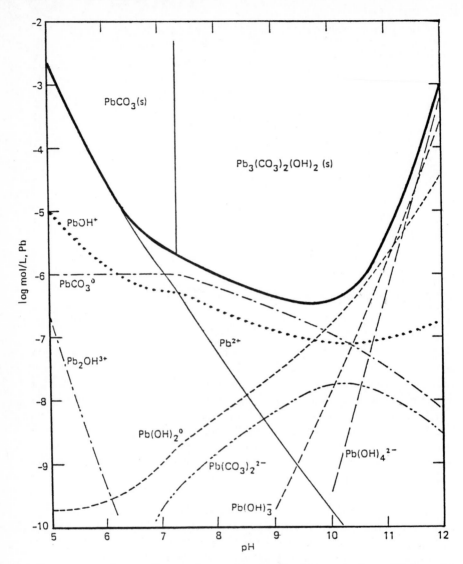

**Figure 17.4** A solubility diagram showing dissolved lead(II) species in equilibrium with lead(II) solids in a pure system containing 3 mg C/L DIC ($2.5 \times 10^{-4}$ mol/L) at 25°C and $I = 0.005$ mol/L using data from Schock and Wagner.

When solubility diagrams are constructed for any metal, the selection of solid and aqueous species must truly represent the system to be modeled, or very erroneous conclusions can result. For example, the aragonite, rather than the calcite, form of calcium carbonate is frequently found in deposits formed in systems having galvanized pipe. Also, the ferric iron deposits formed in mains are frequently a rela-

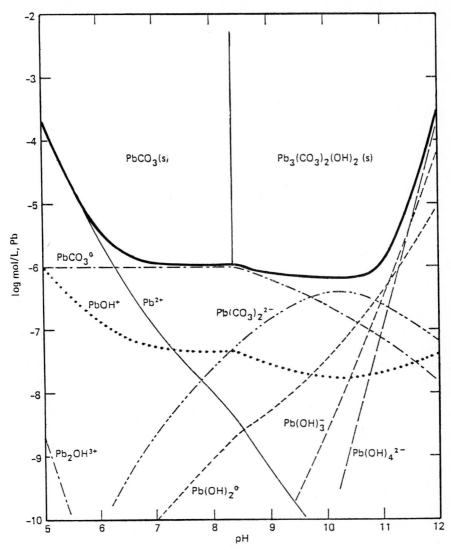

**Figure 17.5**  A solubility diagram showing dissolved lead(II) species in equilibrium with lead(II) solids in a pure system containing 30 mg C/L DIC ($2.5 \times 10^{-3}$ mol/L) at 25°C and $I = 0.005$ mol/L using data from Schock and Wagner.

tively soluble hydroxide or oxyhydroxide form [$Fe(OH)_3$ or $FeOOH$] rather than an ordered form such as hematite ($Fe_2O_3$). To go along with this concept, the equilibrium constants must also be accurate to give realistic concentration estimates, and knowledge of changes in the equilibrium constants with temperature is essential, especially when projections of depositional tendency have to be made into hot or very cold water piping systems. Critical evaluation of data appearing in handbooks and published papers is necessary to avoid using incorrect values, and occasionally review articles or major works by rigorous researchers can be consulted for reliable values.

An important assumption behind the diagrams is that the system must reach thermodynamic equilibrium for the calculations to be truly valid, unless kinetic factors are incorporated into the model. Sometimes improvements can be made in predictions by using metastable species in the calculations, although it is not thermodynamically rigorous to do so.

Stumm and Morgan have discussed the formation of precipitates and the relationship to solubility diagrams in conceptual terms, as illustrated by Fig. 17.6.[6] They define an *active* form of a compound as one that is a very fine crystalline precipitate with a disordered lattice. It is generally the type of precipitate formed incipiently from strongly oversaturated solutions. Such an active precipitate may persist in metastable equilibrium with the solution and may convert ("age") slowly into a more stable *inactive* form. Measurements of the solubility of active forms give solubility products that are higher than those of the inactive forms.[6] The formation of some of the iron hydroxide or oxyhydroxide solids in pipe deposits mentioned previously provide an example of this phenomenon.

Hydroxides and sulfides often occur in amorphous and several crystalline modifications. Amorphous solids may be either active or inactive. Initially formed amorphous precipitates or active forms of unstable crystalline modifications may undergo two kinds of changes

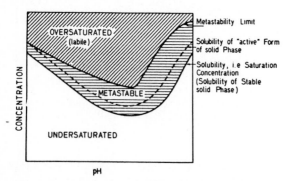

Figure 17.6   Solubility and saturation. A schematic solubility diagram showing concentration ranges versus pH for s upersaturated, metastable, saturated, and undersaturated solutions. *(Source: Aquatic Chemistry. Warner Stumm and J. J. Morgan. Copyright © 1981 Wiley Interscience.)*

during aging. Either the active form of the unstable modification becomes inactive, or a more stable modification is formed. With amorphous compounds deactivation may be accompanied by condensation or dehydration. When several of the processes take place together, nonhomogeneous solids can be formed upon aging. Similar phenomena can occur with basic carbonates, such as a transition from one form to another with changes in pH or DIC over time.

Rather than construct a different detailed diagram for each level of a secondary variable, such as DIC, researchers frequently add lines representing the different levels of this variable to a single diagram, and the aqueous species are omitted. For metals (such as zinc, copper, and lead) that have their solubility greatly influenced by complexation, the expansion of the diagram to include a "third dimension" is often useful. This is becoming much easier to accomplish, given the wide availability of computers of all sizes and sophisticated graphics software.

For a qualitative, conceptual understanding, a three-dimensional (3-D) surface can be constructed that can show multiple trends in a complex system at a glance. Figure 17.7 shows the two "troughs" representing $PbCO_3(s)$ and $Pb_3(CO_3)_2(OH)_2(s)$, the different trend of solubility with DIC for each solid, and the distinct solubility minimum in the system.[10,11]

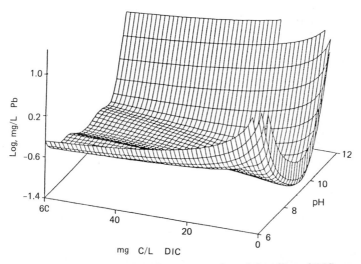

**Figure 17.7**  Three-dimensional representation of the effect of DIC concentration and pH on lead solubility. Ionic strength = 0.005, temperature = 25°C. (*Source: Internal Corrosion of Water Distribution Systems, American Water Works Association Research Foundation, Denver, Colo. 1985*).

To obtain a better quantitative estimate of trends of solubility resulting from interrelationships between two major variables, diagrams such as Fig. 17.8 can be constructed. Operating on the same principles as topographic maps, such "contour diagrams" present a "map review" of surfaces such as Fig. 17.7. The diagrams are derived by interpolating levels of constant concentration within a three-dimensional array of computed solubilities at different combinations of the other two master variables (e.g., pH and DIC). Several different mathematical algorithms are widely used in commercially available computer software. Rapid changes in solubility with respect to a master variable (here, pH or DIC) are shown by closely spaced contour lines. A series of these diagrams at levels of a third master variable (such as orthophosphate concentration, temperature, etc.) can be useful to help display multiple interactions with a minimum of diagrams. They also enable a direct reading of estimated solubilities, without having to guess from an indirect perspective, such as with Fig. 17.7. One problem with this type of diagram, as well as the 3-D surface plots, is that most metals can undergo a change in solubility of three orders of magnitude or even more, over the range of conditions that might be reasonable for potable waters. Thus, logarithmic scales are

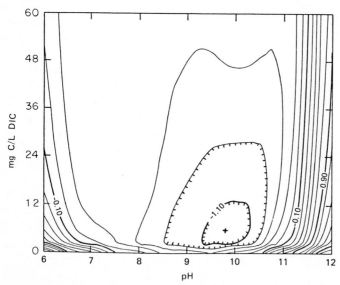

**Figure 17.8** Contour diagram of system described in Fig. 17.7. Contour lines are log (mg/L Pb), contour interval is 0.2. The lowest contour line (−1.10) represents 0.079 mg Pb/L. The system low point is 0.069 mg Pb/L at pH = 9.8 and DIC = 4.8 mg C/L. *(Source: Adapted from Proc. AWWA Water Qual. Tech. Conf., Nashville, Tenn., 1982.)*

often necessary for the metal concentrations, which can be somewhat confusing to read.

## Pourbaix diagrams

Using the Nernst equations for appropriate electrochemical half-reactions, it is possible to construct potential-pH diagrams, which are also called $Eh$-pH or Pourbaix diagrams. These diagrams have been popularized by Pourbaix and his coworkers in the corrosion field,[4,12,13] and by Garrels and Christ in geochemistry.[5] A similar type of diagram uses the concept of electron activity, pE, which is analogous to the concept of pH.[6,7] These diagrams are applicable in drinking water when the metal of interest can exist in different valence states under normal ranges of pH and oxidizing agent concentrations.

The Pourbaix diagrams include the occurrence of different insoluble corrosion products of the dissolved metal that limit the concentration of the free-metal ion. These diagrams mainly give information about thermodynamically stable products under different conditions of electrochemical potential. The position of the boundaries of each region is also a function of the aqueous concentrations (activities) of ions that participate in the half-cell reactions.

Potential-pH diagrams are particularly useful for studying speciation in systems that could contain species of several possible valence states within the range of redox potential normally encompassed by drinking water. Obtaining an accurate estimate of the redox potential of the drinking water is usually an important factor limiting the usefulness of potential-pH diagrams. The diagrams are also useful for gauging the possible reliability of electrochemical corrosion-rate measurement techniques. Measurement methods that rely on the imposition of a potential to the pipe surface may shift the pipe surface into the stability domain of a solid that would not normally form when freely corroding. The imposed potential might also serve to alter the nature of the surface phase, leading to an erroneous identification of the dominating corrosion or passivation reactions as the result of a surface compound analysis.

A potential-pH diagram is related to the solubility-versus-pH plots discussed earlier in the manner shown schematically in Fig. 17.9. If a conventional two-dimensional solubility diagram is considered as a vertical plane, a potential-pH diagram may be thought of as a "slice" that is perpendicular to the solubility plane, which cuts through the plane at a single concentration of the metal. In actuality, potential-pH diagrams are usually computed in terms of activities rather than concentrations, but the difference is usually not important for practical purposes. The activities of all aqueous and solid species must be fixed,

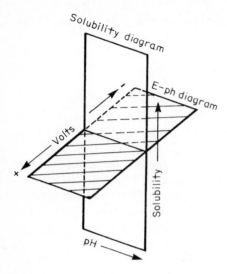

**Figure 17.9** Schematic relationship between potential-pH and conventional solubility versus pH diagrams.

while the electrical potential of the solution and the pH become the master variables. If a given water constituent (such as calcium) does not oxidize or reduce under meaningful physical conditions, no additional useful information is gained beyond that directly available with a solubility diagram by constructing a potential-pH diagram.

Figure 17.10 is an example of a potential-pH diagram for iron in water.[1] This diagram assumes $Fe_2O_3$ and $Fe_3O_4$ to be the solid phases that can control iron solubility, and $Fe^{2+}$ and $Fe^{3+}$ as the only ionic species. The diagram in Fig. 17.10 shows that iron and water are never thermodynamically stable simultaneously because the iron metal field (Fe) falls below the line where water is reduced to $H_2$ gas. At such a low electrode potential the iron will not corrode (i.e., it is immune). The iron potential is reduced to the immune region, for example, by cathodic protection. To accomplish this, the iron must be coupled with another, more easily corrodible material, such as magnesium. At low pH (<5) and intermediate to high potential (approximately −0.5 to 1.3 V) the diagram shows that the stable iron species is $Fe^{2+}$ or $Fe^{3+}$. Corrosion will occur at a high rate under these conditions (i.e., it is active). In the high-potential and high-pH regions, solid products such as $Fe_2O_3(s)$ or $Fe_3O_4(s)$ may form and deposit on the surface of the iron. These solids decrease the corrosion rate by influencing anodic or cathodic reactions. The rate of corrosion of a metal covered with a layer depends on the structure and chemical nature of the layer. The diagram also shows the pH-potential regions for $H_2$—$H_2O$—$O_2$ stability. At very low potential, water is reduced to $H_2$, and at high potential, water is oxidized to $O_2$. The stability of wa-

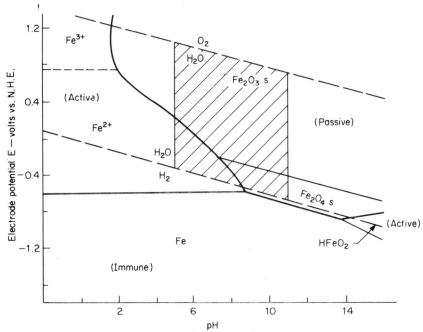

**Figure 17.10**  Potential-pH diagram for the iron-water system at 25°C (considering Fe, $Fe_3O_4$, and $Fe_2O_3$ as solid substances and $[Fe^{2+}]$ or $[Fe^{3+}] = 10^{-6}$ *M*). The cross-hatched region is of the most interest for drinking water.) *(Source: Internal Corrosion of Water Distribution Systems, American Water Works Association Research Foundation, Denver, Colo., 1985.)*

ter thus limits the range over which the potential of a metal can be varied if it is in contact with water.

The same restrictions that apply to solubility diagrams in general, with regard to the necessity of using realistic aqueous and solid species, apply to these diagrams. Also, the selected activities (or concentrations) of the dissolved species should be close to the real situations under study. Because the total activity is fixed across the entire pH range, the phase relationships predicted by the potential-pH diagram may deviate somewhat from those predicted in the previously described solubility diagrams.

Figure 17.11 is of interest to illustrate the relative oxidizing ability (and instability) of chlorine species introduced for water disinfection.[7] The presence of species such as $Cl_2$, $HOCl^0$, or $OCl^-$ will drive the oxidizing state of the water upward toward its stability limit (line A), which will affect the corrosion and speciation of many of the metals and metal surfaces in contact with the water. These are more powerful oxidizing agents than oxygen itself (line A).

The development of any potential-pH diagram depends on the avail-

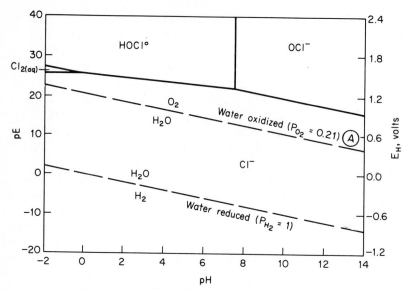

**Figure 17.11** The potential-pH diagram for aqueous chlorine; 25°C, $C_{T,Cl} = 1 \times 10^{-4}$ M. *(Source: Water Chemistry, V. L. Snoeyink and D. Jenkins, Copyright © 1980, John Wiley & Sons, Inc.)*

ability of accurate thermodynamic data for each chemical reaction that will take place and for each solid that will form on the metal surface. Such data are not available for all of the solids that can form on iron when, for example, the aqueous solution contains a variety of inorganic ions (e.g., $CO_3^{2-}$, $Cl^-$, $Ca^{2+}$, etc.) that may become incorporated in the scale. Typical protective scales may contain a wide variety of solid phases, and a potential-pH diagram that would incorporate these would be hopelessly complicated even if the necessary thermodynamic data were available. Further, corrosion scales are not necessarily thermodynamically stable, so they would not appear on the diagram. The diagram is most appropriate for the conditions of the potential and the pH near the pipe surface, when those conditions differ significantly from those of the bulk water phase.[3] Thus, the type of scale that will form cannot necessarily be predicted if the composition of only the bulk water phase is known. Further, thermodynamic considerations alone never can give any information about the velocity of the corrosion process itself. Nevertheless, potential-pH diagrams can still predict or describe corrosion and passivation processes in many water qualities and systems.

Several additional diagrams can be used to illustrate the points about the selection of realistic sets of solids, aqueous species and their activities (concentrations), plus careful consideration of the accuracy of equilibrium constants to substantially improve the utility of the diagrams. Figure 17.12 was constructed with a mixture of solids and aqueous species of iron that is more typical of a drinking water environment than Fig. 17.10.[14] This diagram shows some basis for the frequent observation of a ferric hydroxide outer deposit on cast-iron pipes, and the occurrence of colloidal iron in chlorinated drinking water, or aerated groundwater because of the location of the $Fe(OH)_3(s)$ stability field. In waters having high alkalinities (carbonate concentrations), siderite [$FeCO_3(s)$] can often be found in corrosion deposits on iron pipe.[15,16]

Figures 17.13 to 17.16 demonstrate the necessity of paying close attention to the thermodynamic data and assumptions of the solubility and redox chemistry models behind the diagrams. Figure 17.13 shows a widely referenced potential-pH diagram for the lead-water system at 25°C.[4,13] The theoretical conditions of corrosion, immunity, and passivation that were deduced from Fig. 17.13 are shown by Fig. 17.14 for potable water conditions. The formation of an insoluble lead carbonate passivation film in hard bicarbonate waters was assumed. The carbonate concentration in Fig. 17.14a is 0 mol/L, and in 17.14b it is 1 mol/L. Dissolved lead species activities were apparently assumed to be $10^{-6}$ mol/L (0.21 mg/L).

Figures 17.13 and 17.14a are unrealistic for a drinking water environment, which would virtually always contain carbonate to some extent. Figure 17.14b also suffers from several serious defects. First, the

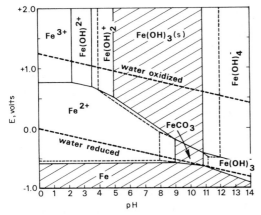

Figure 17.12  Potential-pH diagram of iron in carbonate-containing water at 25°C at $I$ = 0. Stability fields are shown for dissolved iron species activities of 0.1 mg/L (—) and 1.0 mg/L (- - -). Dissolved carbonate species concentrations are 4.8 mg C/L ($4 \times 10^{-4}$ $M$).

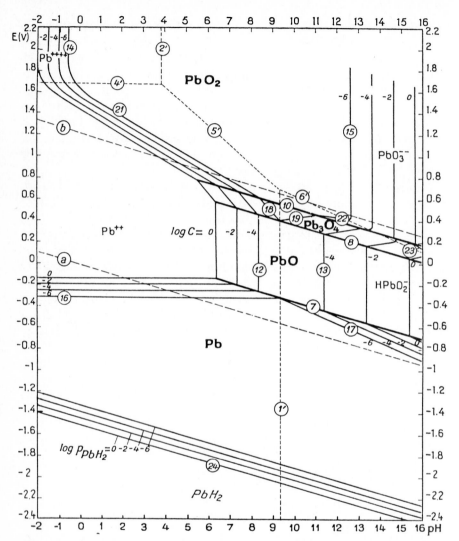

**Figure 17.13**  Potential-pH equilibrium diagram for the system lead-water, at 25°C, from Pourbaix, assuming dissolved lead concentrations (log $C$ = ) of $10^0$, $10^{-2}$, $10^{-4}$, $10^{-6}$ mol/L.[4,13] *(Source: Atlas of Electrochemical Equilibria in Aqueous Solutions, M. Pourbaix, Copyright 1966, Pergamon Press.)*

concentration of carbonate (equal to the activities at zero ionic strength) given is absurdly high for any natural or drinking water. Second, the solid and aqueous species of Pb(II) used to construct Fig. 17.14b and to a large extent, Fig. 17.13 and 17.14a, do not reflect lead solution chemistry as it is now known.[10,11,17] Therefore, they present a misleading picture of the solution behavior of lead.

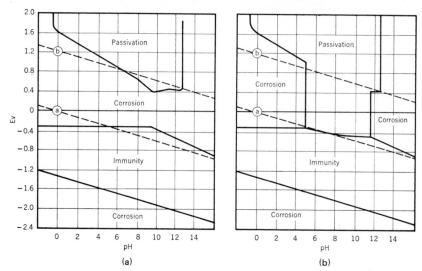

**Figure 17.14**  Theoretical conditions of corrosion, immunity, and passivation for lead at 25°C. (*a*) pure water; (*b*) pure water containing carbonate species concentrations = 1 mol/L. Dissolved lead species concentration = $10^{-6}$ mol/L. (*Source: M. F. Obrecht and M. Pourbaix, "Corrosion of Metals in Potable Water Systems," J. AWWA, vol. 59, no. 8, 1967, p. 977.*)

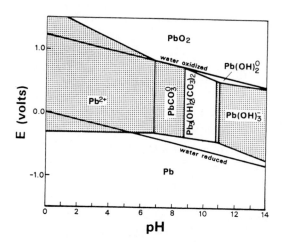

**Figure 17.15**  Potential-pH diagram for the Pb-$H_2$O-$CO_2$ system at 25°C. Areas of passivation and immunity are unstippled. Dissolved lead species activities = 0.05 mg/L. Dissolved carbonate species activities = 2.4 mg C/L ($2 \times 10^{-4}$ mol/L).

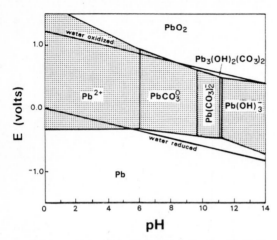

**Figure 17.16**   Potential-pH diagram for the Pb-H$_2$O-CO$_2$ system at 25°C. Areas of passivation and immunity are *unstippled*. Dissolved lead species activities = 0.05 mg/L. Dissolved carbonate species activities = 24 mg C/L ($2 \times 10^{-3}$ mol/L).

Figures 17.15 and 17.16 are revised potential-pH diagrams based on the same species as were used to create Figs. 17.4, 17.5, 17.7, and 17.8. Areas of immunity and passivation are *unstippled*. Note how the area of passivation virtually vanishes as the carbonate level increases from 2.4 (Fig. 17.7) to 24 mg/L (Fig. 17.8), for activities (concentrations) of aqueous lead species set at the current (1990) 0.05 mg/L MCL.

In conclusion, Bockris and Reddy provide several good suggestions and cautions for the use of potential-pH diagrams.[3]

- Potential-pH diagrams can be used to determine whether a corrosion process is thermodynamically possible or not.

- The diagrams provide a compact pictorial summary of the electron transfer, proton transfer, and electron and proton transfer reactions that are thermodynamically favored when a metal is immersed in a solution.

- When a potential-pH diagram indicates that a metal is immune to corrosion, it is so, provided the pH in the close vicinity of the surface is what it is assumed to be.

- Even if a potential-pH diagram indicates that a corrosion process can spontaneously take place, it does not mean that significant corrosion must necessarily be observed. The observability depends on the corrosion reaction rate being appreciable, which cannot be predicted thermodynamically.

## Chemical Reactions and Their Influence on Corrosion

Analysis of corrosion problems is complicated by the variety of chemical reactions that take place. For example, consider the reactions at an iron or steel surface, in a water where oxygen is the only oxidant, and aqueous iron complexation is negligible. First, the primary reaction occurring at the anodic sites is[1]

$$Fe(s) \rightarrow Fe^{2+} + 2e^- \qquad (17.19)$$

The $Fe^{2+}$ may then diffuse into the water, or it may undergo a number of secondary reactions.

$$Fe^{2+} + CO_3^{2-} \rightarrow FeCO_3(s) \text{ (siderite)} \qquad (17.20)$$

$$Fe^{2+} + 2OH^- \rightarrow Fe(OH)_2(s) \qquad (17.21)$$

$$2Fe^{2+} + 1/2\ O_2 + 4OH^- \rightarrow 2FeOOH(s) + H_2O \qquad (17.22)$$

The hydrated ferric oxides that form from reactions similar to that shown in Eq. (17.22) are reddish and, under some conditions, may be transported to the consumer's tap. Tertiary reactions may also occur at the surface. Possibilities are

$$2FeCO_3(s) + 1/2\ O_2 + H_2O \rightarrow 2FeOOH(s) + 2CO_2 \qquad (17.23)$$

$$3FeCO_3(s) + 1/2\ O_2 \rightarrow Fe_3O_4(s) \text{ (magnetite)} + CO_2 \qquad (17.24)$$

Reactions such as these can reduce the rate of oxygen diffusing to the anode, thus contributing to the cause of oxygen concentration cells. Other reactions that affect corrosion may take place, depending on the composition of the water and the type of metal.

Many possible cathodic reactions may occur as well. Perhaps the most common in drinking water distribution systems is the acceptance of electrons by $O_2$.

$$e^- + 1/4\ O_2 + 1/2\ H_2O \rightarrow OH^- \qquad (17.25)$$

This reaction causes an increase in pH near the cathode and triggers the following reactions:

$$OH^- + HCO_3^- \rightarrow CO_3^{2-} + H_2O \qquad (17.26)$$

$$Ca^{2+} + CO_3^{2-} \rightleftharpoons CaCO_3(s) \qquad (17.27)$$

These reactions can cause $CaCO_3(s)$ to precipitate from some waters in which the bulk solutions are undersaturated with this solid, because the pH increase in the vicinity of the cathode could be of sufficient magnitude to cause $CaCO_3(s)$ to precipitate.

The nature of the scales or deposits that form on metals is very important because of the effect that these scales have on the corrosion rate. A very long time, up to 18 months, may be required for the corrosion rate on iron and steel to stabilize because of the complex nature of the scales.[1] A much shorter time may be sufficient for other metals. If a scale reduces the rate of corrosion, it is said to be a *protecting* scale; if it does not, it is called *nonprotecting*.

The importance of scale is also demonstrated by the phenomenon of *erosion corrosion,* observed at points in the distribution system or in domestic plumbing systems where a high flow velocity or an abrupt change in direction of flow exists. The more intense corrosion that often is observed at such locations can be attributed to the abrasive action of the fluid (caused by turbulence, suspended solids, and so forth), which scours away or damages the scale, and to the velocity of flow, which carries away corrosion products before they precipitate and which supplies corrosion reactants more efficiently.[1]

The deposits that form on pipe surfaces may be a mixture of corrosion products that depend both on the type of metal that is corroding and the composition of the water solution [e.g., $FeCO_3(s)$, $Fe_3O_4(s)$, $FeOOH(s)$, $Pb_3(CO_3)_2(OH)_2(s)$, $Zn_5(CO_3)_2(OH)_6(s)$], precipitates that form because of pH changes that accompany corrosion [e.g., $CaCO_3(s)$], precipitates that form because the water entering the system is supersaturated [e.g., $CaCO_3(s)$, $SiO_2(s)$, $Al_2O_3(s)$, $MnO_2(s)$], and precipitates or coatings that form from reaction of components of inhibitors, such as silicates or phosphates that are added to the water prior to transport, with the pipe materials (e.g., lead or iron orthophosphates or silicates). Figure 17.17 shows an example of the complex nature of scale on a cast-iron distribution pipe.[1,15] Changes in water treatment or source water chemistry over time can produce successive layers of new solid phases, remove or change the nature of previously existing deposits, or both. Scales of similar chemical composition can have a significantly different impact on corrosion and metal protection because properties such as uniformity, adherability, and permeability to oxidants can vary, depending on trace impurities, presence of certain organics, temperature of deposition, length of time of formation, and other factors.

Scales that form on pipes may have deleterious effects in addition to the beneficial effect of protecting the metal from rapid corrosion or limiting the levels of toxic metals (such as lead) in solution. Water quality should be controlled so that the scale is protective but as thin

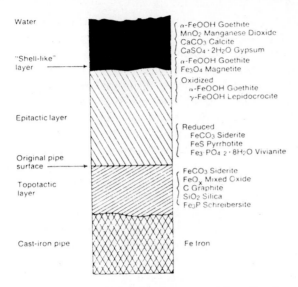

Water

"Shell-like" layer

- α-FeOOH Goethite
- MnO₂ Manganese Dioxide
- CaCO₃ Calcite
- CaSO₄ · 2H₂O Gypsum
- α-FeOOH Goethite
- Fe₃O₄ Magnetite

Epitactic layer

- Oxidized
  - α-FeOOH Goethite
  - γ-FeOOH Lepidocrocite
- Reduced
  - FeCO₃ Siderite
  - FeS Pyrrhotite
  - Fe₃·PO₄ ₂ · 8H₂O Vivianite

Original pipe surface

Topotactic layer

- FeCO₃ Siderite
- FeOₓ Mixed Oxide
- C Graphite
- SiO₂ Silica
- Fe₃P Schreibersite

Cast-iron pipe

Fe Iron

**Figure 17.17**   Schematic of scale on a cast-iron distribution pipe. *(Source: Internal Corrosion of Water Distribution Systems, American Water Works Association Research Foundation, Denver, Colo., 1985.)*

as possible, because as bulk of scale increases the capacity of the main to carry water is reduced. The formation of uneven deposits such as tubercles increases the roughness of the pipe surface, reducing the ability of the mains to carry water, and may provide shelters for the growth of microorganisms.

## Types of Corrosion

Many different types of corrosion exist. The kind of attack depends on the material, the construction of the system, the scale and oxide film formation, and the hydraulic conditions. Corrosion forms range from uniform to intense localized attack. The different forms of corrosion are primarily influenced by the distribution of anodic and cathodic areas over the corroding material. If the areas are microscopic, and very close to each other, corrosion may be relatively uniform over the entire surface. If the areas are scattered, however, and especially if the potential difference is large, pits may form, sometimes with the irregular deposits called *tubercles.*

### Uniform corrosion

According to one model, when uniform corrosion of a single metal occurs, any site on the metal surface may be anodic one instant and cathodic the next. Anodic sites shift or creep about the surface, so the rate of metal loss becomes relatively uniform over the metal surface.

The reasons for development of corrosion cells on such metals are varied. The single metals themselves may be heterogeneous, with possible differences in potential existing between different areas because of differences in crystal structure or imperfections in the metal. Also, the concentrations of oxidants and reductants in solutions may be different, causing momentary differences in potential.

An alternative model for uniform corrosion is that oxidation at a metal surface is accompanied by electron transport through an adherent film (Fig. 17.4).[1] Reduction of oxygen occurs at the film surface, and transport of ions to and away from the oxide film takes place. Electrons are probably not transported through an external portion of the metal. The overall rate of corrosion is controlled by the presence and properties of the film or by transport of reaction products, especially hydroxide ion, away from the film-solution interface. This model has been used for several kinds of metallic corrosion, notably by Ives and Rawson[18] for copper. Its applicability to drinking water systems having chlorine, or chlorine plus oxygen as the oxidizing agents, is somewhat uncertain.

### Galvanic corrosion

Galvanic corrosion occurs when two different types of metals or alloys contact each other, and the elements of the corrosion cell are present. One of the metals serves as the anode, and thus deteriorates, while the other serves as the cathode. Available metals and alloys can be arranged in order of their tendency to be anodic, and the resulting series is called the *galvanic series*. This order, in a potable water environment, depends on the temperature and solution chemistry (which affects the thermodynamic activities of the cell components), as well as the simple relative ordering of the standard electrode potentials of the oxidation or reduction half-cells or simple metal-ion redox reaction couplings.

Depending upon generalizations about the range of chemical characteristics expected for the water in contact with the metal, different "galvanic series" can, and have been, proposed. A particularly relevant and useful one is that presented by Larson,[19] which is excerpted in Table 17.3.

If any two of the metals from different groups in this table are connected in an aqueous environment, the metal appearing first will tend to be the anode and the second will be the cathode. In general, the farther apart the metals in Table 17.3, the greater will be the potential

TABLE 17.3    Empirical Galvanic Series of
Metals and Alloys in Potable Waters[†]

| Corroded end[‡] |
| --- |
| Zinc |
| Aluminum 2S |
| Cadmium |
| Aluminum 17ST |
| Steel or iron |
| Cast iron |
| Lead-tin solders |
| Lead |
| Tin |
| Brasses |
| Copper |
| Bronzes |
| Monel |
| Silver solder |
| Silver |
| Protected end + |

[†]Metals within each of the groupings have relatively similar corrosion potentials.
[‡]Anodic, or least "noble," + cathodic, or most
"noble."
SOURCE: From Larson, Ref. 19.

for corrosion because the potential difference between them will be
greater.

The rate of galvanic corrosion is increased by greater differences in
potential between the two metals. It is increased by large areas of
cathode relative to the area of the anode, and it is generally increased
by closeness of the two metals and by increased mineralization or conductivity of the water. The relative size of the cathode relative to the
anode may be of particular concern in the corrosion of lead-tin soldered joints in copper pipe.

Galvanic corrosion is often a great source of difficulty where brass,
bronze, or copper is in direct contact with aluminum, galvanized iron,
or iron. Copper-bearing metals are cathodic to aluminum, zinc, and
iron, and their underwater contact very often results in corrosion of
the latter metals. Galvanized (zinc-coated) steel is usually more ser-
viceable than steel alone, because the iron exposed at joints and holi-
days is protected at the expense of the zinc.

Proper selection of materials and the order of their use in domestic
hot- and cold-water plumbing systems is critical to the control of corrosion. To prevent galvanic corrosion, for example, only copper tubing
should be used with copper-lined water heaters. Brass valves in con-

tact with steel and galvanized plumbing in waters with high total dissolved solids cause corrosion of the steel and galvanized pipes. Dissolved copper can attack spots on galvanized pipe, thereby causing copper-zinc galvanic cells.[20,21]

### Pitting corrosion

Pitting is a damaging, localized, nonuniform corrosion that forms pits or holes in the pipe surface. It actually takes little metal loss to cause a hole in a pipe wall, and failure can be rapid. Pitting can begin or concentrate at a point of surface imperfections, scratches, or surface deposits. Frequently, pitting is caused by ions of a metal higher in the galvanic series plating out on the pipe surface. For example, steel and galvanized steel are subject to corrosion by small quantities (about 0.01 mg/L) of soluble metals, such as copper, that plate out and cause a galvanic type of corrosion.

Pitting occurs in an environment that offers some but not complete protection. The pit develops at a localized anodic point on the surface and continues by virtue of a large cathodic area surrounding the anode. Chloride ions are particularly notorious for their association with this type of corrosion of steel. Even stainless steel is subject to pitting corrosion with relatively high chloride solutions. Pits may be sharp and deep or shallow and broad, and can occur even without chlorides. In water containing dissolved oxygen, the oxide corrosion products deposit over the site of the pitting action and form tubercules. Pitting-type corrosion may also be associated with galvanic corrosion, concentration corrosion, and crevice corrosion.

### Concentration cell corrosion

Concentration cell corrosion is perhaps the most prevalent type of corrosion, and, because it is difficult to ascertain by field measurement, it is usually deduced by inference. It occurs when differences in the total or the type of mineralization of the environment exist. Corrosion potential is a function of the concentration of aqueous solution species that are involved in the reaction, as well as of the characteristics of the metal. Therefore, if the concentration of such species is different between two parts of the metal, a corrosion reaction can be driven. The corrosion process will always occur in a way that tends to equalize the concentration of the species that affect the corrosion potential.

Differences in acidity (pH), metal-ion concentration, anion concentration, or dissolved oxygen cause differences in the solution potential of the same metal. Differences in temperature can also induce differences in the solution potential of the same metal.

When concentration cell corrosion is caused by dissolved oxygen, it is often referred to as *differential oxygenation corrosion*. For example, the Nernst equation for corrosion of iron with oxygen as the electron acceptor,

$$E = E° + \frac{0.059}{4} \log \frac{\{Fe^{2+}\}^2}{\{O_2(aq)\}\{H^+\}^4} \tag{17.28}$$

can be used to show that if two sites on an iron pipe have the same pH and $Fe^{2+}$ concentration, but the $O_2$ concentration is 5 mg/L at one site and 0.1 mg/L at the other, the potential difference between the two sites will be 25 mV. Some examples of these situations are illustrated in Fig. 17.18. Common areas for differential oxygenation corrosion are between two metal surfaces, for example, under rivets, under washers, or in crevices.

In water containing dissolved oxygen (DO), the corrosion products deposit at the anode, and in the secondary reaction of oxidation of ferrous iron to ferric iron and subsequent hydrolysis, hydrogen ions are formed. This greater acidity at the anode results in a hydrogen-ion concentration cell at this point and increases the rate of corrosion. Concurrently, dissolved oxygen cannot diffuse or penetrate to the anode surface because it first reacts with the ferrous iron; thus an absence of oxygen at the anode occurs. Oxygen can diffuse to the cathode area, however, causing an oxygen concentration cell that also increases the rate of corrosion at the point of absence of oxygen. Likewise, hydroxyl ions accumulate at the cathode area, resulting in drastic reduction in hydrogen-ion concentra-

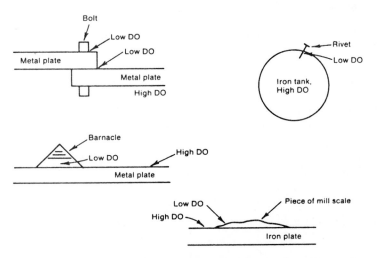

**Figure 17.18**  Examples of differential oxygenation corrosion. *(Source: Internal Corrosion of Water Distribution Systems, American Water Works Association Research Foundation, Denver, Colo., 1985.)*

tion. Therefore, the concentration cell related to the development of hydrogen ions at the anode is enhanced. Although the dissolved oxygen usually stimulates corrosion, the loss in metal takes place at the anode, where no dissolved oxygen exists.

Oxygen concentration cells develop at metal-water interfaces exposed to air, such as in a full water tower, accelerating corrosion a short distance below the surface. The dissolved oxygen concentration is replaced by diffusion from air and remains high at and near the surface, but does not replenish as rapidly at lower depths because of the distance. Therefore, the corrosion takes place at a level slightly below the surface rather than at the surface.

Dirt and debris or local chemical precipitates on a metal surface hinder oxygen diffusion by covering the metal at local areas. Thus, corrosion takes place under the deposit. Clearly, any nonadherent deposition on metal can start a chain of circumstances that result in an oxygen concentration cell.

### Tuberculation

*Tuberculation* occurs when pitting corrosion products build up at the anode next to the pit, as illustrated in Fig. 17.19. In iron or steel pipes, the tubercles are made up of various iron oxides and oxyhydroxides. These tubercles are usually rust colored and soft on the outside and are both harder and darker toward the inside. When copper pipe be-

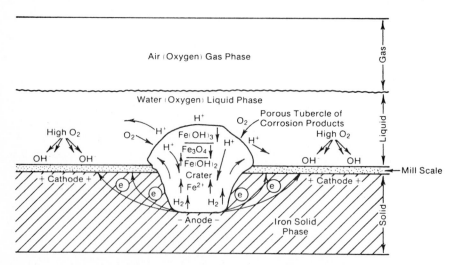

**Figure 17.19** Pitting of iron by tuberculation and oxygen-concentration cells. *(Source: Internal Corrosion of Water Distribution Systems, American Water Works Association Research Foundation, Denver, Colo., 1985.)*

comes pitted, the tubercle buildup is smaller and is green to blue-green.

Tuberculation is observed only when a piece of pipe is removed from the system, because it rarely affects the water quality, although it is possible for some of the tubercles to break loose with changes in flow or when the pipes are hit hard enough to loosen them. This type of corrosion can be suspected, though, when the flow through a pipe is much less than should be expected, as tubercles add to the roughness of a main's interior and reduce the flow. In extreme cases, the flow can be completely stopped by tubercles.

### Crevice corrosion

*Crevice corrosion* is a form of localized corrosion usually caused by changes in acidity, oxygen depletion, dissolved ions, and the absence of an inhibitor. As the name implies, this corrosion occurs in crevices at gaskets, lap joints, rivets, and surface deposits.

### Erosion corrosion

*Erosion corrosion* mechanically removes protective films, such as metal oxides, hydroxycarbonates, and carbonates, that serve as protective barriers against corrosive attack. It can also remove the metal of the pipe itself. Erosion generally results from high flow velocities, turbulence, sudden changes in flow direction, and the abrasive action of suspended materials. Erosion is much worse at sharp bends. Erosion corrosion can be identified by grooves, waves, rounded holes, and valleys it causes on the pipe walls.

### Cavitation corrosion

*Cavitation corrosion* is a type of erosion corrosion. It occurs where water pressure suddenly drops to below vapor pressure, causing water to form water vapor bubbles that collapse with an explosive effect as they move to a region of high pressure. These explosions can blast off protective coatings and even the metal surface itself. Cavitation occurs at high flow velocities immediately following a constriction of the flow or a sudden change in direction. This is most often a problem at pump impellers, partially closed valves, elbows, and reducers.

### Dealloying or selective leaching

*Dealloying* or *selective leaching* is the preferential removal of one or more metals from an alloy in a corrosive medium, such as the removal of zinc from brass (dezincification) or the removal of disseminated lead from brass.[22] This type of corrosion weakens the metals and can lead to pipe failure in severe cases. The stability of brasses (bronzes) in natural waters depends in a complex manner on the dissolved salts, the hardness, the dissolved gases, and on the formation of protective films. Dezincification is common in brasses containing 20 percent or more zinc and is rare in brasses containing less than 15 percent zinc. The occurrence of plug-type dezincification and dezincification at threaded joints suggests that debris and crevices may initiate oxygen concentration cells and result in dezincification.

Soft unstable waters, especially those with a high $CO_2$ content, are particularly aggressive to Muntz metal (60:40 copper-zinc) and yellow brass (approximately 67:33 copper-zinc). The removal of lead from brass faucets can be of concern in new plumbing, and also complicates the interpretation of metal leaching data from copper or galvanized pipe loops or homes having these plumbing materials.[22] Red brass and Admiralty metal are more resistant to dezincification, but they usually contain more lead than yellow brasses.

Lead occurs in lead-tin solder as a disseminated phase and acts principally as a diluent for the tin that actually does the binding with the copper.[23] Factors governing the removal of lead from the soldered joint are presumably similar to those affecting the leaching of lead from brass. Soldered joints in service for a long time frequently show a depletion of lead from the exposed solder.

Selective leaching also applies to the dissolution of asbestos-cement pipe or to the deterioration of cement mortar linings of iron water mains. Highly soluble components, such as free lime, calcium carbonates, and a variety of silicates and aluminosilicates, can be dissolved by aggressive water.[14,24] In some cases, the attack can be so severe as to cause weakening in the walls of the pipe and the dislodging of mats of fibers. The water causing the selective leaching can be either naturally aggressive, in terms of such parameters as pH, alkalinity, hardness, and silica concentration, or artificially aggressive, because of the use of treatment chemicals that sequester calcium or other naturally protective chemical species (e.g., zinc, iron, manganese, etc.).[14,24]

### Graphitization

*Graphitization* is a form of corrosion of cast iron in highly mineralized water or water with a low pH that results in the removal of the iron silicon metal alloy making up one of the phases of the cast-iron microstructure. A black, spongy-appearing, but hard mass of graphite remains. The graphite dispersed in the cast iron serves as the cathode for a large number of small galvanic cells, and the iron-silicon alloy becomes the anode.

Cast-iron products derive their shape from the freezing of a liquid alloy in a mold; thus solidification takes place under different conditions over the cross section of any cast shape. The portion near the mold wall may have a different composition, freezing rate, and structure than the material farther from the mold wall. Less graphite at the surface of the casting is likely; thus the familiar observation that machining of cast iron produces a lower corrosion resistance at the machined areas.

### Stress corrosion

*Stress corrosion* results from tensile stress, usually of external origin, on the metal or alloy. The corrosion usually (but not always) takes place selectively at the microstructure grain boundaries in the metal. Repeated rupture of a protective film on the surface provides a continuously anodic region.

### Microbiologically induced corrosion

*Biological corrosion* results from a reaction between the pipe material and organisms such as bacteria, algae, and fungi. It is an important factor in the taste and odor problems that develop in a system, as well as in the degradation of the piping materials. Controlling such growths is complicated because they can take refuge in many protected areas, such as in mechanical crevices or in accumulations of corrosion products. Bacteria can exist under tubercles, where neither chlorine nor oxygen can destroy them. Mechanical cleaning may be necessary in some systems before control can be accomplished by residual disinfectants. Preventative methods include avoiding dead ends and stagnant water in the system by flushing.

Ainsworth et al.[25] noted that organic carbon appears to be a major factor in controlling the numbers of microorganisms in a distribution system. They noted that the numbers of microorganisms increased as the loss in total organic carbon (TOC), and oxygen, through the distribution system increased. The microbial activity was found to be concentrated in the surface deposits and pipe sediments.

The ways in which bacteria can increase corrosion rates are numerous. Slime growths of nitrifying (and other) organisms may produce acidity and consume oxygen in accordance with Eqs. (17.29) and (17.30).

$$NH_4^+ + 3/2\ O_2 \rightarrow NO_2^- + 2H^+ + H_2O \qquad (17.29)$$

$$NH_4^+ + 2O_2 \rightarrow NO_3^- + 2H^+ + H_2O \qquad (17.30)$$

These reactions can cause oxygen concentration cells that produce localized corrosion and pitting. Lee et al.,[26] for example, showed increased localized corrosion when a water with extensive biological activity was exposed to cast iron, compared to the same water under sterile conditions.

Iron bacteria derive energy from oxidation of ferrous iron to ferric iron. Nuisance conditions often result because the ferric iron precipitates in the gelatinous sheaths of the microbial deposits, and these can be sloughed off and be the cause of red water complaints. Sontheimer et al.[16] noted that they can also interfere with the development of passivating scales and, thus, that the rate of corrosion is higher in their presence than in their absence.

The sulfate-reducing bacteria also play a role. Sulfate can act at the cathode in the place of oxygen,

$$SO_4^{2-} + 8H^+ + 8e^- \rightarrow S^{2-} + 4H_2O \qquad (17.31)$$

and the sulfate-reducing organisms apparently catalyze this reaction. Sulfate reducers have been found in the interior of tubercles and thus may be responsible for maintenance of corrosion under tubercles.[25] Sulfate reducers are often present regardless of whether the supply is aerated, and the activity of sulfate reducers can be controlled by TOC. Thus, reductions in TOC should lessen their activity. Some observations that higher levels of sulfate cause more corrosion[19] may possibly be related to microbial activity.[1]

### Stray-current corrosion

*Stray-current corrosion* is a type of localized corrosion usually caused by the grounding of home appliances or electrical circuits to the water pipes. Corrosion takes place at the anode, the point where the current leaves the metal to the power source or to ground. Stray current corrosion is difficult to diagnose because the point of corrosion does not necessarily occur near the current source. It occurs more often on the outside of pipes, but does show up in house faucets or other valves.

### Physical Factors Affecting Corrosion

Essentially no statement regarding corrosion or the use of a material can be made that does not have an exception.[27] The corrosion of metals and alloys in potable water systems depends on the environmental factors (solution composition) and on the plumbing or fitting material. Concern about the manifestation of corrosion, for example, trace metal dissolution, and contamination of the drinking water, as opposed to plumbing material, degradation or aesthetic concerns, also varies with the material. Therefore, this section will address some of the major factors that contribute to corrosion in potable waters, but the reader must refer to literature that comprehensively describes the relationship for the materials of interest.

Operators and managers of water utilities are obviously concerned with knowing what characteristics of this drinking water determine whether it is corrosive. The answers to this question are important because waterworks personnel can control, to some extent, the characteristics of this drinking water environment.[2]

Those characteristics of drinking water that affect the occurrence and rate of corrosion can be classified as (1) physical, (2) chemical, and (3) biological. In most cases, corrosion is caused or increased by a complex interaction among several factors. Some of the more common characteristics in each group are discussed in this section to familiarize the reader with their potential effects. Controlling corrosion may require changing more than one of these because of their interrelationships.

### Physical characteristics

Flow velocity and temperature are the two main physical characteristics of water that affect corrosion.

**Velocity.**    Flow velocity has seemingly contradictory effects. In waters with protective properties, such as those with scale-forming tendencies, or those containing certain inhibitors, high flow velocities can aid in the formation of protective coatings by transporting the protective material to the surfaces at a higher rate. High flow velocities, however, are usually associated with erosion corrosion or impingement attack in copper pipes in which the protective wall coating or the pipe material itself is mechanically removed. High-velocity waters combined with other corrosive characteristics can rapidly deteriorate pipe materials, without showing much metal pickup in the water because of dilution.

Another way in which high-velocity flow can contribute to corrosion is by increasing the rate at which dissolved oxygen comes in contact with pipe surfaces. Oxygen often plays an important role in determining corrosion rates because it enters into many of the chemical reactions that occur during the corrosion process.

At low velocities, the protective properties of inhibited waters are not used to their best advantage, because the slow movement does not aid the effective diffusion rate of the protective ingredients to the metal surface. Therefore, a water that behaves satisfactorily at medium-to-high velocities may still cause incipient or slow corrosion with accompanying red water problems at low velocities.

Extremely low velocity flows may also cause corrosion in water systems. Stagnant flows in water mains and household plumbing have occasionally been shown to promote tuberculation and pitting, especially in iron pipe, as well as biological growths, so dead ends of lines should be minimized. Red water complaints are frequently associated with stagnant or low-flow situations, which provide ample opportunity for metal pickup with the lengthened contact time.

Proper hydraulic design of distribution and plumbing systems can prevent or minimize erosion corrosion, particularly in household copper lines.

**Temperature.**    The influence of temperature is also often confused by conflicting observations. The confusion can be avoided by recognizing basic equilibria in water chemistry and by remembering that temperature effects are complex and depend on both the water chemistry and the type of plumbing material present in the system.

As a rule of thumb, the rates of chemical reactions tend to double for every 10°C rise in temperature.[1] The electrode potential (the driving force for any corrosion cell) is proportional to the absolute temperature. From either of these standpoints, theory predicts that corrosion rates increase with temperature. Such effects are observed in carefully controlled experiments, but in practice they are less noticeable in everyday experience unless wide differences occur (i.e., hot- versus cold-water systems). In distribution systems temperature fluctuations are somewhat limited, and whatever effect they may have is obscured by other factors.

All other aspects being equal, hot water should be more corrosive than cold. Water that shows no corrosive characteristics in the distribution system can cause severe damage to copper or galvanized iron hot-water heaters at elevated temperatures. In some hot-water systems, however, the effect of the rise in temperature is to turn a non-scaling water into a scaling water, and corrosion is reduced. Also, depending on the design of the system, heating the water may drive off dissolved oxygen and result in less corrosion. If conditions are right for corrosion, however, a hot-water system will corrode more quickly than a cold-water system.

Temperature significantly affects the dissolving of $CaCO_3$. Less $CaCO_3$ dissolves at higher temperatures, which means that $CaCO_3$ tends to come out of solution (precipitate) and form a protective scale more readily at higher temperatures. The protective coating resulting from this precipitation can reduce corrosion in a system, although excessive deposition of $CaCO_3$ can clog hot-water lines. Some other minerals behave similarly, such as calcite, aragonite, calcium sulfate (anhydrite), and many silicates.

Larson has pointed out that the effect of temperature on pH is seldom recognized.[28] For pure water, pH decreases or the $H^+$ concentration increases (as does the $OH^-$ concentration) because the $pK_w$ goes down (increased dissociation) with increasing temperature. But the degree of influence of temperature on pH is a function of the alkalinity (inorganic carbon content) of the water. Increasing concentrations of bicarbonate increasingly buffer or hinder this effect of temperature on pH. Furthermore, the solubility of $CaCO_3$ is commonly assumed to decrease with an increase in temperature. For water of low alkalinity (less than approximately 50 mg $CaCO_3$/L), this is frequently not true. In such water, the higher temperatures decrease the pH at a rate that is greater than the rate of decrease in solubility of $CaCO_3$. This decrease in pH actually increases the solubility of $CaCO_3$, even though the equilibrium constant for $CaCO_3$ dissolution actually decreases. The effect on the saturation state of calcium carbonate is particularly acute for 10 mg $CaCO_3$/L alkalinity, and even for 25 mg $CaCO_3$/L alkalinity, with a temperature change to 40°C (130°F) and, more so, to 55°C (157°F).

Consider water at 15°C (59°F) with pH = 8.71, $[Ca^{2+}]$ = 17.3 mg/L, $[Na^+]$ = 17 mg/L, $[Cl^-]$ = 35 mg/L, and total alkalinity = 30.6 mg $CaCO_3$/L. The water is at equilibrium saturation with calcite. Therefore, it has a saturation index (or Langelier index)[1,7,8,29–32] of 0.00. At this temperature, the solubility constant for the simple dissolution of calcite,

$$CaCO_3(s) \rightleftharpoons Ca^{2+} + CO_3^{2-} \qquad (17.32)$$

is equal to $10^{-8.43}$. If that water is warmed to 55°C, the saturation index drops to $-0.02$, even though the solubility constant decreases to $10^{-8.71}$. This phenomenon occurs because the pH also decreases to 8.16.

Figures 17.20 and 17.21 illustrate how the pH of waters of different alkalinities change when warmed to a temperature of 55°C from 15 or 25°C, respectively. These figures were computed, assuming

- An ionic strength of 0.001 mol/L
- No change in inorganic carbon content
- No redox or hydrolysis reactions that significantly affect the proton balance of the system

These temperature effects explain, in part, the problems of some water supplies, whose alkalinities are too low to buffer the effect of temperature. The low flow velocities in hot-water tanks often aggravate corrosivity by limiting the ability of some chemical inhibitors to be effective.[28]

An additional effect of temperature is that an increase can change the entire nature of the corrosion. For example, water that exhibits

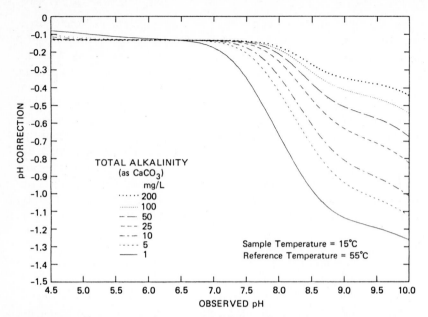

**Figure 17.20** Change in pH for waters of different alkalinities caused by warming a water originally at 15 to 55°C under closed-system conditions, assuming an ionic strength of 0.001. The pH at 55°C is obtained by *adding* the pH correction to the original pH at 15°C (e.g., $pH_{55} - pH_{15} + pH_{corr}$).

pitting at cold temperatures may cause uniform corrosion when hot.[2] Although the total quantity of metal dissolved may increase, the attack is less acute, and the pipe will have a longer life. Another example in which the nature of the corrosion is changed as a result of changes in temperature involves a zinc-iron couple. Normally, the anodic zinc is sacrificed or corroded to prevent iron corrosion. In some waters, the normal potential of the zinc-iron couple may be reversed at temperatures above 46°C (140°F). In other words, the zinc becomes cathodic to the iron, and the corrosion rate of galvanized iron is much higher than would normally be anticipated, a factor that causes problems in many hot-water heaters or piping systems.

By changing solubility constants (as well as pH), temperature changes can also favor the precipitation of different solid phases or transform the identities of corrosion products. These changes may result in either more or less protection for the pipe surface, depending on the materials and water qualities involved.

### Manufacturing-induced characteristics

For several types of common plumbing materials, the manufacturing processes used may play an important role in determining the types of corrosion that will occur and the durability of the pipe or fixtures. Pit-

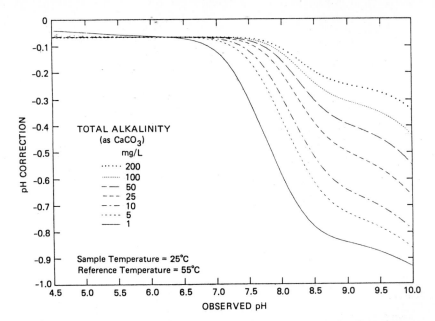

**Figure 17.21**  Change in pH for waters of different alkalinities caused by warming a water originally at 25 to 55°C under closed-system conditions, assuming an ionic strength of 0.001. The pH at 55°C is obtained by *adding* the pH correction to the original pH at 25°C (e.g., $pH_{55} - pH_{25} + pH_{corr}$).

ting of galvanized pipe has been associated with several characteristics of its manufacture, such as thin, improper, or uneven galvinizing coating, poor seam welds, and rough interior finish.[8,33–35] Impurities in the galvinizing dip solution, such as lead and cadmium, could cause concern because of their potential for leaching into the water.[8]

In England, particularly, rapid pitting failures of copper pipe were found to be associated with a carbon film on their interior surfaces.[36] The film was a residue from drawing oil that had become carbonized during annealing. Soft-annealed tubes have been more susceptible to pitting than half-hard tubes, which are more susceptible than hard-temper tubes.[36]

## Chemical Factors Affecting Corrosion

Dissolved substances in water have an important effect on corrosion and corrosion control. This section points out some of the most important factors.

### General

Table 17.4 lists some chemical factors that have an important effect on corrosion or corrosion control. Several of these factors are closely

TABLE 17.4    Chemical Factors Influencing Corrosion and Corrosion Control

| Factor | Effect |
| --- | --- |
| pH | Low pH may increase corrosion rate; high pH may protect pipes and decrease corrosion rates, or could cause dezincification of brasses. |
| Alkalinity-DIC | May help form protective coating, helps control pH changes. Low to moderate alkalinity reduces corrosion of most materials. High alkalinities increase corrosion of copper and lead. |
| DO | Increases rate of many corrosion reactions. |
| Chlorine residual | Increases metallic corrosion, particularly for copper, iron, and steel. |
| TDS | High TDS increases conductivity and corrosion rate. |
| Hardness (Ca and Mg) | Ca may precipitate as $CaCO_3$ and thus provide protection and reduce corrosion rates. May enhance buffering effect in conjunction with alkalinity and pH. |
| Chloride, sulfate | High levels increase corrosion of iron, copper, and galvanized steel. |
| Hydrogen sulfide | Increases corrosion rates. |
| Ammonia | May increase solubility of some metals, such as copper and lead. |
| Polyphosphates | May reduce tuberculation of iron and steel, and provide smooth pipe interior. May enhance iron and steel corrosion at low dosages. Attacks and softens cement linings and A-C pipe. Increases the solubility of lead and copper. Prevents $CaCO_3$ formation and deposition. Sequesters ferrous iron. |
| Silicate, orthophosphate | May form protective films. |
| Natural color, organic matter | May decrease corrosion by coating pipe surfaces. Some organics can complex metals and accelerate corrosion or metal uptake. |
| Iron, zinc, or manganese | May react with compounds on interior of A-C pipe to form protective coating. |
| Copper | Causes pitting in galvanized pipe. |
| Magnesium (and possibly other of trace metals) | May inhibit the precipitation of calcite from $CaCO_3$ on pipe surfaces, and favor the deposition of the more soluble aragonite form of $CaCO_3$. |

related, and a change in one changes another. The most important examples of this are the relationships among pH, carbon dioxide ($CO_2$), DIC, and alkalinity. Although $CO_2$ is frequently considered to be a factor in corrosion, no direct corrosion reactions include $CO_2$.[2] In some cases, the rate of $CO_2$ hydration might influence the bicarbonate concentration and, hence, the buffering ability of the water, but the important corrosion effect usually results from pH. The dissolved $CO_2$ concentration is interrelated with pH and DIC. Knowing all of the complex equations for these calculations is not necessary, but knowing that each of these factors plays some role in corrosion is useful.

Following is a description of some of the corrosion-related effects of the factors listed in Table 17.4. A better understanding of their relationship to one another will aid in understanding corrosion and in choosing corrosion-control methods.

## pH

pH is a measure of the activity of hydrogen ions, $H^+$, present in water. In most potable waters, the activity of hydrogen ion is nearly equal to its concentration. Because $H^+$ is one of the major substances that accepts the electrons given up by a metal when it corrodes, pH is an important factor to measure. At pH values below about 5, both iron and copper corrode rapidly and uniformly. At values higher than 9, both iron and copper are usually protected. Under certain conditions, however, corrosion may be greater at high pH values. Between pH 5 and 9, pitting is likely to occur if no protective film is present. The pH also greatly affects the formation or solubility of protective films, especially for lead, copper, and zinc.

## Alkalinity/DIC

Alkalinity is a measure of the ability of a water to neutralize acids and bases. In most potable waters, alkalinity is mainly described by the relationship

$$\text{Total alkalinity (TALK)} = [HCO_3^-] + 2[CO_3^{2-}] + [OH^-] - [H^+]$$

$$(17.33)$$

where [ ] indicates concentration in moles per liter and total alkalinity is in equivalents/L (equiv/L). The concentrations of bicarbonate and carbonate ions are directly related to the pH of the water, and the DIC concentration because of the dissociation of carbonic acid. Figure 17.22 is a *distribution diagram,* the data in which show the fraction of the DIC that is present in each form. When Eq. (17.33) is valid, total alkalinity is related to DIC in the manner represented by Fig. (17.23). For these calculations, a temperature of 25°C and an ionic strength of 0.005 were assumed. The TALK/DIC relationship is affected by temperature and ionic strength, so using other assumptions would change the slopes of the lines in Fig. 17.23.

When bases other than the carbonates ($HCO_3^-$ and $CO_3^{2-}$) and $OH^-$

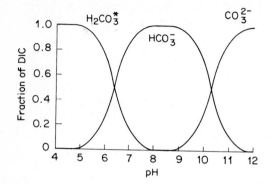

**Figure 17.22** Distribution of dissolved inorganic carbonate species in water at 25°C and $I = 0$.

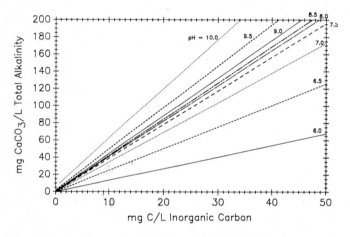

**Figure 17.23**  Relationship between total alkalinity and the total dissolved inorganic carbon concentration for a water at 25°C and an ionic strength of 0.005.

are present in significant quantities, they will consume protons in the alkalinity titration. Then the alkalinity definition must be expanded to accommodate them.

The alkalinity definition of a water containing orthophosphate, hypochlorite, ammonia, silica, and some singly charged organic acid species ($OA^-$) would then be

$$\text{Total alkalinity} = [HCO_3^-] + 2[CO_3^{2-}] + [OCl^-] + 2[HPO_4^{2-}]$$
$$+ [OA^-] + [NH_3] + [H_3SiO_4^-] + [OH^-] - [H^+] \quad (17.34)$$

This equation assumes that the concentrations of $PO_4^{3-}$ and $H_2SiO_4^{2-}$

are negligible, because their dissociation constants are so small that the pH would need to be inordinately high for them to exist to any great degree in potable waters.

Hydrogen ion–consuming complexes of metals, such as $CaHCO_3^+$, $Fe(OH)_2^0$, $Al(OH)_3^0$, $MgCO_3^0$, $Pb(CO_3)_2^{2-}$, and so forth, also contribute to alkalinity, but their concentrations are usually small enough that their contribution can be neglected. If a complex reacts slowly with the acid in an alkalinity titration, it should not be included in equations used to derive DIC from pH and the titration alkalinity.

The ability of a water to provide buffering against pH increases or decreases brought about by corrosion processes or water treatment chemical additions is closely related to the alkalinity (DIC) and pH of the water.[4,6-8,36,37] An example diagram showing this property, the buffer intensity $\beta$, for a water with DIC = $10^{-3}$ mol/L and $SiO_2$ = $10^{-4}$ mol/L is Fig. 17.24. The buffer intensity represents the amount of strong acid or strong base (in equivalents or milliequivalents) that would produce a change of 1 pH unit. Figure 17.24 shows the intrinsic buffering ability of water in the extreme low- and high-pH regions (pH $\lesssim$ 4; pH $\lesssim$ 10). The buffering ability of the carbonate system is added to the water system at pH = $pK_1$ and $pK_2$ (pH 6.35 and 10.33, respectively) proportional to its concentration. Figure 17.24 also shows a minor contribution from dissolved silicic added.

The bicarbonate and carbonate present affect many important reactions in corrosion chemistry, including a water's ability to lay down a protective metallic carbonate coating, such as $CaCO_3$, $FeCO_3$, $Cu_2CO_3(OH)_2$, $Zn_5(CO_3)_2(OH)_6$, or $Pb_3(CO_3)_2(OH)_2$. They also affect the concentration of calcium ions that can be present, which, in turn, affects the dissolving of calcium from cement-lined pipe or from A-C pipe.

Another important effect is the formation of strong soluble complexes with metals such as lead, copper, and zinc.[38] This can acceler-

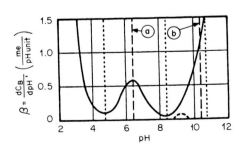

**Figure 17.24** Calculated buffer intensity curve for a $10^{-3}$ $M$ carbonate system and the contribution of $10^{-4}$ $M$ silicate system (dashed curve) at 25°C. Reference lines a and b represent $pk_1$ and $pk_2$ for the carbonic acid system, respectively. *(Source: W. J. Weber, Jr., and W. Stumm, "Mechanism of Hydrogen Ion Buffering in Natural Waters," J. AWWA, vol. 55, no. 12, 1963, p. 1553.)*

ate corrosion or cause high levels of metal pickup given the right pH-alkalinity or pH-DIC conditions.

### Dissolved oxygen

Oxygen is one of the most common and important corrosive agents. In many cases, it is the substance that accepts the electrons given up by the corroding metal, allowing corrosion reactions to continue.

Oxygen reacts with hydrogen gas, $H_2$, released at the cathode, removing hydrogen gas from the cathode and allowing the corrosion reactions to continue. Oxygen also reacts with any ferrous iron ions and converts them to ferric iron. Ferrous iron ions, $Fe^{2+}$, are soluble in water, but ferric iron forms an insoluble hydroxide. Ferric iron accumulates at the point of corrosion, forming a tubercle, or settles out at some point in the pipe and interferes with flow.

When oxygen is present in water, tuberculation or pitting corrosion may take place. The pipes are affected by the pits, by the tubercles, and by the deposits. Red water may also occur if velocities are sufficiently high to cause iron precipitates to be flushed out. In many cases when oxygen is not present, any corrosion of iron is usually noticed by the customer as red water, because the soluble ferrous iron is carried along in the water, and the last reaction happens only after the water leaves the tap and is exposed to the oxygen in the air. In some cases, particularly with iron and copper, oxygen may react with the metal surface to form a protective coating of the metal oxide.

### Chlorine residual

Gaseous chlorine lowers the pH of the water by reacting with the water to form hypochlorous acid, hydrogen ion, and chloride ion. The reaction makes the water potentially more corrosive. In water with low alkalinity, the effect of chlorine on pH is greater because the water has less capacity to resist pH changes. Tests show that the corrosion rate of steel is increased by free-chlorine concentrations greater than 0.4 mg/L.[2] Chlorine is a strong oxidizing agent, which is why it has such good properties as a disinfectant (see Chaps. 13 and 14). A free-chlorine residual is highly aggressive toward copper pipe.[39] The corrosive effect of combined chlorine and the impact of residual free ammonia have not been clearly defined.

### Total dissolved solids (TDS)

A high TDS concentration is usually associated with a high ion concentration in the water that increases conductivity. This increased conductivity in turn increases the water's ability to complete the electrochemical circuit and to conduct a corrosive current. The dissolved solids may also affect the formation of protective films.

## Hardness

Hardness is caused predominantly by the presence of calcium and magnesium ions and is expressed as the equivalent quantity of $CaCO_3$. Hard water is generally less corrosive than soft water if sufficient calcium ions and alkalinity are present at the given pH to form a protective $CaCO_3$ lining or a mixed iron–calcium carbonate film on the pipe walls. High levels of calcium also help stabilize calcium silicate phases in cement-mortar linings and A-C pipes. In spite of common beliefs, significant calcium carbonate films do not usually form on lead, galvanized, or copper cold-water pipes, so they are not primarily the causes of corrosion inhibition in these cases. Calcium may, however, assist in the corrosion buffering reactions at the pipe surface.[8]

## Chloride and sulfate

Chloride and sulfate ions, $Cl^-$ and $SO_4^{2-}$, may cause pitting of metallic pipe by reacting with the metals in solution and causing them to stay soluble, thus preventing the formation of protective metallic oxide films. Chloride is about three times as active as sulfate in this effect. The ratio of the chloride plus the sulfate to the bicarbonate $[(Cl^- + SO_4^{2-})/HCO_3^-]$ has been used by some corrosion experts to estimate the corrosivity of a water. Pitting of copper tubing is strongly related to high concentrations of $Cl^-$ and $SO_4^{2-}$ relative to $HCO_3^-$.

## Hydrogen sulfide (H₂S)

$H_2S$ accelerates corrosion by reacting with the metallic ions to form nonprotective insoluble sulfides.[2] It attacks iron, steel, copper, and galvanized piping to form "black water," even though oxygen is absent. An $H_2S$ attack is often complex, and its effects may begin immediately or may not become apparent for months and then become suddenly severe. $H_2S$ also has been found to attack A-C pipe in some waters, though possibly through microbial interactions.

## Ammonia

Ammonia forms strong aqueous complexes with many metals, particularly copper and probably lead. Thus, ammonia can potentially interfere with the formation of passivating films, or it can increase corrosion rate.

## Silicates and orthophosphates

Silicates and orthophosphates can form protective films that reduce or inhibit corrosion by providing a barrier between the water and the pipe wall. These chemicals are usually added to the water by the utility, although natural levels of silica can be high enough where the silica can react with corrosion by-products on the pipe surface, and form a more protective film. Orthophosphate usually forms insoluble passivating films on the pipe, reacting with the metal of the pipe itself (particularly with lead, iron, and galvanized). Orthophosphate formulations with zinc can decrease the dezincification of brass and can deposit a protective zinc coating (probably basic zinc carbonate or zinc silicate) on the surface of cement or A-C pipe given the proper chemical conditions. The zinc formulation may also have some advantages in some situations with iron, galvanized, or steel pipe.

## Natural color and organic matter

The presence of naturally occurring organic color and other organic substances may affect corrosion in several ways. Some natural organics can react with the metal surface and provide a protective film and reduce corrosion. Others have been shown to react with the corrosion products to increase corrosion. Organics may also complex calcium ions and keep them from forming a protective $CaCO_3$ coating. In some cases, the organics have provided food for organisms growing in the distribution system. This can increase the corrosion rate in instances in which those organisms attack the surface, as discussed in the section on biological characteristics. Which of these instances will occur for any specific water has been impossible to determine, so using color and organic matter as corrosion control methods is not recommended.[2]

## Polyphosphates

Polyphosphates have frequently been used to successfully control tuberculation and restore hydraulic efficiency to transmission mains. Polyphosphates can sometimes alter the type of corrosion form from pitting or concentration cell corrosion to a more uniform type, which causes fewer leaks and aesthetic complaints. Polyphosphates have been used to control the oxidation of ferrous iron dissolved from pipe and the formation of red water. On the other hand, polyphosphates can prevent the deposition of protective calcium-containing films, and enhance the solubility of metals such as lead and copper, interfering with the formation of passivating films. Polyphosphates have been found to attack and substantially soften A-C pipe[24] by accelerating the depletion of calcium and inhibiting the formation of fiber-binding iron or manganese deposits. Although not comprehensively investigated so far, polyphosphates when mixed with orthophosphates may assist in the formation of orthophosphate films by complexing calcium or magnesium in hard waters that otherwise could cause unwanted interfering precipitates.

### Iron, zinc, and manganese

Soluble iron, zinc, and, to some extent, manganese have been shown to play a role in reducing the corrosion rates of A-C pipe.[24] Through a reaction that is not yet fully understood, these metallic compounds may combine with the pipe's cement matrix to form a protective coating on the surface of the pipe. Waters that contain natural amounts of iron have been shown to reduce the rate of A-C pipe corrosion and bind asbestos fibers to the surface. When zinc is added to water in the form of zinc chloride or zinc phosphate, protection from corrosion has been demonstrated by the formation of a hard surface coating of basic zinc carbonate or a zinc silicate.[24]

### Copper

In recirculating water systems, such as the hot-water systems commonly found in industry, hotels, and apartments, with combined copper and galvanized steel piping, some dissolution of the copper piping occurs. This free copper then adsorbs or reacts with the galvanized steel piping, setting up small galvanic cells. These can produce rapid pitting failure of the galvanized steel piping. Similar corrosion has occurred in unprotected galvanized steel pipes in consumers' homes as a result of the addition of even trace amounts of copper, either from copper sulfate as an algicide in distribution reservoirs, from flexible copper tubing connectors to hot-water heaters, or by mixed copper and galvanized plumbing in the same water line.[8,21,33,34]

### Magnesium

Magnesium, and possibly some other trace metals (such as zinc), is known to inhibit the formation of the calcite form of $CaCO_3$. Instead, the aragonite form or some magnesium calcites may be deposited, which are more soluble. Thus, most calcium carbonate precipitation or saturation indices will give erroneous predictions. Whether any difference in the protection against corrosion of these $CaCO_3$ forms occurs is not known.

## Corrosion of Specific Materials

### Cast iron and steel

Dissolved oxygen can either enhance or inhibit the rate of corrosion of steel, depending on the mineral content of the water and the dissolved oxygen concentration.[15] Oxygen is also very important in causing concentration cell corrosion.

The effect of pH is generally through its role in secondary reactions, such as the oxidation of ferrous iron, and on the formation of scales and corrosion products. This effect influences the corrosion rate, as well as iron uptake.

Chlorine can accelerate the rate of attack of iron either by direct increase of the redox potential of the electrolyte that favors the conversion of iron to ferrous and then ferric ions or through a sequence of chemical reactions that produce hydrogen ions, hypochlorous acid, hypochlorite ions, and chloride.[15] The effect of chlorine, like that of oxygen, diminishes after the corroding surface is passivated with corrosion by-products. A secondary effect of chlorine is its significant mitigating impact on microbiological corrosion phenomena. The same is true for other disinfecting agents.

The effect of "natural color" is variable because the nature of organic materials responsible for color varies widely from source to source. These materials can frequently reduce the rate of ferrous iron oxidation, but they can also reduce the rate of calcium carbonate precipitation. No studies have shown an increase in corrosion rate.[15]

Chloride and sulfate can drastically affect the behavior of ferrous materials. Both corrosion rates and iron uptake (into the water from the pipe) have been determined to be increased sharply as the concentration of sodium chloride or sodium sulfate is increased in solution.[15] The work of Larson and Skold[40] has suggested that the ratio, sometimes called the *Larson ratio* (LR),

$$LR = \frac{[Cl^-] + 2[SO_4^{2-}]}{[HCO_3^-]} \tag{17.35}$$

should be less than 5.

Chloride ion has been noted for its role in breaking down passivating films on many ferrous metals and alloys and is one of the main causes of pitting of stainless steels.[15]

If a water contains an appreciable amount of calcium, it may act in conjunction with the pH and bicarbonate concentration to buffer the pH rise from corrosion reactions or to form a carbonate-containing passivating film on the pipe surface. Therefore, the bicarbonate concentration is also an important variable in iron and steel corrosion.

Natural silica has been shown to reduce the corrosion rate of steel, particularly when a layer of scale is already present. Silica is also frequently added as a corrosion inhibitor.

Phosphates have a varied effect, depending on the type of phosphate and other water conditions (such as pH). Sometimes the nature of iron corrosion, such as tuberculation, can be changed through the addition of polyphosphates. High doses can have a "cleansing" action on distribution mains. Frequently, the corrosion rate is not reduced, but the form and manifestation of the corrosion can be made to be less objectionable. Koudelka et al.[41] found that films formed on iron pipe from metaphosphate solutions were actually iron orthophosphates. Corrosion inhibitors of the orthophosphate type, often containing zinc and sometimes blended with silicate solutions, have been found to be useful in suppressing iron uptake and corrosion rates. Keeping a moderate flow rate is also useful in reducing corrosion and enhancing film formation. Extensive research on iron corrosion has been summarized and presented by the American Water Works Association Research Foundation.[15]

### Copper

The performance of copper in potable water systems depends on whether relatively thin and adherent films of corrosion products, cuprous oxide (cuprite, $CuO$), and basic copper carbonate [malachite, $Cu_2(OH)_2CO_3$] can be formed.[42] The conditions to optimize for the formation of the best corrosion-retarded films have not been widely studied.

Copper has only a weak tendency to passivation. Therefore, the effect of unequal aeration is very slight. The influence of copper ion concentration on the potential of copper in solution is very marked. When solution velocities vary over a copper surface, the parts washed by the solution with the higher rate of movement become anodes and not cathodes, as would be the case with iron.[42] Impingement attack of copper by high water velocities is one of the most common problems of copper pipe.

In the presence of oxidants, such as chlorine or dissolved oxygen, complexants such as ammonia accelerate the corrosion and dissolution of copper.[42] Similarly, high concentrations of bicarbonate and carbonate ions at alkaline pH values can also enhance the uptake of copper by aqueous complexation [e.g., $CuHCO_3^+$, $CuCO_3$, $Cu(CO_3)_2^{2-}$].[38] Soft waters, particularly those with an acidic pH and high carbonate content, will dissolve high levels of copper rapidly. Copper is very sensitive to the level of free chlorine in the system,[36,39] with even low levels (such as 0.2 mg/L) affecting the corrosion rate.

Normally, the rate of corrosion of copper in potable water systems is not rapid enough to cause failure of the tubing. Consumer complaints of "green water" and staining, however, can result. The resulting copper concentrations may also exceed drinking water standards.

When such copper dissolution is sufficient to cause complaints, it is usually because a fine dispersion of copper-corrosion products discolor the water. When inspected, the inside surface of the copper tube under such circumstances is characterized by a loosely adhering powdery scale and beneath it or, in areas where no scale is present, by general dissolution or corrosion of the copper.[36] Related to green water is green staining, an annoyance that has largely disappeared from the scene since detergents have replaced soaps. Even a few parts per million of copper in water can react with soap scums and cause green staining of plumbing fixtures.

The most important chemical variables in general uniform copper corrosion are pH and alkalinity, and possibly calcium concentration. The latter is mostly important if the pH and alkalinity are such that a calcium carbonate film is deposited.

Generalized corrosion and green water complaints arise typically when soft waters of slightly acid pH (6 to 7) are involved. The phenomenon clearly occurs below pH 6, but most potable waters are not served at pHs much below this. Waters in the pH range of 6 to 7 typically exhibit high carbon dioxide contents, but this is not the causative factor, as is often stated by some observers.[36] The phenomenon is characterized only by pH because the carbon dioxide content can vary over a wide range at a given pH depending on the alkalinity.

Generally, cations (calcium, magnesium, sodium, and potassium) exert no effect on the rate of corrosion.[36] Anions (chloride, sulfate, and bicarbonate), however, do exert some influence on the rate of corrosion. Chloride in particular is a strong corrosion catalyst. Sulfate is less corrosive than chloride, while bicarbonate generally tends to reduce the corrosivity of chloride and sulfate by an inhibiting action. Quantifying these observations is difficult, but as a first approximation effects caused by equal quantities of bicarbonate and sulfate tend to cancel each other out, and that chloride is at least two or three times as active as sulfate. If tabulated thermodynamic data are accurate, copper orthophosphate compounds are relatively more soluble than lead and zinc orthophosphates, so they would not play as great a role in corrosion inhibition by the formation of insoluble passivation films.[38] Some recent research, however, suggests orthophosphate might provide some corrosion reduction by changing the fundamental form of the anodic reaction.[43] Sodium silicate dosages at pH values over 8 have been helpful in reducing the corrosion rate or pitting attack.

Pitting of copper tubing is most commonly associated with hard well waters.[8,36] Pitting most often occurs in cold-water piping. Usually, pitting is observed when dissolved carbon dioxide exceeds 5 mg/L and dissolved oxygen is 10 to 12 mg/L or more.[8] The water quality parameters typical for water systems having copper pitting problems are not the typical soft, low-pH waters normally associated with corrosion of copper.

Pitting occurs most often in horizontal runs of piping, with the deepest pits concentrated in the bottom of the pipe.[44] One potential explanation is that gravity holds dense solutions of copper salts in the pit, sustaining the corrosion reaction.[45] Pitting attack is most common in new installations, with 80 to 90 percent of the reported failures occurring within the first 2 to 3 years.[8] In extreme cases, pits can break through within only a few months. Pitting occurs in all three standard types of copper tubing (K, L, and M). If unfavorable water quality conditions occur before the protective coat has formed, then serious pitting attack may occur. After 3 or 4 years, the incidence of failure drops significantly, even in systems having serious incidents of pitting.

Thin carbonaceous films, derived from carbonization of the lubricant used in manufacture, have resulted in pitting of copper tubes. Films of manganese oxides, derived by slow deposition from soft moorland waters, can also give rise to troubles with localized attack. Presumably, the presence of an organic chemical at very low concentration has been observed to reduce the corrosion of copper in certain waters. The organic agent responsible for this action has not been isolated, and the way in which it operates is imperfectly understood.[11,36]

Blue staining from water containing high levels of copper can be an irritating aesthetic problem. The staining problem can generally be resolved by adjusting the water quality, usually by increasing the pH, to lower the copper solubility.

### Brasses and bronzes

Although many studies on factors involved in the corrosion and dissolution of brasses and bronzes have been conducted, a clear picture of the relationship to water chemistry has not emerged.[22] Most studies have focused on dezincification problems. Corrosion of these materials results in the failure of some mechanical devices, such as valves and faucets, and also contributes lead, copper, and zinc to the water. This property can confound the interpretation of field studies when the intent is to correlate metal levels at the tap to corrosion rates or the effectiveness of corrosion control treatment programs.[22] Oliphant has studied and discussed several aspects of brass dezincification in water systems,[46] and several water quality properties affecting lead dissolution from brass have been reviewed.[22] Important variables in the performance of brasses and bronzes in potable waters include the pH, flow, temperature, alkalinity, chloride concentration, and the presence of natural inhibitors, as well as the types of alloys used and the manufacturing process.

### Lead and lead-containing solder

To a greater extent than all other commonly used household or utility plumbing materials, the corrosion resistance of lead is governed by the solubility of its corrosion products and their physical properties (permeability to oxidants, surface adhesion). In spite of this apparent simplicity, much experimental work is still needed to elucidate and optimize solutions.[11] Lead dissolution and corrosion rates are relatively fast compared to the concentration of lead in solution that is of concern. Thus, even slightly aggressive water standing in contact with lead pipe or lead-containing materials, for even a brief time, will pick up lead levels that are serious from a public health standpoint.

Passivation of lead usually results from the formation of a surface film composed of cerussite ($PbCO_3$), hydrocerussite [$Pb_3(CO_3)_2(OH)_2$], or plumbonacrite [$Pb_{10}(CO_3)_6(OH)_6O$], or some combination of these forms. Rarely, $PbO$ (litharge or massicot) or $PbO_2$ (plattnerite) have been reported. The formation of these compounds depends primarily on pH, along with different DIC concentrations. These interrelationships are complicated and have been described extensively here and elsewhere.[6,8,11,12,38] The carbonate-containing films are usually off-white, pale bluish grey, or light brown, and are also usually slightly chalky when dry. Therefore, they are frequently mistaken for coatings of calcium carbonate.

Studies have not been performed that adequately isolate the effect of various cations, such as calcium, magnesium, sodium, and so forth, on lead corrosion in potable waters. Their primary effects, however, are likely to be negligible, based on what is currently known about lead corrosion. Generally, studies have shown the effects of sulfate, chloride, and nitrate on lead corrosion in potable waters to be negligible.[11] "Organic acids" have apparently been implicated in increasing lead solubility, and some complexation of dissolved lead by organic ligands has been determined. Some organic materials, however, have been found to coat pipe, reducing corrosion. Therefore, a reasonable prediction cannot be made about the effect of various natural organic substances.

Orthophosphate has been shown in field and laboratory tests to greatly reduce lead solubility through the formation of lead orthophosphate films, but the effectiveness depends on proper control of pH and alkalinity.[11] The rate of formation seems to be slower than the rate of carbonate or hydroxycarbonate film formation, however, so time must be allowed for the reactions to take place. No real information exists on the rate of film formation on "new" versus "used" pipe.

Silicate may also serve to provide some protection against lead solubility, especially at levels exceeding 10 to 20 mg/L.[11] Little is known about the nature of films formed, except that reaction rates may be slow, similar to the case of orthophosphate.

Most polyphosphate treatment chemicals are strong complexing agents for lead and would be expected to either directly enhance lead solubility or to interfere with the formation of calcareous films. Some of the apparently successful applications of polyphosphates to decrease lead corrosion may actually be caused by simultaneous pH adjustment or the reversion of a fraction of the polyphosphate to a protective orthophosphate form.[47]

Studies at Portland, Ore., observed that the use of chloramines for disinfection had the effect of decreasing the rate of lead corrosion compared to the use of free chlorine, although the final pHs of the two alternative treated waters were approximately equal.[8,48] One explanation could be attributed to a lower redox potential in the system with combined chlorine. The addition of ammonia to chlorinated water to generate chloramines, however, tended to increase the concentration of lead by-products in solution.[8] The role of ammonia in possible complexation of metal ions and its effect on corrosion rates are poorly understood.

The leaching of lead from soldered joints, unlike lead pipe dissolution, has a strong galvanic corrosion component. Therefore, the effect of water chemistry and treatment strategies on lead uptake from solder is not necessarily the same as for the pipe. Unfortunately, only a few of studies have attempted to look at the water-solder interaction and the effect of other plumbing practices.[11] Lead leaching from solder is somewhat variable because the lead is contained in the solder essentially as a diluent and does not appear to participate in the alloying with copper that makes the fastening of the joint.[23]

Lyon and Lenihan found that the quality of workmanship and the presence of excess flux on the pipe interior enhanced lead uptake. Though degreasing reduced the lead uptake from this source, higher levels of copper dissolution were then seen.[49] They suggested that well-made joints possessed considerable leaching potential, though a recent study found relatively limited lead leaching from new joints in a 60-ft copper pipe loop.[50]

Oliphant did several electrochemical studies of solder corrosion and concluded that, no matter how small the area of solder exposed, considerable contamination was possible for years.[51] He did not find selective leaching of the lead component of the solder, although a solder residue containing only $SnO_2$ has been found in some copper hotwater systems in Illinois.[52]

Other experiments by Oliphant suggested that lead leaching from solder is not affected by conductivity, carbonate hardness, and both pyro- and orthophosphate.[11,51] Lead leaching was increased by decreasing pH, increasing chloride, and increasing nitrate levels. Leaching rates were decreased by increasing sulfate and silicate concentrations. The effectiveness of sulfate was related to the ratio of sulfate to chloride, with a 2:1 ratio allowing the formation of good crystalline corrosion product layers.[51]

### Galvanized steel

The various aspects of the corrosion of galvanized steel, including the pipe-water chemical reactions and the significance of the manufacturing process of the pipe, have been extensively reviewed by Trussell and Wagner.[33] One of the complexities of this material is that once the zinc layer is corroded away, the pipe behaves as if it were black iron pipe. Therefore, the corrosion potential of the water and corrosion inhibition strategies should take into account the age and probable condition of the galvanized pipe in a system, because chemical effectiveness could differ.

Four aspects of concern arise from galvanized pipe corrosion: uniform corrosion, pitting, metal uptake, and tuberculation.[33] Mechanical failure because of uniform corrosion is rare, but when the zinc corrosion rate is high, failures are often induced by the tuberculation that follows rapid depletion of the zinc layer. The uniform corrosion of zinc depends strongly on pH, which would easily be predicted by considerations of zinc chemistry.[11,14,23,38] The nature of the film formed on the pipe changes in response to many chemical factors, and galvanized pipe corrosion seems to be very sensitive to the type of scale produced. Some are voluminous, chalky, and relatively unprotective, even if metal uptake is not extremely high. Poor film protection can enhance the potential pitting or tuberculation.

Studies have shown increases in the corrosion rate of zinc with increasing carbonate hardness, even in the presence of orthophosphate.[33] This may be a solubility enhancement by carbonate complexation $[ZnHCO_3^+, ZnCO_3^0, Zn(CO_3)_2^{2-}]$,[14,38] or some other factor.

The role of calcium carbonate in galvanized pipe protection, often seen in deposits as aragonite,[33,53] is not clear. Calcium carbonate, along with basic zinc carbonate $[Zn_5(CO_3)_2(OH)_6$, hydrozincite], and forms of zinc hydroxide and zinc oxide appear to be the most significant components of natural scale layers.

Orthophosphate has been demonstrated to be an effective controller of zinc solubility and galvanized material corrosion in several studies.[33] The effectiveness of orthophosphate, as with lead, must depend on the pH, alkalinity, and orthophosphate level in solution.[38]

Silicate has been shown to reduce galvanized pipe corrosion, but most studies have focused on hot-water systems. Zinc silicate compounds are generally of low solubility, however, and should provide some protection in the slightly alkaline pH range.[14,33]

Pitting failure of galvanized pipe is more common in hot-water systems than cold.[33] Major reasons for failure are pipe coating quality, the presence of dissolved copper, and the reversal of electrochemical potential of zinc and iron at high temperature, which have been noted previously.

### A-C and cement-mortar linings

A-C and cement-mortar linings behave very similarly in potable waters. The principal difference is the concern that accompanies A-C pipe deterioration about the input of asbestos fibers into the water. Therefore, two separate issues are involved in A-C pipe corrosion treatment: maintenance of structural integrity and control of asbestos fiber removal.[14,24]

The cement matrix of A-C pipe is a very complicated combination of compounds and phases.[24] Some of the phases are poorly identified or are of indefinite composition. More than 100 compounds and phases important to the chemistry of portland and related cements have been described and identified. Because of solid solution possibilities, probably many more exist. The state of knowledge of the thermodynamic solubilities in water of the individual predominant compounds of the cement is not very advanced.

The main components of portland cement are tricalcium silicate, dicalcium silicate, and tricalcium aluminate, together with smaller amounts of iron and magnesium compounds. Free lime [$Ca(OH)_2$] is also present, but it is regulated to be less than or equal to 1.0 percent by weight in U.S. commercial type II autoclaved A-C pipe.[24]

When mixed with water, the compounds hydrate, and under alkaline conditions this reaction causes the cement to set and to develop strength. Through this hydration process the calcium silicates go into solution, release calcium hydroxide, and precipitate again. The precipitate is chiefly in the form of hydrated monocalcium silicate. Hydrolysis reactions during the setting and hardening of the cement produce further calcium hydroxide.

Groundwater having high mineral content but low pH values shows accelerated rates of pipe leaching.[24] Classically, such water is said to possess *aggressive $CO_2$*, meaning that the actual dissolved carbon dioxide concentrations exceed those that would be in equilibrium with solid calcium carbonate at the observed pH, calcium, alkalinity, and mineral contents. That the dissolved carbon dioxide is the actual reacting chemical species is unlikely, however, although writing chemical reactions in that way has often been convenient.

The mechanism of attack by soft waters of low mineral content generally is believed to proceed as follows. First, the calcium hydroxide is removed. Then, if the water still possesses sufficient acid content, calcium carbonate is either dissolved, or not formed. Because alkaline conditions cannot be maintained within the cement after the initial calcium hydroxide is removed, attack begins on the hydrated calcium silicates. These convert to calcium hydroxide, and the cycle of reactions can continue either until sufficient calcium carbonate is formed and the water is neutralized to the point at which the cycle stops or until no cementitious material remains to bind the aggregates together.[24]

No simple index exists that can predict the behavior and service life of A-C pipe when it is exposed to waters of varying chemistries. Thus, few specific inhibitor formulations or water quality adjustments exist that can be quantitatively predicted to be effective without some field testing and empirical dosage adjustment.

Calcium carbonate saturation indices, described later, are insufficient in themselves to predict whether A-C pipe will be attacked by a given water. Calcium carbonate deposition, however, can be used to prevent attack on A-C pipe. Reasonable field evidence exists that this idea is basically sound,[24] but it has not been investigated in great detail. Some of the success of this approach may be attributed to the mass-action effect of waters with high calcium concentrations and pH values that would tend to stabilize the cement matrix. The primary drawback to calcium carbonate saturation is that in the absence of actual scale formation, or the formation of protective coatings by naturally occurring substances such as iron and manganese, fibers are left exposed at the surface of the pipe, where they could be physically eroded.

Coatings containing iron have been found on many A-C pipe specimens exposed to a variety of aggressive to nonaggressive water qualities ("aggressiveness" being based on their calcium carbonate saturation state).[14,24] Iron coatings formed sometimes have a granular, porous structure that does not necessarily retard calcium leaching from the cement matrix. The iron frequently does help in preventing the exposure of asbestos fibers at the pipe surface.

The iron coatings are often formed not as a continuous film but as crystals of unidentified iron compounds that accumulate on the surface of the pipe. In addition to the protection afforded by naturally occurring iron (as a ferric hydroxide or oxyhydroxide deposit) and manganese(IV) oxide deposit, silica may also be beneficial as an agent that can help maintain the hardness of the pipe.[14,24] This effect has not been studied in detail. Laboratory experiments indicate that high concentrations of silica will inhibit iron hydroxide precipitation, possibly through complexation or by colloid stabilization. If the pipe deterioration can be stabilized without needing the iron precipitate to bind the fibers and help coat the surface, then the usable silica level can be increased substantially. Silica, however, could enhance protection by the formation and adsorption of iron colloids on to the pipe surface.

Orthophosphate, sulfate, and chloride salts of zinc have been found to be useful in preventing softening of A-C pipe specimens and in slowing calcium leaching in numerous laboratory experiments.[14,24] The dosage of zinc necessary depends at least on the pH and dissolved inorganic carbonate concentration of the system. How useful zinc addition is in extending the potential lifetime of newly installed A-C lines, or in increasing the service life of existing mains, has not been established in the short field tests performed thus far. In these experiments, zinc has been found to be the active agent, not its anionic associate. The reaction mechanisms have not been delineated. One possibility is that initially zinc reacts with the water to form a zinc-hydroxycarbonate precipitate. The zinc solid may then react with the pipe surface itself, possibly converting some or all of the coating to a harder zinc-silicate solid phase. Thus, the choice of salt can primarily be based on economic considerations, or whether the system water would ultimately contribute to nutrient loading in waste streams. If the system consists of metallic pipe as well as asbestos-cement, the orthophosphate salt might be preferable in order to lower metallic corrosion.[24]

Particularly destructive to A-C pipe, and probably to all concrete or cement mortar–lined pipes, are waters with low pH (less than approximately 7.5 or 8, unless they contained high calcium, alkalinity, and silicate levels), very high sulfate concentrations, and polyphosphates. Strong sequestering or complexing agents, such as the polyphosphates, have been shown to attack the pipe by enhancing calcium, aluminum, iron, and magnesium leaching from the cement matrix.[24] They could also prevent the formation of protective coatings by metals such as zinc, manganese, iron, and calcium.

## Direct Methods for the Assessment of Corrosion

### Physical inspection

Water plant operators and engineering staff should be aware of the appearance of the interior of the distribution piping and how their water quality is affecting the materials in their system.[53]

Physical inspection is usually the most useful inspection tool to a utility because of the low cost.[2] Both macroscopic (human eye) and microscopic observations of scale on the inside of the pipe are valuable tools in diagnosing the type and extent of corrosion. Macroscopic studies can be used to determine the amount of tuberculation and pitting and the number of crevices. The sample should be examined also for the presence of foreign materials and for corrosion at joints.

Utility personnel should try to obtain pipe sections from the distribution or customer plumbing systems whenever possible, such as when old lines and equipment are replaced. If a scale is not found in the pipe, an examination of the pipe wall can yield valuable information about the type and extent of corrosion and corrosion-product formation (such as tubercles), though it may not indicate the most probable cause. Pipe sections of small diameter may be conveniently examined by sawing them lengthwise.

Examination under a microscope can yield even more information, such as hairline cracks and local corrosion too small to be seen by the unaided eye. Such an examination may provide additional clues to the underlying cause of corrosion by relating the type of corrosion to the metallurgical structure of the pipe.

Photographs of specimens should be taken for comparison with future visual examinations. High magnification photographs should be taken, if possible.

**Rate measurements.**   Rate measurements are another method frequently used to identify and monitor corrosion. The corrosion rate of a material is commonly expressed in mils (0.001/in) penetration per year (mpy). Common methods used to measure corrosion rates include (1) weight-loss methods (coupon testing and loop studies) and (2) electrochemical methods. Weight-loss methods measure corrosion over a period of time. Electrochemical methods measure either instantaneous corrosion rates or rates over a period of time, depending on the method used.

**Coupon weight-loss method.**   Four important criteria for corrosion tests are that (1) the metal sample must be representative of the metal piping, (2) the quality of the water to which the pipe sample is exposed should be the same as that transported in the plumbing system, (3) the flow velocity and stagnation times should be representative of those in the full-scale system, and (4) the duration of the test must allow for development of the pipe scales that have an important effect on corrosion rate and on the quality of the water passing through the pipe system.[54] Estimates of the time (months) required to determine the desired information from tests with pipe inserts are given in Table 17.5. These times are only rough estimates, and specific conditions

TABLE 17.5    Estimated Duration (Months) Required for Corrosion Tests

| Material | Comparison of uniform corrosion rates or metal leaching | Comparison of inhibitors | | Pitting |
| --- | --- | --- | --- | --- |
| | | New pipe | Old pipe | |
| Iron | 12–24 | 3–6 | 12–24 | 12–24 |
| Copper | 3–6 | 1–3 | 3–6 | 12–36 |
| Galvanized iron (zinc) | 3–6 | 1–3 | 6–12 | 12–36 |
| Lead | 6–12 | 2–6 | 6–12 | |
| Asbestos cement | 18–24 | 6–12 | 12–18 | |
| Mortar lining | 24–36 | 12–24 | 24–36 | |

SOURCE: From AWWA, Ref. 15.

may make it advisable to use other times. Corrosion tests for hot-water systems, for example, should require less time. More than one insert should be used, each containing pipe coupons, so that the corrosion rate can be determined at different times. For example, analysis of coupons at 3-month intervals may be desirable if the test duration is to be 12 months or longer. The test duration can then be modified based on the results obtained.

Coupon weight-loss test results do not measure localized corrosion, but are an excellent method for measuring general or uniform corrosion.[2] Coupons are most useful when corrosion rates are high so that weight-loss data can be obtained in a reasonable time.

Following are lists of the advantages and disadvantages of the coupon method:[2]

Advantages

1. Provides information on the amount of material attacked by corrosion over a specified period of time and under specified operating conditions.

2. Coupons can be placed in actual distribution systems for monitoring purposes.

3. The method is relatively inexpensive.

Disadvantages

1. Rate determinations may take a long time (i.e., months, if corrosion rates are moderate or low).

2. The method will not indicate any variations in the corrosion rate that occurred during the test.

3. The specimen or coupon may not be representative of the actual material for which the test is being performed.

4. The reaction between the metal coupon and the water may not be

the same as the reaction at the pipe wall because of friction or flow velocity, because the coupon is placed in the middle of the pipe section.

5. Removing the corrosion products without removing some of the metal may be difficult.

The insertion of flat metal coupons into a pipe is generally unsatisfactory because the altered flow lines are not similar to those at the pipe wall, and thus scale formation and the action of inhibitors on it may not be representative.[54] Also, obtaining coupons that are representative of piping material is sometimes difficult. The use of pipe sections has several advantages relative to flat coupons. Sections of actual pipe material, with the lengths selected based on the objectives of the test, can be inserted into actual plumbing systems or pipe loops that have been designed specifically for the study. Information obtained from these tests will then be more representative of what occurs in actual plumbing systems.

Coupon preparation, handling, and examination are relatively time-consuming and exacting operations that usually require a trained laboratory analyst or corrosion specialist to provide reproducible results.[54] Coupon metal must be carefully degreased, etched, metal-stamped for identification, weighed, and stored in a sealed plastic wrapping in a desiccator to avoid atmospheric corrosion. The edges of a coupon are painted with epoxy, and gloves or tongs are used to handle a coupon so that skin moisture or oil will not mar its surface. Following coupon exposure and examination, the scale must be removed by brushing and using sequestered acids, before weight loss, pitting characteristics, and so forth, can be determined. Again, the coupons must be thoroughly dry and stored in a desiccator. Control blanks are processed along with the exposed coupons in order to determine the weight loss or gain caused by the handling techniques, and these values are used to correct calculated corrosion rates.

Coupon or insert tests for weight loss can be done with a variety of pipe loop configurations, including household installations.[55,56] Several improvements over the original ISWS-ASTM method have recently been developed.[57]

**Loop system weight-loss method.**   Another method for determining water quality effects on materials in the distribution system is the use of a pipe loop or sections of pipe. Either the loop or sections can be used to measure the extent of corrosion and the effect of corrosion control methods. Pipe loop sections can be used also to determine the effects of different water qualities on a specific pipe material. The advantage is that actual pipe is used as the corrosion specimen. The loop may be made from long or short sections of pipe, but for direct monitoring of dissolved constituent levels, longer pipe sections are required.[50]

Water flow through the loop may be either continuous or shut off with a timer part of the time to duplicate the flow pattern of a household. Pipe sections can be removed for weight-loss measurements and then opened for visual examination. This method is called the Illinois State Water Survey (ISWS) method and is an American Society of Testing and Materials (ASTM) standard method (D2688, Method C) and should be followed closely.

Following are lists of the advantages and disadvantages of a loop system:[2]

Advantages

1. Actual pipe is used as the corrosion specimen.
2. Loops can be placed at several points in the distribution system.
3. Loops can be set up in the laboratory to test the corrosive effects of different water qualities on pipe materials.
4. The method provides information on the amount of material attacked by corrosion over a specified period of time and under specified operating conditions.
5. The method is relatively inexpensive, as many corrosive effects can be examined visually.

Disadvantages

1. Determination of corrosive rates can take a long time (i.e., months, if corrosion rates are moderate or low).
2. The method does not indicate variations in the corrosion rate that occur during the test.

The use of "planned interval tests" can enable the determination of changes in corrosion rates during the testing period to some extent, however.[58]

A test to determine the metal uptake rate from the pipe surface in a recirculating loop system has been developed by German researchers and is described in detail elsewhere.[54] The test pipe would normally be obtained from a corrosion test pipe rig through which water of the quality being tested flows during the test. With parallel tests using different pipes and waters of different quality, comparable results between different water qualities can be obtained.

**Electrochemical rate measurements.** Electrochemical rate measurements are based on the electrochemical nature of corrosion of metals in water.[2] An increasing number of these instruments are now on the market. They are relatively expensive, however, and probably not widely used by smaller utilities. They are discussed here for completeness. The primary types of rate-measuring techniques are systems based on either electrical resistance, linear polarization, or galvanic current.

The electrical resistance system uses a low-resistance Kelvin bridge circuit, with all the bridge resistance elements except the slide wire made of the metal tested.[54] Corrosion of the exposed metal element reduces its cross-sectional area and increases the electrical resistance, allowing corrosion to be monitored over time. The corrosion rate can be logged manually or by a recorder to obtain a plot of resistance versus time. By applying the proper probe size factor, one obtains a direct readout, in mils of lost probe thickness. A probe size factor is furnished by the manufacturer for each type of probe.

The electrical resistance instruments are simple, reliable, and relatively economical. The corrosion rate, however, is reported only as uniform corrosion, and the probes provide no information on pitting or pitting rate. Also, information obtained from specimens covered with scales is difficult to interpret.

The linear polarization method is based on the principle that a linear relationship exists between a small amount of electrochemical potential shift or polarization and the corrosion rate.[54] Corrosion rate is measured by the electric current flow associated with a slowly pulsed direct current from an outside source that causes no greater than 20 mV of electrochemical polarization of the metal electrodes. Metal electrodes that also serve as coupons, for loss-of-weight testing and pitting examination, are used. A two-electrode probe system and a three-electrode probe system are currently available. Each of these instruments provides a direct readout of corrosion rates. The three-electrode probe has an advantage when solution conductivity varies appreciably.[54] Standard electrode probes that are about 3/8 in diameter by 1 to 2 in long (9 mm by 25 to 50 mm), depending on metal and service, are available. Larger probes that are placed closer together are needed if the water conductivity is less than about 50 μmhos/cm.

The three-probe instrument uses a null-balance corrosion potential that is then calibrated to read corrosion rates directly in mils per year: Readings are rapid, and both anodic and cathodic reactions can be read on manual instruments or automatically programmed or recorded as continuous measurements. The electrodes are screwed into a plastic terminal holder that has a threaded plug end that is inserted in pipe in a manner similar to insertion of bar coupons. The probes can be used for weight-loss measurements, and a constant to relate the corrosion rate to weight loss is obtained for actual conditions. A generally closer correlation to coupon data occurs if corrosion is uniform and if scale or tubercle buildup is minimal. Nevertheless, the simplicity and economy of these on-line techniques are very useful in an overall corrosion control program.

Both the two- and three-electrode systems indicate when pitting or localized corrosion is taking place.[54] For the two-electrode system, a separate readout is available of a semiquantitative index that is based on the difference in corrosion rates of anodic and cathodic pulses interpreted as a relative change in electrode surface areas. For the three-electrode system, pitting is indicated by the difference in potential between two of the electrodes or by variations of current flow during the polarization pulse. A surface-mounted linear polarization electrode probe has been developed that can provide an exposure more comparable to that of ISWS pipe sleeve coupons.[54]

The galvanic current method utilizes a zero-resistance ammeter and measures corrosion of dissimilar alloys or metals. These instruments are particularly useful in measuring galvanic and ground currents that can sometimes be related to internal corrosion.[54]

### Immersion testing in the laboratory

Immersion tests are a very effective means of making an *initial* appraisal of the effectiveness of inhibitors.[54] These tests use metal coupons to evaluate inhibitors, dosages, and pH control in batch jars of water. The procedures of ASTM Standard 631-72 should be utilized for immersion tests.

Twenty or more tests can be operated concurrently by using glass jars of 3- to 5-gal capacity and suspending four or five coupons in each jar. The coupons can be removed at various times, for evaluation of weight loss and pitting, which indicate the trend toward a steady-state corrosion condition. These tests are usually set up by varying pH of the water from 6.5 to 8.5, as 6.5, 7.0, 7.5, 8.0, 8.5, and then by varying inhibitor doses of polyphosphate, zinc bimetallic phosphates, and zinc orthophosphate from 0.5 to 1.5 times the anticipated control dosage. The water is changed biweekly, and air is bubbled through each jar to provide a continuous saturation of dissolved oxygen and to create a velocity and turbulence past the coupons. If four coupons are used, one is removed each week for evaluation of corrosion penetration, pitting, and scale formation. The water in the tank may also be tested for iron, copper, or whatever pipe material is being tested by the coupon.

With outside and edge surfaces coated by epoxy paint, pipe sleeve coupons (1 in diameter and 2 in long (25 mm by 50 mm)) also work very well in immersion tests. Four sections of pipe are suspended on a string tied to a stick resting on the top of the immersion jar. Each week, one pipe section can be cut from the bottom of the coupon for evaluation. The pH and dissolved oxygen must be monitored daily. The pH is adjusted by adding small amounts of sulfuric or hydrochloric acid or sodium hydroxide.

Immersion tests in the laboratory can be used to evaluate any contemplated change of water quality or water treatment on corrosivity to pipe materials. During the initial week of testing, any pretreatment passivating dosage to be used in the full-system program should be used to provide an accurate simulation of the practice of inhibitor use.

Less elaborate and more direct static immersion tests are possible.[54] Tests of in situ corrosion rates by using resistance or linear polarization instruments are also possible. The wire loop or electrodes of these instruments remain in the immersion jar, but are connected periodically to the portable meter on which readings are taken. Special jar setups for this type of equipment can be obtained from the manufacturers of the electric corrosion meters.

Limitations of all immersion tests are that accurate representation of flow and water chemistry conditions in the field situation are sacrificed for testing simplicity and efficiency. Thus, results of immersion tests might not prove to be accurate simulations when extrapolated to practice at the treatment plant and in the whole distribution system. They should only be used as a screening tool.

### Chemical analysis

Evaluating corrosion control program performance by direct measurement of important chemical constituents is critical when metals involved in drinking water standards are of concern. The concentrations of other constituents that are active agents in the formation of passivation films or in the performance of chemical inhibitors should also be analyzed to evaluate corrosion.

Valuable information about probable corrosion causes can be found by chemically analyzing the corrosion by-product material.[2] Scraping off a portion of the corrosion by-products, dissolving the material in acid, and qualitatively analyzing the solution for the presence of suspected metals or compounds can indicate the type or cause of corrosion. These analyses are relatively quick and inexpensive. If a utility does not have its own laboratory, samples of the pipe sections can be sent to an outside laboratory for analysis. The numerical results of these analyses cannot be quantitatively related to the amount of corrosion occurring because only a portion of the pipe is being analyzed. However, such analyses can give the utility a good overview of the type of corrosion that is taking place.

The compounds for which the samples should be analyzed depend on the type of pipe material in the system and the appearance of the corrosion products. For example, brown or reddish-brown scales should be analyzed for iron and for trace amounts of copper. If found in lead pipe, lead should also be included in the analysis. Greenish mineral deposits should be analyzed for copper. Black scales should be analyzed for iron and copper. Depending on the type of plumbing material, light-colored deposits should be analyzed for calcium, carbonate, silicate and phosphate, in addition to lead, zinc, or whatever is the metal of interest.

Comparing sampling data of scales from various locations within the distribution system can isolate sections of pipe that may be corroding. Increases in levels of metals such as iron or zinc, for instance, indicate potential corrosion occurring in sections of iron and galvanized iron pipe, respectively. The presence of cadmium, a minute contaminant in the zinc alloy used for galvanized pipe, also indicates the probable corrosion of a galvanized iron pipe. Corrosion of cement-lined or A-C pipe is generally accompanied by an increase in both pH and calcium throughout the system, sometimes in conjunction with an elevated asbestos fiber count.

## Microscopic techniques

The field of analysis of pipe scales and inhibitor films can use a variety of sophisticated instrumental analysis techniques, in addition to the traditional wet-chemical procedures, such as those of ASTM (D2331-73). Often, no single technique will provide all the answers in film identification and interpretation of corrosion and corrosion-control data. Judicious use of a combination of methods, however, can give invaluable insight into the problems and solutions.

Optical microscopic techniques have been widely used for the evaluation of A-C pipe corrosion, and have been summarized and described in many articles.[24,59] They have also been employed in the examination of particulates in drinking water[60] and in the examination of films on iron pipe.[25,61] Photomicrographs have also been extremely useful in examining the structure of the alloy and zinc layers on galvanized pipe.[33]

The scanning electron microscope (SEM) has been widely applied to the evaluation of A-C pipe deterioration.[14,24,59] Studies have also attempted to directly relate the morphology of corrosion product crystals of lead carbonate, basic lead carbonate, or a lead phosphate to the protectiveness against corrosion.[62]

SEM shows highly magnified views of the surfaces of materials. Figures 17.25, 17.26, and 17.27 demonstrate the use of this technique in evaluating the corrosion of A-C pipe. Figures 17.25 and 17.26 show

**Figure 17.25** Scanning electron microscope photograph of unused A-C pipe from Seattle (magnification 100×). *(Source: Internal Corrosion of Water Distribution Systems, American Water Works Association Research Foundation, Denver, Colo., 1985.)*

the change in the surface of A-C pipe after exposure to an aggressive water.[24] Note the exposed chrysotile asbestos fibers present in Fig. 17.26. Figure 17.27 is from a USEPA laboratory experiment, in which the pipe surface was protected through the formation of a zinc-containing solid.[14]

The transmission electron microscope (TEM) can be useful to identify individual particles of material by selected area electron diffraction (SAED). This is frequently used to identify chrysotile, amphibole, and other types of asbestos by their characteristic electron diffraction patterns.

### X-ray elemental analysis

For surficial deposits, x-ray fluorescence (XRF) spectrometry is helpful.[63] The sample, or its fusion product, is ground finely into a powder and is compressed into a flat discoidal pellet. The mount is irradiated by an x-ray beam of high energy.

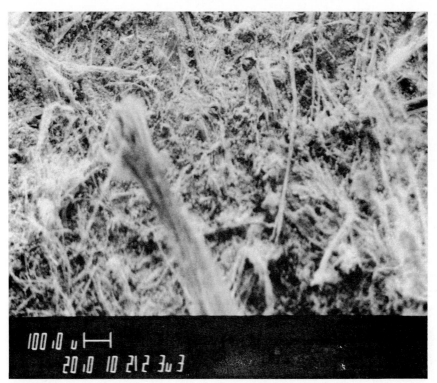

**Figure 17.26**  Scanning electron microscope photograph of A-C pipe from the Seattle distribution system (magnification 100×). *(Source: Internal Corrosion of Water Distribution Systems, American Water Works Association Research Foundation, Denver, Colo., 1985.)*

XRF can be done very quickly in comparison to wet-chemical analysis (including flame or flameless atomic absorption spectrophotometry), because of the relative simplicity of sample preparation and processing. Quantitative analysis can be performed with proper sample preparation precautions and suitable mathematical processing of the raw intensity data. For most major constituents, and for most metals down to the order of 1 percent by weight, the accuracy is comparable to dissolution and analysis techniques.

On the microscale, energy-dispersive x-ray analysis (EDXA) is comparable. Here, an SEM unit is used to excite the generation of the characteristic x rays from the sample. Wavelength-dispersive x-ray analysis (WDXA) is also possible with some electron microscopes, but EDXA is much more widely used today.

Several limitations exist to any chemical analysis performed by EDXA.[59] Only the elements heavier than sodium can be detected. Also, the EDXA is semiquantitative under the sample preparation

**Figure 17.27** Scanning electron microscope photograph of A-C pipe specimen from USEPA laboratory experiment, showing the surface covered by a zinc-containing solid (magnification 300×). *(Source: Internal Corrosion of Water Distribution Systems, American Water Works Association Research Foundation, Denver, Colo., 1985.)*

and analysis conditions normally used, so it does not give the exact concentrations of the elements present on the surface. The relative amounts of the elements present, however, may be important as a means to gain information on the identity of surficial coatings. EDXA analysis cannot identify the compounds present, as would be possible with x-ray diffraction.

For EDXA and SEM analysis, observing representative sections of the pipe is important because the same pipe might vary in color and texture. Often, several sections of the same pipe must be analyzed.

### X-ray diffraction

Another of the sophisticated analytical techniques that can be applied to the analysis of scales and corrosion products is x-ray diffraction (XRD). The fundamental principle behind the method is that when a highly collimated beam of x rays of a particular wavelength strikes a finely ground, randomly oriented sample of material, the x rays are diffracted in a pattern that is characteristic of the crystalline structure of the compound. Mixtures of compounds give a pattern that is a sum of those compounds present. The general method, as applied to corrosion products, was described by Gould.[64] The analyzed patterns are compared to a reference library of known patterns for various solids, minerals, metals, and alloys.

Particularly when used in conjunction with other analytical techniques to reduce the amount of pattern searching necessary (by identifying the combinations of elements present, or probable class of compounds, e.g., phosphates, silicates, oxides, and so forth), XRD provides a tremendous insight into pipe corrosion and corrosion inhibitor effects.

Some examples follow of information provided by XRD that could not be determined by the other methods discussed. Figure 17.28 shows two XRD patterns of corrosion products on galvanized pipe.[33] Interest-

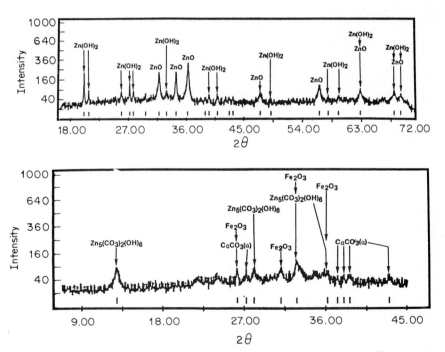

**Figure 17.28**  X-ray diffraction of corrosion products from galvanized pipe.[28] The upper diffractogram was done on pipe exposed to pure, deionized water, and the lower diffractogram was conducted on a water at pH 7.7 with 150 mg/L of alkalinity and a calcium hardness of 100 mg $CaCO_3$/L. Though calcite is the most stable species of calcium carbonate under most conditions of interest to waterworks, aragonite is found instead on this specimen. (*Source: Internal Corrosion of Water Distribution Systems, American Water Works Association Research Foundation, Denver, Colo., 1985.*)

ingly, the aragonite form of $CaCO_3$ was often found instead of calcite when the deposits are associated with galvanized materials. This is possibly caused by some influence of the zinc ion on the form of calcium carbonate crystals that will form, similar to the effect of magnesium ion. Aragonite is more soluble than calcite, so this is important in the modeling of precipitation and corrosion reactions. Conventional wet-chemical methods, as well as EDXA, XRF, or WDXA, cannot differentiate calcite from aragonite, or even compounds like basic zinc carbonate [hydrozincite, $Zn_5(CO_3)_2(OH)_6$] from zinc carbonate (smithsonite, $ZnCO_3$) when a mixture of solids is present. Campbell has pointed out the relative unprotectiveness of smithsonite formed in some very hard groundwater, compared to basic zinc carbonate formed in softer water of higher pH.[65] This differentiation is only possible with XRD analysis.

Figure 17.29 shows a deposit present in a specimen from a lead service line removed from the Chicago, Ill., system.[52] The deposit consists almost entirely of basic lead carbonate [$Pb_3(CO_3)_2(OH)_2$, hydrocerussite] and a small amount of lead carbonate ($PbCO_3$, cerussite). The peaks from lead metal came from pieces of the pipe picked up during scale removal. Notably, peaks from calcite or aragonite are absent, indicating no protection from $CaCO_3$ deposition. This

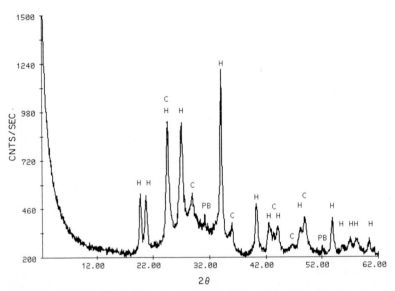

**Figure 17.29** X-ray diffraction pattern from a piece of lead service line from the Chicago, Ill., system. The deposit is a mixture of hydrocerussite [$Pb_3(CO_3)_2(OH)_2$] (P), and cerussite ($PbCO_3$) (C). Two peaks (PB) were caused by lead from the pipe that was inadvertently abraded.

is an excellent example of information not readily available using elemental analysis methods or only SEM.

Figure 17.30 is a recent analysis of a pipe specimen from a Providence, R.I., service line.[52] This is a soft water with a pH of approximately 10 and no history of a lead problem. The deposit on this pipe consists of hydrocerussite [$Pb_3(CO_3)_2(OH)_2$], as would be expected, as well as a significant fraction of another basic lead carbonate, plumbonacrite [$Pb_{10}(CO_3)_6(OH)_6O$], which has hitherto been unreported in corrosion studies. Discoveries such as this suggest that refinement of lead corrosion and solubility models would be desirable.

## Indirect Methods for the Assessment of Corrosion

### Complaint logs and reporting

Usually, customer complaints will be the first evidence of a corrosion problem in a water system.[2] The most common symptoms are listed in Table 17.6, along with their possible causes. The complaints may not always be caused by corrosion. For example, red water may also be caused by iron in the source water that is not removed in treatment. Therefore, in some cases, further investigation is necessary before attributing the complaint to corrosion in the system.

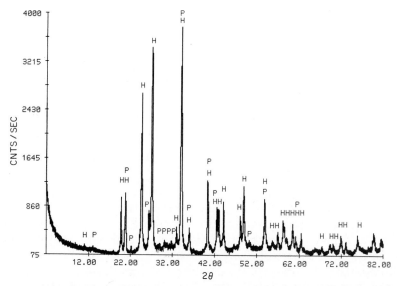

**Figure 17.30**   An x-ray diffraction pattern from a Providence, R.I., lead service line specimen, showing hydrocerussite [$(Pb_3)(CO_3)_2(OH)_2$] (H) and plumbonacrite [$Pb_{10}(CO_3)_6(OH)_6O$] (P).

**TABLE 17.6    Typical Customer Complaints Caused by Corrosion**

| Customer complaint | Possible cause |
|---|---|
| Red water or reddish-brown staining of fixtures and laundry | Corrosion of iron pipes or presence of iron in source water |
| Bluish stains on fixtures | Corrosion of copper lines |
| Black water lines | Sulfide corrosion of copper or iron |
| Foul taste or odors | By-products from microbial activity |
| Loss of pressure | Excessive scaling, tubercle building up from pitting corrosion, leak in system from pitting or other type of corrosion |
| Lack of hot water | Buildup of mineral deposits in hot water system (can be reduced by setting thermostat to under 140°F) |
| Short service life of household plumbing | Rapid deterioration of pipes from pitting or other types of corrosion |

SOURCE: From AWWA, Ref. 2.

Complaints can be a valuable corrosion monitoring tool if records of the complaints are organized. The complaint record should include the customer's name and address, date the complaint was made, and nature of the complaint. The following information should also be recorded:

1. Type of material (copper, galvanized iron, plastic, etc.) used in the customer's system

2. Whether the customer uses home treatment devices prior to consumption (softening, carbon filters, etc.)

3. Whether the complaint is related to the hot-water system and, if so, what type of material is used in the hot-water tank and its associated appurtenances

4. Any follow-up action taken by the utility or customer

These records can be used to monitor changes in water quality because of system or treatment changes.

The development of a complaint map is useful in pinpointing problem areas. The complaint map would be most useful when combined with a materials map that indicates the location, type, age, and use of a particular type of construction material. If complaints are recorded on the same map, the utility can determine if a relationship between complaints and the materials used exists. Sending questionnaires to a random sampling of customers is a useful supplement to the customer complaint records.

A long-term reporting program for pipe and equipment failures can be a valuable tool for monitoring corrosion.[53] Such a program can provide a history of corrosion problems, especially if some means of evaluating failures is in place. Little information, however, is readily available on plumbing failures in homes or buildings. Corrosion of plumbing within buildings is not brought to the attention of the water utility immediately. Few specimens of pipe are available for examination. Plumbers may not recognize corrosion as the cause for failures and attribute the cause to "bad pipe." The water utility should accumulate this information and seek the cooperation of at least one plumbing firm operating within their service area.

Additional information can be gathered for inclusion in the records, such as loss in main capacity, changes in water source, changes in water treatment procedures, and so forth. Sometimes, this information can be more important than weight-loss measurements or chemical analyses of the water supply for the monitoring of corrosion.

Customer complaint records, material service information, and questionnaires are useful monitoring tools that can be used as part of any corrosion monitoring and control program. The low costs associated with keeping a good record of complaints can be well worth the time. The resulting information would indicate the real effect of water quality at the customer's tap and would show the effect of any process changes made as part of a corrosion-control program.

### Corrosion indices

**Indices based on calcium carbonate saturation.**  The desired outcome of corrosion studies is information that will permit the choice of metals and control of water chemistry in a manner that will prevent the occurrence of deleterious effects from corrosion.[1] The extreme complexity of corrosion phenomena has prevented the development of general rules that can be used to predict when corrosion will or will not occur, except in a few limiting cases. Dissolved oxygen, pH, alkalinity, calcium, suspended solids, organic matter, buffer intensity, metal ions that can be reduced in cathodic reactions, total salt concentration, specific anions, such as chloride, sulfate, phosphate, and silicate, biological factors, and temperature have all been shown to have an effect, and many of these have different effects on the corrosion of different metals. Each of the many indices that are available have served some useful purposes, but many have been misapplied, particularly when they are used without recognition of their limitations. Many of the indices are more useful "after the fact," in helping to understand possible causes of the problem rather than being useful in a predictive sense. In general, indices based on calcium carbonate saturation have not been shown to have any significant predictive value for the corrosion and leaching of lead, zinc, and copper from brass, bronze, soldered joints, and their respective pipe materials.

Here, indices will be considered in two broad classifications: those based on the principle of calcium carbonate deposition as a means to provide corrosion control, and those based on relating corrosion to a variety of other water chemistry parameters. Because many indices are essentially the same, only expressed in different ways, only a small subset of the indices that have appeared in the literature will be considered here.

**Langelier saturation index (LSI).** The LSI is the most widely used and misused index in the water treatment and distribution field.[66] The index is based on the effect of pH on the equilibrium solubility of $CaCO_3$. The pH at which a water is saturated with $CaCO_3$ is known as the *pH of saturation,* or $pH_s$. At $pH_s$, a protective $CaCO_3$ scale should neither be deposited nor dissolved. The LSI is defined by the equation

$$LSI = pH - pH_s \tag{17.36}$$

The results of the equation are interpreted as follows:

LSI > 0    Water is supersaturated and tends to precipitate a scale layer of $CaCO_3$.

LSI = 0    Water is saturated (in equilibrium) with $CaCO_3$; a scale layer of $CaCO_3$ is neither precipitated nor dissolved.

LSI < 0    Water is undersaturated, tends to dissolve solid $CaCO_3$.

The minimum information necessary to calculate the LSI is

1. Total alkalinity
2. Calcium
3. Ionic strength, or total dissolved solids and a conversion factor to ionic strength (implied in many tables)
4. pH
5. Temperature
6. $pH_s$

The 1980 amendments to the NIPDWR require all community water supply systems to determine either the LSI or the aggressiveness index (AI) and report these values to the state regulatory agencies.[67] The AI has been critically evaluated by several researchers[14,24,29,53] and has several serious shortcomings as a deposition tendency indicator and index for A-C pipe protection. Therefore, because the AI is only a simplistic approximation of the LSI, only the LSI will be discussed.

The original derivation of the LSI depends on computing the $pH_s$ value, using the analyzed total alkalinity of the water,[29,66] and can be readily derived as follows. The basic definition of alkalinity for a pure system is

$$TALK = 2[CO_3^{2-}] + [HCO_3^-] + [OH^-] - [H^+] \qquad (17.33)$$

From Eq. (17.27), this relationship can be derived,

$$[CO_3^{2-}] = \frac{K_s'}{[Ca^{2+}]} \qquad (17.37)$$

where $K_s'$ represents the solubility constant for calcium carbonate, corrected for temperature and ionic strength.

Combining the second dissociation of carbonic acid with Eq. (17.37) yields

$$[HCO_3^-] = \frac{[H^+]K_s'}{[Ca^{2+}]K_2'} \qquad (17.38)$$

where $K_2'$ is the second dissociation constant for carbonic acid corrected for temperature and ionic strength. Substituting Eqs. (17.37) and (17.38) into Eq. (17.33), plus the relationship

$$[OH^-] = \frac{K_w'}{[H^+]}$$

produces the quadratic equation, after rearrangement,

$$0 = [H^+]^2[1 - \frac{[Ca^{2+}]K_2'}{K_s'}] + [H^+]K_2'[2 - \frac{TALK[Ca^{2+}]}{K_s'}]$$
$$+ \frac{[Ca^{2+}]K_w'K_2'}{K_s'} \qquad (17.39)$$

The hydrogen ion concentration at saturation equilibrium $[H^+]_s$ can be computed using the quadratic formula:[29]

$$[H^+]_s = \frac{-B \pm \sqrt{B^2 - 4AC}}{2A} \qquad (17.40)$$

in which

$$A = 1 - \frac{[Ca^{2+}]K_2'}{K_s'}$$

$$B = K_2'\left[2 - \frac{[Ca^{2+}]TALK}{K_s'}\right]$$

$$C = \frac{K_w'K_2'[Ca^{2+}]}{K_s'}$$

Therefore,

$$pH_s = -\log_{10}[H^+] - \log_{10}\gamma_{H^+} \tag{17.41}$$

where $\gamma_{H^+}$ is the activity coefficient for the hydrogen ion. Equations for the computation of the equilibrium constants and activity coefficients are available in several sources,[6,8,9,29–31] some of which use the best recently evaluated values for the constants.[8,30]

Rossum and Merrill point out several important properties of the computation of LSI.[29] Because it is derived from a quadratic equation, two possible solutions for $[H^+]$ and, therefore, $pH_s$ exist. The existence of two values for $pH_s$ is not generally appreciated. To obtain consistent and continuous values for LSI across the entire range of pH values while still maintaining established conventions (e.g., positive LSI for oversaturated waters), the lesser value of $pH_s$ must always be used.[29] It has also been pointed out that for pH above the bicarbonate-carbonate equivalence point (pH ~ 10.3 to 10.5), the negative root of the quadratic equation applies,[68] and the LSI definition should be reversed[31,68] to LSI = $pH_s$ – pH. This results from the fact that at high pH, more of the alkalinity comes from hydroxide. An important point is that the calculations are only valid when the pH values used were analyzed at that temperature using a temperature-compensated pH meter, or were corrected to that temperature using chemical equations.[29,30,50]

Water that has too little calcium and inorganic carbon to become saturated with $CaCO_3$ at any pH is indicated by an imaginary solution to Eq. (17.39); that is, the quantity $B^2 - 4AC$ in Eq. (17.40) is negative.[29,68]

If alkalinity-contributing species from systems other than carbonate are negligible, complexation can be ignored, and if DIC is conserved during the temperature change (frequently the case in plumbing systems when calcium carbonate does not dissolve or precipitate), the change in pH as temperature is changed can be computed.[50] Necessary measurements are pH, alkalinity (equivalents per liter), and ionic strength. DIC (moles per liter) must be computed from these measurements or directly analyzed. Then the equilibrium constants $K_1'$, $K_2'$, and $K_w'$ are recalculated for the different temperature of interest. The hydrogen ion concentration at the changed temperature

$[H^+]_T$ is determined by solving for the single positive real root of the quadratic equation

$$A[H^+]_T^4 + B[H^+]_T^3 + C[H^+]_T^2 + D[H^+]_T + E = 0 \qquad (17.42)$$

in which

$A = 1$
$B = K_1' + \text{TALK}$
$C = -[(K_w' - K_1'K_2') + K_1' \, (\text{DIC} - \text{TALK})]$
$D = -K_1'\{K_w' + [K_2'(2\text{DIC} - \text{TALK})]\}$
$E = -K_1'K_2'K_w'$

This enables the calculation of the LSI or other saturation or deposition index for a water after it has been warmed or cooled, or it can enable the estimation of the in situ pH of a sample measured in the laboratory.

The LSI can also be derived following a slightly different path.[7,29,30] By this route, $[H^+]_s$ is found by writing the calcium carbonate dissolution equation as

$$\text{CaCO}_3(s) + H^+ \rightleftharpoons \text{Ca}^{2+} + \text{HCO}_3^- \qquad (17.43)$$

which corresponds to the equilibrium constant expression of Eq. (17.38).

Solving for $[H^+]_s$ gives

$$[H^+]_s = \frac{[\text{HCO}_3^-][\text{Ca}^{2+}]K_2'}{K_s'} \qquad (17.44)$$

or

$$\text{pH}_s = -\log\left(\frac{K_2'}{K_s'}\right) - \log[\text{Ca}^{2+}] - \log[\text{HCO}_3^-]$$

Here, analyzed values for calcium and bicarbonate can be substituted into the equation, along with the equilibrium constants corrected for ionic strength and temperature. A difficulty with this approach is that either it must be assumed that $[\text{HCO}_3^-] = \text{TALK}$, which is valid for many waters, or $[\text{HCO}_3^-]$ must be computed from a suitable alkalinity expression, Eq. (17.33) or Eq. (17.34), using appropriate concentrations and constants.

When conversion factors are added, and the temperature and ionic strength corrections are incorporated into Eq. (17.44), Eq. (17.45) can be derived:

$$pH_s = A + B - \log[Ca^{2+}] - \log[HCO_3^-] \qquad (17.45)$$

as is given in the *Federal Register*.[30,67] Corrections to the constant terms in Eq. (17.44) for improvements in equilibrium constant values (constant $A$) and activity coefficient calculation (constant $B$) have been presented[30] to supercede previously published values.[2,67] In this equation, the $Ca^{2+}$ and $HCO_3^-$ concentrations are expressed as milligrams $CaCO_3$ per liter.

The bicarbonate concentration in Eq. (17.45) must be that at the temperature of computation. Though total alkalinity is conserved,[6] that is, it does not change as temperature varies providing there is no gain or loss of DIC, the bicarbonate concentration is not, because $K_w$, $K_1$, and $K_2$ change with the temperature. Therefore, to use equations such as Eq. (17.45), or others using nonconservative species, the necessary concentrations must be adjusted for the temperature change.

Several problems exist with the LSI as a corrosion index:[1,30,38]

1. Complexation of $Ca^{2+}$ and $HCO_3^-$ is not accounted for, although this is possible if the needed analytical data are available. In the presence of polyphosphates, the equations defining $pH_s$ will overestimate calcium carbonate saturation unless correction factors are added to account for the complexation.

2. The crystalline form of $CaCO_3(s)$ has usually been assumed to be calcite. The presence of another form of $CaCO_3(s)$, aragonite, which has a higher solubility, has, however, been observed in several systems. The formation of other forms of $CaCO_3(s)$ may account for some of the observations of substantial supersaturation with respect to calcite.

3. A deposit of $CaCO_3(s)$ does not necessarily aid in preventing corrosion.

4. $CaCO_3$, if present in high enough concentrations, can also be deposited from waters with a negative LSI because of the localized high pH next to the pipe, which is generated by the cathodic reactions.

5. The preoccupation of many with maintaining a positive LSI has led to excessive deposition of $CaCO_3(s)$ and to significant decreases in the capacity of distribution systems to carry water.

Analysis of many protective scales has shown that many types of solids other than $CaCO_3(s)$ are present that provide resistance to corrosion, as reviewed by Sontheimer, Kolle, and Snoeyink for iron and steel,[16] and by many other types of pipe materials. Further, many examples exist of the failure of the LSI to predict corrosivity. Water with a negative LSI can be noncorrosive, although some $Ca^{2+}$ appears to be necessary for the deposit to be protective on iron and steel materials.[26] When corrosion inhibitors are used, or when the dissolution of metals such as lead or copper is of concern, then other water chemistry factors must be considered.

While the LSI predicts whether $CaCO_3(s)$ should precipitate or dissolve, it does not predict how much $CaCO_3(s)$ will precipitate or whether its structure will provide resistance to corrosion. Larson showed, for example, that an LSI of 0.9 is necessary to precipitate 10 mg $CaCO_3$/L at an alkalinity of 50 mg $CaCO_3$/L, but an LSI of only 0.2 is necessary for the same amount of precipitation if the alkalinity is 200 mg/L.[19]

Precipitation of too much $CaCO_3(s)$ could result in reduction of the water-carrying capacity of the main, so the calculation of the quantity of $CaCO_3$ that will precipitate is important. This quantity is also called the *calcium carbonate precipitation potential* (CCPP). Procedures have been given in several sources for these calculations.[8,29,31,32]

The equation for the CCPP is[29]

$$CCPP = 50,000(TALK_i - TALK_{eq}) \qquad (17.46)$$

in units of milligrams of $CaCO_3$ per liter. This is because during $CaCO_3$ precipitation, the equivalents of calcium precipitated must be equal to the equivalents of alkalinity precipitated.[31] The acidity (ACY) of such a system, however, is conserved in the process, so $ACY_i = ACY_{eq}$.[6,7,29,31,32]

The initial acidity of the system may be computed from the relationship[29]

$$ACY_i = \left(\frac{TALK_i + s_i}{t_i}\right)p_i + s_i \qquad (17.47)$$

where

$$p = \frac{2[H^+]_i + K_1'}{K_1'} \qquad (17.48)$$

$$s = [H^+]_i - \frac{K_w'}{[H^+]_i} \qquad (17.49)$$

$$t = \frac{2K_2' + [H^+]_i}{[H^+]_i} \qquad (17.50)$$

The alkalinity after precipitation when equilibrium is reached can then be related to the initial acidity,[29]

$$\text{TALK}_{eq} = \frac{t_{eq}}{p_{eq}}(\text{ACY}_i - s_{eq}) - s_{eq} \tag{17.51}$$

where the terms $t_{eq}$, $p_{eq}$, and $s_{eq}$ correspond to Eqs. (17.48), (17.49), and (17.50), with $[H^+]_i$ replaced by $[H^+]_{eq}$.

The alkalinity at equilibrium can also be related to the initial calcium concentration and alkalinity through the equation

$$2[\text{Ca}^{2+}]_i - \text{TALK}_i = \frac{2K_s'r_{eq}}{\text{TALK}_{eq} + s_{eq}} - \text{TALK}_{eq} \tag{17.52}$$

in which

$$r_{eq} = \frac{[H^+]_{eq} + 2K_2'}{K_2'} \tag{17.53}$$

Combining Eqs. (17.51) and (17.52) gives the relationship[29]

$$2[\text{Ca}^{2+}]_i - \text{TALK}_i = \frac{2K_s'r_{eq}p_{eq}}{t_{eq}(\text{ACY}_i - s_{eq})} - \frac{t_{eq}(\text{ACY}_i - s_{eq})}{p_{eq}} + s_{eq} \tag{17.54}$$

The values for $[\text{Ca}^{2+}]_i$ and $\text{TALK}_i$ are obtained by chemical analysis. $\text{ACY}_i$ can be computed by Eq. (17.47). The terms $p_{eq}$, $r_{eq}$, $s_{eq}$, and $t_{eq}$, are functions of $[H^+]_{eq}$, which must be obtained through an iterative trial-and-error solution of Eq. (17.54).[29] After the calculation of $[H^+]_{eq}$, $\text{TALK}_{eq}$ can then be derived, using Eq. (17.51), and substituted along with $\text{TALK}_i$ into Eq. (17.46).

A positive CCPP denotes oversaturation and the mg $CaCO_3$/L of calcium carbonate that should precipitate. A negative CCPP indicates undersaturation and how much $CaCO_3$ should dissolve.

Several mathematical strategies exist that would be appropriate for solving Eq. (17.54) for $[H^+]_{eq}$ with a programmable calculator or computer.[9,29] The equation can also be solved relatively quickly by using simple manual trial-and-error substitution with PC spreadsheet software. One caution in solving the equation is that the quantity $\text{ACY}_i - s_{eq}$ in Eq. (17.54) must always be greater than zero to give a physically meaningful answer, and it should be constantly checked while solving for $[H^+]_{eq}$.[29]

Note the distinction between $[H^+]_s$ used with the Langelier index and $[H^+]_{eq}$ used with CCPP. The quantity $[H^+]_s$ represents the hydrogen ion concentration if a water of a specific composition was at equilibrium with calcium carbonate. That is different from $[H^+]_{eq}$, which is the final hydrogen ion concentration that would occur after the initial water either precipitated or dissolved $CaCO_3$ to attain saturation equilibrium.

The amount of calcium carbonate (or other solid phase) that would dissolve or precipitate for a given water may also be computed by sophisticated geochemical modeling computer codes, such as PHREEQE[69] or MINTEQ,[70] that can better account for aqueous ion pairing and complexation reactions and for ionic strength changes.

Rossum has discussed the usefulness of the CCPP and its application to some difficult red water problems.[71] He found that the amount of calcium carbonate formed by an increment of hydroxide from the cathode reaction may be an important factor in the prevention of red water. CCPP is the actual degree of calcium carbonate supersaturation (if positive) or undersaturation (if negative) in milligrams per liter. If hydroxide is also expressed in milligrams as calcium carbonate per liter, then CCPP-OH values greater than 1 indicate that more than one molecule of a $CaCO_3$ is formed for each molecule of hydroxide generated. Figure 17.31 is that used by Rossum to illustrate the CCPP-OH relationship for different alkalinities and pH values, assuming calcium concentrations to represent equilibrium with calcite.[71] At high pH values, so much hydroxide is present that small increments of hydroxide have little effect, and at low pH values so much carbonic acid is present that increments of hydroxide form bicarbonate rather than carbonate. At a given pH value, carbonic acid concentrations increase as alkalinity increases.

To simulate the addition of 1 mg $OH^-$ as $CaCO_3$/L, the alkalinity of the water is increased by 1 and the acidity is decreased by 1. From these adjusted values, a new pH value and CCPP can be calculated.[29] If the result is less than 1, red water problems are likely. For many waters in which the Langelier index is close to zero, the value of CCPP-OH may be estimated with sufficient accuracy from Fig. 17.31.

Because $CaCO_3(s)$ is frequently found in protective scales on iron and steel, control of water chemistry using the LSI is usually a reasonable and economical corrosion-control strategy. Even if the films are not $CaCO_3$, these water conditions can often help with the formation of protective scales, such as mixtures of $CaCO_3$, $FeCO_3$, and iron hydroxides or oxides.

In general, the higher the calcium, alkalinity, and pH, the less corrosive a water will be, and higher values are associated with more positive values of LSI.[1] A positive LSI, however, is not necessarily required.

Measurements of the change in pH, calcium, and alkalinity concentrations through segments of distribution systems should be made to monitor the quantity of $CaCO_3(s)$ that precipitates. These measure-

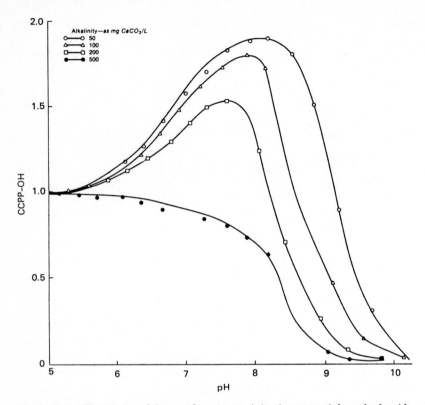

**Figure 17.31**   Change in calcium carbonate precipitation potential per hydroxide versus pH (temperature = 25°C, ionic strength = 0.01, and Langelier index = 0). *(Source: J. R. Rossum, "Dead Ends, Red Water, and Scrap Piles," J. AWWA, vol. 79, no. 7, 1987, p. 113.)*

ments, combined with visual observations of the scale on pipe segments taken from different parts of the system should permit adjustment of the water quality so that enough, but not too much, $CaCO_3(s)$ will precipitate.[1]

The LSI should be used only as one of several pieces of data to indicate corrosivity or noncorrosivity. It should be supplemented with observations of pipe that has been in use, with analytical data on dissolved or particulate corrosion products, consideration of material-specific chemical factors, and with pipe loop studies, if possible.

A comparison of the different scale formation tendencies that can be obtained by using three different approaches illustrates the relative utility of some of the different indices. Consider a water having the following characteristics:

pH 8.35

Temperature 25°C

Ionic strength 0.012

TDS 480 (converted from $I$ using Langelier's estimate[30,66])

Calcium 40 mg/L as Ca or 100 mg/L as $CaCO_3$

Magnesium 25 mg/L

Sodium 120 mg/L

Total alkalinity 100 mg/L (as $CaCO_3$)

Sulfate 150 mg/L

Chloride 163 mg/L

Computations will be made of the LSI[30,31,68] and the CCPP[29,31] using common equations, and these results will be compared to the results of the application of the PHREEQE computerized equilibrium chemical speciation model.[69]

The simplest approach to the calculation of the LSI is through the use of Eq. (17.45). Because the pH is not within approximately 1 pH unit of $pK_1$ or $pK_2$ for carbonic acid (Fig. 17.22), and there is not a significant concentration of other species that could add to alkalinity (such as ammonia, hypochlorite, etc.), essentially all of the alkalinity can be assumed to be bicarbonate. This assumption should be correct within several percent. Inserting these values into the equation, along with the interpolated constants $A$ for calcite and $B$,[30] gives

$$pH_s = 1.8479 + 9.93 - 2 - 2 = 7.778$$

$$LSI = 8.35 - 7.778 = 0.57$$

Computation via the quadratic equation of Rossum and Merrill (Eq. (17.39)) requires several initial computations.[29] The equilibrium constants $K_w'$, $K_2'$, and $K_s'$ must be computed for the temperature and ionic strength of the water system. The ionic strength is most accurately computed from a complete water analysis, as has been shown in many references.[5-9,31,50] Though this approach does not take into account ion pairs and complexes, the impact of ionic strength in the $pH_s$ calculation is usually such that this neglect is inconsequential. In the absence of a complete water analysis, $I$ can be estimated from an empirically determined relationship with the analyzed conductivity or TDS. Ideally, this should be determined for each water supply situation. For some waters, Langelier proposed the approximate relationship[66]

$$I = 2.5 \times 10^5 (\text{TDS})$$

This is widely used today, although other waters usually give relationships that can differ from this one by $\pm$ 50 percent. For consistency, this relationship will be used in this example, and the analyzed TDS will be assumed to be 480 mg/L.

The activity coefficients of monovalent $(f_m)$ and divalent ions $(f_d)$ are computed as

$$\log_{10} f_m = -0.049$$

$$\log_{10} f_d = -0.197$$

from Rossum and Merrill.[29] Then, the ionic strength corrections can be applied to the equilibrium constants $K_2$, $K_w$, and $K_s$, giving

$$K_2' = 7.368 \times 10^{-11}$$

$$K_w' = 1.263 \times 10^{-14}$$

$$K_s' = 9.413 \times 10^{-9}$$

From these and the given analytical input, which converts to the following concentrations:

$$[H^+] = 5.000 \times 10^{-9} \text{ mol/L}$$

$$[Ca^{2+}] = 9.980 \times 10^{-4} \text{ mol/L}$$

$$\text{TALK} = 1.998 \times 10^{-3} \text{ equiv/L}$$

the constant terms for the quadratic equation [Eq. (17.39)] are

$$A = 9.9999 \times 10^{-1}$$

$$B = -1.5463 \times 10^{-8}$$

$$C = 9.8651 \times 10^{-20}$$

Solving the equation via the quadratic formula [Eq. (17.40)] yields $[H^+]_s = 6.3825 \times 10^{-12}$ mol/L and $1.5457 \times 10^{-8}$ mol/L. The latter is the appropriate physically meaningful root for this case. Multiplying the hydrogen ion concentration by its activity coefficient to get the hydrogen ion activity and taking the negative logarithm to get $pH_s$ gives

$$pH_s = -\log([H^+]_s f_m)$$

$$= -\log[(1.5457 \times 10^{-8})(0.893)]$$

$$= 7.86$$

Therefore, LSI = 8.35 − 7.86 = 0.49. The difference between this value and the 0.57 achieved with the abbreviated formula [Eq. (17.44)] is partly the result of differences in equilibrium constants and activity coefficients and partly the result of assuming all TALK to be composed of bicarbonate ion. Using the same equilibrium constants $(K_w', K_s', K_2')$ as Schock,[30] but activity coefficient calculations from Rossum and Merrill,[29] gives LSI = 0.55.

Calculation of the saturation index for calcite using the PHREEQE computer program,[69] which should be close to the LSI and which will be discussed later, gave 0.47. The closeness of this result to that of Eq. (17.39) is somewhat fortuitous and misleading because the PHREEQE model indicates that approximately 13 percent of the calcium assumed to be available in Eqs. (17.39) and (17.45) is actually bound in carbonate, bicarbonate, and sulfate ion pairs and complexes, and almost about 51 percent of the carbonate ion is similarly tied up. Without the ion pairing correction, the SI would be higher by approximately 0.05 unit. If the equilibrium constants and ionic strength corrections were the same across all three methods, there would be a systematic high bias in Eqs. (17.39) and (17.45) relative to the PHREEQE computation given the use of identical thermodynamic constants.

Further modeling with PHREEQE shows a CCPP of 6.5 mg $CaCO_3$/L and ultimate saturation concentrations of

$$[Ca^{2+}] = 8.21 \times 10^{-4} \text{ mol/L} = 32.9 \text{ mg/L}$$

$$TALK = 1.87 \times 10^{-3} \text{ equiv/L} = 93.6 \text{ mg } CaCO_3/L$$

$$[H^+] = 1.36 \times 10^{-8} \text{ mol/L}$$

$$pH_s = 7.92$$

Assuming the same ionic strength, but not correcting for sulfate ion pairing, would give a CCPP = 7.4 mg $CaCO_3$/L.

The CCPP may also be calculated by using Eqs. (17.46), (17.51), and (17.54). The equilibrium constants corrected for ionic strength using the same equations as in the previous example[30,69] give

$$K_1' = 5.5461 \times 10^{-7}$$

$$K_2' = 7.3325 \times 10^{-11}$$

$$K_s' = 8.0980 \times 10^{-9}$$

$$K_w' = 1.2655 \times 10^{-14}$$

Activity coefficients computed for monovalent and divalent ions were computed to be 0.8943 and 0.6396 ($\log f_m = -0.0485$; $\log f_d$

= −0.1941), and for uncharged species,[69] it was 0.9972 ($\log f_d$ = 0.0012). Using these values, the terms for Eqs. (17.47) to (17.50) and their saturation equilibrium analogs were computed to be

$$ACY_i = 1.9712 \times 10^{-3} \text{ equiv/L}$$

$$p = 1.0180$$

$$s = -2.5286 \times 10^{-6}$$

$$t = 1.0294$$

$$p_{eq} = 1.0573$$

$$s_{eq} = -7.8088 \times 10^{-7}$$

$$t_{eq} = 1.0092$$

$$r_{eq} = 2.1861 \times 10^{2}$$

Substituting these quantities into Eq. (17.54) along with analytical values solving by trial and error with an electronic spreadsheet on a personal computer yielded $[H^+]_{eq} = 1.5883 \times 10^{-8}$ mol/L, corresponding to $pH_{eq} = 7.848$. By Eq. (17.51), $TALK_{eq} = 1.8831 \times 10^{-3}$ eq/L. Adding Eq. (17.46) gives the final result:

$$CCPP = 50,000(1.9982 \times 10^{-3} - 1.8831 \times 10^{-3})$$

$$= 5.76 \text{ mg CaCO}_3\text{/L}$$

The agreement with the calculations done by the more complete geochemical equilibrium modeling program[69] is reasonably good, considering the differences in assumptions behind the two approaches.

As higher amounts of carbonate, sulfate, and bicarbonate are present, correction of the LSI or CCPP for ion pairing and complexation becomes more important. However, because of analytical variability and error, and the many complicating factors involved with calcium carbonate film formation, indices such as these must not be interpreted too literally, and they do not have quantitative significance to any great degree of precision.

**Ryznar index.**    The Ryznar (saturation) index (RSI), defined as[2,31,32,72]

$$RSI = 2pH_s - pH \tag{17.55}$$

was developed from empirical observations or corrosion rates and film formation in steel mains and heated water in glass coils. An RSI between 6.5 and 7.0 is considered to be approximately at saturation equilibrium with calcium carbonate. A RSI > 7.0 is interpreted as undersaturated and, therefore, would tend to dissolve any existing solid $CaCO_3$. Waters with RSI < 6.5 would tend to be scale-forming.

This index does not have any particular theoretical justification other than that from mathematical and chemical considerations. It tends to favor waters of higher hardness and alkalinity that would naturally have a greater potential to deposit calcium carbonate if their pH exceeded their $pH_s$. Thus, in that respect it is somewhat consistent with the observations of calcium carbonate deposition potential described by Merrill and Sanks[32] and Loewenthal and Marais.[31] A notable internal inconsistency of the RSI is that the value for saturation equilibrium varies with the $pH_s$ of the water; for example, if $pH_s = 7.0$, the RSI for saturation equilibrium is 7, but if $pH_s = 9$, the RSI for saturation equilibrium is 9. Therefore, the interpretation of the index must be adjusted with the $pH_s$. Although commonly used, the RSI does not offer any tangible advantages to a variety of other methods for computing the calcium carbonate saturation state and deposition potential.

The buffer intensity β or capacity, as discussed in a previous section, has also been suggested as a useful corrosion indicator. The effect of buffer intensity seems to be closely linked with $CaCO_3$ precipitation.

Figure 17.31 suggests that high values of buffer capacity do not necessarily inhibit corrosion,[73] because for any given pH the highest alkalinity has the highest buffer capacity, but it shows the least favorable value of CCPP-OH. Frequently, scanty evidence of a thin, coherent, protective scale of calcium carbonate exists—even for those systems in which a positive Langelier index has been maintained. The benefit of having a CCPP-OH greater than 1 appears to lie in the ability of calcium carbonate to strengthen and harden iron corrosion products so that they remain on the pipe wall and do not cause red water.[71] No assurance can be given that corrosion rates based on the loss of weight of pipe coupons would correspond to those based on red water, because the protective effect of siderite and, to a lesser extent, calcite, depends on some corrosion having already occurred.[71]

The *calcium carbonate buffer intensity* ($\beta^s_{alk}$) has been discussed as another potentially useful measure.[8] This index gives an estimate of the sensitivity of the calcium carbonate saturation of the solution to changes in alkalinity. The index is[8]

$$\beta^s_{alk} = \left[\frac{K_s'}{DIC[Ca^{2+}]K_1'K_2'}\right]\left[\frac{F}{2[H^+] + K_1'}\right]$$

$$\times \left[\frac{DIC\, K_1'([H^+]^2 + 4K_2'[H^+] + K_1'K_2')}{F^2}\right] + \left[\frac{K_w'}{[H^+]^2}\right] + 1 \quad (17.56)$$

where

$$F = [H^+]^2 + [H^+]K_1' + K_1'K_2' \qquad (17.57)$$

and DIC is in mol/L.

Higher values of this buffer intensity are associated with smaller changes in calcium carbonate saturation, for a given external change in alkalinity (for example, by changes in solution composition near an electrochemical corrosion microcell). Therefore, at cathodic sites where hydrogen ion is being drawn from solution, local super-saturations of $CaCO_3$ may be expected,[8] thus reducing the protective effect of calcium carbonate scale formation. A graph of this equation is shown in Fig. 17.32 for a water having 80 mg Ca/L and DIC = 24 mg C/L at 25°C.

In general, scale formation would be enhanced by regions in which $\beta_{alk}^s$ is small, which would be where DIC is larger or $[Ca^{2+}]$ is larger.[8] This approach parallels the advice given by Merrill and Sanks in defining a water with good scale-formation characteristics.[32]

**Other corrosion indices.**  A general *saturation index* (sometimes called a *disequilibrium index*), SI, can be defined for any solid solubility reaction as

$$SI_x = \log_{10}\left(\frac{IAP_x}{K_x}\right) \qquad (17.58)$$

where $IAP_x$ and $K_x$ are the ion activity product and solubility product constant, respectively, for mineral $x$.

For example, for calcium carbonate (calcite) dissolving

$$CaCO_3(s) \rightleftharpoons Ca^{2+} + CO_3^{2-} \qquad (17.27)$$

the expression for $SI_{calcite}$ would be

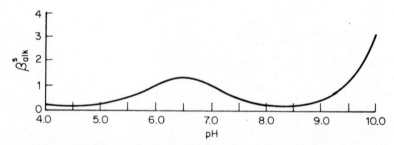

**Figure 17.32**  $CaCO_3$ saturation buffer intensity. *(Source: Water Treatment Principles and Design. James M. Montgomery Consulting Engineers, Inc., Copyright © 1985, John Wiley & Sons, Inc.)*

$$SI_{\text{calcite}} = \log_{10}\left[\frac{\{Ca^{2+}\}\{CO_3^{2-}\}}{K_s}\right] \tag{17.59}$$

The braces represent activities rather than concentrations. For a solid such as the galvanized pipe corrosion product hydrozincite, dissolving in the manner

$$Zn_5(CO_3)_2(OH)_6(s) \rightarrow 5Zn^{2+} + 2CO_3^{2-} + 6OH^- \tag{17.60}$$

then

$$SI_{\text{HZ}} = \log_{10}\left[\frac{\{Zn^{2+}\}^5\{CO_3^{2-}\}^2\{OH^-\}^6}{K_s}\right] \tag{17.61}$$

As with the LSI, SI = 0 is equilibrium saturation. Positive values represent oversaturation, and negative values represent undersaturation. Reactions can be written in other forms, because $K_s$ will change, depending on how the dissolution reaction will be defined. These indices are directly related to the free-energy driving force for the reactions. Indices of these types are useful in attempting to determine if passivating films are being formed on pipe,[14,17] or if a hypothetical final concentration of an added constituent would result in film formation. Very good corrections can be made for the effects of ion pairing of carbonate, sulfate, and so on, ionic strength, and temperature by using computerized equilibrium speciation models.[69,70]

Studies on the effects of $Cl^-$ and $SO_4^{2-}$ on iron and mild steel corrosion found that the effects are primarily related to their concentration relative to $HCO_3^-$.[40] The relationships

$$LI_1 = \frac{2[SO_4^{2-}] + [Cl^-]}{[HCO_3^-]} \tag{17.62}$$

and

$$LI_2 = \frac{[Cl^-]}{[HCO_3^-]} \tag{17.63}$$

where the brackets indicate concentration in mol/L, were suggested by work of Larson and Skold.[40] Observations of corrosion in water of different quality[9] showed the rate of corrosion of mild steel to increase significantly if $LI_2$ was greater than 0.1 to 0.2. Further research is needed to fully establish the effect of $Cl^-$ and $SO_4^{2-}$, however, on iron and even on other materials.

Attempts have been made to develop corrosion indexes by correlating corrosivity to all water quality parameters that might affect corrosion. One such effort by Riddick[1,73] yielded the following index:

$$RI = \frac{75}{ALK}\left[ CO_2 + 0.5(\text{hardness} - ALK) + Cl^- + 2NO_3^-\left(\frac{10}{SiO_2}\right)\frac{(DO + 2)}{DO_{sat}}\right]$$

$$(17.64)$$

where hardness and alkalinity are in milligrams of $CaCO_3$ per liter, $NO_3^-$ is in milligrams of nitrogen per liter, and the remaining parameters are in milligrams per liter. Values of less than 25 indicate noncorrosive water, 26 to 50 indicate moderately corrosive water, 51 to 75 corrosive water, and greater than 75 very corrosive water. This index was developed using data for the soft waters of the northeastern United States and was successfully applied to water from this area but not to higher hardness waters.[73]

Pisigan and Singley conducted a variety of computations and laboratory corrosion tests on mild steel in order to develop a predictive equation for the corrosion rate and to refine the Langelier index to include a correction for significant ion pair and complex species while maintaining a single equation that could be directly calculated.[74] Their suggested equation for the equation to estimate $pH_s$ was

$$pH_s = 11.017 + 0.197\log(TDS) - 0.995\log(Ca^{2+})_T - 0.016\log(Mg^{2+})_T$$

$$- 1.041\log(TALK)_T + 0.021\log(SO_4^{2-})_T \quad (17.65)$$

where the subscript $T$ denotes the total analytical concentration in milligrams per liter (milligrams of $CaCO_3$ per liter for TALK). A statistical analysis of the experimental test systems produced several predictive equations, using 4, 8, and 13 variables. Approximately 98 percent of the variability in the corrosion rates of mild steel (CR) observed in the jar tests could be explained by using the expressions

$$CR(4) = \frac{(TDS)^{0.253}(DO)^{0.820}}{[10^{SI}]^{0.0876}(\text{days})^{0.373}} \quad (17.66)$$

or

$$CR(8) = \frac{(Cl)^{0.509}(SO_4^{2-})^{0.0249}(TALK)^{0.423}(DO)^{0.780}}{(Ca^{2+})^{0.676}(\beta)^{0.0304}(\text{days})^{0.381}(10^{LSI})^{0.107}} \quad (17.67)$$

where $\beta$ is the buffer intensity, days is the exposure time, DO is dissolved oxygen concentration in milligrams per liter, and CR is in mpy. Thus far, neither of these models has been extensively tested by comparison of data to field systems.

An unsuccessful attempt to develop a similar kind of relationship by relating corrosion rates or corrosion product levels to water quality was reported by Neff et al.[56] They found that too many uncontrolled variables existed to enable the extraction and differentiation of chemical relationships from the field data.

## Corrosion-Control Alternatives

The complete elimination of corrosion is difficult, if not impossible. Several ways exist, however, to reduce or inhibit corrosion that are within the capability of most water utilities. Corrosion depends on both the specific water quality and pipe material in a system;[57] a particular method may be successful in one system and not in another.

Corrosion is caused by a reaction between the pipe material and the water in direct contact with each other. Consequently, three basic approaches to corrosion control exist:[2]

1. Modify the water quality so that it is less corrosive to the pipe material.
2. Place a protective barrier or lining between the water and the pipe.
3. Use pipe materials and design the system so that it is not corroded by a given water.

The most common ways of achieving corrosion control are to

1. Properly select system materials and adequate system design.
2. Modify water quality.
3. Use inhibitors.
4. Provide cathodic protection.
5. Use corrosion-resistant linings, coatings, and paints.

### Materials selection

In many cases, corrosion can be reduced by properly selecting system materials and having a good engineering design.[2] Some pipe materials are more corrosion resistant than others in a specific environment. In general, the less reactive the material is with its environment, the more resistant the material is to corrosion. When replacing old lines or putting new lines in service, the utility should select a material that will not corrode in the water it contacts. This provides a limited solution, because few utilities can select materials based on corrosion resistance alone. Usually several alternative materials must be compared and evaluated based on cost, availability, use, ease of installation, and maintenance, as well as resistance to corrosion. In addition, the utility owner may not have control over the selection and installation of the materials for household plumbing. Several guidelines, however, can be used in selecting materials.[2]

Some materials are known to be more corrosion resistant than others in a given environment. For example, a low-pH water that contains high dissolved oxygen levels will cause more corrosion damage in a copper pipe than in a concrete or cement-lined cast iron pipe. Changes in alloy composition within the same general classification (such as "brass") may result in considerable differences in resistance to chemical attack by a given water type. Other guidelines relating water quality to material selection are given in Table 17.1. A good description of the proper selection of materials can be found in *The Prevention and Control of Water-Caused Problems in Building Potable Water Systems,* published by the National Association of Corrosion Engineers.[75]

Compatible materials should be used throughout the system. Two metal pipes having different activities, such as copper and galvanized iron, that come in direct contact with others can set up a galvanic cell and cause corrosion. Particular problems with elevated lead levels resulting from the partial replacement of lead plumbing with copper have been reported in the United Kingdom.[11] As much as possible, systems should be designed to use the same metal throughout or to use metals having a similar position in the galvanic series (Table 17.3). Galvanic corrosion can be avoided by placing dielectric (insulating) couplings between dissimilar metals.

The possible long-term health and economic consequences of replacing materials must be considered. For example, many of the lead-free brasses available have a significant arsenic content. Whether that is leachable under normal conditions of usage has not been studied extensively.

### Engineering considerations

The design of the pipes and structures can be as important as the choice of construction materials. A faulty design may cause severe corrosion, even in materials that may be highly corrosion resistant. Some of the important design considerations include[2]

- Avoiding dead ends and stagnant areas
- Using welds instead of rivets
- Providing adequate drainage where needed
- Selecting an appropriate flow velocity
- Selecting an appropriate metal thickness
- Eliminating shielded areas

- Reducing mechanical stresses
- Avoiding uneven heat distribution
- Avoiding sharp turns and elbows
- Providing adequate insulation
- Choosing a proper shape and geometry for the system
- Providing easy access to the structure for periodic inspection, maintenance, and replacement of damaged parts
- Eliminating grounding of electrical circuits to the system

Many plumbing codes are outdated and allow undesirable situations to exist. Such codes may even create problems. Where such problems exist, working with the responsible government agency to modify outdated codes may be helpful.

### Chemical treatment

In many cases, the easiest and most practical way to make a water noncorrosive is to modify the water quality at the treatment plant. Because of the differences among source waters, the effectiveness of any water quality modification technique will vary widely from one water source to another. Where applicable, however, water quality modification can often result in an economical method of corrosion control.

**pH adjustment.** The adjustment of pH is the most common method of reducing corrosion in water distribution systems. pH plays a critical role in corrosion control for several reasons.

1. Hydrogen ions ($H^+$) act as electron acceptors and enter readily into electrochemical corrosion reactions. Acid waters are generally corrosive because of their high concentration of hydrogen ions. When corrosion takes place below pH 6.5, it is generally uniform corrosion. In the range between pH 6.5 and 8.0, the type of attack is more likely to be pitting.[2]

2. pH is the major factor that determines the solubility of most pipe materials. Most materials used in water distribution systems (copper, zinc, iron, lead, and cement) dissolve more readily at a lower pH. Increasing the pH increases the hydroxide ion ($OH^-$) concentration, which, in turn, decreases the solubility of metals that have relatively insoluble hydroxides, basic carbonates and oxides, including copper, zinc, iron, and lead. When carbonate alkalinity is present, increasing the pH, up to a point, increases the amount of carbonate ion in solution. This may control the solubility of metals that have insoluble ba-

sic carbonates, such as lead and copper. The cement matrix of A-C pipe or cement-lined pipe is also more soluble at a low pH. Increasing the pH is a major factor in limiting the dissolution of the cement binder and thus controlling corrosion in these types of pipes.

3. The relationship between pH and other water quality parameters, such as DIC, alkalinity, carbon dioxide ($CO_2$), and ionic strength, governs the solubility of calcium carbonate ($CaCO_3$), which is commonly used to provide a protective scale on interior pipe surfaces. To deposit this protective scale, the pH of the water must be slightly above the pH of saturation for $CaCO_3$, provided sufficient DIC and calcium are present. pH adjustment alone is often insufficient to control corrosion in waters that are low in carbonate or bicarbonate alkalinity. A protective coating of $CaCO_3$, for instance, will not form unless a sufficient number of carbonate and calcium ions are in the water.

Some metals, notably lead and copper, can form a layer of an insoluble carbonate or, more probably, basic carbonate [e.g., $Pb_3(CO_3)_2(OH)_2$, $Pb_{10}(CO_3)_6(OH)_6O$, $Cu_2(CO_3)(OH)_2$], minimizing corrosion rates and the dissolution of these metals. In very low alkalinity waters, carbonate ion must be added to form these insoluble carbonates. For such waters, soda ash ($Na_2CO_3$) or sodium bicarbonate ($NaHCO_3$) are the preferred chemicals generally used to adjust pH because they also contribute carbonate ($CO_3^{2-}$) or bicarbonate ions ($HCO_3^-$). The number of carbonate ions available is a complex function of pH, temperature, and other water quality parameters. Bicarbonate alkalinity can be converted to carbonate alkalinity by increasing the pH. If carbonate supplementing is necessary to control corrosion in a water system, pH also must be carefully adjusted to ensure that the desired result is obtained. Extreme caution must be employed when using carbonate supplementation for metal solubility control, however. Lead, copper, and zinc form strong aqueous carbonate complexes that increase metal solubility when DIC exceeds a critical level at a given pH.[10,11,14,17,33,38] The exact value of this critical level depends to some degree on many factors, such as ionic strength, temperature, and the presence of competing cations and complex-forming ligands, in addition to the pH. The level must be computed by using the appropriate chemical equilibrium equations to determine the metal solubility for the given water quality.

The proper pH for any given water distribution system is so specific to its water quality and system materials that only general guidance can be provided.[2] If the water contains a moderate amount of carbonate alkalinity and hardness (approximately 40 mg/L as $CaCO_3$ or more of carbonate or bicarbonate alkalinity and calcium hardness), the utility should first calculate the LSI or CCPP to determine at what pH the water is stable with regard to $CaCO_3$. To start, the pH of the water should be adjusted such that the LSI is slightly positive, no more than 0.5 unit above the $pH_s$, or the CCPP should be adjusted to a slightly positive value. If no other evidence is available, such as a good history of the effect of pH on the laying down of a protective coating of $CaCO_3$ or laboratory or field test results, then the LSI and/or CCPP provide a good starting point. Maintenance of the pH above the $pH_s$; that is, a positive LSI, should cause a protective mixed-coating to develop in cast-iron mains. If no coating forms, then the pH should be increased another 0.1 to 0.2 unit, or the CCPP should be raised, until a coating begins to form. Watching the pressure in the system carefully is important because too much scale buildup near the plant could seriously clog the transmission lines. Many systems that employ polyphosphate sequestrants or crystallization poisoners must treat to much higher CCPP levels (or $pH_s$ values) because the water is much more aggressive and the indices do not include the chemical corrections for poisoning or sequestration. This approach is frequently not optimal for the control of lead, copper, or zinc corrosion.

A strong tendency exists to overestimate the accuracy of the calculated values of the LSI. Soft, low-alkalinity waters cannot become supersaturated with $CaCO_3$ regardless of how high the pH is raised.[29,31,32] Raising the pH values greater than about 10.3 is useless because no more carbonate ions can be made available. Excess hydroxide alkalinity is of no value because it does not aid in $CaCO_3$ precipitation.[2] These relationships can be computed, or seen graphically through the use of Caldwell-Lawrence diagrams, such as those found in Chap. 10.

For systems that do not rely on $CaCO_3$ deposition for corrosion control, estimation of the optimal pH is more difficult. If lead or copper corrosion is a problem, adjusting the pH to values of from 8.0 to 8.5 or higher may be required. Practical minimum lead solubility occurs at a pH of about 9.8 in the presence of about 20 to 30 mg/L of alkalinity. For systems with low DIC, pH adjustment coupled with carbonate supplementation may be required to minimize lead corrosion problems. Alternatively, waters having a high concentration of DIC may require removal of $CO_2$ by stripping (see Chap. 5) of softening in addition to pH adjustment.

Orthophosphates and other corrosion inhibitors often require a narrow pH range for maximum effectiveness. If such an inhibitor is used, consideration must be given to adjusting the pH to within the effective range for the material of interest. Several studies have explored the pH-DIC-$PO_4$ relationships relative to the corrosion of lead,[11,38,47] galvanized,[38] copper,[38] and A-C[14,24,59] pipe. Consideration must be given to secondary interactions, such as calcium carbonate or calcium phosphate solubility limitations, that restrict the possible pH range.[50]

Chemicals commonly used for pH adjustment or supplementing recommended carbonate dosages, as well as equipment requirements, are summarized in Table 17.7. The pH should be adjusted after filtration because waters having higher pH values need larger doses of alum for optimal coagulation.[2]

Although pH adjustment can aid in reducing corrosion, it cannot eliminate corrosion in every case. pH adjustment is, however, the least costly and most easily implemented method of achieving some corrosion control, and utilities should use it if at all possible.

**Control of oxygen.**   Oxygen is an important corrosive agent because

- Oxygen can act as an electron acceptor, allowing corrosion to continue.

- Oxygen reacts with hydrogen to depolarize the cathode and, thus, speeds up corrosive reaction rates.

- Oxygen reacts with iron ions to form tubercles and leads to pitting in copper.

If oxygen could be removed from water economically, then the chances of corrosion starting (and the corrosion rate once it had started) would be reduced.[2] Unfortunately, oxygen removal is too expensive for municipal water systems and is not a practical control method. Minimiz-

**TABLE 17.7    Chemicals for pH Adjustment or Carbonate Supplementation**

| pH adjustment chemical | Typical feed rate, mg/L (lb/mil gal) | mg $CaCO_3$/L alkalinity added by 1 mg/L dose† | Equipment required |
|---|---|---|---|
| Lime, as $Ca(OH)_2$ | 1–20 (8–170) | 1.35 | Quicklime slaker, hydrated lime-solution tank, and feed pump with erosion-resistant lining as educator |
| Caustic soda, NaOH (50% solution) | 1–20 (8–170) | 1.25 | Proportioning pump or rotameter |
| Soda ash, $Na_2CO_3$ | 1–40 (8–350) | 0.94 | Solution tank, proportioning pump, or rotameter |
| Sodium bicarbonate, $NaHCO_3$ | 5–30 (40–250) | 0.59 | Solution tank, proportioning pump, or rotameter |

†Caustic soda and lime add only hydroxide alkalinity. Soda ash and sodium bicarbonate add carbonate or bicarbonate alkalinity, depending on pH.
SOURCE: From AWWA, Ref. 2.

ing the addition of oxygen to the source water is possible, however, particularly to groundwater.

Often, aeration is the first step in treating groundwater having high iron, hydrogen sulfide ($H_2S$), or $CO_2$ content (see Chap. 5). Though aeration helps remove these substances from source water, it can also cause more serious corrosion problems by saturating the water with oxygen. In lime–soda softening plants for treating groundwater, the water is often aerated first to save on the cost of lime by eliminating free $CO_2$. Iron is incidentally oxidized and precipitated in this step, but the iron would be removed in the subsequent softening process even if the water were not aerated. The actual result is that dissolved oxygen increases to near saturation, and corrosion problems are increased. Thus, the attempt to save on lime addition may actually end up costing much more in corrosion damage.

Measures that help keep the dissolved oxygen levels as low as possible include sizing well pumps and distribution pumps so as to avoid air entrainment, and using as little aeration as possible when aerating for $H_2S$ or $CO_2$ removal.[2] This can be achieved by bypassing the aerators with part of the source water. Complete elimination of the use of aerators has been possible if enough detention time is available in the reservoir so that enough oxygen can be adsorbed at the surface to oxidize the $H_2S$ or to let the $CO_2$ escape. Dissolved oxygen levels can be kept as low as 0.5 to 2.0 mg/L by this method, low enough, in many cases, to reduce corrosion rates considerably. Stumm[76] and others have shown dissolved oxygen to be an accelerator of corrosion in the absence of deposition of a mixed-carbonate scale.

**Chemical inhibitors.** Corrosion can be controlled by adding to the water chemicals that form a protective film on the surface of a pipe and provide a barrier between the water and the pipe. These chemicals, called *inhibitors,* reduce corrosion or limit metal solubility, but do not totally prevent it. The three types of chemical inhibitors commonly approved for use in potable water systems are chemicals that cause $CaCO_3$ scale formation, inorganic phosphates, and sodium silicate.

The success of any inhibitor in controlling corrosion depends upon three basic requirements.[2] First, starting the treatment at two or three times the normal inhibitor concentration to build up the protective film as fast as possible may be desirable. This minimizes the opportunity for pitting to start before the entire metal surface has been covered by a protective film. Usually several weeks are required for the coating to develop.

Second, the inhibitor usually must be fed continuously and at a sufficiently high concentration. Interruptions in the feed can cause loss of the protective films by its dissolution, and too low concentrations may prevent the formation of a protective film on all parts of the surface. Both interrupted feeding and low dosages can lead to pitting. On the other hand, excessive use of some alkaline inhibitors over a period of time can cause an undesirable buildup of scale, particularly in harder waters. The key to good corrosion inhibitor treatment is feed control.[2]

Third, flow rates must be sufficient to continuously transport the inhibitor to all parts of the metal surface; otherwise an effective protective film will not be formed and maintained. Corrosion will then be free to take place. For example, corrosion inhibitors often cannot reduce corrosion in storage tanks because the water is not flowing and the inhibitor is not fed continuously. To avoid corrosion of tanks, a protective coating, cathodic protection, or both are necessary. Similarly, corrosion inhibitors are not as effective in protecting dead ends as they are in those sections of mains that have a reasonably continuous flow.

Several different types of phosphates are used for corrosion control, including linear and cyclic polyphosphates, orthophosphates, glassy polyphosphates, and bimetallic polyphosphates. Recent developments in corrosion control include the use of zinc along with a polyphosphate or orthophosphate, or blends of orthophosphate and polyphosphates (mainly linear) without zinc.

Low dosages (about 2 to 4 mg/L by product weight) of glassy phosphates, such as sodium hexametaphosphate, have long been used to solve red water problems. In such cases, the addition of glassy phosphates masks the color, and the water appears clear because the iron is tied up as a complex ion. The corrosive symptoms are removed, but the corrosion rates are not reduced. Controlling actual metal loss requires dosages up to 10 times higher (20 to 40 mg/L) of the glassy phosphates. Other glassy phosphates that contain calcium or zinc, as well as sodium, are more effective as corrosion inhibitors. Orthophosphate and zinc + orthophosphate treatment has also been used to eliminate lead pickup, to reduce measured corrosion rates of metals, and limit some aesthetic problems resulting from corrosion (red water, staining).

The choice of a particular type of phosphate to use in a corrosion control program depends on the specific water quality and the material to be protected. Some phosphates work better than others in a given environment. Sometimes they can accelerate corrosion. Laboratory or field tests of one or more phosphate inhibitors usually need to be conducted before long-term use is initiated. Phosphate inhibitors, as noted previously, require particular zones of pH, DIC (or alkalinity), and phosphate level to be effective for corrosion control of most metals. This is often not understood. The complexation or solubility side reactions of calcium, magnesium, and iron (for example) with orthophosphate and polyphosphates can alter the dosages, DIC, and pH required for best performance. Also, limitations of metal levels in wastewater treatment or solubility limitations of basic zinc carbonate might provide limits on the use of zinc-containing phosphate formulations.

Sodium silicate (water glass) has been used for over 50 years to reduce corrosivity.[2] The way in which sodium silicate acts to form a protective film is still not completely understood. Corrosion and red water complaints can effectively be reduced in galvanized iron, yellow brass, and copper plumbing systems in both hot and cold water.[2]

The effectiveness of sodium silicate as a corrosion inhibitor depends on water quality properties such as pH and bicarbonate concentration. As a general rule, feed rates of 2 to 8 mg/L but possibly up to 20 mg/L of sodium silicate may be needed to control corrosion in a system once a protective film is formed. Corrosion control also may be assisted when activated silica is used as a coagulant aid (see Chap. 6). Silicate has been found to be particularly useful in waters having very low hardness and alkalinity and a pH of less than 8.4. It is also more effective under higher-velocity flow conditions.

**Secondary effects of treatment for corrosion control.** A utility must meet a broad spectrum of regulatory and aesthetic water quality goals, aside from those associated with corrosion and corrosion by-products. Therefore, the potential effects of different corrosion control strategies must enter into the determination of the best course to follow.

Adjustment of pH may affect the ability of coagulation to remove turbidity or organic matter effectively, as can the adjustment of alkalinity or the addition of phosphate.[50] The efficiency of disinfection by free chlorine is pH-dependent, so a utility may be forced to consider alternatives, such as the use of monochloramine or changing the location of pH adjustment in the treatment chain (see Chaps. 6 and 14).

The pH will also affect the formation of trihalomethanes (see Chaps. 12 and 14). Consideration must also be given to impacts of changes in water quality on industrial processes, building heating and cooling systems, wastewater loadings of metals (such as zinc), and nutrients (such as phosphates), and on algal or plant growth in open storage reservoirs.

### Cathodic protection

Cathodic protection is an electrical method for preventing corrosion of metallic structures. Metallic corrosion involves contact between a metal and an electrically conductive solution that produces a flow of electrons or current from the metal to the solution. Cathodic protection stops the current by overpowering it with a strong current from some outside source, forcing the metal that is being protected to become a cathode. The metal then has a large excess of electrons and cannot release any of its own.

Two basic methods can be used to apply cathodic protection.[2] One method uses inert electrodes, such as high-silicon cast iron or graphite, that are powered by an external source of direct current. The current impressed on the inert electrodes forces them to act as anodes, minimizing the possibility that the metal surface being protected will become an anode and corrode. The second method uses a sacrificial galvanic anode. Magnesium or zinc anodes produce a galvanic action with iron such that they are sacrificed (or corrode) while the iron structure they are connected to is protected from corrosion. This type of system is common to small hot-water heaters. Another form of sacrificial anode is galvanizing, where zinc is used to coat iron or steel. The zinc becomes the anode and corrodes, protecting the steel that is forced to be the cathode.

Cathodic protection is often used by water utilities to prevent internal corrosion in water storage tanks.[2] It is also an effective method of preventing external corrosion of distribution systems, especially when used as part of an integrated corrosion-control effort that also includes appropriate linings, coatings, and environmental preparation.

### Linings, coatings, and paints

Another technique to keep corrosive water away from the pipe wall is to line the wall with a protective coating. These linings are usually mechanically applied, either when the pipe is manufactured or in the field before it is installed. Some linings can be applied even after the pipe is in service, though this method is much more expensive. The most common pipe linings are coal-tar enamels, epoxy paint, cement mortar, and polyethylene. The use of coatings must be carefully monitored, because they can be the source of several water quality problems,[77] such as support of microbiological growth, taste and odor, and solvent leaching.

Water storage tanks are most commonly lined to protect the inner tank walls from corrosion.[2] Common water tank linings include coal-tar enamels and paints, vinyls, and epoxy. Although coal-tar based products have been widely used in the past for contact with drinking water, concern exists about their use because of the presence of polynuclear aromatic hydrocarbons and other hazardous compounds in coal tar and the potential for their migration in water.

A common method of extending the life of pipe and restoring good corrosion resistance and hydraulic properties is either the controlled deposition of a $CaCO_3$ film[1] or the process of cement-mortar lining. The latter process has recently been comprehensively reviewed.[24,78–80] Either process is also useful in covering exposed asbestos fibers and softened interior surfaces of A-C pipe.

### Costs of corrosion control

The costs of corrosion control may vary considerably among utilities of different sizes and in different regions of the country. Generalizations can be made, however, that guide utilities in estimating the costs for their own circumstances. Equipment costs will vary, depending on size, quality, features, and construction materials. In addition, many site-specific factors will affect the total system costs. These factors include labor costs, quantity of piping and valves needed, construction materials, housing requirements, bulk storage, unloading-conveyance systems, and site preparation needed.

Chemical costs for water quality modification will vary with location, transportation costs, and volume of chemicals purchased. Water utility personnel are advised in all cases to contact water treatment chemical and equipment suppliers in their areas to determine actual costs of an in-place control system.

The following guidelines and discussions of what to consider for estimating costs are essentially those given in the AWWA publication *Corrosion Control for Operators*.[2] A detailed review of all of the costs involved in treating water for corrosion control is currently being undertaken, which will give utilities a good basis for developing their programs[81] and will update this information.

**Sampling and analysis.**   Sampling and analytical costs to monitor and control corrosion will vary among utilities, depending on the number of parameters analyzed, the number of samples collected, the type of materials used in the system, and the type of control program being monitored by the utility.

To comply with the 1980 National Interim Primary Drinking Water Regulation (NIPDWR) amendments,[67] only one or two (if the surface water supplies are used) samples for the following parameters are required:

1. Alkalinity, mg/L as $CaCO_3$

2. pH, as pH units

3. Hardness, mg/L as $CaCO_3$

4. Temperature

5. Total dissolved solids, mg/L

Additional sampling and analysis are required to determine if corrosion is occurring and what materials are being corroded, as will be discussed later. Analyses must also be conducted to monitor the control and performance of inhibitor additions.

**Weight-loss measurements of corrosion rate.**    The main costs of coupon or weight-loss methods are

1. The initial purchase and installation of the coupons
2. Labor costs of setting up the test
3. Dismantling and weighing the coupons after a specified time period
4. The cost of any water quality modifications tested during the test period (such as pH adjustment, reduction of oxygen, pipe lining, or inhibitor treatment)

The costs vary, depending on the number of coupons placed in the system, the number of different materials tested, and whether the utility performs the study in-house or hires an outside consultant to conduct the tests. The direct measurement of corrosion rates with electrochemical monitors and test devices will result in the initial investment in one to several instruments, plus ongoing costs of periodically replacing chart paper, electrodes, and so forth.

### Water Sampling for Corrosion Control

The effects of corrosion may not be evident without monitoring. The effects can be expensive, and in the case of corrosion by-products such as lead, copper, and cadmium they can be injurious to the health of segments of the population.

Corrosion has many causes, and many techniques exist to measure or "cure" corrosion. Corrosion in a system depends on a specific water and the reaction of that water with specific pipe materials; therefore, each utility is faced with a unique set of problems. General methods of measuring and monitoring for corrosion can, however, provide a basis for a sound corrosion-control program for any utility.

The first concern for a utility is to meet all regulatory sampling requirements, in terms of the number of sampling sites, the location and frequency of the samples, and the use of approved and generally accepted analytical methods. Beyond these requirements, monitoring programs may address other questions.

#### Defining the problem

The many factors responsible for variability in lead concentrations that can limit the accurate assessment of exposure levels, treatment performance, and regulatory compliance have recently been reviewed.[82,83] These factors are not unique to lead in most cases. The goal of a sampling program must be to control as many of the analytical, chemical, and physical factors so that proper exposure assessments and decision on the existence of water quality problems or the performance of a treatment program can be made.

The variability represents the actual range of exposures that occur routinely in the population. A monitoring system designed to accurately reflect exposure to lead in drinking water must simultaneously capture that diversity while providing adequately reliable information to critically evaluate exposure.[82] A monitoring program for corrosion-control effectiveness must address similar concerns, although the criteria for sample size based on the desired levels of confidence in the mean values could be different, or a multistage sampling design could be used to obtain information more economically.

Few studies have presented sampling strategies that allow the evaluation of the variability over time of the same sampling site,[83] particularly with respect to conditions before and after the implementation of a corrosion-control program. Even within a water supply system with relatively consistent and uniform water characteristics, successive samplings under equivalent conditions do not necessarily yield the same corrosion by-product concentrations. The design of a monitoring scheme for the determination of treatment effectiveness must take into account the magnitude of these variabilities so that the proper number of sites and frequency of sampling can be chosen. To reduce both random and unidirectional bias, the monitoring program must have standardized procedures for sample collection and analysis, a high level of analytical precision and accuracy, and uniform performance by well-trained and informed personnel.[82] To correctly assess exposure, sampling locations and sampling collection protocols must be established that are likely to coincide with occurrences of elevated by-product levels in drinking water to which members of the service population are likely to be exposed.

Careful thought must be given to the selection of sites within the target groups (for example, lead interior plumbing or service lines, or new construction or faucet replacement when lead is the primary concern). Particular locations should be chosen randomly, insofar as is possible, to restrict numerous possibilities of socioeconomic bias. Additionally, randomization would result in valid statistics for the target groups, because the sampling error of various statistical tests and estimates can be characterized.

The nature of exposure will vary, depending on the type of water system involved (e.g., single-family dwelling, apartment building, office building, schools). Different sampling strategies are essential, and they must be carefully oriented toward the layout of the customers' system. To determine whether the contamination is from a part of the plumbing system (e.g., faucets, soldered joints, parts of a service line, or pigtail connector), small sample volumes may be required to isolate the specific area of plumbing in question, or numerous sequential samples may be required. Some recent guidelines have been developed by the USEPA to isolate locations of contamination in building[84] or school[85] drinking water systems, which are applicable to many other corrosion problems as well. Utilities must also be concerned with monitoring the availability of active inhibitor constituents, such as silica, orthophosphate, polyphosphate, DIC (alkalinity), or pH to ensure that the dosages applied to the treated water are adequate to provide the necessary levels for corrosion control throughout the distribution system.

### Statistical considerations

The number of sites to be monitored for exposure evaluation or corrosion control must be related to the level of constituent or corrosion by-product variability over the proposed range of sites, and the relationship of that variability to the population mean (mean of all sites).[82] A log-normal model for lead levels has been demonstrated, using data from two comprehensive field projects, both before and after treatment.[82,83] The relationship between the logarithm of the standard deviations for repetitive observations at each site versus the logarithm of the mean lead values for each site was reasonably approximated by a simple linear model. This type of model can be used to select the total number of sample sites for given degree of confidence. Further information on estimating the number of samples necessary for monitoring and interpreting treatment programs for lead has recently been developed.[50] A good "before" sampling program must be developed to obtain a "base line" of data. The base line is used to get a handle on the required number and frequency of sampling for the rest of the program and to enable valid comparisons to be made of water characteristics before and after changes in the system (e.g., treatment changes, design changes, source water changes).

### Chemical and physical considerations

Corrosion is affected by the chemical composition of a water, so sampling and chemical analysis of the water can provide valuable corrosion-related information. Some waters tend to be more aggressive or corrosive than others because of the quality of the water. For example, water with a low pH ($<6.0$), low alkalinity ($<40$ mg/L), and high carbon dioxide ($CO_2$) (or DIC) tend to be more corrosive than water with pH > 7.0, high alkalinity, and low $CO_2$. Whether corrosion is occurring in the system, however, depends on the action of the water on the pipe material.[2]

Most utilities routinely analyze their water (1) to ensure that they are providing a safe water to their customers and (2) to meet regulatory requirements. The 1980 amendments to the NIPDWR[67] require all community water supply systems to sample for certain "corrosive characteristics." Table 17.8 summarizes the sampling and analytical requirements of the 1980 amendments.[2]

Water samples should generally be collected at the following locations within the system:

- Entrance to the distribution system
- Various locations in the distribution system prior to household service lines
- In an appropriate number of household service lines throughout the system
- At an appropriate number of customer's taps, when required

Water entering the distribution system at the plant can be conveniently sampled from the clearwell, the storage tank, or a sample tap on a pipe before or after the high-service pump.[2]

**TABLE 17.8     1980 Amendments to the NIPDWR: Sampling and Analytical Requirements (Individual states may add requirements.)**

| Parameters required | Sampling location | Number of samples | |
|---|---|---|---|
| | | Water supply source | Number of samples per year |
| Alkalinity (mg/L as $CaCO_3$) | Sample(s) are to be taken at one representative point as the water enters the distribution system | Groundwater only | One |
| pH (pH units) Hardness (mg/L as $CaCO_3$ Temperature (°C) Total dissolved solids (mg/L) Langelier or aggressive index | | Surface water only or groundwater and surface water | Two samples, taken at different times o fht eyear to account for seasonal variations in furface water supplies, such as mid-summer high temperatures and mid-winter low temperatures, or high-flow and low-flow conditions. |

SOURCE: *Federal Register,* August 1980.

To represent conditions at the customer's tap, "standing" samples should be taken from an interior faucet in which the water has remained for several hours (i.e., overnight). The sample should be collected as soon as the tap is open. If the contribution of the faucet to metal levels is of interest, a small sample must be collected (i.e., 60 to 125 mL) in addition to samples from the rest of the plumbing system. Larger volumes are necessary to include the pipe contribution.[22]

A representative sample from the household service line (between the distribution system and the house itself) can be obtained by collecting a "running" sample from the customer's faucet after letting the tap run only long enough to flush the household lines. Sometimes, the water temperature noticeably decreases when water in the service line reaches the tap. The volume to "waste" before the water from the service line can be most accurately determined by calculating the volume from the size and length of the household lines running from the connection to the sampling tap. By letting the same faucet run for several minutes following the initial temperature change, the running water sample at the tap is most representative of the water recently in the distribution main itself. If a comparison of the sampling results shows a change in the water quality, corrosion may be occurring between the sampling locations.

Many important decisions are likely to be made based on the sampling and chemical analyses performed by a utility. Therefore, care must be taken during the sampling and analysis to obtain the best data. Handling of samples for pH, alkalinity, and $CO_2$ analyses often requires special precautions.[86] Samples should be collected without adding air and with minimal agitation, because air tends to remove $CO_2$ and affects the oxygen content in the sample. To collect a sample without additional air and to minimize exchange of volatile gases (such as $CO_2$ in waters that are frequently out of equilibrium with the atmosphere), fill the sample container to the top so that a meniscus is formed at the opening and no bubbles are present. If possible, the sample bottle should be filled below the surface of the water using tubing so that the water is not contaminated by the faucet material. Cap the sample bottle as soon as possible.

The constituents that should be analyzed in a thorough corrosion-monitoring program depend to a large extent on the materials present in the system's distribution, service, and household plumbing lines. Table 17.9 summarizes parameters recommended to be analyzed in a thorough corrosion-monitoring program. Temperature and pH should be measured in situ (in the field) with proper precautions for atmospheric $CO_2$ exchange.[86]

TABLE 17.9    Recommended Analyses for a Thorough Corrosion Monitoring Program

| | |
|---|---|
| General parameters | |
| In situ measurements | pH, temperature |
| Dissolved gases-oxidants | Oxygen, hydrogen sulfide,† free chlorine, total chlorine residual (if ammonia present or used) |
| Parameters required to calculate $CaCO_3$-based indices | Calcium, total hardness (or magnesium), alkalinity (or DIC), total dissolved solids (or conductivity)‡ |
| Parameters for A-C pipe | Add to general parameters: fiber count, iron, zinc, silica, polyphosphate, manganese |
| Background parameters for metal pipe | Add to general parameters: |
| Iron or steel pipe | Iron |
| Lead pipe or lead-based solder | Lead |
| Copper pipe | Copper, lead |
| Galvanized iron pipe | Zinc, iron, cadmium, lead |
| Brass (faucets and valves) | Zinc, copper, lead |
| Possibly "aggressive" constituents | Chloride, sulfate, total chlorine residual |
| All metal pipes | Polyphosphate§ |
| Potentially "protective" constituents (all metal pipes) | Orthophosphate, silica |

†Usually cannot coexist with oxygen and chlorine, so it should only be analyzed when suspected.

‡If a complete water analysis is done, which would generally just add sodium and potassium to this list, these can be neglected because ionic strength can be directly computed.

§For most potable waters; derived from analyzing total phosphate and subtracting orthophosphate.

Frequency of analysis depends on the extent of the corrosion problems experienced in the system, the degree of variability in source and finished water quality, the type of treatment and corrosion control practiced by the water utility, and cost considerations. When phosphates or silicates are added to the water, samples should be collected at the far reaches of the system and analyzed for polyphosphates, orthophosphates, and sodium silicate, as appropriate. If no residual phosphate or silicate is found, the feed rate should be increased. When calcium carbonate precipitation is practiced, the parameters needed to compute the LSI or the CCPP in the far reaches of the system must be monitored. Continual monitoring of pH and temperature are important because they are so interdependent.

When corrosion-control studies obtain good base-line data and have a comprehensive, well-designed monitoring program in place throughout their corrosion-control effort, the data are invaluable to other utilities and corrosion scientists. They can then work with a larger body of knowledge of drinking water chemistry and treatment, and they can implement future corrosion control and public health protection strategies much more effectively and efficiently.

## Acknowledgments

The author gladly acknowledges the extensive contributions of all of the writers of the AWWA publication *Corrosion Control for Operators,* the AWWARF/DVGW-Forschungsstelle manual *Internal Corrosion of Water Distribution Systems,* and the late Dr. T. E. Larson, whose works were heavily cited and directly included in this chapter. At the time this chapter was written, the author was an Associate Chemist with the Aquatic Chemistry Section, Illinois State Water Survey, Champaign, Ill.

## References

1. *Internal Corrosion of Water Distribution Systems,* AWWARF/DVGW-Forschungsstelle Cooperative Research Report, American Water Works Association Research Foundation, Denver, Colo., 1985, chap. 1.
2. *Corrosion Control for Operators,* AWWA, Denver, Colo., 1986.
3. J. O'M Bockris and A. K. N. Reddy, *Modern Electrochemistry,* Vol. 2, Plenum, New York, 1973.
4. M. Pourbaix, *Lectures on Electrochemical Corrosion,* Plenum, New York, 1973.
5. R. M. Garrels and C. L. Christ, *Solutions, Minerals, and Equilibria,* Harper and Row, New York, 1965.
6. W. Stumm and J. J. Morgan, *Aquatic Chemistry, an Introduction Emphasizing Chemical Equilibria in Natural Waters,* 2d ed., Wiley-Interscience, New York, 1981.
7. V. L. Snoeyink and D. Jenkins, *Water Chemistry,* Wiley, New York, 1980.
8. James M. Montgomery Consulting Engineers, Inc., *Water Treatment Principles & Design,* Wiley-Interscience, New York, 1985.
9. A. J. Bard, *Chemical Equilibrium,* Harper & Row, New York, 1966.
10. M. R. Schock, "Response of Lead Solubility to Dissolved Carbonate in Drinking Water," *J. AWWA,* vol. 72, no. 12, 1980, p. 695; Errata, *J. AWWA,* vol. 73, no. 3, 1981, p. 36.
11. *Internal Corrosion of Water Distribution Systems,* AWWARF/DVGW-Forschungsstelle Cooperative Research Report, American Water Works Association Research Foundation, Denver, Colo., 1986, chap. 4.
12. M. F. Obrecht and M. Pourbaix, "Corrosion of Metals in Potable Water Systems," *J. AWWA,* vol. 59, no. 8, 1967, p. 977.
13. M. Pourbaix, *Atlas of Electrochemical Equilibria in Aqueous Solutions,* Pergamon Press, Oxford; CEBELCOR, Brussels, 1966.
14. M. R. Schock and R. W. Buelow, "The Behavior of Asbestos-Cement Pipe under Various Water Quality Conditions: Part 2. Theoretical Considerations," *J. AWWA,* vol. 73, no. 12, 1981, p. 636.
15. *Internal Corrosion of Water Distribution Systems,* AWWARF/DVGW-Forschungsstelle Cooperative Research Report, American Water Works Association Research Foundation, Denver, Colo., 1986, chap. 2.
16. H. Sontheimer et al., "The Siderite Model of the Formation of Corrosion-Resistant Scales," *J. AWWA,* vol. 73, no. 11, 1981, p. 572.

17. M. R. Schock and M. C. Gardels, "Plumbosolvency Reduction by High pH and Low Carbonate—Solubility Relationships," *J. AWWA,* vol. 75, no. 2, 1983, p. 87.
18. D. J. G. Ives and A. E. Rawson, "Copper Corrosion III: Electrochemical Theory of General Corrosion," *J. Electrochem. Soc.,* vol. 109, 1962, p. 452.
19. T. E. Larson, *Corrosion by Domestic Waters,* Bulletin 59, Illinois State Water Survey, Urbana, Ill., 1975.
20. L. Kenworthy, "The Problem of Copper and Galvanized Iron in the Same Water System," *J. Inst. Metals,* vol. 69, 1943, p. 67.
21. K. P. Fox et al., *Copper-Induced Corrosion of Galvanized Steel Pipe,* U.S. Environmental Protection Agency, Drinking Water Research Division, EPA Rep. 600/2-86-056, Cincinnati, Ohio, 1986.
22. M. R. Schock and C. H. Neff, "Trace Metal Contamination from Brass Fittings," *J. AWWA,* vol. 80, no. 11, 1988, p. 47.
23. J. O. G. Parent et al., "Effects of Intermetallic Formation at the Interface between Copper and Lead-Tin Solder," *J. Mat. Sci.,* 1988, p. 2564.
24. *Internal Corrosion of Water Distribution Systems,* AWWARF/DVGW-Forschungsstelle Cooperative Research Report, American Water Works Association Research Foundation, Denver, Colo., 1986, chap. 6.
25. R. G. Ainsworth et al., The Introduction of New Water Supplies into Old Distribution Systems, Tech. Rep. TR-143, Water Research Centre, Medmenham, England, 1980.
26. S. H. Lee et al., "Biologically Mediated Corrosion and Its Effects on Water Quality in Distribution Systems," *J. AWWA,* vol. 72, no. 11, 1980, p. 636.
27. *Corrosion Basics, An Introduction,* National Association of Corrosion Engineers, Houston, Texas, 1984.
28. American Water Works Association. *Water Quality and Treatment,* 3d ed., McGraw-Hill, New York, 1971.
29. J. R. Rossum and D. T. Merrill, "An Evaluation of the Calcium Carbonate Saturation Indices," *J. AWWA,* vol. 75, no. 2, 1983, p. 95.
30. M. R. Schock, "Temperature and Ionic Strength Corrections to the Langelier Index—Revisited," *J. AWWA,* vol. 76, no. 8, 1984, p. 72.
31. R. E. Loewenthal and G. v.R. Marais, *Carbonate Chemistry of Aquatic Systems: Theory and Applications,* Ann Arbor Science, Ann Arbor, Mich., 1976.
32. D. T. Merrill and R. L. Sanks, "Corrosion Control by Deposition of $CaCO_3$ Films: A Practical Approach for Plant Operators," *J. AWWA,* vol. 69, no. 11, 1977, p. 592, *J. AWWA,* vol. 69, no. 12, 1977, p. 634, *J. AWWA,* vol. 70, no. 1, 1978, p. 12.
33. *Internal Corrosion of Water Distribution Systems,* AWWARF/DVGW-Forschungsstelle Cooperative Research Report, American Water Works Association Research Foundation, Denver, Colo., 1986, chap. 3.
34. G. P. Treweek, et al., "Copper-Induced Corrosion of Galvanized Steel Pipe," *Proc. AWWA Water Qual. Tech. Conf.,* Kansas City, Mo., 1978.
35. K. P. Fox et al., "The Interior Surface of Galvanized Steel Pipe: A Potential Factor in Corrosion Resistance," *J. AWWA,* vol. 76, no. 2, 1983, p. 84.
36. *Internal Corrosion of Water Distribution Systems,* AWWARF/DVGW-Forschungsstelle Cooperative Research Report, American Water Works Association Research Foundation, Denver, Colo., 1986, chap. 5.
37. W. J. Weber, Jr., and W. Stumm, "Mechanism of Hydrogen Ion Buffering in Natural Waters," *J. AWWA,* vol. 55, no. 12, 1963, p. 1553.
38. M. R. Schock, "Treatment or Water Quality Adjustment to Attain MCLs in Metallic Potable Water Plumbing Systems," *Plumbing Materials and Drinking Water Quality: Proc. Seminar,* Cincinnati, Ohio, May 1985, U.S. Environmental Protection Agency, Water Engineering Research Laboratory, Rep. 600/9-85/007, 1985.
39. D. Atlas et al., "The Corrosion of Copper by Chlorinated Drinking Waters," *Water Research,* vol 16:1982, p. 693.
40. T. E. Larson and R. V. Skold, "Corrosion and Tuberculation of Cast Iron," *J. AWWA,* vol. 49, no. 10, 1957, p. 1294.
41. M. Koudelka et al., "On the Nature of Surface Films Formed on Iron in Aggressive and Inhibiting Polyphosphate Solution, *J. Electrochem. Soc.,* vol. 129, 1982, p. 1186.

42. G. Butler and H. C. K. Ison, *Corrosion and Its Prevention in Waters,* Leonard Hill, London, 1966.
43. S. H. Reiber, "Copper Plumbing Surfaces: An Electrochemical Study," *J. AWWA,* vol. 81, no. 7, 1989, p. 114.
44. S. H. Reiber, et al., *Corrosion in Water Distribution Systems of the Pacific Northwest,* Final Report, U.S. Environmental Protection Agency, Cooperative Agreement No. CR-810508, Drinking Water Research Division, Cincinnati, Ohio, NTIS PB87-197521, 1987.
45. H. Cruse and R. D. Pomeroy, "Corrosion of Copper Pipes," *J. AWWA,* vol. 67, no. 8, 1974, p. 479.
46. R. Oliphant, *Dezincification of Potable Water of Domestic Plumbing Fittings: Measurement and Control,* Water Research Centre Technical Report TR-88, Medmenham, England, 1978.
47. M. R. Schock, "Understanding Corrosion Control Strategies for Lead," *J. AWWA,* vol. 81, no. 7, 1989, p. 88.
48. James M. Montgomery Consulting Engineers, Inc., *Internal Corrosion Mitigation Study Final Report,* Bureau of Water Works, Portland, Ore., 1982.
49. T. D. B Lyon and J. M. A. Lenihan, "Corrosion in Solder Jointed Copper Tubes Resulting in Lead Contamination of Drinking Water," *Br. Corros. J.,* vol. 12, 1977, p. 41.
50. *Lead Control Strategies,* American Water Works Association Research Foundation, Denver, Colo., in press.
51. R. Oliphant, *Summary Report on the Contamination of Potable Water by Lead from Soldered Joints,* Water Research Centre External Report 125E, Medmenham, England, 1983.
52. M. R. Schock and C. H. Neff, Unpublished analytical data, Illinois State Water Survey, Champaign, Ill., 1988.
53. M. R. Schock and C. H. Neff, "Chemical Aspects of Internal Corrosion: Theory, Prediction, and Monitoring," *Proc. AWWA Water Qual. Tech. Conf.,* Nashville, Tenn., 1982.
54. *Internal Corrosion of Water Distribution Systems,* AWWARF/DVGW-Forschungsstelle Cooperative Research Report, American Water Works Association Research Foundation, Denver, Colo., 1986, chap. 9.
55. P. M. Temkar et al., "Pipe Loop System for Evaluating Effects of Water Quality Control Chemicals in Water Distribution System," *Proc. AWWA Water Qual. Tech. Conf.,* Baltimore, Md., 1987.
56. C. H. Neff et al., *Relationships between Water Quality and Corrosion of Plumbing Materials in Buildings,* U.S. Environmental Protection Agency, Water Engineering Research Laboratory, EPA Rep. 600/S2-87/036, Cincinnati, Ohio, 1987.
57. S. Reiber et al., "An Improved Method for Corrosion Rate Measurement by Weight Loss," *J. AWWA,* vol. 80, no. 11, 1988, p. 41.
58. D. H. Thompson, General Tests and Principles, in W. H. Ailor (ed.), *Handbook on Corrosion Testing and Evaluation,* Wiley, New York, 1970.
59. M. R. Schock et al., "Evaluation and Control of Asbestos-Cement Pipe Corrosion," *Proc. NACE Corrosion 81,* Toronto, Canada, April, Rep. 600/D-81-067 1981.
60. A. P. Walker, "The Microscopy of Consumer Complaints," *J. Inst. Water Eng. Scient.,* vol. 37; 1983, p. 200.
61. W. Kolle and H. Rosch, "Untersuchungen an Rohrnetz-Inkrustierungen unter Mineralogischen Gesichtspunkten," *Sond. Vom Wasser,* vol. 55, 1980, p. 159.
62. J. H. Colling et al., "The Measurement of Plumbosolvency Propensity to Guide Control of Lead in Tapwaters," *J. Inst. Water Environ. Mgmt.,* vol. 1, 1987, p. 263.
63. M. Rose, "X-Ray Fluorescence Analysis of Waterborne Scale," *Amer. Lab.,* vol. 15, 1983, p. 46.
64. R. W. Gould, "The Application of X-Ray Diffraction to the Identification of Corrosion Products," *Proc. AWWA Water Qual. Tech. Conf.,* Miami Beach, Fla., 1980.
65. H. S. Campbell, "Corrosion, Water Composition and Water Treatment," *Water Treat. Exam.,* vol. 20, 1971, p. 11.
66. W. F. Langelier, "The Analytical Control of Anti-Corrosion Water Treatment," *J. AWWA,* vol. 28, no. 10, 1936, p. 1500.

67. "National Interim Primary Drinking Water Regulations," *Federal Register,* vol. 45, no. 168, 1980, p. 57332.
68. H. K. Miyamoto and M. D. Silbert, "A New Approach to the Langelier Stability Index," *Chem. Eng.,* vol. 89, 1986.
69. D. L. Parkhurst et al., *PHREEQE—A Computer Program for Geochemical Calculations,* U.S. Geological Survey, Water Resources Investigations WRI-80-96, 1980.
70. D. S. Brown and J. D. Allison, *MINTEQA1, An Equilibrium Metal Speciation Model: User's Manual,* U.S. Environmental Protection Agency, Rep. 600/3-87/012, Athens, Ga., 1987.
71. J. R. Rossum, "Dead Ends, Red Water, and Scrap Piles," *J. AWWA,* vol. 79, no. 7, 1987, p. 113.
72. J. W. Ryznar, "A New Index for Determining Amount of Calcium Carbonate Scale Formed by a Water," *J. AWWA,* vol. 36, no. 4, 1944, p. 472.
73. J. E. Singley, "The Search for a Corrosion Index," *J. AWWA,* vol. 73, no. 10, 1981, p. 529.
74. R. A. Pisigan, Jr., and J. E. Singley, "Evaluation of Water Corrosivity Using the Langelier Index and Relative Corrosion Rate Methods," *Mat. Perform.,* vol. 26, 1985.
75. *Prevention and Control of Water-Caused Problems in Building Potable Water Systems,* National Association of Corrosion Engineers, TPC Publication No. 7, Houston, Texas, 1980.
76. W. Stumm, "Calcium Carbonate Deposition at Iron Surfaces," *J. AWWA,* vol. 48, no. 3, 1956, p. 300.
77. K. B. Stinson and K. E. Carns, "Ensuring Water Quality in a Distribution System," *J. Environ. Eng. Div. ASCE,* vol. 109, 1983, p. 289.
78. D. Hasson and M. Karmon, "Novel Process for Lining Water Mains by Controlled Calcite Deposition," *Corros. Prev. Control,* vol. 9, 1984.
79. R. F. McCauley, "Controlled Deposition of Protective Calcite Coatings in Water Mains," *J. AWWA,* vol. 52, no. 11, 1960, p. 1386.
80. R. W. Bonds, "Cement Mortar Linings for Ductile Iron Pipe," *Ductile Iron Pipe News,* Spring/Summer, vol. 8, 1989.
81. *Economics of Internal Corrosion Control,* American Water Works Association Research Foundation, Denver, Colo., 1989.
82. M. R. Schock et al., "The Significance of Sources of Temporal Variability of Lead in Corrosion Evaluation and Monitoring Program Design," *Proc. AWWA Water Qual. Tech. Conf.,* St. Louis, Mo., 1988.
83. M. R. Schock, Causes of Temporal Variability of Lead in Domestic Plumbing Systems, *Environ. Monit. Assessment* (in press).
84. *Suggested Sampling Procedures to Determine Lead in Drinking Water in Buildings Other than Single Family Homes,* U.S. Environmental Protection Agency, Office of Drinking Water, Washington, D.C., 1988.
85. *Lead in School Drinking Water: A Manual for School Officials to Detect, Reduce, or Eliminate Lead in School Drinking Water,* U.S. Environmental Protection Agency, Office of Drinking Water, Washington, D.C., 1988.
86. M. R. Schock and S. C. Schock, "Effect of Container Type on pH and Alkalinity Stability," *Water Research,* vol. 16, 1982, p. 1455.

# Microbiological Quality Control in Distribution Systems

## Edwin E. Geldreich

*Senior Research Microbiologist*
*Drinking Water Research Division*
*U.S. Environmental Protection Agency*
*Cincinnati, Ohio*

The purpose of a water supply distribution system is to deliver to each consumer a safe drinking water that is also adequate in quantity and acceptable in terms of taste, odor, and appearance. Distribution of high-quality drinking water has been a major concern for many centuries. In early times, the primary objective was to provide delivery of adequate amounts of water to centralized fountains and other locations designated for public gathering plus restricted areas of official residences. Aqueducts to carry spring waters were built by ancient Greeks and Romans during their rise to their dominance as centers of civilization. Romans also covered some of their aqueducts to protect the spring waters from interruptions or willful contamination by enemy action.[1] Pipe materials in these early water transport systems were of natural stone, wood, clay, or lead. Distribution lines were simple, generally limited to a main trunk line with few dead-end branches. Water flow was dependent on gravity and the discharge from springs and from mountain stream diversions.[2]

Historically, the initial distribution network of pipes was a response to existing community needs that eventually created a legacy of problems of inadequate supply and low pressure as the population density increased. To resolve the problem of increased water demand along the distribution route, reservoir storage was created. Pressure pumping to move water to far reaches of the supply lines and standpipes was incorporated to afford relief from surges of pressure in pipelines.

In some areas, population growth exceeded the capacity of a water resource so that other sources of water were incorporated and additional treatment plants were built to feed into the distribution network. Another response was to consolidate neighboring water systems and interconnect the associated distribution pipe networks.

## General Considerations for Contamination Prevention

Today (1990), community expansion plans are more fully developed and include the engineering of utility service so that careful consideration is given to meeting future projected water supply needs. Advanced planning provides the opportunity to design the pipe network as a grid with a series of loops to avoid dead ends. The objective is to produce a circulating system capable of supplying high-quality water to all areas yet designed that any section may be isolated for maintenance, repair, or decontamination without interrupting service to all other areas.

To ensure delivery of a high-quality municipal potable water supply to each consumer, management of public water supply systems must be continually vigilant for any intrusions of contamination in the distribution network and the occurrence of microbial degradation. This job is complicated by the very nature of a distribution system: a network of mains, fire hydrants, valves, auxiliary pumping, chlorination substations, storage reservoirs, standpipes and service lines. Following the intrusion of microbial contamination, any of these component parts may serve as a habitat suitable for colonization by certain microorganisms in the surviving flora. The persistence and possible regrowth of organisms in the pipe network is influenced by a variety of environmental conditions that include physical and chemical characteristics of the water, system age, variety of pipe materials, and the availability of sites suitable for colonization.

## Engineering Considerations for Contamination Prevention

Many public water utilities make substantial efforts to expand their distribution networks to keep up with continuing suburban growth. Urban renewal and highway construction projects may at times require the relocation or enlargement of portions of the distribution network. Corrosion, unstable soil, faulting, land subsidence, extreme low temperatures, and other physical stresses often cause line breaks and necessitate repair or replacement of pipe sections. To avoid possible bacteriological contamination of the water supply during these construction projects, a rigorous protective protocol must be followed.

## Distribution system construction practices

The American Water Works Association[3] has developed standards for disinfecting water mains that are used, with variations, by most of the water supply industry. In essence, these recommendations recognize six areas of concern: (a) protection of new pipe sections at the construction site; (b) restriction on the use of joint-packing materials; (c) preliminary flushing of pipe sections; (d) pipe disinfection; (e) final flushing; and (f) bacteriological testing for pipe disinfections.

Pipe sections, fittings, and valves stockpiled in yard areas or at the construction site should be protected from soil, seepages from water or sewer line leaks, storm-water runoff, and habitation by pets and wildlife.[4,5] Each of these contamination sources may deposit significant fecal material in the interior of pipe sections awaiting installation. Septic tank drain fields, subsurface water in areas of poor drainage or high water table, and seasonal or flash flooding may also introduce significant contamination into unprotected pipe sections. Fecal excrement transmitted by contamination sources may become lodged in pipe fittings and valves. Thus, such sites become protected habitats from which coliforms and any associated pathogens in the contaminated material may be shed. Commonsense protective measures include end covers for these pipe materials, drainage of standing water from trenches, and flushing of assembled pipe sections to remove all visible signs of debris and soil.[6,7]

## Pipe joining materials

Gasket seals of pipe joints can be a source of bacterial contamination in new pipes.[8] Annular spaces in joints provide a protected habitat for continued survival and possible multiplication of a variety of bacteria in the distribution network. In these instances, although the heterotrophic plate count (HPC) and any coliform occurrences may be temporarily reduced by main disinfection, bacteria soon become reestablished from the residual population harbored in some joint-packing materials. In this regrowth process, the variety of organisms and dominance of strains change, often restructuring the bacterial flora to a predominant population of *Pseudomonas aeruginosa, Chromobacter* strains, *Enterobacter aerogenes,* or *Klebsiella pneumoniae.* Thus, where the pattern of organisms present is predominantly one bacterial strain, a search for a protective habitat in joint-packing materials or impacted material in pipe sections should be made.[9-14] Nonporous materials such as molded or tubular plastic, rubber, and treated paper products are preferable. Lubricants used in seals must be nonnutritive to avoid bacterial growth in protected joint spaces. Efforts to develop bacteriostatic lubricants have resulted in the inclusion of various quaternary ammonium compounds that minimize contamination from pipe joint spaces.[15]

### Water supply storage reservoirs

Water use in a community varies hourly as a reflection of the activities of the general public and local industries. While industrial uses of potable water are more constant and predictable, expecting water treatment operations to gear production to those frequent and sudden changes in water demand from all consumers is impractical. For this reason, storage reservoirs are an essential element of the distribution network. These water supply reserves supplement water flows in distribution during periods of fluctuating demand on the system, providing storage of water during off-peak periods, equalize operational water pressures, and augment water supply from production wells that must be pumped at a uniform rate. Storage reservoirs also provide a protective reserve of drinking water to guard against discontinuance of water treatment during oil spills in the source water, flooding of well fields, and transmission line or power failure. An important secondary consideration is sufficient storage capacity calculated to be adequate for fire emergencies.

Finished water reservoirs may be located near the beginning of a distribution system, but most often they are situated in suburban areas. Local topography plays an important part in determining the use of low-level or high-level reservoirs. Underground storage basins are usually formed by excavation, while ground-level reservoirs are constructed by earth embankment. Such reservoirs are lined with concrete, Gunite, asphalt, or a plastic sheet over the sides and bottom to prevent or reduce water loss in storage.[16] In earthquake zones, reinforced concrete or a series of flat bed steel compartments are mandatory. Reinforced concrete is often selected because of its minimal rate of deterioration from water contact. Elevated storage tanks and standpipes are constructed of steel with an interior coating applied to prevent corrosion.[17]

Tanks constructed of redwood are common in the western United States, being used by small communities, recreational areas (state and federal), mobile home parks, and motels. Many of these redwood structures are plastic or fiberglass lined to prevent leakage and bacterial colonization in the redwood pores. Investigation of redwood storage reservoirs has revealed that the coliform, *Klebsiella pneumoniae* can colonize such structures. This coliform metabolizes the leached-out wood sugars (cyclitols) from the staves as a source of nutrient.[18,19] The problem is most acute in new redwood tanks and can be controlled by maintaining a free-chlorine residual of 0.2 to 0.4 mg/L until the nutrient supply is leached away with tank usage over a 2-year period.[20] More details are provided in the section on water supply storage.

Care must be taken to prevent potential contamination of the high-quality water entering storage reservoirs and standpipes. One area of concern is in the application of coating compounds over the inner walls of tanks to maintain tank integrity. Organic polymer solvents in bituminous coating materials may not entirely evaporate even after several weeks of ventilation. As a consequence, the water supply in storage may become contaminated from the solvent-charged air and from contact at the side wall. Some of these compounds are assimilable organics that support regrowth of heterotrophic bacteria during warm-water periods.[13,21–23] Liner materials, also used to prevent water loss, may contain bitumen, chlorinated rubber, epoxy resin, or tar-epoxy resin that will eventually be colonized by microbial growth and slime development.[13] PVC film and PVC coating materials are other sources of microbial activity. Nonhardening sealants (containing polyamide and silicone) used in expansion joints should not be overlooked as a possible source of microbial habitation.

Water volumes in large reservoirs mix and interchange slowly with water that is actually distributed to service lines. Standpipes, in contrast, provide a fluctuating storage of water during a down surge, thereby providing surge relief in the system. Abrupt changes in water flow that sometime occur during surge relief can cause the steady-state nature of sediment deposits to become unstable, releasing viable bacteria from biofilm sites into the main flow of water.[24]

Reservoirs of treated water should be covered whenever possible to avoid recontamination of the supply from bird excrements,[25,27] air contaminants, and surface water runoff. The health concern with bird excrement is that this wildlife may be infected with *Salmonella* and protozoans pathogenic to man. Within the wildlife population in every area (as is true for any community of people), constant supply of infected individuals exists that shed pathogenic organisms in their fecal excretions. Sea gulls are scavengers and often are found at landfill locations and waste discharge sites searching for food, which is often contaminated with a variety of pathogens. At night, birds frequently turn inland to aquatic areas, such as source water impoundments and open finished water reservoirs, to roost, thereby introducing pathogens through their fecal excrements.[27]

Air pollution contaminants and surface water runoff can contribute dirt, decaying leaves, lawn fertilizers, and accidental spills to a water supply that is not covered. Such materials increase the productivity of the water by providing support to food-chain organisms and nitrogen-phosphate requirements for algal blooms. This degrades the treated water quality.

Covered distribution system storage structures also are subject to occasional contamination because of air movement in or out of the vents as a result of water movement in the structure. During air transfer, the covered reservoir is exposed to fallout of dust and air pollution contaminants from the inflowing air. Vent ports or conduits from the service reservoir to the open air should be equipped with suitable air filters to safeguard the water quality from airborne contaminants. Birds and rodents may also gain access through air vents that have defective screen protection. Bird or rodent excrement around the vents may enter the water supply and become transported into the distribution system before dilution and residual disinfection is able to dissipate and inactivate the associated organisms.

## Factors Contributing to Microbial Quality Deterioration

Factors contributing to deterioration of microbial quality may be associated with source water quality, treatment processes, or distribution network operation and maintenance. Each of these areas is reviewed in the following sections.

### Source water quality

Bacteria in distributed water mostly originate in the source water. High-quality groundwater may be characterized as containing less than 1 coliform per 100 mL and a heterotrophic bacterial population that is often less than 10 organisms per milliliter, even in waters that reach the regrowth stimulating temperature of 10°C or more.[28] These microbial qualities show little fluctuation, because of groundwater aquifer protection from surface contamination. Some groundwater, however, is not insulated from surface contamination.[29] Agricultural fertilizer runoff can contain nitrate, and improperly isolated landfills may introduce a variety of organics, many of which are biodegradable. In such situations, bacterial populations in the groundwater become excessive, resulting in 1000 to 10,000 heterotrophic bacteria per milliliter. Groundwater containing a high concentration of iron or sulfur compounds provides nutrients for a variety of nuisance bacteria that may become so numerous as to restrict water flow from a well. Where groundwater is poorly protected from contamination by storm-water runoff and wastewater effluents, coliforms and pathogens may be introduced into distributed water unless a disinfection barrier exists.[30]

Surface water sources are subject to a variety of bacterial contaminants introduced by storm-water runoff over the watershed and the upstream discharges of domestic and industrial wastes. While impoundments and lakes provide water volume and buffering capacity to dilute bacterial contamination and thereby reduce density fluctuations, counterproductive factors must be considered. Lake turnovers, decaying algal blooms, and bacterial nutrient conditions deteriorate water quality and introduce a wide range of organisms (some of which may be pathogens) to the source water intake that may pass through marginal treatment processes or improperly operated treatment systems.

### Treatment processes

Water supplies using a single treatment barrier (disinfection) for surface water treatment will not prevent a variety of organisms (algae, protozoan, and multicellular worms and insect larvae) from entering the distribution system.[31] While many of these organisms are not immediately killed by disinfectant concentrations and contact times (C-T values) suited to controlling coliforms and virus,[32] they eventually die because of lack of sunlight (algae) or adverse habitat (multicellular worm and insect larvae). Disinfection is also less effective on a variety of environmental organisms that include spore formers (*Clostridia*), acid-fast bacteria, gram-positive organisms, pigmented bacteria, fungi, yeast, and protozoan cysts. Any of these more resistant organisms may be found in the pipe environment.[33-37]

Filtration is an important treatment barrier for protozoan cysts (*Entamoeba, Giardia, Cryptosporidium*), being more effective than the usual disinfectant concentration and contact times.[38] Improperly operated filtration systems have been responsible for releasing concentrated numbers of entrapped cysts (*Giardia* and *Cryptosporidium*) as a result of improper filter backwashing procedures or filter bypasses and channelization.[39] Filter sand may become infested with nematodes from stream or lake bottom sediments that shed into the process water and the distribution system. While nematodes are not pathogenic, they may harbor viable pathogens ingested from source water or filter media beds.

Properly operated water treatment processes are effective in providing a barrier to coliforms and pathogenic microorganisms reaching the distribution system. This does not, however, preclude the passage of nonpathogenic organisms through the treatment train. Investigation of heterotrophic bacterial populations revealed that a 4 log (99.99 percent) or better reduction can occur through conventional treatment processes (storage, chlorination, coagulation, settling, and rapid sand filtration) for many of these organisms.[37] A less significant reduction of the subpopulation of pigmented organisms in the heterotrophic flora occurs, however, so that these organisms may predominate the residual densities of detectable bacteria after processing and become the dominant strains in distributed water. In one study,[37] yellow and orange pigmented bacteria were consistently present in distributed water at a site 25 mi (40 km) from the water treatment plant. Pink and red strains were frequently found as well. Pigmented bacteria also occurred in waters containing a free-chlorine residual (although in reduced numbers), indicating the possibility that these organisms are either chlorine-resistant or chlorine-tolerant.

The effects that granular activated carbon (GAC) filtration or biological activated carbon treatment have on distributed water quality is largely undocumented. Several coliform species (*Klebsiella, Enterobacter,* and *Citrobacter*) have been found to colonize GAC filters, regrow during warm-water periods, and discharge into the process effluent. Activated carbon particles have also been detected in finished water from several water plants using powdered activated carbon or GAC treatment. Over 17 percent of finished water samples examined from nine water treatment facilities contained activated carbon particle fines colonized with coliform bacteria.[40] These observations confirm that activated carbon fines provide a transport mechanism by which microorganisms penetrate treatment barriers and reach the distribution system. Other mechanisms that could be involved in protected transport of bacteria into the distribution system include aggregates or clumps of organisms released from colonization sites in GAC filtration and by passage of coagulants.

Furthermore, heterotrophic bacterial densities in distributed water from a full-scale treatment train using GAC were found to be significantly higher (Table 18.1) than in water from a similar full-scale treatment train that did not employ GAC.[41] Upon entering the pipe network, persistance and regrowth of these organisms will be influenced by the same factors that also impact disinfectant effectiveness: habitat locations, water temperature, pH, and assimilable organic carbon concentrations.[43,44]

TABLE 18.1 Treated Water Bacterial Populations Following Various Water Treatment Processes Using Standard Plate Medium or R-2A Medium with Extended Incubation Times[†] (organisms/mL)

| Sampling day | Lime-softened water | | | Sand filter effluent | | | GAC adsorber effluent | | |
|---|---|---|---|---|---|---|---|---|---|
| | SPC, 2 days | SPC, 6 days | R-2A, 6 days | SPC, 2 days | SPC, 6 days | R-2A, 6 days | SPC, 2 days | SPC, 6 days | R-2A 6 days |
| Initial | 120 | 350 | 510 | 890 | 1,200 | 1,500 | <1 | 140 | 220 |
| 7 | 31 | 202 | 510 | 820 | 22,000 | 35,000 | 1 | 24,000 | 95,000 |
| 14 | 7 | 7 | 130 | <1 | 1,200 | 9,400 | <1 | 600 | 4,400 |
| 21 | 7 | 18 | 150 | 2200 | 2,500 | 33,000 | <1 | 5,200 | 16,000 |
| 28 | 3 | 39 | 530 | 700 | 7,800 | 67,000 | 1 | 11,000 | 55,000 |
| 35 | <1 | 490 | 330 | 100 | 6,000 | 25,000 | <1 | 12,000 | 74,000 |
| 42 | 70 | 120 | 1700 | 1,200 | 71,000 | 22,700 | N.D. | 56,000 | 52,000 |
| 49 | 9 | 1200 | 23 | 5000 | 41,000 | 3,000 | 80 | 4,200 | 100 |
| 56 | <1 | 10 | <1 | <1 | 700 | 12,000 | N.D. | 1,900 | 50,000 |
| 63 | 29 | 190 | 170 | 170 | 2,000 | 3,000 | N.D. | 5,000 | 48,000 |

[†]Data revised from Symons et al., Ref. 41. All cultures incubated at 35°C; SPC = standard plate count (SPC agar); N.D. = not done

## Distribution network operation and maintenance

Because the public health concern for drinking water microbiological quality has been based solely on limiting total coliform occurrence,[45] the acceptance of new or repaired mains has depended only on a laboratory report that no coliforms are detected in water held in the new pipe sections. A more rigorous check on installed pipe cleanliness would include examination of water in the pipe section for heterotrophic bacterial density in addition to total coliforms.[46,47] The HPC in this situation measures the myriad of soil organisms that could have been introduced into the pipe section during construction or repair. Soil deposits in new pipe sections may not only introduce a variety of heterotrophic bacteria to the distribution network, but they may also create some measure of associated protection from disinfection in the particulates released. Some of the poor disinfection results attributed to chlorine applied in these situations may be traced to excessively dirty line sections or joints.

In Halifax, Nova Scotia, new line construction was found to contain pieces of wood used in construction work embedded in some pipe sections.[48] An environmental *Klebsiella* associated with the wood forms, adjusted to the distribution pipe environment and colonized the exposed wood surfaces. Shearing forces created by water flow velocity introduced this coliform into the passing water, resulting in consistently unsatisfactory test results. The problem was resolved by the addition of more than 5 mg/L lime to the process water that elevated the water pH to 9.1 in the distribution system. At this pH, *Klebsiella* were either inactivated or entrapped in the pipe sediment and the problem was eliminated.

Upon completion of a new pipe line or after emergency repairs are made to a line break, flushing water through the pipe section at a minimum velocity of 10 ft/s (76.2 cm/s) to remove any soil particles is advisable.[49] In lines with diameters of 16 in (4.1 cm) or more, this velocity may not be attainable or may be ineffective so that polypig or foam swab applications should be considered. Following flushing, disinfectant should be introduced into the new sections and the water held for 24 to 48 h to optimize line sanitation. Bacteriological tests for coliforms and the HPC should then be performed. If the results of these tests are satisfactory ($< 1$ coliform/100 mL; $< 500$ HPC/milliliter), the line may be placed in service. If not, the line should again be flushed and refilled with distributed water dosed with 50 mg/L available chlorine. Chlorine levels should not decrease below 25 mg/L during the 24-h holding period before repeat bacteriological testing. In pipes free of extraneous debris, free available chlorine (1 to 2 mg/L), potassium permanganate (2.5 to 4.0 mg/L), or copper sulfate

(5.0 mg/L) have been used to meet coliform requirements.[49-51] Only free available chlorine, however, was found to eliminate large numbers of heterotrophic bacteria.[49]

## Microbial Quality of Distributed Water

Little in-depth information is available in the literature on the identity of all heterotrophic organisms that can be found in water supplies because of the difficulties in isolating and identifying a broad spectrum of organisms with widely differing growth requirements. Previous studies have generally been limited to identification of those organisms that were associated with consumer complaints about taste, odor, and color.[52] Other studies have explored spoilage problems in the food, beverage, cosmetic, and drug industries that use large quantities of potable water in production processes.[53-56] The following section is a brief profile of the organisms generally considered to be nonpathogenic by ingestion that have been found in distribution systems.

### Bacterial profiles

Heterotrophic organisms in water supplies most often originate in the source water, survive the rigors of treatment processes and adapt to the environment of the water distribution network.[57] Trace organic nutrients in the water or concentrated in reservoir and pipe sediment can support a diverse population (Table 18.2) of surviving organisms (heterotrophs). Habitat sites that are successfully colonized almost always invoke the mixed growth of organisms that are attached to each other or to particles, sediments, and porous structures in the pipe tubercles or sediments. The spectrum of organisms may include many gram-negative bacteria (such as coliforms, *Pseudomonas,* etc.) sporeformers, acid-fast bacilli, pigmented organisms, actinomycetes, fungi, and yeast. In addition, various protozoans and nematodes that feed on these microbial populations can also be found in protected areas of slow-flow sections or dead ends.

**Coliforms.**   Total coliform bacteria are used primarily as a measure of water supply treatment effectiveness and as a measure of public health risk. These gram-negative bacteria are occassionally found in water supplies. Data in Table 18.3 show the wide range of coliform species that may be encountered.[57-63] While coliform bacteria are chlorine-sensitive, they may be protected from disinfectant action in particulates originating in source water turbidity, activated carbon fines released from GAC filtration,[64] inorganic sediments in the con-

**TABLE 18.2   Organisms in the Standard Plate Count Population of Three Water Supplies**

| Water plant filter effluent and clearwell | | Distribution water | |
|---|---|---|---|
| Significant categories | Organisms Isolated | Significant categories | Organisms isolated |
| Total coliforms | *Klebsiella pneumoniae*<br>*Enterobacter cloacae*<br>*Erwinia herbicola* | Total coliforms | *Klebsiella pneumoniae*<br>*Enterobacter cloacae*<br>*Erwinia herbicola*<br>*Enterobacter aerogenes*<br>*Escherichia coli*<br>*Aeromonas hydrophila*<br>*Citrobacter freundii* |
| Coliform antagonists | *Pseudomonas fluorescens*<br>*Pseudomonas maltophila*<br>*Flavobacterium* sp. | Coliform antagonists | *Pseudomonas fluorescens*<br>*Pseudomonas maltophila*<br>*Flavobacterium* sp.<br>*Pseudomonas cepacia*<br>*Pseudomonas putida*<br>*Pseudomonas aeruginosa*<br>*Bacillus* sp.<br>*Actinomycetes* sp. |
| Opportunistic pathogens | *Pseudomonas maltophila*<br>*Klebsiella pneumoniae*<br>*Moraxella* sp.<br>*Staphylococcus* (coagulase +) | Opportunistic pathogens | *Pseudomonas Maltophila*<br>*Klebsiella pneumoniae*<br>*Moraxella* sp.<br>*Staphylococcus* (coagulase +)<br>*Pseudomonas aeruginosa*<br>*Klebsiella rhinoscheromatis*<br>*Serratia liquefaciens*<br>*Serratia marcescens* |
| Category not established | *Acinetobacter calcoaceticus*<br>*Neisseria flavescens* | Category not established | *Acinetobacter calcoaceticus*<br>*Streptococcus* sp.<br>*Bacillus* sp.<br>*Corynebacterium* sp.<br>*Micrococcus* sp.<br>*Nitrococcus* sp. |

SOURCE: Data from Geldreich et al., Ref. 57.

TABLE 18.3    Coliforms Identified in 111 Public Water Supply Distribution Systems[†]

| Citrobacter | Escherichia |
|---|---|
| C. freindii | E. coli |
| C. diversus | |
| Enterobacter | Klebsiella |
| Enter. aerogenes | K. pneumoniae |
| Enter. agglomerans | K. rhinoscleromatis |
| Enter. cloacae | K. oxytoca K. ozaenae |

†Published data[57-63] from various distribution systems in six states (United States) and Ontario Province (Canada).

tact basin, and inadequate conditions for disinfection action (contact time, water pH, and temperature). Coliforms may also enter the distribution system through water line breaks, negative water pressure and cross connections.

Coliform colonization of the pipe network may occur in areas where porous sediments accumulate.[31,65] These sediments develop from the action of corrosion or are found in slow-flow and dead-end areas that permit the accumulation of particulates that passed through the treatment works because of improper processing. Such sites are attractive to bacterial colonization because these deposits adsorb trace nutrients from the passing water and provide porous surface areas for bacterial attachment against the flow of water. Among the coliform bacteria, *Klebsiella pneumoniae, Enterobacter aerogenes, Enterobacter cloacae,* and *Citrobacter freundii* are the most successful colonizers. Encapsulation of coliforms provides protection from the effects of chlorine or other disinfectants. Once total coliforms become established in an appropriate habitat, growth can occur and result in occasional sloughing of cells into the flowing water. This condition can persist until either shearing effects of water hydraulics limit colony growth or elevated disinfectant residuals penetrate the protective habitat and inactivate the microbial population.

**Antibiotic-resistant bacteria.** Heterotrophic bacteria in water supply that are resistant to one or more antibiotics may pose a health threat if these strains are opportunistic pathogens or serve as donors of the resistant factor to other bacteria that could be pathogens. Antibiotic-resistant (R factor) bacteria may originate in surface water sources used for public water supplies.[66,67] Polluted waters acquire bacteria with R factors from the fecal wastes of man and domestic animals in wastewater effluents and storm-water runoff, farm pasture lands, and feedlots. Farm animals in particular may receive continuous doses of antibiotics in animal feed and become constant generators of a variety of antibiotic-resistant bacteria. Although treatment processes in-

activate or remove antibiotic-resistant organisms (*Aeromonas,
Hafnia,* and *Enterobacter*) in the source water, a shift of this
transmissable factor to other heterotrophic bacteria (*Pseudomonas/
Alcaligenes* group, *Acinetobacter, Moraxella, Staphylococcus,* and
*Micrococcus*) may occur.[68]

Water supply treatment processes apparently act as a mixing chamber for R factor transfers with surviving organisms acquiring multiple
resistances to different antibiotics. Many of the transformations occur
in the biofilm established on activated carbon and sand filters.[68] The
disinfection process may also have a major impact on the selection of
drug-resistant bacteria. The reason for the common occurrence of
streptomycin resistance among bacteria that survive chlorination is
not known. Multiple antibiotic-resistant bacteria passing through water treatment are more tolerant to metal salts [i.e., $CuCl_2$, $Pb(NO_3)_2$,
and $ZnCl_2$].[68] Examination of bacteria for multiple antibiotic-
resistance from two sites in a distribution system indicate a dynamic
state of fluctuation (16.7 percent R factor organisms at one site and
52.4 percent at the other location). In a typical population of 100
heterotrophic bacteria per milliliter of water from the distribution
system, 40 to 70 of these organisms could be expected to have some
antibiotic-resistance factors.[69] What health risk this represents, particularly when the heterotrophic bacterial population is above the 500
organisms or more per milliliter limit suggested for potable water, is
not yet clearly understood.

**Mycobacteria.** Aquatic mycobacteria (*Mycobacterium avium,* **M.**
*gordonae, M. flavescens, M. fortuitum, M. chelonae,* and *M. phlei*) may
colonize susceptible humans (immunocompromised patients, surgery
cases, individuals on kidney dialysis) by a variety of routes including
water supply.[33,70–71] Densities of environmental mycobacteria in
groundwater supplies ranged from 10 to 500 organisms per 100 mL
and from a few per liter to 60 organisms per 100 mL in water supplies
using surface water. Source waters, open finished water reservoirs,
soil contaminants introduced in groundwater, line repairs, and new
line construction are the major origins of mycobacteria.

The density of mycobacteria (acid-fast bacteria) in surface waters
contaminated with wastewater may approximate $10^4$ organisms per
100 mL. In water treatment, the most significant reductions for these
organisms occur during sand filtration. For example, during an 18-
month study of two water systems, reductions in the concentration of
acid-fast bacteria by rapid sand filtration ranged from 59 to 74 percent. Final disinfection, including the presence of free residual chlorine did not have a statistically significant effect on the residual densities of mycobacteria leaving the treatment plant.[33,72] Even the

presence of a free-chlorine residual at a low pH (5.9 to 7.1) did little to reduce these organisms in the distribution systems. The ability of mycobacteria to survive in water distribution systems may be influenced by the protective waxy nature of the cell wall to resist 1.5 mg free chlorine after 30 min contact time. Some regrowth was noted in the pipe environment at the ends of the distribution lines where chlorine residual disappeared, increased total organic carbon concentrations occurred (providing bacteria nutrients), and pipe sediments and tuberculation accumulated (sites for bacterial attachment). Further amplification of the acid-fast bacteria can be expected in some building plumbing systems and their associated attachments.

**Pigmented bacteria.** A characteristic of some bacteria that may be present in distributed water is the ability to form brightly colored pigments. Little is known about their health significance from ingestion; however, some strains have been the cause of gastroenteritis while others have been associated with pyrogenic reactions and septicemia.[73] Study of the seasonal occurrence of pigmented bacteria in a treated water supply suggests that these bacteria originated from the source water supply at some previous time or from line breaks and repairs to the distribution lines.[37]

More pigmented bacteria were found at a distribution sampling site 25 mi (40 km) from the water treatment plant than in either the river source water, the presedimentation basin, after flocculation, in chlorinated influent to rapid sand filter, or in the filtered water. This observation suggests that pigmented bacteria surviving treatment became adapted to the distribution environment and regrew when conditions were favorable. The proportion of yellow, pigmented bacteria in the source water was lowest in the autumn and highest in the summer. The mean proportion of pigmented bacteria was calculated to be 26 percent of the heterotrophic population. Orange-pigmented bacteria comprise less than 10 percent of the plate count in any season (average 5.8 percent); pink or red organisms, less than 3 percent. Pigmented bacteria such as purple, black, and brown organisms ranged from 0.2 to 1.5 percent.

In contrast, at the distribution sampling site, pigmented bacteria comprised from 65 to nearly 90 percent of the heterotrophic population, depending on the season. Lowest and highest percentages occurred in the winter (5 percent) and autumn (70 percent), respectively. Orange-pigmented bacteria, on the other hand, appeared as 82 percent of the total pigmented population in the winter and only 10 percent of the population in the autumn. Pink bacteria ranged from <1 percent (winter, spring, and fall) to 9 percent (summer). Other pigmented bacteria (purple, black, and brown) were rarely encountered in the distributed water. A mixed population of pigmented bacteria was generally present in the distributed water with surges to dominance among these groups when environmental factors were favorable. The question of whether most pigmented bacteria isolated from distribution systems that contain a free-chlorine residual are chlorine-resistant or chlorine-tolerant is unanswered. Because these organisms are found in drinking water, pigmented bacteria have been reported from water used in hospital therapy machines or on instruments.[74–76]

**Disinfectant-resistant bacteria.**  This bacterial group comprises a wide range of organisms of very limited health significance whose presence is a reflection of disinfection treatment effectiveness or a depiction of the heterotrophic bacterial flora of a stagnant public water supply. Comparison of bacterial genera present in two different public water supplies in southern California by Olson and Hanami[59] indicated that variation in the diversity of the bacterial population can also be related to source water, that is, surface water versus groundwater. The predominant genera in one untreated groundwater supply were *Acinetobacter* and *Pseudomonas. Klebsiella* was also found intermittently in very low numbers in this supply. Chlorinated distributed water from a surface water source contained *Acinetobacter, Pseudomonas/Alcaligenes,* and *Flavobacterium* as predominant genera.

Chlorination of water introduces a strong selective pressure. Ridgway and Olson[77] found that bacteria isolated from a chlorinated surface water distribution system were more resistant to both combined and free forms of chlorine than members of the same genera isolated from an unchlorinated groundwater system, suggesting that selection for more chlorine-tolerant microorganisms occurred. Moreover, the differences in the sensitivities of the two bacterial populations did not appear to be related to substances other than chlorine, because the overall water chemistry of the two public water supplies was similar. The most resistant microorganisms isolated from either water system (gram-positive, spore-forming bacteria, actinomycetes, and some micrococci) were able to survive exposure to 10 mg/L free chlorine for 2 min. The most chlorine-sensitive bacteria isolated from these two water distribution systems (*Corynebacterium/Arthrobacter, Klebsiella, Pseudomonas/Alcaligenes, Flavobacterium/Moraxella, Acinetobacter,* and most gram-positive micrococci) were readily killed by a chlorine concentration of 1.0 mg/L or less. The apparent contradiction in occurrence of genera that are grouped as both chlorine-sensitive and chlorine-resistant reflects variations of strains within genus, physical aggregation of cells, and changes associated with particulate matter.

**Actinomycetes and other related organisms.** These bacteria create some of the taste and odors reported in water supplies. Actinomycetes in source waters of temperate climates may be reduced by approximately 10-fold with storage in holding basins prior to other treatment measures. Some drinking water treatment process configurations that include slow sand filtration or GAC filter adsorbers may support an increase in density and selection of species that enter the distribution network. *Nocardia* strains were predominant in finished water from water treatment trains consisting of aeration and filtration or aeration, sand filtration, ozonation, and activated carbon adsorption. *Micromonospora* made up the greatest percentages in finished water from treatment chains consisting of flocculation, settling and slow sand filtration, or aeration and sand filtration. Some growth may also be expected on PVC-coated walls in finished water reservoirs[78] and in the distribution lines where organic material accumulates in sediments.[79,80] In addition to their presence in cold tap water, thermophilic actinomycetes and mesophilic fungi were found in several hot-water samples in three municipalities in Finland.[81] *Thermoactinomyces vulgaris* was the predominant actinomycete found in 11 of 15 water distribution systems examined. In two studies of treated water supplies in England, chlorination alone was not effective in eliminating *Streptomycetes*. Median densities were two *Streptomycetes* per 100 mL of distributed water. Taste and odor complaints involving *Actinomycetes* often were from waters that had *Streptomycetes* or *Nocardia* counts greater than 10 organisms per 100 mL.

**Fungi.** Although many fungi have been found in the aquatic environment,[82] focus on these organisms in water supply has been limited to their active degradation of gasket and joint materials and association with taste and odor complaints by consumer.[79–81] This focus is somewhat surprising because certain fungi are also the cause of nosocomial infections in compromised individuals, being important agents responsible for allergenic or toxigenic reactions through inhalation of water vapor or by body contact during bathing.

Fungi occur at relatively low densities in ambient waters.[83] Source water storage in temperate climates and chemical coagulation prior to filtration and disinfection has been found to improve the removal efficiency somewhat, but provided no absolute treatment barrier to these organisms. Fungi present in source waters may pass through sand filtration and disinfection treatment processes.[79,80,84] In one study,[81] these organisms were detected in 29 of 32 treated water samples.

Breakthrough in the treatment barriers, soil contamination from line repairs and air passage in storage reservoirs and standpipes are pathways by which fungi enter the distribution network (Table 18.4). Fungi occurrences in distributed water have been observed to be more frequent during summer water temperature conditions. *Aspergillus fumigatus* was the predominant species detected in the distribution system of 15 water supplies in Finland.[81] A variety of fungi (*Cephalosporium* sp., *Verticillium* sp., *Trichodorma sporulosum, Nectria veridescens, Phoma* sp., and *Phialophora* sp.) were identified from water service mains in several water supplies in England.[77,78] Fungi densities in these drinking waters were usually less than 10 organisms per 100 mL.

Water supplies with fungal densities of 10 to 100 organisms per 100 mL frequently receive customer complaints of bad taste or odor.[79] The four most frequently occurring genera of filamentous fungi in two distribution systems (chlorinated and unchlorinated supplies in southern California) were *Penicillium, Sporocybe, Acremonium,* and *Paecilomyces.*[82] In the unchlorinated system, *Penicillium* and *Acremonium* represented approximately 50 percent of 538 colonies identified among 14 genera, while *Sporocybe* and *Penicillium* accounted for 56 percent of 923 fungal strains distributed within 19 genera that occurred in the chlorinated supply. The majority of filamentous fungi appeared to be nonpathogenic saprophytes. The mean

TABLE 18.4    Fungi Densities in Water Supply[†]

| System[‡] | Sampling point | Number of samples | Average density per 100 mL |
|---------|----------------|-------------------|----------------------------|
| SR | Source water | 6 | 2.7 |
| MW |  | 3 | 54.0 |
| BG |  | 8 | 2.0 |
| WH |  | 2 | 8.0 |
| BL |  | 6 | 22.0 |
| BL | Aerated effluent | 3 | 3.0 |
| SR | Finished water | 4 | 3.3 |
| WH |  | 2 | < 1.0 |
| BL |  | 5 | 5.0 |
| BL | Storage tank | 5 | 40.0 |
| SR | Residential tap | 34 | 3.2 |
| MW |  | 65 | 3.8 |
| BG |  | 11 | 2.6 |
| WH |  | 22 | 2.6 |
| BL |  | 20 | 3.7 |
| WH | Fire hydrants | 4 | 16.6 |
| BL |  | 7 | 9.0 |

†Data from five small water systems revised from Rosenzweig et al., Ref. 35.
‡BG, Bradford Glen water system; BL, Brooklawn water system; MW, Marshallton Woods water system; SR, Spring Run water system; WH, Woodbury Heights water system.

density of fungi from the unchlorinated and chlorinated systems were 18 and 34 organisms per 100 mL, respectively. Conidia of *Aspergillus fumigatus, A. niger,* and *Penicillium oxalicum* isolated from distribution systems of three small water supplies (Pennsylvania) showed a greater resistance to chlorine inactivation than yeast that were more resistant than coliforms.[35]

Yeasts are another category of fungi found in the aquatic environment. While chemical coagulation and sedimentation will remove 90 to 99 percent of yeasts from source water and rapid sand filtration will remove at least 90 percent, disinfection is less effective. The resistance of yeast to free available chlorine is primarily a result of the thickness and rigidity of the cell wall, which presents a great permeability barrier to chlorine. Specific species identified in distribution waters include *Candida parapsilosis, C. famata, Cryptococcus laurentis, C. albidus, Rhodotorula glutinis, R. minuta,* and *R. rubra.*[84,85] Densities of yeasts reported in finished drinking water average 1.5 organisms per liter. While these initial densities are low, yeast slowly colonize water pipes and may become more numerous with pipe age.

### Disinfectant stability during water distribution

Stability of disinfectants during water supply distribution is important, particularly to reduce or prevent colonization by surviving organisms and to inactivate bacteria associated with the intrusion of contamination in the pipe network. Microbial colonization may lead to corrosive effects on the distribution system and adverse aesthetic effects such as taste, odor, and appearance problems. Also, the regrowth of health-related opportunistic organisms and their impact on coliform detection should not be dismissed as a trivial problem. Analysis of data (Table 18.5) taken from the National Community Water Supply Study of 969 public water systems[46] revealed that standard plate counts of 10 organisms or less were obtained in over 60 percent of these distribution systems that had a measured chlorine residual of approximately 0.1 to 0.3 mg/L. Protective sediment habitats and selective survival of disinfectant-resistant organisms were the reasons why residual chlorine concentrations greater than 0.3 mg/L chlorine produced no further decreases in the HPC. In an extensive study involving 986 samples taken from the Baltimore and Frederick, Md., distribution systems, the maintenance of a free-chlorine residual was found to be the single most effective measure for maintaining a low standard plate count.[86]

A disinfectant residual in the distribution system can be very effective in the inactivation of pathogens associated with contaminants

TABLE 18.5    Effect of Varying Levels of Residual Chlorine on the Total Plate Count in Potable Water Distribution Systems

| Standard plate count[†] | Residual chlorine, mg/L | | | | | | | |
|---|---|---|---|---|---|---|---|---|
| | 0.0 | 0.01 | 0.1 | 0.2 | 0.3 | 0.4 | 0.5 | 0.6 |
| < 1 | 8.1[‡] | 14.6 | 19.7 | 12.8 | 16.4 | 17.9 | 4.5 | 17.9 |
| 1–10 | 20.4 | 29.2 | 38.2 | 48.9 | 45.5 | 51.3 | 59.1 | 42.9 |
| 11–100 | 37.3 | 33.7 | 28.9 | 26.6 | 23.6 | 23.1 | 31.8 | 28.6 |
| 101–500 | 18.6 | 11.2 | 7.9 | 9.6 | 12.7 | 5.1 | 4.5 | 10.7 |
| 501–1000 | 5.6 | 6.7 | 1.3 | 2.1 | 1.8 | 0 | 0 | 0 |
| > 1000 | 10.0 | 4.5 | 3.9 | 0 | 0 | 2.6 | 0 | 0 |
| Number of samples | 520 | 89 | 76 | 94 | 55 | 39 | 22 | 28 |

†Standard plate count (48 h incubation, 35°C).
‡All values are percent of samples that had the indicated standard plate count.
SOURCE: Data from Geldreich et al., Ref. 46.[46]

seeping into large volumes of high-quality potable water. Obviously, limitations to this protective barrier exist. Chlorine residuals in distributed water often range from 0.1 to 0.3 mg/L and at these concentrations will not provide adequate protection against massive intrusions of gross contamination characterized by odors, color, and milky turbidities. Recent studies conducted in a small abandoned distribution system on a military base indicate that at tap water pH 8, with an initial free-chlorine residual of 0.7 mg/L and wastewater added to levels of up to 1 percent by volume, 3 logs (99.9 percent) or greater bacterial inactivation were obtained within 60 min. Viral inactivation under these conditions was less than 2 logs (<99 percent). In laboratory reservoir experiments, where the residual chlorine is replenished by inflow of fresh uncontaminated chlorinated tap water, greater inactivation was observed at the higher wastewater concentrations tested. Furthermore, a free-chlorine residual was more effective than a combined chlorine residual in the rapid inactivation of microorganisms contained in the contaminant.[86]

Distribution system problems associated with the use of combined chlorine residual or no residual have been documented in several instances.[87–89] In these cases, the use of combined chlorine is characterized by an initial satisfactory phase in which chloramine residuals are easily maintained throughout the system and bacterial counts are very low. Over a period of years, however, problems may develop, including increased bacterial counts, decline in combined chlorine residual, increased taste and odor complaints, and reduced main carrying capacity. Conversion of the system to free-chlorine residual produces an initial increase in consumer complaints of taste and odors resulting from oxidation of accumulated organic material and difficulty in maintaining a free-chlorine concentration at the ends of the distribution system. With application of a systematic main flushing program in these instances, a free-chlorine residual will become established throughout the system, bacterial counts decrease, and taste and odor complaints decline.[90] Of the most common disinfectants (chlorine, chloramines, chlorine dioxide, and ozone) only ozone does not provide a lasting residual because it is highly reactive and therefore disappears very rapidly in the distribution system (see Chap. 12).

When ozone is applied, some nondegradable organics present in the process water will be converted into less complex compounds readily utilized by bacteria. Unless removed, the net result will be more assimilable organic carbon passing into the distribution system. There, they will provide increased nutrient supply for microbial colonization activity during warm-water periods.[91]

## Microbial Colonization Factors

Water mains, storage reservoirs, standpipes, joint connections, fire-plug connections, valves, service lines, and metering devices have the potential to be sites suitable for microbial habitation. No pipe material is immune from potential microbial colonization once suitable attachment sites are established. Given sufficient time, aggressive waters will initiate corrosion of metal pipe surfaces, water characteristics may change the surface structure of asbestos-cement mains, and biological activity may create pitting on the smooth inner surface of plastic pipe materials (see Chap. 17). Noting that not all pipe sections may show evidence of deterioration even after years of active service is important, the reason being the nature of the water chemistry and continuous movement of water under high-velocity conditions.

### Habitat characterizations

The diverse structures in the pipe network and its complexities, which include locations out of the mainstream of continuous moving water, are the most opportune sites for sediments to accumulate, tuberculation to expand, and microbial colonization to develop. Sediment accumulations in these sites are created, in part, from corrosion and corrosion inhibitor additives. Slow-flow areas may also be deposition sites for turbid particles in nonfiltered source water, unstable coagulants, receptors of activated carbon fines, and biological debris.

The corrosion of water mains can be controlled by manipulation of the water chemistry to cause formation of a scale lining on pipe walls,[92,93] but adverse side effects can occur because irregular, porous deposits are more suitable for microbial habitation. Adjusting the calcium carbonate saturation index (Langelier index) for a slight oversaturation does not ensure deposition of a uniform $CaCO_3$ scale.[87,88,94] A uniform $CaCO_3$ scale does not form under conditions of low velocity, high pH values, or high sulfate content.[89] At low velocities, scale is too porous and nonuniform. Thus, in low-flow and dead-end portions of the system, where a minimum flow velocity of 1 ft/s (0.3 m/s) cannot be achieved, scale will be more porous and corrosion will break through the pipe wall. Tuberculation is increased at high pH values because scale formation is less uniform (insufficient calcium and alkalinity present) and less effective in coating the pipe walls.[95,96] Above pH 9, orthosilic acid becomes a factor and changes the chemical nature of the deposits (see Chap. 17). The presence of high sulfate content in the water adversely affects the calculation of a calcium carbonate saturation index, because of the formation of $CaSO_4$. This results in producing false high index (Langelier) values and misjudgment of the treatment application. In water that is slightly oversaturated with $CaCO_3$, selective formation of inorganic precipitate may exist instead of inorganic scale on the pipe walls. The result is a light precipitate that can entrap and protect bacteria. This material will then settle in dead ends or low-flow sections and produce sediment accumulations in finished water reservoirs.

### Turbidity and particulate effects

Turbidity in finished water also contributes to sediment accumulation in the dead ends of the system and in the matrix of porous scale and tubercle formations. Because turbidity is an indirect measurement of the particulate matter in water, it provides no information of the type, number, and size of particles being detected. Turbidity monitoring in the distribution system, however, is a good quality control practice. Values in excess of 5 NTU may signal the need to flush the distribution system and to search for areas of pipe corrosion that must be brought under control.[47]

In general, inorganic particles such as clay and water flocculating agents may trap a variety of organisms. These particles, however, appear to have little, if any, protective effect against disinfection action because of the absence of organic demand substances.[97] Thus, few viable cells are transported in these kinds of particles to potential habitat sites along the distribution system. For example, low turbidity levels of 0.3 to 2.0 NTU caused by alum floc spilling over into the distribution water from the Salem-Beverly, Mass., water plant did not adversely affect the efficiency of chlorination in controlling the standard plate count density, including any coliform breakthrough in treatment.[61] Similarly, a study of the North Berwick (Maine) water supply[98] indicated that a constant decrease in bacterial survival occurred with increased free-chlorine residuals, in spite of plant effluent turbidity levels (1 to 5 NTU) that were consistently higher than the level set by the National Interim Primary Drinking Water Regulations.[45] Chemical characterization of this turbidity indicated it was largely inorganic with traces of tannin, lignins, and a low total organic carbon concentration. The source water had 45 fecal coliforms/100 mL, and was a fast-flowing stream that transversed an area of low human population density along the drainage basin.

In contrast, organic particulates and algal cell masses in seasonal blooms can provide a vehicle of protected transport for microbial entities through some treatment barriers, including disinfection. Organic particulates of concern include fecal cell debris,[99] wastewater solids including aggregates of bacteria and virus,[100] protective mats of algal cells, and activated carbon fines that provide attachment sites for associated heterotrophic bacteria.[31,40,101]

### Passage of microorganisms in macroinvertebrates

Not commonly recognized as a problem in distributed water quality is the occurrence of various larger, more complex, biological organisms including crustaceans (amphipods, copepods, isopods, ostracods), nematodes, flat worms, water mites, and insect larvae such as chironomids.[102–108] While these organisms enter in the source water and most are removed by various treatment processes, some may succeed in becoming established in filter beds, releasing progeny that can successfully survive disinfection and migrate into the distribution system.[109–112] In so doing, these invertebrates may harbor and protect coliforms and pathogenic microorganisms from contact with the disinfectant at concentrations typically present in distribution systems.[114–117] The ingested bacteria may not only survive but multiply within the invertebrate host and be released at a later time when the host cell bursts or disrupts.[118–119] This phenomenon thereby may account for some of the continued release of coliform and other bacteria into the distribution system by passage through treatment barriers. While some bacterial feeders (protozoans, nematodes, etc.) are inactivated immediately by chlorine in the environment of a distribution system, others die slowly or become adapted to the available food sources (biofilms of bacteria) in selected pipe sediments or tubercules or use these sites for attachment and proceed to harvest the organisms passing in the free-flowing waters.

### Key factors in microbial persistance and regrowth

Key factors in the establishment of microbial colonization within the distribution system are a source of nutrients, a protective habitat, and a favorable water temperature for rapid growth.[57,120] Not all bacteria that enter the water supply distribution system persist or are able to adapt to this environment and regrow.[37,121–123] Within the total coliform indicator group, *Klebsiella*, *Citrobacter*, and *Enterobacter* strains are most often noted in distribution systems[48,57,124] and can grow with minimal nutrients. Other bacteria that may regrow include *Pseudomonas*, *Flavobacterium*, *Acinetobacter* and *Arthobacter*. These are especially troublesome because of their potential interference to coliform detection and acknowledged roles as opportunistic pathogens.[57,125–129]

**Bacterial nutrients.**    Nutritive substances containing carbon and nitrogen are introduced in varying concentrations from source waters. Surface waters receive a variety of organics discharged in municipal wastewater effluents, industrial wastes, and agricultural activities. While much of the nutrient content may be removed through conventional treatment, more attention to treatment refinements is needed to further reduce trace organic residuals. Applying GAC adsorption or other equivalent treatment processes for improved trihalomethane precursor control will also minimize the available bacterial nutrients and thereby provide less opportunities for microbial biofilm development and coliform regrowth. For those water utilities that use disinfection as the only treatment process for a relatively clean surface water source, a seasonal threat of organic contributions from natural lignins, algal blooms, and recirculating bottom sediments during lake destratification always exists. These organic materials pass into distribution pipe networks and become an accumulation of nutrients for a wide range of aquatic bacteria.[130]

The basic requirements for nitrogen in the metabolic processes of bacteria can be satisfied in a variety of ways. Watershed conditions involving wastewater effluents released upstream of a water intake, landfill operation in the vicinity of groundwater aquifers poorly protected by the soil barrier, and seasonal application of farm and garden fertilizers over the watershed are often the major contributors of nitrogenous compounds. Water treatment practices must also be carefully controlled to minimize addition of nitrogen. For example, application of ammonia to form chloramines may contribute excess ammonia to water. Trace amounts of metal ions ($Fe^{2+}$, $Fe^{3+}$, $Mg^{2+}$, and others) available in salts, as well as phosphates also appear to be vital to microbial persistence and adaptation to the harsh environment of minimal nutrient concentrations, toxicities of disinfectant residuals, the adversities of a shearing action of flowing water,[131] and the competition for survival from organisms in the microbial flora of drinking water.[52,132,133] The greatest success in reducing the potential for microbial persistence in these nutrients will only be achieved by further reduction of the organic portion of the essential nutrient base.[133]

**Protective habitats in pipe networks.**   Porous sediments and tubercles in the pipe network apparently adsorb nutrients and thereby concentrate these bacterial food sources from the flowing water. Bacteriological colonization found in encrustations, taken from water main sections removed from four municipal water systems during line repairs, demonstrated bacterial densities that ranged from 390 to 760,000 bacteria per milliliter.[133] Variations in bacterial densities will depend on the tubercle sample and fraction examined.[134] Using the scanning electron microscope to locate microcolony sites in tubercles, investigators observed microorganisms predominantly at the surface or near-surface. This is the area where encrustation, water interface, nutrients, and oxygen are constantly present.[31,65]

To colonize surfaces in contact with a flowing stream of water, bacteria must adhere tenaciously. They do so by means of a mass of tangled fibers called a *glycocalyx* that extend from the cell membrane and adhere to surrounding surfaces or other bacteria.[135,136] These glycocalyx adhesions to sediment coatings, tubercles, pipe joints, and rough wall surfaces prevent most of the individual cells of the microcolony from being swept away by the shearing force of flowing water. Because these fiber masses are polysaccharide materials, they may also serve as a protective barrier against the lethal effects of residual disinfectants.

Not all areas of the distribution network are favorable for microbial colonization. High water velocity in smooth pipe sections makes microbial attachment difficult, reduces nutrient adsorption, and provides little or no microbial protection from disinfectant exposure.[122] In contrast, sediment accumulations, tubercle development, and scale formation in slow-flow sections, dead-end, and rough-walled pipe sections and connecting joints are prime sites (Fig. 18.1) for colonization.[28,31,122,133,137–140] Therefore, finding patchiness in colonization along pipe surfaces should be expected.

**Water supply storage.**   Finished water reservoirs, water storage tanks, and standpipes cannot be excluded as potential sites for biofilm development. Operational conditions that may encourage bacterial colonization include prolonged storage time, reduced flow velocities and infrequent cleaning. Water is often not pulled off the bottom of reservoirs, so accumulations of sediments may occur. Sediment buildup in reservoirs is more a problem for those water systems using surface waters and limiting treatment to disinfection. Filtration of

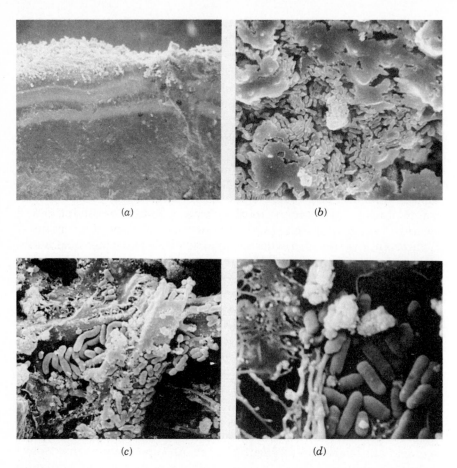

Figure 18.1  (a) Cross section of pipe tubercle showing loose surface material at water interface and compaction of deposits near the pipe wall. Sample magnification 100 × . (b) Bacterial colonization site in porous tubercle encrustation. Electron micrograph magnification 3000 ×. Material supplied from the Cambridge, Mass., distribution system pipe network. (c) Attachment site for bacterial growth in tubercle surface material. Electron micrograph magnification 4000 ×. Material supplied from the Cambridge, Mass., distribution system pipe network. (d) Protected habitat and microbial community in porous encrustation. Electron micrograph magnification 11,000 ×. Material supplied from the Cambridge, Mass., distribution system pipe network.

surface source waters will remove biological materials (algal blooms, vegetation debris) and other suspended solids that otherwise pass through the disinfection process and accumulate in slow-flow pipe sections and finished water storage. These sediments comprise significant amounts of assimilable organic complexes that support bacterial colonization in the distribution network.

Slime or biofilm development in cement structures may develop on a cement mortar with plastic additives, on sealers such as epoxy resin, bitumen, and PVC film, and on areas of cement erosion.[13] Metal structures are also subject to microbial activity in areas of corrosion and at seam and joint construction bonds. Porous materials such as brick and wood used in reservoirs are particularly suited for microbial colonization and may be difficult to dislodge through flushing and disinfection treatment.

The problem with redwood used in some water storage tanks is an example of microbial colonization that impacts on meeting the drinking water coliform standard. Research on coliform occurrences in redwood storage tanks showed that environmental strains of *Klebsiella* colonize the wood tissues of trees. The association begins at embryo fertilization by an insect or windblown pollenization event and the organisms persist in the wood pores, receiving water and nutrients from the xylem and phloem tissues of the tree.[141] New redwood tanks were found to be the source of *Klebsiella* in the water supply and the cause of massive biofilm development over the inner surfaces of the tank.[18] Disinfection and scraping the wood staves was ineffective in eliminating the bacteria because the organisms persist deep inside the wood pores until all the available wood sugars (cyclitols) are leached away or biodegraded, depriving *Klebsiella* of a continuing food source.

As a preventative measure to avoid microbial quality deterioration in all water storage structures, regularly scheduled, systematic inspection for slime development on structural surfaces in contact with the water and for sediment accumulations that serve as protective sites for microorganisms are recommended. These growths and deposits should be removed to reduce sites where taste and odor problems and microbial growth originate.

**Water temperature effects.**    Microbial regrowth is not only keyed to bacterial strains that quickly adjust to limited nutrient sources but also to water temperature elevations. Water temperature above 50°F (10°C) accelerates the growth of adapted organisms with slow generation times. Low water temperatures result in a precarious balance between new cell development and death of old cells. Data available from water systems located in geographical areas of pronounced seasonal temperature changes suggest that regrowth of HPC organisms is more pronounced than among coliforms, and abrupt surges in density may occur in summer.[57,142] Regrowth was not, however, correlated with temperature in southern California water systems,[28] possibly because the water temperatures common to those systems were continually above 50°F (10°C). Nutrient accumulations in particulates during periods of minimal microbial activity in winter may be the key to summer bacterial regrowth occurrences.

## Monitoring Program

Evidence of the reliability of a community water supply must ultimately lie in a record of no disease outbreak in the community. This approach to demonstrating the acceptable quality of a public water supply provides no opportunity to intervene in potential contamination events until waterborne illness is detected. Such a situation is unacceptable to the concepts of public health protection. The logical approach is the development of a monitoring program that will provide a continuing database of treatment effectiveness and early signals of microbial deterioration in the distribution system as indicated by increasing bacterial densities.

Two levels of monitoring a distribution system exist: compliance monitoring and special monitoring designed to better understand system characteristics and performance. Routine compliance monitoring involves the development of a systematic plan to cover all major areas of the distribution network with a sampling strategy that demonstrates the continued delivery of a safe water supply in all service areas. A special monitoring program is designed to understand system characteristics and performance as well as locate problem areas that need corrective action.

### Sampling frequency

Effective monitoring of the microbial quality of drinking water supplies requires careful consideration of sampling frequency and site selection. A number of factors must be taken into consideration if frequency of monitoring is to be optimized.[143–149] Protected groundwater supplies can be expected to be of uniform, high-quality requiring less monitoring than surface water supplies. Treatment barrier penetration is often of increased concern in water supplies using disinfection as the only treatment process, suggesting more frequent monitoring than needed for systems using conventional treatment with multiple processes in series.

Integrity of the distribution system is often an important issue in systems with old pipe networks, cross-connection potential, and flow reversals. Thus every water system is unique in its strengths and weaknesses, and careful customizing of the monitoring frequency is required. Such an approach is complex and difficult to factor into the minimum frequency requirement specified in national regulations.

Several other characteristics exist that could be used to establish a minimum level of monitoring intensity. One possibility is to relate sampling frequency to the volume of water consumed by the population served.[143] This approach has the advantage that the water supply authority will know with reasonable accuracy the volume of water being supplied to particular zones. The weakness to this concept is that the per capita consumption of water will vary seasonally as well as geographically—that is, tropical areas contrasted with temperate regions. Furthermore, very significant portions of the water demand may relate to industrial uses of the public water supply and to lawn watering by homeowners during extended dry periods.

The number of service connections, excluding fire hydrants, is known by the water utility and has received some consideration as a basis for calculating sampling frequency. The inherent weakness with using a count of service connections is that meter connections often do not reflect the number of people using the water supply in multiple-family residences, such as apartment buildings, hospitals, and rest homes, or in the workplace environment of high-rise office buildings and large industries.

Total length of the distribution pipe network has also been proposed as the basis for determining sampling frequency. The basis for this approach is the general assumption that increased length of pipe network will bring increased risk of contamination from multiple residential and commercial service connections and by ground disturbance. Unfortunately, total length of the distribution system can be misleading where local topography requires the use of long distribution lines to reach small clusters of homes. In these situations, a small water utility could have the same amount of total pipeline in its distribution network as that of a medium-sized city with a distribution network configuration of a compact grid of interconnecting pipelines.

Sampling frequency formulas based on population served is the accepted practice in many countries, including the United States.[145,150–153] The basis for this approach is the recognition that as the population increases so will the size and complexity of the system and the potential for distribution network contamination by cross connections and back siphonage.[154] The emphasis is therefore on a demonstration of safe water quality in distribution to the public, with lesser emphasis on monitoring treatment plant effectiveness.

Assuming the concept of sampling based on population served is the best alternative, the next critical consideration is in the establishment of specified numbers of samples required per given population level. While the sampling frequency established in many countries has, in general, proven to be adequate for routine monitoring of most water systems, it is not providing satisfactory monitoring of water supplies serving a population less than 10,000 in the United States.[155–157] In these systems more waterborne outbreaks occur, and often no monitoring occurs for most days in the month. To provide statistically significant data on the bacteriological quality of a small water system would require a minimum of five samples per month, not one per month, or even less frequently as specified in the National Interim Primary Drinking Water Regulations.[45] In small water supplies, the coliform test of finished water quality should be done once per week if occasional coliform occurrences are detected. Monitoring of the distribution system for coliform bacteria should not be ignored just because the pipe network in a small water system is less complicated, with fewer service connections. In small distribution systems, a disinfectant residual measurement will provide confirmation of available disinfectant protection against some contamination events that might occur through cross connections and other potential problems common to pipe network configurations.

Because of laboratory costs associated with additional monitoring, options should be considered to provide relief to small systems yet improve public health protection. For example, for small systems (< 3300 persons) that utilize a disinfected groundwater supply or filter their surface source water prior to disinfection, a triannual sanitary survey of water system operations could be substituted for some of the increased monitoring. This type of survey involves an on-site inspection of the watershed, source water, treatment processes, distribution network, and monitoring database for the public water system by competent personnel. The intent is to determine effectiveness of the water authority to provide continuous safe water to the community.

Another approach has been to use chlorine measurement as a partial substitute for coliform monitoring. Under the National Interim Primary Drinking Water Regulation rule,[45] a water authority could substitute up to 75 percent of bacteriological samples required, provided chlorine residual measurements are taken at points representative of conditions within the distribution system and at a frequency of at least four tests for each substitute coliform analysis. Furthermore, the water authority must maintain no less than 0.2 mg/L free chlorine throughout the distribution system, and daily determinations of chlorine residual are taken. While this approach successfully demonstrates the presence of a disinfectant residual throughout the distribution network, no certainty exists that coliforms might not be present at the same time because of ineffective contact time, particulate protection, and biofilm development.

Using the heterotrophic plate count measurement of less than 500 organisms per milliliter as an alternative to meeting a disinfection requirement (at least 0.2 mg/L) in 95 percent of all distribution system sites tested also has some merit in reducing monitoring requirements. Regrowth of bacteria above the 500 organisms per milliliter limit at these sites in summer would motivate the operator to flush the lines to minimize bacterial growth and perhaps restore a disinfectant residual.

Finally, the presence-absence (P-A) concept is now included in new federal regulations for drinking water.[158] The P-A concept is considered with the frequency of coliform occurrence in a water supply over a specified time span. Such data can be obtained from conventional bacteriological tests by using either the membrane filter or multiple-tube procedure, simply by translating any coliform count or positive tube results into a coliform occurrence. This concept places equal emphasis on all positive samples, regardless of density, with a limit defined by a specific percentage of positive coliform occurrences permitted. New federal regulations (effective December 31, 1990) specify a 5 percent coliform limit for all water supplies. This limit may be relaxed for supplies that have a summer-fall coliform biofilm problem. During this period, monitoring would be intensified and a search for fecal coliform or *E. coli* made to check for any fecal contaminating events. For the public health authority, information on the frequency of coliform occurrences, over the long term, is an important indication of treatment effectiveness and operator skill in providing a continuous supply of safe drinking water.

### Sample site selection

The strategy for sample site selection should be to design a sampling program to monitor all parts of the distribution system over time. Regardless of system size, a percentage of samples should be collected at certain fixed points, such as first customer location, specially constructed sampling stations on the pipe network, pumping stations, storage tanks, and previous problem sites with coliform occurrences. This portion of the sampling program may be described as "representative" sampling—that is, permanent sampling sites chosen by the utility to be representative of water quality variations within the distribution system. Other sites should be selected at random throughout the distribution system, taking care to select taps supplied with water from a service line connected directly with the main rather than to a storage tank. Random points should include areas where a possibility of contamination through cross connections and back siphonage exists, such as hospitals, schools, public buildings, high-rise apartments, hotels, factories, and residential locations.

The sampling tap must be protected from exterior contamination associated with being too close to the sink bottom or to the ground. Contaminated water or soil from the faucet exterior may enter the bottle during the collecting procedure, because placing a bottle underneath a low tap without grazing the neck interior against the outside faucet surface is difficult. Leaking taps that allow water to flow out from around the stem of the valve handle and down the outside of the faucet or taps in which water tends to run up on the outside of the lip are to be avoided as sampling outlets. Aerator, strainer, and hose attachments on the tap must be removed before sampling. These devices can harbor a significant bacterial population if they are not cleaned routinely or replaced when worn or cracked. Whenever an even stream of water cannot be obtained from taps after such devices are removed, a more suitable tap must be sought. Taps whose water flow is not steady should be avoided because temporary fluctuation in line pressure may cause sheets of microbial growth that are lodged in some pipe section or faucet connection to break loose. The chosen cold-water tap should be opened for 2 or 3 min or for sufficient time to permit clearing the service line; a smooth-flowing water stream at moderate pressure without splashing should be obtained. Then, without changing the waterflow, which could dislodge some particles in the faucet, sample collection can proceed.

When glass bottles fitted with ground-glass stoppers are used, a string or paper wedge must be inserted between the bottle and closure before sterilization to facilitate easy opening during sample collection. Upon opening the bottle, discard the string or paper wedge without touching the inner portion of either the bottle or stopper. Reinserting this item into the sample bottle after sample collection will increase the risk of water sample contamination.

Regardless of the type of sample bottle closure used, do not lay the bottle cap down or put it in a pocket. Rather, hold the bottle in one hand and the cap in the other, keeping the bottle cap right side up (threads down) and do not touch the inside of the cap. Likewise, avoid contaminating the sterile bottle with fingers or permitting the faucet to touch the inside of the bottle. The bottle should not be rinsed or wiped out or blown out by the sample collector's breath before use. Such practices may not only contaminate the bottle but may remove the thiosulfate dechlorinating agent. During the filling operation, be careful so splashing drops of water from the ground or sink do not enter into either the bottle or cap. Do not adjust the stream flow while sampling, in order to avoid dislodging some particles in the faucet. Fill the bottle to within 1 in (25 mm) of the bottle top or to the shoulder of the container; cap the bottle immediately. Then turn off the tap.[159,160]

Treating water taps before collecting potable water samples is not necessary if reasonable care is exercised in the choice of sampling tap (clean, free of attachments, and in good repair) and if the water is allowed to flow adequately at a uniform rate before sampling. Superficially passing a flame from a match or an alcohol-soaked cotton applicator over the tap a few times may have a psychological effect on observers, but it will not have a lethal effect on attached bacteria. The application of intense heat with a blow torch may damage the valve-washer seating or create a fire hazard to combustible materials adjacent to the tap. If successive samples from the same tap continue to contain coliforms, however, the tap should be disinfected with a hypochlorite solution to eliminate external contamination as the potential source of these organisms.[161]

This negative position on a protocol for flaming taps before sample collection is supported by several independent studies. Thomas et al.,[162] after a study of 253 samples from farm water supplies, reported that flaming taps before sampling resulted in no significant differences in the multiple-tube test for both total coliforms and fecal coliforms, nor in the standard plate counts incubated at 37° or 22°C. They noted that a tendency for the bacterial content to be lower existed, but the trend was not significant and could have occurred by chance. In a second study involving 527 distribution samples collected without tap flaming from the Chicago public water supply, only two samples (or 0.4 percent) contained coliforms.[163] In a third study, water was flushed from taps located in 76 gasoline service stations in Dayton, Ohio, but again the taps were not flamed or otherwise disinfected.[159] The results showed no coliform positive samples from 40 of the 76 stations, and membrane filter coliform counts in excess of 4/100 mL occurred in only 4 of the 10,916 samples tested.

During a contamination event or when a coliform biofilm is suspected, special sampling sites should be selected, within the pressure zone, in an effort to verify the initial data and isolate the problem area in the pipe network. Isolating the area can be difficult because of flow reversals in the distribution system and the transient nature of some contamination events. Despite these difficulties, some successes in locating the probable site can be achieved by selecting sampling locations upstream and downstream of coliform occurrences, where free-chlorine residual has disappeared or increased turbidity and heterotrophic bacteria population occur.

### Bacterial test selection

Selection of bacterial tests for distributed water quality may take three different approaches: routine testing for compliance with the coliform standard, seasonal survey for microbial regrowth in pipe lines, and special investigative protocols to verify fecal contamination or biofilm development.

Monitoring the effectiveness of water distribution networks to deliver a safe water supply to the public is a high priority. Because a properly designed and operated water treatment system can produce a finished water with an average density of less than one total coliform per 100 mL, water supply throughout the distribution network should also meet this attainable standard. Thus, the major effort in routine monitoring is directed to demonstrate that no posttreatment contamination occurs. The total coliform test has been the traditional bacterial test employed. The original concept was based on the premise that total coliform absence in a drinking water sample was evidence that no fecal contamination was present and, therefore, no risk of pathogen exposure existed, an indirect measure of public health risk. Recent findings suggest this interpretation is not perfect. While occasional reports state that total coliform occurrences do not always parallel some pathogen (virus, *Giardia*) contamination events,[164,165] the mass of data accumulated over the past 80 years clearly demonstrate the continued acceptance of this test as a reliable indication of a bacteriologically safe supply. In this light, the total coliform test is primarily a measure of water supply treatment effectiveness and less significantly a measure of public health risk.

Surrogate indicators such as total coliforms and fecal coliforms are used because some waterborne pathogens are difficult to detect or the available tests may be complex, time-consuming, and often not sufficiently sensitive or selective. Selection of an appropriate surrogate indicator should consider five important aspects: (1) The candidate organisms should be present only in feces at densities far exceeding pathogen concentrations from infected individuals. (2) An unquestionably positive correlation between the indicator and fecal contamination and between the candidate indicator and waters contaminated with fecal excrements should exist. (3) Persistence and regrowth characteristics of the indicator should parallel those of the most persistent waterborne pathogen. (4) Disinfection of water should produce similar kill rates for the indicator and most pathogens. (5) The methodology for detecting the candidate surrogate indicator must be simple, applicable to a variety of waters encountered in water quality monitoring, and defy false positive reactions and interferences from the water flora. Although no candidate surrogate indicators known fit perfectly within all of these ideal requirements for water supply, the total coliform indicator still fulfills most of these specifications. Focusing within the total coliform group reveals fecal coliform and *E. coli* as more exacting evidence of fecal contamination but less desirable if measurement of treatment effectiveness is the first objective.[166]

Coliform detection in water supply samples can be done by any of three different laboratory procedures: the multiple-tube fermentation test, the membrane filter method, or the Colilert analysis.[167,168] Any of these procedures can be configured to provide either a count per 100 mL or a presence-absence determination. The multiple-tube test generally uses a 50- or 100-mL sample portion and requires a minimum of 48 incubation to ensure no coliforms are present. A maximum of 96 h may be necessary to establish a valid total coliform density. With the membrane filter method, larger sample portions (liter or more) can be examined if desired. Test results are available within 24 h. In both the multiple-tube and membrane filter tests, fecal coliform determinations are either an adjunct part of the total coliform examination or are done separately. The Colilert coliform analysis is usually done on 10- or 100-mL sample portions with simultaneous total coliform and *E. coli* results available within 24 h.

As high-quality finished water passes through the distribution network, changes in the general microbial composition occur that do not relate to total coliforms and fecal contamination.[121,169,170] Changes in microbial composition involve the establishment of disinfectant-resistant organisms in the porous infrastructure of tubercles and sediments. As the biofilm of mixed organisms develops, the heterotrophic bacterial density and profile of different bacteria expands to include a variety of gram-negative bacteria, such as pigmented strains, antibiotic-resistant organisms, coliform antagonists, and opportunistic pathogens. In these situations, analyses of potable water for total coliform occurrences will not provide any information on the regrowth of the general bacterial population in areas of corrosion, static water, and slow-flow and dead-end sections of the pipe network. Identifying these areas of water quality deterioration will require a test for heterotrophic bacteria that is accomplished by using either R-2A agar,[171] spread plates,[172] or M-HPC agar[173] in a membrane filtration procedure. Incubation should be at 28°C for 3 to 7 days for optimal recovery on these media. For those laboratories using SPC agar, extending the incubation time to 3 to 5 days is recommended for improving detection of a greater portion of the heterotrophic population.

When total coliform bacteria are detected in the distribution system, immediate testing of repeat samples for coliform occurrence is a universal first response. Any subsequent detection of coliform bacteria should be augmented with a fecal coliform confirmation in EC broth (24 h at 44.5°C) or a colilert *E. coli* determination and a streak plate of M-Kleb agar (24 h at 35°C) to determine if the contamination event is fecal or a *Klebsiella* biofilm occurrence.[174] Supplemental use of commercial biochemical kits for coliform speciation would further verify the specific identity of strains in the contamination event.

## Data interpretation

Bacteriological test results should be promptly reviewed for changes in water quality. When coliforms are detected, the laboratory finding should be verified by procedures described in *Standard Methods.*[167] Concurrently, another sample from the same site should be requested for repeat coliform analyses to confirm that the contamination event still exists. Subsequent coliform positive results should activate a search in treatment protocols and water distribution practices for the cause of contamination. With cause determined, appropriate remedial action must be applied promptly. As a precaution, lines in the affected pressure zone may need an active flushing to remove sediment accumulations that affect diminished disinfection residuals.[49,175]

Fecal coliform or *E. coli* occurrences are of most serious concern because they represent fecal contamination from either a treatment barrier breakthrough, a cross connection or back-siphonage problem that may be intermittent. *E. coli* does not colonize the pipe network; therefore, these occurrences are a signal of immediate danger. When fecal coliforms or *E. coli* are detected in a routine water sample that was positive for total coliforms, the utility should notify the appropriate state agency, take immediate measures to identify and correct the problem, and request a prompt resampling at the site. Repeat sampling should not only be done at the site where a positive sample was reported but also upstream and downstream of the original sample. If any repeat sample contains fecal coliforms or *E. coli,* the system would be out of compliance and a public notification is in order while the utility and water authorities quickly conduct an intensive engineering review of operations and apply the necessary corrective actions.

Other coliforms (*Klebsiella pneumoniae, K. oxytoca, Enterobacter aerogenes, Enterobacter cloacae,* and *Citrobacter freundii*) require less nutrient support than *E. coli* and are more apt to colonize some sediment locations in the network following a brief burst of coliforms through the treatment barriers or into the protected pipeline environment. These coliforms then reappear periodically in the main flow of water, particularly in warm-water periods, and may eventually be detected in samples collected throughout the distribution network. Finding coliform densities ranging from 1 to 150 organisms per 100 mL may be possible, with their occurrence widespread in the distribution system.[176–178] Investigation of past records often shows this problem to be a repeat of some coliform occurrences during past summers, with positive samples ranging from 4 to 62 percent of all bacteriological analyses in the period.[179]

Because coliform colonization in the distribution system may be more an indication of biofilm development in areas of excessive sediment accumulation and rising chlorine demand, determining the need for public notification may require a different approach. Provided no evidence of fecal contamination exists, one approach might be to use coliform frequency of occurrence levels as an appropriate index for establishing a stepwise action plan to avoid sudden increases in taste and odor complaints and increased instability of pipe network sediments. For example, most systems have an annual frequency of coliform occurrence ranging from 1 to 5 percent, with individual positive samples containing 1 to 5 coliforms per 100 mL. When the weekly percent level of coliform occurrences escalates to a level of 5 to 10 percent, coliform densities may range from 1 to 130 organisms per 100 mL. At this time, a sanitary survey of source water, treatment application and distribution system integrity should be performed promptly, a free-chlorine residual established in the pipe network at 0.5 mg/L and the public notified of actions taken. If the weekly percent level of occurrence reaches the next plateau, 10 to 15 percent, a joint meeting of water authorities (utility, state and federal officials) should be held at frequent intervals (weekly or daily, if necessary) to review progress made by the water utility to correct the deficiency, to continue the public advisory on progress with system adjustments, and to monitor local hospitals for waterborne disease outbreaks. Using this approach, issuance of a boil-water order would not be needed unless evidence of fecal coliforms in the water or a waterborne outbreak occurs or the weekly percent coliform frequency ranges beyond 20 percent. The concern with the latter level of coliform biofilm occurrence is one of masking the ability to detect a new fecal contamination event.

The HPC provides a general measure of bacterial water quality in the distribution system and can become an important early sign of excessive microbial growth on distribution system pipe walls and in sediments.[47] Base-line heterotrophic bacteria levels in the treated water can easily be maintained at very low levels, less than 10 organisms/L, with adequate disinfection. When normal background heterotrophic plate counts in the distribution system become greater than $10^3$ organisms/L (a 100-fold increase) and this is confirmed by a second sample, action should be initiated immediately to resolve the microbial regrowth problem.

Another important application for heterotrophic bacterial data on distributed water can be found during periods of adjustments or modification in the disinfection treatment process. During field studies to reduce the production of trihalomethane concentrations in finished water, the point of chlorination for one treatment system was moved and disinfectant residual changed from free chlorine to chloramines.[41] Heterotrophic plate counts were performed on samples from the ends of one section of the distribution pipe network before, during, and after these adjustments. The generated database revealed that immediately following these changes in disinfectant application, the HPC remained in a constant density range for 2 weeks. A gradual decline in the free-chlorine residual to 0.1 mg/L or less occurred at dead-end sections. During warm-water conditions (20 to 25°C), heterotrophic bacterial densities increased suddenly. Within a few more days, some coliform regrowth (12 to 30 organisms/100 mL) began. At this point, free chlorine was restored to the lines, followed by the disappearance of detectable coliforms and later by a decline in the heterotrophic bacterial densities to levels originally encountered. This case history and other similar field experiences demonstrate that sudden changes from the system's normal heterotrophic bacterial densities can serve as an early signal of undesirable quality changes that may precede an unsatisfactory coliform occurrence in distribution water.

## Summary

Drinking water can deteriorate in microbial quality during distribution to the consumer, introducing taste, odor, and, occasionally, waterborne pathogens. Bacteria can be introduced into distributed water by several pathways. Some organisms become established in the pipe environment and eventually form a biofilm community if not effectively controlled by water supply process barriers and treatment to minimize assimilable organic carbon. Corrosion control, effective flushing programs, elimination of static water areas, and maintenance of a disinfectant residual are also important aspects in reducing microbial colonization of the distribution network. Some total coliform bacteria (*Enterobacter, Klebsiella,* and *Citrobacter*) may become part of the biofilm community with periodic releases to the main flow of water during seasonal warm-water periods, reversals of flow and changes in the structure of pipe sediments due to water pH shifts.

The significance of coliform occurrences cannot be ignored because they may indicate a potential pathway for pathogen penetration into the water supply and a pipe environment that is supportive of bacterial colonization. Responses to coliform occurrences may take several different directions, depending on the frequency of occurrences and the detection of fecal coliform or *E. coli.* Because each distribution system is unique, monitoring strategies must be carefully developed to properly characterize distributed water and to provide early warning of any undesirable change in the microbial quality of the public water supply.

## References

1. S. J. Frontinus, *97 A.D. The Water Supply of the City of Rome,* transl., C. Herschel, New England Water Works Association, Boston, Mass., 1973.

2. M. N. Baker, *The Quest for Pure Water,* vol. I, American Water Works Association, Denver, Colo., 1981.
3. *AWWA Standard for Disinfecting Water Mains,* AWWA C651-86, American Water Works Association, Denver, Colo., 1986.
4. R. J. Becker, "Main Disinfection Methods and Objectives. Part I: Use of Liquid Chlorine," *J. AWWA,* vol. 61, no. 2, 1969, p. 79.
5. H. B. Russelman, "Main Disinfection Methods and Objectives, Public Health Viewpoint," *J. AWWA,* vol. 61, no. 2, 1969, p. 82.
6. E. V. Suckling, *The Examination of Waters and Water Supplies,* 5th ed., Blakiston, Philadelphia, 1943.
7. A. R. Davis, "The Distribution System," in L. C. Billings (ed.), *Manual for Water Works Operators,* Texas State Department of Health and the Texas Water and Sanitation Research Foundation, Austin, Texas, 1951.
8. M. Hutchinson, "The Disinfection of New Water Mains," *Chemistry and Industry,* vol. 139, 1971.
9. C. K. Calvert, "Investigation of Main Sterilization," *J. AWWA,* vol. 31, no. 5, 1939, p. 832.
10. G. O. Adams and F. H. Kingsbury, "Experiences with Chlorinating New Water Mains," *J. New Eng. WWA,* vol. 51, 1937, p. 60.
11. E. W. Taylor, *The Examination of Water and Water Supplies,* 7th ed., Little Brown, Boston, 1958.
12. E. W. Taylor, 43rd Report, Metropolitan Water Board, London, 1967–68.
13. D. Schoenen, "Microbial Growth Due to Materials Used in Drinking Water Systems," in H. J. Rehm and G. Reed (eds.), *Biotechnology,* vol. 8, VCH Verlagsgesellschaft, Weinheim, 1986.
14. R. H. N. Schubert, "Das Vorkenmen der aeromonadin in Oberirdischen Gewassern," *Arch. Hyg.,* vol. 156, 1967, p. 688.
15. M. Hutchinson, "WRA Medlube: An Aid to Mains Disinfection," *J. Soc. Water Treat. Exam.,* vol. 23 (part II), 1974, p. 174.
16. F. E. Harem, K. D. Bielman, and J. E. Worth, "Reservoir Linings," *J. AWWA,* vol. 68, no. 5, 1976, p. 238.
17. J. A. Wade, Jr., "Design Guidelines for Distribution Systems as Developed and Used by an Investor-Owned Utility," *J. AWWA,* vol. 66, no. 6, 1974, p. 346.
18. R. J. Seidler, J. E. Morrow, and S. T. Bagley, "*Klebsiella* in Drinking Water Emanating from Redwood Tanks," *Appl. Environ. Microbiol.,* vol. 33, 1977, p. 893.
19. H. W. Talbot, Jr., and R. J. Dwidler, "Gas Chromatographic Analysis of in-situ Cyclitol Utilization by *Klebsiella* Growing in Redwood Extracts," *Appl. Environ. Microbiol.,* vol. 38, 1979, p. 599.
20. H. W. Talbot, Jr., J. E. Morrow, and R. J. Seidler, "Control of Coliform Bacteria in Finished Drinking Water Stored in Redwood Tanks," *J. AWWA,* vol. 71, no. 6, 1979, p. 349.
21. E. Thofern, D. Schoenen, and G. J. Tuschewitzki, "Microbial Surface Colonization and Disinfection Problems," *Off Gesundh.-wes.,* vol. 49, 1987, p. 14.
22. H. Mackle et al., "Koloniezahlerhohung sowie. Geruchs-und Geschmachsbiemtrachtigungen des Trinkwassers durch Losemettelhaltige Auskleidermaterialien," *GWF, Gas-Wasserfach: Wasser/Abwasser,* vol. 129, 1988, p. 22.
23. H. Bernhardt and H. U. Liesen, "Bacterial Growth in Drinking Water Supply Systems Following Bitumenous Corrosion Protection Coatings," *GWF, Gas-Wasservach: Wasser/Abwasser,* vol. 129, 1988, p. 28.
24. J. R. Kroon, M. A. Stoner, and W. A. Hunt, "Water Hammer: Causes and Effects," *J. AWWA,* vol. 76, no. 11, 1984, p. 39.
25. A. J. Alter, "Appearance of Intestinal Wastes in Surface Water Supplies at Ketchikan, Alaska," *Proc. 5th Alaska Sci. Confr. AAAS,* Anchorage, Alaska, 1954.
26. Anonymous. "Ketchikan Laboratory Studies Disclose Gulls are Implicated in Disease Spread," *Alaska's Health,* vol. 11, 1954, p. 1.
27. H. Fennel, D. B. James, and J. Morris, "Pollution of a Storage Reservoir by Roosting Gulls," *J. Soc. Water Treat. Exam.,* vol. 23, 1974, p. 5.
28. B. Olson, *Assessment and Implications of Bacterial Regrowth in Water Distribution*

*Systems,* U. S. Environmental Protection Agency, Rep. 600/52-82-072, Cincinnati, Ohio, 1982.

29. M. J. Allen and E. E. Geldreich, "Bacteriological Criteria for Ground-water Quality," *Ground Water,* vol. 13, 1975, p. 45.

30. G. F. Craun, "A Summary of Waterborne Illness Transmitted Through Contaminated Groundwater," *J. Environ. Health,* vol. 43, 1985, p. 122.

31. M. J. Allen, R. H. Taylor, and E. E. Geldreich, "The Occurrence of Microorganisms in Water Main Encrustations," *J. AWWA,* vol. 72, no. 11, 1980, p. 614.

32. J. C. Hoff, *Inactivation of Microbial Agents by Chemical Disinfectants.* U.S. Environmental Protection Agency, Water Engineering Research Laboratory, Rep. 600/52-86/067, Cincinnati, Ohio, 1986.

33. C. Haas, M. A. Meyer, and M. S. Fuller, "The Ecology of Acid-Fast Organisms in Water Supply, Treatment, and Distribution Systems," *J. AWWA,* vol. 75, no. 3, 1983, p. 139.

34. G. J. Bonde, *Bacterial Indicators of Water Pollution,* Teknisk Forlag, Copenhagen, 1977.

35. W. D. Rosenzweig, H. A. Minnigh, and W. O. Pipes, "Chlorine Demand and Inactivation of Fungal Propugules," *Appl. Environ. Microbiol.,* vol. 45, 1983, p. 182.

36. R. M. Niemi, S. Knuth, and K. Lundstrum, "Actinomycetes and Fungi in Surface Waters and in Potable Water," *Appl. Environ. Microbiol.,* vol. 43, 1982, p. 378.

37. D. J. Reasoner, J. C. Blannon, and E. E. Geldreich, "Nonphotosynthetic Pigmented Bacteria in a Potable Water Treatment and Distribution System," *Appl. Environ. Microbiol.,* vol. 55, 1989, p. 912.

38. G. S. Logsdon, "Comparison of Some Filtration Processes Appropriate for *Giardia* Cyst Removal," *Proc. Calgary Giardia Conf.,* February 1987.

39. A. Amirtharajah and D. P. Wetgstein, "Initial Degradation of Effluent Quality during Filtration," *J. AWWA,* vol. 72, no. 9, 1980, p. 518.

40. A. K. Camper, M. W. LeChevallier, S. C. Broadway, and G. A. McFeters, "Bacteria Associated with Granular Activated Carbon Particles in Drinking Water," *Appl. Environ. Microbiol.,* vol. 52, 1986, p. 434.

41. J. M. Symons et al., *Treatment Techniques for Controlling Trihalomethanes in Drinking Water,* Office of Research and Development, U.S. Environmental Protection Agency, Rep. 600/2-81-156, Cincinnati, Ohio, 1981.

42. C. N. Haas, M. A. Meyer, and M. S. Paller, "Microbial Dynamics in GAC Filtration of Potable Water," *Amer. Soc. Civil Eng. J. Environ. Eng. Div.,* vol. 109, 1983, p. 956.

43. C. N. Haas, M. A. Meyer, and M. S. Paller, "Microbial Alterations in Water Distribution Systems and Their Relationships to Physical-Chemical Characteristics," *J. AWWA,* vol. 75, no. 9, 1983, p. 475.

44. B. E. Rittmann and V. L. Snoeyink, "Achieving Biologically Stable Drinking Water," *J. AWWA,* vol. 76, no. 10, 1984, p. 106.

45. Environmental Protection Agency, *National Interim Primary Drinking Water Regulations,* Rep. 570/9-76-003, Office of Water Supply, Washington, D.C., 1976.

46. E. E. Geldreich, H. D. Nash, D. J. Reasoner, and R. H. Taylor, "The Necessity of Controlling Bacterial Populations in Potable Waters: Community Water Supply," *J. AWWA,* vol. 64, no. 9, 1972, p. 596.

47. Organisms in Water Committee Report, "Microbiological Considerations for Drinking Water Regulation Revisions," *J. AWWA,* vol. 79, no. 5, 1987, p. 81.

48. R. S. Martin, W. H. Gates, R. S. Tobin, D. Grantham, et al., "Factors Affecting Coliform Bacterial Growth in Distribution Systems," *J. AWWA,* vol. 74, no. 1, 1982, p. 34.

49. R. W. Buelow et al., "Disinfection of New Water Mains," *J. AWWA,* vol. 68, no. 6, 1976, p. 283.

50. C. H. H. Harold, 29th Report Director of Water Examination, Metropolitan Water Board, London, England, 1934.

51. J. J. Hamilton, "Potassium Permanganate as a Main Disinfectant," *J. AWWA,* vol. 66, no. 12, 1974, p. 734.

52. M. Hutchinson and J. W. Ridgway, "Microbiological Aspects of Drinking Water

Supplies," in F. A. Skinner and J. M. Schewan (eds.), *Aquatic Microbiology*, Academic Press, London, 1977.

53. S. W. Olson, "The Application of Microbiology to Cosmetic Testing," *J. Soc. Cosmetic Chem.*, vol. 18, 1967, p. 191.

54. S. Tenebaum, "Pseudomonads in Cosmetics," *J. Soc. Cosmetic Chem.*, vol. 18, 1967, p. 797.

55. A. P. Dunnigan, "Microbiological Control of Cosmetic Products," *Federal Regulations and Practical Control Microbiology for Disinfectants, Drugs, and Cosmetics*, Society for Industrial Microbiology Special Publication No. 4, 1969.

56. S. Borgstrom, "Principles of Food Science," in *Food Microbiology and Biochemistry*, vol. 2, Macmillan, London, England, 1978.

57. E. E. Geldreich, H. D. Nash, and D. Spino, "Characterizing Bacterial Populations in Treated Water Supplies: A Progress Report," *Proc. American Water Works Association Water Quality Technology Conf.*, Kansas City, Mo., December 1977.

58. M. W. LeChevallier, R. J. Seidler, and T. M. Evans, "Enumeration and Characterization of Standard Plate Count Bacteria by Chlorinated and Raw Water Supplies," *Appl. Environ. Microbiol.*, vol. 40, 1980, p. 922.

59. B. H. Olson and L. Hanami, "Seasonal Variation of Bacterial Populations in Water Distribution Systems," *Proc. American Water Works Association Water Quality Technology Conf.*, Miami Beach, Fla., 1980.

60. D. S. Herson and H. Victoreen, "Identification of Coliform Antagonists," *Proc. American Water Works Association Water Quality Technology Conf.*, Miami Beach, Fla., 1980.

61. J. K. Reilly and J. Kippin, *Interrelationship of Bacterial Counts with Other Finished Water Quality Parameters within Distribution Systems*, U.S. Environmental Protection Agency, Rep. 0600/52-81-035, Cincinnati, Ohio, 1981.

62. J. A. Clark, C. A. Burger, and L. E. Sabatinos, "Characterization of Indicator Bacteria in Municipal Raw Water, Drinking Water, and New Main Water Samples," *Canad. J. Microbiol.*, vol. 28, 1983, p. 1002.

63. J. T. Staley, *Identification of Unknown Bacteria from Drinking Water*, U. S. Environmental Protection Agency, Project CR-807570010, Cincinnati, Ohio, 1983.

64. A. K. Camper, M. W. LeChevallier, S. C. Broadway, and G. A. McFeters, "Growth and Persistence on Granular Activated Carbon Filters," *Appl. Environ. Microbiol.*, vol. 50, 1985, p. 1378.

65. O. H. Tuovinen, K. S. Button, A. Vuorinen, L. Carison et al., "Bacterial, Chemical and Mineralogical Characteristics of Tubercles in Distribution Pipelines," *J. AWWA*, vol. 72, no. 11, 1980, p. 626.

66. J. L. Armstrong, J. J. Calomiris, and R. J. Seidler, "Selection of Antibiotic-Resistant Standard Plate Count Bacteria during Water Treatment," *Appl. Environ. Microbiol.*, vol. 44, 1982, p. 308.

67. L. Bedard, A. J. Drapeau, S. S. Kasatiya, and R. R. Plaute, "Plasmides de Resistance aux Antibiotiques Ches les Bacteries," *Isolus D'Eaux Potables. Eau Des Quebec*, vol. 15, 1982, p. 59.

68. J. L. Armstrong, J. J. Calomiris, D. S. Shigeno, and R. J. Seidler, "Drug Resistant Bacteria in Drinking Water," *Proc. American Water Works Association Water Quality Technology Conf.*, Seattle, Wash., 1981.

69. H. T. El-Zanfaly, E. A. Kassein, and S. M. Badr-Eldin, "Incidence of Antiobiotic Resistant Bacteria in Drinking Water in Cairo," *Water, Air Soil Pollution*, vol. 32, 1987, p. 123.

70. G. C. du Moulin and K. D. Stottmeier, "Waterborne Mycobacteria: An Increasing Threat to Health," *ASM News*, vol. 52, 1986, p. 525.

71. G. C. du Moulin, I. H. Sherman, D. C. Hoaglin, and K. D. Stottmeier, "*Mycobacterium avinum* Complex, an Emerging Pathogen in Massachusetts," *J. Clin. Microbiol.*, vol. 22, 1985, p. 9.

72. P. A. Pelletier, G. C. du Moulin, and K. D. Stottmeier, "Mycobacteria in Public Water Supplies: Comparative Resistance to Chlorine," *Microbiol. Sci.*, vol. 5, 1988, p. 147.

73. J. M. Quarles, R. C. Belding, T. C. Beaman, and P. Gerhardt, "Hemodialysis Cul-

ture of *Serratia marcescens* in a Goat-Artificial Kidney-Fermentor System," *Infection and Immunity*, vol. 9, 1974, p. 550.

74. M. S. Favero, N. J. Peterson, K. M. Bayer, L. A. Carson, and W. W. Bond, "Microbial Contaminations of Renal Dialysis Systems and Associated Health Risks," *Trans. Amer. Soc. Artificial Internal Organs*, vol. XX-A, 1974, p. 175.

75. M. S. Favero, N. J. Peterson, L. A. Carson, W. W. Bond, and S. H. Hindman, "Gram-Negative Water Bacteria in Hemodialysis Systems," *Health Lab. Sci.*, vol. 12, 1975, p. 321.

76. L. Herman, "Sources of the Slow-Growing Pigmented Water Bacteria," *Health Lab. Sci.*, vol. 13, 1976, p. 5.

77. H. F. Ridgway and B. H. Olson, "Chlorine Resistance Patterns of Bacteria from Two Drinking Water Distribution Systems," *Appl. Environ. Microbiol.*, vol. 44, 1982, p. 972.

78. W. Dott and D. Waschko-Dransmann, "Occurrence and Significance of Actinomycetes in Drinking Water," *Zbl. Bakt. Hyg. I., Abt. Orig. B*, vol. 173, 1981, p. 217.

79. N. P. Burman, "Taste and Odour Due to Stagnation and Local Warming in Long Lengths of Piping," *J. Soc. Water Treat. Exam.*, vol. 14, 1965, p. 125.

80. L. R. Bays, N. P. Burman, and W. M. Lavis, "Taste and Odour in Water Supplies in Great Britain: A Survey of the Present Position and Problems for the Future," *J. Soc. Water Treat. Exam.*, vol. 19, 1970, p. 136.

81. R. M. Niemi, S. Knuth, and K. Lundstrom, "*Actinomycetes* and Fungi in Surface Waters and in Potable Water," *Appl. Environ. Microbiol.*, vol. 43, 1982, p. 378.

82. L. A. Nagy and B. H. Olson, "The Occurrence of Filamentous Fungi in Drinking Water Distribution Systems," *Canad. J. Microbiol.*, vol. 28, 1982, p. 667.

83. W. B. Cooke, "The Fungi of Our Mouldy Earth," in *Beiheft 85 zur Nova Hedwigia*, J. Cramer, Berlin, 1986.

84. F. Hinzelin and J. C. Block, "Yeasts and Filamentous Fungi in Drinking Water," *Environ. Technol. Lett.*, vol. 6, 1985, p. 101.

85. R. S. Engelbrecht and C. N. Haas, "Acid-Fast Bacteria and Yeast as Disinfection Indicators: Enumeration Methodology," *Proc. American Water Works Association Water Quality Technology Conf.*, Kansas City, Mo., 1977.

86. M. C. Snead, V. P. Olivieri, K. Kawata, and C. W. Kruse, "Biological Evaluation of Benefits of Maintaining a Chlorine Residual in Water Supply Systems," *Water Research*, vol. 14, 1980, p. 403.

87. W. F. Langelier, "The Analytical Control of Anti-Corrosion Treatment," *J. AWWA*, vol. 28, no. 10, 1936, p. 1500.

88. R. McCauley, "Controlled Deposition of Protection Calcite Coating in Water Mains," *J. AWWA*, vol. 52, no. 11, 1960, p. 1386.

89. W. Stumm, "The Corrosive Behavior of Water," *Proc. Amer. Soc. Civil Eng.*, vol. 86, No. SA-6, 1960.

90. T. P. Brodeur, J. E. Singley, and J. C. Thurrott, "Effect of a Change to Free Chlorine Residual at Daytona Beach, Florida," *Proc. American Water Works Association Water Quality Technology Conf.*, San Diego, Calif., 1976.

91. L. O. Hiisvirta, "Problems of Disinfection of Surface Water with a High Content of Natural Organic Material," *Water Supply: Rev. J. Int. Water Supply Assoc.*, vol. 4, 1986, p. 53.

92. J. E. Singley, B. A. Beaudet, and P. H. Markey, *Corrosion Manual for Internal Corrosion of Water Distribution Systems*, U.S. Environmental Protection Agency, Rep. 570/9-84-001, Office of Drinking Water, Washington, D.C., 1984.

93. DVGW-Forschungsstelle, AWWA Research Foundation, *Internal Corrosion of Water Distribution Systems*, AWWA Research Foundation, Denver, Colo., 1985.

94. J. Tillmans and O. Heublein, "Investigation of the Carbon Dioxide Which Attacks Calcium Carbonate in Natural Waters," *Gesundh. Ing.*, vol. 35, 1913, p. 669.

95. T. E. Larson and R. V. Skold, "Current Research on Corrosion and Tuberculation of Cast Iron," *J. AWWA*, vol. 50, no. 11, 1958, p. 1429.

96. T. E. Larson, "Deterioration of Water Quality in Distribution Systems," *J. AWWA*, vol. 58, no. 10, 1966, p. 1307.

97. J. C. Hoff, "The Relationships of Turbidity to Disinfection of Potable Water," in

C. W. Hendricks (ed.), in *Evaluation of the Microbiology Standards for Drinking Water* U.S. Environmental Protection Agency, Rep. 570/9-78-002, Washington, D.C., 1978.

98. R. H. Cuillo, H. E. Ferran, Jr., E. E. Whitaker, and H. Leland, *Bacterial Survival in Potable Water with Low Turbidity,* U.S. Environmental Protection Agency Grant R-806329, Nasson College, Springvale, Me., 1983.

99. T. W. Hejkal, F. M. Wellings, P. A. LaRock, and A. L. Lewis, "Survival of Poliovirus within Organic Solids during Chlorination," *Appl. Environ. Microbiol.,* vol. 38, 1979, p. 144.

100. D. M. Foster, M. A. Emerson, C. E. Buck, D. S. Walsh, and O. J. Sproul, "Ozone Inactivation of Cell- and Fecal-Associated Viruses and Bacteria," *J. Water Pollution Control Federation,* vol. 52, 1980, p. 2174.

101. A. K. Camper, S. C. Broadaway, M. W. LeChevallier, and G. A. McFeters, "Operational Variables and the Release of Colonized Granular Activated Carbon Particles in Drinking Water," *J. AWWA,* vol. 79, no. 5, 1987, p. 70.

102. I. C. Small and G. F. Greaves, "A Survey of Animals in Distribution Systems," *J. Soc. Water Treat. Exam.,* vol. 19, 1968, p. 150.

103. K. M. MacKenthun and L. E. Keup, "Biological Problems Encountered in Water Supplies," *J. AWWA,* vol. 62, no. 8, 1970, p. 520.

104. M. H. Gerardi and J. K. Grimm, "Aquatic Invaders," *Water/Eng. Manage.,* vol. 10, 1982, p. 22.

105. S. L. Chang, R. L. Woodward, and P. W. Kabler, "Survey of Free-living Nematodes and Amoebas in Municipal Supplies," *J. AWWA,* vol. 52, no. 5, 1960, p. 613.

106. R. V. Levy, R. D. Cheetham, J. Davis, G. Winer, and F. L. Hart, "Novel Method for Studying the Public Health Significance of Macroinvertebrates Occurring in Potable Water," *Appl. Environ. Microbiol.,* vol. 47, 1984, p. 889.

107. R. V. Levy, F. L. Hart, and R. D. Cheetham, "Occurrence and Public Health Significance of Invertebrates in Drinking Water Systems," *J. AWWA,* vol. 78, no. 9, 1986, p. 105.

108. G. Zrupko, "Examination of Large Volume Samples Taken from the Municipal Water Treatment Plant," *Budapesti Kozegeszsezugy,* vol. 1, 1988, p. 21.

109. N. A. Cobb, "Filter-Bed Nemas: Nematodes of the Slow Sand Filter-Beds of American Cities," *Contr. Sci. Nematology,* vol. 7, 1918, p. 189.

110. M. G. George, "Further Studies on the Nematode Infestation of Surface Water Supplies," *Environmental Health,* vol. 8, 1966, p. 93.

111. A. S. Tombes, A. R. Abernathy, D. M. Welch, and S. A. Lewis, "The Relationship between Rainfall and Nematode Density in Drinking Water," *Water Research,* vol. 13, 1979, p. 619.

112. J. B. Mott and A. D. Harrison, "Nematodes from River Drift and Surface Drinking Water Supplies in Southern Ontario," *Hydrobiologia,* vol. 102, 1983, p. 27.

113. E. B. Shotts, Personal communication on the occurrence of various protozoans (*Tetrahymena, Acanthamoeba, Bodo, Chlamydomonas, Euglena* and *Paramicium*) in sand filters and distribution systems, 1985.

114. S. L. Chang, G. Berg, N. A. Clarke, and P. W. Kabler, "Survival and Protection against Chlorination of Human Enteric Pathogens in Free-living Nematodes Isolated from Water Supplies," *Amer. J. Tropical Medicine Hygiene,* vol. 9, 1960, p. 136.

115. H. W. Tracy, V. M. Camarena, and F. Wing, "Coliform Persistence in Highly Chlorinated Waters," *J. AWWA,* vol. 58, no. 9, 1966, p. 1151.

116. S. M. Smerda, H. J. Jensen, and A. W. Anderson, "Escape of *Salmonellae* from Chlorination during Ingestion by *Pristionchus Theretieri* (Nematoda: Diployasterinae)," *J. Nematology,* vol. 3, 1971, p. 201.

117. D. S. Sarai, "Total and Fecal Coliform Bacteria in Some Aquatic and Other Insects," *Environ. Entomology,* vol. 5, 1976, p. 365.

118. B. S. Fields, E. B. Shotts, J. C. Freeley, G. W. Gorman, and W. T. Martin, "Proliferation of *Legionella* as an Intracellular Parasite of the Ciliated Protozoan, *Tetrahymena pyriformis,*" *Appl. Environ. Microbiol.,* vol. 47, 1984, p. 467.

119. R. L. Tyndall and E. L. Domingue, "Cocultivation of *Legionella* and Free Living Amoebae," *Appl. Environ. Microbiol.*, vol. 44, 1982, p. 954.
120. *Deterioration of Bacteriological Quality of Water during Distribution.* Notes on Water Research No. 6., Water Research Centre, Medmenham, England, 1977.
121. J. Ridgway, R. G. Ainsworth, and R. D. Gwilliam, "Water Quality Changes— Chemical and Microbiological Studies," *Proc. Conf. Water Distribution Systems,* Water Research Centre, Medmenham, England, 1978.
122. H. T. Victoreen, "Water Quality Changes in Distribution," *Proc. Conf. Water Distribution Systems,* Water Research Centre, Medmenham, England, 1978.
123. D. Vander Kooij and B. C. J. Zoeteman, *Water Quality in Distribution Systems,* Special Subject 5, International Water Supply Association Congress, Kyoto, Japan, 1978.
124. D. J. Ptak, W. Ginsburg, and B. F. Willey, "Identification and Incidence of *Klebsiella* in Chlorinated Water Supplies," *J. AWWA,* vol. 65, no. 9, 1973, p. 604.
125. D. Hutchinson, R. H. Weaver, and M. Scherago, "The Incidence and Significance of Microorganisms Antagonistic to *Escherichia coli* in Water," *J. Bact.,* vol. 45, Abstract G34, 1943, p. 29.
126. G. Fischer, "The Antagonistic Effect of Aerobic Sporulating Bacteria on the Coliaerogenes Group," *Zeit. Immun. u. Exp. Ther.,* vol. 107, 1950, p. 16.
127. L. G. Herman, and C. K. Himmelsbach, "Detection and Control of Hospital Sources of *Flavobacteria*," *Hospitals,* vol. 39, 1965, p. 72.
128. A. von Graevenitz, "The Role of Opportunistic Bacteria in Human Disease," *Ann. Rev. Microbiol.,* vol. 31, 1977, p. 447.
129. D. S. Herson, "Identification of Coliform Antagonists," *Proc. American Water Works Association Water Quality Technology Conf.,* Miami Beach, Fla., 1980.
130. J. R. Postgate and J. R. Hunter, "The Survival of Starved Bacteria," *J. General Microbiol.,* vol. 29, 1962, p. 233.
131. M. G. Trulear and W. G. Characklis, *Dynamics of Biofilm Processes.* 34th Ann. Purdue Industrial Waste Conf., West Lafayette, Ind., 1979.
132. J. T. O'Connor, L. Hash, and A. B. Edwards, "Deterioration of Water Quality in Distribution Systems," *J. AWWA,* vol. 67, no. 3, 1975, p. 113.
133. M. J. Allen and E. E. Geldreich, "Distribution Line Sediments and Bacterial Regrowth," *Proc. American Water Works Association Water Quality Technology Conf.,* Kansas City, Mo., 1977.
134. O. H. Tuovinen and J. C. Hsu, "Aerobic and Anaerobic Microorganisms in Tubercles of the Columbus, Ohio, Water Distribution System," *Appl. Environ. Microbiol.,* vol. 44, 1982, p. 761.
135. J. W. Costeron, G. G. Geesey, and K. J. Cheng, "How Bacteria Stick," *Sci. Amer.,* vol. 238, 1978, p. 86.
136. G. Bitton and K. C. Marshall, *Absorption of Microorganisms to Surfaces,* Wiley-Interscience, New York, 1980.
137. H. T. Victoreen, "Bacterial Growth under Optimum Conditions, *J. Maine Public Utilities Assoc.,* vol. 45, 1974, p. 18.
138. H. T. Victoreen, "Water Quality Deterioration in Pipelines." *Proc. American Water Works Association Water Quality Technology Conf.,* Kansas City, Mo., 1977.
139. R. G. Ainsworth, J. Ridgway, and R. D. Gwilliam, "Corrosion Products and Deposits in Iron Mains," *Proc. Conf. Water Distribution Systems,* Paper 8, Water Research Centre, Medmenham, England, 1978.
140. S. H. Lee, J. T. O'Connor, and S. K. Banerji, "Biologically Mediated Corrosion and Its Effects on Water Quality in the Distribution System," *J. AWWA,* vol. 72, no. 11, 1980, p. 637.
141. M. D. Knittel, R. J. Seidler, and L. M. Cabe, "Colonization of the Botanical Environment by *Klebsiella* Isolates of Pathogenic Origin," *Appl. Environ. Microbiol.,* vol. 34, 1977, p. 557.
142. M. J. Howard, "Bacterial Depreciation of Water Quality in Distribution Systems," *J. AWWA,* vol. 32, no. 9, 1940, p. 1501.
143. J. K. Hoskins, "Revising the U.S. Standards for Drinking Water Quality: Some Considerations in the Revision," *J. AWWA,* vol. 33, no. 10, 1941, p. 1804.
144. Technical Subcommittee, "Manual of Recommended Water Sanitation Practice Ac-

companying United States Public Health Service Drinking Water Standards, 1942," *J. AWWA,* vol. 35, no. 2, 1943, p. 135.

145. *International Standards for Drinking Water,* World Health Organization, Geneva, 1971.

146. Safe Drinking Water Committee, *Drinking Water and Health,* vol. 1, National Academy of Sciences, Washington, D.C., 1977.

147. P. S. Berger, and Y. Argaman, *Assessment of Microbiology and Turbidity Standards for Drinking Water,* U.S. Environmental Protection Agency, Rep. 570-9-83-001, Washington, D.C., 1983.

148. Committee on the Challenges of Modern Society (NATO/CCMS), *Drinking Water Microbiology,* in D. O. Cliver and R. A. Newman (eds.), *J. Environ., Pathol., Toxicol. and Onocol.,* vol. 7, no. 516, 1987, p. 1.

149. *Guidelines for Drinking Water Quality,* vol. 1, World Health Organization, Geneva, 1984.

150. Department of Health and Social Security, *The Bacteriological Examination of Water Supplies. Reports on Public Health and Medical Subjects,* vol. 71, London, 1969.

151. Council of the European Communities, "Proposal for a Council Directive Relating to the Quality of Water for Human Consumption," *Official J. Euro. Commun.,* vol. 18, no. C214/2, 1975.

152. Ministry of National Health and Welfare, *Microbiological Quality of Drinking Water,* Health and Welfare, Ottawa, 1977.

153. *European Standards for Drinking Water,* World Health Organization, Geneva, 1970.

154. G. F. Craun, "Impact of the Coliform Standard on the Transmission of Disease," in C. W. Hendricks (ed.), *Evaluation of the Microbiology Standards for Drinking Water,* U.S. Environmental Protection Agency, Rep. 570/9-78-002, Office of Drinking Water, Washington, D.C., 1978.

155. W. O. Pipes, "Monitoring of Microbial Water Quality," in P. S. Berger and Y. Argaman (ed), *Assessment of Microbiology and Turbidity Standards for Drinking Water,* U.S. Environmental Protection Agency, Rep. 570/9-83-001, Office of Drinking Water, Washington, D.C., 1983.

156. N. J. Jacobs, W. L. Ziegler, F. C. Reed, T. A. Stukel, and E. W. Rice, "Comparison of Membrane Filter, Multiple-Fermentation-Tube, and Presence-Absence Techniques for Detecting Total Coliforms in Small Community Water Systems," *Appl. Environ. Microbiol.,* vol. 51, 1986, p. 1007.

157. R. V. Morita, *Sampling Regimes and Bacteriological Tests for Coliform Detection in Groundwater,* U.S. Environmental Protection Agency, Rep. 600/287/083, Cincinnati, Ohio, 1987.

158. "National Primary Drinking Water Regulations; Total Coliforms (Including Fecal Coliforms and *E. coli*)," *Federal Register,* Vol. 54, no. 124, 1989 p. 27544.

159. E. E. Geldreich, *Handbook for Evaluating Water Bacteriological Laboratories,* U.S. Environmental Protection Agency, Rep. 670/9-75-006, Cincinnati, Ohio, 1975.

160. System Water Quality Committee, *Distribution System Bacteriological Sampling and Control Guidelines,* California-Nevada Section, American Water Works Association, 1978.

161. R. W. Buelow, and G. Walton, "Bacteriological Quality vs. Residual Chlorine," *J. AWWA,* vol. 63, no. 1, 1971, p. 28.

162. S. B. Thomas, C. A. Scarlett, W. A. Cuthbert et al., "The Effect of Flaming of Taps before Sampling of the Bacteriological Examination of Farm Water Supplies," *J. Appl. Bacteriol.,* vol. 17, 1975, p. 175.

163. L. J. McCabe, "Trace Metals Content of Drinking Water from a Large System," *Symp. Water Quality in Distribution Systems,* Amer. Chem. Society National Meeting, Minneapolis, Minn., April 1969.

164. P. Payment, M. Trudel, and R. Plante, "Elimination of Viruses and Indicator Bacteria at Each Step of Treatment during Preparation of Drinking Water at Seven Water Treatment Plants," *Appl. Environ. Microbiol.,* vol. 49, 1985 p. 1418.

165. G. F. Craun, "Waterborne Outbreaks of Giardiasis," in S. L. Erlandsen and E. A. Meyer (ed.), *Giardia and Giardiasis* Plenum, New York, 1984.

166. P. W. Kabler and H. F. Clark, "Coliform Group and Fecal Coliform Organisms as Indicators of Pollution in Drinking Water," *J. AWWA,* vol. 52, no. 12, 1960, p. 1577.
167. *Standard Methods for the Examination of Water and Wastewater,* 17th ed., American Public Health Association, Washington, D.C., 1989.
168. S. C. Edberg, M. J. Allen, and D. B. Smith, "National Field Evaluation of a Defined Substrate Method for the Simultaneous Enumeration of Total Coliform and *Escherichia coli* from Drinking Water: Comparison with the Standard Multiple Tube Fermentation Method," *Appl. Environ. Microbiol.,* vol. 54, 1988, p. 1595.
169. H. T. Victoreen, "Control of Water Quality in Transmission and Distribution Mains," *J. AWWA,* vol. no. 6, 1974, p. 369.
170. R. J. Becker, "Bacterial Regrowth within the Distribution System," *Proc. American Water Works Association Water Quality Technology Conf.,* Atlanta, Ga., 1975.
171. D. J. Reasoner and E. E. Geldreich, "A New Medium for the Enumeration and Subculture of Bacteria from Potable Water," *Appl. Environ. Microbiol.,* vol. 49, 1985, p. 1.
172. R. H. Taylor, M. J. Allen, and E. E. Geldreich, "Standard Plate and Spread Plate Methods," *J. AWWA,* vol. 75, no. 1, 1983, p. 35.
173. R. H. Taylor and E. E. Geldreich, "A New Membrane Filter Procedure for Bacterial Counts in Potable Water and Swimming Pool Samples," *J. AWWA,* vol. 71, no. 7, 1979, p. 402.
174. E. E. Geldreich and E. W. Rice, "Occurrence, Significance, and Detection of *Klebsiella* in Water Systems," *J. AWWA,* vol. 79, no. 5, 1987, p. 74.
175. J. F. Rae, "Algae and Bacteria: Dead End Hazard," *Proc. American Water Works Association Water Quality Technology Confr.,* Seattle, Wash., 1981.
176. K. B. Earnhardt, "Chlorine Resistant Coliforms—The Muncie, Indiana, Experience," *Proc. American Water Works Association Water Quality Technology Conf.,* Miami Beach, Fla., December 1980.
177. L. D. Hudson, H. W. Hawkins, and M. Battaglia, "Coliforms in a Water Distribution System: A Remedial Approach," *J. AWWA,* vol. 75, no. 11, 1983, p. 564.
178. J. T. Wierenga, "Recovery of Coliforms in the Presence of a Free Chlorine Residual," *J. AWWA,* vol. 77, no. 11, 1985, p. 83.
179. D. Opheim and D. B. Smith, "Control of Distribution System Coliform Regrowth," *Proc. American Water Works Association Water Quality Technology Conf.,* Philadelphia, 1989.

# Acronyms and Notation

## Acronyms

| | | | |
|---|---|---|---|
| A-C | asbestos-cement | BHC | Bridgeport Hydraulic Company |
| ABS | alkylbenzene sulfonate | BV | bed volumes, i.e., a volume of feedwater equal to the volume of media including void spaces between particles |
| ADIs | acceptable daily intakes | C | disinfectant concentration |
| AI | aggressiveness index | CA | cellulose acetate |
| AIDS | acquired immune deficiency syndrome | CAA | Clean Air Act |
| Alk | alkalinity | CAE | carbon alcohol extract |
| AOC | assimilable organic carbon | CAG | cancer assessment group |
| ALT | alum-to-clay ratio | CAM | carbon adsorption method |
| Alumina · HCl | acid-washed (HCl) activated alumina | CAN | chloroacetonitrile |
| Alumina · HOH | water-washed activated alumina | CCE | carbon chloroform extract |
| ANPRM | Advanced Notice for Proposed Rule Making | CCPP | calcium carbonate precipitation potential |
| ASTM | American Society for Testing and Materials | CDA | cellulose diacetate |
| AWWA | American Water Works Association | CDC | Centers for Disease Control |
| AWWARF | American Water Works Association Research Foundation | CERCLA | Comprehensive Environmental Response and Compensation Liability Act |
| BCAN | bromochloroacetonitrile | CNS | central nervous system |
| BET | Brunauer-Emmett-Teller | COD | chemical oxygen demand |

| | | | |
|---|---|---|---|
| C.P. | compression point | ECH | epichlorohydrin |
| CSTR | completely stirred tank reactor | ED | electrodialysis |
| CT | value time (disinfectant concentration times the contact time) | EDB | ethylenedibromide |
| CU | color units | EDR | electrodialysis reversal |
| CUR | carbon usage rate | EDXA | energy-dispersive x-ray analysis |
| CWA | Clean Water Act | EEC | European Economic Community |
| CWSS | Community Water Supply Survey | equiv | gram equivalent of charged species (equal to $6.023 \times 10^{23}$ charges) |
| DBAN | dibromoacetonitrile | FA | fulvic acid |
| DBCP | dibromochloropropane | FIFRA | Federal Insecticide, Fungicide, and Rodenticide Act |
| DBP | disinfection by-product | FT | fermentation tube |
| dc | direct current | GAC | granular activated carbon |
| DCA | dichloroacetic acid | GI | gastrointestinal |
| DCAN | dichloroacetonitrile | GWSS | Ground Water Supply Survey |
| DCP | dichlorophenol | HA | humic acid |
| 2,4-DCP | 2,4-dichlorophenol | HAV | hepatitis A |
| DDT | dichlorodiphenyltri-chloroethane | HPC | heterotrophic plate count |
| D.E. | diatomaceous earth | HPMC | high-pressure minicolumn |
| DIC | dissolved inorganic carbon | HSDM | homogeneous surface diffusion model |
| DLVO | Derjaguin, Landau, Verwey, and Overbeek model | HWC | Hackensack Water Company |
| DMF | decayed, missing, and filled | IARC | International Agency for Research on Cancer |
| DMP | dimethylphenol | IOC | inorganic chemicals |
| DOC | dissolved organic carbon | ISWS | Illinois State Water Survey |
| DOX | dissolved organic halogen | LR | Larson ratio |
| DWEL | drinking water equivalent level | LSI | Langelier saturation index |
| DWPL | Drinking Water Priority List | MCL | maximum contaminant level |
| EBCT | empty bed contact time, i.e., the hydraulic residence time in an empty bed of the same volume as the media including voids | MCLG | maximum contaminant level goal |

| | | | |
|---|---|---|---|
| MCLs | maximum contaminant levels | NTP | National Toxicology Program |
| MDL | method detection limit | NVTOC | nonvolatile total organic carbon |
| meq | milliequivalent, $10^{-3}$ equivalents | P-A | presence-absence |
| MF | membrane filter | PA | polyamide |
| MIB | 2-methylisoborneol | PAC | powdered activated carbon |
| MIB | methylisoborneol | PACl | polyaluminum chloride |
| MR | macroporous, i.e., a resin with large pores made by incorporating microspheres into a macrosphere (the resin bead) | PAH | polynuclear aromatic hydrocarbon |
| MTD | maximally tolerated dose | PBP | p-bromophenol |
| MTZ | mass transfer zone | PCB | polychlorinated biphenols |
| MW | molecular weight | PCE | perchloroethylene |
| MWC | molecular weight cutoff | PCP | pentachlorophenol |
| NAS | National Academy of Sciences | PDB | paradichlorobenzene |
| NCI | National Cancer Institute | PF | phenol-formaldehyde |
| NES | National Eutrophication Study | PICl | polyiron chloride |
| NF | nanofiltration | PNP | p-nitrophenol |
| NIPDWR | National Interim Primary Drinking Water Regulation | POE | point-of-entry (describes a treatment device located at the point where it can treat all the water entering a house) |
| NOAEL | no observed adverse effect level | POU | point-of-use (describes a treatment device located at the point of use, e.g., under the sink where it treats water for drinking and cooking only) |
| NOM | natural organic matter | PQL | practical quantitations level |
| NOMS | National Organics Monitoring Survey | PTFE | polytetrafluoroethylene |
| NORS | National Organics Reconnaissance Survey | PVA/CA | polyvinylacetate crotonic acid |
| NPDES | National Pollutant Discharge Elimination System | RCRA | Resource Conservation and Recovery Act |
| NSP | National Screening Program | RfD | reference dose |
| NSS | nitrate-sulfate selective [refers to a special SBA resin that prefers nitrate over sulfate at the ionic strength of natural (fresh) waters] | r.m.s. | root mean square |

| | | | |
|---|---|---|---|
| RO | reverse osmosis | TCE | trichloroethene |
| RSI | Ryznar (saturation) index | TCP | trichlorophenol |
| RSSCT | rapid small-scale column test | TCSA | Toxic Substances Control Act |
| RWS | Rural Water Survey | TDS | total dissolved solids |
| SAC | strong-acid cation referring to a cation-exchange resin with sulfonic acid functionality | TEM | transmission electron microscope |
| SAED | selected area electron diffraction | TFC | thin film composite |
| SAR | start action request | THM | trihalomethane |
| SARA | Superfund Amendments and Reauthorization Act | THMFP | trihalomethane formation potential |
| SBA | strong-base anion referring to an anion-exchange resin with quaternary amine functionality | TOC | total organic carbon |
| SCD | streaming current detector | TON | threshold odor number |
| SDI | silt density index | TOX | total organic halogen |
| SDVB | styrene divinylbenzene | UF | ultrafiltration or uncertainty factor |
| SDWA | Safe Drinking Water Act (P.L. 93-523 plus amendments) | USEPA | U.S. Environmental Protection Agency |
| SEM | scanning electron microscope | UV | ultraviolet |
| SFR | service flow rate, i.e., the exhaustion rate for an ion exchanger or adsorbent usually reported in gpm/ft$^3$ or BV/h | VOC | volatile organic contaminant |
| SHMP | sodium hexametaphosphate | WAC | weak-acid cation (refers to a cation-exchange resin typically with carboxylic acid functionality) |
| SI | saturation index | WBA | weak-base anion (refers to an anion exchange resin with primary, secondary, or tertiary amine functionality) |
| SMCL | secondary maximum contaminant level | WDXA | wavelength-dispersive x-ray analysis |
| SOC | synthetic organic chemical | WHO | World Health Organization |
| SPC | standard plate count | WQS | water quality standards |
| SWTR | surface water treatment rule | XRD | x-ray diffraction |
| T | contact time | XRF | x-ray fluorescence |
| TCA | methyl chloroform | ZBC | zero point of charge (the pH at which the net charge on the solid particle is zero) |
| TCAN | trichloroacetonitrile | | |

# Notation

## Chapter 5

| | | | |
|---|---|---|---|
| $a$ | specific interfacial area, m$^{-1}$ or cm$^{-1}$ | $H_m$ | Henry's law, (m$^3$ water)(atm)/mol gas |
| $a_t$ | total packing area, m$^2$/m$^3$ | $H_u$ | Henry's law, unitless |
| $a_w$ | wetted packed area, m$^2$/m$^3$; used as equal to $a$ | HTU | height of one transfer unit, m |
| $A$ | area through which transfer takes place, m$^2$ or cm$^2$ | $J$ | empirical constant for temperature conversion of $H$ |
| $c$ | mole fraction of gas in water, mole of gas per mole of water; or concentration units, kg/m$^3$, mol/L, or mg/L | $k_G$ | interface mass transfer coefficient across the gas interface, m/h or cm/h |
| $C_D$ | coefficient of drag, unitless | $k_L$ | interface mass transfer coefficient across the liquid interface, m/h or cm/h |
| $C_v$ | velocity coefficient from spray nozzle | $K_G$ | overall mass transfer coefficient across the gas interface, m/h |
| $d$ | bubble diameter, cm | $K_L$ | overall mass transfer coefficient across the liquid interface, m/h |
| $d_p$ | nominal packing diameter, cm | $L$ | liquid velocity, m/s, m$^3$/(m$^2$)(s), kg/(m$^2$)(s), or mol/(m$^2$)(s) |
| $D_G$ | diffusion coefficient for gas in air, m$^2$/s | $n, m$ | constants used in the Sherwood-Holloway equations, good only for the units given |
| $D_L$ | diffusion coefficient for gas of interest in water, m$^2$/s (units must be ft$^2$/h for Sherwood-Holloway equations) | NTU | number of transfer units |
| DF | driving force, mg/L | $p$ | mole fraction of gas in air, mole gas per mole air, or concentration units, kg/m$^3$, mol/L, or mg/L |
| $F$ | flux, mass transferred per time per unit area, kg/(h)(m$^2$) | $P$ | pressure of gas in air, atm |
| $g$ | gravity, m/s$^2$ | $P_T$ | total air pressure, usually atmospheric, i.e., 1 atm |
| $G$ | air velocity, m/s, m$^3$/m$^2$s, kg/(m$^2$)(s), or mol/(m$^2$)(s) | $q$ | air-to-water feed ratio, unitless |
| $h$ | height, m | $Q_L$ | volumetric liquid flow rate, m$^3$/s or gpm |
| $H$ | heat absorbed in the evaporation of 1 mol of gas from solution, kcal/kmol | $Q_G$ | volumetric air flow rate, m$^3$/s or ft$^3$/min |
| $H$ | Henry's law, atm | $r_G$ | air density, kg/m$^3$ |
| $H_D$ | Henry's law, (atm)(L)/mg | $r_L$ | liquid density, kg/m$^3$ |

| | | | |
|---|---|---|---|
| $R$ | stripping factor, unitless | $T_d$ | liquid detention time, h |
| $R_c$ | gas constant, 1.987 kcal/kmol | $u_G$ | air viscosity, kg/(m)(s) |
| $R_n$ | Reynolds number, unitless | $u_L$ | liquid viscosity, kg/(m)(s) |
| $s$ | liquid surface tension, N/m = 0.073 at 20°C | $v$ | rise velocity of air bubble, cm/min |
| $s_c$ | critical surface tension of packing material, N/m | $v_d$ | velocity of drop from a spray nozzle, cm/s |
| $t$ | time | $V$ | volume containing the interface through which transfer occurs, $m^3$ or $cm^3$ |
| $t_B$ | bubble rise time, s | $W$ | mass transferred, kg |
| $T$ | temperature, K | $\alpha$ | angle of spray in a spray nozzle |

## Chapter 6

| | | | |
|---|---|---|---|
| $A$ | Hamaker constant for van der Waals attraction | $F_g$ | gravitational force |
| $A_0$ | concentration coefficient for total particulate matter | $G$ | fluid velocity gradient |
| $A_{SH}$ | dimensionless constant for particle size distribution in shear | $G^*$ | optimum velocity gradient |
| $c$ | fluctuating concentration | $h$ | separation gap between particles |
| $c_0'$ | initial mean square concentration fluctuation | $I$ | ionic strength |
| $C$ | coagulant concentration | $I_s$ | intensity of segregation |
| $C_s$ | floc strength coefficient | $J_j$ | total flux of particles |
| $d$ | particle diameter | $k$ | Boltzmann's constant |
| $d_m$ | particle formed by collision of two smaller particles, $d_i$ and $d_j$ | $K$ | empirical coefficient |
| $d_p$ | particle diameter, $\mu$m | $K_A$ | aggregation coefficient |
| $d_s$ | stable floc size | $K_B$ | floc breakup coefficient |
| $dN$ | number of particles per unit fluid volume | $\ell$ | coefficient dependent on breakup mode and size regime of eddies |
| $dq$ | rate of fluid flow | $L_s$ | scale of segregation |
| $dz$ | disk thickness | $n$ | solute concentration |
| $D^x$ | diffusion coefficient for isolated particles | $\dot{n}(d_p)$ | particle size distribution function, number/$(cm^3)(\mu m)$ |
| $E$ | volume of solids per volume of suspension per time | $N$ | collision rate of particles with a central particle |
| $f$ | shape factor | $(N_{ij})_p$ | rate of binary perikinetic collisions among particles $i$ and $j$ |
| $F$ | drag force | $N_m$ | number of concentration of particles of size $d_m$ |

$N_0$    initial concentration of particles at time $t = 0$

$P$    power dissipated

$Q$    flow rate through a reactor

$r$    distance from an assumed datum

$R$    radius of particle

$s$    specific gravity of a particle

$s_p$    surface area of a particle, $\mu m^2$

$t_d$    detention time within a reactor

$t_L$    liquid detention time

$T$    absolute temperature

$u$    fluctuating velocity

$u'$    intensity of turbulence

$U$    instantaneous velocity at a point

$\overline{U}$    time average velocity

$v$    particle velocity

$v_p$    settling velocity of individual particles

$v_s$    settling velocity of whole suspension

$v_u$    nominal upflow velocity

$V$    reactor volume

$V_p$    volume of a particle, $\mu m^3$

$Z$    height of fluid

$\alpha$    fraction of collisions resulting in permanent aggregation, i.e., collision efficiency factor

$\beta$    size distribution coefficient

$\gamma$    $\mu/\rho$ = kinematic viscosity

$\epsilon$    rate of energy dissipation per unit mass of fluid

$\epsilon_o$    porosity of floc

$\eta$    Kolmogoroff microscale of turbulence

$\kappa$    reciprocal thickness of the double layer

$\mu$    dynamic viscosity

$\mu_i$    chemical potential

$\mu_i^{\circ}$    standard state chemical potential

$\rho$    density

$\phi$    floc volume fraction

$\psi$    electrostatic potential energy

$\psi_g$    gravitational potential energy

## Chapter 7

$A$    area

$A_H$    cross-sectional area of channel to liquid flow

$A^*$    plan area of tank with horizontal flow for ideal settlement

$b$    breadth

$C$    concentration

$C_c$    critical concentration when flux is minimum, for Coe-Clevenger method

$C_f$    solids concentration in feed

$C_u$    solids concentration in underflow

$C_0$    initial concentration

$C_2$    concentration after time $t_2$

$C^*$    apparent solids volumetric concentration

$C^+$    blanket concentration at maximum flux

$C_D$    drag coefficient

$d$    diameter

$d_H$    hydraulic diameter

$E$    porosity

$E^+$    value of $E$ when flux $\phi$ has maximum value

$f$    force

$f_b$    buoyancy

$f_d$    drag

$f_g$    external force such as gravity

$F_t$    fraction at time $t$

$F(t)$    fraction of total tracer, in tracer test analysis

Fr    Froude number

$g$    gravitational constant

$G$    velocity gradient

$h$    depth, where $h \leq H$

$U$    velocity at angle $\theta$ to horizontal, in inclined settling

$H$    depth, height

$H_u$    height of mudline after time $t_u$, Eq. (7.52)

$H_0$    initial height of a suspension

$H_1$    height of suspension defined by Eq. (7.50)

$H_2$    height of mudline at time $t_2$

He    Henry's law constant

$k$    constant

$k_1$    constant in Eq. (7.27)

$k_2$    constant in Eq. (7.27)

$k_3$    constant in Eq. (7.42)

$K$    constant, Eq. (7.42)

$l$    length, where $l \leq L$

$L$    length

$L^*$    length of tank for ideal settlement

$L'$    greater than or equal to $L^*$, for particles with velocity $U_t' \leq U_t^*$ to settle

$L_p$    length of surface for settlement, in inclined settling

$m$    fraction as dead space

$n$    power index, dependent on Re

$N$    number of channels, stages

$P$    perimeter

$p$    fraction of plug flow, partial pressure

$q$    constant

$Q$    volumetric flow rate

Re    Reynolds number

$r$    power index

$S_c$    Yao's inclined flow geometry number

$t$    time

$t_u$    time to achieve concentration in underflow, Eq. (7.52)

$t_2$    time of specific value

$T$    retention time

$U$    velocity

$U_\theta$    velocity at angle $\theta$

$U_s$    settling velocity of suspension

$U_t$    terminal settling velocity

$U_t^*$    overflow rate of the ideal settling tank

$U_t'$    less than or equal to $U_t^*$

$U_0$    settling velocity of suspension for concentration extrapolated to

$U_2$    propagation rate of mudline at time $t_2$

$U_2^+$    upward propagation rate of mudline of phase with concentration $C_2$

$V$    volume

$w$    perpendicular spacing, width

$W$    width

$x$    mole fraction

$x_t$    mass fraction of particles with settling velocity $U_t$

$x_t^*$    mass fraction of particles with settling velocity $U_t^*$

$\theta$    angle of inclination

| | | | |
|---|---|---|---|
| $\mu$ | viscosity | $\phi$ | flux |
| $\rho$ | density | $\phi_c$ | maximum flux in Coe-Clevenger method |
| $\rho_g$ | density of gas bubble, in flotation angle of inclination | $\phi_u$ | flux in underflow |
| $\rho_s$ | density of particle | $\Phi$ | shape factor |

## Chapter 8

| | | | |
|---|---|---|---|
| $a$ | grain surface area | $R$ | Reynold's number based on superficial velocity, $d_{eq}V\rho/\mu$ |
| $A1$ | dimensionless grouped parameter in expansion model, Eq.(8.18) | $R'$ | Reynold's number for flow through granular bed, $4(V/\epsilon)r\rho/\mu$ |
| $C$ | particle concentration | $R_1$ | Reynold's number based on interstitial velocity, $V/S_v\mu(1-\epsilon)$ |
| $C_D$ | body feed concentration | $R_{mf}$ | Reynold's number based on minimum fluidization velocity, $d_{eq}V_{mf}\rho/\mu$ |
| $d_c$ | diameter of collector | $S_v$ | specific surface, grain surface area per unit grain volume |
| $d_p$ | diameter of particle in suspension to be filtered | $T$ | absolute temperature, Kelvin |
| $d_{eq}$ | diameter of a sphere of equal volume to an irregular grain | $U$ | mean pipe flow velocity |
| $d_{10}$ | sieve size for which 10 percent by weight of the grains of filter medium are smaller | $UC$ | uniformity coefficient, $d_{60}/d_{10}$ |
| $D$ | pipe diameter | $v$ | grain volume |
| e.s. | effective size = $d_{10}$ size of a filter medium | $V$ | superficial velocity (i.e., approach velocity or empty bed velocity) |
| $f$ | pipe friction factor in Darcy-Weisbach equation | $V_{mf}$ | superficial velocity to achieve fluidization of a granular bed |
| $g$ | acceleration of gravity | $V_s$ | Stoke's settling velocity of a particle |
| $h$ | head loss (column of fluid) | $V'$ | volume of filtrate |
| $k$ | Kozeny's constant | $\alpha,\beta,\delta$ | empirical constants in Ives and Sholji's equation |
| $K$ | Boltzmann's constant | $\epsilon$ | porosity ratio |
| $L$ | depth of filter bed in direction of flow | $\epsilon_o$ | fixed loose bed porosity |
| $N$ | number of grains per unit volume | $\epsilon_{mf}$ | porosity ratio of a bed at a condition of incipient fluidization |
| $p, \Delta p$ | pressure and pressure drop | $\eta$ | single collector efficiency |
| $Q_a$ | airflow rate | $\eta_D$ | single collector efficiency for diffusive transport |
| $r$ | approximate hydraulic radius of a granular bed | $\eta_I$ | single collector efficiency for interception |

| | | | |
|---|---|---|---|
| $\eta_S$ | single collector efficiency for sedimentation transport | $\rho_b$ | bulk density of mixture of water plus grains |
| $\mu$ | absolute viscosity of fluid | $\rho_s$ | mass density of solid grains |
| $\rho$ | mass density of fluid | $\psi$ | sphericity, i.e., ratio of surface area of an equal volume sphere to the surface area of an irregular grain |

## Chapter 9

| | | | |
|---|---|---|---|
| $C$ | total aqueous phases concentration, equiv/L or meq/L | $K_{ij}$ | selectivity coefficient for ion exchange |
| $C_E$ | effluent concentration, equiv/L or mg/L | $K_{N/Cl}$ | selectivity coefficient for nitrate/chloride ion exchange |
| $C_F$ or $C_I$ | feed (influent) concentration to column | $-N(CH_3)_2$ | tertiary amine functional group on an anion resin |
| $q_i$ | resin-phase concentration of ion $i$, equiv/L | $-N(CH_3)_3{}^+$ | quaternary amine functional group on an anion resin |
| $C_i$ | aqueous phase concentration of ion $i$, equiv/L or meq/L | $NO_3$-N | nitrate reported as nitrogen, mg/L |
| $D$ | diameter, ft or m | pCi | picocurie, $3.7 \times 10^{10}$ nuclear disintegrations/s |
| $f_B$ | fraction of water bypassed in an ion-exchange system | $q$ | total ion-exchange capacity of resin, equiv/L |
| $f_F$ | fraction of water fed to the ion-exchange column | $q_n$ | resin-phase concentration of nitrate, equiv/L |
| $h$ | depth of media, ft or m | $\overline{RCOOH}$ | weak-acid resin in the hydrogen form |
| $\{i\}$ | activity of ion $i$ in the aqueous phase, equiv/L | $\overline{(RCOO^-)_2Ca^{2+}}$ | weak-acid resin in the calcium form |
| $\{\bar{i}\}$ | activity of ion $i$ in the resin phase, equiv/L | $\overline{R_3N:}$ | weak-base resin in the free-base form |
| $[i]$ | concentration of ion $i$ in the aqueous phase, mol/L | $\overline{R_3NH^+}$ | protonated form of weak-base resin carrying a positive charge |
| $[\bar{i}]$ | concentration of ion $i$ in the resin phase, mol/L | $\overline{R_3NH^+Cl^-}$ | protonated form of weak-base resin carrying a positive charge balanced by an anion |
| $K$ | thermodynamic equilibrium constant for ion exchange | $-SO_3{}^-$ | sulfonate strong-acid functional group or a resin |

$t_{BW}$    time for backwashing, min

$t_{FR}$    time for fast rinse, min

$t_H$    time to hardness breakthrough, h

$t_R$    time for regeneration, min

$t_S$    out-of-service time for back-washing, regeneration, and rinsing, min

$t_{SR}$    time for slow rinse, min

$V_0$    approach velocity, m/h or gpm/ft$^2$

$X_i$    equivalent fraction of ion $i$ in the aqueous phase, dimensionless

$X_N$    equivalent fraction of nitrate in the aqueous phase, dimensionless

$Y_i$    equivalent fraction of ion $i$ in the resin phase, dimensionless

$Y_N$    equivalent fraction of nitrate in the resin phase, dimensionless

$\alpha_{ij}$    separation factor of $i$ over $j$, dimensionless

$\alpha_{N/Cl}$    separation factor of nitrate over chloride, dimensionless

## Chapter 10

| | | | |
|---|---|---|---|
| $A$ | dosage of additional chemicals | $K_w$ | solubility-product constant for water |
| Acd | acidity | Mg | magnesium carbonate hardness removed |
| Al | alum dose | $(Mg)_e$ | desired magnesium concentration in plant effluent after split-treatment excess lime softening |
| Ca | calcium carbonate hardness removed | $(Mg)_r$ | magnesium concentration in raw water entering the split-treatment excess lime softening process |
| $C_T$ | total carbonic species concentration | $(Mg)_t$ | magnesium concentration in stream from lime treatment process in split-treatment excess lime softening |
| $d[Fe^{2+}]/dt$ | rate of iron(II) oxidation | $[OH^-]$ | hydroxide on concentration |
| Fe | iron dose | $PO_2$ | partial pressure of oxygen, atm |
| $[Fe^{2+}]$ | iron(II) concentration | $Q$ | raw water flow |
| $G$ | velocity gradient | $R$ | reaction rate constant |
| $I$ | ionic strength | $\Delta S$ | dry weight of sludge solids formed |
| $k$ | apparent first-order reaction rate constant | SS | suspended solids |
| $K_1, K_2$ | equilibrium constants for the first and second disassociation of carbonic acid, respectively | $t$ | solution temperature, °C |
| $K_{ap}$ | thermodynamic activity product | $T$ | solution temperature, K |

| | | | |
|---|---|---|---|
| $K_{eq}$ | equilibrium constant | $X$ | fraction of total flow to be bypassed in split-treatment excess lime softening |
| $K_{sp}$ | solubility-product constant | $\alpha_1, \alpha_2, \alpha_3$ | carbonic species ionization fraction |

## Chapter 11

| | | | |
|---|---|---|---|
| $A$ | water permeability constant, $g/(cm^2)(s)(atm)$ | $\overline{mi}$ | molality |
| $B$ | salt permeability constant, cm/s | $\Delta p$ | pressure differential across membrane, atm |
| $C_1$ | solute concentration in the feed | $\pi$ | osmotic pressure, psi |
| $C_2$ | solute concentration in the brine | $T$ | temperature, °C |
| $F_s$ | salt flux, $g/(cm^2)(s)$ | $\Delta\pi$ | osmotic pressure differential across membrane, atm |
| $F_w$ | water flux, $g/(cm^2)(s)$ | | |

## Chapter 12

| | | | |
|---|---|---|---|
| $E$ | potential of reaction at other than standard conditions | $K$ | equilibrium constant |
| $E_a$ | energy of activation | $n$ | number of electrons transferred |
| $E°$ | standard electrode potential | $OH^-$ | hydroxyl ion |
| $F$ | Faraday constant | $pO_2$ | partial pressure of oxygen |
| $\Delta G°$ | standard free energy change of reaction | $R$ | gas constant |
| $HO$ | hydroxyl radical | $T$ | absolute temperature |
| $k$ | rate constant | | |

## Chapter 13

| | | | |
|---|---|---|---|
| $A$ | column surface area | $C_{eff}$ | effluent concentration |
| $b_i$ | Langmuir equation constant for $i$th adsorbate | $C_{e,i}$ | solution concentration of $i$th adsorbate in equilibrium with an adsorbent |
| $BV$ | bed volumes of water, equal to the volume of water treated/volume of GAC in the column | $C_{0,i}$ | initial or influent concentration for $i$th adsorbate |
| $C_l$ | the effluent concentration that represents an average for an entire column run | $d_{SC}/d_{LC}$ | ratio of the diameter of GAC in a small column to that in a large colum |
| $C_B$ | breakthrough concentration; the maximum acceptable concentration from a column | $d_{90}$ | 90 percent size of a filter medium, i.e., 90 percent by weight of the medium is larger than this size |

$EBCT_{SC}/EBCT_{LC}$    ratio of the EBCT of a small column to that of a large column

$f$    fraction ($< C_{eff}/C_0$) or concentration, of adsorbate in the treated, blended effluent of a parallel contactor system

$f_i$    fraction ($C_{eff}/C_0$) or concentration, of adsorbate in the effluent of the $i$th contactor in a parallel adsorber system

$K$    Freundlich equation constant

$L_{critical}$    maximum depth of a column that leads to immediate breakthrough of adsorbate

$L_{MTZ}$    length of the mass transfer zone

$m$    mass of adsorbent

$n$    number of contactors in a parallel adsorber system

$Q$    volumetric flow rate

$q_{e,i}$    quantity or $i$th adsorbate on an adsorbent per unit mass of adsorbent

$q_{max,i}$    Langmuir equation constant for $i$th adsorbate

$(q_e)_0$    mass adsorbed per unit mass of carbon when the equilibrium concentration $C_e$ is equal to the influent concentration $C_0$

$t_{1.01}$    time to reach a solution concentration of 1.01

$V$    volume

$V_B$    volume of water processed by a column, from start to breakthrough

$Y$    volume of water that can be treated per unit volume of activated carbon

$1/n$    Freundlich equation constant

$\theta_n$    the number of bed volumes processed by each adsorber in a parallel system at breakthrough

$\rho_{GAC}$    activated carbon apparent density

## Chapter 14

$A$    ratio of available chlorine in the form of dichloramine to available chlorine in the form of monochloramine

$B$    1-4 $K_{eq}[H^+]$

$C$    disinfectant concentration

$E$    activation energy

$H$    Henry's constant

$k$    rate constant of inactivation

$k'$    rate constant of inactivation corrected for disinfectant concentration

$k_f$    rate of reaction between dissolved $Cl_2$ and chlorite

$k_0$    frequency factor in Ahrrenius relationship

$K_a$    acid dissociation constant

$K_{eq}$    equilibrium constant for dichloramine formation from biomolecular reaction of monochloramine

$K_H$    chlorine hydrolysis equilibrium constant

$n$    coefficient of dilution

$N$    concentration of viable microorganisms at time $t$

| | | | |
|---|---|---|---|
| $N_0$ | concentration of viable microorganisms at time 0 | $r$ | rate of formation of monochloramine |
| $P_{Cl_2}$ | gas phase partial pressure of chlorine | $t$ | time |
| $P_{ClO_2}$ | gas phase partial pressure of chlorine dioxide. | $T$ | absolute temperature |
| $R$ | ideal gas constant | $Z$ | ratio of moles of chlorine (as $Cl_2$) added per mole of ammonia nitrogen present |

## Chapter 17

| | | | |
|---|---|---|---|
| $E$ | cell potential for a redox reaction | $pH_s$ | pH at saturation with $CaCO_3$ |
| $E^\circ$ | standard electrode potential for a redox reaction | $R$ | ideal gas constant, 0.001987 kCal/(deg)(mol) |
| $f_d$ | activity coefficient for any divalent ion, using the Davies equation. | $T$ | absolute temperature, K |
| $f_M$ | activity coefficient for any monovalent ion, using the Davies equation | $z$ | number of electrons transferred in a redox reaction |
| $F$ | Faraday constant, 23.060 kCal/V | $\beta$ | buffer intensity |
| $\Delta G$ | Gibbs free energy of formation | $\gamma_i$ | activity coefficient for ion $i$ |
| $I$ | ionic strength, $I = \frac{1}{2} m_i z_i^2$, where $m_i$ is the molar concentration of dissolved ion $i$ and $z$ is the charge of dissolved ion $i$ | [ ] | concentration of an aqueous species, mol/L |
| $K$ | thermodynamic equilibrium constant for a reaction (activities are used) | { } | activity of an aqueous species |
| $K'$ | equilibrium constant adjusted for ionic strength (concentrations are used) | | |

# Index

## ABOUT THE SPONSORING ORGANIZATION

The American Water Works Association is a 49,200-member
organization dedicated to assessing, promoting, and
recognizing high water quality standards. The Association is
the primary source of information on drinking water for
local, state, and national government and for international
regulatory agencies.